W9-ASU-776

SOCIAL PSYCHOLOGY

Sixth Edition

John D. DeLamater
University of Wisconsin–Madison

Daniel J. Myers
University of Notre Dame

Australia • Brazil • Canada • Mexico • Singapore • Spain
United Kingdom • United States

Social Psychology, **Sixth Edition**
John D. DeLamater and Daniel J. Myers

Senior Sociology Editor: Bob Jucha
Assistant Editor: Kristin Marrs
Editorial Assistant: Katia Krukowski
Technology Project Manager: Dee Dee Zobian
Marketing Manager: Michelle Williams
Marketing Assistant: Jaren Boland
Marketing Communications Manager: Linda Yip
Project Manager, Editorial Production: Cheri Palmer
Creative Director: Rob Hugel
Print Buyer: Doreen Suruki
Permissions Editor: Roberta Broyer
Production Service, Composition, Illustration: G&S Book Services
Photo Researcher: Kathleen Olson
Copy Editor: Kathy Finch
Cover Designer: Yvo Riezebos Design
Cover Printer: Phoenix Color Corp
Printer: R.R. Donnelley/Crawfordsville

Printed in the United States of America
1 2 3 4 5 6 7 10 09 08 07 06

Library of Congress Control Number: 2006926046

ISBN 0-495-09336-X

Thomson Higher Education
10 Davis Drive
Belmont, CA 94002-3098
USA

BRIEF CONTENTS

Chapter 1 *Introduction to Social Psychology* 1

Chapter 2 *Research Methods in Social Psychology* 26

Chapter 3 *Socialization* 52

Chapter 4 *Self and Identity* 82

Chapter 5 *Social Perception and Cognition* 108

Chapter 6 *Attitudes* 141

Chapter 7 *Symbolic Communication and Language* 165

Chapter 8 *Social Influence and Persuasion* 196

Chapter 9 *Self-Presentation and Impression Management* 225

Chapter 10 *Emotions* 250

Chapter 11 *Helping and Altruism* 273

Chapter 12 *Aggression* 299

Chapter 13 *Interpersonal Attraction and Relationships* 325

Chapter 14 *Group Cohesion and Conformity* 354

Chapter 15 *Group Structure and Performance* 378

Chapter 16 *Intergroup Conflict* 408

Chapter 17 *Life Course and Gender Roles* 432

Chapter 18 *Social Structure and Personality* 461

Chapter 19 *Deviant Behavior and Social Reaction* 493

Chapter 20 *Collective Behavior and Social Movements* 524

Glossary 553

References 567

Credits 651

Name Index 653

Subject Index 675

CONTENTS

PREFACE
XII

1

INTRODUCTION TO SOCIAL PSYCHOLOGY
1

Introduction 2

What Is Social Psychology? 3
- *A Formal Definition 3*
- *Core Concerns of Social Psychology 3*
- *Relation to Other Fields 5*

Theoretical Perspectives in Social Psychology 5
- *Role Theory 8*
- *Reinforcement Theory 11*
- *Cognitive Theory 14*
- *Symbolic Interaction Theory 16*
- *Evolutionary Theory 18*
- *A Comparison of Perspectives 20*

Is Social Psychology a Science? 22
- *Characteristics of Science 22*
- *Social Psychology as a Science 23*

Summary 24

List of Key Terms and Concepts 25

2

RESEARCH METHODS IN SOCIAL PSYCHOLOGY
26

Introduction 27
- *Questions About Research Methods 27*

Characteristics of Empirical Research 27
- *Objectives of Research 27*
- *Research Hypotheses 28*
- *Validity of Findings 29*

Research Methods 29
Surveys 30
Field Studies and Naturalistic Observation 37
Archival Research and Content Analysis 39
Experiments 41
Comparison of Research Methods 45
Meta-Analysis 46

Research in Diverse Populations 47

Ethical Issues in Social Psychological Research 48
Potential Sources of Harm 48
Institutional Safeguards 49
Potential Benefits 50

Summary 51

List of Key Terms and Concepts 51

3
SOCIALIZATION
52

Introduction 53

Perspectives on Socialization 54
The Developmental Perspective 54
The Social Learning Perspective 55
The Interpretive Perspective 56
The Impact of Social Structure 57

Agents of Childhood Socialization 57
Family 57
Peers 63
School 64

Processes of Socialization 65
Instrumental Conditioning 65
Observational Learning 68
Internalization 69

Outcomes of Socialization 69
Gender Role 69
Linguistic and Cognitive Competence 71
Moral Development 74
Work Orientations 77

Adult Socialization 78
Role Acquisition 78
Anticipatory Socialization 79
Role Discontinuity 79

Summary 80

List of Key Terms and Concepts 81

4
SELF AND IDENTITY
82

Introduction 83

The Nature and Genesis of Self 85
The Self as Source and Object of Action 85
Self-Differentiation 85
Role Taking 87
The Social Origins of Self 87

Identities: The Self We Know 89
Role Identities 89
Social Identities 90
Research on Self-Concept Formation 91
The Situated Self 93

Identities: The Self We Enact 93
Identities and Behavior 93
Choosing an Identity to Enact 95
Identities as Sources of Consistency 97

The Self in Thought and Feeling 98
Self-Schema 99
Effects of Self-Awareness 99
Effects of Self-Discrepancies 100

Self-Esteem 101
Assessment of Self-Esteem 101
Sources of Self-Esteem 101
Self-Esteem and Behavior 103
Protecting Self-Esteem 105

Summary 106

List of Key Terms and Concepts 107

5
SOCIAL PERCEPTION AND COGNITION
108

Introduction 109

Schemas 110
Types of Schemas 111
Schematic Processing 112
Schemas as Cultural Elements 114

Person Schemas and Group Stereotypes 114
Person Schemas 114
Group Stereotypes 116

Impression Formation 122
Trait Centrality 122
Integrating Information About Others 123
Impressions as Self-Fulfilling Prophecies 125
Heuristics 126
Anchoring and Adjustment 127
Dispositional Versus Situational Attributions 127
Inferring Dispositions From Acts 128
Covariation Model of Attribution 131
Attributions for Success and Failure 132

Bias and Error in Attribution 134
Overattribution to Dispositions 134
Focus of Attention Bias 135
Actor-Observer Difference 136
Motivational Biases 137

Cultural Basis of Attributions 138

Summary 139

List of Key Terms and Concepts 140

6
ATTITUDES
141

Introduction 142

The Nature of Attitudes 142
The Components of an Attitude 142
Attitude Formation 143
The Functions of Attitudes 144

Attitude Organization 144
Attitude Structure 144

Cognitive Consistency 146
Balance Theory 147
Theory of Cognitive Dissonance 149
Is Consistency Inevitable? 151

The Relationship Between Attitudes and Behavior 154
Do Attitudes Predict Behavior? 154
Activation of the Attitude 154
Characteristics of the Attitude 155
Attitude-Behavior Correspondence 157
Situational Constraints 159

The Reasoned Action Model 161
Formal Model 161
Assessment of the Model 163

Summary 164

List of Key Terms and Concepts 164

7
SYMBOLIC COMMUNICATION AND LANGUAGE
165

Introduction 166

Language and Verbal Communication 166
Linguistic Communication 167
The Encoder-Decoder Model 168
The Intentionalist Model 170
The Perspective-Taking Model 172

Nonverbal Communication 175
Types of Nonverbal Communication 175
What's in a Face? 177
Combining Nonverbal and Verbal Communication 178

Social Structure and Communication 180
Gender and Communication 180
Social Stratification and Speech Style 181
Communicating Status and Intimacy 185

Normative Distances for Interaction 188
Normative Distances 188

Conversational Analysis 190
Initiating Conversations 190
Regulating Turn Taking 192
Feedback and Coordination 192

Summary 194

List of Key Terms and Concepts 195

8
SOCIAL INFLUENCE AND PERSUASION
196

Introduction 197
Forms of Social Influence 197

Attitude Change via Persuasion 198
Processing Persuasive Messages 198
Communication-Persuasion Paradigm 199
The Source 200
The Message 203
The Target 208

Compliance with Threats and Promises 210
Effectiveness of Threats and Promises 211
Problems in Using Threats and Promises 215
Bilateral Threat and Negotiation 216

Obedience to Authority 217
Experimental Study of Obedience 218
Factors Affecting Obedience to Authority 220

Resisting Influence and Persuasion 221
Inoculation 221
Forewarning 222
Reactance 222

Summary 222

List of Key Terms and Concepts 224

9
SELF-PRESENTATION AND IMPRESSION MANAGEMENT
225

Introduction 226

Self-Presentation in Everyday Life 227
Definition of the Situation 227
Self-Disclosure 229

Tactical Impression Management 230
Managing Appearances 231
Ingratiation 232
Aligning Actions 235
Altercasting 237

The Downside of Self-Presentation 237
Self-Presentation May Be Hazardous to Your Health 238
Deception May Be Hazardous to Your Relationships 239

Detecting Deceptive Impression Management 239
Ulterior Motives 239
Nonverbal Cues of Deception 240

Ineffective Self-Presentation and Spoiled Identities 242
Embarrassment and Saving Face 242
Cooling-Out and Identity Degradation 244
Stigma 245

Summary 248

List of Key Terms and Concepts 249

10
EMOTIONS
250

Introduction 251

Defining Emotions 251

Classical Ideas About the Origins of Emotion 252

Universal Emotions and Facial Expressions 253
Facial Expressions of Emotion 253
Cultural Differences in Basic Emotions and Emotional Display 257

Social Emotions 260
Cognitive Labeling Theory 261
Five Social Emotions 264
Emotion Work 270
Summary 272
List of Key Terms and Concepts 272

11 HELPING AND ALTRUISM 273

Introduction 274
Motivation to Help Others 275
Egoism and Cost-Reward Motivation 275
Altruism and Empathic Concern 276
Evolution and Helping 277
Characteristics of the Needy That Foster Helping 279
Acquaintanceship and Liking 279
Similarity 280
Deservingness 280
Normative Factors in Helping 282
Norms of Responsibility and Reciprocity 282
Personal Norms and Helping 283
Personal and Situational Factors in Helping 285
Modeling Effects 285
Gender Differences in Helping 286
Good and Bad Moods 286
Guilt and Helping 288
Bystander Intervention in Emergency Situations 288
The Decision to Intervene 289
The Bystander Effect 290
Costs and Emergency Intervention 292
Seeking and Receiving Help 295
Help and Obligation 295
Threats to Self-Esteem 296
Summary 297
List of Key Terms and Concepts 298

12 AGGRESSION 299

Introduction 300
What Is Aggression? 300
Aggression and the Motivation to Harm 300
Aggression as Instinct 301
Frustration-Aggression Hypothesis 301
Aversive Emotional Arousal 303
Social Learning and Aggression 304
Characteristics of Targets That Affect Aggression 304
Gender and Race 305
Attribution for Attack 308
Retaliatory Capacity 308
Situational Impacts on Aggression 309
Reinforcements 309
Modeling 309
Norms 310
Stress 310
Aggressive Cues 311
Reducing Aggressive Behavior 312
Reducing Frustration 312
Punishment to Suppress Aggression 313
Nonaggressive Models 313
Catharsis 313
Aggression in Society 315
Sexual Assault 315
Pornography and Violence 319
Media Violence and Aggression 320
Summary 324
List of Key Terms and Concepts 324

13 INTERPERSONAL ATTRACTION AND RELATIONSHIPS 325

Introduction 326
Who Is Available? 327
Routine Activities 327

Proximity 327
Familiarity 328

Who Is Desirable? 329
Social Norms 329
Physical Attractiveness 330
Exchange Processes 333

The Determinants of Liking 335
Similarity 336
Shared Activities 337
Reciprocal Liking 338

The Growth of Relationships 338
Self-Disclosure 339
Trust 340
Interdependence 341

Love and Loving 343
Liking Versus Loving 343
Passionate Love 343
The Romantic Love Ideal 345
Love as a Story 345

Breaking Up 347
Progress? Chaos? 347
Unequal Outcomes and Instability 347
Differential Commitment and Dissolution 349
Responses to Dissatisfaction 350

Summary 352

List of Key Terms and Concepts 353

14
GROUP COHESION AND CONFORMITY
354

Introduction 355

What Is a Group? 355
Group Cohesion 355
Group Structure and Goals 358
Roles in Groups 358

Status of Group Members 360
Status Characteristics 360
Status Generalization 361
Expectation States Theory 363
Overcoming Status Generalization 364

Conformity to Group Norms 365
Group Norms 365
Conformity 368
Why Conform? 369
Increasing Conformity 371

Minority Influence in Groups 374
Effectiveness of Minority Influence 374
Differences Between Minority and Majority Influence 375

Summary 376

List of Key Terms and Concepts 377

15
GROUP STRUCTURE AND PERFORMANCE
378

Introduction 379

Group Leadership 380
Endorsement of Formal Leaders 381
Revolutionary and Conservative Coalitions 382
Activities of Leaders 383
Contingency Model of Leadership Effectiveness 383

Productivity and Performance 387
The Presence of Others 387
Group Size 388
Group Goals 392

Reward Distribution and Equity 393
Principles Used in Reward Distribution 393
Equity Theory 394
Task Interdependence 395
Responses to Inequity 397

Brainstorming 398
Production Blocking 399

Group Decision Making 400
Groupthink 400
Risky Shift, Cautious Shift, and Group Polarization 404

Summary 406

List of Key Terms and Concepts 407

16
INTERGROUP CONFLICT
408

Introduction 409
Intergroup Conflict 410

Development of Intergroup Conflict 411
Realistic Group Conflict 411
Social Identity 412
Aversive Events and Escalation 415

Persistence of Intergroup Conflict 416
Biased Perception of the Out-Group 417
Changes in Relations Between Conflicting Groups 420

Impact of Conflict on Within-Group Processes 421
Group Cohesion 421
Leadership Militancy 422
Norms and Conformity 422

Resolution of Intergroup Conflict 424
Superordinate Goals 424
Intergroup Contact 425
Mediation and Third-Party Intervention 426
Unilateral Conciliatory Initiatives 429

Summary 430

List of Key Terms and Concepts 431

17
LIFE COURSE AND GENDER ROLES
432

Introduction 433

Components of the Life Course 434
Careers 434
Identities and Self-Esteem 434
Stress and Satisfaction 435

Influences on Life Course Progression 435
Biological Aging 435
Social Age Grading 436
Historical Trends and Events 437

Stages in the Life Course: Age and Gender Roles 439
Stage I: Achieving Independence 440
Stage II: Balancing Family and Work Commitments 442
Stage III: Performing Adult Roles 449
Stage IV: Coping with Loss 451

Historical Variations 455
Women's Work: Gender Role Attitudes and Behavior 455
Impact of Events 457

Summary 459

List of Key Terms and Concepts 460

18
SOCIAL STRUCTURE AND PERSONALITY
461

Introduction 462

Status Attainment 463
Occupational Status 463
Intergenerational Mobility 465
Social Networks 471

Individual Values 471
Occupational Role 473
Education 474

Social Influences on Health 474
Physical Health 475
Mental Health 479

Alienation 488
Self-Estrangement 488
Powerlessness 490

Summary 491

List of Key Terms and Concepts 492

19
DEVIANT BEHAVIOR AND SOCIAL REACTION
493

Introduction 494

The Violation of Norms 494
Norms 494
Anomie Theory 495

Control Theory 498
Differential Association Theory 501
Routine Activities Perspective 503

Reactions to Norm Violations 503
Reactions to Rule Breaking 506
Determinants of the Reaction 506
Consequences of Labeling 509

Labeling and Secondary Deviance 510
Societal Reaction 510
Secondary Deviance 513

Formal Social Controls 514
Formal Labeling and the Creation of Deviance 515
Long-Term Effects of Formal Labeling 521

Summary 522

List of Key Terms and Concepts 523

20
COLLECTIVE BEHAVIOR AND SOCIAL MOVEMENTS
524

Introduction 525

Collective Behavior 525
Crowds 526
Gatherings 530
Underlying Causes of Collective Behavior 532
Precipitating Incidents 535
Empirical Studies of Riots 536

Social Movements 540
The Development of a Movement 541
Social Movement Organizations 545
The Consequences of Social Movements 551

Summary 552

List of Key Terms and Concepts 552

GLOSSARY
553

REFERENCES
567

CREDITS
651

NAME INDEX
653

SUBJECT INDEX
675

PREFACE

ABOUT THIS BOOK

Revising an established text is a balancing act. As authors, we want to preserve the character of the book and at the same time keep it up-to-date, incorporating contemporary work in the discipline and phenomena of current social interest. The sixth edition of *Social Psychology* builds on the strengths of previous editions. Most important, the book covers the full range of phenomena of interest to social psychologists. Not only does it treat intrapsychic processes in detail, but it also provides strong coverage of social interaction and group processes and of larger-scale phenomena, such as intergroup conflict and social movements.

Our goal in writing this book is, as it has always been, to describe contemporary social psychology and to present the theoretical concepts and research findings that make up this broad and fascinating field. We have drawn on work by a wide array of social psychologists, including those with sociological and psychological perspectives. This book emphasizes the impact of social interaction, group membership, and social structure on the social behavior of individuals, but it also covers the intrapsychic processes of cognition, attribution, and learning that underlie social behavior. Throughout the book we have used the results of empirical research—surveys, experiments, observational and qualitative studies, and meta-analyses—to illustrate these processes.

NEW TO THIS EDITION

Persons familiar with previous editions will note that Andrew Michener is no longer credited as an author. Following his retirement from the University of Wisconsin–Madison, he has moved on to other pursuits. For the fifth edition, we welcomed aboard a new co-author, Daniel J. Myers. Dan earned his Ph.D. at the University of Wisconsin–Madison and is a member of the faculty at the University of Notre Dame. Dan regularly teaches social psychology at the undergraduate and graduate levels. He has brought a perspective to the book that is fresh but also compatible with the aims of the text.

In developing this edition, we sought not only to keep the reader abreast of changes within the field of social psychology but also to strengthen our presentation of various topics, based in part on feedback from students and faculty who used previous editions. We have continued to enhance our coverage of diversity. We have revised language to reflect diversity in gender, race, and sexual orientation. We have made a special effort to incorporate research that includes participants who vary on these dimensions, but of course are limited by what is available. We have called attention to these limitations in numerous places.

All of the chapters in this edition have been revised and updated. Descriptions of the most important of these changes follow:

- Chapter 10 (Emotions) is new to this edition. This new chapter allowed us not only to bring together material previously scattered across several chapters—for example, facial expression of emotions and catharsis—but also to add new material on this important topic. There is new coverage of classical nonsociological approaches to emotion as well as to social emotions (pride and shame). There are also two new boxes, on emotional intelligence (Box 10.1) and emotion in social movements (Box 10.2).
- Chapter 3 (Socialization) has been extensively revised to reflect new research. There is an updated discussion of effects of divorce on children, and of use of corporal and psychological punishment by parents. There is also new material discussing resistance by children to adult authority, cultural variation in gender role definitions taught to children, and the diversity of situations in which children live in the United States.
- Chapter 7 (Symbolic Communication and Language) has been revised and updated. The research discussed in the section "Normative Distances for Interaction" has been updated. A new subsection, "The Case of 'Dude'," analyzes the use of this term in conversation and how usage reflects social norms. Material has been added to the chapter on several topics, including the use of ebonics in schools and of silencing during interaction and its relation to social structure.
- Chapter 12 (Aggression) has been updated and includes coverage of several new topics. There is a new box, "Fierce Entanglements," discussing domestic violence and its embeddedness in patterns of relationships. There is discussion of new research on the link between childhood learning and teen aggression and adult abusive behavior, on the role of neighbors in responding to intimate violence, and on neighborhood characteristics that increase and decrease the likelihood of intimate violence. There is coverage of the latest research on the effects of viewing violence shown on television and of violence in video games.
- Chapter 14 (Group Structure and Performance) has been updated and revised. Material has been added on equal status within groups, and the discussion in the section "Why Conform?" has been expanded. There are two new boxes, on emotional group attachments and networks, and on priming and conformity.
- Chapter 17 (Life Course and Gender Roles) has been extensively revised. Statistics have been updated throughout, such as those on female labor force participation and on total hours per day spent by working spouses on various tasks. Statistics and coverage of research on minority populations has been expanded. There is new coverage of the concept of emerging adulthood and of the research on grandparents.
- Chapter 20 (Collective Behavior and Social Movements) features a completely revised discussion of classical theories of the crowd. Criticisms of several concepts have been clarified and expanded. There is new coverage of myths about crowds, of framing in social movements, and of collective identity.

CONTENT AND ORGANIZATION

This book begins with a chapter on theoretical perspectives in social psychology, followed by a chapter on research methods. These first two chapters provide the groundwork for all that follows. The remainder of the book is divided into four substantive sections. Section one focuses on individual social behavior. It includes chapters on socialization, self and identity, social perception and cognition, and attitudes. Section two is concerned with social interaction, the core of social psychology. Each of the chapters in this section discusses how persons interact with others and how they are affected by this interaction. These chapters cover such topics as communication, social influence and persuasion, self-presentation and impression management, emotion, helping and altruism, aggression, and interpersonal attraction. Section three provides extensive coverage of groups. It includes chapters on group cohesion and conformity, group structure and performance, and intergroup conflict. Section four considers the relations between individuals and the wider society. These chapters treat the influence of life course and gender roles; the impact of social structure on the individual, especially on physical and mental health; deviant behavior; and collective behavior and social movements.

EASE OF USE

Because there are many different ways in which an instructor can organize an introductory course in social psychology, each chapter in this book has been written as a self-contained unit. Later chapters do not presume that the student has read earlier ones. This compartmentalization enables instructors to assign chapters in any sequence.

Chapters share a standard format. To make the material interesting and accessible to students, each chapter's introductory section poses four to six focal questions. These questions establish the issues discussed in the chapter. The remainder of the chapter consists of four to six major sections, each addressing one of these issues. A summary at the end of each chapter reviews the key points. Thus, each chapter poses several key questions about a topic and then considers these questions in a framework that enables students to easily learn the major ideas.

In addition, the text includes several learning aids. Tables emphasize the results of important studies. Figures illustrate important social psychological processes. Photographs dramatize essential ideas from the text. Boxes in each chapter highlight interesting or controversial issues and studies and also discuss the applications of social psychological concepts in daily life. Some boxes are identified as "Research Update" (a feature new to this edition); these boxes have been updated by including the latest research. Other boxes are identified as "Test Yourself"; these contain a questionnaire that the student can complete to find out his or her standing on the measure of interest. Key terms appear in bold and are listed alphabetically at the end of each chapter. A glossary of key terms appears at the end of the book.

SUPPLEMENTS

Instructor's Manual with Test Bank, by Shirley Keeton of Fayetteville State University. Prepare for class more quickly and effectively with such resources as learning objectives, lecture outlines, lecture/discussion suggestions, student activities, key terms with page references, Internet links, and InfoTrac® College Edition search terms. A test bank with 35 multiple-choice questions, 10–15 true/false questions, and 3–5 essay questions per chapter, with page references, saves you time creating tests. Each multiple-choice item has the question type (factual, applied, or conceptual) indicated. All questions are labeled as new, modified, or pickup, so instructors know whether the question is new to this edition of the test bank, modified but picked up from the previous edition of the test bank, or picked up straight from the previous edition of the test bank.

ExamView® Computerized Testing. Quickly create customized tests that can be delivered in print or online. ExamView's simple "what you see is what you get" interface allows you to easily generate tests of up to 250 items. (Contains all the Test Bank questions electronically.)

DeLamater/Myers's *Social Psychology* Companion Website (http://sociology.wadsworth.com/Delamater6e). On this site, instructors can access password-protected instructor's manuals and important sociology links. Click on the companion website to find useful learning resources for each chapter of the book. These resources include tutorial practice quizzes that can be scored and emailed to the instructor, and much more!

ACKNOWLEDGMENTS

We extend thanks to reviewers for the sixth edition. They include Pamela M. Hunt, Kent State University; David L. Miller, Western Illinois University; Terri L. Orbuch, Oakland University; Brent Simpson, University of South Carolina; Daphne Pedersen Stevens, University of North Dakota; and Ron Wohlstein, Eastern Illinois University.

Throughout the writing of the various editions of this book, many colleagues have reviewed chapters and provided useful comments and criticisms. We express sincere appreciation to these reviewers of the previous editions: Robert F. Bales, Harvard University; Philip W. Blumstein, University of Washington; Marilyn B. Brewer, University of California at Los Angeles; Peter Burke, University of California at Riverside; Brad Bushman, Iowa State University; Peter L. Callero, Western Oregon State College; Bella DePaulo, University of Virginia; Donna Eder, Indiana University; Nancy Eisenberg, Arizona State University; Glen Elder, Jr., University of North Carolina at Chapel Hill; Gregory Elliott, Brown University; Richard B. Felson, State University of New York–Albany; John H. Fleming, University of Minnesota; Jeremy Freese, University of Wisconsin–Madison; Irene Hanson Frieze, University of Pittsburgh; Jim Fultz, Northern Illinois University; Viktor Gecas, Washington State University; Russell G. Geen, University of Missouri; Christine Grella, University of California at Los Angeles; Allen Grimshaw, Indiana University; Elaine Hatfield, University of Hawaii–Manoa; John Hewitt, University of Massachusetts; George Homans, Harvard University; Judy Howard, University of Washington; Michael Inbar, Hebrew University of Jerusalem; Julia Jacks, University of North Carolina; Dale Jaffe, University of Wisconsin–Milwaukee; Edward Jones, Princeton University; Lewis Killian, University of Massachusetts; Melvin Kohn, National Institute of Mental Health and Johns Hopkins University; Robert Krauss, Columbia University; Marianne LaFrance, Boston College; Robert H. Lee, University of Wisconsin–Madison; David Lundgren, University of Cincinnati; Steven Lybrand, University of Wisconsin–Madison; Patricia MacCorquodale, University of Arizona; Armand Mauss, Washington State University; Douglas Maynard, University of Wisconsin–Madison; William McBroom, University of Montana; John McCarthy, Catholic University of America; Kathleen McKinney, Illinois State University; Clark McPhail, University of Illinois; Norman Miller, University of Southern California; Howard Nixon II, University of Vermont; Pamela Oliver, University of Wisconsin–Madison; Edgar O'Neal, Tulane University; James Orcutt, Florida State University; Daphna Oyserman, University of Michigan; Daniel Perlman, University of Manitoba; Jane Allyn Piliavin, University of Wisconsin–Madison; Michael Ross, University of Waterloo, Ontario; David A. Schroeder, University of Arkansas; Melvin Seeman, University of California at Los Angeles; Diane Shinberg, University of Memphis; Roberta Simmons, University of Minnesota; Lynnell Simonson, University of North Dakota; Douglas Clayton Smith, Western Kentucky University; Lawrence Sneden, California State University–Northridge; Sheldon Stryker, Indiana University; Robert Suchner, Northern Illinois University; James Tedeschi, State University of New York–Albany; Richard Tessler, University of Massachusetts; Elizabeth Thomson, University of Wisconsin–Madison; Henry Walker, Cornell University; Nancy Wisely, Stephen F. Austin State University; Steve Wray, Averett College; Mark P. Zanna, University of Waterloo, Ontario; Morris Zelditch, Jr., Stanford University; and Louis Zurcher, University of Texas.

We also thank the many students who used the previous editions and who provided us with feedback about the book; we applied this feedback to improve the presentation, pace, and style of the new edition.

Finally, we express thanks to the professionals who contributed to the process of turning the manuscript into a book. Bob Jucha, Senior Sociology Editor, managed the book at Wadsworth/Thomson. Katherine Bishop, production coordinator at G&S

Book Services, oversaw the transformation of manuscript into printed pages. Kathleen Olson worked diligently to find illustrative photographs. Our appreciation to them all. Although this book benefited greatly from feedback and criticisms, the authors accept responsibility for any errors that may remain.

1

INTRODUCTION TO SOCIAL PSYCHOLOGY

Introduction

What Is Social Psychology?

A Formal Definition

Core Concerns of Social Psychology

Relation to Other Fields

Theoretical Perspectives in Social Psychology

Role Theory

Reinforcement Theory

Cognitive Theory

Symbolic Interaction Theory

Evolutionary Theory

A Comparison of Perspectives

Is Social Psychology a Science?

Characteristics of Science

Social Psychology as a Science

Summary

List of Key Terms and Concepts

INTRODUCTION

- Why are some people effective leaders and others not?
- What makes people fall in love? What makes them fall out of love?
- Why can people cooperate so easily in some situations but not in others?
- What effects do major life events like getting married, having a child, or losing a job have on physical health, mental health, and self-esteem?
- What causes conflict between groups? Why do some conflicts go past the point where participants can expect to achieve any real gains?
- Why do some people conform to norms and laws and others don't?
- Why do people present different images of themselves in various social situations? What determines the particular images they present?
- What causes harmful or aggressive behavior?
- What causes helpful or altruistic behavior?
- Why are some groups so much better at performing their tasks than others?
- Why are some people more persuasive and influential than others? How do they do it?
- Why do stereotypes of out-groups persist even in the face of information that obviously contradicts them?

Perhaps questions such as these have puzzled you, just as they have perplexed others through the ages. You might wonder about these issues simply because you want to better understand the social world around you. Or you might want answers for practical reasons, such as increasing your effectiveness in day-to-day relations with others.

Answers to questions such as these come from various sources. One such source is personal experience—things we learn from everyday interaction. Answers obtained by this means are often insightful, but they are usually limited in scope and generality, and occasionally they are even misleading. Another source is informal knowledge or advice from others who describe their own experiences to us. Answers obtained by this means are sometimes reliable, sometimes not. A third source is the conclusions reached by philosophers, novelists, poets, and men and women of practical affairs who, over the centuries, have written about these issues. Often their answers have filtered down and become commonsense knowledge. We are told, for instance, that punishment is essential to successful child rearing ("Spare the rod and spoil the child") and that joint effort is an effective way to accomplish large jobs ("Many hands make light work"). These principles reflect certain truths and provide guidelines for action in some cases.

Although commonsense knowledge may have merit, it also has drawbacks, not the least of which is that it often contradicts itself. For example, we hear that people who are similar will like one another ("Birds of a feather flock together") but also that persons who are dissimilar will like each other ("Opposites attract"). We are told that groups are wiser and smarter than individuals ("Two heads are better than one") but also that group work inevitably produces poor results ("Too many cooks spoil the broth"). Each of these contradictory statements may hold true under particular conditions, but without a clear statement of when they apply and when they do not, aphorisms provide little insight into relations among people. They provide even less guidance in situations where we must make decisions. For example, when facing a choice that entails risk, which guideline should we use—"Nothing ventured, nothing gained" or "Better safe than sorry"?

If sources such as personal experience and commonsense knowledge have only limited value, how are we to attain an understanding of social interactions and relations among people? Are we forever restricted to intuition and speculation, or is there a better alternative?

One solution to this problem—the one pursued by social psychologists—is to obtain accurate knowledge about social behavior by applying the methods of science. That is, by making systematic observations of behavior and formulating theories that are subject to testing and potential disconfirmation, we can attain a valid and comprehensive understanding of human social relations.

One goal of this book is to present some of the major findings from systematic research by social psychologists. In this chapter, we lay the foundation for this effort by addressing the following questions:

1. What exactly is social psychology? What are the core concerns of the field of social psychology?
2. What are the broad theoretical perspectives that prevail in social psychology today? What are the strengths and weaknesses of each theory?
3. Is social psychology a science? That is, does social psychology have those properties that are the hallmarks of any scientific field?

WHAT IS SOCIAL PSYCHOLOGY?

There are many ways to answer the question "What is social psychology?" We will offer a formal definition of the field; list in detail the topics investigated by social psychologists; and compare and contrast social psychology with other fields of study.

A Formal Definition

We define **social psychology** as the systematic study of the nature and causes of human social behavior. Note certain features of this definition. First, it states that the main concern of social psychology is human social behavior. This includes many things—the activities of individuals in the presence of others, the processes of social interaction between two or more persons, and the relationships between individuals and the groups to which they belong.

Second, the definition states that social psychology addresses not only the nature of social behavior but also the causes of such behavior. Social psychologists seek to discover the preconditions that cause various social behaviors. Causal relations among variables are important building blocks of theory; and in turn, theory is crucial for the prediction and control of social behavior.

Third, the definition indicates that social psychologists study social behavior in a systematic fashion. They rely on formal research methodologies, including experimentation, structured observation, and sample surveys. A description of the research methods used by social psychologists appears in Chapter 2.

Core Concerns of Social Psychology

Another way to answer the question "What is social psychology?" is to describe the topics that social psychologists actually study. Social psychologists investigate human behavior, of course, but their primary concern is human behavior in a social context. There are four core concerns, or major themes, within social psychology: (1) the impact that one individual has on another; (2) the impact that a group has on its individual members; (3) the impact that individual members have on the groups to which they belong; and (4) the impact that one group has on another group. The four core concerns are shown schematically in Figure 1.1.

Impact of Individuals on Individuals Individuals are affected by others in many ways. In everyday life, communication from others may significantly influence a person's understanding of the social world. Attempts by others at persuasion may change an individual's beliefs about the world and his or her attitudes toward persons, groups, or other objects. Suppose, for example, that Carol tries to persuade Debbie that all nuclear power plants are dangerous and undesirable and, therefore, should be closed. If successful, Carol's persuasion attempt could change Debbie's beliefs and perhaps affect her future actions (picketing nuclear power plants, advocating nonnuclear sources of power, and the like).

Beyond influence and persuasion, the outcomes obtained by individuals in everyday life are often affected by the actions of others. A person caught in an emergency situation, for instance, may be helped by an altruistic bystander. In another situation, one person may be wounded by another's aggressive acts. Social psychologists have investigated the nature and origins of both altruism and aggression, as well as other interpersonal activity such as cooperation and competition.

Also relevant here are various interpersonal sentiments. One individual may develop strong attitudes toward another (liking, disliking, loving, hating) based on who the other is and what he or she does.

FIGURE 1.1 THE CORE CONCERNS OF SOCIAL PSYCHOLOGY

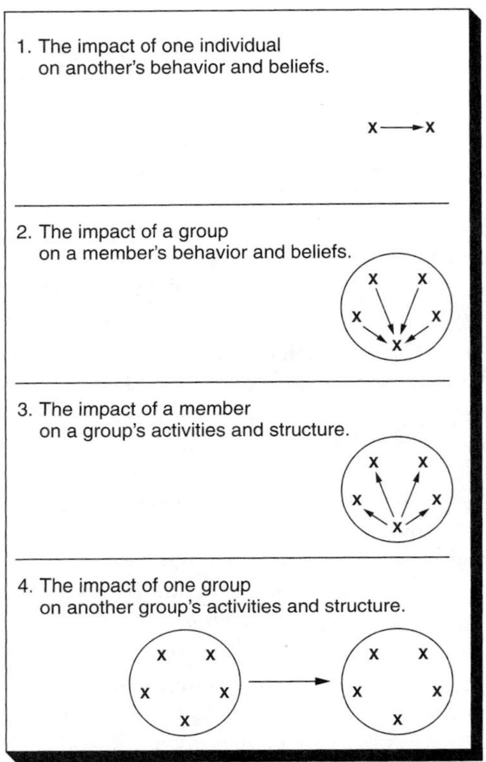

Social psychologists investigate these issues to discover why individuals develop positive attitudes toward some but negative attitudes toward others.

Impact of Groups on Individuals A second concern of social psychology is the impact of a group on the behavior of its individual members. Because individuals belong to many different groups—families, work groups, seminars, and clubs—they spend many hours each week interacting with group members. Groups influence and regulate the behavior of their members, typically by establishing norms or rules. One result of this is conformity, the process by which a group member adjusts his or her behavior to bring it into line with group norms. For example, college fraternities and sororities have norms—some formal and some informal—that stipulate how members should dress, what meetings they should attend, who they can date and who they should avoid, and how they should behave at parties.

Groups also exert substantial long-term influence on their members through socialization, a process that enables groups to regulate what their members learn. Socialization assumes that the members will be adequately trained to play roles in the group and in the larger society. It shapes the knowledge, values, and skills of group members. One outcome of socialization is language skills; another is political and religious beliefs and attitudes; yet another is our conception of self.

Impact of Individuals on Group A third concern of social psychology is the impact of individuals on group processes and products. Just as any group influences the behavior of its members, these members, in turn, may influence the group itself. For instance, individuals contribute to group productivity and group decision making. Moreover, some members may provide leadership, performing functions such as planning, organizing, and controlling, necessary for successful group performance. Without effective leadership, coordination among members will falter and the group will drift or fail. Furthermore, individuals and minority coalitions often innovate change in group structure and procedures. Both leadership and innovation depend on the initiative, insight, and risk-taking ability of individuals.

Impact of Groups on Groups A fourth concern of social psychology is the impact of one group on the activities and structure of another group. Relations between two groups may be friendly or hostile, cooperative or competitive. These relationships, which are based in part on members' identities and may entail group stereotypes, can affect the structure and activities of each group. Of special interest is intergroup conflict, with its accompanying tension and hostility. Violence may flare up, for instance, between two teenage street gangs disputing territorial rights or between racial groups competing for scarce jobs. Conflicts of this type affect the interpersonal relations between groups and within each group. Social psychologists have long studied the emergence, persistence, and resolution of intergroup conflict.

Relation to Other Fields

Social psychology bears a close relationship to several other fields, especially sociology and psychology.

Sociology is the scientific study of human society. It examines social institutions (family, religion, and politics), stratification within society (class structure, race and ethnicity, gender roles), basic social processes (socialization, deviance, social control), and the structure of social units (groups, networks, formal organizations, bureaucracies).

In contrast, psychology is the scientific study of the individual and of individual behavior. Although this behavior may be social in character, it need not be. Psychology addresses such topics as human learning, perception, memory, intelligence, emotion, motivation, and personality.

Social psychology bridges sociology and psychology. In fact, some view it as an interdisciplinary field. Both sociologists and psychologists have contributed to social psychological knowledge. Social psychologists working in the sociological tradition rely primarily on sample surveys and observational techniques to gather data. These investigators are most interested in the relationship between individuals and the groups to which they belong. They emphasize such processes as socialization, conformity and deviation, social interaction, self-presentation, leadership, recruitment to membership, and cooperation and competition. Social psychologists working in the psychological tradition rely heavily on laboratory experimental methodology. Their primary concern is how an individual's behavior and internal states are affected by social stimuli (often other persons). They emphasize such topics as the self, person perception and attribution, attitudes and attitude change, personality differences in social behavior, social learning and modeling, altruism and aggression, and interpersonal attraction.

Thus, sociologically oriented and psychologically oriented social psychologists differ in their outlook and emphasis. As we might expect, this leads them to formulate different theories and to conduct different programs of research. Yet these differences are best viewed as combplementary rather than as conflicting. Social psychology as a field is the richer for the differing contributions of both approaches.

THEORETICAL PERSPECTIVES IN SOCIAL PSYCHOLOGY

Yesterday at work, Warren reported to his boss that he would not be able to complete an important project on schedule. To Warren's surprise, the boss became enraged and told him to complete the task by the following Monday—or else! Warren was not entirely sure what to make of this behavior but he decided to take the threat seriously.

That evening, talking with his girlfriend Alice, Warren announced that he would have to work overtime at the office, so he could not take her to a party on Friday evening as originally planned. Alice immediately got mad at Warren—she definitely wanted to go, and he had promised several times to take her—and threw a paperweight at him. By now, Warren was distressed and a little perplexed.

Reflecting on these two events, Warren noticed that they had some characteristics in common. To explain the behavior of his boss and his girlfriend, he formed a general proposition: "If you fail to deliver on promises and thereby block someone's goals, he or she will get mad at you." He was happy with this simple formulation until the next day when he read an unusual newspaper story: "Man fired from job, then shoots his dog in anger." Warren thought about this event and concluded that his original theory needed some revision. His new version included a chain of propositions: "If someone's goals are blocked, he or she will become frustrated. If someone is frustrated, he or she will become aggressive. If someone is aggressive, he or she will attack either the source of the frustration or a convenient surrogate."

In his own way, Warren had started to informally do the same thing that social psychologists do more elaborately and systematically. Starting from some observations regarding social behavior, Warren

BOX 1.1 Milestones in the History of Social Psychology

Social psychology can trace its historical roots to the philosophers of ancient Greece and even more firmly to the psychological and social theorists of 19th-century Europe. However, the discipline as we know it today is largely a product of efforts by 20th-century researchers and theorists, many of them American (Allport, 1985; Jones, 1985).

The following list summarizes some important milestones in social psychology's development from the beginning of the 20th century up to 1960.

- 1898 Norman Triplett publishes the first social psychological experiment ("The dynamogenic factors in pacemaking and competition"). It investigates social facilitation, a process whereby a person's performance on a familiar task improves in the presence of others performing the same task.
- 1902 Charles Horton Cooley publishes an influential book, *Human Nature and the Social Order,* which presents the idea that the self and society are ultimately the same thing, although viewed at different levels of abstraction.
- 1908 William McDougall and E. H. Ross independently publish the first textbooks in the field. Although both books are titled *Social Psychology,* they differ greatly in content. The book by McDougall (a psychologist) stresses the importance of instincts and innate drives in determining behavior; the book by Ross (a sociologist) discusses groups, crowds, and crazes and emphasizes interpersonal processes such as suggestion and imitation.
- 1918 W. I. Thomas and F. Znaniecki begin their field study of attitudes in immigrant populations (Polish peasants) in Chicago.
- 1922 Morton Prince establishes the first major social psychology journal, the *Journal of Abnormal and Social Psychology* (which in 1965 becomes the *Journal of Personality and Social Psychology*).
- 1924 Floyd Allport, writing from a stimulus-response (behaviorist) perspective, publishes a social psychology textbook that is one of the first systematic treatments of the field. This book advocates the use of the experimental method in social psychology and sets forth a research agenda for the next decade.
- 1928 L. L. Thurstone publishes a path-breaking paper showing how attitudes can be measured.
- 1934 George Herbert Mead, a symbolic interactionist, publishes his seminal work on the self.
- 1934 J. L. Moreno develops sociometry, a system for measuring patterns of social interaction based on individuals' choices regarding whom they would prefer to associate with.
- 1934 Richard T. LaPiere investigates inconsistencies between attitudes (racial prejudice) and related behaviors (discrimination) in a field setting.
- 1936 Muzafer Sherif, by creating social norms in a controlled setting, demonstrates that complex and realistic social situations can be studied experimentally in a laboratory.
- 1936 George Gallup develops methods for conducting public opinion polls and surveys.
- 1937 J. L. Moreno founds *Sociometry,* a journal devoted to research on structure

BOX 1.1 Continued

and process in groups and networks. (The journal is later renamed *Social Psychology Quarterly*.)

- 1939 Kurt Lewin, Ronald Lippitt, and Ralph White, using Lewin's field theory, study group members' reactions to various styles of leadership (autocratic, democratic, and laissez-faire).
- 1943 Theodore Newcomb investigates the effects of social pressures on attitudes held by students at Bennington College.
- 1943 William Foote White uses the technique of participant observation to study the activities of teenage street gangs.
- 1946 Solomon Asch demonstrates that cognitive set can influence the impressions people form of others.
- 1950 Robert Freed Bales develops a categorical framework for systematically observing communication and role differentiation in task groups.
- 1950 George Homans publishes *The Human Group,* a seminal theoretical treatise on group structure and process.
- 1951 Solomon Asch demonstrates some conditions under which individuals in a group will conform to the position of a majority when their beliefs are questioned.
- 1953 Carl Hovland and coworkers at Yale University publish the results of a programmatic study of persuasion and attitude change.
- 1954 Gordon Allport publishes *The Nature of Prejudice,* an important analysis of intergroup prejudice and stereotyping.
- 1957 Leon Festinger proposes the theory of cognitive dissonance, an approach to attitude change based on the idea that people strive for consistency between behavior and attitudes.
- 1958 Fritz Heider publishes *The Psychology of Interpersonal Relations,* which lays the foundation for attribution theory and research.
- 1959 John Thibaut and Harold Kelley publish *The Social Psychology of Groups,* a general theory of social exchange and interpersonal relations.
- 1975 Edward O. Wilson publishes *Sociobiology: The New Synthesis,* which marks a new wave of theorizing about the role of evolution in social and psychological behavior.

After 1960, social psychology continued to expand rapidly and develop as a field. The 1960s saw a large number of laboratory studies of cognitive dissonance and an increased concern with such phenomena as altruism, aggression, and interpersonal attraction. The 1970s witnessed the growth of attribution theory and expanding interest in interpersonal relations. The 1980s saw a renewed concern with the social self and greater emphasis on the cognitive aspects of social behavior. The field grew rapidly, and many new social psychology journals began publication.

Since 1990, social psychologists have continued to investigate social cognition, emotions and the self, language and communication, interpersonal relationships, group performance and decision making, prejudice and intergroup relations, and numerous other topics. In a fundamental sense, the chapters of this book constitute a summary and distillation of the central concerns of social psychology as the field stands today.

attempted to formulate a theory to explain the observed facts. As the term is used here, a **theory** is a set of interrelated propositions that organizes and explains a set of observed phenomena. Theories usually pertain not just to some particular event but to whole classes of events. Moreover, as Warren's example indicates, a theory goes beyond mere observable facts by postulating causal relations among variables. If a theory is valid, it enables its user to explain the phenomena under consideration and to make predictions about events not yet observed.

In social psychology, no single theory explains all phenomena of interest; rather, the field includes many different theories. It is useful to distinguish between middle-range theories and theoretical perspectives. **Middle-range theories** are narrow, focused frameworks that identify the conditions that produce a specific social behavior. They are usually scientific-causal in nature; that is, they are formulated in terms of cause and effect. For example, one middle-range theory tries to explain the processes by which persuasion produces attitude change (Petty & Cacioppo, 1986a, 1986b). Another middle-range theory tries to specify how majorities and minorities within groups differ qualitatively in the ways they influence their targets (Moscovici, 1985a; Nemeth, 1986). Yet another middle-range theory specifies the conditions under which contact between members of different racial and ethnic groups will cause stereotypes to change or disappear (Rothbart & John, 1985). Throughout this book, we will describe many middle-range theories.

In addition to middle-range theories, social psychology includes **theoretical perspectives.** Broader in scope than middle-range theories, theoretical perspectives offer general explanations for a wide array of social behaviors in a variety of situations. These general explanations are rooted in explicit assumptions about human nature. Theoretical perspectives serve an important function for the field of social psychology. By making certain assumptions regarding human nature, a theoretical perspective establishes a vantage point from which we can examine a range of social behaviors. Because any perspective highlights certain features and downplays others, it enables us to more clearly "see" certain aspects or features of social behavior. The fundamental value of any theoretical perspective lies in its applicability across many situations; it provides a frame of reference for interpreting and comparing a wide range of social situations and behaviors.

Social psychology can be organized into a number of distinct theoretical perspectives. The five central perspectives that will guide this textbook are (1) role theory, (2) reinforcement theory, (3) cognitive theory, (4) symbolic interaction theory, and (5) evolutionary theory.

Role Theory

Several months ago, Barbara was invited to participate in a stage production of Molière's comedy *The Learned Women*. She was offered the role of Martine, a kitchen servant dismissed from her job for using poor grammar. Barbara enthusiastically accepted the role and learned her part well. The theater group was scheduled to present the play six times over a period of 3 weeks. Barbara played the role of Martine in the first four shows, but then she got sick. Fortunately, Barbara's understudy was able to substitute as Martine during the final two shows. Barbara's performance was very good, but so was the understudy's. In fact, one reviewer wrote that it was difficult to tell them apart.

Barbara's friend Craig is more interested in football than in theater. Craig plays fullback on the college football team. Although very large and strong, he is a third-string player because he has the unfortunate habit of fumbling the ball, sometimes at the worst possible moment. But Craig believes that with another year's experience and some improvements in his technique, he could perform better than the other fullbacks and win a place in the starting lineup.

Although active in different arenas, Craig and Barbara have something in common: They are both performing roles. When Barbara appears on stage, she performs the role of kitchen servant. When Craig appears on the football field, he performs the role of fullback. In both cases, their behavior is guided by role expectations held by other people. Roles consist of a set of rules (that is, expectations held by others) that function as plans or blueprints and guide behavior.

To run smoothly, a production like this newscast requires that all the participants perform tasks and enact roles specified by their work group.

Barbara's role is defined by very specific expectations. Her part calls for her to say certain lines and perform certain actions at specified points in the plot. There is virtually no room for her to improvise or deviate from her lines. Craig's role is also fairly specific. He has to carry out given assignments—running and blocking—on each of the plays by his team. There is some latitude in exactly how he does these things, but not a great deal. Whenever he misses a block, all the coaches and players know it.

In everyday life, we all perform roles. Anyone who holds a job is performing a role. For instance, unlike Barbara's playacting role as a kitchen servant, an advertising executive's work role does not dictate exactly what lines are to be spoken. But it will certainly specify what goals should be pursued, what tasks must be accomplished, and what performances are required.

The theoretical perspective that best addresses behavior of this type is **role theory** (Biddle, 1979, 1986; Heiss, 1981; Turner, 1990). Role theory holds that a substantial proportion of observable, day-to-day social behavior is simply persons carrying out their roles, much as actors carry out their roles on the stage or ballplayers perform theirs on the field.

Propositions in Role Theory The following propositions are central to the role theory perspective:

1. People spend much of their lives participating as members of groups and organizations.
2. Within these groups, people occupy distinct positions (fullback, advertising executive, police sergeant, and the like).
3. Each of these positions entails a **role,** which is a set of functions performed by the person for the

group. A person's role is defined by expectations (held by other group members) that specify how he or she should perform.

4. Groups often formalize these expectations as **norms,** which are rules specifying how a person should behave, what rewards will result for performance, and what punishments will result for nonperformance.
5. Individuals usually carry out their roles and perform in accordance with the prevailing norms. In other words, people are primarily conformists; they try to meet the expectations held by others.
6. Group members check each individual's performance to determine whether it conforms to the group's norms. If an individual meets the role expectations held by others, then he or she will receive rewards in some form (acceptance, approval, money, and so on). If he or she fails to perform as expected, however, then group members may embarrass, punish, or even expel that individual from the group. The anticipation that others will apply sanctions ensures performance as expected.

Impact of Roles Role theory implies that if we (as analysts) have information about the role expectations for a specified position, we can then predict a significant portion of the behavior of the person occupying that position. According to role theory, to change a person's behavior, it is necessary to change or redefine his or her role. This might be done by changing the role expectations held by others with respect to that person or by shifting that person into an entirely different role (Allen & Van de Vliert, 1982). For example, if the football coach shifted Craig from fullback to tight end, Craig's behavior would change to match the role demands of his new position. Craig himself may experience some strain while adjusting to the new role, but his behavior will change.

Role theory maintains that a person's role determines not only behavior but also beliefs and attitudes. In other words, individuals bring their attitudes into congruence with the expectations that define their roles. A change in role should lead to a change in attitude. One illustration of this effect appears in a classic study of factory workers by Lieberman (1965). In the initial stage of this study, researchers measured the attitudes of workers toward union and management policies in a midwestern home appliance factory. During the following year, a number of these workers changed roles. Some were promoted to the position of foreman, a managerial role; others were elected to the position of shop steward, a union role.

About a year after the initial measurement, the workers' attitudes were reassessed. The attitudes of workers who had become foremen or shop stewards were compared to those of workers who had not changed roles. The recently promoted foremen expressed more positive attitudes than the nonchangers toward the company's management and the company's incentive system, which paid workers in proportion to what they produced. In contrast, the recently elected shop stewards expressed more positive attitudes than the nonchangers toward the union and favored an incentive system based on seniority, not productivity. The most efficient explanation of these results is that the workers' attitudes shifted to fit their new roles, as predicted by role theory.

In general, the roles that people occupy not only channel their behavior but also shape their attitudes. Roles can influence the values that people hold and affect the direction of their personal growth and development. We will discuss these topics in more depth in Chapters 3, 15, and 18.

Limitations of Role Theory Despite its usefulness, role theory has difficulty explaining certain kinds of social behavior. Foremost among these is deviant behavior, which is any behavior that violates or contravenes the norms defining a given role. Most forms of deviant behavior, whether simply refusing to perform as expected or something more serious like committing a crime, disrupt interpersonal relations. Deviant behavior poses a challenge to role theory because it contradicts the assumption that people are essentially conformist—deviant behavior violates the demands of roles. Of course, a certain amount of deviant behavior can be explained by the fact that people are sometimes ignorant of the norms. Deviance may also result whenever people face conflicting or incompatible expectations from several other people (Miles, 1977). In general, however, deviant behavior is an unexplained and problematic exception from the standpoint of role theory. In Chapters 14 and 19, we discuss the conditions that cause deviant behavior and the reactions of others to such behavior.

Even critics of role theory acknowledge that a substantial portion of all social behavior can be explained as conformity to established role expectations. But role theory does not and cannot explain how role expectations came to be what they are in the first place. Nor does it explain when and how role expectations change. Thus, role theory provides only a partial explanation of social behavior.

Reinforcement Theory

Reinforcement theory, another major perspective on social behavior, begins with the premise that social behavior is governed by external events. Its central proposition is that people will be more likely to perform a specific behavior if it is followed directly by the occurrence of something pleasurable or by the removal of something aversive; likewise, people will more likely refrain from performing a particular behavior if it is followed by the occurrence of something aversive or by the removal of something pleasant.

The use of **reinforcement** is illustrated in an early study by Verplanck (1955). The study's point was to show that one person can alter the course of a conversation by the selective use of social approval (a reinforcer). Students conducting the study sought out situations in which each could be alone with another person and conduct a conversation. During the first 10 minutes, the student engaged the other in polite but neutral chitchat; the student was careful neither to support nor to reject the opinions expressed by the other. During this period, the student privately noted the number of opinions expressed by the other and unobtrusively recorded this information by doodling on a piece of paper.

After this initial period, the student shifted behavior and began to express approval whenever the other ventured an opinion. The student indicated approval with reinforcers like "I agree," "That's so," and "You're right," and by smiling and nodding in agreement. The student continued this pattern of reinforcement for 10 minutes, all the while noting the number of opinions expressed by the other.

Next, the student shifted behavior again and suspended reinforcement. Any opinions expressed by the other were met with noncommittal remarks or subtle disagreement. As before, the student noted the number of opinions expressed.

The results of the study showed that during the reward period (when the student expressed approval), their partners expressed opinions at a higher rate than they had during the initial baseline period. Moreover, during the extinction period (when the student suspended approval), about 90 percent expressed opinions at a lower rate than they had during the reward period. Overall, the partners' behavior during the conversation was substantially influenced by social approval.

Some Concepts of Reinforcement Theory Reinforcement theory has a long tradition within psychology. It began at the turn of the century with research by Pavlov and by Thorndike, and evolved through the work of Allport (1924), Hull (1943), and Skinner (1953, 1971). The reinforcement perspective holds that behavior is determined primarily by external events, not by internal states. Thus, the central concepts of reinforcement theory refer to events that are directly observable. Any event that leads to an alteration or change in behavior is called a stimulus. For example, a traffic light that changes to red is a stimulus, as is a wailing tornado siren. The change in behavior induced by a stimulus is called a response. Drivers respond to red lights by stopping; families respond to tornado sirens by rushing for shelter. A reinforcement is any favorable outcome that results from a response; reinforcement strengthens the response—that is, it increases the probability it will be repeated. In Verplanck's (1955) study, the students' social approval was a positive reinforcer that strengthened the partners' response of expressing opinions. Responses that are not reinforced tend to disappear and not be repeated.

Reinforcement is important in some forms of learning, most notably through **conditioning** (Mazur, 1998). In conditioning, a contingency is established between emitting a response and subsequently receiving a reinforcement. If a person emits a particular response and this response is then reinforced, the connection between response and reinforcement is strengthened; that is, the person will more likely emit the same response in the future in hopes of again receiving reinforcement.

A related process, stimulus discrimination, occurs when a person learns the exact conditions under which a response will be reinforced. For example, Karl, a young child, has learned that if his mother rings the dinner bell (a stimulus), he should respond by coming indoors, washing his hands, and sitting in the appropriate place at the table. His mother then puts food on his plate (a reinforcer). He has also learned, however, that if he performs the same response (washing his hands and sitting down at the table) without first hearing the stimulus (dinner bell), his mother merely tells him that he's too early and cannot have food until later. Thus, Karl has learned to discriminate between stimulus conditions (bell vs. no bell), and he knows that reinforcement (food) is obtained only by making the response in the presence of a specific stimulus (bell).

Social Learning Theory Although learning based on reinforcement and conditioning is important, it is not the only form of learning. A central proposition of **social learning theory** (Bandura, 1977) is that one person (the learner) can acquire new responses simply by observing the behavior of another person (the model). This observational learning process, called **imitation,** is distinguished by the fact that the learner neither performs a response nor receives any reinforcement. Many social responses are learned through imitation. For instance, children learn ethnic and regional speech patterns by imitating adult speakers around them.

In imitation, the learner watches the model's behavior and thereby comes to understand how to behave in a similar manner. Learning of this type can occur without any external reinforcement. But the issue of whether the learner will actually perform the behaviors learned through observation may hinge on the consequences that performance has for the learner—that is, on whether the learner receives reinforcement for the performance. A young girl, for example, might observe that her older sister puts on makeup before going out with friends; in fact, if she watches closely enough, she might learn precisely how to apply makeup the right way. But whether the little girl actually puts makeup on herself and wears it around the house may depend heavily on any reinforcements she receives for doing so. If she knows, for instance, that her mother strongly disapproves of little girls wearing cosmetics, she may hesitate to use what she has learned from her big sister.

In sum, social learning theory holds that individuals acquire new responses through conditioning and imitation. Both conditioning and imitation are important processes in socialization, and they help to explain how persons acquire complex social behaviors. Chapter 3 will discuss these processes in more detail.

Social Exchange Theory Another important process based on the principle of reinforcement is social exchange. **Social exchange theory** (Cook, 1987; Homans, 1974; Kelley & Thibaut, 1978) uses the concept of reinforcement to explain stability and change in relations between individuals. This theory assumes that individuals have freedom of choice and often face social situations in which they must choose among alternative actions. Any action provides some rewards and entails some costs. There are many kinds of socially mediated rewards—money, goods, services, prestige or status, approval by others, and the like. The theory posits that individuals are hedonistic—they try to maximize rewards and minimize costs. Consequently, they choose actions that produce good profits (profits = rewards − costs) and avoid actions that produce poor profits.

As its name indicates, social exchange theory views social relationships primarily as exchanges of goods and services among persons. People participate in relationships only if they find that they provide profitable outcomes. An individual judges the attractiveness of a relationship by comparing the profits it provides against the profits available in other, alternative relationships. If a person is participating in a social relationship and receiving certain outcomes, then the level of outcomes available in the best alternative relationship is termed that person's comparison level for alternatives (Thibaut & Kelley, 1959). More concretely, suppose an executive is employed by a food products manufacturer when unexpectedly she is offered an attractive job by a competing firm. The new job entails some additional responsibilities, but it also pays a considerably higher salary and provides

An exchange taking place—a scalper offers tickets to a football fan for a sold-out game. In transactions of this type, the price is often determined through negotiation between buyer and seller.

more benefits. This job offer has the effect of substantially increasing the executive's comparison level for alternatives. In this case, exchange theory predicts that she will leave her job for the new one or possibly use the outside offer as a bargaining chip in dealing with her current employer. She likely will stay with her current employer only if he promotes her to a new position with greater rewards.

Concepts of this type apply not only to work relations but also to personal relations. For instance, studies of heterosexual couples in long-term dating relationships show that rewards and costs can explain whether persons stay in or exit from such relationships (Rusbult, 1983; Rusbult, Johnson, & Morrow, 1986). The results of these studies indicate that individuals are more likely to stay when the partner is physically and personally attractive, when the relationship does not entail undue hassle (such as high monetary costs, broken promises, arguments), and when romantic involvements with attractive outsiders are not readily available. In other words, they are more likely to stay when the rewards are high, the costs are low, and the alternatives are unpromising. Effects of this type are predicted by social exchange theory.

Exchange theory also predicts the conditions under which people try to change or restructure their relationships. A central concept involved is **equity** (Adams, 1963; Walster [Hatfield], Walster, & Berscheid, 1978). A state of equity exists in a relationship when participants feel that the rewards they receive are proportional to the costs they bear. For example, a supervisor may earn more money than a line worker and receive better benefits on the job. But the line worker may nevertheless feel the relationship is equitable because the supervisor bears more responsibility and has a higher level of education.

If, for some reason, a participant feels that the allocation of rewards and costs in a relationship is inequitable, then the relationship is potentially unstable. People find inequity difficult to tolerate—they may feel cheated or exploited and become angry. Social exchange theory predicts that people will try to modify an inequitable relationship. Most likely, they will attempt to reallocate costs and rewards so that equity is established.

Limitations of Reinforcement Theory Despite its usefulness in illuminating why relationships change and how people learn, reinforcement theory has been criticized on various grounds. One criticism is that reinforcement theory portrays individuals primarily as reacting to environmental stimuli rather than as initiating behavior based on imaginative or creative thought. The theory does not account easily for creativity, innovation, or invention. A second criticism is that reinforcement theory largely ignores or downplays other motivations. It characterizes social behavior as hedonistic, with individuals striving to maximize profits from outcomes. Thus it cannot easily explain selfless behavior such as altruism and martyrdom. Despite its limitations, reinforcement theory has enjoyed substantial success in explaining why individuals persist in emitting certain behaviors, how they learn new behaviors, and how they influence the behavior of others through exchange. Ideas from reinforcement and exchange theory are discussed throughout this book, especially in Chapters 3, 8, 11, 12, 13, and 15.

Cognitive Theory

Another theoretical perspective in social psychology is **cognitive theory,** the basic premise of which is that the mental activities of the individual are important determinants of social behavior. These mental activities, called **cognitive processes,** include perception, memory, judgment, problem solving, and decision making. Cognitive theory does not deny the importance of external stimuli, but it maintains that the link between stimulus and response is not mechanical or automatic. Rather, the individual's cognitive processes intervene between external stimuli and behavioral responses. Individuals not only actively interpret the meaning of stimuli but also select the actions to be made in response.

Historically, the cognitive approach to social psychology has been influenced by the ideas of Koffka, Kohler, and other theorists in the gestalt movement of psychology. Central to gestalt psychology is the principle that people respond to configurations of stimuli rather than to a single, discrete stimulus. In other words, people understand the meaning of a stimulus only by viewing it in the context of an entire system of elements (the gestalt) in which it is embedded. A chess master, for example, would not assess the importance of a chess piece on the board without considering its location and strategic capabilities vis-à-vis all the other pieces currently on the board. To comprehend the meaning of any element, we must look at the whole of which it is a part.

Modern cognitive theorists (Fiske & Taylor, 1991; Markus & Zajonc, 1985; Wyer & Srull, 1984) depict humans as active in selecting and interpreting stimuli. According to this view, people do more than react to their environment; they actively structure their world cognitively. First, because they cannot possibly attend to all the complex stimuli that surround them, they select only those stimuli that are important or useful to them and ignore the others. Second, they actively control which categories or concepts they use to interpret the stimuli in the environment. One implication, of course, is that several individuals can form dramatically different impressions of the same complex stimulus in the environment.

Consider, for example, what happens when several people view a vacant house displaying a bright "for rent" sign. When a building contractor passes the house, she pays primary attention to the quality of the house's construction. She sees lumber, bricks, shingles, and glass, and some repairs that need to be made. Another person, a potential renter, sees the house very differently. He notes that it is located close to his job and wonders whether the neighborhood is safe and whether the house is expensive to heat in winter. The realtor trying to rent the house construes it in still different terms—cash flow, occupancy rate, depreciation, mortgage, and amortization. One of the preschool kids living in the neighborhood has yet another view; observing that no person has lived in the house for several months, he is convinced the house is haunted.

Cognitive Structure and Schemas Central to the cognitive perspective is the concept of **cognitive structure,** which refers broadly to any form of organization among cognitions (concepts and beliefs). Because a person's cognitions are interrelated, cognitive theory gives special emphasis to exactly how they are structured and organized in memory and to how they affect a person's judgments.

Social psychologists have proposed that individuals use specific cognitive structures called **schemas** to make sense of complex information about other persons, groups, and situations. The term schema is derived from the Greek word for form, and it refers to the form or basic sketch of what we know about people and things. For example, our schema for "law student" might be a set of traits thought to be characteristic of such persons: intelligent, analytic and logical, argumentative (perhaps even combative), thorough and workmanlike with an eagle eye for details, strategically skillful in interpersonal relations, and (occasionally) committed to seeing justice done. This schema, no doubt, reflects our own experience with lawyers and law students, as well as our conception of which traits are necessary for success in the legal profession. That we hold this schema does not mean we believe that everyone with this set of characteristics is a law student or that every law student will have all of these characteristics. We might be surprised, however, if we met someone who impressed us as unmethodical, illogical, withdrawn, inarticulate, inattentive, sloppy, and not very intelligent, and then later discovered that he was a law student.

Schemas are important in social relations because they help us interpret the environment efficiently. Whenever we encounter a person for the first time, we usually form an impression of what he or she is like. In doing this, we not only observe the person's behavior but also rely on our knowledge of similar persons we have met in the past—that is, we use our schema regarding this type of person. Schemas help us process information by enabling us to recognize which personal characteristics are important in the interaction and which are not. They structure and organize information about the person, and they help us remember information better and process it more quickly. Sometimes they fill gaps in knowledge and enable us to make inferences and judgments about others.

To illustrate further, consider a law school admissions officer who faces the task of deciding which candidates to admit as students. To assist in processing applications, she uses a schema for "strong law student candidate" that is based on traits believed to predict success in law school and beyond. The admissions officer doubtless pays close attention to information regarding candidates that is relevant to her schema for law students, and she most likely ignores or downplays other information. LSAT scores do matter, whereas eye color does not; undergraduate GPA does matter, whereas ability to throw a football does not; and so on.

Schemas are rarely perfect as predictive devices, and the admissions officer probably will make mistakes, admitting some candidates who fail to complete law school and turning down some candidates who would have succeeded. Moreover, another admissions officer with a different schema might admit a different set of students to law school. Schemas also figure centrally in our stereotypes and discriminatory attitudes. If, for example, an admissions officer includes only the race "White" in her schema for successful law students, she will be less likely to admit African Americans. Despite their drawbacks, schemas are more efficient ways to process social information than having no systematic framework at all. Thus, they persist as important cognitive mechanisms even when less than perfect. Schemas will be discussed in more detail in Chapter 5.

Cognitive Consistency One way to study cognitive structure is to observe changes that occur in a person's cognitions when they are under challenge or attack. The changes will reveal facts about the underlying structure or organization of his or her cognitions. An important idea emerging from this approach is the **principle of cognitive consistency** (Heider, 1958; Newcomb, 1968), which maintains that individuals strive to hold ideas that are consistent or congruous with one another, rather than ideas that are inconsistent or incongruous. If a person holds several ideas that are incongruous or inconsistent, then he or she will experience internal conflict. In reaction, he or she will likely change one or more ideas, thereby making them consistent and resolving this conflict.

As an illustration, suppose you hold the following cognitions about your friend Jeff: (1) Jeff has been a good friend for 6 years; (2) you dislike hard drugs and the people who use them; and (3) Jeff has recently started using hard drugs. These cognitions are obviously interrelated, and they are also incongruous with one another. The principle of cognitive consistency predicts that a change in cognitions will occur. That is, you will change either your negative attitude toward hard drugs or your positive attitude toward Jeff—or possibly you will intervene and try to change Jeff's behavior.

Social psychologists have developed several useful theories based on the general notion of cognitive consistency. Among these are balance theory and the theory of cognitive dissonance (see Chapter 6).

Cognitive theory has made many important contributions to social psychology. It treats such diverse phenomena as self-concept (see Chapter 4), perception of persons and attribution of causes (see Chapter 5), attitude change (see Chapter 6), impression management (see Chapter 9), and group stereotypes (see Chapters 5 and 16). In these contexts, cognitive theory has produced many insights and striking predictions regarding individual and social behavior.

Limitations of Cognitive Theory One drawback of cognitive theory is that it simplifies—and sometimes oversimplifies—the way in which people process information, an inherently complex phenomenon. Another drawback is that cognitive phenomena are not directly observable; they must be inferred from what people say and do. This means that compelling and definitive tests of theoretical predictions from cognitive theory are sometimes difficult to conduct. Overall, however, the cognitive perspective is among the more popular and productive approaches in social psychology.

Symbolic Interaction Theory

A fourth perspective in social psychology is **symbolic interaction theory** (Charon, 1995; Stryker, 1980, 1987). Like the cognitive perspective, symbolic interactionism stresses cognitive process (thinking and reasoning), but it places much more emphasis on the interaction between the individual and society. The basic premise of symbolic interactionism is that human nature and social order are products of symbolic communication among people. In this perspective, a person's behavior is constructed through give-and-take during his or her interaction with others. Behavior is not merely a response to stimuli, nor is it merely an expression of inner biological drives, profit maximization, or conformity to roles or norms. Rather, a person's behavior emerges continually through communication and interaction with others.

Negotiating Meanings People can communicate successfully with one another only to the extent that they ascribe similar meanings to objects. An object's meaning for a person depends not so much on the properties of the object itself but on what the person might do with the object. In other words, an object takes on meaning only in relation to a person's plans. A wine merchant, for example, might see a glass bottle as a container for her product; an interior decorator might see it as an attractive vase for some silk flowers; a man in a drunken brawl might see it as a weapon with which to hit his opponent.

Symbolic interaction theory views humans as proactive and goal seeking. People formulate plans of action to achieve their goals. Many plans, of course, can be accomplished only through cooperation with other people. To establish cooperation with others, the meanings of things must be shared and consensual. If the meaning of something is unclear or contested, an agreement must be developed through give-and-take before cooperative action is possible. For example, if a man and a woman have begun to date each other and he invites her up to his apartment, exactly what meaning does this proposed visit have? One way or another, they will have to achieve some agreement about the purpose of the visit before joint action is possible. In symbolic interaction terms, they would need to develop a consensual definition of the situation. The man and woman might achieve this through explicit negotiation or through tacit, nonverbal communication. But without some agreement regarding the definition of the situation, the woman may have difficulty in deciding whether to accept the invitation, the man may find himself behaving in an atypically awkward manner, and cooperative action will be difficult.

Symbolic interactionism portrays social interaction as having a tentative, developing quality. To fit their actions together and achieve consensus, people interacting with one another must continually negotiate new meanings or reaffirm old meanings. Each person formulates plans for action, tries them out, and then adjusts them in light of others' responses. Thus, social interaction always has some degree of unpredictability and indeterminacy.

The Self in Relationship to Others Central to social interaction is the process of role taking, in which an individual imagines how he or she looks from the other person's viewpoint. For example, if an employee is seeking an increase in salary, he might first imagine how his boss would react to one type of request or another. To do this, he might use knowledge gained in past interactions with her and recall what he has heard from others about her reactions to salary requests. By viewing his action from her viewpoint, he may be able to anticipate what type of request would produce the desired effect. If he then actually makes a request of this type and she reacts as expected, his role taking has succeeded. Through the role-taking process, cooperative interaction is established.

Symbolic interaction theory emphasizes that a person can act not only toward others but also toward his or her **self.** That is, an individual can engage in self-perception, self-evaluation, and self-control, just as he or she might perceive, evaluate, and control others. One important component of self is identity, the person's understanding of who he or she is. For an interaction among persons to proceed smoothly, there must be some consensus with respect to the identity of each. In other words, for each person there must be an answer to the question "Who am I in this situation, and who are these other people?" Only by answering this question in some detail can each person understand the implications (meanings) that others have for his or her plan of action.

Sometimes a person's identity is very unusual, and in consequence, interaction becomes awkward, difficult, or even impossible. Consider the old tale by Cervantes of a man, temporarily deranged, who thought he was made of glass (Shibutani, 1961). This man's conception of himself created problems both for him and for others. Whenever people came near, he screamed and implored them to keep away for fear they would shatter him. He refused to eat anything hard and insisted on sleeping only in beds of soft straw. Concerned that loose tiles might fall on him from the rooftops, he walked in the middle of the street. When a wasp stung him in the neck, he did not swat it away because he was afraid of smashing himself to bits. Because glass is transparent and skin is not, he claimed that his body's unusual construction enabled his soul to perceive things more clearly than others, and he offered to assist people perplexed by difficult problems. He gradually developed a reputation for astonishing insight, and many people came to him seeking advice. In the end, a wealthy patron hired a bodyguard to protect him from outlaws and mischievous boys who threw stones at him.

In daily life, of course, we are not likely to meet someone who believes he or she is made of glass. But we might encounter people who believe they are unusually fragile, remarkably strong, superhumanly intelligent, or in contact with the supernatural. Persons with unusual identities can create problems in social interaction, and they make it difficult to achieve consensus. Cooperative action is difficult without such consensus, for people simply do not know how to relate to individuals who insist that they are Superman, Napoleon, Goldilocks, or Jesus Christ.

The self occupies a central place in symbolic interaction theory because social order is hypothesized to rest in part on self-control. Individuals strive to maintain self-respect in their own eyes, but because they are continually engaging in role taking, they see themselves from the viewpoint of the others with whom they interact. To maintain self-respect, they thus must meet the standards of others, at least to some degree.

Of course, an individual will care more about the opinions and standards of some persons than about those of others. The persons whose opinions he or she cares most about are called **significant others.** Typically, these are people who control important rewards or who occupy central positions in groups to which the individual belongs. Because their positive opinions are highly valued, significant others have relatively more influence over the individual's behavior.

In sum, the symbolic interactionist perspective has several strong points. It recognizes the importance of the self in social interaction. It stresses the central role of symbolic communication and language in personality and society. It addresses the processes involved in achieving consensus and cooperation. And it illuminates why people try to maintain face and avoid embarrassment. Many of these topics are discussed in detail in later chapters. The self is discussed in Chapter 4, symbolic communication and language are taken up in Chapter 7, and self-presentation and impression management are treated in Chapter 9.

Limitations of Symbolic Interaction Theory Critics of symbolic interactionism have pointed to various shortcomings. One criticism concerns the balance between rationality and emotion. Some critics argue that this perspective overemphasizes rational, self-conscious thought and de-emphasizes unconscious or emotional states. A second criticism concerns the model of the individual implicit in symbolic interaction theory. The individual is depicted as a specific personality type—an other-directed person who is concerned primarily with maintaining self-respect by meeting others' standards. A third criticism of symbolic interactionism is that it places too much emphasis on consensus and cooperation and therefore neglects or downplays the importance of conflict. The perspective does recognize, however, that interacting people may fail to reach consensus despite their efforts to achieve it. The symbolic interactionist perspective is at its best when analyzing fluid, developing encounters with significant others; it is less useful when analyzing self-interested behavior or principled action.

Evolutionary Theory

When we think of Charles Darwin and evolution, we most often think of the development of physical characteristics. How, for example, did humans develop binocular vision or the ability to walk upright? How did some animals develop an acute sense of smell, whereas others depend for survival on their ability to see at low levels of light? Evolutionary social psychologists do not stop with strictly physical characteristics, however. They extend evolutionary ideas to explain a great deal of social behavior, including altruism, aggression, mate selection, sexual behavior, and even such seemingly arcane topics as why presidents of the United States are taller than the average man (Buss & Kenrick, 1998).

Evolutionary Foundations of Behavior **Evolutionary psychology** locates the roots of social behavior in our genes and, therefore, intimately links the psychological and social to the biological (Buss, 1999; Symons, 1992; Wilson, 1975). In effect, social behavior, or the predisposition toward certain behaviors, is encoded in our genetic material and is passed on through reproduction. In physical evolution, those characteristics that enable the individual to survive and pass on its genetic code are ones that will eventually occur more frequently in the population. For instance, animals whose camouflage coloring allows them to escape predators will be more likely to survive and produce offspring—who will then receive the advantageous coloring from their parents. Animals of the same species whose camouflage coloring is less efficient will be more likely to be caught and killed before they can reproduce. Thus, over time, the camouflaged animals increase in number relative to the others, who will fade from the population over the generations.

The same process, argue evolutionary psychologists, occurs with respect to social behaviors. Consider one area of research that has received a great deal of attention by evolutionary psychologists: mate selection. Psychologists have observed that men strongly value physical attractiveness and youthful appearance in a potential mate, whereas women focus more on the mate's ability to provide resources for herself and their offspring (Buss, 1994). Why does this difference occur? From an evolutionary perspective, it must be that the different strategies differentially enable men and women to produce successful offspring. The source of the difference lies in the span of fertility—men can continue to reproduce nearly their entire lives, whereas women have a much more constricted period to have children. Therefore, men who prefer to mate with women past their childbearing years will not produce offspring. Over time, then, a genetic preference for older women will be eliminated from the population because these men will not reproduce. Men who prefer younger women will reproduce at a much higher rate, and thus this social behavior will dominate men's approach to mating.

Women, on the other hand, are less concerned about their mates' age, because even much older men can produce offspring. Women's concerns about successful reproduction are focused on the resources necessary for a successful pregnancy and for ensuring the proper development of the child. According to Buss and Kendrick (1998), women's solution to this problem has been to select mates who have the resources and willingness to assist during the pregnancy and after. Women who do not prefer such men, or do not have the ability to identify them, will be less likely to have successful pregnancies and child rearing experiences. Therefore, women's preference for resource-providing men will eventually dominate in the population.

Using this basic notion of evolutionary selection, evolutionary psychologists have developed explanations for an extremely wide variety of social behaviors. For example, altruistic or selfless behaviors initially seem to provide a paradox for evolutionary theory. Why would an individual reduce its chances

BOX 1.2 Research Update: Evolutionary Theory and Mate Poaching

When people are searching for mates—either for long-term relationships or for short-term sexual interactions—they must select targets for their advances. One set of individuals who might seem off-limits are those who are already involved in another relationship. When seduction is aimed at someone who is already attached to another, researchers call it "mate poaching," and although we may frown on the idea, in practice, around half of us attempt to poach (Schmitt et al., 2004). But some of us are more likely to poach than others: About 60 percent of men use this mating strategy, while only 40 percent of women try it, and those looking for short-term engagements are more likely to use it than those looking for long-term relationships. Can evolutionary theory help us understand these social patterns?

Recent studies suggest that evolutionary principles are important in explaining mate poaching attempts. First, in a very broad study conducted across 53 different nations, Schmitt and colleagues found that mate poaching occurred commonly in every one of these countries. The fact that poaching exists in such a large variety of starkly different social contexts suggests strongly that it is a universal, genetically encoded behavior. Second, men consistently have different mating strategies than women. Their preferences for mate characteristics are more focused on physical attractiveness and youth, while women are more focused on their potential mate's resources. Evolutionary psychologists believe that men have these preferences because their genetic code will be more successful if they target healthy women who can successfully bear children. Since these women are in high demand, they tend to be in relationships and thus become targets for poaching. In addition, men will be more successful replicating themselves genetically if they broadcast their genetic code broadly. Thus, they are more likely than women to pursue short-term relationships, including short-term attempts to poach desirable women. Because men are more focused on short-term sexual engagement, women who would like to be poached are more successful if they send signals that they are sexually accessible. On the other hand, men who display or devote resources are more likely to be targeted by women poachers who have more limited ability to pass on their genetic code and thus wish to ensure the successful birth and development of their offspring.

For more on mate poaching and evolutionary theory, see Schmitt (2004).

Adapted from Schmitt et al., 2004; Schmitt & Buss, 2001; Schmitt & Shackelford, 2003.

of survival and reproduction by helping others? One answer, as demonstrated in a number of studies, is that individuals are most likely to assist those to whom they are genetically related (Dawkins, 1982). Because individuals share genetic material with those they assist, they are helping to pass on their own genetic code.

Evolution also helps to explain parenting practices. For example, men tend to be somewhat less invested in parenting than women because they invest less in producing offspring—a single sexual act versus 9 months of gestation and giving birth. Adults are also more likely to abuse their stepchildren than their biological children (Lennington, 1981). Again, evolutionary psychologists would argue that this difference can be traced to the fact that parents share genetic material with their biological children but not with their stepchildren (Piliavin & LePore, 1995). These and many other topics will be examined using evolutionary ideas throughout the book, particularly in Chapters 5 (Social Perception and Cognition), 11 (Helping and Altruism), 12 (Aggression), 13 (Interpersonal

Attraction and Relationships), and 17 (Life Course and Gender Roles).

A Unifying Theory? The evolutionary approach to social psychology is not the same kind of theory as cognitive theory or social exchange theory. Rather than attempting to understand the mechanisms that produce specific kinds of social behavior, the evolutionary approach attempts to account for how and why these mechanisms arise in the first place. Evolutionary psychologists believe, therefore, that the evolutionary perspective provides a unifying principle that ties together the many theories about social behavior that have a more specific focus. For example, consider the previous discussions of social exchange theory and social learning theory. Much of the research on social exchange rests on the assumption that people are rational and are trying to maximize profits. Evolutionary theory helps explain this assumption: People attempt to maximize their profits because those resources help them to survive and perpetuate their genes. Of course, most people do not consciously go through life thinking about how they can perpetuate their genes. Instead, the evolutionary process produces behavior that tends to pass itself on to the next generation—those who have a tendency toward behavior that maximizes their resources are the most likely to survive and reproduce.

Similarly, evolutionary processes underlie social learning, because social learning is adaptive. Individuals who have the greatest ability to learn from others will suffer the least from trial-and-error approaches. Therefore, they are more likely to survive and pass that social learning tendency on to their offspring. The ubiquitous nature of evolutionary processes gives this relatively simple notion extremely wide applicability.

Limitations of Evolutionary Theory Although evolutionary theory may appear to be a far-reaching, unifying engine of human social behavior, it is not without its critics (Caporeal, 2001; Rose & Rose, 2000). The most persistent critique accuses evolutionary psychologists of circular reasoning (Kenrick, 1995). Typically, the evolutionary psychologist observes some characteristic of the social world and then constructs an explanation for it based on its supposed contribution to genetic fitness. The logic of the argument then becomes: Why does this behavior occur? Because it improves the odds of passing on one's genes. But how do we know it improves those odds? Because it occurs. This logical trap is, in some sense, unavoidable because we cannot travel back in time to observe the actual evolution of social behavior.

The problem appears most clearly when we consider the possibility of alternative outcomes. For example, we may observe that men are more accepting of casual sex than women. The evolutionary explanation for this difference between men and women is that men can maximize the survival of their genetic material by spreading it as widely as possible. Women, however, need to know who the father of their children is and extract support from him to ensure the successful transmission of their own genes. Suppose, however, that women were actually more accepting of casual sex than men. This could also easily be explained by the evolutionary perspective. Men cannot be certain that a child is theirs, so a strong commitment to a monogamous relationship would help ensure that it is actually their genes that are being passed to a child. Women, on the other hand, are always 100 percent sure that their own genes are passed down to their children, so in terms of genetic fitness, it should not matter to them who is the father. Because these after-the-fact explanations are always easy to construct and difficult to prove, it can be very difficult to judge them against competing arguments. Therefore, although the evolutionary perspective has grown in popularity in recent years, it still has major obstacles to overcome before achieving widespread acceptance as a useful explanation for social behavior.

A Comparison of Perspectives

The five theoretical perspectives discussed here—role theory, reinforcement theory, cognitive theory, symbolic interaction theory, and evolutionary theory—differ with respect to the issues they address. They also differ with respect to the variables they treat as important causes and effects and those that they treat as irrelevant or incidental. In effect, each perspective makes

TABLE 1.1 COMPARISON OF THEORETICAL PERSPECTIVES IN SOCIAL PSYCHOLOGY

	THEORETICAL PERSPECTIVE				
Dimension	**Role Theory**	**Reinforcement Theory**	**Cognitive Theory**	**Symbolic Interaction Theory**	**Evolutionary Theory**
Central Concepts	Role	Stimulus-response; reinforcement	Cognitions; Cognitive structure	Self; role taking	Genetic Fitness
Primary Behavior Explained	Behavior in role	Learning of new responses; exchange processes	Formation and change of beliefs and attitudes	Sequences of acts occurring in interaction	Reproduction and survival
Assumptions About Human Nature	People are conformist and behave in accordance with role expectations	People are hedonistic; their acts are determined by patterns of reinforcement	People are cognitive beings who act on the basis of their cognitions	People are self-monitoring actors who use role taking in interaction	People seek to perpetuate their own genes
Factors Producing Change in Behavior	Shift in role expectations	Change in amount, type, or frequency of reinforcement	State of cognitive inconsistency	Shift in others' standards, in terms of which self-respect is established	Long-term natural selection processes

different assumptions about social behavior and focuses on different aspects of such behavior.

In this section, we compare the various perspectives in terms of four dimensions: (1) the theory's central concepts or focus; (2) the primary social behaviors explained by the theory; (3) the theory's basic assumptions regarding human nature; and (4) the factors that according to the theory produce changes in a person's behavior. Table 1.1 summarizes this comparison by showing the position of each perspective on each of these dimensions.

Central Concepts Each of the theoretical perspectives places primary emphasis on different concepts. Role theory emphasizes roles and norms defined by group members' expectations regarding performance. Reinforcement theory explains observable social behavior in terms of the relationship between stimulus and response and the application of reinforcement. Cognitive theory stresses the importance of schemas and cognitive structure in determining judgments and behavior. Symbolic interaction theory emphasizes the self and role taking as crucial to the process of social interaction. Evolutionary theory is focused on the genetic transmission of behavioral tendencies and the frequency with which these behaviors appear in the population.

Behaviors Explained Although overlapping to some degree, the five theoretical perspectives differ with respect to the behaviors or outcomes they try to explain. Role theory emphasizes role behavior and the attitude change that results from occupying roles. Reinforcement theory focuses on learning and on the impact of rewards and punishments on social interaction. Cognitive theory centers on the mediating effects of a person's beliefs and attitudes on his or her overt response to social stimuli, and it also focuses on factors that produce changes in these beliefs and attitudes. Symbolic interaction theory stresses the sequences of behaviors occurring in interactions among people. Evolutionary theory attempts to generally explain how all social behaviors arise from biological underpinnings, but it has traditionally focused on the behaviors that are most closely linked to reproduction and survival.

Assumptions About Human Nature The five theoretical perspectives differ also in their fundamental assumptions regarding human nature. Role theory, for instance, assumes that people are largely conformist. It views people as acting in accord with the role expectations held by group members. In contrast, reinforcement theory views people's acts—what they learn and how they perform—as determined primarily by patterns of reinforcement. Cognitive theory stresses the ability of people to perceive, interpret, and make decisions about the world. People formulate concepts and develop beliefs, and they act on the basis of these structured cognitions. Symbolic interaction theory assumes that people are conscious, self-monitoring beings who use role taking to achieve their goals through interaction with others. Evolutionary theory assumes that people's behaviors have been shaped by the natural selection process to seek perpetuation of their genetic code.

Change in Behavior The five theoretical perspectives differ in their conception of what produces changes in behavior. Role theory maintains that to change someone's behavior, it is necessary to change the role that he or she occupies. A change in behavior results when the person shifts roles because the new role entails different expectations and demands. Reinforcement theory, in contrast, holds that a change in behavior results from changes in the type, amount, and frequency of reinforcement received. Cognitive theory maintains that a change in behavior results from changes in beliefs and attitudes; it further postulates that these changes in beliefs and attitudes often result from efforts to resolve inconsistency among cognitions. Symbolic interaction theory holds that people try to maintain self-respect by meeting the standards of significant others; the question of which standards are relevant is usually resolved through negotiation. For behavior to change, the standards held by others and accepted as relevant must shift first. A person will detect this shift in standards by role taking and consequently change his or her behavior. Evolutionary theory is not concerned with short-term changes in individual behavior. Instead, it attempts to explain why more individuals come to exhibit certain behavioral tendencies over the generations and why other behaviors become extinct.

IS SOCIAL PSYCHOLOGY A SCIENCE?

As we have noted, social psychology is the systematic study of the nature and causes of human social behavior. The field of social psychology includes not only theoretical perspectives and middle-range theories, but also a large body of facts and empirical generalizations obtained through research. Given these characteristics, we can ask whether social psychology is a science. That is, can we consider social psychology to be a scientific field in the same sense that we consider physics or biology to be scientific?

Characteristics of Science

Any scientific field rests on several basic assumptions. First, scientists assume that a real, external world exists independently of ourselves. This world is subject to investigation by observers. Second, scientists assume that relations in this world are organized in terms of cause and effect. This assumption—which in practice is a working hypothesis—is termed the **principle of determinism.** In its strongest form, this principle holds that there are discoverable causes for all events in a science's domain of interest. Third, scientists assume that knowledge concerning this external world is objective. Facts discovered by one scientist can be checked and verified by others.

In addition to these assumptions, any field that is a science has certain critical characteristics or hallmarks. These include:

1. Any science is based on the observation of facts. No field can be considered a science unless it

includes observation. Thus, so-called "armchair" disciplines without practitioners who make observations cannot be scientific fields.

2. Any science uses an explicit, formal methodology. This methodology is a set of procedures that must be followed by an investigator when establishing something as a known fact. Because any investigator in the field can use this methodology, one scientist's findings can, in principle, be verified by others.
3. Any science involves the accumulation of facts and generalizations. Once relationships are observed to exist in the world, this knowledge is never lost. Of course, facts sometimes undergo a reinterpretation of their meaning, but the essential information is still available.
4. Any science includes a body of theory. This consists of at least one (but more often many) theories that serve to systematize and organize empirical observations. Theory also serves to guide new empirical investigation.
5. After it attains a reasonable level of development, any science provides at least some degree of prediction and control over selected aspects of the environment.

When we hold up a well-developed natural science such as physics or biology against this list, we see immediately that it has all these hallmarks. These sciences are based on observation of the world, have a formal methodology to guide research, have an accumulation of established facts, possess a body of well-developed theory, and provide at least a moderate degree of prediction and control regarding selected aspects of the world.

Social Psychology as a Science

Can social psychology be considered a science? That is, if we hold it up against the criteria just listed, does it measure up? Social psychology certainly has some of the hallmarks of science. Consider the first hallmark—reliance on empirical observation. Social psychology clearly is based on the empirical observation and classification of facts. The field comprises many thousands of empirical studies. Social psychology also meets the second hallmark, for it relies on widely shared methodological procedures for conducting empirical investigation. Among the most widely employed methods in social psychology are experimentation and systematic sample surveys. Chapter 2 will discuss methodology in social psychology.

To a fair degree, social psychology also meets the third hallmark—the accumulation of observed facts. Social psychologists continue to gather facts regarding the conditions under which specific behaviors occur. Of course, more is known about some types of social behavior than about others, but increasingly sophisticated studies have continued to expand the frontiers of knowledge.

Social psychology also measures up fairly well against the fourth criterion—reliance on theory. Although social psychology has no single unified theory covering all phenomena in the field, it does have several theoretical perspectives, as reviewed in this chapter. It also has numerous middle-range theories that make predictions regarding specific types of social behavior under restricted conditions. A large number of empirical studies in social psychology are attempts to test the predictions of middle-range theories. We will encounter many of these middle-range theories in subsequent chapters.

If any problem arises for social psychology as a science, it is primarily with respect to the fifth hallmark, which holds that after attaining a reasonable level of development, any science provides some degree of prediction and control with respect to the phenomena investigated in it. Whether social psychology can accomplish this feat today is unclear.

Although social psychology does fairly well in explaining social behavior (that is, identifying the conditions under which various forms of social behavior occur), it does less well in predicting future events or providing a basis for the control of behavior. Of course, we can point to some successes in prediction and control. For example, political election forecasts based on sample surveys have been fairly accurate in recent years, and programs for the modification of interpersonal behavior based on reinforcement principles have proved effective. Nevertheless, social psychology does not excel in prediction and control. A large percentage of practicing social psychologists believe that, in the coming years, the field's capacity to predict interpersonal behavior will improve to some degree (Lewicki, 1982). At the present time, however, social psychology is no match for the mature physical sciences in this respect.

Part of the problem stems from the nature of social psychological theory. Few theories that make explicit predictions have much generality. They cover only limited ranges of phenomena or apply only under very restrictive and sometimes artificial conditions Theories often fail to make accurate predictions when attempts are made to apply them to new settings.

If the problem ran no deeper than this, we might be optimistic that social psychology will soon predict social behavior with great accuracy. We might conclude, for instance, that social psychology merely needs better, more refined theories. To some degree this is true, but unfortunately the problem is more complex. As noted earlier, any science is based on the principle of determinism; it assumes the world is organized in terms of cause and effect. Science tries to develop laws based on the notion that if X causes Y on one occasion, then X will again cause Y on some similar occasion in the future. Although the assumption of determinism works well for many physical phenomena, it may not work as well for human social behavior. Because human beings are conscious and self-aware, they can exercise some control over their own behavior. Unlike atoms or rocks, they are capable of making decisions and suddenly changing their behavior. They can exercise free will, at least to some degree. In fact, simply knowing about theories of behavior and their predictions can cause individuals to alter their behavior. For this reason, some critics have argued that there will never be anything approaching true "universal laws" describing social behavior. Certainly, it is difficult to reconcile the scientific assumption of determinism with the concept of human free will. Of course, this concern is not unique to social psychology. It besets all the social sciences.

As we have shown, social psychology displays many of the characteristics of the more mature physical sciences. It is based on observation of the social world and relies on a formal methodology to guide research. It has accumulated many descriptive facts regarding human social behavior, and it possesses bodies of formal theory. Nevertheless, it has not yet achieved the same degree of accuracy in prediction as the mature physical sciences. Although it does offer compelling explanations for many types of observed social behavior, social psychology has so far provided only a modest degree of predictability and control of social behavior.

SUMMARY

This chapter considered the fundamental characteristics of social psychology and important theoretical perspectives in the field.

What Is Social Psychology? There are several ways to characterize social psychology. (1) By definition, social psychology is the systematic study of the nature and causes of human social behavior. (2) Social psychology has several core concerns, including the impact of one individual on another individual's behavior and beliefs, the impact of a group on a member's behavior and beliefs, the impact of a member on the group's activities and structure, and the impact of one group on another group's activities and structure. (3) Social psychology has a close relationship with other social sciences, especially sociology and psychology. Although they emphasize different issues and often use different research methods, both psychologists and sociologists have contributed significantly to social psychology.

Theoretical Perspectives in Social Psychology A theoretical perspective is a broad theory based on particular assumptions about human nature that offers explanations for a wide range of social behaviors. This chapter discussed five theoretical perspectives: role theory, reinforcement theory, cognitive theory, symbolic interaction theory, and evolutionary theory. (1) Role theory is based on the premise that people conform to norms defined by the expectations of others. It is most useful in explaining the regular and recurring patterns apparent in day-to-day activity. (2) Reinforcement theory assumes that social behavior is governed by external events, especially rewards and punishments. Reinforcement theory helps to explain not only how people learn but also when social relationships will change. (3) Cognitive theory holds that such processes as perception, memory, and judgment are significant determinants of social behavior. The theory treats ideas and beliefs as organized into structures (schemas) and relies on various principles (such as the principle of cognitive consistency) to explain changes in attitudes and beliefs. Differences in cognitions help to illuminate why individuals may behave differently from one another in a given situation. (4) Symbolic interaction theory holds that human nature and social order are products of communication among people. It stresses the importance of the self, of role taking, and of consensus in social interaction. It is most useful in explaining fluid, contingent encounters among people. (5) Evolutionary theory posits that social behavior is a product of long-term evolutionary adaptation. Behavioral tendencies exist in human beings because these behaviors aided our ancestors in their attempts to survive and reproduce.

Is Social Psychology a Science? To ascertain whether social psychology is a science, we must first identify the five hallmarks that characterize any science. These are (1) scientists engage in empirical observation of the world, (2) use a formal research methodology, (3) accumulate knowledge of facts, (4) develop formal theories to explain facts, and (5) employ these theories to provide some degree of prediction and control. Social psychology meets the first four of these hallmarks, but falls short of meeting the fifth. To date, social psychology has provided only a modest degree of predictability and control over human social behavior.

LIST OF KEY TERMS AND CONCEPTS

cognitive processes (p. 14)
cognitive structure (p. 14)
cognitive theory (p. 14)
conditioning (p. 11)
equity (p. 13)
evolutionary psychology (p. 18)
imitation (p. 12)
middle-range theory (p. 8)
norm (p. 10)
principle of cognitive consistency (p. 15)
principle of determinism (p. 22)
reinforcement (p. 11)
reinforcement theory (p. 11)
role (p. 9)
role theory (p. 9)
schema (p. 14)
self (p. 17)
significant others (p. 17)
social exchange theory (p. 12)
social learning theory (p. 12)
social psychology (p. 3)
symbolic interaction theory (p. 16)
theoretical perspective (p. 8)
theory (p. 8)

2

RESEARCH METHODS IN SOCIAL PSYCHOLOGY

Introduction

Questions About Research Methods

Characteristics of Empirical Research

Objectives of Research

Research Hypotheses

Validity of Findings

Research Methods

Surveys

Field Studies and Naturalistic Observation

Archival Research and Content Analysis

Experiments

Comparison of Research Methods

Meta-Analysis

Research in Diverse Populations

Ethical Issues in Social Psychological Research

Potential Sources of Harm

Institutional Safeguards

Potential Benefits

Summary

List of Key Terms and Concepts

INTRODUCTION

The field of social psychology relies heavily on empirical research, which is the systematic investigation of observable phenomena (behavior, events) in the world. Researchers try to collect information about behavior and events in an accurate and unbiased form. This information, which may be either quantitative or qualitative, enables social psychologists to describe reality in detail and to develop theories about social behavior.

When conducting empirical research, investigators usually employ a **methodology,** which is a set of systematic procedures that guide the collection and analysis of data. In a typical study, investigators first develop a research design. Then they go into a laboratory or field setting and collect the data. Next, they code and analyze the data to test hypotheses and arrive at various conclusions about the behaviors or events under investigation. Throughout this process, investigators follow specific procedures to ensure the validity of the findings.

When investigators report their research to the wider community of social psychologists, they describe not only the results but also the methodology used to obtain the results. By reporting their methods, they make it possible for other investigators to independently verify their findings.

Independent verification of research findings is one of the hallmarks of any science. Suppose, for instance, that an investigator were to report some unanticipated empirical findings that ran contrary to established theory. Other investigators might wish to replicate the study to see whether they can obtain the same findings in other settings with different human subjects. Through this process, investigators with differing perspectives can identify and eliminate biases in the original study. If the results are replicable, they stand a better chance of being accepted by other social psychologists as reliable, general findings.

Questions About Research Methods

In this chapter, we will discuss the research methods used in contemporary social psychology. This discussion will provide a foundation for understanding and evaluating the empirical studies discussed throughout this book. This chapter addresses the following questions:

1. What are the basic goals that underlie social psychological research? What form do research hypotheses assume? What steps can researchers take to ensure the validity of their findings?
2. What are the defining characteristics of research methods such as surveys, naturalistic observation, archival research, and laboratory and field experiments? What are the strengths and weaknesses of each? What is a meta-analysis?
3. What issues are raised when we undertake research on diverse groups within a society or on members of other cultures?
4. What ethical issues are important in the conduct of social psychological research? Which safeguards are available to protect the rights of human subjects or participants? What are the potential benefits to the participants?

CHARACTERISTICS OF EMPIRICAL RESEARCH

In this chapter, we discuss the major research methods used by social psychologists. These methods include surveys, field observation, archival studies, and experiments. Before looking at these in detail, however, we will review some issues common to all forms of empirical research. Specifically, we will consider the objectives that typically underlie empirical research, the nature of the hypotheses that guide research, and the factors that affect the validity of research findings.

Objectives of Research

Investigators conduct social psychological studies for a variety of reasons. Their objectives usually include one or more of the following: describing reality, identifying correlations between variables, testing causal hypotheses, and developing and testing theories.

In some studies, the central objective is simply to describe reality in accurate and precise terms. An investigator may wish to characterize some behavior or describe the features of a social process. Description is often the paramount goal when a researcher investigates a phenomenon about which little or nothing is known. Even when investigating more familiar phenomena, a researcher may wish to ascertain the frequency with which a particular attitude or behavior occurs in a specified group or population. For instance, during election years, researchers routinely conduct public opinion polls to learn how Americans feel about political candidates, issues, and parties. Their goal is to describe public sentiment with great accuracy and precision.

A second objective of research is to ascertain whether a correlation exists between two or more behaviors or attributes. Researchers might conduct a survey, for example, to find out whether growing older is associated with changes in a person's sexual desire (DeLamater & Sill, 2005) or whether how children spend their time is related to their scores on standard achievement tests (Hofferth & Sandberg, 2001). Although a correlation between variables may reflect an underlying causal relation, two variables can be correlated without one causing the other; this will happen, for instance, if both are caused by a third variable. Correlation alone is not sufficient evidence for causation.

A third objective of research, then, is to discover the causes of some behavior or event. When pursing this goal, the researcher first develops a causal hypothesis, which is a statement that differences or changes in one behavior or event produces a difference or change in another behavior or event. For instance, an investigator might hypothesize that studying for an exam in groups will produce higher grades than studying for the exam individually. After specifying the hypothesis, the investigator collects data to test the hypothesis. To support the hypothesis of causality, this test must show that differences or changes in one variable produce differences or changes in another. Moreover, the design of the test must preclude or eliminate plausible alternative (noncausal) interpretations of the data. Frequently, the best way to test a causal hypothesis is by an experiment, a topic discussed in greater detail further on.

A fourth objective of social psychological research is to test existing theories and to develop new ones. A theory is a set of interrelated hypotheses that explains some observable behavior(s) or event(s). Frequently, a theory will serve as a basis for predicting future events. Tests of theories resemble tests of hypotheses, except that several interrelated hypotheses are assessed at once. In some cases, investigators juxtapose theories that make different predictions, and the results of the test may enable them to reject one theory in favor of another.

Research Hypotheses

In broad terms, a **hypothesis** is a conjectural statement of the relation between two or more variables. Many social psychological studies begin with one or more hypotheses. To test whether a hypothesis is correct, an investigator will first ask what observations would be expected if the hypothesis is true; then he or she will take some observations or measures of reality and compare these with what is expected under the hypothesis. If a discrepancy is noted, it constitutes evidence against the hypothesis and may lead to its rejection.

There are various types of hypotheses. Some hypotheses are noncausal in nature; for example, "Variables X and Y are correlated, such that high levels of X occur with low levels of Y" (negative correlation). Noncausal hypotheses make statements about observed relations between variables.

Other hypotheses are explicitly causal in nature. For instance, a causal hypothesis relating two variables might take the form "X causes Y" or "Higher levels of X produce lower levels of Y" or "An increase in X will produce a decrease in Y." Sometimes, of course, causal hypotheses are more explicit and qualified in scope; for example, "If conditions A and B are present, then an increase of 1 unit in X will cause a decrease of 6 units in Y."

Causal hypotheses always include at least two variables—an independent variable and a dependent variable. An **independent variable** is any variable considered to cause or have an effect on some other variable(s). A **dependent variable** is any variable caused by some other variable. The dependent variable changes in response to changes in the independent variable. In the preceding example where X causes Y, X is the independent variable and Y is the dependent variable.

Another important type—the **extraneous variable**—is any variable that is not expressly included in the hypothesis but that nevertheless has a causal impact on the dependent variable. Extraneous variables are widespread in social psychology because most dependent variables of interest have more than one cause.

Validity of Findings

One cannot take for granted that the findings of any given study will have validity. Consider a situation where an investigator is studying deviant behavior. In particular, she is investigating the extent to which cheating occurs on exams by college students. Reasoning that it is more difficult for people monitoring an exam to keep students under surveillance in large classes than in smaller ones, she hypothesizes that a higher rate of cheating will occur on exams in large classes than in small. To test this hypothesis, she collects data on cheating in both large classes and small ones and then analyzes the data. Her results show that more cheating per student occurs in the larger classes. Thus, the data apparently support the investigator's research hypothesis.

A few days later, however, a colleague points out that all the large classes in her study used multiple-choice exams, whereas all the small classes used short-answer and essay exams. The investigator immediately realizes that an extraneous variable (exam format) is confounded with the independent variable (class size) and may be operating as a cause in her data. The apparent support for her research hypothesis (more cheating in large classes) may be nothing more than an artifact. Perhaps the true effect is that more cheating occurs on multiple-choice exams than on essay exams, irrespective of class size.

We say that the findings of a study have **internal validity** if they are free from contamination by extraneous variables. Internal validity is a matter of degree; findings may have high or low internal validity. Obviously, the investigator's findings about the effect of class size on cheating have low internal validity due to the possibly confounding effect of exam format. Internal validity is very important. Without internal validity, a study cannot provide clear, interpretable results.

To achieve results with higher internal validity, the investigator might repeat the study with an improved design. For instance, our investigator might repeat her study with only one exam format (say, multiple choice) in both large and small classes. Then she could test whether class size affects the rate of cheating on multiple-choice exams. By holding constant the extraneous variable (exam format), her new design will have greater internal validity. Better still, she might use a more complex design that includes all four logical possibilities (that is, small class–multiple choice, small class–essay, large class–multiple choice, large class–essay). She could analyze the data from this design to estimate separately the relative impacts of these variables (class size, exam format) on cheating. In effect, this design converts an extraneous variable (exam format) into a second independent variable. Although better, it is not a perfect design, because other extraneous variables could still be operating as causes of cheating—and they may be confounded with class size and exam format.

As important as internal validity is, it is not the only concern of the investigator. Another concern is external validity. **External validity** is the extent to which a causal relationship, once identified in a particular setting with a particular population, can be generalized to other populations, settings, or time periods. Even if an investigator's results have internal validity, they may lack external validity; that is, they may hold only for the specific group and setting studied and not generalize to others. For instance, if the investigator studying cheating and class size conducted her study in a 2-year college, there is no assurance that the findings (whatever they turn out to be) would also apply to students in other settings, such as high schools or 4-year colleges or universities. In general, external validity is important and desirable, because the results of a study often have practical importance only if they generalize beyond the particular setting in which they appeared.

RESEARCH METHODS

Although there are many ways of collecting data about social behavior, most social psychological studies use one or another of four main methods. These methods are surveys, naturalistic observation, archival research based on content analysis, and experiments. We discuss each of these methods in turn.

Surveys

A survey is a procedure for collecting information by asking members of some population a set of questions and recording their responses. The survey technique is very useful for identifying the average or typical response to a question, as well as the distribution of responses within the population. It is also useful for identifying how groups of respondents differ from one another. For instance, Prince-Gibson and Schwartz (1998) used a survey to test a set of hypotheses about gender differences in values. They predicted that men would more strongly value power, achievement, hedonism, and stimulation, whereas women would value benevolence, conformity, tradition, and security. The hypotheses were tested using data from a probability sample of the Israeli Jewish population. Contrary to predictions, there were no significant differences in the mean ratings of the importance of these values given by men and women. Because some research conducted in the 1970s and 1980s did report such differences, the authors concluded that their results suggest that men's and women's values are converging.

© David Frazier

Working from a schedule of questions, this survey interviewer carefully records the answers given by a respondent at a mall.

Purpose of a Survey Investigators often conduct surveys to obtain self-reports from individuals about their own attributes—that is, their attitudes, behavior, and experiences. Information of this type enables investigators to discover the distribution of attributes in the population and to determine whether a relationship exists between two or more attributes of interest.

One form of survey—the public opinion poll—has become very common in the United States. Several organizations specialize in conducting surveys that measure the frequency and strength of favorable or unfavorable attitudes toward public issues, political figures, and candidates for office. These polls play a significant role in American politics, for their findings increasingly influence public policy and the positions taken by political figures (Halberstam, 1979; Ratzan, 1989). Presidential candidates used the results of such polls to guide their decisions during the 2004 election campaign, and they will likely to do so again in 2008.

Investigators also often use surveys to obtain data about various social problems. For instance, government agencies and individual researchers have conducted surveys on pregnancy and contraception use among teenagers (Levine, 2000) and on alcohol and drug use by teenagers (Center on Addiction and Substance Abuse, 2002). Information about the extent of such activities and the people involved in them is requisite to developing effective social policies.

Finally, investigators often conduct surveys with the primary objective of making basic theoretical contributions to social psychology. For instance, many studies of socialization processes and outcomes, psychological well-being, discrimination and prejudice, attitude-behavior relationships, and collective behavior have used survey methods.

Types of Surveys There are two basic types of surveys—those based on interviews and those based on questionnaires. In an **interview survey,** a person serves as an interviewer and records the answers from the respondents. To ensure that each respondent in the study receives the same questions, the interviewer usually works from an interview schedule. This schedule indicates the exact order and wording of questions. In certain studies, however, the interviewer has flexibility in determining the exact order and wording of questions, but he or she is expected to make sure that certain topics are covered. One advantage of using an interview is that the interviewer can adjust the questioning to the respondent. That is, he or she can look for verbal or nonverbal signs that the respondent does not understand a question and repeat or clarify the question as needed (Moore, 2004).

In a **questionnaire survey,** the questions appear on paper, and the respondents read and answer them at their own pace. No interviewer is present. One advantage of questionnaires over interviews is that questionnaires cost less to administer. The cost of a national survey using trained personnel to conduct face-to-face interviews is rather large; it can run as much as $250–$300 or more per completed interview, although this varies with the length of the interview and other factors. In contrast, the same survey using questionnaires mailed to respondents would cost considerably less—maybe as little as $15 per completed form. The major disadvantage of questionnaires lies in the **response rate**—the percentage of people contacted who complete the survey. Whereas an interview study can obtain response rates of 75 to 80 percent or more, mailed questionnaires rarely attain more than a 50 percent response rate. Because a high response rate is very desirable, this is a significant disadvantage for mailed questionnaires.

A compromise between interviews and questionnaires is the telephone interview. This is the standard method used by public opinion polling organizations, such as Gallup and Roper. Investigators are using it in basic research, as well. The telephone interview uses a trained interviewer to ask the questions, but it sacrifices the visual feedback available in a face-to-face interview. It is cheaper (about $60 per completed interview, depending on length) than the face-to-face interview, although it typically involves a somewhat lower response rate (about 65 percent). Many surveys now use computer assisted telephone interviewing (CATI). With CATI, the computer randomly selects and dials telephone numbers. Once a potential respondent is on the line, the interviewer takes over and conducts the interview. He or she reads some questions and enters the answers directly into the computer when the respondent gives them. In listing questions to ask, the computer may alter later questions in light of earlier answers by the respondent.

Measurement Reliability and Validity In surveys, as in any form of research, the quality of measurement is an important consideration. Of primary concern are the reliability and the validity of the instruments. **Reliability** is the extent to which an instrument produces the same results each time it is employed to measure a particular construct under given conditions. A reliable instrument produces consistent results across independent measurements of the same phenomenon. Reliability is a matter of degree; some instruments are highly reliable, whereas others are less so. Obviously, investigators prefer instruments with high reliability and try to avoid those with low reliability.

There are several ways to assess the reliability of an instrument. The first is to see if people's responses to an instrument are consistent across time. In this approach, called the test-retest method, an investigator applies the measuring instrument to the same respondents on two different occasions, and then he or she compares the first responses with the second responses. If the correlation between the first and second responses is high, the instrument has high reliability; if the correlation is low, the instrument has only low reliability.

A second way to assess the reliability of an instrument is to see if people's responses are consistent across items. This approach is called the split-half method. To illustrate, suppose we have a scale of 20 questions measuring psychological well-being. These questions ask the respondent about psychological states, such as how often he or she is sad, nervous, depressed, tense, or irritable and how often he or she has trouble concentrating, working, or sleeping. Assume that we administer all the questions to 300 male respondents. To use the split-half method, we would randomly divide the 20 questions into two groups of 10, calculate a score for each respondent on each group of 10, and compute a correlation between the two scores. A high correlation (if it occurs) provides confirmation that the scale is reliable.

Given that a measure is reliable, the next concern is its validity—that is, does the instrument actually measure the theoretical concept we intend to measure? There are several types of validity, including face validity, criterion validity, and construct validity. First, an instrument has face validity if its content is manifestly similar to the behavior or process of interest. If a researcher wishes to measure the frequency of sexual intercourse, for example, the question "How often do you engage in sexual intercourse?" has face validity.

Second, an instrument has criterion validity if we can use it to predict respondents' standing on some other variable of theoretical or practical interest. Suppose, for example, that an investigator is concerned with traffic safety on the roads and that she develops an instrument to distinguish good drivers from bad drivers. To establish the instrument's predictive validity, she first administers the instrument to young people getting their driver's license and then, several years later, checks their driving records for moving violations. If the drivers' scores on the instrument correlate highly with their level of subsequent violations, the instrument has criterion validity.

Third, an instrument has construct validity if it provides a good measure of the theoretical concept being investigated by the research. In general, an instrument will have construct validity if it measures what people understand the concept to mean and if it relates to other variables as predicted by the theory under consideration. Establishing the construct validity of an instrument can be difficult, especially if the underlying theoretical construct is highly abstract in nature. Suppose, for example, that an investigator's theory includes an abstraction like "intellectual development." The measurement of this concept is somewhat problematic, for there is no readily observable referent, no single behavior or occurrence that the investigator can point to as indicative of intellectual development. The usual method of establishing the construct validity of an instrument is to show that the pattern of correlations between respondents' scores on the instrument and their scores on other variables is what would be expected if the underlying theory holds true.

The Questions The phrasing of questions used in surveys requires close attention by investigators. Subtle differences in the form, wording, and context of survey questions can produce differences in responses (Schwarz, Groves, & Schuman, 1998). Creating good survey questions is as much art as science, but there are certain guidelines that help. First, the more precise and focused a question, the greater will be its reliability and validity. If a question is expressed in vague, ambiguous, abstract, or global terms, respondents may interpret it in different ways, and this in turn will produce uncontrolled variation in responses. A second consideration in formulating survey questions is the exact choice of terms used. It is best to avoid jargon or specialized terminology unless one is interviewing a sample of specialists. Likewise, it is important to adjust questions to the educational and reading level of the respondents. A third consideration is the length of questions. Several studies have shown that questions of moderate length elicit more complete answers than very short ones (Anderson & Silver, 1987; Sudman & Bradburn, 1974). A fourth consideration is whether the topic under investigation is potentially a threatening or embarrassing one (sex, alcohol, drugs, money, and so on). In general, threatening questions requiring quantified answers are better asked by presenting a range of alternative answers (say, 0, 1–5, 6–10) than by asking a question requiring an exact number (Rea & Parker, 1997).

Measuring Attitudes Perhaps the most common purpose of surveys is to measure people's attitudes toward some event, person, or object. Because attitudes are mental states, they cannot be directly observed. Therefore, to find out someone's attitude, we usually ask them.

The most direct way of finding out someone's attitude is to ask a direct question and record the person's answer. This is the way most of us study the attitudes of the people with whom we interact. It is also the technique used by newspaper and television reporters. To make the process more systematic, social psychologists use several methods, including the single-item measure, Likert scales, and semantic differential techniques.

BOX 2.1 The Measurement of Attitudes

Suppose you want to assess attitudes toward premarital sexual behavior. Here are three techniques you could employ.

Single Item

The single item is probably the most common measure of attitudes. An example of this type is:

I think people should wait until they are married to have sex.

____ Yes
____ No
____ Not sure

Likert Scale

The Likert scale consists of a series of statements about the object of interest. The statements may be positive or negative. The respondent indicates how much he or she agrees with each statement. For example:

1. I think people should wait until they are married to have sex.

___ Strongly agree	(+ 2)
___ Agree	(+ 1)
___ Undecided	(0)
___ Disagree	(− 1)
___ Strongly disagree	(− 2)

2. I think having sex before marriage strengthens the marriage.

___ Strongly agree	(− 2)
___ Agree	(− 1)
___ Undecided	(0)
___ Disagree	(+ 1)
___ Strongly disagree	(+ 2)

Semantic Differential Scale

The semantic differential scale consists of a number of dimensions on which the respondent rates the attitude object. For example:

Rate how you feel about premarital sexual intercourse on each of the following dimensions.

good	____ (+ 3)	____ (+ 2)	____ (+ 1)	____ (0)	____ (− 1)	____ (− 2)	____ (− 3)	bad
weak	____ (− 3)	____ (− 2)	____ (− 1)	____ (0)	____ (+ 1)	____ (+ 2)	____ (+ 3)	strong
fast	____ (+ 3)	____ (+ 2)	____ (+ 1)	____ (0)	____ (− 1)	____ (− 2)	____ (− 3)	slow
negative	____ (− 3)	____ (− 2)	____ (− 1)	____ (0)	____ (+ 1)	____ (+ 2)	____ (+ 3)	positive
light	____ (− 3)	____ (− 2)	____ (− 1)	____ (0)	____ (+ 1)	____ (+ 2)	____ (+ 3)	heavy
exciting	____ (+ 3)	____ (+ 2)	____ (+ 1)	____ (0)	____ (− 1)	____ (− 2)	____ (− 3)	boring

Single Items The use of single questions to assess attitudes is very common. The single-item scale usually consists of a direct positive or negative statement, and the respondent indicates whether he or she agrees, disagrees, or is unsure. Such a measure is economical; it takes a minimum of time and space to present. It is also easy to score. The major drawback of the single item is that it is not very precise. Of necessity, it must be general and detects only gross differences in attitude. Using the single-item measure in

Box 2.1, we could separate people into only two groups: those who favor premarital abstinence and everybody else.

Likert Scales Often, we want to know not only how each person feels about the object of interest but also how each respondent's attitude compares with the attitudes of others. The **Likert scale,** a technique based on summated ratings, provides such information (Likert, 1932).

Box 2.1 includes a two-item Likert scale. Each possible response is given a numerical score, indicated in parentheses. We would assess the respondent's attitude by adding his or her scores for both items. For example, suppose you strongly agree with item 1 (+2) and strongly disagree with item 2 (+2). Your score would be +4, indicating strong opposition to premarital intercourse. Your roommate might strongly disagree with the statement that people should wait until they marry (−2) and might also disagree that premarital sex strengthens a marriage (+1). The resulting score of −1 indicates a slightly positive view of premarital intercourse. Finally, someone who strongly disagrees with item 1 (−2) and agrees with item 2 (−1) would get a score of −3 and could be differentiated from a person who received a score of −4.

Typically, a Likert scale includes at least four items. The items should be counterbalanced—that is, some should be positive statements, and others should be negative ones. Our two-item scale in Box 2.1 has this property; one item is positive, and the other is negative. The Likert scale allows us to order respondents fairly precisely; items of this type are commonly used in public opinion polls. Such a scale takes more time to administer, however, and involves a scoring stage as well.

Semantic Differential Scales Like most attitude scales, the single-item and Likert scales measure the denotative or dictionary meanings of the object to the respondent. However, objects also have a connotative meaning, a set of psychological meanings that vary from one respondent to another. For instance, one person may have had very positive experiences with sexual intercourse, whereas another person's experiences may have been very frustrating.

The semantic differential scale (Osgood, Suci, & Tannenbaum, 1957) is a technique for measuring connotative meaning. In using it, an investigator presents the respondents with a series of bipolar adjective scales. Each of these is a scale whose ends are two adjectives having opposite meanings. The respondent rates the attitude object on each scale. After the data are collected, the researcher can analyze them by various statistical techniques. Analyses of such ratings frequently identify three aspects of connotative meaning: evaluation, potency, and activity. Evaluation is measured by adjective pairs such as *good-bad* and *positive-negative;* potency, by *weak-strong* and *light-heavy;* and activity, by *fast-slow* and *exciting-boring.*

The example in Box 2.1 includes two bipolar scales measuring each of the three dimensions. Scores are assigned to each scale from +3 to −3; they are then summed across scales of each type to arrive at evaluation, potency, and activity scores. In the example shown, scores on each dimension could range from −6 (*bad, weak,* and *slow*) to +6 (*good, strong,* and *fast*).

One advantage of the semantic differential technique is that researchers can compare an individual's attitudes on three dimensions, allowing more complex differentiation among respondents. Another advantage is that because the meaning it measures is connotative, it can be used with any object, from a specific person to an entire nation. This technique is also used to assess the meaning of role identities (*mother, doctor*) and role behaviors (*hug, cure;* Heiss, 1979; Smith-Lovin, 1990). Its disadvantages include the fact that it requires more time to administer and to score.

The Sample Suppose a survey researcher wants to ascertain the extent of prejudice toward Blacks among White adults in the United States. These White adults constitute the **population** of interest—that is, the set of all people whose attitudes are of interest to the researcher. It would be virtually impossible—and enormously expensive—to interview all people in the population of White adults, so the researcher instead selects a sample, or representative subset, from that population to interview.

Sample selection is one of the most important aspects of any survey. In some cases, investigators may use a particular sample simply because it is readily available; samples of this type are known as convenience samples. A sample consisting of students taking a class, occasionally used in social science research, is a convenience sample. Convenience samples have a major drawback—they usually lack external validity and do not enable the investigator to generalize the findings to any larger population. For this reason, it is better research practice to select some other type of sample—one that is representative of the underlying population. Only when the sample is representative can the results obtained from it (for example, information regarding racial prejudice obtained from survey respondents) be generalized to the entire population. The nature of the sample, therefore, has a major impact on the external validity of the survey.

Two types of systematic samples are commonly used in social psychological surveys. One is the **simple random sample,** wherein the researcher selects units—usually individuals—from the population such that every unit has an equal probability of being included. To use this technique, the researcher needs a complete list of members of the population. At a university, for example, she might obtain a list of all students from the registrar. At the city or county level, she might use voter registration lists. A frequent problem, especially when the population being studied is large, is the absence of a complete list. Under these circumstances, researchers usually fall back on some substitute, such as a telephone directory. Of course, this will limit the population to which one can generalize, because people who are poor or who move frequently may not have telephones, others may choose not to list their numbers in the directory, and others only have cell phones.

Working from a complete list of the population, the researcher draws a random sample. A common way to do this is to number the people on the list consecutively and then use a table of random numbers to choose people for the sample. Once the researcher has drawn a random sample, she must take steps to ensure that all the members of the sample are interviewed; in other words, the researcher must strive for a high response rate. Without a high response rate, the results of the survey will not be generalizable to the whole population. Bias may result if the people who participate in the study differ in some significant way from those who refuse to participate.

If the population is very large, the investigator may not be able to list all its members and draw a random sample. Under these conditions, researchers frequently employ a **stratified sample.** That is, they divide the population into groups according to important characteristics, select a random sample of groups, and then draw a sample of individuals within each selected group. For example, public opinion polls designed to represent the entire adult population of the United States often use stratified samples. The population is first stratified on the basis of region (Northeast, Midwest, South, Southwest, and West). Next, the population within each region is stratified into urban versus rural areas. Within urban areas, there may be still further stratification by size of urban area. The result will be numerous sampling units—population subgroups of known regional and residential type. Some units are then selected for study in proportion to their frequency in the entire population. Thus, one would sample more urban units from the Northeast than from the South or Midwest; conversely, one would select more rural units in the latter regions. Finally, within each sampling unit, people are selected randomly to serve as respondents. Using this technique, one can represent the adult population of the United States with a sample of 1,500 people and obtain responses accurate within plus or minus 3 percent.

Causal Analysis of Survey Data Social psychologists have long used computers to aid in the descriptive analysis of survey data, and they have often relied on such techniques as cross-tabulations and correlations. In recent years, however, some social psychologists have begun to use more sophisticated, computer-based techniques to aid in the causal interpretation of survey data. Analysis techniques of this type (such as LISREL and path analysis) require the investigator to postulate a pattern of cause-and-effect relations among a set of variables (Bollen, 1989; Jöreskög & Sörbom, 1979). The computer then estimates coefficients of effect from the data. These coefficients indicate the strength of the relationships among the variables, and they provide a test of whether the causal linkages postulated by the theory are indeed present in the data. Using this approach, an analyst can test many alternative hypotheses. Typically, some hypotheses will turn out to be inconsistent with the data, and the analyst can reject these in favor of alternative hypotheses that survive the test. One difficulty with this approach is that for problems involving many variables (say, a dozen or more), there often exist numerous alternative hypotheses that are plausible. Although this process will eliminate many hypotheses, more than one may survive as tenable.

Panel Studies One useful extension of the survey technique is the longitudinal survey or **panel study,** in which a given sample of respondents is surveyed at one point in time and then resurveyed at a later point. For instance, in a panel study, a sample of respondents would be surveyed by telephone interview or mail questionnaire (this is called the first wave of the panel). Then, at some future time (say, 1 year later), the same respondents would be surveyed again (the second wave); the questionnaire items in the second wave will be similar to—or an extension of—those used in the first wave. If desired, the same respondents could be surveyed again at a still later point in time (the third wave), and so on. In principle, there is no upper limit on the number of waves that might be included in a panel study, although there are practical constraints, such as the dollar cost of running the panel and the difficulties in tracking down members of the sample at various times. The waves in a panel study can be spaced either closely together or far apart in time, depending on the study's purpose.

The usual objective of a panel study is to determine whether various outcomes experienced by respondents at later points in time are related to or determined by their experiences, attitudes, and relationships at the earlier points in time. For instance, Orbuch, Veroff, Hassan, and Horrocks (2002) used a panel study with four waves to investigate the risk of divorce over a 14-year period. Initially, both members of 199 White couples and 174 Black couples who had recently married were interviewed. Couples were contacted and re-interviewed in years 3 and 7, and brief follow-up data were collected in year 14. The purpose of the research was to assess the role of social conditions (race, income) and interpersonal processes (positive interaction, frequency and type of conflicts). The results indicated that race and education were related to the risk of divorce. Blacks were twice as likely to be divorced in year 14, and couples in which the wife had more education (12 years or more) were less likely to be divorced. Reports of destructive conflict by husband and wife in earlier waves were related to subsequent risk of divorce. Thus, both social conditions, such as the disadvantaged conditions of some Blacks, and interactional style are related to divorce.

In general, data from a panel study lend themselves somewhat more readily to causal interpretation than data from a simple cross-sectional survey. The waves in the panel study provide a natural temporal ordering among the variables, which usually provides increased clarity when interpreting the results causally.

Strengths of Surveys Surveys can provide, at moderate cost, an accurate and precise description of the characteristics of a specific population. When a social psychological researcher uses measures that are reliable and valid, employs a sampling design that guarantees representativeness, and takes steps to ensure a high response rate, the survey can produce a clear portrait of the attitudes and social characteristics of a population.

Surveys also provide an effective means to study the incidence of various social behaviors. A survey asking people to report their behavior is usually more efficient and cost-effective than observational studies of actual behavior. This is especially true for behavior that occurs only infrequently or in private settings.

Surveys are frequently used to test predictions based on symbolic interaction theory, such as predictions about influences on personal identity and self-esteem. These methods are also used to test hypotheses about attitude structure and function based on cognitive theory.

Weaknesses of Surveys As with any methodology, there are certain drawbacks to the survey technique. Both questionnaires and interviews rely on self-reports by respondents. Under certain conditions, however, self-reports can be invalid sources of information. First, some people may not respond truthfully to questions about themselves. This is not usually a major problem, but it can become troublesome

© Mike Kagan

Observational research involves careful scrutiny of activities in natural settings. Here, a market researcher observes who is eating what in a food court.

if the survey deals with activities that are highly personal, illegal, or otherwise embarrassing to reveal. Second, even when respondents want to report honestly, they may give wrong information due to imperfect recall or poor memory. This can be a nettlesome problem, especially in surveys investigating the past (for example, historical events or childhood). As an illustration, consider the question "When were you last vaccinated?" This may seem simple and straightforward, but it often produces incorrect responses because many people cannot remember the relevant dates. Third, some respondents answering self-report questions have a tendency to fall into a response set. That is, they answer all questions the same way (for example, always agree or disagree) or they give extreme answers too frequently. If many respondents adopt a response set, this will introduce bias into the survey's results.

Field Studies and Naturalistic Observation

Observational research—often termed a **field study**—involves making systematic observations about behavior as it occurs naturally in everyday settings. Typically, the data are collected by one or more researchers who directly observe the activity of people and record information about it. Field studies have been used to investigate many forms of social behavior in their natural settings. For instance, researchers have observed and recorded data about social interaction between judges and attorneys in the courtroom (Maynard, 1983), between teachers and students in the classroom (Galton, 1987), between couples in informal settings (Zimmerman & West, 1975), between working-class boys and girls in grade school (Thorne, 1993), and between street vendors and passersby in

Greenwich Village (Duneier, 2001). Other studies have focused on socialization. Lois (1999) spent 3½ years observing a volunteer search and rescue group, studying the process by which individuals became willing to routinely risk their lives—often in dangerous situations such as blizzards—to save others.

Because field studies investigate social behavior in its natural setting, researchers usually make efforts to minimize or limit the extent to which they intrude on that behavior. In fact, field studies are usually less intrusive than surveys or experiments. Whereas a survey often intrudes on people by asking for self-reports and an experiment involves manipulation of the independent variable(s) and random assignment to treatment, a field study involves nothing more intrusive than recording an observation about the behavior of interest.

Field studies differ in how the observers collect and record information. In some studies, observers watch carefully while the phenomenon of interest is occurring and then make notes about their observations from memory at a later time. The advantage of recording afterward is that the observer is less likely to arouse curiosity, suspicion, or antagonism in the participants. In other studies, the observers may record field notes at the same time that they observe the behavior. For instance, in research on police-citizen encounters (Black, 1980; Lundman, Sykes, & Clark, 1978), trained observers coded the interaction as it occurred. Although taking notes in this manner could potentially be intrusive, it permits more details to be recorded and minimizes any distortion by selective memory on the part of the investigator.

In still other field studies, researchers make audio or video recordings of interactions, and then analyze the tapes later (Whalen & Zimmerman, 1987). Tape recordings may seem a superior alternative to the use of human observers (who may have selective perception), but this is not always the case. The use of recordings maximizes the information obtained, but it can also inadvertently influence behavior if the participants discover that they are being taped.

Participant Observation When the behavior of interest occurs in public settings, such as restaurants, courtrooms, or retail stores, researchers can simply go to the setting and observe the action directly. The researchers do not need to interact with the people being observed or reveal their identities. However, when the behavior of interest is private or restricted in nature (such as intimate sexual activity, use of illegal drugs, or recruiting new members for a cult), observation is usually more difficult. To investigate activities of this type, researchers occasionally use the technique of participant observation. In participant observation, members of the research team not only make systematic observations of others' behavior but also interact with them and play an active role in the ongoing events. Frequently, the fact of being an active participant enables the investigators to approach and observe behavior that otherwise would be inaccessible. In participant observation, researchers usually do not engage in overt coding or any other activity that would disrupt the normal flow of interaction. In some instances, they may even need to use an assumed identity, lest their true identity as investigators disrupt the interaction.

One study (Eder, 1995) used observational techniques combined with participation to investigate adolescent school culture in a midwestern community. To observe interaction patterns and topics of conversation among junior high school students, the investigators participated over an extended period of time in students' lunchroom groups. They identified themselves (truthfully) as being from a nearby university, and they adopted the role of "quiet friend." They did not affiliate with teachers and avoided appearing to be authority figures of any kind. This approach enabled them to establish sufficient rapport and trust with the students that they could ask questions about the students' beliefs regarding gender differences and observe how students' behavior patterns fostered gender inequality.

Unobtrusive Measures Field studies sometimes use unobtrusive measures, which are measurement techniques that do not intrude on the behavior under study and that avoid causing a reaction from the people whose behavior is being studied (Webb, Campbell, Schwartz, & Sechrest, 1981). For example, some unobtrusive measures rely on the physical evidence left behind by people after they have exited from a situation. One illustration is the analysis of inventory records and bar bills to unobtrusively measure the alcohol consumption patterns at various nightclubs and bars (Lex, 1986). Another investigator discovered that the rate at which vinyl floor tiles needed replacement in the Chicago Museum of Science and Industry was a good indicator of the popularity of exhibits.

Strengths and Weaknesses of Field Studies Like any research method, field studies have both strengths and weaknesses. A major strength is that observational techniques allow researchers to study social activity in real-world settings. Careful observation can provide a wealth of information about behavior as it actually occurs in natural settings. These data can be used to investigate ideas about social interaction drawn from role theory or symbolic interaction theory. Moreover, because these techniques are relatively unintrusive, investigators can use them to investigate sensitive or private behaviors—such as drug use or sexual activity—that would be difficult to address through intrusive methods like surveys or experiments.

Many field studies involve only one period of observation, however long that period may be. Burawoy (2003) suggests that a focused revisit to a site can serve several purposes, one of which is to study social change. A revisit to a factory 32 years after the original observational research identified significant changes in the interaction between supervisors and workers. The researchers were able to relate the changes observed in this factory to national trends in labor relations over the 32-year period.

Weaknesses of field studies include their sensitivity to the specific recording methods used. Observations recorded after the fact are often less reliable and valid than those recorded on the spot or those based on audio- or videotaping. Furthermore, the validity of the observations may depend in part on the identities that the investigators publicly project while making their observations; validity may be destroyed if the researchers have been operating covertly and the subjects suddenly discover that they are under observation. Then, too, the external validity of field observation studies can be problematic, because research of this type frequently focuses on only one group or organization, or on a sample of interactions selected for convenience.

In some cases, field investigators do not get informed consent from the people being observed prior to the collection of data. Permission for using the data is sought only after the behavior has been observed or the conversations tape-recorded. Some people construe this as a serious drawback and object to participant observation on ethical grounds. Of course, this concern has to be weighed against the fact that if permission were sought in advance, the behavior under investigation might never occur or might take a different form.

Archival Research and Content Analysis

Although social psychological researchers often prefer to collect original data, it is sometimes possible to test hypotheses and theories by using data that already exist. The term **archival research** denotes the acquisition and analysis (or re-analysis) of information collected previously by others. When archival data of suitable quality exist, a researcher may decide that analyzing them is preferable to collecting and analyzing new data. Archival research usually costs less than alternative methods.

Sources There are many sources of archival data. In the United States, one important source is government agencies. The Census Bureau makes available much of the data it has collected over the years. Census data are a rich source of information about the U.S. population; they often include repeated measures taken at different points in time, which allow an investigator to assess historical trends. The Bureau of Labor Statistics, the Federal Bureau of Investigation, and other agencies also release data to investigators. A second important source of archival data in the United States is the data banks maintained at various large universities. These archives serve as locations where researchers can deposit data they have collected so others can use them. They include, among others, the Interuniversity Consortium for Political and Social Research and the Data Archive on Adolescent Pregnancy and Pregnancy Prevention. A third source of archival data—less used by social psychologists but still important—are formal organizations such as insurance companies and banks. These typically entail over-time data with respect to various measures of financial and economic performance. A fourth source of archival information for research is newspapers. Newspaper articles are a rich source of information about past events. For instance, an investigator wishing to study the reactions of U.S. civilians during a historical event, such as the terrorist acts on September 11, 2001, might use newspapers as a data source. Other types of printed material (for example, corporate annual reports) can also provide archival data usable in research.

Content Analysis In some cases, an investigator relying on newspaper articles, government documents, or annual reports as archival sources can use the information directly as it appears. All the investigator has to do is extract the information and analyze it, usually by computer. In other cases, however, the investigator faces the problem of how to interpret and code the information from the source. Under these circumstances, he or she may use **content analysis,** which involves undertaking a systematic scrutiny of documents or messages to identify specific characteristics and then making inferences based on their occurrence. For example, if newspapers serve as the source, one could use content analysis to code the reportage from newspaper articles into a form suitable for systematic statistical analysis.

Researchers have used content analysis to investigate a wide variety of topics. Some studies, for instance, have analyzed the content of personal advertisements placed in magazines by gay men and lesbians (Bailey, Kim, Hills, & Linsenmeier, 1997). Other studies have addressed such issues as whether the depiction of people with disabilities is distorted in American newspapers (Keller, Hallahan, McShane, Crowley, & Blandford, 1990) and the relationship between the mortality rates associated with a disease, for example, AIDS, and newspaper coverage of that disease (Adelman & Verbrugge, 2000).

When a researcher conducts a content analysis, the first step is to identify the informational unit to be studied—is it the word, the sentence, the paragraph, or the article? The second step is to define the categories into which the units will be sorted. A third step is to code the units in each document into the categories, and the final step is to look for relations within the categorized data.

As an example of content analysis, consider a study of the relationship between rhetorical forms of speech and applause from the audience (Heritage & Greatbatch, 1986). The investigators hypothesized that political speakers will use certain rhetorical forms—for example, a three-element list—to signal the audience when to applaud. The raw data in this study were the texts of 476 speeches delivered by British political leaders at party meetings. The researchers carefully defined the rhetorical devices and identified their use in the speeches. Then they counted the number of times that the speakers used each device and noted whether the audience responded immediately to each use with applause. The results showed that applause was much more likely to occur immediately after the use of certain rhetorical devices (such as a three-element list) than at other points in the speech.

Strengths and Weaknesses of Archival Research One significant advantage of archival research is its comparatively low cost. By reusing existing information, the investigator avoids the cost of collecting new data. A second advantage is that by using information already on hand, an investigator may complete a study more quickly than otherwise. A third advantage is that an investigator can test hypotheses about phenomena that occur over extended periods of time. In some cases, authorities have kept records (such as marriage licenses) for decades or even centuries, and these can serve as a basis for investigating various questions (such as who marries whom).

One major disadvantage of archival research is the lack of control over the type and quality of information. An investigator must work with whatever others have collected. This may or may not include data on all the variables the investigator wishes to study. Moreover, there may be doubts regarding the quality of the original research design or the procedures used for collecting data. A second disadvantage of archival research is that creating a reliable and valid content analysis scheme for use with records can be difficult, especially if the records are complex.

A third disadvantage is that some sets of records contain large amounts of inconsistent or missing information. Obviously, this will hinder the study and limit the validity of any findings.

Experiments

The **experiment** is the most highly controlled of the research methodologies available to social psychologists, and it is a powerful method for establishing causality between variables. For a study to be a true experiment, it must have two specific characteristics:

1. The researcher must manipulate one or more of the independent variables that are hypothesized to have a causal impact on the dependent variable(s) of concern.
2. The researcher must assign the participants randomly to the various treatments—that is, to the different levels of each of the independent variables.

The term **random assignment** denotes the placement of participants in experimental treatments on the basis of chance, as by flipping a coin or using a table of random numbers. Random assignment is desirable because it mitigates the effects of extraneous variables. By using random assignment, the researcher creates groups of participants that are equivalent in all respects except their exposure to different levels of the independent variables. This removes the possibility that these groups will differ systematically on extraneous variables such as intelligence, personality, or motivation. Thus, random assignment enables the investigator to infer that any observed differences between groups on the dependent variable are due only to the effects of the independent variable(s), not to extraneous variables (Aronson, Wilson, & Brewer, 1998).

Whereas researchers manipulate the independent variables in an experiment, they simply measure the dependent variable(s). Experimenters can measure dependent variables in many ways. For example, they can monitor participants' physiological arousal, administer short questionnaires that assess participants' attitudes, record the interactions that occur between participants, or score the participants' performance on tasks. The exact type of measurement used in the experiment will depend on the nature of the dependent variable(s) of interest.

Laboratory and Field Experiments It is useful to distinguish between laboratory experiments and field experiments. Laboratory experiments are those conducted in a laboratory setting, where the investigator can control much of the participants' physical surroundings. In the laboratory, the investigator can determine which stimuli, tasks, information, or situations the participants will face. This control enables the experimenter to manipulate the independent variables, to measure the dependent variables, to hold constant some known extraneous variables, and to implement the random assignment of participants to treatments. For instance, if an investigator is studying the impact of verbal communication on group productivity in a laboratory setting, he may wish to restrict the interaction among participants. To do this, he might limit communication to written notes or verbal messages sent by electronic equipment. This practice would not only eliminate the possibly contaminating influence of nonverbal communication, but it also would permit the content of any messages to be analyzed later by the experimenter.

Field experiments, in contrast with laboratory experiments, are studies where investigators manipulate variables in natural, nonlaboratory settings. Usually, these settings are already familiar to the participants. Investigators have used field experiments to study topics ranging from pay inequity in large bureaucratic organizations to altruistic behavior on street corners and in subway cars. Compared with laboratory experiments, field experiments have the advantage of high external validity. When conducted in natural and uncontrived settings, they usually have greater mundane realism than laboratory experiments. Moreover, participants in field experiments may not be particularly conscious of their status as experimental subjects—a fact that reduces participants' reactivity. The primary weakness of field experiments, of course, is that in natural settings, experimenters sometimes have difficulty manipulating independent variables exactly as they would wish and often have little control over extraneous variables. This means that the internal validity of

Laboratory research enables the investigator to manipulate independent variables and measure behavior in various ways. In this study of aggressive behavior in children, trained observers collect data by viewing the child through a one-way mirror.

field experiments is often lower than in comparable laboratory experiments.

Conduct of Experiments To illustrate how investigators conduct experiments, consider the following laboratory study, which sought to determine the impact of certain independent variables on whether one person will help another in an emergency (Darley & Latané, 1968). The investigators conducted the study at a university in New York City. Male and female students serving as participants came to the laboratory to participate in a discussion of problems they had encountered in adjusting to the university. The experimenters placed each participant in a separate room in the laboratory and instructed them to communicate with other participants via an intercom. The rationale given was that this procedure would permit them to remain anonymous while discussing personal problems.

The independent variable was the number of other persons who the participant believed were participating in the discussion (and who would, therefore, later witness an emergency). Depending on experimental treatment, participants were told there were one, two, or five other participants. Participants were randomly assigned to the various levels of this independent variable.

The discussion proceeded with each participant speaking in turn over the intercom for 2 minutes. Thus, depending on the experimental treatment, the participant heard the voices of one, two, or five others. In reality, the participant was hearing a tape recording of other people, not the voices of actual participants. (This was the real reason for putting participants in separate rooms and having them communicate via intercom.) One of these recorded voices admitted somewhat hesitantly that he was subject to nervous seizures. In his second turn, he started to speak normally, but suddenly his speech became disorganized. Soon, he lapsed into gibberish and choking sounds and then into silence. Evidently, an emergency was occurring. The participant realized that all participants could hear it, although the intercom prevented them from talking to one another.

The dependent variables were whether the participant would leave the room to offer help and how quickly he or she would do so. Participants who elected to help the victim typically came out of their room looking for the victim. The experimenter timed the speed of the participant's response from the beginning of the victim's speech. The results verified the research hypothesis that the greater the number of witnesses, the less likely a participant was to offer help to the victim.

This carefully controlled experiment allowed a straightforward test of the hypothesis. The manipulated independent variable (number of witnesses) and the measured dependent variable (speed of helping response) were unambiguous. Confounds from extraneous variables could be ruled out due to the random assignment of participants to treatments. From these results, we can conclude that the number of witnesses has a causal effect on the speed of helping response.

Note, however, that although the experiment showed the causal effect to hold, it did so only under the conditions prevailing in the laboratory. The causal effect may or may not hold under other conditions. This can be problematic if the conditions that existed in the laboratory setting are uncommon in daily life. (When, for instance, was the last time you discussed personal issues over an intercom with five strangers in other rooms?) Thus, from this study alone, it is not clear whether we can generalize the cause-and-effect findings from the laboratory to everyday, face-to-face situations. The relationship between the number of others present and a person's reaction to an emergency might be different in other situations.

Although this experiment provides some answers regarding intervention in emergencies, it also raises further questions. Why, for instance, should the number of witnesses present affect a person's willingness to help in an emergency? The researchers conducting this study were aware of this question and, based on data from a brief questionnaire administered after the experiment, they proposed that participants in larger groups were slower to help because the responsibility for helping was more diffuse and less focused than in smaller groups. Although this diffusion of responsibility hypothesis is interesting, we must note that this experiment did not demonstrate it to be either true or false. The experiment showed only that under the conditions in the laboratory, the number of witnesses present affected the participants' helping behavior.

Strengths of Experiments The strength of experimental studies lies in their high level of internal validity. This makes experiments especially well suited for testing causal hypotheses. Experiments excel over other methods (surveys, field observation, and so on) in this respect.

Experiments have high internal validity precisely because they control or offset all factors other than the independent variable that might affect the dependent variable. Techniques to accomplish this include (1) randomly assigning participants to treatments, (2) holding constant known extraneous variables, and (3) incorporating extraneous variables as factors in the research design—that is, manipulating them as independent variables, so that they are not confounded with the main independent variables of interest. Another technique is (4) measuring extraneous variables and including them in the data analysis as covariates of the independent variables.

In principle, investigators can design both laboratory experiments and field experiments to have high internal validity. In practice, however, laboratory experiments often have higher internal validity than comparable field experiments. This happens because researchers have more control over extraneous variables in the laboratory than in the field. Field

BOX 2.2 Using Research to Answer Questions

- What makes people fall in love? What makes them fall out of love?
- What causes harmful or aggressive behavior?

In Chapter 1, we suggested that social psychology answers these questions by applying the methods of science. So how might we answer these questions, using the research methods discussed in this chapter?

Consider the questions about love. First, we need to define love. Since love is something people experience, we could begin with a survey. We could ask open-ended questions, such as:

- Have you ever been in love?
- How did you know you were in love?
- What does it feel like to be in love?

We would want to ask these questions of an appropriate sample, so we might choose young adults (college students?). After gathering answers from many respondents, we would study the answers carefully, looking for common themes in the answers to each of the three questions. If we were able to identify certain themes, we could then construct a scale, such as the one in Box 13.2.

Now we can turn to the question of what makes people fall in love. Again, we want to study people's experience. So we might use the method of collecting personal narratives or stories, and conducting a content analysis of the stories. We could request that people "Write a description of the most recent love relationship that you experienced. Describe how you met, what happened in the early days and weeks of the relationship, how you fell in love, and where your relationship is now." Obviously, we would want these narratives from people who are or have recently been in love; we might recruit such people by newspaper ads or flyers on bulletin boards. We would read a sample of the stories and try to develop a set of coding categories that capture the content of the stories. For example, categories for describing how people met could include: school, work, party, bar, music concert, sports event, introduction by friends/relatives, and religious services. Then we would train at least two coders to use our categories and "score" each narrative. Suppose the results showed that one half of the women and one third of the men met through an introduction; what would that suggest about how people fall in love? What if 40 percent of the men and women met at a bar?

experiments, however, often surpass laboratory experiments with respect to external validity.

Experiments have been used to test many causal hypotheses drawn from social exchange theory and cognitive theory. Hundreds of experiments have been conducted in an effort to identify the causes of racial and ethnic prejudice.

Weaknesses of Experiments One weakness of experimentation is that investigators cannot study many social phenomena by this method. Oftentimes, they lack the capacity to manipulate the independent variables of interest or to implement random assignment. Numerous ethical, financial, and practical considerations in everyday life restrict what investigators can manipulate experimentally.

Even when the independent variable(s) can be manipulated, experiments face several threats to internal validity. First, there is the possibility that the experimental manipulation may fail. This might occur, for example, if the participants interpret the manipulation as meaning something other than what the researcher intended. The usual remedy for this problem is to use manipulation checks—measures taken after the manipulation that show whether the participants perceived the manipulation as intended. Use of manipulation checks is routine and widespread in social psychological experiments.

Another threat to the internal validity of experiments is the existence of demand characteristics (also called subject effects). This refers to the possibility that participants may interpret certain subtle cues in the experimental setting as requiring particular responses (Aronson et al., 1998). A subject effect occurs, for instance, when participants bring a stereotyped role expectation or mental set to the experiment and then something in the experimental situation activates that expectation, causing the participants to emit the role-defined behavior. To prevent this, some designs disguise the nature of the research and the research hypothesis by providing a cover story—a plausible albeit false description of its purpose.

Another threat to internal validity is experimenter effects. This refers to the possibility that an experimenter may expect participants to behave in a particular manner (aggressively, cooperatively, and so on) and may unwittingly telegraph these expectations to the participants (Rosenthal, 1966, 1980). The expectations communicated to participants will likely influence their behavior. This can be a serious problem, especially if the expectations conveyed by the experimenter change as a function of the experimental treatment. Should this occur, the expectations communicated may have some implication for the hypothesis under study (such as speciously making the hypothesis appear to be true). People designing an experiment can use several techniques to minimize or eliminate experimenter effects. First, they can restrict the experimenters' contact with the participants and standardize their behavior in the experimental setting. This will limit the opportunities to transmit expectations. Second, they can keep the research personnel "blind" regarding the hypotheses under study and the treatment to which each participant is assigned. If the research personnel do not hold strong expectations, they obviously will not communicate them to the participants. Third, they can use a research design with two or more groups of experimenters, each holding a different hypothesis concerning the variables of the study. Analysis of the data from such a design will show whether experimenter effects are present or absent.

Beyond internal validity, experiments also face problems with external validity. Some experiments take place in settings that seem artificial to participants and have low apparent realism. This is often true of laboratory experiments, although less true of field experiments. One useful distinction is that between mundane realism and experimental realism (Aronson, Ellsworth, Carlsmith, & Gonzales, 1990). Mundane realism is the extent to which the experimental setting appears similar to natural, everyday situations. Experimental realism, in contrast, is the impact the experimental situation creates—that is, the degree to which the participants feel involved in the situation.

Low mundane realism need not imply low experimental realism. A laboratory study can have low mundane realism but high experimental realism. Participants were highly involved, for example, in the previously discussed study where the experimenters staged an emergency in the laboratory. Many participants were nervous and expressed concern when they came out of their room looking for the supposed victim. Most expressed surprise when they later learned that the seizure was simulated, not genuine.

There is no single solution to the problem of establishing high experimental realism. Some investigators use a combination of laboratory experiments and field experiments when investigating a phenomenon. This approach is often successful, for the field experiments provide the mundane realism that the laboratory experiments lack. Other investigators simply note that they are more concerned with experimental realism than with mundane realism. If the situation is real and involving to the participants, they maintain, then the behavior of the participants is real and worthy of study.

Comparison of Research Methods

We have discussed a variety of research methods—surveys, naturalistic observation, archival research, and laboratory and field experiments. Table 2.1 summarizes the strengths and weaknesses of each research method. As this table indicates, no one method

TABLE 2.1 STRENGTHS AND WEAKNESSES OF RESEARCH METHODS

	METHOD				
	Survey	Observational Study	Archival Research	Laboratory Experiment	Field Experiment
Internal Validity	Moderate	Low	Low	High	Moderate
External Validity	Moderate	Moderate	Moderate	Moderate	High
Investigator Control	Moderate	Moderate	Low	High	Moderate
Intrusiveness of Measures	Moderate	Moderate	Low	Moderate	Low
Difficulty of Conducting Study	Moderate	Moderate	Low	Moderate	High
Ethical Problems	Few	Many	Few	Some	Some

Note: Entries in the table indicate the strength of the research methods with respect to the various concerns (validity, control, intrusiveness, and the like).

of empirical investigation is best for all purposes. A method's appropriateness depends on the phenomenon under study and on the research characteristics most important to the investigator.

Surveys, which provide a useful way of obtaining an accurate description of the attributes of some population, usually have at least moderate internal and external validity, and they pose few ethical problems. Field studies relying on observational techniques will tend to have comparatively low internal validity and may confront a variety of ethical issues, but they may still be the best way to investigate previously unexplored social phenomena in their natural settings. Laboratory experiments, which can be especially useful in testing causal hypotheses, are generally high in internal validity, but they may pose some ethical problems (especially if deception is used).

Meta-Analysis

Social psychologists have been conducting empirical research for almost a century. There have been dozens and sometimes hundreds of studies of some phenomena. Unfortunately, the results of different studies on a specific question do not always agree. For instance, some studies show that contact with members of a group produces more positive attitudes (reduces prejudice) toward that group; other studies find that contact has no effect on attitudes. **Meta-analysis** is a technique that allows an investigator to bring order out of this apparent chaos.

Meta-analysis is a statistical technique that allows the researcher to combine the results from all previous studies on a question to determine what, collectively, they say. In conducting a meta-analysis, the researcher performs three steps:

1. The researcher locates all previous studies on the question. Today, this is typically done using computerized searches of libraries and databases.
2. For each study, the investigator computes a statistic that measures how big the difference was, say, between those who did and those who did not interact with members of the group, and what the direction of the difference was (whether those who had contact were more or less prejudiced). This statistic is called *d*. The formula for it is

$$d = \frac{M_c - M_{nc}}{s}$$

where M_c is the mean or average score for the participants who had contact and M_{nc} is the average score for those who did not; s is the standard deviation of the scores of all participants. The standard deviation is a measure of how much variability there is in the scores. The *d* statistic tells us—for this study—how big the difference between the two groups of participants was relative to the variability in scores.

3. The researcher averages all the values of *d* over all the studies that were located. This average *d* value tells what the direction of the difference is in attitudes between those who do and do not have contact with the group and how large the difference is for all the studies combined. A general guide is that a *d* of .20 is a small difference, a *d* of .50 is a moderate difference, and a *d* of .80 is a large difference.

We will include the results of a number of meta-analyses throughout the book.

RESEARCH IN DIVERSE POPULATIONS

For much of the 20th century, the participants in research by social psychologists were often White, often middle class, and often college students. In the past 25 years, there has been increasing interest in studying racial and ethnic minority groups in the United States, and members of other cultures around the world. It is important that research in such groups meet the standards of internal and external validity discussed earlier. This requires that we give careful consideration to the methods we use and be willing to adapt or change them.

Much research is based on theory. The theories and assumptions on which we base studies of diverse groups should take into account the cultural history and present social and economic circumstances of the group(s) being studied. For example, Orbuch et al. (2002), in developing the longitudinal study discussed earlier, assumed that the risk of divorce for Black couples is influenced by past and present social and economic conditions faced by Blacks. The measures must be linguistically equivalent—that is, be worded so that they are understood in the same way by all participants; if the participants speak a different primary language from that of the instrument, a careful process of translation and independent back translation should be employed to produce equivalent instruments. Measures should be standardized or interpreted using data from the population(s) being studied; for example, researchers should not use score distributions obtained from majority samples to interpret the scores of minority populations unless they have been shown to be equivalent. In Orbuch et al.'s research, the measures of positive interaction and of conflict had been used in the earlier waves of the research, and their applicability to both Blacks and Whites had been demonstrated.

If the researcher's intent is to characterize groups or cultures, the samples studied must be representative. If they are not, it should be noted in any reports of the research, and the results should be interpreted accordingly. Whereas the samples in Orbuch et al.'s study were not representative of Blacks or Whites, they did appear to represent the population of couples marrying for the first time in both groups. Finally, the research team should include either researchers who are members of the group(s) or persons who are culturally competent based on supervised training and experience (CNPAAEMI, 2000).

Culture refers to an intersubjective (shared) set of schema, attitudes, and values that members use to perceive and understand the world. When we conduct research, it is important that the results reflect the culture of the group(s) being studied. Some suggest that this requires that quantitative research be supplemented with methods focused on the cultural meanings that group members attribute to the quantitative measures. For example, a study of differences in gender role used scores on the Bem Sex-Role Inventory to compare a sample of European American women with a sample of Women of Color (Landrine, Klonoff, & Brown-Collins, 1995). There were no significant differences between the groups in self-rating on traits such as "feminine," "assertive," and "independent." Following the self-rating items were questions designed to measure the meaning of these words to the respondent. Responses to these questions revealed differences in meaning between the groups. The most common meaning of "assertive" among European American women was "standing up" for themselves, while among Women of Color it meant saying what was on their mind. Thus, understanding differences across groups requires research designs that will capture relevant aspects of the cultures of the groups.

ETHICAL ISSUES IN SOCIAL PSYCHOLOGICAL RESEARCH

As important as the methodological issues are the ethical issues involved in research on humans. There is a consensus among investigators and others affiliated with the scientific community that people who participate in research have certain rights that must be respected. In some cases, protecting those rights requires investigators to limit or modify their research practices.

In the following discussion of ethical issues, we focus first on potential sources of harm to participants. Then we discuss various safeguards, such as risk-benefit analysis and informed consent, to protect participants' rights. Finally, we consider potential benefits to participants in research.

Potential Sources of Harm

Harm to participants in research can take a variety of forms, including physical harm, psychological harm, and harm from breach of confidentiality. We will discuss each of these.

Physical Harm Exposure to physical harm in social psychological research is uncommon. Investigations to measure the effects of stress do sometimes employ an exercise treadmill or tasks where participants immerse one hand in ice water. As a precaution, investigators usually screen prospective participants to exclude those with relevant medical conditions. At the onset of a study, investigators are expected to inform the participants about any risks so that they can decide whether they might be harmed by participating. In studies involving physical stress, investigators typically monitor participants for adverse effects throughout the research.

Psychological Harm A more common risk in social psychological research is psychological harm to participants. This risk is present in studies where participants receive negative information about themselves. For example, a not uncommon experimental manipulation is to give participants false feedback about their physical attractiveness, about others' reactions to them, or about their performance on various tests or tasks. Investigators can use such feedback to raise or lower participants' self-esteem, to induce feelings of acceptance or rejection by others, or to create perceptions of success or failure on important tasks. These manipulations are effective precisely because they do influence the participant's self-perception.

Negative feedback may cause psychological stress or harm, at least temporarily. For this reason, some investigators believe that such techniques should not be employed in research. Others believe, however, that they are acceptable and may be used if alternative, less harmful manipulations are not available. When false feedback is used, an investigator can limit any long-term harmful effects by giving the participants a thorough debriefing after the study, providing the participants with a full description of the study and emphasizing the falsity of the feedback. Debriefing should be done immediately after the study to minimize the time that participants labor under false impressions.

Breach of Confidentiality Confidentiality is another important issue especially in survey and observational research. Interviewers and observers are frequently able to identify participants, and they may recall details regarding the participants' behavior or responses to questions. Were confidentiality to be breached, the effects might be damaging to the participants. This concern arises especially in surveys inquiring about sexual behaviors or other sensitive personal matters. It also arises in observational studies of deviant or criminal activities.

BOX 2.3 Ethical Considerations in Research Design

Before conducting a given study, investigators and members of review boards ask certain ethical questions about the proposed research design and its impact on participants. Among the most commonly asked ethical questions are the following:

1. Is it possible that participants in the study might be harmed physically, for example, by strenuous exercise?
2. Does the study give participants false information about themselves or use any other form of deception?
3. Does the study induce participants to engage in behavior that might threaten their self-respect?
4. If the investigators make audio- or videotapes of the participants, will they obtain permission from the participants to use the tapes as a data source?
5. What steps will the investigators take to preserve the confidentiality of information obtained about the participants?
6. Will the investigators tell potential participants in advance about the foreseeable risks that their participation may entail?
7. Will participants have a chance to ask questions about the study before they consent to participate?
8. Will the investigators inform the participants that they have the right to terminate their participation at any time?
9. At the end of the study, will the investigators fully debrief the participants and tell them about the real nature of the study and its procedures?

One important precaution against breach of confidentiality is to avoid including on the research team any people who are apt to have social contacts with respondents in other settings. Furthermore, some investigators refuse to attach any identifying information such as names and addresses to data after they have been collected. Another approach is to keep any identifying information separate from questionnaires or behavioral records to prevent breaches of confidentiality.

Observational research often deals with a specific group or organization. During their investigation, researchers may gather information about the organization itself and about various members. When these findings are published, the investigators typically refer to the organization by a pseudonym and to members by role only. This practice usually suffices to prevent outsiders from identifying the organization and its members, although it may not prevent members from identifying each other. There are obvious risks to members' positions, reputations, or jobs within the organization if compromising information becomes known to other members.

Box 2.3 lists some of the major ethical questions that apply to many studies.

Institutional Safeguards

As noted earlier, researchers can take various steps to prevent harm to participants. Although many people feel that voluntary self-regulation by researchers suffices to protect the rights and interests of the participants, others feel that some agency other than the researcher should review proposed research designs. Accordingly, many institutions have developed and put into place safeguards against potentially harmful effects of research. The two most important safeguards are conducting a risk-benefit analysis and obtaining informed consent from all participants.

Risk-Benefit Analysis The federal government is a major provider of funds for research in the social and biomedical sciences. Many federal departments and agencies have adopted common criteria for the review of research involving human participants. Under these regulations, investigators and institutions are responsible for minimizing the risks, of whatever type, to participants in research. The rules encourage researchers to develop designs that expose participants to no more than "minimal risk"—meaning risk no greater than that ordinarily encountered in daily life or during the performance of routine physical or psychological examinations or tests (Basic HHS Policy for Protection of Human Research Subjects, 2001).

Furthermore, the regulations require each institution that receives funds from federal agencies to establish an Institutional Review Board responsible for reviewing proposed research involving human participants. The review board (sometimes called a Human Subjects Committee or Research Ethics Committee) assesses the extent to which participants in each proposed study will be placed at risk. As noted earlier, many social psychological studies involve no foreseeable risks to participants, but if the members of the board believe that participants might be harmed—physically, psychologically, or by breach of confidentiality—a detailed assessment must be made. That is, the review board conducts a **risk-benefit analysis,** which weighs potential risks to the participants against anticipated benefits to the participants and the importance of the knowledge that may result from the research. The review board will not approve research involving risk to participants unless it concludes that the risk is reasonable in relation to the benefits.

Informed Consent The other major safeguard against risk is the requirement that investigators obtain informed consent from all individuals, groups, or organizations who participate in research studies. **Informed consent** exists when potential participants or respondents, on being informed by the investigators what their participation will involve, agree willingly to participate in the research. Specifically, six elements are essential to informed consent. (1) The researchers should give potential participants an explanation of the purposes of the research and a brief description of the procedures to be employed; however, they need not and usually do not tell the participants the hypothesis of the research. (2) The investigators should inform participants about any foreseeable risks of participation. (3) The researchers should provide a description of any benefits to the participant or others. (4) The investigators should provide information about which medical or psychological resources, if any, are available to participants who are adversely affected by participation. (5) The researchers should offer to answer questions about the study whenever possible. (6) The researchers should inform potential participants that they have the right to terminate their participation at any time.

In many survey and observational settings, investigators implement informed consent by giving this information to respondents orally. In experiments, especially those involving some risk to participants, investigators usually obtain written consent from each participant.

Potential Benefits

In the process of obtaining informed consent, participants are usually told that they will not benefit directly from the research. Although that is often true, there are exceptions. Field trials of new forms of treatment for physical or psychological problems may directly benefit participants if the new form of treatment proves to be effective. Similarly, participants in some studies may gain insight into themselves and others. For example, a longitudinal study of couples in premarital relationships included measures of how the men and women were affected. Many participants reported that they paid more attention to evaluating their relationship, and those who reported paying more attention reported more satisfaction with their relationship at the end of the year-long study (Hughes & Surra, 2000).

SUMMARY

This chapter discussed the research methods used by social psychologists to investigate social behavior, activity, and events.

Characteristics of Research (1) Objectives of research include describing reality, identifying correlations between variables, testing causal hypotheses, and testing theories. (2) Research is usually guided by a hypothesis, which specifies a causal relationship between two or more variables. (3) Ideally, the findings of empirical research should be high in both internal validity and external validity.

Research Methods Social psychologists rely heavily on four methods—surveys, naturalistic observation, archival research based on content analysis, and experiments. (1) A survey involves systematically asking questions and recording the answers from respondents. Investigators use surveys to gather self-reported information about attitudes and activities. The quality of the data obtained in a survey depends on the reliability and validity of the measures used. (2) Naturalistic observation involves collecting data about naturally occurring events. In a field study, observers view an event or activity as it occurs and then record their observations. (3) Archival research involves the analysis of existing information collected by others. Sources of archival data include the Census Bureau and other federal agencies, data archives, and newspapers. Investigators use content analysis to study textual material such as speeches or reports. (4) An experiment involves the manipulation of independent variables and the random assignment of participants to experimental conditions or treatments. Some experiments are conducted in a laboratory, where the investigator has a high degree of control, whereas others are conducted in natural settings.

Ethical Issues in Research (1) There are several potential sources of harm to participants in research. These include physical harm, psychological harm, and breach of confidentiality. There are various steps that individual investigators can take to prevent or minimize such harm. (2) There are also institutional safeguards against harm. These safeguards require investigators to minimize risks to participants and to obtain informed consent from participants. Institutional review boards monitor research designs to ensure that these conditions are met by investigators. (3) In some cases, participants in research may benefit directly from their participation.

LIST OF KEY TERMS AND CONCEPTS

archival research (p. 39)
content analysis (p. 40)
dependent variable (p. 28)
experiment (p. 41)
external validity (p. 29)
extraneous variable (p. 29)
field study (p. 37)
hypothesis (p. 28)
independent variable (p. 28)
informed consent (p. 50)
internal validity (p. 29)
interview survey (p. 30)
Likert scale (p. 34)
meta-analysis (p. 46)
methodology (p. 27)
panel study (p. 36)
population (p. 34)
questionnaire survey (p. 31)
random assignment (p. 41)
reliability (p. 31)
response rate (p. 31)
risk-benefit analysis (p. 50)
simple random sample (p. 35)
stratified sample (p. 35)

3

SOCIALIZATION

Introduction

Perspectives on Socialization

The Developmental Perspective

The Social Learning Perspective

The Interpretive Perspective

The Impact of Social Structure

Agents of Childhood Socialization

Family

Peers

School

Processes of Socialization

Instrumental Conditioning

Observational Learning

Internalization

Outcomes of Socialization

Gender Role

Linguistic and Cognitive Competence

Moral Development

Work Orientations

Adult Socialization

Role Acquisition

Anticipatory Socialization

Role Discontinuity

Summary

List of Key Terms and Concepts

INTRODUCTION

My 9-year-old son, Levi, is the kind of kid for whom playing in the soccer league means getting to wear shinguards and eating soft ice cream afterward. Not that he minds the game in between. It gives him lots of opportunities for performing pratfalls and planning what he'll say when the teams congratulate one another after it ends. ("Good game." Hand slap. "Great game." Hand slap. "Excellent game." Hand slap.) What Levi isn't particularly interested in is what all those other kids are doing on the field. While the rest of the team is focused on running practice drills and scoring goals, my son seems far more intent on simply going his own way.

"Levi has got to learn how to be part of a team," scolds the mother of one of his friends. I nod soberly as if I agree. But I feel conflicted—caught between knowing what my child ought to do and enjoying what he is actually doing. Because what that other parent doesn't see is that my son has his own way of dealing with team sports. He goofs around, occasionally bursts into song, and has a great time, whether they win or lose. And I can't help secretly approving.

Unfortunately, it isn't fair to force Levi's eccentricities on the rest of his team. His buddies may be too young and klutzy for his antics to be all that disruptive, but their attitudes about the game have been getting more and more serious of late. It's only a matter of time before one of them decks him. . . .

But somehow I can't bring myself to pull him off the team. Not when I feel partially to blame for the way he is. Some of Levi's attitude comes from the fact that he is a creative kid who resists being limited by the pesky rules and regulations that govern scoring goals and winning games. But much of his attitude comes from my tacit approval of his shenanigans. The truth is, I'm proud to be raising a kid who manages to have fun doing something that used to make me miserable.

My first brush with team sports was playing little league softball. Like Levi's, my mind would wander far away from the position I'd been assigned, which was way, way, way out in left field. I lasted about a season, despite having zero aptitude or interest. . . .

The only time I couldn't escape team sports was in phys ed [high school]. As one of the last kids to get picked for any game, I'd think, "Hey, if they think I stink so bad, the last thing they're gonna get is my best. I'll drag 'em all down with me!" Not that I'd try to throw the game on purpose—my natural lack of ability was usually enough to do the job. . . .

I know that, eventually, I'm going to have to act like a responsible parent and steer my son into a solo sport like track or tennis—something with more room for individuality. For now, though, if his fondest wish is to come up with the most creative post-game hand slap ("Superior game"), so be it. (Squier, 2002, p. 10)

This essay reflects on one of the striking features of social life. There is great continuity from one generation to the next—continuity both in physical characteristics and in behavior. Genetic inheritance is one source of continuity. But a major contributor to intergenerational similarity is **socialization,** the ways in which individuals learn and recreate skills, knowledge, values, motives, and roles appropriate to their position in a group or society.

How does an infant become "human"—that is, an effective participant in society? The answer is, through socialization. As we grew from infancy, we interacted continually with others. We learned to speak a language—a prerequisite for participation in society. We learned basic interaction rituals, such as greeting a stranger with a handshake and a loved one with a kiss. We also learned the socially accepted ways to achieve various goals, both material (food, clothing, shelter) and social (respect, love, help of others). As we learned these, we used them; as we used them, we recreated them—adapted them to our particular circumstances.

It is obvious that socialization makes us like most other members of society in important ways. It is not so obvious that socialization also produces our individuality. The self and the capacity to engage in self-oriented acts (discussed in Chapter 4) are a result of socialization.

The first part of this chapter will examine childhood socialization. By childhood, we mean the period from birth to adolescence. Childhood is a social concept, shaped by historical, cultural, and political influences (Elkin & Handel, 1989). In contemporary American society, we define children as immature—in need of training at home and of a formal education. In the last part of this chapter, we will consider the continuing socialization of adults.

The discussion focuses on the following five questions:

1. What are the basic perspectives in the study of socialization?
2. What are the socializing agents in contemporary American society?
3. What are the processes through which socialization occurs?
4. What are the outcomes of socialization in childhood?
5. What is the nature of socialization in adulthood?

Responsiveness to another person develops early in life. By 16 weeks of age, a child smiles in response to a human face. By 28 weeks, a child can distinguish caregivers from strangers.

PERSPECTIVES ON SOCIALIZATION

Which is the more important influence on behavior—nature or nurture, heredity or environment? This question has been especially important to those who study children. Although both influences are important, one view emphasizes biological development (heredity), whereas another emphasizes social learning (environment).

The Developmental Perspective

The human child obviously undergoes a process of maturation. He or she grows physically, develops motor skills in a relatively uniform sequence, and begins to engage in various social behaviors at about the same age as most other children.

Some theorists view socialization as largely dependent on processes of physical and psychological maturation, which are biologically determined. Gesell and Ilg (1943) have documented the sequence in which motor and social skills develop and the ages at which each new ability appears in the average child. They view the development of many social behaviors as primarily due to physical and neurological maturation, not social factors. For example, toilet training requires voluntary control over sphincter muscles and the ability to recognize cues of pressure on the bladder or lower intestine. According to developmental theory, when children around age 2½ develop these skills, they learn by themselves without environmental influences.

Table 3.1 lists the sequences of development of various abilities that have been identified by observational research. The ages shown are approximate; some children will exhibit the behavior at younger ages, whereas others will do so later.

As an example, consider the development of responsiveness to other persons. As early as 4 weeks, many infants respond to close physical contact by relaxing. At 16 weeks, babies can discriminate the human face and usually smile in response. They also show signs of recognizing the voice of their usual caregiver. By 28 weeks, the infant clearly differentiates faces and responds to variations in facial expression. At 1 year, the child shows a variety of emotions

TABLE 3.1 THE PROCESS OF DEVELOPMENT

	16 Weeks	28 Weeks	1 Year	2 Years	3 Years
Visual Activity	Follows objects with eyes; eyes adjust to objects at varying distances	Watches activity intently; hand-eye coordination	Enjoys watching moving objects (like TV picture)	Responds to stimuli in periphery of visual field; looks intently for long periods	
Interpersonal	Smiles at human face; responds to caregiver's voice; demands social attention	Responds to variation in tone of voice; differentiates people (fears strangers)	Engages in responsive play; shows emotions, anxiety; shows definite preferences for some persons	Prefers solitary play; rudimentary concept of ownership	Can play cooperatively with older child; strong desire to please; gender differences in choice of toys, materials
Vocal Activity	Vocalizes pleasure (coos, gurgles, laughs); babbles (strings of syllable-like sounds)	Vocalizes vowels and consonants; tries to imitate sounds	Vocalizes syllables; practices two to eight known words	Vocalizes constantly; names actions; repeats words	Uses three-word sentences; likes novel words
Bodily Movement	Can hold head up; can roll over	Can sit up	Can stand; can climb up and down stairs	Can run; likes large-scale motor activity—push, pull, roll	Motion fluid, smooth; good coordination
Manual Dexterity	Touches objects	Can grasp with one hand; manipulates objects	Manipulates objects serially	Good control of hand and arm	Good fine-motor control—uses fingers, thumb, wrist well

Sources: Adapted from Caplan, 1973; and *The Infant and Child in the Culture of Today* (1943) by Arnold Gesell and Frances L. Ilg. Used with permission of the Gesell Institute of Human Development.

in response to others' behavior. He or she will seek interaction with adults or with siblings by crawling or walking toward them and tugging on clothing. Thus, recognition of, responsiveness to, and orientation toward adults follow a uniform developmental pattern. The ability to interact with others depends in part on the development of visual and auditory discrimination.

The Social Learning Perspective

Whereas the developmental perspective focuses on the unfolding of the child's own abilities, the social learning perspective emphasizes the child's acquisition of cognitive and behavioral skills from the environment. Successful socialization requires that the child acquire considerable information about the world. The child must learn about many physical or natural realities, such as what animals are dangerous and which things are edible. Children also must learn about the social environment. They must learn the language used by people around them to communicate their needs to others. They also need to learn the meanings their caregivers associate with various actions. Children need to learn to identify the kinds of persons encountered in their immediate environment. They need to learn what behaviors they can expect of people, as well as others' expectations for their own behavior.

According to the social learning perspective, socialization is primarily a process of children learning the shared meanings of the groups in which they are reared (Shibutani, 1961). Such variation in meanings gives groups, subcultures, and societies their distinctiveness. Although the content—what is learned—varies from group to group, the processes by which social learning takes place are universal. This viewpoint emphasizes the adaptive nature of socialization. The infant learns the verbal and interpersonal skills necessary to interact successfully with others. Having acquired these skills, children can perpetuate the meanings that distinguish their social groups and even add to or modify these meanings by introducing innovations of their own.

Recent research on socialization has considered both the importance of developmental processes and the influence of social learning. The developmental age of the child obviously determines which acts the child can perform. Infants less than 6 months old cannot walk. All cultures have adapted to these developmental limitations by coordinating the performance expectations placed on children with the maturation of their abilities. However, developmental processes alone are not sufficient for the emergence of complex social behavior. In addition to developmental readiness, social interaction—learning—is necessary for the development of language. This is illustrated by the case of Isabelle, who lived alone with her deaf-mute mother until the age of 6½. When she was discovered, she was unable to make any sound other than a croak. Yet within 2 years after she entered a systematic educational program, her vocabulary numbered more than 1,500 words and she had the linguistic skills of a 6-year-old (Davis, 1947).

Thus, both nature and nurture influence behavior. Developmental processes produce a readiness to perform certain behaviors. The content of these behaviors is determined primarily by social learning—that is, by cultural influences.

The Interpretive Perspective

Socialization occurs primarily through social interaction. Whereas the social learning perspective emphasizes the process of learning—for example, the role of reinforcement in the acquisition of behavior—the interpretive perspective (Corsaro & Fingerson, 2003) focuses on the interaction itself. Drawing on symbolic interaction theory, this perspective views the child's task as the discovery of the meanings common to the social group (such as the family or a soccer team). This process of discovery requires communication with parents, other adults, and other children. Especially important is the child's participation in **cultural routines,** which are recurrent and predictable activities that are basic to day-to-day social life (Corsaro & Fingerson, 2003). Greeting rituals, common games, and mealtime patterns are examples of such cultural routines. These routines provide a sense of security and of belonging to a group. At the same time, their predictability enables children to use them to display their developing cultural knowledge and skills. A good example is Levi, whom we met in the opening essay; he has learned the ritual of complimenting opponents at the end of the game, and is embellishing it in an effort to find "the most creative post-game hand slap."

According to this perspective, development is a process of interpretive reproduction. Children don't simply learn culture. In daily interaction, children use the language and interpretive skills that they are learning or discovering. As they become more proficient in communicating and more knowledgeable about the meanings shared in the family, children attain a deeper understanding of the culture. Children, through interaction, acquire and reproduce the culture.

When children communicate with each other (as in school or at play), they don't simply imitate the acquired culture. They use what they have learned to create their own somewhat unique peer culture.

Children take a traditional game such as hide-and-seek and change the rules to fit their needs and the social context in which they are enacting the game. The changed rules become part of a new routine of hide-and-seek. Thus, from an early age, children are not just imitating culture, but creating it.

The Impact of Social Structure

A fourth perspective emphasizes the influence of social structure. Socialization is not a random process. Teaching new members the rules of the game is too important to be left to chance. Socialization is organized according to the sequence of roles that newcomers to the society ordinarily pass through. In American society, these include familial roles, such as son or daughter, and roles in educational institutions, such as preschooler, elementary school student, and high school student. These are age-linked roles; we expect transitions from one role to another to occur at certain ages. Distinctive socialization outcomes are sought for those who occupy each role. Thus, we expect young children to learn language and basic norms governing such diverse activities as eating, dressing, and bowel and bladder control. Most preschool programs will not enroll a child who has not learned the latter.

Furthermore, social structure designates the persons or organizations responsible for producing desired outcomes. In a complex society such as ours, there is a sequence of roles and a corresponding sequence of socializing agents. From birth through adolescence, the family is primarily responsible for socializing the child. In the opening essay, Levi's dad recognizes that eventually he will "have to act like a responsible parent" and respond to Levi's lack of involvement in soccer. From ages 6 to 12, a child is an elementary school student; we expect elementary school teachers to teach the basics to their students. Next, the adolescent becomes a high school student, with yet another group of agents to further develop his or her knowledge and abilities.

This perspective is sociological; it considers socialization as a product of group life. It calls our attention to the changing content of and responsibility for socialization throughout the individual's life.

AGENTS OF CHILDHOOD SOCIALIZATION

Socialization has four components. It always involves (1) an agent—someone who serves as a source for what is being learned; (2) a learning process; (3) a target—a person who is being socialized; and (4) an outcome—something that is being learned. This section will consider the three primary agents of childhood socialization—family, peers, and school. Later sections will focus on the processes and outcomes of childhood socialization.

Family

At birth, infants are primarily aware of their own bodies. Hunger, thirst, or pain creates unpleasant and perhaps overwhelming bodily tensions. The infant's primary concern is to remove these tensions and satisfy bodily needs. To meet the infant's needs, adult caregivers must learn to read the infant's signals accurately (Ainsworth, 1979). Also, infants begin to perceive their principal caregivers as the source of need satisfaction. These early experiences are truly interactive (Bell, 1979). The adult learns how to care effectively for the infant, and the infant forms a strong emotional attachment to the caregiver.

Is a Mother Necessary? Does it matter who responds to and establishes a caring relationship with the infant? Must there be a single principal caregiver in infancy and childhood for effective socialization to occur?

Psychoanalytic theory (as originally framed by Freud) asserts that an intimate emotional relationship between infant and caregiver (almost always the mother at the time Freud wrote) is essential to healthy personality development. This was one of the first hypotheses to be studied empirically. To examine the effects of the absence of a single, close caregiver on children, researchers have studied infants who were institutionalized. In the earliest reported work, Spitz (1945, 1946) studied an institution in which six nurses cared for 45 infants under 18 months old. The nurses met the infants' basic biological needs. However, they had limited contact with the babies, and there was little evidence of emotional ties between the nurses and the infants. Within 1 year, the infants' scores on developmental tests fell dramatically from an average of 124 to an average of 72. Within 2 years, one third had died, 9 had left, and the 21 who remained in the institution were severely retarded. Recent research on children who lived in orphanages for an average of 16 months following birth found that at age 4½, they had significant difficulty matching facial expressions of emotion with stories, compared to children from control families (Fries & Pollak, 2004). These findings dramatically support the hypothesis that an emotionally responsive caregiver is essential.

Thus, infants need a secure **attachment**—a warm, close relationship with an adult that produces a sense of security and provides stimulation—to develop the interpersonal and cognitive skills needed for proper growth (Ainsworth, 1979). Moreover, being cared for in such a relationship provides the foundation of the infant's sense of self.

For many decades, gender role definitions in American society made mothers primarily responsible for raising children. Fathers' parental responsibility was to work outside the home and provide the income needed by the family. The division of labor in many families conformed to these definitions. As a result, some analysts concluded that a warm, intimate, continuous relationship between a child and its mother is essential to normal child development (Bowlby, 1965). Perhaps only in the mother-infant relation can the child experience the necessary sense of security and emotional warmth. According to this view, other potential caregivers have less emotional interest in the infant and may not be adequate substitutes.

Research on parent-child interaction indicates that if mothers are sensitive to the child's needs and responsive to his or her distress in the first year of life, the child is more likely to develop a secure attachment (Demo & Cox, 2001). This is true in both two-parent and mother-only families. Infants who are securely attached to their mothers in the first 2 years of life evidence less problem behavior and more cooperative behavior from ages 4 to 10. Thus, secure mother-infant attachment is associated with positive outcomes.

There has also been research on the father-infant relationship and attachment at later ages. A meta-analysis of the results of relevant research has reported a weak but significant relationship (De Wolff & Van IJzendoorn, 1997).

Since 1960, gender role definitions have been changing. Married women with children are increasingly working outside the home (see Figure 17.3). The effects of maternal employment on the child is a major continuing public concern.

Effects of Maternal Employment What effects do maternal employment and child care have on children? Some studies have reported negative cognitive and social outcomes in children whose mothers worked during the first year of the child's life; other studies reported positive effects, and some studies found no differences (Perry-Jenkins, Repetti, & Crouter, 2001). Using a large national sample from a longitudinal study, Harvey (1999) found that neither mother's or father's employment was consistently related to child outcomes. The inconsistent results suggest that it is not employment per se that affects the child, but employment in combination or interaction with other variables. For example, a study of 147 employed mothers and their children (MacEwen & Barling, 1991) found that it was the amount of conflict between maternal and work roles that was associated with children's behavior problems and anxiety, not the work itself. It should be noted that much of this research has focused on middle-class families.

Among adolescents, maternal employment is associated with positive outcomes. Youth whose mothers are employed have less stereotyped gender role attitudes and expectations. A study of the likelihood of graduating from high school in a sample of 1,258 young people ages 19 to 23 found that maternal employment during adolescence had a positive effect on school completion. Moreover, maternal employment during childhood had no effect on school completion (Haveman, Wolfe, & Spaulding, 1991).

What about the effects of day care? It depends on the type, quality, and amount of care. A large-scale research project conducted at 10 sites around the United States has followed 1,000 children from birth. At age 4½, children who experienced higher quality

BOX 3.1 Test Yourself: Attachment in Children and Adults

Which of the following best describes your feelings about relationships?

1. I find it relatively easy to get close to others and am comfortable depending on them and having them depend on me. I don't often worry about being abandoned or someone getting too close to me.
2. I am somewhat uncomfortable being close to others; I find it difficult to trust them completely, difficult to allow myself to depend on them. I am nervous when anyone gets too close, and often, love partners want me to be more intimate than I feel comfortable being.
3. I find that others are reluctant to get as close as I would like. I often worry that my partner doesn't really love me or won't want to stay with me. I want to merge completely with another person, and this desire sometimes scares people away (Hazan & Shaver, 1987).

Each of these statements represents one attachment style, an individual's characteristic way of relating to significant others (Hazan & Shaver, 1987). The first describes a secure style, the second an avoidant style, and the third an anxious/ambivalent style.

The roots of the individual's style may be found in childhood. Ainsworth (1979) identified three styles of attachment in caregiver-child interactions. The attachment style of a young child is assessed by observing how the child relates to his or her caregiver when distressed (by, for example, a brief separation in a strange environment). The secure child readily approaches the caregiver and seeks comfort. The avoidant child does not approach the caregiver and appears detached. The anxious/ambivalent child approaches the caregiver and expresses anger or hostility toward him or her. Children as young as 2 behave consistently in one of these ways when distressed.

We bring the style we developed as children into our intimate adult relationships. Surveys of adults (for example, Hazan & Shaver, 1987) have found that about 55 percent describe themselves as secure, 25 percent as avoidant, and 20 percent as anxious/ambivalent. Attachment style influences our responses to other people (Feeney, 1999). It leads us to pay attention to certain aspects of a person (for example, his or her trustworthiness), creates biases in memory (we remember events consistent with our style), and affects how we explain relationship events. A secure person will ignore an event (his partner talking to an attractive person) that would make an anxious person feel jealous. Attachment style also influences relationship quality. Men and women who describe themselves as secure report that their romantic relationships involve interdependence, trust, and commitment (Simpson, 1990). Adults who describe themselves as avoidant say that they do not trust others and are afraid of getting close (Feeney & Noller, 1990). Those who are anxious/ambivalent report intense emotions toward the partner and a desire for deep commitment in a relationship. Since attachment style develops on the basis of childhood experience, analysts assume that it precedes adult relationships. Longitudinal data point to stability in style over time (Feeney, 1999). However, particularly significant relationship experiences may lead to change in style. A secure person who spends a long time with someone who is chronically unfaithful understandably may become anxious.

care and whose care was provided in a center had significantly better cognitive skills and language performance; quality was measured using observers who completed a standardized observational record. Children who received more hours of care between the ages of 3 months and 4½ were given higher ratings on behavior problems (on the 113 item Child Behavior Checklist) by care providers. Twenty-four percent of the sample were children of color; it appears that the results do not vary by ethnicity (NICHD Early Child Care Research Network, 1997a, 1997b, 2002).

Father's Involvement with Children The broadening of maternal role definitions to include work outside the home has been accompanied by changes in expectations for fathers. This new ideology of fatherhood, promoted by television and film, encourages active involvement of fathers in child care and child rearing (Parke, 1996). Many men have adopted these expectations for themselves; some have not. Research finds that married fathers spent significantly more time with their child(ren) each day in 1998 than they did in 1965 (Sayer, Bianchi, & Robinson, 2004). Father's contribution is often through rough-and-tumble play; such play is thought to facilitate the child's development of motor skills. Fathers increasingly also engage in child care and developmental activities. These patterns are found in European American, African American, and Hispanic two-parent families (Parke, 1996).

Several variables influence the extent of fathers' involvement with their children. Maternal attitudes are one important factor; a father is more involved when the mother encourages and supports his participation. Maternal employment is another influence. Husbands of employed women are more involved in child care and in some cases provide full-time care for the child. Finally, a study found that lower levels of stress on the job and greater support from coworkers for being an active father were associated with greater involvement (Volling & Belsky, 1991). Thus, research suggests that work stressors have negative effects on both fathers' and mothers' involvement in child rearing.

TABLE 3.2 CHILDREN'S LIVING ARRANGEMENTS, 2000

Arrangement	Percentage
Working father/nonworking mother	21
Married, both working	41
Male-headed	2.3
Female, previously married	10
Female, never married	5
Cohabiting couple	4.1
Grandparents	6
Unknown	10.6

Source: *Marriage and Family in a Multiracial Society* by Lichter and Qian.

Child Rearing in a Diverse Society There is diversity in the living arrangements of children in the United States today. Table 3.2 indicates the living arrangements of children in 2000. These arrangements vary by race/ethnicity. Compared to Whites, more African American children live with a single mother (35%), and more American Indian children live with a cohabiting couple or grandparents (Lichter & Qian, 2004). Note that there are more than two million children living with a single father (U.S. Census Bureau, 2005).

Studies of socialization have focused on child rearing techniques or parenting styles and their impact on cognitive and social development. Research has consistently found that authoritative parenting—characterized by high levels of warmth combined with control—benefits children. Reliance by parents on this style is associated with greater achievement in school and positive relations with other adults and peers. Authoritarian styles, including physical punishment, and permissive styles are more likely to be associated with poor adjustment in childhood (Demo & Cox, 2001).

Much of this research has involved European American families. In the 1990s, minority researchers challenged the validity of this model for minority families (McLoyd, Cauce, Takeuchi, & Wilson, 2001). White and Black mothers living in poverty are more likely to use physical punishment, partly due to chronic financial stress (Demo & Cox, 2001). Research by Deater-Deckard and Dodge (1997) has suggested that physical discipline is more common in African American families and that they define it as positive parenting. Other research (Chao, 1994) has suggested that Asian American parents rely on providing training and clear and concrete guidelines for behavior, and that this should not be seen as authoritarian. However, longitudinal research with a sample of 3,400 families, each with one or more children, found that parental support, monitoring, and the use of harsh punishment resulted in the same outcomes for children and adolescents in White, African American, and Hispanic families (Amato & Fowler, 2002). We need more research on both the nature and the outcomes of childhood socialization among non–European American cultural groups. In addition to differences from European Americans, it is likely that we will find variation within each group.

With respect to values, White parents emphasize the development of autonomy (Alwin, 1990), which is consistent with the mainstream culture's emphasis on individualism and independence. Minority children are more likely to be socialized to value cooperation and interdependence (Demo & Cox, 2001). African American parents tend to emphasize assertiveness, whereas Mexican American families emphasize family unity and solidarity with the extended family. Asian American parents teach children to value family authority. Thus, as we would expect, socialization in distinctive communities tends to emphasize the values of those communities.

Effects of Divorce Half of all marriages end in divorce; about one half of these divorces involve children under the age of 18. Divorce usually involves several major changes in the life of a child: a change in family structure, a change in residence, a change in the family's financial resources, and perhaps a change of schools. Therefore, it is difficult to isolate the effects of divorce—the change in family structure—independently of these other changes. An additional confounding fact is that divorce is not a one-time crisis; it is a process that begins with marital discord while the couple is living together, continues through physical separation and legal proceedings, and ends, if ever, when those involved have completed the uncoupling process (Amato, 2001).

Research comparing children of divorced with children of married parents has consistently found that the children of divorced parents score lower on measures of academic success (such as grades), psychological adjustment, self-esteem, and long-term health, among other outcomes (Amato, 2001). Some research (for example, Hetherington, 1999) has reported that these deficits were present several years before the divorce, leading to the suggestion that children's problem behaviors cause the discord that leads to divorce. However, if we view the divorce as a process, problems prior to the divorce could be caused by the marital discord. A few studies report positive consequences for some children. Some offspring, especially daughters, develop very positive relationships with custodial mothers (Arditti, 1999).

The view of divorce as a one-time crisis implies that children will show improved function as the time since divorce increases. Some studies (for example, Jekielek, 1998) report that children's well-being does improve over time. On the other hand, longitudinal research finds that the gap in well-being between children of divorced parents and children of intact couples increases (Cherlin, Chase-Lansdale, & McRae, 1998) or remains the same (Sun & Li, 2002) over time. A unique study documents intergenerational effects of divorce. The researchers reported negative effects on subsequent academic achievement, later marital discord, and weak ties to mothers and fathers in both the second and third generations (that is, effects on children and grandchildren) (Amato & Cheadle, 2005).

Although most people acknowledge the undesirability of divorce, it is often justified with the argument that it is less harmful than growing up in a family with chronic marital, social, and perhaps economic problems. Is this true? A longitudinal study in Great Britain followed thousands of children from birth to age 33, enabling researchers to compare adults whose parents divorced when they were 7–16, 17–20, or 21–33 years of age (Furstenberg & Kiernan, 2001). The results show that men and women whose parents

BOX 3.2 The Peer Group

American society is highly segregated by age. Most of us spend most of our time with people of about the same age. This is especially true in childhood and adolescence, because age segregation is the fundamental organizing principle of our schools. Research provides important insights into the nature of peer groups and their significance for socialization.

Among preschool-age children, a major concern is social participation. Kids in American society learn about the role of friend and the expectations associated with that role. Their understanding of this role provides a basis for evaluating their relationships with other children. As children begin to play in groups, maintaining access to the group becomes an issue. Children become concerned with issues of inclusion and exclusion—who is in the group and who is not. These issues remain important ones throughout childhood and into adolescence (Adler & Adler, 1995).

Peer groups reflect the desire of children to gain some control over the social environment and to use that control in concert with other children (Corsaro & Eder, 1995). Children become concerned with gaining control over adult authority, and they learn that a request or plea by several children is more likely to be granted. In elementary school, children develop a strong group identity, which is strengthened by minor rebellions against adult authority. Thorne (1993) observed that in one fourth–fifth grade classroom, most of the students had contraband—small objects such as toy cars and trucks, nail polish, and stuffed animals—which were prohibited by school rules. By keeping these items in desks and by displaying or exchanging them at key moments during class, the kids were displaying resistance, a form of nonconformity challenging the academic regime and rules in the classroom (McFarland, 2004).

Both children and adolescents assert themselves by making fun of and mocking teachers and administrators.

Peer groups play a major role in socializing young persons to gender role norms. As children move through elementary school, they increasingly form groups that are homogeneous by gender. For instance, in one study, Thorne (1993) observed that there is a geography of gender in the school yard. Boys generally were found on the playing fields, whereas girls were concentrated in the areas closer to the building

divorced when they were 7 to 16, compared to men and women whose parents divorced when they were older, completed less schooling and earned higher scores on an index of psychological symptoms; women were more likely to drink heavily as adults. The researchers also found higher rates of early and nonmarital pregnancy among those whose parents had divorced early. All of these results have been reported in studies of persons in the United States (Demo & Acock, 1988; Garfinkel & McLanahan, 1986). Reduced educational attainment and early parenthood and marriage result in a higher rate of poverty among adults raised in single-parent families (McLanahan & Booth, 1989).

An important mechanism producing these effects is the quality of parenting before and following separation. Several studies have reported that divorced custodial parents have fewer rules, use harsher forms of punishment, spend less time with, and engage in less supervision of their children compared to married parents (Astone & McLanahan, 1991). Also, continuing hostility and lack of cooperation between the parents following separation is consistently related to poor outcomes for children and adolescents (Hetherington, 1999). Another mechanism is the economic hardship for both parents and children that follows divorce (McLanahan & Sandefur, 1994).

BOX 3.2 Continued

and in the jungle gyms. Children who violated these gender boundaries risked being teased or even ridiculed. Thorne identified several varieties of **borderwork,** which is "interaction across—yet interaction based on and even strengthening—gender boundaries" (1993, p. 64). One form of borderwork was the chase, which almost always involved a boy chasing a girl or vice versa. Another form was cooties, or treating an individual or group as contaminated, which also was often cross-gender; girls were often identified as the ultimate source of contamination, whereas boys typically were not. Finally, invasion occurred when a group of boys physically occupied the space that girls were using for some activity; Thorne never observed girls invading a boys' game. All of these activities involve the themes of gender and aggression—themes common to heterosexual relationships in American society. There is also the implicit message that boys and their activities are more important than girls and their activities.

In another study, Eder (1995) and her colleagues observed peer relationships in a middle school for 3 years. During the sixth, seventh, and eighth grades, young adolescents shift their focus from gender role norms to norms governing male-female relationships. Boys learn from other boys the "proper" view of girls; in some but not all groups, the prescribed view was that girls were objects of sexual conquest. Girls learn to view boys as potential participants in romantic relationships. Public teasing and ridicule of those who violate norms—common in elementary school—are replaced by gossip and exclusion from the group as sanctions for violations of group norms in middle school.

Eder (1995) also observed that the status hierarchy in the school generally reproduced the class structure of the wider community. Status was accorded to students based on popularity. One became popular by being visible. The most visible students were those on athletic teams and the cheerleader squad. Participating in these activities required money, as they were not funded by the school. Furthermore, the teams and cheerleaders relied on parents to transport them to games, giving an advantage to students who had one parent who did not work or parents whose jobs allowed them to take time off for such activities. Not surprisingly, the popular, visible students were those from middle-class families.

A review of research on low-income families (often single-parent families) concludes that the need for the parent(s) to work long hours in order to earn enough money shifts the burden of family labor onto one or more children, usually girls. This labor includes caring for younger siblings, cooking, and cleaning; it prevents the person providing it from focusing on education and taking advantage of extracurricular and other opportunities, and may funnel her into early childbearing and marriage (Dodson & Dickert, 2004). Very few studies have been done of the effects of divorce in non–European American families. We don't know whether we would find the effects described here in racial and ethnic minority groups.

Peers

As the child grows, his or her peers become increasingly important as socializing agents. The peer group differs from the family on several dimensions. These differences influence the type of interaction and thus the kinds of socialization that occur.

The family consists of persons who differ in status or power, whereas the peer group is composed of status equals. From an early age, the child is taught to treat parents with respect and deference. Failure to do so will probably result in discipline, and the adult will use the incident as an opportunity to instruct the child about the importance of deference (Cahill, 1987; Denzin, 1977). Interaction with peers is more open and spontaneous; the child does not need to be deferential or tactful. Thus, children at the age of 4 bluntly refuse to let children they dislike join their games. With peers, they may say things that adults consider insulting, such as "You're ugly," to another child. This interactional give-and-take is a basic aspect of the friendship process (Corsaro & Fingerson, 2003).

Membership in a particular family is ascribed, whereas peer interactions are voluntary (Gecas, 1990). Thus, peer groups offer children their first experience in exercising choice over whom they relate to. The opportunity to make such choices contributes to the child's sense of social competence and allows interaction with other children who complement the developing identity.

Unlike the child's family, peer groups in early and especially middle childhood (ages 6 to 10) are usually homogeneous in sex and age. A survey of 2,299 children in 3rd through 12th grade measured the extent to which they belonged to tightly knit peer groups, the size of such groups, and whether they were homogeneous by race and gender (Schrum & Creek, 1987). The proportion belonging to a group peaked in sixth grade and then declined. The size of peer groups declined steadily from 3rd through 12th grade. Other research indicates that friendships of 7th- to 12th-grade Black, Hispanic, and White students tend to be homogeneous by race (Quillian & Campbell, 2003).

Peer associations make a major contribution to the development of the child's identity. Children learn the role of friend in interactions with peers, contributing to greater differentiation of the self (Corsaro & Rizzo, 1988). Peer and other relationships outside the family provide a basis for establishing independence; the child ceases to be exclusively involved in the roles of offspring, sibling, grandchild, and cousin. These alternate, nonfamilial identities may provide a basis for actively resisting parental socialization efforts (Stryker & Serpe, 1982). For example, a parent's attempt to enforce certain rules may be resisted by a child whose friends make fun of children who behave that way.

Although peer culture tends to be concerned with the present, it plays an important role in preparing children and adolescents for role transitions. An observational study of Italian preschoolers found that the transition to elementary school was a common topic of discussion and debate (Corsaro & Molinari, 2000).

School

Unlike the peer group, school is intentionally designed to socialize children. In the classroom, there is typically one adult and a group of children of similar age. There is a sharp status distinction between teacher and student. The teacher determines what skills he or she teaches and relies heavily on instrumental learning techniques, with such reinforcers as praise, blame, and privileges to shape student behavior (Gecas, 1990). School is the child's first experience with formal and public evaluation of performance. Every child's behavior and work is evaluated by the same standards, and the judgments are made public to others in the class as well as to parents.

We expect schools to teach reading, writing, and arithmetic, but they do much more than that. Teachers use the rewards at their disposal to reinforce certain personality traits, such as punctuality, perseverance, and tact. Schools teach children which selves are desirable and which are not. Thus, children learn a vocabulary that they are expected to use in evaluating themselves and others (Denzin, 1977). The traits chosen are those thought to facilitate social interaction throughout life in a particular society. In this sense, schools civilize children.

A key feature of social life in the United States is making statements or "claims" about reality and supporting them with evidence. Each of us engages in such discourses many times each day. In legislative arenas and courtrooms, there are multiple perspectives, each with its claims and supporting arguments contending for adherents. Schools, especially public speaking and debate classes and clubs, are the settings in which youth learn and hone these skills (Fine, 2000).

Social comparison has an important influence on the behavior of schoolchildren. Because teachers make public evaluations of the children's work, each child can judge his or her performance relative to others'. These comparisons are especially important to the child because of the homogeneity of the classroom group. Even if the teacher de-emphasizes a child's low score on a spelling test, the child interprets the performance as a poor one relative to those of classmates. A consistent performance will affect a child's image of self as a student.

An observational study of children in kindergarten, first, second, and fourth grades documented the development of social comparison in the classroom (Frey & Ruble, 1985). In kindergarten, comparisons were to personal characteristics—for example, liking ice cream. Comparisons of performance increased sharply in first grade; at first, comparisons were blatant, but they became increasingly subtle in second and fourth grades.

© David Schaefer/Jeroboam

Shaping is a process through which many complex behaviors, such as playing the cello, are learned. Initially, the socializer (teacher or parent) rewards behavior that resembles the desired response. As learning progresses, greater correspondence between the behavior and the desired response is required to earn a reward, such as praise.

PROCESSES OF SOCIALIZATION

How does socialization occur? We will examine three processes that are especially important: instrumental conditioning, observational learning, and internalization.

Instrumental Conditioning

When you got dressed this morning, chances are you put on a shirt or blouse, pants, a dress, or a skirt that had buttons, hooks, or zippers. When you were younger, learning how to master buttons, hooks, zippers, and shoelaces undoubtedly took considerable time, trial and error, and slow progress accompanied by praise from adults. You acquired these skills through **instrumental conditioning,** a process wherein a person learns what response to make in a situation in order to obtain a positive reinforcement or avoid a negative reinforcement. The person's behavior is instrumental in the sense that it determines whether he or she is rewarded or punished.

The most important process in the acquisition of many skills is a type of instrumental learning called shaping (Skinner, 1953, 1957). **Shaping** refers to learning in which an agent initially reinforces any behavior that remotely resembles the desired response and later requires increasing correspondence between the learner's behavior and the desired response before providing reinforcement. Shaping thus involves a series of successive approximations in which the learner's behavior comes closer and closer to resembling the specific response desired by the reinforcing agent.

In socialization, the degree of similarity between desired and observed responses required by the agent depends in part on the learner's past performance. In this sense, shaping is interactive in character. In teaching children to clean their rooms, parents initially reward them for picking up their toys. When children show they can do this consistently, parents may require that the toys be placed on certain shelves as the condition for a reward. Shaping is more likely to succeed if the level of performance required is consistent with the child's abilities. Thus, a 2-year-old may be praised for drawing lines with crayons, whereas a 5-year-old may be expected to draw recognizable objects or figures.

Reinforcement Schedules When shaping behavior, a socializing agent can use either positive reinforcement or negative reinforcement. Positive reinforcers are stimuli whose presentation strengthens the learner's response; positive reinforcers include food, candy, money, or high grades. Negative reinforcers are stimuli whose withdrawal strengthens the response, such as the removal of pain.

In everyday practice, it is rare for a learner to be reinforced each time the desired behavior is performed. Instead, reinforcement is given only some of the time. In fact, it is possible to structure when reinforcements are presented to the learner, using a reinforcement schedule.

There are several possible reinforcement schedules. The fixed-interval schedule involves reinforcing the first correct response after a specified period has elapsed. This schedule produces the fewest correct responses per unit time; if the learner is aware of the length of the interval, he or she will respond only at the beginning of the interval. It is interesting that many schools give examinations at fixed intervals, such as the middle and end of the semester; perhaps that is why many students study only just before an exam. The variable-interval schedule involves reinforcing the first correct response after a variable period. In this case, the individual cannot predict when reinforcement will occur, so he or she responds at a regular rate. Grading a course based on several surprise or "pop" quizzes uses this schedule.

The fixed-ratio schedule provides a reinforcement following a specified number of correct, nonreinforced responses. Paying a worker on a piece rate, such as 5 dollars for every three items produced, uses this pattern. If the reward is sufficient, the rate of behavior may be high. Finally, the variable-ratio schedule provides reinforcement after several nonrewarded responses, with the number of responses between reinforcements varying. This schedule typically produces the highest and most stable rates of response. An excellent illustration is the gambler, who will insert quarters in a slot machine for hours, receiving only occasional, random payoffs.

Punishment By definition, **punishment** is the presentation of a painful or discomforting stimulus (by a socializing agent) that decreases the probability that the preceding behavior (by the learner) will occur. Punishment is one of the major child rearing practices used by parents. The Gallup organization interviewed a nationally representative sample of parents in 1995 (Straus & Stewart, 1999; Straus & Field, 2003). The percentage of parents who reported using corporal punishment—pinching, slapping, spanking, or hitting—during the preceding year varied by the age of the child. The use of corporal punishment was reported by 94 percent of the parents of 3- and 4-year-olds; the prevalence declined steadily from age 5 to age 17. The use of psychological techniques—shouting, name-calling, threatening—was reported by more than 85 percent of parents of children of all ages. The results are displayed in Figure 3.1.

Punishment is obviously widely used in the United States, suggesting that our culture is tolerant of or encourages its use. Corporal punishment was more commonly reported by African American and low-income parents (Straus & Stewart, 1999), while the use of psychological techniques did not vary by race or other sociodemographic characteristics (Straus & Field, 2003).

So does punishment work? Research indicates that it is effective in some circumstances but not in others. One aspect is timing. Punishment is most effective when it occurs in close proximity to the behavior. A verbal reprimand delivered as the child touched the toy was more effective than a prior warning or a reprimand following the action (Aronfreed & Reber, 1965). Also, the effectiveness may be limited to the situation in which it is given. Because punishment is usually administered by a particular person, it may be effective only when that person is present. This probably accounts for the fact that when their parents are absent, children may engage in activities that their parents earlier had punished (Parke, 1969, 1970).

FIGURE 3.1 PERCENTAGE OF PARENTS WHO USE PHYSICAL PUNISHMENT AND PSYCHOLOGICAL PUNISHMENT

The Gallup Organization interviewed a representative sample of 991 parents in 1995. Each parent was asked whether and how often he or she used physical punishment (spanked the bottom; slapped hand, arm, or leg; pinched; shook; hit on the bottom with an object; or slapped head, face, or ears) and psychological punishment (shouted, yelled, or screamed; threatened to hit or spank; swore or cursed; threatened to kick out of the house; or called names, such as dumb or lazy). Most parents reported using both types. The use of physical punishment peaked with 4-year-old children and then declined steadily through age 17. By contrast, the use of psychological punishment was reported to be as common with 17-year-olds as with 1-year-olds (90 percent).

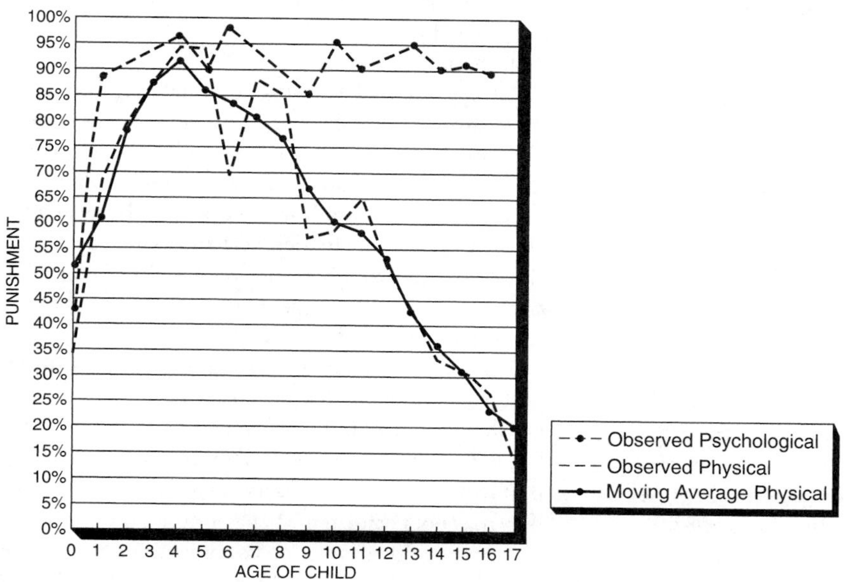

Sources: Straus and Stewart, 1999; Straus and Field, 2003.

Another factor in the effectiveness of punishments is whether they are accompanied by a reason (Parke, 1969). Providing a reason allows the child to generalize the prohibition to a class of acts and situations. Yelling "No!" as a child reaches out to touch the stove may suppress that behavior. Telling the child not to touch it because it is "hot" enables him or her to learn to avoid hot objects as a group. Finally, consistency between the reprimands given by parents and their own behavior makes punishment more effective than if parents do not practice what they preach (Mischel & Liebert, 1966).

What about the long-term consequences of punishment? Clearly parents and caregivers need to control children's behavior. At the same time, they need to recognize that corporal punishment is associated with subsequent antisocial behavior by children. In a study of mothers of 6- to 9-year-old children, the more often the mother reported spanking the child, the higher the level of the child's antisocial behavior (on a six-item scale) 2 years later (Straus, Sugarman, & Giles-Sims, 1997). Also, the use of psychological punishment may be associated with problem behaviors (such as running away, being suspended from school) in adolescence. Punishment should focus on the behavior and not the child, and should be balanced by praise and rewards. Also, the effects of punishment may depend on the social context; research involving African American families found that corporal punishment had negative outcomes in neighborhoods where the use of physical punishment was rare, but not in neighborhoods where its use was common (Simons et al., 2002)

Self-Reinforcement and Self-Efficacy Children learn hundreds, if not thousands, of behaviors through instrumental learning. The performance of some of these behaviors will remain **extrinsically motivated**—that is, they are dependent on whether someone else will reward appropriate behaviors or punish inappropriate ones. However, the performance of other activities becomes **intrinsically motivated**—that is, performed in order to achieve an internal state that the individual finds rewarding (Deci, 1975). Research has demonstrated that external rewards do not always improve performance. Providing a reward for a behavior that is intrinsically motivated, such as drawing, may actually reduce the frequency or quality of the activity (Lepper, Greene, & Nisbett, 1973).

Closely related to the concept of intrinsic motivation is self-reinforcement. As children are socialized, they learn not only specific behaviors but also performance standards. Children learn not only to write but to write neatly. These standards become part of the self; having learned them, the child uses them to judge his or her own behavior and thus becomes capable of **self-reinforcement** (Bandura, 1982b). The child who has drawn a house and comes running up to her father with a big smile saying "Look what I drew," has already judged the drawing as a good one. If her father agrees, her standards and self-evaluation are confirmed.

Successful experiences with an activity over time create a sense of competence at the activity, or self-efficacy (Bandura, 1982b). This, in turn, makes the individual more likely to seek opportunities to engage in that behavior. The greater one's sense of self-efficacy, the more effort one will expend at a task and the greater one's persistence in the face of difficulty. For instance, a young girl who perceives herself as a good basketball player is more likely to try out for a team. Conversely, experiences of failure to perform a task properly, or of the failure of the performance to produce the expected results, create the perception that one is not efficacious. Perceived lack of efficacy is likely to lead to avoidance of the task. A boy who perceives himself as poor at spelling will probably not enter the school spelling bee.

Observational Learning

Children love to play dress-up. Girls put on skirts, step into high-heeled shoes, and totter around the room; boys put on sport coats and drape ties around their necks. Through observing adults, children have learned the patterns of appropriate dress in their society. Similarly, children often learn interactive rituals, such as shaking hands or waving goodbye, by watching others perform the behavior and then doing it on their own.

Observational learning, or modeling, refers to the acquisition of behavior based on the observation of another person's behavior and of its consequences for that person (Shaw & Costanzo, 1982). Many behaviors and skills are learned this way. By watching another person (the model) perform skilled actions, a child can increase his or her own skills. The major advantage of modeling is its greater efficiency compared with trial-and-error learning.

Does observational learning lead directly to the performance of the learned behavior? No; research has shown that there is a difference between learning a behavior and performing it. People can learn how to perform a behavior by observing another person, but they may not perform the behavior until the appropriate opportunity arises. Considerable time may elapse before the observer is in the presence of the eliciting stimulus. A father in the habit of muttering "damn" when he spills something may, much to his chagrin, hear his 3-year-old daughter say "damn" the first time she spills milk. Children may learn through observation many associations between situational characteristics and adult behavior, but they may not perform these behaviors until they occupy adult roles and find themselves in such situations.

Even if the appropriate stimulus occurs, people may not perform behaviors learned through observation. An important influence is the consequences experienced by the model following the model's performance of the behavior. For instance, in one study (Bandura, 1965), nursery school children watched a film in which an adult model punched, kicked, and threw balls at a large, inflated rubber Bobo doll. Three versions of the film were shown to three groups of children. In the first, the model was rewarded for his acts: A second adult appeared and gave the model soft

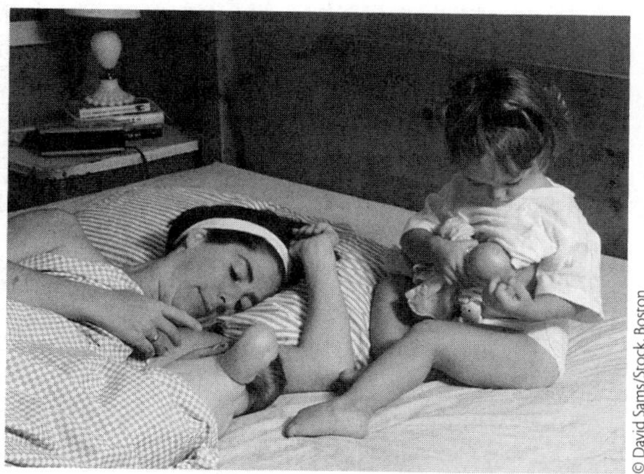

Observational learning is an important process through which children learn appropriate behaviors.

drinks and candy. In the second version, the model was punished: The other adult spanked the model with a magazine. In the third version, there were no rewards or punishments. Later, each child was left alone with various toys, including a Bobo doll. The child's behavior was observed through a one-way mirror. Children who had observed the model who was punished were much less likely to punch and kick the doll than the other children.

Did these other children not learn the aggressive behaviors, or did they learn them by observation but not perform them? To answer this question, the experimenter returned to the room and offered a reward for each act of the model that the child could reproduce. Following this offer, the children in all three groups were equally able to reproduce the acts performed by the model. Thus, a child is less likely to perform an act learned by observation if the model experienced negative consequences.

Whether children learn from observing a model also depends on the characteristics of the model. Children are more likely to imitate high-status and nurturant models than models who are low in status and nurturance (Bandura, 1969). Preschool children given dolls representing peers, older children, and adults consistently chose adult dolls as people they would go to for help and older children as people they would go to for teaching (Lewis & Brooks-Gunn, 1979). Children also are more likely to model themselves after nurturant persons than after cold and impersonal ones. Thus, socialization is much more likely to be effective when the child has a nurturant, loving primary caregiver.

Internalization

Often, we feel a sense of moral obligation to perform some behavior. At other times, we experience a strong internal feeling that a particular behavior is wrong. Usually, we experience guilt if these moral prescriptions or prohibitions are violated.

Internalization is the process by which initially external behavioral standards (for example, those held by parents) become internal and subsequently guide the person's behavior. An action is based on internalized standards when the person engages in it without considering possible rewards or punishments. Various explanations have been offered of the process by which internalization occurs, but all of them agree that children are most likely to internalize the standards held by more powerful or nurturant adult caregivers.

Internalization is an important socializing process. It results in the exercise of self-control. People conform to internal standards even when there is no surveillance of their behavior by others and, therefore, no rewards for their conformity. People who are widely admired for taking political or religious actions that are unpopular—for standing up for their beliefs—often do so because those beliefs are internalized.

OUTCOMES OF SOCIALIZATION

Persons being socialized acquire new skills, knowledge, and behavior. In this section, we discuss some specific outcomes of the socialization process, including gender role, linguistic and cognitive competence, moral development, and orientation toward work.

Gender Role

"Congratulations, you have a girl!" Such a pronouncement by a birth attendant may be the single most important event in a new person's life. The gender assigned to the infant—male or female—has a major influence on the socialization of that child.

Every society has differential expectations regarding the characteristics and behavior of men and women. In our society, men traditionally have been expected to be competent—competitive, logical, able to make decisions easily, ambitious. Women have been expected to be high in warmth and expressiveness—gentle, sensitive, tactful (Broverman, Vogel, Broverman, Clarkson, & Rosenkrantz, 1972). Parents employ these or other expectations as guidelines in socializing their children, and differential treatment begins at birth. Male infants are handled more vigorously and roughly, whereas female infants are given more cuddling (Lamb, 1979). Boys and girls are dressed differently from infancy and may be given different kinds of toys to play with.

Moreover, mothers and fathers differ in the way they interact with infants. Mothers engage in behavior oriented toward fulfilling the child's physical and emotional needs (Baumrind, 1980), whereas fathers engage the child in rough-and-tumble, physically stimulating activity (Walters & Walters, 1980). Fathers also engage sons in more rough-and-tumble play than daughters. These differences are found in European American, African American, and Hispanic families (Parke, 1996). Thus, almost from birth, infants are exposed to models of masculine and feminine behavior. Mothers and fathers differ in their talk to young children; mothers talk more than fathers, and mothers' talk is socioemotional (supportive or negative) whereas fathers' talk is instrumental (Leaper, Anderson, & Sanders, 1998).

By age 2, the child's gender identity—his or her conception of self as male or female—is firmly established (Money & Ehrhardt, 1972). Boys and girls show distinct preferences for different types of play materials and toys by this age. Between the ages of 2 and 3, differences in aggressiveness become evident, with boys displaying more physical and verbal aggression than girls (Hyde, 1984). By age 3, children more frequently choose same-gender peers as playmates; this increases their opportunities to learn gender-appropriate behavior via modeling (Lewis & Brooks-Gunn, 1979). By age 4, the games typically played by boys and girls differ; groups of girls play house, enacting familial roles, whereas groups of boys play cowboys. In middle childhood, gender-segregated play appears to be almost universal (Edwards, Knoche, & Kumru, 2001).

Parents are an important influence on the learning of **gender role**—the behavioral expectations associated with one's gender. Children learn gender-appropriate behaviors by observing their parents' interaction. Children also learn by interacting with parents, who reward behavior consistent with gender roles and punish behavior inconsistent with these roles. The child's earliest experiences relating to members of the other gender occur in interaction with the opposite-gender parent. A woman may be more likely to develop the ability to have warm, psychologically intimate relationships with men if her relationship with her father was of this type (Appleton, 1981).

Obviously, boys are not all alike in our society, and neither are girls. The specific behaviors and characteristics that the child is taught depend partly on the gender role expectations held by the parents. These in turn depend on the network of extended family—grandparents, aunts and uncles, and other relatives—and friends of the family. The expectations held by these people are influenced by the institutions to which they belong, such as churches and work organizations (Stryker & Serpe, 1982). With regard to religion, research suggests that the differences among denominations in socialization techniques and in outcomes such as gender role attitudes have declined in recent decades (Alwin, 1986). The data suggest that church attendance is more influential than the denomination to which one belongs.

Gender role definitions vary by culture. Some research suggests that Latino families teach more traditional expectations for behavior of boys and girls compared to other groups in U.S. society. Other research finds that as education and female labor-force participation increase, such families have more egalitarian views of behavior and decision making (Ginorio, Gutierrez, Cauce, & Acosta, 1995). It is important to remember that "Latino" encompasses people from several different cultural backgrounds, including Mexican American, Puerto Rican, and Cuban. Asian cultures are patriarchal, and parents may socialize female children to restrictive norms designed to serve

Children and adolescents learn gender role expectations through interaction with adults. Meeting his hero may have a lifelong impact on this youth.

the family rather than express their individuality (Root, 1995). Again, "Asian" includes persons of Chinese, Japanese, Korean, and Vietnamese descent; these cultures may differ in the prevailing gender role definitions.

Schools also teach gender roles. Teachers may reward appropriate gender role behavior; they often reinforce aggressive behavior in boys and dependency in girls (Serbin & O'Leary, 1975). A more subtle influence on socialization is the content of the stories that are read and told in preschool and first-grade classes. Many of these stories portray men and women as different. In the past, men were depicted as rulers, adventurers, and explorers; women were wives (Weitzman, Eifler, Hokada, & Ross, 1972). A study of award-winning books for children published in 1995–1999 found men and women equally represented as main characters, but men played a greater variety of roles and were seldom shown engaging in child care, shopping, or housework (Gooden & Gooden, 2001).

A major influence on gender role socialization is the mass media. Researchers analyzing the contents of television programs, television advertising, feature films, and other media report that portrayals of men and women and girls and boys reinforce traditional definitions of gender roles. A content analysis of 175 episodes of 41 animated TV series found that male characters were portrayed as independent, athletic, ambitious, and aggressive, whereas female characters were shown as dependent, emotional, domestic, and romantic (Thompson & Zerbinos, 1995). A study of the fiction in *Seventeen* and *Teen,* the two largest selling magazines for teenage girls, found that the stories reinforced traditional messages (Peirce, 1993). Half of the conflicts were about relationships, and half the female characters relied on someone else to solve their problems. Adult men in the stories were doctors, lawyers, and bankers; adult women were nurses, clerical workers, and secretaries. Perhaps the most stereotyped portrayals are found on MTV. An analysis of 40 music videos found that men engaged in more dominant, aggressive behavior, whereas women engaged in subservient behavior; women were frequently the object of explicit, implicit, and aggressive sexual advances (Sommers-Flanagan, Sommers-Flanagan, & Davis, 1993).

During childhood and adolescence, youth are explicitly taught and rewarded for behavior consistent with gender role norms. They also observe models behaving in a variety of ways. Children do not simply mimic their parents, siblings, or MTV performers. As the interpretive perspective suggests, children learn gender role behaviors and then recreate them, adapting them to their individual social contexts. Williams (2002) refers to this process as trying on gender—experimenting, resisting, and rehearsing ways to be female or male.

Linguistic and Cognitive Competence

Another important outcome of socialization is the ability to interact effectively with others. We shall discuss two specific competencies: language and the ability to cognitively represent the world.

Language Using language to communicate with others is a prerequisite for full participation in social groups (Shibutani, 1961). The child's acquisition of speech reflects both the development of the necessary perceptual and motor skills and the impact of social learning (Bates, O'Connell, & Shore, 1987).

The three main components of language are the sound system (phonology), the words and their associated meanings (lexicon), and the rules for combining words into meaningful utterances (grammar). Young children appear to acquire these in sequence, first mastering meaningful sounds, then learning words, and finally learning sentences. In reality, acquiring speech is a process that involves all three at the same time and continues throughout childhood.

Language acquisition in the first 3 years passes through four stages (Bates et al., 1987). The pre-speech stage lasts for about 10 months and involves speech perception, speech production, and early intentional communication. In the first few weeks of life, infants can perceive all of the speech sounds. They begin producing sounds at 2 to 3 months, and begin producing sounds specific to their parent's language at 4 to 7 months. Speech production involves imitation of the sounds they hear. With regard to intentional communication, observational data indicate that vocal exchanges involving 4-month-old infants and their mothers are patterned (Stevenson, Ver Hoeve, Roach, & Leavitt, 1986). Vocalization by *either* infant or mother was followed by silence, allowing the other to respond. Vocalization by one was likely to be followed by vocalization by the other, a pattern like that found in adult conversation.

The first intentional use of gestures occurs at about 9 months. At this age, infants orient visually to adults rather than to desired objects, such as a cookie. Furthermore, if an initial gesture is not followed by the adult engaging in the desired behavior, the infant will repeat the gesture or try a different gesture.

The second or first word stage occurs at 10 to 14 months and involves the infant's recognition that things have names. The first words produced are usually nouns that name or request specific objects (Marchman, 1991). Obviously, this ability to use names reflects cognitive as well as linguistic development.

At about 18 months, there is a vocabulary burst, with a doubling in a short time of the number of words that are correctly used. The suddenness of this increase suggests that it reflects the maturation of some cognitive abilities. This, in turn, is followed by an increase in the complexity of vocalizations, leading to the first sentence stage at 18 to 22 months. Examples of such sentences include "See truck, Mommy" and "There go one." Such speech is telegraphic—that is, the number of words is greatly reduced relative to adult speech (Brown & Fraser, 1963). At the same time, such utterances are clearly more precise than the single-word utterances of the 1-year-old child.

The fourth stage, grammaticization, occurs at 24 to 30 months. The child's use of language now reflects the fundamentals of grammar. Children at this age frequently overgeneralize, applying rules indiscriminately. For example, they will add an appropriate ending to a novel word although it is incorrect: "He runned." Such usage indicates that the child understands that there are rules. At about the same age, a child puts series of acts in the conventional sequence—for example, undressing a doll, bathing it, drying it, and dressing it. Perhaps both activities reflect the maturation of an underlying ability to order arbitrary units.

An important process in learning to make grammatically correct sentences is speech expansion. That is, adults often respond to children's speech by repeating it in expanded form. In response to "Eve lunch," the mother might say "Eve is eating lunch." One study showed that mothers expanded 30 percent of the utterances of their 2-year-old children (Brown, 1964). Adults probably expand on the child's speech to determine the child's specific meaning. Speech expansion contributes to language acquisition by providing children with a model of how to convey more effectively the meanings they intend.

The next stage of language development is highlighted by the occurrence of private speech, in which children talk loudly to themselves, often for extended periods. Private speech begins at about age 3, increases in frequency until age 5, and disappears by about age 7. Such private talk serves two functions. First, it contributes to the child's developing sense of self. Private speech is addressed to the self as object, and it often includes the application of meanings to the self, such as "I'm a girl." Second, private speech helps the child develop an awareness of the environment. It often consists of naming aspects of the physical and social environment. The repeated use of these names solidifies the child's understanding of the environment. Children also often engage in appropriate actions as they speak, reflecting their developing awareness of the social meanings of objects and persons. Thus, a child may label a doll a "baby" and dress it and feed it.

Gradually, the child begins to engage in dialogues, either with others or with the self. These conversations reflect the ability to adopt a second perspective. Thus, by age 6, when one child wants a toy that another child is using, the first child frequently offers to trade. She knows that the second child will be upset if she merely takes the toy. This movement away from a self-centered view also may reflect maturational changes. Dialogue requires that the child's own speech meshes with that of another.

Language is important in the socialization of gender. A meta-analysis of observational studies of parents' use of language in interaction with their children identified several differences in types of communication between mothers and fathers. For example, mothers were more supportive and less directive compared to fathers. Moreover, mothers and fathers differed in the way they talked to sons, and to daughters (Leaper, Anderson, & Sanders, 1998). Thus children are socialized to gender differences in language use as they observe and interact with their parents/caregivers.

Cognitive Competence Children must develop the ability to represent in their own minds the features of the world around them. This capacity to represent reality mentally is closely related to the development of language.

The child's basic tasks are to learn the regularities of the physical and social environment and to store past experience in a form that can be used in current situations. In a complex society, there are so many physical objects, animals, and people that it is not possible for a child (or an adult) to remember each as a distinct entity. Things must be categorized into inclusive groupings, such as dogs, houses, or girls. A category of objects and the cognitions that the individual has about members of that category (for example, "dog") make up a schema. Collectively, our schemas allow us to make sense of the world around us.

Young children must learn schemas. Learning language is an essential part of the process, because language provides the names around which schemas can develop. It is noteworthy that the first words that children produce are usually nouns that name objects in the child's environment. At first, the child uses a few very general schemas. Some children learn the word "dog" at 12 to 14 months and then apply it to all animals—to dogs, cats, birds, and cows. Only with maturation and experience does the child develop the abstract schema "animals" and learn to discriminate between dogs and cats.

Researchers can study the ability to use schemas by asking children to sort objects, pictures, or words into groups. Young children (ages 6 to 8) rely on visual features, such as color or word length, and sort objects into numerous categories. Older children (ages 10 to 12) increasingly use functional or superordinate categories, such as foods, and sort objects into fewer groups (Olver, 1961; Rigney, 1962). With age, children become increasingly adept at classifying diverse objects and treating them as equivalent.

These skills are very important in social interaction. Only by having the ability to group objects, persons, and situations can one determine how to behave toward them. Person schemas and their associated meanings are especially important to smooth interaction. Even very young children differentiate people by age (Lewis & Brooks-Gunn, 1979). By about 2 years of age, children correctly differentiate babies and adults when shown photographs. By about 5, children employ four categories: little children, big children, parents (ages 13 to 40), and grandparents (age 40-plus).

As children learn to group objects into meaningful schemas, they learn not only the categories but also how others feel about such categories. Children learn not only that Catholics are people who believe in the Trinity of Father, Son, and Holy Spirit but also whether their parents like or dislike Catholics. Thus, children acquire positive and negative attitudes toward the wide range of social objects they come to recognize. The particular schemas and evaluations that children learn are influenced by the social class, religious, ethnic, and other subcultural groupings to which those who socialize them belong.

Moral Development

In this section, we discuss moral development in children and adults. Specifically, we focus on the acquisition of knowledge of social rules and on the process through which children become capable of making moral judgments.

Knowledge of Social Rules To interact effectively with others, people must learn the social rules that govern interaction and in general adhere to them. Beliefs about which behaviors are acceptable and which are unacceptable for specific persons in specific situations are termed **norms.** Without norms, coordinated activity would be very difficult, and we would find it hard or impossible to achieve our goals. Therefore, each group, organization, and society develops rules governing behavior.

Early in life, an American child learns to say "please"; a French child, "s'il vous plait"; and a Serbian child, "molim te." In every case, the child is learning the value of conforming to arbitrary norms governing requests. Learning language trains the child to conform to linguistic norms and serves as a model for the learning of other norms. Gradually, through instrumental and observational learning, the child learns the generality of the relationship between conformity to norms and the ability to interact smoothly with others and achieve one's own goals.

What influences which norms children will learn? The general culture is one influence. All American children learn to cover parts of the body with clothing in public. The position of the family within the society is another influence. Parental expectations reflect social class, religion, and ethnicity. Thus, the norms that children are taught vary from one family to another. Interestingly, parents often hold norms that they apply distinctively to their own children. Mothers and fathers expect certain behaviors of their own sons or daughters but may have different expectations for other people's children (Elkin & Handel, 1989). For instance, they may expect their own children to be more polite than other children in interaction with adults. Parental expectations are not constant over time; they change as the child grows older. Parents expect greater politeness from a 10-year-old than from a 5-year-old. Finally, parents adjust their expectations to the particular child. They consider the child's level of ability and experiences relative to other children; they expect better performance in school from a child who has done well in the past than from one who has had problems in school. In all of these ways, each child is being socialized to a somewhat different set of norms. The outcome is a young person who is both similar to most others from the same social background and unique in certain ways.

In the opening vignette, we met 9-year-old Levi and his father. Dad recognizes that Levi is a child who often does not adhere to "pesky rules," who prefers to go his own way. The father has different expectations for Levi than do the other parents and the coach. He recognizes that Levi is unique, and he encourages this because he did not like team sports as a child. This is an excellent example of the intergenerational transmission of orientations toward norms, and of the adage "Like father, like son."

When children begin to engage in cooperative play at about 4 years of age, they begin to experience normative pressure from peers. The expectations of age-mates differ in two important ways from those of parents. First, children bring different norms from their separate families and, therefore, introduce new expectations. Thus, through their peers, children first become aware there are other ways of behaving. In some cases, peers' expectations conflict with those of parents. For example, many parents do not allow their children to play with toy guns, knives, or swords. Through involvement with their peers, children may become aware that other children routinely play with such toys. As a result, some children will experience normative conflict and discover the need to develop strategies for resolving such conflicts.

A second way that peer group norms differ from parental norms is that the former reflect a child's perspective (Elkin & Handel, 1989). Many parental expectations are oriented toward socializing the child for adult roles. Children react to each other as children

At school, children get their first exposure to universal norms—behavioral expectations that are the same for everyone. Although parents and friends treat the child as an individual, teachers are less likely to do so.

and are not concerned with long-term outcomes. Thus, peers encourage impulsive, spontaneous behavior rather than behavior directed toward long-term goals. Peer group norms emphasize participation in group activities, whereas parental norms may emphasize homework and other educational activities that may contribute to academic achievement.

When children enter school, they are exposed to a third major socializing agent—the teacher. In school, children are exposed to universalistic rules—norms that apply equally to all children. The teacher is much less likely than the parents to make allowances for the unique characteristics of the individual; children must learn to wait their turn, to control impulsive and spontaneous behavior, and to work without a great deal of supervision and support. In this regard, the school is the first of many settings where the individual is treated primarily as a member of the group rather than as a unique individual. As noted in Box 3.2, children may engage in resistance in response to the authority structure in a school.

Resistance sometimes includes physical attacks on school personnel, as in the highly publicized case of a 5-year-old who attacked a teacher in response to an attempt to get the youngster to stop being "silly." Taken to the school office, the child attacked the assistant principal, and was subsequently arrested by the St. Petersburg, FL police (Leary &Tobin, 2005). Many observers felt the arrest was inappropriate. The case highlights the way in which an act of resistance to authority often quickly escalates, with an outcome no one intended.

Thus, school is the setting in which children are first exposed to universalistic norms and the regular use of symbolic rewards, such as grades. Such settings become increasingly common in adolescence and adulthood, in contrast with the individualized character of familial settings.

Moral Judgment We not only learn the norms of our social groups, but we also develop the ability to evaluate behavior in specific situations by applying certain standards. The process through which children become capable of making moral judgments is termed **moral development.** It involves two components: (1) the reasons one adheres to social rules and (2) the bases used to evaluate actions by self or others as good or bad.

How do children evaluate acts as good or bad? One of the first people to study this question in detail was Piaget, the famous Swiss developmental psychologist. Piaget read a child stories in which the central character performed an act that violated social rules. In one story, for example, a young girl, contrary to rules, was playing with scissors and made a hole in her dress. Piaget asked the children to evaluate the behaviors of the characters (that is, to indicate which characters were naughtier) and then to explain their reasons for these judgments. Based on this work, Piaget concluded there were three bases for moral judgments: amount of harm/benefit, actor's intentions, and the application of agreed-upon rules or norms (Piaget, 1965).

Kohlberg (1969) extended Piaget's work by analyzing the reasoning by which people reach moral judgments. He uses stories involving conflict between human needs and social norms or laws. Here is an example:

> In Europe, a woman was near death from cancer. One drug might save her, a form of radium that a druggist in the same town had recently discovered. The druggist was charging $2,000, ten times what the drug cost him to make. The sick woman's husband, Heinz, went to everyone he knew to borrow money, but he could only get together about half of what it cost. He told the druggist that his wife was dying and asked him to sell it cheaper or let him pay later. But the druggist said, "No." The husband got desperate and broke into the man's store to steal the drug for his wife (Kohlberg, 1969).

TABLE 3.3 KOHLBERG'S MODEL OF MORAL DEVELOPMENT

Preconventional Morality	
Moral judgment based on external, physical consequences of acts.	
Stage 1:	Obedience and punishment orientation. Rules are obeyed in order to avoid punishment, trouble.
Stage 2:	Hedonistic orientation. Rules are obeyed in order to obtain rewards for the self.
Conventional Morality	
Moral judgment based on social consequences of acts.	
Stage 3:	"Good boy/nice girl" orientation. Rules are obeyed to please others, avoid disapproval.
Stage 4:	Authority and social-order maintaining orientation. Rules are obeyed to show respect for authorities and maintain social order.
Postconventional Morality	
Moral judgment based on universal moral and ethical principles.	
Stage 5:	Social-contract orientation. Rules are obeyed because they represent the will of the majority, to avoid violation of rights of others.
Stage 6:	Universal ethical principles. Rules are obeyed in order to adhere to one's principles.

Source: Adapted from Kohlberg, 1969, table 6.2.

Respondents are then asked, "Should Heinz have done that? Was Heinz right or wrong? What obligations did Heinz and the druggist have? Should Heinz be punished?"

Kohlberg proposes a developmental model with three levels of moral reasoning, each level involving two stages. This model is summarized in Table 3.3.

Kohlberg argues that the progression from stage 1 to stage 6 is a standard or universal one, and that all children begin at stage 1 and progress through the stages in order. Most adults reason at stages 3 or 4. Few people reach stages 5 or 6. Several studies have shown that such a progression does occur (Kuhn, Langer, Kohlberg, & Haan, 1977). If the progression is universal, then children from different cultures should pass through the same stages in the same order. Again, data suggest that they do (White, Bushnell, & Regnemer, 1978). On the basis of such evidence, Kohlberg claims that this progression is the natural human pattern of moral development. He also believes that attaining higher levels is better or more desirable.

Kohlberg's model is an impressive attempt to specify a universal model of moral development. However, it has limitations. First, like Piaget, Kohlberg locates the determinants of moral judgment within the individual. He does not recognize the influence of the situation. Studies of judgments of aggressive behavior (Berkowitz, Mueller, Schnell, & Pudberg, 1986), of driving while intoxicated (Denton & Krebs, 1990), and of decisions about reward allocation (Kurtines, 1986) have found that both moral stage and type of situation influenced moral judgment.

Second, Kohlberg's model has been criticized as sexist—not applicable to the processes that women use in moral reasoning. Gilligan (1982) identifies two conceptions of morality: a morality of justice and a morality of caring. A justice orientation is concerned with adherence to rules and fairness, whereas a caring orientation is concerned with relationships and meeting the needs of others. Gilligan argues that the former is characteristic of men and is the basis of Kohlberg's model. She believes the latter is more characteristic of women. A meta-analysis of studies testing predictions from the two models indicates that there is a significant but modest tendency for women to base judgments on caring criteria and for men to base judgments on considerations of justice (Jaffee & Hyde, 2000). Several of these experiments suggest that the content of the dilemma has greater influence on the criteria used than does gender; thus, both men and women are flexible in the making of moral judgments (Crandall, Tsang, Goldman, & Pennington, 1999).

Third, Kohlberg shows little interest in the influence of social interaction on moral reasoning. In response to this limitation, Haan (1978) has proposed a model of interpersonal morality. Moral decisions and actions often result from negotiations between people in which the goal is a "moral balance." Participants attempt to balance situational characteristics, such as the options available, with their individual interests to arrive at a decision that allows them to preserve their sense of themselves as moral persons. Haan (1978, 1986) presented moral dilemmas to groups of friends and asked them to decide. In some cases, the decisions were more influenced by individual moral principles; in others, by the group interaction.

Work Orientations

Work is of central importance in social life. In recognition of this, occupation is a major influence on the distribution of economic and other resources. We identify others by their work; its importance is evidenced by the fact that one of the first questions we ask a new acquaintance is "What do you do?"

Most adults want to work at jobs that provide economic and perhaps other rewards. Therefore, it is not surprising that a major part of socialization is the learning of orientations toward work. By the age of 2, the child is aware that adults "go work" and asks why. A common reply is "Mommy goes to work to earn money." A study of 900 elementary school children found that 80 percent of first-graders understood the connection between work and money (Goldstein & Oldham, 1979). The child, in turn, learns that money is needed to obtain food, clothing, and toys. The child of a physician or nurse might be told "Mommy goes to work to help people who are ill." Thus, from an early age the child is taught the social meaning of work.

Occupations vary tremendously in character. One dimension on which jobs differ is closeness of supervision; a self-employed auto mechanic has considerable freedom, whereas an assembly-line worker may be closely supervised. The nature of the work varies; mechanics deal with things, salespeople deal with people, lawyers deal with ideas. Finally, occupations such as lawyer require self-reliance and independent judgment, whereas an assembly-line job does not. So the meaning of work depends on the type of job the individual has.

Adults in different occupations should have different orientations toward work, and these orientations should influence how they socialize their children. Based on this hypothesis, extensive research has been conducted on the differences between social classes in the values transmitted through socialization (Kohn, 1969). Fathers are given a list of traits, including good manners, success, self-control, obedience, and responsibility, and asked to indicate how much they value each for their children. Underlying these specific characteristics, a general dimension—"self-direction versus conformity"—is usually found. Data from fathers of 3- to 15-year-old children indicate that the emphasis on self-direction and reliance on internal standards increases as social class increases. The relationship of values and social class is found not only in samples of American fathers but also in samples of Japanese and Polish fathers (Kohn, Naoi, Schoenbach, Schooler, & Slomczynski, 1990).

These differences in the evaluations of particular traits reflect differences in the conditions of work. In general, middle-class occupations involve the manipulation of people or symbols, and the work is not closely supervised. Thus, these occupational roles require people who are self-directing and who can make judgments based on knowledge and internal standards. Working-class occupations are more routinized and more closely supervised. Thus, they require workers with a conformist orientation. Kohn argues that fathers value those traits in their children that they associate with success in their occupation.

Do the differences in the value parents place on self-direction influence the kinds of activities they encourage their children to participate in? A study of 460 adolescents and their mothers (Morgan, Alwin, & Griffin, 1979) examined how maternal emphasis on self-direction affected the young person's grades in school, choice of curriculum, and participation in extracurricular activities. The researchers reasoned that parents who valued self-direction would encourage their children to take college-preparatory courses, because a college education is a prerequisite to jobs that provide high levels of autonomy. Similarly, they expected mothers who valued self-direction to encourage extracurricular activities, because such activities provide opportunities to develop interpersonal skills. The researchers did not expect differences in grades. The results confirmed all three predictions. Thus, parents who value particular traits in their children do encourage activities that they believe are likely to produce those traits.

By age 16, many adolescents have expectations about jobs they will hold as adults. A longitudinal study in the United Kingdom found that these expectations at 16 were influenced by both parents and teachers; these expectations, in conjunction with amount of education completed, were associated with adult occupational attainment at ages 23, 33, and 42 (Brown, Sessions, & Taylor, 2004). Thus, adolescents' expectations provide a basis for educational and career choices.

ADULT SOCIALIZATION

The process of socialization occurs not only in childhood but continues throughout life (Bush & Simmons, 1990). Its focus changes, however. In childhood, socializing efforts are directed at such basic outcomes as gender role, the acquisition of language, and the learning of social norms. In adolescence, socialization is focused on the acquisition of traits such as independence, responsibility, and the ability to relate to others. In adulthood, it is concerned with equipping the individual to function effectively in adult roles. In this section, we discuss three processes important to adult socialization—role acquisition, anticipatory socialization, and role discontinuity.

Role Acquisition

Throughout our lives, we move out of some roles and into new roles. The major roles we acquire as adults include intimate partner or spouse, parent, work roles, and later, grandparent and retiree. Midlife—the period from age 40 to 60—often involves several role transitions, including marital (divorce, widowhood), parental (as the children leave home), work (entry or exit), and caregiver (for aging parents; Etaugh & Bridges, 2001). Each of these changes involves role acquisition, learning the expectations and skills associated with the new role and entry into the role.

In recognition of the need to train people for new roles, many groups and organizations provide socialization opportunities. Certain agents are given the responsibility for teaching potential or new role occupants the necessary information and skills. Often, this training occurs on the job. The novice checker in a supermarket will work with an experienced clerk at first. Initially, the experienced clerk will perform the work, perhaps instructing the novice while doing so.

Gradually, the roles will be reversed. As the novice becomes more skilled at the mechanics of using the price scanner and interacting with shoppers, he or she will do more and more of the work. Eventually, on demonstrating the requisite level of competence, the novice will be on his or her own. This process can be observed in hundreds of occupational settings. In some cases, there is a formal role designation for those undergoing socialization, such as "trainee."

Alternatively, there may be a separate period of formal training outside the organization before the person occupies the role; this is the basis of many junior or technical college training programs. Many educational programs exist to prepare people for roles they will play in the future. Training programs or schools for beauticians, flight attendants, dental technicians, and truck drivers are only a few examples of this type of socialization.

Anticipatory Socialization

In addition to intentional training before and after a role is acquired, there may be **anticipatory socialization**—activities that provide people with knowledge about, skills for, and values of a role they have not yet assumed. The teenager learning about sexual activity from the boasts of an older friend or from an X-rated movie is undergoing anticipatory socialization. So is the aspiring diplomat or politician who attends closely to the behavior of the President of the United States in a television interview.

Anticipatory socialization is different from explicit training because it is not intentionally designed as role preparation by socialization agents (Clausen, 1968; Heiss, 1990). It can ease the transition into new roles, but it is more effective for some roles than for others (Bush & Simmons, 1990; Thornton & Nardi, 1975). First, anticipatory socialization usually works best for future roles that are highly visible. Socialization of this type usually prepares children more effectively for the parent role than for the spouse role, for example. Children see the parent role directly when interacting with their own parents, whereas important aspects of interaction between spouses occur when children are away or asleep.

Second, anticipatory socialization eases role transition if future roles are presented accurately. The interactions between spouses that children do observe are often intentionally laundered to hide negative feelings and conflict, resulting in poor anticipatory socialization and incorrect expectations about role demands.

Successful anticipatory socialization entails goal setting, planning, and preparation for future roles. Only by setting at least tentative occupational and family goals during our teenage years, for example, can we effectively plan our educational and social lives. Preparation occurs through part-time jobs, special courses, reading, talking with informed individuals, and so on. People also prepare for transitions by trying out elements of their anticipated roles. This is what couples planning marriage are doing when they take joint vacations, live together for a trial period, and share purchases.

Role Discontinuity

The acquisition of a new role does not always proceed smoothly, even if there has been anticipatory socialization. Changing roles can be stressful. A new role can involve meeting new expectations, performing new tasks, and interacting with new types of people. It may necessitate moving to a new location. Furthermore, acquiring a role may involve losses as well as gains. The person may have to leave a prior role to move into the new one.

Entering a new role can be especially difficult when there is **role discontinuity**—that is, when the values and identities associated with a new role contradict those of earlier roles (Benedict, 1938). On entering a discontinuous role, we must revise our former expectations and aspirations. Retirement, for example, often creates role discontinuity. During their working years, career-oriented adults are expected to strive for autonomy and productivity and to build their identities around their work. Retirement introduces contradictory expectations for such people. They must now assume less productive roles and rebuild their identities around these discontinuous expectations (Mortimer & Simmons, 1978).

Role transitions of particular social importance are sometimes marked by rites of passage—public ceremonies or rituals in which the individual's new status is affirmed (Glaser & Strauss, 1971; Van Gennep, 1960). Christenings, bar mitzvahs, graduations, marriage ceremonies, and retirement parties are common rites of passage in American society. They signify both to the individual and to others that the person now has a new identity and that new behaviors, rights, and duties are appropriate. The more radical or drastic the transition, the more important such rituals are. A study of weddings in the Netherlands found that persons who were older, had lived together, or were remarrying were less likely to have elaborate wedding ceremonies (Kalmijn, 2004). These ceremonies also serve as social occasions for giving emotional support, advice, and material aid (presents) to those adopting new roles.

SUMMARY

Socialization is the process through which infants become effective participants in society. It makes us like all other members of society in certain ways (shared language) but distinctive in other ways.

Perspectives on Socialization (1) One approach to the study of socialization emphasizes biological development; it views the emergence of interpersonal responsiveness and the development of speech and of cognitive structure as influenced by maturation. (2) Another approach emphasizes learning and the acquisition of skills from other persons. (3) A third approach emphasizes the child's discovery of cultural routines as he or she participates in them. Society organizes this process by making certain agents responsible for particular types of socialization of specific persons.

Agents of Childhood Socialization There are three major socializing agents in childhood. (1) The family provides the infant with a strong attachment to one or more caregivers. This bond is necessary for the infant to develop interpersonal and cognitive skills. Family composition and social class affect socialization by influencing the amount and kind of interaction between parent and child. Ethnic and racial groups differ in the child rearing techniques they use and in the values they emphasize. (2) Peers provide the child with equal status relationships and are an important influence on the development of self. (3) Schools teach skills—reading, writing, and arithmetic—as well as traits like punctuality and perseverance.

Processes of Socialization Socialization is based on three different processes. (1) Instrumental conditioning—the association of rewards and punishments with particular actions—is a basis for learning both behaviors and performance standards. Studies of the effectiveness of various child rearing techniques indicate that rewards do not always make a desirable behavior more likely to occur and punishments do not always eliminate an undesirable behavior. The use of corporal punishment appears to increase the likelihood of later antisocial behavior. Through instrumental learning, children develop the ability to judge their own behaviors and to engage in self-reinforcement. (2) We learn many behaviors and skills by observation of models. We may not perform these behaviors, however, until we are in the appropriate situation. (3) Socialization also involves internalization—the acquisition of behavioral standards, making them part of the self. This process enables the child to engage in self-control.

Outcomes of Socialization (1) The child gradually learns a gender role—the expectations associated with being male or female. Whether the child is independent or dependent, aggressive or passive, depends on the expectations communicated by parents, kin, and peers. (2) Language skill is another outcome of socialization; it involves learning words and the rules for combining them into meaningful sentences. Related to the learning of language is the development of thought and the ability to group objects and persons into meaningful categories. (3) The learning of social norms involves parents, peers, and teachers as socializing agents. Children learn that conformity to norms facilitates social interaction. Children also develop the ability to make moral judgments. (4) Children acquire motives—dispositions that produce sustained, goal-directed behavior. Orientations toward work are influenced primarily by parents; middle-class families emphasize self-direction, whereas working-class families emphasize conformity.

Adult Socialization Socialization continues throughout life. In adulthood, it involves preparing the person to successfully enact major roles such as intimate partner and parent. (1) Role acquisition is often accompanied by on-the-job training. Also, time spent in the role of trainee may be prerequisite to entry into a new role. (2) Anticipatory socialization involves unintentional role preparation. Its effectiveness depends on the visibility and accuracy of the portrayal of the new role. (3) Role discontinuity makes role acquisition especially difficult. Discontinuity can be eased by anticipatory socialization and by rites of passage.

LIST OF KEY TERMS AND CONCEPTS

anticipatory socialization (p. 79)
attachment (p. 58)
borderwork (p. 63)
cultural routines (p. 56)
extrinsically motivated behavior (p. 68)
gender role (p. 70)
instrumental conditioning (p. 65)
internalization (p. 69)
intrinsically motivated behavior (p. 68)
moral development (p. 76)
norm (p. 74)
observational learning (p. 68)
punishment (p. 66)
role discontinuity (p. 79)
self-reinforcement (p. 68)
shaping (p. 65)
socialization (p. 53)

4

SELF AND IDENTITY

Introduction

The Nature and Genesis of Self

The Self as Source and Object of Action

Self-Differentiation

Role Taking

The Social Origins of Self

Identities: The Self We Know

Role Identities

Social Identities

Research on Self-Concept Formation

The Situated Self

Identities: The Self We Enact

Identities and Behavior

Choosing an Identity to Enact

Identities as Sources of Consistency

The Self in Thought and Feeling

Self-Schema

Effects of Self-Awareness

Effects of Self-Discrepancies

Self-Esteem

Assessment of Self-Esteem

Sources of Self-Esteem

Self-Esteem and Behavior

Protecting Self-Esteem

Summary

List of Key Terms and Concepts

INTRODUCTION

An amnesia victim who walked into a pizza parlor six weeks ago and said "Help me, I don't know who I am," now faces a new identity crisis.

"I've gotten lots of calls saying I'm two entirely different people," said the man, who dubbed himself John Jackson after he was taken to a homeless shelter.

Jackson's ordeal began May 3, when he woke up beside railroad tracks and Interstate 64 near where the borders of West Virginia, Ohio, and Kentucky meet. "I woke up in a field and the first thing I thought was 'It's cold.' And then I wondered, 'Where am I?' And then, 'Who am I?' Then I realized I had no answers and it got very frightening."

"He was bewildered more than anything else," said mission cook Virginia Berry. "He was dressed OK, had on blue jeans. He has a beard but it's well-trimmed and shaped and been taken care of. His hair's that way, too."

The man asked mission staff to call him Jackson because he remembers roads and a baseball park in Jacksonville, Fla. Mission workers asked Jacksonville newspapers to publish the man's photograph and on Friday he was inundated by calls from people claiming to know him.

Jackson did not recognize any of the callers, although some of the details they provided matched the few pieces he has to his puzzle.

One group of people who saw Jackson's photograph told him they believe he is Roy Moses, 33, who frequently vanished for months at a time "on a whim." Moses' stepmother, stepsister, and brother-in-law all talked to Jackson on the telephone "and they seem convinced that's who I am." But Jackson said the family could provide few details, "so I'm not so sure."

Another Jacksonville resident said he recognized Jackson as Michael Shawper, 40, a career Navy man. The caller knew several little known naval bases that Jackson remembers, "and the knowledge of history and geography, that matches. But he said the last time he had seen Shawper was 11 years ago in Guam."

Shawper and Moses both attended the same college in Jacksonville. One has been married two or three times, the other once. "So I'm either an educated responsible man or an educated irresponsible man," Jackson said. "I can't be both, that's for sure. One of them has to be wrong. They both may be wrong."

(*Wisconsin State Journal,* June 12, 1988, p. 4A)

"Who am I?" Few human beings in Western societies live out their lives without pondering this question. Some people pursue the search for self-knowledge and for a meaningful identity eagerly; others pursue it desperately. College students in particular are often preoccupied with discovering who they are. Few, however, have experienced existential uncertainty to the degree faced by John Jackson.

Each of us has unique answers to this question, answers that reflect our **self-schema** or self-concept, the organized structure of cognitions or thoughts we have about ourselves. The self-schema comprises our perceptions of our social identities and personal qualities and our generalizations about the self based on experience.

The contents of self-schema are often assessed by having people answer the question "Who am I?" This test is the focus of Box 4.1. Before you read on, take a few moments and respond to this question yourself in the space provided in the box. For comparison, read the answers of a 9-year-old boy and a female college sophomore to the question "Who am I?" Their responses are listed at the bottom of the box.

Five major questions are addressed in this chapter.

1. What is the self and how does it arise?
2. How do we acquire unique identities—the categories we use to specify who we are? How do we use them to locate ourselves in the world relative to others?
3. How do our identities guide our plans and behavior?
4. How does self-awareness influence the ways we think?
5. Feelings of evaluation inevitably accompany thoughts about ourselves. Where do they come from, and how do they affect our behavior? How do we protect our self-esteem against attack?

BOX 4.1 Test Yourself: Measuring Self-Concepts

In order to study self-concepts, we need ways to measure them. Many methods have been used. For example, one approach asks people to check those adjectives on a list (intelligent, aggressive, trusting, and so on) that describe themselves (Sarbin & Rosenberg, 1955). In another approach (Osgood, Suci, & Tannenbaum, 1957), people rate themselves on pairs of adjectives (strong-weak, good-bad, active-passive): Are they more like one of the adjectives in the pair or more like its opposite? Another technique, developed by Miyamoto and Dornbusch (1956), asks people whether they have more or less of a characteristic (self-confidence, likableness) than members of a particular group (such as fraternities, sororities, and so on). In yet another technique, people sort cards containing descriptive phrases (interested in sports, concerned with achievement) into piles according to how accurately they think the phrases describe them (Stephenson, 1953).

Each of these popular methods provides respondents with a single standard set of categories to use in describing themselves. Using the same categories for all respondents makes it easy to compare the self-concepts of different people. These methods have a weakness, however. They do not reveal the unique dimensions that individuals use in spontaneously thinking about themselves. For this purpose, techniques that ask people simply to describe themselves in their own words are especially effective (Kuhn & McPartland, 1954; McGuire & McGuire, 1982).

Instructions for the "Who Am I?" technique for measuring self-concepts (Gordon, 1968) are provided below. You can try this test yourself.

In the 15 numbered blanks write 15 different answers to the simple question "Who Am I?" Answer as if you were giving the answers to yourself, not to somebody else. Write the answers in the order they occur to you. Don't worry about "logic" or "importance."

I Am

1. ____________
2. ____________
3. ____________
4. ____________
5. ____________
6. ____________
7. ____________
8. ____________
9. ____________
10. ____________
11. ____________
12. ____________
13. ____________
14. ____________
15. ____________

The following responses have been obtained from two persons, Josh and Arlene.

Josh: A 9-year-old male

a boy
do what my mother says, mostly
Louis's little brother
Josh
have big ears
can beat up Andy
play soccer
sometimes a good sport
a skater
make a lot of noise
like to eat
talk good
go to third grade
bad at drawing

Arlene: A female college sophomore

a person
member of the human race
daughter and sister
a student
people-lover
people-watcher
creator of written, drawn, and spoken (things) creations
music enthusiast
enjoyer of nature
partly the sum of my experiences
always changing
lonely
all the characters in the books I read
a small part of the universe, but I can change it
I'm not sure?! (Gordon, 1968)

THE NATURE AND GENESIS OF SELF

The Self as Source and Object of Action

We can behave in a wide variety of ways toward other persons. For example, if Bob is having coffee with Carol, he can perceive her, evaluate her, communicate with her, motivate her to action, attempt to control her, and so on. Note, however, that Bob also can act in the same fashion toward himself—that is, he can engage in self-perception, self-evaluation, self-communication, self-motivation, and self-control. Behavior of this type, in which the individual who acts and the individual toward whom the action is directed are the same, is termed reflexive behavior.

For example, if Bob, a student, has an important term paper due Friday, he engages in the reflexive process of self-control when he pushes himself ("Work on that history paper now"). He engages in self-motivation when he makes a promise to himself ("You can go out for pizza and a movie Friday night"). Both processes are part of the self. To have a self is to have the capacity to engage in reflexive actions—to plan, observe, guide, and respond to our own behavior (Bandura, 1982c; Mead, 1934).

Our understanding of reflexive behavior and the self is drawn from symbolic interaction theory. By definition, the **self** is the individual viewed as both the source and the object of reflexive behavior. Clearly, the self is both active (the source that initiates reflexive behavior) and passive (the object toward whom reflexive behavior is directed). The active aspect of the self is labeled the *I,* and the object of self-action is labeled the *me* (James, 1890; Mead, 1934).

It is useful to think of the self as an ongoing process (Gecas & Burke, 1995). Action involving the self begins with the *I*—with an impulse to act. For example, Bob wants to see Carol. In the next moment, that impulse becomes the object of self-reflection and, hence, part of the *me* ("If I don't work on that paper tonight, I won't get it done on time"). Next, Bob responds actively to this self-awareness, again an I phase ("But I want to see Carol, so I won't write the paper"). This, in turn, becomes the object to be judged, again a me phase ("That would really hurt my grade"). So Bob exercises self-control and sits down at his desk to write. The *I* and *me* phases continue to alternate as every new action (*I*) becomes in the next moment the object of self-scrutiny (*me*). Through these alternating phases of self we plan, act, monitor our actions, and evaluate outcomes (Markus & Wurf, 1987).

Another way to think about self is as the source of agency (Emirbayer & Mische, 1998). While we recognize that social structure influences the person, the symbolic interactionist view of self reminds us that individuals act at times in spontaneous, novel, or deviant ways.

Mead (1934) portrays action as guided by an internal dialogue. People engage in conversations in their minds as they regulate their behavior. They use words and images to symbolize their ideas about themselves, other persons, their own actions, and others' probable responses to them. This description of the internal dialogue suggests there are three capacities human beings must acquire in order to engage successfully in action: They must (1) develop an ability to differentiate themselves from other persons, (2) learn to see themselves and their own actions as if through others' eyes, and (3) learn to use a symbol system or language for inner thought. In this section, we examine how children come to differentiate themselves and how they learn to view themselves from others' perspectives. We also discuss how language learning is intertwined with acquiring these two capacities.

Self-Differentiation

To take the self as the object of action, we must—at a minimum—be able to recognize ourselves. That is, we must distinguish our own faces and bodies from those of others. This may seem elementary, but infants are not born with this ability. At first, they do not even discriminate the boundaries between their own bodies and the environment. Cognitive growth and continuing tactile exploration of their bodies contribute to infants' discovery of their physical uniqueness. So does experience with caregivers who treat them as distinct beings. Studies of when children can

© Elizabeth Crews

To take the self as the object of our action—observing and modifying our own behavior—we must be able to recognize ourselves. Although infants are not born with this ability, they acquire it quickly.

recognize themselves in a mirror suggest that most children are able to discriminate their own image from others' by about 18 months (Bertenthal & Fischer, 1978). Research indicates that children become capable of representing self-other contingencies (for example, "If I do X, she does Y") at 18 to 24 months old (Higgins, 1989).

Children must learn not only to discriminate their physical selves from others, but also to discriminate themselves as a social object. Mastery of language is critical in childrens' efforts to learn the latter (Denzin, 1977). Learning one's own name is one of the earliest and most important steps in acquiring a self. As Allport (1961) put it, "By hearing his name repeatedly the child gradually sees himself as a distinct and recurrent point of reference. The name acquires significance for him in the second year of life. With it comes awareness of independent status in the social group" (p. 115).

A mature sense of self entails recognizing that our thoughts and feelings are our private possessions. Young children often confuse processes that go on in their own minds with external events (Piaget, 1954). They locate their own dreams and nightmares, for example, in the world around them. The distinction between self and nonself sharpens as social experience and cognitive growth bring children to realize that their own private awareness of self is not directly accessible to others. By about age 4, children report that their thinking and knowing goes on inside their heads. Asked further, "Can I see you thinking in there?" they generally answer "No," demonstrating their awareness that self-processes are private (Flavell, Shipstead, & Croft, 1978).

Changes in the way children talk also reveal their dawning realization that the self has access to private information. During their first years of talking, children's speech patterns are the same whether they are talking aloud to themselves or directing their words to others. Gradually, however, they begin to distinguish speech to self from speech to others (Vygotsky, 1962). Speech to self becomes abbreviated until it is virtually incomprehensible to the outside listener, whereas speech to others becomes more elaborated over time. "Cold" suffices for Amy to tell herself she wants to take off her wet socks. But no one else would understand this without access to her private knowledge. When addressing others, Amy would expand her speech to include whatever private information they would need to understand ("Gotta change my wet socks. They're making me cold."). This reflects her growing awareness that each self has its own unique store of knowledge.

Access to private information about the self leads to systematic differences in adults' self-descriptions compared to descriptions of others (McGuire & McGuire, 1986). Descriptions of the self focus on what one does—on physical action and on affective reactions to others. Descriptions of others focus on who the person is—on social interactions and on his or her cognitive reactions. Furthermore, people perceive themselves as more complex than other people (Sande, Goethals, & Radloff, 1988). Did your

TABLE 4.1 SIGNIFICANT OTHERS MENTIONED IN SELF-DESCRIPTIONS, BY AGE

	RATIO OF THE FREQUENCY OF MENTIONING		
Age	Parents versus Teachers	Brothers and Sisters versus Friends and Fellow Students	Nonfamily Members versus Extended Family
7 years	1.7 to 1	1.7 to 1	4 to 1
9 years	1 to 1.4	1 to 1.4	8 to 1
13 years	1 to 1	1 to 1	13 to 1
17 years	1 to 2.3	1 to 2.3	49 to 1

Note: In this study, 560 boys and girls were asked, "Tell us about yourself." The children's responses suggest that their self-definitions in terms of other people tend to shift away from family members with age—from parents to teachers (column 1), from brothers and sisters to friends and fellow students (column 2), from extended family members (cousins, aunts, uncles) to nonfamily members (column 3). For example, 7-year-olds mentioned parents almost twice as often as teachers, and 17-year-olds mentioned teachers more than twice as often as parents (column 1).

Source: Adapted from McGuire and McGuire, 1982.

responses in Box 4.1 reflect these characteristics of self-descriptions?

Role Taking

Recognizing that one is physically and mentally differentiated from others is only one step in the genesis of self. Once we can differentiate ourselves from others, we also can recognize that each person sees the world from a different perspective. The second crucial step in the genesis of self is **role taking**—the process of imaginatively occupying the position of another person and viewing the self and the situation from that person's perspective (Hewitt, 1997).

Role taking is crucial to the genesis of self because through it the child learns to respond reflexively. Imagining others' responses to the self, children acquire the capacity to look at themselves as if from the outside. Recognizing that others see them as objects, children can become objects (*me*) to themselves (Mead, 1934). They can then act toward themselves to praise ("That's a good girl"), to reprimand ("Stop that!"), and to control their own behavior ("Wait your turn").

Long ago, Cooley (1908) noted the close tie between role taking and language skills. One of the earliest signs of role-taking skills is the correct use of the pronouns *you* and *I*. To master the use of these pronouns requires taking the role of the self and of the other simultaneously. Most children firmly grasp the use of *I* and *you* by the middle of their third year (Clark, 1976). This suggests that children are well on their way to effective role taking at this age. Studies indicate that children develop the ability to infer the thoughts and expectations of others between ages 4 and 6 (Higgins, 1989).

The Social Origins of Self

Our self-schema is produced in our social relationships. Throughout life, as we meet new people and enter new groups, our view of self is modified by the feedback we receive from others. This feedback is not an objective reality that we can grasp directly. Rather, we must interpret others' responses in order to figure out how we appear to them. We then incorporate others' imagined views of us into our self-schema.

To dramatize the idea that the origins of self are social, Cooley (1902) coined the term **looking-glass self.** The most important looking glasses for children are their parents and immediate family and, later, their playmates. They are the child's **significant others**—the people whose reflected views have greatest influence on the child's self-concepts. As we grow older, the widening circle of friends and relatives, school teachers, clergy, and fellow workers provides our significant others. The changing images of self we acquire throughout our lives depend on the social relationships we develop (see Table 4.1).

By playing complex games such as baseball, children learn to organize their actions into meaningful roles and to imagine the viewpoints of other players at the same time. Role-taking enables the catcher to coordinate effectively with teammates, for example, to tag a runner out at home plate.

Play and the Game Mead (1934) identified two sequential stages of social experience leading to the emergence of the self in children. He called these stages play and the game. Each stage is characterized by its own form of role taking.

In the **play** stage, young children imitate the activities of people around them. Through such play, children learn to organize different activities into meaningful roles (nurse, doctor, firefighter). For example, using their imaginations, children carry sacks of mail, drop letters into mailboxes, greet homeowners, and learn to label these activities as fitting the role of "mail carrier." At this stage, children take the roles of others one at a time. They do not recognize that each role is intertwined with others. Playing mail carrier, for example, the child does not realize that mail carriers also have bosses to whom they must relate. Nor do children in this stage understand that the same person simultaneously holds several roles—that mail carriers are also parents, store customers, and golf partners.

The **game** stage comes later, when children enter organized activities such as complex games of house, school, and team sports. These activities demand interpersonal coordination because the various roles are differentiated. Role taking at the game stage requires children to imagine the viewpoints of several others at the same time. For Ellen to play shortstop effectively, for example, she must adopt the perspectives of the infielders and of the base runners as she fields the ball and decides where to throw. In the game, children also learn that different roles relate to one another in specified ways. Ellen must understand the specialized functions of each position, the ways the players in different positions coordinate their actions, and the rules that regulate baseball.

The Generalized Other Repeated involvement in organized activities lets children see that their own actions are part of a pattern of interdependent group activity. This experience teaches children that organized groups of people share common perspectives and attitudes. With this new knowledge, children construct a **generalized other**—a conception of the attitudes and expectations held in common by the members of the organized groups with whom they interact. When we imagine what the group expects of us, we are taking the role of the generalized other. We are also concerned with the generalized other when we wonder what people would say or what society's standards demand. As children grow older, they control their own behavior more and more from the perspective of the generalized other. This helps them to resist the influence of impulse or of specific others who just happen to be present at the moment.

Over time, children internalize the attitudes and expectations of the generalized other, incorporating them into their self-concepts. But building up self-concepts involves more than accepting the reflected views of others. We may misperceive or misinterpret the responses that others direct to us, for example, due to our less-than-perfect role-taking skills. Others' responses may themselves be contradictory or inconsistent. Also, we may resist the reflected views we perceive because they conflict with our prior self-concepts or with our direct experience. A boy may reject his peers' view that he is a "sissy," for example, because he previously thought of himself as brave and could still visualize his experience of beating up a bully.

Self-Evaluation The views of ourselves that we perceive from others usually imply positive or negative evaluations. These evaluations also become part of the self we construct. Actions that others judge favorably contribute to positive self-concepts. In contrast, when others disapprove or punish our actions, the self-concepts we derive may be negative.

We also form self-evaluations when reflecting on the adequacy of our role performances—on the extent to which we live up to the standards we aspire to. Our self-evaluations most commonly focus on our competence, self-determination, moral worth, or unity. Self-evaluations also influence the ways we express our role identities. A musician, for example, will pursue opportunities to perform in public more persistently if she sees herself as competent than if she thinks she is never quite good enough. Self-evaluations are so important that the concluding section of this chapter will be devoted to them.

IDENTITIES: THE SELF WE KNOW

In Box 4.1, Arlene described herself as a person, daughter, student, people-lover, and creator of things. This is the self she knows, a self that includes specific identities. **Identities** are the meanings attached to the self by one's self and others (Gecas & Burke, 1995). When we think of our identities, we are actually thinking of various plans of action that we expect to carry out. When Arlene identifies herself as a student, for example, she has in mind that she plans to attend classes, write papers, take exams, and so on. If Arlene does not engage in these behaviors, she will have to relinquish her student identity.

In this section, we consider four questions about the self we know: (1) How do our roles influence the identities we include in our self? (2) How do group memberships influence the self we know? (3) What evidence is there that the self we know is based on the reactions we perceive from others? (4) How do the aspects of self that people note vary from one situation to another?

Role Identities

Each of us occupies numerous positions in society—student, friend, son or daughter, customer. Each of us, therefore, enacts many different social roles. We construct identities by observing our own behavior and the responses of others to us as we enact these roles. For each role we enact, we develop a somewhat different view of who we are—an identity. Because these identities are concepts of self in specific roles, they are called **role identities.** The role identities we develop depend on the social positions available to us in society. As a result, the self we know is linked to society fundamentally through the roles we play. It reflects the structure of our society and our place in it (McCall & Simmons, 1978; Stryker, 1980). Role identities highlight the impact on self of social structure via reciprocal relationships with occupants of complementary roles.

Do societal role expectations strictly dictate the contents of our role identities? Apparently not. Consider, for example, the role expectations for a college instructor. Some instructors deliver lectures, whereas others lead discussions; some encourage questions, whereas others discourage them; some assign papers, and others do not. As this example indicates, role expectations usually leave individuals some room to improvise their own role performances. It is probably more accurate to think of people as "making" their roles—that is, shaping them—rather than as conforming rigidly to role expectations (Turner, 1978). Societal expectations do dictate the goals of role performance; instructors must instruct using means that are consensually agreed on (Burke, 2004).

Several influences affect the way we make the roles we enact. Conventional role expectations in society set a general framework. In the role of student, for example, you must submit assigned work. The person holding the complementary role also has expectations. As a student in Prof. Myers' class, you must write a 15-page lit review and research proposal. Within the boundaries set by these expectations, you can fashion your actual role performances to reflect your personal characteristics and competencies. You can select topics that interest you, highlight your strengths, and cover your weaknesses. You also mold your role performances to impress your audience (say, writing in the style that Prof. Myers prefers). Finally, you adjust your different performances to maintain some consistency among them (say, trying for a level of quality consistent with your other course work). Because each person makes roles in a unique, personal fashion, we each derive somewhat different role identities even if we occupy similar social positions. Consequently, our role identities as student, team player, and so on differ from the role identities of others who also occupy these positions.

In describing themselves, people frequently mention the styles of interpersonal behavior (introverted, cool) that distinguish the way they fashion their unique role performances. People also mention the emotional or psychological styles (optimistic, moody) that characterize these performances. Individual preferences point to specific ways in which people express their role identities. For example, a person who sees herself as a musician expresses this role identity differently depending on whether she prefers Bach or rock. Body image—the aspect of the self we recognize earliest—remains important throughout life. Beyond this, our self extends to include our material possessions, such as our clothing, house, car, records, and so on (James, 1890).

Social Identities

A second source of identities is membership in social categories or groups based on criteria such as gender, nationality, race/ethnicity, sexual preferences, or political affiliation (Howard, 2000). A definition of the self in terms of the defining characteristics of a social group is a **social identity** (Hogg, Terry, & White, 1995; Tajfel & Turner, 1979). Each of us associates certain characteristics with members of specific groups. These characterizations—Chicago Bulls fans are loud, women are emotional—define the group. If you define yourself as a member of the group, these characteristics become standards for your thoughts, feelings, and actions. If your interactions with others, whether members or not, confirm the importance of these attributes, they become part of the self you know. Research indicates that cognitive representations of the self and of the groups to which the person belongs are closely linked (Smith & Henry, 1996). Social Identities highlight the impact on self of social structure via consensually defined social groupings (Deaux & Martin, 2003). Note that one need not interact with other members to identify as a member of the group.

Group members usually perceive the group—and therefore, themselves—in positive terms. They rate traits perceived as typical of the group more favorably than they rate other traits. However, this bias does not lead members to violate social reality as defined by nonmembers (Ellemers, Van Rijswijk, Roefs, & Simons, 1997).

Social groups are often defined in part by reference to other groups. The meaning of being a Young Republican is related to the meaning of being a Young

Socialist and a Young Democrat. The meaning of being a man in American society is closely related to the meaning of being a woman. Thus, when membership in a group becomes a salient basis for self-definition, perceptions of relevant out-groups are also made salient. Often there is an accentuation effect—an emphasis on perceived differences and unfavorable evaluations of the out-group and its members (Hogg et al., 1995). Thus, negative stereotypes directed at persons of a different gender, race, or religion are often closely related to the self-concept of the person who holds them. Research indicates that both in-group favoritism and out-group hostility are reinforced in conversations between group members (Harasty, 1997).

Research on Self-Concept Formation

Two of the key theoretical ideas discussed so far are (1) the formation of the self-schema involves the adoption of role and social identities, and (2) a person's self-concept is shaped by the reactions that he or she receives from significant others during social interaction. Each of these ideas has been the focus of empirical research.

The Adoption of Role and Social Identities Self-schemas are formed in part by adopting identities. The identities available to us depend on the culture. One difference between cultures is whether a culture is individualist or collectivist (Triandis, 1989). Individualist cultures emphasize individual achievement and its associated identities such as president, team captain, idealist, and outstanding player. Collectivist cultures emphasize values that promote the welfare of the group and its associated identities such as son (family), Catholic (religion), Italian (ethnicity), and American. According to research, the self-schemas of persons in individualist cultures (such as the United States) include more individual identities, whereas those of persons in collectivist cultures include more group-linked identities (Triandis, McCusker, & Hui, 1990).

The adoption of a *role* identity involves socialization into the group or organization of which the role is a part. A study of members of a volunteer search and rescue group, Peak, identified three stages of membership: new, peripheral, and core (Lois, 1999) New members were often attracted by the desire to be a hero. To make the transition to (be accepted by others as) a peripheral member, they had to suppress self-oriented attitudes and behavior and acknowledge the importance of the team. They also had to learn survival skills and rescue techniques, demonstrating humility and persistence in the process. To make the transition to core member, they had to accept the roles offered by the team (sometimes very unglamorous ones) and demonstrate that they were skilled by leading training sessions. As members progressed through these stages, they increasingly shared in the sense of "we-ness," and their membership became an important social identity. They ultimately achieved the role of hero by becoming a committed member of the team.

Adopting a *social* identity involves self-categorization—the defining of the self as a member of a social category (Irish American, Black American, Feminist; Stets & Burke, 2000). Whereas enacting a role identity involves behavior conforming to a role, enacting a social identity involves adopting styles of dress, behavior, and thought associated with the social category. Successful adoption may require consensus by other members of the category that you can claim the identity. Whether one identifies with a social category in which one can claim membership depends on how easily one can be identified as a member of that group, for example, by name or skin color (Lau, 1989). It also depends on the general visibility and status of that group or category in society.

Reflected Appraisals The idea that the person bases his or her self-schema on the reactions he or she perceives from others during social interaction is captured by the term reflected appraisal. Studies of this process (Marsh, Barnes, & Hocevar, 1985; Miyamoto & Dornbusch, 1956) typically compare people's self-ratings on various qualities (such as intelligence, self-confidence, physical attractiveness) with the views of themselves that they perceive from others. The studies also compare self-ratings with actual views of others. The results of these studies support the hypothesis that it is the perceived reactions of others rather than their actual reactions that are crucial for self-concept formation (Felson, 1989).

One study of reflected appraisals analyzed perceptions of leadership in small groups (Riley & Burke, 1995). Groups of four persons met and engaged in discussions on four separate occasions. After each discussion, each member rated self and others on leadership identity and leadership performance scales. Self- and others' ratings of self on the identity scale were similar, and both were consistent with scores on the leadership behavior scale. In other words, a shared meaning structure developed among the discussants, and the individual's perception of self was consistent with others' appraisals.

Research has focused on the differential effect of various significant others on one's appraisal of self in particular roles or domains. Felson (1985; Felson & Reed, 1986) has studied the relative influence of parents and peers on the self-perceptions of fourth- through eighth-graders about their academic ability, athletic ability, and physical attractiveness. The results indicate that parents affect self-appraisals in the areas of academic and athletic ability, whereas peers are an important influence on perceived attractiveness. One aspect of attractiveness is weight. Although there is an objective measure of weight (that is, pounds, or pounds in relation to height), it is the social judgment ("too fat," "too thin," or "just right") that is incorporated into the self-concept. A study of adolescent health obtained self-appraisals of weight from 6,500 adolescents, as well as appraisals from their parents and a physician (Levinson, Powell, & Steelman, 1986) These young people were generally unhappy with their weight, with boys judging themselves to be too thin and girls judging themselves to be overweight. For both, parental appraisal was significantly related to the young person's judgment, whereas the physician's rating was not.

A study of married couples with one child examined the relative influence of self-appraisal and partner's appraisal on two types of behavior: caregiving (traditionally female) and breadwinning (traditionally male; Maurer, Pleck, & Rane, 2001). The hypothesis was that own appraisal would be more influential for gender-consistent behavior (male breadwinning, female caregiving), whereas partner's appraisal would be more influential for noncongruent behavior (male caregiving, female breadwinning). The results generally supported the hypothesis. Thus, the appraisals of those presumed to be more knowledgeable about the role were more important.

Typically, a person's self-ratings are related more closely to his or her perceived ratings by others than to the actual ratings by others. Why is this so? Three reasons are especially important. First, others rarely provide full, honest feedback about their reactions to us. Second, the feedback we do receive is often inconsistent and even contradictory. Third, the feedback is frequently ambiguous and difficult to interpret. It may be in the form of gestures (shrugs), facial expressions (smiles), or remarks that can be understood in many different ways ("That's nice"). For these reasons, we may know little about others' actual reactions to us. Instead, we must rely on our perceptions of others' reactions to construct our self-concepts (Schrauger & Schoeneman, 1979).

Evidence that self-concepts are related to the perceived reactions of others does not in itself demonstrate that self-concepts are actually formed in response to these perceived reactions. However, one study (Mannheim, 1966) does suggest such an impact of the perceived reactions of others on self-concepts. The investigators in this study asked college dormitory residents to describe themselves and to report how they thought others viewed them. Several months later, self-concepts were measured again. In the interim, students' self-concepts had moved closer to the views they had originally thought that others held. Change toward the perceived reactions of others had indeed occurred. Similarly, a longitudinal study of delinquent behavior found that parental appraisals of youth as delinquent were associated with subsequent self-appraisals as delinquent; self-appraisal as delinquent was in turn related to delinquent behavior (Matsueda, 1992).

Identity and Multiracial Heritage In a racially diverse society, social identity based on racial heritage is a significant component of self-schema. According to the reflected appraisal model, it is perceived reactions of others that influence self-perception. Also, successful adoption of an identity requires acceptance by others of one's claims. Thus, an important influence on racial identity should be responses of others based on one's appearance. The racial identity of some persons seems obvious; that is, their skin color and physical features fit the social stereotype of what Asians, Blacks, or Whites look like. But the racial identity of others is not obvious. People with ambiguous appearance are frequently asked "What are you?" and may come to hate having to answer that question one more time (Navarro, 2005). To study multiracial identity, Khanna (2004) recruited adults who had one Asian parent and one White parent. She predicted that (apparent) phenotype or appearance (How would others categorize you, Asian or White?) would be the most important influence on racial identity. But what about persons whose phenotype is ambiguous? Khanna predicted that cultural exposure, language proficiency, eating foods, and celebrating holidays would influence identity, that is, identifying oneself as "Asian." Both hypotheses were confirmed. A study of hundreds of Asian and Latino students entering UCLA found that speaking the ethnic language at home and having high school friends of the same ethnicity were the main predictors of strong ethnic identity (Sears, Fu, Henry, & Bui, 2003).

The Situated Self

If we were to describe ourselves on several different occasions, the identities, personal qualities, and self-evaluations mentioned would not remain the same. This is not due to errors of reporting; rather, it demonstrates that the aspects of self that enter our awareness and matter most to us depend on the situation. The **situated self** is the subset of self-concepts chosen from our identities, qualities, and self-evaluations that constitutes the self we know in a particular situation (Hewitt, 1997). Markus and Wurf (1987) refer to the current, active, accessible self-representations as the working self-concept.

The self-concepts most likely to enter the situated self are those distinctive to the setting and relevant to the ongoing activities. Consider a Black woman for whom being Black and being a woman are both important self-concepts. When she interacts with Black men, she is more likely to think of herself as a woman. When she interacts with White women, she is more likely to be aware that she is Black. Similarly, whether gender is part of your situated self depends in part on the gender composition of others present (Cota & Dion, 1986). Male and female college students placed in a group with two students of the opposite gender were more likely to list gender in their self-descriptions than members of all-male or all-female groups. Thus, self-concepts that are distinctive or peculiar to the social setting tend to enter into the situated self (McGuire & McGuire, 1982; see Figure 4.1).

Our activities also determine the self-concepts that constitute the situated self. A job interview, for example, draws attention to your competence; a party makes your body image more salient. The self we experience in our imaginings and in our interactions is always situated, because setting characteristics and activity requirements make particular self-concepts distinctive and relevant.

IDENTITIES: THE SELF WE ENACT

How does the self influence the planning and regulation of social behavior? The general answer to this question is that we are motivated to plan and to perform behaviors that will confirm and reinforce the identities we wish to claim for ourselves (Burke & Reitzes, 1981; Markus & Wurf, 1987). In elaborating on this answer, we will examine three more specific questions: (1) How are behaviors linked to particular identities? (2) Of the different identities available to us, what determines which ones we choose to enact in a situation? (3) How do our identities lend unity and consistency to our behavior?

Identities and Behavior

Each of us makes dozens of decisions every day; most of them influence our behavior. These decisions are influenced by explicit and implicit egotism—that is, giving undue prominence to the self. A study of major life decisions (where to live, choice of career) suggests that these decisions are influenced by our names

FIGURE 4.1 PERCENTAGE OF STUDENTS WHO MENTION A FEATURE SPONTANEOUSLY AS PART OF THEIR SELF-CONCEPT

A group of 252 sixth-graders from 10 classrooms were asked to describe themselves. Students mentioned a particular feature (for example, birthplace) more often if that feature distinguished them from their classmates. Because these are characteristics on which we stand out from our social groups, attracting more notice and social comment, we are more likely to build them into our self-concepts.

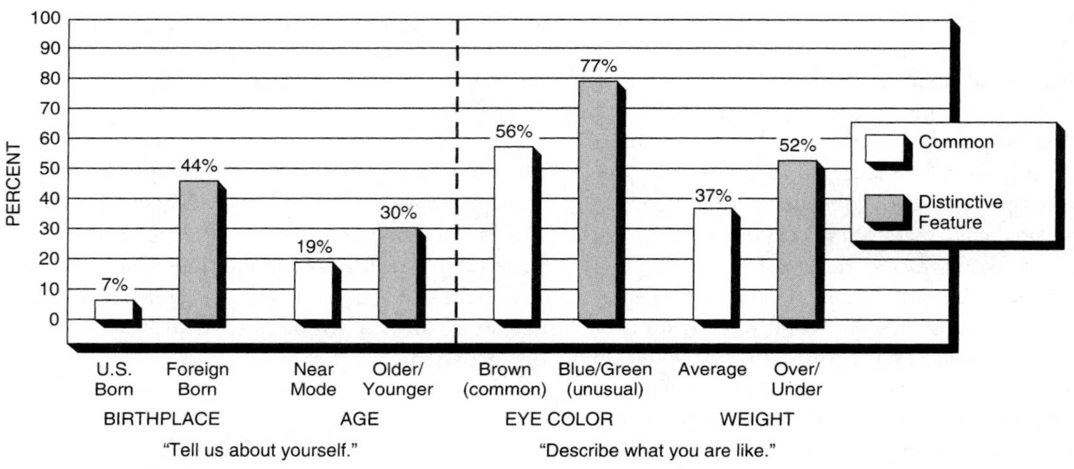

Source: Adapted from "Trait Salience in the Spontaneous Self-Concept" by W. J. McGuire and A. Padawer-Singer, *Journal of Personality and Social Psychology, 33,* 743–754. Copyright © 1976 by the American Psychological Association. Adapted with permission.

(Pelham, Mirenberg, & Jones, 2002). We tend to choose places and occupations with names that resemble our own. According to this study, it is not an accident that Susie sells seashells by the seashore.

Earlier, we noted that self-schemas include both identities and personal qualities. Some people place greater emphasis on one than on the other. For instance, consider responses to the question "Who am I?" Some people emphasize role identities (for example, daughter, psychology major), whereas others list primarily personal qualities (for example, easygoing, friendly). Leary, Wheeler, and Jenkins (1986) predicted that the behavioral preferences of these two groups would differ. In the area of recreation, people who emphasize social aspects would prefer team sports (say, softball, volleyball, basketball), whereas people who emphasize personal qualities would prefer individual sports (swimming, running, aerobics). Furthermore, those whose self-concept is predominantly social should prefer occupations that offer social rewards, such as status and friendship. Those whose self-concept is predominantly personal should prefer jobs that offer rewards such as opportunities for self-expression and personal growth. The results verified both predictions.

The link between identities and behaviors is through their common meanings (Burke & Reitzes, 1981). If members of a group agree on the meanings of particular identities and behaviors, they can regulate their own behavior effectively. They can plan, initiate, and control behavior to generate the meanings that establish the identities they wish to claim. If members do not agree on these meanings, however, people have difficulty establishing their preferred identities. If Roberta sees no connection between competitiveness and femininity, for example, she will have trouble establishing a feminine identity in the eyes of friends who think being feminine means being noncompetitive.

According to **identity control theory** (Burke, 1991), an actor uses the social meaning of his/her identity as a reference point for assessing what is occurring in the situation. The identities of the other actors and elements of the situation also have shared meanings. The behaviors of others and situational elements are evaluated by the actor according to whether they maintain his/her identity. Subsequent behaviors are selected and enacted in order to maintain one's identity in this situation. The (shared) meaning of an identity operates like a thermostat; if reflected appraisals or situational elements are inconsistent with identity, an actor will behave in ways designed to restore it (Smith-Lovin & Robinson, 2005).

Consider a woman whose identity is a "considerate professor." When students hand in assignments on time, her identity is reinforced. Occasionally, when an apparently hardworking student asks for an extension of a due date, it is consistent with her identity as "considerate" to grant the request. But if numerous students ask for extensions for reasons that seem trivial, she may "crack down" and refuse to give an extension to anyone, enacting the "professor" identity.

Since the meanings of role-identity elements, actions, and other identities are widely shared, Burke and other researchers use quantitative techniques to assess them. Adapting the techniques developed by Osgood (see Box 2.1), the meaning of an identity element or action is assessed on the dimensions of affect, evaluation, and potency. Researchers can compare these values across roles or groups or cultures to assess the impact of context on meanings.

Social identities are associated with category or group memberships. There are widely held meanings or stereotypes associated with many categories and groups. Thus, claiming a social identity creates a pressure to accept these stereotypes as self-descriptive. This can have a powerful impact on behavior. We may voluntarily adopt behavior or traits associated with positive stereotypes, such as adopting the food preferences associated with "veganism." But we may be influenced by negative stereotypes as well; stereotype threat refers to a situation in which one is at risk of confirming as self-characteristic a negative stereotype about a group to which one belongs (see Box 5.1). For instance, Blacks may underperform in an academic testing situation because they believe that others stereotype them as "dumb," which creates anxiety that disrupts their performance.

On the other hand, some group members will obviously violate any stereotype of the group. We noted earlier that characteristics that distinguish us from others are more likely to be part of our self-concept. Indeed, research indicates that people are more likely to include in their self-schema areas in which their performance is counterstereotypic (von Hippel, Hawkins, & Schooler, 2001).

Choosing an Identity to Enact

Each of us has many different identities. Each identity suggests its own lines of action. These lines of action are not all compatible, however, nor can they be pursued simultaneously in a single situation. If you are at a family reunion in your parents' home, for example, you might wish to claim an identity as a helpful son or daughter, an aspiring poet, or a witty conversationalist. These identities suggest different, even conflicting ways of relating to the other guests. What influences the decision to enact one rather than another identity? Several factors affect such choices.

The Hierarchy of Identities The many different role identities we enact do not have equal importance for us. Rather, we organize them into a hierarchy according to their **salience**—their relative importance to the self-schema. This hierarchy exerts a major influence on our decision to enact one or another identity (McCall & Simmons, 1978; Stryker, 1980). First, the more salient an identity is to us, the more frequently we choose to perform activities that express that identity (Stryker & Serpe, 1981). Second, the more salient an identity, the more likely we are to perceive that situations offer opportunities to enact that identity. Only a person aspiring to the identity of poet, for example, will perceive a family reunion as a chance to recite his or her poems. Third, we are more active in seeking opportunities to enact salient identities (say, searching for an open-mike poetry reading). Fourth, we conform more with the role expectations attached to the identities that we consider the most important.

What determines whether a particular identity occupies a central or a peripheral position in the salience hierarchy? In general, several factors affect the importance we attach to a role identity: (1) the resources we have invested in constructing the identity (time, effort, and money expended, for example, in learning to be a sculptor); (2) the extrinsic rewards that enacting the identity has brought (for example, purchases by collectors, acclaim by critics); (3) the intrinsic gratifications derived from performing the identity (for example, the sense of competence and aesthetic pleasure obtained when sculpting a human figure); and (4) the amount of self-esteem staked on enacting the identity well (for example, the extent to which a positive self-evaluation has become tied to being a good sculptor). As we engage in interaction and experience greater or lesser success in performing our different identities, their salience shifts.

Social Networks Each of us is part of a network of social relationships. These relationships may stand or fall depending on whether we continue to enact particular role identities. The more numerous and significant the relationships that depend on enacting an identity, the more committed we become to that identity (Callero, 1985). Consider, for example, your role as a student. Chances are that many of your relationships—with roommates, friends, instructors, and perhaps a lover—depend on your continued occupancy of the student role. If you left school, you could lose a major part of your life. Given this high level of commitment, it isn't surprising that for many students, being forced to leave school is traumatic.

The more commitment we have to a role identity, the more important that identity will be in our salience hierarchy. For instance, adults for whom participating in religious activities is crucial for maintaining everyday social relationships rank their religious identity as relatively important compared with their parent, spouse, and worker identities (Stryker & Serpe, 1981). Similarly, the importance rank that undergraduates give to various identities (student, friend, son/daughter, athlete, religious person, and dating partner) depends on the importance to them of the social relationships maintained by enacting each identity (Hoelter, 1983).

Need for Identity Support We are likely to enact those of our identities that most need support because they have recently been challenged. For instance, suppose that someone has recently had difficulty getting a date. That person may now choose actions calculated to elicit responses indicating he or she is an attractive dating partner. We also tend to enact identities likely to bring intrinsic gratifications (such as a sense of accomplishment) and extrinsic rewards (such as praise) that we especially need or miss at the moment. For example, if, after hours of solitary study, you feel a need for relaxed social contact, you might seek gratification by going to a student lounge or union to find someone to chat with.

Situational Opportunities Social situations are restrictive; they let us enact only some identities profitably, not others. Thus, in a particular situation, the identity we choose to enact depends partly on whether the situation offers opportunities for profitable enactment. Regardless of the salience of your identity as poet, if no one wants to listen to your poems, there will be no opportunity to enact that identity.

In a series of studies, Kenrick, McCreath, Govern, King, and Bordin (1990) asked students to rate the extent to which various personal qualities could be displayed in each of six different settings. The traits were adjustment, dominance, intellectual ability, likableness, social control, and social inclination. The students agreed that one can display intellectual ability in academic settings but not in recreational ones. Behaviors expressive of dominance can be displayed in athletic and business settings but not in religious ones. Finally, there are opportunities to display adjustment and social inclination in recreational settings but not in church.

Opportunities to enact an identity depend in part on other persons offering access to the aspiring actor. Offers of access often depend on perceptions of actors or those who control access. In this situation, is it better to be perceived as a specialist or as someone who is versatile at a number of roles? In order to get invited to the party, is it better to have a reputation as the "life of the party" or as a bright, friendly, warm person? Research designed to answer this question looked at the careers of U.S. film actors, specifically at the odds they would get roles in subsequent films (Zuckerman, Kim, Ukanwa, & von Rittman, 2003). Specialization increased the odds that novices would get future roles, but decreased the odds for veterans; when you are relatively unknown, you are more likely to get opportunities if you are known to be good at a specialty. Once you are known, versatility will get you more opportunities than if you are a specialist.

Identities as Sources of Consistency

Although the self includes multiple identities, people usually experience themselves as a unified entity. One reason is the influence of the salience hierarchy. Another reason is that we use several strategies that verify our perceptions of self.

Salience Hierarchy Our most salient identities provide consistent styles of behavior and priorities that lend continuity and unity to our behavior. In this way, the salience hierarchy helps us to construct a unified sense of self from our multiple identities.

The hierarchy of identities influences consistency in three ways. First, the hierarchy provides us with a basis for choosing which situations we should enter and which ones we should avoid. A study of the everyday activities of college students (Emmons, Diener, & Larsen, 1986) found clear patterns of choice and avoidance in each student's interactions; these patterns were consistent with the student's characteristics, such as sociability.

Second, the hierarchy influences the consistency of behavior across different situations. In another study, each person was asked to report the extent to which each of 10 affective states and 10 behavioral responses occurred in various situations over a 30-day period (Emmons & Diener, 1986). The results indicated a significant degree of consistency across situations.

Third, the hierarchy influences consistency in behavior across time. Serpe (1987) studied a sample of 310 first-year college students, collecting data at three points during their first semester in college. The survey measured the salience at each point of five identities: academic ability, athletic/recreational involvement, extracurricular involvement, personal involvement (that is, friendships), and dating. There was a general pattern of stability in salience. Change in salience was more likely for those identities where there was greater opportunity for change, such as dating.

Although the self-concept exhibits consistency over time, it may change (Demo, 1992). Life transitions may change the roles one plays and the situations one encounters. This creates a need to exit from one or more roles, adopt new roles, and change the salience hierarchy. During times such as adolescence and retirement, we are likely to feel a weakened sense of unity and a confusion about how to behave. This has been called an identity crisis (Erikson, 1968). To overcome such confusion, we must reorganize our identity hierarchy, giving greater importance to identities based on our newly available or remaining social positions. A retiree may successfully reorganize the hierarchy, for example, by upgrading identities based on new hobbies (gardener) and on continuing social ties (witty conversationalist).

Self-Verification Strategies We experience ourselves as consistent across time and situations because we employ several strategies that verify our self-perceptions (Banaji & Prentice, 1994).

One set of strategies consists of behaviors that lead to self-confirming feedback from others. First, we engage in selective interaction; we choose as friends, roommates, and intimates people who share our view of self. Second, we display identity cues that elicit identity-confirming behavior from others. In a hospital setting, most people treat a middle-aged man wearing a white coat as a physician. Third, we behave in ways that enhance our identity claims, especially when those claims are challenged. In one study, White students who viewed themselves as unprejudiced were led to believe they were prejudiced toward Blacks. When they were subsequently approached by a Black

© A. Ramey/PhotoEdit, Inc.

The more important an identity is to us, the more consistently we act to express it, regardless of others' reactions. Are any of your identities so important that you would express them by wearing such distinctive clothing as these two women?

panhandler, they gave him more money than did students whose egalitarian identity had not been threatened (Dutton & Lake, 1973).

Another set of strategies involve the processing of feedback from others. As noted in the next section, we often do this in ways that make others' responses to us seem to support our self-concept.

There are limits to the extent to which we engage in self-verifying strategies. There are times when we want accurate feedback about our abilities or about another person's view of our relationship with him or her. When we want such feedback, and we have the necessary cognitive resources (attention, energy), we evaluate feedback from others by comparing it with our self-representations (Swann & Schroeder, 1995). This evaluation may lead to changes in behavior, such as moving toward a goal or a desired identity, or to a change in self-representation.

THE SELF IN THOUGHT AND FEELING

We are often preoccupied by our own thoughts and feelings, as well as by information that is especially relevant to us. At a noisy, crowded party, you can often barely hear the conversation you are directly involved in; but should someone mention your name, even halfway across the room, you are likely to hear it and to immediately shift your attention in that direction.

In this section, we discuss three ways in which the self affects our thoughts and feelings. These include (1) the impact of information's relevance to the self on the processing of that information; (2) ways in which focusing attention on the self influences the relationship between our identities and our behavior;

and (3) the effect of discrepancies in the self on mood. In Chapter 10, we discuss ways in which the self influences the emotions we experience.

Self-Schema

The influence of self on thought occurs through the operation of the self-schema (Greenwald & Pratkanis, 1984). One's self-schema influences cognitive processes in several ways (Markus & Wurf, 1987). The self-schema influences the speed and certainty with which we process information, how we interpret feedback from others, and the storage in and retrieval from memory of information.

The self-schema influences the processing of incoming information. Compare Sara, for whom an athlete identity is important, with Maggie, who does not think about herself as either athletic or nonathletic. Sara will judge more quickly and confidently whether traits like agile, clumsy, muscular, and puny apply to her than will Maggie. Also, she will reject more strongly information purporting to show that she is either more or less coordinated than she had previously thought. In short, people are quicker and more certain when judging and interpreting information related to their important identities or qualities.

Our self-schema influences the way we interpret feedback. When discussing reflected appraisals, we noted that the feedback one receives from others is always incomplete, frequently ambiguous, and sometimes inconsistent. The self-schema determines how we receive and process this feedback. We pay more attention to relevant information and selectively focus on information that confirms our self-concepts, especially for highly salient identities. It also provides us with a basis for interpreting the responses of others to us. Because of the influence of self-schema, we typically perceive more confirmation of our self-concept than actually exists (Swann, 1987).

The self-schema also influences memory. In one study, participants were led to believe that either extroversion or introversion was a desirable trait. They were then asked to remember information about themselves relevant to the trait. In both conditions, participants remembered more information consistent with the trait and remembered it more quickly (Sanitioso, Kunda, & Fong, 1990). Thus, one's memory for events is better the more they relate to the self.

Thus, the important identities and other interrelated self-concepts constituting our self-schema provide a finely tuned set of mental categories that we use to process information.

Effects of Self-Awareness

While eating with friends, reading a book, or participating in conversation, your attention is usually directed toward the objects, people, and events that surround you. But what happens if, on looking up, you discover a photographer, lens focused on you, snapping away? Or what if you suddenly notice your image reflected in a large mirror? In such circumstances, most of us become self-conscious. We enter a state of **self-awareness**—that is, we take the self as the object of our attention and focus on our own appearance, actions, and thoughts. This corresponds to the me phase of action (Mead, 1934).

Numerous circumstances cause people to become self-aware. Mirrors, cameras, and recordings of our own voice cause self-awareness because they directly present the self to us as an object. Unfamiliar situations and blundering in public also cause self-awareness, because they disrupt the smooth flow of action and interaction. When this happens, we must attend to our own behavior more closely, monitoring its appropriateness and bringing it into line with the demands of the situation. In general, anything that reminds us that we are the objects of others' attention will increase our self-awareness.

How does self-awareness influence behavior? When people are highly self-aware, they are more likely to be honest and to more accurately report on their mood state, psychiatric problems, and hospitalizations (Gibbons et al., 1985). In general, people who are self-aware act in ways more consistent with personal and social standards (Wicklund, 1975; Wicklund & Frey, 1980). Their behavior is controlled more consciously by the self. In the absence of self-awareness, behavior is more automatic or habitual. Society gains control over its members through the self-control that individuals exercise when they are self-aware (Shibutani, 1961). This is because the standards to which people conform are largely learned from significant groups in society. Self-awareness is thus often a civilizing influence.

The most widely endorsed theory to explain these effects of self-awareness assumes that attention to self activates the self-schema, which in turn leads to self-evaluation (Gibbons, 1990; Wicklund, 1975). We may evaluate the self at any of three levels: experiential, behavioral, and global.

A focus on experience will heighten one's current affect or mood and elicit behavior associated with that affect; for instance, anger may elicit attack, joy may elicit playfulness. A focus on behavior will lead to the consideration of behavioral standards. The standards may be internal or personal ones. If there is a discrepancy between behavior and internal standards, the person will attempt to align behavior with his or her standard.

Alternatively, the salient standard may be external or social. When social standards and internalized standards correspond, an increase in self-awareness brings our behavior closer to ideals. But what if the social standards conflict with internalized standards, as when groups pressure individuals to change their attitudes or to violate their personal standards? When this happens, the impact of increased self-awareness depends on whether our attention is drawn to the public, social aspects of self or to the private, covert aspects of self (Scheier & Carver, 1981). If we attend to public aspects of self (our public image, our mannerisms), we may respond to group influence and modify our attitude or behavior; however, we will not abandon our internal standard (Gibbons, 1990). On the other hand, if we attend to private aspects of self (our personal values, attitudes, and internalized standards), the effect is the opposite: we guide our behavior more by personal standards and yield less to group influence.

Another possibility is that self-focus may elicit a global evaluation of self. The reference point for this evaluation may be the ideal self. The nature and effects of real-ideal self-discrepancies will be discussed in the next section.

These findings suggest that groups enhance their social control over individual behavior when they expose individuals to conditions—like an attentive audience, unfamiliar circumstances, or socially awkward tasks—that increase awareness of the public self. Interestingly, these are precisely the conditions used so effectively by cults.

Effects of Self-Discrepancies

Research has shown that the relationships between components of one's self-schema influence one's emotional state. There are three components of the self-schema: self as one is (*actual*), as one would like to be (*ideal*), and as one ought to be (*ought*). When we evaluate ourselves, we typically use the ideal self or the ought self as the reference point. When the actual self matches the ideal self, we feel satisfaction or pride. However, when there is a **self-discrepancy**—that is, a component of the actual self is the opposite of a component of the ideal self or the ought self—we experience discomfort (Higgins, 1989).

According to self-discrepancy theory, the two types of discrepancy produce two different emotional states. Someone who has an actual:ideal discrepancy will experience dejection, sadness, or depression. Someone who perceives an actual:ought discrepancy will experience fear, tension, or restlessness. The theory predicts that the larger the discrepancy, the greater the discomfort.

In a study designed to test these hypotheses (Higgins, Klein, & Strauman, 1985), students were asked to list up to 10 attributes each of the actual self, the ideal self, and the ought self. Discrepancy was measured by comparing two lists (say, the actual and the ideal); a self-state listed in both was a match, whereas a self-state listed on one list with its antonym (opposite) listed on the other was a mismatch. The self-discrepancy score was the number of mismatches minus the number of matches. Discomfort was measured by several questionnaires. The results showed that as the actual:ideal discrepancy increased, the frequency and intensity of reported dissatisfaction and depression increased. As the actual:ought discrepancy increased, the frequency and intensity of reported fear and irritability increased.

Self-discrepancy scores also are related to various behaviors. A study of satisfaction with one's body and of eating disorders found that a form of actual:ideal discrepancy was associated with bulimic behaviors, whereas an actual:ought discrepancy was associated with anorexic behaviors (Strauman, Vookles, Berenstein, Chaiken, & Higgins, 1991).

SELF-ESTEEM

Do you have a positive attitude about yourself, or do you feel you do not have much to be proud of? Overall, how capable, successful, significant, and worthy are you? Answers to these questions reflect **self-esteem,** the evaluative component of self-concept (Gecas & Burke, 1995).

This section addresses four questions: (1) How is self-esteem assessed? (2) What are the major sources of self-esteem? (3) How is self-esteem related to behavior? (4) What techniques do we employ to protect our self-esteem?

Assessment of Self-Esteem

Our overall self-esteem depends on (1) which characteristics of self are contingencies of self-esteem, and (2) how we evaluate each of them. Some of our specific role and social identities and personal qualities are important to us; characteristics of self or categories of outcomes on which a person stakes self-esteem are **contingencies** of self-esteem (Crocker & Wolfe, 2001). Others are unimportant. For instance, you may consider yourself an excellent student and a worthy friend, an incompetent athlete and an unreliable employee, and not care about your social identity as Basque French. According to theory, our overall level of self-esteem is the product of these individual evaluations, with each identity weighted according to its salience (Rosenberg, 1965; Sherwood, 1965).

Ordinarily, we are unaware of precisely how we combine and weigh the evaluations of our specific contingencies. If we weigh our positively evaluated identities and traits as more important, we can maintain a high level of overall self-esteem while still admitting to certain weaknesses. If we weigh our negatively evaluated identities heavily, we will have low overall self-esteem even though we have many valuable qualities.

Research on self-esteem often uses explicit measures of global self-esteem—overall judgments of self-worth or self-acceptance. Such measures typically ask for agreement or disagreement with direct statements such as "On the whole, I am satisfied with myself." There also have been attempts to measure implicit self-esteem—the unaware, automatic evaluation of the self—by assessing the person's evaluation of objects and qualities associated with the self (Greenwald & Farnham, 2000).

Sources of Self-Esteem

Why do some of us enjoy high self-esteem whereas others suffer low self-esteem? To help answer this question, consider three major sources of self-esteem—family experience, performance feedback, and social comparisons.

Family Experience As one might expect, parent-child relationships are important for the development of self-esteem. From an extensive study of the family experiences of fifth- and sixth-graders, Coopersmith (1967) concluded that four types of parental behavior promote higher self-esteem: (1) showing acceptance, affection, interest, and involvement in children's affairs; (2) firmly and consistently enforcing clear limits on children's behavior; (3) allowing children latitude within these limits and respecting initiative (such as children setting their own bedtime and participating in making family plans); and (4) favoring noncoercive forms of discipline (such as denying privileges and discussing reasons, rather than punishing physically). Findings from a representative sample of 5,024 New York high school students corroborate these conclusions (Rosenberg, 1965). Note that these results are consistent with our discussion of socialization techniques in Chapter 3.

Family influences on self-esteem confirm the idea that the self-concepts we develop mirror the view of ourselves communicated by significant others. Children who see that their parents love, accept, care about, trust, and reason with them come to think of themselves as worthy of affection, care, trust, and respect. Conversely, children who see that their parents do not love and accept them may develop low self-esteem. A longitudinal study of adolescents found that excessive parental shaming and criticism were associated with low self-esteem and depression (Robertson & Simons, 1989).

Research also suggests that self-esteem is produced by the reciprocal influence of parents and their children on each other (Felson & Zielinski, 1989). Children with higher self-esteem exhibit more self-confidence, competence, and self-control. Such children are probably easier to love, accept, reason with, and trust. Consequently, they are likely to elicit responses from their parents that further promote their self-esteem.

As young people move into adolescence, their overall or global self-esteem becomes linked to the self-evaluations tied to specific role identities. A study of 416 sixth-graders found that evaluations of self as athlete, son/daughter, and student were positively related to global self-esteem (Hoelter, 1986). Also, the number of significant others expands to include friends and teachers in addition to parents. The relative importance of these others appears to vary by gender. A study of 1,367 high school seniors found that the perceived appraisals of friends had the biggest impact on girls' self-esteem, whereas the perceived appraisals of parents had the biggest impact on boys' self-esteem (Hoelter, 1984). For both boys and girls, teachers' appraisals were second in importance.

Both popular (Pipher, 1994) and academic (American Association of University Women, 1990) works have argued that a substantial difference between male and female self-esteem emerges in adolescence. Various causes have been suggested, such as the devaluing of female roles in U.S. society, the development of body consciousness and concern with appearance among girls, and the preferential treatment of boys by teachers. A meta-analysis of studies involving more than 146,000 participants of all ages finds a small difference favoring boys that is larger but not substantial in adolescence (Kling, Hyde, Showers, & Buswell, 1999).

Performance Feedback Everyday feedback about the quality of our performances—our successes and failures—influences our self-esteem. We derive self-esteem from experiencing ourselves as active causal agents who make things happen in the world, who attain goals and overcome obstacles (Franks & Marolla, 1976). In other words, self-esteem is based partly on our sense of efficacy—of competence and power to control events (Bandura, 1982c). People who hold low-power positions (such as clerks, unskilled workers) have fewer opportunities to develop efficacy-based self-esteem because such positions limit their freedom of action. Even so, people seek ways to convert almost any kind of activity into a task against which to test their efficacy and prove their competence (Gecas & Schwalbe, 1983). In this way, they obtain performance feedback useful for building self-esteem.

Social Comparison To interpret whether performances represent success or failure, we must often compare them with our own goals and self-expectations or with the performances of others. Getting a B on a math exam, for example, would raise your sense of math competence if you had hoped for a C at best, but it would shake you if you were counting on an A. The impact of the B on your self-esteem also would vary depending on whether most of your friends got As or Cs.

Social comparison is crucial to self-esteem, because the feelings of competence or worth we derive from a performance depend in large part on whom we are compared with, both by ourselves and by others. Even our personal goals are largely derived from our aspirations to succeed in comparison with people whom we admire. We are most likely to receive evaluative feedback from others in our immediate social context—our family, peers, teachers, and work associates. We are also most likely to compare ourselves with these people and with others who are similar to us (Festinger, 1954; Rosenberg & Simmons, 1972). This reasoning suggests that the self-esteem of minority persons may benefit from being in a consonant environment, that is, one where most people are from the same group; a longitudinal study of a national sample found that as the percentage of Blacks in the

© Brian Bahr/Getty Images

Not everyone can win an Olympic gold medal. But for all of us, an inner sense of self-esteem depends on experiencing ourselves as causal agents who make things happen, overcome obstacles, and attain goals.

college attended increased, postcollege self-esteem increased (St. C. Oates, 2004). A study of adult Chinese in Los Angeles County also found context effects on self-esteem; participation in Chinese culture, e.g., speaking Chinese, eating ethnic foods and celebrating ethnic festivals, was associated with higher self-esteem for persons living in predominantly Chinese neighborhoods, but not for Chinese living in predominantly White neighborhoods (Schnittker, 2002).

A study of job applicants clearly demonstrates the effect of social comparison on self-esteem (Morse & Gergen, 1970). After each applicant had completed a set of forms—including a self-esteem scale—another applicant entered the waiting room. For half the participants, the second applicant wore a dark business suit, carried an attaché case, and communicated an aura of competence. The remaining participants each waited with an applicant who wore a smelly sweatshirt and no socks and appeared dazed. Several minutes later, while still in the presence of the highly impressive or unimpressive competitor, applicants completed additional forms, including a second self-esteem scale. Applicants exposed to the obviously inferior competitor revealed a large increase in self-esteem from the first to the second self-esteem measurement; among participants faced with the impressive competitor, self-esteem dropped substantially.

Losing one's job is generally interpreted as a serious failure in our society. A national survey of American employees reveals that job loss undermined self-esteem, but the size of the drop in self-esteem depended on social comparison (Cohn, 1978). In neighborhoods with little unemployment, persons who lost their jobs suffered a large drop in self-esteem. In neighborhoods where many others were unemployed too, the drop was less. This difference points to the importance of the immediate social context for defining success or failure.

Self-Esteem and Behavior

People with high self-esteem often behave quite differently from those with low self-esteem. At the same time, we should not overestimate the effects of self-esteem (Baumeister, 1998).

Compared with those having low self-esteem, children, teenagers, and adults with higher self-esteem are socially at ease and popular with their peers. They are more confident of their own opinions and judgments and more certain of their perceptions of self (Campbell, 1990). They are more vigorous and assertive in their social relations, more ambitious, and more academically successful. During their school years, persons with higher self-esteem participate more in extracurricular activities, are elected more frequently to leadership roles, show greater interest in public affairs, and have higher occupational aspirations. Persons with high self-esteem achieve higher scores on measures of psychological well-being (Rosenberg, Schooler, Schoenbach, & Rosenberg, 1995). Adults with high self-esteem experience less stress following the death of a spouse and cope with the resulting problems more effectively (Johnson, Lund, & Dimond, 1986).

The picture of people with low self-esteem forms an unhappy contrast. People low in self-esteem tend to be socially anxious and ineffective. They view

BOX 4.2 Minority Status and Self-Esteem

Members of racial, religious, and ethnic minorities may have special problems in developing positive self-esteem. Because of prejudice, minority group members are likely to see a negative image of themselves reflected in appraisals by members of other groups. When they make social comparisons of their own educational, occupational, and economic success with that of the majority, they are likely to compare unfavorably. Therefore, we might assume that members of minority groups will interpret their performances and failures to achieve as evidence of a basic lack of worth and competence—that they will have low self-esteem.

Is this hypothesis true? Hundreds of studies have sought to determine whether minority status undermines self-esteem in America (Porter & Washington, 1993; Wylie, 1979). The vast majority of studies offer little support for the conclusion that minorities (racial, religious, or ethnic) have significantly lower self-esteem. Further research suggests that self-esteem among racial and ethnic minorities has two components. One is **group self-esteem**—how the person feels as a member of a racial or ethnic group. The other is personal self-esteem—how the person feels about the self (Porter & Washington, 1993).

A meta-analysis of data from more than 120 sources found that Blacks score significantly higher than Whites on global measures of personal self-esteem (Gray-Little & Hafdahl, 2000). Reflected appraisals from significant others affect minority group members just as they do majority group members. The self-esteem of Black schoolchildren is strongly related to their perception of what their parents, teachers, and friends think of them. These appraisals are not negative (Rosenberg, 1973, 1990). Living in segregated neighborhoods, minority group children usually see themselves through the unprejudiced eyes of their own group, not the prejudiced eyes of others. The self-esteem of Black adults is related to the quality of their relationships with family and friends and their involvement in religion (Hughes & Demo, 1989).

What about other racial/ethnic groups? A meta-analysis of data from 354 samples of people of all ages, including Hispanics, Asians, and American Indians (Twenge & Crocker, 2002), again found Blacks' mean scores on global measures to be somewhat higher than Whites'; the means of the other three groups were somewhat below the means of Whites.

Group self-esteem, on the other hand, is not associated with reflected appraisals. Among Black Americans, group self-esteem includes Black consciousness, Black racial identity, and support for independent Black politics. High group self-esteem among Blacks is associated with higher education and more frequent contact with Whites, not with relationships with family and friends (Demo & Hughes, 1990). Research indicates that Puerto Ricans, Mexican Americans, and Asian Americans have high levels of group self-esteem (Porter & Washington, 1993). Other data suggest that when members of these groups receive negative feedback from members of other groups, they attribute it to racial prejudice (Crocker, Voelkl, Testa, & Major, 1991). A recent study suggests that many persons are reclaiming American Indian group identity due to Indian political activism and governmental policies that are making resources available (Nagel, 1995).

But what about the effects of social comparisons? Many minority group members are disadvantaged in terms of education, occupation, and income. Minority individuals do compare themselves with the majority, but they often do not blame themselves for their disadvantaged position. Minorities can protect their personal self-esteem by blaming the system of discrimination for their lesser accomplishments. Indeed, minority statuses such as race, religion, and ethnicity show virtually no association with self-esteem (Jacques & Chason, 1977; Rotheram-Borus, 1990). Social failure affects self-esteem only when people attribute it to poor individual achievement (Rosenberg & Pearlin, 1978).

interpersonal relationships as threatening, feel less positively toward others, and are easily hurt by criticism. Lacking confidence in their own judgments and opinions, they yield more readily in the face of opposition. They expect others to reject them and their ideas, and they have little faith in their ability to achieve. In school, they set lower goals for themselves, are less successful academically, less active in the classroom and in extracurricular activities, and less popular. People with lower self-esteem appear more depressed and express more feelings of unhappiness and discouragement. They more frequently manifest symptoms of anxiety, poor adjustment, and psychosomatic illness.

Self-esteem influences our attributions regarding events in our close relationships. College students in dating relationships were recruited to participate in research. Their self-esteem was measured, and then they imagined two scenarios; in one, their partner was in a good mood, in the other, he or she was in a bad mood. When the partner's mood was negative and the cause ambiguous, those with low self-esteem felt more responsible, more rejected, and more hostile (Bellavia & Murray, 2003).

Most of these contrasts are drawn from comparisons between naturally occurring groups of people who report high or low self-esteem. It is, therefore, difficult to determine whether self-esteem causes these behavior differences or vice versa. For example, high self-esteem may enable people to assert their opinions more forcefully and, thus, to convince others. But the experience of influencing others, in turn, may increase self-esteem. Thus, reciprocal influence, rather than causality from self-esteem to behavior, is probably most common (Rosenberg, Schooler, & Schoenbach, 1989).

Protecting Self-Esteem

What grade would you like to get on your next exam in social psychology—an A or a C? Your answer depends in part on whether your self-esteem is high or low. We often think that everybody wants to get positive feedback from others, to have others like them, to be successful—that is, to experience self-enhancement. As noted in the previous section, people with high self-esteem expect to perform well and usually do. People with low self-esteem, on the other hand, expect to perform poorly and usually do. People are motivated to protect their self-esteem whether it is high or low—that is, to experience self-verification in the feedback they receive. Most people have high self-esteem and want self-enhancing feedback. Some people have low self-esteem; to verify their self-evaluation, they want self-derogating feedback.

People use several techniques to maintain their self-esteem. We will examine four of them (McCall & Simmons, 1978).

Manipulating Appraisals We choose to associate with people who share our view of self and avoid people who do not. For example, a study of interaction in a college sorority revealed that women associated most frequently with those they believed saw them as they saw themselves (Backman & Secord, 1962). People with negative self-views seek people who think poorly of them (Swann & Predmore, 1985). Another way to maintain our self-esteem is by interpreting others' appraisals as more favorable or unfavorable than they actually are. For instance, college students took an analogies test and subsequently were given positive, negative, or no feedback about their performance (Jussim, Coleman, & Nassau, 1987). Each student then completed a questionnaire. Students with high self-esteem perceived the feedback—whether positive or negative—as more positive than students with low self-esteem.

Selective Information Processing Another way we protect our self-esteem is by attending more to those occurrences that are consistent with our self-evaluation. In one study, participants high or low in self-esteem performed a task; they were then told either that they succeeded or that they failed at the task. On a later self-rating, all the participants gave biased ratings. High self-esteem participants who succeeded increased their ratings, whereas their low self-esteem counterparts did not. Low self-esteem participants who failed gave themselves lower ratings, whereas high self-esteem participants who failed did not (Schlenker, Weigold, & Hallam, 1990).

Memory also acts to protect self-esteem. People with high self-esteem recall good, responsible, and successful activities more often, whereas those with low self-esteem are more likely to remember bad, irresponsible, and unsuccessful ones.

Selective Social Comparison When we lack objective standards for evaluating ourselves, we engage in social comparison (Festinger, 1954). By carefully selecting others with whom to compare ourselves, we can further protect our self-esteem. We usually compare ourselves with persons who are similar in age, sex, occupation, economic status, abilities, and attitudes (Suls & Miller, 1977; Walsh & Taylor, 1982). We generally rate ourselves more favorably than we rate our friends (Suls, Lemos, & Stewart, 2002.) We tend to avoid comparing ourselves with the class valedictorian, homecoming queen, or star athlete, thereby forestalling a negative self-evaluation.

Once people make a social comparison, they tend to overrate their relative standing (Felson, 1981). This is illustrated by self-ratings obtained from a large sample of American adults (Heiss & Owens, 1972). Only 2 percent rated themselves below average as parents, spouses, sons or daughters, or in the qualities of trustworthiness, intelligence, and willingness to work. These were probably people with low self-esteem.

Selective Commitment to Identities Still another technique to protect self-esteem involves committing ourselves more to those self-concepts that provide feedback consistent with our self-evaluation, downgrading those that provide feedback that challenges it. This protects overall self-esteem because self-evaluation is based most heavily on those identities and personal qualities that are contingencies of self-esteem.

People tend to enhance self-esteem by assigning more importance to those identities (religious, racial, occupational, family) they consider particularly admirable (Hoelter, 1983). They also increase or decrease identification with a social group when the group becomes a greater or lesser potential source of self-esteem (Tesser & Campbell, 1983). In one study, students were part of a group that either succeeded or failed at a task (Snyder, Lassegard, & Ford, 1986). On measures of identification with the group, students belonging to a successful group claimed closer association with the group (that is, basked in the reflected glory) whereas those in an unsuccessful group distanced themselves from the group. Similarly, students are more apt to wear clothing that displays their university affiliation following a football victory than after a defeat. They also identify more with their school when describing victories ("We won") than defeats ("They lost"), thereby enhancing or protecting self-esteem (Cialdini, Borden, Thorne, Walker, & Freeman, 1976).

People who want to verify their low self-esteem behave differently. Low self-esteem participants who were members of a successful group downplayed their connection to the group and minimized their contribution to its success. Low self-esteem participants were more likely to link themselves to the successful group when they were *not* members of it (Brown, Collins, & Schmidt, 1988).

All four techniques for protecting self-esteem described here portray human beings as active processors of social events. People do not accept social evaluations passively or allow self-esteem to be buffeted by the cruelties and kindnesses of the social environment. Nor do successes and failures directly affect self-esteem. The techniques described here testify to human ingenuity in selecting and modifying the meanings of events in the service of self-esteem.

SUMMARY

The self is the individual viewed both as the source and the object of reflexive behavior.

The Nature and Genesis of Self (1) The self is the source of action when we plan, observe, and control our own behavior. The self is the object of action when we think about who we are. (2) Newborn infants lack a sense of self. Later, they come to recognize that they are physically separate from others. As they acquire language, they learn that their own thoughts and feelings are also separate. (3) Through role taking, children come to see themselves through others' eyes. They can then observe, judge, and regulate their own behavior. (4) Children construct their identities based on how they imagine they appear to others. They also develop self-evaluations based on the perceived judgments of others.

Identities: The Self We Know The self we know includes multiple identities. (1) Some identities are linked to the social roles we enact. (2) Some identities are linked to our membership in social groups or categories. These identities may be associated with in-group favoritism and out-group stereotyping. (3) We form self-concepts primarily through learning and adopting role and social identities. The self we know is primarily influenced by the perceived reactions of others. (4) The self we know varies with the situation. We attend most to those aspects of our selves that are distinctive and relevant to the ongoing activity.

Identities: The Self We Enact The self we enact expresses our identities. (1) We choose behaviors to evoke responses from others that will confirm particular identities. To confirm identities successfully, we must share with others our understanding of what these behaviors and identities mean. Adopting these meanings may lead to poorer performance when we experience stereotype threat. (2) We choose which identity to express based on that identity's salience, need for support, and situational opportunities for enacting it. (3) We gain consistency in our behavior over time by striving to enact important identities. We also employ several strategies that lead to verification of our self-conceptions.

The Self in Thought and Feeling The self affects both thought and feeling. (1) We perceive and process information more effectively if it relates to our important identities. We learn and remember information better if it relates to the self. (2) When attention is drawn to the self, we become self-aware and take greater control over our behavior. We then conform more with our personal standards and with salient social standards. (3) Discrepancies between components of the self may cause sadness, depression, fear, or restlessness.

Self-Esteem Self-esteem is the evaluative component of self. Most people try to maintain positive self-esteem. (1) Overall self-esteem depends on the evaluations of our specific role identities. (2) Self-esteem derives from three sources: family experiences of acceptance and discipline, direct feedback on the effectiveness of actions, and comparisons of our own successes and failures with those of others. (3) People with higher self-esteem tend to be more popular, assertive, ambitious, academically successful, better adjusted, and happier. (4) We employ numerous techniques to protect self-esteem. Specifically, we seek reflected appraisals consistent with our self-view, process information selectively, carefully select those with whom we compare ourselves, and attribute greater importance to qualities that provide consistent feedback.

LIST OF KEY TERMS AND CONCEPTS

contingencies (of self-esteem) (p. 101)
game (p. 88)
generalized other (p. 89)
group self-esteem (p. 104)
identity (p. 89)
identity control theory (p. 95)
looking-glass self (p. 87)
play (p. 88)
role identity (p. 89)
role taking (p. 87)
salience (p. 95)
self (p. 85)
self-awareness (p. 99)
self-discrepancy (p. 100)
self-esteem (p. 101)
self-schema (p. 83)
significant others (p. 87)
situated self (p. 93)
social identity (p. 90)

5

SOCIAL PERCEPTION AND COGNITION

Introduction

Schemas

Types of Schemas

Schematic Processing

Schemas as Cultural Elements

Person Schemas and Group Stereotypes

Person Schemas

Group Stereotypes

Impression Formation

Trait Centrality

Integrating Information About Others

Impressions as Self-Fulfilling Prophecies

Heuristics

Anchoring and Adjustment

Dispositional Versus Situational Attributions

Inferring Dispositions From Acts

Covariation Model of Attribution

Attributions for Success and Failure

Bias and Error in Attribution

Overattribution to Dispositions

Focus of Attention Bias

Actor-Observer Difference

Motivational Biases

Cultural Basis of Attributions

Summary

List of Key Terms and Concepts

INTRODUCTION

It is 10 p.m., and the admitting physician at the psychiatric hospital is interviewing a respectable-looking man who has asked for treatment. "You see," the patient says, "I keep hearing voices." After taking a full history, the physician diagnoses the man with schizophrenia and assigns him to an inpatient unit. The physician is well trained and makes the diagnosis with apparent ease. Yet to diagnose correctly someone's mental condition is a difficult problem in social perception. The differences between paranoia, schizophrenia, depression, and normality are not always easy to discern.

A classic study conducted by Rosenhan (1973) demonstrates this problem. Eight pseudo-patients who were actually research investigators gained entry into mental hospitals by claiming to hear voices. During the intake interviews, the pseudo-patients gave true accounts of their backgrounds, life experiences, and present (quite ordinary) psychological condition. They falsified only their names and their complaint of hearing voices. Once in the psychiatric unit, the pseudo-patients stopped simulating symptoms of schizophrenia. They reported that the voices had stopped, talked normally with other patients, and made observations in their notebooks. Although some other patients suspected that the investigators were not really ill, the staff continued to believe they were. Even upon discharge, the pseudo-patients were still diagnosed with schizophrenia, although now it was "schizophrenia in remission."

A man voluntarily checking into a psychiatric hospital may pose a confusing problem for the hospital staff. Is he really "mentally ill" and in need of hospitalization, or is he "healthy"? Is he no longer able to function in the outside world? Or is he merely faking and trying to get a break from his work or his family?

To try to answer these questions, the admitting physician gathers information about the person and classifies it as indicating illness or health. Then the doctor combines these facts to form a general diagnosis (paranoia, schizophrenia, or depression) and determines what treatment the person needs. While performing these actions, the doctor is engaging in **social perception.** Broadly defined, social perception refers to constructing an understanding of the social world from the data we get through our senses. More narrowly defined, social perception refers to the processes by which we form impressions of other people's traits and personalities.

In making her diagnosis, the physician not only forms an impression about the traits and characteristics of the new patient, but she also tries to understand the causes of that person's behavior. She tries, for instance, to figure out whether the patient acts as he does because of some internal dispositions or because of external pressures from the environment. Social psychologists term this process **attribution.** In attribution, we observe others' behavior and then infer backward to causes—intentions, abilities, traits, motives, and situational pressures—that explain why people act as they do.

Social perception and attribution are not passive activities. We do not just register the stimuli that impinge on our senses. Rather, our expectations and cognitive structures influence what we notice and how we interpret it. The intake physician at the psychiatric hospital, for example, does not expect to encounter researchers pretending to be mentally ill. Instead, she expects to meet people who are mentally ill. Even before the interaction begins, the doctor has categorized the patient as mentally ill, and thus she focuses on information relevant to that condition and interprets the information based on the expectation that the patient is a real patient.

Most of the time, the impressions we form of others are sufficiently accurate to permit smooth interaction. After all, few people who are admitted to psychiatric hospitals are researchers faking mental illness. Yet social perception and attribution can be unreliable. Even highly trained observers can misperceive, misjudge, and reach the wrong conclusions.

In February 1999, police officers in New York City were attempting to track down a serial rapist. Sketches of the rapist had been circulated to the police, and so they had some idea what the rapist looked like. Four White officers patrolling the Bronx encountered Amadou Diallo, a Black man, and thought that he resembled the sketches of the rapist. As Diallo was entering his apartment building, the police officers ordered him to stop. Diallo stopped and began to reach for his wallet to produce his identification. The police officers interpreted this action quite differently, however, and, believing he was reaching for a gun, opened fire. They fired a total of 41 shots, and Diallo died immediately. Diallo was not the rapist and had no criminal record—the officers' snap judgments were wrong.

The image of a Black man in a bad neighborhood, reaching into his pocket as he was being stopped by the police, provided too many dangerous cues and the officers reacted immediately. Many have wondered if the police officers would have been slower to act if Diallo had been White. Did race help activate a dangerous image in the police officers' minds and encourage them to respond aggressively? Studies conducted in laboratory settings confirm this type of dynamic. In one study, subjects were asked to act as police officers and decide whether or not to shoot at suspected criminals. The suspected criminals were either holding a gun (in which case the officer should shoot) or were holding a neutral object such as a cell phone (in which case the officer should not shoot). The results showed that the subjects were more likely to mistakenly shoot a suspect holding a cell phone if the suspect was Black. Similarly, they were also more likely to mistakenly hold back from shooting a suspect holding a gun if the suspect was White (Plant, Peruche, & Butz, 2005).

This chapter focuses on these processes of social perception and attribution and addresses the following questions:

1. How do we make sense of the flood of information that surrounds us? How do we categorize that information and use it in social situations?
2. Why do we rely so much on notions about personality and group stereotypes? What problem does this practice solve, and what difficulties does it create?
3. How do we form impressions of others? That is, how do we integrate the diverse or even contradictory information we receive about someone into a coherent, overall impression?
4. How do we ascertain the causes of other people's behavior and interpret the origins of actions we observe? For instance, when we judge someone's behavior, how do we know whether to attribute the behavior to that person's internal dispositions or to the external situation affecting that person?
5. What sorts of errors do we commonly make in judging the behavior of others, and why do we make such errors?

SCHEMAS

The human mind is a sophisticated system for processing information. One of our most basic mental processes is **categorization**—our tendency to perceive stimuli as members of groups or classes rather than as isolated, unique entities. For instance, at the theater, we see a well-groomed woman on stage wearing a short dress and dancing on her toes; rather than viewing her as a novel entity, we immediately categorize her as a "ballerina."

How do we go about assigning people or things to categories? For instance, how do we know the woman should be categorized as a ballerina and not as an "actress" or a "cheerleader"? To categorize some person, we usually compare that person to our prototype of the category. A **prototype** is an abstraction that represents the "typical" or quintessential instance of a class or group—as least to us. Others may have different prototypes for the same category. Usually, prototypes are specified in terms of a set of attributes. For example, the prototype of a "cultured person" may be someone who is knowledgeable about literature, classical music, fine food, and foreign cultures and who indulges these tastes by regularly attending concerts, eating at fine restaurants, and traveling worldwide.

Categorizing people, objects, situations, events, and even the self becomes complicated because the categories we use are not isolated from one another. Rather, they link together and form a structure. For instance, we may think of a person (Jonathan) not only as having various attributes (tall, wealthy) but also as bearing certain relations to other persons or entities (friend of Caroline, stronger than Bill, owner of a Honda). These other persons or entities will themselves have attributes (Caroline: thin, athletic, brunette; Bill: short, fat, mustachioed; Honda: blue, four-door, new). They also have relations with still other persons and entities (Caroline: cousin of Bill, wife of George; Bill: friend of George, owner of a Buick). In this way, we build a cognitive structure consisting of persons, attributes, and relations.

Social psychologists use the term **schema** to denote a well-organized structure of cognitions about some social entity such as a person, group, role, or event. Schemas usually include information about an entity's attributes and about its relations with other entities. To illustrate, suppose that Martha, who is somewhat cynical about politics, has a schema about the role of "member of Congress." In Martha's schema, the member of Congress will claim to insist that he or she serves the needs of his constituents, but will actually vote for the special interests of those who contributed most to his campaign; will run TV advertisements containing half-truths at election time; will spend more time in Washington, DC, than in his home district; will put avoiding scandal above ethics; will vote large pay raises and retirement benefits for himself; and above all, will never do anything to lessen his own power.

Someone else, of course, may hold a less cynical view of politics than Martha and have a different schema about the role of "member of Congress." But, like Martha's, this schema will likely incorporate such elements as the congressional representative's typical activities, relations, motives, and tactics. Whatever their exact content, schemas enable us to organize and remember facts, to make inferences that go beyond the facts immediately available, and to assess new information (Fiske & Linville, 1980; Wilcox & Williams, 1990).

Types of Schemas

There are several distinct types of schemas, including person schemas, self-schemas, group schemas, role schemas, and event schemas (Eckes, 1995; Taylor & Crocker, 1981).

Person schemas are cognitive structures that describe the personalities of others. Person schemas can apply either to specific individuals (such as George W. Bush, Ozzy Osbourne, your mother) or to types of individuals (such as introvert, manic-depressive, sociopath). Person schemas organize our conceptions of others' personalities and enable us to develop expectations about others' behavior.

Self-schemas are structures that organize our conception of our own characteristics (Catrambone & Markus, 1987; Markus, 1977). For instance, if you conceive of yourself as independent (as opposed to dependent), you may see yourself as individualistic, unconventional, and assertive. Then, if you behave in a manner consistent with your self-schema, you may refuse to accept money from your parents, refuse to ask others for help with schoolwork, take a part-time job, or dye your hair an unusual color. Self-schemas are discussed in detail in Chapter 4.

Group schemas, also called **stereotypes,** are schemas regarding the members of a particular social group or social category (Hamilton, 1981). Stereotypes indicate the attributes and behaviors considered typical of members of that group or social category. American culture uses a wide variety of stereotypes about different races (Blacks, Hispanics, Asians), religious groups (Protestants, Catholics, Jews), and ethnic groups (Germans, Irish, Poles, Greeks, Italians).

Role schemas indicate which attributes and behaviors are typical of persons occupying a particular role in a group. Martha's conception of the role of a Congressional representative illustrates a role schema. Role schemas exist for most occupational roles—nurses, cab drivers, store managers, and the like—but they also exist for other kinds of roles in groups: group leader, captain of a sports team. Role schemas are often used to understand and to predict the behaviors of people who occupy roles.

Event schemas (also called scripts) are schemas regarding important, recurring social events (Abelson, 1981; Hue & Erickson, 1991; Schank & Abelson, 1977). In our society, these events include weddings, funerals, graduation ceremonies, job interviews, cocktail parties, and first dates. An event schema specifies the activities that constitute the event, the predetermined order or sequence for these activities, and the persons (or role occupants) participating in the event. Scripts can be revealed by asking people to describe what typically happens during an event. In one study, researchers asked male and female college students to describe the typical sequence of activities on a first date (Rose & Frieze, 1993). There was substantial agreement

TABLE 5.1 CORE ACTIONS OF THE FIRST DATE SCRIPT

Script for Woman	Script for Man
GROOM AND DRESS*	
BE NERVOUS	
Worry about appearance	WORRY ABOUT APPEARANCE
PICK UP DATE (BY MAN)	PICK UP DATE
	MEET PARENTS/ ROOMMATES
Leave	Leave
Confirm plans	Confirm plans
Get to know and evaluate date	Get to know and evaluate date
TALK, JOKE, LAUGH	TALK, JOKE, LAUGH
GO TO MOVIES, SHOW, PARTY	
Eat	EAT
Take date home (by man)	TAKE DATE HOME
Kiss goodnight (by man)	Kiss goodnight
Go home	Go home

*Capital letters indicate the action was mentioned by 50% or more of the participants; lowercase letters indicate the action was mentioned by fewer than 50% of the participants.

Source: Rose and Frieze, 1993.

between male and female respondents, as shown in Table 5.1. Several activities were mentioned by more than half of the participants, including grooming and dressing, picking up the date, and taking the date home. A number of activities were important in both male and female scripts: worrying about appearance, leaving, confirming plans, eating, and going home. Reflecting the impact of gender roles, both men and women agreed that the man would take the initiative in picking up the date, taking the date home, and kissing her goodnight. Notice that irrelevant activities were not mentioned, such as taking a driver's license test or going to the dentist; these are not appropriate for a first date. Note also that the script specifies a sequence or expected order for the various activities—the man will not kiss the woman goodnight before they eat dinner.

Schematic Processing

Why Do We Use Schemas? Although schemas may produce reasonably accurate judgments much of the time, they do not always work. Wouldn't it be better for us to rely less on schemas and so perhaps be able to avoid the kind of tragic mistake the police made with Amadou Diallo? Perhaps, but we come to rely on schemas because they give us a way to efficiently organize, understand, and react to the complex world around us. It is simply impossible to process all the information present in each interaction. We have to find a way to focus on what is most important in defining the situation and the persons involved so that we can respond appropriately. Schemas help us do this in several ways: (1) they influence our capacity to recall information by making certain kinds of facts more salient and easier to remember; (2) they help us process information faster; (3) they guide our inferences and judgments about people and objects; (4) they allow us to reduce ambiguity by providing a way to interpret ambiguous elements in the situation. Once we have applied a schema to the situation, our decisions about how to interact in it become much more straightforward (Mayer, Rapp, & Williams, 1993).

Schematic Memory Human memory is largely reconstructive. That is, we do not usually remember all the precise details of what transpired in a given situation—we are not a video camera instantly recording all the images and sounds. Instead, we typically remember some of what happened—enough to identify the appropriate schema and then rely on that schema to fill in other details. Schemas organize information in memory and, therefore, affect what we remember and what we forget (Hess & Slaughter, 1990; Sherman, Judd, & Park, 1989). When trying to recall something, people often remember better those facts that are consistent with their schemas. For instance, one study (Cohen, 1981) investigated the impact of an occupational role schema on recall. Participants viewed a videotape of a woman celebrating her birthday by having dinner with her husband at home. Half the participants were told the woman was a librarian; the other half were told she was a waitress.

Some characteristics of the woman were consistent with the schema of a librarian: She wore glasses, had spent the day reading, had previously traveled in Europe, and liked classical music. Other characteristics of the woman, however, were consistent with the schema of a waitress: She drank beer, had a bowling ball in the room, ate chocolate birthday cake, and flirted with her husband. Later, when participants tried to recall details of the videotape, they recalled most accurately those facts consistent with the woman's occupational label. That is, participants who thought she was a librarian remembered facts consistent with the librarian schema, whereas those who thought she was a waitress remembered facts consistent with the waitress schema.

What about memory for material inconsistent with schemas? Several studies have tested the recall of three types of information: material consistent with schemas, material contradictory to schemas, and material irrelevant to schemas. The results show that people recall both schema-consistent and schema-contradictory material better than schema-irrelevant material (Cano, Hopkins, & Islam, 1991; Higgins & Bargh, 1987). People recall schema-contradictory material better when the schema itself is concrete (for example, spends money wisely, often tells lies, brags about her accomplishments) rather than abstract (for example, practical, dishonest, egotistical).

Schematic Inference Schemas affect the inferences we make about persons and other social entities (Fiske & Taylor, 1991). That is, they supply missing facts when gaps exist in our knowledge. If we know certain facts about a person but are ignorant about others, we fill in the gaps by inserting suppositions consistent with our schema for that person. For example, knowing your roommate is a nonsmoker, you can infer he will not want to spend time with your new friend who smokes. Of course, the use of schemas can lead to erroneous inferences. If the schema is incomplete or does not correctly mirror reality, some mistakes are likely. For instance, the police officers who confronted Amadou Diallo applied a schema that was incorrect. Their schema for "a Black man who puts his hand in his pocket as he is being confronted by the police" includes the element that the suspect would have a gun in his pocket. From this, they inferred that he would try to shoot at them, and they reacted according to that erroneous inference.

Schemas—especially well-developed schemas—can help us infer new facts. For instance, if a physician diagnoses a patient as having chicken pox, he can make inferences about how the patient contracted the disease, which symptoms might be present, what side effects or complications might arise, and what treatment will be effective. For another person who has no schema regarding this disease, these inferences would be virtually impossible.

Schematic Judgment Schemas can influence our judgments or feelings about persons and other entities. For one thing, the schemas themselves may be organized in terms of evaluative dimensions; this is especially true of person schemas. For another thing, the level of complexity of our schemas affects our evaluations of other persons. Greater schematic complexity leads to less extreme judgments. That is, the greater the complexity of our schemas about groups of people, the less extreme are our evaluations of persons in those groups. This is called the complexity-extremity effect.

For instance, in one study (Linville & Jones, 1980), White college students evaluated a person applying for admission to law school. Depending on treatment, the applicant was either White or Black and had an academic record that was either strong or weak. The results showed an interaction effect between academic record and race. Participants rated a weak Black applicant more negatively than a weak White applicant, but they rated a strong Black applicant more positively than a strong White applicant. Judgments about Black applicants were more extreme—in both directions—than those about White applicants because the participants' schema for their own in-group (White) was more complex than their schema for the out-group (Black). Further research (Linville, 1982) shows that the complexity-extremity effect also holds for other attributes, such as age. College students have less complex schemas for older persons than for persons their own age, so they are more extreme in judgments of older persons.

Drawbacks of Schematic Processing Although schemas provide certain advantages, they also entail some corresponding disadvantages. First, people are overly accepting of information that fits consistently with a schema. In fact, some research suggests that perceivers show a confirmatory bias when collecting new information relevant to schemas (Higgins & Bargh, 1987; Snyder & Swann, 1978). That is, when gathering information, perceivers tend to ask questions that will obtain information supportive of the schemas rather than questions that will obtain information disconfirmatory of the schemas.

Second, when faced with missing information, people fill in gaps in knowledge by adding elements that are consistent with their schemas. Sometimes these added elements turn out to be erroneous or factually incorrect. When this happens, it will, of course, create inaccurate interpretations or inferences about people, groups, or events.

Third, because people are often reluctant to discard or revise their schemas, they occasionally apply schemas to persons or events even when the schemas do not fit the facts very well. Forced misapplication of a schema may lead to incorrect characterization and inferences, and this in turn can produce inappropriate or inflexible responses toward other persons, groups, or events.

Schemas as Cultural Elements

The concept of schema is important in social psychology for several reasons. At the individual level, schemas clearly affect memory, inference, and judgment. However, schemas are also cultural elements (Forgas & Bond, 1985; Harris, Lee, Hensley, & Schoen, 1988). That is, schemas are socially shared patterns of thought that can be communicated intact from one person to another and are often shared by the members of a group. In fact, holding specific schemas may be a characteristic of group membership.

Groups differ in the schemas used by their members. For example, when a professional football player leaves one team and joins another, he has to learn the offensive and defensive scripts (plays) used by his new team. These may or may not resemble those used by his old team. Because schemas shared by group members affect their capacity to work together, differences between members in the schemas used may produce differences in group performance.

Differences in schemas also can affect the relations between groups. If the schemas used by one group differ materially from the schemas used by another, the two groups may have difficulty finding common ground. In one study, for example, two pro-environment groups were discussing whether to form a coalition. One group usually made its decisions by majority vote, whereas the other group made decisions by reaching a consensus after thorough discussion. When observers from the first group attended a long meeting held by the second group, it appeared inefficient and aimless to them; they went back to their own group and recommended against joining the proposed coalition (Lichterman, 1995).

PERSON SCHEMAS AND GROUP STEREOTYPES

Person Schemas

As noted earlier, person schemas are cognitive structures that describe the personalities of other individuals. There are several distinct types of person schemas. Some person schemas are very specific and pertain to particular people. For example, Carolyn married George 3 years ago, and she knew him for 4 years before that. By this time, she has an elaborate schema regarding George, and she can usually predict how he will react to new situations, opportunities, or problems. Similarly, we often have individual schemas for public figures (for instance, Oprah Winfrey—talk show host, actor, advocate for women, Black, extremely wealthy) or for famous historical personages (for instance, Abraham Lincoln—political leader during the Civil War, honest, determined, opposed to slavery, committed to holding the Union together).

Other person schemas are very abstract and focus on the relations among personality traits. A schema of this type is an implicit personality theory—a set of unstated assumptions about which personality traits are correlated with one another (Anderson & Sedikides, 1991; Grant & Holmes, 1981; Sternberg, 1985). These theories often include beliefs about the behaviors that are associated with various personality traits (Skowronski & Carlston, 1989). They are considered implicit because in everyday life we usually do not subject our person schemas to close examination, nor are we explicitly aware of their contents.

Implicit Personality Theories and Mental Maps As do all schemas, implicit personality theories enable us to make inferences that go beyond the available information. Instead of withholding judgment, we use them to flesh out our impressions of a person about whom we have little information. For instance, if we learn someone has a warm personality, we might infer she is also likely to be sociable, popular, good-natured, and so on. If we hear that somebody else is pessimistic, we may infer he is humorless, irritable, and unpopular, even though we lack evidence that he actually has these traits.

We can depict an implicit personality theory as a mental map indicating the way traits are related to one another. Figure 5.1 displays such a mental map. Based on judgments made by college students, this figure shows how various personality traits stand in relation to one another (Rosenberg, Nelson, & Vivekananthan, 1968). Traits thought to be similar are located close together within our mental map, meaning that we expect people who have one trait to have the other. Traits thought to be dissimilar are located far apart, meaning we believe they rarely occur together in one person.

If your mental map resembles the one portrayed in Figure 5.1, you think that people who are wasteful are also likely to be unintelligent and irresponsible (see the lower left part of the map). Similarly, you think that people who are persistent are also likely to be determined and skillful (the upper right part of the map).

As portrayed in some mental maps, personality traits fall along two distinct evaluative dimensions—a social dimension and an intellectual dimension. These dimensions are represented by the lines shown in Figure 5.1. For instance, the traits "warm" and "cold" differ mainly on the social dimension, whereas "frivolous" and "industrious" differ on the intellectual dimension (Rosenberg & Sedlak, 1972). Some traits (such as "important") are good on both the social and the intellectual dimensions; other traits (such as "unreliable") are bad on both. Traits usually tend to be either good on both dimensions or bad on both dimensions, which explains a common bias in impression formation. We tend to judge persons who have several good traits as generally good and persons who have several bad traits as generally bad. And, if we have several bad characteristics associated with an important social marker, such as race, these bad characteristics can be unthinkingly attributed to a person of a particular race. This process of applying stereotypes has been demonstrated in a number of studies that also show how Whites place characteristics that are stereotypically Black with negative characteristics and characteristics that are stereotypically White with positive characteristics (Wittenbrink, Judd, and Park, 2001). Once we have a global impression of someone as, say, generally good, we assume that other positive traits (located nearby in the mental map) also apply. This tendency for our general or overall liking for a person to influence our subsequent assessment of more specific traits of that person is called the halo effect (Lachman & Bass, 1985; Thorndike, 1920). The **halo effect** produces bias in impression formation; it can lead to inaccuracy in our ratings of others' traits and performances (Cooper, 1981; Fisicaro, 1988).

The particular mental map described in Figure 5.1 (from Rosenberg et al., 1968) organizes traits in terms of two dimensions. However, many different researchers have constructed mental maps from trait data (Conley, 1985; Goldberg, 1981). Depending on the statistical techniques used, the maps produced sometimes have more than two dimensions. For instance, McCrae and Costa (1987) uncovered five dimensions—neuroticism, extraversion, openness, agreeableness, and conscientiousness. Three of these (neuroticism, extroversion, agreeableness) correspond roughly to Rosenberg et al.'s social traits, and the others (openness, conscientiousness) correspond to intellectual traits.

Origins of Person Schemas Where do implicit personality theories come from? First, they stem from the distilled wisdom of our socializing agents, who use expressions like "power corrupts," "ignorance is bliss," and "tall, dark, and ______" (fill in the blank). These expressions tell us which traits go together. Second,

FIGURE 5.1 RELATIONSHIPS AMONG ATTRIBUTES: A MENTAL MAP

Each of us has an implicit theory of personality—a theory about which personality attributes tend to go together and which do not. We can represent our theories of personality in the form of a mental map. The closer attributes are located to each other on our mental map, the more we assume that these attributes will appear in the same person. The mental map shown below is based on the mental maps of many American college students.

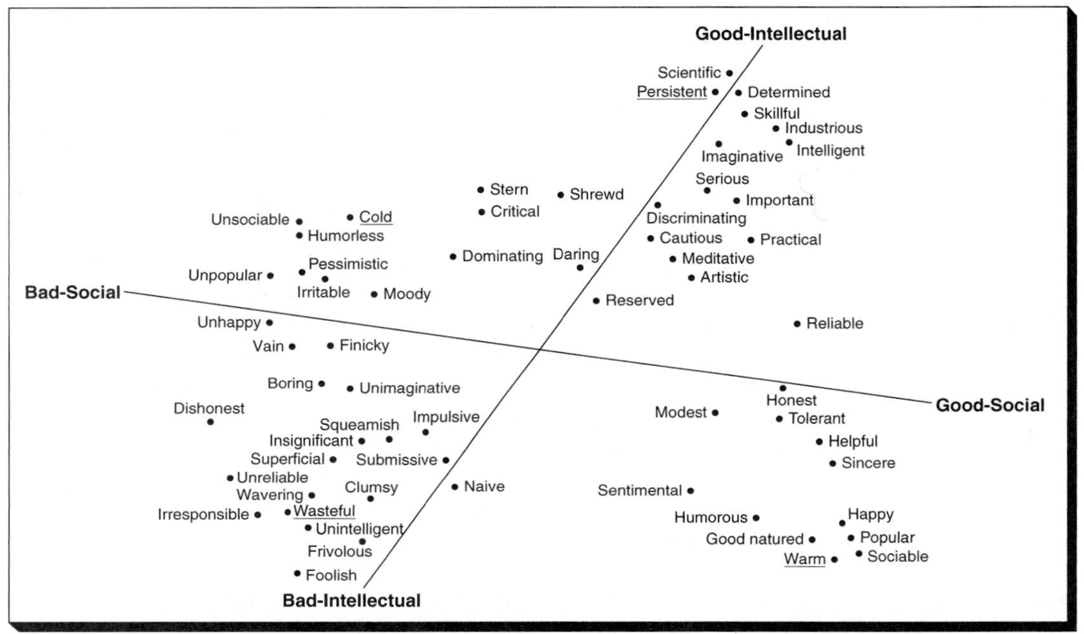

Source: Adapted from "Relationships Among Attributes: A Mental Map" by Rosenberg, Nelson, and Vivekananthan, *Journal of Personality and Social Psychology, 9,* 283–294. Copyright © 1968 by the American Psychological Association. Adapted with permission.

and probably more important, the very meanings of words in our language suggest certain relationships. The word "generous" implies both "helpful" and "sympathetic"; "fickle" implies both "unreliable" and "unpredictable." As a result, when one word applies to a person, we assume that other words that share similar meanings also apply (Shweder, 1977). Third, people we know actually do possess sets of personality characteristics that occur together; we observe their actions, infer the traits underlying these actions, and then formulate our implicit personality theories accordingly (Borkenau & Ostendorf, 1987; Stricker, Jacobs, & Kogan, 1974).

Group Stereotypes

- "The Irish are hot-headed, drunken, and belligerent."
- "Blacks are lazy and unreliable, and aren't good at anything except singing, dancing, and sports."
- "Feminists are left-wing, militant, man-hating radicals."
- "Jocks may be strong, but they're stupid and arrogant."
- "Southerners are rednecked, speech-slurring, barefooted bigots."

- "Republicans are heartless, racist, elitist reactionaries."
- "Lawyers are shrewd, contentious, overpriced troublemakers."

An unfortunate reality in our society is that we have all heard remarks like these—categorical, extreme, inaccurate characterizations. Each of these is an example of a group schema or stereotype. A stereotype is a set of characteristics attributed to all members of some specified group or social category (McCauley, Stitt, & Segal, 1980; Taylor, 1981). Just like other types of schemas, stereotypes simplify the complex social world. Rather than treating each member of a group individually, stereotypes encourage us to think about and treat all feminists, southerners, or lawyers the same way. By helping us to quickly place people into categories, stereotypes let us form impressions of people and predict their behavior with only minimal information: the groups to which they belong.

Stereotypes, however, involve overgeneralization. They lead us to think that all members of a particular group or social category have certain attributes. Although stereotypes might contain a kernel of truth—some members of the stereotyped group may have some of the imputed characteristics—it is almost never the case that all members have those characteristics. For this reason, stereotypes often lead to inaccurate inferences. Consider, for instance, all the persons you know of Irish descent. Perhaps one of them does—as the stereotype suggests—have a quick temper, and maybe another once got into a fistfight. It is certainly false, however, that all your Irish acquaintances spend most of their time fighting, arguing, drinking, and eating potatoes. Moreover, the Irish people you know probably do not get into angry fights anymore often than people who are not Irish.

Although stereotypes are overgeneralizations, we still constantly use them and are often unaware of their impact on our judgments of others (Hepburn & Locksley, 1983). And, although there is nothing inherent in stereotypes that requires them to be negative, many stereotypes do contain negative elements. Of course, some stereotypes are positive ("Asians excel at math"; "Graduate students are hard working"), but many others disparage or diminish the group stereotyped. Stereotypes can have many negative effects, especially when they are used to limit access to important social roles—for example, when an individual applies for a job or for admission to college. Stereotypes can also have less direct effects on members of stereotyped groups through a process called **stereotype threat** (Steele, 1997). When a member of a group believes that there is a real threat of being judged based on group stereotypes, the result can be poorer performance. Box 5.1 explains how stereotype threat reduces the performance of some students on academic tasks and standardized tests.

We can hardly avoid making a snap judgment about the personalities of these individuals, but are we right? Stereotypes enable us to form impressions about people merely by knowing the group to which they belong.

Common Stereotypes As the foregoing examples suggest, in American society, some widely known stereotypes pertain to ethnic, racial, and gender groups. Ethnic (national) stereotypes held by Americans might include, for example, the view that Germans are industrious and technically minded; Italians, passionate; Irish, quick tempered; and Americans, materialistic (Karlins, Coffman, & Walters, 1969). Investigators have studied ethnic, racial, and gender stereotypes for many years, and the results show that the content of stereotypes changes over time (Diekmann, Eagly, Mladinic, & Ferreira, 2005). For instance, few of us now believe—as many once did—that the typical American Indian is a drunk, the typical African American is superstitious, or the typical Chinese American is conservative and inscrutable.

BOX 5.1 Research Update: Stereotype Threat

When people act on their stereotypes, they can produce many negative effects for those who are the subjects of their stereotypes. Members of racial groups may be denied jobs or promotions because the schema the employer holds of their racial group include laziness. Other groups may be denied admissions to selective colleges because admissions officers believe people from that group are lazy or irresponsible. As damaging as these direct uses of stereotypes can be, researchers have recently discovered a second, less direct negative effect of stereotypes called stereotype threat (Steele, 1997, 1999).

Stereotype threat occurs when a member of a group suspects that he or she will be judged based on a common stereotype that is held of that group. For example, one stereotype of women is that they are less proficient at mathematics than men. If a woman enters a situation where her mathematical ability is being judged and she believes that the judgment will be negatively affected by the stereotype about women's mathematical ability, her performance on the exam may deteriorate (Spencer, Steele, & Quinn, 1999). To test for this kind of effect, Steele and Aronson (1995) gave Stanford University students a very difficult test using questions from the Graduate Record Examination in literature. The difficulty of the test provided a stereotype threat for Black students because poor performance would confirm a stereotype that they were not as able as White students. Even though the White and Black students were matched on ability, the Black students scored much lower than the White students. However, when researchers told the students that the test was part of a study to understand how people solved problems and that it did not measure ability, the stereotype threat was removed and the Black and White students did equally well.

Why does performance deteriorate when stereotype threat is present? Isn't it possible that the desire to disprove the stereotype might cause students to try harder and thereby cause them to do even better than they normally would? In a follow-up study, students took the exam on a computer, so that the researchers could time how long the students took with each question. The results showed that under conditions of stereotype threat, Black students were exerting extra effort and were overthinking the questions. They reread questions, changed their answers, and generally became less efficient at taking the test (Steele, 1999). This result also made sense of Steele's finding that stereotype threat affected academically strong students more than academically weak students—for those students who saw academics as an important part of their self-concept, the threat was much more meaningful than for those who cared less about academics (Steele, 1997).

In real life, it may be possible to reduce stereotype threat and to even the playing field. One way of doing this is to convince students who may be experiencing stereotype threat that the test being used is not biased. This is not easy to do given current deeply held beliefs about the unfairness of testing and the pervasiveness of racial stereotypes. However, Cohen, Steele, and Ross (1999) found that they could reduce stereotype threat by informing students that the evaluations of their performance would use very high standards and that they believed the students could perform up to those standards. Such an approach lets the student know that assessment is based on standards rather than stereotypes and that students are not viewed stereotypically.

Stereotypes may not have disappeared over time, but they have changed form (Dovidio & Gaertner, 1996).

Just as stereotypes about ethnic and racial groups are commonly held in our society, so also are stereotypes about gender groups. Usually our first judgment when meeting people involves classifying them as male or female. This classification is likely to activate an elaborate stereotype. This stereotype depicts male persons as more independent, dominant, competent, rational, competitive, assertive, and stable in handling crises. It characterizes female persons as more emotional, sensitive, expressive, gentle, helpful, and patient (Ashmore, 1981; Martin, 1987; Minnigerode & Lee, 1978). Research on the nature of these stereotypes of male and female persons is discussed in Box 5.2. Within gender, stereotypes are linked to titles. For instance, women labeled Ms. are seen as more achieving, more masculine, and less likable than women labeled Mrs. (Dion & Schuller, 1991). In addition to using ethnic, racial, and gender stereotypes, people also stereotype groups defined by occupation, age, political ideology, mental illness, hobbies, school attended, and so on (Milburn, 1987; Miller, 1982).

Origins of Stereotypes How do various stereotypes originate? Some theorists suggest that stereotypes arise out of direct experience with some members of the stereotyped group (Campbell, 1967). We may once have known Italians who were passionate, Blacks who were musical, or Japanese who were polite. We then build a stereotype by generalizing—that is, we infer that all members of a group share the attribute we know to be characteristic of some particular members.

Other theorists (Eagly & Steffen, 1984) suggest that stereotypes derive in part from a biased distribution of group members into social roles. Roles have associated characteristics, and eventually those characteristics are attached to the persons occupying the roles. If a social group is concentrated in roles with negative characteristics, an unflattering stereotype of that group may emerge that ascribe the negative characteristics of the job to members of the group.

Stereotyping may also be a natural outcome of social perception. When people have to process and remember a lot of information about many others, they store this information in terms of group categories rather than in terms of individuals (Taylor, Fiske, Etcoff, & Ruderman, 1978). In trying to remember what went on in a classroom discussion, you may recall that several women spoke and a Black person expressed a strong opinion, although you cannot remember exactly which women spoke or who the Black person was. Because people remember behavior by group category rather than by individual, they attach the behavior to the groups (Rothbart, Fulero, Jensen, Howard, & Birrell, 1978). Remembering that women spoke and a Black person expressed a strong opinion, you might infer that in general, women are talkative and Blacks are opinionated. You would not form these stereotypes if you recalled these attributes as belonging to individuals.

Errors Caused by Stereotypes Because stereotypes are overgeneralizations, they foster various errors in social perception and judgment. First, stereotypes lead us to assume that all members of a group are alike and possess certain traits. Yet individual members of a group obviously differ in many respects. One person wearing a hard hat may shoulder you into the stairwell on a crowded bus; another may offer you his seat. Second, stereotypes lead us to assume that all the members of one group differ from all the members of other groups. Stereotypes of football players and ballet dancers may suggest, for instance, that these groups have nothing in common. But both groups contain individuals who are patient, neurotic, hard-working, intelligent, and so on. In fact, there are ballet dancers who also play football.

Although stereotypes can produce inaccurate inferences and judgments in simple situations, they are especially likely to do so in complex situations. When the judgment to be made is multifaceted and involves a lot of complex data, reliance on stereotypes can prove particularly misleading. If an observer uses a stereotype as a central theme around which to organize information relevant to a decision, he or she may neglect information that is inconsistent with the stereotype (Bodenhausen & Lichtenstein, 1987). Research also indicates that people of higher status have a tendency to use stereotypes more than people of lower status. This seems to occur because people of higher status have more people competing for their attention and thus

BOX 5.2 Gender Schemas and Stereotypes

One of the most consistent research findings on stereotypes is that many people believe men and women have different personality traits. What are the traits believed to be typical of each sex? Where do these sex stereotypes come from?

Studies of sex stereotyping have established a number of characteristics that people associate differentially with men and women. In the accompanying chart, 20 characteristics are listed that studies have found to be consistently associated with men or women. To see how aware you are of these stereotypes, fill out the chart by indicating which of the traits listed are more typical of men and which are more typical of women. Also indicate if you consider each trait as a desirable or an undesirable characteristic.

	Most Typical of		Desirable	
Trait	Men	Women	Yes	No
Independent	____	____	____	____
Aggressive	____	____	____	____
Ambitious	____	____	____	____
Strong	____	____	____	____
Blunt	____	____	____	____
Passive	____	____	____	____
Emotional	____	____	____	____
Easily influenced	____	____	____	____
Talkative	____	____	____	____
Tactful	____	____	____	____
Excitable in minor crises	____	____	____	____
Aware of others' feelings	____	____	____	____
Submissive	____	____	____	____
Strong need for security	____	____	____	____
Feelings easily hurt	____	____	____	____
Self-confident	____	____	____	____
Adventurous	____	____	____	____
Acts as a leader	____	____	____	____
Makes decisions easily	____	____	____	____
Likes math and science	____	____	____	____

Broverman et al. (1972) found that both men and women agreed on the sex stereotypes and on the desirability of each trait. The first five traits listed in the chart were seen as more

have more incentive to use shortcuts, and because they can afford to make more mistakes because of their power (Goodwin, Gubin, Fiske, & Yzerbyt, 2000). This dynamic occurs even when subjects are randomly assigned to higher and lower status roles (Richeson & Ambady, 2003).

Although stereotypes involve overstatement and overgeneralization, they resist change even in the face of concrete evidence that contradicts them. This occurs because people tend to accept information that confirms their stereotypes and to ignore or explain away information that disconfirms them (Lord,

BOX 5.2 Continued

typical of men, whereas the next five were seen as more typical of women. That is, men were seen as more independent, aggressive, ambitious, strong, and blunt; women were seen as more passive, emotional, easily influenced, talkative, and tactful. In general, men were perceived as stronger and more confident than women, and women as weaker and more expressive than men. Subsequent studies have found that these stereotypes have persisted over time (Bergen & Williams, 1991; Deaux & Lewis, 1983).

Broverman et al. (1972) also found that most traits stereotyped as masculine were evaluated as desirable, whereas most traits stereotyped as feminine were evaluated as undesirable. In other words, traits associated with men were usually considered to be better than those associated with women. Did your evaluations of trait desirability favor the male stereotyped traits? If not, you may fit in with a trend among educated respondents toward valuing some traditionally feminine traits (say, emotional) more positively and some traditionally masculine traits (say, ambitious) more negatively (Der-Karabetian & Smith, 1977; Lottes & Kuriloff, 1994; Pleck, 1976). This trend means that even if sex stereotypes persist, women may be evaluated less negatively than before.

Are we more likely to use stereotypes in some situations than in others? Research indicates that the answer is yes. One of the functions of schematic or stereotypic processing is that it "fills in the blanks." Thus, the less information we have about someone, the more likely we are to rely on stereotypes. If we know little about a man other than his sex, we are more likely to assume he will be aggressive and independent. Moreover, we are more likely to apply stereotypes when cues make group membership salient. Thus, if we meet someone whose clothing or grooming emphasizes masculinity or femininity, we are likely to assume that our stereotype fits the person. This has interesting implications for women who work in male-dominated professions. They must decide whether to emphasize their gender difference by wearing dresses and makeup or to de-emphasize the difference by wearing pants and no makeup. Finally, there is a priming effect (Fiske & Taylor, 1991); the more recently a stereotype or schema has been used, the more accessible it is and, therefore, likely to be used again. Gender schemas are among the most frequently activated and therefore are highly accessible.

Gender schemas are deeply ingrained in part because they are taught to children starting at a very young age (Klinger, Hamilton, & Cantrell, 2001). By the time children are 2 years old, they have acquired an understanding of what gender means and can identify people as male or female (Katz, 1986). Cultural differences also play a role. Cultures that emphasize differences between men and women encourage the use of gender schemas—gender is more salient and thus becomes a more trusted perceptual aid (Bem, 1981).

Lepper, & Mackie, 1984; Snyder, 1981; Weber & Crocker, 1983). Suppose, for example, that Stan stereotypes gay men as effeminate, nonathletic, and artistic. If he stumbles into a gay bar, he is especially likely to notice the men in the crowd who fit this description, thereby confirming his stereotype. But how does he construe any rough-looking, athletic men who are there? It is possible that these individuals might challenge his stereotype, but reconstructing schemas is a lot of work, and Stan is more likely to find a way around this challenge. He might scrutinize those who don't fit his stereotype for hidden signs of effeminacy; he might underestimate their number or consider them the exceptions that prove the rule—or even assume they are straight. Through cognitive strategies like these, people explain away contradictory information and preserve their stereotypes.

IMPRESSION FORMATION

Information about other people comes to us from various sources. We may read facts about someone. We may hear something from a third party. We may witness acts by the other. We may interact directly with the other and form an impression of that person based on his or her appearance, dress, speech style, or background. We even infer personality characteristics from people's facial features (Hassin & Trope, 2000; Zebrowitz, Andreoletti, Collins, Lee, & Blumenthal, 1998). Regardless of how we get information about the other, we as perceivers must find a way to integrate these diverse facts into a coherent picture. This process of organizing diverse information into a unified impression of the other person is called impression formation. It is fundamental to person perception.

Trait Centrality

In a classic experiment, Asch (1946) used a straightforward procedure to show that some traits have more impact than others on the impressions we form. Undergraduates in one group received a list of seven traits describing a hypothetical person. These traits were intelligent, skillful, industrious, warm, determined, practical, and cautious. Undergraduates in a second group received the same list of traits, but with one critical difference: The trait "warm" was replaced by "cold." All participants then wrote a brief paragraph indicating their impressions and completed a checklist to rate the stimulus person on such other characteristics as generous, wise, happy, good-natured, humorous, sociable, popular, humane, altruistic, and imaginative.

The findings led to several conclusions. First, the students had no difficulty performing the task. They were able to weave the trait information into a coherent whole and to construct a composite sketch of the stimulus person. Second, substituting the trait "warm" for the trait "cold" produced a large difference in the overall impression formed by the students. When the stimulus person was "warm," the students typically described him as happy, successful, popular, and humorous. But when he was "cold," they described him as self-centered, unsociable, and unhappy. Third, the terms "warm" and "cold" had a larger impact than other traits on the overall impression formed of the stimulus person. This was demonstrated, for instance, by a variation in which the investigator repeated the basic procedure but substituted the pair "polite" and "blunt" in place of "warm" and "cold." Whereas describing the stimulus person as warm rather than cold made a great difference in the impressions formed by the students, describing him as polite rather than blunt made little difference.

We say that a trait has a high level of **trait centrality** when it has a large impact on the overall impression we form of that person. In Asch's study, the warm/cold trait displayed more centrality than the polite/blunt trait because differences in warm/cold produced larger differences in participants' ratings.

A follow-up study (Kelley, 1950) replicated the warm/cold finding in a more realistic setting. Students in sections of a psychology course read trait descriptions of a guest lecturer before he spoke. These descriptions contained adjectives similar to those used by Asch (that is, industrious, critical, practical, determined), but they differed regarding the warm/cold variable. For half the students, the description contained the trait "warm"; for the other half, it contained "cold." The lecturer subsequently arrived at the classroom and led a discussion for about 20 minutes. Afterward, the students were asked to report their impressions of him. The results showed large differences between the impressions formed by those who read he was "warm" and those who read he was "cold." Those who had read he was "cold" rated him as less considerate, sociable, popular, good-natured, humorous, and humane than those who had read he was "warm." Because all students saw the same guest instructor in the classroom, the differences in their impressions could stem only from the use of "warm" or "cold" in the profile they had read.

How could a single trait embedded in a profile have such an impact on impressions of someone's behavior? Several theories have been advanced, but one plausible explanation holds that the students used a schema—a mental map—indicating what traits go with being warm and what traits go with being cold. Looking again at Figure 5.1, we note the locations of the attributes "warm" and "cold" on the map and the nature of the other attributes close by. If the mental maps used by the participants in the Asch (1946) and Kelley (1950) studies resembled Figure 5.1, it becomes immediately clear why they judged the warm person as more sociable, popular, good-natured, and humorous; these traits are close to "warm" and remote from "cold" on the mental map.

Integrating Information About Others

In everyday life, we often receive a lot of information at once. New experiences with others typically add information to impressions we already hold. Some information we receive may be inconsistent with information we already have. If your uncle, whom you view as loving and tolerant, criticizes your cousin for her sloppy clothes and stringy hair, how do you make sense of his behavior? Most likely, you would try to understand how this contradictory information squares with your earlier impression of your uncle. You might conclude, for instance, that your uncle's intolerance stemmed from his love for your cousin and his desire to protect her from others' criticism. In general, when forming an impression of a person, perceivers try to integrate information about many seemingly contradictory attributes to create a unified and coherent impression of that individual (Asch & Zukier, 1984).

Models of Information Integration Many models of how perceivers combine information are based on a key assumption—namely, that the most important aspect of a perceiver's impression of a person is the overall positive or negative evaluation of that person. Although restrictive, this assumption is justified by two considerations. First, empirical studies show that evaluation is the most important dimension on our mental maps of personality traits. Second, this evaluative dimension is crucial when we make practical judgments and decisions. For instance, we may reject a job applicant because we think he is unproductive (negative evaluation) or invite a new acquaintance to a party because we think she is vivacious (positive evaluation).

Assume a theorist wishes to predict the overall evaluation that a perceiver would make of a stranger who has the following traits: sincere, friendly, cautious, and dishonest. To do this, the theorist needs to know several facts. First, the theorist must know how positively or negatively perceivers rate each of these traits. Most college students assign highly positive values to such traits as "sincere" (say, +3), less positive values to "friendly" (+2) and "cautious" (+1), and negative values to "dishonest" (−3; Anderson, 1968).

Next, the theorist needs to know how perceivers combine the trait values to form an overall evaluation of the stranger. Social psychologists have proposed several different models that depict how perceivers combine information on diverse traits to form an overall evaluation of someone. One of these, termed the additive model, postulates that perceivers form an overall evaluation by summing the values of all the single traits. An example of the additive model appears in the top panel of Table 5.2 with four different combinations of traits. A key feature of the additive model is that when we add traits with a positive value, we increase the favorableness of our overall impression (column I vs. column II), whereas when we add traits with a negative value, we decrease its favorableness (column III vs. column IV).

Although the additive model is plausible, theorists also have developed other models. One alternative—termed the averaging model—postulates that perceivers form an overall evaluation by averaging the values of all the single traits (see the bottom panel of Table 5.2). In the averaging model, the impact of new information depends on whether this information is more favorable or less favorable than the overall impression we already have. Thus incorporating a new, mildly positive trait into a strongly positive impression makes the resulting impression less positive (column I vs. column II), whereas incorporating a mildly negative trait to a strongly negative impression makes the resulting impression less negative (column III vs. column IV).

TABLE 5.2 A COMPARISON OF ADDITIVE AND AVERAGING MODELS FOR FORMING IMPRESSIONS

	TRAIT COMBINATIONS							
Model	**I**		**II**		**III**		**IV**	
Additive	sincere	+3	sincere	+3	serious	+1	serious	+1
	friendly	+2	friendly	+2	irresponsible	−3	irresponsible	−3
	tolerant	+1	tolerant	+1	dishonest	−3	dishonest	−3
			cautious	+1			unimaginative	−1
Overall impression:		+6		+7		−5		−6
Averaging	sincere	+3	sincere	+3	serious	+1	serious	+1
	friendly	+2	friendly	+2	irresponsible	−3	irresponsible	−3
	tolerant	+1	tolerant	+1	dishonest	−3	dishonest	−3
			cautious	+1			unimaginative	−1
Overall impression:		+2.00		+1.75		−1.67		−1.50

Although both the additive and averaging models have some appeal, the bulk of the evidence from empirical studies of impression formation supports a third model. This is a refinement of the averaging model called the weighted averaging model (Anderson, 1981). According to this model, perceivers average the values of the traits to form an impression, but they also give more weight to some information and less to other information. For instance, if a stranger is described as tall, dark, handsome, talkative, blue-eyed, armed, and dangerous, the last two traits on this list will overshadow the others and receive more weight in the perceiver's evaluation of the stranger.

Several factors influence the weights that perceivers assign to trait information. First, they give more weight to information from highly credible sources than to information from less credible sources. Second, they weight negative attributes more heavily than positive attributes (Hamilton & Zanna, 1972; Ronis & Lipinski, 1985), perhaps because negative information is distinctive in a world where people usually present socially desirable selves. Third, they attend more to attributes that pertain to the purpose or judgment at hand. Fourth, they will focus more on elements that stand out from the background and less on things that blend in with the overall situation (Nelson & Klutas, 2000). Fifth, they discount information that is very inconsistent with previous impressions or is redundant with what they already know. And finally, they weight first impressions more heavily than subsequent impressions.

First Impressions You have surely noticed the effort that individuals make to create a good impression when interviewing for a new job, entering a new group, or meeting an attractive potential date. This effort reflects the widely held belief that first impressions are especially important and have an enduring impact. In fact, this belief is supported by a body of systematic research. Observers forming an impression of a person give more weight to information received early in a sequence than to information received later. This is called the **primacy effect** (Luchins, 1957).

What accounts for the impact of first impressions? One explanation is that after forming an initial impression of a person, we interpret subsequent information in a way that makes it consistent with our initial impression. Having established that your new roommate is neat and considerate, you interpret the dirty socks on the floor as a sign of temporary forgetfulness rather than as evidence of sloppiness and lack of concern. Thus, the schema into which an observer assimilates new information influences the interpretation of that information (Zanna & Hamilton, 1977).

A second explanation for the primacy effect holds that we attend very carefully to the first bits of information we get about a person, but we pay less attention once we have enough information to make a

This man makes a first impression as strong, stern, and self-controlled. First impressions are hard to change, because people pay less attention to later information. When told this man is fearful, for example, observers are likely to ignore this information, or to interpret it as meaning he shows healthy fear in extremely dangerous situations.

© Roger Allyn Lee/SuperStock

judgment. It is not that we interpret later information differently; we simply use it less. This explanation assumes that whatever information we attend to most has the biggest effect on our impressions (Dreben, Fiske, & Hastie, 1979).

Although primacy effects are commonplace, they do not always occur; sometimes, the direct opposite happens. Under certain conditions, the most recent information we acquire exerts the strongest influence on our impressions—an occurrence known as the **recency effect** (Jones & Goethals, 1971; Steiner & Rain, 1989). A recency effect is likely to occur when so much time has passed that we have largely forgotten our first impression or when we are judging characteristics that change over time, like moods or attitudes. In laboratory settings, investigators can induce a recency effect by asking perceivers to make a separate evaluation after each new piece of information is received (Stewart, 1965).

Although both primacy effects and recency effects can have an impact on the impressions that people form of one another, primacy effects are especially important in everyday life. In one study investigating the relative impact of primacy and recency effects on impression formation (Jones, Rock, Shaver, Goethals, & Ward, 1968), participants observed the performance of a college student on an SAT-type aptitude test. In one condition, the student started successfully on the first few items but then her performance deteriorated steadily. In a second condition, the student started poorly and then gradually improved. In both conditions, the student answered 15 out of 30 test items correctly. After observing one or the other performance, participants rated the student's intelligence and tried to predict how well she would do on the next 30 items. Although the student's overall performance was the same in both conditions (15 of 30 correct), participants rated the student as more intelligent when she started well and then tailed off than when she started poorly and then improved. They also predicted higher scores for the student on the next series when the student started well than when she started poorly. Clearly, participants gave more weight to the student's performance on the first few items—a primacy effect.

Impressions as Self-Fulfilling Prophecies

Whether correct or not, the impressions we form of people influence our behavior toward them. Recall, for instance, the study in which students read that their guest instructor was "warm" or "cold" before meeting him (Kelley, 1950). Not only did the students form different impressions of the instructor, but they also behaved differently toward him. Those who believed the instructor was "warm" participated more in the class discussion than those who believed he was "cold."

When our behavior toward people reflects our impressions of them, we cause them to react in ways that confirm our original impressions. For example, if we ignore someone because we think she is dull, she will probably withdraw and add nothing interesting to the conversation. Because our own actions evoke appropriate reactions from others, our initial impressions—whether correct or incorrect—are often confirmed by the reactions of others. When this happens, our impressions become self-fulfilling prophecies (Darley & Fazio, 1980).

A study of "getting acquainted" conversations between male and female college students shows vividly how impressions may become self-fulfilling (Snyder, Tanke, & Berscheid, 1977). The investigators provided each male student in the study with a folder of information about a female student whom he would subsequently contact by telephone to get acquainted. The folder included a biographical sheet and a photo of the woman. The biographical information was accurate, but (unknown to the male participants) the photo was actually a snapshot of a different woman. The woman shown in the photo was either very attractive or very unattractive (the experimental manipulation). Each male student was asked to form an impression of the woman's personality based on the biographical sheet and the photo. As expected, men who saw the attractive snapshot rated the woman more positively on personality dimensions than those who saw the unattractive snapshot.

When the man then engaged in the "get acquainted" phone conversation, those who believed their partner was attractive spoke with more animation, sociability, and warmth than those who thought their partner was unattractive. In other words, the impressions that the men formed from the snapshots influenced their own behavior. For their part, the women responded over the phone in a more poised, confident, animated, sociable, sexually warm, and outgoing manner when they were speaking with men who thought they were attractive. This occurred even though the women did not realize the men had seen snapshots of any kind. It is easy to understand how the men might interpret the responses of the women as confirmation of their original impressions concerning their partners' sociability. Thus, the prophecy was self-fulfilling.

Heuristics

In most social situations, our impressions could be guided by a number of different schemas. Which schemas will we choose to help us define the situation and the people in it, and which schemas will guide us as we interact with them? Furthermore, there may be some situations for which we have not created an appropriate schema. How do we make decisions on how to characterize these situations? The answer comes in the form of another type of mental shortcut called a heuristic (Tversky & Kahneman, 1974). **Heuristics** provide a quick way of selecting schemas that—although far from infallible—often help us make an effective choice amid considerable uncertainty.

Availability One factor that determines how likely we are to choose a particular schema is how long it has been since we have used that particular schema. If we have recently used a particular schema, it is easier for us to call up that schema for use in the current situation. There are other reasons why certain schemas are more available to us. If, for instance, certain examples of categorizations are easier to remember, then schemas consistent with those examples are more likely to be called up and used. Suppose that you were asked whether there are more words in the English language that begin with the letter R or if there are more words in which the third letter is an R. Most people find it much easier to think of examples of words that begin with R, and thus the ease of producing examples makes it seem as if there are more words that begin with R (Tversky & Kahneman, 1974). These words are more easily available to us, and thus they cause us to overestimate their frequency of occurrence (Manis, Shedler, Jonides, & Nelson, 1993).

Representativeness A second heuristic we often use is called the representativeness heuristic (Tversky & Kahneman, 1974). In this case, we take the few characteristics we know about someone or something and select a schema that matches those characteristics well (Dawes, 1998; Thomsen & Borgida, 1996). Black students who attend the University of Notre Dame sometimes complain that they are assumed to be athletes by White students on campus. These White students are using a representativeness heuristic—they observe their most visible sports teams (football and basketball) and note that the percentage of players on those teams who are Black is much higher than the percentage of students in the overall student population who are Black. Thus, the schema of "athlete" is consistent with being Black, leading White students to the erroneous assumption that Black students must be athletes. People tend to discount statistical information in the face of the representativeness heuristic (Kahneman & Tversky, 1973). Given the size of the Black student body at Notre Dame, those students on the football and basketball teams can only account for a fraction of all Black students.

Anchoring and Adjustment

When faced with making a judgment on something we know very little about, we grasp any cues we can find to help us make a decent guess. Oftentimes, we will use some particular standard as a starting point, then try to determine if we should guess higher or lower than that starting point. Such a starting point is called an anchor, and our modification relative to the anchor is called adjustment (Mussweiler, Strack, & Pfeiffer, 2000; Tversky & Kahneman, 1974). Suppose you were asked on an exam to provide the population of Chicago. If you did not know that population, but you did know the population of New York City, you might use the population of New York as an anchor and, thinking that Chicago must be somewhat smaller than New York, adjust the New York value downward to produce your guess.

When using this heuristic, however, we do not always have meaningful anchors. If a number is in our head for any reason, we are likely to use it as an anchor even if it has nothing whatsoever to do with the situation we are facing (Cadinu & Rothbart, 1996; Wilson, Houston, Etling, & Brekke, 1996). Suppose an employer is conducting an annual evaluation of employees and has the power to give employees a raise of anywhere from 0 to 40 percent depending on their performance. If the boss just attended a retirement party for someone who worked in the firm for 30 years, he or she may unconsciously use this value as an anchor and end up giving relatively high raises. If, on the other hand, the boss just attended the birthday party of a 5-year-old niece, 5 may be used as the anchor, and although the boss may adjust up from 5, the raises are likely to be considerably lower than if 30 were used as the anchor. These kinds of anchoring effects tend to occur even if we are explicitly warned not to allow arbitrary anchors to affect our decisions (Griffin, Gonzalez, & Varey, 2001).

Perhaps most often, we use ourselves as an anchor when judging social situations (Markus, Smith, & Moreland, 1985). We have a tendency to do this even when we know we are unusual. If you are a very generous person who always tips at least 25 percent at a restaurant and are asked if your friend Emily is miserly or charitable, you would be likely to use your own rather unusual behavior as an anchor and report that she is tightfisted because you know that she typically tips "only" 20 percent.

Attribution Theory When we interact with other people, we observe only their actions and the effects those actions have. This is fine as far as it goes, but, as perceivers, we want to know why others act as they do. To figure this out, we must usually make inferences beyond what we observe. For instance, if a woman performs a favor for us, why is she doing it? Is she doing it because she is fundamentally a generous person? Or is she manipulative and pursuing some ulterior motive? Does her social role require her to do it? Have other people pressured her into doing it? To act effectively toward her and to predict her future behavior, we must first figure out why she behaves as she does.

The term attribution refers to the process that an observer uses to infer the causes of another's behavior: "Why did that person act as he or she did?" In attribution, we observe another's behavior and infer backward to its causes—to the intentions, abilities, traits, motives, and situational pressures that explain why people act as they do. Theories of attribution focus on the methods we use to interpret another person's behavior and to infer its sources (Kelley & Michela, 1980; Lipe, 1991; Ross & Fletcher, 1985).

Dispositional Versus Situational Attributions

Fritz Heider (1944, 1958), whose work was an early stimulus to the study of attribution, noted that people in everyday life use commonsense reasoning to understand the causes of others' behavior. They act as "naive scientists" and use something resembling the scientific method in attempting to discern causes of behavior. Heider maintained that whether their interpretations about the causes of behavior are scientifically valid or not, people act on their beliefs. For this reason, social psychologists must study people's commonsense explanations of behavior and events so we can understand their behavior.

The most crucial decision that observers make is whether to attribute a behavior to the internal state(s) of the person who performed it—this is termed a **dispositional attribution**—or to factors in that person's environment—a **situational attribution.** For example, consider the attributions an observer might make when learning that her neighbor is unemployed. She might judge that he is out of work because he is lazy, irresponsible, or lacking in ability. These are dispositional attributions, because they attribute the causes of behavior to his internal states or characteristics. Alternatively, she might attribute his unemployment to the scarcity of jobs in his line of work, to employment discrimination, to the depressed condition of the economy, or to the evils of the capitalist system. These are situational attributions, because they attribute his behavior to external causes.

What determines whether an observer attributes an act to a person's disposition or to the situation? One important consideration is the strength of situational pressures on the person. These pressures may include normative role demands as well as rewards or punishments applied to the person by others in the environment. For example, suppose we see a judge give the death penalty to a criminal. We might infer that the judge is tough (a dispositional attribution). However, suppose we learn that the law in that state requires the death penalty for the criminal's offense. Now we would see the judge not as tough but as responding to role pressures (a situational attribution).

This logic has been formalized as the subtractive rule, which states that when making attributions about personal dispositions, the observer subtracts the perceived impact of situational forces from the personal disposition implied by the behavior itself (Trope & Cohen, 1989; Trope, Cohen, & Maoz, 1988). Thus, considered by itself, the judge's behavior (imposing the death penalty) might imply that she is tough in disposition. The subtractive rule, however, states that the observer must subtract the effect of situational pressures (the state law) from the disposition implied by the behavior itself. When the observer does this, he or she may conclude the judge is not especially tough or overly inclined to impose the death penalty.

For the judge and the death penalty, using the subtractive rule served to weaken the dispositional attribution and strengthen the situational attribution. In certain other cases, however, applying the subtractive rule strengthens or augments the dispositional attribution. This happens, for instance, whenever a person persists in actions that his or her environment discourages or punishes. Suppose that while interacting with political conservatives, a woman expressed liberal sentiments on various issues. If she persists in this despite a negative reaction from the others, we attribute the behavior to her personal disposition. That is, applying the subtractive rule, we subtract the impact of situational forces from the personal disposition implied by the behavior itself. This means we subtract a negative quantity (the negative reaction of others) from the disposition implied by the behavior itself. The net effect is to increase or augment the dispositional attribution. Our conclusion as observers is that the woman must hold strong liberal sentiments.

Inferring Dispositions from Acts

Although Heider's analysis and the subtractive rule are useful in identifying some conditions under which observers make dispositional attributions, they do not explain which specific dispositions they will ascribe to a person. Suppose, for instance, that you are on a city street during the Christmas season and you see a young, well-dressed man walking with a woman. Suddenly, the man stops and tosses several coins into a Salvation Army pot. From this act, what can you infer about the man's dispositions? Is he generous and altruistic? Or is he trying to impress the woman? Or is he perhaps just trying to clear out some nuisance change from his coat pocket?

When we try to infer a person's dispositions, our perspective is much like that of a detective. We can observe only the act (a man gives coins to the Salvation

Army) and the effects of that act (the Salvation Army receives more resources, the woman smiles at the man, the man's pocket is no longer cluttered with coins). From this observed act and its effects, we must infer the man's dispositions.

According to one prominent theory (Jones, 1979; Jones & Davis, 1965), we perform two major steps when inferring personal dispositions. First, we try to deduce the specific intentions that underlie a person's actions. In other words, we try to figure out what the person originally intended to accomplish by performing the act. Second, from these intentions we try to infer what prior personal disposition would cause a person to have such intentions. If we think the man intended to benefit the Salvation Army, for example, we infer the disposition "helpful" or "generous." However, if we think the man had some other intention, such as impressing his girlfriend, we do not infer he has the disposition "helpful." Thus, we attribute a disposition that reflects the presumed intention.

One problem in inferring dispositions from acts, of course, is that any given act may have multiple effects. In order to make confident attributions, perceivers must decide which effect(s) the person intends and which are merely incidental. This is not always easy to do. When the man donated money to the Salvation Army, for example, was his intention to perform a charitable act or to impress the woman accompanying him? His act accomplished both effects. Before making the inference that the man is generous and helpful by disposition, an observer must know which effect(s) the man intended the act to produce.

Several factors influence observers' decisions regarding which effect(s) the person is really pursuing and, hence, what dispositional inference is appropriate. These factors include the commonality of effects, the social desirability of effects, and the normativeness of effects (Jones & Davis, 1965).

Commonality If any given act produced one and only one effect, then inferences of dispositions from acts would always be clear-cut. Because many acts have multiple effects, however, observers attributing specific intentions and dispositions find it informative to observe the actor in situations that involve choices between alternative actions.

Suppose, for example, that a person can engage either in action 1 or in action 2. Action 1, if chosen, will produce effects a, b, and c. Action 2 will produce effects b, c, d, and e. As we can see, two of these effects (b and c) are common to actions 1 and 2. The remaining effects (a, d, and e) are unique to a particular alternative; these are noncommon effects.

Now suppose the person chooses action 2. What can we infer about his intentions and dispositions? The main inference is that although he may or may not have intended to produce effects b and c, he certainly intended either to produce effect d or effect e (or both) or to avoid effect a. The unique (noncommon) effects of acts enable observers to make inferences regarding intentions and dispositions, but the common effects of two or more acts provide little or no basis for inferences (Jones & Davis, 1965).

Thus, observers who wish to discern the specific dispositions of a person try to identify effects that are unique to the action chosen. Research shows that the fewer noncommon effects associated with the chosen alternative, the greater the confidence of observers about their attributions (Ajzen & Holmes, 1976).

Social Desirability In many situations, people engage in particular behaviors because those behaviors are socially desirable. Yet people who perform a socially desirable act show us only that they are "normal" and reveal nothing about their distinctive dispositions. Suppose, for instance, that you observe a guest at a party thank the hostess when leaving. What does this tell you about the guest? Did she really enjoy the party? Or was she merely behaving in a polite, socially desirable fashion? You cannot be sure—either inference could be correct. Now, suppose instead that when leaving, the guest complained loudly to the hostess that she had a miserable time at such a dull party. This would likely tell you more about her, because observers interpret acts low in social desirability as indicators of underlying dispositions (Miller, 1976).

Normative Expectations When inferring dispositions from acts, observers consider the normativeness of behavior. Normativeness is the extent to which we expect the average person to perform a behavior in a particular setting. This includes conformity to social

TABLE 5.3 MEAN RATINGS BY PARTICIPANTS OF INTERVIEWEES

	ROLE			
	Astronaut		Submariner	
Trait Rated	Inner-Directed	Other-Directed	Inner-Directed	Other-Directed
Conformity	13.09	15.91	9.41	12.58
Affiliation	11.12	15.27	8.64	12.00
Candor	9.68	12.42	12.08	10.09

Source: Adapted from Jones, Davis, and Gergen, 1961.

norms and to role expectations in groups (Jones & McGillis, 1976). Actions that conform to norms are uninformative about personal dispositions, whereas actions that violate norms lead to dispositional attributions.

A study by Jones, Davis, and Gergen (1961) illustrates the effect of out-of-role behavior on inferences. Participants listened to a tape-recorded interview of an individual seeking employment either as a submariner or as an astronaut. The first part of the tape provided a description (by the interviewer) of the ideal job applicant. For the submariner (who had to work long hours in cramped quarters with other people), the ideal characteristics were friendliness, cooperativeness, obedience, and gregariousness (an other-directed person). In contrast, for the astronaut (who would travel alone in space), the ideal characteristics were resourcefulness, thoughtfulness, independence, and a capacity to perform without help or the company of others (an inner-directed person).

Next, participants heard a tape of the applicant presenting himself for the job. Depending on the experimental condition, he sought employment either as a submariner or as an astronaut, and he presented himself either as an other-directed person or as an inner-directed person. The important point is that two of these combinations are role appropriate—the other-directed person applying for the submariner job and the inner-directed person applying for the astronaut job. The other two combinations are role inappropriate—the other-directed person applying for the astronaut job and the inner-directed person applying for the submariner job. After listening to the taped interview, the participants rated the applicant on various trait measures. They also indicated how much confidence they had in their own ratings.

Table 5.3 displays the ratings. Participants rated the two applicants whose behavior was role appropriate (the other-directed submariner and the inner-directed astronaut) as conforming and affiliative. Participants had low confidence in their ratings of these applicants. These applicants knew the job requirements and presented themselves accordingly during the interview, so participants could not infer much about them as persons. In particular, participants could not tell whether the applicants were truly what they claimed to be or were merely posing to get the job.

However, when the participants rated the two applicants whose behavior was role inappropriate (the other-directed astronaut and the inner-directed submariner), the results were different. Participants rated the inner-directed submariner as independent (nonconforming) and nonaffiliative. Because his behavior was contrary to role requirements, the participants made strong attributions and reported high confidence in their ratings. In contrast, participants rated the other-directed astronaut as very conforming and affiliative. This candidate's behavior was also contrary to role requirements, and participants again were confident of their ratings.

Note that the participants in this study followed the subtractive rule discussed earlier. In their attributions, participants augmented (increased) personal dispositions as causes for the applicants who were role inappropriate (the other-directed astronaut and the inner-directed submariner). Likewise, they discounted (diminished) personal dispositions as causes for the applicants who were role appropriate (the other-directed submariner and the inner-directed astronaut). Both effects are consistent with the subtractive rule.

Covariation Model of Attribution

Up to this point, we have examined how observers make attributions regarding a person's behavior in a single situation. Sometimes, however, we have multiple observations of a person's behavior. That is, we have information about a person's behavior in a variety of situations or in a given situation vis-à-vis different partners. Multiple observations enable us to make many comparisons, and these, in turn, facilitate causal attribution.

How do perceivers use multiple observations to arrive at a conclusion about the cause(s) of a behavior? Extending Heider's ideas, Kelley (1967, 1973) suggests that when we have multiple observations of behavior, we analyze the information essentially in the same way a scientist would. That is, we try to figure out whether the behavior occurs in the presence or absence of various factors (actors, objects, contexts) that are possible causes. Then to identify the cause(s) of the behavior, we apply the **principle of covariation:** We attribute the behavior to the factor that is both present when the behavior occurs and absent when the behavior fails to occur—the cause that covaries with the behavior.

To illustrate, suppose you are working at your part-time job one afternoon when you hear your boss loudly criticizing another worker, Michael. To what would you attribute your boss's behavior? There are at least three potential causes: the actor (the boss), the object of the behavior (Michael), and the context or setting in which the behavior occurs. For example, you might attribute the loud criticism to your boss's aberrant personality (a characteristic of the actor), to Michael's slothful performance (a characteristic of the object), or to some particular feature of the context.

Kelley (1967) suggests that when using the principle of covariation to determine whether a behavior is caused by the actor, object, or context, we rely on three types of information: consensus, consistency, and distinctiveness.

Consensus refers to whether all actors perform the same behavior or only a few do. For example, do all the other employees at work criticize Michael (high consensus), or is your boss the only person who does so (low consensus)?

Consistency refers to whether the actor behaves in the same way at different times and in different settings. If your boss criticizes Michael on many different occasions, his behavior is high in consistency. If he has never before criticized Michael, his behavior is low in consistency.

Distinctiveness refers to whether the actor behaves differently toward a particular object than toward other objects. If your boss criticizes only Michael and none of the other workers, his behavior is high in distinctiveness. If he criticizes all workers, his behavior toward Michael is low in distinctiveness.

The causal attribution that observers make for a behavior depends on the particular combination of consensus, consistency, and distinctiveness information that people associate with that behavior. To illustrate, Table 5.4 reviews the scenario in which your boss criticizes Michael. The table displays three combinations of information that might be present in this situation. These combinations of information are interesting, because studies have shown they reliably produce different attributions regarding the cause of the behavior (Cheng & Novick, 1990).

As Table 5.4 indicates, observers usually attribute the cause of a behavior to the actor (the boss) when the behavior is low in consensus, low in distinctiveness, and high in consistency. In contrast, observers usually attribute a behavior to the object (Michael) when the behavior is high in consensus, high in distinctiveness, and high in consistency. Finally, observers usually attribute a behavior to the context when consistency is low.

Several studies show that at least in general terms, people use consensus, consistency, and distinctiveness information in the way Kelley theorized (Hewstone & Jaspars, 1987; McArthur, 1972; Pruitt & Insko, 1980), although consensus seems to have a weaker effect on attributions than the other two aspects of covariation (Winschild & Wells, 1997). Of course, in any given

TABLE 5.4 WHY DID THE BOSS CRITICIZE MICHAEL?

Situation: At work today, you observe your boss criticizing and yelling at another employee, Michael.

Question: Why did the boss criticize Michael?

1. Kelley's (1973) model indicates that attributions are made to the actor (boss) when consensus is low, distinctiveness is low, and consistency is high.

   Example: Suppose no other persons criticize Michael (low consensus). The boss criticizes all the other employees (low distinctiveness). The boss criticized Michael last month, last week, and yesterday (high consistency).

   Attribution: The perceiver will likely attribute the behavior (criticism) to the boss. ("The boss is a very critical person.")

2. The model indicates that attributions are made to the stimulus object (Michael) when consensus is high, distinctiveness is high, and consistency is high.

   Example: Suppose everyone at work criticizes Michael (high consensus). The boss does not criticize anyone else at work, only Michael (high distinctiveness). The boss criticized Michael last month, last week, and yesterday (high consistency).

   Attribution: The perceiver will likely attribute the behavior (criticism) to Michael. ("Michael is a lazy, careless worker.")

3. The model indicates that attributions are made to the context or situation when consistency is low.

   Example: Suppose the boss has never criticized Michael before (low consistency).

   Attribution: The perceiver will likely attribute the behavior (criticism) to a particular set of contextual circumstances, rather than to Michael or the boss per se. ("Michael made a remark this morning that the boss misinterpreted.")

situation, the combination of available information may differ from the three possibilities shown in Table 5.4. In such cases, attributions are more complicated, more ambiguous, and less certain. We usually assign less weight to a given cause if other plausible causes are also present (Kelley, 1972; Morris & Larrick, 1995).

Attributions for Success and Failure

For students, football coaches, elected officials, and anyone else whose fate rides on evaluations of their performance, attributions for success and failure are vital. As observers realize, however, attributions of this type are problematic. Whenever someone succeeds at a task, a variety of explanations can be advanced for the outcome. For example, a student who passes a test could credit her own intrinsic ability ("I have a lot of intelligence"), her effort ("I really studied for that exam"), the easiness of the task ("The exam could have been much more difficult"), or even luck ("They just happened to test us on the few articles I read").

These four factors—ability, effort, task difficulty, and luck—are general and apply in many settings. How do observers decide which of these is the "real" cause of success or failure? When observers look at an event and try to figure out the cause of success or failure, they must consider two things. First, they must decide whether the outcome is due to causes within the actor (an internal or dispositional attribution) or due to causes in the environment (an external or situational attribution). Second, they must decide whether the outcome is a stable or an unstable occurrence. That is, they must determine whether the cause is a permanent feature of the actor or the environment or whether it is labile and changing. Only after observers make judgments regarding internality-externality and stability-instability can they reach conclusions regarding the cause(s) of the success or failure.

As various theorists (Heider, 1958; Weiner, 1986; Weiner et al., 1971) have pointed out, the four factors aforementioned—ability, effort, task difficulty, and luck—can be grouped according to internality-externality and stability-instability. Ability, for instance, is usually considered internal and stable. That is, observers usually construe ability or aptitude as a property of the person (not the environment), and they consider it stable because it does not change from moment to moment. In contrast, effort is internal and unstable. Effort or temporary exertion is a property of the person that changes depending on how hard he or she tries. Task difficulty depends on objective task characteristics, so it is external and stable. Luck or

TABLE 5.5 PERCEIVED CAUSES OF SUCCESS AND FAILURE

Degree of Stability	LOCUS OF CONTROL	
	Internal	External
Stable	Ability	Task difficulty
Unstable	Effort	Luck

Source: Adapted from Weiner, Heckhausen, Meyer, and Cook, 1972.

chance is external and unstable. Table 5.5 displays these relations.

Determinants of Attributed Causes Whether observers attribute a performance to internal or external causes depends on how the actor's performance compares with that of others. We usually attribute extreme or unusual performances to internal causes. For example, we would judge a tennis player who wins a major tournament as extraordinarily able or highly motivated. Similarly, we would view a player who turns in an unusually poor performance as weak in ability or unmotivated. In contrast, we usually attribute average or common performances to external causes. If defeat comes to a player halfway through the tournament, we are likely to attribute it to tough competition or perhaps bad luck.

Whether observers attribute a performance to stable or unstable causes depends on how consistent the actor's performance is over time. When performances are very consistent, we attribute the outcome to stable causes. Thus, if a tennis player wins tournaments consistently, we would attribute this success to her great talent (ability) or perhaps to the uniformly low level of her opponents (task difficulty). When performances are very inconsistent, however, we attribute the outcomes to unstable causes rather than stable ones. Suppose, for example, that our tennis player is unbeatable one day and a pushover the next. In this case, we would attribute the outcomes to fluctuations in motivation (effort) or to random external factors such as wind speed, court condition, and so on (luck).

An experimental study (Frieze & Weiner, 1971) clearly illustrates these effects. Participants were given information about an individual's performance (success or failure) at a given task. They also received information about that person's past success rate on the same and similar tasks, as well as information about others' success rates on that task. These data influenced whether the participants viewed the actor's performance as consistent or inconsistent with his past performance and as similar to or different from the performance of others on the same task. Participants then reported their judgments about the impact of internal factors (ability, effort) and external factors (task difficulty, luck) in causing the actor's performance outcome (success or failure) on the immediate task. The results showed that (1) success was more likely to be attributed to internal factors (ability, effort) than was failure; (2) performance similar to that of others was attributed to external factors (task difficulty), whereas performance different from that of others (such as success where others failed or failure where others succeeded) was attributed to internal factors (ability, effort); (3) performance consistent with one's own past record (such as success when one has succeeded in the past or failure when one has failed in the past) was attributed to stable factors (ability, task difficulty), whereas performance inconsistent with one's past was attributed to unstable factors (luck, effort).

Consequences of Attributions Attributions for performance are important because they influence both our emotional reactions to success and failure and our future expectations and aspirations. For instance, if we attribute a poor exam performance to lack of ability, we may despair of future success and give up studying; this is especially likely if we view ability as given and not controllable by us. Alternatively, if we attribute the poor exam performance to lack of effort, we may feel shame or guilt, but we are likely to study harder and expect improvement. If we attribute the poor exam performance to bad luck, we may experience feelings of surprise or bewilderment, but we are not likely to change our study habits, because the situation will not seem controllable; despite this lack of change, we might nevertheless expect improved grades in the future. Finally, if we attribute our poor performance to the difficulty of the exam, we may become angry, but we do not strive for improvement

© Glyn Kirk/Getty Images

In this portrait of failure, a sprinter experiences depression after narrowly losing a race. Although failing is never pleasant, its impact may vary. If we attribute our failure to lack of effort, we are likely to feel guilt or shame and to renew our efforts in the future. If we attribute it to lack of ability, we may despair and quit trying.

(McFarland & Ross, 1982; Valle & Frieze, 1976; Weiner, 1985, 1986).

BIAS AND ERROR IN ATTRIBUTION

According to the picture we have drawn so far, observers scrutinize their environment, gather information, form impressions, and interpret behavior in rational, if sometimes unconscious, ways. In actuality, however, observers often deviate from the logical methods described by attribution theory and fall prey to biases. These biases may lead observers to misinterpret events and to make erroneous judgments. In this section, we will consider several major biases and errors in attribution.

Overattribution to Dispositions

At the time of the Cuban missile crisis, the Cuban leader Fidel Castro was generally unpopular, even feared, in the United States. In an interesting study done shortly after the crisis, Jones and Harris (1967) asked participants to read an essay written by another student. Depending on the experimental condition, the essay either strongly supported the Cuban leader or strongly opposed him. Moreover, the participants received information about the conditions under which the student wrote the essay. They were told either that the essay was written by a student who was assigned by the instructor to take a pro-Castro or anti-Castro stand (no-choice condition) or that the essay was written by a student who was free to choose whichever position he or she wanted to present (choice condition). The participants' task was to infer the writer's true underlying attitude about Castro. In the conditions where the writer had free choice, participants inferred that the content of the essay reflected the writer's true attitude about Castro. That is, they saw the pro-Castro essay as indicating pro-Castro attitudes and the anti-Castro essay as indicating anti-Castro attitudes. In the conditions where the writer was assigned the topic and had no choice, participants still thought the content of the essay reflected the writer's true attitude about Castro, although they were less sure that this was so. Participants made these internal attributions even though it was possible the writer held an opinion directly opposite of that expressed in the essay. In effect, participants overestimated the importance of internal dispositions (attitudes about Castro) and underestimated the importance of situational forces (role obligations) in shaping the essay.

The tendency to overestimate the importance of personal (dispositional) factors and to underestimate situational influences as causes of behavior is so common that it is called the **fundamental attribution error** (Higgins & Bryant, 1982; Ross, 1977; Small & Peterson, 1981) and it has been documented in study after study over the years (for instance, Allison, Mackie, Muller, & Worth, 1993; Jones, 1979; Ross, 2001; Sabini, Siepmann, & Stein, 2001). This error results from a failure by the observer to fully apply the subtractive rule. This tendency was first identified by Heider (1944), who noted that most observers ignore or minimize the impact of role pressures and situational constraints on others and interpret behavior as caused by people's intentions, motives, or attitudes. This bias is especially dangerous when it causes us to overlook the advantages of power built into social roles. For instance, we may incorrectly attribute the successes of the powerful to their superior personal capabilities, or we may incorrectly attribute the failures of persons without power to their personal weaknesses.

FIGURE 5.2 THE FOCUS OF ATTENTION BIAS

This diagram depicts the seating arrangement for speakers and observers in a study investigating the effect of the focus of visual attention on observers' attributions. Arrows indicate visual focus of attention. Following a conversation between both speakers, observers attributed more influence to the speaker they faced than to the other speaker. Observers on the side, however, attributed equal influence to both speakers. This illustrates our tendency to attribute more causal impact to the object of our attention.

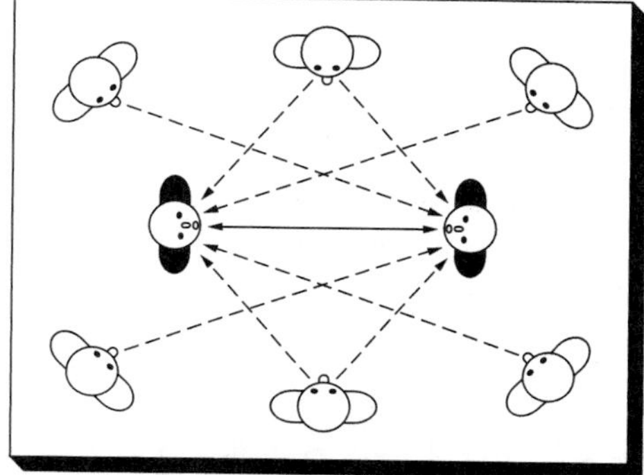

Source: Adapted from "The Focus of Attention Bias" (1978) by S. E. Taylor and S. T. Fiske. Reprinted by permission of the author.

Focus of Attention Bias

A closely related error is the tendency is to overestimate the causal impact of whomever or whatever we focus our attention on; this is called the **focus of attention bias.** A striking demonstration of this bias appears in a study by Taylor and Fiske (1978). The study involved six participants who observed a conversation between two persons (Speaker 1 and Speaker 2). Although all six participants heard the same dialogue, they differed in the focus of their visual attention. Two observers sat behind Speaker 1, facing Speaker 2; two sat behind Speaker 2, facing Speaker 1; and two sat on the sides, equally focused on the two speakers (see Figure 5.2). Measures taken after the conversation showed that observers thought the speaker they faced not only had more influence on the tone and content of the conversation but also had a greater causal impact on the other speaker's behavior. Observers who sat on the sides and were able to focus equally on both speakers attributed equal influence to them.

We perceive the stimuli that are most salient in the environment—those that attract our attention—as most causally influential. Thus, we attribute most causal influence to people who are noisy, colorful, vivid, or in motion. We credit the person who talks the most with exercising the most influence; we blame the person who runs past us when we hear a rock shatter a window. Although salient stimuli may be causally important in some cases, we overestimate their importance (Krull & Dill, 1996; McArthur & Post, 1977).

The focus of attention bias provides an explanation for the fundamental attribution error. The person behaving is the active entity in the environment; therefore, that person is likely to capture our attention. In fact, many of the contextual influences on the actor (for example, things that happened earlier in the day, or ongoing pressure from one's boss or family) may be completely invisible to the observer (Gilbert & Malone, 1995). Because we direct our attention more to people who act than to the context, we attribute more causal importance to people than to their situations.

Actor-Observer Difference

Actors and observers make different attributions for behavior. Observers tend to attribute actors' behavior to the actors' internal characteristics, whereas actors see their own behavior as due more to characteristics of the external situation (Jones & Nisbett, 1972; Watson, 1982). This tendency is known as the **actor-observer difference.** Thus, although other customers in a market may attribute the mix of items in your grocery cart (beer, vegetables, candy bars) to your personal characteristics (hard drinking, vegetarian, chocolate addict), you will probably attribute it to the requirements of your situation (preparing for a party) or the qualities of the items (nutritional value or special treat).

In one demonstration of the actor-observer difference (Nisbett, Caputo, Legant, & Maracek, 1973), male students wrote descriptions explaining why they liked their girlfriends and why they chose their majors. Then, as observers, they explained why their best friend liked his girlfriend and chose his major. When explaining their own actions, the students emphasized external characteristics like the attractive qualities of their girlfriends and the interesting aspects of their majors. However, when explaining their friends' behavior, they downplayed external characteristics and emphasized their friends' internal dispositions (preferences and personalities).

Two explanations for the actor-observer difference in attribution are that actors and observers have different visual perspectives and different access to information.

Visual Perspectives The actor's natural visual perspective is to look at the situation, whereas the observer's natural perspective is to look at the actor. Thus, the actor-observer difference reflects a difference in the focus of attention. Both the actor and the observer attribute more causal influence to what they focus on.

Storms (1973) reasoned that if the actor-observer difference in attributions was due simply to a difference in perspective, it might be possible to reverse the actor-observer difference by making the actor see the behavior from the observer's viewpoint and the observer see the same behavior from the actor's viewpoint. To give each the other's point of view, Storms videotaped a conversation between two people, using two separate cameras. One camera recorded the interaction from the visual perspective of the actor, the other from the perspective of the observer. Storms then showed actors the videotape made from the observer's perspective, and he showed observers the videotape made from the actor's perspective. As predicted, reversing the visual perspectives reversed the actor-observer difference in attribution; finding ways of making individuals more self-aware can therefore reduce the actor-observer bias (Fejfar & Hoyle, 2000).

Information A second explanation for the actor-observer difference is that actors have information about their own past behavior and the context relevant to their behavior that observers lack (Johnson & Boyd, 1995). Thus, for example, observers may assume that certain behaviors are typical of an actor when in fact they are not. This would cause observers to make incorrect dispositional attributions. An observer who sees a clerk return an overpayment to a customer may assume the clerk always behaves this way—resulting in a dispositional attribution of honesty. However, if the clerk knows he has often cheated customers in the past, he would probably not interpret his current

BOX 5.3 Research Update: Detecting Deception and the Fundamental Attribution Error

Social psychological research has demonstrated that, in general, people are very poor lie detectors. Unless they have received special training about what to observe in a potential liar, observers can rarely detect deception any better than random chance. Many reasons we are such poor lie detectors have been offered and demonstrated, but one recent study suggests that the Fundamental Attribution Error may also be contributing to our poor performance.

Researchers induced confederates to lie about their opinion on a controversial social issue of importance to them (such as the death penalty or a ban on smoking in public). If they were able to trick the person interviewing them, they received a bonus payment, thereby increasing their motivation to attempt to lie effectively. Videotapes of these interviews, along with interviews in which the interviewees did not lie, were then shown to other subjects who attempted to determine if the videotaped subject was lying or not.

After viewing the videotapes, the subjects were ask to rate the person on the tape: How trustworthy was he, and was he telling the truth? As the researchers expected, the subject demonstrated a bias toward attributing trustworthiness (a dispositional state) to the subjects, even though they knew that half of them were lying and that they had been randomly assigned to lie. In addition, however, those who were most biased toward the dispositional attribution were also most likely to think that the subjects were telling the truth. Therefore, they did an especially poor job of identifying deception.

The researchers suggest that this dynamic exists because most lies that we encounter in our daily lives are small ones, and therefore, we are better off assuming that people are telling the truth and are trustworthy. If we are wrong, it is of no major significance, and coupled with the fact that detecting lies is a difficult, costly endeavor, it is a more efficient strategy to assume truthfulness and interact with people under that assumption.

Adapted from O'Sullivan, 2003.

behavior as evidence of his honest nature. Consistent with this, research shows that observers who have a low level of acquaintance with the actor tend to form more dispositional attributions and fewer situational attributions than those who have a high level of acquaintance with the actor (Prager & Cutler, 1990).

Even when observers have some information about an actor's past behavior, they often do not know how changes in context influence the actor's behavior. This is because observers usually see an actor only in limited contexts. Suppose that students observe a professor delivering witty, entertaining lectures in class week after week. The professor knows that in other social situations she is shy and withdrawn, but the students do not have an opportunity to see this. As a result, the observers (students) may infer dispositions from apparently consistent behavior that the actor (the professor) knows to be inconsistent across a wider range of contexts.

Motivational Biases

Up to this point, we have considered attribution biases based on cognitive factors. That is, we have traced biases to the types of information that observers have available, acquire, and process. Motivational factors—a person's needs, interests, and goals—are another source of bias in attributions. When events affect a person's self-interests, biased attribution is likely. Specific motives that influence attribution include the desire to defend deep-seated beliefs, to enhance one's self-esteem, to increase one's sense of control over the environment, and to strengthen the favorable impression of oneself that others have.

The desire to defend cherished beliefs and stereotypes may lead observers to engage in biased attribution. Observers may interpret actions that correspond with their stereotypes as caused by the actor's personal dispositions. For instance, they may attribute a female executive's outburst of tears during a crisis to her emotional instability because that corresponds to their stereotype about women. At the same time, people attribute actions that contradict stereotypes to situational causes. If the female executive manages the crisis smoothly, the same people may credit this to the effectiveness of her male assistant. When observers selectively attribute behaviors that contradict stereotypes to situational influences, these behaviors reveal nothing new about the persons who perform them. As a result, the stereotypes persist even in the face of contradictory evidence (Hamilton, 1979).

Motivational biases may also influence attributions for success and failure. People tend to take credit for acts that yield positive outcomes, whereas they deflect blame for bad outcomes and attribute them to external causes (Bradley, 1978; Campbell & Sedikides, 1999; Ross & Fletcher, 1985). This phenomenon, referred to as the **self-serving bias,** is illustrated clearly in a study in which college students were asked to explain the grades they received on three examinations (Bernstein, Stephan, & Davis, 1979). Students who received As and Bs attributed their grades much more to their own effort and ability than to good luck or easy tests. However, students who received Cs, Ds, and Fs attributed their grades largely to bad luck and the difficulty of the tests. Other studies show similar effects (Reifenberg, 1986).

The self-serving bias also appears when athletes report the results of competitions (Lau & Russell, 1980; Ross & Lumsden, 1982). Whereas members of winning teams take credit for winning ("We won"), members of losing teams are more likely to attribute the outcome to an external cause—their opponent ("They won," not "We lost").

Various motives may contribute to this self-serving bias in attributions of performance. For instance, attributing success to personal qualities and failure to external factors enables people to enhance or protect their self-esteem. Regardless of the outcome, they can continue to see themselves as competent and worthy. Moreover, by avoiding the attribution of failure to personal qualities, they maximize their sense of control. This in turn supports the belief that they can master challenges successfully if they choose to apply themselves because they possess the necessary ability. Finally, biased attributions enable people to present a favorable public image and to make a good impression on others.

CULTURAL BASIS OF ATTRIBUTIONS

Because our attributions are linked to schemas, and schemas often contain and reflect cultural elements, culture can play an important role in the attribution process and in producing attribution biases and errors. If a culture bases person and group schemas on race, the members of that culture will be more likely to use those race-based schemas as they interpret behavior—in other words, they are more likely to think that race is the cause of the behavior instead of environmental influences (Hewstone & Jaspars, 1984). Thus, Whites may ascribe Black unemployment to a characteristic (perhaps laziness) their schemas have connected to being Black, rather than to the environment or to economic conditions such as the lack of available jobs in central cities.

Perhaps the attribution issue that has been most studied across cultures is the use of the fundamental attribution error. For many years, the fundamental attribution error was thought to be fundamental in the sense that all people, regardless of culture, were thought to exhibit the tendency toward dispositional attribution. More recently, however, a number of social psychologists have been investigating whether attribution biases and errors are consistent across cultures. In short, it has become plain that attribution errors do vary across cultures, and the main difference has to do with how individualist or collectivist a culture is (Norenzayan & Nisbett, 2000; Triandis, 1995). Individualist cultures emphasize the individual and value individual achievement; collectivist cultures emphasize the welfare of the family, ethnic group, and perhaps work group over the interests of individuals. This difference in emphasis turns out to have a substantial impact on the orientation toward dispositional versus situational attributions for behavior. Individualist cultures focus on the individual—thus, its members are predisposed to use individualist or dispositional attribution. In collectivist cultures, the groups around individuals focus some attention on the context—thus, members of these cultures are more likely to include situational elements in their attributions.

In one study, researchers compared attributions made by students from an individualist society (the United States) with those made by students from a collectivist society (Saudi Arabia). Participants in the study were 163 students recruited from U.S. universities and 162 students from a university in Saudi Arabia (Al-Zahrani & Kaplowitz, 1993). Each student was presented with vignettes describing eight situations—four involving achievement and four involving morality. Students were asked to assign responsibility for the outcome to each of several factors. Consistent with the hypothesis, the results showed that across the eight situations, U.S. students assigned greater responsibility to internal dispositional factors than did Saudi students.

Is this difference really due to people in collectivist cultures appreciating the context more, or does it result from people in collectivist cultures ignoring dispositional information? Either possibility could explain the findings in these cross-cultural studies. Recent studies have demonstrated that the use of dispositional information is quite consistent across cultures and that the difference in the fundamental attribution error comes mainly from collectivists' greater attention to situational cues (Choi, Nisbett, & Norenzayan, 1999; Krull et al., 1999). Researchers approaching this topic have used variants of Jones and Harris's (1967) Castro study described earlier. In these experiments, participants were asked to judge essays like the ones in the Castro study. Deprived of any situational information, people from both collectivist and individualist cultures made the fundamental attribution error. The researchers then made the situational elements of the study more obvious by having the participants write counterattitudinal essays before judging the other essays. This task made the arbitrary nature of the essays more apparent to the participants. Participants from the collectivist cultures used this information to better understand the situational constraints and incorporated this information into their attributions, producing less dispositional attributions. Participants from individualist cultures, however, did not correct their views—they still made dispositional attributions even after writing their own counterattitudinal essays (Choi & Nisbett, 1998; Choi et al., 1999).

SUMMARY

Social perception is the process of using information to construct understandings of the social world and form impressions of people.

Schemas A schema is a well-organized structure of cognitions about some social entity. (1) There are several distinct types of schemas: person schemas, self-schemas, group schemas (stereotypes), role schemas, and event schemas (scripts). (2) Schemas organize information in memory and therefore affect what we remember and what we forget. Moreover, they guide our inferences and judgments about people and objects. (3) Schemas are transmissible cultural elements. They can be taught by one person and learned by another and, therefore, can be shared among members of a social group.

Person Schemas and Group Stereotypes (1) One important type of person schema is an implicit personality theory—a set of assumptions about which personality traits go together with which other traits. These schemas enable us to make inferences about other people's traits. We can depict an implicit personality theory as a mental map. (2) A stereotype is a fixed set of characteristics attributed to all members of a given group. American culture includes stereotypes for ethnic, racial, gender, and occupational groups. Because stereotypes are overgeneralizations, they cause errors in inference; this is especially true in complex situations.

Impression Formation (1) Research on trait centrality using the "warm/cold" variable illustrates how variations in a single trait can produce a large difference in the impression formed by observers of a stimulus person. (2) Perceivers try to combine the bits of information they receive to create a unified impression of others. They average these bits of information after weighting certain types of information more heavily than other types. Information received early usually has a larger impact on impressions than information received later; this is called the primacy effect. (3) Impressions become self-fulfilling prophecies when we behave toward others according to our impressions and evoke corresponding reactions from them. (4) Impressions are informed by schemas that are selected through mental shortcuts called heuristics.

Attribution Theory Through attribution, people infer an action's causes from its effects. (1) One important issue in attribution is locus of causality—dispositional (internal) versus situational (external) attributions. Observers follow the subtractive rule when making attributions to dispositions or situations. (2) To attribute specific dispositions to an actor, observers observe an act and its effects and then try to infer the actor's intention with respect to that act. Observers then attribute the disposition that corresponds best with the actor's inferred intention. (3) Observers who have information about an actor's behaviors in many situations make attributions to the actor, object, or context. The attribution made depends on which of these causes covaries with the behavior in question. Observers assess covariation by considering consensus, consistency, and distinctiveness information. (4) Observers attribute success or failure to four basic causes—ability, effort, task difficulty, and luck. They attribute consistent performances to stable rather than to unstable causes, and they attribute average performances to external rather than internal causes.

Bias and Error in Attribution (1) Observers frequently overestimate personal dispositions as causes of behavior and underestimate situational pressures; this bias is called the fundamental attribution error. (2) Observers also overestimate the causal impact of whatever their attention is focused on. (3) Actors and observers have different attribution tendencies. Actors attribute their own behavior to external forces in the situation, whereas observers attribute the same behavior to the actor's personal dispositions. (4) Motivations—needs, interests, and goals—lead people to make self-serving, biased attributions. People defend deep-seated beliefs by attributing behavior that contradicts their beliefs to situational influences. People defend their self-esteem and sense of control by attributing their failures to external causes and taking personal credit for their successes.

Cultural Basis of Attributions Cultural factors can affect attribution. In comparison to observers in an individualist society, perceivers in a collectivist society are less likely to commit the fundamental attribution error. Although people in both individualist and collectivist societies use dispositional explanations, those in collectivist cultures tend to balance these dispositional factors with situational explanations, whereas those in individualist cultures do not.

LIST OF KEY TERMS AND CONCEPTS

actor-observer difference (p. 136)
attribution (p. 109)
categorization (p. 110)
dispositional attribution (p. 128)
focus of attention bias (p. 135)
fundamental attribution error (p. 135)
halo effect (p. 115)
heuristic (p. 126)
primacy effect (p. 124)
principle of covariation (p. 131)
prototype (p. 110)
recency effect (p. 125)
schema (p. 111)
self-serving bias (p. 138)
situational attribution (p. 128)
social perception (p. 109)
stereotype (p. 111)
stereotype threat (p. 117)
trait centrality (p. 122)

6

ATTITUDES

Introduction

The Nature of Attitudes

The Components of an Attitude

Attitude Formation

The Functions of Attitudes

Attitude Organization

Attitude Structure

Cognitive Consistency

Balance Theory

Theory of Cognitive Dissonance

Is Consistency Inevitable?

The Relationship Between Attitudes and Behavior

Do Attitudes Predict Behavior?

Activation of the Attitude

Characteristics of the Attitude

Attitude-Behavior Correspondence

Situational Constraints

The Reasoned Action Model

Formal Model

Assessment of the Model

Summary

List of Key Terms and Concepts

INTRODUCTION

- "Aerosmith's music is great!"
- "My human sexuality class is really boring."
- "I like my job."
- "Government spending causes inflation."
- "The drinking age law is unfair."
- "Guns don't kill people; people kill people."

What do these statements have in common? Each represents an **attitude**—a predisposition to respond to a particular object in a generally favorable or unfavorable way (Ajzen, 1982). A person's attitudes influence the way in which he or she perceives and responds to the world (Allport, 1935; Thomas & Znaniecki, 1918). For example, attitudes influence attention—the person who likes Aerosmith's music is more likely to notice news stories about the band's activities. Attitudes also influence behavior—the person who opposes the drinking age law is more likely to violate it.

Because attitudes are an important influence on people, they occupy a central place in social psychology. But what exactly is an attitude? What do we mean by a "predisposition to respond"? Furthermore, a particular attitude does not exist in isolation. The person who believes that government spending causes inflation has a whole set of beliefs about the role of government in the economy, and this attitude about spending is related to those other beliefs. And if attitudes influence behavior, perhaps we can change behavior by changing attitudes. How, then, do attitudes change? Politicians, lobbyists, auto manufacturers, and brewers spend billions of dollars every year trying to create favorable attitudes. Even if they succeed, do these attitudes affect our behavior?

In this chapter, we consider three main questions:

1. What is an attitude? Where do attitudes come from, and how are they formed?
2. How are attitudes linked to other attitudes? How does this organization affect attitude change?
3. What is the relationship between attitudes and behavior?

THE NATURE OF ATTITUDES

An attitude exists in a person's mind; it is a mental state. Every attitude is about something, the "object" of the attitude. In this section, we consider the components of an attitude, the sources of attitudes, and their functions.

The Components of an Attitude

Consider the following statement: "My human sexuality class is really boring." This attitude has three components: (1) beliefs or cognitions, (2) an evaluation, and (3) a behavioral predisposition.

Cognition An attitude includes an object label, rules for applying the label, and a set of cognitions or knowledge structures associated with that label (Pratkanis & Greenwald, 1989).The person who doesn't like his or her human sexuality class perceives it as involving certain content, taught by a particular person. Often we cannot prove whether particular beliefs are true or false. For example, economists and government officials disagree on whether government spending causes inflation, with both sides equally convinced they are right.

Evaluation An attitude also has an evaluative or affective component. "It's boring" indicates that the course arouses a mildly unpleasant emotion in the speaker. Stronger negative emotions include dislike, hatred, or even loathing: "I can't stand punk rock." Of course, the evaluation may be positive: "I like Aerosmith's music," or "This food is terrific!" The evaluative component has both a direction (positive or negative) and an intensity (ranging from very weak to very strong). The evaluation component distinguishes an attitude from other types of cognitive elements.

Behavioral Predisposition An attitude involves a predisposition to respond or a behavioral tendency toward the object. "It's boring" implies a tendency to avoid the class. "I like my job" suggests an intention to go to work. People having a specific attitude are inclined to behave in certain ways that are consistent with that attitude.

Relationships Between the Components Cognitive, evaluative, and behavioral components all have the same object, so we would expect them to form a single, relatively consistent whole. However, these three components are distinct; if they were identical, we would not need to distinguish among them. Thus, we should be able to measure each component, and we should be able to find a relationship between them. A survey of women's attitudes toward contraceptives, for example, found that beliefs, feelings, and actions are both distinct and somewhat related (Kothandapani, 1971).

The degree of consistency between components is related to other characteristics of the attitude. Greater consistency between the cognitive and affective components is associated with greater attitude stability and resistance to persuasion (Chaiken & Yates, 1985). Greater consistency is also associated with a stronger relationship between attitude and behavior, as we will discuss later in this chapter.

Attitude Formation

Where do attitudes come from? How are they formed? The answer lies in the processes of social learning or socialization (discussed in Chapter 3). Attitudes may be formed through reinforcement (instrumental conditioning), through associations of stimuli and responses (classical conditioning), or by observing others (observational learning).

We can acquire an attitude toward our classes and jobs through instrumental conditioning—that is, learning based on direct experience with the object. If you experience rewards related to some object, your attitude will be favorable. Thus, if your work provides you with good pay, a sense of accomplishment, and compliments from your coworkers, your attitude toward it will be quite positive. Conversely, if you associate negative emotions or unpleasant outcomes with some object, you will dislike it. For example, repeated exposure to bland, overcooked food leads many students to have a very negative attitude toward cafeteria food.

Only a small portion of our attitudes are based on direct contact with the object, however. We have attitudes about many political figures we have never met. We have attitudes toward members of certain ethnic or religious groups, although we have never been face-to-face with a member of those groups. Attitudes of this type are learned through our interactions with third parties. For one, we learn attitudes from our parents as part of the socialization process. Research shows that children's attitudes toward male-female relations (gender roles), divorce, and politics frequently are similar to those held by their parents (Glass, Bengston, & Dunham, 1986; Thornton, 1984; Sinclair, Dunn, & Lowery, 2005). This influence also involves instrumental learning; parents typically reward their children for adopting the same or similar attitudes.

Friends are another important source of our attitudes. The attitude that the drinking age law is unfair, for example, may be learned through interaction with peers. A classic study of Bennington College women by Newcomb (1943) demonstrated the impact of peers on the political attitudes of college students. Although most of these women grew up in wealthy, politically conservative families, the faculty at Bennington had very liberal political attitudes. The study demonstrated that first-year students who maintained close ties with their families and did not become involved in campus activities remained conservative. Women who became active in the college community and who interacted more frequently with other students gradually became more liberal. Presumably, the students at Bennington rewarded the liberal attitudes of their peers.

We acquire attitudes and prejudices toward a particular group through classical conditioning, in which a stimulus gradually elicits a response through repeated association with other stimuli. Children learn at an early age that "lazy," "dirty," "stupid," and many other characteristics are undesirable. Children themselves are often punished for being dirty or hear adults say, "Don't be stupid!" If they hear their parents (or others) refer to members of a particular group as lazy or stupid, children increasingly associate the group name with the negative reactions initially elicited by these terms. Several experiments have shown that classical conditioning can produce negative attitudes toward groups (Lohr & Staats, 1973; Staats & Staats, 1958).

Another source of attitudes is the media, especially television and films. Here, the mechanism may be observational learning. The media provide interpretive packages or frames about an object that may influence the attitudes of viewers and readers. By portraying events and actors in certain ways, TV news, news magazines, and newspapers can produce cognitive images of a racial group as being volatile, dangerous, or unreasonable that in turn produce negative attitudes (Myers & Caniglia, 2004).

The Functions of Attitudes

We acquire attitudes through learning. But why do we retain them, sometimes for months, years, or even a lifetime? One answer is that they serve at least some important function for us (Katz, 1960; Pratkanis & Greenwald, 1989).

The first is the heuristic or instrumental function. We develop favorable attitudes toward objects that aid or reward us and unfavorable attitudes toward objects that thwart or punish us. Once they are developed, attitudes provide a simple and efficient means of evaluating objects. Businesspeople learn that Republican politicians frequently propose and vote for legislation that benefits business. Learning that a new candidate is a Republican, the businessperson immediately has a favorable attitude.

Second, attitudes serve a schematic or knowledge function—because the world is too complex for us to completely understand, we group people, objects, and events into categories or schemas and develop simplified (stereotyped) attitudes that allow us to treat individuals as members of a category. Our attitudes about that category (object) provide us with meaning—with a basis for making inferences about its members (Bodenhausen & Wyer, 1985). The belief that Blacks are untrustworthy leads some Whites to be guarded in their interaction with Blacks. Reacting to every member of the group in the same way is more efficient, even if less accurate and satisfying, than trying to learn about each person as an individual.

Stereotypes of groups are often associated with intense emotions. A strong like or dislike for members of a specific group is called a **prejudice**. Prejudice and stereotyping go together, with people using their stereotyped beliefs to justify prejudice toward members of the group. (Stereotypes are discussed in Chapter 5.) The emotional component of prejudice can lead to intergroup conflict (see Chapter 16).

Third, attitudes define the self and maintain self-worth. Some attitudes express the individual's basic values and reinforce his or her self-image. Many political conservatives in our society have negative attitudes toward abortion, racial integration, and equal rights for women. Thus, a person whose self-concept includes conservatism may adopt these attitudes because they express that self-image. Some attitudes symbolize a person's identification with or membership in particular groups or subcultures. The attitude "Guns don't kill people; people kill people" is widespread among members of the National Rifle Association. Holding this attitude may be both a prerequisite to acceptance by other group members and a symbol of loyalty to the group.

Finally, some attitudes protect the person from recognizing certain thoughts or feelings that threaten his or her self-image or adjustment. For instance, an individual (say, Tom) may have feelings that he cannot fully acknowledge or accept, such as hostility toward his father. If he recognized this hostility, he would feel very guilty, because such sentiments are contrary to his upbringing. So instead of acknowledging that he hates his father, Tom may direct anger and hatred toward members of a minority group or toward authority figures such as police officers or teachers. Research indicates that experiences that threaten a person's self-esteem, such as failing a test, lead to a more negative evaluation of other groups (Crocker, Thompson, McGraw, & Ingerman, 1987), particularly among people whose self-esteem was initially high.

ATTITUDE ORGANIZATION

Attitude Structure

Have you ever tried to change another person's attitude toward an object (such as a political candidate or a racial group) or a behavior (such as premarital sex)?

If you have, you probably discovered that the person had a counterargument for almost every argument you put forth. She or he probably had several reasons why her or his attitude was correct. This tendency flows from how attitudes are arranged in our minds. Attitudes are usually embedded in a cognitive structure, linked with a variety of other attitudes. We can often find out what other cognitive elements are related to a particular attitude by asking the person why he or she holds that attitude. Consider the following interview:

INTERVIEWER: Why do you think premarital sexual intercourse is bad?

BILL: Because sex outside of marriage is wrong; it is against the teachings of God in the Bible.

INTERVIEWER: Are there any other reasons?

BILL: Well, I think people who have sex before marriage are usually promiscuous, and they could spread AIDS.

INTERVIEWER: Any other reasons?

BILL: Um . . . yeah. They may get pregnant, and teenage pregnancy causes a lot of problems.

This exchange indicates Bill's reasons for his attitude. More than that, it illustrates the two basic dimensions of attitude organization: vertical and horizontal structure (Bem, 1970).

Vertical Structure Bill is opposed to premarital sex because it violates his religious beliefs. Specifically, it violates the biblical injunction against intercourse outside of marriage. Bill accepts this injunction because he views the Bible as a statement of God's teachings. Bill's attitude toward premarital intercourse ultimately rests on his belief in God. The unquestioning acceptance of the credibility of some authority, such as God, is termed a primitive belief (Bem, 1970).

Attitudes are organized hierarchically. Some attitudes (primitive beliefs) are more fundamental than others. The linkages between fundamental beliefs and minor beliefs in cognitive structure are termed vertical. Vertical linkages signify that a minor belief is derived from or dependent on a primitive belief. Such a structure is portrayed in the center of Figure 6.1.

A fundamental or primitive belief, such as a belief in God, is often the basis for a large number of specific or minor beliefs (Bem, 1970). For example, Bill probably is opposed to murder, adultery, and other sins mentioned in the Bible. Changing a primitive belief may result in widespread changes in the person's attitudes. If Bill meets members of the Unification Church, they may attempt to persuade him that the Reverend Moon is the only legitimate religious authority. If Bill is converted, the resulting change in his primitive beliefs will lead to changed attitudes toward many objects, including his family and friends.

Horizontal Structure When the interviewer asked Bill why he was opposed to sex before marriage, Bill also gave two other reasons. One was his belief that people who engage in premarital sexual intercourse are promiscuous and that promiscuity spreads AIDS. The other reason was his belief that premarital sex leads to teenage pregnancy and such undesirable consequences as birth defects. These belief structures are portrayed in the right-hand and left-hand columns of Figure 6.1. When an attitude is linked to more than one set of underlying beliefs—that is, when there are two or more different justifications for it—the linkages are termed horizontal.

An attitude with two or more horizontal linkages, or justifications, is more difficult to change than one based on a single primitive belief. Even if you show Bill statistical evidence that AIDS is not associated with premarital intercourse, his religious beliefs and his concern about teenage pregnancy make it unlikely his attitude will change.

One way to study attitude structure is to interview people and identify vertical and horizontal linkages. A different approach is to study response latency—how long it takes a person to reply to an attitude question (Judd, Drake, Downing, & Krosnick, 1991). What is your attitude toward extramarital sex? What is your attitude toward vegetarians? Chances are it took you longer to retrieve from memory your attitude toward vegetarians. Your attitudes about various types of sexual behavior were primed or activated by our discussion of premarital sex and should be associated with short latencies. The shorter

FIGURE 6.1 THE STRUCTURE OF ATTITUDES

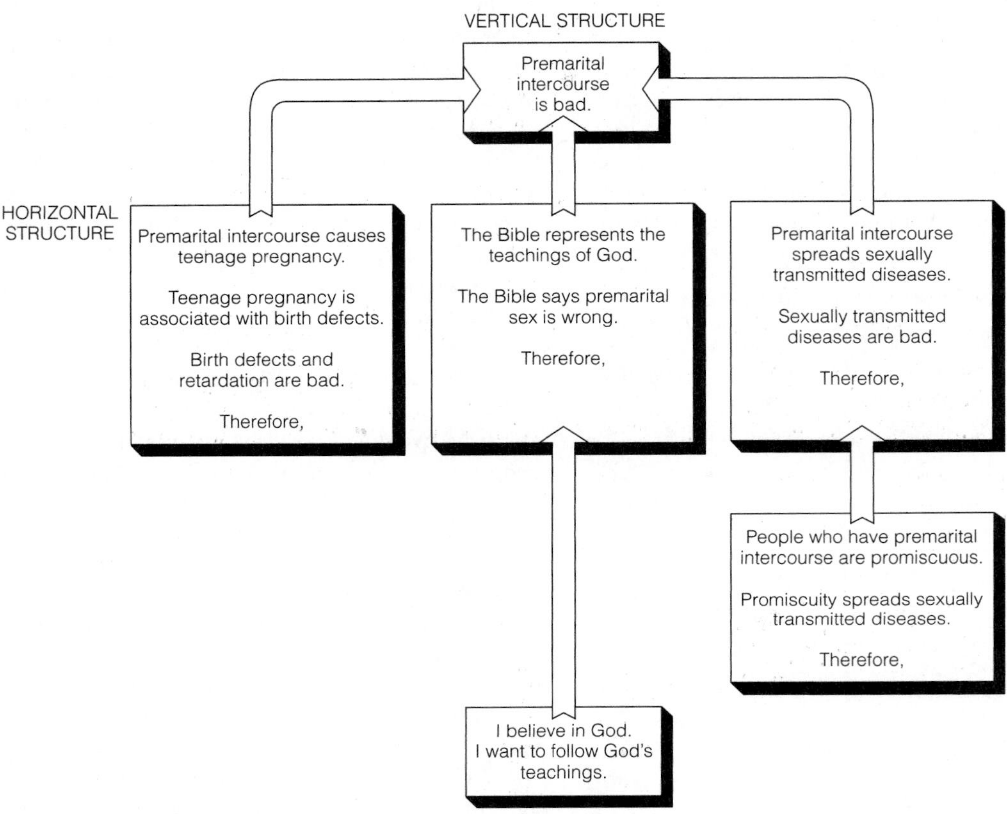

the latency, the closer two attitudes are in a person's attitude structure.

COGNITIVE CONSISTENCY

The elements of a cognitive structure are called **cognitions.** A cognition is an individual's perception of personal attitudes, beliefs, and behaviors. Bill perceives himself as someone who believes in God and follows God's teachings. These two cognitions go together; we are not surprised that Bill perceives both as applying to him. Many of his attitudes are consistent with what he perceives as God's teachings. For example, he has very negative attitudes toward adultery and murder. Given his attitude toward premarital sex, we would expect Bill to abstain from intercourse until he marries. If he does, then his behavior is consistent with his attitudes.

Consistency among a person's cognitions—that is, beliefs and attitudes—is widespread. If you have liberal political values, you probably favor medical assistance programs for people living in poverty. If you value equal rights for all persons, you probably support affirmative action plans. Also, you may try hard to behave in nonsexist ways when you interact with members of the opposite sex. The observation that most people's cognitions are consistent with one another implies that individuals are motivated to maintain that consistency. Several theories of attitude organization are based on this principle. In general, these cognitive consistency theories hypothesize that if an inconsistency develops between cognitive elements, people are motivated to restore harmony between those elements.

Balance Theory

One important consistency theory is **balance theory,** which was formulated by Heider (1958) and elaborated by Rosenberg and Abelson (1960). To see how balance theory works, consider the following statement: "I'm going to vote for Steve Smith; he's in favor of reducing taxes." Balance theory is concerned with cognitive systems like this one. This system contains three elements—the speaker; another person (candidate Steve Smith); and an impersonal object (taxes). According to balance theory, two types of relationships may exist between elements. Sentiment relations refer to sentiments or evaluations directed toward objects and people; a sentiment may be either positive (liking, endorsing) or negative (disliking, opposing). Unit relations refer to the association between elements. For example, a positive unit relation may result from ownership, a social relationship (such as friendship or marriage), or causality. A negative unit relation indicates dissociation, like that between ex-spouses or members of groups with opposing interests. A null unit relation exists when there is no association between elements.

Using these terms, let's analyze our example. We can depict this system as a triangle (see Figure 6.2). Balance theory is concerned with the elements and their interrelations from the speaker's viewpoint. In our first example (Figure 6.2A), the speaker favors reduced taxes, perceives Steve Smith as favoring reduced taxes, and intends to vote for Steve. This system is balanced. By definition, a balanced state is one in which all three sentiment relations are positive or in which one is positive and the other two are negative. Consider another example (Figure 6.2B). Suppose you favor legalizing the possession of marijuana, and candidate Mary Smith wants mandatory prison sentences for its possession. Your cognitions would be balanced if you disliked Mary Smith.

Imbalance and Change According to balance theory, an imbalanced state is one in which two of the relationships between elements are positive and one is negative or in which all three are negative. Consider Judy and Mike, who are seniors in college. They have been going together for 3 years and are in love. Mike is thinking about going to law school. Judy doesn't want him to stay in school after he gets his bachelor's degree. She doesn't want to have to work full-time while he goes to school for 3 more years. Figure 6.2C illustrates the situation from Mike's viewpoint. Mike feels positively toward Judy and toward law school, but Judy is not positive toward law school. Thus there is an imbalance.

In general, an imbalanced situation like this is unpleasant. When people are presented hypothetical triads like those shown in Figure 6.2C and asked to rate each triad, imbalanced triads are rated less pleasant than balanced ones (Price, Harburg, & Newcomb, 1966). Balance theory assumes that people will try to restore balance among their attitudes.

There are three basic ways to do this. First, Mike may change his attitudes so the sign of one of the relations is reversed (Tyler & Sears, 1977). For instance, Mike may decide he does not want to attend law school (Figure 6.2D). Alternatively, Mike may decide he does not love Judy, or he may persuade Judy it is a good idea for him to go to law school. Each of these involves changing one relationship so the system of beliefs contains either zero or two negative relationships.

Second, Mike can restore balance by changing a positive or negative relation to a null relation (Steiner & Rogers, 1963). Mike may decide that Judy doesn't know anything about law school and her attitude toward it is irrelevant. Third, Mike can restore balance by differentiating the attributes of the other person or object (Stroebe, Thompson, Insko, & Reisman, 1970). For instance, Mike may distinguish between major law schools, which require all the time and energy of their students, and less prestigious ones, which require less work. Judy is correct in her belief that they would have to postpone marriage if he went to Yale Law School. However, Mike believes he can go to a local school part-time and also work.

Which technique will a person use to remove the imbalance? Balance is usually restored in whichever

FIGURE 6.2 BALANCED COGNITIVE SYSTEMS AND RESOLUTION OF IMBALANCED SYSTEMS

When the relationships among all three cognitive elements are positive (A), or when one relationship is positive and the other two are negative (B), the cognitive system is balanced. When two relationships are positive and one negative, the cognitive system is imbalanced. In (C), Judy's negative attitude toward law school creates an unpleasant psychological state for Mike. He can resolve the imbalance by deciding he does not want to go to law school (D), by deciding he does not love Judy, or by persuading Judy to like law school.

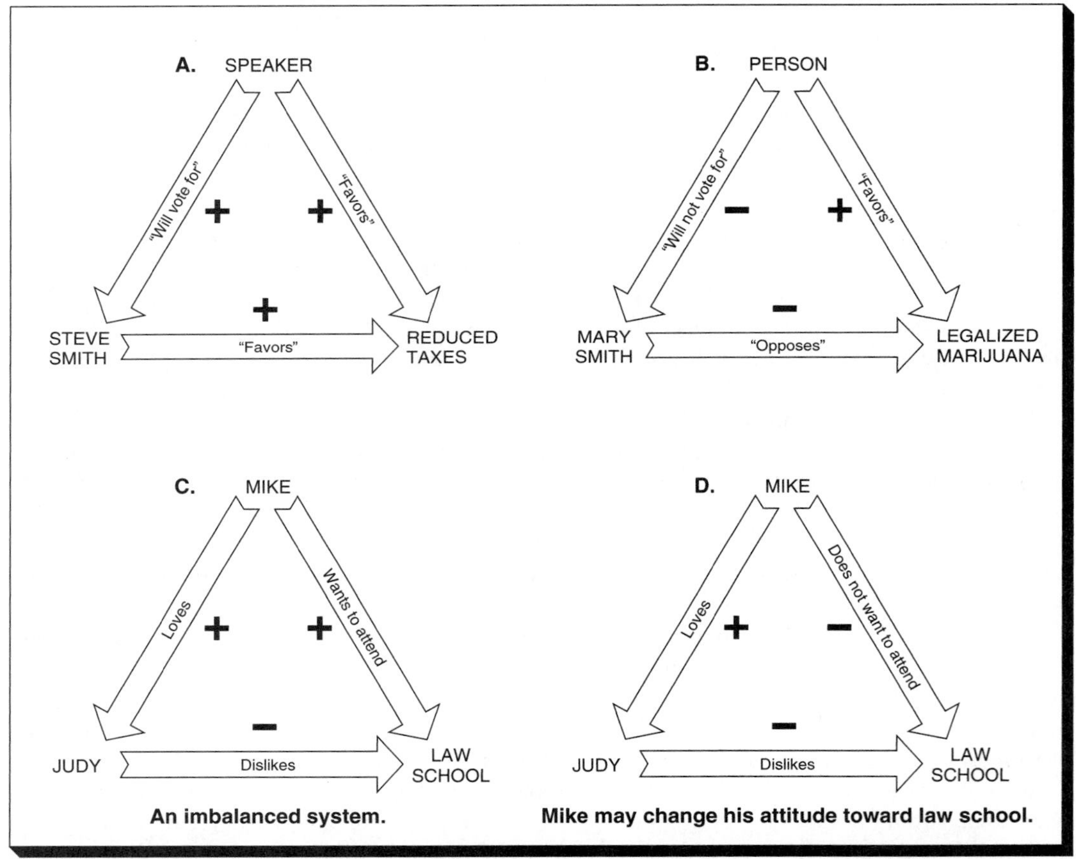

way is easiest (Rosenberg & Abelson, 1960). If one relationship is weaker than the other two, the easiest mode of restoring balance is to change the weaker relationship (Feather, 1967). Because Mike and Judy have been seeing each other for 3 years, it would be very difficult for Mike to change his sentiments toward Judy. It would be easier for him to change his attitude toward law school than to get a new fiancée. However, Mike would prefer to maintain their relationship and go to law school. Therefore, he may attempt to change Judy's attitude, perhaps by differentiating the object (law schools). If this influence attempt fails, Mike will probably change his own attitude toward law school.

Theory of Cognitive Dissonance

Another major consistency theory is the **theory of cognitive dissonance.** Whereas balance theory deals with the relationship among three cognitions, dissonance theory deals with consistency between two or more elements (behaviors and attitudes). There are two situations in which dissonance commonly occurs: (1) after a decision, or (2) when one acts in a way that is inconsistent with his or her beliefs.

Postdecision Dissonance Susan will begin her junior year in college next week. She needs to work part-time to pay for school. After 2 weeks of searching for work, she receives two offers. One is a part-time job doing library research for a faculty member she likes, and it pays $7 per hour with flexible working hours. The other is a job in a restaurant as a cashier that pays $10 per hour but has working hours from 5 p.m. to 9 p.m., Thursdays, Fridays, and Saturdays. Susan has a hard time choosing between these jobs. Both are located near campus, and she thinks she would like either one. Whereas the research job offers flexible hours and easier work, the cashier's job pays more and offers her the opportunity to meet interesting people. In the end, Susan chooses the cashier's job, but she is experiencing dissonance.

Dissonance theory (Festinger, 1957) assumes there are three possible relationships between any two cognitions. Cognitions are consistent or consonant if one naturally or logically follows from the other. They are dissonant when one implies the opposite of the other. The logic involved is psycho logic (Rosenberg & Abelson, 1960)—that is, logic as it appears to the individual, not logic in a formal sense. Two cognitive elements also may be irrelevant; one may have nothing to do with the other. In Susan's case, the decision to take the cashier's position is consonant with its convenient location, the higher pay, and the opportunities to meet people, but it is dissonant with the fact that she will be responsible for hundreds of dollars and has to work weekend nights (see Figure 6.3).

Having made the choice, Susan is experiencing **cognitive dissonance,** a state of psychological tension induced by dissonant relationships between cognitive elements. Although dissonance is a cognitive concept, it has physiological correlates. One study used the galvanic skin response (GSR)—a widely used measure of physiological arousal—as a measure of the tension induced by dissonance (Elkin & Leippe, 1986). As expected, participants who were placed in a dissonance-arousing situation showed an elevated GSR.

Some decisions produce a large amount of cognitive dissonance, others very little. The magnitude of dissonance experienced depends in part on the proportion of elements that are dissonant with a person's decision. In Susan's case, there are three consonant and only two dissonant cognitions, so she will experience moderate dissonance. The magnitude is also influenced by the importance of the elements. She will experience less dissonance if it is not important that she has to work every Friday and Saturday, but more dissonance if an active social life on weekends is important to her.

FIGURE 6.3 POSTDECISIONAL DISSONANCE

Whenever we make a decision, there are some cognitions—attitudes, beliefs, knowledge—that are consonant with that decision, and other cognitions that are dissonant with it. Dissonant cognitions create an unpleasant psychological state that we are motivated to reduce or eliminate. In this example, Susan has chosen a job and is experiencing dissonance. Although three cognitions are consistent with her decision, two other dissonant cognitions are creating psychological tension.

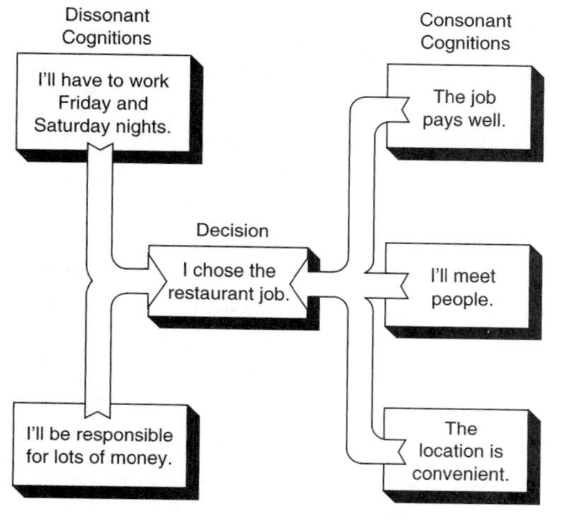

Dissonance is an uncomfortable state. To reduce dissonance, the theory predicts, Susan will change her attitudes. She can either change the cognitive elements themselves or change the importance associated with the elements. It is hard to change cognitions. She chose the restaurant job, and she made a commitment to work weekend nights and to be responsible for large sums of money. Alternatively, Susan can change the relative importance of her cognitions. She can emphasize the importance of one or more of the consonant cognitions and de-emphasize one or more of the dissonant cognitions. Although she has to work to earn money, she can emphasize the fact that the cashier's job pays well. Although she would prefer to be able to go out on weekends, this is less important because the cashier's job will still allow her to meet people.

In a laboratory study of postdecision dissonance, undergraduate women were given a choice between two products, such as a toaster and a coffeemaker. Participants rated the attractiveness of each item before and after their choice. Researchers predicted that to reduce dissonance, the women would minimize the importance of cognitions dissonant with the decision. That is, after the choice was made, the attractiveness of the item chosen would increase and the attractiveness of the item not chosen would decrease. The results verified these hypotheses (Brehm, 1956).

Counterattitudinal Behavior A second circumstance that produces dissonance occurs when a person behaves in a way that is inconsistent with his or her attitudes. Such situations may involve forced compliance, that is, pressures on a person to comply with a request to engage in counterattitudinal behavior (Joule and Azdia, 2003).

Imagine you have volunteered to serve in a psychology experiment. You arrive at the lab and are told you are participating in a study of performance. You are given a pegboard and told to turn each peg exactly one quarter turn. After you have turned the last peg, you are told to start over, to turn each peg another one quarter turn. Later you are told to remove each peg from the pegboard and then to put each peg back. After an hour of such activity, the experimenter indicates that you are finished. The experimenter says, "We are comparing the performance of participants who are briefed in advance with that of others who are not briefed. You did not receive a briefing. The next participant is supposed to be briefed, but my assistant who usually does this couldn't come to work today." He then asks you to help out by telling a waiting participant that the tasks you have just completed were fun and exciting. For your help, he offers you either $1 or $20.

In effect, you are being asked to lie—to say that the boring and monotonous tasks you performed are enjoyable. If you actually tell the next participant the tasks are fun, you may experience cognitive dissonance afterward. Your behavior is inconsistent with your cognition that the tasks are boring. Moreover, lying to the next participant is dissonant with your beliefs about yourself—that you are moral and honest. To reduce dissonance, you can change one of the cognitions. Which one will you change? You cannot change your awareness that you told the next participant the task is fun. The only cognition open to change is your attitude toward the task, which can change in the direction of greater liking for the task.

The theory of cognitive dissonance predicts (1) that you will change your attitudes toward the tasks (like them better), and (2) that the amount of change will depend on the incentive you were paid to tell the lie. Specifically, the theory predicts that greater attitude change will occur when the incentive to tell the lie is low ($1) rather than high ($20), because you will experience greater dissonance under low incentive than you would under high incentive.

These predictions were tested in a classic experiment by Festinger and Carlsmith (1959). In this study, most of the participants agreed to brief the next participant. They told him or her that the tasks were interesting and that they had fun doing them. A secretary then asked each participant to rate the experiment and the tasks. These ratings provided the measures of the dependent variable. As expected, control participants who did not brief anyone and were not offered money rated the tasks as very unenjoyable and did not want to participate in the experiment again.

What about the participants who were paid money to tell a lie? For those receiving $20, the situation was not very dissonant. The money provided ample justification for engaging in counterattitudinal behavior (lying). In the $1 condition, however, the participants experienced more dissonance because they did not have the justification for lying provided by a large amount of money. These participants could not deny they lied, so they reduced dissonance by changing their attitude—that is, by increasing their liking for the task and the experiment. The results of this study confirmed the predictions from dissonance theory. Participants in the high-incentive ($20) condition experienced little dissonance and rated the task and experiment negatively, whereas those in the low-incentive ($1) condition experienced more dissonance and rated the task and experiment positively.

These results reflect the **dissonance effect:** the greater the reward or incentive for engaging in counterattitudinal behavior, the less the resulting attitude change. Sometimes, however, the opposite relationship is observed: the greater the incentive for engaging in counterattitudinal behavior, the greater the resulting attitude change. This is called the **incentive effect.** Different conditions determine which of these two outcomes will occur. Research shows that the dissonance effect is more likely when people choose (or have the illusion of choosing) whether or not to engage in the behavior. In one study (Sherman, 1970), participants were asked to write essays taking a position on current issues that contradicted their own attitudes. They were paid either 50 cents or $2.50 for the essay. In one condition, participants were allowed to choose whether or not to write the essay; in the second condition, students were not given a choice. For those participants given a choice, the results showed a dissonance effect: Participants given little incentive (50 cents) wrote longer, more persuasive essays and showed more attitude change than those given larger incentive ($2.50). For participants given no choice, the results showed an incentive effect: Participants given more incentive ($2.50) wrote longer, more persuasive essays and showed more attitude change than those given a small incentive (50 cents).

Dissonance occurs only in some situations (Wicklund & Brehm, 1976). To experience dissonance, a person must be committed to a belief or course of action (Brehm & Cohen, 1962). Moreover, the person must believe he or she chose to act voluntarily and is thus responsible for the outcome of the decision (Linder, Cooper, & Jones, 1967). This is shown in the case of Susan, who chose the cashier's job. If the owner of the restaurant were Susan's father and he demanded she work for him, she would have had little or no post-decision dissonance.

Is Consistency Inevitable?

If our beliefs and behavior were always consistent, all of our cognitions would be in harmony. Obviously, this is not the case. Practically every adult in the United States accepts that cigarette smoking causes lung cancer, yet millions continue to smoke. Most of us overindulge in a favorite food (pizza, chocolate) or beverage (soda, beer) at least occasionally, although we recognize that it is not healthy to do so. When we do, our behavior is inconsistent with the belief that overindulgence is unhealthy. How is it that people can hold mutually inconsistent cognitions?

For one thing, many of our cognitions never come into contact with one another. In order for people to experience a strain toward consistency, they must perceive the "implicational relationship" between two inconsistent cognitions or attitudes (Lavine, Thomsen, & Gonzales, 1997). Research indicates that thinking about attitude objects, or about the relationship between an attitude and one's values, may lead to a recognition of this relationship. The strength of the perceived implicational relationship determines the strength of the consistency pressure.

A related reason for inconsistency is that some of our behavior is mindless. Because we sometimes do not think about our actions, we are unaware that they are inconsistent with our beliefs (Triandis, 1980).The cigarette smoker often lights up without consciously thinking about the act; sometimes, he is surprised to find a lit cigarette in the ashtray. Thus, the relationship between the act and one's knowledge is often not salient.

A third reason inconsistency occurs is that each belief, attitude, or self-perception is embedded in a larger structure of consistent, related attitudes, beliefs, and self-perceptions. For example, Bill's toward premarital sexual intercourse, discussed earlier, was embedded in a structure of other attitudes and values. Although two attitudes may be inconsistent, each may be related to several other consonant attitudes. To change one or the other would create new inconsistencies. In effect, people tolerate some inconsistencies to avoid others.

People use various strategies for handling the dissonance aroused by messages that are inconsistent with their behavior. Faced with these two ads, a smoker resolves the inconsistencies by denying the risk of cancer and emphasizing the link between smoking and a life of leisure. A nonsmoker resolves the inconsistencies by emphasizing the importance of health and denying that smokers enjoy life more than nonsmokers.

BOX 6.1 Selling with Cognitive Dissonance

Cognitive dissonance is a ubiquitous part of our daily lives. We encounter it almost wherever we go and almost whatever we do. One social interaction where we are very likely to encounter cognitive dissonance is when we encounter a salesperson—particularly one who uses high-pressure sales techniques, such as in automobile sales. Many sales techniques have harnessed the power of cognitive dissonance and use it to increase the chances of convincing the customer to buy. How does this occur?

First, salespeople often make use of a technique called the "foot-in-the-door" (Freedman & Fraser, 1966). In this case, the salesperson attempts to get the customer to agree to some kind of small request and, having established a pattern of compliance, will ask the customer to do bigger things—including purchasing the product. Salespeople might request an appointment at your home, ask you to fill out paperwork, or get you to take a test drive. Once the small request is fulfilled, an inconsistency is produced if you do not go ahead and buy the product (Burger, 1986, 1999). Your refusal to buy causes some dissonance because it is inconsistent with your previous, compliant behavior. Of course this is not always enough to get you to buy, but it can reduce sales resistance.

A second technique, often used by unscrupulous salespeople, is called "low-balling" (Burger & Petty, 1981; Weyant, 1996). In this technique, the salesperson will offer the buyer a very good price on a product. The buyer agrees, and the salesperson sets about to do all the paperwork. Before it is completed, though, the salesperson "discovers" that he or she has made an error and that the price is going to be higher than initially promised. Under these circumstances, the buyer has a tendency to accept the higher price—after all, he or she has already agreed to buy; why should a few more dollars make that much of a difference? Interestingly, though, social psychologists have found that buyers will often pay more than their original upper limit when confronted with the low-balling technique (Cialdini, 1993; Cialdini, Cacioppo, Basset, & Miller, 1978). If you walk into a car dealership knowing that you can buy a certain car for $20,000 elsewhere and are low-balled with an offer of $19,500, you may end up paying $20,500 for the car in the end!

In a third technique that involves consistency, salespeople usually work very hard to get us to like them (Gordon, 1996). In fact, they are often trained in many specific techniques that get buyers to feel like they are friends with the salesperson. It is no surprise that we are more likely to buy things from people we like than from people we do not like, but why does this occur? One reason is that refusing a request from someone we like is inconsistent with our liking them. When a friend asks us to purchase candy for a fund-raiser, it can be difficult to turn him or her down because such behavior is incompatible with the friendship. Salespeople can use this underlying tendency to increase compliance as well (Jones, 1964; Liden & Mitchell, 1988).

Finally, people vary in the strength of their preference for consistency. In one set of experiments, half of the participants did not show this preference (Cialdini, Trost, & Newsom, 1995). It has also been suggested that the motivation toward consistency is stronger in Western (United States, Western Europe) than in Eastern cultures (Japan, India; Markus & Kitayama, 1991). In the end, however, the drive toward consistency and away from dissonance is a powerful influence on how we live our lives, day in and day out (see Box 6.1).

THE RELATIONSHIP BETWEEN ATTITUDES AND BEHAVIOR

Do Attitudes Predict Behavior?

We have seen how behavior can affect our attitudes and how people sometimes change their attitudes when their behavior appears to contradict them. However, most people think of attitudes as the source of behavior. For example, we often assume that when we know a person's attitude toward an object (another person, volleyball, Aerosmith's music), we can predict how that person will behave toward the object. If you know someone enjoys volleyball, you would expect her to accept your invitation to play volleyball with friends. When we are able to predict another person's responses, we can decide how to behave toward that person in order to achieve our own goals. But can we truly predict someone's behavior if we know his or her attitudes?

In 1930, the social scientist Richard LaPiere traveled around the United States by car with a Chinese couple. At that time, there was considerable prejudice against the Chinese, particularly in the West. The three travelers stopped at more than 60 hotels, auto camps, and tourist homes, and more than 180 restaurants. They kept careful notes about how they were treated. In only one place were they denied service. Later, LaPiere sent a questionnaire to each place, asking whether they would accept Chinese guests. He received responses from 128 establishments; 92% of them indicated that they would *not* serve Chinese guests (LaPiere, 1934). Evidently there can be a great discrepancy between what people do and what they say.

Many studies on the topic have found only a modest correlation between attitude and behavior (Wicker, 1969). Correlation (r) is a measure of the relationship between two variables and may range from -1.00 to $+1.00$. If there is no relationship between the variables, the correlation is zero. If two variables increase together (or decrease together), the correlation is positive. A survey of 33 studies of attitudes and behavior found that the average correlation between these two variables is .30 or less—a modest correlation. Several reasons why the relationship is not stronger have been suggested. In this section, we consider four variables that influence the relationship between attitudes and behavior: (1) the activation of the attitude, (2) the characteristics of the attitude, (3) the correspondence between attitude and behavior, and (4) situational constraints on behavior.

Activation of the Attitude

Each of us has thousands of attitudes. Most of the time, a particular attitude is not within our conscious awareness. Moreover, much of our behavior is mindless or spontaneous (Fazio, 1990). We act without thinking—that is, without considering our attitudes. For an attitude to influence behavior, it must be activated—that is, brought from memory into conscious awareness (Zanna & Fazio, 1982).

An attitude is usually activated by exposure of the person to its object, particularly if the attitude was originally formed through direct experience with the object (Fazio, Powell, & Herr, 1983). Earlier sections of this chapter may have activated your attitudes toward many objects, such as Aerosmith's music, African Americans, premarital sexual activity, and cigarette smoking. Thus, one way to activate attitudes is to arrange situations in which persons are exposed to the relevant objects. Soft lighting, a cozy fire, and glasses of wine are all associated with seduction; we often set up these cues in the hope of activating our partner's positive attitudes toward romantic and sexual activity.

Attitudes differ in the ease with which they can be activated—that is, they differ in accessibility. Some attitudes, such as stereotypes, are highly accessible and are activated automatically by the presentation of the object (Devine, 1989). One measure of the degree to which an attitude is accessible is the speed of activation. Attitudes activated instantaneously are by definition highly accessible. Other attitudes are activated more slowly (Fazio, Sanbonmatsu, Powell, & Kardes, 1986). The more accessible an attitude is, the greater its influence on categorizing and judging objects (Smith, Fazio, & Cejka, 1996).

Evidence also indicates that the more accessible an attitude, the more it is likely to guide future behavior. This was shown, for example, in a study of the impact of accessibility on voting in the 1984 presidential election (Fazio & Williams, 1986). In June and July 1984, 245 people were questioned about their attitudes toward presidential candidates Ronald Reagan and Walter Mondale. The latency of the answer—how quickly the person replied to the question about each candidate—was used as a measure of accessibility. After the election, each person was asked whom he or she voted for. The more accessible the attitude—that is, the more quickly the person replied to the original question about the candidate—the more likely the person was to vote for that candidate in November.

Characteristics of the Attitude

The relationship between attitude and behavior is also affected by the nature of the attitude itself. Four characteristics of attitudes that may influence the attitude-behavior relationship are (1) the degree of consistency between the affective (evaluative) and the cognitive components, (2) the extent to which the attitude is grounded in personal experience, (3) the strength of the attitude, and (4) the stability of the attitude over time.

Affective-Cognitive Consistency At the beginning of this chapter, we identified three components of an attitude: cognition, evaluation (affect), and behavioral predisposition. When we consider the relation between attitude and behavior, we are looking at the relationship between the first two components and the third. Not surprisingly, research has shown that the degree of consistency between the affective and cognitive components influences the attitude-behavior relationship. That is, the greater the consistency between cognition and evaluation, the greater the strength of the attitude-behavior relation.

Recall that the cognitive component is a belief about the attitude object (for example, "Capital punishment is necessary to protect society") and the affective component is the emotion associated with the object (for example, "I am strongly in favor of capital punishment"). In this case, there is a high degree of affective-cognitive consistency. Now suppose another person endorses the belief but is opposed to capital punishment. Whose behavior could you confidently predict? The first person is much more likely to write letters to legislators supporting the death penalty and to vote for candidates who advocate its use.

In one experiment, participants' beliefs and evaluations regarding capital punishment were assessed by questionnaires (Chaiken & Yates, 1985). Next, participants who were either high or low in consistency were asked to write two essays—one on the death penalty and one on an unrelated topic. The death penalty essays written by high-consistency participants were much more internally consistent; that is, their attitudes were part of an internally consistent structure. Furthermore, high-consistency participants dealt with discrepant information by discrediting it or minimizing its importance, making their attitudes more resistant to change.

Direct Experience Suppose you have a positive attitude toward an activity based on having done it once, and your roommate has a positive attitude based on hearing you rave about it. Which of you is more likely to accept an invitation to engage in it again?

One study (Regan & Fazio, 1977) provides an answer to this question. The behavior of interest was the proportion of time spent playing with several kinds of puzzles. Participants in the direct-experience condition played with sample puzzles; those in the indirect-experience condition were given only descriptions of the puzzles. Researchers then asked participants to respond to some attitude measures and later gave them an opportunity to play with the puzzles. They discovered that the average correlation between attitude and behavior was much higher for participants who had direct experience than for those who did not.

Attitudes based on direct experience are more predictive of subsequent behavior for several reasons (Fazio & Zanna, 1981). The best predictor of future behavior is past behavior; the more frequently you have played tennis in the past, the more likely you are to play it in the future (Fredricks & Dossett, 1983). An attitude is a summary of a person's past experience; thus, an attitude grounded in direct experience predicts future behavior more accurately. Moreover, direct experience makes more information available about the object itself (Kelman, 1974). In a test of the hypothesis that the attitude-behavior relation will increase as the amount of available information increases, researchers studied three different behaviors,

Although most Americans have attitudes about abortion, only a minority act on their beliefs like these demonstrators. People with strong attitudes, whether pro or con, are more likely to engage in such behavior.

including voting for specific candidates in an election (Davidson, Yantis, Norwood, & Montano, 1985). The results indicated that both the amount of information and direct experience enhanced the relationships.

Strength Suppose you ask two friends which candidate they like in the upcoming presidential election. One replies, "I'm voting for X"; the other hedges a bit, saying, "Well, maybe I'll vote for Y." Which person's behavior do you think you could predict? In general, the greater the strength of an attitude, the more likely it is to influence behavior.

Studies of voting behavior find that many of the errors in predictions occur among those who report indifference to the election—that is, people who have weak or uncertain attitudes (Schuman & Johnson, 1976). In one study, researchers measured people's attitudes before an upcoming election. Participants completed a 15-item Likert scale and indicated how certain they were of each response (Sample & Warland, 1973). For each participant, two measures were constructed—one of attitude and one of certainty. The 243 participants were divided into high- and low-certainty groups. Researchers noted how participants voted in the election 15 days after they had completed the questionnaire. Among participants who were more certain of their attitudinal responses, the attitude-behavior correlation was .47, whereas among those who were less certain, the correlation was .06. The average for all participants was in the usual range (.29).

Attitudes based on direct experience with the object may be held with greater certainty. Certainty is also influenced by whether affect or cognition was involved in the creation of the attitude. Attitudes formed based on affect (for example, fear of snakes) are more certain than attitudes based on cognition (for example, a preference for Hondas based on reading analyses of auto quality; Edwards, 1990).

The relevance of an attitude—the extent to which the issue or object directly affects the person—is an important influence on its strength. Framing an issue in relevant terms (say, tuition increases on your college campus) brings to mind important consequences for you, such as the need for greater income. Framing it in irrelevant terms (say, tuition increases on campuses in Russia) may elicit no thought of personal consequences (Lieberman & Chaiken, 1996). A study of the reactions of 1,300 adults in the Boston area to busing students to achieve racial integration included several measures of relevance: Was busing occurring in the respondent's neighborhood, did she or he have a child in the public schools, and was his or her neighborhood integrated racially? The survey also measured racial attitudes, including tolerance. A much stronger relationship between racial attitude and voting behavior was found among adults for whom busing was a relevant issue (Crano, 1997).

Temporal Stability Most studies attempting to predict behavior from attitudes measure people's attitudes first and their behavior weeks or months later. A modest or small correlation may mean a weak attitude-behavior relationship—or it could mean people's attitudes have changed in the interim period. If the attitude changes after it is measured, the person's behavior may be consistent with his or her present attitude, although it appears inconsistent with our measure of the attitude. Thus, to predict behavior from attitudes, the attitudes must be stable over time.

In general, we would expect that the longer the time between the measurement of attitude and that of behavior, the more likely the attitude will change and the smaller the attitude-behavior relationship will be. In a study designed to test this possibility (Schwartz, 1978), an appeal was mailed to almost 300 students to volunteer as tutors for blind children. Earlier, students had filled out a questionnaire measuring general attitudes toward helping others, including questions about tutoring blind children. Some students had filled out the questionnaire 6 months earlier; some, 3 months earlier; some, both 3 and 6 months earlier; and still others had not seen the questionnaire. The correlation between attitude toward tutoring and actually volunteering was greater over the 3-month period than over the 6-month period. Thus, to avoid problems of temporal instability, the amount of time between the measurement of attitudes and that of behavior should be brief.

Some attitudes evidence a remarkable degree of stability, however. Thornton (1984) studied the attitudes of 458 women toward divorce; their attitudes were measured in 1962, 1977, and 1980. He found substantial stability over the 18-year period, particularly among women who attended church regularly. Marwell, Aiken, and Demerath (1987) studied the political attitudes of 220 White young people who spent the summer of 1964 organizing Blacks in the South to vote. They measured the same attitudes of two thirds of these activists 2 decades later, in 1984. The extreme radical attitudes these people held in 1965 had softened in the intervening 20 years, but in general these people remained liberal and committed to the needs of disadvantaged groups.

Attitude-Behavior Correspondence

Attitudes are more likely to predict behavior when the two are at the same level of specificity (Schuman & Johnson, 1976). For example, suppose you have invited a casual acquaintance to dinner, and you want to plan the menu. You know she is Italian, so she probably likes Italian food. But can you predict with confidence that she will eat green noodles with red clam sauce? Probably not. A favorable attitude toward a type of cuisine does not mean the person will eat every dish of that type.

Many studies have attempted to predict from general attitudes to specific behaviors. For instance, some studies of racial prejudice have tried to predict from people's general attitudes toward African Americans to specific behaviors, such as willingness to have one's photograph taken with particular African Americans in particular settings (Green, 1972). Not surprisingly, the relationship between attitude and behavior was weak. A general attitude is a summary of many feelings either about an object under a variety of conditions or about a whole class of objects. Logically, it should not predict behavior in any particular single situation. But it might predict a composite measure of several relevant behaviors (Weigel & Newman, 1976).

What about predicting a specific behavior, such as whether your Italian guest will eat green noodles and red clam sauce? Just as general attitudes best predict a composite index of behavior, we need a specific measure of attitude to predict a specific behavior. We can think of an attitude and a behavior as having four elements: an action (eating), an object or target (green noodles and red clam sauce), a context (in your home), and a time (tomorrow night). The greater the degree of **correspondence**—that is, the number of elements that are the same in the two measures—the better we can predict behavior from attitudes (Ajzen & Fishbein, 1977).

A study of birth control use by 244 women (Davidson & Jaccard, 1979) demonstrated that attitudinal measures that exhibit correspondence with the behavioral measure are better predictors of behavior. In this study, the behavior of interest was whether women used birth control pills during a particular 2-year period. Attitude was measured in four ways. The measure of the women's general attitude toward birth control had only one element in common with the behavior (object). The correlation between this general measure and behavior was a modest .323, as shown in Figure 6.4. When the attitude measure had two elements in common with the behavior (object and action), the correlation rose to .525. Finally, an attitude measure that included three elements (object, action, and time: "Do you plan to use birth control pills in the next 2 years?") was most highly correlated with the behavioral measure. Thus, attitudinal measures having high correspondence with the behavioral measure are better predictors of behavior than attitudinal measures having low correspondence.

Earlier in this chapter, we mentioned that in LaPiere's (1934) study, most establishments that served the Chinese couple later said they would not. The lack of a relationship between attitude and behavior in LaPiere's study may be due to a lack of correspondence. The behavioral measure was whether a particular Chinese couple (object) was served (action) in a particular restaurant or hotel (context) on a particular day (time). However, LaPiere's questionnaire measuring attitudes simply asked whether Chinese guests would be served. Thus, there was correspondence between the measures on only one element (action), which may account for the discrepancy LaPiere found between attitude and behavior.

FIGURE 6.4 CORRELATIONS OF ATTITUDE MEASURES THAT VARY IN CORRESPONDENCE WITH BEHAVIOR

Every behavior involves a target, action, context, and time. In order to predict behavior from attitude, the measures of attitude and behavior should correspond—that is, involve the same elements. The larger the number of elements in common, the greater the correlation between attitude and behavior. Researchers obtained four measures of attitudes toward birth control from 244 women: (1) general attitude toward birth control; (2) attitude toward birth control pills; (3) attitude toward using pills; and (4) attitude toward using pills in the next 2 years. The behavioral measure was actual use of pills during the 2-year period. Note that as correspondence increased from zero to three elements, the correlation between attitude and behavior also increased.

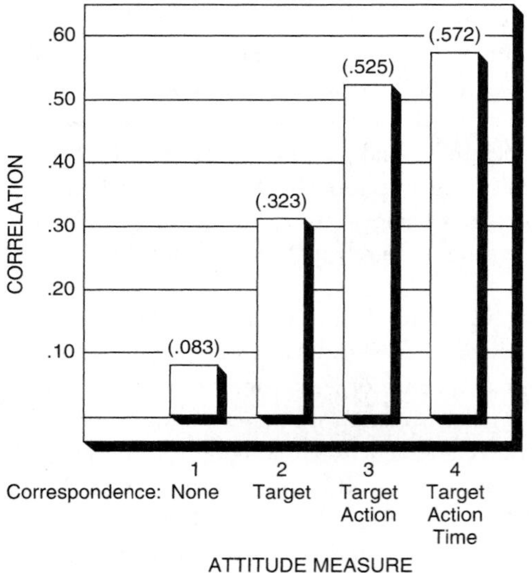

Source: Adapted from Davidson and Jaccard, 1979.

FIGURE 6.5 THE INFLUENCE OF ATTITUDE AND SITUATIONAL CONSTRAINTS ON BEHAVIOR

Our behavior is influenced not only by our attitudes but also by situational constraints, the behavior of others, or the likelihood that others will find out what we do. When the individual has a strongly held attitude and situational influences encourage behavior consistent with that attitude, there will be a strong relationship between attitude and behavior. But when situational influences produce pressure to behave in ways inconsistent with one's attitude or when the attitude is weak, behavior and attitude are less likely to be consistent.

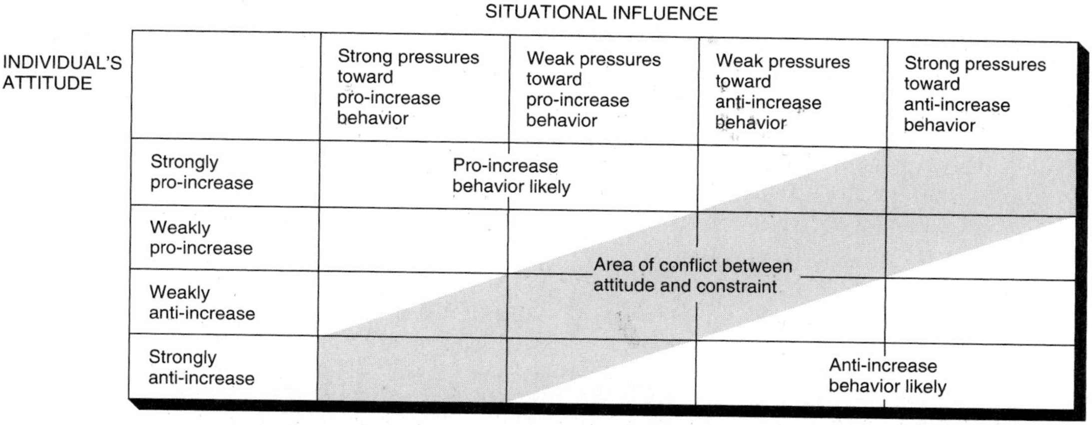

SITUATIONAL INFLUENCE

INDIVIDUAL'S ATTITUDE	Strong pressures toward pro-increase behavior	Weak pressures toward pro-increase behavior	Weak pressures toward anti-increase behavior	Strong pressures toward anti-increase behavior
Strongly pro-increase	Pro-increase behavior likely			
Weakly pro-increase		Area of conflict between attitude and constraint		
Weakly anti-increase				
Strongly anti-increase			Anti-increase behavior likely	

Source: Adapted from "Attitude as an Interactional Concept: Social Constraints and Social Distance as Intervening Variables between Attitudes and Action" (1969) by L. G. Warner and M. L. DeFleur, *American Sociological Review, 34*(2), 168. Used with permission from the American Sociological Association.

Situational Constraints

If you believe tuition increases at your college are necessary to maintain the quality of your education—to retain the best faculty, provide ready access to books, journals, and computers, and so on—and you attend a meeting of Students for Educational Quality, your behavior reflects your attitude. If you are opposed to tuition increases and you participate in a protest march, your behavior is consistent with your attitude.

Suppose, however, that you oppose tuition increases but find yourself in a conversation with your partner's parents in which they express support for the increases. Would you voice your opposition—that is, behave in a manner consistent with your attitudes—or not? Your reaction would probably depend partly on the strength and certainty of your attitude (Pratkanis & Greenwald, 1989). If you are strongly opposed to tuition increases, you may speak your mind. But if you are moderately opposed, you may decide to avoid an argument and behave in a way that is inconsistent with your attitude. In LaPiere's study, for instance, hotel and restaurant employees confronted by a White man and a Chinese couple may have felt compelled to serve them rather than run the risk of creating a scene by refusing to do so.

Situational constraint refers to an influence on behavior due to the likelihood that other persons will learn about the behavior and respond positively or negatively to it. Situational constraints often determine whether our behavior is consistent with our attitudes. In fact, how we behave is frequently a result of the interaction between our attitudes and the constraints present in the situation (Warner & DeFleur, 1969; Klein, Snyder, & Livingston, 2004). This relationship is summarized in Figure 6.5, using attitudes toward college tuition increases as an example. A conversation between someone weakly opposed to

BOX 6.2 Research Update: Modern Versus Old-Fashioned Racism and Sexism

In the first half of the 20th century, overt markers of American bigotry and prejudice against racial minorities were everywhere—particularly in the South, where segregation laws supported separate facilities and practices for African Americans. It was in this social context that the social psychological study of racial prejudice began. Given the widely accepted discriminatory norms and beliefs that Blacks were inferior to Whites, it is no surprise that Whites had little trouble expressing their prejudicial attitudes. Blacks were typically viewed by Whites as being "lazy," "ignorant," and "superstitious," among other negative traits (Katz & Braly, 1933). However, as the civil rights movement began to change the social context through the 1950s and beyond, the attitudes expressed by Whites began to change. As studies have been replicated over time, it has become clear that Whites espouse considerably less negative views of Blacks (Dovidio, Brigham, Johnson, & Gaertner, 1996).

The same kind of changes can be observed in attitudes toward women. The role of women in society has changed dramatically over the years, and individual attitudes about the place of women in the workplace, their responsibilities at home, their right and ability to participate in the governance of the country, and so forth, have all experienced dramatic changes—particularly since the advent of the women's movement in the 1970s (Bolzendahl & Myers, 2002; Oskamp, 1991; Plutzer, 1988; Thorton, Alwin, & Camburn, 1983).

Nevertheless, the meaning of these changes in expressed attitudes has not been completely clear. It is true that when asked in research studies, respondents express less prejudicial views of both African Americans and women. But do they really feel that way, or are they responding to contextual pressures to give more socially acceptable responses? If so, perhaps we have not seen a reduction in prejudice and discrimination at all, but instead have seen a shift from overt and hostile forms of racism and sexism to subtle, less recognizable forms (Benokraitis, 1997; Benokraitis & Feagin, 1986; Fiske, 1998; Nelson, 2002; Nail, Decker, & Harton, 2003).

These subtle forms are often called "modern" racism and sexism and contrasted with "old-fashioned" racism and sexism (McConahay, 1983, 1986; McConahay, Hardee, & Batts, 1981; Swim, Aikin, Hall, & Hunter, 1995). Old-fashioned variants assert that Blacks and women are not equal to White men; they are inferior in intellect, drive, and competence and should therefore play different (that is, lower status) roles in society. Modern variants focus more on denial of prejudice and discrimination—and therefore on a repudiation of programs and interventions designed to address systemic inequality. For example, modern racists assert that discrimination no longer occurs, that Blacks are too pushy in trying to get into places (neighborhoods, jobs, country clubs, and so on) where they are not wanted, and that affirmative action programs and other efforts to help Blacks are unfair to and discriminatory against Whites.

Finding ways to confront and address modern racism and sexism can be a very difficult challenge. Those who hold modern racist and sexist views may not be aware of their negative attitudes toward Blacks and women—or at least, they completely deny these feelings (Dovidio et al., 1996). Thus, because prejudice is expressed very indirectly through opposition to social programs and practices, Whites and men may not realize that the root of their opposition lies in the same kinds of negative feelings that drove acts of old-fashioned racism.

Source: Adapted from Nelson, 2002, chapters 5 and 8.

increases and your partner's parents (weak pressures) would be a situation of conflict for the individual, whereas someone in the same situation who is strongly opposed to the increases is more likely to voice that opposition.

Sometimes we feel constrained by the possibility that others may learn of our behavior. At other times those around us exert direct social influence; they communicate specific expectations about how we should behave. The greater the agreement among others about how we should behave, the greater the situational constraint on persons whose attitudes are inconsistent with the situational norms (Schutte, Kendrick, & Sadalla, 1985). Under these conditions, there is a weaker relationship between attitudes and behavior. Consequently, the less visible our behavior is to others, the more likely it is that our behavior and attitudes will be consistent (Acock & Scott, 1980). With respect to attitudes about race and gender, many scholars who study prejudice have noted a shift in how people express prejudicial attitudes. As the social environment has become less accepting of overt expressions of racism and sexism, people may have responded to the situational constraints by hiding their attitudes and finding different, more subtle ways of expressing prejudice (Gawronski & Strack, 1991; see also Box 6.2).

But what if persons whose opinions we value are not actually present? Several studies have assessed the impact of reference groups on the attitude-behavior relationship. Such research involves measuring participants' attitudes toward some object and then asking them to indicate the positions of various social groups about that object. One survey assessed adults' attitudes toward drinking alcoholic beverages and the degree to which their friends approved of drinking (Rabow, Neuman, & Hernandez, 1987). When attitudes and social support were congruent—that is, when the respondents' and their friends' views were the same—there was a much stronger relation between attitudes and behavior than when attitudes and social support were not congruent. Another study found that the perceived norms of their friends influenced whether women engaged in regular exercise, but only for those who identified strongly with the reference group (Terry & Hogg, 1996).

THE REASONED ACTION MODEL

In the preceding section, we identified several influences on the relationship between a single attitude and behavior. However, at times, an object or situation may elicit multiple attitudes. In these cases, predicting behavior is more difficult. When several attitudes are invoked, the individual often engages in deliberative processing of information (Fazio, 1990). He or she considers the attributes of the object or situation, the relevant attitudes, and the costs and benefits of potential behaviors. One important attempt to specify this process is the **theory of reasoned action,** developed by Fishbein and Ajzen (1975; Ajzen & Fishbein, 1980). This model is based on the assumption that behavior is rational, and it incorporates several factors that have been shown to affect the consistency between attitudes and behavior (see Figure 6.6).

According to the reasoned action model, behavior is determined by behavioral intention. Behavioral intention is primarily influenced by two factors: attitude and subjective norm. Attitude refers to positive or negative feelings about engaging in a behavior. **Subjective norm** is the individual's perception of others' beliefs about whether a behavior is appropriate or not. In other words, subjective norm is one form of situational constraint. The reasoned action model also specifies the determinants of attitude and of subjective norm. Attitude is influenced by a person's beliefs about the likely consequences of the behavior and a person's evaluation—positive or negative—of each of those outcomes. Subjective norm is influenced by a person's beliefs about the reactions of other persons or groups to the behavior and his or her motivation to comply with their expectations.

Formal Model

The model can be summarized using the following expression:

$$\text{Behavior} = \text{Behavioral intention} = \text{Attitude} + \text{Subjective norm}$$

Suppose Bill has two good friends who are followers of Reverend Moon. They have been giving him literature

FIGURE 6.6 THE REASONED ACTION MODEL

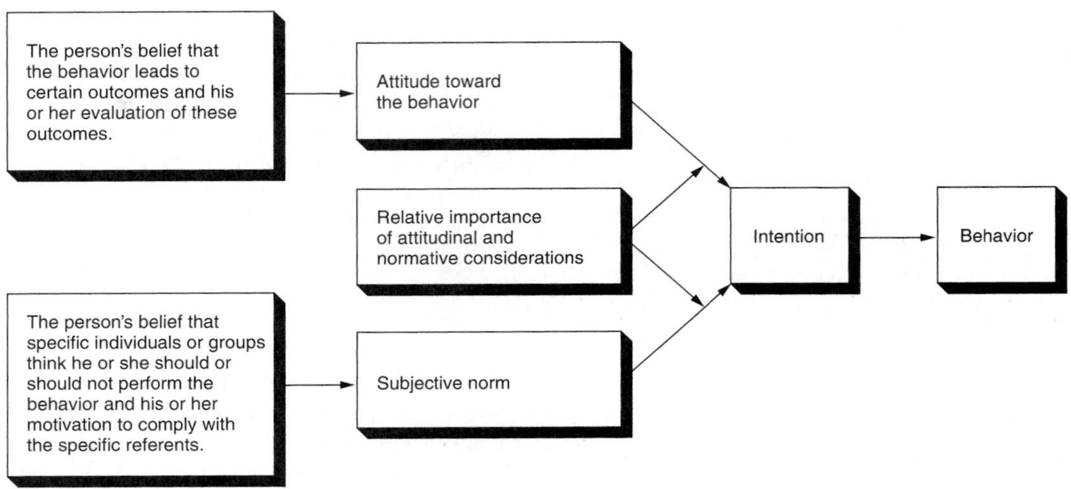

Note: Arrows indicate the direction of influence.

Source: From *Understanding Attitudes and Predicting Social Behavior* by Ajzen and Fishbein. Copyright © 1980 by Prentice-Hall, Inc. Reprinted by permission of Pearson Education, Inc., Upper Saddle River, NJ.

TABLE 6.1 DETERMINING ATTITUDE FROM BELIEFS AND EVALUATIONS

Consequences of Joining the Unification Church	Belief	Evaluation	Product
Gain a sense of purpose	+3	+3	+9
Have one's physical needs provided for	+3	+1	+3
Loss of relationship with Cindy	+2	−2	−4
Loss of some personal freedom	+1	−3	−3
Attitude			**+5**

and encouraging him to join the Unification Church. To predict what Bill will do, we need to know his attitude and his perception of how others will react to his joining the group. Attitude is the sum of beliefs about the likelihood of various consequences of the act multiplied by the evaluation—positive or negative—of each consequence.

Bill has several beliefs about joining Moon's church. If he joins, he will gain a greater sense of purpose in his life, and his physical needs will be met. Also, he will have to end his relationship with Cindy, his girlfriend. Finally, Bill has read that former members claim they had to relinquish their personal freedom when they joined. These beliefs and their evaluations are shown in Table 6.1. Bill is certain that consequences (1) and (2) would occur; hence their value is +3. His evaluation of consequence (1) is very positive (+3), whereas his evaluation of having his physical needs cared for is less positive (+1). He believes it is likely that he will have to give up Cindy (+2), which would be unpleasant (−2). He is skeptical about the claim that he will lose his freedom (+1),

TABLE 6.2 DETERMINING SUBJECTIVE NORM FROM NORMATIVE BELIEFS AND MOTIVATION TO COMPLY

Significant Others	Normative Beliefs	Motivation to Comply	Product
Parents	−3	+2	−6
Friends	+3	+2	+6
Cindy	−2	+3	−6
Sum			**−6**

although he would be very upset if that occurred (−3). To determine Bill's overall attitude, we multiply each belief by its evaluation and sum the products. In this case, the value for the attitude is +5.

Subjective norm is the product of normative beliefs—expectations about how significant others will react—and the motivation to comply with each. For Bill, the significant others are his parents, his peers, and his girlfriend, Cindy. His parents are strongly opposed to his joining (−3), and he is moderately motivated to comply with their views (+2). He is equally motivated to comply with his friends' views (+2), who strongly favor his joining (+3). And he is highly motivated to comply with Cindy, who opposes his joining Moon's church. Thus, the value of the overall subjective norm is −6, as shown in Table 6.2. Behavioral intention is the weighted sum of attitude and subjective norm. Assuming equal weights, the model predicts that Bill will not join the Unification Church; that is, (+5) + (−6) = −1. Although his attitude is positive, the subjective norm is negative to the extent that it will prevent him from joining.

Assessment of the Model

The reasoned action model combines several elements discussed earlier in this chapter. It has been used to predict behaviors such as whether a mother will breastfeed her baby (Manstead, Proffitt, & Smart, 1983). When combined with quantitative measures of the components of attitudes, this model can predict a specific behavior under specific circumstances. For instance, one study attempted to predict weight loss among college women (Schifter & Ajzen, 1985). The participants' subjective intention, attitude, and subjective norm about losing weight were measured. Several other variables were also assessed, including whether or not the participant had a detailed plan regarding weight loss. Six weeks later, the amount of weight actually lost was measured. The amount of weight lost was associated with intention and with having a detailed plan; the intention to lose weight was determined by attitude and subjective norm.

This model has been criticized (Liska, 1984) because it assumes our behavior is determined largely by our intentions. This assumption is not always correct; in some situations, our past behavior may be even more influential than our intentions. For example, whether one has donated blood in the past is a much better predictor of whether he or she will donate blood in the next 4 months than the statement that one intends to do so (Bagozzi, 1981). In effect, much of our behavior is habitual and may not match our conscious intentions. The effect of prior behavior is particularly strong when the stated intention is not compatible with the individual's self-identity (Granberg & Holmberg, 1990). Conversely, a significant relationship between intention and weight loss over an 8-week period was noted among women whose self-schema was consistent with their intention (Kendzierski & Whitaker, 1997).

Also, research suggests that our behavior may be affected not only by intentions but also by whether we have the resources or the ability needed to carry out the intention. Consequently, it has been suggested that an additional variable should be added to the model, **perceived behavioral control** (Ajzen, 1985). A study of intentions to engage in safer sex among 403 undergraduates found that attitude and subjective norm explained substantial variation in the intention to use condoms in the next 3 months. Even more variance was explained when the perception of one's ability to use condoms was added to the analysis (Wulfert & Wan, 1995). The revised model is referred to as the theory of planned behavior.

SUMMARY

The Nature of Attitudes (1) Every attitude has three components: cognition, evaluation, and a behavioral predisposition toward some object. (2) We learn attitudes through reinforcement, through repeated associations of stimuli and responses, and by observing others. (3) Attitudes are useful; they may serve instrumental and knowledge functions, and they define and maintain self.

Attitude Organization An attitude is usually embedded in a larger cognitive structure and is based on one or more fundamental or primitive beliefs. Attitudes derived from primitive beliefs form a vertical structure. When attitudes are supported by more than one underlying belief, they have a horizontal structure that helps the attitude persist even when a primitive belief changes.

Cognitive Consistency Consistency theories assume that when cognitive elements are inconsistent, individuals will be motivated to change their attitudes or behavior to restore harmony. Balance theory assesses the relationships among three cognitive elements and suggests ways to resolve imbalance. Dissonance theory identifies two situations in which inconsistency often occurs: after a choice between alternatives or when people engage in behavior that is inconsistent with their attitudes. The theory also cites two ways to reduce dissonance: by changing one of the elements or by changing the importance of the cognitions involved.

The Relationship Between Attitudes and Behavior The attitude-behavior relationship is influenced by four variables: activation of the attitude, characteristics of the attitude, attitude-behavior correspondence, and situational constraints. (1) For an attitude to influence behavior, it must be activated, and the person must use it as a guide for behavior. (2) The relationship is stronger if affective-cognitive consistency is high and if the attitude is based on direct experience, is strong (relevant), and is stable over time. (3) The relationship is stronger when the measures of attitude and behavior correspond in action, object, context, and time. (4) Situational constraints may facilitate or prevent the expression of attitudes in behavior.

The Reasoned Action Model This model suggests that behavior is determined by behavioral intention. In turn, intention is determined by a person's attitudes and perceptions of social norms. This model allows precise predictions of behavior, and some studies report results consistent with such predictions. However, it may not apply to some types of behavior, such as behavior determined by habit.

LIST OF KEY TERMS AND CONCEPTS

attitude (p. 142)
balance theory (p. 147)
cognitions (p. 146)
cognitive dissonance (p. 149)
correspondence (p. 158)
dissonance effect (p. 151)
incentive effect (p. 151)
perceived behavioral control (p. 163)
prejudice (p. 144)
situational constraint (p. 159)
subjective norm (p. 161)
theory of cognitive dissonance (p. 149)
theory of reasoned action (p. 161)

7

SYMBOLIC COMMUNICATION AND LANGUAGE

Introduction

Language and Verbal Communication

Linguistic Communication

The Encoder-Decoder Model

The Intentionalist Model

The Perspective-Taking Model

Nonverbal Communication

Types of Nonverbal Communication

What's in a Face?

Combining Nonverbal and Verbal Communication

Social Structure and Communications

Gender and Communication

Social Stratification and Speech Style

Communicating Status and Intimacy

Normative Distances for Interaction

Normative Distances

Conversational Analysis

Initiating Conversations

Regulating Turn Taking

Feedback and Coordination

Summary

List of Key Terms and Concepts

INTRODUCTION

Communication is a basic ingredient of every social situation. Imagine playing a game of basketball or buying a pair of shoes without some form of verbal or nonverbal communication. Without communication, interaction breaks down, and the goals of any social encounter are foiled. Indeed, it would be simply impossible to arrange commercial transactions, courtroom trials, birthday parties, or any other social occasion without communication.

Communication is the process whereby people transmit information about their ideas, feelings, and intentions to one another. We communicate through spoken and written words, through voice qualities and physical closeness, through gestures and posture. Often, communication is deliberate: We smile, clasp our beloved in our arms, and whisper, "I love you." Other times, we communicate meanings that are unintentional. A Freudian slip, for instance, may tell our listeners more than we want them to know.

Because people do not share each other's experiences directly, they must convey their ideas and feelings to each other in ways that others will notice and understand. We often do this by means of symbols. **Symbols** are arbitrary forms that are used to refer to ideas, feelings, intentions, or any other object.

Symbols represent our experiences in a way that others can perceive with their sensory organs—through sounds, gestures, pictures, even fragrances. But if we are to interpret symbols as others intend, their meanings must be socially shared. To communicate successfully, we must master the ways for expressing ideas and feelings that are accepted in our community.

Symbols are arbitrary stand-ins for what they represent. A green light could as reasonably stand for "stop" as for "go," the sound *luv* as reasonably for negative as for positive feelings. The arbitrariness of symbols becomes painfully obvious when we travel in foreign countries. We are then likely to discover that the words and even the gestures we take for granted fail to communicate accurately. A North American who makes a circle with thumb and index finger to express satisfaction to a waiter may be in for a rude surprise if he is eating at a restaurant in Ghana, where the waiter may interpret his gesture as a sexual invitation. In Venezuela, it may be interpreted as a sexual insult! The traveler may then have serious difficulties straightening out these misunderstandings because he and the waiter lack a shared language of verbal symbols to discuss them.

Language and nonverbal forms of communication are amazingly complicated. They must be understood and used with flexibility and creativity. Most of us fail on occasion to communicate our ideas and feelings with accuracy or to understand others' communications as well as we might wish. Yet, considering the problems a communicator must solve, most people do surprisingly well. This chapter begins with an examination of language, moves on to nonverbal communication, then analyzes the impacts of communication and social relationships on each other. Finally, this chapter considers the delicate coordination involved in our most common social activity—conversation. This chapter addresses the following questions:

1. What is the nature of language, and how is it used to grasp meanings and intentions?
2. What are the major types of nonverbal communication, and how do they combine with language to convey emotions and ideas?
3. How do social relationships shape communication, and how does it in turn express or modify those relationships?
4. What rules and skills do people employ to maintain a smooth flow of conversation and to avoid disruptive blunders?

LANGUAGE AND VERBAL COMMUNICATION

Although people have created numerous symbol systems (such as mathematics, music, painting), language is the main vehicle of human communication. All people possess a spoken language. There are thousands of different languages in the world (Katzner, 1995). This section addresses several crucial topics regarding the role of language in communication. These include the nature of language, three

Signing by the interpreter parallels the oral message by the speaker. Although sign language lacks the phonetic component, it possesses the morphologic, semantic, and syntactic components of language.

perspectives on how people attain understanding through language use, and the relation between language and thought (see Box 7.1).

Linguistic Communication

Little is known about the origins of language, but humans have possessed complex spoken languages since earliest times (Kiparsky, 1976; Lieberman, 1975). **Spoken language** is a socially acquired system of sound patterns with meanings agreed on by the members of a group. We will examine the basic components of spoken language and some of the advantages of language use.

Basic Components Spoken languages include sounds, words, meanings, and grammatical rules. Consider the following statement of one roommate to another: "Wherewereyoulastnight?" What the listener hears is a string of sounds much like this, rather than the sentence, "Where were you last night?" To understand a string of sounds and to produce an appropriate response, people must recognize the following components: (1) the distinct sounds of which the language is composed (the phonetic component); (2) the combination of sounds into words (the morphologic component); (3) the common meaning of the words (the semantic component); and (4) the conventions for putting words together built into the language (the syntactic component, or grammar). We are rarely conscious of manipulating all these components during conversation, though we do so regularly and with impressive speed.

Unspoken languages, such as Morse code, computer languages, and sign languages lack a phonetic component, although they do possess the remaining components of spoken language. People who use sign languages, for example, use upper body movements to signal words (morphology) with shared meanings (semantics), and they combine these words into sentences according to rules of order (syntax). For a communication system to be considered a language, morphology, semantics, and syntax are all essential. Linguists study these components, seeking to uncover the rules that give structure to language. Social psychologists are more interested in how language fits into social interaction and influences it and in how language expresses and modifies social relationships (Giles, Hewstone, & St. Clair, 1981).

Advantages of Language Use Words—the symbols around which languages are constructed—provide abundant resources with which to represent ideas and feelings. The average adult native speaker of English knows the meanings of some 35,000 words, and actively uses close to 5,000. Because it is a symbol system, language enhances our capacity for social action in several ways.

First, language frees us from the constraints of the here and now. Using words to symbolize objects, events, or relationships, we can communicate about things that happened last week or last year, and we can discuss things that may happen in the future. The ability to do the latter allows us to coordinate our behavior with the activities of others.

Second, language allows us to communicate with others about experiences we do not share directly. You cannot know directly the joy and hope your friend feels at bearing a child, nor her grief and despair at her mother's death. Yet she can convey a good sense of her emotions and concerns to you through words, even in writing, because these shared symbols elicit the same meanings for you both.

Third, language enables us to transmit, preserve, and create culture. Through the spoken and the written word, vast quantities of information pass from person to person and from generation to generation. Language also enhances our ability to go beyond what is already known and to add to the store of cultural ideas and objects. Working with linguistic symbols, people generate theories, design and build new products, and invent social institutions.

We turn now to three models of communication: the encoder-decoder model, the intentionalist model, and the perspective-taking model. We will consider how each model views the communication process and discusses communication accuracy.

The Encoder-Decoder Model

Language is often thought of as a medium of communication that one person uses to transmit information to another. The **encoder-decoder model** views communication as a process in which an idea or feeling is encoded into symbols by a source, transmitted to a receiver, and decoded into the original idea or feeling (Krauss & Fussell, 1996). This process is portrayed in Figure 7.1.

Communication Process According to this model, the basic unit of communication is the message, which has its origin in the desire of the speaker to communicate. A message is constructed when the speaker encodes the information he or she wishes to communicate into a combination of verbal and nonverbal symbols. The message is sent via a channel, whether by face-to-face interaction, telephone, electronic communication, or in writing. The listener must decode the message in order to arrive at the information he or she believes the speaker wanted to communicate.

Communication Accuracy The goal of communication is to accurately transfer the message content from speaker to listener. The speaker hopes to create in the listener the mental image or feeling that the speaker intends to convey. The listener is also motivated to achieve accuracy, in order to coordinate his or her behavior with that of the speaker. **Communication accuracy** refers to the extent to which the message inferred by the listener matches the message intended by the speaker. According to this model, the primary influence on accuracy is codability, which is the extent of interpersonal agreement about what something is called. Codability is partly a function of language. Early research focused on the codability of colors (Lantz & Stefflre, 1964). In this research, one person

FIGURE 7.1 THE ENCODER-DECODER MODEL

According to the encoder-decoder model, communication originates in the speaker's desire to convey an idea or feeling. She encodes the message into a set of symbols and transmits it to the hearer. The hearer decodes the message. The more codable the idea or feeling in the language, the more accurate the communication.

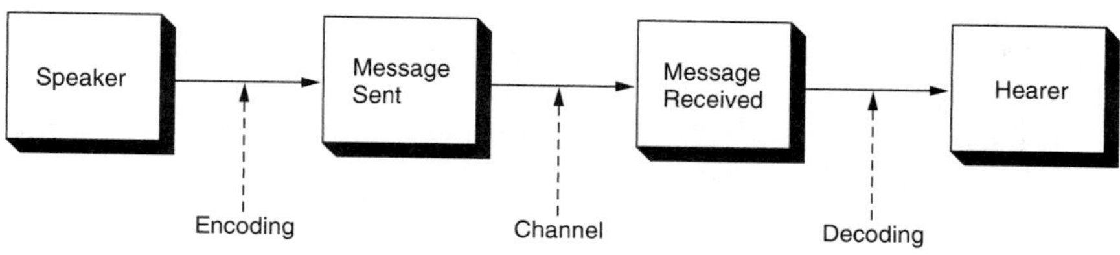

(the encoder) was shown a color and asked to describe that color in words. This verbal message was then sent to a second person (the decoder), who tried to use the verbal description to identify the color intended by the encoder. Some colors are much more easily coded in the English language (fire-engine red) than others (the reddish color of a sunset). By extension, some ideas and feelings are easily expressed in English, whereas others are much more difficult to put into words. In general, messages that are easily coded will be more accurately transmitted.

Codability involves agreement about what something is to be called. It also depends on the extent to which speaker and listener define symbols (such as words or gestures) in the same way. This in turn depends on the language to which each was socialized. Thus, a common cause of miscommunication is differences in language between speaker and listener. This is obvious when we try to converse with someone from a different country. It is less obvious, but perhaps just as important, when we converse with someone of a different race, class, or gender (Maynard & Whalen, 1995).

At times, the processes of encoding and decoding are very deliberate or mindful (Giles & Coupland, 1991). If we are preparing a speech, we may consciously consider alternative ways to phrase a message and alternative gestures to use when communicating. Listening to a speaker, we may pay careful attention to the words used, the speed and volume of the spoken message, the gestures, and the posture of the speaker in order to decide which message is the correct one. We are often mindful of the encoding and decoding process in novel situations or in communicating about novel topics.

Communication is not always a process of consciously translating ideas and feelings into symbols and then transmitting these symbols deliberately in hopes that the listener will interpret them correctly. Much communication occurs without any self-conscious planning. In familiar or routine situations, we often rely on a conversational script—a sequentially organized series of utterances that occur with little or no conscious thought. Thus, when you enter a restaurant, you can interact with the server without much mental effort, because you both follow a conversational script that specifies what each of you should say and in what sequence. Communication accuracy is typically high in situations governed by conversational scripts. When you order food in a fast food outlet, you usually get exactly what you want with minimum effort.

If conversation is scripted, listeners will probably not pay careful attention to the idiosyncratic features of a message. They will tend to remember the generic content of a message but not its unusual characteristics. In a field experiment testing this prediction, students were approached by a stranger who asked for a piece of paper. Prior to the request, one half of the students were asked to pay attention to it; the other half were not forewarned. Later, the forewarned students were more likely to remember the specific words used in the request than the unprepared students (Kitayama & Burnstein, 1988).

The Intentionalist Model

The encoder-decoder model emphasizes messages consisting of symbols whose meaning is widely understood. It directs our attention to the literal meaning of verbal messages. Often, however, messages are not interpreted literally. For example, in most theaters, the feature film is preceded by the message, "Please, silence during the show." But are members of the audience expected to be silent? No. They can laugh if the film is a comedy, boo at the villain, and applaud when the bad guy or gal gets what he or she deserves. Most of us understand this message in terms of its intention: We should not whisper or talk to those seated near us. For this type of communication, we need a different model.

According to the **intentionalist model,** communication involves the exchange of communicative intentions, and messages are merely the means to this end (Krauss & Fussell, 1996). The speaker selects the message she believes is most likely to accomplish her intent. "Please, silence during the show," is intended to keep us from disturbing other members of the audience, and we understand it to mean that.

Communication Process The origin of communication is the speaker's intent to achieve some goal or to have some effect on the hearer. But there is not a fixed, one-to-one relation between words and intended effects, so the speaker can use a variety of messages or utterances to achieve his or her intended effect. For example, imagine you are studying in your living room, and you want your roommate to bring you something to drink. Table 7.1 lists some of the utterances you might use to make the request. Which one would you choose?

TABLE 7.1 "GET ME A DRINK OF LEMONADE."

1. Get me a glass of lemonade.
2. Can you get me some lemonade?
3. Would you get me some lemonade?
4. Would you get me something to drink?
5. Would you mind if I asked you to get me some lemonade?
6. I'm thirsty.
7. Did you buy some lemonade at the store?
8. How is that lemonade we bought?

According to the intentionalist model, decoding the literal meaning of a message is only part of the process of communication. The hearer must also infer the speaker's underlying intention in order to respond appropriately. To the question "How is that lemonade we bought?" a satisfactory response to the literal message is "Good." If the communication is to be successful, however, your roommate needs to infer your intention—that he should bring you a glass of lemonade. Both selecting a message to convey your intention and inferring another's intention from their utterance are carried out according to social conventions.

Communication Accuracy According to this model, accuracy in communication is accuracy in understanding the intentions of the speaker. To achieve accuracy requires more sophisticated processing than merely interpreting the literal meaning of the message. When inferring the speaker's intention, the listener needs to take into account the context, especially (1) the status or role relationship between speaker and listener, and (2) the social context in which the communication occurs. If you and your roommate are lovers, you might choose a less polite form of the request, such as option 2 in Table 7.1, and you would expect a less polite response than if the two of you are simply sharing the residence. If your parents are visiting at the time, your request to them is likely to take a different form, such as option 3.

According to **speech act theory,** utterances both state something and do something (Searle, 1979). In Table 7.1, utterances 1 to 6 state the speaker's desire for a drink (or specifically for lemonade), whereas utterances 7 and 8 do not. But all eight of the utterances perform an action; each has the force of a request. The significance of an utterance is not its literal meaning, but what it contributes to the work of the interaction in which it occurs (Geis, 1995). The use of language to perform actions is rule governed; these rules influence both the creation and the interpretation of speech acts. To achieve accurate communication, both speaker and hearer must be aware of these rules. Miscommunication is caused not only by the lack of shared meaning of symbols, but also by a lack of shared understanding of the rules governing the use of speech to perform actions.

To determine whether the message has achieved the intended effect, the speaker relies on the feedback provided by the listener's reaction. If the reaction indicates that the listener interpreted the message accurately, the speaker may elaborate, change the topic, or end the interaction. However, if the reaction suggests that the listener inferred a meaning different from the intended one, the speaker will often attempt to send the same message, perhaps using different words and gestures. For example, when Jim asked Susan, a coworker, for a date, Susan replied that she liked him as a friend and that she was busy Saturday night. Her intended message was that she was not interested in developing a relationship with Jim. Several days later, Jim tried to give Susan six red roses. Inferring that Jim had not received her intended message, Susan refused the roses and told Jim directly that she was not interested in dating him.

The Cooperative Principle Mutual understanding is a cooperative enterprise. Because language does not convey thoughts and feelings in an unambiguous manner, people must work together to attain a shared understanding of each others' utterances (Goffman, 1983). A speaker must cooperate with a listener by formulating the content of speech acts in a manner that reflects the listener's way of thinking about objects, events, and relationships.

In turn, the listener must cooperate by actively trying to understand. He or she must go beyond the literal meanings of words to infer what the speaker is really saying. A listener must make a creative effort to cope with a speaker's tendency to formulate speech acts indirectly. Without such an effort, a listener would not understand speech acts that leave out words ("Paper come?"), abbreviate familiar terms ("See ya in calc."), and include vague references ("He told him he would come later.").

According to Grice (1975), listeners assume that most talk is based on the **cooperative principle.** That is, listeners ordinarily assume that the speaker is behaving cooperatively by trying to be informative (giving as much information as is necessary and no more), truthful, relevant to the aims of the ongoing conversation, and clear (avoiding ambiguity and wordiness).

The cooperative principle is more than a code of conversational etiquette. It is crucial to the accurate transmission of meaning. Often, a listener can reach a correct understanding of otherwise ambiguous talk only by assuming that the speaker is trying to satisfy this principle. Consider, for example, how the maxim of relevance enables the conversationalists to understand each other in the following exchange:

TONY: I'm exhausted.

CAROLYN: Fred will be back next Monday.

On the surface, Carolyn's statement seems unrelated to Tony's declaration. In some contexts, we might infer that she has changed the subject, indirectly sending the message that she does not care about Tony's physical state. In fact, however, Carolyn is stating that she and Tony won't have to work as hard after their colleague Fred returns to the office next week. But why does she expect that Tony will understand this? Because she expects him to assume that she is adhering to the relevance maxim—that her comment relates to what he said.

The cooperative principle is also crucial for speech forms like sarcasm or understatement to succeed. In sarcasm or understatement, speakers want listeners to recognize that their words mean something quite different from their literal interpretation. One way we signal listeners that we intend our words to imply something different is by obviously violating one or two component maxims of the cooperative principle while holding to the rest. Consider Carrie's sarcastic reply when asked what she thought of the lecturer: "He was so exciting that he came close to keeping most of us awake the first half hour." By flouting the maxim of clarity (responding in an unclear, wordy

BOX 7.1 The Linguistic Relativity Hypothesis

Does the language we speak influence the way we think about and experience the world? The most famous theory on this question—the Sapir-Whorf linguistic relativity hypothesis—holds that language "is not merely a reproducing instrument for voicing ideas, but is itself a shaper of ideas, the program and guide for the individual's mental activity" (Whorf, 1956). Two forms of this hypothesis—strong and weak—have been proposed.

According to the strong form of the linguistic relativity hypothesis, language determines our perceptions of reality, so we cannot perceive or comprehend distinctions that don't exist in our own language. Orwell's description of *Newspeak,* the language developed by the totalitarian rulers in his novel *Nineteen Eighty-Four,* portrays in frightening terms how language restricts thought:

> Don't you see that the whole aim of Newspeak is to narrow the range of thought? In the end we shall make thoughtcrime literally impossible because there will be no words in which to express it. . . . Every year fewer and fewer words, and the range of consciousness always a little smaller. . . . The revolution will be complete when the language is perfect.
>
> —George Orwell, *Nineteen Eighty-Four* (1949; pp. 46–47)

Orwell's description suggests that language determines thought through the words it makes available to people. We cannot talk about objects or ideas for which we lack words. The ways we think about the world are determined by the way our language slices up reality.

The strong form of the linguistic relativity hypothesis has not fared well in research. Consider some of the facts. Some languages have only two basic words ("dark" and "white") to cover the whole spectrum of colors. Yet people from these and all other known language groups can discriminate between and communicate about whatever large numbers of colors they are shown (Heider & Olivier, 1972). Most likely any concept can be expressed in any language, though not with the same degree of ease and efficiency. Before either TV or the word "television" existed, for example, someone undoubtedly referred to the concept of "a device that can transmit pictures and sounds over a distance." When new concepts are encountered, people invent words ("laser") or borrow them

way) while still being informative, truthful, and relevant, Carrie implies that the lecturer was in fact a bore.

The Perspective-Taking Model

A third model is based on symbolic interaction theory (see Chapter 1). It views the process of communication as both creating and reflecting a shared context between speaker and listener. This approach maintains that symbols do not have a meaning that is invariant across situations (see Box 7.1). According to the **perspective-taking model,** communication involves the exchange of messages using symbols whose meaning grows out of the interaction itself.

Communication Process Communication involves the use of verbal and nonverbal symbols whose meaning depends on the shared context created by the participants. The development of this shared context requires reciprocal role taking, in which each participant places himself or herself in the role of the other in

BOX 7.1 Continued

from other languages ("sabotage" from French, "goulash" from Hungarian).

Thus, the strong linguistic relativity hypothesis that language determines thought has found little support. But there is considerable evidence for a weak form of this hypothesis. The weak form of the linguistic relativity hypothesis says that each language facilitates particular forms of thinking because it makes some events and objects more easily codable or symbolizable. In fact, the availability of linguistic symbols for objects or events has been shown to have two clear effects: (1) It improves the efficiency of communication about these objects and events. (2) It enhances success in remembering them.

Counting is difficult for people whose language does not include numbers. The Piraha, a group living in the Amazon, have only two words for numbers, words that mean "one" and "two." When an experimenter lined up several batteries and asked a member of the tribe to match it, the member did well when the line contained two or three, but had a difficult time if there were more than three batteries in the experimenter's line (Gordon, 2004).

Communication efficiency is often improved when the language includes labels that distinguish among similar objects. For example, one tribal group, the Hanunoo, for whom rice is a staple food, have 92 names for rice (Brown, 1965). Each name conveys the shape, color, texture, state, and so on, of a different type of rice. This makes communication easy and precise. To convey the same information in English would be possible, but less efficient.

The availability of linguistic symbols also affects memory for objects and persons. This was shown in a study that involved subjects who spoke English or Chinese (Hoffman, Lau, & Johnson, 1986). This study used English- and Chinese-language descriptions of two people whose traits could be easily labeled in English but not in Chinese and of two other people whose traits could be easily labeled in Chinese but not in English. Three groups of participants read the descriptions: English monolinguals, Chinese-English bilinguals who read in Chinese, and Chinese-English bilinguals who read in English. The participants' memory of the descriptions was assessed. The results showed that memory was much better when the information about the target conformed to labels in the participant's language of processing. These results lend support to the weak form of the linguistic relativity hypothesis.

an attempt to view the situation from the other's perspective. The context created by the ongoing interaction changes from minute to minute; each actor must be attentive to these changes in order to communicate successfully as both speaker and listener.

Communication Accuracy In the perspective-taking model, communication is much more than transmitting and receiving words with fixed, shared meanings. Conversationalists must select and discover the meanings of words through their context. In ordinary social interaction, the meanings of whole sentences and conversations may be ambiguous. Speakers and listeners must jointly work out these meanings as they go along.

Successful communication depends on **intersubjectivity;** each participant needs information about the other's status, view of the situation, and plans or intentions. Strangers rely on social conventions and rules about interpersonal communication. They categorize other participants and use stereotypes as a basis for making inferences about the plans and intentions of the other person(s) who are present in the setting. Notice that this practice perpetuates stereotypes via the self-fulfilling prophecy (see Chapter 19). Persons who know each other can draw on their past experience with each other as a basis for effective communication.

Interpersonal Context According to this model, both the production and the interpretation of communication is heavily influenced by the interpersonal context in which it occurs (Giles & Coupland, 1991). This context influences communication through norms, cognitive representations of prior similar situations, and emotional arousal.

Every social situation includes norms regarding communicative behavior. These norms specify what topics are appropriate and inappropriate for discussion, what language is to be used, and how persons of varying status should be addressed. Depending on these norms, we use one or another of various speech repertoires, ways of communicating the same literal message that vary in words, tone, and so on (Giles & Coupland, 1991). Imagine a man who wishes another person to close a door. To his son in his home, he might say, "Close the door." To his son at work, he might say, "Please close the door, Tom." To an employee, he could say, "Would you close the door?" These different ways of making a request reflect differences in speech rules, which depend on the relationship between speaker and listener and on the setting.

Each new situation evokes representations of prior similar situations and the language one has used or heard in them (Chapman et al., 1992). These conversational histories provide us with the contents of our speech repertoires. Each of us has a set of things we say when we meet a stranger our own age at a party; these are opening lines that in the past have been effective in facilitating conversation with strangers at parties. If, instead, you met the same stranger on a plane, you might use different speech acts.

The processing of messages by listeners is also influenced by these contextual factors. Listeners interpret messages in light of the rules operating in situations, their past experience, and the emotions elicited in them. When speaker and listener have the same understanding of the normative demands, communication should be quite accurate. Similarly, if a situation evokes the same representations and emotions in both, it is likely that the listener will accurately interpret the speaker's message. Communication across group and cultural boundaries is often difficult precisely because speaker and listener differ in their assumptions and experiences, even though they may speak the same language.

The accuracy of indirect or covert communication depends heavily on shared knowledge. In a series of experiments, participants were asked to compose messages—either in writing or on videotape—taking a position they did not believe in. They were also instructed to try to covertly inform the reader or viewer that they did not hold that position. Most of the participants used the device of including false information about themselves in the message. Friends of the subjects who read or viewed the message detected the deception, whereas strangers did not (Fleming, Darley, Hilton, & Kojetin, 1991).

Members of a group share a **linguistic intergroup bias** (Maass & Arcuri, 1992). That is, there are subtle and systematic differences in the language we use to describe events as a function of our group membership and the group to which the actor or target belongs. We describe other members of our own group behaving properly and members of out-groups behaving improperly at very abstract levels (say, "Jim (in-group member) is helpful"; "George (out-group member) is aggressive."). This encourages positive stereotypes of us and negative stereotypes of them. When in-group members behave badly or out-group members behave well, we describe the events concretely ("Jim pushed that guy out of his way"; "George held the door for a woman carrying a baby.") This technique encourages an attribution to the individual rather than to the group. In a study of these processes, participants were asked to write messages describing a man or woman behaving in ways consistent or inconsistent with gender stereotypes. As predicted, stereotype-consistent behaviors were described abstractly, and stereotype-inconsistent behaviors were described concretely. These messages were then given to other participants, who read them and then answered questions about the incidents. As predicted, stereotype-consistent behaviors were attributed to group membership, whereas stereotype-inconsistent behaviors were attributed to individuals (Wigboldus, Semin, & Spears, 2000).

Sociolinguistic Competence To attain mutual understanding, language performance must be appropriate to the social and cultural context. Otherwise, even grammatically acceptable sentences will not make sense. "My mother eats raw termites" is grammatically correct and meaningful; it reflects linguistic competence. But as a serious assertion by a North American, this utterance would draw amazed looks. It expresses an idea that is incongruous with American culture, and listeners would have difficulty interpreting it. In a termite-eating culture, however, the same utterance would be quite sensible. This example shows that successful communication requires **sociolinguistic competence**—knowledge of the implicit rules for generating socially appropriate sentences. Such sentences make sense to listeners, because they fit with the listeners' social knowledge (Hymes, 1974).

Speech that clashes with what is known about the social relationship to which it refers suggests that a speaker is not sociolinguistically competent (Grimshaw, 1990). Speakers are expected to use language that is appropriate to the status of the individuals they are discussing and to their relationship of intimacy. For example, competent speakers would not state seriously, "The janitor ordered the president to turn off the lights in the Oval Office." They know that low-status persons do not "order" those of much higher status; at most, they "hint" or "suggest." Referring to a relationship of true intimacy, sociolinguistically competent speakers would not say, "The lover bullied her beloved." Rather, they would select such socially appropriate verbs as "coaxed" or "persuaded." In short, competent speakers recognize that social and cultural constraints make some statements interpretable and others uninterpretable in a given situation.

Thus, successful communication is a complex undertaking. A speaker must produce a message that has not only an appropriate literal meaning, but also an intention or goal appropriate to the relationship and setting. The message must reflect the present degree of intersubjectivity between speaker and hearer, consistent with the interactional context. The message must also signify the statuses of the participants (Geis, 1995). Given these requirements, it is remarkable that each of us communicates successfully many times each day.

NONVERBAL COMMUNICATION

Have you ever been in a situation in which you tried to communicate without using words? Perhaps you were interacting with someone who was deaf, or someone who was too far away for your words to be heard. Imagine that you are looking out of a window of your third-floor dorm room or apartment. You notice a man on the sidewalk below, dressed immaculately in a three-piece suit, pacing back and forth. He looks up and sees you and immediately begins to gesture. He points to you, then to some other window, and then to his watch. His movements are quick and sharp. His face is tense. What is he trying to communicate to you?

Even without the use of words, most of us can make some inferences about the man's message and emotional state. We do so by interpreting his nonverbal communication. This section examines three questions concerning nonverbal communication: (1) What are the major types of nonverbal communication? (2) What is communicated by the human face? (3) What is gained and what problems arise because nonverbal and verbal communication are combined in ordinary interaction?

Types of Nonverbal Communication

By one estimate, the human face can make some 250,000 different expressions (Birdwhistell, 1970). In addition to facial expressions, nonverbal communication uses many other bodily and gestural cues. Four major types of nonverbal cues (summarized in Table 7.2) are described next.

Paralanguage Speaking involves a great deal more than the production of words. Vocal behavior includes loudness, pitch, speed, emphasis, inflection, breathiness, stretching or clipping of words, pauses,

TABLE 7.2 TYPES OF NONVERBAL COMMUNICATION

Type of Cue	Definition	Examples	Channel
Paralanguage	Vocal but nonverbal behavior involved in speaking	Loudness, speed, pauses in speech	Auditory
Body language (kinesics)	Silent motions of the body	Gestures, facial expressions, eye gaze	Visual
Interpersonal spacing (proxemics)	Positioning of body at varying distances and angles from others	Intimate closeness, facing head-on, looking away, turning one's back	Primarily visual; also touch, smell, and auditory
Choice of personal effects	Selecting and displaying objects that others will associate with you	Clothing, makeup, room decorations	Primarily visual; also auditory and smell

and so on. All the vocal aspects of speech other than words are called **paralanguage.** This includes such highly communicative vocalizations as moaning, sighing, laughing, and even crying. Shrillness of voice and rapid delivery communicate tension and excitement in most situations (Scherer, 1979). Various uses and interpretations of paralinguistic and other nonverbal cues will be examined later in this chapter. For now, see how many distinct meanings you can give to the sentence, "George is on the phone again" by varying the paralinguistic cues you use.

Body Language The silent movement of body parts—scowls, smiles, nods, gazing, gestures, leg movements, postural shifts, caressing, slapping, and so on—all constitute **body language.** Because body language entails movement, it is known as *kinesics* (from the Greek *kinein* meaning "to move"). Whereas paralinguistic cues are auditory, we perceive kinesic cues visually. The body movements of the man in our example were probably particularly useful to you in interpreting his feelings and intentions.

The handshake is a common nonverbal behavior. There are a variety of beliefs about the meaning of a handshake, depending on whether it is firm or limp, dry or damp. Research on how we interpret handshakes involved four trained coders (two men, two women); each shook hands twice with 112 men and women and rated the participants on four measures of personality. The man or woman with a firm handshake was rated by the coders as extroverted and emotionally expressive and given low ratings on shyness. Women who shook hands firmly were also rated as open to new experience (Chaplin, Phillips, Brown, Clanton, & Stein, 2000). Thus, a handshake can make a strong first impression and influence future interactions.

Interpersonal Spacing We also communicate nonverbally by using **interpersonal spacing** cues—positioning ourselves at varying distances and angles from others (for example, standing close or far away, facing head-on or to one side, adopting various postures, and creating barriers with books or other objects). Because proximity is a major means of communication between people, this type of cue is also called proxemics. When there is very close positioning, proxemics can convey information through smell and touch as well.

Choice of Personal Effects Though we usually think of communication as expressed through our bodies, people also communicate nonverbally through the personal effects they select—their choices of clothing, hairstyle, make-up, eyewear (contact lenses), and the like. A uniform, for example, may communicate social status, political opinion, lifestyle, and occupation, revealing a great deal about how its wearer is likely to behave (Joseph & Alex, 1972). You may have made assumptions about the status and lifestyle of the man in our sketch based on the fact that he wore a three-piece suit. The deliberate use of personal

© J. Applewhite/AP/Wide World Photo

The meaning of a gesture can vary greatly from one culture to another. During his inauguration on January 20, 2005, President George W. Bush used a gesture known as "Hook 'em, horns," the salute of the University of Texas Longhorns. What he apparently didn't know was that in Mediterranean cultures, this gesture implies that a man has an unfaithful wife, and in parts of Africa it is used to impose a curse on another person. Photos of his use of the gesture published in those countries understandably upset many of their residents.

effects to communicate impressions is discussed in Chapter 9.

For the most part, nonverbal cues, like language, are learned rather than innate. As a result, the meanings of particular nonverbal cues may vary from culture to culture. Other features of nonverbal communication may have universal meanings, however. These universals are based in our biological nature.

Whether learned or innate, the meaning of a nonverbal behavior depends on the behavioral context in which it occurs. Participants in one study rated individual nonverbal behaviors, and then pairs consisting of one nonverbal and one overt behavior. The ratings involved adjective pairs that measured the evaluation (good-bad), potency (strong-weak), and activity (active-passive) of the behavior or pair. The rating of a nonverbal behavior varied depending on the overt behavior it was paired with (Rashotte, 2002).

What's in a Face?

The face is an important communication channel. Typically, we pay attention to the face of persons with whom we interact. Moreover, the face is capable of many nonverbal behaviors; one dictionary lists 98 behaviors, of which 25 involve the face (Rashotte, 2002). They include barring the teeth, closing one's eyes, frowning, grinning, licking the lips, nodding, raising one's eyebrows, and smiling. The physical features of the face combined with these movements convey a variety of messages, including information about social identities, personality, and emotions. There is extensive information on the latter, which is discussed in Chapter 10.

The physical features of the face, including color, often provide cues to racial or ethnic identity. The features, in combination with grooming, make-up, and jewelry, virtually always indicate gender. Thus, inferences about two important social identities are made the moment we see someone, and these inferences shape our interaction with that person.

Physiognomy, the art of "reading" faces, is based on the belief that personality traits can be inferred from facial features. In research designed to test this, participants were given photographs and descriptions of a target person. The photographs were selected based on ratings by other participants on the confidence, charisma, and dominance of the person in the photo; one half of the photos were of people rated high on these, and the other half were of people rated low on them. Participants were asked to rate the target on 13 personality scales. When the verbal description was ambiguous, the characteristics of the photo significantly influenced ratings (Hassin & Trope, 2000). This research suggests that people do make inferences about personality based on facial features.

Later in this chapter, we discuss research on facial maturity and how it influences interaction.

Combining Nonverbal and Verbal Communication

When we speak on the telephone or shout to a friend in another room, we are limited to communicating through verbal and paralinguistic channels. When we wave to arriving or departing passengers at the airport, we use only the visual channel. Ordinarily, however, communication is multichanneled. Information is conveyed simultaneously through verbal, paralinguistic, kinesic, and proxemic cues.

What is gained and what problems are caused when different communication channels are combined? If they appear to convey consistent information, they reinforce each other, and communication becomes more accurate. But if different channels convey information that is inconsistent, the message may produce confusion or even arouse a suspicion of deception. In this section, we examine some outcomes of apparent consistency and inconsistency among channels.

Reinforcement and Increased Accuracy The multiple cues we receive often seem redundant, each carrying the same message. A smile accompanies a compliment delivered in a warm tone of voice; a scowl accompanies a vehemently shouted threat. But multiple cues are seldom entirely redundant, and they are better viewed as complementary (Poyatos, 1983). The smile and warm tone convey that the compliment is sincere; the scowl and vehement shout imply that the threat will be carried out. Thus, multiple cues convey added information, reduce ambiguity, and increase the accuracy of communication (Krauss, Morrel-Samuels, & Colasante, 1991).

Taken alone, each channel lacks the capacity to carry the entire weight of the messages exchanged in the course of a conversation. By themselves, the verbal aspects of language are insufficient for accurate communication. Paralinguistic and kinesic cues supplement verbal cues by supporting and emphasizing them. The importance of paralinguistic cues is illustrated in a study of students from a Nigerian secondary school and teachers' college (Grayshon, 1980). Although these students took courses in English and knew the verbal language well, they did not know the paralinguistic cues of British native speakers. The students listened to two British recordings with identical verbal content. In one recording, paralinguistic cues indicated that the speaker was giving the listener a brush-off. In the other recording, paralinguistic cues indicated that the speaker was apologizing. Of 251 students, 97 percent failed to perceive any difference in the meanings the speaker was conveying. Failure to distinguish a brush-off from an apology could be disastrous in everyday communication. Accurate understanding requires paralinguistic as well as verbal knowledge.

Our accuracy in interpreting events is greatly enhanced if we have multiple communication cues rather

© Jane Scherr/Jeroboam

Successful communication is a complex process. These two kids are combining language, interpersonal spacing, and body language to accomplish the sharing of a secret.

than verbal information alone. The value of a full set of cues was demonstrated in a study of students' interpretations of various scenes (Archer & Akert, 1977). Students observed scenes of social interaction that were either displayed in a video broadcast or described verbally in a transcript of the video broadcast. Thus students received either full, multichannel communication or verbal cues alone. Afterward, students were asked to answer questions about what was going on in each scene—questions that required going beyond the obvious facts. Observers who received the full set of verbal and nonverbal cues were substantially more accurate in interpreting social interactions. For instance, of those receiving multichannel cues, 56 percent correctly identified which of three women engaged in a conversation had no children; this compared with only 17 percent of those limited to verbal cues. These findings convincingly demonstrate the gain in accuracy from multichannel communication.

Resolving Inconsistency At times, the messages conveyed by different channels appear inconsistent with one another. This makes communication and interaction problematic. What would you do, for example, if your instructor welcomed you during office hours with warm words, a frowning face, and an annoyed tone of voice? You might well react with uncertainty and caution, puzzled by the apparent inconsistency between the verbal and nonverbal cues you were receiving. You would certainly try to figure out the instructor's true feelings and desires, and you might also try to guess why the instructor was sending such confusing cues.

The strategies people use to resolve apparently inconsistent cues depend on their inferences about the reasons for the apparent inconsistency (Zuckerman, DePaulo, & Rosenthal, 1981). Inconsistency could be due to the communicator's ambivalent feelings, to poor communication skills, or to an intention to deceive. A large body of research has compared the relative weight we give to messages in different channels when we do not suspect deception.

In one set of studies, people judged the emotion expressed by actors who posed contradictory verbal, paralinguistic, and facial signals (Mehrabian, 1972). These studies showed that facial cues were most important in determining which feelings are interpreted as true. Paralinguistic cues were second, and verbal cues were a distant third. Later research, exposing receivers to more complete combinations of visual and auditory cues, replicated the finding that people rely more on facial than on paralinguistic cues when the two conflict. This preference for facial cues increases with age from childhood to adulthood, indicating that it is a learned strategy (DePaulo, Rosenthal, Eisenstat, Rogers, & Finkelstein, 1978).

People also use social context to help them judge which channel is more credible (Bugenthal, 1974). They consider whether the facial expression, tone of voice, or verbal content is appropriate to the particular social situation. If people recognize a situation as highly stressful, for example, they rely more on the cues that seem consistent with a stressful context (such as a strained tone of voice) and less on cues that seem to contradict it (such as a happy face or a verbal assertion of calmness). If the emotional expression is ambiguous, situational cues determine the emotion that observers attribute to the person (Carroll & Russell, 1996). For example, a person in a frightening situation displaying an expression of moderate anger was judged to be afraid. In short, people tend to resolve apparent inconsistencies between channels in favor of the channels whose message seems most appropriate to the social context.

SOCIAL STRUCTURE AND COMMUNICATIONS

So far, this chapter has examined the nature of verbal and nonverbal communication, and some consequences of the fact that everyday communication usually combines the two. But how do social relationships shape communication? And how does communication express, maintain, or modify social relationships? These questions pinpoint social psychology's concern with the reciprocal impacts of social structure and communication on each other. This section examines four aspects of these impacts. First, it discusses gender differences in communication. Second, it considers the links between styles of speech and position in the social stratification system. Third, it analyzes the ways in which communication creates and expresses the two central dimensions of relationships—status and intimacy. Finally, it examines the social norms that regulate interaction distances and some of the outcomes when these norms are violated.

Gender and Communication

A fundamental question about how social structure influences communication is whether there are systematic differences between men and women in communication style.

Many empirical studies have been conducted since 1970. Typically, each study focuses on one or two aspects of interaction and compares men and women on it. The most widely studied aspect has been interruptions. Research by Zimmerman and West (1975) reported that in casual conversation of mixed-gender dyads, men interrupted women much more frequently than the reverse. Other research suggested that women's speech involves more frequent use of tag questions ("It's really hot, isn't it?"), hedges ("In my opinion, . . . "), and disclaimers ("I may be wrong but . . . "). These three are often linked and have been said to indicate that women's speech is more tentative than men's. Some studies report that women are more likely to use intensifiers ("It's really hot, isn't it?"). In the nonverbal realm, women smile more often than men and are less likely to look at the other person as they interact.

These and other findings of gender differences (see Box 7.2 for another example) are the basis for the assertion that there are vast differences in style of interaction between men and women. In addition to academic researchers who take this position, it has been popularized in books such as *You Just Don't Understand: Women and Men in Conversation* (Tannen, 1991), and *Men Are From Mars, Women Are From*

Venus (Gray, 1993). The common academic interpretation of these differences is that they reflect the fact that men have greater power than women. Thus, interruptions, declarative statements instead of tentative ones, and speech without intensifiers all reflect the possession of power—that is, the stratification system of the society.

Research on gender differences in communication has gotten more sophisticated in recent years. Instead of descriptive research comparing men and women on a small number of behaviors, researchers now study these processes in specific social contexts. Thus, researchers study how gender and contextual variables such as type of relationship, group task, or authority structure interact to influence communication. For example, studies in the 1970s and 1980s found that when men attempted to change the topic of conversation, they succeeded 96 percent of the time; in contrast, attempts by women succeeded only 36 percent of the time (Fishman, 1983). This was interpreted as reflecting the difference in status of men and women. But if we take a broader look, we see that (1) there are several types of topic shifts in interaction, and (2) any group of three or more people tends to develop an internal status structure that is influenced by the setting, task, and characteristics of the specific people present. A recent study of six-person task-oriented groups found that topic shifts are more sensitive to the internal status structure of the group than to gender (Okamoto & Smith-Lovin, 2001). Moreover, topic shifts often occurred following a lapse in the discussion or an obvious conclusion to the current topic, suggesting that they are not displays of power.

Research on other aspects of communication has reached similar conclusions. A study of nonverbal behavior recruited participants in a company's headquarters; 42 employees each participated in two interactions with another, randomly chosen employee. As a result, the dyads varied in the corporate status of participants. There were 10 all-male, 9 all-female, and 25 mixed-gender dyads. During each interaction, the pair was given two tasks. Interaction was video- and audiotaped, and the tapes were coded by trained observers. The data were analyzed by gender and by corporate status. Some nonverbal behavior varied by gender and some varied by status. The differences associated with status did not correspond to the differences associated with gender. Although women smiled more, there were no differences in smiling by status. There were no stable differences across gender or status, suggesting that the differences observed reflected local or corporate practices and participants' motives (Hall & Friedman, 1999).

In short, men and women do not form two homogeneous groups with respect to communication style (Cameron, 1998). Generalizations about gender and communication require taking into account the context and particular local (group, organizational) communication practices (Eckert & McConnell-Ginet, 1999). It looks like men and women from Mars will communicate differently from men and women from Venus.

Social Stratification and Speech Style

The way we speak both reflects and recreates our social relationships (Giles & Coupland, 1991). Every sociolinguistic community recognizes variation in the way its members talk. One style is usually the preferred or standard style. In addition to this preferred style, there are often other, nonpreferred styles.

Consider an example of each style. As you enter a theater, a young man approaches you. He asks, "Would you please fill out this short survey for me?" Depending on your mood, you might comply with his request. But what if he asked, "Wud ja ansa sum questions?" Many people would be less likely to comply with this request.

The first request employs standard American English. **Standard speech** is defined as characterized by diverse vocabulary, proper pronunciation, correct grammar, and abstract content. It takes into account the listener's perspective. Note the inclusion of "please" in the first request, which indicates that the speaker recognizes that he is asking for a favor. **Nonstandard speech** is defined as characterized by limited vocabulary, improper pronunciation, incorrect grammar, and directness. It is egocentric; the absence of "please" and "for me" in the second request makes it sound like an order, even though it is phrased as a question.

BOX 7.2 Flirting

A distinctive class of communicative behaviors is flirting, or courtship signaling (Birdwhistell, 1970). It is widely believed that men initiate contact, and therefore relationships, with women. Often it is the man who physically approaches and initiates verbal interaction. But a small body of research indicates that women take the initiative, using nonverbal signals, in encouraging the man to initiate verbal interaction.

In her original research, Moore (1985) defined **flirting** as a class of nonverbal facial expressions and behavior exhibited by women that serves to attract the attention and elicit the approach of a man. Of course, men can and do exhibit the same kinds of behaviors when attempting to attract women. Monica Moore has studied female flirting for 20 years. She began by observing women and recording their nonverbal behavior. Observation of 200 White women, judged to be ages 18 to 35, yielded a catalogue of flirtatious behavior. A woman not accompanied by a man was selected at random and observed for at least 30 minutes in settings where there were at least 20 people present, both men and women. Observers recorded every behavior of the focal subject and its consequences. Flirting, or a nonverbal solicitation behavior, was defined as a behavior that resulted in a man's attention within 15 seconds. The resulting catalogue includes 52 behaviors, including facial and head movements (for example, room-encompassing glance, fixed gaze, head toss), gestures (for example, palming—palm facing a man; primping—touching hair or clothing), and posture (for example, lean toward man, touch).

Subsequent research (Moore & Butler, 1989) describes behaviors that attract male attention and those that maintain his attention after interaction begins. Male attention is likely to follow a room-encompassing glance, a smile while looking at him, patting or smoothing the hair, the "lip lick," or a head toss. Male attention is maintained by frequent head nods while he talks, leaning close to him, and touching or brushing part of the body against him.

Flirting is a complex behavior that conveys interest in being approached by another person. These young women are using posture, smiles, and direct gaze to attract the attention of the young men.

Moore provides contextual evidence for the assertion that these behaviors are courtship signals. If these behaviors are intended to attract male attention, we should observe them in contexts where such solicitations are likely, such as singles bars, but not in settings where no men are present, such as women's group meetings.

BOX 7.2 Continued

THE IMPACT OF SOCIAL CONTEXT ON DISPLAY FREQUENCY AND NUMBER OF APPROACHES

	Singles Bar	Snack Bar	Library	Women's Meetings
Number of subjects	10	10	10	10
Total number of displays	706	186	96	47
Mean number of displays	70.6	18.6	9.6	4.7
Mean number of categories used	12.8	7.5	4.0	2.1
Number of approaches to the subject by a man	38	4	4	0
Number of approaches to a man by the subject	11	4	1	0

Note: The tabulated data are for a 60-minute observation interval. Asymmetry in display frequency: $x^2 = 25.079$, $df = 3$, $p < 0.001$; asymmetry in number of categories used: $x^2 = 23.099$, $df = 3$, $p < 0.001$.

Source: Adapted from Moore, 1985.

She studied 10 women in each of four social settings: singles bar, university snack bar, university library, and women's center meeting. Again, focal sampling was employed; a woman was observed only if at least 25 people were present and she was not accompanied by a man. The display of courtship signals was clearly context specific (see table on following page). Women were much more likely to engage in these behaviors in the singles bar, and least likely to engage in them at women's center meetings.

A recent study (Moore, 1995) looks at the display of nonverbal solicitation behaviors by adolescent girls. Observations were made at school events, swimming pools, and shopping malls, when there were at least 20 boys and girls present. One hundred girls judged to be between 13 and 16 were randomly selected. Each girl was observed for at least 30 minutes.

Overall, girls exhibited the same behaviors, though much less frequently than adult women in similar settings. Also, the behaviors often had an exaggerated and playful quality. These observations are consistent with the conclusion that young women are learning to flirt and rehearsing these behaviors in appropriate contexts. This is another example of trying on gender, as discussed in Chapter 3.

Moore's research program clearly documents female influence on the initiation of relationships. Flirting is an important component of the development of relationships. The selection of a superior mate—one with desirable qualities—is important to both men and women. Traditional gender role norms specify the man as the initiator of interaction. Flirting allows a woman to influence who approaches her. It also signals the man that his overture will be welcome, reducing his anxiety about making an approach.

In the United States, as in many other countries, speech style is associated with social status (Giles & Coupland, 1991). The use of standard speech is associated with high socioeconomic status and with power. People in positions of economic and political power are usually very articulate and grammatically correct in their public statements. In contrast, the use of nonstandard speech is associated with low socioeconomic status and low power.

Speech style is also influenced by the interpersonal context. In informal conversations with others of equal status, such as at some parties or in bars, we often use nonstandard speech, regardless of our socioeconomic status. In more formal settings, especially public ones, we usually shift to standard speech. Thus, our choice of standard or nonstandard speech gives listeners information about how we perceive the situation.

Studies in a variety of cultures have found systematic differences in how people evaluate speakers using standard and nonstandard speech. In one study, students in Kentucky listened to tape recordings of young men and women describing themselves. Four of the recordings, two by men and two by women, were of speakers with "standard" American accents. Four others, identical in content, were of speakers with Kentucky accents. On the average, students gave the standard speakers high ratings on status and the nonstandard speakers low ratings on status (Luhman, 1990).

Nonstandard speech involves limited vocabulary, is rooted in the present, and does not allow for elaboration and qualification of ideas. As a result, some analysts advocate so-called deficit theories, which claim that people who use nonstandard speech are less capable of abstract and complex thought. These theories also claim that nonstandard speech styles are typical of lower class, Black, and other culturally disadvantaged groups in America, Great Britain, and other societies. Combining these two claims, deficit theorists argue that the children from disadvantaged groups perform poorly in school because their restricted language makes them cognitively inferior. Their poor academic performance in turn leads to unemployment and poverty in later life.

The strongest criticism of deficit theories has come from Labov (1972). Based on interviews in natural environments, he demonstrates that "Black English," which has been described as nonstandard speech, is every bit as rich and subtle as standard English. Black English differs from standard English mainly in surface details like pronunciation ("ax" = *ask*) and grammatical forms ("He be busy" = *He's always busy*). Nonstandard speech may appear impoverished because nonstandard speakers feel less relaxed in the social contexts where they are typically observed (such as schools or interviews), and so they limit their speech. Social researchers or other "outsiders" who observe nonstandard speakers may also inhibit their language (Grimshaw, 1973). When interviewed by a member of their own race, for instance, Black job applicants used longer sentences and richer vocabularies and employed words more creatively (Ledvinka, 1971). Overall, speech differences between groups have not been shown to reflect differences in cognitive ability (Thorlundsson, 1987), and deficit theories have not received much empirical support.

In 1996, the Oakland, California, school board decided that Black English is not a nonstandard language, but a distinctive language, **ebonics** (ebony phonics). The board decided to make classes available in ebonics in the hope that it would improve Black students' comprehension of schoolwork. Several organizations now recognize the legitimacy and cultural value of this language spoken by many African Americans. Some teachers use ebonic communication techniques in an effort to enhance the comprehension and learning of African American students (Bohn, 2003). A field researcher in one high school in Washington, D.C., observed that for many African American students, ebonics was the preferred speech style; they associated Standard English with White, majority culture, and its history of oppressing Blacks (Fordham, 1999). For these students, Standard English is the nonstandard vernacular, and they dissed those Blacks who used it. These students "leased" Standard English, that is, used it when they had to while in school, from 9:00 a.m. to 3:00 p.m., but not outside the school building. Their refusal to adopt standard English is a conscious form of resistance to what they perceive as the White-dominated school structure.

Communicating Status and Intimacy

The two central dimensions of social relationships are status and intimacy. Status is concerned with the exercise of power and control. Intimacy is concerned with the expression of affiliation and affection that creates social solidarity (Kemper, 1973). Verbal and nonverbal communication express and maintain particular levels of intimacy and relative status in relationships. Moreover, through communication we may challenge existing levels of intimacy and relative status and negotiate new ones (Scotton, 1983).

Communication can signal our view of a relationship only if we recognize which communication behaviors are appropriate for an expected level of intimacy or status, and which are inappropriate. The following examples suggest that we easily recognize when communication behaviors are inappropriate. What if you,

- Repeatedly addressed your mother as Mrs. _____?
- Used vulgar slang during a job interview?
- Draped your arm on your professor's shoulder as he or she explained how to improve your test answers?
- Looked away each time your beloved gazed into your eyes?

Each of these communication behaviors would probably make you uncomfortable, and they would doubtlessly cause others to think you inept, disturbed, or hostile. Each behavior expresses levels of intimacy or relative status easily recognized as inappropriate to the relationship. In the following section, we survey systematically how specific communication behaviors express, maintain, and change status and intimacy in relationships.

Status Forms of address clearly communicate relative status in relationships. Inferiors use formal address (title and last name) for their superiors (for example, "When is the exam, Professor Levine?"), whereas superiors address inferiors with familiar forms (first name or nickname; for example, "On Friday, Daphne"). Status equals use the same form of address with one another. Both use either formal (Ms./Mr./Mrs.) or familiar forms (Carol/Bill), depending on the degree of intimacy between them (Brown, 1965). When status differences are ambiguous, individuals may even avoid addressing each other directly. They shy away from choosing an address form because it might grant too much or too little status.

A shift in forms of address signals a change in social relationships, or at least an attempted change. During the French Revolution, in order to promote equality and fraternity, the revolutionaries demanded that everyone use only the familiar (*tu*) and not the formal (*vous*) form of the second-person pronoun, regardless of past status differences. Presidential candidates try to reduce their differences with voters by inviting the use of familiar names (John, George-Dubya). In cases where there is a clear status difference between people, the right to initiate the use of the more familiar or equal forms of address belongs to the superior (for example, "Why don't you drop that 'Doctor' stuff?"). This principle also applies to other communication behaviors. It is the higher status person who usually initiates changes toward more familiar behaviors such as greater eye contact, physical proximity, touch, or self-disclosure.

We each have a speech repertoire, different pronunciations, dialects, and a varied vocabulary from which to choose when speaking. Our choices of language to use with other people express a view of our relative status and may influence our relationships. People usually make language choices smoothly, easily expressing status differences appropriate to the situation (Gumperz, 1976; Stiles, Orth, Scherwitz, Hennrikus, & Vallbona, 1984). Teachers in a Norwegian town, for instance, were observed to lecture to their students in the standard language (Blom & Gumperz, 1972). When they wished to encourage student discussion, however, they switched to the local dialect, thereby reducing status differences. Note how your teachers also switch to more informal language when trying to promote student participation.

An experiment involving groups composed of a manager and two workers studied the effect of authority and gender composition of the group on verbal and nonverbal communication (Johnson, 1994). The researcher created a simulated retail store; the manager gave instructions to the subordinates and monitored their work for 30 minutes. The interaction was coded as it occurred. Authority affected verbal behavior; subordinates talked less, were less directive, and gave less feedback compared to superiors, regardless of gender. Gender affected the nonverbal behaviors of smiling and laughing; women in all-female groups smiled more than men in all-male groups.

Paralinguistic cues also communicate and reinforce status in relationships. People of higher status interrupt their partners more during conversations and talk more themselves. Inferiors grant status to others by not interrupting and by responding with "Mm-hmm" more frequently at appropriate intervals to indicate they are following the conversation. Among status equals, these paralinguistic behaviors are distributed more equally (Krauss & Chiu, 1998).

An experimental study of influence in small three-person groups systematically varied the paralanguage of one member (Ridgeway, 1987). This member, a confederate, was most influential when she spoke rapidly, in a confident tone, and gave quick responses. She was less influential when she behaved dominantly (that is, spoke loudly, gave orders) or submissively (that is, spoke softly, in a pleading tone). A subsequent study found that a person who spoke in a task-oriented style (that is, rapid speech, upright posture, eye contact) or a social style (that is, moderate volume, relaxed posture) was more influential (Carli, LaFleur, & Loeber, 1995). Persons who spoke using dominant or submissive paralanguage were less influential. Thus, engaging in the paralinguistic behaviors appropriate to the statuses of group members—in these experiments equals—enhances one's influence; engaging in behaviors inappropriate to one's status (say, like a superior toward equals) reduces one's influence.

Body language also serves to express status. When status is unequal, people of higher status tend to adopt relatively relaxed postures with their arms and legs in asymmetrical positions. Those of lower status stand or sit in more tense and symmetrical positions. The amount of time we spend looking at our partner, and the timing, also indicate status. Higher status persons look more when speaking than when listening, whereas lower status persons look more when listening than when speaking. Overall, inferiors look more at their partners, but they are also first to break the gaze between partners. Finally, superiors are much more likely to intrude physically on inferiors by touching or pointing at them (Dovidio & Ellyson, 1982; LaFrance & Mayo, 1978; Leffler, Gillespie, & Conaty, 1982).

Recent research has focused on the impact of facial maturity on the status and intimacy of one's interactions. In one study, judges rated 114 people (50 men, 64 women) on the extent to which each had a "baby face" (Berry & Landry, 1997). These 114 people kept diaries of their social encounters for 1 week, yielding records for 5,106 interactions. Men who were judged to have babyish faces reported less influence on (that is, lower status) and greater intimacy of their interactions with women compared to men with mature faces. Facial maturity was not related to variation in the interactions reported by women. Another study found a "baby-faced" overgeneralization effect (Zebrowitz, Voinescu, & Collins, 1996). Both men and women with babyish faces were perceived as more honest by observers.

An important phenomenon that both expresses and produces status differences is silencing. In many interactions, being silent is not a passive state reflecting the absence of a desire to communicate. The silence of one or more of the actors may reflect an active state produced by the ongoing interaction. A common form of silencing involves not replying to a comment or question addressed to you, which may silence the other person. Bodily movement may contribute to silencing, as when you turn away from someone and pick up the TV remote, or leave the room. Silencing can be an especially complex process when it occurs in a group setting, as illustrated by a detailed analysis of the silencing of one student during a classroom discussion (Leander, 2002). In response to a teacher's question about equal rights for women, one woman, Chelle, sitting in the back of the room says quietly, "No, we don't have equal rights." A young man in front of her gestures with his thumb over his shoulder and says, "We got somebody back here who says they don't have equal rights." The young man, by gesturing rather than looking at Chelle, and by invoking we-they (in-group vs. out-group identities) is attempting to silence her. Both students are sitting near friends, and the friends become engaged in the conversation, so that interacting groups are now attempting to control the discourse. This contest by the groups is facilitating by seating arrangements, and participants turn their bodies and direct their gaze in ways that signal alignment. Other techniques employed included speaking over a member of the other group, and ridicule of an example given by one of the women. Thus, silencing involves language, gaze, gesture, bodily orientation, and symbolic invocation of group ties within the setting.

Intimacy Communication also expresses another central dimension of relationships—intimacy. One way we signal intimacy or solidarity is by addressing each other with first names. The exchange of title and last names is common for strangers. In other languages, speakers express intimacy by their choice of familiar versus formal second-person pronouns. As noted earlier, the French can choose between the familiar *tu* or the formal *vous;* the Spanish have *tu* or *usted;* the Germans *du* or *sie;* and so on.

Our choice of language is another way to express intimacy. For example, the residents of a Norwegian town were found to use the formal version of their language with strangers and the local dialect with friends. They spoke the formal language when transacting official business in government offices, then switched to dialect for a personal chat with the clerk after completing their business (Blom & Gumperz, 1972). The use of slang gives strong expression to in-group intimacy and solidarity. Through slang, group members assert their own shared social identity and express their alienation from and rejection of the out-group of slang illiterates.

The intimacy of a relationship is clearly reflected in and reinforced by the content of conversation. As a relationship becomes more intimate, we disclose more personal information about ourselves. Intimacy is also conveyed by conversational style. In one study (Hornstein, 1985), telephone conversations were recorded and later analyzed; the conversations were between strangers, acquaintances, and friends. Compared to strangers, friends used more implicit openings ("Hi," or "Hi. It's me."), raised more topics, and were more responsive to the other conversationalist (for example, asked more questions). Friends also used more complex closings (for example, making concrete arrangements for the next contact). Conversations of acquaintances were more like those of strangers.

The **theory of speech accommodation** (Beebe & Giles, 1984; Giles, 1980) illustrates an important way that people use verbal and paralinguistic behavior to express intimacy or liking. According to this theory, people express or reject intimacy by adjusting their speech behavior during interaction to converge with or diverge from their partner's. To express liking or evoke approval, they make their own speech behavior more similar to their partner's. To reject intimacy or communicate disapproval, they accentuate the differences between their own speech and their partner's.

Adjustments of paralinguistic behavior demonstrate speech accommodation during conversations (Taylor & Royer, 1980; Thakerar, Giles, & Cheshire, 1982). Individuals who wish to express liking tend to shift their own pronunciation, speech rate, vocal intensity, pause lengths, and utterance lengths during conversation to match those of their partner. Individuals who wish to communicate disapproval modify these vocal behaviors in ways that make them diverge more from their partner's. Among bilinguals, speech accommodation may also determine the choice of language (Bourhis, Giles, Leyens, & Tajfel, 1979). To increase intimacy, bilinguals choose the language they believe their partner would prefer to speak. To reject intimacy, they choose their partner's less preferred language.

If greater intimacy leads to accommodation, can accommodation lead to greater intimacy? Research suggests that extreme accommodation, in the form of mimicry, leads to behaviors associated with greater intimacy. Using the methodology of the field experiment, 60 groups of customers in a restaurant were randomly assigned to one of two conditions. In one condition, a waitress literally repeated the orders of her customers; in the other, she merely acknowledged the orders by saying "okay" or "coming up."

Customers whose orders were mimicked were more generous, giving significantly larger tips than those in the other condition (van Baaren, Holland, Steenaert, & van Knippenberg, 2003). In a related experiment conducted in a laboratory, the experimenter mimicked the posture (bodily orientation, positions of arms and legs) of one half of the participants during a 6-minute interaction; those whom she mimicked were more likely to help her later when she dropped some pens (van Baaren, Holland, Kawakami, & van Knippenberg, 2004).

Accommodation is evident even in very subtle paralinguistic cues. Using audiotapes of interviews by talk-show host Larry King of 25 guests (stars, athletes, politicians), analyses indicated voice convergence between partners (Gregory & Webster, 1996). Lower status persons accommodated their voices to higher status persons. Moreover, student ratings of the status of Larry King and of his guests were correlated with the voice characteristics that showed convergence.

The ways we express intimacy through body language and interpersonal spacing are well recognized. For instance, research supports the folklore that lovers gaze more into each other's eyes (Rubin, 1970). In fact, we tend to interpret a high level of eye contact from others as a sign of intimacy. We communicate liking by assuming moderately relaxed postures, moving closer and leaning toward others, orienting ourselves face to face, and touching them (Mehrabian, 1972). Increasing emotional intimacy is often accompanied by increasing body engagement, from an arm around the shoulders to a full embrace (Gurevitch, 1990). There is an important qualification to these generalizations, however. Mutual gaze, close distance, and touch reflect intimacy and promote it only when the interaction has a positive cast. If the interaction is generally negative—if the setting is competitive, the verbal content unpleasant, or the past relationship antagonistic—these same nonverbal behaviors intensify negative feelings (Schiffenbauer & Schiavo, 1976).

The Case of "Dude" Let's apply the themes in this section to a concrete case. Language is continually evolving; some words and phrases fall into disuse (remember "valley girl"?) while new ones appear, like "Dude." Think about the last time you used "Dude" in conversation; who were you talking to, and what was the context? Research using diaries, surveys of students, and analysis of conversations yields a snapshot of its use (Kiesling, 2004). "Dude" is used primarily by young men in conversation with other young men, suggesting that it is a marker of youth and masculinity. Further, men rarely use the term in conversation with parents and professors; its use indexes a relationship between persons of equal status. In terms of intimacy, it occurs in conversations involving friends, but not close friends; this suggests to the researcher that "Dude" is used to indicate a "cool solidarity," an effortless interaction with other men. Like many terms that are adopted widely by youth, "Dude" has many uses, as a greeting ("What's up, Dude?"), an exclamation ("Dude!"), to one-up someone ("That's lame, Dude."), and to express agreement. Thus, like all use of speech, the use of this term is governed by sociolinguistic norms, and reflects group membership and the status and intimacy of the relationship between the conversationalists.

NORMATIVE DISTANCES FOR INTERACTION

American and Northern European tourists in Cairo are often surprised to see men touching and staring intently into each other's eyes as they converse in public. Surprise may turn to discomfort if the tourist engages an Arab man in conversation. Bathed in the warmth of his breath, the tourist may feel sexually threatened. In our own communities, in contrast, we are rarely made uncomfortable by the overly close approach of another. People apparently know the norms for interaction distances in their own cultures and they conform to them. What are these norms, and what happens when they are violated?

Normative Distances

Edward Hall (1966) has described four spatial zones that are normatively prescribed for interaction among middle-class Americans. Each zone is considered appropriate for particular types of activities and relationships. Public distance (12–25 feet) is prescribed for interaction in formal encounters, lectures, trials, and other public events. At this distance, communication is often one way, sensory stimulation is very weak, people speak loudly, and they choose language carefully. Social distance (4–12 feet) is prescribed for many casual social and business transactions. Here, sensory stimulation is low. People speak at normal volume, do not touch one another, and use frequent eye contact to maintain smooth communication. Personal distance (1½–4 feet) is prescribed for interaction among friends and relatives. Here, people speak softly, touch one another, and receive substantial sensory stimulation by sight, sound, and smell. Intimate distance (0–18 inches) is prescribed for giving comfort, making love, and aggressing physically. This distance provides intense stimulation from touch, smell, breath, and body heat. It signals unmistakable involvement.

Many studies support the idea that people know and conform to the normatively prescribed distances for particular kinds of encounters (LaFrance & Mayo, 1978). When we compare different cultural and social groups, both similarities and differences in distance norms emerge. All cultures prescribe closer distances for friends than for strangers, for example. The specific distances for preferred interactions vary widely, however. Latin Americans, Arabs, Greeks, and French use shorter interaction distances than Americans, British, Swiss, and Swedes (Sommer, 1969). Women tend to interact with one another at closer distances than men do in various cultures (Sussman & Rosenfeld, 1982). Social class may also influence interpersonal spacing. In Canadian school yards, lower class primary school children were observed to interact at closer distances than middle-class children, regardless of race (Scherer, 1974).

Differences in distance norms may cause discomfort in cross-cultural interaction. People from different countries or social classes may have difficulty in interpreting the amount of intimacy implied by each other's interpersonal spacing and in finding mutually comfortable interaction distances. Cross-cultural training in nonverbal communication can reduce such discomfort. For instance, Englishmen were liked more by Arabs with whom they interacted when the Englishmen had been trained to behave nonverbally like Arabs—to stand closer, smile more, look more, and touch more (Collett, 1971).

Two aspects of interpersonal spacing that clearly influence and reflect status are physical distance and the amount of space each person occupies. Equal status individuals jointly determine comfortable interaction distances and tend to occupy approximately equal amounts of space with their bodies and with the possessions that surround them. When status is unequal, superiors tend to control interaction distances, keeping greater physical distance than equals would choose. Superiors also claim more direct space with their bodies and possessions than inferiors (Gifford, 1982; Hayduk, 1978; Leffler et al., 1982).

Violations of Personal Space What happens when people violate distance norms by coming too close? In particular, what do we do when strangers intrude on our personal space?

The earliest systematic examination of this question included two parallel studies (Felipe & Sommer, 1966). In one, strangers approached lone male patients in mental hospitals to a point only 6 inches away. In the other, strangers sat down 12 inches away from lone female students in a university library. The mental patients and the female students who were approached left the scene much more quickly than the other patients and students who were not approached. After only 2 minutes, 30 percent of the patients who were intruded on had fled, compared with none of the others. Among the students, 70 percent of those whose space was violated had fled by the end of 30 minutes, compared with only 13 percent of the others. The results of this study are shown in Figure 7.2.

Flight is not the only response to space violation. People may also protect their privacy by turning their backs on intruders, leaning away, and placing barriers such as books, purses, or elbows in the intervening space. Only rarely do people react verbally to space violations (Patterson, Mullens, & Romano, 1971). The reaction to violations depends in part on the setting in which they occur. Whereas violations of space norms at library tables lead to flight, violations in library aisles lead to the person spending more time in the aisle (Ruback, 1987). Similarly, intrusion into the space of someone using a public telephone is

FIGURE 7.2 REACTIONS TO VIOLATIONS OF PERSONAL SPACE

How do people react when strangers violate norms of interpersonal distance and intrude on their personal space? A common reaction is illustrated here. Strangers sat down 12 inches away from lone female students in a library or approached lone male patients in a mental hospital to within 6 inches. Those who were approached left the scene much more quickly than control subjects who were not approached. Violations of personal space often produce flight.

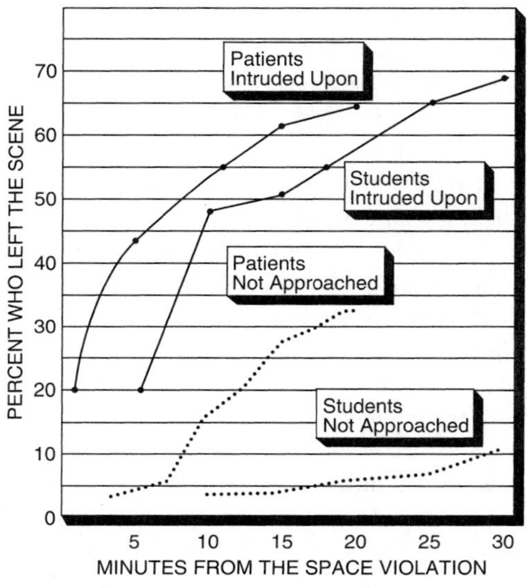

Source: Adapted from "Invasions of Personal Space" (1966) by N. J. Felipe and R. Sommer, *Social Problems*, *14*(2), 206–214.

associated with the caller spending more time on the phone (Ruback, Pape, & Doriot, 1989). It is possible that when you are looking for a book or talking on the phone, a violation of distance norms is distracting, so it takes you longer to complete your task.

Violating personal space is uncomfortable for intruders too. Individuals required to pass through the personal space of others who are engaged in conversation feel more awkward and display more unpleasant facial expressions than individuals who merely pass nearby (Efran & Cheyne, 1974). Due to their discomfort, people will avoid intruding. They will drink from a water fountain much less frequently, for instance, if someone is standing within a foot of it (Baum, Riess, & O'Hara, 1974).

Staring is a powerful way to violate another's privacy without direct physical intrusion. Staring by strangers elicits avoidance responses, indicating that it is experienced as an intense negative stimulus. When stared at by strangers, for instance, pedestrians cross the street faster, and drivers speed away from intersections more quickly (Ellsworth, Carlsmith, & Henson, 1972; Greenbaum & Rosenfeld, 1978).

CONVERSATIONAL ANALYSIS

Although conversation is a regular daily activity, we all have trouble communicating at times. The list of what can go wrong is long and painful: inability to get started, irritating interruptions, awkward silences, failure to give others a chance to talk, failure to notice that listeners are bored or have lost interest, changing topics inappropriately, assuming incorrectly that others understand, and so on. This section examines the ways people avoid these embarrassing and annoying blunders. To maintain smooth-flowing conversation requires knowledge of certain rules and communication skills that are often taken for granted. We will discuss some of the rules and skills that are crucial for initiating conversations, regulating turn taking, and coordinating conversation through verbal and nonverbal feedback.

Initiating Conversations

Conversations must be initiated with an attention-getting device—a summons to interaction. Greetings, questions, or the ringing of a telephone can serve as the summons. But conversations do not get underway until potential partners signal that they are attending and willing to converse. Eye contact is the crucial nonverbal signal of availability for face-to-face interaction. Goffman (1963) suggests that eye contact places a person under an obligation to interact: When a waitress permits eye contact, she places herself under the power of the eye-catcher.

The most common verbal lead into conversation is a **summons-answer sequence** (Schegloff, 1968). Response to a summons ("Jack, you home?" "Yeah.")

© Katherine Karnow/Corbis

Despite the crowded circumstances, the people at this bus stop are maintaining some privacy. Strangers feel uncomfortable when they must intrude on each other's personal space. To overcome this discomfort, they studiously ignore each other, avoiding touch, eye contact, and verbal exchanges.

indicates availability. More importantly, this response initiates the mutual obligation to speak and to listen that produces conversational turn taking. The summoner is expected to provide the first topic—a conversational rule that little children exasperatingly overlook. Our reactions when people violate the summons-answer sequence demonstrate its widespread acceptance as an obligatory rule. When people ignore a summons, we conclude either that they are intentionally insulting us, socially incompetent, or psychologically absent (sleeping, drunk, or crazy).

Telephone conversations exhibit a common sequential organization. Consider the following conversation between a caller and a recipient:

0. (ring)
1. Recipient: Hello?
2. Caller: This is John.
3. R: Hi.
4. C: How are you?
5. R: Fine. How are you?
6. C: Good. Listen, I'm calling about . . .

The conversation begins with a summons-answer sequence (lines 0, 1). This is followed by an identification-recognition sequence (lines 2, 3); in this example, the recipient knows that the caller, John, recognizes his voice, so he does not state his name. Next, there is a trading of "how are you" sequences (lines 4–6). Finally, at line 6, John states the reason for the call. This organization is found in many types of telephone calls. However, in an emergency, when seconds count, the organization is quite different

(Whalen & Zimmerman, 1987). Consider the following example:

0. (ring)
1. R: Mid-City Emergency.
2. C: Um, yeah. There's a fire in my garage.
3. R: What's your address?

Notice that the opening sequence is shortened; both the greeting and the "how are you" sequences are omitted. In emergency calls, the reason for the call is stated sooner. Note also that the recognition element of the identification-recognition sequence is moved forward, to line 1. Both of these changes facilitate communication in an anonymous, urgent situation. However, if the dispatcher answers a call and the caller says, "This is John," that signals an ordinary call. Thus, the organization of conversation clearly reflects situational contingencies.

Regulating Turn Taking

A pervasive rule of conversation is to avoid bumping into someone verbally. To regulate turn taking, people use many verbal and nonverbal cues, singly and together, with varying degrees of success (Duncan & Fiske, 1977; Kendon, Harris, & Key, 1975; Sacks, Schegloff, & Jefferson, 1978).

Signaling Turns Speakers indicate their willingness to yield the floor by looking directly at a listener with a sustained gaze toward the end of an utterance. People also signal readiness to give over the speaking role by pausing and by stretching the final syllable of their speech in a drawl, terminating hand gestures, dropping voice volume, and tacking relatively meaningless expressions (such as "You know") onto the end of their utterances. Listeners indicate their desire to talk by inhaling audibly as if preparing to speak. They also tense and move their hands, shift their head away from the speaker, and emit especially loud vocal signs of interest (such as "Yeah," "M-hmn").

Speakers retain their turn by avoiding eye contact with listeners, tensing their hands and gesticulating, and increasing voice volume to overpower others when simultaneous speech occurs. People who persist in these behaviors are soon viewed by others as egocentric and domineering. They have violated an implicit social rule: "It's all right to hold a conversation, but you should let go of it now and then" (Richard Armour).

Verbal content and grammatical form of speech also provide important cues for turn taking. People usually exchange turns at the end of a meaningful speech act, after an idea has been completed. The first priority for the next turn goes to any person explicitly addressed by the current speaker with a question, complaint, or other invitation to talk. People expect turn changes to occur after almost every question, but not necessarily after other pauses in conversation (Hanni, 1980). It is difficult to exchange turns without using questions. When speakers in one study were permitted to use all methods except questions for signaling their desire to gain or relinquish the floor, the length of each speaking turn virtually doubled (Kent, Davis, & Shapiro, 1978).

Turn Allocation Much of our conversation takes place in settings where turn taking is more organized than in spontaneous conversations. In class discussions, meetings, interviews, and therapy sessions, for example, responsibility for allocating turns tends to be controlled by one person, and turns are often allocated in advance. Prior allocation of turns reduces strains that arise from people either competing for speaking time or avoiding their responsibilities to speak. Allocation of turns also increases the efficiency of talk. It can arrange a distribution of turns that best fits the task or situation—a precisely equal distribution (as in a formal debate) or just one speaker (as in a football huddle).

Feedback and Coordination

We engage in conversation to attain interpersonal goals—to inform, persuade, impress, control, and so on. To do this effectively, we must assess how what we say is affecting our partner's interest and understanding as we go along. Both verbal and nonverbal feedback help conversationalists in making this assessment. Through feedback, conversationalists coordinate what they are saying to each other from moment to moment. The responses called **back channel feedback** are especially important for regulating speech as it is happening. These are the small vocal and visual comments that a listener makes while a speaker is talking, without taking over the speaking turn. They include such responses as "Yeah," "M-hmn," short clarifying questions (such as "What?" "Huh?"), brief repetitions of the speaker's words or completions of his or her utterances, head nods, and brief smiles. When conversations are proceeding smoothly, the fine rhythmic body movements of listeners (such as swaying, rocking, blinking) are precisely synchronized with the speech sounds of speakers who address them (Condon & Ogston, 1967). These automatic listener movements are another source of feedback that indicates to speakers whether they are being properly tracked and understood (Kendon, 1970).

Both the presence (or absence) and the timing of back channel feedback influence speakers. In smooth conversation, listeners time their signs of interest, agreement, or understanding to occur at the end of long utterances, or when the speaker turns his or her head toward them. When speakers are denied feedback, the quality of their speech deteriorates. They become less coherent and communicate less accurately. Their speech becomes more wordy, less organized, and more poorly fitted to the situation (Bavelas, Coates, & Johnson, 2000). Lack of feedback causes such deterioration because it prevents speakers from learning several things about their partners. They cannot discern whether their partners (1) have relevant prior knowledge they need not repeat; (2) understand already so they can wrap up the point or abbreviate; (3) have misinformation they should correct; (4) feel confused so they should backtrack and clarify; or (5) feel bored so they should stop talking or change topics.

Alerted to the possible loss of listener attention and involvement by the absence of feedback, speakers employ attention-getting devices to evoke feedback. One such attention-getting device is the phrase "You know." Speakers frequently insert "You know" into long speaking turns immediately prior to or following pauses if their partner seems to be ignoring their invitation to provide feedback or to accept a speaking turn (Fishman, 1980).

Another device a speaker can use to regain the attention of another participant is to ask her or him a question. Such displays of uncertainty (for example, "What was the name of that guy on the Oprah Winfrey Show?") restructure the interaction by getting listeners more involved (Goodwin, 1987). If the speaker shifts his gaze to a specific person as he asks the question, it will draw that person into the conversation.

The fact that feedback influences the quality of speech has another interesting consequence. Listeners who frequently provide their conversational partners with feedback also understand their partner's communication more fully and accurately. Through their feedback, active listeners help shape the conversation to fit their own needs. The information needed varies on several dimensions, one of which is precision; recall that the cooperative principle assumes the actors provide relevant precision. In responding to an invitation, it may be sufficient, if exaggerated, to say, "I don't have any money", but in bankruptcy court, counsel or the judge will want greater precision. When we fail to provide relevant precision, we will be challenged by an alert listener (Drew, 2003). This reinforces a central theme of this chapter: Accurate communication is a shared social accomplishment.

Feedback is important not only in conversations, but also in formal lectures. Lecturers usually monitor members of the audience for feedback. If listeners are looking at the speaker attentively and nodding their heads in agreement, the lecturer infers that her message is understood. On the other hand, quizzical or out-of-focus expressions suggest failure to understand. Similarly, members of the audience use feedback from the lecturer to regulate their own behavior; a penetrating look from the speaker may be sufficient to end a whispered conversation between listeners.

An important form of feedback in many lectures is applause. Speakers may want applause for a variety of reasons, not just ego gratification. Sometimes, lecturers subtly signal the audience when to applaud; audiences watch for such signals in order to maintain their involvement. For instance, an analysis of 42 hours of recorded political speeches suggests that there is a narrow range of message content that stimulates applause (Heritage & Greatbatch, 1986). Attacks on political opponents, foreign persons, and collectivities;

statements of support for one's own positions, record, or party; and commendations of individuals or groups generate applause. When these messages are framed within particular rhetorical devices, applause is from two to eight times more likely. For example:

SPEAKER: Governments will argue [pause]
that resources are not available [short pause]
to help disabled people. [long pause]
The fact is that too much is spent on the
munitions of war, [long pause]
and too little is spent [applause begins]
on the munitions of peace.

In this example, the speaker uses the rhetorical device of contrast or antithesis. Using this device, the speaker's point is made twice. Audiences can anticipate the completion point of the statement by mentally matching the second half with the first. This rhetorical device is an "invitation to applaud," and in the example, the audience begins to applaud even before the speaker completes the second half.

SUMMARY

Communication is the process whereby people transmit information about their ideas and feelings to one another.

Language and Verbal Communication Language is the main vehicle of human communication. (1) All spoken languages consist of sounds that are combined into words with arbitrary meanings and put together according to grammatical rules. (2) According to the encoder-decoder model, communication involves the encoding and sending of a message by a speaker, and the decoding of the message by a listener. Accuracy depends on the codability of the idea or feeling being communicated. (3) In contrast, the intentionalist model argues that communication involves the speaker's desire to affect the listener, or the transmission of an intention. The context of the communication influences how messages are sent and interpreted. (4) The perspective-taking model argues that communication requires intersubjectivity—the shared context created by speaker and listener. Thus, communication is a complex undertaking; to attain mutual understanding, conversationalists must express their message in ways listeners can interpret, take account of others' current knowledge, and actively work to decipher meanings.

Nonverbal Communication A great deal of information is communicated nonverbally during interaction. (1) Four major types of nonverbal communication are paralanguage, body language, interpersonal spacing, and choice of personal effects. (2) The face is an important channel of communication; it provides information that observers use to infer social identities and personal characteristics. (3) Information is usually conveyed simultaneously through nonverbal and verbal channels. Multiple cues may add information to each other, reduce ambiguity, and increase accuracy. But if cues appear inconsistent, people must determine which cues reveal the speaker's true intentions.

Social Structure and Communication The ways we communicate with others reflect and influence our relationships with them. (1) Gender is related to communication style; its impact depends on the interpersonal, group, or organizational context. (2) In every society, speech that adheres to rules governing vocabulary, pronunciation, and grammar is preferred or standard. Its use is associated with high status or power and is evaluated favorably by listeners. Nonstandard speech is often used by lower status persons and evaluated negatively. (3) We express, maintain, or challenge the levels of relative status and intimacy in our relationships through our verbal and nonverbal behavior. Status and intimacy influence and are influenced by forms of address, choice of dialect or language, interruptions, matching of speech styles, gestures, eye contact, posture, and interaction distances. (4) The appropriate interaction distances for particular types of activities and relationships are normatively prescribed. These distances vary from one culture to another. When strangers violate distance norms, people flee the scene or use other devices to protect their privacy.

Conversational Analysis Smooth conversation depends on conversational rules and communication skills that are often taken for granted. (1) Conversations are initiated by a summons to interaction. They get underway only if potential partners signal availability, usually through eye contact or verbal response. (2) Conversationalists avoid verbal collisions by taking turns. They signal either a willingness to yield the floor or a desire to talk through verbal and nonverbal cues. In some situations, turns are allocated in advance. (3) Effective conversationalists assess their partner's understanding and interest as they go along through vocal and visual feedback. If feedback is absent or poorly timed, the quality of communication deteriorates. An effective speech also involves coordination between speaker and audience; the timing of applause is a joint accomplishment.

LIST OF KEY TERMS AND CONCEPTS

back channel feedback (p. 193)
body language (p. 176)
communication (p. 166)
communication accuracy (p. 168)
cooperative principle (p. 171)
ebonics (p. 184)
encoder-decoder model (p. 168)
flirting (p. 182)
intentionalist model (p. 170)
interpersonal spacing (p. 176)
intersubjectivity (p. 173)
linguistic intergroup bias (p. 174)
nonstandard speech (p. 181)
paralanguage (p. 176)
perspective-taking model (p. 172)
sociolinguistic competence (p. 175)
speech act theory (p. 170)
spoken language (p. 167)
standard speech (p. 181)
summons-answer sequence (p. 190)
symbols (p. 166)
theory of speech accommodation (p. 187)

8

SOCIAL INFLUENCE AND PERSUASION

Introduction

Forms of Social Influence

Attitude Change via Persuasion

Processing Persuasive Messages

Communication-Persuasion Paradigm

The Source

The Message

The Target

Compliance with Threats and Promises

Effectiveness of Threats and Promises

Problems in Using Threats and Promises

Bilateral Threat and Negotiation

Obedience to Authority

Experimental Study of Obedience

Factors Affecting Obedience to Authority

Resisting Influence and Persuasion

Inoculation

Forewarning

Reactance

Summary

List of Key Terms and Concepts

INTRODUCTION

Consider some examples of social influence:

- In front of her house, Julie is met by Erika, a neighbor. Erika has heard that a waste management company plans to open a new landfill only 1/8 mile from their neighborhood. Trying to mobilize opposition, Erika argues that the landfill would pose dangers to health and diminish land values. She asks Julie to attend a meeting and sign a petition against the landfill. Somewhat alarmed by developments, Julie finds Erika's view persuasive, and she agrees to sign.
- One evening, the owner of a 24-hour convenience store is suddenly confronted by a man wearing a ski mask and brandishing a pistol. The man threatens, "Hand over your money or I'll blow you away!" Facing a choice between two undesirable alternatives—losing his money or his life—the victim opens the cash register and hands over the money.
- During a military action in Afghanistan, a U.S. commander orders a platoon of men to attack a series of caves where terrorists are thought to be hiding. The danger involved is great. Night has fallen, the entire area is covered with antipersonnel mines, and the enemy has been firing on the troops from the hills. Despite these obstacles, the troops move out as ordered.

These stories illustrate various forms of social influence. By definition, **social influence** occurs when one person (the **source**) engages in some behavior (such as persuading, threatening or promising, or issuing orders) that causes another person (the **target**) to behave differently from how he or she would otherwise behave. In the preceding illustrations, the sources were Erika (persuading Julie), the thief (threatening his victim), and the infantry commander (ordering his troops to attack).

Various outcomes can result when social influence is attempted. In some cases, the influencing source may produce **attitude change**—a change in the target's beliefs and attitudes about some issue, person, or situation. Attitude change is a fairly common result of social influence. In other cases, however, the source may not really care about changing the target's attitudes but only about securing compliance. **Compliance** occurs when the target's behavior conforms to the source's requests or demands. Some social influence attempts, of course, produce both attitude change and compliance.

Moreover, we must recognize that many social influence attempts prove ineffective, producing little or no change in the target. Orders issued by direct authority frequently obtain compliance, but other times their targets may respond with defiance or open revolt. Because influence attempts vary in their degree of success, one concern of this chapter is to discern the conditions under which influence attempts are more effective.

Forms of Social Influence

Influence attempts can be either open or covertly manipulative (Tedeschi, Schlenker, & Lindskold, 1972). In open influence, the attempt is readily apparent to the target. The target understands that someone is trying to change his or her attitudes or behavior. In manipulative influence, the attempt is hidden from the target. Examples of manipulative influence include ingratiation and tactical self-presentation. We will discuss manipulative influence in Chapter 9. This chapter focuses on open influence.

There are many forms of open influence. Among the more important forms are (1) the use of persuasive communication to change the target's attitudes or beliefs, (2) the use of threats or promises to gain compliance, and (3) the use of orders based on legitimate authority to gain compliance.

Consider, first, the resources involved in persuasion. When attempting to persuade, the source uses information to change the target's attitudes and beliefs about some issue, person, or situation. Certain types of information are more useful than others in bringing about persuasion. For instance, a persuasion attempt is more likely to succeed if the source can introduce facts not already known to the target; likewise, success is more likely if the source can advance compelling and valid arguments not previously considered by the target. Having the right type of information is important when attempting persuasion.

Influence attempted by means of threats or promises is based on punishments and rewards rather than of information. If a threat is to produce compliance, the target must believe that the source controls whether the threatened punishment will be imposed or not. The same is true for influence based on promises, except that it involves the control of rewards rather than punishments. If the target believes that the source has no real control over the punishments or rewards involved, then the threat or promise is unlikely to succeed.

Influence through the use of orders from an authority or officeholder is based on the target accepting the authority's legitimacy. Influence of this type is especially common within formal groups or organizations. When attempting influence by invoking legitimate authority, the source makes demands on the target that are vested in his or her role within the group. Such an attempt will succeed only if the target believes that the source actually holds a position of authority and has the right to issue orders of the kind involved in the influence attempt.

Because influence attempts can vary greatly in their degree of success, and because we all use social influence in our relationships with others, it is important to understand the conditions under which influence attempts are effective. Specifically, in this chapter we will address the following questions:

1. What factors determine whether a communication will succeed in persuading a target to change his or her beliefs or attitudes? In what ways, for instance, do the characteristics of the source and the target and the properties of the message itself determine whether the persuasion attempt will be effective?
2. Under what conditions do threats and promises prove successful in gaining compliance from the target? When two interacting persons can threaten and punish one another, under what conditions will threats intensify and conflict escalate?
3. When a person in authority issues an order, under what conditions are targets likely to obey it?
4. How can persons resist persuasion attempts and maintain their original attitudes?

ATTITUDE CHANGE VIA PERSUASION

Day in and day out, others bombard us with messages that seek to persuade. As an example, consider what happens to Steve Maxwell on a typical day. Early in the morning, Steve's clock radio comes on. Before Steve can get out of bed, a cheerful announcer is trying to sell him a new mouthwash. On the way to work, one of the people in his car pool attempts to persuade him to vote for a particular candidate in the upcoming election. At lunch, a friend mentions her plans to attend a concert the following weekend and urges him to come along. In mid-afternoon, he listens to an argument from a coworker who wants to change some paperwork procedures in the office. When he arrives home in the evening, Steve opens his mail. One letter is a carefully worded appeal from a charitable organization asking him to volunteer his time. Other letters are junk mail fliers asking for money. Later that night, when Steve is watching television, advertisers bombard him endlessly with ads for their products—laundry soaps, light beers, anti-dandruff shampoos, and imported sports cars.

All these messages received by Steve have something in common: They seek to persuade. **Persuasion** may be defined as changing the beliefs, attitudes, or behaviors of a target through the use of information or argument. Persuasion is widespread in social interaction and assumes many different forms (McGuire, 1985). In this section, we will consider various facets of message-based persuasion. First, we will examine how targets process messages. Next, we will discuss the communication-persuasion paradigm. Finally, we will consider the characteristics of sources, messages, and targets that affect the persuasiveness of a message.

Processing Persuasive Messages

When a target receives a message advocating a position different from what he or she believes, the target can respond in various ways. Of course, the target may accept the message and change his or her attitudes or beliefs. But this does not always happen and a variety of other reactions are possible.

Reactions to Persuasive Messages One obvious reaction is that the target might simply reject the message or refuse to listen further. For instance, rather than heed the automobile advertisement or the beer commercial, the television viewer can switch channels. Another second response to a persuasive communication is to derogate the source, and thereby render the argument invalid. Instead of changing attitudes, the target might dismiss the communicator as poorly informed, excessively partisan, or illogical. A third possible response is to listen to the message but to suspend judgment on the issue. If the target knows some contrary facts not discussed in the message, he or she might decide to seek additional information before concluding that one viewpoint or another is correct. A fourth possible response is to distort the message. In this case, the target misperceives or misconstrues the content of the message (bringing it, perhaps, into line with his or her own cognitive schemas). The message might produce an attitude change, although not the one the source intended. A fifth possible response is to attempt counterpersuasion. That is, the target might react by arguing back and trying to change the source's own beliefs and attitudes. Obviously, counterpersuasion is not possible with messages transmitted via TV or other mass media, but people frequently use it in face-to-face situations.

Central and Peripheral Routes The **elaboration likelihood model,** a theory advanced by Petty and Cacioppo and their coworkers (Petty & Cacioppo, 1986a, 1986b; Petty, Cacioppo, Strathman, & Priester, 1994), attempts to explain the processes by which messages produce attitude change. The elaboration likelihood model holds that there are two basic routes through which a message may alter a target's existing attitudes—the central route and the peripheral route.

Persuasion via the central route occurs when a target actively scrutinizes the arguments contained in a persuasive message, interprets and evaluates them (taking into account what he or she already knows), and then integrates them into a coherent position. Petty and Cacioppo use the term elaboration to describe the process whereby the target thinks through the implications of the arguments contained in the message. In elaboration, attitude change occurs only when the arguments are strong, internally coherent, consistent with known facts, and so on. Information held by the target before receiving the message, as well as new information in the message itself, comes into play during elaboration.

In contrast, persuasion via the peripheral route occurs when, instead of elaborating the content of the message, the target pays attention primarily to extraneous cues linked to the message. Among these cues are the characteristics of the source (expertise, trustworthiness, likableness), superficial characteristics of the message (message length), or characteristics of the situation (the response of other audience members). If a target processes the message via the peripheral route, he or she does not carefully think about the message but instead appraises peripheral cues and uses them as a basis for accepting or rejecting the message.

Thus, the elaboration likelihood model states that what matters most about a message is not just its properties, but whether it undergoes elaboration. If a given message undergoes elaboration, attitude change by the target will depend largely on the strength and quality of its arguments (central route factors). On the other hand, if the message does not undergo elaboration, attitude change by the target will depend on such factors as the identity and credibility of the source or the reaction of other audience members (peripheral route factors). Attitudes established via the central route tend to be more strongly held than those established via the peripheral route. That is, attitudes obtained via the central route tend to be longer lasting and more resistant to change, because the target has thought through the issue in more detail (Chaiken, 1980; Petty et al., 1994).

Communication-Persuasion Paradigm

Consider the question, "Who says what to whom with what effect?" This question is one way of organizing modern research on persuasion. In this question, the

FIGURE 8.1 THE COMMUNICATION-PERSUASION PARADIGM

Source →	*Message* →	*Target* →	*Effect*
expertise	discrepancy	intelligence	change attitude
trustworthiness	fear appeal	involvement	reject message
attractiveness	1-sided or 2-sided	forewarned	counterargue
			suspend judgment
			derogate source

"who" refers to the source of a persuasive message, the "whom" refers to the target, and the "what" refers to the content of the message. The phrase "with what effect" refers to the various responses of the target to the message. These elements (source, message, target, response) are fundamental components of the **communication-persuasion paradigm.** Figure 8.1 displays this paradigm and shows how these components interrelate. First, the properties of the source can affect how the target audience will construe the message. For instance, characteristics such as the expertise and trustworthiness of the source can affect whether a target changes attitudes or not. Second, the properties of the message itself can have a significant impact on its persuasiveness. For instance, whether a message carries a fear appeal or presents only one-sided arguments can affect whether a persuasion attempt is successful. Third, the characteristics of the target also enter the picture. For instance, what a target already believes about an issue, as well as the extent of the target person's involvement in the issue and commitment to a position, can affect whether a message leads to attitude change or merely to rejection.

In the discussion that follows, we will consider these factors in detail—properties of the source, the message, and the target—and examine their impact on attitude change.

The Source

Suppose we ask 25 persons selected at random to read a persuasive communication (such as a newspaper editorial) that advocates a position on a nutrition-related topic. We tell this group that the message came from a Nobel Prize-winning biologist. At the same time, we ask 25 other persons to read the same message, but we tell this group that it came from a cook at a local fast food establishment. Subsequently, we ask both groups to indicate their attitude toward the position advocated in the message. Which group of persons will be more persuaded by the communication?

Most likely, the persons who read the message ascribed to the prize-winning biologist will be more persuaded than those who read the message ascribed to the fast food cook.

Why should the source's identity make any difference? The identity of the source provides the target with information above and beyond the content of the message itself. Because some sources are more credible than others, the target may pay attention to the source's identity when assessing whether to believe the message. The term **communicator credibility** denotes the extent to which the communicator is perceived by the target as a believable source of information. Note that the communicator's credibility is "in the eye of the "beholder"—a given source may be credible for some audiences but not for others.

Many factors influence the extent to which a source is credible. Two of—these, the source's expertise and the source's trustworthiness, are of special importance.

Expertise Generally, a message from a source having a high level of expertise relevant to the issue will bring about greater attitude change than a similar message from a source having a lower level of expertise (Chebat, Filiatrault, & Perrien, 1990; Hass, 1981; Maddux & Rogers, 1980). This may occur because targets may be more accepting and less critical of messages from high-expertise sources.

The impact of source expertise is illustrated by a study in which participants were exposed to a message advocating the passage of a consumer protection bill by the U.S. Senate (Sternthal, Dholakia, & Leavitt, 1978). For some participants, the message was ascribed to a lawyer who was educated at Harvard and had extensive experience in consumer issues (a high-credibility source). For other participants, the message was ascribed to a citizen interested in consumer affairs (a low-credibility source). Afterward, participants indicated their reaction to the message. In general, participants who initially opposed the position advocated in the message expressed more unquestioning agreement and less counterargument when the message came from the high-credibility source than when it came from the low-credibility source.

Even if a persuasion attempt from a low-credibility source at first fails, there can sometimes be a sleeper effect in which the target eventually forgets the source of the argument, but still remembers the argument. In this case, the target can later be persuaded, but only if they forget that they originally considered the source to be noncredible (Kumkale & Albarracin, 2004).

The source's expertise interacts with the target's involvement and knowledge in determining attitude change. When the target has little involvement or prior knowledge on a given issue, messages from highly expert sources produce more attitude change than those from less expert sources. But the more involving the issue or the more knowledge the target has about the issue, the less likely it is that communicator expertise will make much difference in persuasion (Rhine & Severance, 1970). When involvement and knowledge are high, the target is more likely to engage in detailed processing and elaboration, so the content of the message itself becomes the overriding determinant of attitude change (Petty & Cacioppo, 1979a, 1979b; Stiff, 1986).

As an automobile owner listens to the message from the garage mechanic, she assesses not only the quality of the argument but also the credibility of the communicator. He may have expertise, but can he be trusted?

Trustworthiness Although expertise is an important factor in communicator credibility, it is not the only one. Under some conditions, a source can be highly expert but still not very credible. As an example, suppose that your car is running poorly, so you take it into a garage for a tune-up. A mechanic you have never met before inspects your car. He identifies several problems, one of which involves major repair work on the engine. The mechanic offers to complete this work for $870 and claims that your car will soon fall apart without it. The mechanic may be an expert, but can you accept his word that the expensive repair is necessary? How much does he stand to gain if you believe his message?

As this example shows, the target pays attention not only to a communicator's expertise but also to his or her motives. If the message appears highly self-serving and beneficial to the source, the recipient may distrust the source and discount the message (Hass, 1981). In contrast, communicators who argue against their own vested interests seem especially candid and trustworthy. For example, suppose that an employee of a local business told you that you should not purchase a product made by her company but one made by a Japanese competitor. Her remarks would probably be unexpected, but they would have more impact than if she had argued for purchasing her own American-made model. Even if you normally prefer to buy products made in the United States, you might think twice in this case. A source who violates our initial expectations by arguing against her own vested interest will therefore be especially persuasive (Eagly, Wood, & Chaiken, 1978; Walster [Hatfield], Aronson, & Abrahams, 1966).

Trustworthiness also depends on the source's identity, because this carries information about the source's goals and values. A source perceived as having goals similar to the audience will be more persuasive than one perceived as having dissimilar goals (Berscheid, 1966; Cantor, Alfonso, & Zillmann, 1976). For example, a given policy proposal will be received differently by conservative Republicans depending on whether it was advanced by Dick Cheney or by Hillary Clinton. The political identity of the source reveals much about his or her underlying goals and values, and these, in turn, affect perceived trustworthiness.

Attractiveness and Likability The physical attractiveness of the source can also affect the extent to which a message is persuasive. Advertisers regularly select attractive individuals as spokespersons for their products, as we can see in television and magazine advertisements. Because it is rewarding to look at attractive spokespersons, these advertising messages receive more attention than they otherwise would. Higher source attractiveness leads us to give greater attention to the message, and higher levels of attention facilitate greater persuasion (Chaiken, 1986). Moreover, because physical attractiveness leads to liking, we like attractive persons more and, thus, are sometimes more positively disposed to accept products or positions they advocate (Eagly & Chaiken, 1975; Horai, Naccari, & Fatoullah, 1974; Burger et al., 2001). Whatever the source of likability (similarity, attractiveness, or simple contact), it tends to increase persuasive influence because individuals wish to maintain and enhance relationships with those they like (Cialdini, 2001; Roskos-Ewoldsen, Bichsel, & Hoffman, 2002).

A source's attractiveness can have greater effect when combined with other factors. In one study investigating the impact of persuasive advertisements for suntanning oil, the participants received a message that depending on treatment contained either strong or weak arguments and came from either an attractive or an unattractive female spokesperson (DeBono & Telesca, 1990). Results showed that in general, the attractive source was more persuasive than the unattractive one. But the attractive source was especially persuasive when the message arguments were strong rather than weak. When the arguments were weak, attractiveness made very little difference in persuasion.

Effect of Multiple Sources Factors other than the source's expertise and trustworthiness can affect whether a message is persuasive. **Social impact theory** (Jackson, 1987; Latané, 1981; Sedikides & Jackson, 1990), which is a general framework applicable to both persuasion and obedience, states that the impact of an influence attempt is a direct function of strength (that is, social status or power), immediacy (that is, physical or psychological distance), and number of influencing sources. A target will be more influenced when the sources are strong (rather than weak), when the sources are physically close (rather than remote), and when the sources are numerous (rather than few).

Although not all the predictions from social impact theory have yet been fully tested (Jackson, 1986; Mullen, 1985), there is support for one of the theory's more interesting predictions—the one regarding the impact of multiple sources. The theory predicts that a message will be more persuasive when a target receives it from multiple sources than from a single source. Consistent with this prediction, several studies have shown that a message presented by several different sources is more persuasive than the same message presented by a single source (Harkins & Petty, 1981b, 1987; Wolf & Bugaj, 1990; Wolf & Latané, 1983). This is especially true when the arguments presented in the message are strong rather than weak. Strong messages coming from multiple sources receive greater scrutiny and foster more issue-relevant thinking by the target, which leads to attitude change; however, weak messages from multiple sources may receive added scrutiny but will produce no extra attitude change (Harkins & Petty, 1981a).

Certain qualifications apply to this multiple-source effect. First, for multiple sources to have more impact than a single source, the target must perceive the multiple sources to be independent of one another. If the target believes that the sources colluded in sending their messages, the added impact of multiple sources will vanish, and the communication will have no more effect than if it came from a single source (Harkins & Petty, 1983).

Second, there is an upper limit to increases in persuasion from the multiple-source effect (Tanford & Penrod, 1984). Adding more and more sources will increase persuasion, but only up to a point. For instance, a message from three independent sources will be more persuasive than the same message from a single source, but a message coming from, say, 13 sources may not be appreciably more persuasive than the same message coming from 11 sources.

The Message

Persuasive communications differ dramatically in their content. Some messages contain arguments that are highly factual and rational, whereas others contain emotional appeals that motivate action by arousing fear or greed. Messages differ in their detail and complexity (simple versus complex arguments), their strength of presentation (strong versus weak arguments), and their balance of presentation (one-sided versus two-sided arguments). These properties affect how a person will scrutinize, interpret, and elaborate a message. For this reason, they pertain more to the central route than to the peripheral route. In this section, we will discuss the impact of these properties on persuasion.

Message Discrepancy Suppose a woman told you that Elizabeth II, the Queen of England, is 5 feet 4 inches tall. Would you believe her? What if she said 5 feet 10 inches tall—would you believe that? How about 6 feet 3 inches? Or 7 feet 6 inches? You may not know how tall the queen actually is, but you probably have a rough idea. Although you might believe 5 feet 10 inches, you would probably doubt 6 feet 3 inches and certainly doubt 7 feet 6 inches. The message asserting that the queen is 7 feet 6 inches tall is highly discrepant from your beliefs.

By definition, a **discrepant message** is one advocating a position that is different from what the target believes. Discrepancy is a matter of degree; some messages are highly discrepant, others less so. To cause a change in beliefs and attitudes, a message must be at least somewhat discrepant from the target's current position; otherwise, it would just reaffirm what the target already believes. Up to a certain point, greater levels of message discrepancy will lead to greater change in attitudes (Jaccard, 1981). A message that is moderately discrepant will be more effective in changing a target's beliefs and attitudes than a message that is only slightly discrepant. Of course, it is possible for a message to be so discrepant that the target will simply dismiss it. To say that the Queen of England is 7 feet 6 inches tall is just not believable.

There is an important interaction between message discrepancy and source credibility. Sources with high credibility produce maximum attitude change at higher levels of discrepancy than do sources with low credibility. Thus, a target is more likely to accept a highly discrepant message from a high-credibility source than from a low-credibility source. Highly discrepant messages from a low-credibility source are ineffective, because the target will quickly derogate the source. Figure 8.2 summarizes the joint impact of message discrepancy and communicator credibility on attitude change.

Many empirical studies report findings consistent with the relationships shown in Figure 8.2 (Aronson, Turner, & Carlsmith, 1963; Fink, Kaplowitz, & Bauer, 1983; Rhine & Severance, 1970). In one experiment, for instance, participants were given a written message on the number of hours of sleep that people need each night to function effectively (Bochner & Insko, 1966). In some cases, the message was attributed to a Nobel Prize-winning physiologist (high credibility), whereas in other cases it was attributed to a YMCA director (medium credibility). The arguments contained in the message were identical for all participants, with one important exception. In some cases the message proposed that people need 8 hours of sleep

FIGURE 8.2 ATTITUDE CHANGE AS A FUNCTION OF COMMUNICATOR CREDIBILITY AND MESSAGE DISCREPANCY

These three curves summarize the relationship between message discrepancy and attitude change conditional on the credibility of a source. Note that messages from low-credibility communicators produce maximum attitude change at moderate levels of discrepancy, whereas messages from high-credibility communicators produce maximum attitude change at high levels of discrepancy.

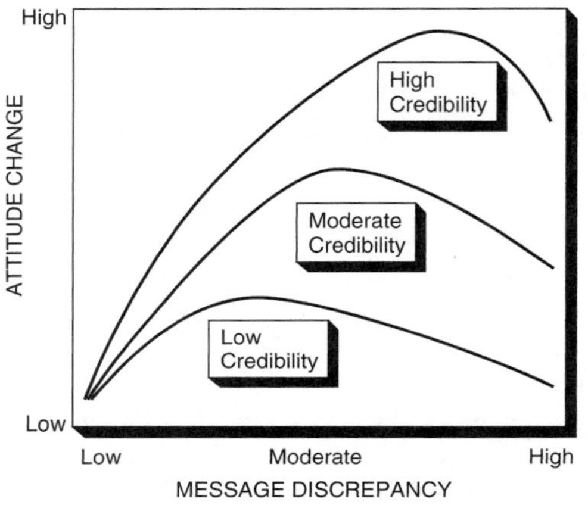

per night; in others, the message proposed 7 hours; in others, 6 hours; and so on down to zero hours of sleep per night. Most participants began the experiment believing that people need approximately 8 hours of sleep each night. Therefore, these messages differed in level of discrepancy.

The results of this study show that the more discrepant the position advocated by the high-credibility source (the Nobel Prize winner), the greater the amount of attitude change. Only when this source argued for the most extreme position (zero hours of sleep) did the participants refuse to believe the message. The same pattern appeared for the medium-credibility source (the YMCA director), except that his effectiveness peaked out at moderate levels of discrepancy (3 hours of sleep per night). For very extreme positions (2 hours of sleep or less), the medium-credibility source was less effective. Thus, this study demonstrates that sources with higher credibility produce greater amounts of attitude change at higher levels of discrepancy.

Fear Arousal Most messages intended to persuade incorporate either rational appeals or emotional appeals. Rational appeals are factual in nature; they present specific, verifiable evidence to support claims. Rational appeals frequently address a need already felt by the audience and provide the missing solution; that is, these messages are drive reducing. Emotional appeals, in contrast, try to arouse basic drives and to stimulate a need where none was present. These messages are drive creating.

Perhaps the most common emotional appeals are those involving fear. Fear-arousing messages are especially useful when the source is trying to motivate the target to take some specific action. A political candidate, for example, may warn that if voters elect her opponent to office, the nation will become embroiled in international conflict. Likewise, in an antismoking advertisement on TV, a victim dying of throat cancer and emphysema warns young persons that if they start smoking cigarettes, they may end up as diseased victims themselves. In each of these cases, the source is using a fear-arousing communication. Messages of this type direct the target's attention to some negative or undesired outcome that is likely to occur unless the target takes certain actions advocated by the source (Higbee, 1969; Ruiter et al., 2003).

Some studies have shown that communications arousing high levels of fear produce more change in attitude than communications arousing low levels of fear (Dembroski, Lasater, & Ramires, 1978; Leventhal, 1970). If a message arouses fear and the targets believe that attending to the message will show them how to cope with this fear, then they may analyze the message carefully and change their attitudes via the central route (Petty, 1995). Fear-arousing communications have been effective in persuading people to do many things, including reducing their cigarette smoking, driving more safely, improving their dental hygiene practices, changing their attitudes toward Communist China, and so on (Insko, Arkoff, & Insko, 1965; Leventhal, 1970; Leventhal & Singer, 1966).

Some studies suggest, however, that fear-arousing messages can fail if they are too strong and create

© Bill Aron/PhotoEdit, Inc.

This billboard is part of a media campaign to sell perfume. To define the intended market for the product, the ad uses the notion of male sexual dominance and traditional gender identities.

too much fear. If people feel very threatened, they may become defensive and deny the reality or the importance of the threat, rather than think rationally about the issue (Johnson, 1991; Lieberman & Chaiken, 1992). In this sense, a message arousing moderate fear may prove more effective than one arousing extremely high fear.

The impact of fear-arousing communications is shown clearly by a study in which college students received messages advocating inoculations against tetanus (Dabbs & Leventhal, 1966). These messages described tetanus as easy to catch and as producing serious, even fatal consequences. The message also indicated that inoculation against tetanus, which could be obtained easily, provided effective protection against the disease. Depending on experimental treatment, the participants received either high-fear, low-fear, or control communications. In the high-fear condition, the messages described tetanus in extremely vivid terms, thereby creating a high level of fear and apprehension. In the low-fear condition, the messages described tetanus in less detailed terms, thereby creating no more than low to moderate fear. In the control condition, the message provided little detail about the disease, thereby arousing no fear.

To determine the message's effectiveness, the students were asked whether they thought it was important to get a tetanus inoculation and whether they actually intended to get one. The responses showed that students exposed to the high-fear message had stronger intentions to get shots than those exposed to the other messages. Moreover, records kept at the university health service indicated that students receiving the high-fear message were more likely to obtain inoculations during the following month than were students receiving the other messages.

This study demonstrates that fear-arousing messages can change attitudes. In general, however, fear-arousing messages are effective only when certain conditions are met. First, the message must assert that if

BOX 8.1 Persuasion via Mass Media

Although many of the persuasive messages we receive daily come to us through face-to-face interaction with other persons, a significant portion of them arrive via the mass media. The term **mass media** refers to those channels of communication that enable a source to reach a large audience. Whereas face-to-face communication typically can reach only a small audience, the mass media can potentially influence many people. The most influential mass medium in the United States is television, followed by newspapers, radio, and magazines (Atkin, 1981).

Not everyone has equal exposure to the mass media. For example, in the United States, women view more television than men. Children and retirees view more television than do adolescents and working adults. Viewing is negatively related to the level of education, income, and occupational status (Comstock, Chaffee, Katzman, McCombs, & Roberts, 1978; Newspaper Advertising Bureau, 1980).

Consideration of the mass media immediately raises a central question: To what extent are communications transmitted by the mass media effective in changing the beliefs and attitudes of large numbers of people? We will look at this issue from the standpoint of media campaigns.

Media Campaigns

A **media campaign** is a systematic attempt by a source to use the mass media to change the attitudes and beliefs of a select target audience. Media campaigns are common in the industrialized world. They are used by advertisers to sell new products or services and by political parties to sway voters' sentiments. They are also used by public officials to change citizens' behavior through public service announcements that attempt to stop drunk driving, to encourage people to try to quit smoking, to get people to vote on election day, and so on (Cummings, Sciandra, Davis, & Rimer, 1989; Farhar-Pilgrim & Shoemaker, 1981; Solomon, 1982).

Each year, advertisers spend tens of billions of dollars on media campaigns. Nevertheless, most media campaigns do not produce large amounts of attitude change. In general, messages sent via the mass media have only a small impact on their target audience's attitudes (Barber & Grichting, 1990; Bauer, 1964; Finkel, 1993). Consider, for example, what occurs during presidential campaigns. Soon after the political conventions nominate their candidates in midsummer, most Americans know how they intend to vote in the upcoming November election. Although the parties spend millions of dollars on political advertising during the fall campaign, they will not change many voters' attitudes. In most presidential elections, only about 7 to 10 percent of the voters switch their preferences regarding candidates during the campaign. It is difficult for campaigners to change political attitudes via the mass media (Klapper, 1960; Lazarsfeld, Berelson, & Gaudet, 1948).

Of course, from another perspective, a shift of 7 to 10 percent may be very significant. Many professional advertisers and politicians are satisfied if their media messages shift public opinion a few percentage points in the intended direction. In a close political race, a gain of 1 or 2 percent might be sufficient to win the election. Thus, even though a media campaign might be a disappointment in terms of producing widespread attitude change, it may simultaneously be a success from the standpoint of getting a candidate elected (Mendelsohn, 1973). Even a small amount of attitude change may be sufficient to justify the cost of the media campaign in a close election.

Why are media campaigns usually able to produce only small amounts of attitude change? There are several reasons. First, there is the phenomenon of selective exposure. Many messages do not reach the audience they are intended to influence, because audience members attend mostly to those sources with which

BOX 8.1 Continued

they agree. Instead of reaching persons who disagree with the message—and whose opinions might therefore be changeable—many media communications are received by persons who already agree with the message and whose opinions will therefore be reinforced, not changed. In media exposure, persons are affected more by messages supporting than not supporting their preexisting attitudes (Klitzner, Gruenewald, & Bamberger, 1991; Sears & Freedman, 1967). Some have argued, however, that selective exposure is being reduced with the advance of modern communication technology. People are now exposed to a greater variety of viewpoints through television news programming and advertising (Mutz & Martin, 2001), although there is still a tendency for the ads to reinforce previously held attitudes (Ansolabehere & Iyengar, 1994).

Second, even if the intended targets receive messages from the media, they may reject them or derogate the source. Recipients of media communications are certainly not passive, and the impact of a message depends heavily on the uses and gratifications that the audience can obtain from the information (Dervin, 1981; Swanson, 1979). For example, in selling consumer products, media persuasion is more effective when the target's involvement with the decision is low and when he or she perceives relatively small differences between alternative products. In contrast, the impact of the media will be slight when target involvement with the decision is high and the differences between products appear clear-cut (Chaffee, 1981; Ray, 1973).

Third, even when a target finds a media message compelling, he or she may be subject to counterpressures that inhibit attitude change (Atkin, 1981). Some of these pressures come from social groups such as family, friends, and coworkers; these groups may exert influence that nullifies the media's impact. Moreover, targets are exposed to conflicting persuasive communications and cross-pressures transmitted via the media. For example, beer advertisements would probably be enormously successful if only one manufacturer advertised its product. But because many brands advertise, media messages offset one another.

Other Effects of Media Campaigns

Although media campaigns do not usually cause a massive change in attitudes, they do exert other impacts on audiences. First, they are effective in strengthening pre-existing attitudes. In other words, they reinforce and buttress preferences already held by the target audience (Ansolabehere & Iyengar, 1994). Televised debates between presidential candidates, for example, usually strengthen existing attitudes rather than change them (Kraus, 1962; Sears & Whitney, 1973).

In addition to strengthening preexisting attitudes, mass media are successful in creating attitudes toward objects that previously were unknown or unimportant to the audience. Advertisers use media campaigns to cultivate positive attitudes toward newly introduced products (breakfast cereals, new-design toys for children). Political parties also use media campaigns to create positive attitudes toward new, little-known candidates running for office. Today, former Presidents Jimmy Carter and Bill Clinton are well known to the American public. But when both began running for the presidency, the situation was very different. Both had been the governors of southern states, but both were almost entirely unknown outside the South. Carter had even appeared on the game show *What's My Line?* and none of the celebrity panel was able to guess who he was. To win the Democratic nomination for president, both candidates launched massive media campaigns to make themselves recognizable to Americans. At the same time the candidates were introducing themselves, they also worked hard to establish a positive public image of themselves.

the target does not change behavior, he or she will suffer serious negative consequences. Second, the message must show convincingly that these negative consequences are highly probable. Third, the message must recommend a specific course of action that, if adopted, will enable the target to avoid the negative consequences. A message that predicts negative consequences but fails to assure the target that he or she can avoid them by taking specific action will produce little attitude change. Instead, it will leave the target feeling that the negative consequences are inevitable regardless of what he or she may do (Job, 1988; Maddux & Rogers, 1983; Patterson & Neufeld, 1987).

One-Sided Versus Two-Sided Messages When a source uses rational rather than emotional appeals, other message characteristics also come into play. One such characteristic is the number of viewpoints, or sides, represented in the message. A one-sided message emphasizes only those facts that explicitly support the position advocated by the source. A two-sided message, in contrast, presents not only the position advocated by the source but also opposing viewpoints. For example, if a man used a one-sided message to persuade his wife to spend their vacation at the seashore, he would mention only the reasons for going to the seashore. If he used a two-sided message, he would mention both the reasons for going and the reasons for not going. Of course, if he really preferred the seashore, he might also try to refute or discredit the reasons for not going.

Which is more effective, a one-sided message or a two-sided message? The answer depends heavily on the nature of the target audience. One-sided messages have the advantage of being uncomplicated and easy to grasp. They are more effective when the audience already agrees with the source; they also tend to be effective when the audience does not know much about the issue, for they keep the audience blind to opposing viewpoints. Two-sided messages are more complex, but they have the advantage of making the source appear less biased and more trustworthy. They are more effective when the audience initially opposes the source's viewpoint or knows a lot about the alternative positions (Karlins & Abelson, 1970; Sawyer, 1973).

The Target

So far, we have discussed how the characteristics of the source and the content of the message affect persuasion. Yet it is also true that the characteristics of the target play a role in persuasion. One important target characteristic that affects persuasion is the degree to which the target is involved with the issue. Moreover, the persuasion attempt can be affected by personality factors, such as how much the target person likes thinking things through. Finally, we consider the role of how focused or distracted the target is during the persuasion attempt.

Involvement with the Issue One important attribute of targets is the extent of their involvement with a particular issue (Johnson & Eagly, 1989; Petty & Cacioppo, 1990). Suppose, for example, that someone advocates a fundamental change at your college, such as increasing the number of comprehensive exams required for graduation. The proposed change would take effect in September of the next year. Many students would be very involved with this issue, because the change would affect the difficulty of completing their degrees. Now, suppose the source advocated that the change take place 10 years in the future rather than next September. Current students would probably have little interest in this proposal, because they will finish college long before any changes take effect.

Involvement with the issue fundamentally affects the way the target processes a message. When highly involved, a target will want to scrutinize the message closely and think carefully about its content. Strong arguments will likely produce substantial attitude change, whereas weak arguments will produce little or no attitude change. In contrast, the target who is uninvolved will have less motivation to think carefully about the message. If any change in attitude occurs, it will result more from peripheral factors, such as source expertise or trustworthiness, than from the arguments themselves (Chaiken, 1980; Leippe & Elkin, 1987; Petty, Cacioppo, & Heesacker, 1981).

In one study, a message similar to the one just described was presented to a group of college students (Petty, Cacioppo, & Goldman, 1981). The message proposed that college seniors be required to take a

FIGURE 8.3 THE EFFECTS OF PERSONAL INVOLVEMENT ON PERSUASION

In this study, students received a message advocating that college seniors be required to take a comprehensive exam prior to graduation. Half the students were told that the new policy would take effect next year (high involvement), whereas the others were told the policy would take effect 10 years later (low involvement). Results show that students in the high-involvement condition were affected primarily by the strength of the arguments rather than by the expertise of the source, whereas students in the low-involvement condition were affected primarily by the expertise of the source rather than by the strength of the arguments.

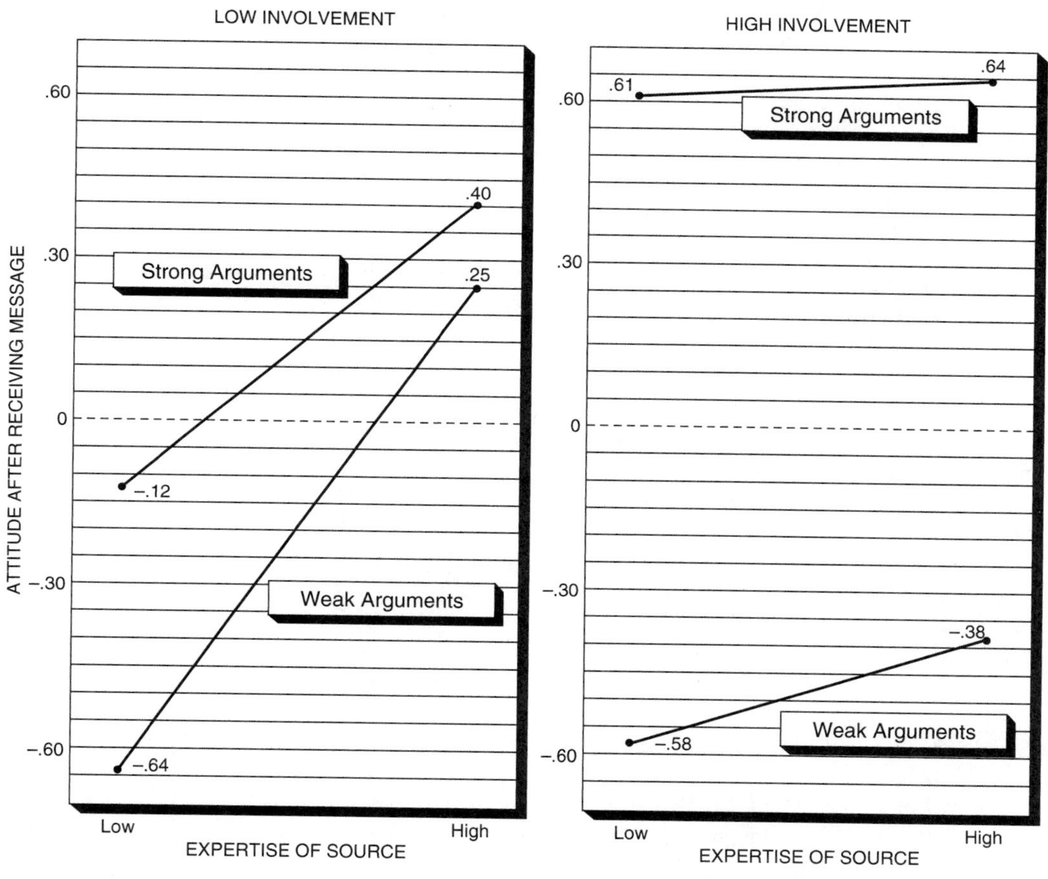

Source: Adapted from "Effects of Personal Involvement on Persuasion" by Petty, Cacioppo, and Goldman, *Journal of Personality and Social Psychology, 41,* 164–169.

comprehensive exam before graduation. Three independent variables were manipulated in this study. The first variable was personal involvement with the issue. Half the participants were told that the new policy would take effect next year at their college (high involvement), whereas the other half were told that the policy would take effect 10 years in the future (low involvement). The second variable was the strength of the message's argument. Half the participants received eight strong and cogent arguments in favor of the proposal; the other participants received eight weak and specious arguments. The third variable was the expertise of the source. Half of the participants were told that the source of the message was a professor of education at Princeton University (high-expertise source); the other half were told that the source was a student at a local high school (low-expertise source).

The results of this study appear in Figure 8.3. In the high-involvement condition, the target's attitude toward comprehensive exams was determined primarily by the strength of the arguments. Strong arguments produced significantly more attitude change than weak ones. The expertise of the source had no significant impact on attitude change. In the low-involvement condition, attitudes were determined primarily by the source's expertise; the high-expertise source produced more attitude change than the low-expertise source. The strength of the arguments had little effect on this group.

Thus, the target's involvement with the issue moderated which factor was the primary determinant of attitude change. For participants with high involvement, the strength of the argument was more important than source expertise (a peripheral factor) because participants cared about the issue. For those with low involvement, source expertise was more important because the participants had little motivation to scrutinize the arguments. Similar findings have been reported more recently by Chaiken and Maheswaran (1994).

Need for Cognition Beyond involvement with the issue, there are some personality traits that affect the persuasion process. In particular, how much an individual enjoys puzzling through problems and thinking about issues plays an important role in persuasion attempts. Those who do enjoy these thinking tasks are said to have a high need for cognition (Cacioppo, Petty, Feinstein, & Jarvis, 1996) and are motivated to examine arguments more carefully and thoroughly than those who have a low need for cognition. Thus, they are more likely to use the central route when forming attitudes and impressions and more likely to ignore the peripheral cues. When facing an audience of people (for example, college professors) with a high need for cognition, one would be wise to pay careful attention to constructing a solid set of arguments that will stand up to the scrutiny of central route elaboration.

Distraction Even people with a high need for cognition who are strongly involved in an issue will sometimes have trouble paying attention to arguments. This can occur because the audience is distracted by any number of things—perhaps they aren't feeling well, perhaps there is street noise that makes it hard to hear, perhaps the speaker has an annoying habit that bothers the listener, and so on. Anything that prevents the target from giving full attention to the argument will affect the persuasion attempt. Given the discussion so far, it will be no surprise to learn that those who are distracted are more likely to use peripheral cues when forming their opinions. The distracting element in the environment prevents them from fully engaging and appreciating the details of the argument, and therefore they fall back on peripheral indicators such as the attractiveness of the speaker (Petty & Brock, 1981; Petty, Wells, & Brock, 1976).

COMPLIANCE WITH THREATS AND PROMISES

As important as attitude change is, it is not the only outcome of social influence. Another important outcome is compliance—that is, behavioral conformity by the target to the source's requests or demands. When considering compliance, the fundamental concern is about producing a particular behavior from the target, irrespective of whether the target's beliefs and attitudes change or not. Of course, in some cases, compliance can be obtained indirectly by changing attitudes—if the source can change what the target believes, he or she also may change how the target behaves. But often persuasion is not required to change behavior. French and Raven (1959; Raven, 1992) proposed that there are six kinds of social power that can be used to induce compliance—some of which require actual persuasion and some do not (see Box 8.2). In this section, we examine two in more detail: threats and promises.

Consider a home owner, Richard Sorenson, who lives in an area of Michigan where it snows heavily. One cold day in January, a snowstorm dumps 12 inches of snow on his driveway and sidewalk. Although Richard has been the person in his household who has always shoveled the snow, he believes his teenage son is now old enough to take on the task and has been considering the best way to shift the responsibility. He could approach his son and say, "I'll give you $20 if you shovel the snow out of the driveway." This would be an attempt to gain compliance in the form of a promise: Richard promises to pay $20 in return for a specified performance. Or he could use a threat: "Shovel the snow or I won't let you use the car for a week." Here compliance is demanded or Richard will levy a penalty.

Influence based on promises and threats differs from persuasion attempts in a fundamental way. When using persuasion, the source tries to change the way a target views the situation. Sorenson, for example, might have attempted to persuade his son that shoveling snow is enormous fun or that clearing out the driveway would make him a good son. These appeals, if successful, would change how the boy looks at the situation, but they would not actually change the situation itself. This contrasts with the use of promises or threats, where the source does restructure the situation. By promising to pay money for a clear driveway or threatening punishment if it is not done, Sorenson has added a new reinforcement contingency to the situation—money in return for snow removal, or the inconvenience of walking if it is not removed. In both cases, he hopes the looming reinforcement will induce his son to comply, but which approach will be most effective?

Effectiveness of Threats and Promises

A **threat** is a communication from one person (the source) to another (the target) that takes the general form, "If you don't do X (which I want), then I will do Y (which you don't want)" (Boulding, 1981; Tedeschi, Schlenker, & Lindskold, 1972). For example, an employer might say to her employee, "If you don't complete this project before the deadline, I'll cut your salary." If the employee needs his salary to keep food on the table and has no other job prospects, he will take the threat seriously and do his best to comply with the demand.

When a source issues a threat, the sanction threatened can be virtually anything—a physical beating, the loss of a job, a monetary fine, the loss of love. The important point is that for a threat to be effective, the target must want to avoid the sanction. If the employee threatened by his boss happens to have a new job lined up elsewhere, he will not care whether his boss intends to cut his salary, and the threat will have little impact.

In the context of compliance, a **promise** is similar to a threat, except that it involves contingent rewards, not punishments. A person using a promise says, "If you do X (which I want), then I will do Y (which you want)." Notice that a promise involves a reward controlled by the source. Richard Sorenson promises a payment of $20 if his teenage son clears the driveway and sidewalk. People frequently use promises in exchanges, both monetary and nonmonetary.

By issuing a promise, the source creates a set of options for the target. Suppose, for example, that the source makes the promise, "I'll give you $20 if you clear the snow from the driveway." In response, the target can (1) comply with the source's request and clear the driveway, (2) refuse to comply and let the matter drop, or (3) make a counteroffer (such as "How about $30? It's a long driveway, and the snow is very deep"). In similar fashion, a threat creates a choice for the target. Once a threat is issued, the target can (1) comply with the threat, (2) refuse to comply, or (3) issue a counterthreat (Boulding, 1981).

The range of possible responses to threats and promises raises a fundamental question: Under what conditions will threats and promises be successful in gaining compliance, and under what conditions will they fail? Certain characteristics of threats and promises, such as their magnitude and credibility, affect the probability that the target will comply.

Magnitude of Threats and Promises In promises, the greater the magnitude of the reward offered by a source, the greater the probability of compliance by the target (Lindskold & Tedeschi, 1971). For example, a

BOX 8.2 Social Power and Compliance

Suppose a high school student did not do well on her last set of exams, and her father wants to try to influence her to study harder for her winter finals. The father can choose from a number of different tactics to try to produce compliance. These tactics can be organized by the type of power the father might use to influence the daughter. According to a model forwarded by French and Raven (1959; Raven, 1992), there are six major social bases of power that can be used in such a situation.

1. *Promise of Reward.* One way of inducing compliance is to promise to provide a reward if the target performs the desired behavior. The father might tell his daughter, "If you spend 2 hours a day studying for the next 2 weeks, I will buy you a new stereo." Oftentimes, explicit agreements about behavior and rewards are made, but other times they are more subtle, such as when we work hard to gain approval from our parents, even though we have never explicitly agreed on such an arrangement.
2. *Coercion Through Threat.* In contrast to the reward strategy, the father might use the threat of a negative outcome to induce compliance. "If you don't do better on your exams next time, I will cancel the ski trip you have been planning." As with rewards, the threats do not necessarily have to be explicit in order to be effective.
3. *Referent Power.* **Referent power** uses our desire to be accepted by members of valued social groups. When we seek acceptance, we may be more likely to comply with the demands of the group or we may try to become more similar to the group by imitating the behavior of its members (see Chapter 14). To use referent power, the father could identify people whom his daughter admires and then point out how studious those people are: "Your older sister spends at least 2 hours a day studying."
4. *Legitimate Power.* The social positions people occupy often supply them with power over other individuals, and this hierarchical

factory supervisor might obtain compliance from a worker by offering a large incentive: "If you are willing to work the late shift next month, I'll approve your request for 4 extra days of vacation in September." The worker's reaction might be less accommodating, however, if his supervisor offered only a trivial incentive: "If you work the late shift next month, I'll let you take your coffee break 5 minutes early today."

A similar principle holds for threats. Compliance with threats varies directly with the magnitude of the punishment involved. Other factors being equal, targets will dismiss threats that entail trivial consequences, but they will more likely comply with threats that entail large and serious consequences (Miranne & Gray, 1987).

Credibility of Threats and Promises Suppose you own a little puppy that often runs wild. Your not-so-nice neighbor hates dogs. One day, he issues a threat: "If you don't keep your dog off my property, I will call the city dogcatcher." This threat is troublesome, because your dog romps on his property frequently. But would your neighbor really do what he says, or is he merely bluffing? You will comply and tether your dog if the threat is real, but you do not want to comply if it is merely a bluff. Unfortunately, there is no surefire, risk-free way to find out whether the threat is credible. The only true way to test your neighbor's credibility is to call his bluff—that is, to refuse to comply. Then, if your neighbor was merely bluffing, that fact will quickly become evident. Of course, if he was

BOX 8.2 Continued

arrangement is often accepted by both the higher power and lower power persons involved (see Chapter 15). Bosses have the power to tell employees what to do, parents have the power to tell children what to do, and police officers have the power to tell motorists what to do. When authority is accepted as a right associated with a social role, it is called **legitimate power.** The father in our example could invoke legitimate power by saying "I'm the parent, and one of my jobs as a parent is to make sure you study. So, get to work!" If the daughter accepts the traditional authority arrangement, she will head off to study, even if she does not really want to.

5. *Information.* Sometimes we can actually change people's attitudes about the behavior we want them to exhibit, and then the behavior change will follow in order to produce consistency with the attitude. One way of doing this is to provide information about the effects of the behavior. "The grades you have now are not going to be high enough for you to get into college. The average grades of entering students at State College last year were in the B+ range. You currently only have a C+ average."
6. *Expertise.* Information can play a less direct role in compliance as well. There are many times in life when we do not need to know all information about the behavior as long as we think the person telling us what to do is an expert. We assume that because the person is an expert, she knows what she is talking about, and thus we will comply with her request. When a doctor prescribes drugs, we usually take them even if we don't know exactly how they work because we can rely on the expertise of the doctor. In the case of the father and the high school student, he might refer to an expert on studying who claims that an additional 2 hours of studying per day will raise a student's GPA by a full letter grade.

not bluffing, you will suffer the consequences. Retrieving your puppy from the dogcatcher may cost some money and entail some anxious moments.

Bluffing or not, any threatener wants the target to believe the threat is credible and to comply with his demand. He does not want the target to call his bluff. After all, a successful threat is one that obtains compliance without actually having to be carried out. If the target refuses to comply, the threatener must either admit that he is bluffing or incur the costs of carrying out the threat.

To judge the credibility of a threat, targets gauge the cost to the source of carrying out the threat. Threats that cost a lot to carry out are less credible than those costing less. Targets also estimate the credibility of a threat from the social identity of the source. A threat involving physical violence, for example, will be more credible if it comes from a karate expert wearing a black belt than if it comes from the proverbial 97-pound weakling.

The SEV Model A threat's **subjective expected value (SEV)** is a measure of the pressure that the target feels from the threat. The level of SEV depends on several factors. SEV increases as the threat's credibility increases and as the magnitude of punishment threatened increases (Stafford, Gray, Menke, & Ward, 1986; Tedeschi, Bonoma, & Schlenker, 1972). When both credibility and punishment magnitude are high, the SEV of the threat will be high; consequently, the target

A robber holds a store clerk at gunpoint. Targets are more likely to comply when threats are both large and credible.

will feel a lot of pressure to comply. When both credibility and punishment magnitude are low, SEV is low; consequently, the target will feel little or no pressure to comply. When one is high and the other is low (as, for example, when magnitude is high but credibility is low), the threat will have a moderate-to-low SEV.

The SEV model of threat effectiveness predicts that a target's compliance with a threat depends on two factors: the SEV of the threat and the cost to the target of complying with the threat (Tedeschi, Bonoma, & Schlenker, 1972). That is, when deciding whether to comply with the threat, a target will estimate both the SEV of the threat and the cost of complying. These factors will have opposite effects on compliance. The probability of compliance will increase directly as a function of SEV but will decrease as a function of the cost to the target of complying.

Various empirical studies have supported the SEV model of threat effectiveness (Faley & Tedeschi, 1971; Grasmick & Bryjak, 1980; Stafford et al., 1986). For example, in one laboratory study by Horai and Tedeschi (1969), several pairs of persons played a game repeatedly over 150 trials. These pairs were composed so that one member was a naive participant, whereas the other was a confederate of the researchers programmed to follow a specific strategy. Both the participant and the confederate were required to make certain choices that determined how points would be allocated during each trial. The object of the game was to gain as many points as possible.

In this game, threat capability was unilateral; that is, the confederate could issue threats but the participant could not. To influence the participant's choice and gain extra points, the confederate occasionally made threats in this form: "If you do not pick Choice 1 on the next trial, I will take [some number of] points away from your counter." The magnitude of the punishment threatened by the confederate was either high

FIGURE 8.4 COMPLIANCE WITH THREATS

Compliance is a function of both the magnitude of the punishment threatened and the credibility of the threat. The greater the punishment threatened, the greater the compliance with the threat. Likewise, the greater the credibility of the threat, the greater the compliance.

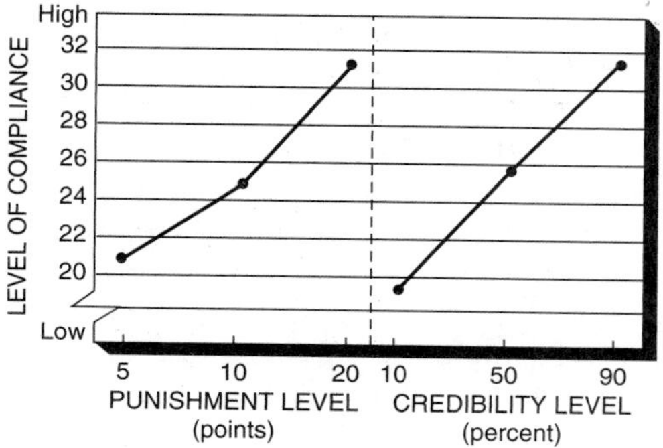

Source: Adapted from "Compliance with Threats" by Tedeschi and Horai, *Journal of Personality and Social Psychology, 12,* 164–169. Copyright © 1969 by the American Psychological Association. Adapted with permission.

(20 points), medium (10 points), or low (5 points), depending on experimental treatment. The credibility of these threats was manipulated by varying the frequency with which the confederate carried them out if the participant did not comply. In the high-credibility condition, the confederate carried out threats 90 percent of the time; in the medium-credibility condition, 50 percent of the time; and in the low-credibility condition, 10 percent of the time.

The results of this study, shown in Figure 8.4, demonstrated that both the magnitude of punishment and the credibility of the threat affected the participant's level of compliance. Compliance increased as a direct function of both variables. Compliance was greatest under the highest magnitude of punishment and the highest level of threat credibility. The results of this study are consistent with the SEV model.

Although we have described the SEV model as it pertains to threats, it also applies to promises. Of course, the relevant variables in this case are the magnitude of the reward and the credibility of the promise. The results of one study showed that targets were more influenced when the reward promised was large rather than small and when the source had credibility in following through on the promises (Lindskold et al., 1970). Consistent with the SEV view, these conclusions hold true only when the reward promised for compliance is greater than the reward(s) that might be gained from refusing to comply.

Problems in Using Threats and Promises

Threats and promises pose certain problems for their user. First, the source using threats or promises must maintain surveillance to determine whether the target has complied. This typically entails some cost to the source in the form of time and trouble to monitor the target's behavior. The problems involved in maintaining surveillance can be especially troublesome if the source uses threats rather than promises. When the source uses threats, the target not only may fail to comply but also may attempt to conceal his or her noncompliance and thereby avoid punishment (Ring & Kelley, 1963). If the target engages in concealment, then adequate surveillance and detection may entail even greater costs for the source.

Another problem is that threats often cause resentment or hostility by the target toward the influencing source (Zipf, 1960). In some cases, these negative interpersonal sentiments may linger long after the threat is finished. Promises usually do not entail this difficulty, because they are based on rewards and not on punishments. A target generally will develop more positive feelings toward someone who uses promises than toward someone who uses threats (Rubin & Lewecki, 1973). It is interesting to note that some kinds of contingent reinforcement can be presented to the target as either a threat or a promise. For example, if your boss would like you to take on a new task this year, she could promise you a big raise at the end of the year. Or, she could threaten you with a very small raise if you don't take it on. Costs can often be viewed as negative rewards, and thus the presentation of the contingency can be very important. Therefore, when a source enjoys a good relationship with the target and wants to maintain it, he or she may prefer to present outcomes as promises and to avoid using threats.

Bilateral Threat and Negotiation

So far, we have considered various means of social influence, including persuasion, threats, and promises. We have relied on the simplifying assumption that influence flows in only one direction—from the source to the target. But in many real-life situations, both persons are able to exercise influence. That is, when person A (the source) attempts to influence person B (the target), person B (now the source) may respond by attempting to counterinfluence person A (now the target). Two-way influence can surely involve persuasion or promises, but it is especially interesting when it entails threats.

Negotiation Consider a situation where (1) two individuals have strongly opposing interests, and (2) each person has the capacity to issue threats and inflict punishments on the other. This situation carries the potential for open conflict, but if the individuals involved are circumspect, they may hesitate to use threats and try instead to negotiate an acceptable solution. **Negotiation** is defined as communication between opposing sides in a conflict wherein the parties make offers and counteroffers and a solution occurs only when the parties reach an agreement (Pruitt & Carnevale, 1993). At first, the negotiators might attempt to find a straightforward compromise, such as one based on the split-the-difference principle. This compromise would yield a solution that is midway between person A's most preferred outcome and person B's most preferred outcome. If a solution like this is not acceptable to the negotiators, they may attempt something more creative, like an integrative solution, which is a compromise whereby the two parties make tradeoffs on issues according to their relative importance to each side (Neale & Bazerman, 1991; Pruitt, 1983). Specifically, each side makes its largest concessions on issues that are unimportant to it but important to the other side; the result is a solution that provides greater total benefits to negotiators than a split-the-difference compromise.

Although the notion of an integrative solution is appealing, such solutions are surprisingly difficult to achieve in practice, in part because there are barriers to identifying them (Ross & Ward, 1995; Thompson, 1990). For one thing, participants in a negotiation frequently fail to discover one another's true interests. And the more that the participants have at stake in a negotiation, the more biased their perceptions of one another. They may come to view one another as opponents and distrust any proposals made by the other side (Thompson, 1995). If the participants overlook interests held in common or fail to discover one another's true interests, integrative solutions will prove elusive.

Deterrence and Conflict Spiral If attempts at negotiation fail, the situation—as perceived by the participants—now has a different, more ominous character. The situation has become one where (1) two individuals have strongly opposing interests, (2) each person has the capacity to issue threats and inflict punishments on the other, and (3) negotiations have shown that there is little realistic possibility for compromise between them. Circumstances like these can become quite volatile and erupt into open conflict. If one or both participants believe that compromise is impossible, the situation may change quickly from one of simple negotiation to one of **bilateral threat.** A bilateral threat situation is one where both participants issue threats and inflict damage on one another.

Bilateral threat situations involve a tension between the temptation to use punishment against the other and the fear of having punishment used against oneself. Most bilateral threat situations end up in one of two outcomes (Lawler, 1986; Lawler & Ford, 1995). On one hand, they can end in a state of **mutual deterrence,** which is a kind of armed standoff. If both participants especially want to avoid suffering damage inflicted by the other, a state of mutual deterrence will exist and the participants will refrain from initiating threats and punishments (Bacharach & Lawler, 1981; Michener & Cohen, 1973). Alternatively, bilateral threat situations can devolve into a **conflict spiral,** which is a form of escalation. In a conflict spiral, the participants insist on compliance to their demands, even though they realize the other can inflict damage. They exchange successively larger threats and counterthreats to bolster their demands and end up inflicting more and more damage on each other (Deutsch & Krauss, 1962; Lawler, Ford, & Blegen, 1988; Smith & Anderson, 1975).

Power Differences and Deterrence One factor of special interest in bilateral threat situations is the relative strength or power of the participants. Several studies have investigated bilateral situations involving asymmetrical power relations—that is, situations where both persons have threat and punishment capability, but one person has more than the other. Power differences have an effect on the participants' use of threat, on concessions made, and on deterrence.

First, situations of unequal power are less stable and less likely to result in mutual deterrence than situations of equal power. If the participants have equal coercive power, the situation is less likely to result in a conflict spiral with threats and punishments than if the participants have unequal power (Lawler et al., 1988). Moreover, given equal power, participants are less likely to engage in damage tactics when they are both strong rather than weak, because both will fear retaliation by the other (Hornstein, 1965; Lawler, 1986; Lawler & Bacharach, 1987).

In situations of unequal power, the high-power negotiators tend to behave exploitatively, whereas those with low power behave more submissively (Rubin & Brown, 1975). Given unequal power, the high-power negotiators usually maintain firm aspirations, adopt tougher tactics, and prefer not to make concessions. Not only do they avoid initiating concessions, they may refuse to match concessions from the low-power negotiator (Michener, Vaske, Schleifer, Plazewski, & Chapman, 1975; Smith & Leginski, 1970). In contrast, low-power negotiators frequently match any concessions made by the other.

In unequal bilateral situations, the low-power negotiator will often hesitate to use threats or to behave aggressively, because he or she fears retaliation from the high-power opponent (who can, after all, inflict a lot of damage). The deterrence effect is unmistakable: as the punishment magnitude of the high-power person increases, the aggressive behavior of the low-power person decreases (Michener & Cohen, 1973).

Despite this deterrence effect, situations of unequal power are still more likely to result in escalation than situations of equal power. Persons with lower but still significant levels of power often refuse to comply with demands from stronger negotiators. Under these conditions, even a single punitive act may be sufficient to trigger a conflict spiral. Negotiators, in fact, often overmatch or exceed the other's threats at low levels of conflict, an unhappy tendency that can lead to conflict spirals (Youngs, 1986). One threat provokes another of larger size, making an agreement ever more difficult to achieve.

OBEDIENCE TO AUTHORITY

So far, we have discussed attitude change through persuasion and through compliance based on threats and promises. Important though these are, they are not the only forms of social influence used in everyday life. We have all witnessed situations where—without the use of threat, promises, or persuasion—one person issues an order and another person complies. For example, a baseball umpire tosses an unruly manager out of the game and orders him to leave the field; the manager, after showing his resentment by throwing his cap to the ground and kicking first base, grudgingly complies. The umpire does not attempt to persuade the manager to leave voluntarily; he simply issues an order directing the manager to leave. Compliance in this case is based on the fact that both the umpire and the manager are participating in a larger social system (two ball clubs playing a game) in which behavior is regulated by rules and roles. The capacity of the source (the umpire) to influence the target (the manager) stems from the rights conferred by their roles within the game. Under the rules of the game, the umpire has the right to throw the manager out of the game for disruptive behavior.

When persons occupy roles within a group, organization, or larger social system, they accept certain rights and obligations vis-à-vis other members in that social unit. Typically, these rights and obligations give one person authority over another with respect to certain acts and performances. **Authority** refers to the capacity of one member to issue orders to others—that is, to direct or regulate the behavior of other members by invoking rights that are vested in his or her role. When the umpire tosses the manager out of the game, the basis of his power is legitimate authority.

The lines of authority become salient as a military commander gives orders to his troops prior to a mission.

Orders by police officers, decisions by judges, directives by corporate executives, and exhortations by clergymen—all these entail authority and the invocation of norms. A source can exercise authority only when, by virtue of the role that he or she occupies in a social group, others accept his or her right to prescribe behavior regarding the issue at hand (Kelman & Hamilton, 1989; Raven & Kruglanski, 1970). In exercising authority, the source invokes a norm and thereby obliges the target(s) to comply. The greater the number of persons the source can directly or indirectly influence in this manner and the wider the range of behaviors over which the source has jurisdiction, the greater his or her authority within the group (Michener & Burt, 1974; Zelditch, 1972).

In this section, we will discuss influence based on legitimate authority. First, we will consider some interesting experimental studies of destructive obedience. These studies illustrate the extremes to which authorities can push behavior. Second, we will consider some factors that determine whether a target will comply or refuse to comply with an authority's directives.

Experimental Study of Obedience

We note that obedience to authority frequently produces beneficial results, for it facilitates coordination among persons in groups or collective settings. Civil order hinges on obedience to orders from police officers and judicial officials, and effective performance in work settings often depends on following the directives of bosses or employers.

Yet if obedience to authority is unquestioning, it can sometimes produce disquieting or undesirable outcomes. For instance, in one study, hospital nurses received orders from doctors to administer a drug to a patient. The order came by telephone, and the nurses involved did not previously know the doctors giving the order. The drug was one not often used in the hospital; hence, it was not very familiar to the nurses. The dosage prescribed was heavy and substantially exceeded the maximum listed on the package. The results showed that nearly all the nurses in this study were nevertheless ready to follow orders and administer the drug at the prescribed dosage (Hofling, Brotz, Dalrymple, Graves, & Pierce, 1966). Of course, the conditions in this study were very favorable for obedience; under different conditions, obedience rates will not be so high. Subsequent research has indicated, for instance, that when nurses are more familiar with the medicine involved and when they are able to consult freely with their colleagues, the rates of obedience are considerably lower (Rank & Jacobson, 1977).

In some cases, obedience to authority can produce very negative consequences—especially if the orders involve actions that are morally questionable or reprehensible. History provides many examples, such as the My Lai massacre during the Vietnam War, where soldiers obeyed Lieutenant Calley's orders to kill innocent villagers. Then there is the activity of the Third Reich of Nazi Germany during the 1930s and 1940s, which produced the Holocaust. In complying with the dictates of Hitler's authoritarian government, some German citizens committed acts that most people consider morally unconscionable—beatings, confiscation of property, torture, and murder of millions of people. This may seem like madness, but Hannah Arendt (1965) has argued that most participants in the Holocaust were not psychotics or sadists who enjoyed committing mass murder but ordinary individuals exposed to powerful social pressures.

To explore the limits of obedience to legitimate authority, Stanley Milgram carried out a program of experimental research in a laboratory setting (Milgram, 1965a, 1974, 1976; Miller, Collins, & Brief, 1995). Milgram created a hierarchy in which one person (the experimenter, who assumed the role of authority) directed another person (the participant) to engage in actions that hurt a third person (a confederate, who played the role of victim). The primary goal of this research was to understand the conditions under which participants would follow morally questionable orders to hurt the confederate.

At the outset, Milgram (1963) recruited 40 adult men to serve as participants. These men, contacted through newspaper advertisements, were adults (ages 20–50) with diverse occupations (labor, blue-collar, white-collar, and professional). When a participant arrived for the experiment, he found that another person (a gentle, 47-year-old male accountant) had also responded to the newspaper advertisement. This person, though ostensibly another participant, was actually a confederate of the experimenter. The participants were told by the experimenter that the purpose of the research was to study the effects of punishment (that is, electric shock) on learning. One of the participants was to occupy the role of learner, whereas the other was to occupy the role of teacher. Participants drew lots to determine their roles; unknown to the participant, the drawing was rigged so that the confederate was selected as the learner. The confederate was then taken into the adjacent room and strapped into an "electric chair," and electrodes were attached to his wrist. He mentioned that he had some heart trouble and expressed concern that the shock might prove dangerous. The experimenter, who was dressed in a lab coat, replied that the shock would be painful but would not cause permanent damage.

The participant and the confederate then participated in a paired-associates learning task. The participant, in the role of teacher, read pairs of words over an intercom system to the confederate in the adjacent room, and the confederate was supposed to memorize these. After reciting the entire list of paired words, the participant then tested the amount learned by the confederate. Going through the list again, he read aloud the first word of each pair and four alternatives for the second word of the pair, like a multiple-choice exam. The confederate's task was to select the correct alternative response for each item.

Consistent with the cover story that they were investigating the effects of punishment on learning, the experimenter ordered the participant to shock the learner whenever he made an incorrect response.

This shock was to be administered by means of an electric generator that had 30 voltage levels, ranging from 15 to 450 volts. The participant was directed to set the first shock at the lowest level (15 volts) and then, with each successive error, increase to the next higher voltage. That is, the participant was to increase the voltages from 15 to 30 to 45, and so on up to the 450-volt maximum. On the shock generator, the lowest voltage level (15 volts) was labeled *slight shock;* a higher level (135 volts) read *strong shock;* higher still (375 volts) read *danger: severe shock;* the highest level (450 volts) was ominously marked *XXX.* In actuality, this equipment was a dummy generator, and the confederate never received any shock, but its appearance was quite convincing to participants.

Soon after the session began, it became apparent that the confederate was a slow learner. Although he got a few answers right, his responses were incorrect on most trials. The participants reacted by administering ever higher levels of shock, as they had been ordered to do. When the shock level reached 75 volts, the confederate (who was still in the adjacent room) grunted loudly. At 120 volts, he shouted that the shocks were becoming painful. At 150 volts, he demanded to be released from the experiment ("Get me out of here! I won't be in the experiment anymore! I refuse to go on!"). At 270 volts, his response to the shock was an agonized scream. (Actually, the shouts and screams that participants heard from the adjacent room came from tape recordings; this made the learner's response uniform for all participants.)

Whenever a participant expressed concern or dismay about the procedure, the experimenter urged him to persist ("The experiment must continue"; "You have no other choice, you must go on"). At the 300-volt level, the confederate shouted in desperation that he wanted to be released from the electric chair and would not provide any further answers to the test. In reaction, the experimenter directed the participant to treat any refusal to answer as an incorrect response. At the 315-volt level, the learner gave out a violent scream. At the 330-volt level, he fell completely silent, and from that point on nothing more was heard from him. Stoically, the experimenter directed the participant to continue toward the 450-volt maximum, even though the learner did not respond.

The basic question addressed by this study was, "What percentage of the participants would continue to administer shocks up to the 450-volt maximum?" The results showed that of the 40 participants, 26 (65 percent) continued to the end of the shock series (450 volts). Although they could have refused to proceed, not a single participant stopped before administering 300 volts. Despite the tortured reaction of the confederate, most participants followed the experimenter's orders.

Understandably, this situation was very stressful for the participants, and many felt some concern for the learner's welfare. As the shock level rose, the participants grew increasingly worried and agitated. Some began to sweat or laugh nervously, and many pleaded with the experimenter to check the learner's condition or to end the study immediately. A few participants became so distressed that they refused to follow the experimenter's orders. The overall level of compliance in this study, however, was quite high, reflecting the enormous impact of directives from a legitimate authority.

Factors Affecting Obedience to Authority

As Milgram's results show, persons in authority usually obtain compliance with their orders, especially when these are accepted as legitimate or backed by potential force. Nevertheless, orders from an authority can set off a complex process, which can lead to various responses (Blass, 1991). Compliance does not always occur, and subordinate members sometimes defy orders from an authority. Although most participants in Milgram's research obeyed orders, some refused to comply. Other studies have reported similar effects: Obedience is the most common response to authority, but defiance occurs in some cases (Martin & Sell, 1986; Michener & Burt, 1975). This raises a basic question: Under what conditions will people comply with authority, and under what conditions will they refuse? What factors affect the probability that group members will comply with authority?

Certain factors affecting compliance are straightforward. For instance, it matters whether the person issuing orders uses an overt display of symbols, such as wearing a uniform with authoritative insignia (Bushman, 1988). Other things being equal, a direct display of authority symbols will increase compliance. In one study (Sedikides & Jackson, 1990), visitors at the bird exhibit of the Bronx Zoo were approached by a person who told them not to touch the handrail of the exhibit. They were significantly more likely to obey this directive when it came from a person dressed in a zookeeper uniform than when it came from a person dressed in casual clothes. The use of authoritative symbols may also have played a part in the Milgram studies, where the experimenter wore a white lab coat.

Another factor that matters is whether the person in authority can back up his or her demands with punishment in the event of noncompliance. Although this was not an explicit factor in Milgram's studies, other research has manipulated the magnitude of the potential punishment wielded by the authority, and the results support the view that greater punishment magnitude leads to higher levels of compliance (Michener & Burt, 1975).

Milgram (1974) extended his basic experiment to study some other factors that affect compliance with orders. For instance, one variation manipulated the degree of surveillance by the experimenter over the participant (Milgram 1965a, 1974). In one condition, the experimenter sat a few feet away from the participant during the experiment, maintaining direct surveillance; in another condition, after giving basic instructions, the experimenter departed from the laboratory and issued orders by telephone from a remote location. The results show that the number of obedient participants was almost three times greater in the face-to-face condition than in the order-by-telephone condition. In other words, obedience was greater when participants were under direct surveillance than under remote surveillance. During the telephone conversations, some participants specifically assured the experimenter that they were raising the shock level when in actuality they were using only the lowest shock and nothing more. This tactic enabled them to ease their conscience while at the same time avoiding a direct confrontation with authority.

In another variation, Milgram (1974) manipulated the participant's physical proximity to the victim. The findings showed that bringing the victim closer to the participant—and therefore increasing the participant's awareness of the learner's suffering—substantially reduced the participant's willingness to administer shock. In the extreme case, when the victim was seated right next to the participant, obedience decreased substantially. Tilker (1970) reported similar results and also showed that expressly making participants totally responsible for their own actions rendered them less likely to administer shocks to the learner.

Obedience to authority is also affected by the participant's position in a larger chain of command. Kilham and Mann (1974) used a Milgram-like situation in which one participant (the executant) actually pushed the buttons to administer shock, while another participant (the transmitter) simply conveyed the orders from the experimenter. The results showed that obedience rates were approximately twice as high among transmitters as among executants. In other words, persons positioned closer to the authority but farther from the unhappy task of throwing the switch were more obedient.

RESISTING INFLUENCE AND PERSUASION

Up to this point in the chapter, we have focused on the factors that produce attitude change and compliance with the wishes of the communicator. But we are not simply hapless victims of the persuasion and compliance efforts of other people. Social psychologists have identified a number of factors that enhance our ability to resist attitude change. In this section, we discuss four major contributors to persuasion resistance: inoculation, forewarning, reactance, and selective avoidance.

Inoculation

Interested in how persons develop resistance to persuasion, McGuire (1964) proposed that a target can be inoculated against persuasion. He specified various **attitude inoculation** treatments that would enable target persons to defend their beliefs against persuasion attempts. One such treatment, called a refutational defense, is analogous to medical inoculation, in which a patient receives a small dose of a pathogen so that he or she can develop antibodies. The refutational defense consists of giving the target (1) information that is discrepant with their beliefs and (2) arguments that counter the discrepant information and that support their original beliefs. By exposing a target to weak attacks and allowing the target to refute them, this inoculation builds up the target's resistance and prepares the target to resist stronger attacks on their attitudes in the future.

Some research (McGuire & Papageorgis, 1961) has demonstrated the effectiveness of a refutational defense against persuasion attempts. College students received messages attacking three commonly held beliefs or "cultural truisms." Two days before the attack, the students had received an inoculation treatment to foster their resistance to persuasion. For one truism, they received a refutational defense. For a second truism, they received a different immunization treatment called a supportive defense—information containing elaborate arguments in favor of the truism. For the third truism, they received no defense. Following exposure to attacks on their attitudes, students rated the extent of their agreement with each of the truisms. The results show that the refutational defense provided the highest level of resistance to persuasion. The supportive defense provided less resistance, and when no defense was present, there was still less resistance to persuasion.

Forewarning

A second aid to resisting influence is simply warning people that they are about to be exposed to a persuasion attempt. It is not necessary to be provided with information to refute the arguments for this effect to occur—if we are warned that our attitudes will be coming under attack, we begin to develop our own counterarguments (Freedman & Sears, 1965). The more advance notice people have that the persuasion attempt is coming, the more time they have to develop counterarguments and the more resistant they are to the persuasion attempt (Chen, Reardon, Rea, & Moore, 1992; Petty & Cacioppo, 1979a).

There is a qualification for the forewarning process to work, however: The targets of the persuasion attempt must care about and be psychologically involved in the issue. If they do care about the issue, then they are motivated by the warning to defend their position. If, on the other hand, they do not care about the issue, the forewarning may have little effect and, in some instances, can even produce greater attitude change (Apsler & Sears, 1968).

Reactance

Sometimes, persuasion attempts can go too far. When trying to convince people to change their attitudes, we become too heavy-handed and actually produce a reaction in the opposite direction we intended. This phenomenon is called **reactance,** and it occurs when the target of the persuasion attempt begins to feel that their independence and freedom are being threatened (Brehm, 1966). Feeling the need to reassert control, the targets will behave in a way counter to the persuasion attempt in order to demonstrate their independence. Reactance effects have been demonstrated in studies of anti-drinking, anti-smoking, and anti-graffiti persuasion attempts; physician's advice; and warning labels on television programming (for example, Bensley & Wu, 1991; Bushman & Stack, 1996; Pennebaker & Sanders, 1976).

SUMMARY

Social influence occurs when behavior by one person (the source) causes another person (the target) to change an opinion or to perform an action he or she would not otherwise perform. Important forms of open influence include persuasion, use of threats and promises, and exercise of legitimate authority.

Attitude Change via Persuasion Persuasion is a widely used form of social influence intended to produce attitude change. (1) There are many possible reactions to persuasive messages. Whereas some attempts at persuasion succeed, others merely lead to rejection of the message, derogation of the source, or suspension of judgment. Persuasive messages can change attitudes and beliefs through either the central route or the peripheral route. (2) The communication-persuasion paradigm points to many factors—properties of the source, the message, and the target—that affect whether a message will change beliefs and attitudes. (3) Certain attributes of the source affect a message's impact. Sources who are credible (that is, highly expert and trustworthy) are more persuasive than sources who are not. Attractive sources are more persuasive than unattractive ones, especially if message arguments are strong. A message coming from multiple, independent sources will have more impact than the same message from a single source. (4) Message characteristics also determine a message's effectiveness. Highly discrepant messages are more persuasive when they come from a source having high credibility. Fear-arousing messages are most effective when they specify a course of action that can avert impending negative consequences. One-sided messages have more impact than two-sided messages when the target already agrees with the speaker's viewpoint or is not well informed. (5) Attributes of the target also determine a message's effectiveness. Targets who are highly involved with an issue, who like thinking issues through in detail, and who are not distracted tend to scrutinize messages closely and are more influenced by the strength of the arguments than by peripheral factors.

Compliance with Threats and Promises Threats and promises are influence techniques used to achieve compliance (not attitude change) from the target. In using threats and promises, the source alters the environment of the target by directly manipulating reward contingencies. (1) The effectiveness of a threat depends on both the magnitude of the punishment involved and the probability that it will be carried out. Greater compliance results from high magnitude and high probability. Similar effects hold true for promises, although these involve rewards rather than punishments. (2) The use of threats and promises poses several problems for the source. For threats to be effective, the source must maintain surveillance over the target. Furthermore, threats often arouse resentment or hostility toward the source and therefore disrupt relationships. (3) When influence is bilateral rather than unilateral, participants may attempt to resolve their differences through negotiation. If successful, negotiation may lead to an integrative solution; solutions of this type, however, are difficult to achieve. If each party has the capacity to inflict punishment, bargaining may escalate into a conflict spiral. If the two parties have very unequal coercive power, the weaker one will probably hesitate to issue threats and will more likely comply with the demands of the stronger one. But bilateral situations are volatile, in part because threats by one party often provoke threats by the other.

Obedience to Authority Authority refers to the capacity of one group member to issue orders or make requests of other members by invoking rights vested in his or her role. (1) Research on obedience to authority shows that participants will comply with orders to administer extreme levels of electric shock to an innocent victim. (2) Obedience to authority is more likely to occur when the authority is dressed in uniform, when the authority can back up orders with punishments, when participants are under direct surveillance by the person issuing orders, when participants are distant from rather than close to the victim, and when participants are transmitters rather than executants of a command.

Resisting Influence and Persuasion Resistance to persuasion attempts can be increased through inoculation processes, in which targets are exposed to some of the source's arguments before the persuasion attempt occurs and provided with counterarguments. Persuasion can also be reduced by warning the target that a persuasion attempt is going to occur. Finally, if a persuasion attempt is too heavy-handed, targets may feel their freedom is threatened and attempt to re-establish their independence by defying the persuasion attempt.

LIST OF KEY TERMS AND CONCEPTS

attitude change (p. 197)
attitude inoculation (p. 221)
authority (p. 217)
bilateral threat (p. 216)
communication-persuasion paradigm (p. 200)
communicator credibility (p. 200)
compliance (p. 197)
conflict spiral (p. 216)
discrepant message (p. 203)
elaboration likelihood model (p. 199)
legitimate power (p. 213)
mass media (p. 206)
media campaign (p. 206)
mutual deterrence (p. 216)
negotiation (p. 216)
persuasion (p. 198)
promise (p. 211)
reactance (p. 222)
referent power (p. 212)
social impact theory (p. 202)
social influence (p. 197)
source (p. 197)
subjective expected value (SEV) (p. 213)
target (p. 197)
threat (p. 211)

9

SELF-PRESENTATION AND IMPRESSION MANAGEMENT

Introduction

Self-Presentation in Everyday Life

Definition of the Situation

Self-Disclosure

Tactical Impression Management

Managing Appearances

Ingratiation

Aligning Actions

Altercasting

The Downside of Self-Presentation

Self-Presentation May Be Hazardous to Your Health

Deception May Be Hazardous to Your Relationships

Detecting Deceptive Impression Management

Ulterior Motives

Nonverbal Cues of Deception

Ineffective Self-Presentation and Spoiled Identities

Embarrassment and Saving Face

Cooling-Out and Identity Degradation

Stigma

Summary

List of Key Terms and Concepts

INTRODUCTION

Strolling down the aisle of the Exhibition Hall at the State Fair, you notice the man in the next booth. He sees you at the same time and says, "Come on up. We're going to do it for you one more time." As you get closer, you see that he is surrounded by bowls of salsa and of coleslaw and piles of vegetables. On the table in front of him is a hard-plastic, hand-operated food processor—the Quick Chopper 2000.

"Let me show you how to work these real quick, all right? You guys seen these on TV before? Cool. You didn't see me on TV, did you, 'America's Most Wanted' Saturday? OK. Now the blades are the best part."

He makes it look effortless. He chops tomatoes, green peppers, and onions, all the while keeping up a steady banter. "Folks," he calls out, "come on up here. Help me get a crowd together. Sir, come on up here. You don't have to buy a thing, sir. Nobody else has." Other potential customers approach the booth.

He finishes the onions. "And then salt it to taste. This is my Daddy's recipe, by the way. He's from Cuba. My mother's from Iceland. I'm an Ice Cube. What can I tell you? That's cool." A woman reaches into her purse. "Did you want to go ahead and get that now, ma'am? OK. Cash, check, or charge? Folks, come on up here. Grab him by the hand. Hi there. I'll get your change, ma'am."

It looks easy. But it isn't. Bill Daniels and other product demonstrators who work State Fairs spend weeks in training before they hit the stage. They are learning the art of "retailtainment"—how to run the demonstration, take the money, run the credit cards, keep talking the whole time, roll over the audience, and start another demonstration smoothly. Much harder than it looks, but very rewarding if you are good at it; you can earn $70,000 per year working long weekends. The successful ones have learned the art of tactical impression management and are making it pay (National Public Radio, 2002).

Although few of us make our living by creating such a finely tuned impression, we all present particular images of who we are. When we shout or whisper, dress up or dress down, smile or frown, we actively influence the impressions others form of us. In fact, presenting some image of ourselves to others is an inherent aspect of all social interaction. The term **self-presentation** refers to the processes by which individuals attempt to control the impressions that others form of them in social interaction (Leary, 1995; Leary & Kowalski, 1990; Schlenker & Weigold, 1992). The individuals involved may be self-aware of these processes or not.

For certain purposes, it is useful to distinguish between authentic self-presentation, ideal self-presentation, and tactical self-presentation (Baumeister, 1982; Kozielecki, 1984; Swann, 1987). In authentic self-presentation, our goal is to create an image of ourselves in the eyes of others that is consistent with the way we view ourselves (our real self). In ideal self-presentation, our goal is to establish a public image of ourselves that is consistent with what we wish we were (our ideal self). In tactical self-presentation, our concern is to establish a public image of ourselves that is consistent with what others want or expect us to be. We may do this, for instance, by claiming to have some attributes they value, even if we really do not have them.

Persons engaging in tactical self-presentation usually have some ulterior motive(s) in mind. In some cases, they want others to view them positively because it will enable them to get some reward(s) that others control. Bill Daniels, for example, is earning money to support his lifestyle. In other cases, they are trying to pass as specific kinds of persons in hopes of gaining access to individuals and situations that are otherwise unavailable. If an undercover narcotics agent is trying to set up a sting, for example, he may first need to infiltrate the drug operation, create the impression that he is an experienced drug runner, and gain the confidence of the bad guys. In tactical self-presentation, a person cares only about the impact of the image he or she is presenting to others, not about whether that image is consistent with his or her real self or ideal self. When a person uses self-presentation tactics calculated to manipulate the impressions formed of him or her by others, we say that he or she is engaging in **tactical impression management.**

Of course, there are hybrid situations in which a person uses several forms of self-presentation simultaneously. For instance, a woman might try to remain largely authentic in self-presentation (that is, giving others a correct impression of her) but also try to hide a few little flaws (so that others form a positive impression of her).

This chapter considers the ways in which people actively determine how others perceive them. It addresses the following questions:

1. What content is conveyed through self-presentation in everyday life? What factors—both personal and situational—affect self-disclosure between persons?
2. What impression management tactics can we use when we want to claim a particular identity such as "overworked employee," "attractive date," or "competent student"? What factors influence our choice to use one impression management tactic rather than another?
3. Is there a downside to the use of self-presentation strategies? How might they harm your health and your relationships?
4. To what extent can people detect when others are using impression management tactics against them? What cues reveal that an impression manager is trying to deceive them?
5. What are some of the consequences when people try but ultimately fail to project the social identities they desire?

SELF-PRESENTATION IN EVERYDAY LIFE

In this section, we discuss self-presentation in everyday situations. Our primary concern is authentic self-presentation, although we must recognize that many processes in authentic self-presentation also apply to tactical self-presentation. In everyday settings, people routinely project specific social identities, and they must take care that others understand and accept their identity claims. For example, when a temporarily out-of-work individual meets a potential employer during a job interview, she may naturally strive to create a positive first impression and claim the identity of "productive worker." However, she has to be careful to create an authentic impression and not to claim too much. If she is hired, it would be quite difficult to maintain a false image for very long when she has to perform on the job.

To project a social identity successfully, an individual must share with others some understandings about the situation in which they are participating. Successful self-presentation involves efforts (1) to establish a workable definition of the situation and (2) to disclose information about the self that is consistent with the claimed identity. We discuss each of these topics in this section.

Definition of the Situation

For interaction to be successful, participants in a situation must share some understandings about their social reality. Symbolic interaction theory (Blumer, 1962; Charon, 1995; Stryker, 1980) holds that for social interaction to proceed smoothly, people must somehow achieve a shared **definition of the situation**—an agreement about who they are, what their goals are, what actions are proper, and what their behaviors mean. Persons can establish a definition of the situation by various means. In some interactions, they can establish a shared definition by actively negotiating the meaning of events (McCall & Simmons, 1978; Stryker & Gottlieb, 1981). In other interactions, people may invoke pre-existing event schemas to provide a definition of the situation. Event schemas are particularly relevant when the event is of a common or recurring type, such as weddings, job interviews, funerals, first dates, and the like.

To establish a definition of the situation, people must agree on the answers to several questions: (1) What type of social occasion is at hand? That is, what is the frame of the interaction? (2) What identities do the participants claim, and what identities will they grant one another? We consider these issues in turn.

Frames The first requirement in defining the situation is for people to agree regarding the type of social occasion in which they are participating. Is it a wedding? A family reunion? A job interview? The type of social occasion that people recognize themselves to be in is called the frame of the interaction (Goffman,

1974; Manning & Cullum-Swan, 1992). More strictly, a **frame** is a set of widely understood rules or conventions pertaining to a transient but repetitive social situation that indicates which roles should be enacted and which behaviors are proper. When people recognize a social occasion to be a wedding, for example, they immediately expect that a bride, a groom, and someone authorized to perform the ceremony will be present. They also know that the other guests attending are mostly friends and relatives of the couple and that it is acceptable—indeed, appropriate—to kiss the bride.

Participants usually know the frame of interaction in advance, or else they discover it quickly once interaction commences. Sometimes, however, there will be conflict and they must negotiate the frame of interaction. When parents send their wayward teenage daughter to a physician for a talk, for example, the discussion may begin with subtle negotiations about whether this is a psychiatric interview or merely a friendly chat. Once established, the frame limits the potential meanings that any particular action can have (Gonos, 1977). If the persons involved define the situation as a psychiatric interview, for example, any jokes the teenager tells may end up being interpreted as symptoms of illness, not as inconsequential banter.

© Mug Shots/Corbis

Even if he has done nothing wrong, this teenager had best dramatize his innocence by presenting himself to this police officer with polite deference. True identities may not be self-evident because perceivers are biased by the stereotypes and expectations they bring to a situation.

Identities Another issue in defining a situation is for people to agree on the identities they will grant one another and, relatedly, on the roles they will enact. That is, people must agree on the type of person they will treat each other as being (Baumeister, 1998). The frame places limits on the identities that any person might claim. For example, a teenager in a psychiatric interview cannot easily claim an identity as a "normal, well-adjusted kid." And employers would find it incongruous and bizarre if a young woman tried to claim the identity of "blushing bride" in a job interview.

Each person participating in an interaction has a **situated identity**—a conception of who he or she is in relation to the other people involved in the situation (Alexander & Rudd, 1984; Alexander & Wiley, 1981). Identities are "situated" in the sense that they pertain to the particular situation. For instance, the identity projected by a person while discussing a film ("insightful critic") differs from the identity projected by the same person when asking for a small loan ("reliable friend"). Situated identities usually facilitate smooth interaction. For this reason, people sometimes adopt particular situated identities in public settings even though they may not accept them privately (Muedeking, 1992). To avoid unpleasant arguments, for example, you might relate to your friend as if she were an insightful or reliable person even though privately you believe she is not.

Much of the time, our identities are not self-evident to others because their perceptions of us filter through the person schemas and stereotypes they bring to a situation. These schemas bias the identities they perceive and grant to us. Thus, even if the identity claimed by us is authentic—in the sense of being consistent with our self-concept—we may need to highlight or dramatize it (Goffman, 1959b). For instance, consider some adolescents who are innocent of any wrongdoing. If they display their usual nonchalant, defiant image when stopped by police, they risk being detained or arrested. They are more likely to avoid arrest if they dramatize their innocence by presenting a polite, deferential demeanor (Piliavin & Briar, 1964).

Self-Disclosure

A primary means we use to make authentic identity claims is to reveal certain facts about ourselves. When we first meet someone, we usually discuss only safe or superficial topics and reveal rather little about ourselves. Eventually, however, as we get to know the other better, we disclose more revealing and intimate details about ourselves. This might include information about our needs, attitudes, experiences, aspirations, and fears (Archer, 1980). This process of revealing personal aspects of one's feelings and behavior to others is termed **self-disclosure** (Derlega, Metts, Petronio, & Margulis, 1993; Jourard, 1971).

Self-disclosure is usually bilateral or reciprocal. There is a widely accepted social norm that one person should respond to another's disclosures with disclosures at a similar level of intimacy (Rotenberg & Mann, 1986; Taylor & Belgrave, 1986). This is termed the norm of reciprocity in disclosure. Most people follow it, although strict reciprocity in disclosure is more common in new relationships or developing friendships than in established ones where people already know a lot about one another (Davis, 1976; Won-Doornink, 1979). Furthermore, we are more likely to reveal more personal information to those we initially like and find attractive (Collins & Miller, 1994).

Men and women differ somewhat in self-disclosure. There is evidence that overall, women disclose more than men, although this difference is rather small (Dindia & Allen, 1992). This difference is conditioned by certain factors, one of which is the sex of the person to whom the disclosures are made. When the partner to whom disclosures are made is female, women disclose more than men, but when the partner to whom disclosures are made is male, women do not disclose more than men. Another conditioning factor is the nature of the relationship between the discloser and the partner. When the person to whom disclosures are made is a stranger, there is rather little difference between men and women in the amount of disclosure. But when the person to whom disclosures are made has an established relationship with the discloser—such as friend, spouse, or parent—women disclose more than men (Dindia & Allen, 1992).

Some men find it more difficult than women to reveal information about themselves and their relationships (Davidson & Duberman, 1982; Hacker, 1981). Some men view excessive disclosure as a sign of weakness (Cunningham, 1981). Moreover, when men and women engage in self-disclosure about their personal shortcomings, they discuss different things. Men tend to discuss such topics as unwarranted risk taking or excessively aggressive behavior, whereas women discuss such topics as immature behavior or flaws in their appearance (Derlega, Durham, Gockel, & Sholis, 1981).

Self-disclosure usually leads to liking and social approval from others. People who reveal a lot of information about themselves tend to be liked more than people who disclose at lower levels (Collins & Miller, 1994). This holds especially true if the content of the self-disclosure complements what their partner has revealed (Daher & Banikiotes, 1976; Davis & Perkowitz, 1979).

Although self-disclosure usually produces liking, there is such a thing as revealing too much about oneself. Self-disclosure that violates the audience's normative expectations may actually produce dislike. For instance, self-disclosure that is too intimate for the depth of the relationship (such as a new acquaintance describing the details of her latest bladder infection) will not strengthen the friendship and may just create the impression that the discloser is indiscreet or maladjusted (Cozby, 1972). Likewise, self-disclosure that reveals negatively valued attributes (such as a person discussing his prison record for felonious assault) or profound dissimilarities with the partner (such as a believer revealing his strong religious commitment to a nonbeliever) may produce disliking (Derlega & Grzelak, 1979).

Perhaps not surprisingly, the level of self-disclosure is related to loneliness. Young adults low in self-disclosure feel more lonely and isolated than those high in self-disclosure (Mahon, 1982). Lonely persons tend to have fewer skills in self-presentation and are less effective in making themselves known to others than are nonlonely persons (Solano, Batten, & Parish,

© Oscar Palmquist/Lightwave

These chess players seem to be building trust and liking through reciprocal self-disclosure, an important process in authentic self-presentation.

1982). The self-disclosure style of lonely persons may impair the normal development of social relations.

TACTICAL IMPRESSION MANAGEMENT

As we noted previously, self-presentation is inherent in social situations. Most people strive to create images of themselves that are authentic or true—that is, consistent with their own self-concept.

Nevertheless, under certain conditions, individuals may try to present themselves in such a way as to create false, exaggerated, or misleading images in the eyes of others. The process of creating false images is called tactical impression management.

There are various reasons we might engage in tactical impression management (Jones & Pittman, 1982; Tetlock & Manstead, 1985). One major reason is to make others like us more than they would otherwise (ingratiation). Other reasons for impression management are to make others fear us (intimidation), respect our abilities (self-promotion), respect our morals (exemplification), or feel sorry for us (supplication).

One aspect of the self that often requires management is the expression of emotion. The frame of a situation defines some emotions as appropriate and others as inappropriate. Service workers such as airline flight attendants and servers are required to be polite to customers and to conceal anger, even if the customer is being unreasonable or insulting (Hochschild, 1983). Professional hockey players, on the other hand, are required to act aggressively on the ice and even attack an opponent if provoked. An important part of the socialization into some professions involves learning to manage emotions; for instance, mortuary science students must learn to suppress negative reactions to dead bodies, bodily fluids, and disfigurement (Cahill, 1999). Some situations, such as the loss of a spouse, a job, or some other salient identity or resource, may elicit very strong emotions that the person has difficulty managing. One reaction to such loss is aggression directed at others (see Chapter 11). Alternatively, the person may seek professional help from a therapist, counselor, or support group. Support groups frequently provide a redefinition of the event (for instance, a divorce is an opportunity to start over; "turn your scar into a star") and an identity for the person that encourages emotions that are consistent with the group's ideology (Francis, 1997).

In this section, we examine some of the tactics used in impression management. In particular, we discuss managing appearances, ingratiation, aligning actions, and altercasting.

Managing Appearances

People often try to plan and control their appearance. As used here, the term appearance refers to everything about a person that others can observe. This includes clothes, grooming, overt habits such as smoking or chewing gum, choice and arrangement of personal possessions, verbal communication (accents, vocabulary), and nonverbal communication. Through the appearances we present, we show others the kind of persons we are and the lines of action we intend to pursue (DePaulo, 1992; Stone, 1962).

Physical Appearance and Props Many everyday decisions regarding appearance—which clothes to wear, how to arrange our hair, whether to shave, and so on—stem from our desire to claim certain identities. In some situations, we arrange our clothing and accessories to achieve certain effects. This would be true, for example, if we were attending a dance or going on a date. It is also true when we go to a job interview, as illustrated in a study of female job applicants (von Baeyer, Sherk, & Zanna, 1981). Some applicants in this study were led to believe that their (male) interviewer felt that the ideal female employee should conform closely to the traditional female stereotype (passive, gentle, and so on); other applicants were led to believe that he felt the ideal female employee should be nontraditional (independent, assertive, and so on). The results showed that applicants managed their physical appearance to match their interviewer's stereotyped expectations. Those expecting to meet the traditionalist wore more makeup and used more accessories, such as earrings, than those planning to meet the nontraditional interviewer.

An important aspect of personal appearance is the location and visibility of hair on the body. U.S. social norms dictate groomed hair on the heads of both men and women unless one is bald; hair on parts of the body such as underarms and legs is expected on men but not on women. In fact, women who do not shave these areas are subject to harassment and ridicule (Hawkins, 2004). A woman may refuse to shave as a matter of principle, of not yielding to an arbitrary grooming norm, and may want this act of independence to be visible to others. But other women react with "How do you expect to get a boyfriend looking like that?" and a man may pointedly ask "Are you a lesbian?" Thus, this nonconformity of appearance leads others to question the woman's sexual orientation.

Visible tattoos as a type of personal adornment are becoming increasingly popular; one source estimates that 7 million people in the United States have tattoos, most of them teenagers and young adults (Knutson, 2002). Several studies have compared college students with and without tattoos on various measures; in these studies 12 to 33 percent of the participants report having one or more tattoos (perhaps high if students with tattoos are especially likely to participate). Those with tattoos do not differ in personality characteristics or reported childhood experiences from those without (Forbes, 2001). Men and women with tattoos do report significantly more risk-taking behavior and greater use of alcohol and drugs (Drews, Allison, & Probst, 2000). Studies of students' reactions to tattoos find that both men and women have significantly more negative reactions to a woman with a visible tattoo (Degelman & Price, 2002); also, participants with more conservative gender attitudes rank her more negatively (Hawkes, Senn, & Thom, 2004).

The impression an individual makes on others depends not only on clothes, makeup, and grooming, but also on props in the environment. The impression a woman makes on her friends and acquaintances, for instance, will depend in part on the props she uses—the big pile of books she places on her desk, the music she selects for her CD player, the wine she serves, and the like. A study of the impact of cleanliness of an apartment on perceptions of the resident found that persons (both male and female) with dirty apartments were given significantly lower ratings on

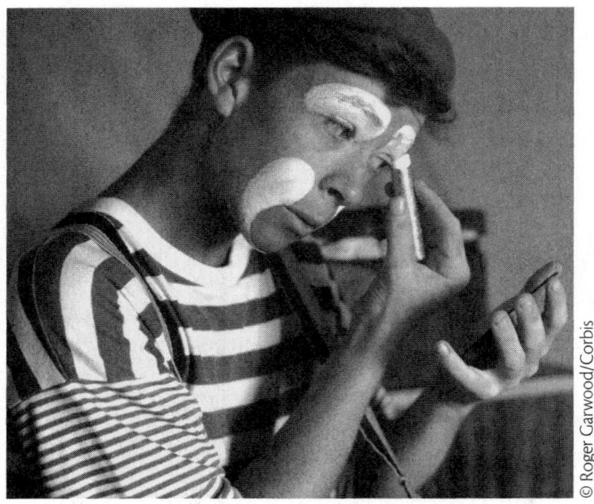

© Roger Garwood/Corbis

Physical appearance is important in impression management. If impression management is to be successful, one's appearance in the eyes of the audience must be consistent with the identity one claims. If he lacked the makeup and costume of a clown, this performer would have a hard time convincing even young children that he really is a clown.

agreeableness, conscientiousness, and intelligence, and higher ratings on openness and neuroticism. Ratings did not vary by the gender of the rater (Harris & Sachau, 2005). Thus, others make inferences about one's character and interests from the props that surround her.

Regions Goffman (1959b) draws a parallel between a theater's front and back stages and the regions we use in managing appearances. He uses the term **front regions** to denote settings in which people carry out interaction performances and exert efforts to maintain appropriate appearances vis-à-vis others. One example of a front region is a restaurant's dining room, where waiters smile and courteously offer to help customers. **Back regions** are settings inaccessible to outsiders in which people knowingly violate the appearances they present in front regions. Behind the kitchen doors, the same waiters shout, slop food on plates, and even mimic their customers. In general, persons use back regions to prepare, rehearse, and rehash the performances that occur in front regions.

Front and back regions are often separated by physical or locational barriers to perception, like the restaurant's kitchen door. These barriers facilitate impression management, because they block access of outsiders to the violations of images that occur in back regions. Any breakdown in these barriers will undermine the ability of persons to manage appearances. In recent years, for example, one such breakdown has occurred regarding national political figures. Because the mass media are pervasive, they sometimes catch presidents, senators, and corporate officers off guard. National figures are shown expressing views and performing actions they would strongly prefer to keep hidden from the public. American presidents find it difficult to project a heroic identity when the media publicize one choking on a pretzel, another entering a building to be questioned about sexual improprieties, and a third nearly collapsing while jogging. It was easier to be a hero in the days of Jefferson or Lincoln, before the invention of electronic media that penetrate the barriers between front and back regions (Meyrowitz, 1985).

Ingratiation

Most people want to be liked by others. Not only do we find it inherently pleasant, but being liked may gain us a promotion or a better grade, and it may save us from being fired or flunked. How do we persuade others to like us? Whereas much of the time we are authentic and sincere in our relations with others, occasionally we may resort to **ingratiation**—attempts to increase a target person's liking for us (Wortman & Linsenmeier, 1977). The original theory (Jones, 1964) included the assumption that these attempts are conscious, but subsequent work has broadened the definition to include attempts that occur automatically due to social learning (Jones & Wortman, 1973).

Certain preconditions make ingratiation more likely. Individuals may try to ingratiate themselves when they depend on the target person for certain benefits and believe or assume that the target person is more likely to grant those benefits to someone he or she likes. Moreover, people are more likely to use ingratiation tactics when the target is not constrained by regulations and can therefore exercise his or her discretion in distributing rewards (Jones, Gergen,

Gumpert, & Thibaut, 1965). In organizational settings, when roles are ambiguous, so that members are uncertain whether they are doing a good job, they may engage in ingratiation in an effort to ensure that they are perceived as competent and to receive rewards (Kacmar, Carlson, & Bratton, 2004).

There are a number of ingratiation tactics. Three of them are intended to increase the other's liking for an actor, that is, are other-focused. These are opinion conformity (that is, pretending to share the target person's views on important issues), other enhancement (that is, outright flattery or complimenting of the target person), and supplication (that is, convincing others you are deserving).

Opinion Conformity Faced with a target person who has discretionary power, an ingratiator may try to curry favor by expressing insincere agreement on important issues. This tactic, termed opinion conformity, is often successful because people tend to like others who hold opinions similar to their own (Byrne, 1971). Of course, obvious or excessive opinion conformity on issue after issue would quickly arouse a target's suspicion, so a clever ingratiator will mix conformity on important issues with disagreement on unimportant issues.

Opinion conformity sometimes requires us to tailor the content of the opinions we express to match a target person's general values rather than any specific opinions he or she may hold. Such matching of opinions to values was illustrated by an experiment in which participants were working for a supervisor (Jones et al., 1965). The supervisor was a target for ingratiation because, although failing at a task, participants needed a positive rating from the supervisor to earn money. Some participants were led to believe that the supervisor valued solidarity (that is, friendliness and social compatibility), whereas others believed that the supervisor valued productivity (that is, efficiency and independence). When given an opportunity to express opinions publicly, participants shifted their opinions to reflect their supervisor's values. This experiment investigated ingratiation directed upward in a hierarchy (from a worker to a high-status supervisor). Indeed, there is some evidence that persons tend to show more ingratiation responses of all kinds toward their boss than toward a stranger or a friend (Bohra & Pandey, 1984). However, a meta-analysis of 69 studies (Gordon, 1996) indicates that ingratiation attempts directed upward (that is, toward persons of higher status) are less likely to be effective in promoting liking than are ingratiation attempts directed laterally (that is, toward persons of equal status) or downward (that is, toward persons of lower status). High-status targets, aware that others may have a motive to ingratiate, may be somewhat more vigilant than equal- or low-status targets.

Other Enhancement A second ingratiation tactic is other enhancement—that is, using flattery on the target person. To be effective, flattery cannot be careless or indiscriminate. More than two centuries ago, Lord Chesterfield (1774) stated that people are best flattered in those areas where they wish to excel but are unsure of themselves. This hypothesis was tested in a study similar to the one discussed earlier, in which female participants were told that their supervisor valued either efficiency or sociability (Michener, Plazewski, & Vaske, 1979). The supervisor was a target for ingratiation because the participants' earnings depended on the evaluations they received from her. Before the supervisor made these evaluations, the participants had an opportunity to flatter her. The experimenter asked them to rate the supervisor's efficiency and sociability and indicated that the supervisor would see the ratings. The results showed that the supervisor's assumed values channeled the form of flattery the participants used. Those who believed the supervisor valued efficiency publicly rated her higher on efficiency than on sociability, whereas those who believed she valued sociability publicly rated her higher on sociability than on efficiency. Thus, the participants were discriminating in their use of praise; moreover, they avoided extreme ratings that might suggest insincerity.

Ingratiation works. Research shows that targets of flattery are more likely to believe it—and to like the flatterer—than observers (Gordon, 1996). A set of experiments were conducted to identify which of several plausible reasons—vanity of the target, reduced ability to make accurate attributions, or the desire to like the other person—account for the target's reactions. The results suggest that it is the target's vanity; people

BOX 9.1 Research Update: Playing Dumb

"Playing dumb" is an ingratiation tactic used with some frequency in interaction. When playing dumb, impression managers pretend to be less intelligent or knowledgeable than they really are. By playing dumb, they present themselves as inferior, thereby giving the target person a sense of superiority. Thus, playing dumb is a form of other enhancement.

Although popular belief and early research suggested that women are more prone than men to playing dumb (Wallin, 1950), a national survey of American adults indicated just the opposite (Gove, Hughes, & Geerken, 1980). Significantly more men than women said that they had pretended at least once to be less intelligent or knowledgeable than they really were. Men reported playing dumb more often than women in most of the situations examined, including work.

Interestingly, men tend to play dumb more than women with their boss and coworkers, whereas women tend to play dumb more than men with their spouse. Gove et al. (1980) suggested that this reflected a cultural expectation that a wife should refrain from displaying superior knowledge that might challenge her husband's assumed superiority, at least on certain issues. One wonders what a contemporary survey would find.

What leads people to play dumb? The data indicate that people who use this technique tend to be young, highly educated men (Gove et al., 1980). In contemporary American society, these persons are likely to hold lower status positions in competitive occupations where knowledge is valued (junior executives, law clerks, graduate students, and the like). Because of their youth, many of these people are located at the bottom of an occupational ladder they aspire to climb.

These people interact with superiors in settings where intelligence and knowledge are prized. Under these circumstances, a person's relatively low status may require deferring to superiors despite one's own knowledge and ability. People may stand to gain by hiding any intellectual superiority they feel—that is, by playing dumb.

Alternatively, playing dumb may be a defensive tactic, used to avoid action (Ashforth & Lee, 1990). The actor may attempt to avoid acting by pleading ignorance or lack of ability. Again, this may be more common in a highly structured, competitive organization, where midlevel personnel are motivated to avoid irritating others in order to enhance their long-term prospects. In the early years of the 21st century, playing dumb has become a common tactic for avoiding responsibility for corporate misconduct and fraud (Steffy, 2005). On trial for fraudulent financial practices, a CEO pleads ignorance, arguing that he didn't know what the CFO (chief financial officer) or the auditors were doing. In this case, the actor is playing dumb in an effort to avoid significant penalties rather than to enhance other's liking for him/her.

like to be evaluated positively (Vonk, 2002). Other enhancement can also take forms other than flattery; one example, playing dumb, is discussed in Box 9.1.

Supplication A third other-focused tactic is **supplication**—convincing a target person that you are needy and deserving (Baumeister, 1998). This is the tactic that roadside panhandlers use. By dressing in ragged clothes, they convey their need for money; by holding a sign that suggests a good use of the money ("Vet needs money to support kids"), they attempt to convey that they are deserving. Students sometimes use this tactic in attempts to get an instructor to change a grade, "But I studied *really* hard and I knew a *lot* more than was on the exam." Whereas some people choose to use this tactic, others are forced to do so, for example, to get benefit payments from government or charitable agencies. In the latter case, the supplicant may feel embarrassed or angry, and will have to manage his or her emotional display.

Selective Self-Presentation The fourth, self-focused ingratiation tactic is selective self-presentation, which involves the explicit presentation or description of one's own attributes to increase the likelihood of being judged attractive by the target. There are two distinct forms of selective self-presentation: self-enhancement or self promotion (Baumeister, 1998) and self-deprecation. When using self-enhancement, a person advertises his or her strengths, virtues, and admirable qualities. If successful, this tactic creates a positive public identity and gains liking by others. A field study of job interviews in a campus placement office assessed the degree to which each applicant (61 men, 58 women, 91% White) used opinion conformity and self-promotion during the interview; the interviewer's perception of the applicant's fit to the job was assessed following the interview. The results indicated that opinion conformity enhanced perceived fit and influenced hiring recommendations, whereas self-promotion had little effect (Higgins & Judge, 2004).

In contrast, when using self-deprecation, a person makes only humble or modest claims. Self-deprecation can be an effective way to increase others' approval and liking, especially when it aligns the ingratiator with such important cultural values as honesty and objectivity in self-appraisal.

Although often effective, the tactic of selectively emphasizing our admirable qualities can be risky. This is especially true if the target knows enough about us to suspect we are boasting or if uncontrollable future events could prove our claims invalid. Wise ingratiators, therefore, use self-enhancing descriptions only when these risks are minimal—that is, when the target person does not know them well and has no way to check their future performances (Frey, 1978; Schlenker, 1975).

Due to the risks inherent in self-enhancement, the opposite approach—self-deprecation or modest self-presentation—is often a safer tactic. To be effective, however, self-deprecation must be used in moderation. Excessively harsh and vigorous public self-criticism may gain expressions of support from others, but these expressions run the risk that others may actually believe them and form a negative private evaluation of the person using them (Powers & Zuroff, 1988). A more effective form of self-deprecation is an assured, matter-of-fact modesty that understates or downplays one's substantial achievements. In one experiment, members of a group were asked to evaluate other members following the group's success or failure at a task (Forsyth, Berger, & Mitchell, 1981). Group members reported greater liking for those who took blame for the group's failure or credited others for the group's success (self-deprecation) than for those who blamed others for failure and claimed credit themselves for the group's success (self-enhancement). These results suggest that when observers have evidence about someone's performance—whether favorable or unfavorable—self-deprecation can be an effective tactic of ingratiation.

Aligning Actions

During interaction, occasional failures of impression management are inevitable. Others may sometimes catch us performing actions that violate group norms (such as missing an appointment) or contravene laws (such as running a red light). Such actions potentially undermine the social identities we have been claiming and disrupt smooth interaction. When this occurs, people engage in **aligning actions**—attempts to define their apparently questionable conduct as actually in line with cultural norms. Aligning actions repair cherished social identities, restore meaning to the situation, and re-establish smooth interaction (Hunter, 1984; Spencer, 1987). In this section, we discuss two important types of aligning actions—disclaimers and accounts.

Disclaimers When people anticipate that their impending actions will disrupt smooth social interaction, invite criticism, or threaten their established identity, they often employ disclaimers. A **disclaimer** is a verbal assertion intended to ward off any negative implications of impending actions by defining these actions as irrelevant to one's established identity (Bennett, 1990; Hewitt & Stokes, 1975). By using disclaimers, they suggest that although the impending acts ordinarily imply a negative identity, theirs is an extraordinary case. For example, before making a bigoted remark, a person may point to her extraordinary credentials (for example, "My best friend is Hispanic, but . . . "). Disclaimers are also used prior to acts that would otherwise undermine one's identity as moral (for example, "I know I'm breaking the rules, but . . . ") or as mentally competent (for example, "This may seem crazy to you, but . . . "). These disclaimers emphasize that although the speakers are aware the act could threaten their identity, they are appealing to a higher morality or to a superior competence.

Still other disclaimers plead for a suspension of judgment until the whole event is clear: "Please hear me out before you jump to conclusions." When individuals are not certain how others will react to new information or suggestions, they are more likely to preface their actions with hedging remarks (such as "I'm no expert, but . . . " or "I could be wrong, but . . . "). Such remarks proclaim in advance that possible mistakes or failures should not reflect on one's crucial identities.

Although disclaimers can be helpful in smoothing interaction, they tend to lose their effectiveness when they are overused. One study of disclaimers in dyadic interaction found that whereas stimulus persons who used disclaimers to a limited degree were not evaluated negatively by observers, others who used a large number of disclaimers were viewed as unrealistic and were evaluated more negatively (Bell, Zahn, & Hopper, 1984).

Accounts After individuals have engaged in disruptive behavior, they may try to repair the damage by using accounts. **Accounts** are the explanations people offer to mitigate responsibility after they have performed acts that threaten their social identities (Harvey, Weber, & Orbuch, 1990; Scott & Lyman, 1968; Semin & Manstead, 1983). There are two main types of accounts: excuses and justifications. Excuses reduce or deny one's responsibility for the unsuitable behavior by citing uncontrollable events (for example, "My car broke down"), coercive external pressures (for example, "She made me do it"), or compelling internal pressures (for example, "I suddenly felt dizzy and couldn't concentrate on the exam"). Presenting an excuse reduces the observer's tendency to hold the individual responsible or to make negative inferences about his or her character (Riordan, Marlin, & Kellogg, 1983; Weiner, Amirkhan, Folkes, & Verette, 1987). Excuses also preserve the individual's self-image and reduce the stress associated with failure (Snyder, Higgins, & Stucky, 1983). Justifications admit responsibility for the unsuitable behavior but also try to define the behavior as appropriate under the circumstances (for example, "Sure I hit him, but he hit me first") or as prompted by praiseworthy motives (for example, "It was for his own good"). Justifications reduce the perceived wrongness of the behavior.

Persons are more likely to accept accounts when the content appears truthful and conforms with the explanations commonly used for such behavior (Riordan et al., 1983). Accounts are honored more readily when the individual who gives them is trustworthy, penitent, and of superior status, and when the identity violation is not serious (Blumstein, 1974). Thus, we are more likely to accept a psychiatrist's quiet explanation that he struck an elderly mental patient because she kept shouting and would not talk with him than to accept a delinquent's defiant use of the same excuse to explain why he struck an elderly woman.

A staple of public life in many countries is the political scandal, allegations that a politician has engaged in some improper or illegal behavior. The politician typically either denies the allegation outright ("I did not have sex with that women"), or offers an excuse ("I did not know that my housekeeper was in the United States illegally") or a justification ("I did accept $200,000 from that group. I did so on the advice of my lawyer that it was legal."). How effective are these responses? Does their effectiveness vary depending on the transgression and the politician's gender? To answer these questions, researchers prepared four newspaper stories involving hiring an illegal alien, engaging in sex with a superior, accepting illegal gifts, and engaging in sex with a subordinate. Within each, gender was varied. Within each, the politician's response was denial, an excuse, or a justification. The results (mean ratings by undergraduates and adults from the community) show that denials and justifications were associated with more favorable ratings than were excuses. Contrary to predictions, respondents did not judge women politicians more harshly than men for the same offence. However, respondents did judge more harshly persons whose offense was consistent with gender stereotypes, men accepting illegal contributions and having sex with a subordinate, and women hiring an illegal alien and having sex with a superior (Smith, Powers, & Suarez, 2005). Thus, had Bill Clinton been a woman, he might have escaped impeachment!

Altercasting

The tactics discussed so far illustrate how people claim and protect identities. The actions of one person in an encounter will place limits on who the others can claim to be. Therefore, to gain an advantage in the interaction, we might try to impose identities on others that complement the identities we claim for ourselves. We might also pressure others to enact roles that mesh with the roles we want for ourselves. **Altercasting** is the use of tactics to impose roles and identities on others. Through altercasting, we place others in situated identities and roles that are to our advantage (Weinstein & Deutschberger, 1963).

In general, altercasting involves treating others as if they already have the identities and roles that we wish to impose on them. Teachers engage in altercasting when they tell a student, "I know you can do better than that." This remark pressures the student to live up to an imposed identity of competence. Altercasting can entail carefully planned duplicity. An employer may invite subordinates to dinner, for example, casting them as close friends in hopes of learning employee secrets.

People frequently use altercasting to put someone on the defensive. "Explain to the voters why you can't control the runaway national debt," says the challenger, altercasting the incumbent official as incompetent in dealing with the economy. Should the incumbent rise to her own defense, she admits that the charge merits discussion and that the negative identity may be correct. Should she remain silent, she implies acceptance of the altercast identity. Putting one's rivals on the defensive is an effective technique, because a negative identity is difficult to escape.

When bargaining over identities, persons often use altercasting to achieve an edge. A study of identity bargaining between members of dating couples illustrates the use of altercasting (Blumstein, 1975). The researchers instructed women to claim an identity of "healthily assertive" by altercasting their dates into a more submissive identity. The women did this by making critical remarks to their dates, such as, "Must you insist on making all the decisions?" Some men conceded the assertive identity claimed by their dates by presenting a self consistent with the altercast. (They might have said, for example, "Sorry I've been so pushy. Whatever you say goes.") Other men rejected their date's assertive identity and pressured her to return to a submissive identity (for example, "You always liked me to make the decisions before. What's up?"). One of the factors determining whether people resist altercasting is dominance, a personality variable. Men who had earlier rated dominance as an important part of their self-concept resisted their date's altercasting, whereas men who had rated dominance as unimportant were more prone to accept the submissive identity. In general, the men responded to altercasting by conceding identities that were unimportant to their overall self-concept but by retaining identities that were more central.

THE DOWNSIDE OF SELF-PRESENTATION

So far, we have discussed the role of self-presentation in facilitating smooth interaction and the use of impression management tactics to gain benefits from others. However, this is only one part of the picture. A complete analysis requires us to consider potential harmful effects of these practices. In this section, we

© Dynamic Graphics, Inc./Jupiterimages

These young people are sunbathing, hoping to get a golden tan that will enhance their attractiveness to others. The downside is that exposure to ultraviolet radiation is a major cause of skin cancer.

discuss two issues: (1) the relationship between self-presentation and risky behavior, and (2) the consequences of tactical self-presentation in romantic relationships.

Self-Presentation May Be Hazardous to Your Health

Leary and his colleagues (Leary, Tchividjian, & Kraxberger, 1994; Martin, Leary, & Rejeski, 2000) have written about the connection between concern with how others perceive you and risky behavior. We usually want others to evaluate us favorably and support the social identities we claim in interactions. We want to avoid failures in self-presentation because they are painful and because they tarnish others' image of us. These motives lead to a number of behaviors that jeopardize our physical health.

Teen pregnancy and sexually transmitted infections (STIs) are major public health problems and can be traumatic or life-threatening to those affected by them. There are 800,000 pregnancies among teens and four million new cases of STIs among people under 25 in the United States each year. Most of these could be prevented by the correct and consistent use of condoms. Why don't sexually active young people, many of whom are aware of these risks, use condoms? Research indicates that self-presentational concerns are a major reason (Leary et al., 1994). Some men and women are afraid to buy condoms because others will infer they are sexually active. Some are afraid to produce a condom during sexual interaction, for fear they will appear prepared (gasp!) for sexual activity. Some are afraid to suggest condom use, because it might suggest that they are unfaithful or that they think their partner is unfaithful.

Consider skin cancer. The incidence of skin cancer increases every year in the United States. A major cause is excessive exposure to ultraviolet radiation. Many people intentionally expose themselves to this radiation by sunbathing. Why? To enhance others' impressions of their attractiveness. Research indicates that people who are concerned with others' impres- sions or high in body consciousness are mor likely to sunbathe or use tanning facilities (Leary al., 1994). An experiment confirmed that conc with others' impression is leading to health risk avior (Martin & Leary, 1999). aviors result in part

Numerous other risk able impression on oth- from the desire to make ng and eating disorders; ers, including excessi use; and excessive use of alcohol, tobacco, ns may engage in risky behav- steroids by athl pted by their friends: Discussing ior in order king, one woman said, "There are why she I'll cross the line just so I won't lose many n, 2002, p. D1). Numerous teens die friesly a result of showing off, whether by driv- or diving into shallow water.

Deception May Be Hazardous to Your Relationships

Many of us engage in selective self-presentation—that is, accentuating our positive features and withholding information or avoiding issues that might create negative impressions. Research indicates that we are most likely to engage in these practices in our romantic relationships. Obviously, we engage in these practices in an effort to preserve the relationship and to avoid costly interactions, such as conflict with or punishment by our partner. A study of 128 heterosexual couples found that many men and women reported using "misleading communication" with their partners for such purposes (Cole, 2001). However, they also reported using these practices when they perceived that their partner was using these tactics. And people who reported using deception or who perceived that their partner was dishonest reported lower levels of satisfaction with and commitment to their relationships.

One of the processes at work in this situation is the norm of reciprocity. Just as there is reciprocity in self-disclosure, there is reciprocity in withholding information and intentionally using misleading communication (that is, lying) in close relationships. These behaviors, motivated by a desire to preserve the relationship, can lead to a downward spiral and the eventual dissolution of the relationship.

DETECTING DECEPTIVE IMPRESSION MANAGEMENT

Up to this point, we have discussed various techniques used by impression managers to project identities and control relationships. Now we will shift our focus to the person (target) toward whom impression management tactics are directed. Impression managers try to create a deceitful, false image. This image may or may not be challenged by the person targeted. In some cases, the target will accept the false image because he or she has little to gain by questioning the sincerity of the impression manager. For instance, funeral directors strive to convey an air of sympathy and concern although they usually did not know the deceased person. Mourning relatives realize that the sympathy is superficial, but they ask very few questions because they would only be more upset to discover the mortician's true feelings of boredom and indifference.

In other cases, however, the accurate detection of deception is crucial for protecting our own interests. In attempting to win a contract, for example, builders may claim to be reliable businesspeople and skilled artisans even when they are total frauds. For the homeowner about to make a down payment, it is literally worth thousands of dollars to determine whether the builder's hearty handshake belongs to a responsible contractor or to a fly-by-night operator.

How do people go about trying to unmask the impression manager? In general, they attend to two major types of information: (1) the ulterior motives the other person may have for an action, and (2) the nonverbal cues that accompany the action. In this section, we discuss both of these cues.

Ulterior Motives

The recognition that another person has a strong ulterior motive for his or her behavior usually colors an interaction. For example, when a used car salesman tells us that a battered vehicle with sagging springs was driven only on Sundays by his aged aunt, his ulterior motive is transparent, and we are certain to suspect deceit. In such a case, we will probably discount what the salesman says about any car on his lot or even refuse to do business with him.

When ulterior motives become apparent to a target, they undermine the success of tactical impression management. For instance, in one study (Dickoff, 1961), an experimenter expressed praise for the performances of participants under two different conditions. In one condition, when the experimenter had no apparent ulterior motive, this flattery worked. The participants liked her better the more she praised them. In the second condition, however, when the experimenter had an obvious ulterior motive (she wanted them to volunteer for another experiment), participants discounted her remarks, and the flattery did not increase their liking for her.

Ironically, the very conditions that increase the temptation to use ingratiation tactics also make the target more vigilant. As noted earlier, ingratiators are especially prone to use such tactics as flattery or opinion conformity when the target person controls important rewards and can use discretion in distributing them. Unhappily for ingratiators, these same conditions alert the target to be vigilant and to expect deception. This state of affairs, termed the ingratiator's dilemma, means that ingratiators must be doubly careful to conceal their ulterior motives and avoid detection under conditions of high dependency. As documented by Gordon (1996) based on a meta-analysis, ingratiation attempts that are transparent tend to be relatively ineffective, sometimes to the point of backfiring. Ingratiators usually understand this, and indeed, there is some evidence that ingratiators avoid using tactics such as opinion conformity under conditions of blatant power inequality; they are more likely to use them under conditions that are less likely to alert the target (Kauffman & Steiner, 1968).

Nonverbal Cues of Deception

It is widely believed that nonverbal cues such as the avoidance of eye contact and nervous gestures are telltale signs of deception. One hundred years ago, Sigmund Freud wrote, "He who has eyes to see and ears to hear may convince himself that no mortal can keep a secret. If his lips are silent, he chatters with his fingertips; betrayal oozes out of him at every pore" (1905, p. 78). Although this view is a bit overblown, research indicates that nonverbal cues do provide a basis for detecting deception at a rate somewhat better than chance (DePaulo, Zuckerman, & Rosenthal, 1980; Kraut, 1980).

Cues Indicating Deception

When people interact face to face, they send messages through both verbal and nonverbal channels. People transmit meanings not only by words (verbal expressions), but also by facial expressions, bodily gestures, and [illegible] The multichannel nature of communication [illegible] problems for impression managers, because the feelings transmitted through some of these channels are more controllable than those transmitted through others. For instance, if an impression manager is trying to deceive a target, he or she may tell a lie verbally but then inadvertently reveal his or her true intentions or emotions through nonverbal channels. The term nonverbal leakage denotes the inadvertent communication of true intentions or emotions through nonverbal channels (Ekman & Friesen, 1969, 1974).

An impression manager will generally have a high level of control over his or her verbal expression (choice of words) and a fair amount of control over facial expressions (smiles, frowns, and so on). The deceiver may have less control, however, over body movements (arms, hands, legs, feet) and over voice quality and vocal inflections (the pitch and waver of his or her voice). The nonverbal channels that are least controllable—voice quality and body movements—are the ones that leak the most information (Blanck & Rosenthal, 1982; DePaulo & Rosenthal, 1979).

Several studies have demonstrated that the fundamental pitch of the voice is higher when someone is lying than when telling the truth (DePaulo, Stone, & Lassiter, 1985; Ekman, Friesen, & Scherer, 1976). The difference is fairly small—individuals cannot discriminate just by listening—but electronic analysis of vocal signals can reliably detect when an impression manager is lying. Other vocal cues associated with deception include speech hesitation (liars hesitate more), speech errors (liars stutter and stammer more), and response length (liars give shorter answers; DePaulo et al., 1985; Zuckerman et al., 1981).

Certain facial and body cues are also associated with deception. Tipoffs regarding deception include eye pupil dilation (liars show more dilation) and blinking (liars blink more); another tipoff is self-directed gestures (liars touch themselves more; DePaulo et al., 1985). The musculature of a smile is slightly different when people are lying than when they are telling the truth. Lying smiles contain a trace of muscular activity usually associated with expression of disgust, fear, or sadness (Ekman, Friesen, & O'Sullivan, 1988). Using a high-tech heat detection camera, researchers found that people who are lying get hot around the eyes (Pavlidis et al., 2002). In the popular mind and in the media, the lie detector is often associated with detecting deception, via the sensors that supposedly monitor pulse, breathing, and sweating. But the polygrapher rarely looks at the machine's output; he is busy listening for the verbal cues and watching for the behavioral changes listed here (Editorial, 2004).

Accuracy of Detection Most of us rarely concern ourselves with the possibility of deception as we interact with others. But the events of September 11, 2001, led us to realize that in some situations, the costs of undetected deception are high indeed. As a result, there is much greater interest in the question, "How good are observers at detecting acts of deception?" Although some people believe they can always detect deception when it is used by an impression manager, research suggests the contrary. The results of most experiments reveal that observers are not especially adept at correctly identifying when others are lying. Rates of detection are generally somewhat better than chance but not especially good in absolute terms (Ekman & O'Sullivan, 1991; Zuckerman et al., 1981). This occurs in part because observers often use the wrong cues or do not rely on the most useful kinds of information in judging whether someone is lying.

Difficulty in liar detection is illustrated by a study in which travelers at an airport in New York were asked to participate in a mock inspection procedure (Kraut & Poe, 1980). Some of these travelers were given "contraband" to smuggle past inspection, whereas others carried only their own legitimate luggage. All participants were instructed to present themselves as honest persons. As motivation, the researchers offered travelers prizes up to $100 for appearing honest. Later, professional customs inspectors and lay judges watched videotaped playbacks of each of the travelers and tried to decide which travelers ought to be searched. The results showed that both the customs inspectors and the inexperienced judges failed to identify a substantial proportion of the travelers who were smuggling contraband. The rate of detection, even by the customs inspectors, was no better than chance. Interestingly, however, the professional inspectors and the inexperienced judges agreed on which travelers should be searched. That is, the inspectors and the lay judges used the same (invalid) behavioral cues as indicative of deception. Travelers were more likely to be selected for search if they were young and lower class, appeared nervous, hesitated before answering questions, gave short answers, avoided eye contact, and shifted their posture frequently. Unfortunately for the inspectors, these cues were imperfect indicators of deception. The results of this experiment remind us of the difficulties facing immigration and customs officials in airports and at border crossings.

Why aren't observers better at detecting deception? First, nonverbal behaviors that do reveal deception—such as high vocal pitch and short response length—are imperfect indicators. They do arise from deception, but they can also result from conditions unrelated to deception, such as excitement or anxiety. In such circumstances, the innocent will appear guilty, and observers will make mistakes in detection. Second, there are certain cues that are commonly believed to reveal deception but that actually do not (DePaulo et al., 1985). These cues include speech rate (people think liars talk slower), smiling (people think liars smile less), gaze (people think liars engage in less eye contact), and postural shifts (people think liars shift more). If observers rely heavily on these cues, they will make mistakes in detection. Third, certain skilled impression managers are able to give near-flawless performances when deceiving. One study (Riggio & Friedman, 1983) finds evidence that certain people can give off what seem to be honest emotional cues (such as facial animation, some exhibitionism, few nervous behaviors) even when they are deceiving. If an impression manager has this capability, he or she will appear innocent, again causing mistakes in detection by observers. Fourth, we note that face-to-face interaction is a two-way street; impression managers not only exhibit behavior, but they also observe the reactions of their audiences. The feedback from audiences in face-to-face situations is fairly rich, and it often provides impression managers with a clear indication whether their attempts at deception are succeeding. If they are not succeeding, they may be able to adjust or fine-tune their deceptive communications to be more convincing.

The picture is not entirely bleak, however. First, members of some groups are accurate at catching liars. Law enforcement officers, judges, and professional psychologists were shown videotapes of the head and shoulders of 10 persons; each person was speaking about an issue he felt strongly about, and half of them were lying about their position. Federal officers (most from the CIA) attained an accuracy score of 73, while sheriffs, federal judges, and clinical psychologists interested in deception attained scores of 67 to 62. Law enforcement officers and academic psychologists attained the lowest scores (Ekman, O'Sullivan, & Frank, 1999). Second, observers' success in detecting deception can be increased by special discrimination training (Zuckerman, Koestner, & Alton, 1984). Moreover, success in detecting deception can be affected by the instructions given to observers. For instance, one study (DePaulo, Lassiter, & Stone, 1982) varied the instructions given to observers in face-to-face interaction. When given instructions to pay particular attention to auditory cues and to downplay visual cues, observers were more successful in discriminating truth from deception than when they were given instructions to pay attention to both visual and auditory cues. By emphasizing auditory and downplaying visual cues, observers more fully attended those cues that are least under an impression manager's control, such as voice quality. In general, lack of attention to verbal content and paralinguistic cues seriously impairs the ability to detect deception (Geller, 1977; Littlepage & Pineault, 1978). Hopefully, the events of September 11, 2001, have led to changes in the training of officials who are supposed to detect deception—training based on the research results described here.

INEFFECTIVE SELF-PRESENTATION AND SPOILED IDENTITIES

Social interaction is a perilous undertaking, for it is easily disrupted by challenges to identity. Some of us may recover when a challenge occurs, but others will be permanently saddled with spoiled identities. In this section, we discuss what happens when tactical impression management fails. First, we consider embarrassment—a spontaneous reaction to sudden or transitory challenges to our identities. Second, we examine cooling-out and identity degradation, which are deliberate actions aimed at destroying or debasing the identities of persons who fail repeatedly. Third, we analyze the fate of those afflicted with stigmas—physical, moral, or social handicaps that may spoil their identities permanently.

Embarrassment and Saving Face

Embarrassment is the feeling we experience when the public identity we claim in an encounter is discredited (Edelmann, 1987; Semin & Manstead, 1982, 1983). Many people describe it as an uncomfortable feeling of exposure, mortification, awkwardness, and chagrin (Miller, 1992; Parrott & Smith, 1991). It may entail such physiological symptoms as blushing, increased heart rate, and increased temperature (Edelmann & Iwawaki, 1987).

Whereas we experience embarrassment when our own identity is discredited, we also experience embarrassment when the identities of people with whom we are interacting are discredited (Miller, 1987). In this sense, embarrassment is contagious. It may be more acute when our own adequate performance serves as a frame of reference that highlights the inadequacy of others' performances (Bennett & Dewberry, 1989). We feel embarrassment at others' spoiled identities because we have been duped about the assumptions on which we built our interaction

We can read the embarrassment on the face of President Bill Clinton as he faces questioning by media representatives during his impeachment. People experience embarrassment when an important social identity they claim for themselves or accept in others is discredited.

with them, including our unwarranted acceptance of their identity claims (Edelmann, 1985; Goffman, 1967). For example, someone who claims to be an outstanding ballplayer will experience embarrassment when he drops the first three routine fly balls to center field, but the manager who let him play in a crucial game also will feel embarrassment and chagrin for accepting the ballplayer's claim of competence.

Sources of Embarrassment In several studies, investigators have analyzed hundreds of cases of embarrassment to ascertain the conditions that produce this feeling (Gross & Stone, 1970; Miller, 1992; Sharkey & Stafford, 1990). The results show that any of several conditions can precipitate embarrassment. To begin with, people feel embarrassed if it becomes publicly apparent that they lack the skills to perform in a manner consistent with the identity they claim. This is the plight, for example, of the math professor who suddenly discovers that he cannot solve the demonstration problem he has written on the chalkboard. Closely related to lack of skill is cognitive shortcoming, such as forgetfulness. A host's inability to remember others' names during introductions at a dinner party can cause embarrassment for all concerned.

Another condition that precipitates embarrassment is violation of privacy norms. If one person barges unaware into a place where he or she does not belong (such as a bathroom occupied by another), both persons are likely to experience embarrassment at the violation of privacy. The sudden and unexpected conversion of a back region into a front region is embarrassing for those whose identities are tarnished or discredited.

A further condition that often precipitates embarrassment is awkwardness or lack of poise. A person can lose poise if he or she trips, stumbles, spills coffee, or miscoordinates physically with others. Loss of control of equipment (a dentist dropping her drill), of clothing (a speaker splitting his pants), or of one's own body (trembling, burping, or worse) also will destroy poise. In general, poise vanishes and embarrassment increases whenever we lose control over those aspects of our self-presentation that we ordinarily manage routinely.

A study of Japanese undergraduates (Higuchi & Fukada, 2002) found that the causes of embarrassment include disruption of social interaction, fear of negative evaluation by others, inconsistency with self-image, and loss of self-esteem. The first two were rated as most important when the event occurred in the presence of others (criticism by an instructor during class, falling in public), and the last two as most important when the individual considered a prior event in private (failing to support a friend, failing an examination). In an experiment, male and female university students viewed slides of nudes and erotic couples either alone or with two strangers. Participants self-reported greater embarrassment when others were present, but careful analysis of videotapes showed fewer instances of nonverbal indicators of embarrassment, such as face touches and downward gazes, in the public condition (Costa, Dinsbach, & Manstead, 2001).

It may be that we try to control nonverbal indicators in the presence of others.

Responses to Embarrassment A continuing state of embarrassment is uncomfortable for everyone involved. For this reason, it is usually in everyone's interest to restore face—that is, to eliminate the conditions causing embarrassment. The major responsibility for restoring face lies with the person whose actions produced the embarrassment, but interaction partners frequently try to help the embarrassed person restore face (Levin & Arluke, 1982). For instance, if a party guest slips and falls while demonstrating his dancing prowess, his partner might help him save face by remarking that the floor tiles seem newly waxed and very slippery. Mutual commitment to supporting each other's social identities is a basic rule of social interaction (Goffman, 1967).

To restore face, the embarrassed person will often apologize, provide an account, or otherwise realign his or her actions with the normative order (Knapp, Stafford, & Daly, 1986; Metts & Cupach, 1989). When providing accounts, people will either make excuses that minimize their responsibility or offer justifications that define their behavior as acceptable under the circumstances. If the interaction partners accept these accounts—and partners have been known to accept the lamest excuses rather than endure continuing embarrassment—a proper identity is restored. If accounts are unavailable or insufficient, the embarrassed individual may offer an apology for the discrediting behavior and admit that his or her behavior was wrong. In this way, the person reaffirms threatened norms and reassures others that he or she will not violate those norms again. Research suggests that blushing is particularly important in restoring the normative order. Observers rated videotapes of a public gaffe; an actor who visibly blushed following the incident was judged less negatively, as less responsible, and as more trustworthy than an actor who did not blush (Jong, 1999). The results suggest that blushing communicates to others that the actor is attached to the social rules in question despite the violation.

When our behavior discredits a particular, narrow identity, we can sometimes restore face through an exaggerated reassertion of that identity. A man whose masculine identity is threatened by behavior suggesting he is infantile, for example, might try to reassert his courage and strength. In a test of this hypothesis (Holmes, 1971), some male participants were asked to suck on a rubber nipple, a pacifier, and a breast shield—all embarrassing experiences. Other participants were asked to touch surfaces such as sandpaper and cloth. The men were next asked how intense an electric shock they would be willing to endure later in the experiment. Men who had faced the embarrassing experiences indicated willingness to endure more intense shocks than men who had faced no threat to their masculine identity. By taking the intense shocks, the embarrassed men could present themselves as tough and courageous, thereby reasserting their threatened masculinity.

Sometimes people embarrass others intentionally and make no effort to help them to save face. In such circumstances, the embarrassed persons are likely to react aggressively. They may vigorously attack the judgment or character of those who embarrassed them. Some research indicates that an aggressive response to embarrassment is more likely between status unequals than between status equals (Sueda & Wiseman, 1992). Alternatively, the embarrassed persons may assert that the task on which they failed is worthless or absurd (Modigliani, 1971). Finally, they may retaliate against those who embarrassed them intentionally. Retaliation not only reasserts an image of strength and achieves revenge, but it also forestalls future embarrassment by showing resolve to punish those who discredit us. In these ways, embarrassment may lead to interpersonal aggression.

Cooling-Out and Identity Degradation

When people repeatedly or glaringly fail to meet performance standards or to present appropriate identities, others cease to help them save face. Instead, they may act deliberately to modify the offenders' identities or to remove them from their positions in interaction. Failing students are dropped from school, unreliable employees are let go, tiresome suitors are rebuffed, people with schizophrenia are institutionalized.

Persons will modify an offender's identity either by cooling-out (Goffman, 1952) or by degradation (Garfinkel, 1956), depending on the social conditions surrounding the failure.

The term **cooling-out** refers to gently persuading a person whose performance is unsuitable to accept a less desirable, though still reasonable, alternative identity. A counselor at a community college may cool-out a weak student by advising him to switch from premed to an easier major, for example, or by recommending that he seek employment after completing community college rather than transfer to a university. Persons engaged in cooling-out seek to persuade offenders, not to force them. Cooling-out actions usually protect the privacy of offenders, console them, and try to reduce their distress. Thus, the counselor meets privately with the student, emphasizes the attractiveness of the alternative, listens sympathetically to the student's concerns, and leaves the final choice up to him.

The process of destroying the offender's identity and transforming him or her into a lower social type is termed **identity degradation.** Degradation establishes the offender as a nonperson—an individual who cannot be trusted to perform as a normal member of the social group because of reprehensible motives. This is the fate of a political dissident who is fired from her job, declared a threat to society, and relegated to isolation in a prison or work camp.

Identity degradation imposes a severe loss on the offender, so it usually is done forcibly. Identity degradation often involves a dramatic ceremony—such as a criminal trial, sanity hearing, or impeachment proceeding—in which a denouncer acts in the name of the larger society or the law (Scheff, 1966). In such ceremonies, persons who had previously been treated as free, competent citizens are brought before a group or individual legally empowered to determine their "true" identity. They are then denounced for serious offenses against the moral order. If the degradation succeeds, offenders are forced to give up their former identities and to take on new ones like "criminal," "insane," or "dishonorably discharged."

Two social conditions strongly influence the choice between cooling-out and degradation: (1) the offender's prior relationships with others, and (2) the availability of alternative identities (Ball, 1976). Cooling-out is more likely when the offender has had prior relations of empathy and solidarity with others and when alternative identity options are available. For example, during a breakup, lovers who have been close in the past can cool-out their partners by offering to remain friends. Identity degradation is more likely when prior relationships entailed little intimacy or when respectable alternative identities are not readily available. Thus, strangers found guilty of sexually molesting children are degraded and transformed into immoral, subhuman creatures.

Stigma

A **stigma** is a characteristic widely viewed as an insurmountable handicap that prevents competent or morally trustworthy behavior (Goffman, 1963; Jones et al., 1984). There are several different types of stigma. First, there are physical challenges and deformities—missing or paralyzed limbs, ugly scars, blindness, deafness. Second, there are character defects—dishonesty, unnatural passions, psychological derangements, treacherous beliefs. These may be inferred from a known record of crime, imprisonment, sexual abuse of children, mental illness, or radical political activity, for example. Third, there are characteristics such as race, sex, and religion that—in particular segments of society—are believed to contaminate or morally debilitate all members of a group.

Once recognized during interaction, stigmas spoil the identities of the persons having them. Stigmas operate via reflected appraisals; "normals" (nonstigmatized persons) convey expectations and negative evaluations of the stigmatized person (Kaufman & Johnson, 2004). No matter what their other attributes, stigmatized individuals are likely to find that others will not view them as fully competent or moral. As a result, social interaction between normal and stigmatized persons is shaky and uncomfortable.

Sources of Discomfort Discomfort arises during interaction between "normals" and stigmatized individuals because both are uncertain which behavior is appropriate. "Normals" may fear, for example, that if they show direct sympathy or interest in the stigmatized person's condition, they will be intrusive (for example, "Is it difficult to write with your artificial hand?"). Yet if they ignore the defect, they may make impossible demands (for example, "Would you help me move the refrigerator?"). To avoid being hurt, stigmatized individuals may vacillate between shamefaced withdrawal (avoiding social contact) and aggressive bravado ("I can do anything anyone else can!").

Another source of discomfort for "normals" is the threat, a sense of anxiety or even danger that they experience during interaction with stigmatized individuals (Blascovich, Mendes, Hunter, Lickel, & Kowai-Bell, 2001). "Normals" may fear that associating with a stigmatized person may discredit them (for example, "If I befriend a convicted criminal, people may wonder about my trustworthiness"). In recent times, this problem has arisen frequently with respect to AIDS, which is a heavily stigmatized condition due in part to its association with drug use and homosexuality as well as the lingering fear of transmission. Persons with AIDS experience the stigma, of course; but beyond that, the compassionate confidants who provide care and social support for persons with AIDS often encounter social difficulties as well. The stigma associated with AIDS results in some of these confidants being rejected by their friends and family (Jankowski, Videka-Sherman, & Laquidara-Dickinson, 1996).

Effects on Behavior and Perceptions "Normals" react toward stigmatized persons with an attitude of ambivalence (Katz, 1981; Katz, Wackenhut, & Glass, 1986). Toward a person with quadriplegia, for instance, "normals" have feelings of aversion and revulsion but also of sympathy and compassion. This ambivalence creates a tendency toward behavioral instability, in which extremely positive or extremely negative responses may occur toward the stigmatized person, depending on the specific situation.

When interacting with stigmatized individuals, "normals" alter their usual behavior. They gesture less than usual, refrain from expressing opinions that reflect their actual beliefs, maintain less eye contact, and terminate the encounters sooner (Edelmann, Evans, Pegg, & Tremain, 1983). Moreover, "normals" speak faster in interactions with stigmatized persons than in interactions with other "normals," ask fewer questions, agree less, make more directive remarks, and allow the stigmatized persons fewer opportunities to speak (Bord, 1976). By limiting the responses of the stigmatized person, "normals" reduce uncertainty and diminish their own discomfort. Negative messages are likely to be expressed nonverbally; normals often monitor their speech and try to restrain or suppress negative remarks, but nonverbal leakage may carry the day (Hebl & Dovidio, 2005).

For their part, stigmatized persons also have difficulty interacting with "normals." Remarkably, the mere belief that we have a stigma—even when we do not—leads us to perceive others as relating to us negatively. In a dramatic demonstration of this principle (Kleck & Strenta, 1980), some female participants were led to believe that a woman with whom they would interact had learned that they had a mild allergy (a nonstigmatizing attribute). Other female participants believed that the woman would view them as disfigured due to an authentic-looking scar that had been applied to their faces with stage makeup (a stigmatizing attribute). In fact, the interaction partner had no knowledge of either attribute. In the allergy condition, the partner had received no medical information whatsoever. In the scar condition, there was actually no scar to be seen, because the experimenter had surreptitiously removed the scar just before the discussion.

After a 6-minute discussion with the interaction partner, the participants described their partners' behavior and attitudes. Those participants who believed they had a facial scar remarked more frequently that their partners had stared at them. They also perceived their partners as more tense, more patronizing, and less attracted to them than the nonstigmatized participants did. Judges who viewed videotapes of the interaction perceived none of these differences. This is not surprising, as the partner knew nothing about either disability. However, these results show that people who believe they are stigmatized perceive others as relating negatively to them. This occurs even if the others are not, in fact, doing anything negative or irregular. These findings are illustrated in Figure 9.1.

When people believe they are stigmatized, they tend not only to perceive the social world differently but also to behave differently. In one study, for

FIGURE 9.1 PERCEPTIONS OF INTERACTION PARTNERS BY STIGMATIZED AND NONSTIGMATIZED INDIVIDUALS

In this study, some female students were led to believe that a large facial scar stigmatized them in the eyes of their female interaction partner. Others were led to believe their partner knew they had a mild allergy—a nonstigmatized characteristic. In fact, interaction partners were unaware of either of these characteristics. Nonetheless, students who believed they were stigmatized perceived their partners as substantially more tense and patronizing and as less attracted to them. This suggests that the mere belief that we are stigmatized leads us to perceive others as behaving negatively toward us.

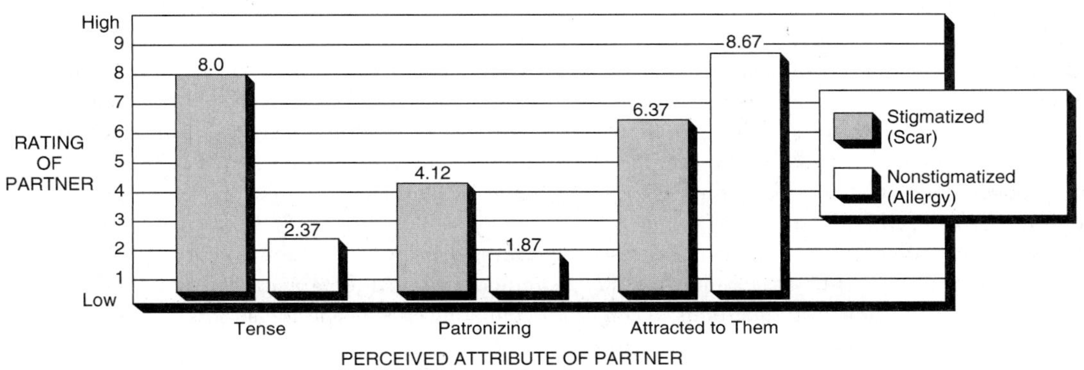

Source: Adapted from "Perceptions of the Impact of Negatively Valued Physical Characteristics on Social Interaction" by Kleck and Strenta, *Journal of Personality and Social Psychology, 39,* 861–873. Copyright © 1980 by the American Psychological Association. Adapted with permission.

instance, one group of mental patients believed that the person with whom they were interacting knew their psychiatric history, whereas another group thought their stigma was safely hidden (Farina, Gliha, Boudreau, Allen, & Sherman, 1971). Patients in the first group performed more poorly on a cooperative test and found the task more difficult. Moreover, outside observers of the interaction perceived these patients to be more anxious, more tense, and less well adjusted.

Coping Strategies Stigmatized persons adopt various strategies to avoid awkwardness in their interactions with "normals" and to establish the most favorable identities possible (Gramling & Forsyth, 1987). Persons who are handicapped or physically challenged often must choose between engaging in interaction (thereby disclosing their stigma) or withdrawing from interaction (concealing their stigma; Lennon, Link, Marbach, & Dohrenwend, 1989). A stutterer, for instance, may refrain from introducing himself to strangers; were he to introduce himself, he could do so only at the risk of stumbling over his own name and drawing attention to his stigma (Petrunik & Shearing, 1983). People whose speech reveals their stigmatized foreign origin or lack of education face a similar dilemma when meeting strangers.

In interaction, stigmatized persons often try to induce "normals" to behave tactfully toward them and to build relationships around the aspects of their selves that are not discredited. Their strategies depend on whether their stigma can be defined as temporary—such as a broken leg on the mend or a passing bout of depression—or whether it must be accepted as permanent—such as blindness or stigmatized racial identity (Levitin, 1975). Persons who are temporarily stigmatized focus attention on their handicap, recounting how it befell them, detailing their favorable prognosis, and encouraging others to talk about their own past injuries. In contrast, people who are permanently stigmatized often try to focus attention on attributes unrelated to their stigma

(Davis, 1961). They often use props to highlight aspects of the self that are unblemished, such as proclaiming their intellectual interests (say, by carrying a heavy book), their political involvements (say, by wearing campaign buttons), or their hobbies (say, by toting a knitting bag).

In cases where a stigma does not force excessive dependency, permanently stigmatized individuals often try to strike a deal with "normals": They will behave in a nondemanding and nondisruptive manner in exchange for being treated as trustworthy human beings despite their stigma. Under this arrangement, they are expected to cultivate a cheerful manner, avoid bitterness and self-pity, and treat their stigma as a minor problem with which they are coping successfully (Hastorf, Wildfogel, & Cassman, 1979).

Everyone gains some benefit from handling stigmas in this way. Stigmatized persons avoid the constant embarrassment of indelicate questions, inconsiderateness, and awkward offers of help. They gain some acceptance and enjoy relatively satisfying interaction in most encounters. "Normals" gain because this resolution assuages the ambivalence they feel toward stigmatized persons and spares them the true pain the stigmatized person suffers.

Some persons are stigmatized because of some individual characteristic—birth defect, illness, disfigurement due to an accident, or history of deviant or criminal behavior. They often rely on these strategies. Others are stigmatized because they are members of certain groups; that is, because of a shared social identity—mental retardation, schizophrenia, obesity, or racial/ethnic minority status (Crocker & Major, 1989). In these cases, stereotypes that are widely shared by both stigmatized and stigmatizers shape the attitudes and behavior of members of both groups, including the identities claimed in interaction (Renfrow, 2004). These persons have an additional coping strategy; they can attribute the stigma they experience to the prejudiced attitudes of others and base their self-perception on traits on which they rank well. They may also seek out relationships with others who share the stigma in an effort to experience positive reflected appraisals (Kaufman & Johnson 2004).

Another coping strategy is passing, distancing oneself from the stigmatized identity by hiding information (Renfrow, 2004). The person may hide the identity from normals while cultivating discreet associations with others who share the stigma; this will prevent negative appraisal by normals and provide the person with positive appraisals by the others. Millions of college students have used this strategy to gain access to bars and alcoholic beverages before they reach the legal drinking age! Or the person may distance the self from other stigmatized persons and associate with normals or withdraw from interaction (the closeted GLBT person); the latter strategy may result in great psychological distress. The central emotion in passing is fear; fear of the consequences of being identified by normals as stigmatized leads to passing, and fear of discovery dominates the passing experience.

Persons may attempt to cope with stigma by seeking therapy (Kaufman & Johnson, 2004). Physical and occupational therapy may reduce or remove the debilitating effects of accidents, loss of limbs, or loss of abilities. Psychological or behavioral therapy may change the beliefs and behavior that accompany mental retardation, mental illness, or unnatural passions. A final strategy is to join a social movement intended to change the perceptions and stereotypes of normals.

SUMMARY

Self-presentation refers to our attempts—both conscious and unconscious—to control the images we project of ourselves in social interaction. Some self-presentation is authentic, but some may be tactical.

Self-Presentation in Everyday Life Successful presentation of self requires efforts to control how others define the interaction situation and accord identities to participants. (1) In defining the situation, people negotiate the type of social occasion considered to be at hand and the identities they will grant each other. (2) Self-disclosure is a process through which we not only make identity claims but also promote friendship and liking. Self-disclosure is usually two-sided and gradual, and it follows a norm of reciprocity.

Tactical Impression Management People employ various tactics to manipulate the impressions that others form of them. (1) They manage appearances (clothes, habits, possessions, and so on) to dramatize the kind of person they claim to be. (2) They ingratiate themselves with others through such tactics as opinion conformity, other enhancement, and selective presentation of their admirable qualities. (3) When caught performing socially unacceptable actions, people try to repair their identities through aligning actions, which are attempts to align their questionable conduct with cultural norms. They explain their motives, disclaim the implications of their conduct, or offer accounts that excuse or justify their actions. (4) They altercast others, imposing roles and identities that mesh with the identities they claim for themselves.

The Downside of Self-Presentation (1) Self-presentational motives such as the desire to be liked by or obtain rewards controlled by others may lead to behavior that is risky to your health, such as unprotected sexual intercourse or alcohol or drug abuse. (2) The desire to maintain romantic relationships may lead to withholding information from or lying to your partner; people who report deceiving their partner or who perceive their partner as deceptive report reduced commitment to their relationship.

Detecting Deceptive Impression Management Observers attend to two major types of information in detecting deceitful impression management. (1) They assess others' possible ulterior motives. If a large difference in power is present, an impression manager's ulterior motives may become transparent to the target, making tactics like ingratiation difficult. (2) They scrutinize others' nonverbal behavior. Although detection of deceit is difficult, observers are more accurate when they concentrate on leaky cues, such as tone of voice, and discrepancies between messages transmitted through different channels. Some professionals are quite accurate in detecting deception.

Ineffective Self-Presentation and Spoiled Identities Self-presentational failures lead to several consequences. (1) People experience embarrassment when their identity is discredited. Interaction partners usually help the embarrassed person to restore an acceptable identity. Otherwise, embarrassed persons tend to reassert their identity in an exaggerated manner or to attack those who discredited them. (2) Repeated or glaring failures in self-presentation lead others to modify the offender's identity through deliberate actions. They may try to cool-out offenders by persuading them to accept less desirable alternative identities, or they may degrade offenders' identities and transform them into lesser social types. (3) Many physical, moral, and social handicaps stigmatize individuals and permanently spoil their identities. Interaction between stigmatized and "normal" persons is marked by ambivalence and is frequently awkward and uncomfortable. In general, "normals" pressure stigmatized individuals to accept inferior identities, whereas stigmatized individuals seek to build relationships around the aspects of their selves that are not discredited. Some persons with stigma attempt to pass to avoid the negative reflected appraisals they would receive from "normals."

LIST OF KEY TERMS AND CONCEPTS

accounts (p. 236)
aligning actions (p. 235)
altercasting (p. 237)
back region (p. 232)
cooling-out (p. 245)
definition of the situation (p. 227)
disclaimers (p. 235)
embarrassment (p. 242)
frame (p. 228)
front region (p. 232)
identity degradation (p. 245)
ingratiation (p. 232)
self-disclosure (p. 229)
self-presentation (p. 226)
situated identity (p. 228)
stigma (p. 245)
supplication (p. 234)
tactical impression management (p. 226)

10

EMOTIONS

Introduction

Defining Emotions

Classical Ideas About the Origins of Emotion

Universal Emotions and Facial Expressions

Facial Expressions of Emotion

Cultural Differences in Basic Emotions and Emotional Display

Social Emotions

Cognitive Labeling Theory

Five Social Emotions

Emotion Work

Summary

List of Key Terms and Concepts

INTRODUCTION

Rob Schwartz is a kindergarten teacher in a suburb of Columbus, Ohio, facing a Monday morning of teaching 25 energetic 5-year-olds for 3 hours. Rob is a very dedicated teacher and always strives to give his best to his students. He is usually energized for class and follows well-developed lesson plans. His students love him and look forward to school every day.

Unfortunately, Rob is coming off a very difficult weekend. On Friday afternoon, he found out he was denied a loan for a new house he wanted to buy, and on Saturday he was in a serious automobile accident. He was not hurt, but his car was damaged beyond repair and was towed to the junkyard. To top it all off, his favorite football team lost its final game of the season and thus was eliminated from the playoffs. On Monday morning, Rob got up in his shabby apartment, rode the bus to work, and proceeded to pay off a bet he had made on the football game with another teacher. Needless to say, he was not in a very good mood when he headed down the hall to his classroom.

But Rob still wanted to do a good job with the children in his class, so he decided to hide his feelings and put on a cheerful face. He bounced into the room acting as though absolutely nothing was wrong, and in fact that he was just as happy as he could be. The students in the class thought he was in a great mood, as did the teacher's aide who helped in the class. They all reacted very positively toward him and everyone had a great day in the classroom. As things progressed, Rob himself began to feeling much better and in the end thought that this was one of the best days he'd ever had in the classroom.

In the space of a few days, Rob has experienced a wide range of emotions. He started off the weekend feeling fine, until he experienced a severe disappointment when he found out about his loan. When he was hit in his car, he first felt fear and then anger toward the other driver. Later, he started watching the football game with excited anticipation, rode a roller coaster of ups and downs as the game progressed, and came close to tears when the game was over. By Monday morning, he felt positively depressed, but then managed to recover—very much by his own doing—during the course of teaching that morning.

How can we explain all of Rob's emotional experiences and the changes from one emotional state to another? During the football game, for example, Rob vacillates quickly between anger, sadness, disgust, joy, and satisfaction, all because of a few images on a television screen. On Monday morning, he is able to consciously choose behaviors that end up actually changing his mood. What is involved, both on an individual level and a social level, that produces all of these different emotional states and the changes from one to another? Given that Rob is not an inherently unstable individual, we need to understand the social context of his interactions to understand his emotional states.

Therefore, in this chapter we'll consider several key questions about emotions that can help us understand emotion stories like Rob's:

1. What exactly are emotions and psychological emotional states?
2. Where do emotions come from? What are the physiological, psychological, and social bases of emotions?
3. What kinds of emotions and expression of them are universal human traits, and which are specific to specific social or cultural contexts?
4. How do social contexts produce emotions?
5. How do we psychologically and socially control the expression of emotions? Beyond limiting how we express them, how can we control our emotions, either by suppressing or producing them?

DEFINING EMOTIONS

To talk about emotions, we first need to know exactly what we mean by that term—and emotions are not particularly easy to define. In addition, there are other related terms that are partially synonymous with emotion, including sentiment, affect, mood, and of course, feelings (Smith-Lovin, 1995). All of these terms are used in common everyday language to refer to emotions, and they are also used by social psychologists in specific ways, so we first need to sort them out from one another.

Affect is usually considered the most general label that encompasses virtually any kind of subjective positive or negative evaluation of some other object. It includes short-term reactions like the anger you might feel after being the target of an insult, and longer term orientations such as the pleasant association many people have with the Christmas season. Affect can usually be described not just in terms of direction, but also in term of strength and the level of activity associated with it (Osgood, Suci, & Tannenbaum, 1957).

Emotions are usually treated as a subset of affect. Usually, **emotions** are thought of as short-lived reactions to a stimulus outside of the individual that involve both physiological and cognitive reactions. Of course, this definition is not enough to differentiate emotions from other human stimulus-response reactions. Sociologist Peggy Thoits (1990) expands the definition of emotions to include (1) situational stimulus, (2) physiological changes, (3) expressive gesturing of some kind, and most importantly, (4) a label to identify a cluster of the first three. Being slapped by one's girlfriend, plus elevated temperature and heart rate, plus a furrowed brow and a clenched fist equals anger. Being slapped by one's girlfriend, plus elevated heart rate, plus laughter and smiling equals humorous teasing. Various culturally defined combinations produce what we typically think of, and experience, as an emotion.

Emotions are also intimately related to goals. Whether we are frustrated by a blocked goal, joyful upon achieving one, anxious as we approach one, or motivated by an emotion to pursue a goal (such as fleeing from a fearful situation), emotions and goals are intimately entwined (Frijda & Mequita, 1994).

Sentiment is very close to emotion in its meaning, but when social psychologists use the term sentiment, they emphasize the social parts of the emotional response. Rather than focusing on the automatic physiological portions of emotions, sentiments are steeped in the social characteristics of the situation. Early social psychologists used sentiment to refer to the components of human responses that separate them from analogous responses that animals would have (Cooley, 1909). Accordingly, sentiment relies not just on the responses of the individual to the stimulus, but also on how that stimulus is understood by other human beings. In later years, as social psychologists have come to increasingly accept that social elements are a key piece of emotions, the idea of sentiment has become less distinguishable from that of emotion (Stets, 2003). Instead, the term sentiment is now often used to distinguish immediate emotional responses from longer term emotional states such as love, grief, and jealously (Gordon, 1990). These sentiments are highly social in their construction and can endure for days, weeks, and even years after the initial event that triggered them.

Moods have a more enduring quality than emotions. Whereas we might experience one emotion and move on to another in a matter of seconds or minutes (Ekman, 1992), a **mood** is a general psychological condition that characterizes our experience and emotional orientation for hours or even days. Moods are considerably less specific than emotions. When an emotion occurs, we have a very good sense of where it is directed and what stimulus caused it. Moods, however, are diffuse and can be widely directed at anything that comes in our path—whether or not it has anything to do with the origin of the bad mood.

Returning to the experiences of the kindergarten teacher, Rob, almost all of the events described involve affect—some were stronger, such as when he was in the car accident, and some were weaker, such as when he paid his gambling debt. He also experienced many short-term emotional states with a great deal of social content (such as losing the loan) and others with much less social content (such as the initial fear and shock when hit by another car). These events accumulated and by the time he got to work on Monday, he had probably been experiencing a morose mood for quite a number of hours.

CLASSICAL IDEAS ABOUT THE ORIGINS OF EMOTION

Where do emotions come from? Some of the earliest scientific work on emotions focuses largely on the nonsocial origins of emotions. Although these perspectives are now viewed as quite incomplete, they do provide some of the essential building blocks used in later understandings of emotions, and thus require a brief review.

The beginnings of the study of emotion can be found in the work of Charles Darwin, especially in

From *The Expression of the Emotions in Man and Animals*, Charles Darwin, p. 118; drawing by Mr. Wood.

Charles Darwin's analysis of emotional expression focused on the similarities in expression across cultures and species. He thought that some elements of human emotional expression (such as those related to anger) were related through evolution to such practices of other species as baring of teeth and biting in battles with other animals.

his important book, *The Expression of the Emotions in Man and Animals.* Darwin was motivated to write this book because he thought that if humans and other animals had common ancestors, there ought to be some similarities in emotional expression as well. At the time he wrote, Darwin was battling those who believed that humans had all kinds of unique characteristics not shared by other animals—and this included facial expressions. His book pointed out fascinating continuities in the emotional expressions made by humans and by other animals, such as monkeys and dogs, and also found some facial expressions and gestures that he thought were universal across cultures (Darwin, 1998).

The theory Darwin developed about emotional expression was based on these similarities across cultures and species. If some emotions and expressions of them were universal, then they must be genetically encoded. And if they are genetically encoded, then they must have value that enhances genetic survival. Thus emotional gestures, for example, were residual expressions related to threat or sexual attraction. The act of gritting and showing our teeth when angry could be a derivative of the act of biting in a battle for survival.

While Darwin focused on the expression of emotions, other early thinkers were more concerned about the internal sources of emotions. James (1890) and Lange (1922) separately proposed a biological approach to emotions that focused heavily on the physiological responses to stimulus. In this model, the stimulus is followed by the biological reaction, and *then* the individual cognitively processes the physiological sensation and interprets it as an emotion. The physiological component of the emotional experience (for example, increased blood pressure) helps us to identify the emotion (anger) rather than anger causing the physiological change.

Sigmund Freud (1905) found the sources of emotions (principally anxiety and guilt) in repressed childhood sexual desires. Although many of Freud's ideas have been discredited over time, he did contribute to the study of emotions in several ways. First, he drew attention to the ways that emotions could develop unconsciously. Individuals can hide their feelings from themselves and yet still have these emotions affect the way they think and act. Second, he drew attention to the role that an individual's past experience plays in helping to understand the meaning of emotions and the physical sensations that accompany them. Recognizing the past context of emotional reactions is an important step toward understanding the social forces that shape and define emotions.

UNIVERSAL EMOTIONS AND FACIAL EXPRESSIONS

The classic perspectives discussed above point to one of the more persistent questions in the study of emotions: Are some emotions genetically encoded in human biology? If so, are emotions universal human experiences that are similar across cultural and geographic boundaries, and over historical epochs? When we feel anger, for example, are we feeling the same physiological and psychological sensations that people experienced hundreds and thousands of years ago or experience in other parts of the world? Darwin's work suggests that at least some (but not all, he believed) emotions are universal—not just among humans, but also across some different species.

Facial Expressions of Emotion

As questions about the universality of emotions have developed over time, scholars have focused a great deal of attention on facial expression of emotion.

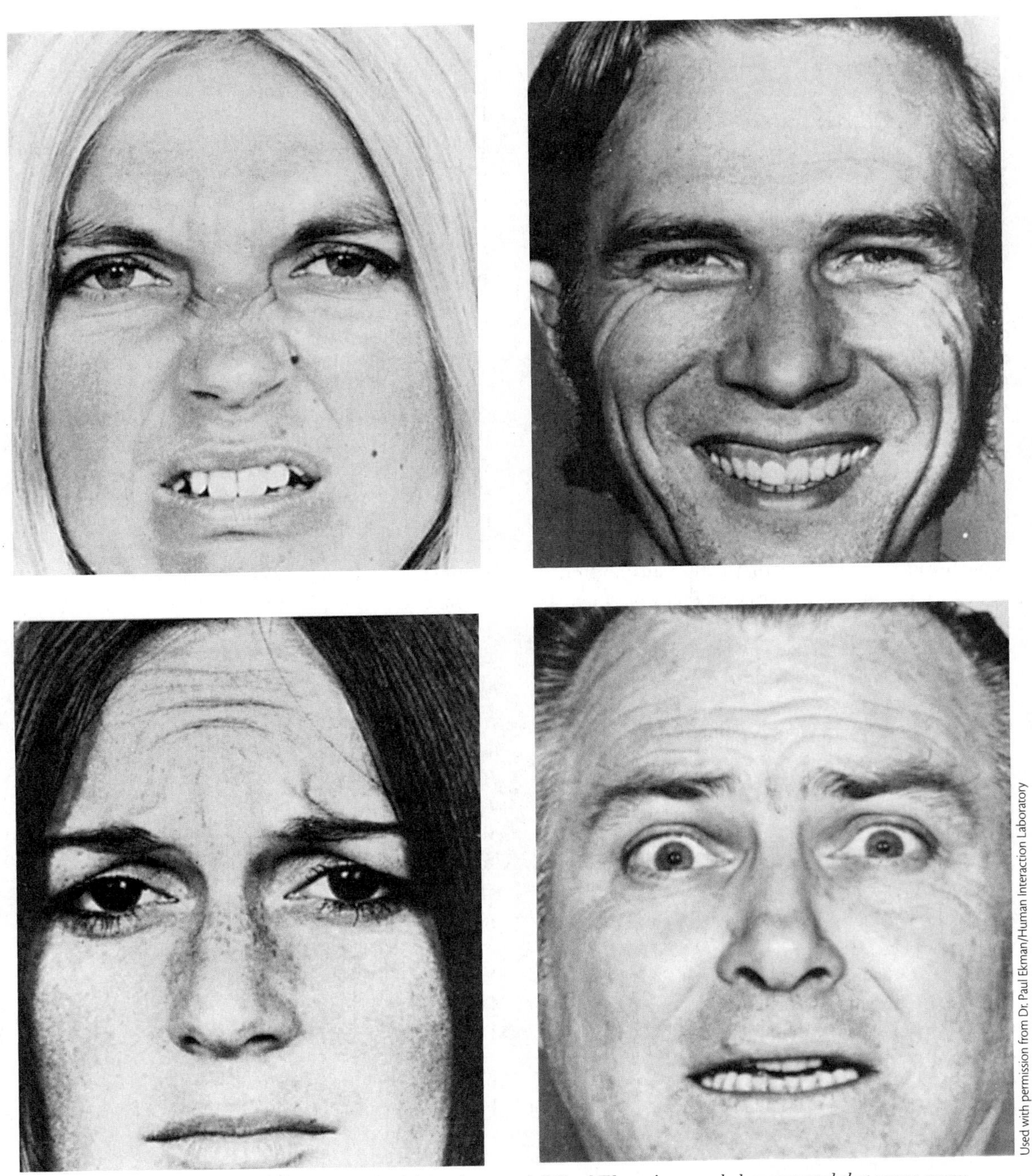

Can you identify the emotions displayed in each of these photographs? Paul Ekman's research demonstrated that across many cultures, people easily identified these faces as portraying disgust, happiness, sadness, and fear.

TABLE 10.1 SINGLE-EMOTION JUDGMENT TASK: PERCENTAGE OF PARTICIPANTS WITHIN EACH CULTURE WHO CHOSE THE PREDICTED EMOTION

Nation	Happiness	Surprise	Sadness	Fear	Disgust	Anger
Estonia	90	94	86	91	71	67
Germany	93	87	83	86	61	71
Greece	93	91	80	74	77	77
Hong Kong	92	91	91	84	65	73
Italy	97	92	81	82	89	72
Japan	90	94	87	65	60	67
Scotland	98	88	86	86	79	84
Sumatra	69	78	91	70	70	70
Turkey	87	90	76	76	74	79
United States	95	92	92	84	86	81

Source: Adapted from "Single Judgment Emotion Task" by P. Ekman and M. Friesan, *Journal of Personality and Social Psychology, 53*, 712–717. Copyright © 1987 by the American Psychological Association. Adapted with permission.

Some of the facial expressions we make are largely involuntary—we perform them without thought or conscious effort in certain circumstances. If we are trying to hide our emotions, we often try to look away or hide our face because we know we are involuntarily revealing our feelings (Goffman, 1959b). Other emotions are consciously displayed to emphasize a point, to appear friendly and welcoming, to threaten others, and so forth.

Social psychologists have used this distinction between involuntary and voluntary facial displays to study the issue of emotional universality: If involuntary facial expressions are (1) produced by the same emotional state across individuals and are (2) identified by many observers as meaning the same thing, then we have reason to believe that they are universal expressions of emotions (Ekman, Sorenson, & Friesen, 1969). Furthermore, if these emotional expressions are consistent across all world cultures, then Darwin's view is supported.

Paul Ekman has worked on this very problem for many years. In his initial studies, he took photographs of thousands of people portraying six fundamental emotions: happiness, sadness, surprise, fear, anger, and disgust (Ekman & Friesen, 1975). Each of these emotions has a particular configuration of facial muscles, and these configurations are crucial for the expression of each emotion. The lower face, for example, is very important in identifying happiness, but fear cannot easily be identified without seeing the area around the eyes (Boucher & Ekman, 1975; Ekman, Friesen, & Tomkins, 1971).

To test the universality of emotional expression, these photographs were shown to individuals from different cultural groups who were asked to identify what emotion the person in the photograph was feeling (Ekman & Friesen, 1975; Ekman et al., 1987). All six emotions were recognized at very high rates (see Table 10.1) strongly suggesting that these facial expressions represented a set of primary emotions across cultural groups.

At the time of Ekman's studies, however, Western media had permeated much of the world and critics wondered if exposure to Western television and movies had taught respondents to recognize these emotional expressions. Subsequent research, however, has verified Ekman's claims. First, some social psychologist have carefully studied the facial expressions of people who were born blind—and thus could not learn from other people how to express emotions. These studies demonstrate common expressions of many emotions—indeed, blind people still smile, laugh, and frown in much the same way as sighted people do (Eibl-Eibesfeldt, 1979).

Ekman himself took up the challenge as well and struck out to find a cultural group that was uncontaminated by Western media and influence. He located a group in New Guinea called the Fore, who lived in primitive conditions, without television or print media. Again, Ekman asked the subjects to identify the emotions expressed in various photographs, and again the respondents identified the same emotion 80–90 percent of the time. Second, he asked the Fore to act out different scenarios, such as, "Your child has died and you are sad," or "You are angry and about to fight." After returning to the United States, Ekman asked college students to look at the photographs he had taken of the Fore acting out these scenarios. Once again, they were able to accurately identify the expressed emotion, although they were considerably less accurate when identifying the fear scenario (Ekman, 1984, 1993).

Although most of the research about the facial expressions of emotions seems to support the notion that some emotions are universal across cultures, the research has its limitations as well. For instance, the methods used to conduct research on facial expressions have been criticized for a number of shortcomings. One critique suggests that experimental studies study expressions in highly contrived circumstances and that results are in some ways elicited from the participants (Zajonc, 1998; Zajonc & McIntosh, 1992). Many studies of facial expressions provide the labels that can be used to identify the emotions displayed in the pictures. This results in a task that is more like a matching or multiple-choice test, which is likely to end up with much higher degrees of agreement than if the participants could give any emotional description they wish to the facial expression. Subsequent research has tried to address these concerns and in general has successfully supported Ekman's findings (see Haidt & Keltner, 1999).

There are two important limitations to Ekman's approach:

1. Just because subjects who see a certain facial expression associate it with a particular emotion, it does not mean that the particular emotion is always expressed this way, or even expressed on the face at all. For instance, we might recognize a smiling face as expressing happiness, but we do not smile every time we are happy. Indeed, individuals in situations that should elicit happiness (for example, when one's favored team wins a game) often do not display smiles (Fernandez-Dols & Ruiz-Belda, 1995). Furthermore, some people are very good at faking their emotional displays and can smile very convincingly, even when they are extremely angry or sad. Recall our story about Rob and how he convinced his coworkers and his students that he was in a very good mood—this dynamic is common and even astute observers can easily be fooled. We will explore the selective display of emotions in more detail later in the chapter.
2. Ekman's studies of emotional universality examined a very limited set of emotions. The six core emotions (happiness, sadness, anger, fear, disgust, and surprise) appear repeatedly in these studies and are thought to be the most universally experienced and expressed. But there is no universal agreement that these are the core emotions. Social psychologists have used different methods to identify the most "basic" of emotions and typically cluster them into only five categories: love, joy, anger, sadness, and fear (Shaver, Schwartz, Kirson, & O'Connor, 1987). Compared to Ekman's list, this one is missing surprise and disgust, but adds love. Ekman's approach focuses on those emotions that have uniquely identifiable facial expressions whereas Shaver uses a method in which subjects generate lists of emotions in a free-form way. Using this method, surprise, for example, is rarely generated by the subjects.

In addition to some emotions not being universal, perhaps there are others that are missing from the list. Using the emotional display approach, Keltner and colleagues have demonstrated that there is a unique way of expressing embarrassment that is widely recognized by others (Haidt & Keltner, 1999; Keltner, 1995). When we are embarrassed, we tend to avert our gaze, move our heads down, and touch our faces. Perhaps there are other emotions that also have universal recognition.

Cultural Differences in Basic Emotions and Emotional Display

There also may be cultural differences in basic emotions, emotions being central parts of personal and social experience within a culture in the same way that joy, anger, and sadness are core to our social experience. Shaver, Wu, and Schwartz (1992) discerned a unique category called "sad love" in Chinese culture that combined notions about unrequited love and nostalgia. In addition, these emotions may not be distinct from each other in some cultures as they are in ours. In China, joy and love are not differentiable from each other as they are in the United States.

But even if some basic emotions are invariant across cultures, meaning that they lead us all to use certain combinations of facial muscles in a particular way, there are strong cultural influences that can suppress, exaggerate, or change the display of these emotions. As we interact in daily life, certain situations call for the display of a particular emotion or demand that we do not show others. In most situations, attending a funeral requires the expression of sadness and suppression of any levity or laughter. Typically, a reunion with a cherished brother or sister that you haven't seen in 10 years requires the expression of excitement and joy. But if this reunion occurs at a funeral, it is usually required that the excitement and joy be suppressed. The requirement for solemn sadness at a funeral is not a cultural universal, however, and we must learn the norms of our culture in order to produce the proper emotional displays in each social situation. The impact of situation on emotional display has been documented in several studies. One study carefully examined the facial expressions of Olympic gold medalists and looked at how the specifics of the social situation changed the medalists' expressions of happiness during the medal awards ceremony (Fernandez-Dols & Ruiz-Belda, 1995). Although medalists were judged by observers of the videotaped ceremonies to be happy throughout, they suppressed their smiles throughout most of the ceremony. They allowed the smile to emerge only when they were interacting with others.

One set of cultural norms about emotional expression norms deal with how we must modify our facial expressions to make them fit the social situations. These norms are called **display rules** (Ekman, 1972). Display rules are typically learned in childhood and become habits that automatically control facial muscles. Display rules may require modifying facial expressions of emotion in one of several ways. They may require (1) greater intensity in the expression of an emotion, (2) less intensity in the expression of an emotion, (3) complete neutralization of the emotional expression, or (4) masking one emotion with a different one. If cultures vary in the intensity of emotional displays that are considered appropriate, we would expect observers to have a hard time assessing the intensity of an emotional display by someone from another culture. In the study by Ekman et al. (1987), using observers from 10 cultures, each observer was asked to rate how intensely the person in the photograph was experiencing happiness, anger, and so on. Whereas judgments of the emotion itself showed high levels of agreement across cultures, the judgments of intensity showed much lower levels of agreement from one culture to another. In an experiment with women as participants, facial displays were found to be influenced not only by the intensity of the emotional experience and the situation, but also by the relationship between the woman and the audience (Hess, Banse, & Kappas, 1995). The most intense display occurred when the emotion was intense, the display norm allowed display, and the other participant was a friend.

Cultures can vary more generally in their demands about emotional display, in addition to providing guidance for displays in specific kind of situations. For example, some cultures have a much more disapproving stance toward anger in general. In other cultures, anger and the expression of it is seen as a normal part of healthy social interaction. When anger is viewed more negatively, as in Eskimo culture (see Briggs, 1970), whatever natural expression of anger may exist can be moderated in such a way that people from other cultural groups, who are used to a more open and obvious expression of anger, might not even recognize that there is a problem.

One of the most important cultural differences that affects the expression of emotions is how collectivist or individualistic the culture is. Individualistic cultures, like in the United States, are usually more focused on the individual person as a key social

TABLE 10.2 EMOTIONAL EXPRESSION IN COLLECTIVIST AND INDIVIDUALISTIC CULTURES

Emotional Element	Collectivist Orientation	Individualist Orientation	Cultures Compared	Example Study
Estimate of emotional intensity	Report low intensity	Report high intensity	Japan, U.S.	Matsumoto, 1987
Emotional content in vocal cues	Higher attention to vocal cues	Higher attention to word meanings	Japan, U.S.	Ishii, Reyes, & Kitayama, 2003
Expressing happiness	Expression of happiness promotes connection to others	Expression of happiness reflects individual achievement	Japan, U.S.	Mesquita & Karasawa, 2002
Shame and guilt	Shame is of central importance	Guilt is more important, shame develops later	China, Dutch	Hazen & Shaver, 1992; Wellman, Phillips, & Rodriguez, 2000
Anger	Anger is dangerous and must be suppressed	Anger is healthy and promotes assertiveness	Eskimo, Western	Briggs, 1970
Grief	Elaborate mourning rituals	Brief mourning period followed by a return to normal life	Many non-Western, U.S., and Europe	See Braesicke et al., 2005

Source: This table and the accompanying discussion are adapted from portions of Parkinson, Fischer, and Manstead, 2005.

unit. Individuals have their own goals, accomplishments, and behaviors that stand apart from group membership. Collectivist cultures are more focused on groups as the sources of identify. For example, individuals are seen in terms of their membership in their family and are much more affected by the interconnected behaviors, accomplishments, and failures of others in their groups. They are less independent and more interdependent in their social relationships and identities.

How does collectivism or individualism affect the expression of emotion? Table 10.2 summarizes some of the many findings about emotional expression in these two types of cultures. Most often, the studies compare American participants, who are usually highly individualistic, and Japanese participants, who are usually highly collectivist. It is important to remember, however, that there are many cultural contexts that are more and less collectivist and individualistic than these two, and that individuals within each country vary a great deal in terms of their orientation toward collectivism and individualism.

To understand some of these differences, it is important to remember that collectivistic cultures are more concerned about disrupting social interaction. When social relations, as opposed to individual states, are at the center of identity and interaction, ensuring smooth, trouble-free exchange with others becomes more important. To avoid negative disruptions in social interaction, members of collectivist cultures such as Japan are more likely to suppress emotional expression and also to downplay the intensity of emotional reaction they observe in others' faces (Matsumoto, 1987; Matsumoto & Ekman, 1989). Thus, collectivistic cultures have not only more subdued display rules, but also have decoding norms that reduce the interpretation of emotional intensity.

This does not mean, however, that collectivist cultures are less sensitive to emotional expression. In fact, experimental research comparing Japanese and Americans showed that the Japanese were more sensitive to indirect emotional cues. In this study, the experimenters listened to words that were pleasant or unpleasant in meaning and were read in either a pleasant or unpleasant manner. Thus the research subjects were presented with a curious mix of words with unpleasant meaning but were conveyed using positive vocal cues. Likewise, they were also presented with pleasant words conveyed using negative emotional cues. The results demonstrated that the Japanese participants were much more attentive to the vocal cues, whereas the Americans were more attentive to the words themselves (Ishii, Reyes, & Kitayama, 2003). Thus, while collectivist cultures might work at reducing the expression and interpretation of negative emotions, they are actually more sensitive to them than those from individualistic cultures.

Collectivist and individualist cultures also differ in their approaches to happiness. What might those in individualist cultures be most likely to express happiness about? The answer, of course, is individual achievements and attributes. If the individual is good at something or has a characteristic that causes a positive reaction, then that individual has reason to be happy and to express it. Members of collectivist cultures express happiness in situations that emphasize their connectedness with others in their identity groups. When they feel close to, and consonant with, other people, this is the primary reason to experience and display happiness. This difference was demonstrated in a study by Mequita and Karasawa (2002) in what is called a diary study. During the course of 1 week, American and Japanese students recorded information about their emotions every 3 hours. Japanese students, reflecting their collectivist orientation, reported pleasant feelings mainly when they experienced positive interdependent engagement with others. The American students reported positive feeling with respect to both interdependent events and independent events (such as having control or mastery).

Another important emotional experience that differentiates collectivist and individualist cultures is shame. We will discuss shame in more detail below, but shame plays a substantially different role—in fact, a much more central role—in some cultures. Collectivist cultures emphasize the relationship of the individual to other people. Therefore, shame results from the negative evaluation that others have of the individual. The interpersonal sanctions that result are most important in the collectivist cultures. In contrast, individualistic cultures tend to focus more on guilt—failing to meet one's own standards, which is internally evaluated. Two studies suggest the relative importance of shame in different cultures. In one study, the researchers found that 95 percent of Chinese mothers reported that their children understood the concept of shame by the age of 3 (Shaver, Wu, & Schwartz, 1992). In another study, none of a group of Dutch 7-year-olds knew what shame was (Wellman, Phillips, & Rodriguez, 2000).

Anger is a very important emotion because it is so volatile. When people are angry, they can be extremely forceful, can lose control, and even become highly destructive. When people express anger, they can make others angry—especially the targets of their anger. The expression of anger can lead to retribution and may set off a spiral of escalating anger and destruction. Therefore, anger is an emotion that we must find ways to manage, control, and appropriately direct. Sigmund Freud believed that managing anger was one of the central functions of society, in that society could devise ways for people to express their aggression in less destructive ways, rather than allowing it to build up and cause serious problems.

But how might collectivist and individualist cultures differently approach anger? Your first guess might be that collectivist cultures try to avoid the expression of anger. Given that it has such potential to upset and damage social interactions with others, we should expect that collectivists would work very hard to avoid experiencing and acting on anger. Briggs's (1970) study of an Eskimo culture provides a dramatic example of this very principle. Briggs determined that she was studying a highly interdependent, collectivist culture. The group she studied depended heavily on the relationships among those in the group for survival. They were very isolated and often faced difficult conditions. Combined, these factors led them to emphasize group harmony very heavily, and anger was seen as a major threat to the stability and existence of the group. Western, individualist cultures tend to associate anger more with assertiveness and individual rights (Averill, 1980; Braesicke et al., 2005). The expression of anger is often encouraged so that others know how we feel, as a tool to rectify an unjust situation, and to blow off steam before the pressure builds to a catastrophic climax.

The final emotion we will discuss in the context of collectivist and independent cultural contexts is grief. As we look across cultures, we can witness an incredible range in the expression of grief and mourning. In almost all cultures, mourning a death involves sadness and tears. Given the wide variety of funeral and bereavement practices, it can be difficult to draw clear

Cultural norms strongly influence the expression of emotions. The expression of bereavement, for example, varies widely, ranging from solitary reflection to group wailing practices. Here Iraqi mourners beat their chests in a Shiite grieving ritual in Baghdad.

distinctions among the practices of collectivist and individualistic cultures. Some of the most dramatic cultural mourning acts are associated with social groups in which there are strong interdependent relationships. These practices can range from elaborate wailing episodes to the suicide of a widow or widower. Individualist societies view these kinds of practices as extremely strange and instead carry a belief that although mourning and grief after a death is normal, there are limits. Mourners who cannot get over a death are considered unstable and can end up being isolated until they can recover.

As we look across cultures and observe the different ways that emotions are expressed, we come away with a complicated picture of basic emotions. For the most part, social psychologists agree that some emotions are universal and that they are biologically connected to distinctive facial expressions, and yet the expression of even the most primary emotions is heavily influenced by cultural norms and can vary a great deal from place to place. If people are to communicate emotions effectively in everyday interaction, they must learn and use the display rules of their own culture. In addition, some emotions are far more conditioned on social processes than others, and sociologists who study emotion are much more focused on these kinds of social-emotion processes than on the biological links. We will turn our attention to these social emotions in the following section.

SOCIAL EMOTIONS

When we think about social emotions, we have to start by breaking away from the notion that emotions are simply natural reactions to things that happen in the world around us. As we have seen, some aspects of some emotions may be biologically hardwired. But others are much more a matter of interpretation. When we experience a physiological reaction that is part of an emotion, before we can decide which emotion is involved, we have to interpret the physical sensation in its social context.

Consider the feeling of being slightly nauseous and having sweaty palms. There are many potential interpretations of this physical symptom. Perhaps you are a first-year medical student feeling disgusted upon seeing your first cadaver. Maybe you are grief stricken at the unexpected death of a cherished pet. Then again, perhaps you are excited about an upcoming date that you have been looking forward to. Or perhaps you are nervously heading into an exam you did not study enough for. When we think more closely about emotions, we find that they are often just plausible explanations for our physiological reactions, and how we interpret them depends very much on what is happening in our social world. To the degree that we are actively experiencing, interpreting, and constructing our social world, we are also interpreting and constructing our emotions.

Cognitive Labeling Theory

One theory that tries to understand the emergence of social emotions is called **cognitive labeling theory** (Schachter, 1964). This theory proposes that emotional experience is the result of the following three-step sequence:

1. An event in the environment produces a physiological reaction.
2. We notice the physiological reaction and search for an appropriate explanation.
3. By examining situational cues ("What was happening when I reacted?"), we find an emotional label (joy, disgust) for the reaction.

The theory further assumes that physical arousal is a general state. We are somehow stimulated by the environment to change from a quiescent state to a situation of arousal and that this state of arousal does not substantially differ from one emotional state to another. One arousal state is not physiologically distinguishable from another and therefore virtually any emotion can be attached to the arousal state. Which emotion becomes attributed to the arousal depends on the context.

Many social psychologists have demonstrated that our understanding of physical arousal is indeed fairly easy to manipulate. For example, in one early study of cognitive labeling theory (Schachter & Singer, 1962), researchers gave students an injection of epinephrine, a drug that produces mild physiological arousal. They informed one group of students that this injection would probably cause them to experience a pounding heart, flushed face, and trembling. They told a second group nothing about the drug's effects. All students then waited with a confederate who, though appearing to be another student, was actually employed by the researchers. Depending on the experimental treatment, the confederate behaved either euphorically (for example, shooting crumpled paper at a waste basket, flying paper airplanes, playing with a hula hoop) or angrily (for example, reacting with hostility to items on a questionnaire and finally tearing it up).

According to the theory, students in the informed group would not need to seek an explanation for their arousal, because they knew their symptoms were drug induced. Students in the uninformed group, however, lacked an adequate explanation for their symptoms and thus would need to search the environment for cues to help them label their feelings. The results confirmed these predictions. Students in the uninformed group adopted the label for their arousal suggested by their environment. That is, those who waited with the euphoric confederate described themselves as happy, whereas those who waited with the angry confederate described themselves as angry. The self-descriptions of the informed group, on the other hand, were largely unaffected by the confederate's behavior.

Numerous later studies have expanded these findings to additional emotions (Kelley & Michela, 1980). They show that people who are unaware of the true cause of their physiological arousal can be induced to view themselves as anxious, guilty, amused, or sexually excited by placing them in environments that suggest these emotions (Dutton & Aron, 1974; Zillman, 1978). As the theory predicts, environmental conditions strongly influence people's labeling of their physiological arousal.

Research suggests that the emotional label sometimes even precedes the awareness of arousal (Leventhal, 1984; Pennebaker, 1980). Our social context suggests that we should be experiencing a particular emotion and only then search our bodily sensations for signs that will verify our belief. If environmental cues give us reason to believe we are angry, we attend to our flushed face and racing heart and verify our

BOX 10.1 Research Update: Emotions and Social Movements

So much of the research on emotions is concerned with the nature of emotions and the sources of emotions. Psychologists of all types have been heavily focused on how emotions emerge—both from the body and from the social context. But there is different set of questions that use emotions not as a dependent, or outcome, variable, but rather look at the effects of emotions on the individual, on other people, and on the social environment. These approaches treat emotions as independent variables—the sources as opposed to the outcomes. Most researchers recognize, of course, that the experience and expression of emotions are part of complicated cycles in which felt emotions are expressed and cause changes in the social context. These changes introduce feedback in the system that can result in new emotions being experienced and introduced to the environment or in changes in the original emotions.

Some researchers who examine political and social activism have come to see emotions as a key element that drives individuals to participate. Oftentimes, activists face very difficult challenges as they pursue their goals, and the emotions they experience can play a very important part in keeping them involved or causing them to abandon their activism. In Doug McAdam's (1988) study of college students who traveled to the South to register Black voters during the Civil Rights Movement, it was apparent that the volunteers had transformative emotional experiences—ranging from fear that had to be overcome, to anger that strengthened their resolve, to love for their fellow activists.

More recently, sociologist Erika Summers-Effler (2005) has been studying the emotional experiences of those who live and work in Catholic worker houses. These individuals are engaged in very challenging and highly involved activism in service to the poor in inner cities. They essentially give their lives to the Catholic worker movement, abandoning their possessions and former lives to live in the inner-city environment with those they serve. The Catholic worker house hosts "guests," who essentially are anyone who wishes to live at the house and typically are those who would otherwise be homeless. In addition to hosting the guests in the house, the Catholic workers also provide meals for people in the neighborhood, maintain a clothing pantry for those in need in the neighborhood, and provide after-school tutoring. The challenges are many and the support systems are sparse, resulting in difficult daily challenges that can stretch the workers to their limits.

Given these kinds of extreme challenges, Catholic workers need intangible support in order to continue with their work—they need emotional outlets and emotional rituals that give

anger. If the cues suggest we are happy, we attend to our feelings of alertness and trembling and confirm our happiness. At any given time, our physiological state may afford evidence to support several emotional labels. Once the emotional label is applied, it can induce further physiological arousal that provides additional confirmation of the emotional label we have applied.

As the study of emotions has developed, researchers have come to better understand the centrality of the social situation in defining emotions. As cognitive labeling theory posits and the empirical research around it demonstrates, our immediate emotional reactions are products of internal physical processes that must also be recognized and interpreted in light of the present social context. This is true not only of immediate emotional reactions that are tied to arousal, but also of complex, enduring feelings like love and jealousy. These emotional states last even after physical arousal has passed (Gordon, 1990). Each is

BOX 10.1 Continued

them the fortitude to continue on and maintain their commitments. Summers-Effler documents a number of critical emotional processes that provide this kind of support. For one, she emphasizes the importance of laughter in the community. Humor, she found, allowed individuals to face difficult situations and maintain a sense of cohesion within the group. Joint laughter allowed individuals to escape any shame they might have felt or any sense of failure by recognizing the absurd in the situation without giving it to it. This story illustrates that process:

> Lynn said that the ice cream was in the basement because the refrigerator was broken. I asked, half seriously, if they were waiting for God to get them a new refrigerator. Everyone thought this was really funny; most were doubled over in laughter. The laughter was irresistible, contagious, I found myself laughing as well . . . Finally, Lynn stopped laughing long enough to say that she was debating whether to call the repair man or not, and that she had had a few discussions with St. Francis about it . . . she was immediately consumed with laughter again (pp. 143–144).

A second key process related to emotion was its relationship to rituals in the community. The Catholic worker environment was full of both emotion and ritual—rituals both formal and informal. What Summers-Effler found was that emotion and ritual combined to produce positive outcomes for the workers—in particular, greater solidarity. What seemed to occur was that negative emotion was processed by the ritual and the result was a strengthening of bonds among the group. One example was the weekly community gathering:

> Once a week the extended community, mostly white middle class supporters but also some guests and people from the neighborhood, participate in a community dinner at the Catholic worker house . . . during this weekly ritual, which often involves the formal ritual of liturgy, the extended group reaffirms the Catholic workers as sacred symbols of the community. Because the Catholic workers are the center of attention and praise, these weekly dinners are an emotional boost to the Catholic workers from their extended community.

Thus, we can see some ways that emotions produce important social outcomes, and therefore their role as inputs into social situations should be considered as carefully as are questions about how emotions emerge from social situations.

Source: Excerpted from Summers-Effler, 2005.

a pattern of sensations, emotions, actions, and cultural beliefs that are appropriate to a social relationship. Sentiments such as grief, loyalty, envy, and patriotism develop around our attachments to family, friends, fellow workers, and country.

Sentiments reflect the nature of our social relationships and the changes in them. Grief and nostalgia reflect social losses. Jealousy and envy reflect problems over control of possessions. Anger and resentment reflect betrayal of commitments. We label our feelings with the culturally appropriate sentiment to make sense of our diverse emotional responses. For example, a husband's joy in his wife's presence, his sorrow in her absence, his anger when she is criticized, and his fear when threatened with losing her make sense if he labels his feeling "love." Like simpler emotions, sentiments are produced by cognitive labeling. In choosing a sentiment label, however, we consider all the information we have about our enduring relationship, not just the immediate social context.

Five Social Emotions

While many emotions have both social and nonsocial components, there are certain emotions that simply cannot be understood or even defined without reference to the social world. These emotions are called **social emotions** and they are defined as emotions that (1) involve an awareness of oneself in the social context, (2) emerge out of interaction with at least one other actor, and (3) are often experienced in reference to some kind of societal standard (Barrett, 1995; Stets, 2003).

To understand this rather complex definition, think about the notion of empathy. To experience emotional empathy, you first have to be aware that you have some kind of connection to a person that is experiencing an emotion. For example, when we feel empathy for someone who is feeling pain, we are at least partially recognizing that we are involved, and perhaps even responsible for easing the person's pain. (The notion that helping can reduce empathic suffering is explored in greater detail in Chapter 11.) Second, there must be someone else in the social environment who is experiencing some kind of emotion in order for us to experience empathy. By definition, then, empathy is social. Third, when one experiences empathy, he or she is reminded of society's standards. If the object of our empathy is feeling pain, we may evaluate, based on societal standards, our own obligation to intervene to ease their pain. If their pain is minimal and the risk of harm to ourselves is great, we are not obligated to intervene and therefore may experience minimal empathy.

In this section, we will examine the social bases of five emotions to illustrate how important social interaction is in experiencing and defining some of our most important social processes. There are many other social emotions, but here we will focus on guilt, shame, jealousy, empathy, and love.

Guilt We feel guilt when we judge that we have done something that we should not have. Guilt is therefore, inherently evaluative. We are not necessarily a bad person if we have done something that causes us guilt, but we are certainly less good than we could be. Where do the standards come from that we use to judge ourselves? They come from others in our social groups or environment. Guilt therefore involves a self-reflexive judgment in which we see ourselves through other people's eyes. When we feel guilt, we are engaged in an appraisal of ourselves using standards that we may have accepted, but that were constructed by others.

Guilt also implies action. Just like so many social psychological processes, guilt involves an uncomfortable feeling. Individuals who feel guilt need to do something to eliminate it so that they can return to a more pleasant psychological state. They attempt to engage in some kind of reparative activity so that they can be forgiven by someone who has power to release them from guilt (Stets, 2003). Thus, not only is the negative emotion itself inherently social, so is the method of dealing with it. Returning for a moment to our story about Rob, the kindergarten teacher, suppose in his frustration he had blown up at his teacher's aide for not getting the children settled for story time. After yelling at his aide, Rob may feel guilty for the unnecessary outburst, which made the aide feel underappreciated; the aide may feel guilty for not following classroom procedures properly; and if the children observed the outburst, they might also feel guilty for not getting into the story-time circle quickly enough. In each instance of guilt, there is a social standard of behavior (for Rob, for the aide, and for the children) that has been violated. Each guilty party can reduce or eliminate the guilt by, for example, apologizing to the offended party and receiving forgiveness.

Where do we most often see guilt in social interaction? Guilt has a number of social functions, the first of which is socialization (Hoffman, 1998). When a parent disapproves of a child's behavior, the result is a feeling of discomfort. If the parent can then induce the child to realize that the source of the discomfort is the child's own behavior, then the child will feel a sense of guilt. The child can then be taught to avoid the guilt by avoiding the behavior.

Closely related, people can use guilt as a method to get what they want from others. If they can induce others to feel guilty about doing or not doing something, they may be able to change the target's behavior (Stets, 2003). Panhandlers use various methods of making their targets feel guilty for refusing to help them. By pointing out the disparities between the target and their own difficult circumstances, they can induce guilt and inspire giving.

Finally, although we often think that guilt has a negative impact on social relationships, social psychologists have found that it actually functions to support and strengthen relationships. It does this in two ways. First, it distributes the negative consequences of a bad social interaction. If I do something that hurts you, you are bearing all of the costs of the social exchange. But, if I feel guilty about it, I also have to bear some costs—so guilt evens out the suffering to some degree (Baumeister, Stillwell, & Heatherton, 1994). Second, when we feel guilt and act to correct whatever problem caused the guilt, we are sending a powerful message to those who we harmed—we are telling them that we value the relationship and that we don't want our own poor behavior to damage it. If Rob apologizes to his classroom aide, his actions may actually endear him more to her than if he had never yelled at her in the first place.

Shame A counterpart to the emotion of guilt is shame. Although both emotions share certain characteristics—a person feeling either of these emotions has committed some kind of offense against another—shame is a much deeper and longer lasting state than guilt. When we feel guilty, it is typically tied to a single incident that is easy to identify and oftentimes has an easily identifiable response that will relieve the guilt. If, for example, you steal something and feel guilty, you can return what you stole or replace it.

Shame, on the other hand, is not so much about an incident or transgression as it is about how you evaluate yourself as a person. When you feel shame, there is a deep sense of the self, not as someone who has just done something wrong, but as someone who is a bad person. There is something wrong with your intrinsic character, and it is not something that can be easily rectified (Babcock & Sabini, 1991). As a result, the response to shame is not as simple as repairing the damage done to the social relationship. Most of the time the more likely response to shame is to flee. Shamed individuals want to run away from the situation, hide from everyone, or just disappear completely (Barrett, 1995; Tangney, 1995). Given that shame threatens the very core of an individual's self-regard, it is a very intense emotional experience—much more so than that of guilt.

Because shame is such an intense emotion, people are all the more motivated to try to escape it. One way of reducing shame is to escape the blame for the problems that caused it. Those who have been shamed sometimes try to accomplish this by blaming others. The result can be a volatile "shame-anger cycle" (Scheff & Retzinger, 1991) that escalates tension and shame. This model suggests that one person, let's call her Karla, insults her husband, Leo, telling him that he is a worthless human being, as evidenced by his repeated failure to hold down a job. If Leo were to accept this appraisal, he would become ashamed. Escaping shame is a powerful motivator, so Leo responds by getting angry and insulting Karla with a retaliatory slur that implies she is the reason he has lost his job. He claims that he had to leave work early many times to deal with their son, because she was an incompetent mother. Each partner now feels shame and sees the other person as the source of the problem. Shame leads to more anger through a retaliatory cycle. Oftentimes this is exactly the kind of interaction that leads to spousal assault.

Jealousy Jealousy is also an inherently social emotion. Jealousy is a negative emotional reaction we feel when something good happens to someone else (Ben Ze'ev, 2000). Because an individual cannot, by definition, experience jealousy without the participation of someone else, the emotional state requires the participation of the social context.

In addition, jealousy is often the result of a triad of relationships in which the jealous person has lost a significant relationship with a second person. That second person has formed a relationship with a third person, who is the object of the jealousy. A good example of this dynamic is the breakup of a romantically involved couple. If the girlfriend, for example, leaves the relationship and starts dating a new partner, the boyfriend who is left behind may feel jealous of the new boyfriend. But this kind of dynamic can be observed in virtually any kind of paired relationship. Social and developmental psychologists have documented jealousy directed from one toddler to another when a mother shifts her attention from playing with one child to playing with another (for example, Draghi-Lorenz, Reddy, & Costall, 2001).

If social emotions exist for a reason, what could be the functions of jealousy? We typically think of jealousy as being a destructive force that we must struggle to control. But if jealousy is really born of social forces and is not completely natural, it must have some useful purpose. One function may be to draw oneself back into a social interaction. When we express jealousy, we are signaling to others in the environment that we have been left out, that we want to be included, and that they are in some way responsible for helping to reintegrate us into the social relationship. A child that expresses jealousy when his mother is attending to some other child is sending the message that he wants to interact with her. If she responds, as parents often do, by increasing her interaction with him, picking him up, or drawing him into activities with the other child, then the expression of jealousy has been rewarded, and thereby reinforced.

Jealousy can also signal to others in the environment that the jealous person has some kind of claim on the object in question (be it a person or a physical item). If, for example, my daughter were to express jealousy that my son is getting to play with a favored toy, it would indicate that she is making a claim on the toy: She has a right to play with it, she deserves more time with the toy and he deserves less, and she wants to exercise that claim by playing with the toy. Thus jealousy and envy help to establish turf boundaries around objects and around people.

Empathy In many social emotions, it is necessary to place oneself in the role of others in order to experience the emotion (Shott, 1979). Pride, for example, can occur only if we can see ourselves and our own accomplishments from the perspective of others. It is how others will evaluate us that drives our sense of pride. When we accomplish an important goal—an A on a major paper, a successful recital, or an athletic victory—we may role-take and imagine the positive reactions of family and close friends. Likewise, shame and guilt also require a certain amount of role-taking. When we fail at an important task or engage in behavior that violates group norms, we may imagine others' reactions and then feel shame or guilt. In each of these cases, viewing one's own behavior from the perspective of others arouses emotion.

Parents experience strong empathic responses to the emotional states of their children. They find it easy both to recognize distress in their offspring and to put themselves in the place of their children.

Empathy is driven by role-taking in a slightly different way. Here it is not the evaluation that others may make that is central, but rather it is what others are emotionally experiencing themselves (Davis, 1996). Empathic role-taking emotions are evoked by imagining yourself in the position of another. Hearing a survivor recount the devastation caused by a tornado or the injuries and deaths caused by a bomb elicits sadness and perhaps tears in the listener. Reading a tale of great achievement or heroism may elicit feelings of joy and accomplishment. Empathic role-taking may be the source of motivation to prevent future suffering or to accomplish a great feat.

The ability to experience an empathic emotion depends on the situation and on the characteristics of the observer. In a study of empathic embarrassment, groups of women observed another woman perform either a very embarrassing or an ambiguous task (Marcus, Wilson, & Miller, 1996). Most observers reported feeling embarrassed as they observed another perform the embarrassing task. Some observers of the innocuous task also reported feeling embarrassed, suggesting that people vary in their readiness to perceive emotion in others. Indeed, individuals vary a great deal in their ability to read, understand, and respond to the emotional cues and behavior of others. Those who are skilled in this process are said to have high emotional intelligence. Box 10.2 explores the notion of emotional intelligence and allows you to test yourself for emotional intelligence.

In earlier years, many people believed that women had greater empathy than men. The notion of "women's intuition" suggests that women are better able to understand and identify with the emotional states of others. The idea has been extended to support the notion that mothers are better parents because they are better at reading children's internal states of discomfort—this in turn motivates arguments that women should be primary caretakers and should attend to home and children rather than have careers outside the home. Most recent research on the topic, however, shows that men and women do not differ with respect to their empathic abilities (Stets, 2003). Rather, there is a difference in how men and women are socialized to use empathy. Graham and Ickes (1997) pointed out that when tested for empathy, men and women did equally well—except when they were told that their empathy was being tested. The researchers concluded from this that because of gender socialization, women were more motivated to prove that they were empathic. If being a good mother is an important past, present, or future identity, a test that could show you to be deficient on a core aspect of mothering is threatening—causing increased attention to the empathy tasks. Men, on the other hand, do not perceive empathy as being central to their identity and thus are not motivated to try harder when told their empathy is being tested.

As with our other social emotions, empathy has important social functions (Davis, 1996). It promotes altruism and prosocial behavior (see Chapter 11). It assists in developing strong interpersonal relationships because one person is better able to address the needs and concerns of another. And it inhibits aggressive behavior. As we will see in Chapter 12, if someone is hurting someone else or thinking about hurting someone else, he or she is more likely to terminate the attack if he or she can empathically feel the other person's pain.

Love We have discussed many negative emotions in this chapter, but these are not the only social emotions. It is fitting therefore to end our discussion of social emotions with a brief discussion of a positive one: love. Despite our immediate sense that love is a happy, positive emotion, it is also important to remember that it is not always so. In fact, some Asian cultures actually identify a separate emotion that might be called "sad love," which usually involves some unrequited feelings of loss. Nevertheless, most people experience love as a positive, mood-enhancing emotion.

We will discuss love in greater detail in Chapter 13 (Interpersonal Attraction and Relationships), and we will see that there are different kinds of love that are experienced across a wide range of intensity. It is a complex emotion, or, more appropriately, a set of related emotions. Love is, however, an inherently social emotion. It is plain that love involves at least two people—one who loves and one who is the object of the love. The type of love involved, and the experience of it, depends very heavily on the relationship to the object of love and the reaction to any expression of love. Friends have different experiences of love than spouses, teenage boyfriends and girlfriends have different experiences of love than a couple that has been married for 30 years. Even within a single loving relationship, there can be significant asymmetry. A parent, for example, shares love with a child, but the adult's experience of love is considerably different than the child's. The differences in all of these experiences of love are not derived from some kind of natural reaction, but rather from the different expectations associated with the role of parent, child, boyfriend, girlfriend, husband, wife, best friend, and so on. Parents have responsibilities for the growth, development, and well-being of their children that indicates a caretaking element in their

BOX 10.2 Test Yourself: What's Your Emotional IQ?

Think you're smart? Think you're street smart? Maybe you are, but will those smarts help negotiate all social situations? Maybe not, say social psychologists who study emotional intelligence. In 1995, Daniel Goleman published a best-selling book called *Emotional Intelligence: Why It Can Matter More Than IQ,* which summarized and extended earlier work done by psychologists John Mayer and Peter Salovey on the notion of **emotional intelligence.** These authors claim that an individual's ability to understand the emotional content in social interactions constituted a unique dimension of intelligence that was substantially different from the different kinds of intelligences measured by IQ that had been previously discussed in the psychological literature on intelligence.

According to Mayer and Salovey, emotional intelligence consists of four capacities:

1. being able to accurately perceive emotions (one's own and others)
2. being able to use emotional information in rational thinking
3. being able to understand the meaning of emotions
4. being able to manage emotions (both one's own and those of others)

These four capacities have been tied to success and failure in a number of social environments including family and marital relationships, managing employees and coworkers, intergroup relations, and even personal health (Goleman, 1995).

Although the idea of emotional intelligence is relatively new, a number of measures have been developed to assess it (see Brackett & Mayer, 2003; Lopes, Salovey, Cote, & Beers, 2005). One is called the Emotional Competence Inventory (ECI) developed by the Hay Group. As with most other measures, the ECI is too long to be reproduced here, but the Hay Group has provided some sample items that give a preliminary sense of how emotional intelligence is typically measured. These 10 items do not represent a comprehensive assessment of emotional intelligence. Those interested in a comprehensive assessment can visit the Hay Group's website at http://ei.haygroup.com/.

1. You are on an airplane that suddenly hits extremely bad turbulence and begins rocking from side to side. What do you do?
   A. Continue to read your book or magazine, or watch the movie, trying to pay little attention to the turbulence.
   B. Become vigilant for an emergency, carefully monitoring the stewardesses and reading the emergency instructions card.
   C. A little of both a and b.
   D. Not sure—never noticed.
2. You are in a meeting when a colleague takes credit for work that you have done. What do you do?
   A. Immediately and publicly confront the colleague over the ownership of your work.
   B. After the meeting, take the colleague aside and tell her that you would appreciate in the future that she credits you when speaking about your work.
   C. Nothing, it's not a good idea to embarrass colleagues in public.
   D. After the colleague speaks, publicly thank her for referencing your work and give the group more specific detail about what you were trying to accomplish.
3. You are a customer service representative and have just gotten an extremely angry client on the phone. What do you do?
   A. Hang up. It doesn't pay to take abuse from anyone.
   B. Listen to the client and rephrase what you gather he is feeling.
   C. Explain to the client that he is being unfair, that you are only trying to do your job, and you would appreciate it if he wouldn't get in the way of this.
   D. Tell the client you understand how frustrating this must be for him, and offer a specific thing you can do to help him get his problem resolved.
4. You are a college student who had hoped to get an A in a course that was important for your future career aspirations. You have just found out you got a C− on the midterm. What do you do?
   A. Sketch out a specific plan for ways to improve your grade and resolve to follow through.
   B. Decide you do not have what it takes to make it in that career.

BOX 10.2 Continued

C. Tell yourself it really doesn't matter how much you do in the course; concentrate instead on other classes where your grades are higher.

D. Go see the professor and try to talk her into giving you a better grade.

5. You are a manager in an organization that is trying to encourage respect for racial and ethnic diversity. You overhear someone telling a racist joke. What do you do?

   A. Ignore it—the best way to deal with these things is not to react.

   B. Call the person into your office and explain that their behavior is inappropriate and is grounds for disciplinary action if repeated.

   C. Speak up on the spot, saying that such jokes are inappropriate and will not be tolerated in your organization.

   D. Suggest to the person telling the joke that he go through a diversity training program.

6. You are an insurance salesman calling on prospective clients. You have left the last 15 clients empty-handed. What do you do?

   A. Call it a day and go home early to miss rush-hour traffic.

   B. Try something new in the next call, and keep plugging away.

   C. List your strengths and weaknesses to identify what may be undermining your ability to sell.

   D. Sharpen up your resume.

7. You are trying to calm down a colleague who has worked herself into a fury because the driver of another car has cut dangerously close in front of her. What do you do?

   A. Tell her to forget about it—she's OK now and it is no big deal.

   B. Put on one of her favorite tapes and try to distract her.

   C. Join her in criticizing the other driver.

   D. Tell her about a time something like this happened to you, and how angry you felt, until you saw the other driver was on the way to the hospital.

8. A discussion between you and your partner has escalated into a shouting match. You are both upset and in the heat of the argument, start making personal attacks which neither of you really mean. What is the best thing to do?

   A. Agree to take a 20-minute break before continuing the discussion.

   B. Go silent, regardless of what your partner says.

   C. Say you are sorry, and ask your partner to apologize too.

   D. Stop for a moment, collect your thoughts, then restate your side of the case as precisely as possible.

9. You have been given the task of managing a team that has been unable to come up with a creative solution to a work problem. What is the first thing that you do?

   A. Draw up an agenda, call a meeting, and allot a specific period of time to discuss each item.

   B. Organize an off-site meeting aimed specifically at encouraging the team to get to know each other better.

   C. Begin by asking each person individually for ideas about how to solve the problem.

   D. Start out with a brainstorming session, encouraging each person to say whatever comes to mind, no matter how wild.

10. You have recently been assigned a young manager in your team, and have noticed that he appears to be unable to make the simplest of decisions without seeking advice from you. What do you do?

   A. Accept that he "does not have what it take to succeed around here" and find others in your team to take on his tasks.

   B. Get an HR manager to talk to him about where he sees his future in the organization.

   C. Purposely give him lots of complex decisions to make so that he will become more confident in the role.

   D. Engineer an ongoing series of challenging but manageable experiences for him, and make yourself available to act as his mentor.

Scoring: Question 1. A-10 points, B-10 points, C-10 points; 2. B-5, D-10; 3. B-5, D-10; 4. A-10, C-5; 5. B-5, C-10, D-5; 6. B-10, C-5; 7. C-5, D-10; 8. A-10; 9. B-10, D-5; 10. B-5, D-10.

concept of parental love, an element that their children do not share. Other role demands produce very different conceptions of love, not only in different types of relationships, but also across cultures.

Emotion Work

When a person enters into a social situation, he or she can make some decisions about what kind of emotions to display. In some situations, it might advantageous to suppress anger and deal with the other in the environment as politely as possible. In other cases, such as those in which a person needs to be particularly assertive, it might be advantageous to reveal one's anger or even enhance the display of it beyond what one really feels—a coach criticizing his team during halftime for a lazy performance might fall into this category.

In fact, social psychologists often argue that we are always acting in social situations, since we cannot possibly express every nuance of every feeling we experience. We are always making choices to suppress and exaggerate the performance of some emotions. But this kind of work extends beyond mere acting. In some situations, we actually attempt to change our actually feeling, not just our behaviors. Have you ever tried to psych yourself up for a performance, sporting event, or examination? Have you ever forced yourself to have a good time at a party even though you were tired? Maybe you have tried to feel grateful for a gift that you really didn't like or displayed a stiff upper lip despite severe disappointment. These are all instances of **emotion work**—attempts to change the intensity or quality of our feelings to bring them into line with the requirements of the occasion (Hochschild, 1983).

Emotion work occurs because we are subject to **feeling rules**—rules that dictate what people with our role identities ought to feel in a given situation. If we are receiving a gift, the feeling rule is that we should feel grateful. If we are at a party, we should be having fun. Social psychologists have attempted to identify feeling rules by presenting scenarios to subjects. Many times, there was a high degree of agreement about which emotions should be present in certain social situations—evidence that there indeed are emotional expectation or feeling rules (Heiss & Calhan, 1995). If our feelings are not in line with the current feeling rules and we were to express our true feelings, we would be violating the norms of the situation. Therefore, we try not only to express the right feeling, but we also try to feel the right feelings.

Arlie Hochschild conducted a very interesting study of flight attendants and bill collectors to examine this phenomenon (1983). Here we will focus on the flight attendants. Generally speaking, airline flight attendants are expected to feel calm and cheerful as they interact with passengers. But suppose a group of flight attendants has been working for 10 hours, serving hundreds of people on three different flights. Fatigue and irritation may be the main feelings they are experiencing. If so, they must then work directly on their own emotions to evoke feelings of cheerfulness and to suppress feelings of irritation. In fact, Hochschild showed that this kind of emotion work was a fundamental requirement of the job and that flight attendants were trained to perform it. Table 10.3 lists some of the emotion work activities that flight attendants were trained to perform.

What is most interesting about emotion work is that when individuals act in accordance with feeling rules, they often begin to actually change their internal feelings to bring them in line with the way they are acting. Flight attendants, who appear to be cheerful even if they are actually fatigued and irritable, tend to actually feel more cheerful as they continue the act. Recall our story about Rob the kindergarten teacher at the beginning of the chapter. Although he didn't feel like it at all, feeling rules demanded that he act energetic and cheerful in front of his class. The longer he did this, the more cheerful he actually became. This may occur because of cognitive consistency effects (see Chapter 6), in which we are motivated to bring our attitudes in line with our behaviors. It can also occur because when we act cheerfully, people in turn interact with us in the way they would with a cheerful person. Their behavior reflects positively on us and can enhance our mood.

As individuals attempt to manage their emotions, they can engage in a number of strategies. One method used to evoke suitable emotions and suppress unsuitable ones is simply to adopt an appropriate physical posture. Slumping over in a chair is unlikely to evoke feeling of cheerfulness, but standing up straight and

TABLE 10.3 EMOTION WORK TASKS FOR FLIGHT ATTENDANTS

Emotion Work Category	Detailed Description
Relax and smile	Present a calm and cheerful demeanor. "It's incredible how much we have to smile, but there it is. We know that, but we're still doing it, and you would too."
Consider passengers as friends and family	Protect and comfort passengers as you would your own family.
Do not engage in ridicule	No matter how the passengers behave or what they demand, flight attendants may not ridicule them or their requests.
Never appear alarmed or frightened	No matter what the situation, flight attendants must suppress fear and panic. "Even though I am an honest person, I have learned not to allow my face to mirror my alarm or my fright."
Sincerity	Act from the heart. Manage others' emotions by remaining calm.
Do not blame passengers for anything	Even if it is their own fault.
Never display anger	Training programs emphasize strategies for reducing anger, including role taking as passengers.

Source: Adapted from Hochchilds, 1983.

© Mark Peterson/Corbis

A job requirement for flight attendants is to display upbeat, positive, energetic personas. Research has found that these portrayals of emotions can change the actual emotions the person is feeling under the surface of his or her performance.

walking briskly can. Other methods include shaping our facial expressions, breathing quickly or deeply, and imagining a situation that produces the required feeling. To recapture some energy and become more upbeat, flight attendants may take a deep breath to relax and reduce irritation, imagining how good it will feel to be home tonight. Actively pursuing these strategies allows us to gain some control over our feelings and to project images necessary for a particular social situation.

We should note, however, that the demands for emotion work are not evenly distributed. Those who have higher status in a social situation have more freedom in expressing their emotions. A boss or a parent can easily express anger or irritability in circumstances in which lower level employees and children are expected to follow the feeling rules more closely. In effect, a person's status allows different feeling rules. Emotion work can also become a commodity, in that people who are proficient at it can be rewarded. For example, employers can demand emotion work from their employees as a part of their job requirements. Such is the case too, in no uncertain terms, for flight attendants.

SUMMARY

Defining Emotions Affect is a general label that encompasses any kind of evaluation of an object. Affect varies in direction, intensity, and activity. Emotions are short-lived reactions to stimuli involving cognitive reactions, physiological reactions, and expressive gesturing, and is also a label we associate with a cluster of the first three elements. Sentiment refers to the social components of emotional situations or to longer term emotional states. Moods are general psychological conditions that can last for days or even weeks.

Classical Ideas About the Origins of Emotion Charles Darwin believed that some emotions and their expression were universal, not only among human beings, but also across species. These ideas were developed to support his theories of evolution. James and Lange developed a physiological notion of emotion in which physical changes occurred first and then were cognitively processed and interpreted as emotion. Freud focused on how emotions could be developed unconsciously and how past experience affected the experience of emotion.

Universal Emotions and Facial Expressions Facial expressions are often directly tied to certain emotional experiences. Ekman and colleagues have demonstrated that facial expressions of happiness, sadness, surprise, fear, anger, and disgust are all readily recognized by almost all cultures, thereby suggesting that they are universal, basic emotions. But, there are also significant differences in the experience and expression of emotions across cultures. Some exaggerate or suppress basic emotions via complex display rules that may make emotions harder or easier to recognize. Collectivist cultures process and display emotion in ways that protect and reinforcement social bonds. Individualist cultures display emotions in ways that broadcast individual states and draw attention to the individual as the key social unit.

Social Emotions Many emotions cannot be defined or experienced without reference to the social context in which they exist. Cognitive labeling theory includes situational elements as a key factor in experiencing emotion; studies of generic arousal and how it is experienced on an emotional level support the notion that emotional experience is dependent on the social context. Guilt and shame require the judgments of others. Jealousy requires another person to compare oneself to. Empathy requires another person who is experiencing an emotion, and love requires an object of affection. Social environments also define feeling rules, which dictate which emotions are appropriate for particular roles in that social context. Individuals often engage in emotion work to produce the necessary emotional portrayal. They can do this by acting or by actively manipulating their feelings.

LIST OF KEY TERMS AND CONCEPTS

affect (p. 251)
cognitive labeling theory (p. 261)
display rules (p. 257)
emotion work (p. 270)
emotional intelligence (p. 268)
emotions (p. 252)
feeling rules (p. 270)
mood (p. 252)
sentiment (p. 252)
social emotions (p. 264)

11

HELPING AND ALTRUISM

Introduction

Motivation to Help Others

Egoism and Cost-Reward Motivation

Altruism and Empathic Concern

Evolution and Helping

Characteristics of the Needy That Foster Helping

Acquaintanceship and Liking

Similarity

Deservingness

Normative Factors in Helping

Norms of Responsibility and Reciprocity

Personal Norms and Helping

Personal and Situational Factors in Helping

Modeling Effects

Gender Differences in Helping

Good and Bad Moods

Guilt and Helping

Bystander Intervention in Emergency Situations

The Decision to Intervene

The Bystander Effect

Costs and Emergency Intervention

Seeking and Receiving Help

Help and Obligation

Threats to Self-Esteem

Summary

List of Key Terms and Concepts

INTRODUCTION

Jennifer Beyer, age 22, was driving along Old River Road in Appleton, Wisconsin, on a cold day in February. She was on the way to visit a friend, but when a soaking wet child flagged her down, she pulled over immediately. Shivering and frightened, Jeff Lasrewski hurriedly explained that he and his friend, 9-year-old Colin Deeg, had been playing on the frozen Fox River when the ice gave way. Jeff had managed to climb back onto the ice and make it to shore, but Colin was still in the water.

Starting down the river bank, Jennifer saw Colin splashing in the frigid water. At the point where many others would have stopped due to the great personal risk, she went onto the frozen river to rescue him. Inching her way onto the ice, she tried to use her scarf to pull Colin out, but the ice cracked and she plunged into the water. At this point, Colin was still conscious but fading fast. In the meantime, Jeff reached another adult, Cyndy Graf, who quickly dialed 911 for help and then ran to the river.

Jennifer grabbed Colin to keep him from going under and tried to get him out of the water. This proved impossible, however. Colin soon passed out and the weight of his wet clothes made him too heavy to push onto land. Jennifer's limbs were numb with cold by the time police and fire teams reached the river with rescue equipment, but she had kept Colin's head above water and prevented him from drowning. Officers performed CPR on Colin, then rushed the pair to nearby St. Elizabeth's Hospital, where doctors used a bypass machine to warm Colin's blood, which had dropped in temperature to 78 degrees. Jennifer was treated for hypothermia. A week later, Colin was doing fine.

Jennifer Beyer's story is extraordinary for its valor and heroism, but everyday life is filled with smaller tales of people helping others in need. Individuals help others in many ways. They may give someone a ride, help change a flat tire, donate blood, make contributions to charity, return lost items to their owners, assist victims of accidents, and so on.

Of course, the mere fact that someone needs help does not mean that others will rush to give aid. Humans are capable of vastly different responses to persons in need. Although Jennifer Beyer went onto the ice to rescue Colin Deeg, many others would not have taken that risk. Some will not even stop to help a stranded motorist or make contributions to charitable causes. Thus, a challenge for social psychologists is to explain variations in helping behavior. When will people help others, when will they refuse to help, and why? Drawing on research and theory, this chapter addresses the following questions:

1. What motivates one person to help another? How do such factors as cost-reward and empathic concern for others affect helping and altruistic behavior?
2. How do characteristics of the person needing help influence help giving by others?
3. What impact do cultural factors such as norms and roles have on helping behavior?
4. How do characteristics of the situation influence helping? How do the emotional states of the helper (guilt, mood) affect helping behavior?
5. In emergency situations, which factors determine whether bystanders will intervene and offer help?
6. When help is given, which factors determine the recipient's reactions?

To address these questions, we must first define some important terms. Social psychologists use the term **prosocial behavior** to refer to a broad category of actions that are considered by society as being beneficial to others and as having positive social consequences. A wide variety of specific behaviors can qualify as prosocial, including donation to charity, intervention in emergencies, cooperation, sharing, volunteering, sacrifice, and the like. Prosocial behavior often contrasts with antisocial behavior, which refers to activity that is aggressive, violent, destructive, or criminal.

One important type of prosocial behavior is helping. By **helping,** we mean any action that has the consequence of providing some benefit to or improving the well-being of another person. This definition, which is intentionally broad, has certain properties. First, it does not place the intent of the helper at issue. There is no requirement, for instance, that the helper intend to benefit another person with his or her action.

Second, the definition does not explicitly address the issue of whether the helper can also benefit from giving help. There is no requirement, for instance, that to qualify as helping behavior, an action must benefit another person more than oneself. Under this definition, helping behavior may involve selfish or egoistic motives.

Another type of prosocial behavior is altruism. **Altruism** is helping that is intended to provide aid to someone else without expectation of any reward (other than the good feeling that may result). Under this definition, whether an act is altruistic or not does depend on the intentions of the helper—that is, the helper must intend to benefit the other (Piliavin & Charng, 1990; Schroeder, Penner, Dovidio, & Piliavin, 1995; Simmons, 1991).

MOTIVATION TO HELP OTHERS

What motivates one person to help another? There are at least three major views on this issue, which are rooted in different conceptions of human nature. The first view depicts humans as egoistic or selfish beings, concerned primarily with their own gratification. Although this view acknowledges that helping behavior occurs with considerable frequency, it treats helping as always originating from some ulterior, self-serving consideration. For instance, a woman might help another with a shopping chore to get admiration and approval from the other, to avoid feelings of guilt or shame, to obligate the other to her, or to bolster her own self-esteem. Social psychologists often refer to helping behavior motivated by a person's sense of self-gratification as **egoism.**

The second view depicts humans as rather more generous and unselfish beings, capable of real concern for the welfare of others. In this view, at least some helping behaviors originate primarily because humans are able and willing to extend themselves unselfishly to benefit other persons in need. For instance, a bystander may rush to rescue an accident victim in order to relieve the victim's pain and anguish; help like this can be based on altruistic concerns, without any selfish motives.

The third view, driven by evolutionary psychology, sees helping behavior as an evolved trait that helps individuals ensure they will pass on their genes to the next generation. If, for example, a mother sacrifices herself to save her daughter, the mother's genes can be passed on through the daughter's offspring.

In this section, we look at these three views in more detail. First, we consider the view that humans are primarily egoistic, and we look at the role of reinforcement (rewards and costs) in helping. Then we consider the view that humans can behave altruistically, and we look at the role of empathic concern for others in helping. Finally, we explore the evolutionary viewpoint and consider how helping behavior can enhance genetic fitness.

Egoism and Cost-Reward Motivation

Although some helping may be altruistic, much of it is egoistic in nature. Some reinforcement theorists maintain that human nature is basically selfish and that individuals are motivated to maximize their own net gains (Gelfand & Hartmann, 1982). There is little doubt that considerations of reward and cost influence decisions to give or withhold help. Some theorists have suggested that individuals will generally not give help unless they perceive that the rewards to themselves for helping will outweigh the costs (Lynch & Cohen, 1978; Piliavin, Dovidio, Gaertner, & Clark, 1981).

Rewards for Helping The rewards that motivate potential helpers are many and varied. They may include such things as thanks from the victim, admiration and approval from others, financial rewards and prizes, and recognition for competence. In general, it is clear that individuals do receive rewards such as status enhancement (Bienenstock & Bianchi, 2004) through helping others, and if they anticipate these rewards, they will help more (Kerber, 1984). The form of help that a potential helper will provide may depend on the specific rewards that he or she seeks, and these in turn may depend on his or her own needs. This is illustrated by a study in which undergraduates were invited to volunteer their assistance in projects such as studying unusual states of consciousness (ESP and hypnosis) or

© B. P/Taxi/Getty Images

Who will offer help to this homeless man? Personal attitudes often drive this kind of altruistic behavior, but situational cues that remind us of the social responsibility norm are also important.

counseling troubled high school students (Gergen, Gergen, & Meter, 1972). Those who enjoyed novelty volunteered more frequently to help with the project on unusual states of consciousness. Those who liked close social relationships volunteered more frequently to help troubled high school students. Thus, the rewards people sought through helping matched their own personal needs.

Costs of Helping and Not Helping Every helping act imposes costs on the helper. These costs may include exposure to danger, loss of time, financial costs, expenditure of effort, or exposure to repulsive people and objects. In general, the greater these costs, the less likely persons are to help (Kerber, 1984; Shotland & Stebbins, 1983). There are also some costs to potential helpers for not helping, such as public disapproval by others, embarrassment and loss of face, and condemnation by the victim. Students may donate blood, for example, to avoid condemnation for failing to help their fraternity meet its quota. In certain places, for example, bystanders who fail to help victims not only incur disapproval but may even face criminal prosecution. Some helping is simply compliance with others' demands to avoid socially imposed costs.

Altruism and Empathic Concern

People often react to the distress of others on an emotional level and offer help in response. The term **empathy** refers to the vicarious experience of an emotion that is congruent with or possibly identical to the emotion that another person is experiencing (Barnett, 1987; Eisenberg & Miller, 1987).

Empathy is heightened by various situational factors. Believing that a victim is similar to oneself—whether in race, sex, personality, or attitudes—promotes empathy. In one study, for example, students observed another student receiving painful shocks when he lost at roulette. Students who had been told the victim had a personality similar to their own became more physically aroused than those who thought the victim had a personality different from their own. Those who were more aroused also donated more of their own money to help the victim (Krebs, 1975). Empathy can also be heightened by situational norms or role expectations that expressly induce role-taking (Shott, 1979). When observers actively try to imagine how another person feels rather than simply observing how that person reacts, both empathy and helping behavior increase (Harvey, Yarkin, Lightner, & Tolin, 1980; Toi & Batson, 1982).

There is considerable evidence that feelings of empathy for a person in need will lead to helping behavior. Numerous studies have shown that greater empathy leads to greater help giving (Batson, Duncan, Ackerman, Buckley, & Birch, 1981; Dovidio, Allen, & Schroeder, 1990; Eisenberg & Miller, 1987; Fultz, Batson, Fortenbach, McCarthy, & Varney, 1986).

Empathy-Altruism Model The issue of motivational links between empathy and helping has been addressed by the **empathy-altruism model,** developed by Batson and his coworkers (Batson, 1987, 1991; Batson & Oleson, 1991). This model proposes that adults can experience two distinct states of emotional arousal while witnessing another's suffering: distress and empathy. Distress involves unpleasant emotions such as shock, alarm, worry, and upset at seeing another person suffer. Empathy entails such emotions as compassion, concern, warmth, and tenderness toward the other (Batson & Coke, 1981). These states of emotional arousal give rise to different motivations.

More specifically, suppose that a bystander witnesses another person suffering. If the bystander experiences distress at seeing the other suffer, he or she may be motivated to reduce this unpleasant emotion. By helping the person in need, the bystander will not only assuage the victim's suffering but will also reduce his or her own distress. Notice that although this help benefits the victim, the bystander's fundamental goal in helping is to reduce his or her own distress (that is, an egoistic motive, not an altruistic one). This contrasts with the situation in which a bystander experiences empathy when witnessing the suffering of another. Empathy entails other-oriented feelings such as compassion, tenderness, sympathy, and the like. Feelings of this type may cause the bystander to help the victim, but the point to notice is that this help is motivated fundamentally by a desire to reduce the other's distress (that is, an altruistic motive, not an egoistic one). Thus, the arousal states of distress and empathy give rise to very different motives, but either one of them can lead to helping.

The empathy-altruism model has received support from various experiments. Typically, the participants in these studies witness a person in distress and must decide whether to offer help or not. The independent variables in these studies are the level of empathy and the ease of escape from the situation. For instance, in one experiment (Batson, O'Quin, Fultz, Vanderplas, & Isen, 1983), participants served as observers who watched another participant (in the role of a worker) over closed-circuit television. The worker performed a sequence of 10 digit-recall trials and received electric shocks at random intervals; when these shocks were applied, the worker reacted with expressions of pain.

Some participants were instructed that they needed to watch only the first two trials and were then free to leave if they wanted (easy escape condition). Other participants were instructed to stay and watch all 10 trials (difficult escape condition). After the second trial, participants completed a brief questionnaire that contained adjective scales measuring their personal feelings regarding the situation of the worker. Some of these adjectives reflected personal distress (alarmed, grieved, upset, worried, disturbed), whereas other adjectives reflected empathy (sympathetic, moved, compassionate). Later, each participant's average score on the distress items was subtracted from his or her average score on the empathy items to create a measure of the participant's predominant emotional response to the situation (ranging from empathy to distress).

After completing the questionnaire, the participants learned that the worker had had a traumatic experience with shocks as a child, making them extremely difficult to bear. Naturally, this news was disturbing, and the experimenter gave the participants an opportunity to help the worker by taking the remaining shock trials in her place.

The results showed that the participant's willingness to take the shocks depended on the predominant emotional response (empathy or distress) and on the ease of escape (easy or difficult). When empathy was high, the frequency of helping behavior was high irrespective of whether escape was easy or difficult. However, when distress was high, the frequency of helping behavior dropped off substantially when escape was easy; participants left the situation rather than taking the shocks themselves. This pattern of behavior is consistent with the empathy-altruism model.

Evolution and Helping

As discussed in Chapter 1, evolutionary theories rely on the Darwinian principle of survival of the fittest. The basic notion driving the theory is that any genetically determined physical attribute or trait that helps an individual survive will be passed on to the next generation. Because the individual having the attribute is more likely to survive than others lacking it, he or she will tend to have more offspring than those who do not possess the trait, and these offspring will tend to carry the attribute. The offspring themselves will have a greater tendency to survive, pass on the trait to their own offspring, and so on. Eventually, individuals with the attribute will become more numerous than those without. When extended to include behavior (Pinker, 1997), this evolutionary principle is easy to use to explain selfish or aggressive behavior. For instance, in many species, the strongest and most aggressive animals occupy the top positions in the group's social hierarchy. To fight for position in this hierarchy is adaptive in a Darwinian sense, for it gives the animal control over food, shelter, and other resources needed to survive, as well as access to mating partners. Another form of aggressive behavior, the killing of the young (infanticide), is common in some animal species. When a male langur monkey takes control of another male's harem, for instance, he usually proceeds to murder all the infants sired by his competitor (Hrdy, 1977, 1982), thereby increasing the relative number of his own offspring.

However, animal behavior is not limited to the aggressive elimination of competitors. In fact, helping behaviors and even altruistic, self-sacrificing behaviors are common. Ground squirrels, for instance, frequently sound alarm calls when a predator approaches. These calls warn other squirrels of the threat, but they also draw the attention of the predator to the individual sounding the alarm—thereby increasing the chances of that individual being killed (Sherman, 1980). Many other animal species exhibit similar warning behaviors. Other animals sacrifice themselves to predators to protect the larger group. When a termite hive is attacked by intruders, soldier termites will place themselves in front of the other termites. Many soldier termites sacrifice themselves so that the other termites may live and the nest survive (Wilson, 1971).

At first, these patterns of self-sacrificing behavior seem to run counter to evolutionary theory. The result of altruism among animals is that those who are most helpful will be the least likely to survive. This means they will be less likely to have offspring and may not have any at all. How then, could the altruistic tendency persist generations later? The same question can be posed, or course, with respect to humans.

Genetic Fitness and Helping Evolutionary psychology and a related theoretical perspective called sociobiology (Archer, 1991; Buss, 1999; Ketelaar & Ellis, 2000; Wilson, 1975, 1978) have constructed a response to the problem of altruism and have assembled evidence that supports their view (Buss & Kenrick, 1998; Krebs & Miller, 1985). To understand how helping can make sense in an evolutionary context, it is important to appreciate that the perpetuation of genes is important, rather than the perpetuation of physical attributes (Dawkins, 1976). In this view, the "fittest" animal is the one that passes on its genes to subsequent generations. This can happen either by the animal itself producing offspring or by the animal's close relatives, such as brothers, sisters, and cousins (who share many of its genes), producing offspring.

It is true that altruistic behavior will not have survival value for an individual. But altruistic acts can increase the survival of one's genes if those altruistic acts are directed toward others who share the same genes—a phenomenon called **kin selection** (Hamilton, 1964; Meyer, 2000). If an individual helps his or her close relatives, that act increases the chances that those relatives will survive and eventually have offspring. As the relatives share many genes with the altruistic individual, the reproduction of these relatives passes many of the altruist's own genes to the next generation (Krebs & Miller, 1985; Ridley & Dawkins, 1981). Consider a mother bird that sacrifices herself to save the lives of her eight babies. Each of the babies carries half of the genes of the mother; thus, between them, they have four times as many of the mother's genes as she does herself. Thus, although individual survival is compromised by altruistic behavior, genetic survival may be enhanced.

Furthermore, some sociobiologists have argued that altruistic behavior is perpetuated because of reciprocation. If all the animals in a group engage in helping behavior, they will all be better off in the long run (Trivers, 1971). If, for example, the animals all take turns playing the role of sentry and warning the group of approaching predators, many more members of that group will survive and reproduce than if none of them had warned the group.

Evidence About the Evolutionary Perspective Although it is very difficult to separate the effects of evolution on behavior from those of socialization, the evolutionary perspective implies some very interesting propositions. For one, animals should be most altruistic toward those that most closely resemble them genetically—that is, they should help immediate family members more than distant cousins, and distant cousins more than outsiders or strangers (Burnstein, Crandall, & Kitayama, 1994; Rushton, Russell, & Wells, 1984). Second, parents will tend to behave altruistically toward healthy offspring (who are likely to survive and pass on their genes) but less altruistically toward sick or unhealthy offspring (who are likely to die before reproducing; Dovidio, Piliavin, Gaertner, Schroeder, & Clark, 1991). Third, helping behavior should only favor those genetically related to the helper if they can still reproduce. Thus, helping behavior should be targeted more toward young women than to older women who are past the age of menopause (Kruger, 2001). Fourth, mothers will be more altruistic toward their offspring than fathers. Male animals have the biological capacity to sire many children, whereas females in many species can have only a small number of offspring. Thus, males can perpetuate their genes without much altruism toward any one child, whereas females will try to ensure their genetic survival by helping each of their smaller number of offspring to a greater extent. Finally, if prosocial or altruistic behavior is part of our genetic makeup, then the tendency to behave altruistically ought to be fairly stable over a person's lifetime and fairly impervious to environmental stimulus (Eisenberg et al., 2002).

Generally speaking, these evolutionary propositions have found support in the studies that have been conducted. However, there have also been exceptions, and there are alternative explanations for these patterns that do not depend on evolutionary mechanisms (Buss & Kenrick, 1998; Caporeal, 2001; Dovidio et al., 1991). For example, one study (Sime, 1983) examined the behavior of people in a fire emergency and found that they were much more likely to endanger themselves by searching for family members than by searching for friends. Rather than attributing this behavior to genetic kin selection, however, we may just as likely assume that people would sacrifice more to save someone they love than someone who is simply an acquaintance.

Although interesting, the sociobiological perspective is controversial, especially as applied to humans. For example, critics have questioned whether altruism is genetically transmitted (Buck & Ginsburg, 1991; Kitcher, 1985). If animals and humans behaved in accord with the sociobiological model, they would help only close relatives and rarely or never help those who are genetically unrelated. Yet we know that humans often help others who are—unrelated—even total strangers. Some critics argue that to explain altruism of this type, it is necessary to rely on cultural constructs (for instance, norms that indicate when help is appropriate, institutions such as religion that define certain unrelated others as appropriate recipients of help). At best, evolution is therefore an incomplete explanation for altruism.

CHARACTERISTICS OF THE NEEDY THAT FOSTER HELPING

When in need, some people have a much better chance of receiving help than others. Our willingness to help needy persons depends on various factors. Important among these factors are whether we know them and like them, whether they are similar to or different from us, and whether we consider them truly deserving of help.

Acquaintanceship and Liking

We are especially inclined to help people whom we know and to whom we feel close. Studies of reactions following natural disasters, for example, indicate that whereas people generally become very helpful toward others, they tend to give aid first to needy family members, then to friends and neighbors, and lastly to strangers (Dynes & Quarantelli, 1980; Form & Nosow, 1958). Even a brief acquaintanceship is sufficient to make us more likely to help someone (Pearce, 1980). Relationships like these increase helping because they involve relatively stronger normative obligations, more intense emotion and empathy, and greater costs if we fail to help.

We are more likely to help someone we like than to help someone we do not like. This effect occurs whether our positive feelings about the other are based on his or her physical appearance, personal characteristics, or friendly behavior (Kelley & Byrne, 1976; Mallozzi, McDermott, & Kayson, 1990). Moreover, we are more likely to help someone who likes us than to help someone who does not.

Similarity

In general, we are more likely to help others who are similar to ourselves than to help others who are dissimilar (Dovidio, 1984). That is, we are more likely to help those who resemble us in race, attitudes, political ideologies, and even in mode of dress. For instance, with respect to race, several studies have reported that in situations where refusing to help may be justified on nonracist grounds, Whites are more likely to help other Whites than to help Blacks (Benson, Karabenick, & Lerner, 1973; Dovidio & Gaertner, 1981).

A series of field studies demonstrated that similarity of opinions and political ideologies increases helping (Hornstein, 1978). In these studies, New York pedestrians came across "lost" wallets or letters that had been planted by researchers in conspicuous places. These objects contained information indicating the original owner's views on the Arab-Israeli conflict, on worthy or unpopular organizations, or on trivial opinion items. The owner's views on these topics either resembled or differed from the views known to characterize the neighborhoods in which the objects were dropped. Persons finding the wallets or letters took steps to return them to their owner much more frequently when the owner's views were similar to their own.

Even similarity in something as superficial as mode of dress can increase helping. In one study, for example, a confederate of the researchers dressed either in jeans and a work shirt or in conventional sports clothes and asked college students for a coin to make a telephone call. Students complied with the requests more often when the confederate's mode of dress was similar to their own than when it was different (Emswiller, Deaux, & Willis, 1971).

Deservingness

Suppose you received a call asking you to help elderly people who had just suffered a sharp reduction in income after losing their jobs. Would it matter whether they lost their jobs because they were caught stealing and lying or because their work program was being phased out? A study of Wisconsin homemakers who received such a call showed that respondents were more likely to help if the elderly people had become dependent because their program was cut than because they had been caught stealing (Schwartz & Fleishman, 1978).

What matters in this situation is the potential helpers' causal attribution regarding the origin of need. Potential helpers respond more when the needy person's dependency is caused by circumstances beyond his or her control. Such people are true "innocent victims" who deserve help.

In contrast, needs caused by a person's own actions, misdeeds, or failings elicit little help (Bryan & Davenport, 1968; Frey & Gaertner, 1986). For instance, one study found that students were less sympathetic and less likely to help a person who developed AIDS through promiscuous sexual contact than through a blood transfusion (Weiner, Perry, & Magnusson, 1988). Need viewed as stemming from illegitimate sources undermines helping by inhibiting empathic concern, blocking our sense of normative obligation, and increasing the possibility of condemnation rather than social approval for helping (Brickman et al., 1982).

Even in emergencies, potential helpers are influenced by whether they consider a victim deserving. Consider responses to an emergency staged by experimenters in the New York subway (Piliavin, Rodin, & Piliavin, 1969). Shortly after the subway train left the station, a young man (a confederate) collapsed to the

BOX 11.1 Research Update: Group Boundaries and Emergency Helping

Similarities along the lines of important social statuses like race can be important determinants of helping behavior. Likewise, any kind of strongly held group identity can affect our willingness to help others. When individuals are lined up on one side or another of a long-standing feud (for example, the deadly feud between the Hatfield and McCoy families, see Chapter 16), it is easy to see why members of the groups might be more likely to help those in their own group than those in the out-group. But what about other kinds of group identities that are typically considered less important—for example, an identity as a fan of a particular sport teams, or even less salient, an identity as a fan of a particular sport, like baseball? Can similarity based on these kinds of identities prevent or encourage helping behavior?

A group of researchers set out to explore this question by using a choreographed accident in which a confederate fell down and feigned a painful injury in the presence of a subject. Prior to the accident, the subjects had taken a survey questionnaire about their favorite soccer team, thereby priming their identity as fans of their favorite team. Then, the subjects were directed to walk to a different location for the second part of the soccer study in which they were supposed to watch a video about soccer teams. Along the way, they passed the confederate, who fell and pretended to be hurt. The outcome of interest was if the subjects stopped to help and, if so, how much help was offered.

The experimental manipulation in the experiment was simple: The confederate either wore a shirt identified with the subjects' favorite team, a shirt identified with the main rival of the subjects' favorite team, or wore a neutral shirt that did not identify with any team at all. The results were surprisingly stark. Those confederates wearing the favorite team's shirt received help from the subjects over 90 percent of the time, while those wearing a plain shirt or the rival team's shirt received help less than one-third of the time. There were no differences between the plain-shirt situation and the rival-team situation, suggesting that there was a positive inclination to help those the subjects identified with rather than a disinclination to help their rivals.

In the second experiment, these same researchers attempted to examine a more diffuse identity: that of soccer fans in general, rather than those of a particular team. This experiment followed a similar procedure as the first, except that instead of priming the subjects' identity about their favorite team, the researchers primed the subjects to think about their identity as soccer fans in general. They did this by telling the subjects that there are a few troublemakers among soccer fans who got into drunken brawls and thereby gave soccer fans a bad name. However, there are also many positive aspects about being soccer fans, and the purpose of the research was to examine these positive aspects. After hearing this information, the subjects filled out a survey about being soccer fans, how they had become soccer fans, what they shared in common with other soccer fans, and so forth.

After the survey, the subjects were sent to the second location, as in the first experiment, and encountered the false accident, again with the confederate wearing one of three shirts (their favorite team, the rival, or the neutral shirt). Again, the primed identity had a strong effect on helping, but the pattern was different. If the confederate was wearing either soccer shirt, help was received about 75 percent of the time. If the confederate wore the generic shirt, help was received less than 25 percent of the time.

Source: Adapted from Levine, Prosser, Evans, and Reicher, 2005.

floor of the car and lay staring at the ceiling during the 7½ minute trip to the next station. In one experimental condition, the man carried a cane and appeared crippled. In another condition, he carried a liquor bottle and reeked of whiskey. Bystanders helped the seemingly crippled victim quickly, usually leaping to his aid within seconds, but they often left the apparent drunk lying on the floor for several minutes. Much of this difference in response probably reflects the fact that many people—rightly or wrongly—blame drunks for their own plight.

NORMATIVE FACTORS IN HELPING

Would you intervene in a heated argument between a man and a woman you believe are married? In one experiment (Shotland & Straw, 1976), participants unexpectedly witnessed a realistic fight between a man and a woman in an elevator. The man attacked the woman, shaking her violently, while she struggled and resisted. In one treatment, the man and woman were depicted as strangers; the woman screamed, "Get away from me! I don't know you!" In the other treatment, they were depicted as married; the woman screamed, "Get away from me! I don't know why I ever married you!" This simple variation greatly affected the participants' propensity to help; whereas 65 percent of the subjects intervened in the stranger fight, less than 20 percent intervened in the married fight.

This difference may have been due, in part, to the participants' perception of a greater likelihood of injury to the woman in the stranger fight than in the married fight. However, it may also have been due in part to normative expectations. The participants who witnessed the married fight said they hesitated to take any action because they were not sure their help was wanted. Almost all the participants who did not intervene said they felt the fight was "none of my business."

Clearly, "wife" and "husband" are social roles, and the relations between wives and husbands (and outsiders) are regulated by some widely understood norms. One of these is that except in the case of physical abuse, outsiders should basically mind their own business and let married couples resolve disputes as they will. When the woman in the elevator identified herself as the man's wife, this norm suddenly became relevant and changed the meaning of intervention. To intervene in the fight would be an intrusion on the marital relation and might invite reprisals from the husband, the wife, or both. In fact, participants who thought the attacker was the woman's husband believed that he was more likely to attack them if they intervened than did participants who believed that the attacker was a stranger.

Norms of Responsibility and Reciprocity

Cultural norms mandate helping as appropriate under some conditions, and they define it as inappropriate under others. When mandated as appropriate, helping becomes an approved behavior, supported by social sanctions. Here we discuss the responsibility norm and the reciprocity norm, broad social norms that indicate when helping is appropriate.

Social Responsibility Norm The **social responsibility norm** is a general norm stating that individuals should help others who are dependent on them. People often mention their sense of what they "ought to do"—their internalized standards—when asked why they offer to help (Berkowitz, 1972). For example, Simmons (1991) reports the words of a bone marrow donor prior to giving: "This is a life and death situation and you must do anything you can to help that person, whether it is family, friends, or [someone] unknown" (p. 14). The word "must" in this statement suggests that a norm is operative.

Applicable in many situations, the social responsibility norm is readily activated. Some research suggests that simply informing individuals that another person—even a stranger—is dependent on them is enough to elicit help (Berkowitz, Klanderman, & Harris, 1964). Recognize, however, that there are stronger and weaker versions of the social responsibility norm. Whereas the norm that we must help dependent kin or needy friends is widely held, the belief that we must help needy strangers or unknown persons is not so universally accepted. Although the awareness of a stranger's dependency will sometimes elicit help, it does not always do so. Speeding passersby, for example, frequently disregard stranded motorists they notice on the roadside. Bystanders watch, apparently fascinated but immobile, during rapes and other assaults. Thousands of people reject charity appeals every day.

Some theorists have suggested that the social responsibility norm effectively motivates helping only when people are expressly reminded of it. In a test of this hypothesis (Darley & Batson, 1973), theological students were asked to prepare a talk on the parable of the Good Samaritan. On the way to record their talk, the students passed a man slumped in a doorway. Although these students were presumably thinking about the virtues of altruism, they helped the stranger only slightly more than a similar group of students who had prepared a talk on an unrelated topic (careers). A second variable—being in a hurry—had a much stronger impact on the amount of help offered. Students who were in a hurry offered much less help than those who were not. These findings suggest that the social responsibility norm is a fairly weak source of motivation to help and is easily negated by the costs of helping.

The Norm of Reciprocity Another cultural standard, the **norm of reciprocity,** states that people should (1) help those who have helped them and (2) not help those who have denied them help for no legitimate reason (Schroeder et al., 1995; Trivers, 1983). This norm applies to a person who has previously received some benefit from another. Small kindnesses that create the conditions for reciprocity are a common feature of family, friendship, and work relationships. The reciprocity norm is found in different cultures around the world (Gergen, Ellsworth, Maslach, & Siepel, 1975).

People report that the reciprocity norm influences their behavior, and behavioral studies have demonstrated that people are inclined to help those who helped them earlier (Bar-Tal, 1976; Wilke & Lanzetta, 1982). Reciprocity is especially likely when the person expects to see the helper again (Carnevale, Pruitt, & Carrington, 1982). People try to match the amount of help they give to the quantity they received earlier, and they are less likely to ask for help when they believe they will not be able to repay the aid in some form (Fisher, Nadler, & Whitcher-Alagna, 1982; Nadler, Mayseless, Peri, & Chemerinski, 1985). By matching benefits, people maintain equity in their relationships and avoid becoming overly indebted to others.

People do not reciprocate every benefit they receive, however. Whether we feel obligated to reciprocate depends in part on the intentions we attribute to the person who helped us. We feel more obligated to reciprocate if we perceive that the original help was given voluntarily rather than coerced and that it was chosen consciously rather than accidental (Gergen et al., 1975; Greenberg & Frisch, 1972).

Personal Norms and Helping

Although broad norms like social responsibility and reciprocity undoubtedly affect helping behavior, they are, by themselves, inadequate bases from which to predict the occurrence of helping behavior with precision. There are several reasons for this. First, given the wide variety of contingencies that people encounter, these norms are simply too general to dictate our behavior with any precision in all cases. Second, these norms are not accepted to the same degree by everyone in society; some persons internalize them to a greater extent than others. Third, the social norms that apply to any given situation occasionally conflict with one another; the social responsibility norm may obligate us to help an abused wife, for example, but the widely accepted norm against meddling in others' marriages tells us not to intervene.

In response to these criticisms, a different type of normative theory has been developed by social psychologists (Schwartz & Howard, 1981, 1984). This theory explains not only the conditions under which norms are likely to motivate helping but also individual differences in helping in particular situations. Instead of dealing with broad social norms, this theory focuses on personal norms—feelings of moral obligation to perform specific actions that stem from an individual's internalized system of values.

To investigate whether feelings of moral obligation motivate helping, investigators use a survey questionnaire to measure respondents' personal norms.

FIGURE 11.1 VOLUNTEERING AS A FUNCTION OF PERSONAL NORMS AND RESPONSIBILITY DENIAL

In a survey on social issues, university students indicated how much of a moral obligation they would feel (personal norm) to read texts to blind children. Three months later, the director of the Institute for the Blind wrote to the students, requesting that they volunteer time for just this purpose. Students who rarely denied responsibility for the consequences of their acts (low responsibility denial) behaved in a manner consistent with their personal norms: The stronger their moral obligation, the more they volunteered. Students moderate in responsibility denial showed weak consistency between personal norms and behavior. Students high in responsibility denial showed no consistency between personal norms and behavior. These findings indicate that the impact of our personal norms on helping behavior depends on whether we accept or deny our own responsibility.

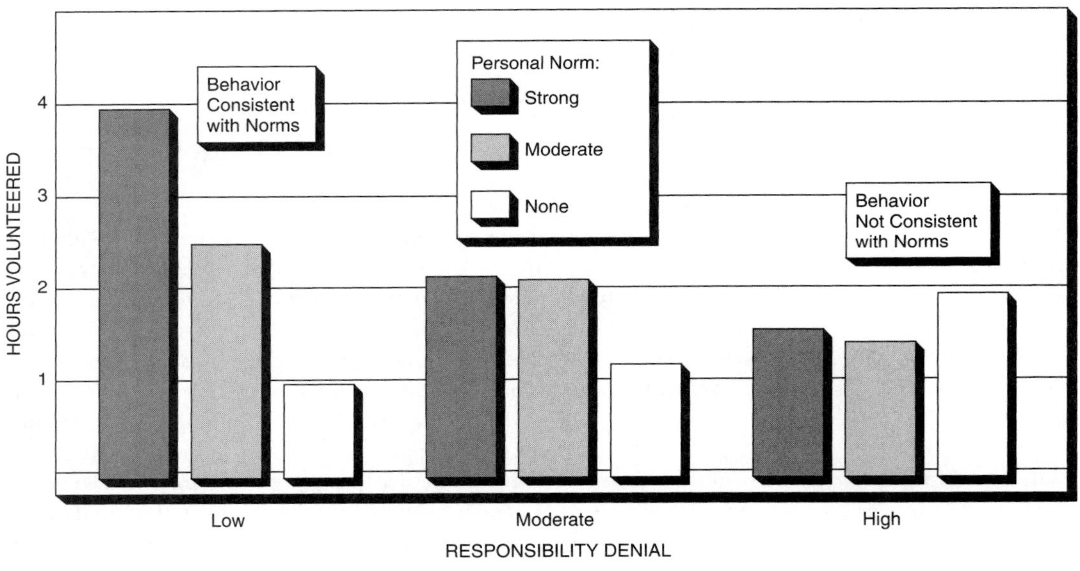

Source: Adapted from "Explanations of the Moderating Effect of Responsibility Denial on the Personal Norm-Behavior Relationship" (1980) by S. H. Schwartz and J. A. Howard, *Social Psychology Quarterly, 43*(4), 445. Used with permission from the American Sociological Association.

For example, a survey on medical transplants might ask, "If a stranger needed a bone marrow transplant and you were a suitable donor, would you feel a moral obligation to donate bone marrow?" This survey would then be followed by an apparently unrelated encounter with a representative of an organization who would ask these individuals for help. In various studies, individuals' personal norms have predicted differences in their willingness to donate bone marrow or blood, to tutor blind children, to work for increased welfare payments for the needy (Schwartz & Howard, 1982), and to participate in community recycling programs (Hopper & Nielsen, 1991).

Helping is most likely to occur when conditions simultaneously foster the activation of personal norms and suppress any defenses that might neutralize personal norms. For example, in one study (Schwartz & Howard, 1980), personal norms predicted quite accurately how much time college students would volunteer to tutor blind children. Among students who accepted responsibility for their actions, the hours volunteered depended on the strength of the students' personal norms. Among students who tended to deny such responsibility, however, there was no relationship between personal norms and volunteering. The results of this study are displayed in Figure 11.1. Other

Rescuers carry an injured climber down a mountain to safety. Many people satisfy their own needs, such as the need for approval or for excitement, by helping others.

studies (Schwartz & Howard, 1984) have revealed that the relationship between personal norms and helping depends on conditions that support norm activation (for example, the tendency to notice others' suffering) and suppress defensive denial of norms (for example, the focus of responsibility on the potential helper).

PERSONAL AND SITUATIONAL FACTORS IN HELPING

Helping behavior is influenced not only by normative factors, but also by situational and personal factors. In this section, we consider how an important situational factor—the presence of models—influences helping. We also consider the impact of mood states in helping. We focus on how various transient emotional states—whether somebody is in a good mood or a bad mood or whether he or she feels guilty about something—affect the tendency to give or withhold help.

Modeling Effects

An important factor that promotes helping is the presence of someone else who is helping—that is, a behavioral model. For instance, one study found that the presence of a model increased adults' willingness to donate blood (Rushton & Campbell, 1977). Other studies have shown that the presence of a model increases motorists' tendency to help a stranger change a flat tire and shoppers' tendency to donate coins to the Salvation Army kettle during the Christmas season (Bryan & Test, 1967).

Not surprisingly, modeling effects hold for children as well as for adults. Studies have consistently shown that children behave more generously toward others when they are exposed to generous models than when they are exposed to selfish models (Lipscomb, Larrieu, McAllister, & Bregman, 1982). In one study, for instance, children exposed to models who behaved inconsistently (once generously, once selfishly) donated less than children exposed to a consistently generous model but more than children exposed to a consistently selfish model (Lipscomb, McAllister, & Bregman, 1985).

The presence of a behavioral model tends to increase helping for several reasons. First, a model demonstrates what kinds of actions are possible or effective in the situation. Others who previously did not know how to help can emulate the model. Second, a helping model conveys the message that to offer help is appropriate in the particular situation. A model may increase the salience of the social responsibility norm; once aware of this norm, others may decide to help. Third, a model provides information about the costs and risks involved in helping—a consideration that is especially important in situations involving danger. By offering help under conditions of danger or potential damage to self, models demonstrate to others that the risks incurred are tolerable or justified.

Gender Differences in Helping

It is not possible to maintain that men are always more helpful than women or vice versa, because so much depends on the type of situation and the relevant role demands. Research findings do indicate that men are more likely than women to intervene and offer assistance in emergency situations that entail danger (Eagly & Crowley, 1986). For instance, in one study the investigators interviewed people who had been publicly recognized as heroes by the state of California (that is, persons who had intervened to protect someone during a dangerous criminal act, such as a mugging or bank robbery). It turned out that all but one of these persons was a man (Huston, Ruggiero, Connor, & Geis, 1981). Acting heroically and dealing with risk and danger is often considered part of the traditional male role. Then, too, men may perceive the costs of dealing with danger as lower than women do, because men are physically stronger and more likely to have relevant abilities, such as self-defense training (Huston et al., 1981).

Studies of helping show that people tend most to offer the types of help that are consistent with their gender role expectations (Eagly & Crowley, 1986; Piliavin & Unger, 1985). Thus, if men are more likely than women to help in heroic or chivalrous situations entailing risk, women are more likely than men to help in situations requiring nurturance, caretaking, and emotional support. For instance, women tend more than men to care for children and aging parents on a day-to-day basis, an important help-giving function (Brody, 1990). Women are also more likely than men to provide their friends with personal favors, emotional support, and informational counseling about personal or psychological problems (Eisenberg & Fabes, 1991; Otten, Penner, & Waugh, 1988).

Good and Bad Moods

A mood is a transitory feeling, such as being happy and elated or being frustrated and depressed. A person's mood state can influence whether he or she gives help to another. In this section, we examine the effects on helping of both good mood and bad mood.

Good Mood and Helping There is substantial evidence that when individuals are in a good mood, they are more likely to help others than when they are in a neutral mood (Salovey, Mayer, & Rosenhan, 1991). Good moods promote both spontaneous helping and compliance with requests for help. But moods can change quickly, and as a person's good mood fades, helping also quickly drops back to normal (Isen, Clark, & Schwartz, 1976).

Almost every experience that puts us in a good mood increases the chance that we will help others. For instance, suburban schoolteachers who learned they scored well on a battery of tests donated more to a school library fund than teachers who received no feedback on their performance (Isen, 1970). People were more likely to help a stranded caller who had dialed the wrong number from a pay phone if they had

TABLE 11.1 THE EFFECTS OF GOOD MOOD ON HELPING

	PERCENTAGE WHO HELPED		
	Picking up Dropped Papers	Mailing a Lost Letter	Making a Phone Call for a Stranger
Neutral mood	4%	10%	12%
Good mood	88%	88%	83%

Sources: Adapted from Isen, Clark, and Schwartz, 1976; Levin and Isen, 1975; Isen and Levin, 1972.

just received a small gift than if they had received no gift (Isen & Levin, 1972). Other mood-enhancing experiences that have been shown to increase helping include recalling happy experiences, reading statements describing pleasant feelings, hearing good news on the radio, listening to soothing music, and enjoying good weather. Some powerful effects of good mood on helping are shown in Table 11.1.

There are several reasons why being in a good mood increases our propensity to help others (Carlson, Charlin, & Miller, 1988). First, people in a good mood are less preoccupied with themselves and less concerned with their own problems. This allows them to focus more attention on the needs and problems of others, which—through empathy—often leads to helping. Second, people in a good mood often feel relatively fortunate compared to others who are deprived. They recognize that their good fortune is out of balance with others' needs; to restore a more just balance, they use their resources to help others (Rosenhan, Salovey, & Hargis, 1981). Third, people in a good mood tend to see the world in a positive light, and they want to maintain the warm glow of happiness. Thus, if by offering help that benefits others, they can maintain and even increase their own positive feelings, they will do so. On the other hand, people in a good mood may tend to avoid forms of helping that involve unpleasant or embarrassing activities, for these may interrupt or destroy the good mood (Cunningham, Steinberg, & Grev, 1980; Isen & Simmons, 1978).

Bad Mood and Helping Bad mood—feeling sad or depressed—can have rather complex effects on helping. The results of studies show that under some conditions, bad mood (when contrasted with neutral mood) inhibits helping; under other conditions, however, it promotes helping (Rosenhan et al., 1981; Carlson & Miller, 1987).

Bad mood can suppress helping for several reasons. First, it has an impact on the salience of others' needs. People in a bad mood are often concerned about their own problems and therefore less likely to notice others' needs than are people in a neutral mood. When others' needs do not grab the attention of a potential helper because of bad mood, help is less likely to be given (Aderman & Berkowitz, 1983; Rogers, Miller, Mayer, & Duvall, 1982). Second, people in a bad mood often see themselves as less fortunate than others. Feeling relatively impoverished and underbenefited, they may resist using their own resources to help others, lest they become even more disadvantaged (Rosenhan et al., 1981). The net effect is that the feelings of relative impoverishment associated with bad mood inhibit helping, especially when the costs of helping are large.

Despite these effects, bad mood can sometimes increase helping. One explanation how this can come about is provided by the negative-state relief hypothesis (Cialdini, Kendrick, & Baumann, 1982; Cialdini et al., 1987). This hypothesis starts with the following assumptions: (1) Persons experiencing unpleasant feelings, such as sadness, will be motivated to reduce them, and (2) many persons have learned from childhood that helping others will improve their own mood, often through the receipt of thanks or praise. The hypothesis predicts that persons of this type, when in a bad mood, will help others primarily as a means to boost their own spirits. This is an egoistic motive for helping, not an altruistic one, because persons are offering help primarily to relieve sadness in themselves rather than to relieve suffering in others.

One implication of the negative-state relief hypothesis is that if people in a bad mood are relieved by some event or act other than giving help, they will not volunteer to help as much as others whose bad mood has not been relieved. The results from several studies are consistent with this prediction (Cialdini, Darby, & Vincent, 1973; Schaller & Cialdini, 1988). Another implication of this hypothesis is that negative moods will motivate help giving only if people believe that their moods will be improved by helping. One study on this point manipulated whether participants did or did not believe that their (sad) mood could be changed or improved by helping; the results were consistent with expectations from the negative-state relief hypothesis (Manucia, Baumann, & Cialdini, 1984).

Guilt and Helping

Guilt is a negative emotional state aroused when we transgress against others or do something we consider wrong (Tangney, 1992). For instance, we may feel guilty if we accidentally break someone's camera or spill ink on a person's clothes. To study the impact of guilt on helping, investigators in various studies have induced participants to perform actions that transgress against others (such as killing a laboratory animal, giving painful electric shocks, damaging expensive machinery) and that consequently produce guilt. After the transgression, participants are presented with an opportunity to help the other; the forms of help in these studies include picking up scattered papers, volunteering to participate in further experiments, donating blood, making telephone calls for an ecology group, and the like (Rosenhan et al., 1981). Studies of this type generally show that individuals are more likely to help others when they believe they have harmed these persons in some way (and feel guilty about it) than when not (Salovey et al., 1991).

Guilt from transgressions of this type has led to helping regardless of whether the transgressions were intentional or unintentional, public or private, in the laboratory or in the field. Such transgressions have also produced help both spontaneously and in response to requests, and benefiting either the victim of the transgression or an unrelated third party.

We can explain these findings in part by assuming that helping allows transgressors to relieve their feelings of guilt. For instance, by giving help to the injured party, a transgressor may be able to directly compensate or make up for the transgression. This would assuage the transgressor's guilt and improve his or her mood.

One problematic finding, however, is that helpfulness induced by guilt is often directed not just toward the injured victim but toward unrelated third parties who may need help. This raises the possibility that helping under conditions of guilt can involve something more than just direct compensation to victims. In this regard, the image reparation hypothesis (Cunningham et al., 1980) maintains that help given by a transgressor is not necessarily directed at undoing the harm done. Rather, the prosocial response serves to repair the transgressor's tarnished self-image and restore his or her self-esteem. Most of us do not see ourselves as abusive or mean, so giving help to someone—anyone—after a transgression can restore our image as positive, worthwhile, decent persons. Note that this is essentially an egoistic, not an altruistic motive for giving help. Although the image reparation hypothesis cannot be taken as a total explanation for helping under conditions of guilt, it identifies an important process and is consistent with a body of research findings (Cunningham, Barbee, & Pike, 1990).

BYSTANDER INTERVENTION IN EMERGENCY SITUATIONS

Some of the earliest and most interesting social psychological research on helping was inspired by the tragic murder of a young woman named Catherine (Kitty) Genovese. Shortly before 3:20 a.m. on March 13, 1964, in the borough of Queens in New York City, Kitty Genovese parked her red Fiat in the lot of the Long Island Railroad station. Although she lived in Kew Gardens, a quiet middle-class residential area, Kitty must have sensed something wrong. Instead of taking the short route home, she started running along well-lighted Lefferts Boulevard. She didn't get far before an assailant attacked her.

Milton Hatch awoke at the first scream. Staring from his apartment window, he saw a woman kneeling on the sidewalk directly across the street and a small man standing over her. "Help me! Help me! Oh God, he's stabbed me!" she cried. Leaning out his window, Hatch shouted, "Let that girl alone!" As other windows opened and lights went on, the assailant fled in his car. No one called the police.

With many eyes now following her, Kitty dragged herself along the street—but not fast enough. More than 10 minutes passed before the neighbors saw her assailant reappear, hunting for her. When he stabbed her a second time, she screamed, "I'm dying! I'm dying!" Still no one called the police. Emil Power would have, had his wife not insisted someone else must already have done so.

The third, fatal attack occurred in the vestibule of a building a few doors from Kitty's own entrance. Onlookers saw the assailant push open the building door, though few could hear the weak cry that greeted him. Finally, at 3:55—35 minutes after Kitty's first scream—Harold Klein, who lived at the top of the stairs where Kitty was murdered, called the police. The first patrol car arrived within 2 minutes, but by then it was too late (Seedman & Hellman, 1975).

The tragic story quickly became front-page news in New York and across the country. When it was subsequently discovered that a total of 38 people had witnessed the stalking and stabbing, the editorials asked, "Why had no one offered help or called the police sooner?" A parade of experts attributed the failure to urban apathy, to the depersonalizing environment of New York, to people's unwillingness to get involved.

These explanations may be partly true, but they don't cut to the heart of the matter. We know that bystanders sometimes help in emergencies, and sometimes not. So, for social psychologists, the more fundamental questions raised by Kitty's murder are, "Under what conditions will bystanders and witnesses intervene in an emergency and give help?" and, "Why do people help in some emergency situations but not in others?" In this section, we will consider this issue in detail and look at various factors that influence whether a bystander will help a victim.

The Decision to Intervene

The term **bystander intervention** denotes a (quick) response by a person witnessing an emergency to help another who is endangered by events. Whether and how to intervene in an emergency is a complex decision, because providing emergency assistance often places the helper in considerable danger.

A theory proposed by Latané and Darley (1970) maintains that bystanders (potential helpers) go through a specific decision making sequence prior to actually giving help in emergencies. In order to provide help, bystanders must go through a series of five steps in order (see Figure 11.2). If any of these steps fail, the decision making process ends and the bystander does not provide assistance.

The first critical step in the process is that the bystander must notice the situation. If bystanders are not aware of a situation that may require their assistance, they obviously will not intervene—the bystanders will simply proceed with whatever they were doing before the incident occurred. Some studies have manipulated how preoccupied potential helpers were and, not surprisingly, those who were more caught up in their own thoughts were less likely to notice the emergency situation and, therefore, less likely to respond (Darley & Batson, 1973).

Once the bystander has noticed the situation, he or she must interpret the situation as an emergency. Most emergency situations are quite ambiguous, and the failure to interpret them as emergencies will produce inaction among bystanders. When we hear a crash in the hallway, do we interpret this as typical work-related noise, or do we think that someone has been hurt? When we hear students shouting at night on campus, do we think something is wrong, or do we interpret it as drunken revelry? According to Shotland and Huston (1979), there are five characteristics of a situation that increase the chances we will interpret it as an emergency: (1) The incident is sudden and unexpected, (2) there is a clear threat of harm, (3) the harm to the victim will increase unless

FIGURE 11.2 DECISIONS LEADING TO INTERVENTION IN AN EMERGENCY

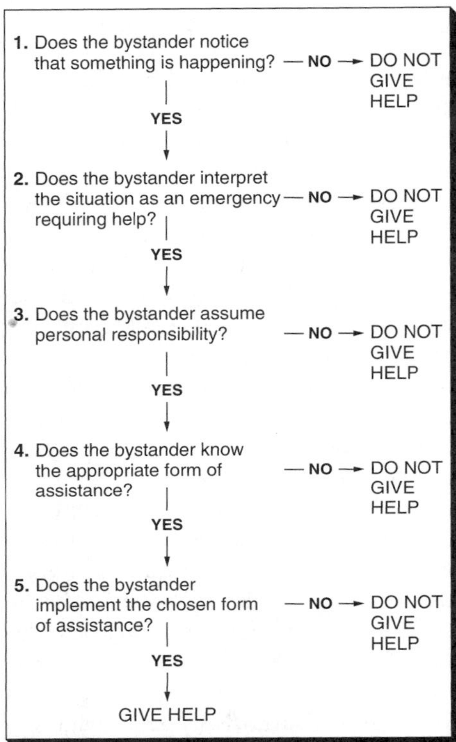

help is provided, (4) the victim cannot address the problem without assistance from someone else, and (5) there is something the bystander can do to help.

Third, the bystander must take personal responsibility. One famous study created a situation at a beach where the researchers staged the theft of a radio while its owner was swimming. Most people—about 80 percent—did nothing to try to stop the thief or to intervene in any manner. However, when the owner of the radio asked the person next to her to keep an eye on the radio while she was swimming, almost all of them confronted the person stealing the radio (Moriarty, 1975). Once they had taken on the responsibility for the radio, they were much more likely to act to help the victim. If bystanders interpret the situation to be "none of their business," they will not respond.

Fourth, the bystander must know how to help. Sometimes, the assistance required is something very simple, like dialing 911 for assistance. Other times, the situation is more complex. If we do not know how to perform CPR, we are less likely to try to resuscitate a collapsed individual. When witnessing an epileptic seizure, most people have no idea how to respond, and so they do nothing. People with medical training are much more likely to attempt to provide assistance at accident sites than those without medical knowhow (Cramer, McMaster, Bartell, & Dragna, 1988).

Finally, the bystander must make the decision to act. Even if all of the first four conditions are fulfilled, people often will hesitate to act because they are afraid of negative consequences to themselves. Typically, people engage in some kind of risk calculation before they act in emergency situations (Fritzsche, Finkelstein, & Penner, 2000). For example, we are often hesitant to break up a fight between other individuals because we are afraid of getting hurt—or even that the two combatants will turn on us.

The Bystander Effect

In emergency situations, potential helpers are influenced by their relations with other bystanders (Dovidio, 1984; Latané & Darley, 1970). This influence is apparent at each step in the decision making process (see Figure 11.2). To investigate the nature of bystander influence, researchers have conducted a variety of laboratory studies that entail simulated emergencies of one kind or another. For instance, in one experiment (Latané & Rodin, 1969), participants heard a loud crash from the room next door, followed by a woman screaming, "Oh my God, my foot! I . . . I . . . can't move it. Oh my ankle. I . . . can't get this . . . thing off me." In another experiment (Darley & Latané, 1968), individuals participating in a discussion over an intercom suddenly heard someone in their group begin to choke, gasp, and call for help, apparently gripped by an epileptic seizure.

In each experiment, the number of people who were supposedly present when the emergency

FIGURE 11.3 THE BYSTANDER EFFECT

Students who were engaged in a discussion via intercom of their adjustment to college life heard one participant begin to choke, then gasp and call for help, as if he were undergoing a serious nervous seizure. Students intervened to help the victim most quickly and most often when they believed they were the lone bystander to witness the emergency. More than 90% of lone bystanders helped within the first 90 seconds after the seizure. Among those who believed other bystanders were present, however, fewer than 90% intervened, even after 4 minutes. The bystander effect refers to the fact that the greater the number of bystanders in an emergency, the less likely any one bystander will help.

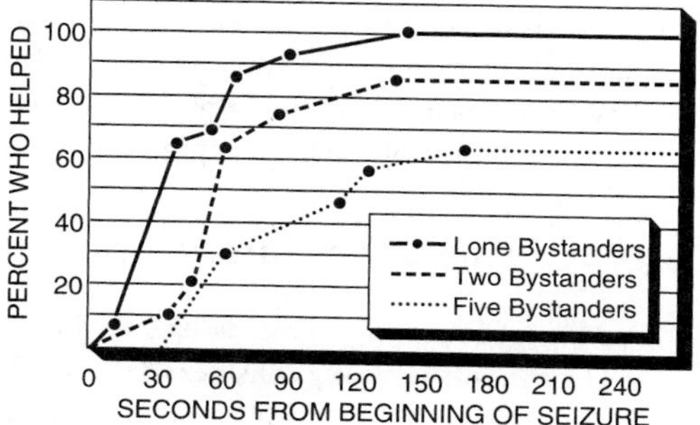

Source: Adapted from "Bystander Intervention in Emergencies: Diffusion of Responsibility" by Darley and Latane, *Journal of Personality and Social Psychology, 8*, 377–383. Copyright © 1968 by the American Psychological Association. Adapted with permission.

occurred was varied. Participants believed either that they were alone with the victim or that one or more bystanders were present. Time and again, the same finding emerged: As the number of bystanders increased, the likelihood that any one of them would help decreased (Latané & Nida, 1981). Bystanders helped most often and most quickly when they were alone with the victim. The knowledge that other potential helpers were present inhibited intervention in an emergency.

Social psychologists use the term **bystander effect** to refer to the finding that as the number of bystanders increases, the likelihood that any one bystander will help a victim decreases. This effect is illustrated in Figure 11.3, using data from the epileptic seizure experiment.

Theorists have identified several distinct processes that contribute to the bystander effect. These include social influence regarding the interpretation of the situation, evaluation apprehension, and diffusion of responsibility (Latané, Nida, & Wilson, 1981; Piliavin et al., 1981). Each of these processes affects specific steps in the decision making sequence.

Interpreting the Situation Emergencies and other situations requiring help are often ambiguous, at least initially. For instance, is a choking, gasping student in real trouble? When faced with ambiguity, people look to the reactions of others for cues about what is going on. The reactions of others can influence three of the steps in the decision making sequence leading to helping (see Figure 11.2). If others appear calm, the bystander may decide that nothing special is happening (step 1) or that whatever is happening requires no help (step 2). Likewise, the failure of others to act may influence the bystander to decide that there is no appropriate way to help (step 4).

Bystanders often try to appear calm, avoiding overt signs of worry until they see whether others are alarmed or not. Through such cautiousness, they unintentionally encourage each other to define the situation as not problematic. In that way, they inhibit each other from helping. The larger the number of apparently unruffled bystanders, the stronger their inhibiting influence is on one another. However, consistent with this explanation, increasing the number of bystanders does not inhibit individual helping under certain conditions, such as (1) when observation reveals that others are indeed alarmed (Darley, Teger, & Lewis, 1973), and (2) when the need for help is so unambiguous that others' reactions are unnecessary to define the situation (Clark & Word, 1972).

Evaluation Apprehension Bystanders are not only interested in others' reactions; they also realize that the other bystanders are an audience for their own reactions. As a result, bystanders may feel **evaluation apprehension**—concern about what others expect of them and how others will evaluate their behavior. Evaluation apprehension can either inhibit or promote helping. On the one hand, evaluation apprehension inhibits helping when bystanders fear that others will view their intervention as foolish or wrong. When they see that other witnesses to an emergency are not reacting (as in the Kitty Genovese case), they may infer that the others see no need to intervene, or even oppose intervention. In the decision making sequence, evaluation apprehension mainly affects step 4 (choosing a way to react) and step 5 (deciding whether to implement the chosen course of action).

On the other hand, evaluation apprehension promotes helping if there are no cues to suggest that the other witnesses oppose intervention. Bystanders then tend to assume that others approve of intervention. In three laboratory studies demonstrating this effect, bystanders witnessed a convulsive nervous seizure or a violent assault (Schwartz & Gottlieb, 1976, 1980). The knowledge that an audience of other bystanders was present led to increased helping when the audiovisual system prevented each bystander from learning how others were reacting.

Diffusion of Responsibility When one and only one bystander witnesses an emergency, the responsibility to intervene is focused wholly on that individual. But when there are multiple bystanders, the responsibility to intervene is shared, as is the blame if the victim is not helped. Hence, a witness is less likely to intervene when other bystanders are present ("Why should I help? Let someone else do it"). This process wherein a bystander does not take action to help because other persons share the responsibility for intervening is called **diffusion of responsibility.**

In the decision sequence (see Figure 11.2), diffusion of responsibility operates primarily at step 3 (bystander decides whether he or she has the responsibility to act). When multiple witnesses are present, bystanders sometimes limit their own responsibility by assuming that others have already taken action. Of the 38 witnesses to Kitty Genovese's murder, many claimed they thought someone else must surely have called the police.

Diffusion of responsibility occurs only when a bystander believes that the other witnesses are capable of helping. We diffuse responsibility less to witnesses who are too far away to take effective action or too young to cope with the emergency (Bickman, 1971; Ross, 1971). The tendency to diffuse responsibility is particularly strong if a bystander feels less competent than others who are present. Bystanders helped less, for example, when one of the other witnesses to a seizure was a premed student with experience working in an emergency ward (Pantin & Carver, 1982; Schwartz & Clausen, 1970).

Despite the presence of others having a strong direct effect on helping, bystander intervention can also be reduced merely by thinking about other people—even if they are not there. Researchers call this effect the implicit bystander effect (Garcia, Weaver, Moskowitz, & Darley, 2002). In these cases, if individuals are primed to think about themselves as part of a group of friends and then confront a situation that calls for their help, they are less likely to help than if they have not been primed. This implicit bystander effect points out the importance of emphasizing each person's individual contribution in a helping situation.

Costs and Emergency Intervention

When deciding whether to offer assistance in an emergency, bystanders typically consider the costs involved. Whereas some interventions involve little or no cost, others are more costly, and help is presumably easier to give when the cost of doing so is low.

One theory, the **arousal/cost-reward** model of helping (Dovidio et al., 1991; Piliavin et al., 1981), proposes that bystanders weigh the needs of the victim and their own needs and goals, and then decide whether helping is too costly in the circumstances. Bystanders often take into account several kinds of costs to themselves in emergency situations. First, bystanders consider the cost of giving direct help. This includes the costs to them if they offer help—lost time, exposure to danger, expenditure of effort, exposure to disgusting experiences, and the like. Second, bystanders consider the cost of not giving help. Costs borne by the bystanders if the victim receives no help include the burden of unpleasant emotional arousal while witnessing another's suffering and the costs associated with one's personal failure to act in the face of another's need (self-blame, possible blame from others, embarrassment, and the like).

Intervention as a Function of Cost The cost approach to helping hypothesizes that the responses of bystanders in an emergency situation will depend on

BOX 11.2 Research Update: Bystander Intervention and Social Control: Littering and Graffiti

Many experiments about the bystander effect have been conducted examining the intervention in a personal emergency—for example, when someone needs immediate emergency medical intervention. But there are many other more mundane opportunities for prosocial intervention or helping, and if the bystander effect is robust, it ought to occur in those less dramatic circumstances as well. We are all well acquainted with social norms against littering. Children are taught from a young age that we have a personal responsibility to avoid this kind of environmental pollution. Likewise, there is a general social condemnation of graffiti—another kind of behavior that fouls the environment. Given that litter and graffiti are both normatively negative, intervening to prevent, confront, or at least express disapproval of littering or graffiti artists is a opportunity to exhibit prosocial behavior. Are these kinds of social control behaviors subject to the bystander effect? French social psychologists Peggy Chekroun and Markus Brauer set out to find out if individuals would be more likely to intervene to address acts of littering and graffiti when alone or when others were also present.

The researchers also tested the idea that the bystander effect is conditional on personal relevance. That is, they hypothesized that increased numbers of bystanders would reduce social control acts only if the counternormative act was not personally relevant to the observer. Thus, the researchers created four experimental conditions: low personal relevance with one participant; low personal relevance with multiple participants; high personal relevance with one participant; and high personal relevance with multiple participants.

The experiment proceeded first by assessing the personal relevance of two different situations. The first was the drawing of graffiti in a shopping mall elevator. The second was littering by throwing down a plastic bottle in a neighborhood park. As the researchers expected, the littering incident has much more personal relevance to bystanders in the park than the graffiti did to elevator riders in the mall. Armed with this information, the experimenters sent confederates into the elevators and the parks when different numbers of bystanders were nearby in the park or in the elevators. The confederates made eye contact with each bystander after conducting the deviant act and recorded the reactions immediately after leaving the scene. Bystander responses ranged on this scale: (0) no social control, (1) angry look, (2) loud audible sigh, (3) comment made to another person, (4) polite comment to the offender, (5) aggressive comment to the offender, (6) personal insult to the offender.

The results of the experiment showed three things. First, there were about 33 percent more social control behaviors exhibited in the personally relevant condition, and they were more likely to be stronger responses (littering in the park). Second, increased numbers of bystanders did inhibit bystanders from exerting social control. Third, the bystander effect only occurred in the elevator situation—the low personal relevance condition. Thus, the researchers concluded that bystander effects for social control behaviors can be mitigated if the situation is personally relevant to the bystanders.

Source: Adapted from Chekroun and Brauer, 2002.

TABLE 11.2 PREDICTED EFFECTS OF COST FACTORS ON HELPING

Cost of Not Helping the Victim	COST OF DIRECTLY HELPING THE VICTIM: Low	High
High	Direct intervention by bystander	Indirect intervention or Redefinition of the situation, disparagement of the victim, etc.*
Low	Variable reactions (largely a function of perceived norms in the situation)	Leaving the situation, ignoring, denial, etc.

This table displays theoretically predicted reactions of an observer to an emergency as a function of costs to the observer and costs to the victim if no help is given.

* The response of redefining the situation or disparaging the victim lowers the cost of giving no help to the victim; this may lead to subsequent responses like leaving the situation, ignoring, denial, etc.

Source: Adapted from Piliavin, Piliavin, and Rodin, 1975.

whether the cost to themselves of giving direct help is high or low and whether the cost to themselves of not giving help is high or low (Piliavin, Piliavin, & Rodin, 1975). Table 11.2 displays the predicted responses of bystanders as a function of cost. The model predicts that direct intervention will occur primarily when the bystander incurs low costs for giving direct help and high costs for not helping the victim. Other modes of response (indirect helping, redefinition of the situation, leaving the scene, and so on) will occur under the other cost combinations.

Evidence Regarding Cost Various studies have documented that cost has an impact on help giving. First, some results support the proposition that the greater the cost to self of giving direct help, the less likely one is to help (Darley & Batson, 1973; Shotland & Straw, 1976). This was demonstrated, for instance, in a study conducted in the New York City subway (Allen, 1972). Aboard a subway car, a bewildered-looking man asked the participant (a passenger) whether the train was going uptown or downtown. The man in the neighboring seat—a muscular type reading a bodybuilding magazine—responded quickly but gave an obviously wrong answer. Both the bewildered man and the bodybuilder were confederates. The participant could help by correcting this misinformation, but only at the risk of challenging the bodybuilder. Whether the participants helped or not depended on how threatening the bodybuilder appeared to be. Threat was manipulated by varying his reaction to an incident a minute before. When the bodybuilder had previously threatened physical harm to a person who had stumbled over his outstretched feet, only 16 percent of the participants helped. When the bodybuilder had only insulted and embarrassed the stumbler, 28 percent helped. When the bodybuilder had given no reaction to the stumbler, 52 percent helped. Thus, the greater the anticipated cost of antagonizing the misinforming bodybuilder, the less likely people were to help the bewildered man.

Furthermore, studies show that the greater the costs to self of not helping the victim, the more likely one is to help (Berkowitz, 1978a; Gottlieb & Carver, 1980). For instance, Staub (1974) conducted a study with a simulated health emergency on a city sidewalk. A male confederate faked a bad knee (that is, he grabbed his knee, fell to the sidewalk, and failed to stand up). The study varied the ease with which bystanders could escape from the emergency scene. In one treatment, the emergency happened on the other side of the street, so escape was easy, and the cost of not helping was fairly low. In the other treatment, the emergency happened right in front of the bystander on the same side of the street, so escape was difficult, and the cost of not helping was higher. Victims received considerably more help when escape was difficult than when it was easy.

© David Maialetti/Philadelphia Daily News

High potential costs and diffusion of responsibility inhibit bystander intervention in this fight. The man on the ground clearly needs help, but bystanders watch without getting involved. They feel little responsibility for the wounded man and wish to avoid entanglement in the fight, still in progress.

SEEKING AND RECEIVING HELP

Up to this point, we have focused primarily on giving help rather than on receiving it. Yet recipients' reactions to receiving help—and related phenomena, such as people's willingness to seek help in the first place—are important topics that also deserve attention. How does it feel to receive help from other persons? Help not only relieves need but also demonstrates to the recipient that someone cares. We might expect, therefore, that recipients will feel gratitude and joy—and of course they frequently do. In other cases, however, help elicits a different emotional reaction from recipients—resentment, hostility, and anxiety. In this section, we will look at the various reactions to help and the reasons for them.

Help and Obligation

Although our society has some norms—such as the norms of reciprocity and responsibility—that mandate giving help when needed, it also has norms of adult independence and self-reliance that tell us to avoid asking for help unless it is really necessary. When help is sought and received, resources (such as labor and materials) are transferred from one person to another. If the norm of reciprocity is salient in the situation, the person receiving help may feel obligated or indebted to the helper (Greenberg & Westcott, 1983).

In consequence, needy persons (in nonemergency situations) sometimes experience a dilemma. On the one hand, they can ask for help and possibly endure some embarrassment or social obligation; on the other hand, they can suffer through the difficulties of trying to solve their problems on their own (Gross & McMullen, 1983). In cases where the recipient has the opportunity and ability to reciprocate, there may be no problem. But in cases where this transfer of resources is strictly one-way from the helper to the needy, it may create a lingering sense of indebtedness in the needy toward the helper. Several studies have shown that if a person lacks the capacity to repay or to reciprocate, he or she is less likely to ask for help than otherwise (Nadler, 1991; Wills, 1992); this is especially so if the person needing help has high self-esteem. Persons tend to develop resentment and negative sentiments toward a benefactor they cannot repay (Clark, Gotay, & Mills, 1974; Gross & Latané, 1974).

Moreover, help that goes beyond what is absolutely necessary may be resented by the recipient (Schwartz & Ames, 1977). We are often wary of people who are overly generous or extend help beyond our ability to reciprocate. Accepting substantial generosity obligates us to submit to our benefactors' wishes, at least to some degree. Gifts we cannot reciprocate threaten our freedom of action (Greenberg, 1980).

Threats to Self-Esteem

In studying people's reactions to receiving help, theorists have proposed that an important determinant of whether help is appreciated or resented is the extent to which it undermines the recipient's self-esteem (Nadler, 1991; Nadler & Fisher, 1986; Shell & Eisenberg, 1992). As we have noted, to receive help is a mixed blessing. Although it provides relief, it can also impair a recipient's self-esteem and sense of self-reliance. Of course, helpers usually intend just the opposite. The avowed purpose of welfare, for instance, has been to aid impoverished individuals and to help families escape hunger while they establish themselves as self-supporting. The purpose of foreign aid is to enable nations to overcome crises and to develop their own resources. Yet welfare, foreign aid, and other forms of assistance are sometimes given reluctantly or in ways that do not promote these outcomes. Intentionally or otherwise, helpers may communicate the message that those who need and accept help are inferior in status and ability because they fail to display the self-reliance and achievement admired in Western societies (DePaulo & Fisher, 1980; Rosen, 1984). When help is couched in these terms, the recipients' acceptance of help may diminish their self-esteem and even change their self-concept.

Ego Centrality Help is especially threatening to self-esteem—and hence less likely to be sought or accepted gratefully by persons in need—when it implies inferiority in intelligence, competence, morality, or other qualities that are central to a recipient's self-concept. Help is less threatening—and hence more likely to be sought—when it does not imply any major inadequacy with respect to central personal attributes (Nadler, 1987; Nadler & Fisher, 1984a). For example, help can be nonthreatening if need is explicitly attributed to uncontrollable or chance factors—like a drought, epidemic, or unprovoked attack—or if the aid is defined as enabling one to overcome a trivial inadequacy in experience or effort.

Similarity of Help Provider Surveys regarding help seeking for personal and psychological problems indicate that we are most likely to ask either our friends or people who are similar to us for assistance. Wills (1992) indicates that persons looking for help of this type are several times more likely to seek it from friends, acquaintances, or family members than from professionals or strangers. Generally, help seeking is more common among close friends and people in communal relationships than among strangers and people in exchange-based relationships.

Nevertheless, the helper's similarity to the recipient is a complex factor in help giving and help seeking. Help that implies an important inadequacy is often more threatening to our self-esteem when we receive it from those who are similar to us in attitudes or background than from those who are dissimilar (Nadler, 1987; Nadler & Fisher, 1984a). Similarity can aggravate recipients' self-evaluations, because similar helpers are relevant targets for self-comparison (say, "If we are both alike, why do I need help while you can give it?"). People who accept aid from helpers similar to themselves on a task central to their self-concept report lower self-esteem, less self-confidence, and more personal threat than when they accept aid from dissimilar helpers (DePaulo, Nadler & Fisher, 1983; Nadler, Fisher, & Ben-Itzhak, 1983).

Threat to Self-Esteem as a Motivator Is the fact that aid threatens recipients' pride and self-esteem wholly undesirable? Not if we consider the long-term consequences (Nadler & Fisher, 1984b). If recipients experience aid as nonthreatening, they will feel favorably toward themselves and the helper, but they will have little motivation to change. Consequently, they will invest little in developing self-reliance and will just continue to seek help in the future. Nations become dependent satellites; individuals become helpless parasites. In contrast, aid that threatens the self may generate negative feelings toward the helper as well as the self, but it can motivate the recipients to change. This motivation promotes self-reliance and drives individuals to reestablish their independence.

SUMMARY

Helping is behavior intended to benefit others. Altruism, a specific kind of helping, is voluntary behavior intended to benefit another with no expectation of external reward.

Motivation to Help Others There are several motivations for helping. (1) Egoistic considerations of cost and reward often motivate helping. People tend to help more when it leads to rewards, such as social approval, material gain, and satisfaction of personal needs. People often refuse to help when it entails costs, such as danger and effort. They also avoid costs imposed for not helping, such as public disapproval and self-condemnation. (2) Altruistic behavior is often mediated by empathic arousal in response to others' distress. Our similarity to victims increases our empathy. Empathic concern motivates altruism—helping to benefit others with no expectation of reward. (3) Evolutionary psychologists posit that altruistic behavior is selectively targeted to increase the chances that our own genes will be passed on to subsequent generations. Thus, we are most likely to help those who are closely related to us.

Characteristics of the Needy That Foster Helping The characteristics of the needy influence whether others will give help. (1) People tend to help those to whom they are related or with whom they are well acquainted. They also tend to give more help to those they like than to those they do not like. (2) Needy persons have a better chance of receiving help if they are similar rather than dissimilar to the potential help giver. This holds for similarity in a wide variety of attributes, including appearance (race, dress) and attitudes (political opinions). (3) Needy persons have a better chance of receiving help if they are seen as deserving and not having caused their own plight.

Normative Factors in Helping (1) Norms define when helping is the appropriate thing to do. The social responsibility norm directs us to help whoever is dependent on us. The norm of reciprocity tells us to reciprocate intentional benefits we receive. (2) Personal norms, based on internalized values, motivate help when we notice another's need and feel responsible for relieving it.

Personal and Situational Factors in Helping (1) The presence of a model (that is, a person actively helping) will increase the probability that others will also offer to help. (2) Consistent with traditional gender roles, men are more likely than women to help in heroic or chivalrous situations entailing risk, whereas women are more likely than men to help in situations requiring nurturance, caretaking, and emotional support. (3) Good moods promote helping because they reduce self-preoccupation, increase the attention paid to others' needs, and increase one's feelings of relative good fortune. Bad moods inhibit helping because they increase self-preoccupation and increase feelings of being relatively underbenefited. However, a bad mood may increase helping when it is perceived as a route to improving one's own mood. (4) Guilt promotes helping because helping boosts transgressors' self-esteem and relieves guilt.

Bystander Intervention in Emergency Situations (1) Prior to actually giving help in emergencies, bystanders go through a decision sequence. A potential helper must notice that something is happening, interpret the situation as an emergency, decide that he or she has the responsibility to act, know or recognize the appropriate form of assistance, and decide to implement the chosen behavior. (2) Bystanders to an emergency influence each others' behavior in three ways. First, the reactions of other bystanders affect whether a bystander interprets the situation as one that requires help. Second, the expectations of other bystanders create evaluation apprehension, which may either inhibit or enhance helping. Third, the presence of other bystanders fosters diffusion of responsibility, which reduces helping. (3) Bystanders' responses to an emergency will depend on whether the costs to themselves of giving help are high or low and on whether the costs to themselves if the victim receives no help are high or low. Direct intervention will occur primarily when the bystander incurs low costs for helping and high costs for not helping.

Seeking and Receiving Help (1) From the standpoint of the recipient, help is a mixed blessing. Although providing benefits, it may also cause the recipient to become indebted to the helper, especially if the recipient lacks the capacity or opportunity to reciprocate. In such a case, a recipient may develop resentment or other negative sentiments toward the helper. (2) An important determinant of whether help is appreciated or resented is the extent to which it undermines the recipient's self-esteem. If helpers communicate the message that recipients are inferior in status and competence, the net result may be a reduction in recipients' self-esteem. Help can increase recipients' dependency and weaken their future self-reliance.

LIST OF KEY TERMS AND CONCEPTS

altruism (p. 275)
arousal/cost-reward model (p. 292)
bystander effect (p. 291)
bystander intervention (p. 289)
diffusion of responsibility (p. 292)
egoism (p. 275)
empathy (p. 276)
empathy-altruism model (p. 277)
evaluation apprehension (p. 291)
helping (p. 274)
kin selection (p. 278)
norm of reciprocity (p. 283)
prosocial behavior (p. 274)
social responsibility norm (p. 282)

12

AGGRESSION

Introduction

What Is Aggression?

Aggression and the Motivation to Harm

Aggression as Instinct

Frustration-Aggression Hypothesis

Aversive Emotional Arousal

Social Learning and Aggression

Characteristics of Targets That Affect Aggression

Gender and Race

Attribution for Attack

Retaliatory Capacity

Situational Impacts on Aggression

Reinforcements

Modeling

Norms

Stress

Aggressive Cues

Reducing Aggressive Behavior

Reducing Frustration

Punishment to Suppress Aggression

Nonaggressive Models

Catharsis

Aggression in Society

Sexual Assault

Pornography and Violence

Media Violence and Aggression

Summary

List of Key Terms and Concepts

INTRODUCTION

- Denise Farmer, a 40-year-old mother of two living in Chicago, got up and dressed for work. At 7:00 a.m., she left her apartment and walked down the stairs. According to police, one or more attackers were waiting at the foot of the stairs. The assailant(s) stabbed her more than 20 times; four of the thrusts penetrated her heart and killed her. Another resident of the building found Farmer, her pockets turned out and empty.
- On Monday, March 5, 2001, Charles "Andy" Williams, 15, went to his school in Santee, CA, carrying a gun owned by his father. When he arrived, he opened fire. In the next 6 minutes, he killed Bryan Zuckor, 14, and Randy Gordon, 14, and wounded 13 others, including 11 students and a teacher. When police arrived, he surrendered.
- Kevin Singer and his girlfriend, Carrie Young, were in Kevin's apartment talking to Carrie's brother Steven. An argument began, and Kevin picked up a hammer and a knife. Steven grabbed the hand with the hammer, and Kevin stabbed Steven in the back. Steven Young died from a combination of knife wounds and blows to the head, according to the coroner.

These incidents portray in stark relief a person's ability to inflict pain and death on other human beings. How can we account for such incidents and for the much more common and less extreme forms of aggression—harassment, abuse, assault—that occur several times each minute in the United States? These phenomena are the focus of this chapter.

What Is Aggression?

Defining aggression seems a simple enough task: Aggression is any behavior that hurts another. But this definition considers only the observable consequences of behavior, and ignores the actor's intentions. Hence, it often leads to absurd conclusions. Under this definition, for instance, we would consider a surgeon an aggressor if a heart transplant patient died on the operating table despite heroic efforts to preserve the patient's life.

Because intentions are clearly important in defining an act as aggression (Krebs, 1982), we will use the following definition: **Aggression** is any behavior intended to harm another person that the target person wants to avoid. According to this definition, a bungled assassination is an act of aggression; it involves intended harm that the target surely would wish to avoid. Heart surgery—approved by the patient and intended to improve his or her health—is clearly not aggression, even if the patient dies. The intended harm may be physical, psychological, or social (for example, harm to the target's reputation).

Drawing on research and theory, this chapter addresses the following questions:

1. What motivates people to aggress against others?
2. How do the characteristics of the target influence aggression?
3. How do the characteristics of the situation influence aggression?
4. How can we reduce the frequency of aggressive behavior in society?
5. What influences the incidence of interpersonal aggression—abuse, assault, sexual assault, and murder—in our society?

AGGRESSION AND THE MOTIVATION TO HARM

As the examples in the introduction show, human beings have a remarkable capacity to harm others. Our first question concerns the motivation for human aggression. Why do people turn against others? There are at least four possible answers: (1) people are instinctively aggressive; (2) people become aggressive in response to events that are frustrating; (3) people aggress against others as a result of aversive emotion; and (4) people learn to use aggression as an effective means of obtaining what they want. We consider each of these answers in turn.

Aggression as Instinct

The best known proponent of the theory that aggression is an instinct was Sigmund Freud (1930, 1950). In Freud's view, from the moment of conception, we carry within us both an urge to create and an urge to destroy. The innate urge to destroy, or **death instinct,** is as natural as our need to breathe. This instinct constantly generates hostile impulses that demand release. We release these hostile impulses by aggressing against others, by turning violently against ourselves (suicide), or by suffering internal distress (physical or mental illness).

Many studies of animal behavior provide evidence that aggression is instinctive. According to Lorenz (1966, 1974), the aggressive instinct has evolved because it contributed to an animal's survival. Animals motivated to fight succeed better in protecting their territory, obtaining desirable mates, and defending their young. Through evolution, animals have also developed an instinct to inhibit their aggression once their opponents signal submission. Humans have no such instinct, however, so in this sense, humans are more dangerous and destructive than animals.

Instinct theories postulate that the urge to harm others is part of our genetic inheritance. As a result, the proponents of such theories are pessimistic about the possibility of controlling human aggression. At best, they believe, aggression can be channeled into approved competitive activities such as athletics, academics, or business. Social rules that govern the expression of aggression are designed to prevent competition from degenerating into destructiveness. Quite often, however, socially approved competition stimulates aggression: Football players start throwing punches, soccer fans riot violently, and businesspeople destroy competitors or cheat the public through ruthless practices. If aggression is instinctive, we should not be surprised that it is always with us.

Despite the popularity of instinct theories of aggression, most social psychologists find them neither persuasive nor particularly useful. Generalizing findings about animal behavior to human behavior is hazardous. Moreover, cross-cultural studies suggest that human aggression lacks two characteristics that are typical of instinctive behavior in animals—universality and periodicity. The need to eat and breathe, for example, are universal to all members of a species. They are also periodic, for they rise after deprivation and fall when satisfied. Aggression, in contrast, is not universal in humans. It pervades some individuals and societies but is virtually absent in others. Moreover, human aggression is not periodic. The occurrence of human aggression is largely governed by specific social circumstances. Aggressive behavior does not increase when people have not aggressed for a long time or decrease after they have recently aggressed. Thus, our biological makeup provides only the capacity for aggression, not an inevitable urge to aggress. We must look elsewhere to explain why particular people harm others in particular circumstances.

Frustration-Aggression Hypothesis

The second possible explanation of aggressive behavior is that aggression is an internal state that is elicited by certain events. The most famous view of aggression as an elicited drive is the **frustration-aggression hypothesis** (Dollard, Doob, Miller, Mowrer, & Sears, 1939). This hypothesis makes two bold assertions. First, every frustration leads to some form of aggression. Second, every aggressive act is due to some prior frustration. In contrast to instinct theories, this hypothesis states that aggression is instigated by external, environmental events.

In an early experiment (Barker, Dembo, & Lewin, 1941), researchers showed children a room full of attractive toys. They allowed some children to play with the toys immediately. They made other children wait about 20 minutes, looking at the toys, before they allowed them into the room. The children who waited behaved much more destructively when given a chance to play, smashing the toys on the floor and against the walls. Here, aggression is a direct response to **frustration**—that is, to the blocking of a goal-directed activity. By blocking the children's access to the tempting toys, the researchers frustrated them. This in turn elicited an aggressive drive that the children expressed by destroying the researchers' toys.

Several decades of research have led to modifications of the original hypothesis (Berkowitz, 1978b). First, studies have shown that frustration does not always produce aggressive responses (Zillman, 1979). Although motivated to behave aggressively, individuals may restrain themselves due to fear of punishment. Being laid off is a frustrating experience. Researchers predicted that small increases in layoffs would lead to violence by those laid off. Large increases, however, will lead to reduced violence, because those still working will be afraid of being laid off (Catalano, Novaco, & McConnell, 1997). Data from San Francisco supported these predictions. Also, frustration sometimes leads to different responses, such as despair, depression, or withdrawal.

Second, research indicates that aggression can occur without prior frustration (Berkowitz, 1989). The ruthless businessperson or scientist may attempt to destroy competitors due to the desire for wealth and fame, even though they have not blocked her goal-directed activity.

The frustration-aggression hypothesis implies that the nature of the frustration influences the intensity of the resulting aggression. Two factors that intensify aggression are the strength and the arbitrariness of frustration.

Strength of Frustration The more we desire a goal and the closer we are to achieving it, the more frustrated and aroused we become if blocked. If someone cuts ahead of us as we reach the front of a long line, for example, our frustration will be especially strong. According to theory, this strong frustration should lead to aggressiveness.

A field experiment based on this idea demonstrated that stronger frustration elicits more aggression (Harris, 1974). Researchers directed a confederate to cut ahead of people in lines at theaters, restaurants, and grocery checkout counters. The confederate cut in front of either the 2nd or the 12th person in line. Observers recorded the reactions of the person. As predicted, people at the front of the line responded far more aggressively. They made more than twice as many abusive remarks to the intruder than people at the back of the line.

Cases of "road rage," in which one driver engages in aggressive or violent behavior toward another, often exemplify the frustration-aggression hypothesis. One motorist engages in a behavior—driving excessively slowly, taking the only parking space, or cutting in front of you—that causes frustration, blocking your attempt to achieve a goal, such as arriving on time for an appointment. This frustration may lead to aggression. Researchers distinguished between driver aggression—honking, yelling, making obscene gestures—and driver violence—chasing the other car or its driver, throwing objects, or shooting at him or her. Mild frustration causes the former, whereas stronger frustration causes the latter. A measure of the extent to which persons who commuted by car (average 93 minutes per day) found such events frustrating was positively correlated with reports of road rage. Of interest is the finding that men and women were equally likely to report driver aggression in response to such incidents, whereas only men who were highly frustrated reported driver violence (Hennessy & Wiesenthal, 2001).

Arbitrariness of Frustration People's perceptions of the reasons for frustration markedly influence the degree of hostility they feel. People are apt to feel more hostile when they believe that the frustration is arbitrary, unprovoked, or illegitimate than when they attribute it to a reasonable, accidental, or legitimate cause. As a result, arbitrary or illegitimate frustration elicits more aggression.

In a study demonstrating this principle, researchers asked students to make appeals for a charity over the telephone (Kulick & Brown, 1979). The students were frustrated by refusals from all the potential donors (in reality, confederates). In the legitimate frustration condition, potential donors offered good reasons for refusing (such as "I just lost my job"). In the illegitimate frustration condition, they offered weak, arbitrary reasons (such as "charities are a rip-off"). As shown in Figure 12.1, individuals exposed to illegitimate frustration were more aroused than those exposed to legitimate frustration. They also directed more verbal aggression against the potential donors.

FIGURE 12.1 EFFECTS OF LEGITIMACY OF FRUSTRATION ON AGGRESSIVE RESPONSES

Students were frustrated when potential donors whom they telephoned all refused their appeal for a charity. Half the students heard legitimate reasons for refusal, whereas the others heard illegitimate and arbitrary reasons. Those exposed to illegitimate frustration showed a higher level of angry arousal and aggression than those exposed to legitimate frustration. They slammed the receiver down harder when hanging up and expressed more verbal aggression toward potential donors during their conversations. These findings demonstrate that the perceived reasons for frustration influence angry arousal and aggression.

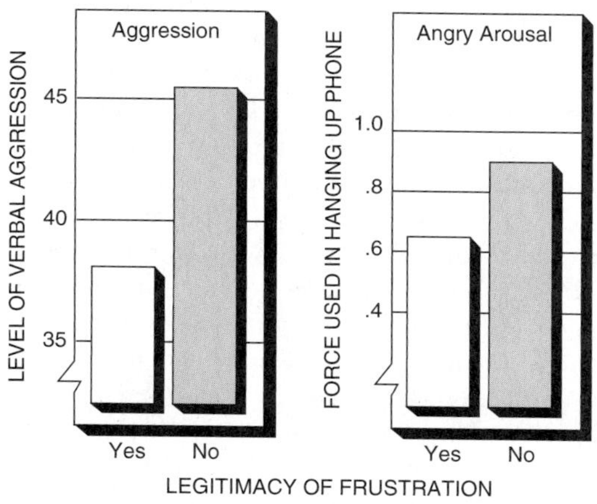

Source: Adapted from "Frustration, Attribution of Blame, and Aggression" by Kulick and Brown, *Journal of Experimental Social Psychology, 15*, 183–194. Copyright © 1979, with permission from Elsevier.

A young boy reacts aggressively to a physical attack by another. An attack typically produces anger, which in turn leads to an aggressive response.

Aversive Emotional Arousal

In the 7 decades since the original statement of the frustration-aggression hypothesis, research has identified several other causes of aggression. In one study, community residents and university students were asked what events upset or angered them (Averill, 1982). Some replied that legitimate actions by others and unavoidable accidents triggered aggressive reactions. What makes you aggressive? Chances are that insults—especially those involving traits that you value, perhaps your intelligence, honesty, ethnicity, or attractiveness—would be on your list. Physical pain also can produce aggression.

Verbal and physical attacks may arouse us and elicit an aggressive response. Repeated insults and bullying by classmates may have contributed to the shootings committed by Charles Williams.

Accidents, insults, and attacks all arouse **aversive affect**—negative affect that people seek to reduce or eliminate (Berkowitz, 1989). Often, this affect is in the form of anger, but it can be pain or other types of discomfort. (For example, we will discuss evidence that high temperature produces discomfort and aggression.) Often, the resulting aggression is instrumental—that is, intended to reduce or eliminate the cause of the affect. Turning on an air conditioner, slapping someone who insults you, or shooting an attacker are instrumental actions.

Aggression resulting from aversive affect is called affective aggression, in contrast to aggression due to hostile thought or cognition. In one experiment, participants either experienced extreme temperatures or viewed pictures of weapons (Anderson, Anderson, & Denser, 1996). The former expressed increased anger and hostile attitudes; the latter did not. A series of experiments found that highly aggressive responses to harsh criticism were more likely among participants who believe aggression will make them feel better.

Social Learning and Aggression

Social learning theories provide a fourth explanation for aggressive behavior. Two processes by which aggression can be learned are imitation and reinforcement.

Imitation Some people, perhaps many, learn aggressive behavior by observing others commit aggressive acts and then imitating them. An experiment conducted by Bandura, Ross, and Ross (1961) illustrates such learning. Children observed an adult playing with toys. In one condition, the man played with Tinkertoys for 1 minute. Then he played with a 5-foot-tall, inflated rubber Bobo doll. He engaged in aggressive behavior toward the doll, including punching and kicking it and sitting on it. These actions, accompanied by shouted aggressive words and phrases, continued for 9 minutes. In the other condition, the man played with the Tinkertoys for the entire 10 minutes. Later, each child was intentionally frustrated. Then the child was left alone in a room with various toys, including a smaller Bobo doll. The children who had observed the aggressive model were much more aggressive toward the doll than those who had observed the nonaggressive model. They engaged in aggressive behavior such as kicking the doll and made comments similar to those they had observed.

Many children learn aggressive behavior at home. We noted in Chapter 3 (see Figure 3.1) that more than 90 percent of parents in the United States report using physical punishment to discipline their children. Children who are spanked or slapped for transgressions are learning that if someone's behavior breaks rules or makes you angry, it is okay to punish them physically. A longitudinal study of 717 boys found that boys who experienced harsh parenting practices at ages 10 to 12 were more likely to be involved in violent dating relationships at age 16 (Lavoie et al., 2002). More serious aggressive behavior within the family—child abuse, spouse abuse, or sibling abuse—can be explained by social learning theory. People who abuse their spouses or children often themselves grew up in families in which they either witnessed or were the targets of abuse (Gelles & Cornell, 1990). Growing up in a family in which some members abuse others teaches the child that it is acceptable to engage in physical aggression. It also teaches that occupants of certain roles, such as husbands or children, are appropriate targets for aggression.

Reinforcement Often people behave aggressively because they anticipate that the aggressive act will be rewarding to them. Muggers may attack a person to take his or her money. One child knocks down another to obtain the toy he or she desires. Students destroy library materials to improve their own chances—and worsen others' chances—of doing well on exams. These and other aggressive acts provide rewards to their perpetrators. According to social learning theory, the expectation of reward is a major motivation for aggression (Bandura, 1973). Social learning theory holds that aggressive responses are acquired and maintained—like any other social behavior—through experiences of reward. It appears, for instance, that Denise Farmer was killed for her money.

If the expectation of a reward motivates a person to aggress, which aggressive responses—if any—will he or she perform in a particular situation? The answer depends on two factors: (1) the range of aggressive responses the person has acquired, and (2) the cost-reward consequences the person anticipates for performing these responses. A person may be skilled, for example, in using a switchblade knife, a Molotov cocktail, or a sarcastic comment to harm others. People also consider the likely consequences of enacting particular aggressive behaviors in a particular situation. They try to calculate which actions will produce the rewards they seek, and at what cost. These considerations largely determine which aggressive acts, if any, people perform.

CHARACTERISTICS OF TARGETS THAT AFFECT AGGRESSION

In the preceding section, we discussed four potential sources of the motivation to aggress. Once aroused, such motives incline us toward aggressive behavior. Whether aggression occurs, however, also depends on

TABLE 12.1 MARITAL VIOLENCE BY HUSBANDS AND WIVES IN 1985

	INCIDENCE RATE (PERCENT REPORTING EACH ACT)		FREQUENCY (NUMBER OF INCIDENTS IN 1985)			
			Mean		Median	
Violent Behavior	Husband	Wife	H	W	H	W
1. Threw something at spouse	2.8	4.3	3.7	2.7	1.5	1.0
2. Pushed, grabbed, or shoved spouse	9.3	8.9	2.9	3.1	2.0	2.0
3. Slapped spouse	2.9	4.1	2.8	2.7	1.0	1.0
4. Kicked, bit, or hit with fist	1.5	2.4	3.9	2.9	1.5	1.0
5. Hit or tried to hit spouse with something	1.7	3.0	3.6	3.3	1.2	1.1
6. Beat up spouse	.8	.4	4.2	5.7	2.0	2.0
7. Threatened spouse with knife or gun	.4	.6	4.3	2.0	1.8	1.1
8. Used a knife or gun	.2	.2	18.6	12.9	1.5	4.0
Overall violence (1–8)	11.3	12.1	5.4	6.1	1.5	2.5
Wife-beating/husband-beating (4–8)	3.0	4.4	5.2	5.4	1.5	1.5

Source: Gelles and Strauss, 1988.

the characteristics of the **target**—the person toward whom the aggressive behavior is directed. In this section, we discuss three target characteristics: gender and race, attributions for an aggressor's attack, and retaliatory capacity.

Gender and Race

Aggression does not occur at random. If it did, we would observe aggressive behaviors by all kinds of people directed at targets of both genders, all ethnic groups, and all ages. In fact, aggression is patterned. First, aggressive behavior usually involves two people of the same race or ethnicity. This is true of aggression within the family—whether it involves child, sibling, spouse, or elder abuse—as most families are ethnically homogeneous. It is also true of violent crime—that is, assault, sexual assault, and murder.

The relationship between aggression and gender depends on the type of aggressive behavior. In cases of abuse within the family, men and women are about equally likely to be the targets. Boys and girls are equally likely to be abused by a parent. Wives abuse their husbands as often as husbands abuse their wives (Gelles & Strauss, 1988; see Table 12.1). The relationship between Sylvia and Michael described in Box 12.1 clearly indicates the interactional nature of domestic violence. Sylvia and Michael are both offenders and victims in their sometimes violent relationship.

In cases of violence involving current or former spouses, cohabitors, or intimates, women are the victims of 74 percent of the murders and 85 percent of the assaults and sexual assaults (Greenfeld et al., 1998). These patterns are found among Blacks, Hispanics, and Whites (Rennison, 2001). The rate of violence directed at women within the family is the same among Blacks, Hispanics, and Whites; it may be lower among Asian and Pacific Islanders (Johnson & Ferraro, 2001). Again, we see that men and women are equally likely to engage in aggressive behavior, but only men engage in violence.

Violence outside the family also involves primarily targets of one gender or the other. More than

BOX 12.1 The Faces of Domestic Violence

When Michael Wilkes left Sumter, SC, two decades ago, he was trying to escape what he perceived to be his bloodstained fate. He was only 21, but he was already in trouble—in petty trouble with the law, in big trouble at home. He didn't want to end up like his father: a career criminal, a wife beater, and dead by his wife's hand. So, baby-faced and jittery, Michael boarded a Greyhound for New York, fleeing an urge to exact vengeance on the stepmother who killed his father. He was running, too, from a failed, violence-ridden marriage of his own.

At first, things went well. Michael quickly found a job as a gofer for an art studio in Manhattan. He worked hard, the owner took him under his wing, and within 2 years he had settled into a cozy basement apartment in Queens. That's where he met Sylvia. Sylvia, the landlady's daughter, was voluptuous and dark-skinned, with fine features and twinkly eyes. She thought that Michael was "adorable and nice," and she was impressed by his cooking, especially his barbecue sauce. It was only a matter of time before they got involved. Michael's plans to re-create himself suddenly became more complicated.

Michael had succeeded in starting over as an industrious working man. But he thought that his hostility toward women was something he could not choose simply to rise above. He just felt it in him; if he got passionate about a woman, he was prepared to be betrayed, and his guard went up. And with Sylvia, the passion was intense. "I loved her to death," Michael says. It's a phrase that a man given to battering women probably shouldn't use.

Right from the start, Michael found himself falling into familiar patterns with Sylvia. "The distrust of a woman—I had it deep," he says. "I physically abused my first wife—smacks, punches, kicks. And then I turned around and did it again with Sylvia. The least little thing, I would fight her. I would hurt her. And she didn't deserve none of it." Speaking now as a sober-minded 41-year-old, after all he has been through and more precisely all he has put others through, Michael is trying hard to shoulder full responsibility for his actions.

Sylvia, however, argues that the dynamic was mutual all along. Michael wasn't the only one who had issues, she says. When they met, she had just escaped from a violent relationship that deteriorated to the point where the man was stalking her, armed with a knife. She was defensive and her fuse was short. "It's inaccurate to say only that Michael would beat me," Sylvia says, more forgiving of Michael than he is of himself. "He did. But we would beat each other. We would destroy the house. It became kind of dangerous for both of us. I didn't know who was going to kill who."

95 percent of reported cases of rape or sexual assault involve a male offender and a female victim. Sexual assaults represent about 6 percent of the violent crime reported in the United States. More than 80 percent of violent crime involves **aggravated assault**—an attack by one person on another with the intent of causing bodily injury. These assaults overwhelmingly involve men as both offender and victim. Most murders also involve two men.

These patterns indicate that the display of aggression is channeled by social beliefs and norms. Observing violence within one's family teaches a child that violence directed at children or spouses is acceptable. Similarly, beliefs and norms in American society encourage men to direct sexual aggression toward women. Men in our society frequently compete with each other for various rewards, such as influence over each other, status in a group, the companionship of a

BOX 12.1 Continued

It was a complex situation, murkier than the black-and-white portrayal of domestic violence that currently guides public policy. In that view, there's a batterer and a victim; the batterer is an ogre molded—misshapen—by patriarchal society; the victim, a mouse made helpless by it. There is only one happy ending: The batterer is punished, the victim liberated.

But Sylvia did not see Michael as a monster. She saw him as the product of a lousy childhood. She also saw him as a good provider and, in time, as the father of their two daughters. Nor did she see herself as defenseless but rather as the beneficiary of a good upbringing, as a self-reliant working woman and as someone who stood her ground. She never wanted Michael locked up; she wanted him to change. She wanted to rehabilitate her family, not to break it up. And in that way, Sylvia—like so many other women who refuse to call themselves victims—is a formidable challenge to doctrinaire thinking about the nature of domestic violence and how to combat it.

As the feminist movement grew in the 1970s, advocates for women struggled to build a network of shelters for battered women and to get domestic violence redefined as a serious crime. They lobbied for new state laws that would remove the police's discretion and mandate arrests for domestic violence. And they succeeded. Over the last two decades, and especially in the last 10 years, mandatory arrests have become a linchpin of the government's effort to address the issue; they are seen as a way to protect women, punish offenders, deter future violence, and send a message that spousal abuse won't be tolerated.

Now, though, a growing number of professionals are questioning the effectiveness of the mandatory arrest policies that advocates fought so long and hard for. Making more arrests and ensnaring more couples in the criminal justice system has not yet proved itself as a policy of deterrence, they say. And arrests sometimes backfire, especially in inner-city neighborhoods, causing unintended problems for some of the women that society is trying to protect. It would seem, they argue, that we are ignoring human nature, putting principle above the lives involved, and creating an unproductive antagonism between the system and some victims. Many battered women, for instance, don't want their men arrested or put away. The questioners, who include academics, crime experts, Black feminists, and social workers, are wondering aloud if we have come to rely too much on the law to solve a problem that defies easy solutions.

Source: "Fierce Entanglements" by Deborah Sontag from the *New York Times,* November 17, 2002.

woman, or other symbols of success. These competitions often lead to insults that provoke anger or direct physical challenges. There are norms in some groups that require men to defend themselves in such situations. The norms regarding interpersonal violence vary by subculture or culture. Observers have often described the American South as having norms that require men to defend themselves against insults. An experiment was designed to examine whether there is such a "culture of honor" in the South. Men from either the North or the South were intentionally bumped by another man (a confederate of the experimenters), who then called the participant an "asshole." Men who grew up in the North generally did not react. Men raised in the South were more upset, more primed for aggression (as shown by elevated testosterone levels), and more likely to engage in aggressive behavior (Cohen, Nisbett, Bowdle, & Schwarz, 1996).

It is likely that the incident caused aversive affect in men from the South.

Attribution for Attack

Direct attacks, both verbal and physical, typically produce an aggressive reaction (Geen, 1968; White & Gruber, 1982). Nevertheless, we withhold retaliation when we perceive that an attack was not intended to harm us. We are unlikely to respond aggressively, for example, if we see that a man who has smashed his grocery cart into us was trying to save a child from falling. Aggression following an attack is both more probable and stronger when we attribute the attack to the actor's intentions rather than to accidental or legitimate external pressures (Dyck & Rule, 1978).

Research indicates that attributions for physical abuse by one's spouse play a key role in determining the victim's response. In one study of 70 women, those living with their violent partner sometimes blamed themselves for the abuse. They attributed it to their incompetence, unattractiveness, or talking back to the partner. Others blamed situational factors such as stress. On the other hand, women who had left their abusive partner blamed him for the abuse (Andrews & Brewin, 1990).

An important influence on attributions is whether the attacker apologizes. An apology often states or implies that the harm done to us by another was unintentional. In one study, an experimenter made mistakes that caused the participant to fail at a task. When the experimenter apologized, the participants refrained from aggression against her. However, as the severity of the harm increased, the effect of the apology decreased (Ohbuchi, Kameda, & Agarie, 1989).

Retaliatory Capacity

Earlier, we identified four potential sources of the motivation to harm someone. Two involve emotions—frustration and aversive affect. When we experience intense frustration or anger, we usually are not concerned with the possible consequences of aggressive behavior. However, when we experience less intense emotions or are motivated by the expectation of reward, we may assess the likely consequences of an aggressive act. A consequence of particular importance is retaliation by the target.

Research suggests that the threat of retaliation reduces aggressive behavior. In one experiment, male and female participants were told to deliver electric shocks to another person; they could select the intensity of the shock. In one condition, researchers told the participants that after they had delivered shocks, the experiment would be over. In another condition, researchers told the participants that after they had delivered shocks, they would change places with the other person; that is, the other person could retaliate. Participants in the latter condition delivered significantly less intense shocks than participants in the former condition (Rogers, 1980). The finding that expected retaliation reduces aggressive behavior is consistent with social reinforcement theory; anticipated punishment leads to the inhibition of a behavior.

The fact that those who provoke us often have the power to retaliate gives rise to the idea of **displaced aggression,** defined as aggression toward a target that exceeds what is justified by provocation by that target; instigated by a different source, the aggression is displaced onto a less powerful or more available target (Marcus-Newhall, Pedersen, Carlson, & Miller, 2000). Displaced aggression is an everyday explanation for aggression directed toward partners, children, or pets. We have all heard the story of someone who has a bad day at work coming home and taking it out on her partner, daughter, or gerbil. But does it really occur? Contemporary social psychology textbooks rarely mention it, according to a content analysis. A meta-analysis of the research literature, based on 49 experimental studies, found that there is substantial evidence that displaced aggression does occur (Marcus-Newhall et al., 2000). Additional analyses indicated that the more negative the insult, attack, or frustration, and the more similar the instigator and the target, the greater the displaced aggression. (See the discussion of spillover from work to home of stress and frustration in Chapter 18.)

SITUATIONAL IMPACTS ON AGGRESSION

We turn now to characteristics of the situation in which the motive to aggress is aroused. We discuss five features: potential rewards, presence of models, norms, stress, and aggressive cues.

Reinforcements

Three rewards that promote aggression are direct material benefits, social approval, and attention. The material benefits that armed robbers, mafiosi, and young bullies obtain by using violence support their aggression. If material benefits are reduced—say, by vigorous law enforcement or by training the bullied children in karate—this type of aggressive violence will decline.

Despite the general condemnation of aggression, social approval is a second common reward for specific aggressive acts. Virtually every society has norms that approve aggression against particular targets in particular circumstances. We honor soldiers for shooting the enemy in war. We praise children for defending their siblings in a fight. Most of us, on occasion, urge friends to respond aggressively to insults or exploitation.

Attention is a third type of reward for aggressive acts. The teenager who aggressively breaks school rules basks in the spotlight of attention from peers even as he is reproached by school authorities. If we ignored aggressive behavior and rewarded cooperation with attention and praise, would this reduce aggressive acts? This is, in fact, what researchers found in a study of aggression among 27 male nursery-school children (Brown & Elliott, 1965).

Modeling

A second situational factor that promotes aggression is the presence of behavioral models. We noted earlier that aggressive behavior is often learned by observing and then imitating a model, as demonstrated by Bandura's research (Bandura et al., 1961). In specific situations, a model's aggressive behavior may encourage others to behave in similar ways.

In 1992, four Los Angeles police officers were on trial, charged with abusing a Black suspect, Rodney King. On May 6, 1992, a California jury returned verdicts of not guilty on all but one of the charges. Within 2 hours, a group of Black youth assembled at the intersection of Normandie and Florence Avenues; some began throwing rocks and bottles at passing cars. The crowd grew; some participants stopped cars and trucks, pulled the drivers out of their vehicles, and beat them. The aggressive acts of each participant served as a model for others.

Aggressive models provide three types of information that influence observers. First, through their actions, models demonstrate specific aggressive acts that are possible in a situation. The idea of stopping a car, for example, never crossed the minds of most participants until they saw a model do it. The model's acts identified available opportunities for aggressive action.

Second, models provide information about the appropriateness of aggression—about whether it is normatively acceptable in the setting. The behavior of the initial participants in the Los Angeles riot signaled that violence was appropriate. The live television coverage of the riot provided by several local stations unwittingly transmitted this message to tens of thousands of other Angelenos. The result of this process is termed an emergent norm (see Chapter 20).

Later, after the looting started, television announcers identified the exact locations, undoubtedly attracting additional looters. Research on the location and timing of riots by Black Americans between 1961 and 1968 demonstrates that riots in one city tended to spread (diffuse) to other cities in response to mass media coverage (Myers, 1997). Furthermore, prior riots in one city are associated with subsequent riots in the same city (Olzak, Shanahan, & McEneaney, 1996).

Third, models provide information about the consequences of acting aggressively. Observers see whether the model succeeds in attaining goals—whether the behavior is punished or rewarded. Observers are more likely to imitate aggressive behaviors that yield reward and avoid punishment. Hesitation by the police in the first 4 hours of the riot gave participants the impression that smashing cars and beating motorists would go unpunished. Had the police immediately surrounded the initial rioters at Florence and Normandie, observers would probably not have joined in, although they might have erupted in another way.

A successful aggressive model conveys the message that aggression is acceptable in a particular situation. This message matters little when observers are not motivated to do harm. But people who feel provoked and are suppressing their inclination to aggress—like thousands in Los Angeles that afternoon—often lose their inhibitions after observing an aggressive model. They are the most likely to imitate aggression. The aggressive motivation experienced by these men and women is a consequence of decades of racial discrimination and years of harsh treatment of Blacks by a White-dominated police force.

Five years later, in 1997, 85 percent of the 1,000 businesses damaged or destroyed in the riot had been rebuilt. But the underlying tensions remain, making future outbreaks likely (Purdum, 1997).

Norms

Just as there is a positive norm of reciprocity (see Chapter 11), there is a negative norm of reciprocity. This norm—"an eye for an eye, a tooth for a tooth"—justifies retaliation for attacks. In a national survey, more than 60 percent of American men considered it proper to respond to an attack on one's family, property, or self by killing the attacker (Blumenthal, Kahn, Andrews, & Head, 1972). Milder attacks call for milder retaliation. Data from a representative sample and a sample of ex-offenders found that endorsing "an eye for an eye," and similar statements, was related to frequency of violent behavior reported in the past year (Markowitz & Felson, 1998).

The negative reciprocity norm requires that the retaliation be proportionate to the provocation. Numerous experiments indicate that people match the level of their retaliation to the level of the attack (Taylor, 1967). In the heat of anger, however, we are likely to overestimate the strength of another's provocation and to underestimate the intensity of our own response. When angry, we are also more likely to misinterpret responses that have no aggressive intent as intentional provocation. Thus, even when people strive to match retaliation to provocation, aggression may escalate.

A study of 444 assaults against police officers revealed that escalation of retaliation due to mutual misunderstanding was the most common factor leading to violence (Toch, 1969). Typically, the police officer began with a routine request for information. The person confronted interpreted the officer's request as threatening, arbitrary, and unfair, and refused to comply. The officer interpreted this noncompliance as an attack on his own authority and reacted by declaring the suspect under arrest. Angered further by the officer's seemingly illegitimate assertion of power, the suspect retaliated with verbal insults and obscenities. From there the incident escalated quickly. The officer angrily grabbed the suspect, who retaliated by attacking physically. This sequence illustrates how a confrontation can spiral into violent aggression even when the angry participants feel they are merely matching their opponents' level of attack.

In an experimental study, two participants engaged in a competitive reaction time task; after each trial, the faster person could direct a noxious blast of noise at the slower person (Bushman, Baumeister, & Stack, 1999). The experiment was rigged so that the participant received the noise on one half of the trials (randomly selected) and could deliver noise on the other half. Over time, the participant increasingly matched the noise level delivered to him—clear evidence of reciprocity. This kind of spiral clearly occurred in the incident involving Kevin and Steven described in the introduction.

Stress

Stress increases the likelihood of aggressive behavior. There are several sources of stress within couples that may lead to intimate violence. Analyses of survey data from a large representative sample found that these stressors include a short relationship duration (that is, they don't know each other well), a mismatch in gender role definitions (one has traditional views, the other modern ones), substance abuse, and large number of children. These are related to intimate violence in part through their relationship to more frequent disagreements, and a more heated disagreement style (that is, disagreements escalate). (DeMaris, Benson, Fox, Hill, & Van Wyk, 2003). This process of escalation is captured by the colloquial phrase, "throwing everything including the kitchen sink" during a fight; as the argument/fight continues, each person reintroduces past grievances.

Social stressors, such as chronic unemployment and the experience of discrimination, are related to aggression because of their effects on frustration and anger. A study of the impact of economic distress on violence in married and cohabiting couples found that objective indicators such as reported household income were negatively related to violence; as household income increased, the frequency of physical violence decreased. However, regardless of actual income, when either partner wished that the other worked more hours (earned more money), this discrepancy between desire and reality was positively related to physical violence (Fox, Benson, DeMaris, & Van Wyk, 2002). This is excellent evidence of the importance of social psychological processes in physical aggression.

Analysis of data from a national survey replicated the finding that intimate violence was linked to economic distress; in addition, violence was related to neighborhood economic characteristics. Intimate violence occurred even more often and was more severe in economically disadvantaged neighborhoods (Benson & Fox, 2004). Results also indicated that African Americans and Whites with the same economic characteristics had the same rates of intimate violence. Other research shows that neighborhood characteristics also can reduce intimate violence. Research linking survey data, census data, and homicide data for the city of Chicago found that, even in disadvantaged neighborhoods, if residents shared a sense of collective efficacy (for example, "people in this neighborhood can be trusted") and a sense that neighbors could be counted on, rates of intimate violence were lower (Browning, 2003).

Moreover, situational stressors can produce high levels of aggression. Several studies have shown that temperature is related to the occurrence of violent crime. Analysis of data from 260 cities for the year 1980 found that the higher the average temperature, the higher the rate of assault, sexual assault, and murder (Anderson, 1987). A study of 102 incidents of collective violence or riots found that these events were much more likely to occur on days when the air temperature was between 71 and 90 degrees than when it was cooler (Baron & Ransberger, 1978). When it is hot, people are more irritable and thus more likely to respond to provocation with aggression. Interestingly, the increase in rates of violence peaks at temperatures around 75 degrees; higher temperatures lead to reduced violence, perhaps because people want to escape the heat and break off interaction with others (Cohn & Rotton, 1997).

Aggressive Cues

Whether the motivation to harm actually leads to aggressive acts depends in part on the presence of suitable aggressive cues in the environment (Berkowitz, 1989). These cues may intensify the aggressive motivation or lower inhibitions.

Aggressive cues can intensify the arousal that an angry person experiences. Seeing a gun or a violent film, for example, may further arouse a frustrated person and make him or her more aggressive. People who have been frustrated respond more aggressively when in the presence of a gun than in the presence of neutral objects (Carlson, Marcus-Newhall, & Miller, 1990); this so-called **weapons effect** occurs when people are already aroused. The effect involves cognitive priming; the sight of a weapon makes more accessible or primes aggression-related concepts or scripts for behavior (Anderson, Benjamin, & Bartholow, 1998). Thus, if family members become angry, the presence of a handgun or knife in the home creates the possibility of a weapons effect.

Aggressive cues may reduce the inhibitions that normally prevent aggressive behavior. As noted earlier, viewing aggressive behavior that is not punished increases the likelihood that an individual will behave aggressively.

© Spencer Grant/Stock, Boston

One way to reduce interpersonal aggression is to provide alternative methods of conflict resolution. In many communities, police officers are trained to use special techniques to resolve disputes, such as this one involving neighbors.

Aggressive cues also affect aggression by a process of ruminative thought (Marcus-Newhall et al., 2000). Such cues prime other thoughts and emotions related to aggression, the target, and the situation. As these become accessible, they may lead to increased anger, desire to aggress, or ideas about ways to aggress against the target. This process is perhaps characteristic of people who are "slow to burn"—who display aggression hours or days after the triggering incident.

REDUCING AGGRESSIVE BEHAVIOR

Aggressive behavior is often costly to individuals as well as to the groups and the society to which they belong. For this reason, there is great interest in identifying ways to reduce the frequency of aggressive behavior. Four methods that hold some promise are reducing frustrations, punishing aggressive behavior, providing nonaggressive models, and providing opportunities for catharsis.

Reducing Frustration

We noted that one cause of aggression is frustration, and that the stronger the frustration, the more likely aggression is to occur. Thus, one way to reduce aggression is to reduce the frequency or strength of frustration. A major source of frustration in American society is inadequate resources. Many cases of robbery, assault, and murder are motivated simply by desire for money or property; this appears to fit Denise Farmer's case. Two studies compared the homicide rates of American cities and states (Land, McCall, & Cohen, 1990) and nations (Gartner, 1990) in an attempt to identify the determinants of homicide. Both studies found that economic deprivation was the best predictor. Frustration and, therefore, aggression could be reduced if everyone had access to life's necessities (for example, through guaranteed income and universal health insurance programs).

Many of the frustrations we experience arise from conflicts with other people. The resulting anger often leads to aggressive attacks directed either at the target or at others perceived as similar to the target. Thus, another way to reduce aggressive behavior is to provide people with alternative means of resolving interpersonal conflicts. Some people take their conflicts to lawyers; others call the police. Recent innovations in dispute resolution involve the increasing use of professionally trained mediators and the training of selected community members in conflict resolution techniques. Conflict resolution programs using peers as mediators have been introduced in many high schools in recent years, partly in response to mass shootings in schools such as the one in Santee High School in 2001.

Punishment to Suppress Aggression

Because punishment is widely used to control aggression, we might assume that punishment or the threat of punishment are effective deterrents. In fact, however, threats are effective in eliminating aggression only under certain narrowly defined conditions (Baron, 1977). For threats to inhibit aggression, the anticipated punishment must be great and the probability that it will occur very high. Even so, threatened punishment is largely ineffective when potential aggressors are extremely angry and when they believe they will gain by being aggressive.

For actual punishment (in contrast to mere threats) to control aggression, certain other conditions must be met (Baron, 1977): (1) The punishment must follow the aggressive act promptly; (2) it must be seen as the logical outcome of that act; and (3) it must not violate legitimate social norms. Unless these conditions are met, people perceive punishment as unjustified, and they respond with anger.

The criminal justice system often fails to meet these conditions. The probability that any single criminal act will be punished is low, simply because most criminals are not caught. Even when criminals are caught, punishment rarely follows the crime promptly. Moreover, few criminals see the punishment as a logical or legitimate outcome of their act. Finally, they often have much to gain through their aggression. A longitudinal study of a sample of 1,497 adult offenders found that the perception of risk of sanctions by the system was not related to criminal activity. The most significant predictor of crime was the perception of opportunities to gain economically by breaking the law (Piliavin et al., 1986). As a result, the criminal justice system is not very effective in deterring criminal aggression.

Nonaggressive Models

Just as aggressive models may increase aggression, nonaggressive models may reduce it. Mahatma Gandhi, who led the movement to free India of British colonialism, used pacifist tactics that have since been imitated by protesters around the world. Laboratory research has also demonstrated the restraining influence of nonaggressive models. In one study (Baron & Kepner, 1970), participants observed an aggressive model deliver many more shocks to a confederate than required by the task. Other participants observed a nonaggressive model who gave the minimum number of shocks required. A control group observed no model. The results showed that the participants who observed the nonaggressive model displayed less subsequent aggression than the participants in the control group. The participants who observed the aggressive model displayed more subsequent aggression than the control group. Other research shows that nonaggressive models not only reduce aggression but can also offset the influence of aggressive models (Baron, 1971).

Catharsis

Infuriated by a day of catering to the whims of her boss, Ruth turned on her teenage son as he drove her home. "Why must you drive like a maniac?" she snapped. Ralph was stunned. He was driving a sedate 35 miles per hour and had done nothing to provoke his mother's aggression.

Did Ruth feel better after venting her anger on Ralph? Many people believe that letting off steam is better than bottling up hostility. A very old psychological concept, catharsis, captures this idea (Aristotle, *Poetics,* Book 6). **Catharsis** is the reduction of aggressive arousal brought about by performing aggressive acts. The catharsis hypothesis states that we can purge ourselves of hostile emotions by intensely experiencing these emotions while performing aggression. A broader view of catharsis suggests that by observing aggression as an involved spectator to drama, television, or sports, we also release aggressive emotions.

Numerous studies support the basic catharsis hypothesis: Aggressive acts directed against the source of anger do reduce physiological arousal (Geen & Quanty, 1977). We usually feel relieved after letting off steam against a tormentor. However, often we cannot direct aggression at the source. Even when we can, aggression may not bring catharsis because the fear of retaliation keeps us aroused. Aggression against a tormentor is also unlikely to reduce arousal under two other conditions: (1) if we feel our aggression is inappropriate to the situation and will make us look foolish or (2) if our internalized values oppose the aggression and make us feel guilty about it. Thus, catharsis results from aggression directed at the source only under limited conditions.

A second hypothesis that extends the original catharsis idea asserts that catharsis reduces subsequent aggression (Dollard et al., 1939; Freud, 1950). That is, once people have acted aggressively and released their anger, they are less inclined to future aggression. This hypothesis underlies advice such as "Put on the gloves and get the fight over with once and for all." Yet, with few exceptions, research has shown that performing aggressive acts will increase future aggression, not reduce it. This is true whether the initial aggression is a verbal attack, a physical attack, or even aggressive play. One particularly telling study, detailed in Figure 12.2, shows a clear catharsis effect after retaliating against a tormentor—followed by increased aggression against that same tormentor (Geen, Stonner, & Shope, 1975). Recent research using a similar experimental design reports the same results (Bushman et al., 1999).

Initial aggression promotes further aggression in several ways. First, initial aggressive acts produce **disinhibition**—the reduction of ordinary internal controls against socially disapproved behavior. Disinhibition is reflected in the reports of murderers and soldiers who commented that killing was difficult the first time but became easier thereafter. Second, initial aggressive acts serve to arouse our anger even further. Third, they give us experience in harming others. Finally, the catharsis that follows aggression and the pleasure of thwarting our tormentor

FIGURE 12.2 AGGRESSION, CATHARSIS, AND SUBSEQUENT AGGRESSION

Participants in a learning experiment who were antagonized by a confederate became physically aroused. Some participants had an opportunity to retaliate immediately, whereas others did not. Those who retaliated experienced catharsis—a sharp drop in arousal indicated by a drop in blood pressure. Those who did not retaliate remained aroused (left). Later, both groups of participants had a chance to attack the confederate (right). Those who had retaliated earlier delivered more intense shocks to the confederate than those who had not retaliated earlier. Thus the group that experienced catharsis earlier subsequently behaved more aggressively. These findings contradict the idea that releasing pent-up anger through retaliation reduces future aggression against a target. Instead, as this study showed, catharsis increases subsequent aggression.

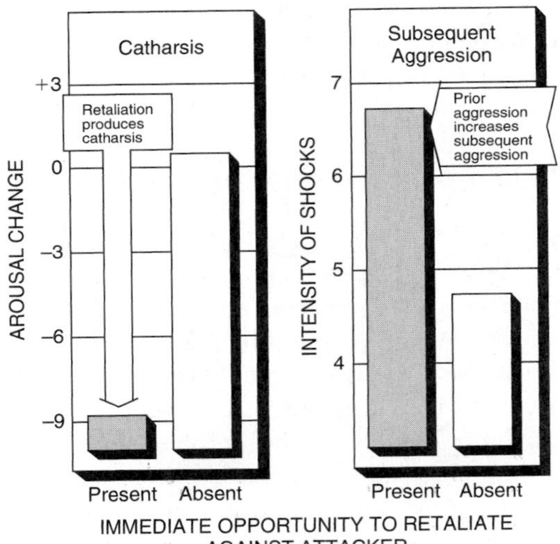

Source: Adapted from "Aggression, Catharsis, and Subsequent Aggression," by Green, Stonner, and Shope, *Journal of Personality and Social Psychology, 31,* 721–726. Copyright © 1980 by the American Psychological Association. Adapted with permission.

reward aggression, and rewarded behaviors are repeated more frequently.

AGGRESSION IN SOCIETY

In recent years, there has been increasing recognition that aggressive behavior is at the heart of several major social problems. This awareness is due, in part, to the widespread publicity given to certain incidents, such as the murderous assault by Charles Williams on his classmates. But fortunately, mass murders are rare. Much more common are other types of interpersonal violence, in which one person directs physical aggression toward another with the intent to injure or kill the target. This section discusses three specific aspects of interpersonal violence. First, it looks at the causes and consequences of sexual assault. Next, it examines the question of whether pornography contributes to sexual assault. Finally, it discusses whether television programming and video games contribute to violence.

Sexual Assault

Sexual assault refers to sexual contact or intercourse without consent, accomplished by the threat or the use of force. The greater the force used or the resulting injury, the more severe the assault. Most cases involve a male offender and a female victim from the same racial or ethnic group.

In some cases, the offender is motivated by sexual desire. In other cases, however, the offender's intent is to dominate, humiliate, or injure the victim. Sexual assault is one form of sexual aggression; sexual aggression is really a continuum, ranging from the use of bribes, through verbal pressure, the intentional use of alcohol or drugs, physical force, and kidnapping, to sexual murder (Global Forum for Health Research, 2002). One study of sexual coercion collected data from 165 men and 131 women who were new members of fraternities and sororities. Men were as likely as women to report being coerced into unwanted sexual contact, but only women reported being physically forced to do so (Larimer, Lydum, Anderson, & Turner, 1999).

In this section, we consider three aspects of sexual assault. These are characteristics of society that encourage assault, characteristics of those who commit assault, and characteristics of the victims.

Rape-Prone Society What causes sexual aggression? There are several answers to this question. One is that a specific set of cultural beliefs and practices creates conditions that encourage rape. In a rape-prone society, the sexual assault of women by men is allowed or overlooked (Sanday, 1981). Rape-prone societies share several characteristics. First, there are high levels of interpersonal violence. Second, women are viewed as property and are generally subservient to men. Third, men and women are regularly separated, for example, during religious rituals. The United States is a rape-prone society. Rates of violent crime are high. Until recently, men dominated women politically, economically, and sexually. Also, there is a continuing separation of men and women, for example, in athletic programs and in many workplaces (see Chapter 18). Sexual assault has been studied in many societies, including Australia, Bangladesh, India, Indonesia, and South Africa. Many societies that are characterized by these cultural beliefs and practices have substantial rates of sexual violence against women (Global Forum for Health Research, 2002).

Perpetrators of Sexual Assault Of course, individuals rather than societies commit rape. A second approach to determining what causes sexual assault is to identify the characteristics of men that may be related to their aggressive behavior. Research suggests that some men are sexually aggressive—that is, they rely on aggressive behaviors in their relationships with women (Malamuth, Heavey, & Linz, 1993). These men attain high scores on measures of the desire to dominate women and of hostility toward women. They have a variety of attitudes that facilitate aggression toward women (Malamuth, 1984). These attitudes include various rape myths, such as the belief that women secretly desire to be raped and enjoy it, that victims cause rape, and that other men are prone to rape (see Box 12.2). These men are sexually aroused by portrayals of rape. In laboratory studies, such men

BOX 12.2 Test Yourself: Rape Myths

Among the causes of sexual aggression are cultural beliefs that encourage rape. These beliefs are **rape myths**—prejudicial, stereotyped, and false beliefs about rape, rape victims, and persons who commit rape (Burt, 1980). Examples of these myths are "Only bad girls get raped," "Women ask for it," and "Rapists are sex-starved, insane, or both." These beliefs create a climate that encourages sexual assault and is suspicious of and hostile toward victims.

An attitude scale that is widely used to assess these beliefs is the *Rape Myth Acceptance Scale,* developed by Burt (1980). Some of the items that make up the scale are reproduced here.

Read each statement and circle the appropriate response: Strongly Agree (StA), Agree (A), Slightly Agree (SlA), Don't know (?), Slightly Disagree (SlD), Disagree (D), or Strongly Disagree (StD).

1. A woman who goes to the home or apartment of a man on their first date implies she is ready to have sex.

   StA A SlA ? SlD D StD

2. Any healthy woman can successfully resist a rape if she wants to.

   StA A SlA ? SlD D StD

3. When women go around braless or wearing short skirts and tight tops, they are just asking for trouble.

   StA A SlA ? SlD D StD

4. If a girl engages in necking or petting and she lets things get out of hand, it is her own fault if her partner forces sex on her.

   StA A SlA ? SlD D StD

5. If a woman gets drunk at a party and has intercourse with a man she's just met there, she should be considered "fair game" to other males at the party.

   StA A SlA ? SlD D StD

Answer the following questions by selecting the best answer.

6. What percentage of women who report a rape would you say are lying because they are angry and want to get back at the man they accuse?

   Almost all About 3/4 About half
   About 1/4 Almost none

A person comes to you and claims they were raped. How likely would you be to believe their statement if the person were,

7. Your best friend?

   Always Frequently Sometimes Rarely Never

are more likely to aggress against a woman who has mildly insulted or rejected them (Malamuth, 1983).

The tendency of men to be sexually aggressive is stable over time. Researchers collected data on 423 young men, including measures of hostility toward women and attitudes supportive of violence. Ten years later, they reinterviewed 176 of the men and 91 of their female partners (Malamuth, Linz, Heavey, Barnes, & Acker, 1995). The characteristics measured 10 years earlier predicted which men were sexually aggressive toward their partners, as reported both by the men and by their partners.

Other research focuses specifically on the tendency of a man to use aggression to obtain sexual gratification. This tendency is measured by asking men whether they have ever used verbal threats or physical force to obtain sexual activity with women (Koss & Oros, 1982). Men who report the use of threat or force to obtain intercourse are termed sexually assaultative. Men who report the use of threat or force to obtain

BOX 12.2 Continued

8. A neighborhood woman?

Always Frequently Sometimes Rarely Never

9. A Black woman?

Always Frequently Sometimes Rarely Never

10. A White woman?

Always Frequently Sometimes Rarely Never

As you can see, the scale is made up of two types of items. One type—items 1 through 5—asks the respondent whether rape is justified in particular situations. Agreement with an item indicates acceptance of that rape myth. The other type—items 6 through 10—asks the respondent whether he or she believes the claims of women who have been raped. For item 6, the belief that women who report rape are lying is considered a rape myth. For items 7 through 10, the belief that particular types of women (for example, Black women) lie about rape is a rape myth.

A high score on this scale indicates that the respondent believes most sexual activity is a direct result of the woman's behavior and not due to threat or force. A male respondent with a high score believes that if he has sexual intercourse with a woman who comes home with him on the first date, it is not rape even if she offers some resistance. This is one of the dangers of rape myths: They provide scripts that legitimize sexual activity to which the woman may not have overtly consented. The other type of rape myth—that claims of rape are not true—creates an environment in which such claims are not believed and, therefore, sexual assault is not punished.

There have been many studies of the correlates of endorsing rape myths. One review summarizes the findings of 72 studies (Anderson, Cooper, & Okamura, 1997). Men, older persons, and persons from lower socioeconomic class backgrounds are more likely to hold such attitudes. Acceptance of rape was associated with traditional beliefs about gender roles, an adversarial view of male-female relationships, and conservative political beliefs. These results are consistent with the theory that rape myth acceptance is the result of socialization to gender-typed, conservative beliefs.

other types of intimacy (for example, breast fondling) are intermediate in the use of sexual aggression (see Table 12.2). Koss and Leonard (1984) report that a high level of sexually aggressive behavior is associated with the acceptance of rape myths and with the belief that sexual aggression is normal. The use of coercion or violence can also be viewed as an attempt to maintain or gain control in a relationship (Johnson & Ferraro, 2001). Sexual violence may reflect the fear that one's control in a relationship—over sexual behavior or over the partner generally—is threatened. A study of motivational factors found that men who had been sexually assaultative were angry toward women and had feelings of low power or inadequacy (Lisak & Roth, 1988).

This research suggests that men who commit sexual assault have learned a script for heterosexual interactions that includes the use of force (Huesmann, 1986). This script suggests the use of verbal abuse or physical force to exercise influence over or obtain

TABLE 12.2 A TYPOLOGY OF SEXUALLY ASSAULTATIVE MEN

Type	Behavior(s)	Percentage
Sexually nonaggressive	Consensual sexual activity	59.0%
Sexually coercive	Accomplished intercourse by a) threatening to end the relationship; or b) after "continual discussion and argument"; or c) saying things they didn't mean.	22.4%
Sexually abusive	Accomplished kissing or petting by use of physical force; or threatened or used force in an unsuccessful attempt to obtain intercourse.	4.9%
Sexually assaultative	Accomplished vaginal, oral, or anal intercourse by threat or use of force.	4.3%

Sample: 1,846 male college students

Source: Adapted from "Sexually Assaultive Men" by Koss and Leonard in *Pornography and Sexual Aggression*, Malamuth and Donnerstein (eds.). Copyright © 1984, with permission from Elsevier.

sexual gratification from a woman. Once learned, it is used to regulate behavior in various situations. Research suggests that it is learned in childhood (Jacobson & Gottman, 1998). It is most likely to be learned when the child observes aggression frequently, is reinforced for aggressive behavior, and is the object of aggression.

Victims of Sexual Assault The victims of sexual assault are primarily women between the ages of 15 and 24. Some women—probably a minority of all victims—are assaulted by men they do not know. Assaults by strangers often occur outdoors, in parks or deserted parking lots, or in the victim's residence. The offenders in these cases are often opportunistic, attacking any woman who is available or appears to be vulnerable.

More often, women are assaulted by someone they know. This may be a man they are dating (date rape) or a neighbor or coworker (acquaintance rape). The victims in most cases of date rape are young, single women—often high school or college students. The couple goes out on a date. After dinner or a movie or dancing, they drive to a secluded spot or to his or her residence. The man initiates physical intimacy, perhaps at first with the woman's cooperation. At some point, the woman indicates that she does not wish to continue, but the man persists and uses force.

Several factors contribute to the occurrence of sexual assault. One is alcohol, which is a factor in up to one half of all sexual assaults (Abbey, Ross, McDuffie, & McAuslan, 1996); it lowers internal inhibitions that might otherwise prevent aggression. Alcohol also influences judgment; both surveys and experiments provide evidence that people are more likely to engage in risky behavior—including entering a risky situation—when they have been drinking. In fact, some men use alcohol or drugs intentionally to make a woman more likely to voluntarily take such risks (Abbey, McAuslan, Zawacki, Clinton, & Buck, 2001).

Another factor in sexual assault is misinterpreted verbal or nonverbal messages. The woman may engage in some behavior that the man incorrectly interprets as a sexual invitation (Bondurant & Donat, 1999). Cultural beliefs are a third factor. According to one survey of 14- to 17-year-olds, both male and female teenagers believe that a man is justified in forcing a woman to have intercourse if she gets him sexually excited, leads him on, or has dated him for a long time (Goodchilds & Zellman, 1984).

Many women who experience forced, nonconsensual sexual activity do not perceive the experience as rape (Kahn, Mathie, & Torgler, 1994). This may be because their experience—being assaulted by someone they know during a date after some sexual foreplay—does not match their script for rape: a violent attack by a stranger. Researchers asked women to write a description "of events before, during, and after a rape." Women who reported that they had been forced to have sex but replied "no" to the question, "Have you been raped?" were more likely to describe rape as an attack by a stranger than women who answered "yes." Similarly, women's script for a loving relationship is one of equality and romance (which encourages us to overlook our partner's bad behavior), with an emphasis on male rather than female sexual drive (Lloyd & Emery, 2000). An experience of aggression does not fit this script, and so may be ignored. This is a good example of the power of scripts to shape experience.

Pornography and Violence

Earlier, we suggested that some men learn scripts that encourage sexual aggression. One possible source of such scripts is growing up in an abusive family. Another is viewing or reading pornography.

On January 24, 1989, Florida prison officials executed convicted serial murderer Ted Bundy. Before his execution, Bundy admitted killing at least 24 young women. Authorities believe he killed many others. Bundy claimed that he was addicted to violent pornography and that it had contributed to the murders he committed. Such claims create great interest in the connection between pornography and violence. Considerable research has been conducted on this link.

Nonaggressive Pornography Various studies have shown that the effect of pornography on behavior depends on what the pornography portrays. Pornography that explicitly depicts adults engaging in consenting sexual activity is termed **nonaggressive pornography** or erotica. Reading or viewing nonaggressive pornography creates sexual arousal (Byrne & Kelley, 1984), usually through the mechanism of cognitive and imaginative processing. Nonaggressive pornography by itself does not produce aggression toward women (Donnerstein, 1984). However, when the viewer's inhibitions are lowered, it may do so. Research has also demonstrated that when anger or frustration is induced in men, and they then view nonaggressive pornographic images, aggressive behavior toward a woman may result (Donnerstein & Barrett, 1978). The mechanism is thought to be transfer of arousal: The sexual arousal that results from viewing pornography is added to the arousal induced by the anger, resulting in aggression.

In the past 20 years, many nationally distributed R-rated movies have contained scenes of apparently consensual sexual activity that are degrading or humiliating to a woman. What is the effect of viewing such scenes? In one experiment, men and women saw either scenes from *9½ Weeks* and *Showgirls* or scenes from an animated film. They subsequently read and evaluated a magazine story about a date rape or a stranger rape. Men who saw the sexual videotape were more likely to say that the victim of the date rape enjoyed it and "got what she wanted" (Milburn, Mather, & Conrad, 2000).

Aggressive Pornography The term **aggressive pornography** refers to explicit depictions of sexual activity in which force is threatened or used to coerce a woman to engage in sex (Malamuth, 1984). Unlike erotica, aggressive pornography has lasting effects on both attitudes and behavior.

In a study of its effects on attitudes (Donnerstein, 1984), men viewed one of three films featuring aggression, nonaggressive sexual activity, or aggressive sexual activity. Following the film, the participants completed several attitude scales, including one that measures acceptance of rape myths (see Box 12.2). Men who saw the films depicting aggression or aggressive sexual activity scored higher on the rape myth acceptance scale than men who saw the film depicting nonaggressive sexual activity. These men also indicated greater willingness to use force to obtain sex. The fact that both films depicting aggression affected attitudes more than the nonaggressive film suggests that it is aggression rather than explicit portrayals of sex that influences attitudes toward sexual aggression. In another study, Malamuth and Check (1981) found that viewers of films that portrayed sexual aggression as having positive consequences showed greater acceptance of rape myths and of violence toward women. Finally, repeated exposure to sexually violent films is associated with more calloused attitudes toward female victims of domestic abuse (Mullen & Linz, 1995).

The mechanism by which viewing these films influences attitudes may be priming (Malamuth, 1984). Priming refers to the process whereby exposure to a stimulus makes psychologically related cognitions more accessible. Exposure to aggressive pornography makes cognitions related to sex and aggression more accessible, and these cognitions influence later judgments. The effect of aggressive pornography on attitudes is indirect (Malamuth & Briere, 1986). For most people, viewing such portrayals does not produce sexual arousal. (In fact, it may produce the aversive emotion of disgust.) It does, however, affect thought patterns in ways that make many viewers more tolerant of sexual aggression. If other influences occur, such as peer encouragement or hostility toward a woman, men with these attitudes may commit sexual assault (Malamuth, 1989).

Exposure to aggressive pornography also influences behavior, especially aggression toward women. This effect is evident in a study by Donnerstein and Berkowitz (1981). Male participants were either angered or treated neutrally by a male or female confederate. They next viewed one of four films: a neutral film, a nonaggressive pornographic film, or one of two aggressive pornographic films. In the latter films, a young woman is shoved around, tied up, stripped, and raped. In one version, she finds the experience disgusting, whereas in the other she is smiling at the end. Following the film, the men were given an opportunity to aggress against the male or female confederate by delivering electric shocks. The films did not affect aggression toward the male confederate. Participants who saw the aggressive films delivered more intense electric shocks to the female confederate.

The fact that aggressive pornography produces aggressive behavior reflects three influences: sexual arousal, aggressive cues, and reduced inhibitions. Some men experience high levels of arousal in response to such portrayals. Moreover, such pornography portrays women as targets of aggression. In the experiment conducted by Donnerstein and Berkowitz, the film created an association in the viewer's mind between the victim in the film and the woman who angered him, suggesting aggression toward the latter. Note that aggressive films led to increased violence toward the female confederate and not the male confederate, a finding consistent with this interpretation. Finally, these films also may reduce inhibitions to aggression, by suggesting that aggression directed toward women has positive outcomes.

One important question is whether we can generalize from the results of laboratory research to natural settings. Does the viewing of aggressive pornography in nonlaboratory settings contribute to violence against women? Some anecdotal evidence comes from sex offenders and therapists who work with them. These persons describe many actual sexual offenses that closely mimic activities depicted in pornographic materials available to the offender before the offense (Court, 1984). Another form of evidence comes from researchers who have studied the correlation between the availability of pornography and rates of violent crime. For instance, Baron and Straus (1984) obtained circulation figures for eight "sex magazines" (including *Playboy, Chic,* and *Hustler*) and the rates of assault, rape, and murder reported to police for each state. Of several variables included in their study, the circulation index (number of magazines sold per 100,000 residents) was the strongest predictor of rates of rape.

Media Violence and Aggression

Violent Television and Aggression Evelyn Wagler was carrying a 2-gallon can of gasoline back to her stalled car. She was cornered by six young men who forced her to douse herself with the fuel. Then one of the men tossed a lighted match. She burned to death. Two nights earlier, a similar murder had been depicted on national television.

Violence pervades television: stabbings, shootings, poisonings, and beatings. Not just humans, but even cartoon characters torment each other in astonishingly creative ways. Both heroes and villains perform aggression on television. There are three to five violent incidents per hour during prime time,

© Rick Kopstein

By age 18, the average American child will see 200,000 violent acts on television. Such exposure directly contributes to aggressive behavior by the viewer.

and 20 to 25 violent incidents per hour during Saturday morning programs for children (American Psychological Association, 1993).

Do these depictions of violence encourage viewers to behave aggressively? Is our society more violent because of violence shown by the media? These are vexing questions, especially when the average American child spends more time watching television than engaging in any other waking activity (Huston & Wright, 1982). By age 18, the average American child is likely to have seen about 200,000 violent acts on television, including 40,000 homicides ("Violence in our culture," 1991). The aforementioned incident suggests that in some cases, media violence induces aggression. Yet establishing a causal link between watching television and subsequent aggression has been difficult and highly controversial (Milavsky, Kessler, Stipp, & Rubens, 1983).

There are five processes that explain why exposure to media violence might increase aggressive behavior (Huesmann & Moise, 1996):

1. *Imitation.* Viewers learn specific techniques of aggression from media models. Social learning through imitation evidently played a part in the violent attack on Evelyn Wagler.
2. *Cognitive priming.* Portrayals of violence activate aggressive thoughts and pro-aggression attitudes.

As noted in Chapter 6, the activation of an attitude increases the likelihood that it will be expressed in behavior.

3. *Legitimation/justification.* Exposure to violence that successfully attains goals and has positive outcomes (for example, punishes wrongdoers) legitimizes aggression and makes it more acceptable.
4. *Desensitization.* After observing violence repeatedly, viewers become less sensitive to aggression. This makes them less reluctant to hurt others and less inclined to ease others' suffering.
5. *Arousal.* Viewing violence on television produces excitement and physiological arousal, which may amplify aggressive responses in situations that would otherwise elicit milder anger.

Research has shown that all of these processes operate in linking media violence to aggression (Murray & Kippax, 1979).

To investigate these processes, researchers have conducted hundreds of laboratory and field experiments. Children, adolescents, and adults have been shown portrayals of aggression, both live and on film. Some of these scenes have been specially prepared, whereas others have been taken directly from popular television shows. With rare exceptions, these experiments show that observing violence increases subsequent aggression by viewers (Comstock, 1984; Friedrich-Cofer & Huston, 1986). Moreover, these results have been found in experiments with boys and girls of all ages, races, social classes, and levels of intelligence, in many countries (Huesmann & Moise, 1996).

These experiments demonstrate a causal link between viewing violence and aggression. However, our ability to generalize from these experiments to the effects of media violence on society has been challenged. For one thing, few of us watch several violent scenes within a brief period of time. For another, few of us watch movies or television programs in research settings. To complement the experimental approach, researchers have turned to studies of the correlation between exposure to violence on television and aggressive behavior in everyday settings.

Many studies have been carried out with thousands of participants of widely varying age, social class, and ethnicity. These studies consistently report moderate positive correlations between television viewing and aggressive behavior (Friedrich-Cofer & Huston, 1986). In one correlational study (Singer & Singer, 1981), researchers measured the television viewing experiences and spontaneous aggression of preschoolers several times over a period of 1 year. Consistently, aggressive behavior during this period was linked to heavy viewing of aggressive action-adventure programs or cartoons. The level of aggressiveness these children exhibited 4 to 5 years later also correlated positively with the amount of violent television they had viewed as preschoolers. Another study (Johnson, Cohen, Smailes, Kasen, & Brook, 2002) recruited 707 families in 1975 with a child between ages 1 and 10. Data were collected on television viewing by the children during 1991–1993, and on aggressive behavior through 2000. Time spent watching television at age 14 was positively related to aggressive acts against others reported at age 16 and age 22 (see Figure 12.3).

The correlations in these studies may reflect a causal impact of viewing television violence on subsequent aggression, but correlations do not prove causality. The correlation also might suggest that aggressive children prefer violent programs. The growing body of evidence suggests that the relationship between aggression and television viewing is circular (Friedrich-Cofer & Huston, 1986). Because aggressive children are relatively unpopular with their peers, they spend more time watching television. This exposes them to more violence, teaches them aggressive scripts and behaviors, and reassures them that their behavior is appropriate. When they then try to enact these scripts in interactions with others, they become even more unpopular and are driven back to television—and the vicious cycle continues (Huesmann, 1986; Singer & Singer, 1983). In a related study, researchers measured the aggressiveness of 210 men and 210 women, and then gave them descriptions of several films (Bushman, 1995). Men and women with high aggressiveness scores were more likely to choose a violent film than persons who scored low on aggression.

A meta-analysis of all of the available research reports that virtually every study—whether cross-sectional ($n = 86$), longitudinal ($n = 46$), or experi-

FIGURE 12.3 TELEVISION VIEWING AND SUBSEQUENT AGGRESSION

Families with a child ages 1 to 10 were recruited in 1975. Television viewing by the child was measured during 1991–1993, and aggressive behavior was measured by self-report in the year 2000, and by checking the records of New York State and the FBI for arrests and criminal charges. The results showed that television viewing at age 14 was associated with aggressive acts toward others at ages 16 or 22. Notice that the relationship was much stronger for boys. Also, the effect of viewing television was greater on boys and girls with a history of aggressive actions toward others prior to age 14.

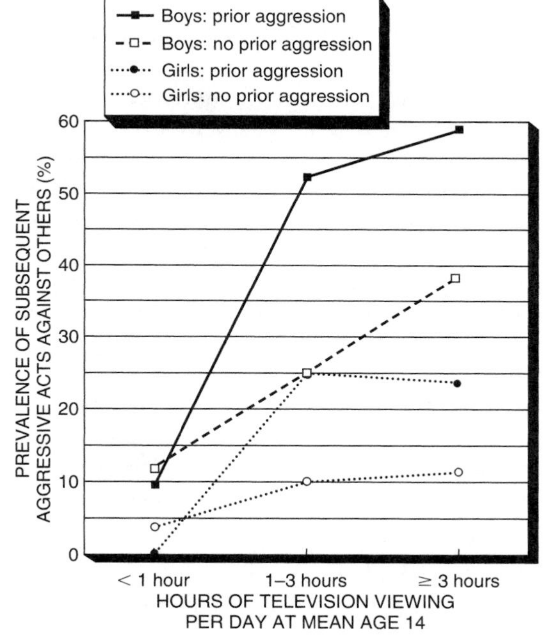

Source: Adapted from Johnson, Cohen, Smailes, Kasen, and Brook, 2002.

mental ($n = 152$)—finds a significant relationship between exposure to violence via the media and aggression (Anderson & Bushman, 2002). There is no question that violence in the mass media contributes to violence in society. The question is, "What should be done about violence on television?" Comprehensive rating systems and V-chips that enable parents to exercise some control over what children watch are important steps. There are several changes that the television industry could make as well (Kunkel, 1997). First, the number of violent incidents portrayed on both prime-time and Saturday morning programs for children should be reduced. Second, more portrayals of violent acts should include their negative consequences, such as pain and suffering for victims, family members, and the community. Third, programs should feature more alternatives to the use of violence in resolving interpersonal conflicts, such as problem-focused discussion and mediation. Finally, when violence is portrayed, antiviolence themes should be included; for example, perpetrators could be shown as remorseful about the need to resort to violence. Changes such as these will not eliminate violence in interpersonal relationships. But they will help.

Violent Video Games and Aggression Research suggests also a relationship between exposure to violence in video games and aggression. First, many video games portray interpersonal violence. A content analysis of 33 popular Nintendo and Sega games found that 80 percent of them involved aggression or violence as part of the strategy. About half encouraged violence directed at people, and 21 percent included violence directed at women (Dietz, 1998). Playing violent video games involves two theoretical principles discussed earlier—observational learning and reinforcement. Players are learning behavior patterns or scripts that include violence as a means of achieving higher scores, and are rewarded when they do (Funk, Flores, Buchman, & Germann, 1999). Second, there is a relationship between playing video games and aggression. A survey of college students asked participants to name their five favorite games and to rate how often they played each one and how violent it was. Both men and women who reported playing violent video games were more likely to report engaging in eight types of aggressive behavior. The relationship was stronger for participants who had high scores on a measure of aggressive personality, and for men. In a subsequent experiment, participants played a violent or a nonviolent video game for 30 minutes and then were asked to list their thoughts; those who played the violent game listed more aggressive thoughts, reflecting a priming effect. Participants returned 1 week later and played one of the games for 15 minutes. They then played the competitive reaction time task described earlier (Bushman et al., 1999). Those who played the violent game reacted more aggressively to noise blasts supposedly administered by their partner (Anderson & Dill, 2000).

SUMMARY

Aggression is behavior intended to harm another person that the target person wants to avoid.

Aggression and the Motivation to Harm There are four main theories regarding the motivation for aggression. (1) Aggression is based on a biological instinct that generates hostile impulses demanding release. (2) Aggression is a drive elicited by frustration. (3) Aggressive behavior is a response to aversive emotional arousal, such as anger. (4) Aggression is a learned behavior motivated by rewards.

Characteristics of Targets that Affect Aggression Once aggressive motivation has been aroused, target characteristics influence whether aggressive behavior occurs. (1) Aggression is more likely if the target is of the same race or ethnicity. The target's gender also influences the response. (2) When we are attacked, our response is influenced by the attributions we make about the attacker's intentions. (3) We are less likely to engage in aggression toward a target whom we perceive as capable of retaliating. We may, however, engage in displaced aggression.

Situational Impacts on Aggression Situational conditions are important influences on aggressive behavior. (1) Rewards that encourage aggression include material benefits, social approval, and attention. (2) Aggressive models provide information about available options, normative appropriateness, and consequences of aggressive acts. (3) The negative reciprocity norm encourages aggressive behavior in certain situations. (4) Aggressive behavior is more likely when stressors, such as high temperature, are present. (5) Aggressive behavior is more likely in the presence of aggressive cues, especially weapons.

Reducing Aggressive Behavior (1) Reducing frustration levels by guaranteeing everyone the basic necessities would reduce the aggression motivated by desire for these rewards. (2) Punishment is effective in controlling aggression only when it follows the aggressive act promptly, is seen as the logical outcome of that act, and does not violate social norms. (3) Nonaggressive models reduce the likelihood of aggression and can offset the effect of aggressive models. (4) Catharsis—the reduction of angry arousal—may follow aggressive acts, but such acts still may promote later aggression.

Aggression in Society Interpersonal violence is a serious problem in American society. (1) Sexual assault is facilitated by societal characteristics, such as male domination of women, and by scripts that encourage male aggression toward women. (2) Nonaggressive pornography has little effect on attitudes or behavior. Exposing men to aggressive pornography in laboratory settings is associated with greater acceptance of rape myths and aggression toward women. (3) Experimental research shows that observing violence on film and television increases subsequent aggression. Correlational studies find a positive relationship between viewing violent programs and aggressive behavior in everyday settings. (4) Research indicates that playing violent video games is associated with aggressive behavior toward others.

LIST OF KEY TERMS AND CONCEPTS

aggravated assault (p. 306)
aggression (p. 300)
aggressive pornography (p. 319)
aversive affect (p. 303)
catharsis (p. 314)
death instinct (p. 301)
disinhibition (p. 314)
displaced aggression (p. 308)
frustration (p. 301)
frustration-aggression hypothesis (p. 301)
nonaggressive pornography (p. 319)
rape myths (p. 316)
sexual assault (p. 315)
target (p. 305)
weapons effect (p. 311)

13

INTERPERSONAL ATTRACTION AND RELATIONSHIPS

Introduction

Who Is Available?

Routine Activities

Proximity

Familiarity

Who Is Desirable?

Social Norms

Physical Attractiveness

Exchange Processes

The Determinants of Liking

Similarity

Shared Activities

Reciprocal Liking

The Growth of Relationships

Self-Disclosure

Trust

Interdependence

Love and Loving

Liking Versus Loving

Passionate Love

The Romantic Love Ideal

Love as a Story

Breaking Up

Progress? Chaos?

Unequal Outcomes and Instability

Differential Commitment and Dissolution

Responses to Dissatisfaction

Summary

List of Key Terms and Concepts

INTRODUCTION

Dan was looking forward to the new semester. Now that he was a junior, he would be taking more interesting classes. He walked into the lecture hall and found a seat halfway down the aisle. As he looked toward the front, he noticed a very pretty young woman removing her coat; as he watched, she sat down in the front row.

Dan noticed her at every class; she always sat in the same seat. One morning, he passed up his usual spot and sat down next to her.

"Hi," he said. "You must like this class. You never miss it."

"I do, but it sure is a lot of work."

As they talked, they discovered they were from the same city and both were economics majors. When the professor announced the first exam, Dan asked Sally if she wanted to study for it with him. They worked together for several hours the night before the exam, along with Sally's roommate. Dan and Sally did very well on the exam.

The next week, he took her to a film at a campus theater. The week after, she asked him to a party at her dormitory. That night, as they were walking back to her room, Sally told Dan that her roommate's parents had just separated and that her roommate was severely depressed. Dan replied that he knew how she felt because his older brother had just left his wife. Because it was late, they agreed to meet the next morning for breakfast. They spent all day Sunday talking about love, marriage, parents, and their hopes for the future. By the end of the semester, Sally and Dan were seeing each other two or three times a week.

At its outset, the relationship between Dan and Sally was based on **interpersonal attraction**—a positive attitude held by one person toward another person. Over time, however, the development of their relationship involved increasing interdependence and increasing intimacy, or pair relatedness (see Figure 13.1).

The development and outcome of personal relationships involves several stages. This chapter discusses each of these stages. Specifically, it considers the following questions:

1. Who is available? What determines with whom we come into contact?
2. Who is desirable? Of those available, what determines with whom we attempt to establish relationships?
3. What are the determinants of attraction or liking?
4. How do friendship and love develop between two people?

FIGURE 13.1 LEVELS OF PAIR RELATEDNESS

A relationship between two persons develops through a series of stages, or levels. Before it begins, the two are unrelated. The next level occurs when one person becomes aware of the other. The third level involves interaction. Relationships that develop beyond the surface-contact stage involve progressively greater mutuality—disclosure of personal information, trust, and interdependence.

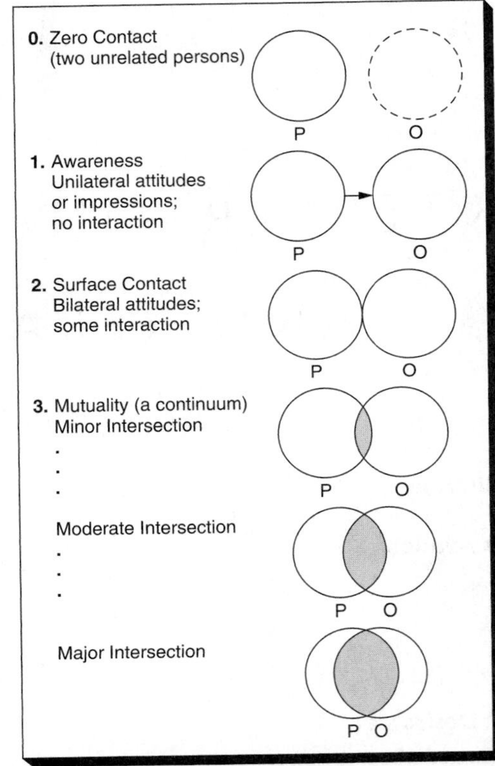

Source: Adapted from Levinger, 1974.

5. What is love?
6. What determines whether love thrives or dies?

WHO IS AVAILABLE?

Dozens or thousands of persons may go to school or live or work where you do. Most of them remain strangers—persons with whom you have no contact. Those persons with whom we come into contact, no matter how fleeting, constitute the field of **availables**—the pool of potential friends and lovers (Kerckhoff, 1974). What determines who is available? Is it mere chance that George rather than Bill is your roommate, or that Dan met Sally rather than Heather? The answer, of course, is no.

Two basic influences determine who is available. First, institutional structures influence our personal encounters. The admissions office of your school, the faculty committees that decide degree requirements, and the scheduling office all influence whether Dan and Sally enroll in the same class. Second, individuals' personal characteristics influence their choice of activities. Dan chose to take the economics class where he met Sally because of an interest in that field and a desire to go to graduate school in business. Thus, institutional and personal characteristics together determine who is available.

Given a set of persons who are available, how do we make contact with one or two of these persons? Three influences progressively narrow our choices: routine activities, proximity, and familiarity.

Routine Activities

Much of our life consists of a routine of activities that we repeat daily or weekly. We attend the same classes and sit in the same seats, eat in the same places at the same tables, shop in the same stores, ride the same bus, and work with the same people. These activities provide opportunities to interact with some availables but not with others. More importantly, the activity provides a focus for our initial interactions. We rarely establish a relationship by saying "Let's be friends" at a first meeting. To do so is risky, because the other person may decide to exploit us. Or that person may reject such an opening, which may damage our self-esteem. Instead, we begin by talking about something shared—a class, an ethnic background, a school, or the weather.

Most relationships begin in the context of routine activities. A study of college students found that relationships began with a meeting in a class, a dorm, or at work (36 percent); with an introduction by a third person (38 percent); or at parties (18 percent) or bars (14 percent). A study of 3,342 adults ages 18 to 59 asked how respondents met their sexual partners (Laumann, Gagnon, Michael, & Michaels, 1994). One-third reported that they were introduced by a friend, and another third said they were introduced by family members or coworkers. Thus, social networks play an important role in the development of relationships. Studies of the friendship patterns of city dwellers have found that friends are selected from relatives, coworkers, and neighbors (Fischer, 1984). Thus, routine activities and social networks are important influences on the development of relationships.

Proximity

Although routine activities bring us into the same classroom, dining hall, or workplace, we are not equally likely to meet every person who is present. Rather, we are more likely to develop a relationship with someone who is in close physical proximity to us.

In classroom settings, seating patterns are an important influence on the development of friendships. One study (Byrne, 1961a) varied the seating arrangements for three classes of about 25 students each. In one class, they remained in the same seats for the entire semester (14 weeks). In the second class, they were assigned new seats halfway through the semester. In the third class, they were assigned new seats every 3½ weeks. The relationships between students were assessed at the beginning and at the end of the semester. Few relationships developed among the students in the class where seats were changed every 3½ weeks. In the other two classes, students in neighboring seats became acquainted in greater numbers than students in nonneighboring seats. Moreover, the

© PhotoDisc/Getty Images

When we think about where people meet available partners, we often picture the singles bar. However, one study of heterosexual relationships found that relatively few people met their partners at a bar. Much more common were meetings in classes, dorms, or workplaces.

relationships were closer in the class where seat assignments were not changed.

Similar positive associations between physical proximity and friendship have been found in a variety of natural settings, including dormitories (Priest & Sawyer, 1967), married student housing projects (Festinger, Schachter, & Back, 1950), and business offices (Schutte & Light, 1978).

We are more likely to develop friendships with persons in close proximity because such relationships provide interpersonal rewards at the lowest cost. First, interaction is easier with those who are close by. It costs less time and energy to interact with the person sitting next to you than with someone on the other side of the room. A second factor is the influence of social norms. In situations where people are physically close or interact frequently—such as in dormitories, classes, and offices—we are expected to behave in a polite, friendly way. Polite, friendly behavior provides increased rewards in these interactions. Failure to adhere to these norms might result in disapproval from others, which increases costs.

Familiarity

As time passes, people who take the same classes, live in the same apartment building, or do their laundry in the same place become familiar with each other. Having seen a person several times, sooner or later we will smile or nod. Repeated exposure to the same novel stimulus is sufficient to produce a positive attitude toward it; this is called the **mere exposure effect** (Zajonc, 1968). In other words, familiarity breeds liking, not contempt. This effect is highly general and has been demonstrated for a wide variety of stimuli—such as music, visual art, and comic strips—under many different conditions (Harrison, 1977).

Does mere exposure produce attraction? The answer appears to be yes. In one experiment, female undergraduates were asked to participate in an experiment on their sense of taste. They entered a series of booths in pairs and rated the taste of various liquids. The schedule was set up so that two participants shared the same booth either once, twice, five times, ten times, or not at all. At the end of the experiment, each woman rated how much she liked each of the other participants. As predicted, the more frequently a woman had been in the same booth with another participant, the higher the rating (Saegert, Swap, & Zajonc, 1973).

WHO IS DESIRABLE?

We come into contact with many potential partners, but contact by itself does not ensure the development of a relationship. Whether a relationship of some type actually develops between two persons depends on whether each is attracted to the other. Initial attraction is influenced by social norms, physical attractiveness, and processes of interpersonal exchange. If the attraction is mutual, the interaction that occurs is governed by scripts.

Social Norms

Each culture specifies the types of relationships that people may have. For each type, norms specify what kinds of people are allowed to have such a relationship. These norms tell us which persons are appropriate as friends, lovers, and mentors. In U.S. society, there is a **norm of homogamy**—a norm requiring that friends, lovers, and spouses be similar in age, race, religion, and socioeconomic status (Kerckhoff, 1974). Research shows that homogamy is characteristic of all types of social relationships from acquaintance to intimate (McPherson, Smith-Lovin, & Cook, 2001). Interviews with 832 students attending the same (all White) high school obtained data on their romantic/sexual relationships (Bearman, Moody, & Stovel, 2004). The students' relationships were homophilous on IQ, family SES, getting drunk, sexual activity, and college plans. A survey of 3,342 adults assessed the extent to which partners in relationships were similar on the following dimensions (Laumann et al., 1994): 75 to 83 percent were homophilous (similar) by age, 82 to 87 percent by education, 88 to 93 percent by race/ethnicity, and 53 to 72 percent by religion. Differences on one or more of these dimensions make a person less appropriate as an intimate partner and more appropriate for some other kind of relationship. Thus, a person who is much older but of the same social class and ethnicity may be appropriate as a mentor—someone who can provide advice about how to manage your career. Potential dates are single persons of the opposite sex who are of similar age, class, ethnicity, and religion.

Norms that define appropriateness influence the development of relationships in several ways. First, each of us uses norms to monitor our own behavior. We hesitate to establish a relationship with someone who is defined by norms as an inappropriate partner. Thus, a low-status person is unlikely to approach a high-status person as a potential friend. For example, the law clerk who just joined a firm would not discuss his hobbies with the senior partner (unless she asked). Second, if one person attempts to initiate a relationship with someone who is defined by norms as inappropriate, the other person will probably refuse to reciprocate. If the clerk did launch into an extended description of the joys of restoring antique model trains, the senior partner would probably end the interaction. Third, even if the other person is willing to interact, third parties often enforce the norms that prohibit the relationship (Kerckhoff, 1974). Another member of the firm might later chide the clerk for presuming that the senior partner cared about his personal interests.

As we move into the 21st century, interracial relationships continue to be rare in the United States. A study of adolescent friendships found that "best friends" are typically of the same race/ethnicity, particularly among Whites (92 percent) and Blacks (85 percent), compared to Hispanics (51 percent) and Asians (48 percent) (Kao & Joyner, 2004). In 2003, only 3½ percent of married couples are interracial; of these, less than one third are Black/White, and the rest are White/other. The norm of homogamy remains especially strong on this dimension (Blackwell & Lichter, 2004). Research on interracial romantic relationships

© Ronnie Kaufman/Corbis

© Gerhard Steiner/Corbis

According to the matching hypothesis, people seek partners whose level of social desirability is about equal to their own. We frequently encounter couples who are matched—that is, who are similar in age, race, ethnicity, social class, and physical attractiveness.

found that non-White males reported more disapproval from their White female partners' family and friends than any other race/gender combination (Miller, Olson, & Fazio, 2004).

Physical Attractiveness

In addition to social norms that define who is appropriate, individuals also have personal preferences regarding desirability. Someone may be normatively appropriate but still not appeal to you. Physical attractiveness can have a significant impact on desirability.

Impact of Physical Attractiveness A great deal of evidence shows that given a choice of more than one potential partner, individuals will prefer the one who is more physically attractive (Hendrick & Hendrick, 1992). A study of 752 first-year college students, for example, demonstrates that most individuals prefer more attractive persons as dates (Walster [Hatfield], Aronson, Abrahams, & Rottman, 1966). As part of the study, students were invited to attend a dance. Before the dance, each student's physical attractiveness was secretly rated by four people, and each student completed a questionnaire. Although the students were told they would be paired by the computer, in fact, men and women were paired randomly. At the dance, during the intermission, students filled out a questionnaire that measured their impression of their date.

This study tested the **matching hypothesis**—the idea that each of us looks for someone who is of approximately the same level of social desirability. The researchers predicted that the students whose dates matched their own level of attractiveness would like their date most. Those whose dates were very different in attractiveness were expected to rate their date as less desirable. Contrary to the hypothesis, in this situation, students preferred a more attractive date, regardless of their own attractiveness. The matching hypothesis is supported by analyses of singles ads (see Box 13.1).

How can we explain the significance of attractiveness? One factor is simply esthetic; generally, we prefer what is beautiful. Although beauty is, to a degree,

BOX 13.1 Let's Make a Deal

> Reasonably good-looking professional man, 34, wishes to meet attractive, professional, sensitive woman, 24–30, to share interest in music, hockey, people, and conversation. Reply to Box 3010.
>
> Attractive Gemini, 36, blonde, blue eyes, 5′7″, 125 lbs., tanned, loves sports, playing piano, guitar. Interested in meeting man 26–46, physically fit, financially secure, witty. Reply to Box 2407.

Dozens of ads like these can be found in most daily newspapers in the United States. If we look at these ads in terms of what they offer and what they seek, we can see that people are looking for a complementary exchange.

One study analyzed 800 such ads to identify patterns in the revelations and stipulations contained in them (Harrison & Saeed, 1977). The ads were taken from a national weekly over a 6-month period. The results of the study showed that most women offered physical attractiveness and were seeking financial security and an older companion. Conversely, most men offered financial security and were seeking an attractive, younger woman. Thus, each group offered complementary resources.

Also, people sought partners whose level of social desirability was approximately equal to their own. This supports the matching hypothesis (Berscheid, Dion, Walster [Hatfield], & Walster, 1971). The researchers calculated an overall desirability score for both the seeker and the person being sought. Attractiveness, financial security, and sincerity were each worth one point on the seeker's index, whereas seeking each of these was worth one point on the other index. Desirability scores ranged from 0 to 3. There was a significant correlation between the level of desirability offered and the level of desirability sought. Moreover, most persons who stated they were physically attractive were seeking good-looking partners—evidence of the importance of physical attractiveness.

A study of 400 ads from two weekly newspapers replicated the finding that women offered physical attractiveness and were seeking status (security), whereas men offered status and were seeking attractiveness (Koestner & Wheeler, 1988). Moreover, women offered low body weight and were seeking tall men, whereas men offered height and were seeking light women. Finally, women offered instrumental traits—independent, competent, bright—valued by men, whereas men offered expressive traits—warm, gentle, emotional—valued by women. Advertisers seem to understand attraction, offering precisely the traits sought by the opposite sex.

Singles ads have been appearing in newspapers for more than 30 years. Have they changed over time? Research comparing ads published in 1986 and 1991 suggests they have not (Willis & Carlson, 1993). In both years, differences between ads placed by men and by women replicated the results summarized earlier.

Research comparing ads placed by gay men and lesbians with those placed by heterosexuals (Gonzales & Meyers, 1993) finds that gay men mention physical characteristics more often and lesbians mention such traits less often than heterosexuals. Gay men also explicitly mention sex more often than the other groups. Finally, heterosexuals more frequently mention sincerity and financial security, suggesting an orientation toward long-term relationships.

In addition to advertisements, people can use telephone messages, computer bulletin boards, and videodating services in their search for partners. What do the people who use these mechanisms want to accomplish? A study of the clients of a videodating service provides some answers (Woll & Young, 1989). The men believed that the women were looking for financial security and tried to convey in their video that they were secure. The women believed that the men were looking for physically attractive and independent women, so they tailored their self-presentation accordingly. Both men and women said that they tried to create a video that would appeal to one special person, Ms. or Mr. Right.

Did they succeed? Most said "no."

"in the eye of the beholder," cultural standards influence our esthetic judgments. A study of female facial beauty found substantial agreement among male college students about which features are attractive (Cunningham, 1986). These men rated such features as large eyes, small nose, and small chin as more attractive than small eyes, large nose, and large chin. What male features do women find attractive? Female college students rated men with large eyes, prominent cheekbones, and a large chin as more attractive (Cunningham, Barbee, & Pike, 1990). Research has also found a high level of agreement among men that certain female body shapes are more appealing than others (Wiggins, Wiggins, & Conger, 1968) and agreement among women about which male body shapes are attractive (Beck, Ward-Hull, & McLear, 1976).

A second factor is that we anticipate more rewards when we associate with attractive persons. A man accompanied by an extremely attractive woman receives more attention and prestige from other persons than if he is seen with an unattractive woman, and vice versa (Sigall & Landy, 1973).

The Attractiveness Stereotype A third factor is the **attractiveness stereotype**—the belief that "what is beautiful is good" (Dion, Berscheid, & Walster [Hatfield], 1972). We assume that an attractive person possesses other desirable qualities. Research consistently finds that we infer that physically attractive people possess more favorable personality traits and are more likely to experience successful outcomes in their personal and social lives (Berscheid & Reis, 1998).

There are limits to the influence of this stereotype. A meta-analysis of more than 70 studies found that attractiveness has a moderate influence on judgments of social competence—how sensitive, kind, and interesting a person is (Eagly, Ashmore, Makhijani, & Longo, 1991). It has less influence on judgments of adjustment and intelligence, and no influence on judgments of integrity or concern for others. Also, the influence of attractiveness on judgments of intellectual competence is reduced when other information about the person's competence is available (Jackson, Hunter, & Hodge, 1995).

When we believe another person possesses certain qualities, those beliefs influence our behavior toward that person. Our actions may then lead him or her to behave in ways that are consistent with our beliefs (see Chapter 5). In one experiment, men were shown photographs of either an attractive or an unattractive woman. They were then asked to interact with that woman via intercom for 10 minutes. The woman was actually a student volunteer. Each conversation was tape-recorded and rated by judges. Women who were perceived as attractive by the men were rated as behaving in a more friendly, likable, and sociable way compared with women who were perceived as unattractive. This happened in part because the men gave the target person opportunities to act in ways that would confirm their expectations based on the attractiveness stereotype (Snyder et al., 1977).

Judgments of attractiveness seem to be based on several dimensions. College students were asked to sort photographs of 95 female fashion models. Analyses suggested that both men and women distinguished three dimensions in their judgments—sexy, cute (youthful), and trendy (up to date in clothing and grooming; Ashmore, Solomon, & Longo, 1996).

Each of us knows that physically attractive people may receive preferential treatment. As a result, we spend tremendous amounts of time and money trying to increase our own attractiveness to others. Men and women purchase clothing, jewelry, perfumes, colognes, and hair color products in an effort to enhance their physical attractiveness. Our choice of products reflects current standards of what looks good. Increasingly, people are using cosmetic surgery to enhance their appearance. Plastic surgeons can lift your eyelids; pin your ears; fill your wrinkles; reshape your nose, jaw, or chin; enlarge your breasts, pectorals, or penis; and suck the fat from your abdomen, thighs, or ankles ("Body Rebuilding," 1997). So strong is the motivation to be attractive that some people have toxic substances (Botox) injected into their bodies in the hope that it will remove wrinkles!

Not everyone prefers attractive persons. Many of us were taught—and some of us believe—that "Beauty is only skin deep," and "You can't judge a book by its cover." What kinds of people are influenced by another's attractiveness, and what kinds of people "read the book" before making a judgment? Research suggests that people who hold traditional attitudes toward men and women are those whose judgments are much more likely to be influenced by beauty (Touhey, 1979).

Evolutionary Perspective on Attractiveness According to the evolutionary perspective, men and women have an evolved disposition to mate with healthy individuals, so that they will produce healthy offspring, who will in turn successfully mate and pass on their genetic code. According to this view, facial and bodily physical attractiveness are markers for physical and hormonal health (Thornhill & Grammar, 1999). Thus, we prefer young, attractive partners because they have high reproductive potential.

Research based on this perspective argues that women and men face different adaptive problems in short-term (casual) mating compared to long-term mating and reproduction. These differences lead to different strategies or behaviors designed to solve these problems. In short-term mating, a woman may choose a partner who offers her immediate resources such as food or money (dinner?). In long-term mating, she will prefer a partner who appears willing and able to provide resources for the indefinite future (marriage?). A man may choose a sexually available woman for a short-term liaison and avoid such women when looking for a long-term mate. A study of mating strategies found that physical attractiveness and possession of resources were judged important in selecting a long-term mate, whereas sexual availability and giving gifts were judged more important in selecting a partner for a "one-night stand" (Schmitt & Buss, 1996). Moreover, both men and women are more selective when choosing a partner for a long-term relationship (Stewart, Stinnett, & Rosenfeld, 2000). It is not surprising, according to this perspective, that singles ads emphasize attractiveness and resources. It is also worth noting that mate-selection criteria do not vary much by age, in the range from age 20 to age 60 (Buunk, Dijkstra, Fetchenhauer, & Kenrick, 2002).

Attractiveness Isn't Everything Physical attractiveness may have a major influence on our judgments of others because it is readily observable. When we meet someone for the first time, one characteristic we can assess quickly is his or her attractiveness. If other relevant information is available, it might reduce or eliminate the impact of attractiveness on our judgments. In fact, an analysis of 70 studies found that when perceivers have other personal information about the target person, the effect of the attractiveness stereotype is smaller (Eagly et al., 1991).

Exchange Processes

How do we move from the stage of awareness of another person to the stage of contact? Recall that in our introduction, Dan noticed Sally at every lecture. Because she was young and not wearing a wedding ring, Dan hoped that she was available. She was certainly desirable—she was very pretty and seemed like a friendly person. What factors did Dan consider when deciding whether to initiate contact? One important factor in this decision is the availability and desirability of alternative relationships (Backman, 1990). Thus, before Dan chose to initiate contact with Sally, he probably considered whether there was anyone else who might be a better choice.

Choosing Friends We can view each actual or potential relationship—whether involving a friend, coworker, roommate, or date—as promising rewards but entailing costs. Rewards are the pleasures or gratifications we derive from a relationship. These might include a gain in knowledge, enhanced self-esteem, satisfaction of emotional needs, or sexual gratification. Costs are the negative aspects of a relationship, such as physical or mental effort, embarrassment, and anxiety.

Exchange theory proposes that this is, in fact, the way people view their interactions (Blau, 1964; Homans, 1974). People evaluate interactions and relationships in terms of the rewards and costs that each is likely to entail. They calculate likely outcomes by subtracting the anticipated costs from the anticipated rewards. If the expected outcome is positive, people are inclined to initiate or maintain the relationship. If the expected outcome is negative, they are unlikely to initiate a new relationship or to stay in an ongoing relationship. Dan anticipated that dating Sally would be rewarding; she would be fun to do things with, and others would be impressed that he was dating such an attractive woman. At the same time, he anticipated that Sally would expect him to be committed to her and that he would have to spend money to take her to movies, plays, and restaurants.

What standards can we use to evaluate the outcomes of a relationship? Two standards have been proposed (Kelley & Thibaut, 1978; Thibaut & Kelley, 1959). One is the **comparison level** (CL), the level of outcomes expected based on the average of a person's experience in past relevant relationships. Each relationship is evaluated as to whether it is above or below that person's CL—that is, better or worse than the average of past relevant relationships. Relationships that fall above a person's CL are satisfying, whereas those that fall below it are unsatisfying.

If this were the only standard, we would always initiate relationships that appeared to promise outcomes better than those we already experienced and avoid relationships that appeared to promise poorer outcomes. Sometimes, however, we use a second standard. The **comparison level for alternatives** (CL_{alt}) is the lowest level of outcomes a person will accept in light of the available alternatives. A person's CL_{alt} varies depending on the outcomes that he or she believes can be obtained from the best of the available alternative relationships. The use of CL_{alt} explains why we may sometimes turn down opportunities that appear promising or why we may remain in a relationship even though we feel that the other person is getting all the benefits.

Whether a person initiates a new relationship or not will depend on both the CL and the CL_{alt}. An individual usually avoids relationships whose anticipated outcomes fall below the CL. If a potential relationship appears likely to yield outcomes above a person's CL, then initiation will depend on whether the outcomes are expected to exceed the CL_{alt}. Dan believed that a relationship with Sally would be very satisfying. He was casually dating another woman, and that relationship was not gratifying. Thus, the potential relationship with Sally was above both CL and CL_{alt}, leading Dan to initiate contact.

Whereas CL is an absolute, relatively unchanging standard, several factors influence a person's CL_{alt}. These factors include the extent to which routine activities provide opportunities to meet people, the size of the pool of eligible persons, and one's skills in initiating relationships.

Making Contact Once we decide to initiate interaction, the next step is to make contact. Sometimes we use technology, such as the telephone or e-mail. Often, we arrange to get physically close to the person. At parties and in bars, people often circulate, which brings them into physical proximity with many of the other guests.

Once in proximity, a stranger attracted to another person wants to communicate interest without making a commitment to interaction. In initial opposite-sex encounters, the problem can be resolved by using ambiguous cues. The gender that has "more to lose" (the woman) will try to control the interaction; to do so, she will initially use nonverbal cues that the man may not consciously perceive (Grammar, Kruck, Juette, & Fink, 2000). Researchers observed 45 male-female pairs of strangers (ages 18 to 23) left alone in a waiting room. Questionnaires completed later were used to assess each participant's interest in the other person. Women interested in the man were more likely to display several cues, including short glances, coy smiles, and primping (adjusting clothing without a visible reason). Men interested in the woman were more likely to speak to her in the first 3 minutes (a direct cue); if she responded with head nods, his rate of speech increased. This pattern is probably repeated many times every day in airplanes, on trains and buses, and in classes and waiting rooms. (See Box 7.2, Flirting.)

Scripts The development of relationships is influenced by an event schema or script. A script specifies (1) the definition of the situation (a date, job interview, or sexual encounter), (2) the identities of the social actors involved, and (3) the range and sequence of permissible behaviors (see Chapter 5).

The initiation of a relationship requires an opening line. Often, it is about some feature of the situation. At the beginning of this chapter, Dan initiated the conversation by commenting that Sally never missed the class. Two people waiting to participate in a psychology experiment may begin talking by speculating about the purpose of the experiment. The weather is a widely used topic for openings. The opening line often includes an identification display—a signal that we believe the other person is a potential partner in a specific kind of relationship

(Schiffrin, 1977). When Dan commented about Sally's attitude toward the class, it conveyed an interest in friendship. A different message would have been sent if he had asked whether she knew the woman sitting next to her. The person who is approached, in turn, decides whether she is interested in that type of relationship. If she is, she engages in an **access display**—a signal that further interaction is permissible. Thus, Sally responded warmly toDan's opening line, encouraging continued conversation.

Once initiated, scripts specify the permissible next steps. American society—or at least the subculture of college students—is characterized by a specific script for "first dates" (Rose & Frieze, 1993). When asked to describe "actions that a woman (man) would typically take" on a first date, both men and women identified a core action sequence: dress, be nervous, pick up date, leave (meeting place), confirm plans, get to know, evaluate, talk, laugh, joke, eat, attempt to make out/accept or reject, take date home, kiss, go home. In general, both men and women ascribed a proactive role to the man and a reactive role to the woman.

Actual first dates, of course, are characterized by departures from the script. A study of college students focused on the extent to which the roles of men and women are changing (Lottes, 1993). Both men and women were asked about the extent to which they had experienced the woman's initiating a date, initiating sexual intimacy, and paying for a date. Increasing proportions of women are engaging in these traditionally male activities.

How do we learn these scripts, and the departures from them? One source is the mass media. Both men and women learn about relationships and how to handle them from popular magazines. A study of magazines oriented toward women (*Cosmopolitan, Glamour,* and *Self*) and men (*Playboy, Penthouse,* and *GQ*) found that they portrayed relationships in similar terms (Duran & Prusank, 1997). The dominant focus in both types of magazines was sexual relationships. In women's magazines (January 1990 to December 1991), the themes were (1) women are less skilled at and more anxious about sex, and (2) sex is enjoyed most in caring relationships. In men's magazines during the same period, the themes were (1) men are under attack in sexual relationships, and (2) men have natural virility and strong sexual appetites. Also, the articles in women's magazines portrayed men as incompetent about relationships.

The United States is one of a small number of societies in which love is widely included in the script for getting married. In many other societies, marriages reflect political and economic influences, not romance.

THE DETERMINANTS OF LIKING

Once two people make contact and begin to interact, several factors will determine the extent to which each person will like the other. Three of these factors are considered in this section: similarity, shared activities, and reciprocal liking.

Similarity

How important is similarity? Do "birds of a feather flock together?" Or is it more the case that "opposites attract?" These two aphorisms about the determinants of liking are inconsistent and provide opposing predictions. A good deal of research has been devoted to finding out which one is more accurate. The evidence indicates that birds of a feather do flock together; that is, we are attracted to people who are similar to ourselves. Probably the most important kind of similarity is **attitudinal similarity**—the sharing of beliefs, opinions, likes, and dislikes.

Attitudinal Similarity A widely employed technique for studying attitudinal similarity is the attraction-to-a-stranger paradigm, initially developed by Byrne (1961b). Potential participants fill out an attitude questionnaire that measures their beliefs about various topics, such as life on a college campus. Later, participants receive information about a stranger as part of a seemingly unrelated study. The information they receive describes the stranger's personality or social background and may include a photograph. They also are given a copy of the same questionnaire they completed earlier, ostensibly filled out by the stranger. In fact, the stranger's questionnaire is completed by the experimenter, who systematically varies the degree to which the stranger's supposed responses match the participant's responses. After seeing the stranger's questionnaire, the participants are asked how much they like or dislike the stranger and how much they would enjoy working with that person.

In most cases, the participant's attraction to the stranger is positively associated with the percentage of attitude statements by the stranger that agree with the participant's own attitudes (Byrne & Nelson, 1965; Gonzales, Davis, Loney, Lukens, & Junghans, 1983). We rarely agree with our friends about everything; what matters is that we agree on a high proportion of issues. This relationship between similarity of attitudes and liking is very general; it has been replicated in studies using both men and women as participants and strangers under a variety of conditions (Berscheid & Walster [Hatfield], 1978).

In the attraction-to-a-stranger paradigm, the participant forms an impression of a stranger without any interaction. This allows researchers to determine the precise relationship between similarity and liking. But what do you think the relationship would be if two people were allowed to interact? Would similarity have as strong an effect?

A study attempting to answer this question arranged dates for 44 couples (Byrne, Ervin, & Lamberth, 1970). Researchers distributed a 50-item questionnaire measuring attitudes and personality to a large sample of undergraduates. From these questionnaires, they selected 24 male-female couples whose answers were very similar (66 to 74 percent identical) and 20 couples whose answers were not similar (24 to 40 percent identical). Each couple was introduced, told they had been matched by a computer, and asked to spend the next 30 minutes together at the student union; they were even offered free sodas. The experimenter rated each participant's attractiveness before he or she left on the date. When they returned, the couple rated each other's sexual attractiveness, desirability as a date and marriage partner, and indicated how much they liked each other. The experimenter also recorded the physical distance between the two as they stood in front of his desk.

The results of this experiment showed that both attitudinal similarity and physical attractiveness influenced liking. Partners who were attractive and who held highly similar attitudes were rated as more likable. Moreover, similar partners were rated as more intelligent and more desirable as a date and marriage partner. The couples high in similarity stood closer together after their date than the couples low in similarity—another indication that similarity creates liking.

At the end of the semester, 74 of the 88 participants in this study were contacted and asked whether they (1) could remember their date's name, (2) had talked to their date since their first meeting, (3) had dated their partner, or (4) wanted to date their partner. Participants in the high attractiveness/high similarity condition were more likely to remember their partner's name, to report having talked to their partner, and to report wanting to date their partner than those in the low attractiveness/low similarity condition.

The story of Dan and Sally at the beginning of this chapter illustrates the importance of similarity in the early stages of a relationship. After their initial meeting, they discovered they had several things in common. They were from the same city. They had chosen the same major and held similar beliefs about their field and about how useful a bachelor's degree would be in that field. Each also found the other attractive; like the participants in the high attraction/high similarity condition, Dan and Sally continued to talk after their first meeting.

Why Is Similarity Important? Why does attitudinal similarity produce liking? One reason is the desire for consistency between our attitudes and perceptions. The other reason focuses on our preference for rewarding experiences.

Most people desire cognitive consistency—consistency between attitudes and perceptions of whom and what we like and dislike. If you have positive attitudes toward certain objects and discover that another person has favorable attitudes toward the same objects, your cognitions will be consistent if you like that person (Newcomb, 1971). When Dan discovered that Sally had a positive attitude toward his major, his desire for consistency produced a positive attitude toward Sally. Our desire for consistency attracts us to persons who hold the same attitudes toward important objects.

Another reason we like persons with attitudes similar to our own is because our interaction with them provides three kinds of reinforcement. First, interacting with persons who share similar attitudes usually leads to positive outcomes (Lott & Lott, 1974). At the beginning of this chapter, Dan anticipated that he and Sally would get along well because they shared similar likes and dislikes.

Second, similarity validates our own view of the world. We all want to evaluate and verify our attitudes and beliefs against some standard. Sometimes physical reality provides objective criteria for our beliefs. But often there is no physical standard, and so we must compare our attitudes with those of others (Festinger, 1954). Persons who hold similar attitudes provide us with support for our own opinions, which allows us to deal with the world more confidently (Byrne, 1971). Such support is particularly important in areas, such as political attitudes, where we realize that others hold attitudes dissimilar to our own (Rosenbaum, 1986).

Research indicates that similarity in mood is also an important influence on attraction. In the attraction-to-a-stranger paradigm, nondepressed participants prefer nondepressed strangers (Rosenblatt & Greenberg, 1988). In an experiment, male and female students interacted with a depressed or nondepressed person of the same sex. People in homogeneous pairs (both depressed or both nondepressed) were more satisfied with the interaction than people in mixed pairs (Locke & Horowitz, 1990). In another study, researchers measured the depression levels of people and of their best friends. Depressed people had best friends who were also depressed (Rosenblatt & Greenberg, 1991).

Third, we like others who share similar attitudes because we expect that they will like us. In one experiment, college students were given information about a stranger's attitudes and the stranger's evaluation of them (Condon & Crano, 1988). The participants' perceptions of the strangers' similarity to and evaluation of them were also assessed. The students were attracted to strangers whom they perceived as evaluating them positively, and that accounted for the influence of similar attitudes.

Shared Activities

As people interact, they share activities. Recall that after Sally and Dan met, they began to sit together in class and to discuss course work. When the professor announced the first exam, Dan invited Sally to study for it with him. Sally's roommate was also in the class; the three of them reviewed the material together the night before the exam. Sally and Dan both got A's on the exam, and each felt that studying together helped. The next week, they went to a movie together. Several days later, Sally invited Dan to a party.

Shared activities provide opportunities for each person to experience reinforcement. Some of these reinforcements come from the other person; Sally finds Dan's interest in her very reinforcing. Often the other person is associated with a positive experience, which leads us to like the other person (Byrne & Clore, 1970). Getting an A on the examination was a very positive experience for both Dan and Sally. The association of the other person with that experience led to increased liking for the other.

Thus, as a relationship develops, the sharing of activities contributes to increased liking. This was shown in a study in which pairs of friends of the same sex both filled out attitude questionnaires and listed their preferences for various activities (Werner & Parmelee, 1979). The duration of the friendships averaged 5 years. The results of the study showed similarity between friends in both activity preferences and attitudes. A study of romantic relationships found that sharing of tasks or activities was a strong predictor of liking (Stafford & Canary, 1991). Thus, participation in mutually satisfying activities is a strong influence on the development and maintenance of relationships. The results of a series of five studies indicate that participation in novel and arousing activities rather than mundane and trivial pursuits is associated with relationship quality (Aron, Norman, Aron, McKenna, & Heyman, 2000). As Dan and Sally got to know each other, their shared experiences—studying, seeing movies, going to parties—became the basis for their relationship, supplementing the effect of similar attitudes.

If shared activities are important, what happens when a couple does not do things together? This is not an idle question; many students and working adults are separated from their partner by distance, and may see each other (share activities) only occasionally. And what about secret relationships, where contact is limited by the need to prevent others from learning about it? Some research suggests that secrecy will be associated with greater attraction (Wegner, Lane, & Dimitri, 1994). But if shared activities are important, absence will prevent the development of shared reinforcements and may increase costs, such as a sense of burden (a sense that the relationship is difficult to coordinate, requires a great deal of work, energy). A series of studies found that participants in secret relationships reported greater burden, less satisfaction, and lower relationship quality than persons in open ones (Foster & Campbell, 2005). People in secret relationships also reported significantly less love for their partner. These results strengthen the conclusion that shared activities make an important difference.

Reciprocal Liking

One of the most consistent research findings is the strong positive relationship between our liking someone and the perception that the other person will like us in return (Backman, 1990). In most relationships, we expect reciprocity of attraction; the greater the liking of one person for the other, the greater the other person's liking will be in return. But will the degree of reciprocity increase over time as partners have greater opportunities to interact? To answer this question, one study obtained liking ratings from 48 persons (32 men and 16 women) who had been acquainted for 1, 2, 4, 6, or 8 weeks (Kenney & La Voie, 1982). The results showed a positive correlation between each person's liking rating and the other's, and the reciprocity of attraction increased somewhat with the duration of the acquaintance. Some participants in this study were roommates rather than friends; they would be expected to like each other due to the proximity effect. When roommate pairs were eliminated from the results, the correlation between liking ratings increased substantially.

THE GROWTH OF RELATIONSHIPS

We have traced the development of relationships from the stage of zero contact through awareness (who is available) and surface contact (who is desirable) to mutuality (liking). At the beginning of this chapter, Dan and Sally met, discovered that they had similar attitudes and interests, and shared pleasant experiences—such as doing well on an examination, going to a movie, and later, to a party.

Many of our relationships remain at the "minor" level of mutuality (see Figure 13.1). We have numerous acquaintances, neighbors, and coworkers whom we like and interact with regularly but to whom we do not feel especially close. A few of our relationships grow closer; they proceed through "moderate" to "major" mutuality. Three aspects of this continued growth of relationships are examined in this section: self-disclosure, trust, and interdependence. As the degree of mutuality increases between friends, roommates, and coworkers, self-disclosure, trust, and interdependence also will increase.

Self-Disclosure

Recall that when Dan and Sally returned from the party, Sally told Dan that her roommate's parents had just separated and that her roommate was very depressed. Sally said that she didn't know how to help her roommate—that she felt unable to deal with the situation. At this point, Sally was engaging in self-disclosure—the act of revealing personal information about oneself to another person. Self-disclosure usually increases over time in a relationship. Initially, people reveal things about themselves that are not especially intimate and that they believe the other will readily accept. Over time, they disclose increasingly intimate details about their beliefs or behavior, including information they are less certain the other will accept (Backman, 1990).

Self-disclosure increases as a relationship grows. In one study, same-sex pairs of previously unacquainted college students were brought into a laboratory setting and asked to get acquainted (Davis, 1976). They were given a list of 72 topics. Each topic had been rated earlier by other students on a scale of intimacy from 1 to 11. The participants were asked to select topics from this list and to take turns talking about each topic for at least 1 minute while their partner remained silent. The interaction continued until each partner had spoken on 12 of the 72 topics. The results showed that the intimacy of the topic selected increased steadily from the 1st to the 12th topic chosen. The average intimacy of topics discussed by each couple increased from 3.9 to 5.4 over the 12 disclosures. Research also indicates that greater self-disclosure during a 10-minute conversation was associated with an increase in positive affect—happiness, excitement—and attraction to the partner (Vittengl & Holt, 2000).

When Sally told Dan about the situation with her roommate, Dan replied that he knew how she felt, because his older brother had just separated from his wife. This exchange reflects reciprocity in self-disclosure; as one person reveals an intimate detail, the other person usually discloses information at about the same level of intimacy (Altman & Taylor, 1973). In the Davis (1976) study, each participant selected a topic at the same level of intimacy as the preceding one or at the next level of intimacy. However, reciprocity decreases as a relationship develops. In one study, a researcher recruited students to be participants and asked each to bring an acquaintance, a friend, or a best friend to the laboratory (Won-Doornink, 1985). Each dyad was given a list of topics that varied in degree of intimacy. Each was instructed to take at least four turns choosing and discussing a topic. Each conversation was tape-recorded and later analyzed for evidence of reciprocity. The association between the stage of the relationship and the reciprocity of intimate disclosures was curvilinear; that is, there was greater reciprocity of intimate disclosures between friends than between acquaintances but less reciprocity between best friends than among friends (see Figure 13.2).

Not all people divulge increasingly personal information as you get to know them. You have probably known people who were very open—who readily disclosed information about themselves—and others who said little about themselves. In this regard, we often think of men as less likely to discuss their feelings than women. However, research has shown that self-disclosure depends not only on gender but also on the nature of the relationship. In casual relationships (with men or women), men are less likely to disclose personal information than women (Reis, Senchak, & Solomon, 1985). In dating couples, the amount of disclosure is related more to gender role orientation than to gender. Men and women with traditional gender role orientations disclose less to their partners than those with egalitarian gender role orientations (Rubin, Hill, Peplau, & Dunkel-Scheker, 1980). Traditional gender roles are more segregated, with each person responsible for certain tasks, whereas egalitarian orientations emphasize sharing. An emphasis on joint activity leads to greater self-disclosure. In intimate

FIGURE 13.2 THE RELATIONSHIP BETWEEN RECIPROCITY AND INTIMACY

Reciprocity—picking a topic of conversation that is as intimate as the last topic introduced by your partner—is the process by which relationships become more intimate. The extent of reciprocity depends on the intimacy of the topic and the stage of the relationship. Students talked with an acquaintance (early stage), friend (middle stage), or best friend (advanced stage). With topics that were not intimate (such as the weather), reciprocity declined steadily as the stage increased. With intimate topics, in contrast, reciprocity was greatest at the middle stage, less at the advanced stage, and least at the early stage of a relationship.

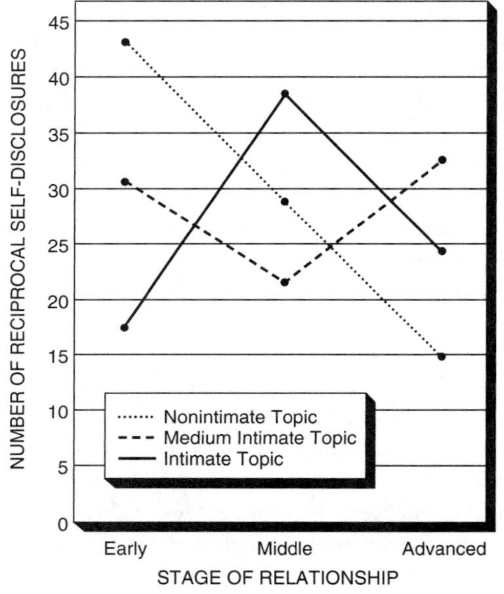

Source: Adapted from Won-Doornink, 1985, figure 4.

heterosexual relationships, men and women do not differ in the degree of self-disclosure (Hatfield, 1982).

The work discussed so far is concerned with the intimacy of self-disclosure. But self-disclosure is a complex phenomenon and has several characteristics including content (information about self or about the relationship), amount, and emotional tone (Bradford, Feeney, & Campbell, 2002). Using diaries, researchers gathered data on 1,908 conversations lasting more than 10 minutes from both members of the couple. The results indicated that disclosures could be scored on these dimensions and that there were differences in intimacy and amount of disclosure related to the partner's attachment style.

Trust

Why did Dan confide in Sally that his brother had just left his wife? Perhaps he was offering reciprocity in self-disclosure. Because Sally had confided in Dan, she expected him to reciprocate. But had he been suspicious of Sally's motives, he might not have. This suggests the importance of trust in the development of a relationship.

When we **trust** someone, we believe that person is both honest and benevolent (Larzelere & Huston, 1980). We believe that the person tells us the truth—or at least does not lie to us—and that his or her intentions toward us are positive. One measure of interpersonal trust is the interpersonal trust scale reproduced in Table 13.1. The questions focus on whether the other person is selfish, honest, sincere, fair, or considerate. We are more likely to disclose personal information to someone we trust. How much do you trust *your* partner? Answer the questions on the scale and determine your score. Higher scores indicate greater trust.

To study the relationship between trust and self-disclosure, researchers recruited men and women from university classes, from a list of people who had recently obtained marriage licenses, and by calling persons randomly selected from the telephone directory. Each person was asked to complete a questionnaire concerning his or her spouse or current or most recent date. The survey included the interpersonal trust scale in Table 13.1. Researchers averaged the trust scores for seven types of relationships, as shown in Figure 13.3. Note that as the relationship becomes more exclusive, trust scores increase significantly. Is there a relationship between trust and self-disclosure? Each person was also asked how much he or she had disclosed to the partner in each of six areas—religion, family, emotions, relationships with others, school or work, and marriage. Trust scores were positively correlated with self-disclosure—that is, the more the person trusted the partner, the greater the degree of self-disclosure.

TABLE 13.1 INTERPERSONAL TRUST SCALE

	Strongly Agree	Agree	Slightly Agree	?	Slightly Disagree	Disagree	Strongly Disagree
1. My partner is primarily interested in his or her own welfare.	______	______	______	–	______	______	______
2. There are times when my partner cannot be trusted.	______	______	______	–	______	______	______
3. My partner is perfectly honest and truthful with me.	______	______	______	–	______	______	______
4. I feel I can trust my partner completely.	______	______	______	–	______	______	______
5. My partner is truly sincere in his or her promises.	______	______	______	–	______	______	______
6. I feel my partner does not show me enough consideration.	______	______	______	–	______	______	______
7. My partner treats me fairly and justly.	______	______	______	–	______	______	______
8. I feel my partner can be counted on to help me.	______	______	______	–	______	______	______

Note: For items 1, 2, and 6, Strongly Agree = 1, Agree = 2, Slightly Agree = 3, and so on. For items 3, 4, 5, 7, and 8, the scoring is reversed.

Source: Adapted from "The Dyadic Trust Scale: Toward Understanding Interpersonal Trust in Close Relationships," by Larzelere and Huston, *Journal of Marriage and the Family, 42*(3). Copyrighted 1980 by the National Council on Family Relations, 3989 Central Ave. NE, Suite 550, Minneapolis, MN 55421. Reprinted by permission.

Other research on interpersonal trust suggests that in addition to honesty and benevolence, reliability is an important aspect of trust. We are more likely to trust someone who we feel is reliable—on whom we can count (Johnson-George & Swap, 1982)—and predictable (Rempel, Holmes, & Zanna, 1985).

Interdependence

Earlier in this chapter, we noted that people evaluate potential and actual relationships in terms of the outcomes (rewards minus costs) they expect to receive. Dan initiated contact with Sally because he anticipated that he would experience positive outcomes. Sally encouraged the development of a relationship because she, also, expected the rewards to exceed the costs. As their relationship developed, each discovered that the relationship was rewarding. Consequently, they increased the time and energy devoted to their relationship and decreased their involvement in alternative relationships. As their relationship became increasingly mutual, Sally and Dan became increasingly dependent on each other for various rewards (Backman, 1990). The result is strong, frequent, and diverse interdependence (Kelley et al., 1983).

Increasing reliance on one person for gratifications and decreasing reliance on others is called **dyadic withdrawal** (Slater, 1963). One study of 750 men and women illustrates the extent to which such withdrawal occurs. Students identified the intensity of their current heterosexual relationships, then listed the names of persons whose opinions they considered important. They also indicated how important each person's opinions were and how much they had disclosed to that person (Johnson & Leslie, 1982). As predicted, the more intimate his or her current heterosexual relationship, the smaller the number of friends listed by the respondent; there was no difference in the number of relatives listed. Furthermore, as the

FIGURE 13.3 AVERAGE INTERPERSONAL TRUST SCORES FOR SEVEN TYPES OF HETEROSEXUAL RELATIONSHIPS

Trust involves two components: the belief that a person is honest and that his or her intentions are benevolent. More than 300 persons completed the interpersonal trust scale (see Table 13.1) for their current or most recent heterosexual partner. Results showed a strong relationship between the degree of intimacy in a relationship and the degree of trust.

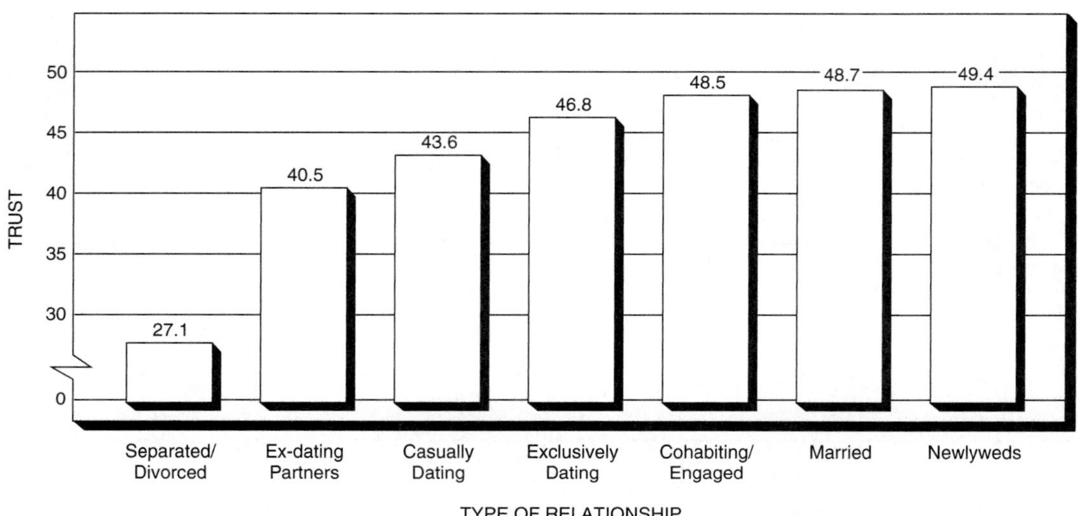

Source: Adapted from "The Dyadic Trust Scale: Toward Understanding Interpersonal Trust in Close Relationships," by Larzelere and Huston, *Journal of Marriage and the Family, 42*(3). Copyrighted 1980 by the National Council on Family Relations, 3989 Central Ave. NE, Suite 550, Minneapolis, MN 55421. Reprinted by permission.

degree of involvement increased, the proportion of mutual friends of the couple also increased (Milardo, 1982). Other studies have found that as heterosexual relationships become more intimate, each partner spends less time interacting with friends, whereas interaction with relatives may increase (Surra, 1990).

Interdependence evolves out of the process of negotiation (Backman, 1990). Each person offers various potential rewards to the partner; the partner accepts some and rejects others. As the relationship develops, the exchanges stabilize. Shared activities are an important potential source of rewards. Each person has activity preferences. As the relationship develops, the couple must blend their separate preferences into joint activities. A study of dating couples found that men liked sex, games, and sports better than women, whereas women preferred companionship, entertainment, and cultural activities (Surra & Longstreth, 1990). Some couples achieved a blend by taking turns, alternately engaging in activities preferred by each. Others cooperated, engaging in activities they both liked, such as preparing food and running errands. Some couples experienced continuing conflict over what to do.

A potential reward in many relationships is sexual gratification. As relationships develop and become more mutual, physical intimacy increases as well. The couple negotiates the extent of sexual intimacy, with the woman's preferences having a greater effect on the outcome (Lear, 1997). How important is sexual gratification in dating relationships? A study of 149 couples assessed the importance of various rewards in relationships of increasing intimacy (preferred date, going steady, engaged, living together, and married). Among intimate couples, sexual gratification was much more likely to be cited as a major basis for the relationship (Centers, 1975). Other surveys indicate that the more emotionally intimate a couple is, the more likely they are to engage in sexual intimacy (Christopher & Roosa, 1991).

LOVE AND LOVING

It is fair to say that what we feel for our friends, roommates, coworkers, and some of the people we date is attraction. But is that all we feel? Occasionally, at least, we experience something more intense than a positive attitude toward others. Sometimes we feel and even say "I love you."

How does loving differ from liking? Much of the research in social psychology on attraction or liking is summarized earlier in this chapter. By contrast, there has been much less research on love. Three views of love are considered in this section: the distinction between liking and loving, passionate love, and romantic love.

Liking Versus Loving

One of the first empirical studies of love distinguished between liking and loving (Rubin, 1970). Love is something more than intense liking; it is the attachment to and caring about another person (Rubin, 1974). Attachment involves a powerful desire to be with and be cared about by another person. Caring involves making the satisfaction of another person's needs as significant as the satisfaction of your own.

Based on this distinction, Rubin developed scales to measure both liking and love. The liking scale evaluates one's dating partner, lover, or spouse on various dimensions, including adjustment, maturity, responsibility, and likability. The love scale measures attachment to and caring for one's partner, and intimacy (self-disclosure). These scales were completed by each member of 182 dating couples, both for her or his partner and best friend of the same sex (Rubin, 1970). The results showed a high degree of internal consistency within each scale and a low correlation between scales. Thus, the two scales do measure different things.

If the distinction between liking and loving is valid, how do you think you would rate a dating partner and your best friend on these scales? Rubin predicted high scores on both liking and love scales for the dating partner, lover, or spouse and a high liking score but lower love score for the (platonic) friend. The average scores of the 182 couples confirmed these predictions. Research work by Davis (1985) also distinguishes between friendship and love. Friendship involves several qualities, including trust, understanding, and mutual assistance. Love involves all of these plus caring (giving the utmost to and being an advocate for the other) and passion (obsessive thought, sexual desire).

Passionate Love

Love certainly involves attachment and caring. But is that all? What about the agony of jealousy and the ecstasy of being loved by another person? An alternative view of love emphasizes emotions such as these. It focuses on **passionate love**—a state of intense physiological arousal and intense longing for union with another (Hatfield & Walster, 1978).

Cognitive and emotional factors interact to produce passionate love. Each of us learns about love from parents, friends, movies, and popular music. We learn whom it is appropriate to fall in love with, how it feels, and how we should behave when we are in love. We experience an emotion only when we are physiologically aroused. Thus, we experience passionate love when we experience intense arousal and the circumstances fit the cultural definitions we have learned.

Passionate love has three components: cognitive, emotional, and behavioral (Hatfield & Sprecher, 1986). The cognitive components include a preoccupation with the loved one, an idealization of the person or the relationship, and a desire to know the other and be known by him or her. Emotional components include physiological arousal, sexual attraction, and desire for union. Behavioral elements include serving the other and maintaining physical closeness to him or her. A scale designed to measure passionate love is reproduced in Box 13.2. Notice that each item deals with one of these components.

Research in the United States using the passionate love scale finds that the items are closely related; that is,

BOX 13.2 Test Yourself: Passionate Love

Think of the person you love most passionately right now. If you are not in love right now, think of the last person you loved passionately. If you have never been in love, think of the person you came closest to caring for in that way. Keep that person in mind as you complete this questionnaire. (The person you choose should be of the opposite gender if you are heterosexual and of the same gender if you are gay or lesbian, or either if . . .) Try to record how you felt at the time when your feelings were the most intense.

Use the following scale to answer each item.

1	2	3	4	5	6	7	8	9
Not at all true				Moderately true				Definitely true

1. I would feel deep despair if ________ left me.
2. Sometimes I feel I can't control my thoughts; they are obsessively on ________.
3. I feel happy when I am doing something to make ________ happy.
4. I would rather be with ________ than anyone else.
5. I'd get jealous if I thought ________ were falling in love with someone else.
6. I yearn to know all about ________.
7. I want ________ physically, emotionally, and mentally.
8. I have an endless appetite for affection from ________.
9. For me, ________ is the perfect romantic partner.
10. I sense my body responding when ________ touches me.
11. ________ always seems to be on my mind.
12. I want ________ to know me—my thoughts, my fears, and my hopes.
13. I eagerly look for signs indicating ________'s desire for me.
14. I possess a powerful attraction for ________.
15. I get extremely depressed when things don't go right in my relationship with ________.

Source: Adapted from "Scale for Determining Passionate Love" by Hatfield and Sprecher, *Journal of Adolescence, 9,* 383–410. Copyright © 1986, with permission from Elsevier.

all of them measure a single factor (Hendrick & Hendrick, 1989). A study of 60 men and 60 women found that scores on the scale are related to the stage of the relationship. Passionate love increases substantially from the early stage of dating to the stage of an exclusive relationship. It does not increase further as the relationship moves from exclusively dating to living together or becoming engaged (Hatfield & Sprecher, 1986). A study of 197 couples at various stages of courtship, including recently married, found that passionate love did decline as the length of the relationship increased (Sprecher & Regan, 1998).

Passionate love is associated with other intense emotions. When our love is reciprocated and we experience closeness or psychological union with the other person, we experience fulfillment, joy, and ecstasy. Conversely, positive emotional experiences—excitement, sexual excitement—can enhance passionate love. The study of 197 couples found that passionate love and sexual desire were positively related (Sprecher & Regan, 1998). Interestingly, sexual desire and sexual activity are *not* related (Regan, 2000). Unrequited love, on the other hand, is often associated with jealousy, anxiety, or despair. Loss of a love can be emotionally devastating.

An important question is whether passionate love is universal or only found in some (Western?) cultures. Analyses of answers to the 15-item scale using data from nine cultural groups ($n = 1{,}809$) identified a common-factor structure; the six dimensions included commitment/affection, security/insecurity, and self-/other-centered (Landis & O'Shea III, 2000). More detailed analyses of the responses separating men and women identified variation in the relative importance of the factors across culture by gender groups.

The Romantic Love Ideal

The studies and theories of love discussed so far assume that love consists of a particular set of feelings and behaviors. Furthermore, most of us assume that we will experience this emotion toward a member of the opposite sex at least once in our lives. But these are very culture-bound assumptions. There are societies in which the state or experience we call love is unheard of. In fact, U.S. society is almost alone in accepting love as a major basis for marriage.

In U.S. society, we are socialized to accept a set of beliefs about love—beliefs that guide much of our behavior. The following five beliefs are known collectively as the **romantic love ideal:**

1. True love can strike without prior interaction ("love at first sight").
2. For each of us, there is only one other person who will inspire true love.
3. True love can overcome any obstacle ("Love conquers all").
4. Our beloved is (nearly) perfect.
5. We should follow our feelings—that is, we should base our choice of partners on love rather than on other, more rational considerations (Lantz, Keyes, & Schultz, 1975).

Researchers have developed a scale to measure the extent to which individuals hold these beliefs (Sprecher & Metts, 1989). When the scale was completed by a sample of 730 undergraduates, the results indicated that the first four beliefs are held by many young people. Interestingly, male students are more likely to hold these beliefs than female students.

Research suggests that the fourth belief, idealization of the partner, is an important influence on relationship satisfaction. Two studies of Dutch adults found that many of them believed that their relationship was better than the relationships of others, and that this belief was associated with reported happiness (Buunk & van der Eijnden, 1997). In another study, researchers asked the members of dating (98) and married (60) couples to rate themselves, their partner, and their ideal partner on 21 interpersonal characteristics (Murray, Holmes, & Griffin, 1996a). Analyses indicated that the participant's ratings of the partner were more similar to the ratings of the self and the ideal partner than to the partner's self-ratings. Furthermore, people who idealized their partner and whose partner idealized them were happier. A longitudinal study found that over a 1-year period, partners came to share the individual's idealized image of him or her (Murray, Holmes, & Griffin, 1996b).

The romantic love ideal has not always been popular in the United States. A group of researchers conducted an analysis of best-selling magazines published during four historical periods (Lantz, Keyes, & Schultz, 1975; Lantz, Schultz, & O'Hara, 1977). They counted the number of times the magazines mentioned one or more of the five beliefs that make up the romantic love ideal. The number of times the ideal was discussed increased steadily over time, as shown in Figure 13.4. These findings suggest that American acceptance of the romantic love ideal occurred gradually from 1741 to 1865. The romantic love ideal first really came into its own about the time of the Civil War.

Love as a Story

When we think of love, our thoughts often turn to the great love stories: Romeo and Juliet, Cinderella and the Prince (Julia Roberts and Richard Gere), King Edward VIII and Wallis Simpson, and *Pygmalion/My Fair Lady*. According to Sternberg (1998), these stories are much more than entertainment. They shape our beliefs about love and relationships, and our beliefs in turn influence our behavior.

> Zach and Tammy have been married 28 years. Their friends have been predicting divorce since the day they were married. They fight almost constantly. Tammy threatens to leave Zach; he tells her that nothing would make him happier. They lived happily ever after.

FIGURE 13.4 OCCURRENCE OF THE ROMANTIC LOVE IDEAL IN MAJOR AMERICAN MAGAZINES, 1741–1865

One way to measure the influence of the romantic love ideal on American society is to determine the frequency with which it is mentioned in popular magazines. A team of researchers selected a sample of the best-selling magazines from four historical periods and counted the number of times each of the five romantic ideals was mentioned—including (1) idealization of the beloved, (2) love at first sight, (3) love conquers all, (4) there is one and only one for each of us, and (5) we should follow our hearts. They discovered that the number of times the ideals were mentioned increased more than 300% from 1741–94 to 1850–65.

Source: Adapted from "Occurrence of the Romantic-Love Ideal in Major American Magazines, 1741–1865" (1975) by J. Lantz, J. Keyes, and M. Schultz, *American Sociological Review, 40*(1). Used with permission from the American Sociological Association.

> Valerie and Leonard had a perfect marriage. They told each other and all of their friends that they did. Their children say they never fought. Leonard met someone at his office and left Valerie. They are divorced. (Adapted from Sternberg, 1998)

Wait a minute! Aren't those endings reversed? Zach and Tammy should be divorced, and Valerie and Leonard should be living happily ever after, right? If love is merely the interaction between two people—how they communicate and behave—you're right; the stories have the wrong endings. But there is more to love than interaction; what matters is how each partner interprets the interaction. To make sense out of what happens in our relationships, we rely on our love stories.

A **love story** is a story (script) about what love should be like; it has characters, plot, and theme. There are two central characters in every love story, and they play roles that complement each other. The plot details the kinds of events that occur in the relationship. The theme is central; it provides the meaning of the events that make up the plot, and it gives direction to the behavior of the principals. The love story guiding Zach and Tammy's relationship is the "War" story. Each views love as war; that is, a good relationship involves constant fighting. The two central characters are warriors, doing battle, fighting for what they believe. The plot consists of arguments, fights, threats to leave—in short, battles. The theme is that love is war; one may win or lose particular battles, but the war continues. Zach and Tammy's relationship endures because they share this view, and it fits their temperaments. Can you imagine how long a wimp would last in a relationship with either of them?

According to this view, falling in love occurs when you meet someone with whom you can create a relationship that fits your love story. Furthermore, we are satisfied with relationships in which we and our partner match the characters in our story (Beall & Sternberg, 1995). Valerie and Leonard's marriage looked great on the surface, but it didn't fit Leonard's love story. He left when he met his "true love"—that is, a woman who could play the complementary role in his primary love story.

Where do our love stories come from? Many of them have their origins in the culture—in folk tales, literature, theater, films, and television programs. The cultural context interacts with our own personal experience and characteristics to create the stories that each of us has (Sternberg, 1996). As we experience relationships, our stories evolve, taking account of unexpected events. Each person has more than one story; the stories often form a hierarchy. One of Leonard's stories was "House and Home"; home was the center of the relationship, and he (in his role of Caretaker) showered attention on the house and kids—not on Valerie. But when he met Sharon with her aloof air, ambiguous past, and dark glasses, he was hooked; she elicited the "Love Is a Mystery" story that was more salient to Leonard. He could not explain why he left Valerie and the kids; like most of us, he was not consciously aware of his love stories. It should be obvious from these examples that love stories derive their power from the fact that they are self-fulfilling. We create in our relationships events according to the plot and then interpret those events according to the theme. Our love relationships are literally social constructions. Because our love stories are self-confirming, they can be very difficult to change.

Sternberg and his colleagues have identified five categories of love stories found in U.S. culture, and several specific stories within each category. They have also developed a series of statements that reflect the themes in each story. People who agree with the statements, "I think fights actually make a relationship more vital," and, "I actually like to fight with my partner" are likely to hold the "War" story. Sternberg and Hojjat studied samples of 43 and 55 couples (Sternberg, 1998). They found that couples generally held similar stories. The more discrepant the stories of the partners, the less happy the couple was. Some stories were associated with high satisfaction—for example, the "Garden" story, in which love is a garden that needs ongoing cultivation. Two stories associated with low satisfaction were the "Business" story (especially the version in which the roles are Employer and Employee), and the "Horror" story, in which the roles are Terrorizer and Victim.

Love stories, or implicit theories of relationships (Franiuk, Cohen, & Pomerantz, 2002), are stable over time. Persons who believe there is a one and only love, or soul mate, for them believe that finding the right person is the key to a satisfying relationship; people who believe that a successful relationship requires continuing work, Sternberg's "love is a garden," believe that hard work is the key. So men and women in the first group emphasize the partner's characteristics in assessing their satisfaction with the relationship; if they decide their partner is not Ms. or Mr. Right, they may leave in search of the true love. Gardeners view the relationship as a work in progress and place less emphasis on the partner in assessing their satisfaction; if unhappy, they work harder and apply more water and fertilizer.

BREAKING UP

Progress? Chaos?

You may have noticed that much of the work we have reviewed assumes or implies that intimate relationships develop or progress in a linear way. We meet, disclose, trust, disclose more, trust more, become sexually intimate, become interdependent, fall in love, and (hopefully) live happily ever after. This linear model underlies much of the work on relationships and relationship stability/instability. There is, however, an alternative model that may be (more) appropriate: chaos theory (Weigel & Murray, 2000). Chaos theory suggests that relationships do not develop in a steady linear progression. Instead, relationships may shift suddenly or spontaneously; they may go up (get better) or down (get worse). A small event (say, a missed phone call) may have a major impact; a traumatic event (say, a diagnosis of cancer) may have little or no effect. As a result, it may be impossible to predict the future of an individual relationship. We are just beginning to explore the implications of this model of intimacy.

Whether linear or not, few relationships last forever. Roommates who once did everything together lose touch after they finish school. Two women who were once best friends gradually stop talking. Couples fall out of love, and break up. What causes the dissolution of relationships? Research suggests two answers: unequal outcomes and unequal commitment.

Unequal Outcomes and Instability

Earlier, this chapter discussed the importance of outcomes in establishing and maintaining relationships. Our decision to initiate a relationship is based on what we expect to get out of it. In ongoing relationships, we can assess our actual outcomes; we can evaluate the rewards we are obtaining relative to the costs of maintaining the relationship. A survey of college students examined the impact of several factors on satisfaction with a relationship; one factor was the value of overall outcomes compared with a person's comparison level (CL; Michaels, Edwards, & Acock, 1984). In an analysis of the reports of men and women involved in exclusive relationships, the outcomes being experienced were most closely related to satisfaction with the relationship. Several other studies report the same results (Surra, 1990).

The comparison level for alternatives (CL_{alt}) is also an important standard used in evaluating outcomes. Are the outcomes from this relationship better than those obtainable from the best available alternative? One dimension on which people may evaluate relationships is physical appearance. A relationship with a physically attractive person may be rewarding. Two people who are equally attractive physically will experience similar outcomes on this dimension. What about two people who differ in attractiveness? The less attractive person will benefit from associating with the more attractive one, whereas the more attractive person will experience less positive outcomes. Because attractiveness is a valued and highly visible asset, the more physically attractive person is likely to find alternative relationships available and to expect some of them to yield more positive outcomes.

This reasoning was tested in a study of 123 dating couples. Photographs of each person in the study were rated by five men and five women for physical attractiveness, and a relative attractiveness score was calculated for each member of each couple. Both men and women who were more attractive than their partners reported having more friends of the opposite sex—that is, alternatives—than men and women who were not more attractive than their partners. Follow-up data collected 9 months later indicated that dating couples who were rated as similar in attractiveness were more likely to be still dating each other (White, 1980). These results are consistent with the hypothesis that persons experiencing outcomes below CL_{alt} are more likely to terminate the relationship. Another study of 120 couples asked each to rate the relationship on 16 dimensions (Attridge, Berscheid, & Simpson, 1995). Six months later, the researchers determined whether the couple was still together. The best predictor of whether a couple broke up was the assessment of the "weak link"—the person who gave the relationship lower ratings than the partner.

But not everyone compares their current outcomes with those available in alternative relationships. Individuals in White's study who were committed—that is, cohabiting, engaged, or married—did not vary in the number of alternatives they reported. Also, their relative attractiveness was not related to whether they were still in the relationship 9 months later. Persons who are committed to each other may be more concerned with equity than with alternatives.

Equity theory (Walster [Hatfield], Berscheid, & Walster, 1973) postulates that each of us compares the rewards we receive from a relationship to our costs or contributions. In general, we expect to get more out of the relationship if we put more into it. Thus, we compare our outcomes (rewards minus costs) to the outcomes our partner is receiving. The theory predicts that **equitable relationships**—in which the outcomes are equivalent—will be stable, whereas inequitable ones will be unstable.

This prediction was tested in a study involving 537 college students who were dating someone at the time (Walster [Hatfield], Walster, & Traupmann, 1978). Each student read a list of things that someone might contribute to a relationship, including good looks, intelligence, loving, understanding, and helping the other make decisions. Each student also read a list of potential consequences of a relationship, including various personal, emotional, and day-to-day rewards and frustrations. Each student was then asked to evaluate the contributions he or she made to the relationship, the contributions the partner made, the things he or she received, and the things the partner received. Each evaluation was made using an 8-point scale that ranged from extremely positive (+4) to extremely negative (−4). The researchers calculated the person's overall outcomes by dividing the rating of consequences by the rating of contributions. They calculated the perceived outcomes of the partner by dividing the rating of the consequences the partner received by the rating of the contributions the partner was making. By comparing the person's outcomes with the perceived partner's outcomes, the researchers determined whether the relationship was perceived as equitable.

Students were interviewed 14 weeks later to assess the stability of their relationships. Stability was determined by whether they were still dating their partner and by how long they had been going together (or how long they had gone together). The results clearly demonstrated that inequitable relationships were unstable. The less equitable the relationship was at the start, the less likely the couple was to be still dating 14 weeks later. Furthermore, students who perceived that their outcomes did not equal their partner's outcomes reported that their relationships were of shorter duration.

Differential Commitment and Dissolution

Are outcomes (rewards minus costs) the only thing we consider when deciding whether to continue a relationship or not? What about emotional attachment or involvement? We often continue a relationship because we have developed an emotional commitment to the person and feel a sense of loyalty to and responsibility for that person's welfare. The importance of commitment is illustrated by the results of a survey of 234 college students (Simpson, 1987). Each student was involved in a dating relationship and answered questions about 10 aspects of the relationship. Three months later, each respondent was recontacted to determine whether he or she was still dating the partner. The characteristics that were most closely related to stability included length and exclusivity of the relationship and having engaged in sexual intimacy; all three are aspects of commitment. A review of research on premarital relationships concludes that commitment—the person's intent to remain in the relationship—is consistently related to stability (Cate, Levin, & Richmond, 2002).

When both persons are equally committed, the relationship may be quite stable. But if one person is less involved than the other, the relationship may break up. The importance of equal degrees of involvement is illustrated in another study. Couples were recruited from four colleges and universities in the Boston area (Hill, Rubin, & Peplau, 1976). Each member of 231 couples filled out an initial questionnaire and completed three follow-up questionnaires 6 months, 1 year, and 2 years later. At the time the initial data were collected, couples had been dating an average of 8 months; most were dating exclusively, and 10 percent were engaged. Two years later, researchers were able to determine the status of 221 of the couples. Some were still together, whereas others had broken up.

What distinguished the couples who were together 2 years later from those who had broken up? Couples who were more involved initially—who were dating exclusively, who rated themselves as very close, who said they were in love, and who estimated a high probability that they would get married—were more likely to be together 2 years later. Of the couples who reported equal involvement initially, only 23 percent broke up in the following 2 years. But of the couples who reported unequal involvement initially, 54 percent were no longer seeing each other 2 years later.

Earlier in this chapter, we discussed the importance of similarity in establishing relationships. How important is similarity in determining whether a relationship persists over time? Among the 221 couples, couples who stayed together and couples who broke up were initially similar in their gender role attitudes, approval of premarital sexuality, importance of religion, and the number of children they wanted. Thus, although attitudinal similarity seems to determine the formation of relationships, it does not distinguish couples whose relationships persist from those whose relationships dissolve.

Not surprisingly, the breakup of a couple was usually initiated by the person who was less involved. Of those whose relationships ended, 85 percent reported that one person wanted to break up more than the other. There was also a distinct pattern in the timing of breakups; they were much more likely to occur in May-June, September, and December-January. This suggests that factors outside the relationship—such as graduation, moving, and arriving at school—led one person to initiate the breakup. Such changes, or life course transitions (see Chapter 17), are likely to increase the costs, such as the difficulty of meeting, of continuing a relationship.

A cultural event that focuses attention on intimate relationships is Valentine's Day; it comes with a script of what we should do for our romantic partner, in the hope of nurturing the garden. Does it work? Researchers surveyed samples of students in relationships twice, 2 weeks apart, in September, November, February (one week before and one week after Valentine's Day), and April. Couples participating in February were *more* likely to break up than those surveyed in the other 2-week periods. Data analyses suggest that Valentine's Day serves as a catalyst, resulting in the dissolution of relationships that are declining in quality (Morse & Neuberg, 2004). Apparently, the emphasis on what intimate relationships should be like provides a standard that leads participants in declining relationships to decide to quit, and perhaps to look for alternatives.

The dissolution of a relationship is often painful. But breaking up is not necessarily undesirable. It can be thought of as part of a filtering process through which people who are not suited to each other terminate their relationships.

Responses to Dissatisfaction

Not all relationships that involve unequal outcomes or differential commitment break up. What makes the difference? The answer is, in part, the person's reaction to these situations. The level of outcomes a person experiences and his or her commitment to the relationship are the main influences on satisfaction with that relationship (Bui, Peplau, & Hill, 1996; Rusbult et al., 1986). A study of 60 students and 36 married couples found that an important influence on satisfaction is the perception that your partner supports your attempts to achieve goals that are important to you (Brunstein, Dangelmayer, & Schultheiss, 1996). As long as the person is satisfied, whatever the level of rewards or commitment, he or she will want to continue the relationship. People who are satisfied are more likely to engage in **accommodation**—to respond to potentially destructive acts by the partner in a constructive way (Rusbult, Verette, Whitney, Slovik, & Lipkus, 1991). A study of Black and White married couples over 14 years found that reports of frequent conflict and of using insults, name-calling, and shouting in response to conflict (in other words, not engaging in accommodation), predicted subsequent divorce (Orbuch et al., 2002).

An individual in an unsatisfactory relationship has four basic alternatives (Goodwin, 1991; Rusbult, Zembrodt, & Gunn, 1982): exit (termination), voice (discuss with your partner), loyalty (grin and bear it), and neglect (stay in the relationship but not contribute much). Which of these alternatives the person selects depends on the anticipated costs of breaking up, the availability of alternative relationships, and the level of reward obtained from the relationship in the past.

To assess the costs of breaking up, the individual weighs the costs of an unsatisfactory relationship against the costs of ending that relationship. There are three types of barriers or costs to leaving a relationship: material, symbolic, and affectual (Levinger, 1976). Material costs are especially significant for partners who have pooled their financial resources. Breaking up will require agreeing on who gets what, and it may produce a lower standard of living for each person. Symbolic costs include the reactions of others. A survey of 254 persons, 123 of whom were in relationships, measured the perception of friends' and family members' support for the relationship and commitment to it (Cox, Wexler, Rusbult, & Gaines, 1997). Persons who perceived more support were more committed, in both dating and married couples. Will close friends and family members support or criticize the termination of the relationship? A longitudinal study of dating couples found that lower levels of support by friends for the relationship was associated with later termination of it (Felmlee, Sprecher, & Bassin, 1990). Affectual costs involve changes in one's relationships with others. Breaking up may cause the loss of friends and reduce or eliminate contact with relatives; that is, it may result in loneliness (see Box 13.3). A study of married persons asked each to name "the most important factors keeping you together"; the most frequently mentioned barriers were children (31 percent of respondents), religion (13 percent), and financial need (6 percent) (Previti & Amato, 2003).

A second factor in this assessment is the availability of alternatives. The absence of an attractive alternative may lead the individual to maintain an unrewarding relationship, whereas the appearance of an attractive alternative may trigger the dissatisfied person to dissolve the relationship.

We noted at the beginning of this chapter that two factors influence who is available—personal characteristics and institutional structure. With regard to

BOX 13.3 Are You Lonely Tonight?

Did you feel lonely when you first entered school here? If you did, you weren't alone. People entering a college or university are likely to feel lonely for the first several weeks or months (Cutrona, 1982). In fact, most people have experienced loneliness sometime during their lives.

Loneliness is an unpleasant, subjective experience that results from the lack of social relationships satisfying in either quantity or quality (Perlman, 1988). Loneliness is different from being alone or social isolation. Social isolation is an objective situation, whereas loneliness is a subjective, internal experience. You can feel lonely in the midst of a family reunion, and you can be alone in your room and yet feel connected to others.

Loneliness is different from shyness. Shyness is a personality trait that reflects characteristics of the person rather than the state of one's social ties. Shyness is defined as "discomfort and inhibition in the presence of others" (Jones, Briggs, & Smith, 1986). A study of several measures of shyness found that the common element in these measures is distress in and avoidance of interpersonal situations. When shy people interact with others, they are afraid they are being evaluated by the other person and are more likely to think they are making a negative impression on the other person (Asendorpf, 1987).

There are two types of loneliness (Weiss, 1973), which differ in their cause. One is social loneliness, which results from a lack of social relationships or ties to others. Several studies have found that people with few or no friends and few or no family ties are more likely to feel lonely (Stokes, 1985). Thus, events that disrupt ties to social networks can cause loneliness (Marangoni & Ickes, 1989).

The other type is emotional loneliness, which results from the lack of emotionally intimate relationships. One study of adolescents found a strong association between self-disclosure and loneliness; greater self-disclosure to others was associated with reduced loneliness (Davis & Franzoi, 1986). Thus, shyness can cause loneliness by inhibiting self-disclosure. There is evidence that loneliness in men is the result of having few or no relationships with others, whereas in women it is the result of having no intimate relationships (Stokes & Levin, 1986). Clearly, loneliness is tied to the state of one's interpersonal relationships.

Because loneliness is related to the number and quality of interpersonal relationships, we can predict that people in some circumstances are more likely to experience it. First, people undergoing a major social transition are generally at greater risk of loneliness. The transition from school to work may be accompanied by feelings of loneliness, especially when this transition involves a geographic move. Second, living arrangements are related to feeling lonely. A study of 554 adult men and women found that living alone was the most important determinant of these feelings (de Jong-Gierveld, 1987). Third, one's relationship status is important. Earlier in this chapter, we described the increasing self-disclosure and interdependence that accompanies the development of romantic relationships; people who report that interdependence developed faster and broader and who feel a strong sense of "we-ness" in their relationship are less likely to report loneliness (Flora & Segrin, 2000). Conversely, people who have recently gone through the termination of an intimate relationship—through breaking up, divorce, or death of a partner—may be especially vulnerable to loneliness.

the first, people who are in relationships perceive opposite-sex persons of the same age as less physically attractive than do people who are not in relationships (Simpson, Gangestad, & Lerum, 1990). This devaluation of potential partners contributes to relationship maintenance. However, a longitudinal study found that the perceived quality of alternative partners increased among persons whose relationships subsequently ended (Johnson & Rusbult, 1989). With regard to institutional structures, the sex ratio in a community determines the number of eligible partners. Research combining survey data with census data for the area where the respondent lived found that the risk of divorce is greatest in areas where husbands or wives encounter numerous alternatives (South, Trent, & Shen, 2001).

A third factor is the level of rewards experienced before the relationship became dissatisfying. If the relationship was particularly rewarding in the past, the individual is less likely to decide to terminate it.

How important are each of these three factors? That is, which factors are most important in determining whether a dissatisfied person responds by discussing the situation with his or her partner, waiting for things to improve, neglecting the partner, or terminating the relationship? In one study, participants were given short stories describing relationships in which these three factors varied. They were asked what they would do in each situation (Rusbult et al., 1982). The results showed that the lower the prior satisfaction—that is, the less satisfied and the less positive their feelings and caring for their partner—the more likely they were to neglect or terminate the relationship. The less the investment—that is, the degree of disclosure and how much a person stands to lose—the more likely participants were to engage in neglect or termination. Finally, the presence of attractive alternatives increased the probability of terminating the relationship. A later study of ongoing relationships yielded the same results (Rusbult, 1983).

A study of the stability of the relationships of 167 couples over a 15-year period also found that satisfaction, level of investments, and quality of alternatives predicted commitment. Relationships in which commitment was high were more likely to endure (Bui et al., 1996).

SUMMARY

Interpersonal attraction is a positive attitude held by one person toward another person. It is the basis for the development, maintenance, and dissolution of close personal relationships.

Who Is Available? Institutional structures and personal characteristics influence who are available to us as potential friends, roommates, coworkers, and lovers. Three factors influence whom we select from this pool. (1) Our daily routines make some persons more accessible. (2) Proximity makes it more rewarding and less costly to interact with some people rather than others. (3) Familiarity produces a positive attitude toward those with whom we repeatedly come into contact.

Who Is Desirable? Among the available candidates, we choose based on several criteria. (1) Social norms tell us what kinds of people are appropriate as friends, lovers, and mentors. (2) We prefer a more physically attractive person, both for esthetic reasons and because we expect rewards from associating with that person. Attractiveness is more influential when we have no other information about a person. (3) We choose based on our expectations about the rewards and costs of potential relationships. We choose to develop those relationships whose outcomes we expect will exceed both comparison level (CL) and comparison level for alternatives (CL_{alt}). We implement our choices by making contact, using an opening line that often indicates the kind of relationship in which we are interested.

The Determinants of Liking Many relationships—between friends, roommates, coworkers, or lovers—involve liking. The extent to which we like someone is determined by three factors. (1) The major influence is the degree to which two people have similar attitudes. The greater the proportion of similar attitudes, the more they like each other. Similarity produces liking because we prefer cognitive consistency and because we expect interaction with similar others to be reinforcing. (2) Shared activities become an important influence on our liking for another person as we spend time with them. (3) We like those who like us; as we experience positive feedback from another, it increases our liking for them.

The Growth of Relationships As relationships grow, they change on three dimensions. (1) There may be a gradual increase in the disclosure of intimate information about the self. Self-disclosure is usually reciprocal, with each person revealing something about themselves in response to revelations by the other. (2) Trust in the other person—a belief in his or her honesty, benevolence, and reliability—also increases as relationships develop. (3) Interdependence for various gratifications also increases, often accompanied by a decline in reliance on and number of relationships with others.

Love and Loving (1) Whereas liking refers to a positive attitude toward an object, love involves attachment to and caring for another person. Love also may involve passion—a state of intense physiological arousal and intense absorption in the other. (2) The experience of passionate love involves cognitive, emotional, and behavioral elements. (3) The concept of love does not exist in all societies; the romantic love ideal emerged gradually in the United States and came into its own about the time of the Civil War. (4) Love stories shape our beliefs about love and relationships, and our beliefs influence how we behave in and interpret our relationships.

Breaking Up There are three major influences on whether a relationship dissolves. (1) Breaking up may result if one person feels that outcomes (rewards minus costs) are inadequate. A person may evaluate present outcomes against what could be obtained from an alternative relationship. Alternatively, a person may look at the outcomes the partner is experiencing and assess whether the relationship is equitable. (2) The degree of commitment to a relationship is an important influence on whether it continues. Someone who feels a low level of emotional attachment to and concern for his or her partner is more likely to break up with that person. (3) Responses to dissatisfaction with a relationship include exit, voice, loyalty, or neglect. Which response occurs depends on the anticipated economic and emotional costs, the availability of attractive alternatives, and the level of prior satisfaction in the relationship.

LIST OF KEY TERMS AND CONCEPTS

access display (p. 335)
accommodation (p. 350)
attitudinal similarity (p. 336)
attractiveness stereotype (p. 332)
availables (p. 327)
comparison level (p. 334)
comparison level for alternatives (p. 334)
dyadic withdrawal (p. 341)
equitable relationship (p. 348)
interpersonal attraction (p. 326)
loneliness (p. 351)
love story (p. 346)
matching hypothesis (p. 330)
mere exposure effect (p. 328)
norm of homogamy (p. 329)
passionate love (p. 343)
romantic love ideal (p. 345)
trust (p. 340)

14

GROUP COHESION AND CONFORMITY

Introduction

What Is a Group?

Group Cohesion

Group Structure and Goals

Roles in Groups

Status of Group Members

Status Characteristics

Status Generalization

Expectation States Theory

Overcoming Status Generalization

Conformity to Group Norms

Group Norms

Conformity

Why Conform?

Increasing Conformity

Minority Influence in Groups

Effectiveness of Minority Influence

Differences Between Minority and Majority Influence

Summary

List of Key Terms and Concepts

INTRODUCTION

Groups are everywhere. We all participate in them, spending a significant portion of our days engaging in group activities. Families, work groups, sports teams, street gangs, classes and seminars, therapy and rehabilitation groups, classical quartets and rock groups, small military units, neighborhood clubs, church groups—these are only some of the groups we encounter.

Groups are important because they provide social support, a cultural framework to guide performance, and rewards and resources of all kinds. Without them, most individuals would be lost. The character Robinson Crusoe was depicted as living on an island, isolated and alone; he managed to survive, but his plight was less than ideal, and he was sustained primarily by the dream of rejoining his fellow humans. Without groups, most individuals would be isolated, unloved, disoriented, relatively unproductive, and very possibly hungry.

WHAT IS A GROUP?

We all have a notion of a "group," but our commonsense notion is too broad for use in social psychology. In the next three chapters, the term **group** will specifically mean a social unit that consists of two or more persons and that has all of the following attributes:

1. Membership. To be a member, a person must think of himself or herself as belonging to the group and must also be recognized by other members as belonging to the group (Lickel et al., 2000).
2. Interaction among members. Group members interact—they communicate with one another and influence one another.
3. Goals shared by members. Group members are interdependent with respect to goal attainment. Progress by one member toward his or her objectives makes it more likely that another member will also reach his or her objectives.
4. Shared Norms. Group members hold a set of expectations (that is, norms or rules) that place limits on members' behavior and guide action.

As this definition suggests, groups are not just collections of individuals; rather, they are organized systems in which the relations among individuals are structured and patterned. Not all social units involving two or more persons are groups. Persons in a theater crowd escaping in panic from a fire would not constitute a group. Although there may be some communication among the individuals in the crowd, there are no explicit normative expectations or a sense of shared membership among those present. Likewise, a commercial transaction between yourself and a store clerk selling you a bag of groceries would not qualify as a group interaction, because there is no common goal or explicit basis for group membership.

Questions About Groups This chapter focuses on the forces that unify a group and define the behavior of the members of the group. It addresses the following questions:

1. What factors hold a group together as a unit? That is, what produces cohesion—or the lack of it—in groups?
2. How does a group define its structure and goals? What determines the relative status of group members as they claim different roles within the group?
3. What are group norms, and in what ways do they regulate the behavior of members?
4. How do groups influence their members to conform to the wishes of the majority in the group?
5. Under what conditions can a minority within the group successfully influence the majority?

Group Cohesion

A recreational baseball club—the Jaguars—has a long record of championships in its city league. The Jaguar players take pride in their performance and are very committed to their team. At practice and during games, this team is a model of enthusiasm and coordination. On the rare occasion when they have a losing streak, all the team members voluntarily hold extra practice sessions to sharpen their skills and teamwork.

The players like each other, and they enjoy playing together and celebrating their victories. Although they do not always agree on strategy, the Jaguars resolve their differences quickly. Several of the players consider their teammates best friends, and they often spend time together off the field. The Jaguars team rarely loses any of its players—not even its second-stringers.

The players of another team in the league—the Penguins—provide a very different story. Less than a model of competence, the Penguins have finished in last place for the last three seasons. Occasionally, the Penguins have to forfeit a game because they cannot even field a team of nine players. The team is not a high priority for the players—they are often busy with other activities, and they often miss practice. This may be related to the fact that the players seldom run into one another outside of team activities. The Penguins' planning session last spring dissolved into chaos when the players could not agree on how to pay for some new equipment. The friction was so bad that there is doubt whether the team will even participate in the league next year.

The Jaguars and the Penguins differ in a number of respects. For one thing, the Jaguars win a lot more than the Penguins. But the teams also differ notably in their members' willingness to participate. The Jaguar players care about their membership on the team and want to participate, whereas the Penguin players seem to care much less. The Jaguars have a stronger grip on members' loyalty than the Penguins, and the team is bonded together more firmly—the Jaguars have a higher level of group cohesion than the Penguins.

Group cohesion refers to the extent to which members of a group desire to remain in that group and resist leaving it (Balkwell, 1994; Cartwright, 1968). A highly cohesive group will in general maintain a firm hold over its members' time, energy, loyalty, and commitment. Cohesive groups are marked by strong ties among members, a positive emotional feeling about membership, and by a tendency for members to perceive events in similar terms (Bollen & Hoyle 1990; Braaten, 1991; Evans & Jarvis, 1980). Because members of a cohesive group desire to belong, the interactions among them will typically have a positive, upbeat character and reflect a "we" feeling.

The Nature of Group Cohesion People may have very different motives for joining and staying in groups. Some may belong to a group because they like the tasks they perform in the group, because they enjoy interacting with the other members, because the group reflects their own values, or because the group helps them get something they want (such as prestige, money, future opportunities, protection, or social contacts). These differing motives lead to different levels and kinds of cohesion among the members (Cota, Evans, Dion, Kilik, & Longman, 1995; Hogg & Haines, 1996; Mullen & Copper, 1994; Tziner, 1982).

One of the fundamental types of group cohesion is social cohesion. A group has social cohesion if its members stay in the group primarily because they like one another as persons and desire to interact with one another (Aiken, 1992; Lott & Lott, 1965). Other things being equal, social cohesion will be greater when group members are similar. Similarity increases liking; therefore, groups whose members have similar education, ethnicity, and status, and hold similar attitudes will have greater social cohesion.

The other major type of group cohesion is task cohesion. When a group has high task cohesion, its members remain together primarily because they are heavily involved with the group's task(s). Task cohesion will be greater if members find the group's task(s) intrinsically valuable, interesting, and challenging. It will also be greater if the group's objectives (and the related tasks) are clearly defined (Raven & Rietsema, 1957). Groups that succeed at achieving their goals (like the Jaguars) often have higher task cohesion than groups that fail repeatedly (like the Penguins).

Consequences of Group Cohesion What difference does it make whether a group is highly cohesive or not? First, cohesion affects the amount of interaction among group members. Given the opportunity, members of highly cohesive groups communicate more with one another than do members of less cohesive groups. This holds true for a wide variety of groups, ranging from student organizations to industrial training groups (Moran, 1966). However, cohesion affects not only the amount of interaction but also its quality. Interaction among members in highly cohesive groups is usually friendlier, more cooperative, and

BOX 14.1 Research Update: The Ties That Bind: Attachments to a Sorority

Why are people strongly attached to groups? Social psychologists have identified a number of different factors that strengthen cohesion among group members. Recent work on the problem has produced a new proposition: Emotional cohesion develops from the network structure in a group. That is, patterns of who knows whom, and who interacts with whom, produce emotional commitments to the group and the ensuing effects of cohesion (conformity to group norms, productivity of the group, and so on).

To be emotionally attached to the group, individuals must feel that they are full members of the group—that they belong to the group—and they must feel good about being members of the group—belonging to it makes them happy. Being emotionally attached implies that having to detach from the group produces negative emotional costs. If someone separates from a group in which he or she does not feel very connected it would not produce much negative emotion because the individual does not feel connected in the first place. Likewise, if someone is a member of a group and is not happy about it (perhaps it is a group that is constantly arguing), they will not suffer much from separating.

Sociologists Pamela Paxton and James Moody believed that differences in an individual's involvement and position in the network of individuals inside the group would produce differences in the sense of belonging and the individual's emotional satisfaction with the group.

To test these ideas, Paxton and Moody studied the relationships among the members of a sorority at a college in the southern United States. Each member of the sorority was asked who in the sorority was her best friend, who she went out with socially on a regular basis, and who she would confide in. As expected, some members were named more often than others, and those people were considered more central to the sorority network.

The researchers found, not surprisingly, that sorority sisters that were more central to the network felt more belonging to the group and were happier with their membership. But the researchers also located several subgroups or cliques within the larger sorority. They found that people who were more central to those subgroups were lower on their emotional attachment to the group. In terms of group cohesion, then, relationships with others can increase cohesion, but if cohesion among a subgroup becomes strong enough, it can damage the commitment of its members to the larger group.

Source: Adapted from Paxton and Moody, 2003.

entails more attempts to reach agreements and to improve coordination (Shaw & Shaw, 1962).

Members of high-cohesion groups also have more influence on each other than members of low-cohesion groups (Lott & Lott, 1965). Not only do members of high-cohesion groups try to influence each other more, they are also more likely to be successful. This also means that members of highly cohesive groups conform more to the expectations of their fellow members than do members of less cohesive groups (Sakurai, 1975; Wyer, 1966). Members of groups with high cohesion care about belonging and want their group to perform well, so they use their influence to bring about coordination and consensus in the group. The preponderance of evidence in group research indicates that cohesion tends to increase the productivity and performance of groups, though not invariably (Evans & Dion, 1991; Gully, Devine, & Whitney, 1995; Mullen & Copper, 1994). However, this finding depends on the type of cohesion holding the group together. Task cohesion (that is, members' commitment to the group's task) has a significant effect on group productivity, but other forms of cohesion (such as social cohesion and group pride) have little or no effect on productivity (Mullen & Copper, 1994). Perhaps members of socially cohesive groups prefer to spend their time socializing rather than producing.

Group Structure and Goals

In addition to cohesion, groups can also be characterized by their goals and by the structure they adopt in pursuit of their goals. A **group goal** is an outcome viewed by group members as desirable and important to attain. These goals can differ in terms of specificity, ranging from general statements about what the group does and why it exists to more specific targets and tasks that the group members attempt to achieve along the way to its larger goals.

Group Goals and Individual Goals Although individual and group goals can be related, they are not always the same, and these differences can be very important for the functioning of the group.

Most groups function best when there is substantial similarity or isomorphism between group goals and the individual goals of its members. The term **goal isomorphism** refers to a state where group goals and individual goals are similar in the sense that actions leading to group goals also lead to the attainment of individual goals. High isomorphism benefits the group, because members are motivated to pursue group goals and to contribute resources and effort to the group (Sniezek & May, 1990). To heighten isomorphism, groups can recruit selectively—that is, admit as members only persons who strongly support the group's main goal(s). Also, a group can try to increase acceptance of group goals through socialization and training. Furthermore, a group can increase its members' awareness that they belong to the group and make their identity as members more salient (Mackie & Goethals, 1987). This may be done by increasing the proximity of members to one another, increasing the experiences they share in common, increasing the amount of social contact and communication among them, or using a common designation to label them (Dion, 1979; Turner, 1981).

Roles in Groups

As groups pursue their goals, the members of the group take on different roles. Usually, it is not efficient for all members of a group to try to perform the same tasks. Instead, the group engages in a division of labor in which members are assigned different tasks. In addition to the assignment of tasks, members of the group also differ in the informal roles they take on in the group. Some group members will exercise more influence over the group (even if they have no official authority in the group), some will talk more than others, some will become close friends with each other, and so on.

From one perspective, a role is a set of functions that a member performs for the group. From another perspective, a role is a cluster of rules or expectations indicating the set of duties to be performed by a member occupying a given position within a group. Group members hold role expectations regarding one another's performance, and they feel justified in making demands on each other. For instance, the members of a sales group expect the salespeople to contact potential customers, to identify customers' needs, and to offer customers the products that meet these needs. If, for some reason, one of the salespeople suddenly stopped contacting customers, other members of the sales group would view it as a violation of role expectations and would doubtless take action to correct the situation.

Roles in Group Communication Sometimes, roles in groups are well defined and have been operating for long periods of time even as group membership changes. On the other hand, groups often form only for temporary purposes and face the problem of developing norms, roles, and goals as part of their group tasks. Consider five undergraduates meeting for the first time to discuss a case for their human relations course. The case they must analyze involves a delinquent juvenile who has committed a serious crime but who comes from an underprivileged background. The group members have been instructed to read a summary of the boy's history, discuss it, and reach a collective decision about his case. At one extreme, they might decide that he should be punished severely for the crime, whereas at the other extreme they might decide he should be treated leniently. Because the case is complex and ambiguous, there is no obvious best decision.

All the members of this group have very similar social attributes—that is, they are of the same age, same race, same gender, and same basic background. The group has not been assigned a formal leader, so the five members enter the group on an equal footing.

Interaction in groups such as this one—newly formed, homogeneous, problem-solving groups working on human relations problems—have been extensively studied by investigators (Bales, 1965; Slater, 1955). When members of such groups start to discuss their problem, certain things typically happen. For instance, the initial equality among members disappears, and distinctions quickly arise among them. Some members participate more than others and exercise more influence regarding the group's decision. Members develop different expectations about their own and others' activities in the group. Most often, one or more members start to exercise leadership within the group. Role differences begin to emerge among the members.

One finding that regularly emerges in studies of these kinds of group situations is that the members do not participate equally in a discussion. The most talkative person in a problem-solving group typically initiates 40 to 45 percent of all communicative acts. The second most active person initiates approximately 20 to 30 percent of the communication. This pattern is apparent in Table 14.1, which summarizes the percentage of acts initiated by each member for groups ranging from three to eight members. As the size of the group increases, the most talkative person still initiates a consistently large percentage of communicative acts, whereas the less talkative individuals are crowded out almost completely (Bales, 1970).

Another finding concerns the stability of participation: The group member who initiates the most communication during the beginning minutes of interaction is very likely to continue doing so throughout the life of the group. If a problem-solving group meets for several sessions, for example, the member who ranked highest in participation during the first session is likely to rank highest during subsequent sessions. In general, the ranking of group members in terms of participation is stable over time (Fisek, 1974).

TABLE 14.1 PERCENTAGE OF TOTAL ACTS INITIATED BY EACH GROUP MEMBER AS A FUNCTION OF GROUP SIZE

Member Number	Group Size 3	4	5	6	7	8
1	44	32	47	43	43	40
2	33	29	22	19	15	17
3	23	23	15	14	12	13
4		16	10	11	10	10
5			6	8	9	9
6				5	6	6
7					5	4
8						3

Note: Data are based on a total of 134,421 acts observed in 167 groups consisting of 3 to 8 members.

Source: Adapted from Bales, 1970, pp. 467–474.

Task Specialists and Social-Emotional Specialists

Group members differ not just with respect to the amount they contribute to the group's communication processes, they also differ with respect to the kind of communication they offer.

One study (Bales, 1953) investigated groups of five men who met to discuss a problem in a laboratory setting. Group interaction was scored by observers, and then, at the end of the discussion period, members filled out questionnaires and rated each other. Items included such questions as "Who had the best ideas in the group?" "Who did the most to guide the group discussion?" and "Which group member was the most likable?" Typically, there was high agreement among group members in their answers regarding ideas and guidance but less agreement in their answers regarding liking. This study's results showed a high correlation between participation and members' perceptions of who offered the most guidance and the best ideas. In short, the person initiating the most communicative acts was perceived as the group's task leader. Results of this type have been observed in other studies as well (Reynolds, 1984; Sorrentino & Field, 1986).

But this is not the entire picture. Often, the person initiating the most acts was not the best-liked member; indeed, he was sometimes the least-liked member. More typically, the second highest initiator was the best-liked member. Why does this occur?

In general, the highest initiator is someone who drives the group toward the attainment of its goals. A high proportion of the acts initiated by this person are task oriented. For this reason, we can call the high initiator the group's **task specialist.** In the effort to get things done, however, the task specialist tends to be pushy and, in some instances, openly antagonistic. He or she makes the most impact on the group's opinion, but this aggressive behavior often creates tension. So it remains for some other member, the **social-emotional specialist,** to ease the tension and soothe hurt feelings in the group. The acts initiated by this person are likely to be acts showing tension, releasing tension, and encouraging solidarity. The social-emotional specialist is the one who exercises tact or tells a joke at just the right moment. This person helps to release tension and maintain good spirits within the group. Not surprisingly, the social-emotional specialist is often the best-liked member of the group.

Thus, in problem-solving groups, there are two basic functions—getting things done and keeping relations pleasant—that are typically performed by different members. When group members divide up functions in this manner, we say that **role differentiation** has occurred in the group. Although this type of role differentiation occurs frequently, it is not inevitable. In some groups, a single member performs both the task-oriented and the social-emotional functions. Observations suggest that for problem-solving groups in laboratory settings, a single member successfully performs both functions in about 20 to 30 percent of the cases (Lewis, 1972). For groups in natural, nonlaboratory settings, the incidence of combined roles may be higher (Rees & Segal, 1984).

STATUS OF GROUP MEMBERS

So far, we have considered interactions in newly formed problem-solving groups that consist of very similar members. Despite this initial equality among members, role and status differences quickly emerge in groups. Members differ in their rate of participation, their influence over group decisions, and the types of acts they contribute.

But what about interactions in newly formed groups whose members are not identical in social attributes? In particular, what about groups composed of members who differ in gender, race, age, education, and occupation? We encounter such groups every day—PTAs, student committees, neighborhood associations, juries, church groups, and so on. In this section, we discuss how within-group status differences emerge in groups based on the characteristics of their members.

Status Characteristics

A **status characteristic** is any social attribute of a person around which evaluations and beliefs about that person come to be organized. We often take status characteristics into consideration when establishing expectations regarding others in groups. To illustrate, suppose that members of a new group are meeting for the first time. They have not had prior contact, so they do not know one another. In the course of the meeting, they start to develop beliefs and expectations about each other; this may include expectations about who will be able to contribute most toward the group's task and who will be able to provide leadership.

In forming these expectations, members initially may not have much information to go on. Usually, however, they will know or observe such attributes as the age, gender, and race of the other members, so they use these as shortcuts to infer what they want to know. The term diffuse status characteristics refers to attributes that provide an indirect indication of a member's level of ability on the group's task. Attributes such as age, gender, race, ethnicity, education, and physical attractiveness serve as diffuse status characteristics in many contexts (Berger, Cohen, & Zelditch, 1972; Jackson et al., 1995). In addition to these indicators, members may take note of others' specific status characteristics—attributes that more directly and precisely indicate someone's level of ability on the task to be performed by the group. For example, a player's height could be a specific status characteristic on a basketball

© Tony Freeman/PhotoEdit, Inc.

The status hierarchy in this church is apparent, as a priest, bishop, altar boy, and members of the choir participate in a service. Each person's function in the recessional, as well as his or her physical location, depends on formal status.

team; prior experience in writing advertising copy would be a specific status characteristic in a work group developing the advertising campaign.

In many cases, specific status characteristics provide a more secure basis for inference than diffuse status characteristics. Expectations and evaluations based on diffuse status characteristics tend to reflect prevailing cultural stereotypes. In the United States, for example, despite a general ideology of equality, many persons consider it preferable to be male rather than female, White rather than Black, adult rather than juvenile, and white-collar rather than blue-collar. Although we may not think of ourselves as holding such views, we are nevertheless very sensitive to attributes such as age, gender, and race, and use them to make judgments in group settings (Berger, Rosenholtz, & Zelditch, 1980; Ellard & Bates, 1990; Meeker, 1981).

Status Generalization

Status characteristics can significantly affect interactions among members in newly forming groups. Studies show that persons with high standing on status characteristics are accorded more respect and esteem than other members, and they are chosen more frequently as leaders. Their contributions to group problem solving are evaluated more positively, they are given more chance to participate in discussions, and they exert more influence over group decisions (Balkwell, 1991; Berger et al., 1972; Webster & Foschi, 1988). The tendency for members' status characteristics to affect group structure and interaction is called **status generalization.**

When status generalization occurs, a member's status outside a group affects his or her status inside that group. That is, the members who hold higher status in society at large will tend to hold higher status in the group (Cohen & Zhou, 1991).

Consider the impact of status generalization in a jury. The members of a jury, who usually differ in status characteristics, must discuss the facts of a case and reach a verdict. A study by Strodtbeck, Simon, & Hawkins (1965) demonstrated that status characteristics such as occupation and gender affected jury deliberations. Because the investigators were precluded by law from observing actual jury deliberations, they studied mock juries composed as authentically as possible. These juries, selected from the voter registration list in an urban area, consisted of men and women who had occupations of varying status, including proprietors, clerical workers, and skilled and unskilled laborers. During the study, they listened to a tape recording of a trial, after which they were instructed to do everything a real jury does—select a foreman, deliberate on the case, and reach a verdict.

From the recording of the deliberations, it was possible to determine which jury members talked the most. Table 14.2 shows the rates of participation by jury members as a function of their occupation and

TABLE 14.2 RATES OF PARTICIPATION DURING DELIBERATION BY OCCUPATION AND SEX OF JUROR

	OCCUPATION				
Sex	**Proprietor**	**Clerical**	**Skilled**	**Laborer**	**Combined Average**
Male	12.9	10.8	7.9	7.5	9.6
Female	9.1	7.8	4.8	4.6	6.6
Combined average	11.8	9.2	7.1	6.4	8.5

Note: Entries in this table are percentage rates of participation. Because there were 12 persons on a jury, the "average" juror would theoretically have a rate of participation of 8.3%. This can be used as a frame of reference against which to compare tabled values. (The "combined-combined" value in the table is 8.5 rather than 8.3 because 26 of 588 jurors in the study were not satisfactorily classified by occupation and were omitted from the analysis.)

Source: Adapted from Strodtbeck, Simon, and Hawkins, 1965.

TABLE 14.3 AVERAGE VOTES RECEIVED AS "HELPFUL JUROR" BY OCCUPATION AND SEX OF JUROR

	OCCUPATION				
Sex	**Proprietor**	**Clerical**	**Skilled**	**Laborer**	**Combined Average**
Male	6.8	4.2	3.9	2.7	4.3
Female	3.2	2.7	2.0	1.5	2.3
Combined average	6.0	3.4	3.5	2.3	3.6

Source: Adapted from Strodtbeck, Simon, and Hawkins, 1965.

gender. The results indicate that men initiated more interaction than women in the mock juries. The data also show the impact of occupational status: The higher the occupational status, the greater the rate of participation. This held for both men and women.

The questionnaire completed by jurors at the end of the session provided information on their perceptions of one another. Table 14.3 shows the votes received by the jury members for being "most helpful in reaching the verdict." This measure reflected the amount of influence each member had over the group decision as perceived by other members. The findings are very similar to those on the rates of participation. On average, male jurors were perceived as more helpful than female jurors, and jurors of high occupational status were perceived as more helpful than those of lower occupational status.

Overall, this jury study revealed the typical impact of status generalization: Persons with higher standing in terms of gender and occupation became the group members with the higher status inside the group.

Although the findings in this study seem clear-cut, the interpretation in terms of status generalization is open to criticism. A critic might argue that a person's status inside a group is not a function of his or her status outside but is instead caused by the same qualities or personal traits that determine that person's status outside the group. One might hypothesize, for instance, that people of high intelligence translate their intelligence into both high occupational status and better contributions to the group. If this were the case, a person's standing inside a group would not be caused by his or her external occupational status; rather, both internal and external status would be caused by a third factor—intelligence.

To check this possible confound, several studies have manipulated status characteristics experimentally. One of these studies (Moore, 1968) investigated small, two-person groups of female participants. Both women were shown a series of large figures made up of smaller black-and-white rectangles. Their task was to judge which of the two colors—black or white—covered the greater area in each of the large figures.

This task was difficult because the areas were in fact approximately equal, making the figures ambiguous. The participants, who were seated so that they could not see or talk with each other, signaled their preliminary judgments by pressing buttons on consoles in front of them. Each participant's answer was revealed to the other through a system of lights. The participants knew that after seeing each other's initial judgments, they would have an opportunity to make a final judgment. Because their goal was to make accurate final judgments, participants had been told that they should weigh their own answers against the answers of their partners. The participants did not know, however, that the connections on the consoles were, in fact, controlled by the experimenter, who could manipulate how often one participant perceived that her partner disagreed with her. Between their first judgment and their final judgment, the participants were free to reverse their decisions when their partners appeared to disagree with them.

All the participants in this experiment were junior college students. As a manipulation of status, one half of the participants were told that their partner was a high school student (low-status partner), whereas the other half were told that their partner was from Stanford University (high-status partner). The results show that the women who believed their partner to be of higher status changed their answers on the judgmental task more often than those who thought themselves to have higher status than their partner. The random assignment of participants to experimental treatments eliminated the possibility that participants differed systematically in intelligence or ability on the judgment task. These findings and other like experiments support the case for status generalization.

Expectation States Theory

Why does status generalization occur? That is, why do the status characteristics of group members affect the interaction occurring within the group? One answer to this question comes from **expectation states theory** (Berger, Conner, & Fisek, 1974; Berger, Webster, Ridgeway, & Rosenholtz, 1986; Knottnerus, 1988). This theory proposes that at the outset of interaction in a task group, members form expectations regarding one another's potential performance. These expectations, which are formed through basic attribution processes, affect subsequent interaction among members. In particular, members are more likely to defer to and accept influence from those they expect to perform well than from those they expect to perform poorly.

Because members form these expectations in the early stages of group interaction, they often base them on limited, sketchy information. Status characteristics such as age, race, gender, and occupation are readily apparent or easily learned, so they influence the expectations that members form regarding one another. These status characteristics are most likely to influence performance expectations when group members (1) have no prior history of interaction, (2) have no information about one another except for their status characteristics, and (3) have no special experience or information about the group's task.

Burden of Proof It might seem reasonable that when forming performance expectations from status characteristics, group members would insist on proof that the characteristics are relevant to the task at hand. In fact, however, it works the other way around. When developing performance expectations from status characteristics, group members behave as if the burden of proof is placed on demonstrating that these characteristics are not relevant to the task at hand rather than on demonstrating that they are relevant. If these characteristics (such as race or gender) are not explicitly demonstrated as irrelevant to the task (such as planning a weekend trip), they often will be treated as if they are. This phenomenon has been confirmed in several studies (Freese, 1976; Webster & Driskell, 1978; Wilke, Van Knippenberg, & Bruins, 1986). Furthermore, when statuses suggests equality (for example, if two group members are both men), this information tends to be ignored and does not reduce the difference in performance expectations between the participants (Walker & Simpson, 2000).

Multiple Status Characteristics As already noted, each member in a newly forming group possesses a variety of status characteristics. How, then, do members combine information on two or more status characteristics to form performance expectations regarding a given person? What performance expectations result

© John Neubauer/PhotoEdit, Inc.

A jury listens carefully as a lawyer presents his case. Although formally of equal status, some jury members exercise more influence than others in determining the verdict. Characteristics such as occupation, gender, and race can affect the patterns of influence among jurors during deliberations.

if that person's standing on one status characteristic is inconsistent with his or her standing on another?

Consider, for example, the case of a female physician. Such a person has inconsistent status characteristics. On the one hand, she is a physician, and this is a high-status characteristic. On the other hand, she is a woman, and some studies show that this is perceived as lower status than being a man (Berger et al., 1980; Ridgeway, 1982). How will group members respond to her? Clearly, there are several possibilities. They might ignore one of the status characteristics and focus on the other; that is, they might respond to her only as a physician or only as a woman. Alternatively, they might combine both status characteristics to form some kind of intermediate expectation. The results of several studies show that in these cases, members generally form an intermediate expectation rather than focusing on only one or another of these status characteristics (Markovsky, Smith, & Berger, 1984; Pugh & Wahrman, 1983; Zelditch, Lauderdale, & Stublarec, 1980). Therefore, confronted with a female physician and lacking other specific information, group members will most likely consider her more competent than female nonphysicians but less competent than male physicians.

Overcoming Status Generalization

Because group members often treat diffuse status characteristics as relevant to performance expectations even when in actuality they are not, status generalization can work to an individual's disadvantage (Forsyth, 1999). In a mixed setting with both men and women, for example, the women may find that they are not permitted to influence the group's decision significantly even though they are as qualified as—and may be more qualified than—men with respect to the problem under discussion. Without a clear demonstration that gender is irrelevant to performance, verbal protests regarding gender equality may be to no avail (Pugh & Wahrman, 1983). Likewise, in an interracial interaction between Blacks and Whites, the Blacks may feel that they are treated as low-status minority members. Because irrelevant diffuse status characteristics can so easily place someone at a disadvantage, researchers have asked whether status generalization can be overcome or eliminated in face-to-face interaction.

Some researchers have suggested that the best way to overcome status generalization is by direct methods—that is, by raising the expectations of lower status persons regarding their own performance on group tasks, so that they can in turn force a change in other people's expectations regarding their performance. Unfortunately, this approach does not work very well. In one study, for example, Blacks at a northern university were trained in assertiveness techniques and then participated with Whites in biracial groups. The training raised Blacks' expectations for themselves, and they behaved in an assertive and confident manner. But because the Whites' expectations regarding the Blacks' performance were not affected by the training, what ensued was not smooth interaction but a status struggle. The Whites thought the Blacks were arrogant and overreaching in light of their "limited ability," whereas the Blacks viewed the Whites as engaging in racist bigotry (Katz, 1970).

To overcome status generalization, one must change not only the expectations held by low-status group members but also those held by high-status group members. Demonstrating the right to high status through successful performance is usually more effective than merely claiming or asserting it (Freese & Cohen, 1973; Martin & Sell, 1985). A key to overcoming status generalization is to satisfy the burden-of-proof requirement by supplying group members with information that contradicts performance expectations inferred from a diffuse status characteristic (Berger et al., 1980).

For instance, in one study (Cohen & Roper, 1972), investigators taught Black junior high school students how to build a radio, then showed them how to teach another pupil to build a radio. This created two specific status characteristics inconsistent with students' conception of race. Investigators then had the Blacks train White pupils to build a radio, thereby establishing the relative superiority of the Blacks on this task. Finally, they informed some of the students that the skills involved in building the radio and teaching others to build it were relevant to another, entirely different task. This new task was a decision-making game, which the boys subsequently played. The results of this study indicated that this pattern of training modified the performance expectations held by both Black and White boys. The change in expectations produced a significant increase in equality between Blacks and Whites, as indicated by who exercised influence over decisions when the boys played the game. The overall conclusion is that status generalization can be overcome, provided that the expectations of both low-status and high-status persons are modified simultaneously. Similar findings appear in related studies (Riordan & Ruggiero, 1980).

CONFORMITY TO GROUP NORMS

Group Norms

A **norm** is a rule or standard that specifies how group members are expected to behave under given circumstances (Hechter & Opp, 2001). Most groups develop a variety of norms that regulate the activities of their members. A factory work group may have norms specifying what time of the day workers are expected to start on the assembly line and how much they are expected to produce; norms of this type will obviously have an impact on the level of productivity achieved by the work group. A group of college admissions officers may have norms that regulate how judgments are made by the officers; the nature of these norms will indirectly affect which applicants will be admitted and which not. A family may have norms regulating who washes the dishes or mows the lawn, as well as norms specifying who can and cannot have sexual relations with whom (for example, the incest taboo). A street gang may have norms regulating various actions of its members. For instance, one study of gangs of

Group norms can extend to any aspect of behavior, including dress. Bikers have a different dress code from corporate executives, but conformity is high within each group.

adolescent boys in the Southwest found that they had norms governing what kinds of information members could reveal to outsiders (for example, parents and police) and how members should behave during street fights with rival gangs (Sherif & Sherif, 1964).

Functions of Norms Norms serve a number of important functions for groups. First, they foster coordination between members in pursuit of group goals. Because norms usually reflect a group's fundamental value system, they prescribe behaviors that foster the attainment of important goals. When members conform to group norms, they know what to expect of each other, facilitating coordination among them.

Second, norms provide a cognitive frame of reference through which group members interpret and judge their environment. That is, norms provide a basis for distinguishing good from bad, important from unimportant, tenable from untenable. They are especially useful in novel or ambiguous situations, where they serve as pointers on how to behave. Because they are anchored in the group's values and culture, norms bring predictability and coherence to group activities.

Third, norms define and enhance the common identity of group members. This is especially true when group norms require members to behave differently from persons outside the group. Thus, norms that prescribe distinctive dress (for example, clothing or hairstyles) or distinctive speech patterns (for example, dialects or vocabulary) will differentiate group members from nonmembers. These norms demarcate group boundaries and reinforce the group's distinctive identity.

Types of Norms One useful approach to understanding norms is the **return potential model** (Jackson, 1965), which treats norms as having two dimensions. The behavior dimension specifies the frequency or amount of behavior regulated by the group norm, whereas the evaluation dimension refers to the response to that behavior by other members (positive or negative).

Suppose a group of editors working on a magazine has a norm about how much work each editor does, such as editing so many pages of manuscript per day. Under this norm, members encourage and reward more work done by each editor—the more pages a member edits, the more he or she is rewarded. Any member who produces at a very low level on this behavior dimension will receive negative evaluation from other group members. He or she may be criticized or perhaps even face a monetary fine.

But the group may also have norms that do not continually increase with additional performance: There can often be situations where members can produce too much. Perhaps editing too many pages per day will make the other members of the group look bad, and so a norm is created under which a member is rewarded most for producing in the middle range.

A norm of this type was evident in a classic early study of an industrial work group (Roethlisberger & Dickson, 1939). A group of 14 men, known as the Bank Wiring Group, assembled banks of telephone switching equipment for the Western Electric Company. The management of the company had recently instituted a wage incentive plan rewarding higher levels of productivity by individual workers. Managers believed that this plan would bring about an increase in the productivity of the Bank Wiring Group. Surprisingly, however, the workers continued to produce at the same rate as before, despite the opportunity to increase their earnings. The workers feared that if they began to produce at higher levels, the company might eventually switch the pay schedule and require still higher levels of productivity for a given wage. Therefore, the group enforced its own production norm on individual members. If a man produced too much, he was ridiculed by other members as a "rate buster." If he produced too little, he was disparaged as a "chiseler." If an offender did not respond to these verbal reprimands, he was soon subjected to another form of sanction termed "binging," whereby other members would punch him in the arm. Although this may seem unusual, groups often develop their own special form of sanctions. What matters is not the form, but whether the sanction is effective in regulating the behavior of its members.

The norms we have discussed up to this point involve both positive and negative evaluation. Other norms, however, may pertain to behaviors that group members consider simply undesirable. For instance, group members may tolerate low levels of an undesirable behavior, but their reaction becomes more negative as the behavior becomes more frequent or intense. These kinds of norms might apply to such behaviors as talking too much, pestering others, making a nuisance of oneself, and so on.

Another characteristic of a norm is its intensity. The intensity of a norm reflects the strength of group feelings—whether approval or disapproval—regarding that behavior and the level of sanctions or rewards the group will levy for the behavior. Norms that affect the attainment of important group goals often have high intensity, whereas norms about matters of personal taste (such as style of dress or manner of speaking) have lower intensity.

FIGURE 14.1 JUDGMENTAL TASK EMPLOYED IN ASCH CONFORMITY STUDIES

In the Asch paradigm, naive participants are shown one standard line and three comparison lines. The task is to judge which of the three comparison lines is closest in length to the standard line. By itself, this task appears easy. However, participants are surrounded by other persons (supposedly also naive participants, but actually experimental confederates) who publicly announce erroneous judgments regarding the match between lines. Such a situation imposes pressure on the participant to conform to their erroneous judgments.

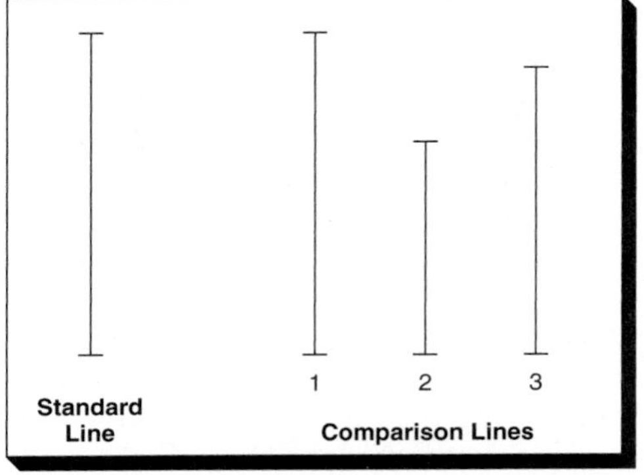

Source: Adapted from Asch, 1952.

Conformity

Norms do not mean much unless the group can somehow oblige its members to live up to its norms. When an individual adheres to group norms and standards, it is called **conformity.** A great proportion of the behavior that we witness in daily life involves conformity to one group norm or another. Group members often change their behavior expressly so that it will conform to group norms.

The Asch Conformity Paradigm In groups, influence flows in many directions—members influence other members and are influenced in turn. Of special importance, however, is the influence exercised by the group's majority over the behavior of individual members. Social psychologists use the term **majority influence** to refer to the processes by which a group's majority pressures an individual member to conform or to adopt a specific position on some issue. Majority influence is important, for it gives a group integrity and continuity over time. Of course, the amount of influence exerted by the majority on individual members varies from group to group. In some cases, it is slight to moderate, whereas in others it is large.

The impact of majority influence on individual group members was illustrated in a series of classic experiments by Asch (1951, 1955, 1957). Using a laboratory setting, Asch created a situation in which an individual was confronted by a majority that agreed unanimously on a factual matter (spatial judgments) but was obviously in error. These studies showed that within limits, groups can pressure their members to change their judgments and conform with the majority's position even when that position is obviously incorrect.

In the basic Asch experiment, a group of eight persons participated in an investigation of "visual discrimination." In fact, all but one of these persons were confederates working for the experimenter. The remaining individual was a naive participant. In front of the experimental room, large cards displayed a standard line and three comparison lines, as shown in Figure 14.1. The participant's objective was to decide which of the three comparison lines was closest in length to the standard line.

The task seemed simple and straightforward, for one of the comparison lines was the same length as the standard line, whereas the other two were very different. The group repeated this task 18 times, using a different set of lines each time. On each trial, the standard line matched one of the three comparison lines. During each trial of the experiment, the confederates announced their judgments publicly, one after another. The participant also announced his or her opinion publicly. The group was seated so that the confederates responded prior to the real participant.

Although this task seemed easy, it turned out to be a difficult experience for the naive participant. On 6 of the 18 trials, the confederates gave a correct response, but on the other 12 trials, the confederates responded incorrectly. The erroneous responses by the confederates put the naive participants in a trying social position. On the one hand, they knew the correct response based on their own perception of the lines. On the other hand, they heard all the other persons (whom they believed to be sincere) unanimously announcing a different and incorrect judgment on 12 trials.

The purpose of the study was to observe how the naive participants would behave during the 12 critical trials. The results indicated that the incorrect opinion expressed by the majority strongly influenced the judgments announced by the naive participants. In the 12 critical trials, nearly one third of the responses by participants were incorrect (Asch, 1957). This compares with an error rate of less than 1 percent in a control condition in which no confederates were present and participants recorded their judgments privately on paper. Whereas about one quarter of the participants showed no conformity and remained independent throughout, the remainder conformed, at least to some degree. One third of the participants conformed on 50 percent or more of the critical trials.

Interviews conducted after the experiment revealed that most of the participants were quite aware of the discrepancy between the majority's judgments and their own. They felt puzzled and under pressure and tried to figure out what might be happening. Some wondered whether they had misunderstood the experimental instructions; others began to look for other explanations or to question their eyesight. Even those participants who did not conform to the majority felt some apprehension, but they eventually decided that the problem rested more with the majority than with themselves. The interviews indicated that the conformity by participants in this study was of a particular type. To a fair extent, it involved public compliance without private acceptance: Although many participants conformed publicly, they did not believe or accept the majority's judgment privately. In effect, they viewed public compliance as the best choice in a difficult situation.

Why Conform?

Normative Influence The occurrence of majority influence and conformity in groups can be explained generally by the fact that individual members are dependent on the majority, cognitively, socially, and for utilitarian reasons as well. For one thing, members seek information about social reality, and they depend on the majority to validate their understanding of and opinions about the group and the world. For another, individual members want to obtain various rewards and benefits—not the least of which is the acceptance of their continuing membership in the group—and they depend heavily on the majority for these outcomes.

The dependence of group members on the majority thus leads to the exercise of influence by the majority in groups because it can withhold these outcomes from the minority. Influence by the majority can take various forms. Many analyses distinguish between normative influence and informational influence (Cialdini & Trost, 1998; Deutsch & Gerard, 1955; Kaplan, 1987; Turner, 1991). **Normative influence** occurs when a member conforms to expectations held by others (that is, to norms) in order to receive the social rewards or avoid the punishments that are contingent on meeting these expectations (Janes & Olson, 2000). Being liked and accepted by other members is one important reward in normative influence. To exercise influence of this type, a group will need to maintain at least some degree of surveillance over the behavior of its members. The impact of normative influence is heightened, for instance, when members respond publicly rather than anonymously (Insko, Drenan,

Solomon, Smith, & Wade, 1983; Insko, Smith, Alicke, Wade, & Taylor, 1985).

A closely related perspective also views conformity to social norms as a utility-seeking activity, but less in terms of directly avoiding punishments or currying favor, but rather as a means to stabilize relationships and enhance the predictability of behavior in the group. Theorists suggest that conforming to norms and enforcing them produces more easily understandable relationships among people, and that these relationships are therefore easier to use for exchange purposes (Horne, 2004). If, for example, Bob wishes to sell an item on eBay (an Internet auction site), he must follow the prescribed norms about describing his product accurately, adequately packaging his product for shipping, and charging reasonable shipping and handling costs. He is motivated to conform to the standards of the eBay community, not just because he will be sanctioned (by negative testimonials and perhaps even losing his account) if he does not, but also because following the norms enhances the trading system for everyone involved. It makes buying and selling behavior predictable, comfortable, and easy to manage.

Informational Influence **Informational influence** occurs when a group member accepts information from others as valid evidence about reality. This type of influence occurs especially when members need to reduce uncertainty—as in situations that involve ambiguity or that entail an absence of objective standards to guide judgment (Baron, Vandello, & Brunsman, 1996). More concretely, informational influence often occurs in situations where members are trying to solve a complex problem unfamiliar to them (Kaplan & Miller, 1987); members considered more expert or knowledgeable are especially likely to exercise informational influence during such tasks. It also occurs frequently in crisis situations when members must act immediately but lack knowledge about the appropriate action. The common element in all these situations is that the group exerts influence on individual members by providing information that defines reality and serves as a basis for making judgments or decisions.

With respect to the Asch line judgment task, it seems that normative influence was operating prominently in the situation. Of those participants who conformed in the Asch experiment, many did so to avoid being embarrassed, ridiculed, or laughed at by the majority. They were seeking acceptance by the majority (or at least they were seeking to avoid outright public rejection). It is hard to argue that Asch's majorities exercised informational influence to any great extent; after all, most persons have some skill in comparing line lengths and are not very dependent on others for this. Moreover, in one variation, Asch retested his participants on the same stimuli with the majority group no longer present, and they gave correct answers; their experience of judging lines in the presence of the majority did not permanently alter their understanding of the lines' lengths.

Although informational influence was not very central in the Asch situation, we should not underestimate its importance in other situations. A famous study by Sherif (1935, 1936)—conducted years before Asch did his line judgment research—dramatically illustrates the impact of informational influence under conditions of uncertainty. Sherif's study used a physical phenomenon known as the autokinetic effect (meaning "moves by itself"). The autokinetic effect occurs when a person stares at a stationary pinpoint of light located at a distance in a completely dark room. For most people, this light will appear to move in an erratic fashion, even though the light is not actually moving at all. Sherif used the autokinetic effect as a basis for studying informational influence in groups. First, he placed participants in a laboratory setting by themselves and asked them to estimate how far the light moved. In making these judgments, the participants were literally in the dark—they had no external frame of reference. From their individual estimates, the researcher was able to determine a stable range for each participant. Participants differed quite a bit in this respect. Whereas some thought the light was moving only an inch or 2, others believed it was moving as much as 8 or 10 inches.

Shortly thereafter, Sherif put the same participants together in groups of three and placed them back in the autokinetic situation. Although the estimates the participants had made when alone were different, the estimates they made in groups converged on a common standard. This change in judgments by members provides evidence for the operation of informational influence. Lacking an external frame of reference and being uncertain about their own judgment, group members began to use one another's estimates as a basis for defining reality. Each group established its own arbitrary standard, and members used this as a frame of reference. This process of norm formation can be quite subtle; in fact, other research (Hood & Sherif, 1962) has shown that participants involved in an autokinetic experiment are often unaware that their judgments are being influenced by other members.

One interesting finding from Sherif's original study emerged when, a week or two after their initial exposure, the participants were again placed alone in the autokinetic situation. The results showed that the participants used the acquired group norm as the framework for their new, individual judgments. Although not all studies have found evidence of such enduring norm internalization, at least one study retested individual participants in the autokinetic task a year after their initial exposure to the group norm and found evidence that the group norm still influenced the participants' judgments despite the passage of time (Rohrer, Baron, Hoffman, & Swander, 1954).

Increasing Conformity

When discussing the Asch line judgment study, we saw that pressure from the majority can substantially influence the behavior of individual group members. But an individual's tendency to conform will be greater under some conditions than under others. Investigators have tried to identify factors that have important effects on the amount of conformity occurring in groups. We will review some of these factors here (see also Box 14.2).

Size of the Majority Consider again the Asch situation in which a single participant is confronted by a majority. If the majority is unanimous—that is, if all the members of the majority are united in their position—then the size of the majority will have an impact on the behavior of the participant. As the size of the unanimous majority increases, the amount of conformity by participants increases (Asch, 1955; Rosenberg, 1961). For example, a participant confronted by one other person in an Asch-type situation will conform very little; he or she will answer independently and correctly on nearly all trials. However, when confronted by two persons, the participant will experience more pressure and will agree with the majority's erroneous answer more of the time. Confronted by three persons, the participant will conform at a still higher rate. In his early studies, Asch (1951) found that conformity to unanimous false judgments increased with majority size up to three members and then remained essentially constant beyond that point. Although some research (Bond & Smith, 1996; Gerard, Wilhelmy, & Conolley, 1968) has questioned the exact point at which the effect of majority size begins to level off, there does seem to be some point at which additional persons do not further increase conformity.

Unanimity What happens when the group's majority is not unanimous? Basically, lack of unanimity among majority members has a liberating effect on the behavior by participants. A participant will be less likely to conform if a member breaks away from the majority (Gorfein, 1964; Morris & Miller, 1975). One explanation for this is that the member who abandons the majority provides validation and social support for the participant. In an Asch experiment, for example, if one or several members abandon the majority and announce correct judgments, their behavior will reaffirm the participant's own perception of reality and reduce his or her tendency to conform to the majority.

Beyond this, however, any breach in the majority—whether it provides social support or not—will reduce the pressure on the participant to conform (Allen & Levine, 1971; Levine, Saxe, & Ranelli, 1975). In one study (Allen & Levine, 1969), individuals participated in groups of five persons, four of whom were confederates. The participants made judgments on a variety of items. These included visual tasks similar to those used by Asch as well as informational items (for example, "In thousands of miles, how far is it from San Francisco to New York?") and opinion items for which there were no correct answers ("Agree or disagree: 'Most young people get too much education'"). Depending on the experimental condition, participants were confronted with either a unanimous majority of

BOX 14.2 Research Update: Conformity and Priming

The Asch conformity paradigm and the Sherif autokinetic-effect experiments were provocative demonstrations of conformity effects. But what kinds of experiments have been done more recently to better understand the conditions under which these kinds of effects operate? One study used an experimental procedure somewhere in between the Asch and Sherif approaches to determine if priming could influence conformity. These researchers, Rachel Pendry and Rachel Carrick, asked their subjects to count the number of beeps they heard—a task not as ambiguous as the Sherif autokinetic effect, but still subject to considerable error by many subjects. Each time, the subjects actually heard 100 beeps, but the confederates (as in the Asch experiments) were instructed to lie and report between 120 and 125 beeps.

What was most interesting about this experiment, though, was not that the subjects often conformed and reported much higher than 100 beeps. Rather, it was that the experimenters were able to manipulate how much the subjects conformed through a process of priming. To achieve this manipulation, the researchers exposed their subjects to either a "punk" stimulus (representing anarchy and nonconformity), an "accountant" stimulus (representing the neat and orderly conformist), or no stimulus (the control condition). To prime the individuals, they were shown a picture and induced to read text that described the person in the text as either a punk rocker or an accountant.

The results showed that in the beep-counting task, the accountant-primed subjects conformed to the confederates' estimates of the number of beeps the most of the three groups. The group that received no prime conformed, but less than the accountant-primed group. The punk-primed group essentially did not conform at all. Their estimates were not significantly different than subjects who performed the beep-counting task in isolation (and thus had no conformity pressure).

If this simple priming effect extends beyond the lab, conformity in real life could be manipulated using conforming role models. Some situations call for high levels of conformity, such as in a military unit, and simple exposure to conformists could increase the uniformity of behavior. Likewise, exposure to stereotypical nonconformists could increase individualized thinking and behaviors.

Source: Adapted from Pendry and Carrick, 2001.

four persons (control condition), a majority of three persons and a fourth person who broke from the majority and gave the correct answer (social support condition), or a majority of three persons and a fourth person who broke from the majority but gave an answer even more erroneous than that of the majority (extreme erroneous dissent condition).

The results of this study are shown in Figure 14.2. The control condition, which involved a unanimous majority, produced a high level of conformity. The social support condition, in which the dissenter joined the participant, produced significantly less conformity than the control condition. Even the extreme erroneous dissent condition, in which the dissenter gave an answer that was more extreme and incorrect than the majority's, produced significantly less conformity. Thus, any breach in the majority reduced conformity. Participants in the various conditions, however, had very different impressions of dissenters. Under the social support condition, the participants held a positive impression of the dissenter, whereas under the extreme erroneous dissent condition, the participants

FIGURE 14.2 CONFORMITY SCORES AND DISSENT FROM THE MAJORITY

In an Asch-type experiment, participants conformed less when one member broke from the majority and gave the same answer as the participant (the social support condition) than when faced with a unanimous majority (the control condition). Conformity was also lower when one member broke from the majority but gave an answer even more erroneous than the majority's (the extreme erroneous dissent condition). Thus, a breach in the majority of any kind had a liberating effect on the participants and reduced conformity.

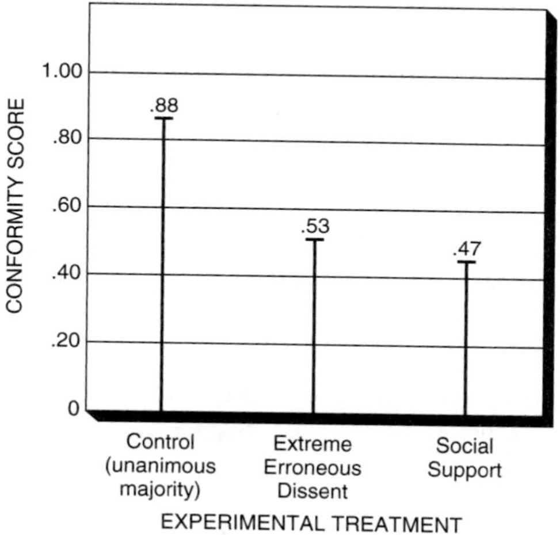

Source: Adapted from Allen and Levine, 1969.

had a negative impression and viewed the dissenter as unlikable, stupid, insincere, and badly adjusted. Either way, the breach in the majority had the same effect: It called into question the correctness of the majority's position and reduced the participant's tendency to conform.

Attraction to the Group Members who are highly attracted to a group will conform more to group norms than members who are less attracted to it (Kiesler & Kiesler, 1969; Mehrabian & Ksionzky, 1970). One explanation for this is that when individuals are attracted to a group, they also wish to be accepted personally by its members. Because acceptance and friendship are strengthened when members hold similar attitudes and standards, individuals who are highly attracted to a group conform more to the views held by the other members (Feather & Armstrong, 1967; McLeod, Price, & Harburg, 1966). However, attraction to a group will increase conformity only if that conformity leads to acceptance by others in the group (Walker & Heyns, 1962).

Commitment to Future Interaction Members are more likely to conform to group norms when they anticipate that their relationship with the group will be permanent or enduring, as opposed to short term. This was demonstrated by a study in which groups of five persons were asked to discuss such problems as urban affairs, international relations, pollution, and population (Lewis, Langan, & Hollander, 1972). Some of the participants were led to believe that they would review these problems with group members again in the future, whereas others were not given this expectation. Those who anticipated future interaction conformed more to the majority's opinion than those who did not anticipate future interaction. Returning to our eBay seller, if he were making his last transaction before leaving the system, he might be more tempted to cheat the buyer than if he anticipated more interaction—not just with that buyer, but with the rest of the eBay community.

Commitment to future interaction affects conformity whether members are attracted to a group or not (Forsyth, 1999). For instance, one investigation showed that even if members do not especially like the others in a group, they will conform at a high level provided they are already committed to continuing in the group and cannot readily leave. Such persons are likely to experience distress or dissonance at having to interact with others they do not like, and they may resolve this problem by bringing their own attitudes and beliefs into line with group standards (Kiesler & Corbin, 1965; Kiesler, Zanna, & DeSalvo, 1966).

Competence Another factor affecting conformity is an individual member's level of expertise relative to that of other members. If members who are skilled at the group's task differ from the majority's view, they will resist pressure to the degree that they believe themselves to be more competent than the other group members (Ettinger, Marino, Endler, Geller, & Natziuk, 1971). Interestingly, the extent to which a person believes that he or she is competent may be more important than the actual level of competence (Stang, 1972). Persons who, in fact, are not competent will still resist conformity pressure if they believe they have more skill than the other members. This is because group members who believe themselves competent rely less on the judgments of others; when confronted by a majority, they usually try to persuade other members to change their positions.

MINORITY INFLUENCE IN GROUPS

Although influence exercised by the majority over individual group members is important, this should not blind us to the activities of minorities within groups. In some instances, a small subgroup may disagree with the majority on some issue and advocate a different position. A dissenting minority is a coalition of group members that holds a viewpoint different from the majority on some important issue(s) and that presses for the adoption of its position by the group. If the members of a dissenting minority are able to persuade or induce majority members to accept their viewpoint, they are said to exercise **minority influence** (DeDreu & DeVries, 2001).

Attempts at influence by minority individuals or coalitions prove effective in some cases, but in many others they do not (Kalven & Zeisel, 1966). Throughout history, we encounter many instances of minority groups or individuals in scientific or religious contexts advocating viewpoints that were met initially with resistance or derision. Some of these minorities eventually succeeded in persuading or converting others, but many other minorities have been less successful and their messages are now forgotten or lost in oblivion. Viewpoints advocated by minorities often encounter resistance (or at least nonacceptance) from others, and sometimes they are even considered to be deviant or seditious positions, having the potential to impair the group's established way of life. This raises a basic question: Under what conditions are attempts to exercise influence by a minority coalition likely to be successful? We will turn to this question now.

Effectiveness of Minority Influence

According to the conversion theory of minority influence (Moscovici, 1980, 1985b), when minority coalitions exercise influence, the process involved is usually conversion, not compliance. In other words, minority coalitions try to win over the other members by changing their underlying attitudes and beliefs rather than by compelling their compliance. To achieve this end, minority coalitions typically use persuasion or exemplification; this contrasts with majority coalitions, which more often use coercive pressure or (veiled) threats to get their way.

More specifically, this theory holds that for a minority coalition to be successful in exercising influence, it must usually do certain things. First, the minority will need to create conflict and disrupt the established order, thereby producing doubt and uncertainty in the minds of other group members. Sometimes this can be accomplished merely by stubbornly resisting pressure to conform to the majority's view. In doing this, the minority will make itself visible and salient, focusing attention on itself. Second, the minority will need to offer a constructive alternative—a coherent point of view different from that of the majority—and indicate that it has strong confidence in the correctness of this viewpoint. Third, it must signal its intention not to compromise or abandon this view, with the implication that if other members wish to re-establish consensus and stability within the group, they must shift their own position toward that of the minority.

Consistency of the Minority A central hypothesis of conversion theory is that a minority coalition has a greater chance of influencing the judgments or opinions of others if it takes a distinctive position and holds it consistently in the face of pressure. By holding the position consistently, the minority coalition can demonstrate its commitment to and confidence in the position. This hypothesis has received support in empirical studies.

For instance, one study (Moscovici, Lage, & Naffrechoux, 1969) used a color perception paradigm (the "blue-green paradigm") to show that even if a minority coalition lacks power or status, it can influence the judgments of other members by maintaining a consistent position over time. The groups in this study consisted of six members; four of these were naive participants, whereas the other two were confederates who constituted the minority coalition. These groups performed a simple color perception task. The members sat in front of a screen that displayed a set of six blue slides differing in luminance. They saw these slides in six different orders; each slide was shown for 15 seconds. The members judged the color of each slide and announced their judgments aloud. In fact, all the slides used were blue. The minority coalition, however, voiced some judgments that departed from this standard. In one experimental condition, the two confederates in the minority coalition responded consistently and always labeled the blue slides "green." In another experimental condition, the two confederates responded inconsistently—they labeled the slides "green" 24 times and "blue" 12 times.

The results show that the behavior of the minority coalition influenced the judgments by the naive participants. In the consistent minority condition, 8.42 percent of all answers from naive participants were "green," and as many as 32 percent of the naive participants reported seeing a green slide at least once during the session. In the inconsistent minority condition, only 1.25 percent of the responses from the naive participants were "green." (These results contrast with the error rate in control groups consisting of six naive participants, which was virtually zero. The control participants always saw the slides as blue.) Thus, overall, the consistent minority exerted more influence than the inconsistent minority, as predicted by conversion theory. Results similar to these have been found in subsequent studies (Moscovici & Lage, 1976; Nemeth, Swedlund, & Kanki, 1974; Nemeth, Wachtler, & Endicott, 1977).

Behavioral Style Another factor that affects the success of a dissenting minority is the extent to which its negotiating style is flexible rather than rigid. Minorities are most influential when their position is consistent but their behavioral style is not rigid—that is, when their style of presentation is flexible and multifaceted. A dissenting minority using this style is more likely to be successful than one that adopts a rigid, hard-line, single-note approach (Mugny, 1982, 1984; Papastamou & Mugny, 1985). A flexible negotiating style tends to connote competence and honesty, whereas a rigid negotiation style, with its attendant refusal to make any concessions, is more likely to cause others to perceive the minority's consistency as dogmatic and, hence, as idiosyncratic.

Size of the Minority Another factor that affects a minority coalition's capacity to influence others is its size. In general, minority coalitions having many members can exert more influence than those having just a few members. For instance, one study found that a minority coalition consisting of two members is more influential than a minority of one (Arbuthnot & Wayner, 1982). In another study, the size of the majority was fixed at six persons, whereas the size of the minority coalition was varied experimentally. The results indicated that a minority of three or four persons exerted more influence than a minority of one or two (Nemeth et al., 1977). Reviewing findings across studies, Wood, Lundgren, Ouellette, Busceme, and Blackstone (1994) reported that a minority coalition's influence increases as a function of its size. This effect was especially apparent on measures of public change and direct private change.

Differences Between Minority and Majority Influence

One current controversy is whether minority influence (innovation) and majority influence (conformity) entail processes that are fundamentally similar or different. Some theorists have hypothesized that the influence exercised by a minority and the influence exercised by a majority involve the same underlying process (Doms, 1984; Latané & Wolf, 1981; Tanford & Penrod, 1984). In this view, minority and majority influence may differ in quantity (with majorities exercising more) but not in quality. Several studies have reported results consistent with this single-process view (Doms & Van Avermaet, 1980; Personnaz, 1981; Wolf, 1985).

Nevertheless, the single-process view is not universally accepted. Some other theorists (Maass, West, & Cialdini, 1987; Moscovici, 1985a, 1985b) have proposed a dual-process view, which maintains that majorities and minorities differ qualitatively in the ways they influence their targets. Under this theory, majority influence is based on a social comparison process, in which a member compares his or her own response with that of the majority in an attempt to achieve consensus, whereas minority influence is based on a validation process, in which a member tries to comprehend the minority's viewpoint in an attempt to figure out what is valid or real about the world. Majority influence is largely normative in nature and involves compliance based on the target's desire to gain social rewards and approval; minority influence is largely informational and, hence, involves conversion (that is, private agreement and belief change).

This dual-process model is controversial, and some writers have concluded that the evidence supporting it is weak (Kruglanski & Mackie, 1990). Other work, however, provides at least support for it. A recent meta-analysis of studies addressing this issue (Wood et al., 1994) concluded that majorities and minorities have somewhat different impacts on targets' public and private responses to influence. Majorities tend to exert more influence than minorities on measures of public change and direct private change, whereas minorities exert equal or greater influence than majorities on measures of indirect private change.

Also relevant to the dual-process model, some research shows that persons confronted by a dissenting minority tend to think more carefully and more creatively about the issue than they otherwise would (Nemeth, 1995; Nemeth & Kwan, 1985, 1987; Smith, Tindale, & Dugoni, 1996). Although they may not be persuaded to adopt precisely the position advocated by the minority, they do think in more divergent terms, attend to more aspects of the situation, and seek novel solutions (Maass et al., 1987). Thus, the effectiveness of a dissenting minority is perhaps best measured in terms of its capacity to change opinions and viewpoints, not its capacity to force behavioral compliance.

SUMMARY

What Is a Group? A group is a social unit that consists of two or more persons and that has certain defining attributes, including recognized membership, interaction among members, shared goals and objectives, and norms that guide members' behavior. A cohesive group is one that can strongly attract and hold its members. Two important types of cohesion are social cohesion and task cohesion and the level of a group's cohesion affects the interaction among members. Members in highly cohesive groups communicate more than those in less cohesive groups; they also exert more influence over one another, and their interaction is friendlier and more cooperative. A group goal is a desirable outcome that members strive collectively to bring about. Group goals differ from individual goals—outcomes desired by members for themselves.

Status of Group Members Members' positions in most groups differ with respect to roles (required performances) and status (rank or prestige in the group). A member's status characteristics outside the group can substantially determine his or her status inside the group. Studies of juror interaction, for instance, have shown that characteristics such as occupation and gender affect which jury members participate the most, exert the most influence, and hold positions of leadership within the jury. The tendency for members' status characteristics to affect group structure and interaction is called status generalization. Expectation states theory holds that group members develop expectations regarding one another's performance. Status generalization in this form is most likely to occur when members have no prior history of interaction, no experience with the group's task, and no explicit proof to the contrary. To overcome or eliminate status generalization, one must change not only the expectations of low-status members but also those of high-status members.

Conformity to Group Norms A norm is a rule or standard that specifies how group members are expected to behave under given circumstances. Group norms coordinate activity among members, provide a frame of reference that enables members to interpret their environment, and define the common identity of group members. Conformity means adherence by an individual to group norms and expectations. (1) The Asch conformity paradigm uses a simple visual discrimination task to investigate conditions that produce conformity by individuals to the majority's judgment. (2) Group majorities can use both normative influence and informational influence to exert pressure on individual members. Sherif's autokinetic effect studies illustrate the impact of informational influence on group members. (3) Many factors affect the amount of conformity in Asch-type situations. Conformity increases with group size up to three, and it is greater when the majority is unanimous than when it is not. Conformity is also greater when members are highly attracted to a group and when conformity leads to liking and acceptance by other members. Commitment to future interaction affects conformity; conformity is greater when members believe that their relationship with the group will be relatively permanent. Task competence affects conformity; members who oppose the majority's view will resist conformity pressures to the extent that they believe themselves to be more competent than other members.

Minority Influence Minority influence refers to efforts by a dissenting minority to persuade majority members to accept a new viewpoint and adopt a new position. (1) A minority will be more influential if it maintains its position consistently over time, adopts a flexible negotiating style, has many members, and consists of members with an in-group identity similar to that of the majority. (2) Although controversial, there is some evidence that minorities and majorities exert different types of influence. Whereas a majority can often compel members to comply, a minority must rely primarily on persuasion to change others' viewpoints.

LIST OF KEY TERMS AND CONCEPTS

conformity (p. 368)
expectation states theory (p. 363)
goal isomorphism (p. 358)
group (p. 355)
group cohesion (p. 356)
group goal (p. 358)
informational influence (p. 370)
majority influence (p. 368)
minority influence (p. 374)
norm (p. 365)
normative influence (p. 369)
return potential model (p. 367)
role differentiation (p. 360)
social-emotional specialist (p. 360)
status characteristic (p. 360)
status generalization (p. 361)
task specialist (p. 360)

15

GROUP STRUCTURE AND PERFORMANCE

Introduction

Group Leadership

Endorsement of Formal Leaders

Revolutionary and Conservative Coalitions

Activities of Leaders

Contingency Model of Leadership Effectiveness

Productivity and Performance

The Presence of Others

Group Size

Group Goals

Reward Distribution and Equity

Principles Used in Reward Distribution

Equity Theory

Task Interdependence

Responses to Inequity

Brainstorming

Production Blocking

Group Decision Making

Groupthink

Risky Shift, Cautious Shift, and Group Polarization

Summary

List of Key Terms and Concepts

INTRODUCTION

Groups in society exist for many different purposes. A work group in a clothing factory strives to meet its production quota for swimsuit apparel; an airline crew seeks to transport passengers to their destinations on time; a police detective squad works around the clock to crack an urgent murder case; a professional ball club tries to win games and take the league championship; a jury deliberates a case carefully with the intent of reaching a correct verdict. Not every group, of course, pursues explicit goals or produces something tangible. Some groups—a Wednesday-night poker club, for example—exist largely for simple pleasure and sociability. No matter how purposeful a group is, it exhibits some kind of structure that is determined in part by its goals. The structure of the group can define leaders, set a hierarchy, determine how rewards are distributed, and have a major impact on whether the group achieves its goals or not.

Consider, as an example, a sales group working for a small company in California that manufactures laboratory equipment. The sales group consists of 11 people—one vice president, two managers, and eight salespersons. Their primary goal is to sell equipment to industrial laboratories in the northern and southern regions of the state.

This group has a hierarchical structure, which requires that different members of the group contribute in different ways to the group's operations. The most influential person in the group is the woman who serves as the vice president. She has a lot of experience in the industry and, over the years, has amassed an outstanding record as a salesperson. In the role of vice president, she is responsible for maintaining the group's sales performance. She establishes sales goals and closely monitors the group's performance. Furthermore, she communicates with executives in other divisions about which products are selling and why. Any major changes made with respect to sales must receive her approval.

Next in importance within the group are the two managers—one responsible for the northern region and the other for the southern region. Both have excellent skills and extensive experience in sales, and as a result are well paid—although less than the vice president.

The eight salespersons are less central than the managers, but they do perform important activities for the group. They are responsible for making direct contact with the customers in their area and for providing products that meet their customers' needs. Younger and less experienced than the managers, the salespersons are paid partly on salary and partly on commission, and they earn less than the managers.

The positions in this group are diagrammed in Figure 15.1. This kind of diagram not only shows the structure of the group, but also conveys differences in status, influence, rewards, and leadership.

Often, groups are structured in specific ways in order to help the group achieve certain outcomes, and social psychologists have long studied various characteristics of groups that may help or inhibit the group as it tries to achieve its goals. **Group productivity** refers to a group's output gauged relative to something else, such as the level of resources used by the group or the group's targeted objective (Gilbert, 1978; Pritchard & Watson, 1992). A task group is considered highly efficient if it generates a lot of output from a small amount of input. For example, if a clothiers' work group produces a large number of garments per day and requires only a small number of workers to do it, that group is more efficient than another group that produces the same number of garments per day but requires more workers to do it. In contrast, a group is considered highly effective if its output meets or exceeds its goal or targeted objective. For example, if a group of college students working on a project for a political science class has a goal of getting an A on the project, but ends up with a C, it obviously has a low level of effectiveness.

From everyday observation, we know groups differ greatly in their levels of productivity. Even when groups are identical in size and similar in talent, they may perform at significantly different levels. Some groups produce more than others, win competitions more frequently, achieve better solutions to problems,

FIGURE 15.1 STATUS STRUCTURE IN A WORK GROUP

This work group has a three-level status hierarchy. The high-status member (vice president) exercises the most control and receives the greatest rewards. Reporting to the vice president are two intermediate-status members (managers). The eight low-status members (salespersons) exercise the least control and receive the least rewards.

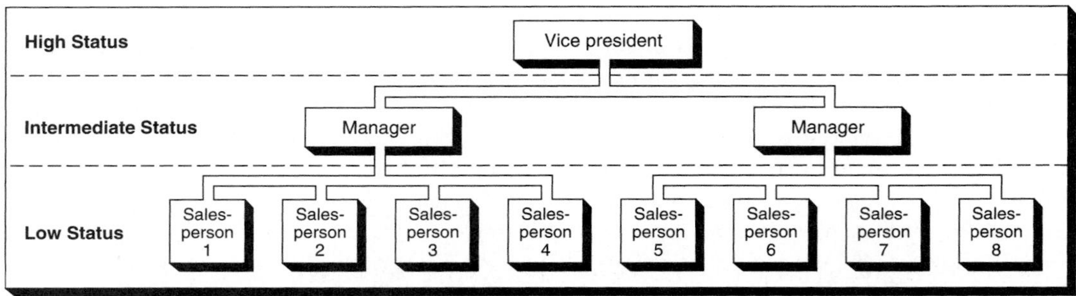

work at a higher speed, commit fewer errors, and create less waste.

Questions About Group Structure and Performance

This chapter focuses on the various factors that influence a group's performance and productivity, including the role of leadership in a group.

1. How do some members of the group become formal and informal leaders? How does the style of leadership in a group affect productivity? What circumstances can lead to the leader being replaced?
2. To what extent is individual performance affected by the presence of others? How do such group characteristics as size, group goals, and task characteristics affect group productivity?
3. How do groups distribute rewards to their members? How do group members react if there is a mismatch between contributions and rewards?
4. How do group processes affect the quality of group decisions? In what ways do conformity pressures and group discussion affect the character of group decisions and the risk-taking propensity of members?

GROUP LEADERSHIP

Many groups that we interact within are informal groups, which are not very structured or regulated. Role definitions—who is supposed to do what—and group goals are often loosely defined or even nonexistent. Formal groups, on the other hand, have much more structured interaction, explicit role definitions and authority, and tasks that are well specified. Formal groups frequently exist within larger organizations or bureaucracies, and they often endure for extended periods, even years.

One of the most important features of formal groups is that the activities of the group are formally regulated by its leadership—that is, the leaders have some degree of **authority** over the behavior of the members (Pescosolido, 2001). In formal groups such as the laboratory equipment sales group, members differ with respect to the status and authority they hold, and these differences usually persist over time. The persons who exercised authority yesterday (the vice president and two managers) are the same persons who exercise it today. Yet the pattern of status and authority in formal groups cannot be taken for granted. Even when it is formally defined, some members of the group may informally challenge the leaders, and at other times the formal leadership may be changed. Persons of high status wishing to remain in control must take steps to ensure they have the endorsement of other group members, and they must build alliances that will support their authority in the event that a challenge arises.

Endorsement of Formal Leaders

Role systems in formal groups are structured in various ways. We often see formal groups organized in a hierarchy or pyramid structure, but others can have a more diffuse set of links among the members. The sales group was structured hierarchically into three levels (eight salespersons who reported to two managers, who in turn reported to one vice president). The person at the top of such a pyramid holds high status and typically performs various leadership functions for the group. These leadership functions—planning, organizing, and controlling the activities of group members—are essential if a group is to achieve its goals (Bass, 1990).

Groups usually maintain a pyramid structure like this through a tacit exchange between high-status members, who fulfill leadership functions, and the other group members, who accept direction from the high-status members (Hollander, 1985; Hollander & Julian, 1970). If the group achieves its goals and low-status members are rewarded, they in turn reciprocate by giving rewards, benefits, and privileges to the high-status group members. Thus, in return for providing effective leadership, high-status members receive rewards and benefits and, moreover, they receive support for their actions (plans, decisions, orders) and reaffirmation of their right to occupy high-status positions in the group.

This support is essential if the status and authority structure within a group is to remain stable. Without it, formal leaders are usually ineffective and vulnerable to removal from office (Walker, Thomas, & Zelditch, 1986).

Support for a formal leader can come from two sources within a group (Zelditch & Walker, 1984). When it comes from members having still higher status in the group, it is usually referred to as authorization. When it comes from members having equal or lower status, it is termed endorsement. We can view **endorsement** as an attitude held by a lower status member indicating the extent to which he or she supports the group's formal leader (Hollander & Julian, 1970; Michener & Tausig, 1971).

Lower status group members are not always unanimous in their attitudes regarding persons in position of authority. Moreover, the endorsement accorded a high-status member often fluctuates over time. With a high level of endorsement from members, a person may feel confident when exercising vigorous leadership; with a low level of endorsement, however, exercising leadership may be difficult or impossible.

Why do some formal leaders receive high levels of endorsement from members whereas others receive only low levels? Formal leaders generally receive endorsement in proportion to the benefits they deliver to group members. Thus, the extent to which a group achieves its goals is one factor affecting endorsement. A distinction must be drawn, however, between initial failure and repeated failure on group objectives. When a group fails initially, members may actually increase their endorsement and rally around the leaders in hopes of improving the situation (Hollander, Fallon, & Edwards, 1977). Repeated failure, however, usually indicates that something is fundamentally wrong. If the responsibility for failure is attributed to faulty leadership, endorsement usually declines. In this case, group members may attempt to remove high-status members from their positions or redistribute authority within the group (Michener & Lawler, 1975; Wit, Wilke, & Van Dijk, 1989).

Another determinant of endorsement is the level of consideration a formal leader shows toward other members. For instance, if a high-status member controls the allocation of rewards and decides to share these rewards equitably, this demonstrates that person's concern for the welfare of other group members, and it will result in strong endorsement. In contrast, if a high-status member—even one who is very competent—behaves selfishly and inequitably when allocating rewards, that behavior tends to diminish endorsement (Michener & Lawler, 1975; Wit & Wilke, 1988). Group members are sensitive not only to the level of rewards they receive but also to the procedures used by a high-status member in allocating rewards. Formal leaders receive higher levels of endorsement if they allocate rewards by procedures viewed as fair rather than by procedures viewed as unfair or arbitrary (Tyler & Caine, 1981; Tyler & Folger, 1980).

Revolutionary and Conservative Coalitions

Lack of endorsement poses serious problems for an established high-status member, and it can render precarious his or her position in the group. Under certain conditions, low levels of endorsement lead to the formation of a **revolutionary coalition,** which is an alliance of medium- and low-status members who oppose the existing leadership (Crosbie, 1975). By combining forces, the revolutionary coalition hopes to remove the current leadership, as in a mutiny, revolt, or coup d'état.

Mobilization of Revolutionary Coalitions Revolutionary coalitions are usually difficult to mobilize. Low-status members must first have reason to end their allegiance to the established status order and then agree to take joint action to overthrow that order. In November of 2002, football players at the University of Arizona formed a revolutionary coalition against their head coach, John Mackovic. The players had developed a strong sense of discontent resulting from the verbal abuse of Mackovic and the sorry state of the football program at Arizona. The low-status members must also feel that the existing channels of appeal cannot or will not provide a remedy for their discontent, and they must be willing to face the risk of punishment if the mutiny fails. Faced with serious grievances against their coach, the Arizona players decided to risk further abuse from their coach and met with the president of the university in an attempt to have Mackovic removed. For a revolutionary coalition to succeed, the members involved usually must find a way to mobilize without alerting the leaders they hope to dislodge; this may entail secret meetings, covert communications, and surreptitious planning. All these factors make mobilization difficult.

Certain conditions increase the likelihood of a revolutionary coalition. In essence, these are the same conditions that produce low levels of endorsement—for example, repeated failure to achieve group goals (Michener & Lawler, 1971). Inequitable treatment of group members by high-status members may also encourage the formation of a revolutionary coalition. If an authority shows favoritism or selfishness in allocating rewards, group members are more likely to chafe under the inequity and form a revolutionary coalition than if the authority is more evenhanded in allocating rewards (Michener & Lyons, 1972; Ross, Thibaut, & Evenbeck, 1971; Webster & Smith, 1978). Likewise, perceived indiscretion or unethical personal or professional behavior by an authority may provoke an overthrow (Deluga, 1987).

Characteristics of the group members may also play a role. Commonalities among members increase members' expectations of support from other members, which in turn increases the probability of each individual joining the revolutionary coalition (Lawler, 1975a, 1975b).

Reactions to Revolutionary Coalitions High-status members have a lot to lose by the emergence of a revolutionary coalition; once they discover an upheaval in the making, they usually take action to suppress the coalition and bolster their own position. One counterstrategy is simply to threaten to punish insurgent members. If the threats are sufficiently large and credible, this may effectively quash the revolt.

Another useful counterstrategy is to mobilize a **conservative coalition**—an alliance of members who support the existing order. Conservative coalitions usually consist of members who view the status differences in the group as legitimate and who—consistent with expectation states theory—hold performance expectations based on these status differences (Wilke, 1985).

A third counterstrategy is to co-opt the revolutionaries. By definition, **co-optation** is a strategic attempt to weaken the bond among potential revolutionaries by singling out one or several lower status members for favored treatment. For instance, a high-status member might offer several low-status members the prospect of future promotion. This would increase these members' investment in the existing status order, thereby making it more difficult for other revolutionaries to influence them or to recruit them as coalition members (Lawler, 1983; Lawler, Youngs, & Lesh, 1978).

Activities of Leaders

Leaders who are in power and are able to maintain it perform a number of very important functions in groups. Their primary purpose, of course, is to guide the group toward its goals, but the actions they take as the leader can have strong effects on how well the group works toward its goals. By definition, **leadership** is the process whereby one group member (the leader) influences and coordinates the behavior of other members in pursuit of group goals (Yukl, 1981). In return for support from the others, a leader provides guidance, specialized skills, and environmental contacts that help the group attain its goals.

A person serving as a leader fulfills certain functions necessary for successful group performance. These functions include planning, organizing, and controlling the activity of group members (Bass, 1990). In formal groups, where roles are organized in a status hierarchy, leadership functions are typically fulfilled by high-status members. These members have both the right and the responsibility to provide leadership for the group. In informal groups, however, these functions may be fulfilled by one or several persons who emerge during the interaction.

What are the specific activities of leaders in groups? Leaders usually do some or all of the following: (1) formulate a clear conception of the group's goals and communicate this conception to group members; (2) develop specific strategies for the attainment of group goals; (3) specify role assignments and establish standards of productivity for members; (4) facilitate communication among members; (5) recruit new members and train members in crucial skills; (6) interact with members personally to maintain good relations; (7) provide persuasion, rewards, and punishments to encourage the members; (8) monitor the group's progress and take corrective steps if necessary; (9) resolve conflict between members; and (10) serve as a representative of the group to outside agencies and organizations.

As this list suggests, leadership often involves a tacit exchange between the leader and the other group members. By fulfilling the planning, organizing, and controlling functions in a group, the leader helps move the group toward the attainment of its goals. In return, the leader receives support for continued control, as well as special rewards and privileges. Leadership of this type, based on an exchange between the leader and group members, is termed **transactional leadership** (Hollander, 1985; Hollander & Julian, 1970; Homans, 1974).

In some circumstances, however, leaders do more than merely mediate rewards and goal attainment for group members in exchange for power and privilege. They strengthen group productivity by changing the way members view their group, its opportunities, and its mission. Some leaders foster high levels of group productivity by conveying an extraordinary sense of mission to group members, arousing new ways of thinking within the group, and stimulating new learning by members (Conger, Kanungo, & Menon, 2000; Judge & Bono, 2000). Leadership in this form, termed **transformational leadership,** often creates structural change and institutionalizes new practices within the group and, thereby, strengthens group productivity (Hater & Bass, 1988; Kohl, Steers, & Terborg, 1995). Transformational leadership (also called charismatic leadership) is particularly effective in situations that are volatile and unpredictable, because members of the group are highly committed to supporting the leader (Waldman, Ramirez, House, & Puranam, 2001).

Contingency Model of Leadership Effectiveness

Depending on the situation and on their personalities, leaders use different techniques to plan, organize, and control group activities. For example, some adopt an authoritarian leadership style whereas others use a democratic leadership style. A leader's effectiveness results from a combination of the leadership style and the characteristics of the group and the situation in which it is working. This basic notion underlies the **contingency model of leadership effectiveness** (Fiedler, 1978a, 1978b, 1981).

Leadership Style There are four elements of the contingency model of leadership effectiveness. The first is that leadership style can be characterized as either relationship oriented or task oriented. When assessing leadership style, one approach asks the leader to recall people whom he or she has worked with in the past and identify the one that was the most difficult to work with. Next, the leader is asked to rate this person on dimensions such as pleasant-unpleasant, helpful-frustrating, cooperative-uncooperative, and efficient-inefficient. Some leaders rate this person negatively across the board, whereas others find positive qualities even for those persons they strongly prefer not to work with. These tendencies indicate two different leadership styles—those who rate others with consistently negative scores tend to be more task oriented in their leadership style, whereas those who give a combination of positive and negative ratings tend to be more relationship oriented—caring primarily about establishing congenial interpersonal relations (Fiedler, 1978a; Rice, Marwick, Chemers, & Bentley, 1982).

Situational Factors The other three elements of the contingency model of leadership effectiveness are characteristics of the situation. In order of importance, these are (1) the leader's personal relations with other group members (good or poor), (2) the degree of structure in the group's task (structured or unstructured), and (3) the leader's position of power in the group (strong or weak).

In the contingency model, a leader's personal relations with other members is thought to be the most important factor determining the leader's influence. A leader whom the other members trust, like, and respect is more likely to be able to influence the group than one with poor rapport. The second factor, task structure, refers to how clearly defined the task requirements of a group are. The clearer the group's goals and the path to achieving them, the more favorable the situation is for the leader. The third factor, position power, has to do with a leader's formal authority over group members and the degree to which the leader can use rewards and punishments. A leader who wields more power is in a more favorable situation.

These three factors combine to produce eight different kinds of group environments, as illustrated in Figure 15.2. From a leader's standpoint, a situation is favorable if it enables him or her to exercise a great deal of influence. Thus, a situation involving good leader-member relations, a highly structured task, and strong leader power is the most favorable (situation I in the figure). At the other extreme, a situation involving poor leader-member relations, an unstructured task, and weak leader power is the most unfavorable (situation VIII).

Predictions According to the contingency model of leadership effectiveness, situations and leadership styles combine to produce different levels of effectiveness. Task-oriented leaders are most effective in situations that are either highly favorable or highly unfavorable, but less so in moderate conditions. In contrast, relationship-oriented leaders are effective in moderately favorable situations, but less so on the extremes (see Figure 15.2).

Why does this occur? When conditions are extremely unfavorable, the group needs strong task leadership to get its job done. Task-oriented leaders are willing to overlook interpersonal conflicts to concentrate on the task. Relationship-oriented leaders fail to get the job done because they expend too much energy smoothing over interpersonal problems.

However, when conditions for leadership are intermediate in favorableness, the relationship-oriented person will do a better job. Some situational factors are positive and others negative; thus, interpersonal problems may come to the surface that could slow down progress toward the goal. The relationship-oriented leader will be better able to address these problems.

Finally, when conditions are highly favorable, the task orientation again emerges as most effective. Under these circumstances, task-oriented leaders feel they can relax because goal attainment is assured. They actually change their orientation and turn their attention to interpersonal relations because they don't have

FIGURE 15.2 LEADERSHIP EFFECTIVENESS AS A FUNCTION OF LEADERSHIP STYLE AND SITUATIONAL FACTORS

The contingency model predicts that group performance depends on both leadership style and situational factors. According to this model, task-oriented leaders are most effective in situations that are either highly favorable (that is, involving good leader-member relations, a structured task, and strong leader power) or highly unfavorable (that is, involving poor leader-member relations, an unstructured task, and weak leader power). In contrast, relationship-oriented leaders are most effective in situations that are of medium favorableness.

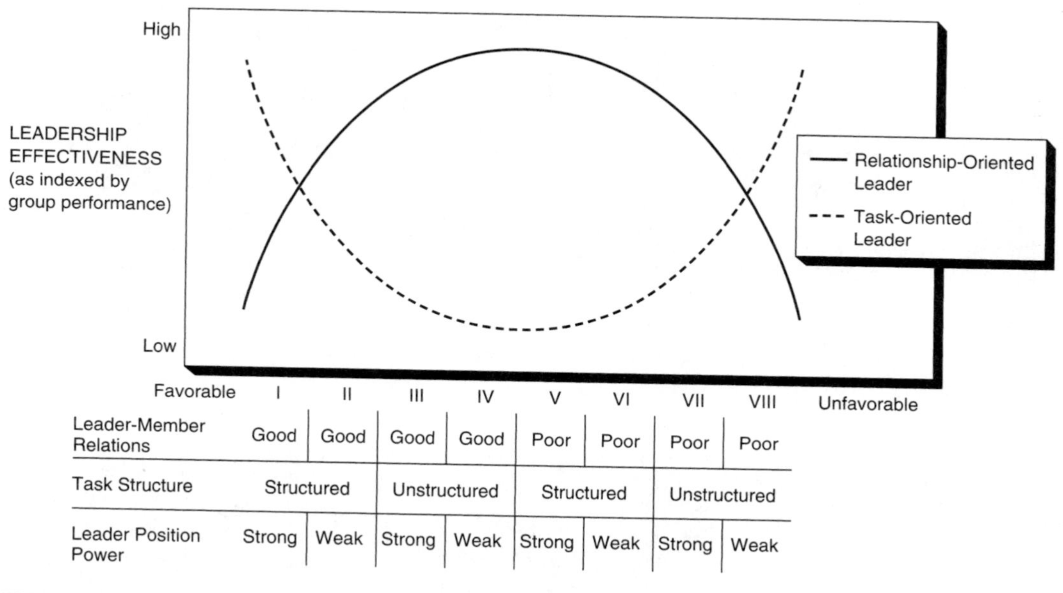

	I	II	III	IV	V	VI	VII	VIII
Leader-Member Relations	Good	Good	Good	Good	Poor	Poor	Poor	Poor
Task Structure	Structured		Unstructured		Structured		Unstructured	
Leader Position Power	Strong	Weak	Strong	Weak	Strong	Weak	Strong	Weak

Source: Adapted from Fiedler, 1978a.

to worry about getting the job done. Under these same conditions, however, relationship-oriented leaders are relatively ineffective. They often start looking for things to do and become bossy and less concerned with the feelings and opinions of their coworkers, which diminishes their effectiveness (Fiedler, 1978a; Larson & Rowland, 1973).

Tests of the Contingency Model Although there are some exceptions, research generally supports the contingency model's predictions (Chemers, 1983; Peters, Hartke, & Pohlmann, 1985; Strube & Garcia, 1981). One study (Chemers & Skrzypek, 1972) used West Point cadets as participants and manipulated all three dimensions of situational favorableness. Groups of four men were assembled on the basis of information about preexisting relations among members. In half of the groups, the leader-member relations were good, whereas in the other half they were poor. Each group performed one structured task (converting blueprints from metric units to inches) and one unstructured task (discussing an issue and making policy recommendations). In half of the groups, the leader had strong power—that is, members believed that the leader would assess their performance and that this would affect their standing at the academy. In the other half, the leader had only weak power—members believed he would have little effect on their standing. The

© Phyllis Graber Jensen/Stock, Boston

Courtesy of Catherine Comet, Conductor of the Grand Rapids, Michigan, Symphony and Shaw Concerts, Inc.

A drill instructor uses a strict authoritarian style to direct his troops. Would the drill instructor's leadership style be as effective in another setting, such as conducting a symphony orchestra?

predictions of the contingency model occurred just as the researchers had expected. Similar findings have emerged in studies of college and elementary school students (Hardy, 1975; Hardy, Sack, & Harpine, 1973) and naval personnel (Fiedler, 1966).

The contingency model predicts that group performance depends on both leadership style and situational factors. According to this model, task-oriented leaders are most effective in situations that are either highly favorable (that is, involving good leader-member relations, a structured task, and strong leader power) or highly unfavorable (that is, involving poor leader-member relations, an unstructured task, and weak leader power). In contrast, relationship-oriented leaders are most effective in situations that are of medium favorableness.

PRODUCTIVITY AND PERFORMANCE

When confronted with the tasks of everyday life, individuals can take them on by themselves or can band together in groups in an attempt to be more productive. But is working in a group likely to produce more output or help people accomplish their goals more efficiently? Sometimes, it seems that the presence of others does indeed help the individuals in the group move toward their goals. When tasks can be divided up so that some individuals can specialize, the increased level of efficiency makes the group more than the sum of its parts. On the other hand, having too many individuals involved can also hinder performance—when group members get in each other's way and find coordinating their activities more trouble than it is worth. Given that group productivity is a matter of considerable importance in everyday life, several questions have become important social psychological research topics. First, is individual performance enhanced in group settings? Second, why do some groups perform better than other, seemingly similar groups, and what group properties determine the level of productivity?" Structural properties of groups, such as size, goal specification, and reward allocation can all have a substantial impact on the quality and quantity of group productivity.

The Presence of Others

The first social psychological experiment published examined the phenomenon of **social facilitation** (Triplett, 1898). Norman Triplett had observed that in cycling races, when cyclists raced head-to-head on the same track, they were likely to produce faster times than when they raced alone on the track against the clock. Triplett believed that this was due to the social facilitation effect, in which the mere presence of the other competitors caused individuals to perform better. You probably have experienced this effect yourself: Imagine yourself out for a jog and you are coming to the end of your run. You have been pushing hard and are losing steam—having spent most of your energy in the early part of the run. Unexpectedly, you see another runner turn onto the street near you. Immediately, you feel a surge of energy, improve your form, and proceed to beat the other runner to the corner where the two of you part ways.

Triplett systematized a test of social facilitation in a laboratory setting by having subjects reel in fishing line as quickly as they could. In some trials they were alone, and in others they were in the presence of another subject doing the same task. As Triplett expected, those in the presence of others tended to reel in the line more quickly. The social facilitation research program has been a very active one. Social psychologists have confirmed that the process is a very general one—it occurs not just when people are doing the same task, but also just when others are observing as passive bystanders (Gates, 1924). More importantly, the effect has been observed in animal species other than humans, including dogs, frogs, and even cockroaches (for example, Zajonc, Heingartner, & Herman, 1969)!

But the effect is not completely ubiquitous. Some experiments have actually found the opposite effect in certain conditions: The presence of others actually damages the performance of individuals on certain tasks (for example, Allport, 1920). What makes the difference? The difficulty of the task for the individuals interacts with social facilitation to produce the different effects. If the task is simple or is one that the individuals are accomplished in performing, then social presence facilitates performance. The presence of others emotionally and cognitively arouses the subjects

BOX 15.1 Types of Group Tasks

Several investigators have attempted to develop classification schemes for group tasks (Hackman & Morris, 1975; Laughlin, 1980; Steiner, 1972). Recently, McGrath (1984) proposed a taxonomy of group tasks that is quite general and consistent with other approaches. McGrath suggests that group tasks can be divided usefully into the following eight types:

1. Planning Tasks. In tasks of this type, a group's major goal is to generate an action-oriented plan to achieve some objectives on which the members have already agreed. The group must consider alternative paths to the goal, contemplate the constraints imposed by resource availability, and develop a viable program. Examples include planning a military maneuver or mapping out an advertising campaign for a new product.
2. Creativity Tasks. Tasks of this type require a group to generate new, original ideas. Usually, group interaction emphasizes creativity or entails brainstorming to generate new ideas, alternatives, or images. Examples include brainstorming to develop new product concepts or new comedy routines.
3. Intellective Tasks. Tasks of this type require a group to solve a problem for which there is—or is believed to be—a correct answer. To solve problems of this type, group members typically must consider many alternative solutions and discard unsuitable ones. The group must strive to discover, select, or compute the right answer. Once uncovered, the answer usually seems intuitively right and compelling. Examples include arithmetic or logical reasoning problems, puzzles, cryptograms, bus routing problems, and computer programs.
4. Decision-Making Tasks. Tasks of this type require a group to solve a problem for which there is no inherently right answer. Through interaction, the group must consider a variety of alternatives and reach a consensus regarding which alternative is preferred. Examples include jury deliberations, investment decisions, and decision tasks used in risky shift and group polarization studies.
5. Cognitive Conflict Tasks. In tasks of this type, group members hold varying viewpoints and strive to resolve their differences. Conflicts of this type occur not because members have different interests and goals but because they disagree about the use of

and then they focus more on their performance and can get the job done easier, quicker, and with higher quality. If the task is difficult or unfamiliar, the arousal can make the subjects nervous or apprehensive about performance, thereby detracting from their ability to do a good job (Zajonc, 1965). In other words, an audience is bad for a beginner, but good for a well-practiced professional. Thus, all other things being equal, individuals' performance in a group may suffer if they are novices at the task assigned to them, but may be facilitated if they have expertise in that task.

Group Size

Beyond the mere presence of others, how does the size of a group affect its level of productivity? Are large groups more productive than smaller ones? Large groups generally have the advantage of greater resources (such as information and skills), which may lead to more productivity. On the other hand, large groups often have more difficulty than small ones in establishing adequate coordination among their members. To understand the effects of group size, it is

BOX 15.1 Continued

information. That is, they disagree regarding which information is relevant to the group's goals and how the information should be weighted and combined. A resolution of these differences usually entails discussion and negotiation among members. Examples include some jury tasks and cognitive conflict tasks used in social judgment studies.

6. Mixed Motive Tasks. In tasks of this type, group members face an underlying conflict of interest with respect to conditions for reward. To resolve this conflict, group members must negotiate and bargain with one another. Examples include labor-management wage negotiations, prisoner's dilemma studies, and reward allocation tasks.
7. Contests/Battles. In tasks of this type, group members compete as a unit against an external opponent or enemy. The outcome is interpreted in terms of a winner and a loser, with corresponding payoffs. Examples include competition between sports teams, battles between military combat groups, and other winner-take-all conflicts.
8. Performances/Psychomotor Tasks. Tasks of this type require group members to exercise manual or psychomotor skills to bring about desired results. Much of the heavy work of the world—lifting, connecting, extruding, digging, pushing—falls into this category. These tasks can be subclassified in a myriad of ways: by the type of material being worked on, the type of activity involved, and the intended product from the activity. In general, when performing these psychomotor tasks, groups strive to meet objective or absolute standards of excellence. Examples include laying a pipeline, loading a ship to meet a deadline, or achieving high output on an assembly line.

Underlying the eight types of tasks proposed by McGrath are four general processes: generating ideas and alternatives (task types 1 and 2), choosing among options (types 3 and 4), negotiating resolutions to conflict (types 5 and 6), and executing manual and psychomotor tasks (types 7 and 8). Note also that four task types are basically cognitive or conceptual in nature (types 2, 3, 4, and 5), whereas the other four are behavioral or action oriented (types 1, 6, 7, and 8). Similarly, four task types are essentially cooperative in nature (types 1, 2, 3, and 8), whereas the other four involve at least some degree of conflict (types 4, 5, 6, and 7).

important to consider the type of task facing the group (see Box 15.1).

Additive Tasks In a classic study published before World War I, a French agricultural engineer named Ringelmann investigated the effects of group size on tasks where individual group members' contributions are summed to produce the total group output (Kravitz & Martin, 1986; Moede, 1927; Ringelmann, 1913). Ringelmann asked individuals to pull as hard as possible on a rope that was attached to heavy weights, and he used a gauge to measure the amount of pull exerted. The participants in this study worked either by themselves or as members of a group. The results showed that although a group of eight men could pull harder than a smaller group or a single individual, the average contribution of members declined as the group's size increased (see Figure 15.3).

Several explanations have been advanced for the Ringelmann effect. One suggests that losses arise from coordination loss. In Ringelmann's rope-pulling task, the possibility of unintentionally pulling in slightly

FIGURE 15.3 PULLING POWER PER PERSON BY GROUP SIZE IN THE RINGELMANN ROPE-PULLING TASK

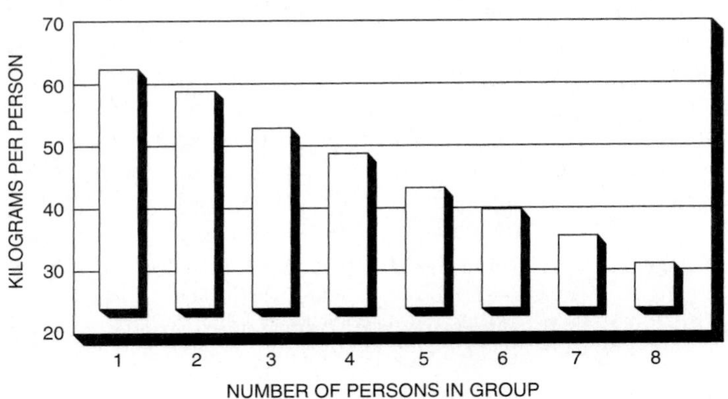

Source: Reconstructed from Ringelmann, 1913; Moede, 1927; Kravitz and Martin, 1986.

different directions or failing to pull at precisely the same time increases with group size (Ringelmann, 1913, p. 9). Several studies have confirmed that coordination loss does occur on tasks of this type (Ingham, Levinger, Graves, & Peckham, 1974; Latané, Williams, & Harkins, 1979).

Nevertheless, coordination loss does not account for all of the decline produced by group size (Harkins & Petty, 1982). Thus, theorists have pointed to another factor. Motivation loss refers to a circumstance where group members reduce their effort and slack off, thereby producing less than they otherwise would. In recent times, social psychologists have looked extensively at **social loafing**—a form of motivation loss that occurs when there is no clear way to know how much individual members are contributing to the group product (Hardy & Latané, 1986; Kerr & Bruun, 1981; Williams, Harkins, & Latané, 1981). Social loafing is surprisingly widespread, and many studies have documented a falloff in effort by group members on additive tasks (Harkins & Szymanski, 1987; Karau & Williams, 2001; Latané et al., 1979).

A number of factors can increase or decrease social loafing (Karau & Williams, 1993, 1995). Individuals are less likely to slack off when they perceive that their contributions to the group are unique and indispensable rather than merely redundant with the inputs from other members (Kerr & Bruun, 1981, 1983). Another factor that affects social loafing is the valence of the group's task. When a group's members are working on an involving task—one that is interesting, meaningful, and challenging—they are less likely to loaf than when they are working on an uninvolving task (Brickner, Harkins, & Ostrom, 1986; Harkins & Petty, 1982; Zaccaro, 1984). Closely related to this is the impact of members' attraction to the group itself on their tendency to loaf. Individuals are less likely to engage in social loafing when they strongly value the group and their membership in it than when they do not (Karau & Williams, 1993).

Disjunctive and Conjunctive Tasks Group size affects group productivity not only on additive tasks, but also on disjunctive and conjunctive tasks. **Disjunctive tasks** are those in which the group's output depends solely on its strongest or most able member. In other words, only the best performance is considered when judging the output of the group. **Conjunctive tasks** are those in which the group's productivity depends on its weakest member. According to a theory

© Jean-Claude Lejeune/Stock, Boston

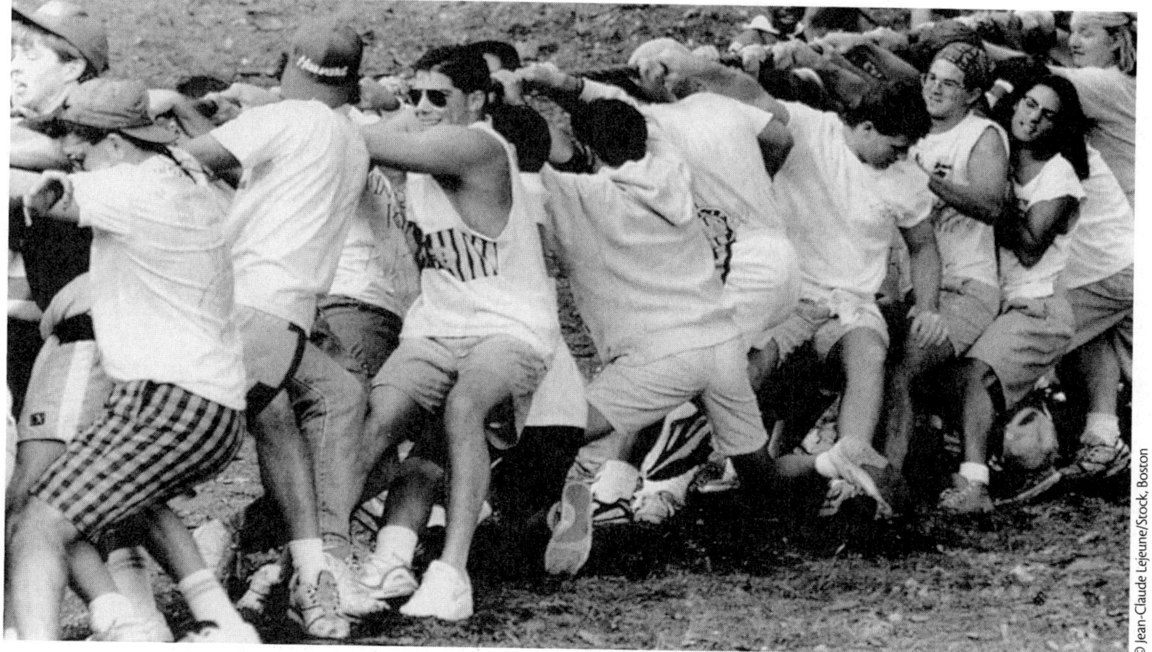

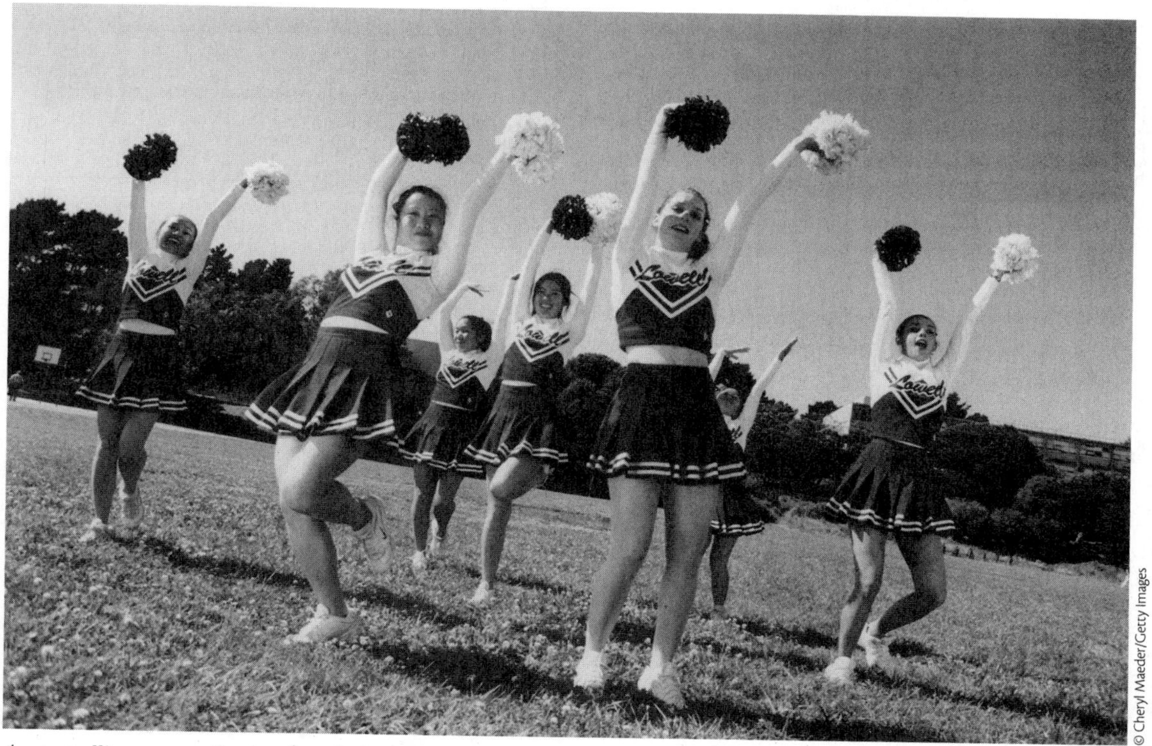

© Cheryl Maeder/Getty Images

A rope-pulling contest often involves the Ringelmann effect because people get in one another's way (faulty coordination). In contrast, a motivational loss can occur in a task like cheering due to social loafing.

proposed by Steiner (1972), the impact of group size on disjunctive tasks is different from that on conjunctive tasks. Steiner reasoned that if a task is disjunctive, then productivity will increase with group size, but if a task is conjunctive, increasing group size will decrease productivity.

The reasoning behind these predictions is relatively simple. If we assume that groups are composed at random, a large group has more chance than a small group of containing members of very high ability. In consequence, as group size increases, the chances of someone giving a good performance increase—improving performance on a disjunctive task. The converse is true for conjunctive tasks. If groups are composed at random, a large group is more likely to have very weak members than a small group. Consequently, large groups should perform less well than small ones on conjunctive tasks.

One study that investigated the effects of group size on conjunctive and disjunctive tasks was conducted by Frank and Anderson (1971). Each group worked on a series of tasks that required group members to generate ideas or images (for example, "Write three points pro and con on the issue of legalizing gambling"). Some groups received disjunctive instructions: As soon as any member had completed the task, the group could move on to the next task. Other groups received conjunctive instructions, which stated the group was not permitted to move to the next task until every member had completed the task. Thus, a group in the disjunctive condition could proceed at the speed of its most competent member, whereas a group in the conjunctive condition could move no faster than its slowest member.

The findings from this study are shown in Figure 15.4. The observed results generally support the aforementioned predictions. On disjunctive tasks, large groups performed better than small ones. On conjunctive tasks, large groups performed worse than small ones—although this difference was rather small. The findings of other studies have also provided some support for these predictions, especially with respect to disjunctive tasks (Bray, Kerr, & Atkin, 1978; Littlepage, 1991; Wittenbaum, Vaughan, & Stasser, 1998).

FIGURE 15.4 EFFECTS OF DISJUNCTIVE AND CONJUNCTIVE TASKS ON GROUP PRODUCTIVITY

On disjunctive tasks—tasks in which a group's performance depends on the performance of its best member—large groups were shown to perform better than small ones. However, on conjunctive tasks—tasks in which a group's performance is restricted by the performance of its worst member—large groups performed slightly less well than small ones.

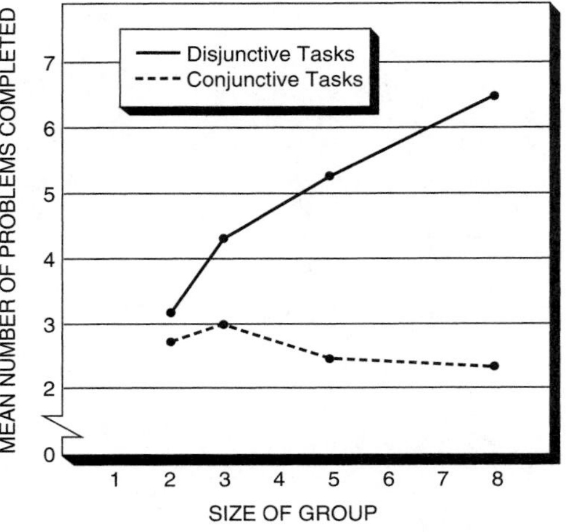

Source: Adapted from "Effects of Task and Group Size Upon Group Productivity and Member Satisfaction" (1971) by Frank and Anderson, *Sociometry, 39,* 135–149. Used with permission from the American Sociological Association.

Group Goals

The goals established by a group have an important impact on the level of productivity by the group's members (Zander, 1985). In general, we can say that if a group establishes explicit, demanding objectives with respect to the group's performance, and if the group's members are highly committed to those objectives, then the group will perform at a higher level than if it does not do these things (O'Leary-Kelly, Martocchio, & Frink, 1994; Pritchard, Jones, Roth, Stuebing, & Ekeberg, 1988). This phenomenon is often referred to as the **group goal effect** on productivity.

The magnitude of the goal effect on productivity can be large. For instance, one study (Gowen, 1986) found that setting explicit individual goals led to a 19 percent increase in group performance relative to not setting goals; and setting group goals led to a 12 percent increase in group performance relative to not setting goals. The combination of compatible individual goals and group goals led to a 31 percent increase in group productivity.

Why does establishing clear goals lead to higher productivity? First, setting a goal tends to strengthen members' work efforts. With an explicit target, they have a focus for their efforts; this channels their performance, resulting in higher productivity. Second, setting an explicit goal often makes the need for planning clear to the group. Planning by the group often leads to greater efficiency, hence to more total productivity. Third, the presence of an explicit and demanding goal often leads to higher levels of cooperation among members. This is sometimes reflected in a high level of morale-building communication among members ("We can do it!"), the effect of which is to heighten members' enthusiasm and sense of group efficacy. Finally, establishing an explicit goal often reduces members' concern for aspects of performance unrelated to the goal. This leads to less wasted motion and, hence, higher productivity (Weldon & Weingart, 1993).

REWARD DISTRIBUTION AND EQUITY

When members contribute to the attainment of group goals, they typically receive rewards such as money and approval. Although there is often a direct relation between the amount of the contribution and the amount of the benefits, this is not always the case. The distribution of rewards within a group concerns all members because it raises questions of fairness and because a group's level of productivity can be influenced by the distribution of rewards. It goes without saying that not all groups allocate their earnings equally to their members. In many groups, certain members are rewarded more than others, depending on such factors as differences in contribution, seniority of membership, degree of task interdependence, and the like (Chen & Church, 1993). When members are not all rewarded equally, we say that a condition of differential rewarding exists in the group.

Members often make judgments regarding what level of rewards they should receive for participating in a group. Reflecting a concern with fairness or equity, these judgments take into account such factors as the level of their own contributions relative to those of other members.

Principles Used in Reward Distribution

There are many criteria—called **distributive justice principles**—that group members can use to judge the fairness of the distribution of rewards (Deutsch, 1985; Elliott & Meeker, 1986; Saito, 1988). Three of the most commonly used are the equality principle, the equity principle, and the relative needs principle. When group members use the equality principle, they distribute rewards equally among members, regardless of members' contributions. When group members follow the equity principle, they distribute rewards in proportion to members' contributions. When they follow the relative needs principle, they distribute rewards according to members' personal needs, regardless of contributions (Lamm & Schwinger, 1980).

When allocating rewards among members, a group may rely exclusively on one of these justice principles or may apply several of them simultaneously. These justice principles are often contradictory in the sense that they lead to different distributions of rewards, but what really matters is the relative importance (or weighting) accorded each principle. Not surprisingly, their relative importance varies from group to group and from situation to situation. For instance, the equality principle often prevails in situations where members are concerned with solidarity and wish to avoid conflict (Leventhal, Michaels, & Sanford, 1972). It also prevails in cultural settings that are relationship oriented, rather than economically oriented (Mannix, Neale, & Northcraft, 1995). There is some evidence that women favor the equality principle over the equity principle to a greater extent than men do (Leventhal & Lane, 1970; Watts, Messe, & Vallacher,

© Rose Skytta/Jeroboam

Equity is relevant in marital relationships just as in work relationships. For this couple, equity means sharing the housework. What would happen to this relationship if either the husband or the wife insisted on doing less around the house?

1982). Friends are more likely to follow the equality norm than strangers (Austin, 1980), and members of small (three-person) groups are more likely to use it than members of large (12-person) groups (Allison, McQueen, & Schaerfl, 1992).

In contrast, the equity principle is often used in work situations, where many persons want their share of rewards to reflect the importance of their contribution. The needs principle is frequently salient in close or intimate relationships involving friends, lovers, and relatives. However, this principle has also been invoked in other contexts. Karl Marx, for example, advocated the adoption of the needs principle in communist societies, where individuals would contribute according to their abilities and receive according to their needs.

In addition to their concerns about the distribution of rewards, people are also concerned about the process that produces the distribution (Brockner et al., 1994). Are the procedures applied uniformly to all people involved? Is the information used to determine who deserves what accurate? Is there any bias that affects the distribution? These are all concerns about procedural justice, although they do not contribute as much to our evaluation of fairness as the actual payoffs (Magner, Johnson, Sobery, & Welker, 2000).

Equity Theory

The equity criterion for reward allocation is used by many groups. A state of **equity** exists when members receive rewards in proportion to the contributions they make to the group. For example, in an industrial work group in the United States, a worker normally would expect to receive better outcomes (salary, benefits) than others if his or her job required higher skill, more hours per week, and so on. Likewise, a wife would probably feel some inequity if she contributes more to the family than her husband but receives little help or love in return. As these examples suggest, equity judgments are made when one group member compares his or her own outcomes and inputs against those of another member.

In an effort to formalize the nature of these comparisons, theorists have developed equity theory (Brockner & Wiesenfeld, 1996; Greenberg & Cohen, 1982; Homans, 1974; Walster, Walster, & Berscheid, 1978). The fundamental idea underlying equity can be expressed in terms of the following equation:

$$\frac{\text{Person A's outcomes}}{\text{Person A's inputs}} = \frac{\text{Person B's outcomes}}{\text{Person B's inputs}}$$

This equation states that equity exists when the ratio of person A's outcomes to inputs is equal to the ratio of person B's outcomes to inputs. What matters is not merely the level of outcomes or inputs but the equality between the ratios of outcomes to inputs.

To make this more concrete, consider the case of two women employed by the same company. One of the women (person A) receives a high outcome—a salary of $100,000 a year, 4 weeks of paid vacation, reserved parking in the company's lot, and a fancy corner office with thick rugs and a nice view. The other woman (person B) is about the same age but receives a lesser outcome—a salary of $40,000 per year, no paid vacation, no reserved parking, and a cramped, noisy office with no windows.

Will persons A and B feel that this distribution of rewards is equitable? If their inputs to the company are identical, then the arrangement will almost certainly be experienced as distressing, especially by person B. For example, if both A and B work a 40-hour week, have only high school educations, and have approximately equal experience, there is little basis for paying person A more than person B. Person B will probably feel angry because the reward distribution is inequitable, and person A may feel uncomfortable or guilty.

Suppose instead that person A's inputs are much greater than B's. Say that person A works a 60-hour week, holds an advanced degree such as an M.B.A., and has 12 more years of relevant experience than person B. Suppose, also, that person A's job involves a high level of stress because it entails the risk of serious failure and financial loss for the company. In this event, A not only has greater "investments" (that is, education and experience) but she is also bearing greater immediate "costs" (60 hours of work a week, plus high stress). Under these conditions, both A and B may feel that their outcomes, although not equal, are nevertheless equitable.

This example highlights the basic difference between an equitable and an inequitable relationship, but precise calculations regarding equity can be difficult to make in everyday life. Two persons may view the same situation in different ways. They may disagree, for instance, over how to evaluate particular inputs and outcomes. If one worker holds an advanced university degree, whereas another worker has 7 years' more seniority, who is contributing the more important input? Moreover, persons may disagree over exactly which inputs and outcomes are to be included in the equity calculation. Should a college education be considered as an input even if the degree earned has nothing to do with the job at hand? Should seniority be rewarded even when experience does not cause workers to do a better job?

Task Interdependence

In addition to the characteristics of workers and the costs they pay to the group, some elements of the situation can also enter into group members' assessments of their value to the group. In particular, if there is a high degree of task interdependence among members (Miller & Komorita, 1995), individual group members may make other group members change their ideas about how important their contributions are relative to others in the group. A high degree of task interdependence exists when the task demands a high level of communication, coordination, and mutual performance monitoring among members for successful completion. In other words, the more the outcome depends on contributions from every member of the group, the greater the task interdependence. When task interdependence is high, unequal rewards may cause some members to feel that smaller rewards are not fair, given that the task cannot be completed successfully without their contribution—no matter how small it is relative to others.

The effects of task interdependence and differential rewarding on group productivity were demonstrated vividly in a study by Miller and Hamblin (1963). In this study, three group members were placed in separate booths and their task was to determine which one of 13 numbers had been selected by the experimenter. Each group member was informed of four numbers that had not been selected by the experimenter. These sets of four numbers were different for each member; thus, if the three members shared their information, they would immediately see the correct answer. The group's productivity was measured in terms of time: The faster it discovered the correct answer, the higher the score it received.

Two conditions were manipulated in the study. One was the degree of task interdependence. In the high-interdependence condition, guessing was discouraged by a substantial penalty. Group members, therefore, were encouraged by the penalty to share information to solve the problem. In the low-interdependence condition, there was no restriction on guessing. To solve the problem, individuals could merely continue to guess until they hit the correct solution.

The other manipulation in the study was the degree of differential rewarding among group members. Rewards for group members were based both on their own efforts and on those of the group as a whole. Each group started with 90 points, and 1 point was subtracted for every second that elapsed before all three members had solved the problem. For example, if the group took 60 seconds to reach a solution, it scored 30 points. The group's points were then allocated among its members. In one condition (equality), each member received one third of the group reward. In a second condition (medium differential rewarding), the member who first solved the problem received one half of the group's points, the member who finished second received one third, and the one who finished third received one sixth. In a third condition (high differential rewarding), the member who first solved the problem received two thirds of the group's points, the member who finished second received one third, and the one who finished third received none.

FIGURE 15.5 GROUP PRODUCTIVITY AS A FUNCTION OF INTERDEPENDENCE AND DEGREE OF DIFFERENTIAL REWARDING

Under low task interdependence, differential rewarding (that is, unequal reward distribution) had little effect on group performance. However, under high task interdependence, a high level of differential rewarding induced low group productivity as well as competition among members, and a low level of differential rewarding induced high productivity and cooperation.

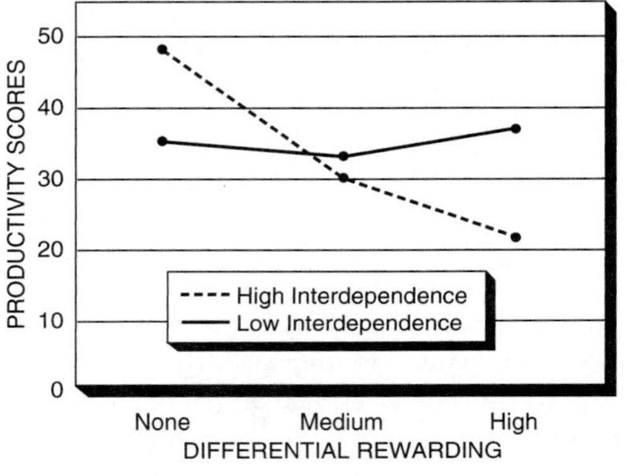

Source: Adapted from Miller and Hamblin, 1963.

The results of this study are displayed in Figure 15.5. Under conditions of low interdependence (where all three individuals could freely guess at the answer), differential rewarding did not significantly affect group productivity. But under conditions of high interdependence (where guessing was prohibited and members had to share information), higher levels of differential rewarding sharply reduced the group's level of productivity.

To interpret these results, note that under conditions of high task interdependence and no differential rewarding, members realized that their best strategy for achieving higher rewards was to cooperate with each other. They shared information quickly, and, as a result, the group got a high productivity score. Under conditions of high task interdependence and high differential rewarding, however, members realized they could achieve high rewards for themselves only by outperforming others in their group. This created a competitive atmosphere in which members used a blocking strategy that highlighted their own importance to solving the problem. For example, one member might try to obtain information from others without yielding information in return, thereby hindering others' performance. This would diminish coordination, slow the group's progress, and inevitably result in poor group productivity.

Responses to Inequity

Inequity produces not only emotional reactions of distress (anger, guilt) and reduced productivity and commitment to the group but also direct attempts to change the conditions that produce it. There are two distinct types of inequity: underreward and overreward. Underreward occurs when a person's outcomes are too low relative to his or her inputs; overreward occurs when a person's outcomes are too high relative to his or her inputs.

Responses to Underreward Persons who are underrewarded typically become dissatisfied or angry (Austin & Walster, 1974; Cropanzano, 1993; Scher, 1997; Sweeney, 1990). The greater the degree of underreward, the greater their dissatisfaction and desire to reestablish equity. To illustrate, suppose that person A is underrewarded in a relationship. In this case, the inequity may be expressed as follows:

$$\frac{\text{Person A's outcomes}}{\text{Person A's inputs}} < \frac{\text{Person B's outcomes}}{\text{Person B's inputs}}$$

Given that A is underrewarded, equity could be restored by (1) increasing the outcomes received by A, (2) reducing the inputs from A, (3) reducing the outcomes received by B, or (4) increasing the inputs from B.

Studies show that most underrewarded persons take direct steps to reduce inequity. If the situation permits, such a person usually would attempt to reduce inequity by increasing his or her own outcomes. For example, this person might aggressively seek a pay raise. If an industrial employee paid on a piecework basis feels underrewarded, he or she might reduce the quality of effort on each piece in order to increase the total number of pieces produced per hour. This would increase the worker's outcomes without increasing his or her input (Andrews, 1967; Lawler & O'Gara, 1967), but at the same time it would reduce group productivity. More recently, feelings of underreward have been connected to workers' stealing from their employers and work-related sabotage (Greenberg & Scott, 1996).

Responses to Overreward What happens when a person receives more than his or her fair share in a relationship? Will he or she be content just to enjoy the benefits? Although overreward is apparently less troubling to individuals than underreward (Greenberg, 1996), it can still create feelings of inequity, often in the form of guilt rather than anger (Perry, 1993; Sweeney, 1990). A person who feels guilty about overreward may attempt to rectify the inequity (Austin & Walster, 1974).

Research findings show that in some situations, overrewarded persons sacrifice some of their rewards to increase those of others. However, the extent of the redistribution often will not be complete, and equity may be only partially restored (Leventhal, Weiss, & Long, 1969). There is some evidence that overrewarded members prefer to restore equity by increasing their inputs. For example, in a work situation, overrewarded members can strive to produce more or better products as a means of reducing inequity (Goodman & Friedman, 1971; Patrick & Jackson, 1991); this enables them to restore equity without sacrificing any of the outcomes they receive.

This process was investigated in a classic study in which students were hired to work as proofreaders (Adams & Jacobsen, 1964). In one condition, participants were told that they were not really qualified for the job—due to inadequate experience and poor test scores—but that they would nevertheless be paid the same rate as professional proofreaders (30 cents per page). In a second condition, participants were told that due to their lack of qualifications, they would be paid a reduced rate (20 cents per page). In a third condition, participants were told that they had adequate experience and ability for the job and that they would be paid the full rate (30 cents per page). Thus, the participants in the first condition viewed themselves as overrewarded, whereas those in the second and third conditions saw their pay as equitable. Measures of the quality of the students' work showed that the overrewarded students caught significantly more errors than the equitably paid students. In fact, the overrewarded students were so vigilant that they often challenged the accuracy of material that was correct. These results indicate that the overrewarded students increased their inputs, thereby restoring equity. Similar findings appear in related studies (Adams & Rosenbaum, 1962; Goodman & Friedman, 1969).

Other Responses to Inequity We have noted that inequity—both underreward and overreward—can be resolved by changing outcomes or inputs. There are other ways of coping with inequity, however. Although these methods do not rectify the conditions producing inequity, they make the inequity itself more tolerable. For instance, a person who is underrewarded might choose a different person for comparison. Person A, instead of comparing her inputs and outcomes with those of person B, might compare them with person C. By adjusting her frame of reference, person A would see the situation in a different light and experience less inequity.

Another way to cope with inequity is to withdraw from the situation entirely. An individual might choose to resign from his or her job, for example, rather than continue to perform under severe inequity. One study found evidence that among blue-collar workers, those experiencing inequity (underreward) are somewhat more likely to quit the job, be absent, or report sick than those not experiencing inequity (Van Yperen, Hagedoorn, & Geurts, 1996). If it is extreme, inequity can be so distressing that it produces behavior that seems economically irrational. For instance, one study found that workers were often willing to accept an alternative position that paid less in order to remove themselves from severe inequity (Schmitt & Marwell, 1972).

Still another way to reduce inequity is through perceptual distortion or bias. By distorting inputs and outcomes, equity can be reestablished in a psychological sense. For example, an employer might convince himself that his overworked and underpaid secretary is, in fact, being treated equitably by minimizing the secretary's inputs ("You wouldn't believe how incompetent she is!") or exaggerating her outcomes ("Work gives her a chance to socialize with all her friends"). Among other things, this spares the executive any need to reduce the secretary's workload or increase her salary (Austin & Hatfield, 1980). Bias of this type is especially convenient for those who are overrewarded. If a person is receiving more than he or she deserves, the person might simply conclude that his or her inputs are greater than he or she originally believed; this eliminates any need to reallocate rewards (Gergen, Morse, & Bode, 1974).

BRAINSTORMING

In everyday life, many cognitive products—solutions, decisions, verdicts, and judgments—are generated by groups. So prevalent is this activity that some persons have even suggested that most of the truly important decisions and judgments in society are made by groups, not by individuals. Whether that is true or not, groups surely are a major source of important decisions and ideas. In the remainder of this chapter, we will discuss various cognitive products of groups. Here, we focus on groups generating one special type of cognitive output, namely, novel or innovative ideas.

Suppose an advertising executive has the responsibility to develop a strong media campaign to promote a new product, such as a toothpaste that not only cleans but also strengthens teeth. Should she ask her staff members to work individually on this task, or should she bring them together in a group and ask them to work jointly on the task? A few decades ago, an advertising executive named Alex Osborn (1963) proposed that groups are better at generating creative new ideas than individuals working alone. In particular, Osborn maintained that groups employing a specific procedure called **brainstorming** will be able to generate a large number of high-quality, novel ideas in a brief period.

Brainstorming requires that a group follow set procedures when generating new ideas. Given a specific problem, such as developing an advertising campaign for toothpaste, their objective would be to think up as many different advertising themes or slogans as possible. All ideas are recorded or written down. During the idea generation session, group members should do the following:

1. Express any idea that comes to mind, no matter how wild or unrealistic. In fact, the wilder the idea, the better.
2. Withhold criticism and defer judgment on all ideas until later. Members should not even criticize their own ideas; instead, they should let their ideas flow freely without precensoring them.
3. Generate as many ideas as possible—the more, the better. This improves the odds of getting some good ones.
4. Try to build on ideas suggested by others—generate new ideas by combining, modifying, or extending the ideas of others.

Although Osborn claimed considerable success with the brainstorming technique, systematic empirical studies do not suggest that the technique actually does what it is supposed to do (Paulus, Larey, & Dzindolet, 2001). For instance, one early study (Taylor, Berry, & Block, 1958) asked four-person brainstorming groups to work on various problems. The results showed that the number of novel suggestions generated by groups was greater than the number a single individual could produce. At first, this might seem to support Osborn's notion, but it ignores the fact that the four-person groups involved four times the person power. To take this into account, the investigators compared the productivity of the four-person brainstorming groups with the combined productivity of four persons who had worked alone. Surprisingly, persons working alone produced almost twice as many original ideas as the four-person brainstorming groups. Another study (Bouchard & Hare, 1970) used groups of five, seven, or nine participants and asked them to discuss the following problem: What would happen if people had an extra thumb? The participants wrote down as many implications of this anatomical change as they could imagine. The answers generated by these face-to-face groups were compared with the pooled answers of individuals who had worked alone. Once again, the results showed that those working alone produced more solutions—and this difference increased as the size of the groups increased. Moreover, the quality and the originality of the ideas produced by brainstorming groups is, on average, no higher than that of ideas produced by individuals working alone—and it is sometimes worse (Bouchard, Barsaloux, & Drauden, 1974).

Production Blocking

Why aren't brainstorming groups more effective? One prominent answer is that **production blocking** limits the ideas generated by brainstorming groups (Diehl & Stroebe, 1987, 1991; Lamm & Trommsdorff, 1973). Production blocking occurs when members of a brainstorming group are unable to express their ideas due to turn taking among group members. Because most individuals follow the norm that persons should take turns when talking, members in brainstorming sessions often wait for others to stop speaking before they voice their own ideas—slowing the expression of new ideas. Moreover, because suggestions from others can be distracting, members may have to concentrate on remembering or "rehearsing" their own ideas while others are speaking; this will further limit the group's productivity because members who are rehearsing cannot turn their attention to generating more new ideas.

Production blocking is illustrated in a study by Diehl and Stroebe (1987). This study used four-person groups on a brainstorming task and manipulated whether or not participants could speak while others were speaking. In one condition, the four persons participated in an actual, interacting brainstorming group, where suggestions were tape-recorded; the members were placed in separate rooms, but they could hear one another over an intercom; a signaling system of lights indicated when others were speaking, and members could speak into their own microphones only when others' were not speaking (that is, when others' lights were not on). In a second condition, the setup was similar, except that participants could not actually hear each others' voices over the intercom. In a third condition, the participants could not hear one another over the intercom; they could see the lights, but were told to ignore these; and they were free to express their ideas into the microphone whenever they wanted. The results indicated that group production of ideas in the first two conditions (in which the participants could speak only when the light indicated that others were not speaking) was much lower than the production of ideas in the third condition (in which the participants could express their ideas whenever they wanted, irrespective of whether others were speaking). The interference caused by the requirement that members speak in turn (when lights permitted) blocked the production of ideas and reduced total group output.

GROUP DECISION MAKING

Beyond the matter of idea generation or brainstorming, another important group process is decision making. Typically, when a group makes a decision, its members consider the merits of several mutually exclusive options and then select one in preference to the others. Reduced to its essentials, group decision making is fairly easy and involves several basic steps (Janis & Mann, 1977). To make a decision effectively, group members need to define a set of possible options, gather all the relevant information about these options, share this information among themselves, carefully assess all the potential consequences of each option under consideration, and then calculate the overall value of each option. Once this is done, it remains only to select the most attractive option as the group's choice.

In practice, however, group decision making is not always so easy or straightforward, for the decision making process can go awry in various ways. Information regarding certain options may prove hard to obtain, leading to incomplete or inadequate consideration of these options. Even if the individual members do have all the relevant information, they may fail to share it fully with one another (Stasser, 1992; Stasser & Titus, 1987). If members hold different values, they may disagree regarding which options are most attractive; this can spawn arguments and block consensus within the group. Then, too, conformity pressures within the group may impel members to abbreviate or short-circuit the deliberation processes; if this happens, group discussion may lead to ill-considered or unrealistic decisions.

Groupthink

Aberrations in decision making can plague any group—even those at the highest levels of business and government. The history of American foreign policy provides many examples. The decisions by the United States to invade the Cuban Bay of Pigs, to cross the 38th parallel in the Korean conflict, and to escalate the Vietnam War were all made by committees. The infamous Bay of Pigs invasion, for example, was planned by a small group of top government officials immediately after President John Kennedy took office in 1961. The group included what some considered the nation's "best and brightest": McGeorge Bundy, Dean Rusk, Robert McNamara, Douglas Dillon, Robert Kennedy, Arthur Schlesinger, Jr., and President Kennedy himself, with representatives of the Pentagon and the Central Intelligence Agency (CIA). This group decided to invade Cuba in April 1961, using a small band of 1,400 Cuban exiles as troops. The invasion was to be staged at the Bay of Pigs and assisted covertly by the U.S. Navy and Air Force and the CIA. As it turned out, the invasion was poorly conceived. The material and reserve ammunition on which the exiles were depending never arrived, because Castro's air force sank the supply ships. The exiles were promptly surrounded by 20,000 well-equipped Cuban soldiers, and within 3 days, virtually all had been captured or killed. The United States suffered a humiliating defeat in the eyes of the world, and Castro's communist government became more strongly entrenched on the Caribbean island.

How could it happen? How could a group of such capable and experienced men make a decision that turned out so poorly? In a post hoc analysis, Janis (1982) suggests that a process termed groupthink may have produced the defective decision. **Groupthink** refers to a faulty mode of thinking by group members in which their desire to realistically evaluate alternative courses of action is overwhelmed by pressures for unanimity within the group. That is, concerned that they not disrupt apparent group consensus, the group members neglect to appraise alternatives critically and to weigh the pros and cons carefully. Once groupthink sets in, the typical result is an ill-considered decision.

Symptoms of Groupthink How can one detect whether groupthink is present in a group discussion? Janis (1982) suggests certain symptoms that indicate when groupthink is operating. These include the following:

1. Illusions of invulnerability. Group members may think that they are invulnerable and cannot fail and, therefore, display excessive optimism and take excessive risks.
2. Illusions of morality. Members may display an unquestioned belief in the group's inherent superior morality, and this may incline them to ignore the ethical consequences of their decisions.
3. Collective rationalization. Members may discount warnings that, if heeded, would cause them to reconsider their assumptions.
4. Stereotyping of the adversary. Especially in the political sphere, the group may develop a stereotyped view of enemy leaders as too evil to warrant genuine attempts to negotiate or as too weak to mount effective counteractions.
5. Self-censorship. Members may engage in self-censorship of any deviation from the apparent group consensus, with each member inclining to minimize the importance of his or her own doubts.
6. Pressure on dissenters. The majority may exert direct pressure on any member who dissents or argues against any of the group's stereotypes, illusions, or commitments.
7. Mindguarding. There may emerge in the group some self-appointed "mind guards"—members who protect against information that might shatter the complacency about the effectiveness and morality of the group's decisions.
8. Apparent unanimity. Despite their personal doubts, members may share an illusion that unanimity regarding the decision exists within the group.

Janis suggests that some of these symptoms were present during the decision making process for the Bay of Pigs invasion. For example, there was an assumed air of consensus that caused members of the decision making group to ignore some glaring defects in their plan. Although several of Kennedy's senior advisors had strong doubts about the planning, the group's atmosphere inhibited them from voicing criticism. Several members emerged as "mind guards" within the group; they suppressed opposing views by arguing that the decision to invade had already been made and that everyone should help the president instead of distracting him with dissension. Open inquiry and clearheaded exploration were discouraged. Even the contingency planning was unrealistic. For instance, if the exiles failed in their primary military objective at the Bay of Pigs, they were supposed to join the anti-Castro guerrillas known to be operating in the Escambray Mountains. Apparently, no one was troubled by the fact that 80 miles of impassable swamp and jungle stood between the guerrillas in the mountains and the exiles.

Groupthink might not be such a concern—except for the recognition that it can occur and recur in many groups. Janis notes that the Bay of Pigs invasion is not the only fiasco in which groupthink was implicated. He suggests groupthink was also involved in other high-level government decisions, including the decision to cross the 38th parallel and invade North Korea during the Korean War, the failure to defend Pearl Harbor adequately on the eve of World War II, the decision to escalate the Vietnam War, and the decision to engage in the Watergate cover-up. (See Box 15.2 for a more recent application of groupthink ideas to the United States' decision to invade Iraq.)

Causes of Groupthink According to Janis's theory, groupthink is more likely to occur in high-cohesion groups than in low-cohesion groups. It is caused by various factors, including homogeneity of members, insulation of the group from its environment, lack of clear-cut rules to guide decision making behavior within the group, and high levels of group stress. Another contributing factor is promotional leadership—that is, a leader who actively promotes his or her own favored solution to the problem facing the group, to the neglect of other possible solutions. According to Janis, each of these factors contributes to groupthink,

BOX 15.2 Groupthink and the War in Iraq

The incidents of groupthink in governmental decision making clearly did not end with the publication of Irving Janis's analysis. Even though governmental decision-making bodies have instituted procedures intended to avoid the emergence of groupthink, it is still a common occurrence and has been implicated in a variety of poor decisions ranging from invasions to the launch of the space shuttle, Challenger, which exploded in midflight (Moorhead, Ference, & Neck 1991). Most recently, analysts have determined that groupthink made a major contribution to President George Bush's decision to go to war against the nation of Iraq in 2003 and more generally to the processing of intelligence information about the production of weapons of mass destruction (WMD) in Iraq. After the invasion, it was discovered that WMD and production facilities did not exist in Iraq, and President Bush was forced to admit that the intelligence information, and the government's decision making based on it, were faulty.

The U.S. Senate Intelligence Committee concluded in its review of the decision to invade that groupthink was an important factor and that the administration had circumvented important safeguards specifically constructed to avoid groupthink: "The Intelligence Community suffered from a collective presumption that Iraq had an active a growing weapons of mass destruction program. This 'group think' dynamic led Intelligence Community analysts, collectors and managers to both interpret ambiguous information as conclusively indicative of a WMD program as well as ignore or minimize evidence that Iraq did not have active and expanding weapons of mass destruction programs. This presumption was so strong that formalized Intelligence Community mechanisms established to challenge assumptions and group think were not utilized" (Select Committee on Intelligence, 2004).

Psychologists Alexander Rinehart and Philip Dunwoody (2005) conducted a more detailed analysis of the decision to invade and demonstrate in their research how each of the elements of groupthink displayed in Figure 15.6 was played out in the decision. Below are listed some of Rinehart and Dunwoody's observations related to groupthink symptoms and the Iraq decision.

and their simultaneous occurrence makes groupthink very probable. The groupthink framework is displayed in Figure 15.6.

Laboratory and case study research on groupthink supports some of Janis's hypotheses (Park, 1990; Paulus, 1998). Several studies have shown that promotional leadership contributes to the occurrence of groupthink symptoms in business decisions (Leana, 1985; Moorhead & Montanari, 1986), personnel decisions (Flowers, 1977), and foreign policy decisions (McCauley, 1989). Other studies have shown that groupthink symptoms are stronger when a group is insulated from its environment—and therefore gets little information or criticism—than when it is not insulated (Hensley & Griffin, 1986; Manz & Sims, 1982; Moorhead & Montanari, 1986).

Although cohesion is important in Janis's theory, there is only limited support for the hypothesis that group cohesion contributes to groupthink (Aldag & Fuller, 1993; Michener & Wasserman, 1995; Park, 1990). Whereas a few studies support this hypothesis (Callaway & Esser, 1984; Courtright, 1978), many others simply do not (Flowers, 1977; Fodor & Smith, 1982; Leana, 1985; Moorhead & Montanari, 1986). This

BOX 15.2 Continued

Groupthink Symptoms

Overestimation of the Group

Illusion of invulnerability: Donald Rumsfeld and Paul Wolfowitz argue that, after victory in Afghanistan, Iraq can be beaten.

Illusions of morality: Saddam's regime labeled part of the "axis of evil" by President Bush in 2002 State of the Union address.

Closemindedness

Collective Rationalization: In a presidential briefing regarding Iraq's chemical and biological weapons production, Paul O'Neill interrupted the discussion stating, "I've seen a lot of factories around the world that look a lot like this one. What makes us suspect that this one is producing chemical or biological agents for weapons?" The response was that we had "no confirming intelligence as to the materials being produced," but discussion immediately continued toward military options in Iraq. The Senate Intelligence Committee report details a large number of rationalizations present within the intelligence community. Others have documented that Vice President Dick Cheney was part of these sessions where negative evidence was routinely ignored.

Pressures Toward Uniformity

Pressure on Dissenters: Paul O'Neill later noted that the administration favored loyalty over analysis: "Hard-eyed analysis would be painted as disloyalty."

Self-Censorship: ". . . how do you follow the lead of a President if you're sure he's on the wrong path and there's no process to hash out differences?"

Mindguarding: Paul O'Neill said, "Everybody played their parts: literally. For this President, cabinet meetings and the many midsize to large meetings he attended were carefully scripted. Before most meetings, a cabinet secretary's chief of staff would receive a note from someone on the senior staff in the White House. The note instructed the cabinet secretary when he was supposed to speak, about what, and how long."

Apparent Unanimity: Most moderate members of Bush's initial cabinet have been replaced.

See Rinehart and Dunwoody (2005), Suskin (2004), and Select Committee on Intelligence (2004) for additional information.

Source: Adapted from Rinehart and Dunwoody, 2005.

inconsistency may arise in part because the levels of cohesion that can be induced in laboratory studies are low compared to what can arise in real-life situations.

Avoiding Groupthink If groupthink produces poor decisions and outcomes, how can one guard against it? There are several ways to prevent groupthink from occurring (Janis, 1982). Basically, these methods increase the probability that a group will obtain all the information relevant to a decision and then evaluate that information with care. First, a group's leader should encourage dissent and call on each member to express any objections and doubts. Second, a leader should be impartial and not announce a preference for any particular option or plan. By describing a problem rather than recommending a solution, a leader can foster an atmosphere of open inquiry and impartial exploration. Third, a group should divide itself into several independent subgroups, each working on the same problem and carrying out its deliberation independently. This will prevent the premature development of consensus in the main group. Fourth, after a tentative consensus has been reached, a group should hold a "second chance" meeting, at which each member can

FIGURE 15.6 JANIS'S MODEL OF GROUPTHINK

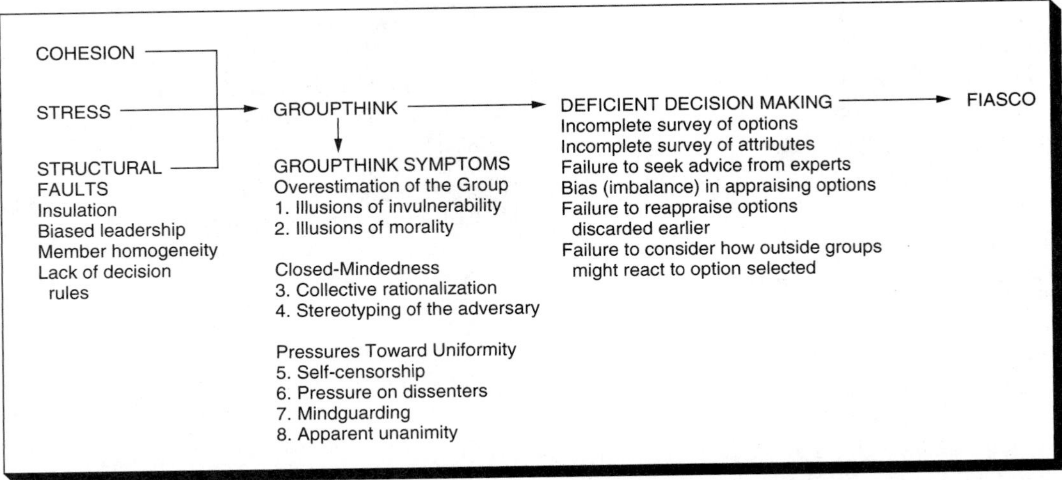

express any remaining doubts before a firm decision is taken. Fifth, a group can appoint a "devil's advocate" who is responsible for challenging the positions of other group members even if he or she really agrees with them (Hirt & Markmann, 1995). The net result of these steps will be a better, more realistic decision.

Risky Shift, Cautious Shift, and Group Polarization

Even when group decision making is not dysfunctional and follows a fairly rational course, it can still produce surprising consequences. For instance, there is some evidence that discussion in groups causes individuals to favor courses of action that are riskier than what they would choose if they made the decision alone. This was demonstrated by Stoner (1968), in a study where the participants responded individually to a series of 12 problems called choice dilemmas. In each problem, the participants were asked to advise a fictional character how much risk he or she should assume. The following item illustrates this task:

> Mr. A, an electrical engineer who is married and has one child, has been working for a large electronics corporation since graduating from college 5 years ago. He is assured of a lifetime job with a modest, although adequate, salary and liberal pension benefits upon retirement. On the other hand, it is very unlikely that his salary will increase much before he retires. While attending a convention, Mr. A is offered a job with a small, newly founded company that has a highly uncertain future. The new job would pay more to start and would offer the possibility of a share in the ownership if the company survived the competition with the larger firms.
>
> Imagine that you are advising Mr. A. Listed below are several probabilities or odds of the new company proving financially sound. Please check the lowest probability that you would consider acceptable to make it worthwhile for Mr. A to take the new job.
>
> ____ The chances are 1 in 10 that the company will prove financially sound.
>
> ____ The chances are 3 in 10 that the company will prove financially sound.
>
> ____ The chances are 5 in 10 that the company will prove financially sound.

____ The chances are 7 in 10 that the company will prove financially sound.

____ The chances are 9 in 10 that the company will prove financially sound.

____ Place a check here if you think Mr. A should not take the new job no matter what the probabilities.

(Kogan & Wallach, 1964)

After working as individuals and responding to 12 different choice dilemma items, the participants assembled in groups of six and discussed each item until they reached a unanimous decision. The participants were then separated and asked again to review each item and indicate an individual decision. The basic finding was that the group decisions following discussion were, on the average, riskier than the decisions made by individual members prior to the discussion. Moreover, the responses made individually after participating in the group were also riskier on average than the responses prior to discussion. This tendency to advocate more risk following a group discussion is termed a **risky shift.** This phenomenon has been observed in many studies (Cartwright, 1971; Dion, Baron, & Miller, 1970).

Other studies using similar tasks, however, have revealed something directly opposite to the risky shift. On certain issues when members are cautious or risk avoidant, group discussion actually causes members to become even more cautious than they were initially (Fraser, Gouge, & Billig, 1971; Stoner, 1968; Turner, Wetherell, & Hogg, 1989). This move away from risk following a group discussion is termed a **cautious shift.** So, although group discussion leads to more extreme decisions, these are not necessarily riskier decisions.

Both risky shift and cautious shift are forms of an underlying phenomenon called **group polarization** (Levine & Moreland, 1998). Polarization occurs when group members shift their opinions toward a position that is similar to but more extreme than their opinions before group discussion. Thus, if members favored a moderately risky position prior to a group discussion, polarization would occur if they shifted toward greater risk following the discussion. Likewise, if they initially favored a moderately cautious position, polarization would occur if they shifted in the direction of even greater caution after the group discussion (Myers & Lamm, 1976).

The tendency for group discussion to create polarization is quite general. That is, discussion produces polarization not only on decisions involving risk, but also on judgments and attitudes in general. Polarization has been observed with respect to political attitudes (Paicheler & Bouchet, 1973), jury verdicts (Isozaki, 1984; Myers & Kaplan, 1976), satisfaction with new consumer products (Johnson & Andrews, 1971), judgments of physical dimensions (Vidmar, 1974), ethical decisions (Horne & Long, 1972), perceptions of other persons (Myers, 1975), and interpersonal bargaining and negotiating (Lamm & Sauer, 1974).

Why does group discussion lead to polarization? That is, what causes group members to shift their risk-taking responses toward an extreme position? Several basic explanations have been proposed (Isenberg, 1986; Myers & Lamm, 1976). According to one theory, group polarization results from a process of social comparison (Goethals & Zanna, 1979; Jellison & Riskind, 1970). This theory suggests that people often value opinions more extreme than those they personally advocate. They fail to adopt these ideal (extreme) positions as their own, because they fear being labeled as extremist or deviant. However, during a group discussion in which members compare their positions, these persons may discover that other members hold opinions closer to their ideal position than they had realized. This motivates the moderate members to adopt more extreme positions. The overall result is a polarization of opinions.

Although controversial, the social comparison theory has been supported by various studies. The major source of support stems from demonstrations that exposure to simple information about other group members' positions can by itself, without any persuasive argumentation, produce polarization effects (Baron & Roper, 1976; Blascovich, Ginsburg, & Veach, 1975; Myers, Bruggink, Kersting, & Schlosser, 1980).

A second theory explains group polarization as the result of persuasive argumentation (Burnstein, 1982; Burnstein & Vinokur, 1973). According to this view, group polarization occurs whenever the preponderance of compelling arguments advanced during a group discussion favors a position more extreme than that held initially by the average member. Discussion within a group serves to persuade members who—because they had been unaware of the arguments—initially chose relatively moderate positions. After discussion, the moderate members shift their opinions in the direction of the most compelling—and relatively extreme—arguments. This produces group polarization.

Some research supports the persuasive argumentation theory. First, studies have shown that polarization does result from the exchange of persuasive arguments. In particular, the greater the number of novel and valid arguments exchanged during discussion, the greater the influence of those arguments on group members and the greater the polarization (Kaplan, 1977; Vinokur & Burnstein, 1978). Second, the greater the proportion of arguments favoring a particular point of view, the greater the shift of opinion in its direction (Ebbesen & Bowers, 1974; Lamm, 1988; Madsen, 1978). Thus, participants who are exposed to mostly risky arguments become more risk taking, whereas those who hear mostly cautious arguments become more risk avoidant.

Overall, then, there is support for both the social comparison and the persuasive argumentation theories. Both of these processes occur in combination to produce polarization, although the effects of persuasive argumentation tend to be stronger than those of social comparison (Isenberg, 1986).

SUMMARY

This chapter has discussed group structure and group performance.

Group Leadership (1) In return for helping the group achieve its objectives, a leader receives endorsement as well as special rewards and privileges. Endorsement declines if the group fails repeatedly to achieve its goal(s), if the leader is judged incompetent, or if the level of consideration the leader shows toward members is low. (2) Without endorsement, members may attempt to oust an established leader by forming a revolutionary coalition. Revolutionary coalitions are not always successful, because they may be thwarted by conservative coalitions or by a leader's use of cooptation. (3) Leaders who have authority in groups perform specific roles, including planning, organizing, and controlling the group's activities. Transactional leadership entails an exchange between a leader and group members, whereas transformational leadership fosters high levels of group productivity by conveying an extraordinary sense of mission to group members and arousing new ways of thinking. (4) The contingency model of leadership effectiveness maintains that group productivity is a function not only of leadership style (task orientation versus relationship orientation) but also of the situation in which a leader performs.

Group Productivity (1) The mere presence of others facilitates performance when the task is simple or practiced. It inhibits performance when the task is difficult or novel. (2) Group size interacts with task type to affect productivity. On disjunctive tasks, group productivity increases directly with group size; on conjunctive tasks, it decreases with group size. On additive tasks, the Ringelmann effect occurs—that is, total productivity increases directly with group size, but relative productivity per member decreases. (3) If a group establishes explicit, demanding objectives with respect to the group's performance and if the group's members are highly committed to those objectives, then the group will perform at a higher level than if it does not do these things.

Equity and Reward Distribution (1) Groups use various distributive justice principles when determining the level of rewards that members receive. These include equality, equity, and relative need. (2) A state of equity exists when members receive rewards in proportion to the contributions they make to the group. (3) Reward structure also affects productivity. Under low task interdependence, differential rewarding has little effect on group productivity. Under high task interdependence, greater differentials in rewards reduce group productivity. (4) If inequity occurs within a group, members typically react with distress (anger for underreward or guilt for overreward) and initiate efforts to restore equity.

Brainstorming (1) Despite some expectations to the contrary, brainstorming groups generate fewer novel ideas than the same number of individuals working alone. Moreover, the quality and the originality of the ideas produced by brainstorming groups is, on average, no higher than that of ideas produced by individuals working alone. (2) The main reason for the unfavorable level of productivity in brainstorming groups is the phenomenon of production blocking, in which participants are unable to express their ideas due to turn taking among group members.

Group Decision Making Although many decisions made by groups are good ones, the process of group decision making entails potential hazards that can lead to poor or inferior choices. (1) One factor affecting decisions is groupthink—a mode of thinking that occurs when pressures for unanimity overwhelm members' motivation to realistically appraise alternative actions. Groupthink is especially likely to occur in groups that are highly cohesive and have leaders who strongly promote their own opinion. Groupthink can be prevented or reduced if group leaders not only strive to obtain all information relevant to a decision but also encourage an atmosphere of impartial exploration of alternatives. (2) Group members in decision making groups often shift toward a more extreme position following a group discussion; this phenomenon is termed group polarization. Underlying group polarization are several distinct processes, including social comparison and persuasive argumentation.

LIST OF KEY TERMS AND CONCEPTS

authority (p. 380)
brainstorming (p. 398)
cautious shift (p. 405)
conjunctive task (p. 390)
conservative coalition (p. 382)
contingency model of leadership effectiveness (p. 384)
co-optation (p. 382)
disjunctive task (p. 390)
distributive justice principle (p. 393)
endorsement (p. 381)
equity (p. 394)
group goal effect (p. 392)
group polarization (p. 405)
group productivity (p. 379)
groupthink (p. 400)
leadership (p. 383)
production blocking (p. 399)
revolutionary coalition (p. 382)
risky shift (p. 405)
social facilitation (p. 387)
social loafing (p. 390)
transactional leadership (p. 383)
transformational leadership (p. 383)

16

INTERGROUP CONFLICT

Introduction

Intergroup Conflict

Development of Intergroup Conflict

Realistic Group Conflict

Social Identity

Aversive Events and Escalation

Persistence of Intergroup Conflict

Biased Perception of the Out-Group

Changes in Relations Between Conflicting Groups

Impact of Conflict on Within-Group Processes

Group Cohesion

Leadership Militancy

Norms and Conformity

Resolution of Intergroup Conflict

Superordinate Goals

Intergroup Contact

Mediation and Third-Party Intervention

Unilateral Conciliatory Initiatives

Summary

List of Key Terms and Concepts

INTRODUCTION

Kanawha County in West Virginia includes the state capital of Charleston and several smaller communities. Some years ago, the Kanawha County School Board was considering whether to adopt a new set of 325 progressive language arts textbooks or instead to stick with traditional textbooks of the type already in use. The conflict began in the spring, instigated by the wife of a fundamentalist minister who was a member of the school board. She opposed adopting the new books and argued that they were excessively liberal in viewpoint. At the April 11 meeting of the school board, she objected to the method of textbook selection. At the June 2 meeting, she again objected to the new books and argued that the texts did not support a traditional, fundamentalist conception of God, the Bible, and religion. On June 23, she spoke in opposition to the books to the congregation of a local Baptist church. These events were covered by the local news media.

Other people, concerned about the larger issue of what curriculum should be taught in the schools, joined the board member's crusade. On June 27, more than a thousand textbook protestors appeared at the regularly scheduled school board meeting. Nevertheless, after hours of testimony, the board voted formally to adopt the disputed books.

During July and August, the fundamentalist protestors organized their ranks and developed strategies of action. Several distinct groups organized themselves, including the Concerned Citizens of Kanawha County (a large coalition of church congregations), the Businessmen and Professional People's Alliance for Better Textbooks, and the Christian American Parents. These groups held marches and rallies, circulated petitions, appealed to elected officials, and planned a boycott of the school system for September.

Supporters of the new textbooks also got organized through the Kanawha County School Board and a group called Citizens for Quality Education. Further support for the new textbooks came from liberal organizations, such as the American Civil Liberties Union and the National Association for the Advancement of Colored People, and from teachers and school administrators.

The Kanawha County textbook controversy began with disagreement over the selection of textbooks. It escalated into a classroom boycott by students and parents, destructive attacks on the Board's building, and even the creation of alternative schools.

On September 3, the new school term started for some 45,000 students. Protesting parents withheld their children (10,000 by some estimates) from the schools and prevented school and city buses from operating. On September 12, the school board closed the schools for a 3-day cooling-off period. The board also removed the controversial textbooks from classrooms pending review by a special citizens' committee.

During this period, violence broke out. Police received reports of random gunfire and sniping near the schools. Vandalism of school property was commonplace. The fundamentalist protestors demanded that members of the school board and the superintendent resign and that the liberal textbooks be banned permanently. On October 28, the citizens' review committee recommended that the school board restore all but 35 of the 325 controversial books to the classroom.

Some of the protestors, extremely displeased by this recommendation, reacted with force and dynamited the county board building on the night of October 30. The explosion partially destroyed the building.

On November 9, the school board voted 4-1 to reintroduce most of the textbooks into the classrooms. In response, the protestors filed a formal complaint with civil authorities, and the police arrested four board members and the school superintendent on November 15 for "contributing to the delinquency of minors." During the remainder of the school term, many parents continued to withhold their children from the public schools. Some enrolled their children in newly created, private Christian schools that stressed a fundamentalist curriculum.

By the end of the year, the protestors appeared to have won several concessions from the school board. The board issued new guidelines for the selection of textbooks, and it planned to open some "alternative" elementary schools with a more traditional curriculum the following semester. Although the vehemence of the protest subsided, the anger of the protestors lingered on (Page & Clelland, 1978).

Intergroup Conflict

The Kanawha County textbook controversy is an instance of **intergroup conflict**—a situation in which groups take antagonistic actions toward each other to influence an outcome each considers important. The major participants in the textbook controversy were organized groups—including the Concerned Citizens of Kanawha County, the Businessmen and Professional People's Alliance for Better Textbooks, and the Christian American Parents on one side, and the Kanawha County School Board and the Citizens for Quality Education on the other.

The textbook controversy illustrates how easily intergroup conflict can expand and escalate. The issues at stake often become larger and more apparent as the conflict progresses. Members of opposing groups develop antagonistic attitudes toward each other; distrust and hostility grow. Actions by participants become more damaging and destructive. Groups solidify their commitment to various positions, and the conflict becomes increasingly difficult to resolve.

Conflict between organized groups typically involves competing values, beliefs, and norms. For instance, a protesting mother who blocks a school bus may, in the eyes of the school board, be performing an unlawful and deviant act. But she will not view the act as deviant, because she is conforming to a different set of norms—the fundamentalist norms of the antitextbook coalition. In intergroup conflict, behavior considered appropriate by members of one group is often considered unacceptable by members of another. Thus, the conflict is rooted not merely in individual behavior but in the different values and belief systems.

The term intergroup conflict is often used in two distinct ways. First, we use it when referring to conflict between organized groups—each group consisting of members who interact with one another, who have well-defined role relationships, and who have interdependent goals. Second, we also use intergroup conflict to refer to what might be better described as conflict between persons belonging to different social categories. Although not necessarily members of organized groups, these people perceive themselves as members of the same social category and are involved emotionally in this common definition of themselves. For instance, conflict between members of ethnic or racial categories (such as neighborhood conflict between Blacks and Hispanics in Miami) is usually considered intergroup conflict, even though the individuals involved may not belong to organized groups. Throughout this chapter, we will use the term intergroup conflict to refer both types of conflict.

Intergroup conflict is widespread, and many instances are reported every day on TV and in the newspapers. Street fights between gangs, strife between different religious groups such as the Catholics and Protestants in Northern Ireland, confrontations between the Ku Klux Klan and the Black community, strikes by organized labor against management, economic competition and rivalry among ethnic groups, and long-standing family feuds—all of these are instances of intergroup conflict.

In this chapter, we will address the following questions:

1. What causes intergroup conflict to emerge and to escalate?

2. What factors sustain intergroup conflict? When conflict persists over a long time, what mechanisms support its persistence?
3. What effect does intergroup conflict have on relationships among the members of each group? How does conflict with another group change the structure of a group and the way its members relate to each other?
4. How can intergroup conflict be de-escalated or resolved?

DEVELOPMENT OF INTERGROUP CONFLICT

There are several basic origins of intergroup conflict. Overt conflict can develop because groups have an underlying opposition of interests. When this opposition prevents them from achieving their goals simultaneously, it can lead to antagonism and friction, and eventually to open conflict. Likewise, conflict can develop because members of one group view themselves as different in important ways from members of another group and act in a discriminatory way toward the other group. Or conflict may occur because one group suddenly threatens or deprives another group and thereby provokes an aggressive reaction. These factors are not mutually exclusive; in fact, they often work together to cause open conflict between groups (Taylor & Moghaddam, 1987, chapter 10). In this section, we consider each of these factors in intergroup conflict.

Realistic Group Conflict

Years ago, Muzafer Sherif and his colleagues conducted an important study of intergroup conflict at Robbers Cave State Park in Oklahoma (Sherif, 1966; Sherif, Harvey, White, Hood, & Sherif, 1961; Sherif & Sherif, 1982). The participants in this experiment were well-adjusted, academically successful, White, middle-class American boys, ages 11 and 12. These boys attended a 2-week experimental summer camp and participated in camp activities, unaware that their behavior was under systematic observation. The research objective was to investigate how an underlying opposition of interest can lead to overt intergroup conflict. Therefore, the boys were divided into two groups, named the Eagles and the Rattlers.

The experiment progressed in several stages. The first stage, which lasted about a week, was designed to produce cohesion within each of the groups. The boys arrived at the camp on two separate buses and settled into cabins located a considerable distance apart. By design, contact within each group was high, but contact between the two groups was minimal.

The boys within each group engaged in various activities, many of which required cooperative effort for achievement. They camped out, cooked, worked on improving swimming holes, transported canoes over rough terrain to the water, and played various games. As they worked together, the boys in each group pooled their efforts, organized duties, and divided tasks of work and play. Eventually, the boys identified more and more with their own groups, and each unit developed a high degree of group cohesion and solidarity.

Next, the experimenters began the second stage, in which conflict was induced between the groups. Specifically, the camp staff arranged a tournament of games, including baseball, touch football, tug-of-war, and a treasure hunt. In this tournament, prizes were awarded only to the victorious group. Thus, one group could only attain its goal at the expense of the other.

The tournament started in the spirit of good sportsmanship, but as it progressed, the positive feelings faded. The good sportsmanship cheer that customarily follows a game, "2-4-6-8, who do we appreciate," turned into "2-4-6-8, who do we appreci-HATE." Intergroup hostility intensified, and members of each group began to refer to their rivals as "sneaks" and "cheats." After suffering a stinging defeat in one game, the Eagles burned a banner left behind by the Rattlers. When the Rattlers discovered this "desecration," they confronted the Eagles, and a fistfight nearly broke out. The next morning, the Rattlers seized the Eagles' flag. Name-calling, threats, physical scuffling, and cabin raids by the opposing groups became increasingly frequent. When asked by the experimenters to rate each other's characters, a large proportion of the boys in each group gave negative ratings to all the boys in the other group. When the tournament was finally over, the two groups refused to have anything to do with each other.

In later stages of the study, when the level of intergroup antagonism was high, the experimenters tried various strategies for reducing strife. Several of these techniques failed, but the experimenters did succeed in reducing conflict by introducing important goals that required cooperation between groups for attainment (Sherif et al., 1961).

This study is a classic illustration of **realistic group conflict theory,** a well-established theory that provides one explanation for the development of intergroup conflict. The basic propositions of realistic group conflict theory are (1) when groups are pursuing objectives in which a gain by one group necessarily results in a loss by the other, they have what is called an opposition of interest; (2) this opposition of interest causes members of each group to experience frustration and to develop antagonistic attitudes toward the other group; (3) as members of one group develop negative attitudes and unfavorable perceptions about members of the other group, they become more strongly identified with and attached to their own group; (4) as solidarity and cohesion within each group increases, the likelihood of overt conflict between groups increases, and even a very slight provocation can trigger direct action by one group against another.

The pattern of conflict in the relationship between the Eagles and the Rattlers is consistent with this theory. So is the conflict between fundamentalists and liberals in Kanawha County regarding the content of textbooks and the social values taught in public schools. Intergroup conflict stemming from an underlying opposition of interest is also apparent in the everyday struggle for economic survival, such as the competition between ethnic groups for access to jobs, housing, and schooling (Bobo, 1983, 1999, 2000; Olzak, 1992).

Social Identity

Beyond opposition of interests, another factor in intergroup conflict is how strongly members identify with their own group. When interests are opposed, strong group identification can intensify conflict between groups. But even when an underlying opposition of interest is not present, strong group identification can, by itself, produce biased behavior toward out-groups.

In-Group Identification and Ethnocentrism Many years ago, Sumner (1906) observed that people have a fundamental tendency to like their own group (the in-group) and to dislike competing or opposing groups (the out-groups). He hypothesized that members who strongly identify with the in-group are especially prone to hold positive attitudes toward the in-group and to hold negative attitudes toward out-groups. Sumner's term for this phenomenon was **ethnocentrism**—the tendency to regard one's own group as the center of everything and as superior to out-groups. Ethnocentrism involves a pervasive and rigid distinction between the in-group and one or more out-groups. It entails stereotyped positive imagery and favorable attitudes regarding the in-group, combined with stereotyped negative imagery and hostile attitudes regarding the out-groups.

Table 16.1 presents a summary of the in-group and out-group orientations in ethnocentrism. These include seeing the in-group as superior and the out-group as inferior, viewing the in-group as strong and the out-group as weak, and construing the in-group as honest and peaceful and the out-group as treacherous and hostile (LeVine & Campbell, 1972; Wilder, 1981). Although ethnocentrism often plays a part in intergroup conflict, not all facets of ethnocentrism appear in every conflict. In some instances, only some of the orientations listed in Table 16.1 occur (Brewer, 1986; Brewer & Campbell, 1976).

The complexity of an individual's group identity also affects relations with an out-group. An individual does not belong to just one in-group, and in fact, usually has multiple group memberships (based on nationality, religion, recreational pursuits, occupations, and so forth). When individuals see their various in-groups as having a great deal of overlap, they are said to have a low identity complexity. Those who see their in-groups as relatively distinct from one another have high identity complexity. Research shows that those with low identity complexity have lower levels of tolerance for out-groups and diversity (Brewer & Pierce,

TABLE 16.1 ETHNOCENTRIC ORIENTATIONS TOWARD THE IN-GROUP AND THE OUT-GROUP

Member Orientations Toward the In-Group	Member Orientations Toward the Out-Group
See themselves as virtuous and superior	See the out-group as contemptible, immoral, and inferior
See their own standards of value as universal and intrinsically true	Reject out-group values
See themselves as strong	See the out-group as weak
Maintain cooperative relations with other in-group members	Refuse to cooperate with the out-group
Obey authorities within the in-group	Disobey authorities in the out-group
Demonstrate a willingness to retain membership in the in-group	Reject membership in the out-group
Trust in-group members	Distrust and fear out-group members
Show positive affect toward other in-group members	Show negative affect and hate toward out-group members
Take credit for in-group successes	Blame the out-group for in-group troubles

Source: Adapted with modifications from LeVine and Campbell, 1972.

2005), which can intensify ethnocentrism and intergroup conflict.

Ethnocentric attitudes not only cause in-group members to devalue and demean out-group members, they also lead to discrimination. The term **discrimination** refers to overt acts that treat members of certain out-groups in an unfair or disadvantageous manner. In circumstances entailing competition or direct opposition of interest between groups, ethnocentric attitudes will often produce discriminatory responses toward the out-group. More striking, even when an underlying opposition of interest is not present, the mere categorization of persons as belonging to an out-group can lead to discriminatory responses by in-group members.

Discrimination in the Minimal Intergroup Situation

The simple process of social categorization—placing people into arbitrarily defined groups that have no important meaning—is sufficient to produce intergroup discrimination. This effect was demonstrated by studies using an experimental paradigm called the minimal intergroup situation (Tajfel, 1982b; Tajfel & Billig, 1974). An example of this kind of experiment was conducted with English schoolboys, ages 14 through 16, who thought the experiment was a study of visual perception (Tajfel, Billig, Bundy, & Flament, 1971). In the first stage of the experiment, eight boys gathered in a lecture hall and were shown slides of clusters of dots. The exposure time for each slide was brief (½ second or less), and the boys wrote down their estimates of the number of dots in each cluster. While these estimates were being scored for accuracy, the experimenter told the participants that on this task some people consistently overestimate the number of dots (overestimators), whereas others consistently underestimate the number (underestimators). After completing the estimation task, the boys were placed in separate rooms. Each boy was assigned a code number to keep his personal identity unknown to others, and then he was informed either that he was an overestimator or that he was an underestimator. In fact, unknown to the boys, this assignment to a category was done at random and had nothing to do with their actual judgments. The experimenter wanted simply to divide the participants into two categories on an arbitrary basis.

In the second stage of the study, each boy performed a task that involved allocating points to others. In this task, each boy had to allocate points between pairs of other boys participating in the study. They received no specific directions on how to award these points, but they were told that at the end of the experiment, each of them would receive money based on the number of points allotted to them by the other participants. In this task, each boy was given information about the code number and the category membership (overestimator or underestimator) of the others to whom he could allocate points. Of course, this information was not especially revealing or helpful, because no boy knew who had which code numbers and, therefore, did not know the identities of those to whom he was allocating points.

Given this situation, the fairest allocation might be simply to give an equal number of points to each of the others. Yet the results show that the allocations made were not equal—the boys heavily favored their own in-group. That is, boys who thought of themselves as overestimators gave relatively more money to other overestimators, whereas boys who thought of themselves as underestimators gave more money to other underestimators. This effect was widespread—approximately 70 percent of the participants showed a bias favoring their own group.

The outcomes of the experiment were particularly remarkable because of the conditions under which point allocation occurred. First, the allocation exercise had no value for the boys themselves because they were distributing money to other people, not to themselves. Second, personal characteristics or affinities were not involved because the participants did not know the identities of the other boys in their own group. Third, the distinction between the groups (overestimator vs. underestimator) had no great importance or special meaning. Fourth, there was no communication or interaction—either within a group or between groups. Finally, there was neither an opposition of interest nor any previous hostility between the groups.

Other studies in this basic paradigm have extended these results. Some studies have used complete strangers as participants and formed groups using the most trivial criteria imaginable. In one instance, participants were openly assigned to categories at random based on a coin toss. The results still show the same pattern: Participants discriminate in favor of their own in-group and against the out-group. This bias is reflected both in attitudinal and evaluation measures and in the allocation of money and other rewards (Brewer, 1979; Brewer & Brown, 1998; Oakes & Turner, 1980; Tajfel, 1981).

Bias can also be increased simply by renaming a group in a less inclusive way. In one study, researchers surveyed residents of two nearby towns, one in Scotland and one in England. If the researchers referred to both towns as "British," stereotypes of the out-group and favoritism toward the in-group were markedly less than if they referred to one town as "Scottish" and the other as "English" (Hopkins & Moore, 2001).

Social Identity Theory Because the participants' discriminatory behavior in these studies had no direct utilitarian value, it is very hard to explain the results in terms of opposing interests between groups. Thus, realistic group conflict theory does not offer a convincing explanation for discrimination in these cases. A more satisfactory explanation comes from the **social identity theory of intergroup behavior,** developed by Tajfel and others (Tajfel, 1981, 1982a; Tajfel & Turner, 1986). This theory starts by assuming that individuals want to hold a positive self-concept. According to this view, the self-concept has two components—a personal identity and a social identity—and improving the evaluation of either of these can improve one's self-concept. The social identity component depends primarily on the groups or social categories to which one belongs, and the evaluation of one's own group is determined in part by a comparison with other groups. Thus, positive social identity depends on whether the comparisons made between one's in-group and some relevant out-groups are favorable or not.

The desire to maintain a positive self-concept, then, creates pressures to evaluate one's own group positively. Thus, in the minimal intergroup situation, when an individual is assigned to a group, he or she ends up thinking of that group (the in-group) as better than the other (the out-group) and as a result will have higher personal self-esteem (Aberson, Healy, & Romero, 2000; Rubin & Hewstone, 1998). He or she will also engage in actions to support this idea, such as allocating money to members of the in-group.

For groups in natural settings, Tajfel and others suggest that individuals will respond to a negative or unsatisfactory social identity in any of several ways. First, individuals might react by leaving an existing group and joining some group that is more positively evaluated. Second, individuals might try to hide their identity as members of a particular group (gay men and lesbians passing as heterosexual, Blacks passing as White in South Africa, and so on). Third, individuals might engage in social protest and collective

© David McNew/Getty Images

A police barricade defines the point of confrontation between gay-pride marchers and Christian fundamentalist counterprotestors. Strongly held social identities elicit group stereotypes, intensifying the underlying conflict.

action to elevate the status or welfare of the group to which they belong. This alternative, of course, may provoke overt conflict with other groups.

Aversive Events and Escalation

Sometimes members of one group hold antagonistic attitudes toward members of another group but refrain from engaging in open conflict. By staying away from one another as much as possible, they can minimize confrontation. Still, if underlying antagonism is present, a single provocative incident may be enough to set off open conflict. For example, in Long Island, New York, an unexpected defeat in a crucial high school basketball game led to an argument between some fans of the competing teams. This argument quickly escalated into a serious brawl involving most of the people attending the game. The fight started in the gym but then spread into the streets. These fans had different racial identities—those supporting the losing home team were largely Black, whereas those supporting the visitors were mostly White. During the fight, people were beaten and several school buses were overturned. Squads of police arrived and eventually brought the brawl to an end. But the conflict among groups of teenagers from the two communities continued for weeks, punctuated with outbreaks of street violence.

As this example shows, a single aversive event can provoke open hostilities between groups (Berkowitz, 1972; Konecni, 1979). An **aversive event** is a behavioral episode caused by or attributed to an out-group that entails undesirable outcomes for members of an in-group. The unexpected loss of the basketball game was an aversive event for fans of the home team, and it triggered conflict extending far beyond the hardwood court. Although aversive events can assume many forms, they always involve outcomes that people would prefer to avoid, and they include such things as being physically or verbally attacked, being slighted or humiliated, or being subjected to a loss of income or property.

As another illustration of how aversive events can trigger overt hostility, consider again the Kanawha County textbook controversy. In Kanawha County, a long-standing but submerged difference existed between fundamentalist church groups and the more liberal school board. But it took an aversive event—the decision to adopt new, liberal textbooks—to galvanize the fundamentalists into action. Once the protests began, groups supporting the different ideologies and interests mobilized quickly, and the conflict escalated.

The idea that aversive events trigger overt intergroup conflict is based on the general frustration-aggression hypothesis (see Chapter 12). This hypothesis holds that frustration leads to annoyance or anger, which can quickly turn into aggression if situational conditions are conducive (Berkowitz, 1989; Gustafson, 1989). The hypothesis is pertinent not only at the individual level, but also at the group level: If provoked by an aversive event seen to be caused by an out-group, an in-group will mobilize and attack the out-group. This response is most likely to happen when an underlying opposition of interest exists between groups, when easily identifiable characteristics (such as language, religion, or skin color of members) serve as a basis for differentiation between groups, and when members of one group already hold antagonistic attitudes and negative stereotypes regarding the other.

PERSISTENCE OF INTERGROUP CONFLICT

Probably the most famous family feud in the history of the United States was the long-enduring conflict between the Hatfields and the McCoys. In the early days, these mountain people lived peacefully on opposite sides of a narrow river: the Hatfields in West Virginia and the McCoys in Kentucky. The feud began one day in 1873 when Floyd Hatfield drove a razorback sow and her piglets into his pigsty on the McCoy side of the river. The pigs settled in comfortably, but trouble broke out a few days later when Randolph McCoy, Floyd's brother-in-law, came up to the pigsty and accused Floyd of stealing the pigs. A furious argument broke out, and the dispute eventually wound up in a backwoods court. During the trial, witness upon witness went to the stand to testify regarding the ownership of the pigs. All witnesses named Hatfield swore that Floyd owned the pigs, whereas all those named McCoy pointed to Randolph as the rightful owner. When the trial ended, Floyd Hatfield retained possession of the animals.

This was only the beginning, however. The McCoys were furious regarding the trial's outcome. And their familial affection toward the Hatfields quickly turned to hatred. The feud grew and deepened in intensity. Time passed, and the pigs were forgotten, but the entire McCoy family joined the fight against the Hatfield clan. Several months later, a McCoy murdered Ellison Hatfield; in retaliation, the Hatfields shot three McCoy boys. Hatred intensified, and the vicious conflict escalated. Eventually, the civil authorities of West Virginia and Kentucky became involved as they attempted to maintain order and protect human rights. According to one estimate, more than 100 people lost their lives during the long feud. It was not until 1928, when Tennis Hatfield and Uncle Jim McCoy shook hands in public, that the conflict ended (Jones, 1948).

The Hatfield-McCoy feud is an interesting and puzzling event. It began as a simple disagreement over the ownership of some pigs and escalated into a conflict that lasted 55 years and cost many lives. The feud illustrates a fundamental point about intergroup conflict, namely, that processes internal to a conflict can cause it to persist over time. Even without outside intervention or provocation, intergroup conflicts are often self-sustaining and feed on themselves.

What are some processes that cause intergroup conflict to persist? It is possible to identify several, but perhaps most important are the biased perception of

This clinic has been the scene of demonstrations both for and against abortion. Demonstrators on both sides are likely to hold strong negative stereotypes and attitudes regarding members of the other group.

out-group members and the changes in the structure of the relationship between adversaries.

Biased Perception of the Out-Group

In intergroup conflict, it is not uncommon for members of an in-group to harbor unrealistic impressions regarding out-group members. When in-group members hold mistaken perceptions of the out-group, disputes become increasingly difficult to resolve. Mistaken impressions arise from certain biases inherent in group perception, including the illusion that the out-group is homogeneous, an excessive reliance on stereotypes, errors in causal attribution, and incorrect evaluation of in-group performance relative to that of the out-group.

Out-Group Homogeneity There is a tendency for in-group members to overestimate the degree of similarity or homogeneity among out-group members (Linville, Fischer, & Salovey, 1989; Quattrone, 1986). Research findings indicate that individuals usually perceive less variability among members of the out-group than among members of their own group (Mullen & Hu, 1989; Rothbart, Dawes, & Park, 1984). In other words, although in-group members perceive and appreciate the diversity within the in-group, they tend to perceive the out-group members as "all alike." This is referred to as the **illusion of out-group homogeneity.**

This perceptual bias is quite general and widespread. It has been observed in men's and women's perceptions of each other (Park & Rothbart, 1982), of students attending rival universities (Quattrone & Jones, 1980), of young and elderly persons (Brewer & Lui, 1984), and of people in different occupations (Brauer, 2001).

Quattrone (1986) has suggested that out-group members are perceived as more homogeneous than in-group members because the perceivers have limited contact with out-group members, whereas they have fuller contact with in-group members. This makes it less likely that the perceivers will have a chance to see or appreciate the extent to which out-group members differ from one another in important ways.

Group Stereotypes and Images In-group members often make use of stereotypes (see Chapter 5) of the out-group. Although stereotypes do have certain virtues (for instance, they make it possible to process information more quickly), reliance on them can foster mistaken impressions of the out-group and its members. For one thing, stereotypes often exaggerate or accentuate the differences between an in-group and an out-group; they make the groups seem to differ from one another to a greater extent than they really do (Eiser, 1984). Moreover, many stereotypes are depreciatory, and they often ascribe negatively valued traits or characteristics to out-group members.

Individuals are also likely to overestimate similarities between themselves and their in-groups relative to themselves and their out-groups. Thus, the stereotypes created of out-groups tend to attribute characteristics to its members that are opposed to the individual's view of self and in-group. This in turn helps create unrealistic contrasts between in-groups and out-groups, exaggerating differences and promoting intergroup conflict (Riketta, 2005).

Images of the out-group can turn very negative when groups are perceived to have a large conflict of interests. One study of Israeli adults (primarily persons who were secular or traditional Jews) measured their impressions of an out-group (ultraorthodox Jews), whose growing movement into residential neighborhoods was viewed as a threat. The results indicated that the more threatening the respondents considered the out-group to be, the more likely they were to view the out-group as deficient in moral virtue—unfriendly, dishonest, and untrustworthy. Sentiments such as these produced support for overt aggression toward the out-group (Struch & Schwartz, 1989). Other investigators have reported a tendency during conflict for each side to perceive itself as relatively peaceful and cooperative and to perceive the adversary as aggressive and competitive (Bronfenbrenner, 1961).

Another important characteristic of stereotypes is that they tend to have low schematic complexity—that is, they are oversimplified and unrealistic. The in-group's stereotype of the out-group is usually less complex—sometimes far less complex—than the in-group's image of itself. This occurs in part because in-group members have less information about the out-group than they do about their own group (Linville & Jones, 1980). Due to this lower schematic complexity, in-group members are at risk of neglecting or misinterpreting new information about the activities of the out-group, especially if it is inconsistent with the stereotype. For instance, if a peaceful overture by the out-group is difficult to understand in light of the in-group's stereotype of the out-group, the in-group members may incorrectly reinterpret the action as a veiled threat and react with hostility.

Ultimate Attribution Error Several studies have revealed a perceptual bias that Pettigrew (1979) has called the **ultimate attribution error.** When a member of our own in-group behaves in a positive or desirable manner, we are likely to attribute that behavior to the member's internal, stable characteristics (such as positive personality dispositions). If that same person behaves in a negative or undesirable manner, we will tend to discount it and attribute it to external, unstable factors (she was operating under unusual stress or having a bad day). However, when perceiving a member of an out-group, we display the opposite bias. Positive behaviors by out-group members are attributed to unstable, external factors, such as situational pressures or luck. Negative behaviors are attributed to stable, internal factors, such as undesirable personal traits or dispositions. In other words, an in-group observer will blame the out-group for negative outcomes but will not give it credit for positive outcomes (Cooper & Fazio, 1986; Hewstone, 1990; Taylor & Jaggi, 1974). This attribution bias helps maintain favorable stereotypes of the in-group and unfavorable stereotypes of the out-group.

One study illustrating this attributional bias was based on the recurring intergroup violence in Northern Ireland (Hunter, Stringer, & Watson, 1991). This study examined how real instances of violence were interpreted by persons identifying with different groups in the conflict. Catholic and Protestant university students were shown newsreel footage depicting

BOX 16.1 Research Update: Morality and Attitudes Toward Out-Groups

Intergroup conflict is common in our world today. We need only to pick up a newspaper or watch a single episode of the evening news to see multiple examples, both recently developed and long-standing. But while intergroup conflict is easy to find, it is not ubiquitous. Many times, intergroup interaction is not conflictual, but cooperative. And, even when conflicts do exist, they vary substantially in their severity and tractability. Likewise, the in-group and out-group biases that contribute so much to mobilizing and maintaining conflict vary considerably across circumstances, groups, and individuals. We discuss several factors that can limit or prevent intergroup conflict in this chapter, but an additional factor, moral identity, is now being investigated by social psychologists interested in in-group/out-group dynamics.

Karl Aquino and Americus Reed propose that moral identity varies from person to person such that some have their self-concept highly organized around moral traits, while for others, these moral traits are less important. Some traits that we typically associate with morality include caring, compassionate, generous, honest, and so forth. If these traits are relatively important to a person's self-concept, they will react differently to out-groups and to potential intergroup conflict than will those whose moral self-identity is less central.

In one study, Reed and Aquino conducted four different examinations of moral identity and intergroup attitudes and behaviors. Each of these studies demonstrates that higher moral identity produced more acceptance, support, and tolerance toward out-groups. The first study showed that subjects who had high moral identity were more concerned about the welfare of out-groups and were more likely to exchange resources (both material exchanges, such as money, and symbolic exchanges like a smile or greeting) with out-groups and strangers.

In the second study, subjects were given information about a post-9/11 humanitarian charity to benefit those suffering in Afghanistan. Those with high moral identity showed more willingness to contribute to the charity and rated the cause as more worthy of support.

The third study gave American subjects three $1 bills and asked them to distribute them to two different charities—one benefiting Americans and one benefiting Afghanis. The three bills forced the subject to allocate the money unequally and once again, those with a high moral identity were more likely to favor the out-group than were those with a low moral identity.

Finally, the researchers tested whether moral identity affected individuals when they had the opportunity to either forgive an out-group that had caused harm or to exact revenge. Subjects were shown pictures of the World Trade Center bombing and told that a bombing raid to punish those responsible was about to occur, a bombing raid in which some innocent civilians would die. They were then asked to indicate how many such civilian deaths would be morally acceptable in this situation. They were then asked how morally acceptable each of the following would be: "(a) Use any means necessary to kill those responsible for these acts; (b) forgive those responsible, meaning negative emotions like hatred and anger should be replaced with positive emotions like compassion and love." As the researchers predicted, those with high moral identity reported smaller numbers of acceptable civilian deaths, lower acceptability of the "any means necessary" response, and higher acceptability of the forgiveness response.

For more on the notion of moral identity, see Aquino & Reed (2002).

Source: Adapted from Reed and Aquino, 2004.

scenes of in-group and out-group violence. One film showed a Protestant attack on mourners at a Catholic funeral, and the other showed a Catholic attack on a car containing two Protestant soldiers. Participants were asked to provide explanations for the attackers' actions in these films, and the investigators then classified their explanations into internal and external attributions. The results show that the participants ascribed negative behavior by their own in-group to external causes, whereas they ascribed negative behavior by the out-group to internal causes. In other words, both Catholics and Protestants were more likely to attribute out-group violence than in-group violence to internal causes (such as strongly held, hostile attitudes). Such attributions tend to maintain each side's negative view regarding the character and motives of the other side.

Biased Evaluation of Group Performance Another common bias is that in-group members rate the performance of their own group more favorably than that of the out-group, even when there is no objective basis for this difference (Hinkle & Schopler, 1986). This bias increases as the distinction between in-group and out-group becomes more salient (Brewer, 1979). One illustration of this bias appeared in the Robbers Cave study discussed earlier. When antagonism between the Eagles and the Rattlers was at its peak, the investigators arranged for the boys to participate in a bean-collecting contest. They scattered beans on the ground, and the boys collected as many as possible in 1 minute. Each boy stored his beans in a sack with a narrow opening, so he could not check the number of beans in it. Later, the experimenters projected a picture of the beans gathered by each boy on a screen in a large room. Boys from both groups tried to estimate the number of beans in each boy's collection. The projection time was very short and precluded counting. In reality, the experimenters projected the same number of beans (35) each time, although in different arrangements. The boys' estimates revealed a strong in-group bias; they overestimated the number of beans collected by members of their own group and underestimated the number collected by the out-group.

Bias in the evaluation of group performance can produce a variety of consequences. This bias can serve as a positive motivational device that strengthens the in-group's effort, boosts group morale, and helps members avoid complacency (Worchel, Lind, & Kaufman, 1975). On the other hand, overvaluation of an in-group's relative performance can lead to faulty decision making or groupthink (Janis, 1982). Overestimation of its own capacity relative to that of an adversary may cause the in-group to become overconfident and, hence, too willing to continue a fight that realistically should be abandoned or settled.

Changes in Relations Between Conflicting Groups

Once a conflict is under way, changes occur in the relationship between the conflicting groups. Often, the issues under dispute will expand in number, the relationship between groups will become more polarized, and intergroup communication may break down. These changes will, in turn, make it more difficult to resolve the conflict.

Expansion of the Issues When groups are disputing a particular matter, they often compound the conflict by introducing additional issues. This process of expansion usually moves from very specific issues to more general ones. The Kanawha County textbook controversy provides a good illustration. Initially, participants in this conflict were concerned with the issue of which textbooks should be adopted by the school system. This quickly expanded to the larger issue of what curriculum should be taught. Soon, the conflict broadened further to include the issue of who should serve on the school board and make decisions for the school system. Opposing sides also raised the larger umbrella issue of which broad viewpoint—liberal or traditional—should be taught in the public schools. The conflict then expanded yet further to include another issue: Should schools teaching the "wrong" viewpoint be permitted to continue operating without interference, or should they be shut down by force? Some extremists went even further and questioned if legitimacy should be withdrawn entirely from the public school system and invested instead in a new system of alternate schools teaching the fundamentalist viewpoint.

Expansion of issues occurs for several reasons. First, as relations between conflicting groups deteriorate, issues that members may have previously suppressed or ignored come to the fore. The conflicting parties will feel less inhibition about raising these issues, and they may even view the overt conflict as a good occasion to get even or settle accounts with the out-group (Ikle, 1971). Second, if one group takes aggressive steps against another, those actions become issues themselves. In the Kanawha County conflict, certain actions by the participants—blocking the passage of school buses, bringing charges leading to the arrest of school board members, and dynamiting the county board's building—were highly provocative. These actions became issues in themselves and pulled new participants into the fray.

Polarization and Communication Breakdown As a conflict expands, the relationship between adversaries typically becomes more polarized. Members' identifications with their own groups will intensify. Their attitudes will harden, growing more negative toward the out-group. Any positive bonds that may have existed between the members of different groups before the conflict will begin to weaken and disintegrate. Instead of having both positive and negative ties with one another, the groups will be left with only negative ones. As a result, they will become increasingly polarized and have fewer reservations about using coercive tactics.

Moreover, as the conflict expands, patterns of communication will change. Often there will be a communication breakdown—that is, a sharp decrease in communication between the conflicting groups (Dube-Simard, 1983; Giles & Coupland, 1991). In some cases, a communication breakdown takes the form of an actual decrease in the quantity or frequency of messages sent and received between conflicting groups. In other cases, it takes a different form, wherein groups continue to send and receive messages, but the impact of these messages is nil. Once a conflict is underway, stereotypes exert considerable influence over how an in-group interprets messages received from the out-group (Hewstone & Giles, 1986).

A communication breakdown will, of course, cause the in-group and the out-group to become even more isolated from each other, and this in turn makes it more difficult for participants to resolve or de-escalate the conflict. Members of the opposing groups will have fewer opportunities to communicate openly about issues. People on one side will lack accurate information about the plans and desires of the other side. Fresh proposals and new ideas will not be communicated. Even if one group sincerely wants to resolve or de-escalate the conflict, it may have a hard time signaling that intention. Efforts by one side to reduce the conflict may be interpreted by the other side as a trick or trap. Overall, polarization and the accompanying communication breakdown make the conflict more difficult to resolve.

IMPACT OF CONFLICT ON WITHIN-GROUP PROCESSES

Intergroup conflict also produces changes in the internal structure of the groups participating in the conflict. Once a struggle has begun, each group in the conflict undergoes changes that tend to promote escalation or make conflict resolution more difficult. Three main changes that can occur are increased group cohesion, increased militancy of group leaders, and an alteration of norms in the group.

Group Cohesion

Theorists have long recognized that overt conflict and external threats can produce changes in the internal structure of groups (Coser, 1967). One change demonstrated in a number of studies is that when a group engages in conflict against another group or is threatened by another group, it will become more cohesive (Dion, 1979; Ryen & Kahn, 1975; Worchel & Norvell, 1980). During conflict, a group's boundaries will become more firmly etched, and its members will generally show higher levels of loyalty and commitment to it. In the Robbers Cave study, for instance, as the conflict between the Eagles and the Rattlers escalated, various indicators of group cohesion—such as cooperativeness and friendship choice among group members—rose to high levels.

Of course, there are some limits to this effect. If a group is embroiled in a conflict in which it cannot possibly prevail, members may give up all hope. When this occurs, cohesion can decline, and some members may leave the group. But under conditions where success is still possible, in-group cohesion will usually increase when conflict develops with another group.

Why does intergroup conflict lead to higher cohesion? First, as the conflict escalates, a group's cause becomes more significant to its members, and they increase their commitment to it. Second, intergroup conflict frequently entails threats; if an out-group issues a threat, that action quickly identifies the out-group as an enemy. Having a common enemy heightens perceived similarity among in-group members and increases cohesion (Holmes & Grant, 1979). In one study (Samuels, 1970), researchers varied the degree of cooperation and competition within groups separately from the presence or absence of competition between groups. The results indicated that groups facing intergroup competition were more cohesive than those not facing intergroup competition, regardless of whether members within a given group related cooperatively or competitively with each other. Thus, the common antagonism of group members toward an opposing group overshadowed any friction among themselves.

What are the consequences of heightened in-group cohesion during intergroup conflict? As noted previously (see Chapter 14), if a group is cohesive, its members will desire to remain in it and resist leaving it. A highly cohesive group will, in general, maintain a firm hold over its members' time, energy, loyalty, and commitment. Conformity and cooperation tend to be greater in high-cohesion groups than in low-cohesion groups (Sakurai, 1975). For this reason, cohesive groups are capable of taking well-coordinated action in pursuit of their goals. In the context of intergroup conflict, high-cohesion groups are often more vigorous and contentious than low-cohesion groups.

Leadership Militancy

The activities of group leaders under conditions of intergroup conflict are somewhat different from those of leaders under conditions of peace. Under conflict, leaders have to direct the charge against the adversary. They plan the group's strategic moves, obtain resources needed for the conflict, coordinate members' actions, and serve as spokespersons in negotiations with the adversary. How well these activities are performed will have an important impact on a group's success or failure in intergroup conflict.

It is not uncommon for groups embroiled in heavy conflict to change leaders. If the campaign against an opposing group is not progressing well, rivals for leadership may emerge within the in-group. Frequently, these rivals will be angrier, more radical, and more militant than the existing leaders. This challenge from rivals will place the existing leaders under pressure. To defend against this threat, they may react by adopting a harder line and taking stronger action against the out-group. Although existing leaders do not always react to rivals in this manner, they are especially prone to do so when their own position within the group is insecure or precarious (Rabbie & Bekkers, 1978). In this manner, the competition for leadership within a group will increase group militancy and intensify the level of conflict with outside groups (Kriesberg, 1973).

The impact of competition for leadership is illustrated by a study of civil rights leaders in 15 U.S. cities in the mid-1960s (McWorter & Crain, 1967). At that time, civil rights organizations were trying to bring about societal changes favorable to Blacks and other minorities. The researchers interviewed civil rights leaders to determine the extent to which there was rivalry for leadership within civil rights groups. The results showed that organizations with higher levels of rivalry also were more militant. Militancy was apparent both in the extremity of attitudinal responses and in the frequency of civil rights demonstrations in the cities. Thus, a process internal to these civil rights groups—rivalry for leadership leading to greater militancy—was connected to intensified conflict between groups.

Norms and Conformity

Intergroup conflict not only increases group cohesion and leadership militancy, but it also changes group norms and goals. Once serious intergroup hostilities have begun, group members will grow concerned

Conflict with other groups increases group cohesion in many ways. What changes in this combat unit might result as they enter into armed confrontation with Iraqis?

with winning (or surviving) the conflict. Some behaviors and activities that the group considered valuable prior to the conflict may now seem useless or even detrimental to success in the conflict; if this happens, the group will reorder goal priorities and favor those behaviors that can help it win the conflict.

As part of this, the group may reassess the importance of various tasks and make corresponding changes in members' role definitions and task assignments. This can result in a redistribution of status and rewards among members that—if judged by preconflict standards—would not appear fair. The re-allocation of tasks may impose an unequal sharing of costs and hardships, and it may not reflect members' seniority or past contributions. Changes such as these can increase tensions among members within the group (Leventhal, 1979). But if the conflict is intense, concerns about group effectiveness and survival will overshadow concerns about equity and fairness.

Under severe conflict, the members will increase their demands on one another for conformity to group norms and standards. The group needs enhanced coordination and task performance to achieve success in the conflict. There will also be pressure to adopt the group's negative attitudes and stereotypes regarding the adversary—a form of "right thinking." The importance of loyalty to the in-group will increase, and members will increasingly expect one another to display a distrusting, competitive orientation toward the out-group. Those who do not will be trusted less by the group and may even be ostracized or ejected from the group.

These conformity pressures may well impinge on the rights and liberties of individual members. Yet the group will care less about these rights than it did before the conflict, and there will be less tolerance of dissent (Korten, 1962). If internal dissent does occur, the majority will likely react by suppressing it or by forcing the dissidents out of the group, especially if they suspect them of sympathizing with the adversary or engaging in behavior that jeopardizes the group's chance of victory.

RESOLUTION OF INTERGROUP CONFLICT

If a conflict expands beyond rational bounds, as occurred in the Kanawha County textbook controversy and the Hatfield-McCoy family feud, it can consume vast amounts of human energy, time, and other resources. Conflicts often begin as small disagreements and then grow in scope and intensity, pulling in new participants. Because intergroup conflicts are potentially dangerous and costly, many scholars have investigated means of stopping them in the early stages, before they spiral out of control.

This problem is surprisingly complex. One cannot, for example, resolve intergroup conflict merely by "reversing" the processes that initially caused it. It is often impossible to eliminate the underlying opposition of interest, to diminish the ethnocentric identification with the in-group, or to forestall aversive events. Nevertheless, investigators and practitioners have developed various techniques to reduce or resolve intergroup conflict. In this section, we discuss four of them.

Superordinate Goals

One of the most effective techniques for resolving intergroup conflict is to develop what are called superordinate goals. A **superordinate goal** is an objective held in common by all groups in a conflict that cannot be achieved by any one group without the supportive efforts of the others. Research findings indicate that once introduced, superordinate goals usually reduce in-group bias and intergroup conflict (Bettencourt, Brewer, Croak, & Miller, 1992; Gaertner et al., 1999; Sherif et al., 1961).

The Robbers Cave study provides a clear illustration of how superordinate goals can reduce conflict. When the conflict between the Eagles and the Rattlers was at its peak, the researchers introduced several goals that involved important shared needs. First, the researchers arranged for the system that supplied water to both groups to break down. To find the source of the problem and restore water to the camp, the two groups of boys had to work together. Next, the food delivery truck became stuck along the roadway. If the boys were to eat, they had to work together to free the heavy vehicle and push it up a steep grade. By inducing some cooperation between the groups, the superordinate goal structure also reduced hostility (Sherif et al., 1961).

The impact of superordinate goals on conflict reduction is not usually immediate, but gradual and cumulative. The results are stronger when several goals are introduced one after another, rather than a single goal. Because superordinate goals are cumulative in effect, they have greater impact when they are massed (Blake, Shepard, & Mouton, 1964; Sherif et al., 1961).

Why does this work? First, superordinate goals serve as a basis for restructuring the relationship between groups. Superordinate goals create cooperative interdependence between the in-group and the out-group. By changing a hostile win-lose situation into one of collaborative problem solving, with the possibility of a win-win outcome, a superordinate goal reduces friction between groups. The activities of out-group members will become valued by in-group members because members of one group are contributing to outcomes desired by the other.

Second, the introduction of a superordinate goal often increases interaction between in-group and out-group members. Increased contact by itself is generally not sufficient to ensure a reduction in intergroup bias or hostility. But if some of the interaction with the out-group members is personalized (rather than just task oriented), or if it provides information that reduces stereotyping, the superordinate goal will reduce bias and hostility (Bettencourt et al., 1992; Brewer & Miller, 1984; Worchel, 1986).

Third, the introduction of a superordinate goal can generate a new, superordinate social identity shared by all members. The superordinate goal reduces the sharp distinction between the in-group ("us") and the out-group ("them"), and a new common identity applying to all members of both groups is created. One theory of recategorization, termed the common in-group identity model, proposes that when persons belonging to separate social groups come to view themselves as members of a single social unit or category, their attitudes toward one another will become more positive (Dovidio, Gaertner, Isen, & Lowrance, 1995; Gaertner, Dovidio, Anastasio,

Bachman, & Rust, 1993; Gaertner, Mann, Murrell, & Dovidio, 1989). Former out-group members will increase in attractiveness, and the favoritism that in-group members originally afforded their own group will now be extended to the whole collective.

Intergroup Contact

Some theorists have proposed that an increase in contact and communication between members of opposing groups will reduce intergroup conflict. Increased contact should lessen stereotypes and reduce bias and consequently lessen antagonism between groups. This concept, called the **intergroup contact hypothesis,** has been investigated primarily with respect to different racial and ethnic groups (Amir, 1976; Cook, 1985; Stephan, 1987).

Support for this hypothesis is mixed. Research findings show that although intergroup contact reduces prejudice and conflict between groups in some cases, it does not do so in others (Brewer & Kramer, 1985; Herek & Capitanio, 1996; Hewstone & Brown, 1986; Pettigrew, 1997; Riordan, 1978). School desegregation, for instance, has increased contact between Black and White children, but this has not always produced positive changes in intergroup relations (Cook, 1984; Gerard, 1983). In some instances, increasing the level of intergroup contact can actually increase conflict (Brewer, 1986). Given this state of affairs, investigators have focused their attention on identifying the conditions under which intergroup contact leads to reduced bias and conflict, and the conditions under which it does not.

Sustained Close Contact Contact between members of different groups is more likely to bring about a reduction in prejudice and conflict if the contact is sustained and personal rather than brief and superficial (Amir, 1976; Brown & Turner, 1981). Some early experiments on interracial contact found that it required interaction over as much as a 20-day period between prejudiced Whites and a Black coworker to bring about change in the Whites' attitudes toward the coworker (Cook, 1985). Laboratory studies have also shown that contact involving repeated sessions with an out-group is more effective in decreasing intergroup bias than contact involving just a single session (Wilder & Thompson, 1980).

The extent of closeness or personalization of intergroup contact will also affect the extent to which attitudes are changed and stereotyping reduced. Low levels of intimacy will have little effect on intergroup prejudice and stereotyping (Segal, 1965).

There are several reasons why sustained close contact tends to reduce prejudice and stereotyping. First, cognitive dissonance may produce attitude change. If individuals with negative attitudes find themselves subject to situational pressures, and if they consequently engage in positive actions toward members of an out-group, then their behavior will be inconsistent with their attitudes, which may create a state of cognitive dissonance. The theory of cognitive dissonance predicts that these persons will end up changing their attitudes (becoming more positive toward the out-group) as a means of justifying their new behavior to themselves.

Second, during close contact, members of different groups may engage in self-disclosure. Higher levels of self-disclosure generally promote interpersonal liking, provided that the attributes revealed by one person are viewed positively—or at least not negatively—by the other (Collins & Miller, 1994). If members of different groups discover through self-disclosure that they hold certain fundamental values in common, liking may be enhanced.

Third, sustained close contact between members of different groups can serve to break down stereotypes. Of course, contact with a single representative or "token" member of an out-group is usually not sufficient to change group stereotypes, because that person can too easily be viewed as an exception who is not representative of the out-group (Weber & Crocker, 1983). But close contact with multiple members of an out-group sustained over time may provide enough contrary information to compel a change in old stereotypes.

Equal-Status Contact Intergroup contact is more likely to reduce prejudice when in-group and out-group members occupy positions of equal status than when they occupy positions of unequal status (Riordan, 1978; Robinson & Preston, 1976). One early demonstration of equal-status contact comes from a classic study conducted in the military during World War II (Mannheimer & Williams, 1949). At that time, the U.S. Army was still largely segregated by race; only a few companies were integrated. This study showed that White soldiers changed their attitudes toward Black soldiers after the two racial groups fought in combat as equals, side by side. When asked how they felt about their company including Black as well as White platoons, only 7 percent of the White soldiers from integrated units reacted negatively. In contrast, 62 percent of the soldiers in completely segregated White companies reacted negatively to the prospect of having Black platoons in their unit.

Equal-status contact has been effective in reducing prejudice in other situations as well. For instance, it has been a factor in reducing prejudice among Black and White children at interracial summer camps (Clore, Bray, Itkin, & Murphy, 1978), and in interracial housing situations (Hamilton & Bishop, 1976). See Box 16.2 for more on this point.

To see why equal-status contact is important, consider what happens when contact is not based on equal status (Cohen, 1984). When status is unequal, members of a higher status group may refuse to accept influence or to learn from a lower status group. They can justify this to themselves on the grounds that the lower status group has less skill or experience. With one side unwilling to accept influence, expectations of lesser competence will appear to be supported, and stereotypes will be all the more difficult to overcome. To have any impact, the lower status group will need repeatedly to demonstrate to the higher status group that it is as good as the other in relevant respects. For all these reasons, intergroup contact is more effective in reducing prejudice and conflict if members of the different groups enter a situation on an equal footing.

Institutionally Supported Contact Intergroup contact is more likely to reduce stereotyping and create favorable attitudes if it is backed by social norms that promote equality among groups (Adlerfer, 1982; Cohen, 1980; Williams, 1977). If the norms support openness, friendliness, and mutual respect, the contact has a greater chance of changing attitudes and reducing prejudice than if they do not.

Institutionally supported intergroup contacts—that is, contacts sanctioned by an outside authority or by established customs—are more likely to produce positive changes than unsupported contacts. Without institutional support, members of an in-group may be reluctant to interact with outsiders because they feel it is deviant or simply inappropriate. In the presence of institutional support, however, they may view contact between groups as appropriate, expected, and worthwhile. For instance, with respect to desegregation in elementary schools, there is evidence that students are more highly motivated and learn more in classes conducted by teachers (that is, authority figures) who support rather than oppose desegregation (Epstein, 1985).

In sum, intergroup contact tends to reduce conflict when it is anchored by institutional or authoritative support, when it is based on equal rather than unequal status, and when it is personal rather than superficial in character.

Mediation and Third-Party Intervention

Disputes of certain types—such as strife between organized labor and management or conflict between community groups—are sometimes most easily resolved through the intervention of third parties, such as mediators or arbitrators. A **mediator** is any third party who serves as a go-between and who helps groups in conflict to identify issues and to agree on some resolution. Typically, mediators are independent third parties. Mediators generally serve as advisors rather than as decision makers in the dispute. In contrast to a mediator, an arbitrator is a neutral third party who has the power to decide how a conflict will be resolved. An **arbitrator** listens to arguments from conflicting parties and then makes a decision that is binding on the conflicting groups. This differs from mediation, where a resolution may or may not be attained and, even if attained, may not be binding. Mediators or arbitrators are found in divorce and small-claims courts, labor-management negotiations, community and neighborhood disputes, and other situations including international disputes.

Under what conditions is mediation most effective in resolving conflicts? In research on mediation,

BOX 16.2 Equal-Status Contact and Interracial Attitudes

According to various theorists (Allport, 1954; Amir, 1969), contact between ethnic and racial groups is likely to reduce prejudice when members of different groups have equal status during the encounter. In U.S. society, however, equal-status contacts may be more the exception than the rule. Because Blacks historically have not enjoyed the same economic position as Whites, interracial contacts often involve status inequality instead of status equality. Nevertheless, any contacts that are structured in terms of equal status have the potential to change interracial attitudes. One demonstration of this occurred in an experimental interracial summer camp for children (Clore et al., 1978).

This camp was structured to foster equal-status contact among children, ages 8 to 12. Half of the children were White, and half were Black. Moreover, half of the counselors and administrative staff were White, and half were Black. Living assignments in the camp ensured that each unit was half White and half Black. The living situation provided an opportunity for intimate acquaintanceships rather than the casual associations typical of many integrated social settings. All children had equal privileges and duties around the camp. Moreover, counselors provided tasks—such as fire building and cooking—that required cooperative efforts among the children.

Approximately 200 children attended the camp during the summer. They attended in groups of 40, with each group staying for one week. Interracial attitudes were measured in several ways. First, researchers asked the children how they felt toward persons of the other race; the children responded in terms of evaluative scales, such as good-bad, clean-dirty, pleasant-unpleasant, valuable-worthless, and so on. Second, researchers assessed the extent of interracial liking and friendship by means of games that required the children to indicate their interpersonal choices. For instance, in the "name game," children designated the others they knew well by circling names on a card with a pencil. At the end of the week, children indicated the names of three other campers whose telephone numbers and addresses they wanted to have.

The results of this study showed that a change in attitude occurred for girls but not for boys. Boys began camp sessions with neutral attitudes toward persons of the other race and did not have much room for change. Girls began with negative attitudes and shifted in a positive direction as a function of the interracial contact. Similarly, there was a significant increase in cross-race interpersonal choices from the beginning of the camp to the end. This increase was more pronounced for girls than for boys. Overall, the results of this study suggest that prolonged, intimate, equal-status contact across races can lead to change in interracial attitudes.

one feature stands out: The greater the magnitude of the conflict and the worse the parties' relationship, the dimmer the prospects that mediation will prove successful (Carnevale & Pegnetter, 1985; Kressel & Pruitt, 1985). Conflicts that have endured a long time, that include a wide range of disputed issues, or that entail destructive or violent tactics will be difficult to mediate.

Mediation has a better chance of success when each of the conflicting groups trusts the mediator. Without trust, a mediator will find it difficult to obtain access to the groups and to conduct candid discussions with their members. Of course, in some conflicts, persons trusted by all sides may be difficult to find.

Mediators can engage in various actions that improve the chances of resolving the conflict. One of the most crucial actions is to facilitate communication between the conflicting groups (Donohue, Allen, &

A mediator tries to settle a running dispute between gangs in California. Efforts by mediators are more likely to be successful when the mediator is trusted by the conflicting groups.

Burrell, 1988; Thoennes & Pearson, 1985). Some mediators prefer to meet individually with each side to explore the issues in the dispute; later, they bring the conflicting groups together to discuss possible resolutions. Other mediators engage in "shuttle diplomacy" and carry messages between groups. In cases where the parties in a conflict might otherwise refuse to reply to messages from the other side, mediators can encourage a response.

Another action by mediators that aids conflict reduction is to diagnose the conflict carefully. Diagnosis enables the mediator to offer insight and clarification regarding the roots of the conflict; it also enables the mediator to separate the issues and to suggest trade-offs that facilitate resolution (Carnevale & Pegnetter, 1985; Thoennes & Pearson, 1985). In some cases, the mediator can help the parties design their own resolution to the conflict; this may better suit the participants' needs than an imposed settlement. In other cases, where the dispute involves as lot of tension and hostility, the mediator may need to exert some pressure on one party or another.

Although aggressive or assertive tactics may seem alien to good mediation, mediators frequently do pressure the parties to accept proposals on specific issues (Hiltrop, 1985; Kressel & Pruitt, 1985; Zartman & Touval, 1985). One tactic used by mediators is threatening to quit unless the conflicting parties begin to make concessions. Research on collective bargaining in Great Britain indicates that in the context of rancorous disputes, use of the threaten-to-quit tactic by a mediator increased the chances of a settlement being reached (Hiltrop, 1985). Mediators are most likely to use assertive tactics when they think there is little likelihood of reaching agreement without it and when they do not have a strong concern for the parties' own aspirations (Carnevale & Henry, 1989).

Even at its best, mediation is successful in resolving conflicts only some of the time. For instance, mediation of labor-management disputes produces successful settlements about 50 percent of the time (Hiltrop, 1985; Kressel & Pruitt, 1985; Lewin, Feuille, & Kochan, 1977, Chapter 5). Even though it is not completely successful in all cases, mediation is still an important element in the toolkit of conflict resolution techniques.

Unilateral Conciliatory Initiatives

The approaches discussed up to this point—superordinate goals, intergroup contact, and mediation by third parties—will often suffice to reduce bias or resolve conflict. But if a conflict has already escalated or reached a stalemate, the opposing sides may not be open to further intergroup contact or mediation. In such a case, any group wishing to reduce tensions can do little except take steps unilaterally. Resolving intergroup conflict by making unilateral conciliatory initiatives is a difficult undertaking; even if one group does make sincere conciliatory moves, the other may react with hostility or demand total capitulation. Still, the use of conciliatory initiatives can often be successful if done through a specific sequence of actions.

One approach to conflict reduction through unilateral initiatives is a strategy called **GRIT,** which stands for "Graduated and Reciprocated Initiatives in Tension Reduction" (Osgood, 1962, 1979, 1980). Originally formulated during the cold war between the United States and the Soviet Union, GRIT arose from a desire to reduce tension between the superpowers (Granberg, 1978). Despite its origins, the GRIT strategy is general in character; it can be applied not only to conflicts between nations but also to conflicts between smaller organizations and groups.

GRIT assumes that each side in a conflict has an interest in reducing tension and that each would like to reallocate resources from the conflict to better purposes. At base, GRIT is a kind of "tit-for-tat" strategy involving de-escalatory moves. One side initiates de-escalatory steps in the hope that the other side will reciprocate; if reciprocation occurs, the initiator then quickly responds with still further de-escalation, and so on. More specifically, the principles of the GRIT strategy are as follows:

1. To begin, the group using the strategy (that is, the in-group) issues a public statement that describes its plan to reduce intergroup tension through subsequent actions.
2. Prior to making any unilateral move, the in-group should publicly announce the move and label it as part of the overall strategy.
3. With each announcement of a unilateral move, the in-group should explicitly invite reciprocation in some form by the out-group.
4. To bolster credibility, each unilateral move should be carried out by the in-group in accordance with the announced timetable.
5. Initiatives by the in-group should be continued for some time, even without reciprocation by the out-group. This entails risk, but it also intensifies pressure on the out-group to reciprocate.
6. The initiatives by the in-group should be clear-cut, unambiguous, and open to verification by the out-group.
7. The in-group should not unilaterally disarm and should retain the capacity to retaliate if the out-group reacts with hostility to the in-group's initiatives. Ideally, the initiatives should be sufficiently risky that they are vulnerable to exploitation, but not so risky that they seriously impair the capacity of the in-group to retaliate against an attack from the out-group.
8. Further initiatives by the in-group should be graduated to match the responses by the out-group. If the out-group responds in a friendly manner, the in-group should react by taking further risk. If the out-group responds in a hostile manner, the in-group should react with retaliation but avoid escalation.
9. Unilateral moves by the in-group should be diversified in form, so that all they have in common is their conciliatory nature. This will demonstrate the in-group's underlying peaceful intentions, and it will also illustrate to the out-group that a variety of moves could be made in reciprocation.

It will also help to avoid developing a large gap in the in-group's defenses.

10. The out-group should be rewarded for cooperating, with the level of reward depending on the level of cooperation (Lindskold, 1986; Osgood, 1980).

The GRIT strategy tries not only to change the out-group's perceptions of the initiating group, but also to change its behavior through the principles of reinforcement. If the out-group reciprocates positively, it is rewarded with further conciliatory initiatives; but if the out-group responds with aggressive action, it is punished with retaliation.

How effective is the GRIT strategy in practice? At the level of international relations, GRIT has not been fully tried or tested. During the cold war, however, there were several events that provided partial tests of the GRIT principles. Notable among these was the unilateral initiative in June 1963 by U.S. President John Kennedy to halt all atmospheric nuclear testing. Kennedy began by unilaterally halting U.S. testing and followed this up with some further concessions on other issues, such as the status of the Hungarian delegation at the United Nations. These concessions were reciprocated by Premier Khrushchev of the Soviet Union. This led immediately to new activity regarding a test ban treaty, which had been stalled for a long period. Kennedy's GRIT-like strategy proved successful and, within just a few months, the players agreed on a treaty (Etzioni, 1967).

Further evidence regarding the effectiveness of GRIT comes from simulation studies of conflict and from experimental games. These studies show that many of the principles in the GRIT strategy do contribute to tension reduction. For instance, the act of announcing intentions in advance (principles 1 and 2) is more likely to elicit reciprocal cooperation from the out-group than not announcing them (Lindskold, Han, & Betz, 1986). Conciliatory moves without such an opening announcement may be incorrectly construed by the out-group as indicating weakness or passivity and therefore lead to exploitation (Oskamp, 1971). Repeated truthful announcements that one intends to cooperate on the next move produces greater reciprocation from the out-group than does the same rate of cooperation unaccompanied by announcements (Lindskold & Finch, 1981; Voissem & Sistrunk, 1971).

Some studies show that carrying out initiatives as announced (principle 4) increases the credibility of the initiator (Ayers, Nacci, & Tedeschi, 1973; Schlenker, Helm, & Tedeschi, 1973), whereas failure to carry out initiatives reduces credibility and diminishes reciprocation (Gahagan & Tedeschi, 1968). Furthermore, maintaining the capacity to retaliate (principle 7) is important, because if one side makes conciliatory moves but fails to maintain the capacity to retaliate, it will quickly weaken itself and create an imbalance of power. Cooperation drops off when a power imbalance exists (Aronoff & Tedeschi, 1968; Michener & Cohen, 1973). Studies regarding reciprocity (principle 8) show that GRIT with communication is superior to simple tit-for-tat responses (Han & Lindskold, 1983).

GRIT also has limitations. In particular, it is much less effective when the disputing groups have very unequal power. When groups differ in power, initiatives from the strong group are likely to be met with concessions by the weaker one, but initiatives from the weaker group are not likely to be met with concessions by the strong one (Lindskold & Aronoff, 1980; Michener et al., 1975). The GRIT strategy works best when the conciliatory initiatives come from a group having equal or superior power in the conflict.

SUMMARY

Intergroup conflict is a circumstance in which groups engage in antagonistic actions toward one another to control some outcome important to them.

Development of Intergroup Conflict Intergroup conflict has several origins. (1) Groups often have opposing interests that prevent them from achieving their goals simultaneously, leading to friction, hostility, and overt conflict. (2) A high level of in-group identification, accompanied by ethnocentric attitudes, may create discrimination between groups, which escalates conflict. (3) One group, by threatening or depriving another, may create an aversive event that turns submerged antagonism into overt conflict.

Persistence of Intergroup Conflict Although some conflicts between groups dissipate quickly, others last for a long time. Several mechanisms support the persistence of intergroup conflict. (1) Perception of the out-group by in-group members is often biased. This bias, caused by insufficient information regarding the out-group and excessive reliance on stereotypes, produces an incorrect understanding of the characteristics and intentions of out-group members and an overestimation of in-group capabilities. (2) Once a conflict escalates, changes occur in the relationship between the opposing sides. The number of issues under dispute will often expand, and the number of parties in the conflict may increase. Relations between adversaries will polarize, communication across groups will decline, and trust will become harder to establish.

Impact of Conflict on Within-Group Processes Intergroup conflict changes the internal structure of the in-group. (1) Conflict increases the level of cohesion of the in-group, as members increase their commitment and unite to face a common adversary. (2) Conflict may produce rivalry for leadership among in-group members, and this rivalry can produce more militant leadership. (3) Conflict often changes the normative structure of the in-group. Standards of fairness may shift as the in-group reorders its priorities to achieve a strategic advantage in the conflict. Conflict often increases the pressure on in-group members to conform and lessens the majority's tolerance of dissenters.

Resolution of Intergroup Conflict Several techniques can be used to reduce intergroup conflict. (1) One is to introduce superordinate goals into the conflict. Because goals of this type can be achieved only through the joint efforts of opposing sides, they promote cooperative behavior and serve as a basis for restructuring the relationship between groups. (2) Another technique is to increase intergroup contact. This approach is more effective in reducing bias and conflict when contact is sustained, close, based on equal status, and supported institutionally. (3) Another technique is to use third parties as mediators in the dispute. Mediators improve the communication between conflicting parties; they help analyze the dispute and develop possible resolutions. The use of mediators is most effective when the conflict is not extreme and when the mediator is trusted by all parties involved. (4) A further technique is to use unilateral conciliatory initiatives (such as the GRIT strategy) to reduce conflict. Under GRIT, one side takes a firm but conciliatory stance and initiates de-escalatory steps in the hope that the other side will reciprocate. The GRIT approach works best when the initiating group is at least as strong as its opponent.

LIST OF KEY TERMS AND CONCEPTS

arbitrator (p. 426)
aversive event (p. 415)
discrimination (p. 413)
ethnocentrism (p. 412)
GRIT (p. 429)
illusion of out-group homogeneity (p. 417)
intergroup conflict (p. 410)
intergroup contact hypothesis (p. 425)
mediator (p. 426)
realistic group conflict theory (p. 412)
social identity theory of intergroup behavior (p. 414)
superordinate goal (p. 424)
ultimate attribution error (p. 418)

17

LIFE COURSE AND GENDER ROLES

Introduction

Components of the Life Course

Careers

Identities and Self-Esteem

Stress and Satisfaction

Influences on Life Course Progression

Biological Aging

Social Age Grading

Historical Trends and Events

Stages in the Life Course: Age and Gender Roles

Stage I: Achieving Independence

Stage II: Balancing Family and Work Commitments

Stage III: Performing Adult Roles

Stage IV: Coping with Loss

Historical Variations

Women's Work: Gender Role Attitudes and Behavior

Impact of Events

Summary

List of Key Terms and Concepts

INTRODUCTION

"I still can't get over Liz," said Sally. "I sat next to her in almost every class for 3 years, and still, I hardly recognized her. Put on some weight since high school, and dyed her hair. But mostly it was the defeated look on her face. When she and Hank announced they were getting married, they were the happiest couple ever. But that lasted long enough for a baby. Then there were years of underpaid jobs. She works part-time in sporting goods at Sears now. Had to take that job when her real estate work collapsed in the recession."

Jim had stopped listening. How could he get excited about Sally's Lincoln High School reunion and people he'd never met? But Sally's mind kept racing. A lot had happened in 20 years.

John—Still larger than life. Football coach at the old school, and assistant principal too. Must be a fantastic model for the tough kids he works with. That scholarship to Indiana was the break he needed.

Frank—Hard to believe he's in a mental hospital! He started okay as an engineer. Severely burned in a helicopter crash, and then hooked on pain killers. Just fell apart. And we voted him Most Likely to Succeed.

Andrea—Thinking about a career in politics. She didn't start college until her last kid entered school. Now she's an urban planner in the mayor's office. Couldn't stop saying how she feels like a totally new person.

Tom—Head nurse at Westside Hospital's emergency ward. Quite a surprise. Last I heard, he was a car salesman. Started his nursing career at 28. Got the idea while lying in the hospital for a year after a car accident.

Julie—Right on that one, voting her Most Ambitious. Finished Yale Law, clerked for the New York Supreme Court, and just promoted to senior partner with Wine and Zysblat. Raised two kids at the same time. Having a husband who writes novels at home made life easier. Says she was lucky things were opening up for women just when she came along.

Linda—Too bad she quit journalism school to put her husband through med school. She was a great yearbook editor. Still, says she enjoys writing stories as a stringer for the *News*. Leaves time for family and travel.

Sally's reminiscences show how different lives can be—and how unpredictable. When we think about people like Liz or Frank or Tom, change seems to be the rule. There is change throughout life for all of us. But there is continuity too. Julie's string of accomplishments is based on her continuing ambition, hard work, and competence. John is back at Lincoln High—once a football hero, now the football coach. Though not a journalist as she planned, Linda writes occasionally for a paper, and she may yet develop a serious career in journalism. Even Frank had started on the predicted path to success before his tragic helicopter crash.

As we look into the future, we cannot project with any certainty what will happen to us. But people's lives have patterns that allow us to make sense of them. Each of us will experience a life characterized both by continuity and by change. This chapter examines the **life course**—the individual's progression through a series of age-linked social roles embedded in social institutions (Elder & O'Rand, 1995), and the important influences that shape the life course that one experiences.

Our examination of the life course is organized around four broad questions:

1. What are the major components of the life course?
2. What are the major influences on progression through the life course? That is, what causes people's careers to follow the paths they do?
3. What are the typical courses of life for men and women in American society? What happens when people depart from the typical patterns?
4. In what ways do historical trends and events modify the typical life course pattern?

COMPONENTS OF THE LIFE COURSE

Lives are too complex to study in all their aspects. Consequently, we will focus on the three main components of the life course: (1) careers, (2) identity and self-esteem, and (3) stress and satisfaction. By examining these components, we trace the continuities and changes that occur in what we do through the life course.

Careers

A **career** is a sequence of roles—each with its own set of activities—that a person enacts during his or her lifetime. Our most important careers are in three major social domains: family and friends, education, and work. The idea of a career comes from the work world, where it refers to the sequence of jobs held. Liz's work career, for example, consisted of a sequence of jobs as waitress, checkout clerk, clothing salesperson, real estate agent, and sporting-goods salesperson.

The careers of one person differ from those of another in three ways—in the roles that make up the careers, the order in which the roles are performed, and the timing and duration of role-related activities. For example, one woman's family career may consist of roles as infant, child, adolescent, spouse, parent, grandparent, and widow. Another woman's family career may include roles as stepsister and divorcee but exclude the parent role. A man's career might include the roles of infant, child, adolescent, partner, and uncle. The order of roles also may vary. "Parent" before "spouse" has very different consequences from "spouse" before "parent." Furthermore, the timing of career events is important. Having a first child at 36 has different life consequences than having a first child at 18. Finally, the duration of enacting a role may vary. For example, some couples end their marriages before the wedding champagne has gone flat, whereas others go on to celebrate their golden wedding anniversary.

Societies provide structured career paths that shape the options available to individuals. The cultural norms, social expectations, and laws that organize life in a society make various career options more or less attractive, accessible, and necessary. In the United States, for example, educational careers are socially structured so that virtually everyone attends kindergarten, elementary school, and at least a few years of high school. Thereafter, educational options are more diverse—night school, technical and vocational school, apprenticeship, community college, university, and so on. But individual choice among these options is also socially constrained. The norms and expectations of our families and peer groups strongly influence our educational careers.

A person's total life course consists of intertwined careers in the worlds of work, family, and education (Elder, 1975). The shape of the life course derives from the contents of these careers, from the way they mesh with each other, and from their interweaving with those of family members. Sally's classmates, Julie and Andrea, enacted similar career roles: Both finished college, held full-time jobs, married, and raised children. Yet the courses of their lives were very different. Julie juggled these roles simultaneously, helped by a husband who was able to work at home. Andrea waited until her children were attending school before continuing her education and then adding an occupational role. The different content, order, timing, and duration of intertwining careers make each person's life course unique.

Identities and Self-Esteem

As we engage in career roles, we observe our own performances and other people's reactions to us. Using these observations, we construct role identities—conceptions of the self in specific roles. The role identities available to us depend on the career paths we are following. When Liz's work in real estate collapsed, she got a job in sales at Sears; she was qualified to sell sporting goods because of her prior work experience.

As we enact major roles, especially familial and occupational roles, we evaluate our performances and thereby gain or lose self-esteem—one's sense of how good and worthy one is. Self-esteem is influenced by one's achievements; Julie has high self-esteem as a consequence of being a senior partner in a prestigious law firm. Self-esteem is also influenced by the feedback one receives from others.

Identities and self-esteem are crucial guides to behavior, as discussed in Chapter 4. We therefore consider identities and self-esteem as the second component of the life course.

Stress and Satisfaction

Performing career activities often produces positive feelings, such as satisfaction, and negative feelings, including stress. These feelings reflect how we experience the quality of our lives. Thus, stress and satisfaction are the third component of the life course.

Changes in career roles, such as having a baby, adopting a child, or changing jobs, place emotional and physical demands on the person. Life events, such as moving or serious conflict with a parent or lover, may have similar effects. At times, the demands made on a person exceed the individual's ability to cope with them; such a discrepancy is called **stress** (Dohrenwend, 1961). People who are under stress often experience psychological (anxiety, tension, depression) and physical (fatigue, headaches, illness) consequences (Wickrama, Lorenz, Conger, & Elder, 1997).

These feelings vary in their intensity in response to life course events. Levels of stress, for example, change as career roles become more or less demanding (parenting roles become increasingly demanding as children enter adolescence), as different careers compete with each other (family versus occupational demands), and as unanticipated setbacks occur (one's employer goes bankrupt). Levels of satisfaction vary as career rewards change (salary increases or cuts) and as we cope more or less successfully with career demands (meeting sales quotas, passing exams) or with life events (a heart attack or a skiing accident).

The extent to which particular events or transitions are stressful depends on several factors. First, the more extensive the changes associated with the event, the greater the stress. For example, a change in employment that requires a move to an unfamiliar city is more stressful than the same new job located across town. Second, the availability of social support—in the form of advice and emotional and material aid—increases our ability to cope successfully with change. To help their members, families reallocate their resources and reorganize their activities. Thus, parents lend money to young couples, and older adults provide care for their grandchildren so their children can work.

Personal resources and competence influence how one copes with stress. Coping successfully with earlier transitions prepares individuals for later transitions. Men who develop strong ego identities in young adulthood perceive events later in their lives as less negative (Sammon, Reznikoff, & Geisinger, 1985). Conversely, early experiences of failure reduce an individual's sense of competence and may leave people with a sense of helplessness when they face later events and transitions.

INFLUENCES ON LIFE COURSE PROGRESSION

At the beginning of this chapter, we noted many events that had an important impact on the lives of Sally's classmates: loss of a job due to economic recession, a helicopter crash, a car accident, having a baby, and graduating from a prestigious law school. These are **life events**—episodes that mark transition points in our lives and involve changes in roles. They provoke coping and readjustment (Hultsch & Plemons, 1979). For many young people, for example, the move from home to college is a life event marking a transition from adolescence to young adulthood. This move initiates a period during which students work out new behavior patterns and revise their self-expectations and priorities.

There are three major influences on the life course: (1) biological aging, (2) social age grading, and (3) historical trends and events. These influences act on us through specific life events (Brim & Ryff, 1980). Some life events are carefully planned—a trip to Europe, for example. Other events, no less important, occur by chance—like meeting one's future spouse in an Amsterdam hostel (Bandura, 1982a).

Biological Aging

Throughout the life cycle, we undergo biological changes in body size and structure, in the brain and central nervous system, in the endocrine system, in our susceptibility to various diseases, and in the acuity of our sight, hearing, taste, and so on. These changes are rapid and dramatic in childhood. Their pace slows considerably after adolescence, picking up again in old age. Even in the middle years, however, biological changes may have substantial effects. The shifting hormone levels associated with menstrual periods in women and with aging in men and women, for example, are thought by many to affect mood and behavior (Sommer, 2001).

Biological aging is inevitable and irreversible, but it is only loosely related to chronological age. Puberty may come at any time between 8 and 17, for example, and serious decline in the functioning of body organs may begin before age 40 or after age 85. The neurons of the brain die off steadily throughout life and do not regenerate. Yet intellectual functioning—long assumed to be determined early in life and to decline with aging—is now known to be capable of increase over the life course. Even in old age, mental abilities can improve with opportunities for learning and practice (Baltes & Willis, 1982).

Biologically based capacities and characteristics limit what we can do. Their impact on the life course depends, however, on the social significance we give them. How does the first appearance of gray hair affect careers, identities, and stress, for instance? For some, this biological event is a painful source of stress. It elicits dismay, sets off thoughts about mortality, and instigates desperate attempts to straighten out family relations and to make a mark in the world before it is too late. Others take gray hair as a sign to stop worrying about trying to look young, to start basing their priorities on their own values, and to demand respect for their experience. Similarly, the impact of other biological changes on the life course—such as the growth spurt during adolescence, or menopause in middle age—also depends on the social significance given them.

Social Age Grading

Which members of a society should raise children, and which should be cared for by others? Who should attend school, and who should work full-time? Who should be single, and who should marry? Age is the primary criterion that every known society uses to assign people to such activities and roles (Riley, 1987). Throughout life, individuals move through a sequence of age-graded social roles. Each role consists of a set of expected behaviors, opportunities, and constraints. Movement through these roles shapes the course of life.

Each society prescribes a customary sequence of age-graded activities and roles. In American society, many people expect a young person to finish school before he or she enters a long-term relationship. Many people expect a person to marry before she or he has or adopts a child. There are also expectations about the ages at which these role transitions should occur. These expectations vary by race and ethnicity; Hispanic adolescent girls expect to marry and have a first child at younger ages (22, 23) than Whites (23, 24) or Blacks (24, 24; East, 1998). These age norms serve as a basis for planning, as prods to action, and as brakes against moving too fast (Neugarten & Datan, 1973).

Pressure to make the expected transitions between roles at the appropriate times means that the life course consists of a series of normative life stages. A **normative life stage** is a discrete period in the life course during which individuals are expected to perform the set of activities associated with a distinct age-related role. The order of the stages is prescribed, and people try to shape their own lives to fit socially approved career paths. Moreover, people perceive deviations from expected career paths as undesirable.

Not everyone experiences major transitions in the socially approved progression. Consider the transition to adulthood; the normative order of events is leaving school, performing military service, getting a job, and getting married. Analyzing data about the high school class of 1972 collected between 1972 and 1980, researchers found that half of the men and women experienced a sequence that violated this "normal" path (Rindfuss, Swicegood, & Rosenfeld, 1987). Common violations included entering military service before one finished school and returning to school after a period of full-time employment.

In some cases, violating the age norms associated with a transition has lasting consequences. The transition to marriage is expected to occur between the ages

Violating the age norms associated with a major transition, such as the transition to parenthood, may have lasting consequences. Having a baby at the age of 16 may force a young woman to leave school and limit her to a succession of poorly paid jobs.

of 19 and 25. Research consistently finds that making this transition earlier than usual has long-term effects on marital as well as occupational careers. A survey of 63,000 adults allowed researchers to compare men who married as adolescents with men of similar age who married as adults (Teti, Lamb, & Elster, 1987). Because the sample included people of all ages, the researchers could study the careers of men who married 20, 30, and 40 years earlier. Men who married as adolescents completed fewer years of education, held lower status jobs, and earned less income. Furthermore, the marriages of those who married early were less stable. These effects were evident 40 years after marriage. Early marriage has similar effects on women. Women who marry before age 20 experience reduced educational and occupational attainment, and are more likely to get divorced (Teti & Lamb, 1989).

Movement from one life stage to another involves a **normative transition**—socially expected changes made by all or most members of a defined population (Cowan, 1991). Although most members undergo this institutional passage, each individual's experience of it may be different, reflecting his or her past experience. Normative transitions are often marked by a ceremony, such as a bar mitzvah, graduation, commitment ceremony, wedding, or baby shower. But the actual transition is a process that may occur over a period of weeks or months. This process involves both a restructuring of the person's cognitive and emotional makeup and of his or her social relationships.

Transitions from one life stage to another influence a person in three ways. First, they change the roles available for building identities. The transition to adulthood brings major changes in roles. Those who marry or have their first child begin to view themselves as spouses and parents, responsible for others. Second, transitions modify the privileges and responsibilities of persons. Age largely determines whether we can legally drive a car, be employed full-time, or serve in the military. Third, role transitions change the nature of socialization experiences. The content of socialization shifts from teaching basic values and motivations in childhood, to developing skills in adolescence, to transmitting role-related norms for behavior in adulthood (Lutfey & Mortimer, 2003). The power differences between socializee and socializing agents also diminish as we age and move into higher education and occupational organizations. As a result, adults are more able to resist socialization than children (Mortimer & Simmons, 1978).

Historical Trends and Events

Recall that Sally's classmate Julie attributed her rapid rise to senior law partner to lucky historical timing. Julie applied to Yale Law School shortly after the barriers to women had been broken, and she sought a job just when affirmative action came into vogue at the major law firms. Sally's friend Liz attributed her setback as a real estate broker to an economic recession coupled with high interest rates that crippled the housing market. As the experiences of Julie and Liz illustrate, historical trends and events are another major influence on the life course. The lives of individuals are shaped by trends that extend across historical periods (such as increasing equality of the sexes and improved nutrition) and by events that occur at particular points in history (such as recessions, wars, earthquakes, and tsunamis).

Birth Cohorts To aid in understanding how historical events and trends influence the life courses of individuals, social scientists have developed the concept of cohorts (Ryder, 1965). A **birth cohort** is a group of people who were born during the same period. The period could be 1 year or several years, depending on the issue under study. What is most important about a birth cohort is that its members are all approximately the same age when they encounter particular historical events. The birth cohort of 1950, for example, was in college at the height of the protests against the Vietnam War in 1967 and 1968. Many of these young people were profoundly influenced by those events. Members of the birth cohort of 1970 don't remember Vietnam protests, but they were in college when the protests against racism erupted in 1987 and 1988. Some of them were profoundly influenced by those events.

A person's membership in a specific birth cohort locates that person historically in two ways. First, it points to the trends and events the person is likely to have encountered. Second, it indicates approximately where an individual is located in the sequence of normative life stages when historical events occur. Life stage location is crucial because historical events or trends have different impacts on individuals who are in different life stages.

To illustrate, consider the effects of the economic collapse of several large corporations in 2001 and 2002. Enron and Arthur Andersen virtually collapsed; several other firms went out of business; and K-Mart, Tyco, and others downsized. Tens of thousands of workers and managers were laid off. Some people in their 50s found it impossible to get new jobs, perhaps due to age discrimination, and experienced prolonged unemployment. Some persons in their 30s and 40s returned to school and subsequently entered new fields. Workers who survived were left with insecurity and increased workloads. Persons just finishing college—the birth cohort of 1980—found fewer employment opportunities than those who graduated in 1995. Of course, not all members of a cohort experience historical events in the same way. Members of the class of 2002 who majored in liberal arts faced more limited opportunities than those earning professional degrees.

Placement in a birth cohort also affects access to opportunities. Members of large birth cohorts, for example, are likely to be disadvantaged throughout life. They begin their education in overpopulated classrooms. They then must compete for scarce openings in professional schools and crowded job markets. As they age, they face reduced retirement benefits because their numbers threaten to overwhelm the Social Security system. Table 17.1 presents examples of how the same historical events affect members of different cohorts in distinct ways. These historically different experiences mold the unique values, ideologies, personalities, and behavior patterns that characterize each cohort through the life course. Within each cohort, there are differences too. For example, the war in Iraq led to a father's or mother's absence for some children but not for others.

Cohorts and Social Change Due to the differences in their experiences, each birth cohort ages in a unique way. Each cohort has its own set of collective experiences and opportunities. As a result, cohorts differ in their career patterns, attitudes, values, and self-concepts. As cohorts age, they succeed one another in filling the social positions in the family and in political, economic, and cultural institutions. Power is transferred from members of older cohorts with their historically based outlooks to members of younger cohorts with different outlooks. In this way, the succession of cohorts produces social change. It also causes intergenerational conflict about issues on which successive cohorts disagree (Elder, 1975).

Occasionally, a major event or trend occurs that is profoundly discontinuous with the past; examples include Operation Desert Storm in 1991, the attacks on the World Trade Center and the Pentagon on September 11, 2001, and the wars in Iraq and Afghanistan that began in 2003. Cohorts that are in late

TABLE 17.1 HISTORY AND LIFE STAGE

	COHORT OF 1960–1965		COHORT OF 1985–1990	
Trend or Event	**Life Stage When Event Occurred**	**Some Life Course Implications of the Event**	**Life Stage When Event Occurred**	**Some Life Course Implications of the Event**
Women's movement (1972–1978)	Adolescence	For girls, increased opportunities in education, athletics. Less gender segregation.	—	—
Recession (1980–1982)	Young adulthood	Prolonged education, delayed marriage. Blue-collar, minority unemployment.	—	—
Economic expansion (1992–2000)	Adulthood	Increased employment, income. Improved standard of living.	Childhood	Raised in dual-career family. Greater opportunities for girls.
Terrorist attacks (09/11/01)	Middle adulthood	Increased awareness of family, reordered priorities. Anxiety about health, safety.	Adolescence	Shaken sense of security, uncertainty about the future. Increased stress.
War in Iraq		Increased political awareness, belt-tightening due to unemployment, recession.	Young adulthood	Reduced opportunities due to budget deficits, experience death, or injury of friend.

adolescence or early adulthood when such events occur may be profoundly affected by them and, in consequence, may develop a generational identity—a strong identification with their own generation and a sense of difference from older and younger cohorts (Stewart, 2002). This identity may shape their lives, influencing their choice of work, political views, and family relationships.

In this section, we have provided an overview of changes during the life course. Based on this discussion, it is useful to think of ourselves as living simultaneously in three types of time, each deriving from a different source of change. As we age biologically, we move through developmental time in our own biological life cycle. As we pass through the intertwined sequence of roles in our society, we move through social time. And as we respond to the historical events that impinge on our lives, we move together with our cohort through historical time.

We have emphasized the changes that occur as individuals progress through the life course. However, there is also stability. Normative transitions usually involve choices, and individuals usually make choices that are compatible with pre-existing values, selves, and dispositions (Elder & O'Rand, 1995). More than 90 percent of all Americans experience the normative transition of marriage. Most persons choose when and whom they marry. Longitudinal research indicates that we choose a spouse who is compatible with our own personality, thus promoting stability over time (Caspi & Herbener, 1990).

STAGES IN THE LIFE COURSE: AGE AND GENDER ROLES

People in every society experience a standard sequence of normative life stages. In this section, we examine some typical life course patterns in American society and some important variations on these patterns. We consider experiences that distinguish men and women as well as experiences common to both genders.

TABLE 17.2 POSTCHILDHOOD LIFE STAGES

Stage	Major Challenge	Conventional Labels	Age Range
I	Achieving independence	Youth, Late adolescence	16–23
II	Balancing family and work commitments	Young adulthood	18–40
III	Performing adult roles	Adulthood, Maturity, Middle age	35–70
IV	Coping with loss	Late maturity, Old age	60–90

Note: The overlap in ages between the stages shows that age is only a rough indicator of life stage.

Every society expects certain role behaviors, values, and attributes of men and others of women (Hyde, 2006). Differences in expectations for men and women are least pronounced at very young ages, increase through adolescence, and become even sharper in young adulthood. Before children are 10, for example, similar amounts of self-reliance, obedience, and leadership are usually expected of both girls and boys. By age 18, however, men and women typically face sex-differentiated expectations for these and other attributes as well as for choices of college major, occupation, and lifestyle. Expectations remain strongly differentiated by gender throughout adulthood. With retirement and old age, expectations for men and women become more similar again.

The four postchildhood life stages are shown in Table 17.2. Each stage is labeled according to the major social task or challenge that characterizes it. The labels point to the fact that these stages are socially derived, rather than the direct products of biological or cognitive development. In discussing each stage, we will focus on three major components of the life course—careers, identity and self-esteem, and stress and satisfaction. We also note the normative transitions inherent in moving through these stages and consider how people cope with the problems posed by these transitions.

Stage I: Achieving Independence

Most college students are in the stage of achieving independence. This is a period of transition from lives centered psychologically and economically on parents to lives in which we stand on our own. This stage challenges us to disengage from our parents and take responsibility for ourselves. For both men and women, achieving independence means acting in ways that are competent, persevering, and task oriented.

Several major social transitions are typically associated with this life stage: leaving the family home, finishing school, entering the workforce, developing a committed relationship, getting married, and becoming financially independent. In the 1960s, 65 percent of men and 77 percent of women had reached adulthood (by this definition, by age 30); in 2000, only 31 percent of the men and 46 percent of the women had done so (Furstenberg, Kennedy, McLoyd, Rumbaut, & Setersten, 2004). Due to changes that we will detail in this chapter, these changes are spread out over a longer period of time, they occur in various orders, and some persons do not achieve financial independence or enter a committed relationship. In an effort to capture these changes, Arnett (2004) refers to this as emerging adulthood, emphasizing the process rather than the outcome. Individuals are freer now to choose when and whether to make each transition, but many are constrained in these choices by the requirements of getting a formal education and preparing for an occupation. Individuals who make transitions too early or too late pay a price. Being out of step can lead to lost income and lost occupational and marital opportunities (Hogan, 1981), as noted earlier.

Individuals not only choose when to make these transitions, but they make choices about the amount and type of education they complete, the occupation they pursue, and who they partner with. We might expect that women and men who are socially competent as adolescents will make wiser choices. Wise choices should, in turn, result in more stable occupational careers and long-term relationships. Researchers in the

San Francisco area collected data on more than 500 children and adolescents between 1928 and 1931. Follow-up data were obtained from 281 of them at ages 53 to 62. Those who got higher scores on measures of competence in adolescence indeed had more stable lives throughout young and middle adulthood (Clausen, 1991). Also, wise choices in adolescence—for example, entering the military after completing high school—can counter the effects of deprivation in childhood or early adolescence (Caspi, 1992).

Careers During this stage, individuals explore the fit between their personal abilities and interests and their education and work options. Following high school, many youths join the workforce in entry-level jobs (as stock clerks, workers in fast food restaurants, servers, and so on). Others, especially young Black men in large cities, are frustrated in their search for jobs (Holzer & Offner, 2001). They experience sporadic employment or unemployment during this life stage. Still others enter military service. Sixty-two percent of male and 68 percent of female high school graduates attend college (National Center for Education Statistics, 2004); by race, 64 percent of White, 55 percent of Black, and 51 percent of Hispanic high school graduates go to college. About one half of those attending college work as well (Sweet & Bumpass, 1987). Some young people try out a variety of jobs during this life stage, acquiring skills and preferences that eventually lead to more permanent employment.

College students benefit from institutional support in their movement toward independence. College provides a partially protected environment in which one can develop self-reliance with the support of peers. College also instills in many students the motivation to pursue long-range, socially approved goals—such as studying 8 years to earn an advanced degree—even if this is not personally attractive. Such motivation is very helpful in the struggle for occupational success (Becker, 1964). The academic interests of students often reflect the gender typing of subject areas (physics and engineering for men, humanities and social work for women). Sex-differentiated preferences for subject areas arise in middle school, when boys or girls see particular subjects as useful in their future occupational choices (Eccles, 1987).

One ultimate goal for most men and women is to establish a long-term, intimate relationship. In American society, dating is the mechanism by which potential partners get to know each other. As discussed in Chapter 13, the conventions of dating encourage men to take the initiative and women to be more passive and dependent. If followed, this pattern fosters a sense of autonomy among men, but it inhibits the achievement of independence among women.

Identities and Self-Esteem A central challenge of this life stage is to solidify a personal identity—to develop a firm sense of continuity and direction in one's life (Erikson, 1968). Because men and women face different adult role expectations, they tend to build their identities on different bases. A survey of 1,267 never-married men and women ages 19 to 25 measured the importance of marital, parental, and career identities. As predicted, women gave greater importance to the parental identity; however, men and women placed equal importance on both career and marital identities (Kerpelman & Schvaneveldt, 1999).

Most men do not perceive their occupational career as dependent on future family roles. In contrast, many women consider their anticipated roles as wives and mothers in constructing their identities (Hyde, 2006). Young women's identity commitments may be more tentative and ambiguous than men's because the exact nature of a woman's future roles depends in part on the decisions she makes about familial roles. On the other hand, in recent years an increasing number of young women hold nontraditional views of gender roles—particularly women whose mothers held a nontraditional gender role ideology when they were young (Moen, Erickson, & Dempster-McClain, 1997). These women anchor their identities in their own vocational aspirations.

Studies of self-evaluation during this life stage show that self-esteem may drop after high school as youths struggle to achieve independence. Illustrating this drop, a sample of male college undergraduates from Michigan rated themselves as less competent, successful, active, and strong in their senior year than they did in their freshman year (Mortimer et al., 1982). Self-esteem rises later, when people successfully adapt to family and work roles. For the Michigan sample,

TABLE 17.3 PERCENTAGE OF MEN AND WOMEN SINGLE BY AGE GROUP IN 1970, 1980, 1991, AND 2003

	WOMEN				MEN			
Age	1970	1980	1991	2003	1970	1980	1991	2003
18–19	75.6	82.8	90.4	94.5	92.8	94.3	96.6	98.3
20–24	35.8	50.2	64.1	74.4	54.7	68.8	79.7	86.0
25–29	10.5	20.9	32.3	40.3	19.1	33.3	46.7	54.6
30–34	6.2	9.5	18.7	22.7	9.4	15.9	27.3	33.1

Source: U.S. Bureau of the Census, 1992, 1996, 2005.

self-ratings rose over the 10 years following college graduation.

Stress and Satisfaction Both men and women experience this period of achieving independence as stressful. Women are more likely than men to feel frightened, to feel that life is hard, and to feel financially insecure (Campbell, Converse, & Rodgers, 1976). These sentiments may well reflect the fact that some young women perceive having less control than men over the important directions their lives are taking. Once they set their life directions, however, the stress that women feel is usually reduced. Women may seriously pursue occupational goals at this time; their experience of stress may be similar to that of career-oriented men. If they enter traditionally male occupations and professions, they may experience social disapproval.

According to data collected 30 years ago, both overall life satisfaction and satisfaction with specific aspects of life tend to be lower for women and men during this life stage than during most later stages (Campbell et al., 1976; Gould, 1978; Lowenthal, Thurnher, & Chiriboga, 1975). Individuals at this stage are unsure of their objectives, of their abilities, and of their futures. Aspirations for occupational and relationship success may be high, but individuals worry about translating their dreams into realities. As people make firm career commitments and solidify their identities—and thereby move into the next stage—life satisfaction increases.

Stage II: Balancing Family and Work Commitments

During their late 20s and 30s, most men and women hold stable worker, partner/spouse, and parent roles. The central challenge of this stage is to establish oneself firmly in these roles—to forgo other options and commit one's energy, time, and self-definition to a particular job and to a particular partner. Priorities also must be set for work and family roles. Women often give first priority to their roles as partner and mother, men to their work roles (Bielby & Bielby, 1989).

Careers Both familial and occupational roles are developed during this stage. Continuing a trend dating to the 19th century, in 2000 more than 90 percent of Americans were married by age 50. Young adults are marrying later now than in recent decades (see Table 17.3). In 1980, half of the women had married by age 22.5, and half of the men had married by age 24.5. In 2003, half of the women were married by age 27, and half of the men were married by age 31. By ethnicity, rates of marriage by age 30 for Hispanic, White, and Asian women are the same (Lichter & Qian, 2004). Blacks postpone marriage longer than adults of other races, or remain single; contributing factors include high unemployment rates and more time devoted to education (Glick, 1997). Furthermore, Black women are less likely to ever marry than White women; this is partly due to the lack of employed Black men (Raley, 1996). This imbalance advantages employed Black men; research in Chicago, IL, found that some employed Black men were able to experience the benefits of long-term relationships without making a long-term commitment (Laumann, Ellingson, Mahay, Paik, & Youm, 2004).

Survey data clarify the nature of this delay in marriage. Three-fourths of the decline in the proportion of people marrying in their 20s is due to the increasing number of couples who are living together. This trend includes White, Black, Hispanic, Asian, and American Indian couples (Lichter & Qian, 2004). Young people today set up such living arrangements at the same ages at which members of earlier cohorts married (Bumpass, Sweet, & Cherlin, 1991). In recent years, one-third of young adults live with a member of the opposite sex for 6 months or more before marriage (Cherlin, 1981). Forty percent of these couples break up; 50 percent of them marry within 5 years. Finally, remaining single for an extended period has now become a viable lifestyle (Bernard, 1981). These changes have changed the composition of the population; among persons of ages 18 to 34, there are substantially fewer couples with children (Gauthier & Furstenberg, 2002). A key influence on delaying marriage and childbirth is the availability of safe and effective birth control; an analysis of data from 1960 through 2000 suggests that widespread adoption of the pill by women was associated with an increase in advanced education and increasing rates of entry into professions (Goldin & Katz, 2002).

Divorce rates reveal that many young adults have difficulties establishing firm family commitments. If recent trends continue, about half of those who marry this year will eventually divorce, most within the first 10 years of marriage. Analyses indicate a strong cohort effect on divorce rates; the divorce rate among persons who married between 1960 and 1964 is much lower than the divorce rate among persons who married between 1975 and 1979 (Morgan & Rindfuss, 1985). A longitudinal study of marriages found that experiencing certain problems substantially increases the likelihood of divorce. These problems include infidelity, spending money in ways the spouse considers foolish, and drinking and drug use; each increases the risk of divorce by 20 percent (Amato & Rogers, 1997). Also, adult children of divorced parents and adults who never lived with a father perceive greater likelihood of divorce than adults from two-parent families (Webster, Orbuch, & House, 1995). These adults were more likely to report patterns of marital interaction (arguing, shouting, and hitting) that strain a relationship. Marital discord seems to be transmitted from parents to their children; a study of 297 married couples and their married children found that parental reports of behaviors such as being domineering, critical, moody, getting angry, and not talking to the spouse predicted discord in their children's marriages (Amato & Booth, 2001).

The most striking transition of this stage is the transition to parenthood. For about 13 percent of couples, the first child arrives within 7 months of marriage. For 26 percent of married couples, the first child arrives between 8 and 23 months into the marriage. The probability of having a child slowly declines as the length of marriage increases beyond 2 years (Sweet & Bumpass, 1987). Newly married couples need to develop priorities, a shared lifestyle, and commitments to each other. When the first child arrives, it brings the challenge of developing a parental role and a division of labor regarding child care. If the first child arrives shortly after marriage, couples are forced to make both transitions—to marriage and to parenthood—at virtually the same time. A longitudinal study of African American and European American couples who had a child within the first 2 years of marriage compared them with couples who remained childless (Crohan, 1996). New parents reported more frequent conflicts after the transition than before, and they reported lower marital happiness than childless couples.

For men, parental and occupational roles are relatively independent. Most men work full time throughout stage II, barring problems of unemployment or disability. A study comparing parents and nonparents of similar age and marital duration found no differences in career orientation or job characteristics between men who had children and men who did not (Waite, Haggstrom, & Kanouse, 1986). On the other hand, in 1993, 1.1 million fathers were their preschool child(ren)'s primary caregivers (Casper, 1997). Some fathers do so because they want to be involved in their children's lives; others do so because of their wives' greater earning capacity, and others because they are ill/disabled (Fields, 2003). A study of fathers of children ages 3 to 5 found that the man's fathering style was related to his age when the child was born; the parenting of men who were younger than 26 at the birth was not related to aspects of the marital and career context, whereas the parenting of men older than 29 when the child was born was (Neville & Parke, 1997). It may be that men in their 30s are more focused on careers.

For many women, parental and occupational roles are intertwined. Most women work full-time until they have their first child, stay home to care for their young child(ren), and then return to work as the youngest child grows out of infancy. Statistics for women's employment in 2003 reflected this pattern. More than 69 percent of all married women were in the labor force, but only 60 percent of women with children under age 6. Regular jobs outside the home were held by 77 percent of the women with children ages 6 to 17. Among Black wives, 70 percent with a child under 6 and 86 percent with a child ages 6 to 17 worked (U.S. Bureau of the Census, 2005). Many women with young children choose to stay home. But women are also constrained to stay home because their housework load typically doubles with the birth of a child and because child care is both difficult to arrange and expensive. In some Mexican American families, men have traditional views about the responsibility for child care and expect their wives to stay home and care for the children (Gowan & Treviño, 1998).

With the arrival of the first child comes the need to develop a division of home-related work between husband and wife. Studies of household work consistently find that women spend more time performing it than do men. Defining housework as "cooking, cleaning, grocery shopping, doing laundry and dishes, doing repairs, paying bills, making arrangements, and caring for children," men reported performing 42 percent of the work, compared to 68 percent reported by women (Bird, 1999). Furthermore, there are differences in the kind of tasks performed. Within married couple households, women perform "core" housework—cooking, cleaning, dishwashing, laundry—whereas men report doing shopping and repairs; cooking and cleaning constitute two-thirds of the hours spent doing housework (Bianchi, Milkie, Sayer, & Robinson, 2000). Also, working mothers spend 1 to 2.4 hours per day on child care, compared with 0.5 to 1.3 hours by working fathers. Interestingly, working husbands and wives view the woman's often substantial contribution to family income as minimal, whereas they view the man's often minimal contribution to child care as substantial (Thompson & Walker, 1989).

What influences how much time men spend on household work? Three alternative explanations have been offered. The first is the gender perspective. According to this view, the gender role attitudes learned by men and women during socialization determine the division of household work. Several studies report supportive results. For example, nontraditional attitudes, particularly of the husband, are positively related to men's participation in housework (Starrels, 1994). Similarly, wives with more egalitarian ideology do less housework, but the wife's ideology has no effect on the husband's housework hours (Bianchi et al., 2000). The second perspective is exchange theory, specifically, the relative resources hypothesis. According to this view, in working couples, relative contributions to the family's economic situation determine housework. Thus, husbands typically do less housework because men typically earn more than women. A review of the research reports that wives who have greater income than their husbands spend fewer hours per week in housework (Shelton & John, 1996). Recent research finds that this relationship is not linear; wives who earn substantially more than their husbands do relatively more housework, perhaps to compensate for challenging his masculinity (Bittman, England, Sayer, Folbre, & Matheson, 2003). At the extreme, "gender trumps money." The third explanation emphasizes the time that each has available; the person who does fewer hours of paid work spends more hours doing housework (South & Spitze, 1994). The division of household labor is influenced by context. One contextual factor is the family life cycle; although wives consistently perform more housework than their husbands, the gap is larger when there are young children at home, as illustrated in Figure 17.1 (U.S. Bureau of Labor Statistics, 2005). The cultural context is also im-

FIGURE 17.1 TOTAL HOURS PER DAY SPENT BY WORKING MEN AND WOMEN IN HOUSEWORK, CHILDCARE, AND EMPLOYMENT

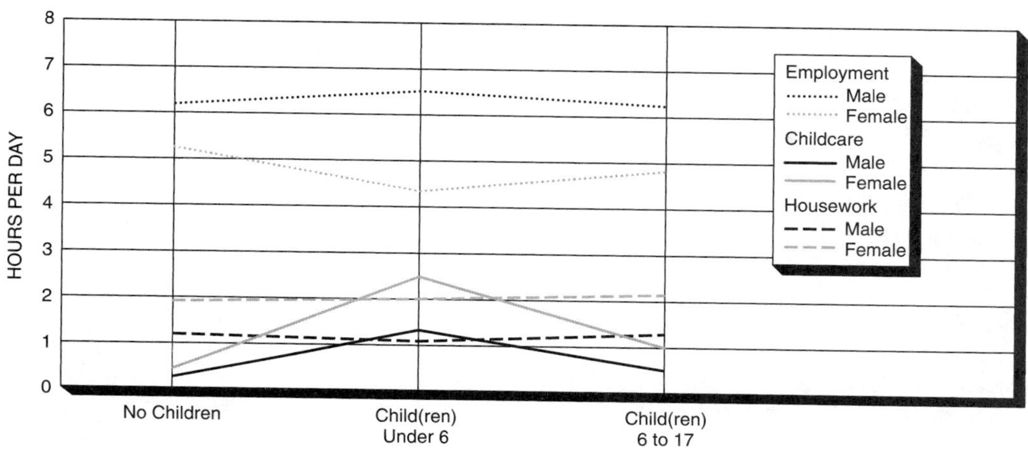

Note: Data collected via diary method, number of hours spent in the activity from 4:00 a.m. yesterday to 4:00 a.m. today (day of interview).

Source: U.S. Bureau of Labor Statistics, online at www.bls.gov/news.release/atus.t06.htm. Last modified January 12, 2005.

portant; the socialization explanation presumes that the division of labor reflects the culture's gender norms (South & Spitze, 1994; Zvonkovic, Greaves, Schmiege, & Hull, 1996). A survey of married couples in 22 countries found that the extent to which husbands performed "core" housework was strongly related to a measure of gender empowerment in the country (including percentage of parliamentary seats held by women).

The findings with regard to the relationship between race and the household division of labor are inconsistent (Shelton & John, 1996). One study of dual earner African American couples found that mothers spent more time caring for their infant and doing housework and that the father participated in both (Hossain & Roopnarine, 1993). The father's involvement in both did not vary by the number of hours the mother worked. In another study, women were divided into three categories: homemakers, wage earners, and career wives. Among Whites, husbands of both wage-earning and career wives participated more in housework—in effect, supporting wives who worked. Among Blacks, the husband's participation did not vary by whether his wife worked or was pursuing a career (Orbuch & Custer, 1995).

Identities and Self-Esteem Men committed to a particular line of work become increasingly caught up in their occupational careers during this stage. They build their identities around their performance at work and around the future advancement they anticipate (Maines & Hardesty, 1987). They are likely to think of themselves as "rising executives," "budding craftsmen," "maturing scholars," or "salesmen on the move." Success in their occupational careers brings a rise in self-esteem during this stage. In contrast, men who fail to establish themselves in a work career by the end of this stage experience confusion about their identities and have relatively low self-esteem (Van Maanen, 1976).

Many women become increasingly committed to their families during this stage. Even if they are employed, married women tend to build their identities primarily around their relationships with their husbands and children (Bielby & Bielby, 1989). They are likely to think of themselves as wives and mothers—as persons who try to provide a pleasant, supportive home for their children and husbands. Women who enter business and the professions often do construct firm identities around their occupational careers. They are sometimes hindered in this,

BOX 17.1 Gender Differences in Communication

Two of the best-selling books of the 1990s, *You Just Don't Understand* (Tannen, 1990) and *Men Are from Mars, Women Are from Venus* (Gray, 1992), proclaim that there are important differences in the way men and women communicate. According to Tannen, men and women have different goals in conversation. Men intend to exert control, maintain their independence, and enhance their status; women want to establish and maintain relationships. Men engage in conversational dominance, women in conversational maintenance. (Does this sound like the gender stereotypes discussed in Box 5.2?)

One assertion about gender differences in communication points to interruptions. Interest in this feature has its origins in a now classic study by Zimmerman and West (1975), which involved recording conversations in public places. They recorded conversations involving two men, two women, and mixed-gender pairs. The results showed that interruptions—that is, simultaneous speech that penetrated the speaker's utterance—were equally distributed in conversations involving two men or two women, but that men were much more likely to interrupt women in the mixed pairs. The researchers interpreted this as men asserting control over the conversation. Several subsequent studies reported the same findings—that men interrupt more than women, and men interrupt women more than women interrupt men—reinforcing the original interpretation. However, more recent analyses have challenged that interpretation. Interruptions actually play several different roles in conversation: request for clarification, expression of agreement, disagreement, making fun of the speaker, or changing the topic. Whereas the last three may represent attempts at dominance, the first two do not. Research incorporating such distinctions shows that a majority of interruptions in mixed-gender pairs are of the first two types (Aries, 1996).

Lakoff (1979) called attention to the greater use by women of tag questions—statements that are between an assertion ("male" speech style) and a question. For example, "Richard is here, isn't he?" Lakoff and others argue that tag questions express a lack of confidence in the speaker—a desire to avoid commitment to a statement and potential conflict. Empirical results with regard to gender differences in the use of tag questions are conflicting: In some studies, women use them more; in other studies, men use them more; and in some studies there are no differences. Again, if we look at the functions of tag questions in conversation, we see that there are

however, by the treatment they receive as a minority in the upper reaches of large organizations. Women are more often excluded from informal male peer networks and find it harder to be taken seriously than men (Kanter, 1976). One area where women are breaking through the "glass ceiling" is higher education; in 2002, 22 percent of the nation's college presidents were women, and women held the top positions at 11 major universities (Kantrowitz & Wingert, 2002). Differences in communicative style between men and women are further explored in Box 17.1.

It is evidently more difficult to build a sense of self-esteem on work as a homemaker than on work outside the home (Mackie, 1983). In fact, when we talk about "work," we are usually referring to activities for which people get paid. Some devalue housework because it is unpaid labor (Daniels, 1987). There are no clear criteria for judging the quality of homemaking and no raises or promotions to signify a job well done. In contrast, jobs outside the home usually provide supportive social contacts, and the regular paycheck indicates that other people value one's work. Thus, even low-prestige jobs can bolster self-esteem.

BOX 17.1 Continued

several: They may express uncertainty, but they also may express solidarity ("You were really sad about losing her, weren't you?") or politeness ("Sit down, won't you?"). Again, a closer look suggests that it is too simplistic to interpret the use of tag questions as an indication of lack of confidence, regardless of the gender of the person using them (Aries, 1996).

Another oft-discussed difference is in the use of back channel feedback—small vocal comments a listener makes while a speaker is talking (see Chapter 7). Women use less intrusive responses than men to indicate attention or agreement during conversation. Women prefer head nods and "M-hmn" rather than the more assertive "Yeah" or "Right." Women also make more such responses than men. Again, research shows that gender interacts with other variables. Back channel responses occur more often in cooperative than in competitive interactions, carry different meaning depending on whether they are inserted in the middle (showing active attention) or at the end of a long utterance (indicating an end to the topic). Back channel comments are not consistently associated with power or dominance.

There are also gender differences in nonverbal behavior. Men tend to signal dominance through freer staring, pointing, and walking slightly ahead of the women they are with. Women are more likely to avert or lower their eyes and move out of a man's way when they are passing him (LaFrance & Mayo, 1978; Leffler, Gillespie, & Conaty, 1982). However, when men are in subordinate positions to women, they avert their eyes or move out of her way. Thus, the gender difference is really a difference in the numbers of men and women who occupy superordinate positions. An observational study of 799 instances of intentional touch found that in public situations—at shopping malls, outdoors on a college campus—men are more likely to touch women. In greeting or leave-taking situations—at bus stations and airports—there was no asymmetry by gender (Major, Schmidlin, & Williams, 1990).

Thus, a comprehensive review of the literature on gender differences in communication leads to the conclusion that speech patterns, conversational style, and nonverbal behavior vary not only by gender but by characteristics of the context, such as the goals of the interaction and the roles of the participants. Anyone is capable of displaying "masculine" or "feminine" styles of communication when it is appropriate.

Stress and Satisfaction Marriage brings a drop in feelings of stress for most women but an increase in stress for men. Newly married husbands are more likely than their wives to feel rushed, to feel life is hard, and to worry about paying their bills (Campbell et al., 1976). More recent data suggest that these differences are getting smaller (Lee, Seccombe, & Shehan, 1991).

The birth of a first child increases stress to the highest level in the life course for both men and women. Strain between spouses increases. One study followed couples who were having their first child from late pregnancy until 9 months after the baby was born (Belsky, Lang, & Rovine, 1985). Marital quality declined significantly over the period, particularly as reported by wives. The level of stress associated with the transition to motherhood is influenced by the degree to which the birth disrupts the wife's relationship with the husband (Stemp, Turner, & Noh, 1986). Also, wives with a nontraditional gender role ideology are more distressed by an unequal distribution of child care responsibilities (Ross & Van Willigen, 1996). On the other hand, the transition is less stressful if the new mother's social network includes other parents (McCannell, 1988) and if she perceives

FIGURE 17.2 MARITAL HAPPINESS AND MARITAL DURATION

A national, longitudinal, five-wave panel study involved telephone interviews with married persons in 1980, 1983, 1988, 1992, and 1997. The initial sample consisted of 2,034 respondents, with 78 percent to 89 percent being reinterviewed at the four follow up points. Analyses of the data collected in 1980 found that marital happiness was associated in a curvilinear way with marital life course; couples were happiest immediately after marriage and much later in life. However, analyses of the data from all five waves identified that pattern displayed here. The relationship was essentially unchanged by including controls for children in the household, income, home ownership, and retirement.

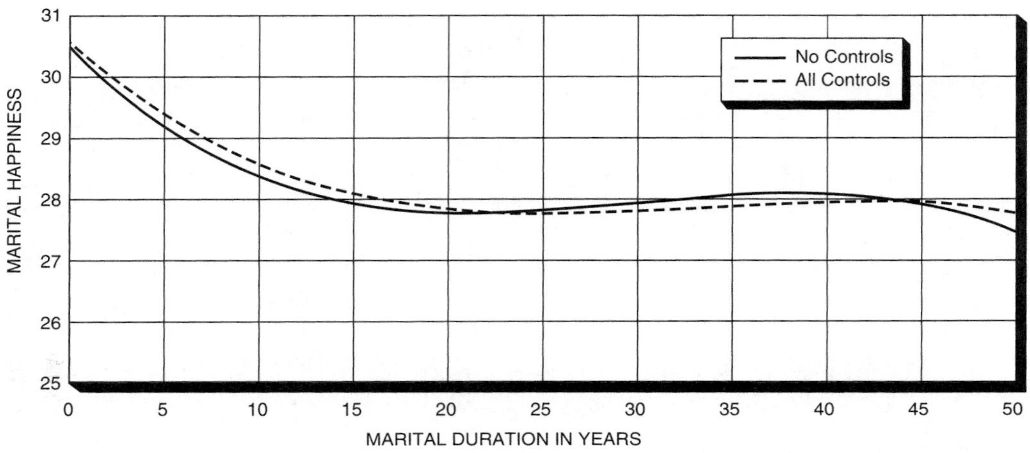

Source: Adapted from "Marital Happiness, Marital Duration, and the U-Shaped Curve: Evidence from a Five-Wave Panel Study" by VanLaningham, Johnson, and Amato, *Social Forces, 78,* 1313–1341. Copyright © 2001, with permission.

network members as concerned and caring. We can see that intimate relationships are an important buffer during stressful transitions.

Levels of satisfaction are only partly determined by stress; hence, patterns of satisfaction or marital happiness differ somewhat across time. Analyses of cross-sectional data report variation across the life course. Young married persons report high satisfaction. The arrival of children reduces satisfaction, especially among women. When young children are present, there is less satisfaction with standards of living, with savings, with housing, and with the marital relationship. Older men and women report increasing satisfaction (Campbell et al., 1976). However, analyses of longitudinal data suggest that happiness does not increase in later years, that in fact marital satisfaction declines as the length of marriage increases (see Figure 17.2).

In dual-earner families, satisfaction is influenced by the time spent together; couples who spend more time talking, eating meals, and having fun together report higher levels of satisfaction with the marriage (Kingston & Nock, 1987). On the other hand, couples in which the wife works more than 40 hours per week outside the home are characterized by higher levels of marital instability (Booth, Johnson, White, & Edwards, 1984), especially if the wife has a nontraditional gender ideology (Greenstein, 1995). Work and family demands are obviously intertwined and must be balanced if satisfaction is to be kept high and stress kept low.

The majority of Black married couples are well educated, own their own home, and are raising their biological children. With regard to global satisfaction, married Black men and women are more satisfied than nonmarried ones (Taylor, Chatters, Tucker, &

Lewis, 1990; Zollar & Williams, 1987). The distinction between the man as provider and the woman as homemaker is less rigid in Black families than in White ones. This is probably because Black women have historically been more likely to work outside the home due to economic necessity. In fact, one study found that Black men whose wives are homemakers have poorer psychological well-being than those married to working and career-oriented women (Orbuch & Custer, 1995). The former have the burden of providing sufficient financial resources for their families with no help from their wives. Also, among Blacks, men who report doing most of the housework are less satisfied with their family life (Broman, 1991). Research on perceived marital quality found that financial strain (lack of sufficient income) was associated with lower quality (Cutrona et al., 2003). In contrast to the research on White couples, years married was not significantly related to perceived quality. The other significant influence was social context; as neighborhood economic disadvantage increased, marital quality increased.

Stage III: Performing Adult Roles

What is the major challenge around which people of your parents' age organize their lives? For most adults, the major challenge from their late 30s into their 60s is to put their lives to a useful purpose—to make a meaningful social contribution (Erikson, 1968). Of course, few people talk about making "meaningful social contributions." Instead, people try to be good workers, parents, and spouses; that is, they try to meet high standards for performance in the adult roles to which they are committed.

Careers Most men and childless career women spend the first part of this stage working their way up the occupational ladder. Devoting themselves primarily to their jobs, they seek the increased responsibility, respect, and financial rewards available in their occupations.

The occupational experiences of women and men are likely to be different. More than half of all employed women are in clerical (secretarial, typing, data entry, and so on), retail sales, and service occupations. Men are much more likely to be craftsmen (carpenters, plumbers, electricians, and so on) or machinery operators (Reskin & Hartmann, 1986; U.S. Bureau of the Census, 2000). Although almost equal percentages of working women and working men are in the professions, women are more often in the lower status professions, such as teaching, nursing, and social and recreational services. These occupational differences produce differences in the prestige and income that men and women derive from employment.

At some point in their 40s or 50s, many workers recognize that their occupational life has reached a plateau that may extend to retirement. Taking stock of where they are, these workers often become less concerned with their own achievements and more interested in promoting others (Gould, 1978; Lowenthal et al., 1975). Thus lawyers, machinists, managers, or researchers who earlier enjoyed the guidance of mentors may now take pleasure in guiding younger associates themselves.

Workers who want greater self-determination may respond to the occupational plateau by pursuing new experiences and challenges, often with the object of increasing their income. Some seek jobs similar to their current ones in firms or institutions where the path to further advancement is still open (such as leaving one automobile company for another, or a local government job for one in Washington). Others return to school or study on their own in order to launch new careers (such as switching from teaching to stock brokerage, or from banking to real estate). Starting one's own business, moonlighting at a second job, and turning a hobby into a money-making venture are other self-determining responses at this life stage.

Some employed mothers devote themselves primarily to their families when their children are young. This causes them to fall behind men in their work careers (Reskin & Hartmann, 1986). Thus, women who seek jobs in stage III, after several years at home, are often at a disadvantage; they have no established record of reliable employment, nor have they developed or maintained marketable skills. Consequently, they must often settle for jobs that are below their educational level and less than fulfilling (Sewell, Hauser, & Wolf, 1980). As a result, these women may pursue occupational advancement and personal achievement most intensively during their 40s and 50s, after their children become teenagers. Paradoxically, this may be just when their husbands are beginning to feel less driven or to perceive restricted opportunities for further advancement.

During stage III, another type of family role may become important—the role of sibling. Sibling relationships may be the longest of all relationships, with a long history of shared experience; if they are close in age, siblings may be experiencing life transitions at about the same time. Brothers and sisters may provide emotional support and direct services throughout one's life (Goetting, 1986). Geographic proximity and emotional closeness are the main influences on how often siblings see each other (Lee, Mancini, & Maxwell, 1990). In midlife, the need to care for elderly parents may lead to increased sibling interaction and to cooperation or conflict. A study of 50 pairs of sisters noted distinct styles of participation in parental care, ranging from routinely providing care to providing no assistance at all (Matthews & Rosner, 1988). The results of interviews with the women suggest that birth order and geographic proximity are important influences on caring style; oldest siblings and those who live closest to the parent(s) usually provide the routine care.

Identities and Self-Esteem It is widely assumed that men continue to anchor their identities in their occupational roles throughout this midlife stage. The results of two surveys of national samples of men and women challenge this assumption (Pleck, 1985). On measures of role involvement, including items such as, "My main satisfaction in life comes from my work," employed men and women report higher levels of involvement in family roles than in work roles. Moreover, satisfaction with family was more closely associated with overall well-being than satisfaction with work.

Some women and men anchor their identities primarily in their family roles. For these parents, loss of the parental role when children leave home may therefore weaken their sense of identity temporarily and undermine their self-esteem. But the ensuing period of freedom is usually accompanied by an expanding sense of competence, maturity, and self-assurance, especially for women (Etaugh & Bridges, 2001).

Menopause was also once thought to weaken women's self-assurance and self-esteem. Perhaps this was true when menopause signified the end of childbearing—the loss of what was considered women's most important capacity. But menopause no longer has this social meaning in the United States, because women end childbearing earlier today and cultivate alternative roles around which to build identities. In other cultures, women often have different menopausal experiences; for instance, women of high caste in India report few negative physical symptoms such as hot flashes, suggesting that menopause is socially constructed rather than a universal biological phenomenon (Etaugh & Bridges, 2001). In some non-Western societies, postmenopausal women gain freedom from social and other restrictions placed on them during their reproductive years (Sommer, 2001).

Stress and Satisfaction For both married men and married women, stage III is a period of declining psychological stress. For parents, the following types of stress are greatest when children are under 6, and they decline steadily as children grow older: feeling tied down, feeling life is hard, worrying about having a nervous breakdown, and—especially—worrying about finances (Campbell et al., 1976). There is a marked increase in marital happiness when the youngest child leaves home (White & Edwards, 1990). Interviews with both members of 60 couples married at least 20 years explored four dimensions of marital relationships following the departure of the youngest child; the sample included White (57 percent), African American (23 percent), and Mexican American (20 percent) couples. Compared to their child-rearing years, couples reported reduced conflict, a continuing decline in satisfaction with the sexual relationship, no change in psychological intimacy (already high), and increased marital satisfaction. There were few differences by cultural background (Mackey & O'Brien, 1999). Reduced work responsibilities are also associated with an increase in marital satisfaction in later life (Orbuch, House, Mero, & Webster, 1996).

In addition to the characteristics of one's roles, personality is an important source of well-being at midlife. A longitudinal study of women found that quality of work and family roles were positively

© Bob Daemmrich/The Image Works

Grandchildren are a major source of satisfaction in later life. As Americans live longer, men and women will be able to enjoy their grandchildren for a longer span of time.

associated with life satisfaction at age 48. Further, generativity—concern with teaching others—was also related to satisfaction (Vandewater, Ostrove, & Stewart, 1997).

A common source of stress at this stage is the awareness that one is aging, brought on by physical changes such as thinning hair, sagging and wrinkling skin, the appearance of age spots, and increasing weight. In our youth-oriented society, these changes are greeted with either grudging acceptance—and perhaps declining self-esteem—or the effort to forestall such changes with cosmetic surgery, increased exercise, and perhaps lifestyle changes. Unfortunately, there is a "double standard of aging"; these changes decrease the attractiveness and desirability of older women but not of older men (Etaugh & Bridges, 2001). The more important her appearance has been to a woman, the greater the negative impact of these changes.

As the midlife stage progresses, two other sources of stress gradually increase: physical illness, and the death of parents or close friends. In one study, women's reports of deteriorating health (eyesight, hearing, teeth, hair, and so on) increased greatly after age 45 (Rossi, 1980). The death of a parent or close friend can be stressful because it increases one's own sense of mortality, deprives one of important roles as child or friend, may increase fears about health, and may impose added financial burdens.

A major source of stress is caring for someone who is physically or mentally ill for a prolonged period. A family may be providing such care for an adult child, an ailing parent, or a sibling of one of the adults. The primary care provider usually experiences increased stress, impaired relationships with other family members, and a decrease in social activities (Dura & Kiecolt-Glaser, 1991). Providing long-term care also can result in severe financial strain on the family.

Stage IV: Coping with Loss

Most of us will enter the final life stage during our 60s. Retirement or the onset of a major physical disability are the key markers of the transition into this stage. The central challenge of this stage is to cope with a series of practically unavoidable losses: loss of one's occupational role through retirement, loss of significant relationships through death, and eventually loss of health, energy, and independence. Despite the severity of these losses, most older people say they are satisfied with life and that they are coping well (Harris, 1981).

Most people who reach age 65 in the United States today will spend as many years in this final stage of life as they spent in childhood and adolescence. White men who reach 65 can expect to live another 17 years, White women another 20. Black men who reach 65 will likely live another 14 years, Black women another 18 years (U.S. Bureau of the Census, 2005).

Through many of these years, older people can actively enjoy a wide range of activities, because most remain reasonably healthy at least until age 75. Yet older people as a group are often mistakenly believed to be narrow-minded, unteachable, not very bright, uninterested in sex, and not good at getting things done (Harris, 1975). Thus, in addition to coping with losses, older people must often cope with **ageism**—prejudice and discrimination against them based on negative beliefs about aging. In fact, research suggests that the adult brain is more flexible than the stereotype asserts (Linkenhoker & Knudsen, 2002). Furthermore, elderly persons in good health with an available partner remain sexually active well into their seventies (American Association of Retired Persons, 1999).

Careers Most older people maintain strong primary relationships. These include ties to their adult children. As children grow into adulthood, parents gradually give up responsibility for their offspring's personal care, work, and financial status; the relationship evolves toward status equality (Blieszner & Mancini, 1987). Parents and children usually continue to have strong emotional ties, communicate regularly, and may provide various kinds of help to each other. Parents are selective in providing aid, giving it to the offspring who needs it—for example, temporary financial support in the event of unemployment (Aldous, 1987).

More than half of all elderly persons live with their spouses, and marriage continues to be their most important social tie. But the marital relationship often becomes more egalitarian at this stage. Women typically continue the shift toward greater assertiveness and independence begun in the preceding life stage, and men continue their shift toward greater nurturance and expressiveness (Guttman, 1977). Retirement adds to equality by reducing differences between spouses' activities. Retirees spend more time in housework—both in their own and in their partner's domains—than employed spouses (Dorfmann, 2002). Even among the elderly, gender roles make a difference; couples with more egalitarian gender role attitudes practice more flexible allocation of housework (Szinovacz, 2000).

Death of one's spouse/partner is the most severe trauma that older people confront. Women are typically widowed, because they have longer life expectancies than their husbands. Half of all married women lose their husbands by age 65. Black women are widowed earlier than White women. Widowed women often do not remarry, partly due to the limited number of single older men (Etaugh & Bridges, 2001). When husbands outlive their wives, they usually become widowers only after age 85. Death of a spouse/partner causes many types of loss. It severs the deepest of emotional bonds, takes away the main companion in day-to-day activities, frustrates the fulfillment of sexual needs, removes the key significant other for affirming one's identity, and—especially for women—produces economic loss.

How well a widow copes with her husband's death depends on several factors. First, the death of a spouse is more stressful if it is sudden and unexpected, if other family members or friends die within weeks of the spouse's death, or if the surviving spouse is in poor health prior to the death (Sanders, 1988). Furthermore, a woman whose identity and lifestyle were built on her marital role will experience greater disorganization when her husband dies (Lopata, 1988). Finally, women with strong social support networks cope with bereavement more successfully. Continuing interaction with married sisters seems to be a particularly important contributor to a widow's well-being (O'Bryant, 1988), and support from children, especially daughters, is also associated with enhanced well-being (Etaugh & Bridges, 2001).

Although there are a few tragic exceptions, older Americans—married or not—are rarely abandoned by their families. In fact, more than half see at least one of their adult children almost every day, and the vast majority have contact with a child or sibling every week (Harris, 1981; Shanas, 1979). In the 21st century, contacts between parents and distant children are made easy by email. The image of elderly persons as isolated in institutions such as nursing homes is inaccurate. Only about one in four is likely to spend any part of this life stage in an institution (Tobin, 1980). More than three quarters of all elderly persons live in independent households (U.S. Bureau of the Census, 2000). A few older people live with their adult children; those who do are typically single and are unable to maintain an independent residence due to poor health or low income.

A major source of social contacts and satisfaction are grandchildren (Dorfmann, 2002). Grandparents often feel greater freedom interacting with their grandchildren than they did with their own children, and so interactions are more enjoyable. Further, grandchildren represent the continuity of generations, which is very important to many older adults. As Americans live longer, more men and women will be able to enjoy their grandchildren for a longer span of time.

Occupations, the second main career line, end for most older people with retirement between ages 60 and 70. Ideally, retirement would be a gradual process loosely linked to aging, as occupational abilities and inclinations diminish only gradually. People who have control over giving up their occupational careers (for example, top management and self-employed persons) do indeed withdraw more gradually (Hochschild, 1975). Poor health is a major cause of early retirement; aging Black men and women are often in poorer health than Whites, and so retire early (Etaugh & Bridges, 2001). Retirement brings losses; income, prestige, a sense of competence and usefulness, and social contacts may all decline. Retired persons may need to fill free time, develop new everyday routines, and—if in a relationship—adjust to spending more time with their partners.

The main fear of retirees is that they will be cut off from social participation. Does this happen? Usually, people continue the level of social involvement they developed in their preretirement years. Those who were constantly busy and involved with people find new outlets in voluntary activities, hobbies, and social visits. Individuals who were uninvolved remain so. Few withdraw further (Palmore, 1981).

Reduced financial resources do, however, disrupt social participation, especially among retirees from the working class (Robson, 1982). Retirees who must worry about finances hesitate to spend money traveling to visit friends and entertaining them. Nor can they buy "proper" clothes or tickets for social events.

Identities and Self-Esteem Retirement, loss of a partner, and declining health deprive people of many of the central roles and relationships around which their identities have been built. Considering these losses, identity change in this stage is less than one might expect (Atchley, 1980). The relative stability of identities occurs because older people continue to think of themselves in terms of their former roles. Though retired, individuals still think of themselves as nurses, accountants, or musicians, for example. Though widowed, they remain "John's wife" or "Sarah's husband" in their own eyes. Interactions with siblings often reinforce continuing identification with roles occupied earlier in life, for example, through reminiscences (Goetting, 1986). Identities are also preserved because many personal qualities remain stable. Individuals are likely to see themselves as unchanged in their honesty, outspokenness, religiousness, and so on.

Neither widowhood nor retirement alone do serious damage to self-esteem (Atchley, 1980). Several aspects of aging are, however, associated with loss of self-esteem. Whatever deprives older people of their independence and of control over their own lives—such as ill health or falling into poverty—weakens their self-esteem. Moving into an institution or family residence where one becomes highly dependent also undermines self-esteem if caretakers make all the decisions for an older person. These well-meaning actions communicate the assumption that the older person is mentally and physically incompetent.

Stress and Satisfaction An important determinant of satisfaction levels in older persons is continued participation in activities. A survey of 618 persons, average age 65, assessed participation in eight clusters

FIGURE 17.3 FOUR PROFILES OF AGING AMONG TERMAN MEN

Using longitudinal data on 412 men, researchers created scores on five domains of life: life satisfaction (with family, friendships, cultural life, and service to society), vitality (energy and happiness at last survey), family engagement (contact with children, siblings, and relatives), occupational success (degree to which they reached their goals), and civic involvement (participation in nine different activities). Using scores on the five domains, the men were clustered; the four profiles depicted here emerged from the analysis.

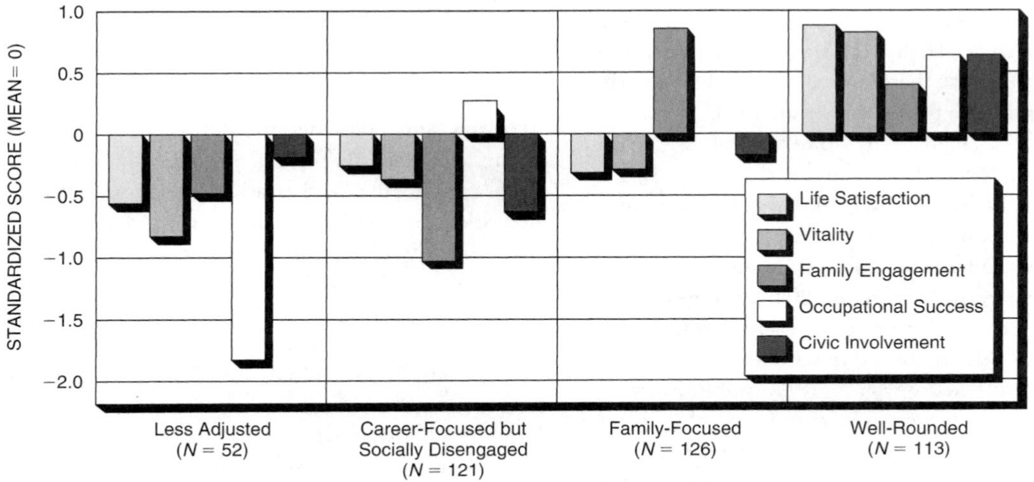

Source: Crosnoe and Elder, 2002.

of activities. Those who reported participation in community service and social activities had higher levels of life satisfaction (Harlow & Cantor, 1996). Participation in social activities was especially important for retirees, and more so for men than women.

Loss of one's spouse is a major source of stress for older persons (Zantra, Reich, & Guarnaccia, 1990). Becoming widowed often results in declines in mental and physical health and in income (Brubaker, 1990). Financial worries are a source of stress for retirees living on fixed incomes and those who have experienced a recent divorce. Elderly parents whose adult children experience serious problems—for example, impaired physical or mental health, abuse of alcohol—experience increased stress (Pillemer & Suitor, 1991).

Physical health is a primary predictor of well-being among older persons (Brubaker, 1990). Stress from failing health rises throughout this life stage, especially after age 75. Older people become anxious and depressed when they experience reduced energy levels, lack of motivation, memory loss, a slowdown in their ability to process information, and chronic or acute diseases. Depression induced by ill health is often misdiagnosed as mental deterioration and confusion. As a result, depressed older people often fail to receive the treatment that could restore them to alertness and satisfactory functioning.

In addition to considering the specifics of aging, we can ask a more global question: What characteristics are associated with successful aging? The longitudinal data from the Stanford Terman study provides data covering more than 50 years of life (Crosnoe & Elder, 2002). Using the data on 412 men, researchers identified four aging profiles; the profiles and the component scores on five dimensions are illustrated in Figure 17.3. The profiles are "less adjusted," career-focused/socially disengaged," "family-focused," and "well-rounded." Less adjusted men (13 percent) were characterized by below-average scores on all dimensions, including life satisfaction, vitality (energy and happiness), family engagement, lifetime occupational success, and civic involvement; men in this group had less stable marital histories and were more likely to have experienced two traumatic events, military service in World War II and death of a spouse or child earlier in life. Career-focused men (29 percent) were high on occupational success, but low in family and civic engagement; they reported the most unstable marital histories and long-term alcohol problems. Family-focused men (31 percent) were highly engaged in the family but average in other respects; they had completed less education and reported poorer physical health. The well-rounded group (27 percent) had completed significantly greater education and attained a substantial socioeconomic status that provided benefits in several areas; they had stable marriages and reported healthy lifestyles. Thus, the quality of life in stage 4 reflects both earlier life course experiences and current circumstances.

HISTORICAL VARIATIONS

Throughout this chapter, we have based our description of stages in the life course on recent findings and on projected future trends. But unique historical events—wars, depressions, medical innovations—change life courses. And historical trends—fluctuating birth and divorce rates, rising education, varying patterns of women's work—also influence the life courses of individuals born in particular historical periods.

No one can predict with confidence the future changes that will result from historical trends and events. What can be done is to examine how major events and trends have influenced life courses in the past. Two examples will be presented: the historical trend toward greater involvement of women in the occupational world, and the effects of historical events on different cohorts of high school graduates. The goals of this section are (1) to emphasize the influence of historical trends on the typical life course, and (2) to illustrate how to analyze the links between historical events and the life course.

Women's Work: Gender Role Attitudes and Behavior

There has been a substantial increase in the percentage of women who work outside the home in the United States since 1960. We will consider the role of attitudes and of economic changes in this trend.

Gender Role Attitudes In the past four decades, attitudes toward women's roles in the world outside the family have changed dramatically. The historical trend in attitudes has been away from the traditional division of labor (paid occupations for men and homemaking for women) to a more egalitarian view.

Consider the following statements. Do you agree with them?

1. It is much better for everyone involved if the man is the achiever outside the home and the woman takes care of home and family.
2. Women should take care of running their homes and leave the running of the country up to men.
3. Most men are better suited emotionally for politics than are most women.

These are typical of attitude statements included in one or more large-scale surveys of adults during the 1970s, 1980s, and 1990s. In the 1970s, two-thirds or more of the people surveyed agreed with the first statement, and one third agreed with the second and third statements. However, by 1998, only one-third agreed with statement 1, and statements 2 and 3 were endorsed by only 15 and 21 percent, respectively (Davis, Smith, & Marsden, 2000). This shift from traditional to egalitarian gender role attitudes has been quite strong among women. Hispanic women are often characterized as having more traditional gender role attitudes. However, young, well-educated, working Latinas have more egalitarian attitudes, similar to White women (Ginorio, Gutierrez, Cauce, & Acosta, 1995). Many Asian women struggle with conflicts between traditional attitudes common in their cultures

Many elderly people participate in organized activities, such as this exercise group. As long as they stay healthy and economically independent, most elderly people maintain their social involvements, activities, and self-esteem.

and the more egalitarian attitudes found in the United States (Root, 1995).

Workforce Participation This historical trend is not limited to attitudes. Women's actual participation in the workforce has been on the increase for almost a century. Figure 17.4 shows the percentage of women employed outside the home since 1960. The proportion of married women who are employed grew steadily from 1960 to 1995; since 1995, employment levels have remained stable or declined slightly. Among young single women, the employment level, already very high in 1960, has remained high. The proportion of women who work during pregnancy and who return to work while their child is still an infant has also grown steadily over this time period (Sweet & Bumpass, 1987). In 1999, Black women, controlling for age and family status, were more likely to be employed outside the home than White women (U.S. Bureau of the Census, 2000). Overall, Hispanic women were less likely to be employed than Whites; rates for Asian women vary considerably, from 59 percent for South Asian women to 77 percent for Filipinas (Cotter, Hermsen, & Vanneman, 2004).

Why have women joined the workforce in ever greater numbers throughout the 20th century? Has the spread of egalitarian attitudes been an important source of influence? Probably not. The idea that wives and mothers should not work except in cases of extreme need was widely held until the 1940s. Yet women's employment increased steadily between 1900 and 1940. The change in gender role attitudes occurred largely in the 1970s, yet women's employment rose rapidly during the two decades preceding these attitude changes. It therefore seems likely that gender role attitude changes have not been a cause of the increased employment of women but a response to it—an acceptance of what more and more women are, in fact, doing.

What, then, are the causes? Perhaps most convincing is the argument that the types of industries and occupations that traditionally demand female labor are the ones that have expanded most rapidly in this century. Light industries like electronics, pharmaceu-

FIGURE 17.4 WOMEN'S EMPLOYMENT: 1960–2003

The percentage of married women who are employed rose steadily from 1960 to 1995. Young single women have maintained virtually the same high level of employment throughout this period. Among married women, the level of employment rose slowly for those with no children and more rapidly among those with children under age 17. Since 1995, employment levels have been stable, or declined slightly.

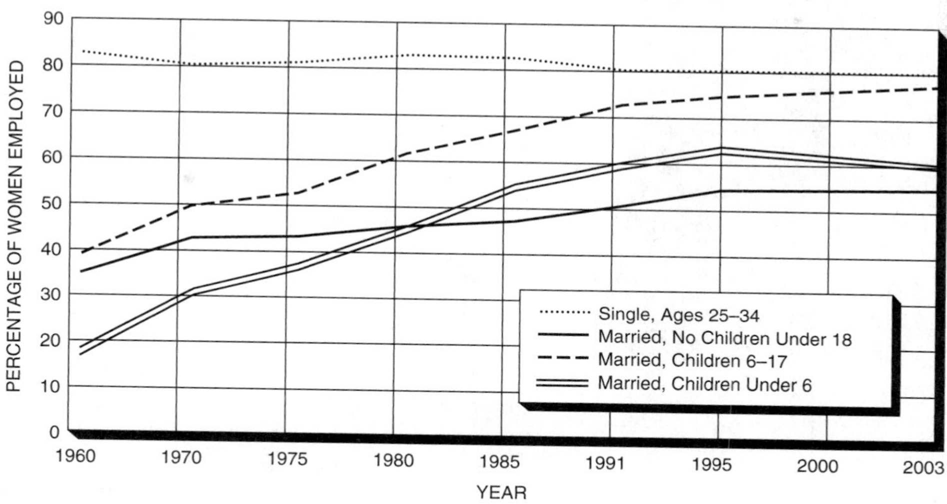

Sources: U.S. Bureau of the Census, 1996, 2001, 2004; U.S. Bureau of Labor Statistics, 1989.

ticals, and food processing have grown rapidly, for example, and service jobs in education, health services, and secretarial and clerical work have multiplied. Many of these occupations were so strongly segregated by sex that men were reluctant to enter them (Oppenheimer, 1970). Moreover, male labor has been scarce during much of the century due to rapidly expanding industry and commerce. The majority of the slack was taken up by a large pool of unemployed married women. These women could be pulled into the workforce at a lower wage because they were often supplementing their family income.

The changes noted in the preceding paragraph led to increased job opportunities for women. Other factors influenced women's desire to work outside the home. One of these was continuing inflation and rising interest rates; in many families, two incomes became necessary to make ends meet. Other factors that may have promoted the increased employment of women include rising divorce rates, falling birth rates, rising education levels, and the invention of labor-saving devices for the home. None of these factors alone can explain the continuing rise in the employment of women over the whole century. However, at one time or another, each of these factors probably strengthened the historic trend, along with changes in gender role attitudes.

The specific changes in women's work behavior demonstrate that the timing of a person's birth in history greatly influences the course of his or her life. Whether you join the workforce depends in part on historical trends during your lifetime. So does the likelihood that you will get a college education, marry, have children, divorce, die young or old, and so on.

Impact of Events

Life course researchers are also interested in the impact of events on those who experience them. One dimension of impact is the magnitude of the event—that is, the number of people who are affected. The

© PhotoDisc/Getty Images

Some parents are able to blend work roles and family roles by working at home. As further advances occur in telecommunication, more women and men may choose this option.

events of September 11, 2001, in the United States affected millions of people across the United States and in other parts of the world. The closing of a school affects hundreds of people in the community where the school is located.

How an event affects people depends on the life stage at which it is experienced. One model of this relationship is displayed in Table 17.4. In one sense, events have the greatest impact on children, by influencing their basic values and attitudes. The effects of an event on adolescents and young adults may be on one or more of their identities and on the social and economic opportunities they experience. A helicopter accident had a profound effect on the opportunities of Frank (whom we met in the introduction), leaving him partially paralyzed. Events may affect an adult's behavior, but they are unlikely to influence his or her identity or basic values. On the other hand, for those at midlife, some events, such as a major illness or the loss of a job, may create new identities and opportunities.

The impact of an event may also vary depending on the person's location in the social structure—that is, class, gender, and race. Consider the closing of a high school in Oak Valley, a prosperous midwestern community. In the mid-1960s, the community and the school were racially integrated; about 50 percent of the students were African American. As the civil rights movement gathered momentum in the United States, it affected the identities and behavior of some of the students; some African American students adopted distinctive dress and grooming patterns. The principal of the high school responded by imposing a dress code prohibiting facial hair; some students, parents, and faculty interpreted his action as racist. There was a walkout by African American students and their supporters, public protests, and some parents demanded action by the Oak Valley school board. Eventually, the Board decided to close the school (Stewart, 2002).

TABLE 17.4 THE IMPACT OF SOCIAL EVENTS ON THE PERSON

Life Stage When Event Is Experienced	Focus of Impact of Event
Childhood	Values and attitudes
Adolescence, young adulthood	Identities, opportunities
Adulthood	Behavior, opportunities
Later adulthood	New life choices, revised identity

Source: Adapted from Stewart, 2000.

A team of researchers has been studying two cohorts of persons who were students at the school: members of the classes of 1955, 1956, and 1957, and of the classes of 1968 and 1969 (Stewart, 2000; Stewart, Henderson-King, Henderson-King, and Winter, 2000). The research involves three methods—ethnographic observation, surveys, and in-depth interviews with selected persons. The team is interested in how the social structure—that is, race, class, and gender—shaped the students' lives in interaction with their experiences at the high school. Note that these people went to the same school in the same neighborhood, and many knew each other. The researchers could talk to each participant about the same people and events, being attentive to differences from one person to another in interpretation and experience.

Many of the graduates still live in Oak Valley. The researchers also read newspapers and other documents from the 1950s and 1960s and interviewed people who were teachers, administrators, ministers, and other community members during this period.

The 1950s graduates, asked 45 years later about the significance of events in their lives, rated past events like World War II high in meaning to them personally. They viewed their years in high school as idyllic; both African and European Americans described the school as a successful "melting pot," where differences were accepted and there was no conflict. There also were no major differences in the descriptions of men and women. In contrast, the 1960s cohort rated then-current events such as the civil rights and women's movements as highly meaningful personally. Reflecting the significance of race, African Americans rated the civil rights movement as much more meaningful than did European Americans. Both Blacks and Whites described Midwest High in terms of the diversity of students and teachers. Probing deeper, differences by race reappeared; African Americans discussed discrimination, racism, and the dress code, whereas European Americans discussed their fear of violence.

Turning to gender, African American men spoke of the school with pride and noted the power of the community in the response to the dress code. These men successfully resisted a code they viewed as racist. One said, "My experience left little to be desired." African American women spoke of the good teachers and the friends they made, but also about their limited social life as Black women and about racism. One said the worst thing about high school was "not being accepted or even noticed by many students." White men discussed the diversity of the student body; they also sometimes pointed to a breakdown of authority in the school. One said the worst thing was "getting beat up a couple of times." Like Black women, White women discussed friendships, but they also discussed the breakdown of authority, recalling instances of sexual harassment.

Thus, the social structure interacts with events to determine their impact on persons. Carrying out an intensive study of specific events, such as the imposition of the dress code and subsequent events at Midwest High, makes us aware that the same events may be perceived very differently depending on the perceiver's race and gender.

SUMMARY

This chapter has discussed the life course and gender roles in American society.

Components of the Life Course This chapter focuses on three components of the life course. (1) The life course consists of careers—sequences of roles and associated activities. The principal careers involve work, family, and friends. (2) As we engage in career roles, we develop role identities, and evaluations of our performance contribute to self-esteem. (3) The emotional reactions we have to career and life events include feelings of stress and of satisfaction.

Influences on Life Course Progression There are three major influences on progression through the life course. (1) The biological growth and decline of body and brain set limits on what we can do. The effects of biological developments on the life course, however, depend on the social meanings we give them. (2) Each society has a customary, normative sequence of age-graded roles and activities. This normative sequence largely determines the bases for building identities, the responsibilities and privileges, and the socialization experiences available to individuals of different ages. (3) Historical trends and events modify an individual's life course. The impact of a historical event depends on the person's life stage when the event occurs.

Stages in the Life Course: Age and Gender Roles Four broad life stages characterize the life course beyond childhood. (1) Achieving independence (age 16 to 25) entails several crucial transitions in education, work, and family life. Stress is high and satisfaction with life relatively low during this stage. Young men emphasize jobs in building their identities; young women emphasize family ties. (2) Balancing family and work commitments (age 18 to 40) is a major life task. The patterns and timing of family and work careers have undergone major changes in recent decades. For many men and some women, identity and self-esteem are tied to occupational success; for many women and some men, the family is central. Satisfaction rises with marriage, but the birth of a first child increases stress and reduces satisfaction. An important source of stress is the unequal division of the household labor between men and women. (3) Performing adult roles competently (age 35 to 70) leads to an occupational plateau for many employed adults, who then seek other challenges. Adults who have been family oriented turn more to occupational advancement or to community involvements as family demands recede. Self-esteem is supported by occupational achievement for both women and men. Stress eases during this stage because resources grow faster than needs. (4) Coping with loss (age 60 to 90) is the key challenge for older persons. Most cope well with retirement and even with the death of a spouse, unless or until they experience financial difficulties, failing health, or reduced social participation. As long as older persons retain independence, their self-esteem remains stable and their identities change little.

Historical Variations The historical timing of one's birth influences the life course through all stages. (1) Over the past 40 years, women's participation in the workforce has increased dramatically, and attitudes toward women's employment have become much more favorable. It is likely that the changes in attitudes reflect the changes in labor force participation, rather than the reverse. The likelihood that women will experience pressures and opportunities to work outside the home is now greater at every life stage. (2) Events also influence the life course of those affected by them. The impact of an event depends on its scope and on the life stage and social structural location of the persons influenced by it.

LIST OF KEY TERMS AND CONCEPTS

ageism (p. 452)
birth cohort (p. 438)
career (p. 434)
life course (p. 433)
life event (p. 435)
normative life stage (p. 436)
normative transition (p. 437)
stress (p. 435)

18

SOCIAL STRUCTURE AND PERSONALITY

Introduction

Status Attainment

Occupational Status

Intergenerational Mobility

Social Networks

Individual Values

Occupational Role

Education

Social Influences on Health

Physical Health

Mental Health

Alienation

Self-Estrangement

Powerlessness

Summary

List of Key Terms and Concepts

INTRODUCTION

Fred is 38, married, the father of two children, and sells pacemakers and artificial joints to hospitals. He travels 2 or 3 days a week and works at home the rest of the time in his $250,000 house in the suburbs. He earns almost $100,000 a year. Because his income is based entirely on commission, Fred worries about his sales falling off; but on the whole, he is satisfied with his life. His values are conservative, and he votes for Republican candidates.

Larry is also 38 and has a wife and two children. He runs a service station, works 6 days a week from early morning until 6 or 7 p.m. Larry and his family live in a small, three-bedroom house. Last year, he made about $48,000. He worries a lot about money and has been very tense the past year. He has liberal values and usually votes for Democratic candidates.

Marie is 39. She is head nurse in a hospital pediatric ward. Last year, her salary was $45,500. Although she enjoys her patients, she hates all the paperwork and the personnel problems. Some of her values are conservative, whereas others are liberal; she considers herself an Independent.

Fred, Larry, and Marie are three very different people. Each has a different occupation, which produces differences in income and lifestyle. They differ in their values—in what they believe is important—and in the amount of stress they feel.

Where do these differences come from? Often, they are the result of one's location in society. Every person occupies a **position**—a designated location in a social system (Biddle & Thomas, 1966). The ordered and persisting relationships among these positions in a social system make up the **social structure** (House, 1981a).

This chapter considers the impact of social structure on the individual. There are three ways in which social structure influences a person's life. First, every person occupies one or more positions in the social structure. Each position carries a set of expectations about the behavior of the occupant of that position, called a **role** (Rommetveit, 1955). Role expectations are anticipations of how a person will behave based on the knowledge of his or her position. Through socialization and personal experience, each of us knows the role expectations associated with our positions (Heiss, 1990). For example, Fred enacts several roles, including salesman, husband, and father. The expectations associated with these roles are a major influence on his behavior.

A second way that social structure influences the individual is through **social networks**—the sets of relationships associated with the various positions a person occupies. Each of us is woven into several networks, including those involving coworkers, family, and friends. Two examples of social networks are depicted in Figure 18.1. Both networks center on a single individual, Mary. The network on the left depicts the pattern of relationships between Mary and her friends that have evolved out of their shared experience. For example, the ties between Mary, Margie, and John developed because they took classes together, whereas Mary got to know Kathy at work. Mary described this network when asked to name the people to whom she is closest, excluding relatives. When asked to include relatives, Mary described the network on the right of Figure 18.1. The most striking difference is the greater number of direct ties between the persons in the network on the right. This network is dense—most of the persons in it know each other independently of their ties to Mary (Milardo, 1988). In both networks, a tie between Mary and another person reflects a **primary relationship**—one that is personal, emotionally involving, and of long duration. Such relationships have a substantial effect on one's behavior and self-image (Cooley, 1902).

A third way that social structure influences the individual is through **status**—the social ranking of a person's position. In every society, some positions are accorded greater prestige than others. Differences in ranking indicate a person's relative standing—his or her status—in the social structure. Each of us occupies several positions of differing status. In the United States, occupational status is especially influential. It is the major determinant of income, which has a substantial effect on one's lifestyle. One of the obvious differences between Fred and Larry, for instance, is their annual income.

This chapter focuses on how social structure influences the individual via roles, social networks, and status. Specifically, it considers four questions:

FIGURE 18.1 SOCIAL NETWORKS

These two networks are focused on Mary. When asked to name her best friends, excluding relatives, Mary described the network on the left. John is her boyfriend, whereas Mark is John's roommate. Mary met Chuck through John, and since then she and Chuck have become friends. Mary's three best women friends are Kathy (a co-worker), Leslie (whose parents are friends of Mary's parents), and Margie (a classmate). Margie has recently married Bob. This is a loosely knit network, because few of Mary's friends are close to each other. The network on the right represents the people to whom Mary feels closest including her relatives. Notice that only Margie and John remain from the friendship network. Mary's intimates now include Gregg (her brother) and Amy (Gregg's girlfriend). They also include her mother (Karen), stepfather (Bob), and her father (Jerry). This network is dense, because most of its members are close to Mary and to each other.

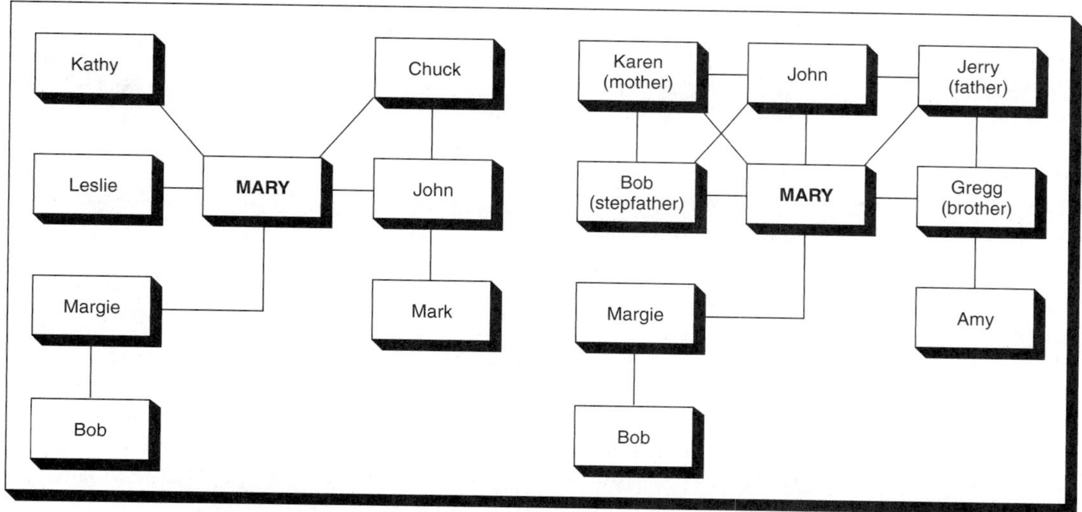

1. How does location in society affect educational and occupational achievement?
2. How does social location influence people's values?
3. How does social location influence a person's physical and mental health?
4. How does social location influence a person's sense of belonging in society, or the lack thereof?

STATUS ATTAINMENT

The individual's relative standing or status in the social structure is perhaps the single most important influence on his or her life. Status determines access to resources—to money and to influence over others. In the United States, occupation is the main determinant of status. This section considers the nature of occupational status, the determinants of the status that particular individuals attain or achieve, and the impact of social networks on the attainment of status.

Occupational Status

Occupational status is a key component of social standing and a major determinant of income and lifestyle. Fred is a sales representative for a company that makes artificial hip and elbow joints, pacemakers, and other medical equipment. These items are in great demand, and few companies make them. Fred sells a single pacemaker for $3,200 and keeps half of the money as his commission. He needs to be on the road only 2 or 3 days per week to earn $100,000 a year. He has a beautiful suburban home and two cars. Larry, by contrast, owns a service station. He works from morning until night pumping gas and repairing cars. His station is in a good location, but his overhead is high; he earned only $48,000 last year, and he worries that this year that figure will be lower. Larry and his family live in a smaller, older house and have a 6-year-old car.

The benefits that Fred and Larry receive from their occupational statuses are clearly different. First, Fred earns twice as much money as Larry. This determines the quality of housing, clothing, and medical care his family receives. Fred also has much greater control over his own time. Within limits, he can choose which days he works and how much he works; this, in turn, affects the time he can spend with family and friends. Larry doesn't have much free time. Finally, Fred receives a great deal of respect from the people he works with. He controls a scarce resource, so doctors and hospital personnel generally treat him well. Larry, however, deals with people who are usually preoccupied or angry because their cars are not running properly.

In addition to these tangible benefits, occupational status is associated with prestige. Several surveys in the United States have found that there is widespread agreement about the prestige ranking of specific occupations. In these studies, respondents typically are given a list of occupations and asked to rate each occupation in terms of its "general standing" or "social standing." The average rating is often used as a measure of relative prestige. The prestige scores for the United States shown in Table 18.1 were taken from an occupational prestige scale of 0–100 (Nakao & Treas, 1994). Surprisingly, there is considerable agreement across diverse societies in the average ranking of occupations. Even adults in China give rankings similar to those displayed in Table 18.1 (Lin & Xie, 1988).

The social structure of the United States can be viewed as consisting of several groups or social classes. A social class consists of persons who share a common status in the society. There are various views regarding the nature of social classes in the United States. One view of social class emphasizes occupational prestige in conjunction with income and education in defining class boundaries. This approach ordinarily classifies people into upper upper, upper lower, upper middle, lower middle, working, and lower classes (Coleman & Neugarten, 1971). A very different approach emphasizes the control, or lack thereof, an individual has over his or her work and coworkers as the main determinant of class standing (Wright, Costello, Hachen, & Sprague, 1982).

TABLE 18.1 OCCUPATIONAL PRESTIGE IN THE UNITED STATES

Occupation	Nakao-Treas Prestige Score
Physician	86
Lawyer	75
College or university professor	74
Registered nurse	66
Electrical engineer	64
Elementary school teacher	64
Police officer	60
Social worker	52
Dental hygienist	52
Office manager	51
Electrician	51
Housewife	51
Office secretary	46
Data entry keyer	41
Farmer	40
Auto mechanic	40
Beautician	36
Assembly-line worker	35
Housekeeper (private home)	34
Precision assembler	31
Truck driver	30
Cashier	29
Waitress/waiter	28
Garbage collector	28
Hotel chambermaid	20
Househusband	14

Source: Hauser and Warren, 1997.

Intergenerational Mobility

When a person moves from an occupation lower in prestige and income to one higher in prestige and income, he or she is experiencing **upward mobility.** To what extent is upward mobility possible in the United States? On the one hand, we have the Horatio Alger rags-to-riches imagery in our culture: Anyone who is determined and works hard can achieve economic success. This imagery is fueled by stories about the astonishing success of the woman who founded Mary Kay Cosmetics, Bill Gates' success as founder and head of Microsoft, Tracy Reese' rise to the top of the fashion industry in just 8 years, and so on. On the other hand, some argue that America is not an open society—that our eventual occupational and economic achievements are fixed at birth by our parents' social class, our ethnicity, and our gender. To be sure, every city has families who have been wealthy for generations and families who have been poor for as long. This suggests that the United States is characterized by castes—groups whose members are prevented from changing their social status.

These two views of upward mobility in American society are concerned with intergenerational mobility—the extent of change in social status from one generation to the next. To measure intergenerational mobility, we compare the social status of persons with that of their parents. If the rags-to-riches image is accurate, we should find that a large number of adults attain a social status significantly higher than their parents'. If the caste society image is correct, we should find little or no upward mobility.

What are the influences on upward (intergenerational) mobility in American society? In this section, we consider the impact of three factors: socioeconomic background, gender, and occupational segregation.

Socioeconomic Background Occupational attainment in American society rests heavily on educational achievement. To be a doctor, dental assistant, computer programmer, lawyer, or business executive, one needs the required education. To become a registered nurse, Marie (whom we met in the introduction) had to complete nursing school. Fred, our medical equipment salesman, earned a bachelor's degree in business.

Beyond education, what other factors influence occupational attainment? To answer this question effectively, we need to trace the occupational careers of individuals over their life course. Such longitudinal data are available from a research project begun in the 1950s (Sewell & Hauser, 1980). In 1957, all high school seniors in Wisconsin were surveyed about their post–high school plans. From this population, a random sample of 10,317 was selected for continuing study. In 1964, researchers obtained information from students' parents about post–high school education, military service, marital status, and current occupation. Later, they obtained information about the students' earnings and about the colleges or universities they attended. In 1975, 97 percent of the original sample were located, and most were interviewed by telephone. The interview focused on post–high school education, work history, and family characteristics. The data from this study enabled researchers to trace the impact of the characteristics of high school seniors on subsequent education, occupation, earnings, and work experience.

Figure 18.2 presents a diagram of the relationships found among the variables studied. The arrows indicate causal influences. Variables are arranged from left to right to reflect the order in which they affect the person through time. These results indicate that children from more affluent homes have greater ability, higher aspirations, and receive more education. Children with higher ability get better grades, which reward them for their academic work and reinforce their aspirations. Children who do well are also encouraged by significant others, such as teachers and relatives, which also contributes to their high aspirations. These children are likely to choose courses that will prepare them for college. They are likely to spend more time on academic pursuits and less time on dating and social activities (Jessor, Costa, Jessor, & Donovan, 1983). As a result, they are likely to continue their education beyond high school and perhaps beyond college. Finally, high ability, encouragement of significant others, and high educational attainment lead to greater occupational status and earnings.

Note that socioeconomic background and grades have an indirect effect on occupational status and a direct effect on educational attainment. This does not

FIGURE 18.2 THE DETERMINANTS OF OCCUPATIONAL STATUS ATTAINMENT

This figure summarizes the influences that determine educational and occupational status over the life course. Socioeconomic background (parents' education, occupation, and income) influences ability, aspirations, and educational attainment. Ability influences grades, which, in turn, affect encouragement from significant others and aspirations for educational attainment. Occupational status is affected by education and also by ability, aspirations, and significant others.

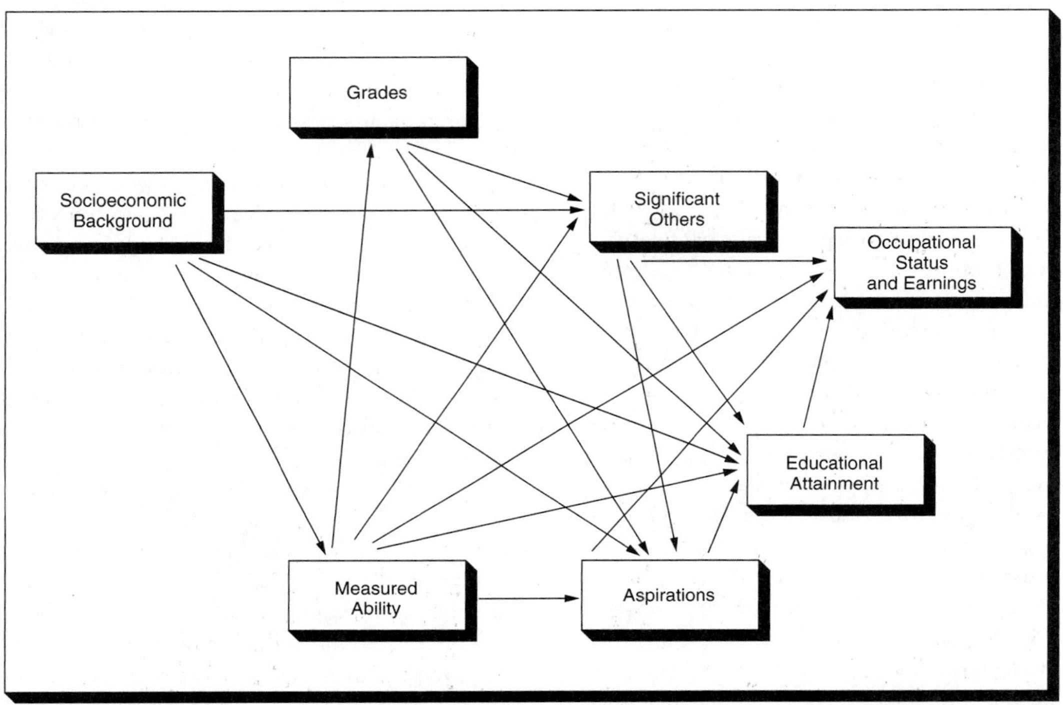

Source: Adapted from William H. Sewell and Robert M. Hauser, "The Wisconsin Longitudinal Study of Social and Psychological Factors in Aspirations and Achievements." *Research in Sociology of Education and Socialization,* Vol. 1, 1980, pp. 55–99. Used by permission of the author.

mean that parental socioeconomic status and an individual's grades are unrelated to occupational status. Rather, it indicates that status and grades influence occupational attainment through other variables—like aspirations—that have a direct impact on occupational attainment (Sewell & Hauser, 1975).

In the research summarized in Figure 18.2, the family characteristics studied are mothers' and fathers' socioeconomic standing—education, occupation, and income. How is it that variables such as your father's education and your mother's income influence your educational attainment? Parents often use their resources to create a home environment that facilitates doing well in school (Teachman, 1987). Thus, they provide such aids as a quiet place to study and encyclopedias. Moreover, they may provide cultural enrichment activities, such as attending concerts and arts events (DiMaggio & Mohr, 1985). A study of the daily activities of children ages 3 to 11 found that children of highly educated parents spent more time reading and studying and less time watching TV (Bianchi & Robinson, 1997).

One review of the research on the intergenerational transmission of poverty concludes that children

TABLE 18.2 EDUCATIONAL ATTAINMENT BY ETHNICITY AND GENDER, 2003

	WHITE		BLACK		ASIAN		HISPANIC	
	Male %	Female %	Male %	Female %	Male %	Female %	Male %	Female %
High school graduate or more	84.5	85.7	79.6	80.3	89.5	86.0	56.3	57.8
College graduate or more	29.4	25.9	16.7	17.8	53.9	46.1	11.2	11.6

Source: U.S. Census Bureau, *Statistical Abstract of the United States: 2004–2005*, Table 212.

raised in poor families will complete fewer years of school, are less likely to attend college, and are more likely to be poor as adults (Corcoran, 1995). On the other hand, many Black families, though not wealthy, do give their children the motivation and the skills to succeed in school (McAdoo, 1997).

Family structure also plays a role in the attainment process. A study of a national sample of 30- to 59-year-old men and women compared those raised in an original two-parent family with those raised in other family structures. Those raised in original families earned more as adults (Powell & Parcell, 1997). Among Blacks, the presence of two parents, both employed outside the home, is essential to mobility (McAdoo, 1997).

The experiences of Fred and Larry clearly reflect the importance of these processes. Fred's parents were upper middle class; they sent him to preschool at age 4 and encouraged him to learn to read. Larry's parents were working class; they encouraged him to get out and play and not to waste time reading. Fred did well in school; his grades were always high. Larry struggled with his schoolwork, especially math. By 8th grade, Fred had an excellent record, and his teachers gave him lots of encouragement; Larry's teachers, on the other hand, didn't pay much attention to him. Fred worked hard in high school, got good grades, and, with the support of his teachers and family, went to the university. After finishing high school, Larry went into the Army, where he learned vehicle mechanics. When Fred finished college, he got a job in a medical equipment firm. Ten years after he graduated from high school, Fred was selling $200,000 worth of equipment per year and earning 20 percent commissions. After he finished his military service, Larry went to work in a gas station. Ten years after Larry graduated from high school, he was earning $26,000 per year working in a gas station.

Thus, there is upward mobility in American society, and one's socioeconomic background does not fix one's occupational attainment and earnings. Through greater education, many persons achieve an occupational status and income larger than would be expected based solely on their background. Thus, America is not a caste society. At the same time, one's socioeconomic background is not irrelevant to one's educational and occupational attainment. Not everyone can be a doctor, lawyer, or engineer. Opportunities for upward mobility are not unlimited.

Education The research summarized in the preceding section clearly indicates the importance of education in determining one's occupational attainment and adult status. Thus, differences in the amount of education completed will result in differences in status. In U.S. society, there are large disparities in educational attainment by ethnicity. The percentage of all persons who have completed high school (and more) and college (and more) are displayed in Table 18.2. Looking at the data, Asian Americans are the most educated (almost one-half complete college), followed by Whites (27.6 percent), Blacks (17.3 percent) and Hispanics (11.4 percent). Not surprisingly, these differences translate into differences in occupations, which in turn create differences in income and lifestyle.

Gender Is the process of status attainment different for men and women? According to the data obtained on Wisconsin high school students, the determinants

© Tim Barnwell/Stock, Boston

© PhotoDisc/Getty Images

Segregation of occupations by gender is widespread in the United States; for example, most water department repair workers are men, whereas most nurses and unit clerks in hospitals are women. This segregation has serious consequences for a woman's earnings and occupational prestige.

of occupational status as depicted in Figure 18.2 are the same for both men and women, although the size of some relationships varies. Using a prestige scale of 0–100, the first jobs held by women were, on the average, 6 points higher than the first jobs held by men. Women's first jobs were concentrated within a narrow range of prestige, whereas there was much greater variation in the prestige scores of first jobs held by men (Sewell et al., 1980). Table 18.1 reveals how this occurred. The first jobs that women held included registered nurse, schoolteacher, social worker, dental assistant, and secretary. The prestige scores of these jobs range from 66 to 46. In contrast, men's first jobs ranged from physician (86) to garbage collector (28).

When the researchers looked at 1975 occupations, they found that men had gained an average of 9 points in status in the 18-year period since their graduation from high school. Women, on the other hand, had actually lost status; the average prestige of current occupations for women was 2 points lower than the average prestige of their first jobs. Men experience upward mobility because they work continuously. Moreover, they are in occupations with possibilities of promotion and advancement. Women's work careers are often interrupted by marriage and by raising children; when they return to work, they often take up the same job. Thus, women are often unable to build up enough continuous experience to gain promotions. Moving also interferes with women's advancement, especially when the purpose of the move is to further the man's career (Shihadeh, 1991). Advancement is also more limited in occupations held largely by women. The top positions in schools, social work, airlines, and sales are more often held by men than by women. So the occupational status achieved by men and women differs over the course of their careers.

These differences are evident in the lives of Fred, Larry, and Marie, who were introduced at the beginning of this chapter. After college, Fred began in sales (prestige score 49), and his income increased substantially every year. If he wanted, he could move up in the company to regional sales manager, national sales manager, and perhaps vice president of sales. Larry has moved from gas station attendant (prestige score 21) to owner of a service station (prestige score 44). Like Fred, Marie went to college and earned a bachelor's degree. Her first job involved working on a surgical unit in a large hospital (prestige score 66). As head nurse in pediatrics, she works days now, gets weekends off, and earns more, but her occupation is unchanged. She could move up to director of nursing, but she is unlikely to strive for this, because the added responsibility isn't balanced by added pay.

Occupational Segregation In the preceding section, we saw that the influence of factors such as socioeconomic background and ability on occupational attainment is similar for men and women. At the same time, working men and women are not proportionately distributed across occupational categories. Look back at Table 18.1. As you look at each occupation in the list, which gender comes to mind? Chances are that when you think of engineers, carpenters, or auto mechanics, you picture men performing those jobs. In 2003, of those employed in these occupations, 92 percent, 98 percent, and 98 percent, respectively, were men (U.S. Bureau of the Census, 2005). Similarly, when you think of registered nurse or dental hygienist, you picture women in these roles; in 2003, of those employed in these occupations, 92 percent and 99 percent, respectively, were women. Many occupations consist overwhelmingly of either men or women; there is substantial occupational level segregation by gender (Reskin & Padavic, 1994). As occupation is the basis of prestige and a major determinant of income, this segregation has serious consequences.

There are several processes that perpetuate occupational segregation. Our awareness of the gender composition of occupations through our daily experience and via media portrayals influences our aspirations. Further, cultural beliefs about gender differences in skills and abilities, for example, that girls are less skilled than boys at math, influence educational decisions and career choices (Correll, 2004) in ways that maintain occupational segregation (men become engineers) (see Box 18.1).

Direct experience with occupational segregation begins in adolescence. Data from a sample of 3,101 tenth- and 11th-grade students in suburban high schools provide concrete evidence (Greenberger & Steinberg, 1983). Adolescents' first jobs are segregated by sex, with girls earning a lower hourly wage. These differences reflect differential opportunity; employers hire primarily girls or primarily boys for a particular job (for example, newspaper carrier, fast food sales), and they pay boys more. Performing different roles results in differences in the skills developed by gender.

Adults often experience gender segregation in the workplace. In a survey of 290 organizations with a total of more than 50,000 employees, the results indicated that men and women rarely perform similar work in a single organization; when they do, they usually have different job titles (Bielby & Baron, 1986). There is little evidence that employers' practices in this regard are a rational response to differences between men and women. Jobs held by men in one organization are held by women in other organizations. What happens when a person of the other gender enters a segregated occupation? That person may experience stereotyping and harassment by coworkers or supervisors, and leave the job. On the other hand, the person may experience unusual rewards and rapid advancement. Elementary school teachers are primarily women (81.7 percent in 2003); research comparing the outcomes of men and women elementary teachers found that men were more likely than women to be promoted to administrative positions (Cognard-Black, 2004). Either way, the dominance of one gender in the occupation is maintained.

Differences in work performed or in job titles often result in large differences in pay. In 2003, the median weekly earnings of White women (over 25) employed full-time was $567, whereas the median for White men (over 25) was $715 (U.S. Bureau of the Census, 2005). Thus, the median annual earnings of a White woman was $29,484, whereas the median for a White man was $37,180—a difference of $7,700.

BOX 18.1 Can Girls Do Math? Cultural Beliefs and Occupational Segregation

The power of cultural beliefs was convincingly demonstrated by the controversy in early 2005 over remarks made by the president of Harvard University, Lawrence Summers. Speaking at a meeting of the National Bureau of Economic Research on January 14th, Summers suggested three reasons why women are underrepresented "in high-end scientific professions. One is what I would call . . . the high-powered job hypothesis. The second is what I would call different availability of aptitude at the high end, and the third is what I would call different socialization and patterns of discrimination in a search" (Summers, 2005). In an elaboration of the second point, he suggested that, among those with the very highest abilities in science and math, the ratio of men to women was probably 5:1. Summers is restating a widely held cultural belief in the United States, that girls/women are inferior at math and, by extension, will not succeed in occupations that require high mathematical ability. A professor of biology at MIT, Dr. Nancy Hopkins, walked out on Summers, and later discussed his remarks with a reporter.

One reason why many reacted to his remarks with indignation is that these beliefs become a self-fulfilling prophecy. Many teachers and guidance counselors (and many parents?) believe girls are inferior at math and so they caution girls not to take advanced math or science courses. This, of course, restricts the number of girls who take the courses needed to later pursue degrees in these fields. This belief is held by many boys and girls, and influences the way they interact with each other. Furthermore, a girl who believes she is less skilled will not work as hard or be as persistent as a girl who believes she can do it, making it less likely that she will succeed. If we really want to encourage members of a group to enter an occupational field in larger numbers, we need to stop constraining them by voicing and acting upon such beliefs. It is especially upsetting when someone of Summers status, president of one of the oldest universities in the United States, voices them.

More importantly, Summers is wrong. There have been hundreds of studies of differences in mathematical and related skills and abilities, at all age levels. These studies have been the focus of two large-scale meta-analyses; one of them used data from more than 100 studies, testing three million persons (Hyde, Fenema, & Lamon, 1990). Overall, the average effect size was $d = -.05$; that is, it was basically zero. When the analysis was done by age, the results indicated that there is no gender difference in math performance in elementary or middle school, but a significant difference emerges in high school. These results flatly reject the hypothesis that there are innate differences in ability, and are entirely consistent with the hypothesis that girls and boys are being socialized to believe there is a difference, and that that cultural belief increasingly affects math performance as girls progress through the school system. It is not differential ability, but differential socialization that results in few women in top positions in science.

Summers' remarks were the subject of many news articles, opinion columns, and segments on national television programs. Many of these were highly critical of his remarks. A letter signed by 81 distinguished persons, including astronaut Sally Ride, was published in *Science*. It called on Summers and others to put in place institutional policies and procedures "that enable women and other underrepresented groups to step beyond the historical barriers in science and engineering" (Muller et al., 2005). For these remarks and other matters, members of the Faculty of Arts and Sciences at Harvard adopted a motion on March 15, 2005, stating that they lacked confidence in his leadership; this is the first time in history that Harvard faculty have adopted such a motion. Hopefully, another result of this very public controversy will be to hasten the disappearance of the cultural belief that girls can't do math.

There are also differences in earnings by race and ethnicity. In 2003, the median annual salary of Black women (over 25) was $25,532, whereas the salary of Hispanic women was $21,320. Thus, Black women earned almost $4,000 less than White women, and the White-Hispanic difference was $8,164. The median salary of Black men was $28,912, and of Hispanic men $24,128. The White-Black difference was $8,268 and the White-Hispanic gap was $13,052 (U.S. Bureau of the Census, 2005). A substantial proportion of these gaps is due to discrimination, not to differences in measures of workers' skills.

Social Networks

We have seen that socioeconomic background, ability, educational attainment, and earlier jobs influence occupational attainment over the life course. In part, this is because differences in experiences create differences in an individual's aspirations and abilities to cope with the occupational world. Varied experiences also move people into different social networks. This exposes them to varied social contacts, which have an important effect on their upward mobility. This section considers some of the ways in which position in social networks affects the person.

Networks provide channels for the flow of information, including information about job opportunities. What types of networks are likely to provide information on finding new jobs? You might think it is networks characterized by strong ties, such as families or peer groups. Surprisingly, employment opportunities are often found through networks characterized by weak ties—relationships involving infrequent interaction and little closeness or emotional depth (Marsden & Campbell, 1984). Those to whom our ties are weak are involved in different groups and activities than we are. Consequently, they will be exposed to information that is different from the information we and our friends already have. For this reason, new information is more likely to come via a weak tie than via a strong one. In one study, of those who found jobs through contacts, only 17 percent were obtained through strong ties (Granovetter, 1973).

A study of the hiring process in a mid-sized high-tech organization gathered information on all 35,229 applicants over a 10-year period (1985–1994). The results indicated that there were only small differences in hiring by gender, and they were accounted for entirely by age and education. For ethnic minorities, on the other hand, some of the differences were accounted for by referral method; members of minority groups lacked access to the informal networks that were associated with success in getting hired (Petersen, Saporta, & Seidel, 2000).

We noted earlier that women are less likely than men to experience upward occupational mobility during their careers. Might this occur because men and women differ in their access to networks that carry job information? Our ties to networks grow out of the activities we share with others. The organizations we belong to are a major setting for such activities (Feld, 1981). The larger the organization, the larger the potential number of weak rather than strong ties. If men belong to larger organizations than women, they would have more weak ties and, hence, better access to information useful in finding jobs.

To examine this possibility, 1,799 adults were asked the name and size of each organization to which they belonged (Miller-McPherson & Smith-Lovin, 1982). On the average, men belonged to organizations such as business and professional groups and labor unions, whereas women were more likely to belong to smaller charitable, church, neighborhood, and community groups. Moreover, job-related contacts are more likely to develop in business, professional, and union groups. Findings showed that men had an average of 170 job-related potential contacts, whereas women had an average of fewer than 35. Apparently, men are in networks that allow greater access to information about and opportunities for advancement.

Social networks also contribute to mobility within one's workplace. A longitudinal study of employees in one high-tech firm found that having a large network of informal ties was associated with promotions and salary increases (Podolny & Baron, 1997).

INDIVIDUAL VALUES

Last year, Fred, Larry, and Marie were each approached by a labor union organizer. Fred, the sales representative, was approached by a member of Retail

Clerks International. The organizer explained that under a union contract, Fred would spend fewer days on the road and would receive a travel allowance from his employer. Larry was approached by a representative of the Teamsters. The organizer sympathized with the problems of independent service station owners and urged Larry to let the Teamsters represent his interests in dealing with his supplier. Marie was approached by the president of United Health Care Workers; she was promised higher wages and greater respect from physicians if she would join.

Fred flatly rejected the invitation, believing that a union contract would limit his freedom and perhaps reduce his income. Larry's reaction was mixed. On the one hand, he feels he is at the mercy of "big oil." On the other hand, he is also a self-employed businessman; like Fred, he doesn't want to join a labor organization that might limit his ability to determine his prices and the pace at which he works. Marie reacted very favorably to her invitation and began to attend union meetings "to see what they are like." She felt that a union might lead to higher pay and might force the hospital to give her more freedom in determining the pace at which she worked.

In making their decisions, Fred, Larry, and Marie used their personal **values,** which are enduring beliefs that certain patterns of behavior or end states are preferable to others (Rokeach, 1973). All three were concerned with protecting or enhancing their freedom and income. These values provided criteria for making decisions. Thus, each person weighed the potential effect of joining a union on freedom and income. Fred felt that the effect on both would be negative. Larry was sure that union membership would limit his freedom but uncertain about its effect on his income. Marie perceived a potential gain in both freedom and income, so she decided to explore union membership.

An important theory of values has been developed by Schwartz (1992, 1994). He identifies 10 motivationally distinct values; these are portrayed in Figure 18.3. One study assessed the values of 999 adults, 52 percent of them women. There were no differences by gender in either the structure of values or the mean ratings of the importance of each value. The rated importance of specific values was related to age; older persons gave higher ratings to tradition and benevolence,

FIGURE 18.3 THE STRUCTURE OF INDIVIDUAL VALUES

A theory of values developed by Schwartz (1992) identifies 10 motivationally distinct types of values. Each value is defined in terms of its central goal. The theory also specifies a structure of relationships among the values. Values that lead to actions that conflict with each other are located opposite each other; complementary values are located close to each other. Thus, actions that provide hedonistic rewards often conflict with social norms and traditions; actions that conform to social norms enhance security. Finally, the values can be thought of as lying along two dimensions, from self-enhancement to self-transcendence, and from conservation to openness to change.

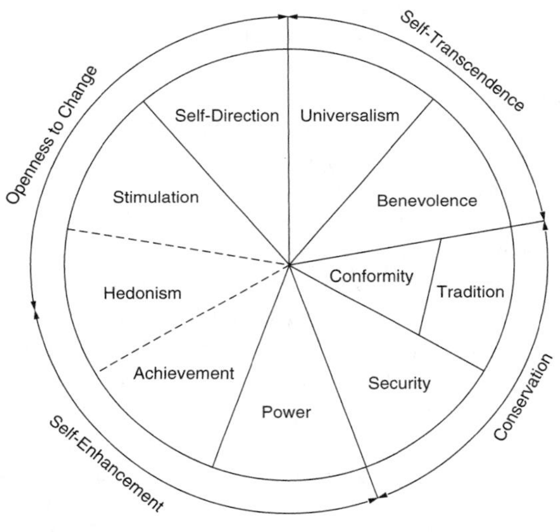

Source: From "Value Priorities and Gender" (1998) by Prince-Gibson and Schwartz, *Social Psychology Quarterly, 61,* 49–67. Used with permission from the American Sociological Association.

and lower ratings to achievement and hedonism. Education was also related to importance; as education increased, persons gave lower ratings to tradition, conformity, and power, and higher ratings to stimulation, self-direction, and universalism (Prince-Gibson & Schwartz, 1998).

Because values are general, they can provide integration or coherence across the many roles an individual plays (Hitlin, 2003). Although general, they influence many specific attitudes, choices, and behaviors as well. For example, values are related to our attitudes toward public policy. Thus, the importance one places on personal property and on social equality is related to one's attitudes toward paying higher taxes to help the poor (Tetlock, 1986). Those who place greater value on property will oppose higher taxes, whereas those who place greater importance on equality will favor increasing taxes to help the poor. Those who feel these values are equally important should find it hard to decide.

Values are related to choices. A study of university students assessed their values and asked them to respond to 10 hypothetical scenarios. Each scenario required a choice between two options, each representing a different value. The respondents' choices were consistent with their values (Feather, 1995).

Values are related to behavior. A survey of shoppers at supermarkets and natural food stores assessed several individual values. Shoppers who valued self-fulfillment and self-respect reported shopping more frequently and spending more money at natural food stores than those who valued sense of belonging and security (Homer & Kahle, 1988).

How do value systems arise? They are influenced by our location in the social structure. This section examines two aspects of social position that affect individual values—occupational role and education.

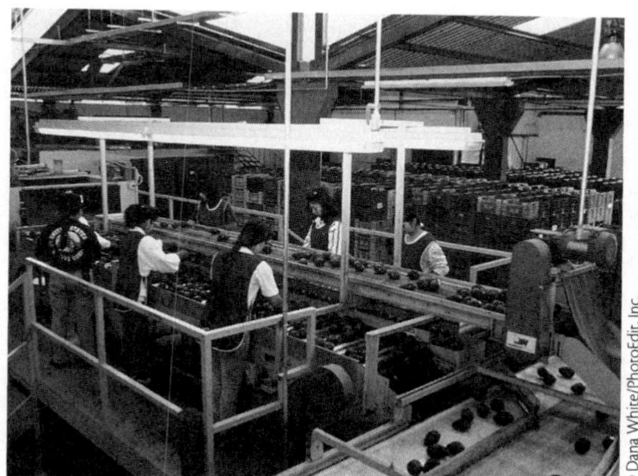

© Dana White/PhotoEdit, Inc.

Workers on an assembly line often experience alienation. Assembly-line jobs are monotonous, do not allow workers to exercise initiative, and give them no influence over working conditions.

Occupational Role

We spend up to half of our waking hours at work, so it is not surprising that our work influences our values. But occupational experiences vary tremendously. To determine their effect on values, we must identify the basic differences between occupations. Three important characteristics have been suggested (Kohn, 1969). The first is closeness of supervision—the extent to which the worker is under the direct surveillance and control of a supervisor. As a traveling salesman, Fred is rarely under close supervision, whereas Marie's work is supervised by the director of nursing and various physicians. The second occupational characteristic is routinization of work—the extent to which tasks are repetitive and predictable. Much of Larry's work is routine—pumping gas, tuning engines, relining brakes. But Larry's work is not highly predictable. From one day to the next, he never knows what kind of auto breakdown he will encounter or what unusual request some customer may make. The third characteristic is substantive complexity of the work—how complicated the work tasks are. Work with people is usually more complex than work with data or work with objects. Marie's occupation as a nurse is especially complex because she must constantly cope with the problems posed by doctors, patients, and families.

All three of these characteristics were measured in several studies of employed men to determine the impact of occupational role on values and personality (Kohn & Schooler, 1983). The results of these studies show a relationship between particular occupational characteristics and particular values: Men whose jobs were less closely supervised, less routine, and more complex placed especially high value on responsibility, good sense, and curiosity. Men whose work was closely supervised, routine, and not complex were more likely to value conformity. Thus, the occupational conditions that encourage self-direction—less supervised, nonroutine, complex tasks—are associated with valuing individual qualities that facilitate adjustment and success in a self-directed environment—responsibility, curiosity, and good sense. Occupational conditions that encourage adherence to a prescribed routine—close supervision and routine and simple tasks, such as bolting bumpers on new cars—are associated with qualities that facilitate success in that environment, such as neatness and obedience. This pattern has emerged in studies of employed men and women (Miller, Schooler, Kohn, & Miller, 1979) and in studies conducted in several countries including the United States, Japan, and Poland (Slomczynski, Miller, & Kohn, 1981).

Early studies of the relationships between workers' values and their occupational conditions revealed that workers exposed to particular conditions hold particular values. However, these studies were unable to determine with certainty whether adjustment to occupational conditions actually *caused* people to value particular qualities. Perhaps men who value curiosity and desire responsibility select occupations that allow them to exercise these traits (Kohn & Schooler, 1973). In attempting to identify the causal order, researchers compared the men's values and occupational conditions in 1974 with their values and occupational conditions 10 years earlier (Kohn & Schooler, 1982). What they found indicated causal effects in *both* directions between values and occupational conditions. Men who had valued self-direction highly in 1964 were more likely to be in work roles that were more complex, less routine, and less closely supervised 10 years later. Thus, values influenced job selection. At the same time, men who were in occupations that allowed or required self-direction in 1964 placed greater value on responsibility, curiosity, and good sense in 1974. Thus, their earlier job conditions influenced their later values.

A recent review of research on cross-national psychological differences concludes that opportunities for self-direction in one's work are consistently associated with differences in values (Schooler, 1996).

Education

Are differences in education also related to differences in an individual's values? The research by Kohn and his colleagues described in the preceding section demonstrated that men in jobs that are not closely supervised and that are nonroutine and substantively complex value self-direction, whereas men in jobs with the opposite characteristics value conformity. Education is associated with the value one places on these characteristics; the higher one's education, the greater the value placed on self-direction.

Substantively complex occupations involve working independently with people, objects, or data. Such work requires intellectual flexibility, the ability to evaluate information or situations, and the ability to solve problems. These abilities should be related to educational attainment, so education should be related to intellectual flexibility. Analyses of data from a sample of 3,101 men indicate that as education increases, so does intellectual flexibility (Kohn & Schooler, 1973). Thus, education influences both the value placed on self-direction and the abilities needed for success in substantively complex occupations.

In fact, it is possible to identify variation among students in self-direction. In one study, researchers assessed the complexity of a student's course work, of his or her most recent term paper or project, and of his or her extracurricular activities (Miller, Kohn, & Schooler, 1986). In a sample of students from seventh grade through the fourth year of college, the greater the substantive complexity of the student's work, the greater the value he or she placed on self-direction. Thus, the exercise of self-direction in school or at work increases the value one places on self-direction.

SOCIAL INFLUENCES ON HEALTH

Most of us attribute diseases to biological rather than social factors. But the transmission of disease obviously depends on people's interactions, and our physical susceptibility to disease is influenced by our lifestyles. This is true, for instance, with a disease such as AIDS. Similarly, our mental health is influenced by our relationships with relatives, friends, lovers, professors, supervisors, and so on. Thus, social position affects both physical and mental health. This section examines the impact of occupation, gender, marital role, and social class on physical health. It also considers the relationship between these factors and mental health.

Physical Health

Occupational Roles What do the physician addicted to alcohol, the executive with an ulcer, the coal miner with black lung disease (chronic obstructive pulmonary disease), and the factory worker with chronic back pain all have in common? The answer is, a health problem that may be due largely to their occupational role.

Occupational roles affect physical health in two ways. First, some occupations directly expose workers to health hazards. Miners who are exposed to coal dust, workers exposed to chemical fumes, and workers who process grain often suffer damage to lung tissue. Waitstaff, bartenders, and kitchen workers exposed to cigarette smoke may develop lung cancer. Workers exposed to various toxic chemicals may die of bladder cancer. Occupational conditions caused 60,300 deaths and 862,200 illnesses in 1992, about 165 deaths and 2,300-plus illnesses per day (Leigh, Markowitz, Fahs, Shin, & Landrigan, 1997). The highest death and injury rates are in construction and transportation (U.S. Bureau of Labor Statistics, 2004).

Second, many occupational roles expose individuals to stresses that affect physical health indirectly. Each of the roles we play carries a set of obligations or duties. Meeting these demands requires time, energy, and resources. When these demands exceed the person's perceived ability to meet them, the result is stress (Lazarus, 1991).

Stress is related to the number of physical ailments we experience. In one study, researchers asked adults to report the amount of stress in their lives and their health problems 20 times over a 6-month period. Those who reported higher levels of stress also reported more health problems, including sore throats, headaches, and flu. Increased stress was associated with more ailments on the same day and on subsequent days (DeLongis, Folkman, & Lazarus, 1988). The connection between stress and physical health may be the body's immune system. A study of college students found that as their reports of stress increased, the concentration of antibodies in their saliva decreased (Jemmott & Magloire, 1988). The lower the level of antibodies, the more susceptible one is to illness.

Stress is often temporary. A move from one apartment to another, for example, will be stressful during the weeks when all the arrangements are being made and during the move itself. As one becomes settled, however, the demands will decline. Also, the ability to respond to the demands of moving may increase as one learns how to cope with packing, disconnecting the old phone and getting a new one, and so on. An important, until now unrecognized, stress may be a "broken heart," or what physicians refer to as stress cardiomyopathy. Data from small samples indicate that sudden emotional stress releases massive amounts of hormones that stun the heart, creating symptoms that mimic heart attack—chest pains, shortness of breath, and heart failure. However, electrocardiograms and ultrasound indicate patterns quite distinct from those of patients who have had a heart attack (Ebert, 2005).

On the other hand, stress may be continuous. Chronic stress may lead to physical illness. Excellent evidence of this link comes from a longitudinal study of two samples of adults (140 and 190 persons, respectively) employed at a large company (Maddi, Bartone, & Puccetti, 1987). Each person's level of stress was assessed by a carefully designed measure of stressful life events. One or 2 years later, each person completed a questionnaire regarding illness that included both mild (for example, influenza) and serious (for example, heart attack) conditions. There were strong associations between the level of stress experienced initially and reported illness 1 or 2 years later.

Many people spend energy, time, and money jogging, playing tennis, or working out. Does it do any good? There is evidence that people who are physically fit are less likely to experience stress-related illness. One study of students obtained self-reports of time spent per week in each of 14 fitness activities. The researchers also assessed fitness directly, measuring blood pressure, aerobic capacity, and endurance. Higher levels of self-reported fitness were associated with higher levels of health. Greater fitness as measured directly was associated with fewer visits to the student health center (Brown, 1991b). However, whereas stress is related to health, it is not related to self-reported fitness (Roth, Wiebe, Fillingian, & Shay, 1989). Fitness does not reduce the amount of stress one experiences, but it does reduce illness. A review of literature, including randomized clinical trials of physical-activity interventions, concludes that persons who engage in regular physical activity are less likely to experience several diseases and have a better quality of life (Penedo & Dahn, 2005).

The most widely studied relationship between job characteristics and physical health is the impact of occupational stress on coronary heart disease. As one's workload increases—including perceived demands on one's time, number of hours worked, and feelings of responsibility—so does the incidence of coronary heart disease (House, 1974). Heart attacks are associated with a high level of serum cholesterol in the blood. Several studies report that the level of serum cholesterol rises among persons under high work-related stress (Sales, 1969). This suggests another tangible link between role demands and physical health.

An important aspect of jobs that are associated with an increased risk of heart attack is the lack of control over work pace and task demands (Karasek et al., 1988). Occupations associated with the highest risk include cooks, waitstaff, assembly-line operators, and gas station attendants. These jobs are characterized by high demand—heavy workload and rapid pace—over which the worker has little or no control. Cashiers and waiters are four to five times more likely to have a heart attack than foresters or civil engineers. One study assessed the contribution of lack of control on the job to coronary heart disease, controlling for other factors, including individual risk factors and the availability of social support (Marmot, Bosma, Hemingway, Brunner, & Stansfeld, 1997). Longitudinal data were obtained from 7,372 employed men and women at three points in time. A reported lack of control at time 1 was associated with increased incidence of self-reported chest pain and angina and doctor-diagnosed angina or heart attack at times 2 and 3.

Gender Roles Who is more likely to experience coronary heart disease, cirrhosis of the liver, or lung cancer—men or women? You probably picked men, and if you did, you are right. Men are twice as likely as women to die from these conditions. Although there is evidence that genetic and hormonal factors play a role, traditional role expectations for men and women and occupational role segregation in our society are also significant factors.

Some gender differences in health are associated with reproductive roles (Macintyre, Hunt, & Sweeting, 1996). Health problems related to reproduction, such as premenstrual syndrome and pregnancy-related conditions are most likely among women of childbearing age. Men exposed to various chemicals, heat, or radiation may experience reduced sperm production, abnormal sperm types, or impaired sperm transfer (National Institute of Occupational Safety and Health, 2002). Hormonal changes at menopause effect the physical health of some women, causing osteoporosis and an increased likelihood of broken bones. Older men are very likely to experience enlarged prostate glands, and may die of prostate cancer.

The top four causes of death are the same for men and women in the United States: heart disease, cerebrovascular disease, cancer, and chronic respiratory disease. However, age-adjusted death rates indicate that males are at greater risk (about 1.6 to 1.0) of all except cerebrovascular disease (National Center for Health Statistics, 2004). One analysis suggests that the greater likelihood of male death from these illnesses is related to smoking (Case & Paxson, 2004). There is also variation by ethnicity. Blacks have the highest age-adjusted death rates of heart disease, cerebrovascular disease, and cancer, while Hispanics have the lowest rates of these three, with Whites intermediate.

Role overload, in which the demands of one's role(s) exceed the amount of time, energy, and other resources the person has, is associated with coronary heart disease. Professionals such as physicians, lawyers, accountants, and so on are especially vulnerable to overload; until recently, the persons holding these positions have been primarily men. Other studies have shown that heart attacks are correlated with certain personality traits known as coronary prone behavior patterns (Jenkins, Rosenmann, & Zyzanski, 1974). People who exhibit these behavior patterns are work oriented, aggressive, competitive, and impatient. They often initiate two or more tasks simultaneously (Kurmeyer & Biggers, 1988). Men are much more likely to be characterized by this behavior pattern than women.

Men are more prone than women to have cirrhosis of the liver because they are more likely than women (1.7 to 1) to be heavy drinkers (Schoenborn, 2004). Until recently, men were much more likely to smoke cigarettes and, therefore, more likely than women to contract lung cancer and emphysema. They are also more likely to die in auto accidents, both because of higher rates of driving under the influence of alcohol and because of poor driving habits (Waldron, 1976).

Clearly, certain behaviors increase the risk of illness or death. The transmission of these health risk behaviors, such as not eating an adequate diet, smoking, and drinking, was the focus of a study of 330 teenagers and their parents. The results showed that the father's lifestyle affected only boys and the mother's lifestyle affected only girls (Wickrama, Conger, Wallace, & Elder, 1999). Thus, health risk behaviors are learned as part of gender role socialization.

These generalizations highlight overall differences between men and women, but we need to recognize that gender roles vary by culture and subculture—and that gender roles are changing (Watkins & Whaley, 2000); for example, increasing numbers of women are smoking, and so the gender gap in deaths due to lung cancer will gradually get smaller.

Marital Roles Marriage is associated with physical health. Married men and women are less likely to report conditions such as back pain and headaches, and limitations on activity (Schoenborn, 2004). They experience fewer acute and chronic health conditions (Ross, Mirowsky, & Goldsteen, 1990). On a variety of indicators, widowed persons are more likely than the divorced or separated to experience health problems. These patterns are found in Whites, Blacks, and Hispanics.

Why is it that being married protects people against illness and accidents? The most likely explanation is that married persons are less likely to engage in behaviors that expose them to illness and accidents. They probably eat and sleep better than unmarried persons. They are less likely to smoke and drink (Schoenborn, 2004). They may take fewer risks, reducing the likelihood they will be involved in accidents. Finally, they may be more likely to seek medical care when ill (Verbrugge, 1979).

This explanation suggests that the health advantage of married people is the result of living with another person. To test this interpretation, data were analyzed from a national sample of women. Measures of illness included the number of days spent in bed and the number of doctor visits in the past year. Women who lived with another adult reported no more illness than married women, regardless of whether they were single, separated, divorced, or widowed (Anson, 1989). When another adult is present, she or he can provide emotional support, help identify illness early, and provide care that encourages rapid recovery.

Is merely being married sufficient to reduce risk? Perhaps. But being happily married is even more beneficial. According to one study, married men and women who were satisfied with their marital roles reported better physical health than those who were dissatisfied (Wickrama, Conger, Lorenz, & Matthews, 1995). On the other hand, marital conflict has been shown to have direct negative influences on cardiovascular, immune, and other physiological mechanisms. Marital stress also has indirect consequences for health by increasing depression and by negatively affecting health behaviors (Kiecolt-Glaser & Newton, 2001).

In contemporary U.S. society, one-half of marriages end in divorce, and most divorced persons remarry. How do these transitions—which are often stressful—affect health? A longitudinal study of the relationship between marital status and mortality in a sample of 12,484 people shows that the longer one is married, the greater one's life expectancy. Women especially benefit from not being single. On the other hand, any transition—to (re)marriage, divorce, or widowhood—increases the risk of death (Brockmann & Klein, 2002).

Social Class We noted earlier that status has a major impact on lifestyle. One aspect of this impact is the effect of class on physical health.

A model of the influences on health is presented in Figure 18.4 (Williams, 1990). In this model, socioeconomic status is one of three influences on health. Whether education, occupation, or income is used as

FIGURE 18.4 A MODEL OF THE INFLUENCES ON HEALTH

There are several influences on a person's physical and mental health. Biomedical, socioeconomic (such as occupation), and demographic (including age, race, gender, and marital status) factors all influence health, both directly and indirectly. All influence such immediate psychosocial factors as health practices and stress. Socioeconomic status is the major influence on the amount and quality of medical care available to the person: The availability of medical care, in conjunction with the other factors, influences how physically and mentally healthy one is.

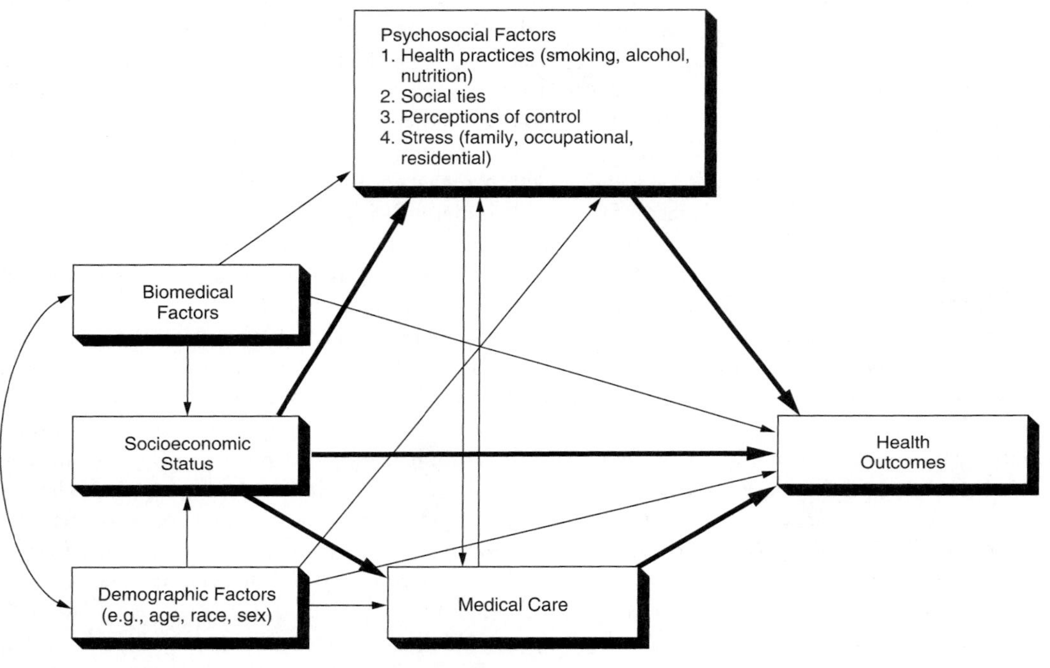

Source: From "Socioeconomic Differentials in Health: A Review and Redirection" (1990) by David R. Williams, *Social Psychology Quarterly*, 53(2), 82. Used with permission from the American Sociological Association.

the indicator of status, lower socioeconomic status groups in the United States experience higher death rates. A study of 18,733 deaths in 1986 found that social class was strongly correlated with mortality (Rogers, 1995). The highest mortality was found among persons who were single and poor. Controlling for class, there were few differences by race. Rates of chronic illness are higher among lower status persons, as are rates of various disabilities. Rates of infant mortality are also negatively related to social class; the rate of mortality among Black infants is twice as high as the rate for White infants.

Several factors have been identified as causes of the negative relationship between class and health. First, persons with higher status are more likely to be employed full-time, to have subjectively rewarding jobs, and to have higher income. Second, higher status persons are more likely to have a sense of control on the job and of control over their lives and health. Finally, higher status persons are less likely to engage in health risk behaviors—smoking and heavy drinking—and more likely to eat properly, exercise, and use health care services. In other words, social class is associated with several of the factors discussed earlier. Analyses of data from two national probability samples found that full-time employment, sense of control on the job, and lifestyle were all related to self-reported health (Ross & Wu, 1995). A study of mortality differences between men and women found that differences in death rates by class (income, occupation) were as large for women as for men (Koskinen & Martelin, 1994). The exception was among married women, whose death rates varied little by class—that is, being married compensates for the health risks experienced by working-class single women.

Class differences in physical health vary across the life course (House et al., 1994). Differences in health associated with class are small in early adulthood (25 to 44), then increase with age (45 to 54) until late in life, when they become small again (75 and older). The large differences in later adulthood reflect greater exposure of lower class persons to risk factors such as occupational health hazards, lack of control on the job, excessive alcohol consumption, and number of stressful life events.

In sum, the relationship between social status and physical health is complex. Occupational roles may expose men and women directly to health risks, causing illness and death, or to the stressful effects of lack of control on the job. Gender differences in reproductive roles lead to differences in health risks. Role overload is associated with some occupations and differentially affects men and women due to occupational segregation. Being married or living with another adult reduces one's health risk. Social class is also associated with differences in mortality, due to its association with occupation and lifestyle.

Mental Health

At the beginning of this chapter, we introduced Larry, who owns a service station, works long hours, and earns about $48,000 per year. He has two children, a large mortgage on his home, and he has trouble making ends meet. He comes home from work every day exhausted. He worries about the economy and whether or not there will be another energy crisis, leading to inadequate supplies of gasoline, or an oil glut, leading to gasoline price wars. Either one would ruin his business, because more than one-half of his income is from gasoline sales.

Like many Americans, Larry finds that his life situation is very demanding. His customers expect him to do high-quality repair work at low prices, his wife expects him to support the family and spend time with her, and his children want more toys and electronics than he can afford. At times, the demands made on him by others exceed his ability to cope with them, causing psychological stress. People who are under stress often become tense and anxious, are troubled by poor appetite, or experience insomnia. A widely used questionnaire designed to measure stress-related symptoms is reproduced in Box 18.2. Many of the items on this scale measure behavior, thoughts, and feelings associated with depression.

Stress is a major influence on mental health. Short-term stressors, such as final exam week or an approaching deadline, may produce a temporary increase in stress-related symptoms or depression. As soon as the exams are over or the deadline passes, mood may return to normal. Long-term stressors, such as the continuing economic worries that Larry experiences, may produce impaired psychological functioning. Neuroses, schizophrenia, and affective disorders such as depression are among the mental illnesses associated with severe stress. The experience of stress and impaired psychological functioning varies by occupation, by gender, by marital and work roles, by membership in social networks, and by social class. Some particularly stressful events have an impact on large numbers of people simultaneously. Box 18.3 provides an example of such a stressful event.

Occupational Roles Work-related stress not only affects physical health but also can affect mental health. We noted earlier that occupations that involve heavy workloads and in which workers have little or no control over their work pace are stressful. Physicians are under high levels of stress; they are expected to heal their patients, to be available to others whenever needed, to be compassionate, and to be cost-effective. These expectations are internalized by most physicians, and they may be exacerbated by personal traits such as perfectionism. The high rates of suicide, depression, substance abuse, and marital problems

BOX 18.2 Test Yourself: How Do You Respond to Stress?

Stress is a discrepancy between the demands on a person and his or her ability to successfully respond to those demands. Individuals under stress experience a variety of physical and psychological symptoms. A widely used measure of these symptoms is reproduced here. Read the instructions and complete the scale.

Here is a list of ways you might have felt or behaved in *the past week*. Read each of the following statements and then circle the appropriate number to the right of the statement to indicate how often you have felt this way *during the past week*.

During the past week:	Rarely or none of the time (less than 1 day)	Some or a little of the time (1–2 days)	Occasionally or moderate amount of the time (3–4 days)	Most or all of the time (5–7 days)
1. I was bothered by things that usually don't bother me.	0	1	2	3
2. I did not feel like eating; my appetite was poor.	0	1	2	3
3. I felt I could not shake off the blues even with help from my family and friends.	0	1	2	3
4. I felt that I was just as good as other people.	0	1	2	3
5. I had trouble keeping my mind on what I was doing.	0	1	2	3
6. I felt depressed.	0	1	2	3
7. I felt that everything I did was an effort.	0	1	2	3
8. I felt hopeful about the future.	0	1	2	3
9. I thought my life had been a failure.	0	1	2	3
10. I felt fearful.	0	1	2	3
11. My sleep was restless.	0	1	2	3
12. I was happy.	0	1	2	3
13. I talked less than usual.	0	1	2	3
14. I felt lonely.	0	1	2	3
15. People were unfriendly.	0	1	2	3
16. I enjoyed life.	0	1	2	3
17. I had crying spells.	0	1	2	3
18. I felt sad.	0	1	2	3
19. I felt that people disliked me.	0	1	2	3
20. I could not "get going."	0	1	2	3

These questions measure depression, a common response to stress. This is the CES-D (Center for Epidemiological Studies-Depression) scale.

You can determine your score by adding up the numbers you circled (0, 1, 2, or 3) for all of the items except items 4, 8, 12, and 16. Notice that these four items refer to positive feelings, whereas the other items refer to negative ones. To score items 4, 8, 12, and 16, give yourself 0 if you circled 3, 1 if you circled 2, 2 if you circled 1, and 3 if you circled 0. Total scores on the scale may range from 0 to 60. If your score is more than 16 points, you could be diagnosed as depressed.

among physicians appear to result from these stresses (Miller & Megowen, 2000).

Beyond the stresses associated with specific occupations, one's economic circumstances are an important source of stress (Voydanoff, 1990). Economic hardship—insufficient income to meet basic needs—is stressful. Interviews with more than 2,000 adults showed that not having enough money to provide food, clothing, and medical care for self or family was the major variable associated with depression scores (Pearlin & Johnson, 1977).

Economic uncertainty—concern over one's prospects of finding or keeping a job—is also stressful. A study of 7,095 workers found that the level of unemployment in an industry as a whole is associated with the level of distress experienced by employed workers in that industry (Reynolds, 1997). The relationship was stronger for workers in complex, rewarding jobs. Unemployment is especially debilitating (Vinokur, Price, & Caplan, 1996). It is associated with anxiety, depression, and admission to mental hospitals (Voydanoff, 1990), and with an increased risk of death (Voss, Nylen, Floderus, Diderichsen, & Terry, 2004). In addition to economic rewards, one's work is a highly salient role identity for many people. The meaning the individual attaches to work influences its importance to psychological well-being (Simon, 1997). The impact of unemployment is greater on men than on women, probably because the work role identity is more salient for men.

In some instances, the stress associated with unemployment leads to violence. Each year, there are incidents in which a fired employee returns to the workplace with guns or rifles, often seeking the person believed responsible for his or her being fired. These incidents can lead to injuries and deaths, and to suicide by the former employee.

Gender Roles Adult women in the United States have somewhat poorer mental health than men. On measures of distress, women attain significantly higher scores than men. For example, Macintyre, Hunt, and Sweeting (1996) found that women report greater malaise (sleep problems, difficulty concentrating, worry, and fatigue) than men at all ages. Are women under greater stress? Or are they more likely to report symptoms of distress than men? One study assessed the frequency of experiencing emotions such as anger, sadness, and happiness and the levels of distress in men and women (Mirowsky & Ross, 1995). The results indicated that women expressed emotions more freely than men, but this did not fully account for the differences in stress scores. Overall, women experienced stress about 30 percent more often than men.

In an attempt to identify sources of chronic stress in women's lives, researchers asked female students, professionals, and mothers to write items representing stressful situations for women. A factor analysis yielded 5 factors, including fear of being unattractive, of victimization, and of failing to be nurturing (Watkins & Whaley, 2000). These sources of stress may be unique to women.

There also may be a gender difference in how people respond to stress. Whereas men experiencing high stress report higher rates of substance use and abuse, women experiencing high stress report higher rates of impaired psychological functioning. In a survey of 3,131 adults, stress was measured by the number of life events experienced by the person, or by someone close to him or her, in the prior 6 months. Respondents were also asked questions about participation in various behaviors and psychological functioning. Men experiencing stress were more likely to report alcohol or drug use or dependence, whereas women experiencing stress reported increased anxiety and emotional disorders (Armeli, Carney, Tennen, Affleck, & O'Neil, 2000).

Marital Roles Just as married men and women are physically healthier than people who are not married, they are characterized by greater psychological well-being and less depression than single, separated, divorced, or widowed persons (Ross et al., 1990). Again, it appears that it is the presence of another adult in the residence—rather than being married per se—that is associated with being healthy.

Of course, it might be that people who have higher levels of well-being are more likely to marry or cohabit, whereas persons with lower levels remain single. A longitudinal study of 18- to 24-year-old men and women, some of whom married whereas others remained single, found that marriage did improve

BOX 18.3 The Impact of September 11, 2001, on Psychological Functioning

Much of the research on the impact of stress on health and psychological functioning is concerned with "everyday" stressors. These include stressful events such as moving, starting a new job, changes in relationships, and conflict with family or coworkers. They also include chronic stressors, such as prolonged major illness or the burden of caring for a family member with a disability. The research summarized in this chapter indicates that these relatively common stressors affect both physical and mental health, in some cases seriously. Occasionally, a large group or a population experiences extreme situations—events that Dohrenwend (2000) terms fateful. **Fateful events** share several characteristics; they are beyond the individual's control, unpredictable, often life-threatening, often large in magnitude, and they disrupt people's usual activities. Such events have consequences far beyond those of everyday stressors.

One such event in the United States was the terrorist attacks on September 11, 2001, on the World Trade Center and the Pentagon. More than 2,800 people were killed, affecting hundreds of workplaces and thousands of families. There was the immediate stress of the deaths and destruction, the continuing stress for persons in New York and Washington due to the disruption of their lives, and the continuing stress for most Americans due to the uncertainty about further attacks. Because the event and its consequences were uncontrollable, thousands of people experienced fear, anger, and rage that could not be channeled into effective action.

A wide variety of responses are seen in adults following such traumatic events. In addition to the emotions just listed, they include disbelief, irritability, anxiety, depression, sleep disturbances, and increases in alcohol and other substance use. For most individuals, these acute traumatic stress symptoms resolve over time. For some, however, the intensity and duration of the symptoms justify the diagnosis of post-traumatic stress disorder (PTSD; Norwood, Ursano, & Fullerton, 2002).

Excellent data on the impact of the terrorist attacks comes from the longitudinal National Tragedy Survey conducted by the National Opinion Research Center (NORC) at the University of Chicago (Rasinski, Berktold, Smith, & Albertson, 2002).

Two weeks after the attacks, a probability sample of 1,013 Americans and a probability sample of 406 residents of New York City were interviewed. Four to 6 months later, in January to March 2002, re-interviews were completed with 805 Americans (79 percent of the original sample) and 296 residents of New York City (73 percent).

To begin, consider national pride. One goal of the terrorist attacks was to demoralize the American population. However, measures of pride taken 2 weeks after September 11 either remained the same or increased compared to previous NORC surveys. For example, 97 percent of those surveyed said they would rather be a citizen of America than of any other country. On the other hand, confidence in major institutions, including the executive branch, Congress, and banks and financial institutions fell by 7 to 13 percent; the exception was confidence in the U.S. military, which remained at pre-attack levels (81 percent).

The interview included questions assessing the experience of stress. In the first survey, respondents were asked which of 15 symptoms they had experienced following the attacks. Nine symptoms were reported by 20 percent or more of the national sample; the five most common were crying (60.3 percent), trouble sleeping (51.2 percent), feeling nervous or tense (49.9 percent), feeling dazed and numb (45.7 percent), and feeling more tired than usual (37.5 percent). In the sample of New York City residents,

BOX 18.3 Continued

12 symptoms were reported by 20 percent or more; the five most common were crying (74.1 percent), feeling nervous or tense (62.5 percent), trouble sleeping (59.4 percent), feeling more tired than usual (47.6 percent), and not feeling like eating (46.4 percent). Not surprisingly, New York City residents were somewhat more likely to report specific symptoms. Clearly, the impact of the attacks was substantial, measured by the numbers who experienced symptoms of stress.

At the re-interview 4 to 6 months later, reports of symptoms experienced in the past 2 weeks were generally lower than in the initial survey. In the national sample, the numbers reporting crying declined 40 percent, those reporting trouble sleeping declined 20 percent, feeling tense and nervous 22 percent, and feeling dazed and numb 33 percent. Comparable declines were observed in the reports of New York City residents. These longitudinal results indicate a trend toward recovery—that is, reduced incidence of stress. Declines were greatest among those who reported that they knew someone who had been hurt or killed in the attacks. African Americans, on the average, reported fewer symptoms in September, but showed little decline in the follow-up interviews. Slower recovery was also observed in persons with less than a high school education and with family incomes less than $40,000 per year. The follow-up interview included a standard measure of PTSD. Among the New York City residents, 15 percent scored in the range indicative of the disorder. In the national sample, the percentage scoring in this range was 8 percent. The highest PTSD scores were observed among those with poor general health, less education, and less income.

These results suggest that those of vulnerable social status have more or longer lasting adverse reactions to fateful events. Research on a broad variety of disasters, including natural disasters such as earthquakes (Seplaki, Goldman, Weinstein, & Lin, 2003) and hurricanes, finds that persons of low SES in that society (whether the country is rich or poor), people who are socially isolated, and people who are directly affected by the disaster report higher levels of depressive symptoms. Several hypotheses have been offered to account for the increased vulnerability of low SES persons. First, such persons may be more likely to have suffered as a result of the event due to poor housing or inadequate public services. Second, economic assets and education can enhance recovery because the individual has access to resources. Third, persons who are of low SES may also be socially isolated, lacking the social support that family and friends can provide.

In response to the rise in reports of symptoms and of widespread PTSD after traumatic events, debriefing has become standard clinical practice. Debriefing involves group sessions in which those affected by a fateful event are encouraged to process their emotional reactions. Following many recent traumatic events, such as 9/11 and the Columbine High School shootings, large numbers of persons have participated in debriefings. Do they help persons cope with traumatic events? An expert panel assembled by the National Institutes of Mental Health (2002), having reviewed the empirical evidence, concludes that early interventions such as debriefing do not reduce the risk of later or continuing disorder. Research, such as the surveys following the 9/11attacks, show that most people are resilient; relying on their own resources, social support networks, and community services, they will recover. Psychological services should be available for those whose personal resources are not sufficient, who do not recover on their own.

well-being (Horowitz, White, & Howell-White, 1996). Another study with data from a national sample compared persons who were stably married with those who experienced separation or divorce. Persons who experienced a loss reported increased symptoms on the CES-D (see Box 18.2); the effect was greater on persons who believed marriage should be a lifelong commitment (Simon & Marcussen, 1999).

The greater well-being of married persons reflects the beneficial effects of social support. A spouse can provide social support—care, advice, and aid in times of stress—and emotional support. Do husbands and wives share equally in receiving these benefits? Apparently not. Married men are characterized by better mental health than married women (Kurdek, 1991).

A study of a representative sample of more than 13,000 adults assessed the relationship between roles and mental health in four ethnic groups (Jackson, 1997). Occupying the spousal role was associated with greater well-being among Blacks, Mexican Americans, Puerto Ricans, and non-Hispanic Whites. Occupying other family roles, especially sibling, was related to better mental health in all groups except Puerto Ricans.

Some researchers focus on the interrelations of work and family roles. Of particular interest is work-family conflict—the extent to which the demands associated with one role are incompatible with the other. One common circumstance is **spillover,** in which the stress experienced at work or in the family is carried into the other domain (Bolger, DeLongis, Kessler, & Wethington, 1989). A study of air traffic controllers and their wives documented the impact of work stress on marital interaction (Repetti, 1989). As the controller's daily workload increased (larger number of planes handled, poorer visibility), wives reported that the men were more withdrawn at home. A longitudinal study of 166 married couples obtained completed diaries from both partners for either 28 or 42 consecutive days. Each person reported on stressors at work (too much to do, arguments with coworkers) and at home (too much to do, arguments with spouse, arguments with child). For both husbands and wives, increased stress at work was associated with increased stress at home (Bolger et al., 1989).

Stress associated with marital roles can influence work role performance. The research by Bolger et al. (1989) also found that for husbands, increased stress at home was associated with increased stress at work. Forthofer, Markman, Cox, Stanley, and Kessler (1996) analyzed data from a study of 8,098 persons ages 15 to 54; they focused on 1,431 employed married men and 1,138 employed married women. The results indicated that problems within the marriage were related to the number of days the person was "unable to function" in the preceding month among both men and women.

If marital strains can lead to reduced performance at work, can positive experiences at home enhance work experience? Two studies suggest that the answer is "yes." Barnett (1994) studied 300 full-time employed women in dual-earner couples. Positive experiences in the roles of partner or parent buffered the effects of negative job experiences on distress. A longitudinal study found that increases/decreases in marital satisfaction are related to increases/decreases in work satisfaction, among employed men and women, but not the reverse (Rogers & May, 2003).

Do married employed women and married employed men respond differently to work-family conflict? Apparently they do not. A study of 200 dual-earner couples found that high levels of parental stress and of occupational stress had the same effect on reports of depression symptoms by husbands and wives (Windle & Dumenci, 1997).

Work-family conflict can affect the quality of the marital relationship by influencing the couple's interaction. The study of air traffic controllers discussed earlier found that as the stress experienced at work increased, husbands became more withdrawn at home. Other research demonstrates the impact of conflict and the resulting distress on two dimensions of marital interaction—hostility and warmth. As distress increases, both the person and the spouse report greater hostility and less warmth (Matthews, Conger, & Wickrama, 1996). Work-family conflict also is related to alcohol consumption. A study of employed adults ages 35 to 65 found that higher levels of marital disagreement (regarding spending money or household tasks) and stress at work (too much to do, conflicting demands) were each related to reports of problem drinking. Positive spillover from family (talking at home, expressions of love) to work was associated with less frequent reports of problem drinking; interestingly, both positive and negative spillover from work to family were associated with greater problem drinking (Grzywacz & Marks, 2000).

Another type of spillover involves the worker using behavior patterns acquired or reinforced at work in interactions with family members. This possibility was explored in a study that linked occupational conditions to the use of violence by a man against his female partner. The results showed that men in violent (for example, law enforcement) and dangerous (for example, construction) occupations were more likely to engage in violence directed at the partner (Melzer, 2002). This may reflect a spillover of the stress associated with these occupation, or the violence supportive attitudes learned on the job.

How do people cope with work-family conflict? They use one or more of several strategies. In one study, wives reported the use of planning and cognitive restructuring—changing their definition of the situation; for example, deciding the house does not need to be cleaned every week. Husbands reported restructuring and withdrawing from interaction (Padden & Buehler, 1995). In a study of 221 managers, both men and women reported the use of prioritizing, reducing their personal standards (restructuring), asking others for help, and ending involvement in one or more roles (for example, in community organizations; Kirchmeyer, 1993).

Thus, the relationships between occupational, gender, and marital roles and psychological well-being are complex (Ross et al., 1990). The demands of work roles may lead to distress; this is especially likely when work demands are high and not under the person's control. Economic hardship and unemployment cause distress. Men typically have somewhat better mental health than women, in part because men react to stress by drug and alcohol use whereas women respond psychologically. Married men report greater well-being than married women, apparently because wives are more likely to experience family-related strains (Mennino, Rubin, & Brayfield, 2005). Stresses experienced at work can spill over and affect marital relationships; conversely, strain at home can produce losses at work. In any case, the social roles one occupies are major influences on mental health.

Social Networks Up to this point, we have reviewed evidence showing that our relationships with others—that is, our membership in social networks—can be major sources of stress. At the same time, social networks can serve as an important resource in coping with stress (Wellman & Worley, 1990). Network members provide us with social support and with help during stressful events and teach us strategies for coping with stressors.

First, a network of close friends and kin eases the impact of stressful events by providing various types of support (Cooke, Rossmann, McCubbin, & Patterson, 1988; House, 1981). One type is emotional support—letting us know that they care for and are concerned about us. Emotional support is an important buffer for negative psychological states like depression (Harlow & Cantor, 1995). A second type is esteem support—providing us with positive feedback about our abilities and worth as a person. A poor grade, for example, is less stressful if our friends let us know they think we are a good student. Informational support from others prepares us to avoid problems or to handle them when they arise. Advice from friends on how to handle job interviews, for example, improves our ability to cope with this situation. Finally, network members provide each other with instrumental support—money, labor, and time. Research shows that people who report poor well-being tend to seek out others who can provide the type of support they need (Harlow & Cantor, 1995).

The presence or absence of support is a major determinant of the impact of a stressful life event. A study of 882 women seeking an abortion obtained longitudinal data from 615 of them. Before the abortion, each woman rated the degree to which she received positive (expressed concern, offered help) and negative (argued, criticized) support from her partner, mother, and friends. Perceptions of positive support from each source was associated with greater well-being following the abortion (Major, Zubek, Cooper, Cozarelli, & Richards, 1997).

Research has documented the impact of supportive relationships on the individual's ability to cope

Family members help us to cope with stressful events, such as the death of a relative or close friend. They are an important source of emotional support and may help by temporarily taking over some of our role responsibilities.

with stress. A longitudinal study of a representative sample of 900 adults focused on the relationship between social network membership and physical health (Seeman, Seeman, & Sayles, 1985). Persons who reported in the initial interview that they had instrumental support available (that is, persons who, when ill, had others who would call, express concern, and offer help) were in better physical health 1 year later. Another longitudinal study assessed the impact of family support on mental health (Aldwin & Revenson, 1987). The sample consisted of 245 men and 248 women from randomly selected families in an urban area. The availability of support from one's family at the time of the initial survey was associated with better psychological adjustment one year later. Other research indicates that individuals with family support are more likely to cope with stressful events by using active strategies rather than avoidance or withdrawal strategies (Holohan & Moos, 1990). A longitudinal study of the relation between coping strategies and mental health found that people who used active strategies at the time of the initial survey reported fewer psychological symptoms on the second survey (Aldwin & Revenson, 1987). Finally, using active behavioral coping strategies is associated with shorter duration of several types of stressful events (Harnish, Aseltine, & Gore, 2000).

A second way social networks reduce stress is by teaching us strategies for coping with stressful events or crises when they occur. When members of a group are all subjected to similar stressors, the group may develop coping strategies. A study of interns and residents in a hospital found that they were subjected to long hours of demanding work in often poor facilities (Mizrahi, 1984). These physicians coped with stress by minimizing the time spent with each patient, by limiting interaction with patients to "relevant topics," and by treating patients as nonpersons—for example, by focusing exclusively on their illness. These strategies were passed on from experienced group members to new ones.

Several types of relationships can provide support, including primary kin (parents, siblings, and adult children), secondary kin, and friends and neighbors. The kind of support provided depends on the type of relationship. Persons to whom we have strong ties provide emotional support and companionship. Primary kin provide us with financial aid and services, whereas friends and neighbors give us services and emotional support (Wellman & Worley, 1990). Research indicates that, among Blacks, kin primarily provide services such as transportation and child care, whereas among Whites, kin are more likely to provide financial support; however, the data suggest that this difference may have more to do with social class than with race (Sarkisian & Gerstel, 2004). Also, Black women are more likely to engage in reciprocal exchanges of services, White women to engage in reciprocal exchanges of emotional support. Research on provision of support among low-income families suggests that these exchanges of services provide an important resource for coping with daily demands (Henley, Danziger, & Offer, 2005).

Recognizing the significance of supportive relationships, an innovative approach, social network mapping, is being used to assess the support available to organ transplant recipients (Lewis, Winsett, Cetingok, Martin, & Hathaway, 2000). The map both increases the person's awareness of the resources available and enables health professionals and social workers to work with the person more effectively.

Social Class The lower a person's socioeconomic status, the greater the amount of stress reported (Mirowsky & Ross, 1986). According to data from interviews with a representative sample of U.S. adults, 8.3 percent of the poor, 5.3 percent of the near poor, and 2 percent of the nonpoor are characterized by serious psychological distress (National Center for Health Statistics, 2004). Education, occupation, and income are the principal measures of socioeconomic status. Does low standing on each of these components contribute to stress independently? Or is stress related to only one or two of these components?

An analysis of data from surveys of eight quite diverse samples (Kessler, 1982) shows a consistent pattern: low education, low occupational attainment, and low income contribute separately to stress. The relative importance of these three components as sources of stress is different for men and women. For men, income appears most important; for women (employed or not) education appears to be the most important component. Occupational attainment is the least important determinant of stress for both genders.

How does education affect stress? Earlier in this chapter we summarized research that shows that people with more financial resources, more control over their work, and partners who provide support are in better physical and mental health. Research shows that people who are well educated have lower levels of distress, primarily because of paid work and financial resources (Ross & Van Willigen, 1997). The evidence suggests that the relationship between social class and health also reflects lifestyle differences; persons higher in socioeconomic status are more likely to have a healthy diet, exercise regularly, and get adequate sleep, and less likely to smoke (Mulatu & Schooler, 2002). Note that greater income enables one to live a healthier lifestyle.

In the United States, a large percentage of the lower class are Black. As a result, one might expect Blacks to have poorer mental health than Whites. However, the results of research comparing the psychological functioning of Blacks and Whites are inconsistent; whereas some studies find higher average symptom scores among Blacks, others do not (Vega & Rumbaut, 1991).

Further analyses have sought to identify the causes of the negative relationship between social status and stress. Are lower class persons exposed to greater stress, or are they simply less able to cope effectively with stressful events? The answer is, both (Kessler & Cleary, 1980). On the one hand, lower class persons are more likely to experience economic hardship—not having enough money to provide adequate food, clothing, and medical care (Pearlin & Radabaugh, 1976). They also experience higher rates of a variety of physical illnesses (Syme & Berkman, 1976).

Both economic hardship and illness increase the stress that an individual experiences. Furthermore, persons who are low in income, education, and occupational attainment lack the resources that would enable them to cope with these stresses effectively. Low income reduces their ability to cope with illness. Moreover, low-status persons are less likely to have a sense of control over their environment, and they have less access to political power or influence. For this reason, they are less likely to attempt to change stressful conditions or events.

Resilience to economic adversity—that is, no increase in distress scores in response to hardship—is provided by several resources. A longitudinal study of 558 rural youth and their families began when the youth were in 7th grade. Resilience among the parents was associated with marital support, effective problem-solving skills, and a sense of mastery. Resilience among the youth was promoted by nurturance (support) by parents and support by older siblings (providing warmth, not drinking alcohol) and friends (Conger & Conger, 2002).

If stress increases as socioeconomic status decreases, we would expect persons lower in status to have poorer mental health. Research over the past 30 years has consistently confirmed this expectation; there is a strong correlation between social class and serious mental disorders (Eaton, 1980). This correlation has been found in studies conducted in numerous countries (Dohrenwend & Dohrenwend, 1974). In general, persons in the lowest socioeconomic class have the highest rates of mental illness.

The differences by social class in rates of mental disorders are due in part to differences in stress. Persons in low-status occupations are more likely to experience lack of control over work. They may also experience economic uncertainty due to risk of layoff or seasonal variations in employment opportunities. This stress is likely to spill over into family interaction patterns, causing familial relations to become an added source of stress rather than a buffer. Research indicates that the shift from adequate to inadequate employment, involuntary part-time work, or low-wage jobs is associated with increased depression, and the shift to unemployment is related to even higher depression scores (Dooley, Prause, & Ham-Rowbottom, 2000). Unemployment is especially stressful, and rates of unemployment are highest among the least educated (U.S. Bureau of the Census, 2000, Table 678). Finally, the members of one's social networks have fewer economic and emotional resources.

ALIENATION

Jim dragged himself out of bed and headed for the shower. As the water poured over him, he thought, "Thursday . . . another 10-hour shift . . . if the line doesn't shut down, I'll bolt 500 bumpers . . . sick of car frames . . . I'd rather do almost anything else . . . if only I'd finished high school . . . damn the money! . . . Let 'em take the job and shove it . . . but what else pays a guy who quit school $17.36 an hour?"

Jim is experiencing **alienation**—the sense that one is uninvolved in the social world or lacks control over it. Several types of alienation have been identified (Seeman, 1975). Two will be discussed here: self-estrangement and powerlessness.

Self-Estrangement

Jim's hatred for his job reflects **self-estrangement**—the awareness that he is engaging in activities that are not rewarding in themselves. Work is an important part of one's waking hours. When work is meaningless, the individual perceives the self as devoting time and energy to something unrewarding—that is, something "alien." Whereas social background and individual characteristics have some influence, alienation from work is primarily determined by the occupational and organizational conditions of work (Mortimer & Lorence, 1995).

What makes a job intrinsically rewarding? Perhaps the most important feature is autonomy. Work that requires the individual to use judgment, exercise initiative, and surmount obstacles contributes to self-respect and a sense of mastery. A second feature is variety in the tasks that the person performs. Jim has no autonomy; his job does not allow him to exercise judgment or initiative. It also has no variety; it is monotonous and boring.

Four features of industrial technology produce self-estrangement. First, self-estrangement will be higher if the worker has no connection with the finished product itself. Second, it will be higher if the worker has no control over company policies. Third, it will be higher if the worker has little influence over the conditions of employment—over which days, which hours, or how long he or she works. Finally, it will be higher if the worker has no control over the work process—for example, the speed with which he or she must perform tasks (Blauner, 1964). Notice that alienation, like stress, is caused by lack of control over the conditions of work.

These features are especially characteristic of assembly-line work, in which each person performs the same highly specialized task many times per day. Thus, workers on assembly lines should be more likely to experience self-estrangement than other workers. A study testing this hypothesis (Blauner, 1964) compared assembly-line workers in textile and automobile plants with skilled printers and chemical industry technicians. As expected, assembly-line workers were more alienated than skilled workers who had jobs that were more varied and involved the exercise of judgment and initiative.

Work in bureaucratic organizations—like large insurance companies or government agencies—may also produce self-estrangement. In many bureaucratic organizations, workers have little or no control over the work process and do not participate in organizational decision making. Thus, workers at the lowest levels of such organizations should experience self-estrangement or dissatisfaction with their work. Conversely, workers who are involved in decision making should be less alienated. A survey of 8,000 employees in 100 companies located in the United States or Japan found that workers involved in participatory decision-making structures had higher commitment to their work (Lincoln & Kalleberg, 1985). Such workers were willing to work harder and were proud to be employed by and wanted to remain with the company.

More generally, the extent to which workers are alienated depends on the system of production in which they work. Hodson (1996) identifies five systems: craft, where each worker produces a product; direct supervision; assembly line; bureaucratic; and participatory. A review of studies of all five types of workplaces reveals a U-shaped relationship between workers' attitudes and the system of production (see Figure 18.5). Both craft and participatory systems are associated with high job satisfaction and pride in one's work; direct supervision is the most alienating system. Results such as these have led many large firms, such as General Motors, to introduce participatory systems.

As noted earlier, individual characteristics do influence reactions to work (Mortimer & Lorence, 1995). Surveys indicate that job satisfaction and involvement are most stable among workers ages 30 to 45. Women

FIGURE 18.5 SYSTEMS OF PRODUCTION AND ALIENATION FROM WORK

The way in which work is organized, or the system of production, is a major influence on people's attitudes toward their work. Historically, five different systems of production have been used: craft, where each worker has considerable autonomy; direct supervision, where another person monitors one's work; assembly line, where the work activity is determined by the organization and speed of the line; bureaucratic, where many aspects of work are governed by impersonal rules; and participatory organization, where teams of managers and workers make decisions.

A review of the research literature suggests that the pride workers have in their work and their job satisfaction vary depending on the system in which they work. The graph displays these variations, indicating that although participative systems are less alienating than the assembly lines and bureaucratic forms they replaced, they are not as satisfying as the craft system of production.

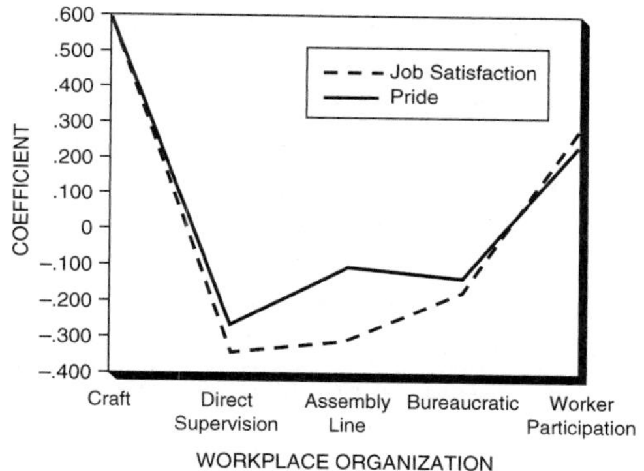

Source: From "Dignity in the Workplace Under Participative Management" (1996) by Randy Hodson, *American Sociological Review, 61*(5), 730. Used with permission from the American Sociological Association.

are as committed to work as men, though they place greater emphasis on the quality of interpersonal relations in the workplace. Among those holding comparable jobs, Blacks are as committed as Whites to their jobs and employers.

According to the theory developed more than a century ago by Karl Marx (Bottomore, 1964), whether a person will experience self-estrangement is determined by his or her relation to the means of production. The most alienated employees are hypothesized to be those who have no autonomy, who do not have the freedom to solve nonroutine problems, and who have no subordinates. Marx referred to such workers as the proletariat. In contemporary society, assembly-line workers, salesclerks, file clerks, and laborers are all in occupations that have these characteristics. A survey of 1,499 working adults found that 46 percent were in jobs of this type (Wright et al., 1982). Several studies have found that men whose jobs are characterized by lack of autonomy and complexity typically have high scores on measures of self-estrangement (Kohn, 1976) and low scores on measures of job involvement (Lorence & Mortimer, 1985).

There is some evidence that the characteristics of the work environment influence psychological well-being. One researcher assessed the common environment of office workers by averaging the ratings of all of those employed in each of 37 branch offices; each worker's own ratings were used as a measure of his or her immediate environment (Repetti, 1987). Workers who rated his or her branch more positively (on interpersonal climate and support and respect from coworkers) reported lower levels of anxiety and depression. Aggregate ratings by the workers of the environment in the branch were also related to anxiety and depression scores.

The graffiti gracing buildings and public telephones in some American cities are responses by youth to alienation. Spray-painted messages reflect the lack of control over their lives that many youth feel.

Powerlessness

Consider the facts that vandalism is widespread in certain sections of large cities, that many middle-class and upper class adults do not vote in presidential elections, and that some people on welfare make no effort to find a job. These facts all have something in common. They reflect, at least in part, people's sense of

powerlessness—the sense of having little or no control over events.

Powerlessness is a generalized orientation toward the social world. People who feel powerless believe they have no influence on political affairs and world events; this is different from feeling a lack of control over events in day-to-day life. A typical measure of powerlessness includes items like "This world is run by a few people in power and there is not much the little guy can do about it." Agreement with such statements indicates powerlessness. Most people's scores on measures of powerlessness are quite stable over a period of many years (Neal & Groat, 1974). There is some evidence that a sense of powerlessness develops during childhood (Seeman, 1975). Interestingly, a sense of powerlessness is *not* associated with social class—that is, income, occupation, or education.

Statements that measure powerlessness, such as "People like me have no say," and "Politicians don't care what I think," were included in several surveys between 1952 and 1980. Analysis of patterns of agreement with these items shows that powerlessness or political alienation declined from 1952 to 1960, rose steadily from 1960 to 1976, and then declined (Rahn & Mason, 1987). The increase in the 1960s and 1970s was associated with increased concern about such political and social issues as civil rights for Blacks, the war in Vietnam, and the Watergate political scandal. Thus, fluctuations in powerlessness reflect, at least in part, events in the larger society.

Although the sense of powerlessness is found in all classes, upper and lower classes may have different means of expressing it. Middle-class and upper class persons may be more likely to stay home on election day or to feel apathetic about political affairs or organizations that influence public policy. Lower class persons may be more likely to have a hostile attitude toward city officials and to vandalize city buses, subway trains, and businesses in their neighborhoods. Thus, how an individual expresses frustration over his or her lack of influence on the world may depend on his or her social position.

SUMMARY

This chapter considers the impact of social structure on four areas of a person's life: achievement, values, physical and mental health, and sense of belonging in society. Social structure influences the individual through the expectations associated with one's roles, the social networks to which one belongs, and the status associated with one's positions.

Status Attainment An individual's status determines his or her access to resources—to money, lifestyle, and influence over others. Three generalizations can be made about status in the United States. (1) An individual's status is closely tied to his or her occupation. (2) Occupational attainment is influenced directly by the individual's educational level and ability and indirectly by socioeconomic background. Among women, occupational status and income is limited by gender segregation. (3) Information about job opportunities is often obtained via social networks, especially those characterized by weak ties.

Individual Values Two aspects of the individual's position in society influence his or her values. (1) Particular values are reliably associated with certain occupational role characteristics. Men and women whose jobs are closely supervised, routine, and not complex value conformity, whereas those whose jobs are less closely supervised, less routine, and more complex value self-direction. (2) A formal education influences values. Higher education is associated with placing greater value on self-direction and with greater intellectual flexibility.

Social Influences on Health Physical health is influenced by occupation, gender, marital roles, and social class. (1) Occupational roles determine the health hazards that individuals are exposed to and whether they experience role overload. (2) The traditional role expectations for men and women make men more vulnerable than women to illnesses such as coronary heart disease. (3) Marriage protects both men and women from illness and premature death. (4) Members of lower status groups experience higher rates of illness, disability, and death.

Mental health is also influenced by social factors. (1) Economic hardship, uncertainty, and unemployment are associated with poor mental health. (2) Women have somewhat poorer mental health than men. (3) Marriage is associated with reduced stress for both men and women; working adults may experience spillover of stress from work into family relationships. (4) Social networks are an important resource in coping with stress; they provide the person with emotional, esteem, and informational support, as well as instrumental aid. (5) Lower class persons report greater stress and experience a higher incidence of mental illness.

Alienation Two types of alienation are self-estrangement and powerlessness. (1) Self-estrangement is associated with work roles that do not allow workers a sense of autonomy, such as assembly-line jobs. (2) Powerlessness is a generalized sense that one has little or no control over the world.

LIST OF KEY TERMS AND CONCEPTS

alienation (p. 488)
fateful events (p. 482)
position (p. 462)
powerlessness (p. 490)
primary relationship (p. 462)
role (p. 462)
role overload (p. 476)
self-estrangement (p. 488)
social network (p. 462)
social structure (p. 462)
spillover (p. 484)
status (p. 462)
upward mobility (p. 465)
values (p. 472)

19

DEVIANT BEHAVIOR AND SOCIAL REACTION

Introduction

The Violation of Norms

Norms

Anomie Theory

Control Theory

Differential Association Theory

Routine Activities Perspective

Reactions to Norm Violations

Reactions to Rule Breaking

Determinants of the Reaction

Consequences of Labeling

Labeling and Secondary Deviance

Societal Reaction

Secondary Deviance

Formal Social Controls

Formal Labeling and the Creation of Deviance

Long-Term Effects of Formal Labeling

Summary

List of Key Terms and Concepts

INTRODUCTION

Virginia and Susan wandered through the department store, stopping briefly to look at blouses and then going to the jewelry counter. Each looked at several bracelets and necklaces. Susan kept returning to a 24-karat gold bracelet with several jade stones, priced at $199.50. Finally, she picked it up, glanced quickly around her, and dropped the bracelet into her shopping bag.

The only other shopper in the vicinity, a well-dressed man in his 40s, saw Susan take the bracelet. He looked around the store, spotted a security guard, and walked toward him. Virginia stammered to Susan, "I, uh, I don't think we should do this."

"Oh, it's okay. Nothing will happen," Susan replied, before walking quickly out of the store. Moments later, Virginia followed her. As Susan entered the mall, the security guard stepped up to her, took her by the elbow, and said, "Come with me, please."

Shoplifting episodes like this one occur dozens of times every day in the United States. Shoplifting is one of many types of **deviant behavior**—behavior that violates the norms that apply in a given situation. In addition to crime, deviance includes cheating, substance use or abuse, fraud, corruption, delinquent behavior, harassment, and behavior considered symptomatic of mental illness.

There are two major reasons why social psychologists study deviant behavior, one theoretical and one practical. First, social norms and conformity are the basic means by which the orderly social interaction necessary to maintain society is achieved. By studying nonconformity, we learn about the processes that produce social order. For example, we might conclude that Susan took the bracelet because there were no store employees nearby, suggesting the importance of surveillance in maintaining order. Second, social psychologists study deviant behavior to better understand its causes. Deviant behaviors such as alcoholism, drug addiction, and crime are perceived as serious threats to society. Once we understand its causes, we may be able to develop better programs that reduce or eliminate deviance or that help people change their deviant behavior.

This chapter addresses four fundamental questions:

1. What are the causes of deviant behavior?
2. How important for deviant behavior is the reaction of observers? That is, does someone have to react to behavior in particular ways for it to be considered deviant?
3. Why do some people engage in deviance regularly? Why do they adopt a lifestyle that involves participation in deviant activities?
4. What determines how authorities and agents of social control deal with incidents of deviance? Is their reaction influenced by the deviant person's gender, social status, or other characteristics of the situation?

THE VIOLATION OF NORMS

When we read or hear that someone is accused of murder, or embezzling money from a bank, or engaging in illegal accounting practices, we often ask "why?" In Susan's case, we would ask, "Why did she take that bracelet?" In this section, we consider first the meaning of norms and then look at several theories about the causes of deviant behavior. These include anomie theory, control theory, differential association theory, and routine activities theory.

Norms

Most people would regard Susan's behavior in the department store as deviant because it violated social norms. Specifically, she violated laws that define taking merchandise from stores without paying for it as a criminal act. Thus, deviance is a social construction; whether a behavior is deviant or not depends on the norms or expectations for behavior in the situation in which it occurs.

In any situation, our behavior is governed by norms derived from several sources (Suttles, 1968). First, there are purely "local" and group norms. Thus, roommates and families develop norms about what personal topics can and cannot be discussed. Second, there are subcultural norms that apply to large numbers of persons who share some characteristic. For example, there are racial or ethnic group norms governing the behavior of Blacks or persons of Polish descent that do not apply to other Americans. A subculture that is particularly relevant to the discussion of deviance is the subculture of violence, which will be discussed later. Third, there are societal norms, such as those requiring certain types of dress or those limiting sexual activity to certain relationships and situations. Thus, the norms that govern our daily behavior have a variety of origins, including family and friends; socioeconomic, religious, or ethnic subcultures; and the society in general.

The repercussions of deviant behavior depend on which type of norm an individual violates. Violations of local norms may be of concern only to a certain group. Failing to do the dishes when it is your turn may result in your roommate being angry, although your friends may not care about that deviance. Subcultural norms are often held in common by most of those with whom we interact, whether they are friends, family members, or coworkers. Violations of these norms may affect most of one's day-to-day interactions. Violations of societal norms may subject a person to action by formal agencies of control, such as the police or the courts. Earlier in this book, we discussed the violation of local norms (see Chapter 9) and group norms (see Chapter 14). In this chapter, we focus on the violation of societal norms and on reactions to norm violations.

Anomie Theory

The **anomie theory** of deviance (Merton, 1957) suggests that deviance arises when people striving to achieve culturally valued goals, such as wealth, find that they do not have any legitimate way to attain these goals. These people then break the rules, often in an attempt to attain these goals illegitimately.

Anomie Every society provides its members with goals to aspire to. If the members of a society value religion, they are likely to socialize their youth and adults to aspire to salvation. If the members value power, they will teach people to seek positions in which they can dominate others. U.S. culture extols wealth as the appropriate goal for most members of society. In every society, there are also norms that define acceptable ways of striving for goals, called **legitimate means.** In the United States, legitimate means for attaining wealth include education, working hard at a job to earn money, starting a business, and making wise investments.

A person socialized into U.S. society will most likely desire material wealth and will strive to succeed in a desirable occupation—to become a teacher, nurse, business executive, doctor, or the like. The legitimate means of attaining these goals are to obtain a formal education and to climb the ladder of occupational prestige. The person who has access to these means—who can afford to go to college and has the accepted skin color, ethnic background, and gender—can attain these socially desirable goals.

What about those who do not have access to the legitimate means? As Americans, these people will desire material wealth like everyone else, but they will be blocked in their strivings. Because of the way society is structured, certain members are denied access to legitimate means. Government decisions regarding budgeting and building or closing schools determine the availability of education to individuals. Similarly, certain members of society are denied access to jobs. Not only individual characteristics—such as lack of education—but also social factors—such as the profitability of making steel in Ohio—determine who is unemployed.

A person who strives to attain a legitimate goal but is denied access to legitimate means will experience anomie—a state that reduces commitment to norms or the pursuit of goals. There are four ways a person may respond to anomie; each is a distinct type of deviance. First, an individual may reject the goals, and give up trying to achieve success, but continue to conform to social norms. This adaptation is termed ritualism. The poorly paid stock clerk who never misses a day of work in 45 years is a ritualist. He is deviant because he has given up the struggle for success. Second, the individual might reject both the goals and the

© PhotoDisc/Getty Images

Most Americans are socialized to strive for economic success. But some people do not have access to legitimate employment, so they seek wealth by alternative, sometimes illegal means, such as prostitution.

means, withdrawing from active participation in society by retreatism. This may take the form of drinking, drug use, withdrawal into mental illness, or other kinds of escape. Third, one might remain committed to the goals but turn to disapproved or illegal ways of achieving success. This adaptation is termed innovation. Earning a living as a burglar, prostitute, or loan shark is an innovative means of attaining wealth. Finally, one might attempt to overthrow the existing system and create different goals and means through rebellion.

Shoplifting is a form of innovation. Like other types of economic crime, it represents a rejection of the normatively prescribed means (paying for what you want) while continuing to strive for the goal (possessing merchandise). According to anomie theory, Susan, the shoplifter, has been socialized to desire wealth but does not have access to a well-paying job due to her poor education. As a result, she steals what she wants because she does not have the money to pay for it.

Another influence on an individual's adaptation is access to deviant roles. Using a means of goal achievement—whether legitimate or illegitimate—requires access to two structures (Cloward, 1959). The first is a **learning structure**—an environment in which an individual can learn the information and skills required. A shoplifter needs to learn how to conceal objects quickly, how to spot plainclothes detectives, and so forth. The second is an **opportunity structure**—an environment in which an individual has opportunities to play a role, which usually requires the assistance of those in complementary roles. Anomie theory assumes that anyone can be an innovator—through shoplifting, prostitution, or professional theft. But not everyone has access to the special knowledge and skills needed to succeed as a prostitute (Heyl, 1977) or a black market banker (Weigand, 1994). Just as access to legitimate means to achieve goals is limited, so is access to illegitimate means. Only those who have both the learning and opportunity structures necessary to become a shoplifter, prostitute, or embezzler can use these alternative routes to success (Coleman, 1987).

The opportunities for deviance available to a person depend on age, sex, kinship, ethnicity, and social class (Cloward, 1959). These characteristics, with the possible exception of class, are beyond the individual's control. Thus, prostitution in our society primarily involves young, physically attractive persons. People who do not have access to the learning or opportunity structures necessary for deviance are double failures; they can succeed neither through legitimate nor through illegitimate means. Double failure often produces retreatism. Drug addicts, alcoholics, and mentally ill persons may be losers in both the conventional and criminal worlds.

Anomie and Social Class Anomie theory emphasizes access to education and employment. Those who have access to both should not engage in deviant behavior. Those who do not have access to one or both should experience anomie and are likely to engage in deviance. A survey of 1,614 youths ages 15 to 18 measured commitment to success goals ("making a lot of money") and perceived access to college education (Farnworth & Leiber, 1989). Respondents who said they wanted to make a lot of money but did not expect to complete college were much more likely to report delinquent behavior.

One measure of access to legitimate means is the unemployment rate. According to the theory, as unemployment increases, rates of deviance also should increase. One study analyzed the relationship between unemployment rates and crime rates in the United States for each year of the 1948–1985 period (Devine, Sheley, & Smith, 1988). There was a strong relationship; as unemployment increases, so does crime. The relationship is stronger for economic crime (burglary) than for violent crime (murder). Another study looked at the relationship between job availability (unemployment) and job quality (pay per hour) and arrest rates by state from 1977 through 1980 (Allan & Steffensmeier, 1989). The results indicate that unemployment was strongly associated with juvenile arrest rates. Job quality was highly correlated with adult arrest rates. Evidence of a direct connection between unemployment and economic crime comes from a longitudinal study in which ex-addicts, ex-offenders, and "drop-out" youth reported their legal and illegal income for up to 3 years; as the unemployment rate in the city increased, youth reported greater income from illegal activities (Uggen & Thompson, 2003).

Two studies suggest that it is relative rather than absolute socioeconomic standing that determines whether one experiences anomie. A study of arrest rates for burglary and robbery from 1957 to 1990 found that as income inequality among Blacks increased, so did Black arrest rates (LaFree & Drass, 1996). Similarly, an analysis of the number of Latinos murdered in 1980 found that the degree of income inequality among Latinos was an important factor (Martinez, 1996). Thus, it is one's economic standing relative to similar others that matters, not one's standing in the society as a whole.

The relationship between socioeconomic disadvantage and crime is not limited to minority groups. A study of 124 central cities with substantial African American populations found that, in cities where the economic circumstances (such as rates of home ownership) of Blacks and Whites were similar, murder rates were similar as well (Boardman, Finch, Ellison, Williams, & Jackson, 2001). Research has also assessed the effect of neighborhoods on psychological distress; when exposure to discrimination and poverty are the same, there are no racial differences in mental health (Schulz et al., 2000). That is, Blacks and Whites living in similarly disadvantaged neighborhoods experience similar levels of mental illness. Another study used a composite score of disadvantage, including the percentage of households below poverty and male unemployment rates. Neighborhood disadvantage was associated with increased exposure to social stressors (illness, criminal victimization) and psychological distress (feeling sad, anxious, hopeless); all three were associated with drug use, especially among those with the lowest incomes. These results are consistent with anomie theory.

Anomie theory directs our attention to the importance of social class. Because lower class members are more frequently excluded from quality education and jobs, the theory predicts that they will commit more crimes. Data collected by police departments and the FBI generally confirm this prediction, showing that a disproportionate number of those arrested for crimes are poor, minority men. This has led some to conclude that crime and social class are inversely related—that the highest crime rates are found in the lower social strata (Cloward, 1959).

However, there is a class bias built into the official statistics on crime. Not all illegitimate economic activities are included in these statistics. Whereas data on burglary, robbery, and larceny are compiled by police departments, data on income tax evasion, price fixing, and insider trading are not. Police and FBI statistics are much more likely to include "street" crimes than the kinds of economic crimes committed by the wealthy, corporate executives, and stockbrokers. The latter are called white-collar crime—activities that violate norms of trust, usually for personal gain (Shapiro, 1990). To embezzle or misappropriate funds or engage in insider trading of stocks, one needs access to a position of trust. Such positions usually are filled by middle-class and upper class persons. These crimes are facilitated by the social organization of trust; the acts of trustees are invisible, hidden in a network of often electronic connections between organizations. Thus, although specific crimes may vary by class, illegitimate economic activity may be common to all classes.

General Strain Theory One limitation of anomie theory is that it does not specify the mechanism by which the lack of access to legitimate means produces delinquent or criminal behavior. One attempt to do so is Agnew's general strain theory (Agnew, 1992; Agnew & White, 1992). Agnew proposes that emotion connects the experience of strain with deviant behavior; strain elicits negative affective states—frustration, anger, or fear—that create the motivation to act. These actions may be deviant or criminal. Such actions include crimes that provide access to the goal (robbery, burglary, selling drugs), aggression against people perceived as responsible for the strain (abuse, assault), or drug and alcohol use to escape the emotions. The role of emotion can explain incidents such as an angry former employee returning to the workplace and killing supervisor(s) and former coworkers.

A longitudinal study of high school youth provides data to test the theory. Youth in the 9th, 10th, and 11th grades in three primarily White suburban communities were interviewed three times over a 2-year period. The research measured life stress and family conflict, anger and anxiety, and aggressive delinquency (damaging property, carrying a weapon, fighting), nonaggressive delinquency (stealing, joyriding, running away), and marijuana use. The results indicated that life stressors and family conflict were related to delinquency and marijuana use. As predicted by the theory, family conflict was related to anger, and anger was related to engaging in aggressive delinquency. However, anger was not related to nonaggressive delinquency or marijuana use, and anxiety was not related to any of the three types of behaviors (Aseltine, Gore, & Gordon, 2000). Thus, the results provide only modest support for the key predictions.

A longitudinal study of youth living in Dade County, FL yielded a sample with substantial numbers of African Americans, Hispanics, and Whites, allowing a test of the extension of the theory to these minorities. To the extent that there are differences by racial/ethnic group in criminal behavior, general strain theory suggests that these are caused by differences between groups in strain. The research included three measures of strain: recent life events (in the preceding 12 months), chronic stressors (for example, unemployment, relationship, child care, residence), and lifetime major events (for example, abandonment, school failure, divorce, physical or sexual assault). It also included measures of social support. Using data from 898 young men, analyses indicated that strain as measured by recent life events was related to criminal activity, and that greater involvement of African Americans in crime was associated with greater exposure to major lifetime events (Eitle & Turner, 2003).

Control Theory

If you were asked why you don't shoplift clothing from stores, you might reply, "Because my parents (or lover, or friends) would kill me if they found out." According to **control theory,** social ties influence our tendency to engage in deviant behavior. We often conform to social norms because we are sensitive to the wishes and expectations of others. This sensitivity creates a bond between the individual and other persons. The stronger the bond is, the less likely the individual is to engage in deviant behavior.

There are four components of the social bond (Hirschi, 1969). The first is attachment—ties of affection and respect for others. Attachment to parents is especially important, because they are the primary socializing agents of a child. A strong attachment to them leads the child to internalize social norms. The second component is commitment to long-term educational and occupational goals. Someone who aspires to go to law school is unlikely to commit a crime, because a criminal record would be an obstacle to a career in law. The third component is involvement. People who are involved in sports, scouts, church groups, and other conventional activities simply have less time to engage in deviance. The fourth component is belief—a respect for the law and for persons in positions of authority.

We can apply control theory to the shoplifting incident described in the introduction. Susan does not

© Catherine Karnow/Woodfin Camp

These urban Boy Scouts are nailing planks together as part of a work project. Participation in such group projects increases attachment to and involvement in conventional society, reducing the likelihood of delinquency.

feel attached to law-abiding adults; therefore, she was not concerned about their reactions to her behavior. Nor did she seem deterred by commitment when she said, "Nothing will happen." Susan's deviant act reflects the absence of a bond with conventional society.

The relationship between delinquency and the four components of the social bond has been the focus of numerous studies. Several studies have found a relationship between a lack of attachment and delinquency; young people from homes characterized by a lack of parental supervision, communication, and support report more delinquent behavior (Hoffman, 2002; Hundleby & Mercer, 1987; Messner & Krohn, 1990). Attachment to school, measured by grades, is also associated with delinquency. Boys and girls who do well in school are less likely to be delinquent. Regarding commitment to long-term goals, research indicates that youths who are committed to educational and career goals are less likely to engage in property crimes such as robbery and theft (Johnson, 1979; Shover, Novland, James, & Thornton, 1979). Findings relevant to the third component, involvement, are mixed. Whereas involvement in studying and homework is negatively associated with reported delinquency, participation in athletics, hobbies, and work is unrelated to reported delinquency. Involvement in religion, as reflected in frequent church attendance and rating religion as important in one's life, is associated with reduced delinquency (Sloane & Potrin, 1986). Finally, evidence suggests that conventional beliefs reduce the frequency of delinquent behavior (Gardner & Shoemaker, 1989).

Control theory asserts that attachment to parents leads to reduced delinquency. Implicitly, the theory assumes that parents do not encourage delinquent behavior. Although this assumption may be correct in most instances, there are exceptions. Studies suggest that some parents encourage some delinquent behaviors. Adolescent drinking is associated with parental alcohol consumption; parents who are heavy drinkers are more likely than nondrinking parents to have adolescents who are heavy drinkers (Barnes, Farrell, & Cairns, 1986). Some parents explicitly teach their children how to shoplift, commit burglaries, and steal cars and trucks (Butterfield, 2002). These are cases in which crime really does "run in the family." In these instances, parental attachment leads to increased delinquency and crime.

Does a lack of attachment to parents in childhood relate to adult deviant behavior? Yes. Research consistently shows that children who are physically and sexually abused are more likely to be involved as adults in violent and property crime, prostitution, and alcohol and substance abuse (Macmillan, 2001).

The strength of adult social bonds is also related to adult criminal behavior. One study assessed month-to-month variations in circumstances that could strengthen or weaken the bond, and related this variation to the occurrence of criminal behavior (Horney, Osgood, & Marshall, 1995). The circumstances were starting/stopping school, starting/stopping work, and starting/stopping living with a girlfriend or wife. Interviews were conducted with 658 men in prison who had committed felonies. Increases in criminal behavior were closely related to changes that reduced the men's bond to others—stopping school or work, and stopping living with a girlfriend or wife. A study of Swedish men found that alcohol use and unemployment often preceded suicide (Norstrom, 1995). Both alcohol use and job loss reduce one's social integration—that is, social ties to other persons.

One longitudinal study indicates that the strength of the social bond influences whether adults engage in deviant behavior (Sampson & Laub, 1990). The researchers studied 500 boys ages 10 to 17 who

FIGURE 19.1 THE RELATIONSHIP BETWEEN AGE AND CRIME

Involvement in criminal behavior is not constant across the life course. People of a certain age are much more likely to commit some crimes and not others. Youth ages 10 to 20 are especially likely to commit burglary; half of all those arrested for burglary are under 18. Fraud, in contrast, is much more likely to be committed by persons ages 20 to 40. Those arrested for gambling are as likely to be 50 as 20.

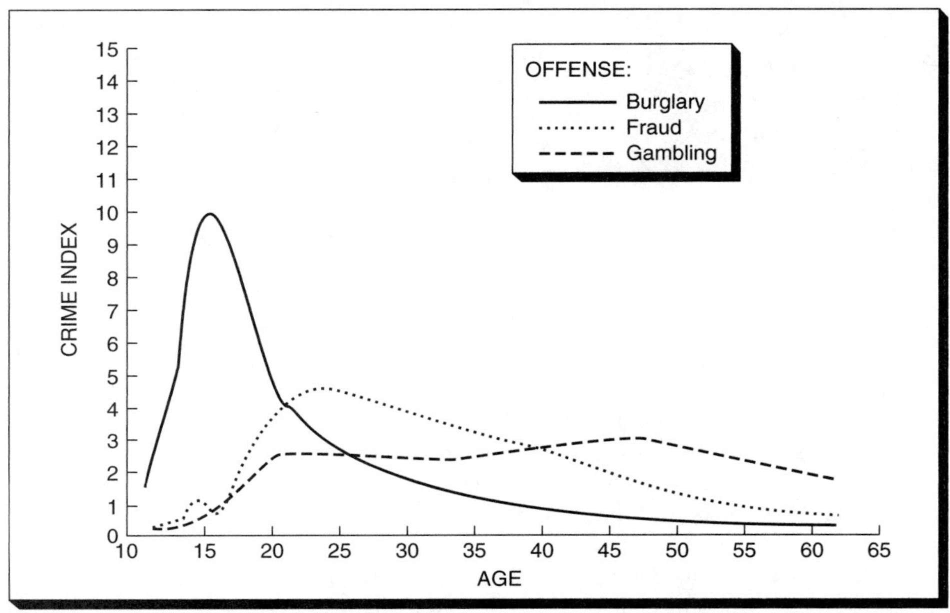

Source: Steffensmeier et al., 1989, figure 1.

were in a correctional school and 500 boys of the same age from public school. Each boy was followed until he was 32. Generally, strong ties to social institutions were associated with reduced rates of crime, alcohol abuse, gambling, and divorce. In adolescence, the important attachments were to family and school. In young adults, the influential ties were to school, work, or marriage. In later adulthood, the important ties were to work, marriage, and parenthood.

What about women? Would work, marriage, and parenthood be associated with desistance from crime among women who were serious delinquents in adolescence? A study comparing women and men who were in institutions during adolescence found that neither job stability nor marriage was associated with adult desistance. In narratives of their lives, women were more likely than men to describe their children and religious transformation as the forces for change in their lives (Giordano, Cernkovich, & Rudolph, 2002). Moreover, it was not the fact of having children or a good job that was important; it was a transformation in the woman's identity or her thinking about those aspects of her life. These results are consistent with symbolic interaction theory and its emphasis on meaning constructed by the person in interaction with others.

A study of the relationship between age and crime found large variation by age in the type of crime committed (Steffensmeier, Allan, Haver, & Streifel, 1989; see Figure 19.1) Typical adolescent offenses include vandalism, auto theft, and burglary. Persons of ages 18 to 28 are more likely to be involved in drug violations and homicide. Middle-age persons are more frequently involved in gambling offenses.

Differential Association Theory

Are all types of deviance explained by the absence of a social bond? Perhaps not. Sometimes people deviate from one set of norms because they are being influenced by a contradictory set of norms. U.S. society is composed of many groups with different values, norms, and behavior patterns. With respect to many behaviors, there is no single, society-wide norm. An adolescent's use of marijuana may deviate from her parents' norms, for example, but it may conform to her friends' norms. Here, the deviance involved in marijuana use reflects a conflict between the norms of two groups rather than an insensitivity to the expectations of others. In fact, the use of marijuana may reflect a high degree of sensitivity to the expectations of one's peers.

This view of deviance is the basis of **differential association theory,** developed by Sutherland. He argued that although the law provides a uniform standard for deviance, one group may define a behavior as deviant, whereas another group defines it as desirable. Shoplifting, for example, is legally defined as a crime. Some groups believe it is wrong because (1) it leads to increased prices, which hurts everyone; (2) it violates the moral principle against stealing; and (3) it constitutes lawbreaking. Other groups, in contrast, believe shoplifting is acceptable because (1) businesses deserve to have things taken because they overcharge; (2) the loss is covered by insurance; and (3) the shoplifter won't be caught. Susan's comment, "It's okay. Nothing will happen," reflects the latter belief.

Attitudes about behaviors are learned through associations with others, usually in primary group settings. People learn motives, drives, and techniques of engaging in specific behaviors. What they learn depends on whom they interact with—that is, on their differential associations. Whether someone engages in a specific behavior depends on how frequently he or she is exposed to attitudes and beliefs that are favorable toward that behavior.

The principle of differential association states that a "person becomes delinquent because of an excess of definitions favorable to violation of the law over definitions unfavorable to violation of the law" (Sutherland, Cressey, & Luckenbill, 1992). Studies designed to test this principle typically ask individuals questions about their attitudes toward a specific behavior and about their participation in that behavior. One study revealed that the number of definitions favorable to delinquency accurately predicted which young men reported delinquent behavior (Matsueda, 1982). The larger the number of definitions a youth endorsed, the larger the number of delinquent acts he reported having committed in the preceding year. A subsequent study found that associating with delinquent peers was also related to delinquent behavior (Heimer & Matsueda, 1994).

Certain groups within the United States hold a set of beliefs that justify the use of physical aggression in certain situations. This set of beliefs is referred to as the subculture of violence. Within this subculture, violence is considered appropriate when used as a means of self-defense and protection of one's home. A review of state laws governing spouse abuse, corporal punishment, and capital punishment found that southern states have laws that are more accepting of violence (Cohen, 1996). Several studies report a relationship between these beliefs and behavior. Felson, Liska, South, and McNulty (1994) studied young men in 87 high schools. The young men were asked whether aggressive responses were appropriate in three situations involving insults or threats. Those young men who endorsed the use of violence were much more likely to report involvement in eight types of interpersonal violence, including striking a parent or teacher, fighting, and using weapons in disputes. Endorsing the use of violence was also associated with delinquency within the school, including cheating, tardiness, and truancy.

The theory of differential association does not specify the process by which people learn criminal or deviant behavior. For this reason, Burgess and Akers (1966) developed a modified theory of differential association. This modified version emphasizes the influence of positive and negative reinforcement on the acquisition of behavior. Much of this reinforcement comes from friends and associates. Thus, if beliefs are learned through interaction with others, then people whose attitudes are favorable toward a behavior should have friends who also have favorable attitudes toward that behavior. Alternatively, people whose attitudes are opposed to an activity should have friends who share those negative views.

A survey of 3,056 high school students was conducted to test these hypotheses (Akers, Krohn, Lanza-Kaduce, & Radosevich, 1979). In particular, it assessed the relationship between differential association, reinforcement, and adolescents' drinking behavior and marijuana use. Differential association was measured by three questions: "How many of your (1) best friends, (2) friends you spend the most time with, and (3) friends you have known longest smoke marijuana and/or drink?" The survey also assessed students' definitions of drug and alcohol laws. Both social reinforcement (whether the adolescent expected praise or punishment for use from parents and peers) and nonsocial reinforcement (whether the effects of substance use were positive or negative) were measured. The findings of this survey showed that differential association was closely related to the use of alcohol or drugs. The larger the number of friends who drank or smoked marijuana, the more likely the student was to drink alcohol or smoke marijuana. Reinforcement was also related to behavior; those who used a substance reported that it had positive effects. The students' definitions were also related to those with whom they associated; if their friends drank or used marijuana, they were more likely to have positive attitudes toward the behavior and negative attitudes toward laws defining that behavior as criminal. Finally, students' attitudes were consistent with their behavior. Those who opposed marijuana use and supported the marijuana laws were much less likely to use that substance.

A similar study (Akers, LaGreca, Cochran, & Sellers, 1989) focused on drinking among older persons. Interviews were conducted with 1,410 people ages 60 and over. The measures used were the same or similar to those used with adolescents. The results were essentially the same. The drinking behavior of persons 60 and older was related to the drinking behavior of spouse, family, or friends, reinforcements, and an individual's attitudes toward drinking.

Survey data collected at one point in time often cannot be used to test hypotheses about cause-effect relationships. However, survey data collected from the same people at two or more times can be. Stein, Newcomb, and Bentler (1987) analyzed data from 654 young people who were surveyed three times at 4-year intervals that began when they were in junior high school. The measures included peer drug use, adult drug use, and community approval of drug use. The results showed that adolescents who believed that both peers and adults were using drugs were more likely to become drug users. Thus, association with persons who use alcohol and drugs, especially in primary relationships, is one cause of substance use by adolescents. In addition to being modeled by peers or family, deviant behaviors can sometimes be provoked by media coverage, as detailed in Box 19.1.

Recall from Chapter 18 that an important characteristic of social networks is density—the extent to which each member of the network or group knows the other members. Networks that are dense should have more influence on their members' behavior; if all of your friends drink alcohol, it will be hard for you to "just say no." A study of a nationally representative sample of 7th to 12th graders found that peers' delinquency has a stronger association with an adolescent's delinquency when the friendship network is dense (Haynie, 2001).

Because each person usually associates with several groups, the consistency or inconsistency in definitions across groups is also an important influence on behavior (Krohn, 1986). Network multiplexity refers to the degree to which individuals who interact in one context also interact in other contexts. When you interact with the same people at church, at school, on the athletic field, and at parties, multiplexity is high. When you interact with different people in each of these settings, multiplexity is low. When multiplexity is high, the definitions of an activity will be consistent across groups; when it is low, definitions may be inconsistent across groups. Thus, differential associations should have the greatest impact on attitudes and beliefs when multiplexity is high. A survey of 1,435 high school students measured the extent to which individuals interacted with parents and with the same peers in each of several activities (Krohn, Massey, & Zielinski, 1988). Students who participated jointly with parents and peers in various activities were less likely to smoke cigarettes.

Routine Activities Perspective

So far, we have considered characteristics of the person (motivation, beliefs) and of his associations with others (parents, friends). These have been shown to be related to delinquency, assault, murder, burglary, economic crimes, suicide, and alcohol and drug use. The **routine activities perspective** focuses on a third class of influences—how these behaviors emerge from the routines of everyday life (Felson, 1994).

Each instance of deviant behavior requires the convergence of the elements necessary for the behavior to occur. Crimes such as burglary, larceny, or robbery require the convergence of an offender and a likely target (residence, store, or person) and the absence of some guardian who could intervene. In the illustration at the beginning of the chapter, the shoplifting incident involves such a convergence: Susan, the bracelet, and the absence of a clerk or security guard. Illegal consumption requires two offenders (seller and user), a substance, and a setting with no guardian; "crack houses" provide the latter in many large cities. Without such convergence, deviance will not occur. We can understand another aspect of deviance if we analyze everyday activity from the perspective of how it facilitates or prevents such convergences. This perspective calls our attention to the contributions of situations to behavior.

One class of situations that facilitates deviance is unstructured socializing with peers in the absence of an authority figure (Osgood, Wilson, O'Malley, Bachman, & Johnston, 1996). The presence of peers makes it likely that definitions will be shared—including definitions favorable to particular forms of deviance. The absence of an authority figure or guardian reduces the likelihood of punishment for deviance. Lack of structure makes time available for deviance. What situations have these characteristics? They include joyriding in a car with friends, going to parties, and "hanging out" with friends. Data from a longitudinal study of a national sample of 1,200 persons ages 18 to 26 allowed researchers to relate involvement in these situations to deviance. Frequency of participation in them was related to alcohol and marijuana use, dangerous driving, and criminal behavior. Changes across five waves of data collection in an individual's participation in these activities were related to changes in his or her involvement in deviance.

Researchers have consistently noted that men are much more likely to commit criminal acts than women. This is not only true of street crime but also of economic crimes involving violation of trust, such as insider trading. The routine activities perspective explains this as due to gender role socialization, which teaches women different norms and definitions; to lack of access to tutelage in various forms of deviance; and to restrictions on activities that keep women out of certain settings (Steffensmeier & Allan, 1996). Thus, few women commit either burglary or insider trading because of their lack of access in everyday life to the apprenticeships where one learns these behaviors. Similarly, we are not surprised that the corporate executives of Enron, WorldCom, and other companies who committed fraud in the period 1995–2003 were men; the "glass ceiling" prevents women from occupying such roles.

The anomie, control, differential association, and routine activities perspectives are not incompatible. Anomie theory suggests that culturally valued goals and the opportunities available to achieve these goals are major influences on behavior. Opportunities to learn and occupy particular roles are influenced by age, social class, gender, race, and ethnic background—that is, by the structuring of everyday life based on these variables. According to control theory, we are also influenced by our attachments to others and our commitment to attaining success. Our position in the social structure and our attachments to parents and peers determine our differential associations—the kinds of groups to which we belong. Within these groups, we learn definitions favorable to particular behaviors, and we learn that we face sanctions when we choose behaviors that group members define as deviant.

REACTIONS TO NORM VIOLATIONS

When we think of murder, robbery, or sexual assault, we think of cases we have read about or heard of on radio or television. We frequently refer to police and

BOX 19.1 The Power of Suggestion

Rape, robbery, murder, and other types of deviant behavior receive a substantial amount of coverage in newspapers and on radio and television. One function of publicizing deviance is to remind us of norms—to tell us what we should not do (Erikson, 1964). But is this the only consequence? Could the publicity given particular deviant activities increase the frequency with which they occur? In some cases, the answer appears to be yes.

A study of the relationship between the publicity given to suicides and suicide rates suggests that the two are positively correlated (Phillips, 1974). This study identified every time a suicide was publicized in three major U.S. daily newspapers during the years 1947–1968. Next, the researchers calculated the number of expected suicides for the following month by averaging the suicide rates for that same month from the year before and the year after. For example, the researchers noted that the suicide of a Ku Klux Klan leader on November 1, 1965, was widely publicized. They then obtained the expected number of suicides (1,652) by averaging the total number of suicides for November 1964 (1,639) and November 1966 (1,665). In fact, there were 1,710 suicides in November 1965; the difference between the observed and the expected rates (58) could be due to suggestion via the mass media.

The results of this study showed that suicides increased in the month following reports of a suicide in major daily papers. Moreover, the more publicity a story was given—as measured by the number of days the story was on the front page—the larger the rise in suicides was. If a story was published locally—in Chicago but not in New York, for example—the rise in suicides occurred only in the area where it was publicized.

Why should such publicity lead other persons to kill themselves? There must be some factor that predisposes a small number of persons to take their own lives following a publicized suicide. That predisposing factor may be anomie. According to this theory, suicide is a form of retreatism—of withdrawal from the struggle for success. Persons who don't have access to legitimate means are looking for some way to adapt to their situation. Publicity given to a suicide may suggest a solution to their problem.

When we think of suicide, we think of shooting oneself, taking an overdose of a drug, or jumping off a building. We distinguish suicide from accidents, in which we presume the person did not intend to harm himself or herself. But the critical difference is the person's intent, not the event itself. Some apparent accidents may be suicides. For example, when a car

FBI statistics as measures of the number of crimes that have occurred in our city or county. Our knowledge of alcohol or drug abuse depends on knowing or hearing about persons who engage in these behaviors. All of these instances of deviance share another important characteristic as well: In every case, the behavior was discovered by someone who called it to the attention of others.

Does it matter that these instances involve both an action (by a person) and a reaction (by a victim or an observer)? Isn't an act equally deviant whether others find out about it or not? Let's go back to Susan's theft of the bracelet. Suppose Susan had left the store without being stopped by the security guard. In that case, she and Virginia would have known she had taken the bracelet, but she would not have faced sanctions from others. She would not have experienced the embarrassment of being confronted by a store guard and accused of a crime. Moreover, she would have had a beautiful bracelet. But the fact is

BOX 19.1 Continued

hits a bridge abutment well away from the pavement on a clear day with no evidence of mechanical malfunction, this may be suicide.

If some auto accidents are, in fact, suicides, we should observe an increase in motor vehicle accidents following newspaper stories about a suicide. In fact, data from newspapers and motor vehicle deaths in San Francisco and Los Angeles verify this hypothesis (Phillips, 1979). Statistics show a marked increase in the number of deaths due to automobile accidents 2 and 3 days after a suicide is publicized—especially accidents involving a single vehicle. In the Detroit metropolitan area, an analysis of motor vehicle fatalities for the years 1973–1976 revealed an average increase in fatalities of 35 to 40 percent the third day after a suicide story appeared in the daily papers (Bollen & Phillips, 1981). Again, the more publicity, the greater the increase. Finally, if the person whose suicide is publicized was young, deaths of young drivers increase; whereas if the person killing himself was older, the increase in fatalities involves more older drivers.

Does an increase in suicide follow any publicized suicide, or are some suicides more likely to be imitated than others? Stack (1987) studied instances in which celebrities killed themselves. Each was classified according to whether the person was an entertainer, political figure, artist, member of the economic elite, or villain. The results showed that when entertainers and politicians took their own lives, there was an increase in suicide rates. Suicides by artists, members of the elite, and villains were not followed by an increase. Moreover, the findings suggest that the effect of publicized suicide is gender and race specific. Suicide by a male celebrity was followed by an increase in the number of men who killed themselves but not in the number of women who took their lives, and vice versa. Similarly, an increase in suicides by Whites followed a publicized case involving a White celebrity, whereas rates among Blacks were unaffected. The fact that the effects of publicized suicide are age, gender, and race specific is consistent with the concept of imitation.

A very different explanation of suicide is that it reflects a lack of social integration. The imitation explanation would be strengthened if we find clustering of suicides after controlling for the effects of variables such as marital and residential stability. Such an analysis was performed of suicide rates for U.S. counties, 1989–1991. Interestingly, the variance in suicide rates in the western third of the United States was explained by integration; in the nonwest, clustering remained, strengthening the argument that suicide involves imitation (Baller & Richardson, 2002).

that she was stopped by the guard. She will be questioned, the police will be called, and she may be arrested. Thus, the consequences of committing a deviant act are quite different when certain reactions follow. Whether a rule-violation becomes "known about" depends in part on the actor's social and economic resources (Jackson-Jacobs, 2004). Consider two crack users, both 21-year-old-men. Joe lives in the ghetto; he works at a car wash, and his earnings support his habit. When he isn't working, he hangs out with other users. Sometimes he commits burglary to get more money. He shoots up in a "crack house," the only place where he can get some privacy and escape surveillance. He is at risk of assault by fellow users, who know he is a user; they want his money or his drugs. Joseph is a junior at a university. He lives in an apartment near campus, where he can use heroin with little risk. His part-time job supplies the money. His schedule of classes and work make it relatively easy for him to restrict his use to leisure times in his apartment, with carefully selected friends. Joseph's legitimate and sufficient income, control over his life, and access to private space make it unlikely he will be arrested.

This reasoning is the basis of **labeling theory**—the view that reactions to a norm violation are a critical element in deviance. Only after an act is discovered and labeled "deviant" is the act recognized as such. If the same act is not discovered and labeled, it is not deviant (Becker, 1963).

If deviance depends on the reactions of others to an act rather than on the act itself, the key social psychological question becomes, "Why do particular audiences choose to label an act as deviant, whereas other audiences may not?" Labeling theory is an attempt to understand how and why acts are labeled deviant. In the case of the stolen bracelet, labeling analysts would not be concerned with Susan's behavior. Rather, they would be interested in the responses to Susan's act by Virginia, the male customer, and the security guard. Only if an observer challenges Susan's behavior or alerts a store employee does the act of taking the bracelet become deviant.

Reactions to Rule Breaking

Labeling theorists refer to behavior that violates norms as **rule breaking,** to emphasize that the act by itself is not deviant. Most rule violations are "secret," in the sense that no one other than the actor (and on occasion, the actor's accomplices) is aware of them. Many cases of theft and tax evasion, many violations of drug laws, and some burglaries are never detected. These activities can be carried out by a single person. Other acts, such as robberies, assaults, and various sexual activities, involve other people who will know about them, but who may not label the act as deviant.

How will members of an audience respond to a rule violation? It depends on the circumstances, but studies suggest that very often people ignore it. When wives of men hospitalized for psychiatric treatment were asked how they reacted to their husband's bizarre behavior, for example, they often replied that they had not considered their husbands ill or in need of help (Yarrow, Schwartz, Murphy, & Deasy, 1955). People react to isolated episodes of unusual behavior in one of four ways. A common response is denial, in which the person simply does not recognize that a rule violation occurred. In one study, denial was typically the first response of women to their husband's excessive drinking (Jackson, 1954). A second response is normalization, in which the observer recognizes that the act occurred but defines it as normal or common. Thus, wives often reacted to excessive drinking as normal, assuming that many men drink a lot. Third, the person may recognize the act as a rule violation but excuse it, attributing its occurrence to situational or transient factors; this reaction is attenuation. Thus, some wives of men who were later hospitalized believed that the episodes of bizarre behavior were caused by unusually high levels of stress or by physical illness. Finally, people may respond to the rule violation by balancing it, recognizing it as a violation but de-emphasizing its significance in light of the actor's good qualities.

The man who witnessed Susan's behavior looked around, spotted a security guard, and reported the act. In doing so, he labeled the actor. Labeling involves a redefinition of the actor's social status; the man placed Susan into the category of "shoplifter" or "thief." The security guard, in turn, probably defined Susan as a "typical shoplifter." Although labeling is triggered by a behavior, it results in a redefinition or typing of the actor. As we shall see, this has a major impact on people's perceptions of and behavior toward the actor.

Determinants of the Reaction

What determines how an observer reacts to rule breaking? Reactions depend on three aspects of the rule violation, including the nature of the actor, the audience, and the situation.

Actor Characteristics Reactions to a type of deviance, such as mental illness, depend on the specific behavior. Given a vignette describing a person who meets the diagnostic criteria for mental illness, adults are more unwilling to interact with (for example, have as a neighbor or coworker) someone who is drug (72

The reactions of others to rule-breaking behavior depend on the characteristics of the actor. The dress and grooming of this shoplifter make it less likely that someone observing his behavior will report it.

percent) or alcohol (56 percent) dependent than someone who has schizophrenia (48 percent) or depression (37 percent). People who view these behaviors as caused by stress are less likely to reject interaction with the person. The belief most closely associated with rejection is the belief that such persons are dangerous (Martin, Pescosolido, & Tuch, 2000).

Reaction to a rule violation often depends on who performs the act. First, people are more tolerant of rule breaking by family members than by strangers. The research cited earlier revealed extraordinary tolerance of spouses for bizarre, disruptive, and even physically abusive behavior. Many of us probably know of a family attempting to care for a member whose behavior creates problems for them. Second, people are more tolerant of rule violations by persons who make positive contributions in other ways. In small groups, tolerance is greater for persons who contribute to the achievement of group goals (Hollander & Julian, 1970). We seem to tolerate deviance when we are dependent on the person committing the act—perhaps because if we punish the actor, it will be costly for us. Third, we are less tolerant if the person has a history of rule breaking (Whitt & Meile, 1985).

Does gender affect reactions to deviant behavior? It depends on the behavior. An ingenious field experiment suggests that it does not affect responses to shoplifting. With the cooperation of stores, shoplifting events were staged near customers who could see the event. The experiment was conducted in a small grocery store, a large supermarket, and a large discount department store. The gender of the shoplifter, the appearance of the shoplifter, and the gender of the observer were varied. Neither the shoplifter's nor the customer's gender had an effect on the frequency with which the customer reported the apparent theft (Steffensmeier & Terry, 1973). Gender does affect reactions to persons who are mentally ill. People are more willing to interact with a woman who is described as having schizophrenia, depression, or drug or alcohol dependence than with a man described the same way. Also, women are perceived as less dangerous than men, which partly explains the greater tolerance (Schnittker, 2000). We are less likely to label women than men for violations of criminal law (Haskell & Yablonsky, 1983). Research indicates that women are less likely to be kept in jail between arraignment and trial and that they receive more lenient sentences than men. One explanation for this differential treatment is that women are subject to greater informal control by family members and friends and so are treated more leniently in the courts. A study of the influences on pretrial release and sentence severity found that both men and women with families received more lenient treatment; the effect was stronger for women (Daly, 1987).

On the other hand, research suggests that psychiatrists are more likely to label women as having a personality disorder than men (Dixon, Gordon, & Khomusi, 1995). Case histories were prepared that included symptoms of clinical disorders (as defined in the *Diagnostic and Statistical Manual of Mental Disorders,* third edition; *DSM-III;* American Psychiatric Association, 1981) and personality disorders (*DSM-III,* Axis II). Personality disorders are generally less serious and more ambiguously defined than clinical disorders. The case histories were identical except for gender: male, female, or unspecified. The psychiatrists' diagnoses of clinical disorders were not influenced by gender, but they were more likely to diagnose women as having personality disorders than men with the same symptoms.

Audience Characteristics The reaction to a violation of rules also depends on who witnesses it. Because groups vary in their norms, audiences vary in their expectations. People enjoying a city park on a warm day will react quite differently to a nude man walking through the park than will a group of people in a nudist park. Recognizing this variation in reaction, people who contemplate breaking the rules—by smoking marijuana, drinking in public, or jaywalking, for example—often make sure no one is around who will punish them.

People who violate rules often belong to groups. What influences whether other group members will ignore or punish a violation? One variable is the cohesiveness of the group. A laboratory experiment found that members of a cohesive group are more likely to reward a member who punished a deviant person than members of groups low in cohesion (Horne, 2001). As a result, members were more likely to punish violators. The effect was found in all-male groups and all-female groups.

Social identity theory (see Chapter 4) suggests that the group membership of the deviant person and the audience both influence reactions. We are motivated to maintain a positive in-group identity, and one means we employ is to maximize the differences we perceive between our group and other groups. Thus, we negatively judge members of our in-group who deviate, especially if the deviation is negative. We judge favorably an out-group member who deviates from his or her group's norms. A laboratory experiment tested these predictions; the results supported them (Abrams, Marques, Brown, & Henson, 2000).

An important influence on whether a witness will label a rule violation is the level of concern in the community about the behavior. Citizens who are concerned about drug use as a social problem are probably more alert for signs of drug sales and use and are more likely to label someone as a drug user. A major determinant of the level of concern is the amount of activity by politicians, service providers, and the mass media calling attention to the problem (Beckett, 1994).

Officials who routinely deal with suspects react very differently to suspected offenders than do most citizens. One study focused on officials working in a court-affiliated unit who evaluated suspected murderers following arrest. These officials had a stereotyped image of the type of person who commits murder (Swigert & Farrell, 1977). When lower class male members of ethnic minorities committed murder, these officials believed that it was in response to a threat to their masculinity. For example, if an Italian American truck driver was arrested for murder, they were likely to assume that he had killed the other man in response to verbal insults. This labeling based on a stereotype had important consequences. Suspects who fit this stereotype were less likely to be defended by a private attorney, more likely to be denied bail, more likely to plead guilty, and more likely to be convicted on more severe charges.

Consider the example of a student with a drinking problem seeking help at a university counseling center. The treatment will depend on how counselors view student "troubles." One study found that the staff of a university clinic believed that students' problems could be classified into one of the following categories: problems in studying, choosing a career, achieving sexual intimacy, or handling personal finances; conflict with family or friends; and stress arising from sociopolitical activities. When a student came to the clinic because of excessive drinking, the therapist first decided which of these categories applied to this person's troubles; that is, which type of problem was causing this student to drink excessively. How the problem was defined in turn determined what the therapist did to try to help the student (Kahne & Schwartz, 1978).

Situational Characteristics Whether a behavior is construed as normal or labeled as deviant also depends on the definition of the situation in which the behavior occurs. Marijuana and alcohol use, for example, are much more acceptable at a party than at work (Orcutt, 1975). Various sexual activities expected between spouses in the privacy of their home would elicit condemnation if performed in a public park.

Consider so-called gang violence. In some major cities, incidents in which teenage gangs assault each other are common. News media, police, and other outsiders often refer to such incidents as "gang wars." These events often occur in the neighborhoods where the gang members live. How do their parents, relatives, and friends react to such incidents? According to a study of one Chicano community, it depends on the situation (Horowitz, 1987). Young men are expected to protect their families, women, and masculinity. When violence results from a challenge to honor, the community generally tolerates it. On the other hand, if the violence disrupts a community affair, such as a dance or a wedding, it is not tolerated.

We often rely on the behavior of others to help us define situations. Our reaction to a rule violation may be influenced by the reactions of other members of the audience. The influence of the reactions of others is demonstrated in a field experiment of intrusions into waiting lines (Milgram, Liberty, Toledo, & Wackenhut, 1986). Members of the research team intruded into 129 waiting lines with an average length of six persons. One or two confederates approached the line and stepped between the third and fourth person. In some cases, other confederates served as buffers; they occupied the fourth and fifth positions and did not react to the intrusion. When the buffers were present, others in the line were much less likely to react verbally or nonverbally to the intrusion.

A good deal of research suggests that interpersonal violence—especially assaults and murders—often involves two young men and is triggered by a verbal insult (Katz, 1988). But whether a remark is an insult is a matter of social definition. Not surprisingly, fights are more likely to erupt following a remark when there is a male audience and when the men have been drinking (Felson, 1994). A remark is less likely to lead to a fight if the audience includes women.

Consequences of Labeling

Assume that an audience defines an act as deviant. What are the consequences for the actor and the audience? We will consider four possible outcomes.

Institutionalization of Deviance In some cases, individuals who label a behavior as deviant may decide that it is in their own interest for the person to continue the behavior. They may, in fact, reward that person for the deviant behavior. If you learn that a good friend is selling drugs, you may decide to use this man as a source and purchase drugs from him. Over time, your expectations will change; you will come to expect him to sell drugs. If your drug-selling friend then decides to stop dealing, you may treat him as a rule breaker. Illegal activities by stockbrokers are likely to be ignored or encouraged by other employees and supervisors when all of them benefit economically from the activity (Zey, 1993). The process by which members of a group come to expect and support deviance by another member over time is called **institutionalization of deviance** (Dentler & Erikson, 1959).

Consider the following sworn statement by a former Enron employee, Timothy N. Belden:

> I was Director of Enron's California energy trading desk . . . [We] marketed and supplied electricity to Californian wholesale customers. . . . Beginning in approximately 1998, and ending in approximately 2001, I and other individuals in Enron agreed to devise and implement a series of fraudulent schemes through these markets. We designed the schemes to obtain increased revenue for Enron. . . .
>
> We exported and then imported amounts of electricity generated within California in order to receive higher, out-of-state prices from the [ISO] when it purchased "out of market." We scheduled energy that we did not have, or did not intend to supply. (United States of America *v.* Timothy Belden, U.S. District Court, Northern District of California, Doc. CR 02-0313 MJJ, October 17, 2002)

Note the repeated use of "we"; Belden and other Enron employees supported (and expected the support of) each other as they engaged in these deviant activities. It is possible that support for their criminal activities extended to the highest levels of the corporation.

Backtracking Even when an audience reacts favorably to a rule violation, the actor may decide to discontinue the behavior. This second consequence of labeling is called backtracking. It may occur after the actor learns that others label his or her act as deviant. Though some audiences react favorably, the actor may wish to avoid the reaction of those who would not react favorably—and the resulting punishment. Many teenagers try substances like marijuana once or twice. Although their friends may encourage its continued use, some youths backtrack because they want to avoid their parents' or others' negative reactions.

Effective Social Control An audience that reacts negatively to rule breaking and threatens or attempts to punish the actor may force the actor to give up further involvement in the deviant activity. This third consequence of labeling is known as effective social control. This reaction is common among friends or family members, who often threaten to end their association with an actor who continues to engage in deviance. Similarly, they may threaten to break off their relationship if the person does not seek professional help. In these instances, the satisfaction of the actor's needs is contingent on changing his or her behavior. Members of the audience also may insist that the actor renounce aspects of his or her life that they see as contributing to future deviance (Sagarin, 1975). If excessive drinking is due to job-related stressors, for example, family members may demand that the person find a different type of employment. Displays of remorse may also lead to reduced punishment for an offense (Robinson, Smith-Lovin, & Tsoudis, 1994).

Unanticipated Deviance Still another possibility is that the individual may engage in further or unanticipated deviance. Note the use of the term "unanticipated." Negative reactions by members of an audience are intended to terminate rule-breaking activity. However, such reactions may, in fact, produce further deviance. This occurs when the audience's response sets in motion a process that leads the actor to greater involvement in deviance. This process and its outcomes are the focus of the next section.

LABELING AND SECONDARY DEVIANCE

Labeling a person as deviant may set in motion a process that has important effects on the individual. The process of societal reaction produces changes in the behavior of others toward the labeled individual and may lead to corresponding changes in his or her self-image. A frequent consequence of the process is involvement in secondary deviance and participation in a deviant subculture. In this section, we consider this process in detail.

Societal Reaction

Earlier in this chapter, we mentioned that labeling is a process of redefining a person. By categorizing a person as a particular kind of deviant, we place that person in a stigmatized social status (see Chapter 9). The deviant person (addict, pimp, thief) is defined as undesirable—not acceptable in conventional society—and frequently treated as inferior. There are two important consequences of stigma: status loss and social discrimination (Link & Phelan, 2001). The loss of status causes a gradual change in self-conception; the person comes to perceive the self as a type of deviant. Discriminatory behavior by others not only affects one's self-concept but constrains one's behavior and opportunities.

Changes in the Behavior of Others When we learn that someone is an alcoholic, a convicted rapist, or mentally ill, our perceptions and behavior toward that person change. For example, if we learn that someone has a drinking problem, we may respond to his or her request for a drink with "Do you think you should?" or "Why don't you wait?" to convey our objection. We may avoid jokes about drinking in the person's presence, and we may stop inviting him or her to parties or gatherings where alcohol will be served.

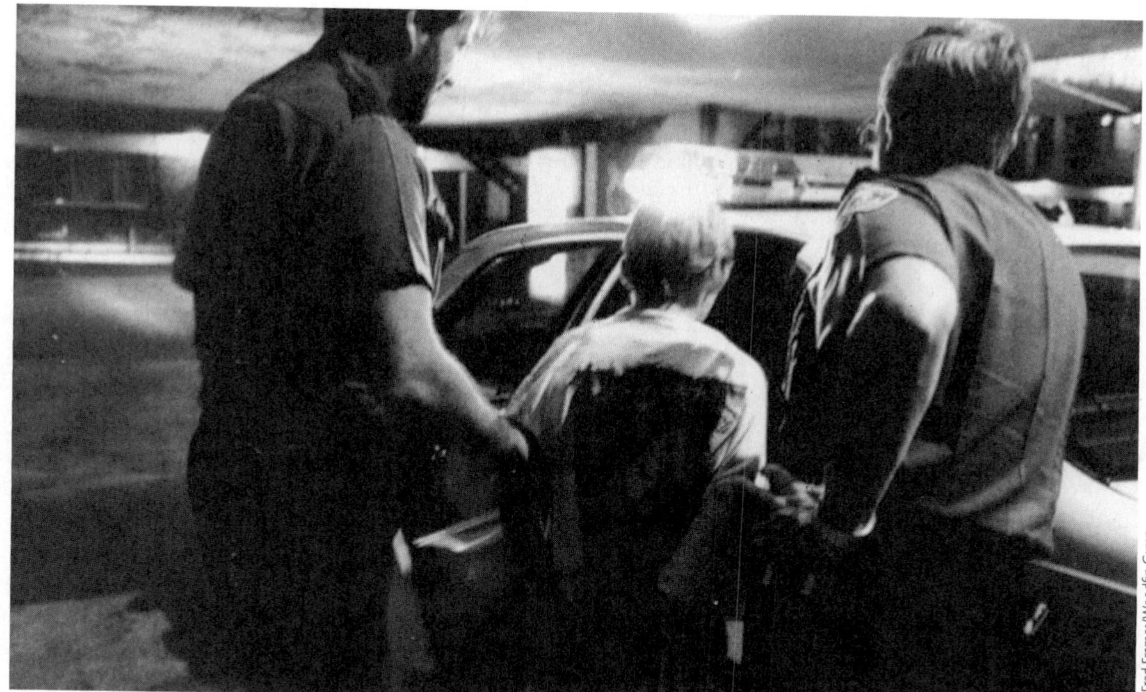

Being caught in a deviant act has important consequences. This youth may experience disrupted schooling and separation from family and friends as a result of being arrested.

A more severe behavioral reaction involves withdrawal from the stigmatized person (Kitsuse, 1964). For instance, the labeled shoplifter, alcoholic, or gay man may be fired from his or her job. Behavioral withdrawal may occur because of hostility toward the deviant person, or it may reflect a sincere desire to help the person. For example, the employer who fires an alcoholic may do so because he dislikes alcoholics or because he believes that relief from work obligations will reduce the stress that may be causing the drinking problem.

Paradoxically, our reaction to deviance may produce additional rule breaking by the labeled person. We expect people who are psychologically disturbed to be irritable or unpredictable, so we avoid them to avoid an unpleasant interaction. The other person may sense that he or she is being avoided and respond with anger or distrust. This anger may cause coworkers to gossip behind his or her back; he or she may respond with suspicion and become paranoid. When members of an audience behave toward a person according to a label and cause the person to respond in ways that confirm the label, they have produced a **self-fulfilling prophecy** (Merton, 1957). Lemert (1962) documents a case in which such a sequence led to a man's hospitalization for paranoia.

Self-Perception of the Deviant Another consequence of stigmatized social status is that it changes the deviant person's self-image. A person labeled deviant often incorporates the label into his or her identity. This redefinition of oneself is due partly to feedback from others who treat the person as deviant. Moreover, the new self-image may be reinforced by the individual's own behavior. Repeated participation in shoplifting, for example, may lead Susan to define herself as a thief.

Redefinition is facilitated by the social programs and agencies that deal with specific types of deviant persons. Such agencies pressure persons to

FIGURE 19.2 THE RELATIONSHIP OF SELF-REJECTION TO DEVIANT BEHAVIOR

A person who engages in deviant behavior anticipates that others will reject him or her, which, in turn, can lead to self-rejection. A longitudinal study collected data from junior high school students three times, each 1 year apart. At time 1, reported participation in deviance was positively related to self-rejection (feeling one is no good, a failure, rejected by parents and teachers). Self-rejection at time 1 was associated with more favorable dispositions (attitudes) toward deviance but a decreased likelihood of associating with other deviants 1 year later (at time 2). Favorable dispositions and deviant associations at time 2, as well as deviance at time 1, were related to increased deviance—theft, gang violence, drug use, and truancy—at time 3.

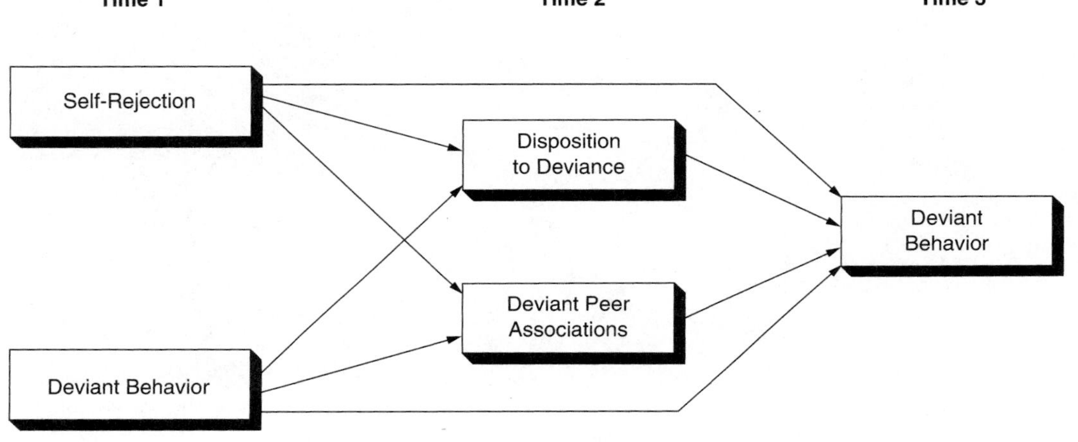

Source: From "Deviant Peers and Deviant Behavior: Further Elaboration of a Model" (1987) by Kaplan, Johnson, and Bailey, *Social Psychology Quarterly, 50*(3), 281. Used with permission from the American Sociological Association.

acknowledge that they are deviant. Admitting that one is a thief will often lead police and prosecutors to go easy on a shoplifter, especially if it is a first offense. Failure to acknowledge this may lead to a long prison sentence. Admitting that one is mentally ill is often a prerequisite for psychiatric treatment (Goffman, 1959a). Mental health professionals often believe that a patient cannot be helped until the individual recognizes his or her problem. Employees of an agency that provided jobs for unemployed persons viewed their clients' employment problems as partly the result of individual failure (Miller, 1991). To receive agency services, clients had to agree with this view and change their behavior accordingly.

Thus, the deviant person experiences numerous pressures to accept a stigmatized identity. Acceptance of a stigmatized identity has important effects on self-perception. Everyone has beliefs about what people think of specific types of deviant persons. Accepting a label such as "thief," "drunk," or "crazy" leads a person to expect that others will stigmatize and reject him or her, which in turn produces self-rejection. Self-rejection makes subsequent deviance more likely (Kaplan, Martin, & Johnson, 1986). In a study of junior high school students, data were collected three times at 1-year intervals. Self-rejection (that is, feeling that one is no good, a failure, rejected by parents and teachers) was related to more favorable dispositions (definitions) toward deviance and to an increased likelihood of associating with deviant persons 1 year later. A high disposition and associations with deviant peers were related to increased deviance—theft, gang violence, drug use, and truancy—1 year later (Kaplan, Johnson, & Bailey, 1987).

Figure 19.2 summarizes these relationships. Delinquent behavior, in turn, is associated with reduced self-esteem (McCarthy & Hoge, 1984).

In short, labeling may set in motion a cycle in which changes in the labeled person's behavior produce changes in other people's behavior, which in turn changes the deviant person's self-image and subsequent behavior. Self-fulfilling prophecies can also be positive. One study assessed the expectations of 98 sixth-grade math teachers for their students (N = 1,539; Madon, Jussim, & Eccles, 1997). Teachers' expectations (positive or negative) predicted performance much better for students who were low achievers. Also, teachers' overestimates—that is, positive expectations—predicted actual achievement better than their underestimates. Perhaps positive expectations inspire underachievers.

Although more attention has been given to situations in which others label the person, some persons become committed to deviance without such labeling. For example, some persons voluntarily seek psychiatric treatment; some of these cases reflect self-labeling (Thoits, 1985). People know that others view certain behaviors as symptoms of mental illness. If they observe themselves engaging in those behaviors, they may label themselves as mentally ill.

Secondary Deviance

A frequent outcome of the societal reaction process is **secondary deviance,** in which a person engages increasingly in deviant behavior as an adjustment to others' reactions (Lemert, 1951). Usually, the individual becomes openly and actively involved in the deviant role, adopting the clothes, speech, and mannerisms associated with it. For example, initially, a person with a drinking problem may drink only at night and on weekends to prevent his or her drinking from interfering with work. Once the person adopts the role of "heavy drinker" or "alcoholic," however, he or she may drink continuously. The first time a woman engages in commercial sex, she may do so tentatively and anxiously, and the interaction may be awkward. As she continues, it is likely that she will begin to dress and talk like others who engage in such work, and become comfortable with engaging in sexual activity with several clients in one evening.

As an individual becomes openly and regularly involved in deviance, he or she may increasingly associate with others who routinely engage in the same or related activity. The individual may join a **deviant subculture**—a group of people whose norms encourage participation in the deviance and who regard positively those who engage in it. Subcultures provide not only acceptance but also the opportunity to enact deviant roles. Through a deviant subculture, the would-be drug dealer or prostitute can gain access to customers more readily.

Subcultural groups are an attractive alternative for deviant persons for two reasons. First, these people are often forced out of nondeviant relationships and groups through others' reactions. As family and friends progressively break off relationships with them, they are compelled to seek acceptance elsewhere. Second, membership in subcultural groups may result from the deviant person's desire to associate with persons who are similar and who can provide them with feelings of social acceptance and self-worth (Cohen, 1966). Deviant persons are no different from others in their need for positive reflected appraisals.

Deviant subcultures help persons cope with the stigma associated with deviant status. We have already noted that deviant persons are often treated with disrespect and sanctioned by others for their activity. Such treatment threatens self-esteem and produces fear of additional sanctions. Subcultures help the deviant person cope with these feelings. They provide a vocabulary of motives—beliefs that explain and justify the individual's participation in the behavior.

The norms and belief systems of subcultures support a positive self-conception. In the early 1970s, a prostitutes' rights group, COYOTE (Cast Off Your Old, Tired Ethics), emerged in San Francisco. Although it did not obtain the legalization of prostitution, it did enhance the self-images of its members (Weitzer, 1991). Many people think that nudists are exhibitionists who take off their clothes to get sexual kicks. Nudists, on the other hand, consider themselves morally respectable and hold several beliefs designed to enhance that claim: (1) Nudity and sexuality are unrelated, (2) there is nothing shameful about

FIGURE 19.3 IMAGES OF THE DEVIANT

We often have very negative images of many types of deviant persons. For example, many people view marijuana users as dirty, unkempt dropouts, like the person on the left. Because these images are widely shared, persons who engage in some form of deviant behavior are usually aware that others look down on them. To counter this stigma, deviants attempt to create a positive self-image, which is reinforced by members of deviant subcultures. The marijuana user views himself as clean, cool, and in touch, like the person on the right. It is easier to view oneself as "normal" when others support that view.

the human body, (3) nudity promotes a feeling of freedom and natural pleasure, and (4) nude exposure to the sun promotes physical, mental, and spiritual well-being. There are also specific norms—"no staring," "no sex talk," and "no body contact"—designed to sustain these general beliefs (Weinberg, 1976). The contrast between the stigmatized image of the deviant person and the deviant person's self-image is illustrated in Figure 19.3. The belief systems of deviant subcultures provide the social support the person needs to maintain a positive self-image.

Joining a deviant subculture often stabilizes participation in one form of deviance. It also may lead to involvement in additional forms of deviant behavior. For instance, many prostitutes become drug users through participation in a subculture.

FORMAL SOCIAL CONTROLS

So far, this chapter has been concerned with **informal social control**—the reactions of family, friends, and acquaintances to rule violations by individuals. Informal controls are probably the major influence on an individual's behavior. In modern societies, however, there are often elaborate systems set up specifically to process rule breakers. Collectively, these are called **formal social controls**—agencies given responsibility for dealing with violations of rules or laws. Typically, the rules enforced are written, and, in some cases, punishments also may be specified. The most prominent system of formal social control in our society is

© The Boston Globe

Deviant subcultures create opportunities for people to enact roles not acceptable elsewhere in society. This nudist camp provides a place where people can undress without attracting attention or being arrested.

the criminal justice system, which includes police, courts, jails, and prisons. A second system of formal social control is the juvenile justice system, which includes juvenile officers, social workers, probation officers, courts, and treatment or detention facilities. A third system of formal social control deals with mental illness. It includes mental health professionals, commitment procedures, and institutions for the mentally ill and mentally impaired.

Formal Labeling and the Creation of Deviance

Most of us think of formal agencies as reactive—as simply processing individuals who have already committed crimes or who are mentally retarded or in need of psychiatric treatment. But these agencies do much more than take care of persons already known to be deviant. It can be argued that the function of formal social control agencies is to select members of society and identify or certify them as deviant (Erikson, 1964).

In the 1990s, crime control became big business in U.S. society. Federal and state governments provided funds to hire thousands of additional police officers, sheriff's deputies, and federal agents. Many states built new prisons. Additional officers and new prisons require large investments in new equipment. It has been suggested that there is a crime control industry, with many people lobbying for its preservation and growth (Chambliss, 1994). More officers and prisons lead to more arrests and further increases in prison populations. Is this expansion due to real increases in crime? No. Crime has not increased substantially in the past 25 years. In fact, in 1997, crime rates were declining. What has increased is political rhetoric on and mass media attention to a stable level of crime, leading the public to perceive an increase. Politicians have used this perception as a basis for fear campaigns to enlist support for the expansion of formal control systems.

Functions of Labeling Of what value is labeling people as "criminals," "delinquents," or "mentally ill"? There are three functions of labeling persons as deviant: (1) to provide concrete examples of undesirable behavior, (2) to provide scapegoats for the release of tensions, and (3) to unify the group or society.

First, the public identification of deviance provides concrete examples of how we should not behave (Cohen, 1966). When someone is actually apprehended and sanctioned for deviance, the norms of society are made starkly clear. For instance, the arrest of someone for shoplifting dramatizes the possible consequences of taking things that do not belong to us. The controversy and publicity in 1998–1999 surrounding the claims of sexual harassment of women by President Bill Clinton heightened awareness of these behaviors.

According to the **deterrence hypothesis,** the arrest and punishment of some individuals for violations of the law deters other persons from committing the same violations. To what extent does general deterrence really affect people's behavior? Most analysts agree that the objective possibility of arrest and punishment does not deter people from breaking the law. Rather, conformity is based on people's perceptions of the likelihood and severity of punishment. Thus, youths who perceive a higher probability that they will be caught and that the punishment will be severe are less likely to engage in delinquent behavior (Jensen, Erickson, & Gibbs, 1978). Similarly, a study of theft of company property by employees found that

© Bob Daemmrich

A man convicted of a crime talks to high school students about the nature and consequences of his deviance. By publicizing the penalties for crime, such programs attempt to deter others from breaking the law.

those who perceived greater certainty and severity of organizational sanctions for theft were less likely to have stolen property (Hollinger & Clark, 1983).

For the punishment of some offenders to deter others, the punishment must be publicized. In recent years, executions of murderers have been widely publicized. Does this publicity deter murder? Specifically, does coverage of executions on the evening news on network television lead to a reduction in homicide rates? A study of news coverage and homicide rates from 1976 through 1987 found no relationship (Bailey, 1990).

Perceived certainty of sanctions generally has a much greater effect on persons who have low levels of moral commitment (Silberman, 1976). People whose morals define a behavior as wrong are not as affected by the threat of punishment. For example, personal moral beliefs are a more important influence on whether adults use marijuana than the fear of legal sanctions (Meier & Johnson, 1977). Adults who believe that the use of marijuana is wrong do not use it, regardless of their perception of the likelihood that they will be sanctioned for its use.

A second function of the public identification of deviant persons is to provide a scapegoat for the release of tension. Many people face threats to the stability and security of their daily lives. Some fear the possibility that they will be victimized by aggressive behavior or the criminal activity of others. The existence of such threats arouses tension. Persons identified publicly as deviant persons provide a focus for these fears and insecurities. Thus, the publicly identified deviant person becomes the concrete threat we can deal with decisively.

This scapegoating process is illustrated among the Puritans, who came to New England in the 1600s to establish a community based on a specific Christian theology. As time passed, groups within the community periodically challenged the ministers' claims that they were the sole interpreters of the theology. Furthermore, the community faced the threat of Indian attacks and the problems of daily survival in a harsh environment. In 1692, a group of young women began to behave in such bizarre ways as screaming, convulsing, crawling on all fours, and barking like dogs. The community focused attention on these women. The physicians defined them as "witches," representatives of Satan, and the entire community banded together in search of others who were under the "devil's influence." The community imprisoned many persons suspected of sorcery and sent 22 persons to their deaths. Thus, the witch hunt provided a scapegoat—an outlet for people's fears and anxieties (Erikson, 1966).

A third function of the public identification of deviant persons is to increase the cohesion and solidarity of society. Nothing unites the members of a group like a common enemy (Cohen, 1966). Deviant persons, in this context, are "internal enemies"—persons whose behavior threatens the morale and efficiency of a group. Should the solidarity of the group be threatened, it can be restored by identifying one member as deviant and imposing appropriate sanctions. Suppose you are given the case study of a boy with a history of delinquency who is to be sentenced for a minor crime. You are asked to discuss the case with three other persons and decide what should be done. One member of the group argues for extreme discipline, whereas you and the other two favor leniency. Suddenly, an expert in criminal justice, who has been sitting quietly in the corner, announces that your group should not be allowed to reach a decision. How might you deal with this threat to the group's existence? The reasoning just outlined suggests that the person who took the extreme position will be identified as the cause of the group's poor performance and that the other members will try to exclude him or her from future group meetings. A laboratory study used exactly this setup, contrasting the reaction of threatened groups to the person taking the extreme position with the reaction of nonthreatened groups. In the former condition, the person taking the extreme position was more likely to be stigmatized and rejected (Lauderdale, 1976).

Thus, controlled amounts of deviant behavior serve important functions. If deviance is useful, we might expect control agencies to "create" deviance when the functions it serves are needed. In fact, the number of persons who are publicly identified as deviant seems to reflect the levels of stress and integration in society (Scott, 1976). When integration declines, there is an increased probability of deviance. Eventually, the level and severity of deviance may reach a point where citizens will demand a "crackdown." Social control agencies will step up their activity, increasing the number of publicly identified deviant persons. This, in turn, will increase solidarity and lower stress, leading to an increase in the amount of informal control and a reduction in deviance.

The Process of Labeling Labeling is not a simple, one-step procedure for formal agencies. The processing of rule breakers usually involves a sequence of decisions. At each step, someone has to decide whether to terminate the process or to pass the rule breaker on to the next step. Figure 19.4 shows the sequence of steps involved in processing criminal defendants.

Each of the control agents—police officers, prosecutors, and judges—has to make many decisions every day. Like anyone else, they develop cognitive schemas and rules that simplify their decision making. A very common police-citizen encounter occurs when an officer stops a motorist who has been drinking. What determines whether a driver who has been drinking is labeled a "drunken driver?" Officers on the street have to rely on a variety of subjective data, as the breathalyzer or blood or urine test may only be available at the police station. Research suggests that police officers develop a series of informal guidelines that they use in deciding whether to arrest the motorist. In one study of 195 police encounters with persons who had been drinking, arrests were more likely if the encounter occurred downtown and if the citizen was disrespectful (Lundman, 1974).

Prosecutors also develop informal rules that govern their decisions. For example, in one large midwestern city, taking an object worth less than $100 is a misdemeanor, and conviction normally results in a fine. The theft of a more valuable object is a felony and results in a prison sentence. Because felony theft cases require much more time and effort, the prosecutor has charged most persons arrested for shoplifting with

FIGURE 19.4 FORMAL SOCIAL CONTROL: PROCESSING CRIMINAL DEFENDANTS

Formal social control often involves several control agents, each of whom makes one or more decisions. The first step in the criminal justice system is an encounter with a law enforcement officer. If you are arrested, the case is passed to a prosecutor, who decides whether to prosecute. If your case goes to court, the judge or jury decides whether you are guilty. Finally, the judge renders a sentence. These decision makers are influenced by their own personal attitudes, cognitive schemas, role expectations, and the attitudes of others regarding their decisions. Much research is devoted to the social psychological aspects of decision making in the criminal justice system.

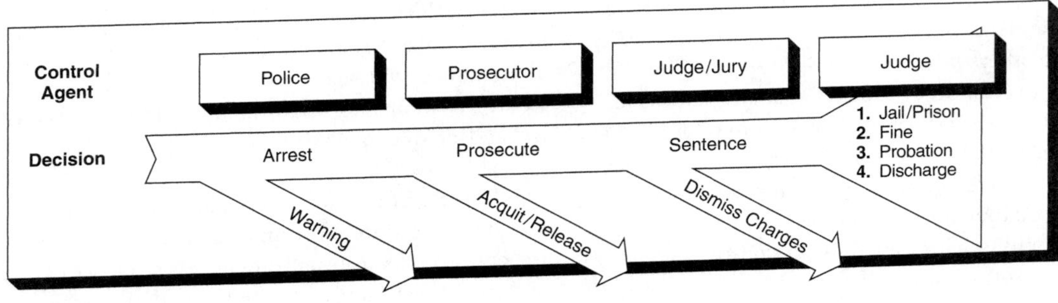

misdemeanors, even if they have taken jewelry worth hundreds of dollars.

In many jurisdictions, probation officers are asked to prepare a presentencing report and to recommend a sentence for the convicted person. Research indicates that these officers have a set of typologies or schemas into which they sort persons (Lurigis & Carroll, 1985). Semistructured interviews with probation officers in one community identified 10 schemas, including burglar, addict, gang member, welfare fraud, and conman. Each schema was associated with beliefs about the motive for the crime and the appropriate treatment and prognosis. When officers were asked to evaluate sample cases, those fitting a schema were evaluated more quickly and confidently. More experienced officers were more likely to use similar schemas (Drass & Spencer, 1987).

Each step in this process involves interaction between professionals and the alleged rule breaker—and often members of his or her family. The professional's goal is to have the rule breaker and other laypersons accept the label. Research on the labeling of children with developmental disabilities suggests that family members are more likely to accept a diagnosis if the professional elicits family members' schemas and frames the diagnosis/label in those terms (Gill & Maynard, 1995).

Biases in Social Control Not all persons who violate the rules are labeled. Most social control agencies process only some of those who engage in rule-breaking behavior. In the study of police encounters with drunken persons, only 31 percent were arrested (Lundman, 1974). In some cases, control agents may be influenced by the demeanor of the rule breaker, by the agent's schema, or by where the violation occurs. This leads one to ask whether systematic biases exist in the social control system.

It has been suggested that control agents are more likely to label those people who have the least power to resist their certification as deviant (Quinney, 1970). This hypothesis predicts that lower class persons and members of racial and ethnic minorities are more likely to be certified as deviant than upper class, middle-class, and White persons. This hypothesis offers a radically different explanation for the correlation between crime and social class. Earlier in this chapter, we suggested that crime rates are higher for lower class persons because they do not have access to nondeviant means of economic success. Here we are

© Jeremy Woodhouse/SuperStock

Whether a police officer gives a citizen a traffic ticket or not depends partly on the demeanor of the citizen. Officers are more likely to ticket or arrest hostile, argumentative persons than polite and submissive ones.

suggesting that crime rates are higher among lower class persons because they are more likely to be arrested, prosecuted, and found guilty, even though the underlying rate of deviant activity may not vary as a function of social class.

Does social class or race influence how an individual is treated by control agents? One way to answer this question is by studying police-citizen encounters through the ride-along method, in which trained observers ride in squad cars and systematically record data about police-citizen encounters. In the largest study of this kind, observers rode with some officers on all shifts every day for 7 weeks. Data were collected in Boston, Washington, and Chicago and included 5,713 encounters. There was no evidence that Blacks were more likely to be arrested than Whites. Rather, arrests were more likely when a third party demanded an arrest, when the evidence was strong, and when the crime was serious (Black, 1980). A study of how police officers managed violent encounters between citizens found that arrest was more likely if the incident involved White persons; two men instead of one woman and one man, or two women; or if one person acted abusively toward the officer (Smith, 1987). On the other hand, research that included ride-alongs in Washington, DC, suggests that at least in that city, Blacks are subjected to more intense police surveillance than other racial/ethnic groups (Chambliss, 1994).

What about decisions by prosecutors? Do they entail discrimination based on race or class? Prosecutors are generally motivated to maximize the ratio of convictions to trials. This may be one criterion that citizens use in evaluating the performance of a district attorney. Prosecutors develop beliefs about which cases are "strong"—that is, likely to result in conviction. A study of a random sample of 980 defendants charged with felonies found that prosecutors are more likely to prosecute cases involving serious crimes where the evidence is strong and the defendant has a serious prior police record. Race was not generally influential (Myers & Hagan, 1979).

Does the social class of an arrested person influence how he or she is treated by the courts? Several studies of the handling of juvenile cases report little evidence of class or race bias. A study of cases in Denver and Memphis found that the seriousness of the offense and the youth's prior record were the major determinants of the sentence given (Cohen & Kluegel, 1978). Two longitudinal studies, of 9,945 boys in Philadelphia (Thornberry & Christenson, 1984) and of cases in Florida (Henretta, Frazier, & Bishop, 1986), found that the most important influence on the disposition of a charge was the disposition imposed for a prior offense or offenses.

A common practice in adult criminal cases is plea bargaining, in which a prosecutor and a defendant's lawyer negotiate a plea to avoid the time and expense of a trial. A single action frequently violates several laws. For instance, if a driver who has been drinking runs a red light and hits a pedestrian who later dies, that incident involves at last three crimes: drunken driving, failure to obey a signal, and vehicular manslaughter. These offenses vary in seriousness and, thus, in their associated sentences. The prosecutor may offer not to indict the driver for manslaughter if a plea of guilty is entered to a drunken driving charge. The attorney may accept the offer, provided the prosecutor also recommends a suspended sentence.

Are the members of certain groups more likely to be tried or to get bigger reductions in sentences? An analysis of charge reduction or plea bargaining in a sample of 1,435 criminal defendants found that women and Whites received slightly more favorable reductions than men and Blacks (Bernstein, Kick, Leung, & Schulz, 1977). Another study of 1,213 men charged with felonies found that the characteristics of an offense—especially the seriousness of the crime and the strength of the evidence—were most important in determining the disposition. The outcomes of the cases were not related to age, ethnicity, or employment status (Bernstein, Kelly, & Doyle, 1977). A study of 296 women who killed another person found that whereas they were all initially charged with murder, in two thirds of the cases, the charge was reduced to manslaughter or a lesser offense (Mann, 1996). Women in southern cities and women who killed men were less likely to have the charge(s) reduced and received more severe sentences if convicted.

Among the persons convicted, do we find a class or racial bias in the length of the sentences given? One study focused on the sentences received by 10,488 persons in three southern states: North Carolina, South Carolina, and Florida (Chiricos & Waldo, 1975). The researchers examined sentences for 17 different offenses and found no relationship between socioeconomic status or race and sentence length. Again, the individual's prior record was the principal variable related to sentence length. A study of a random sample of 16,798 felons convicted during the years 1976–1982 in Georgia looked at racial differences in sentencing (Myers & Talarico, 1986). In general, the seriousness of a crime was the principal influence on the sentence length. Another study of the influence of race on sentencing analyzed federal court proceedings for the years 1993–1996. The sentences given male defendants varied by race/ethnicity. For offenses of the same seriousness, there were small to moderate effects, with Whites receiving shorter sentences and Hispanics receiving longer ones; Blacks received intermediate sentences (Steffensmeier & Demuth, 2000).

Earlier, we discussed white-collar crime, which is often committed by middle-class and upper class persons. Are white-collar offenders more likely to receive lenient sentences? A study of persons charged with embezzlement and tax, lending, credit, postal, and wire fraud found that within this group, high-status persons were no less likely to be imprisoned or to receive shorter sentences (Benson & Walker, 1988). The significant influences were the total amount of dollars involved and how widespread the offenses were. Blacks did receive longer sentences than Whites. It is sometimes argued that judges are lenient on high-status offenders because they suffer serious informal sanctions, such as the loss of a job. A study of the likelihood of job loss and the influence of job loss on sentence severity found no relationship (Benson, 1989). However, class position did influence job loss; high-status offenders and those whose frauds were larger in scale were less likely to lose their jobs.

Long-Term Effects of Formal Labeling

How long does the official label of deviant stick to a person? Can it be shaken? In contrast with the trial or hearing in which a person is formally certified as deviant, there is no formal ceremony terminating one's deviant status (Erikson, 1964). People are simply released from prison or mental hospital, or the final day of their probation passes—with no fanfare. Does the individual regain his former status upon release, or does deviant status in our society tend to be for life?

Some argue that ex-convicts, ex-patients, and others who have been labeled as deviant face continuing pressures from family, friends, employers, and coworkers that prevent them from readjusting to normal life. Such pressures constitute a reminder of their former deviant status.

One domain in which persons who have been officially labeled might face discrimination is employment. In the United States, a good job is essential to health and welfare as an adult. Labeling theory asserts that contact with authorities will reduce one's occupational attainment, independently of the rule breaking or deviant behavior. Longitudinal data allowed a study of the effect of contact with authorities (suspended from school, stopped by police, charged, convicted, sentenced, jailed) at ages 15 to 23 on status (employed, status, income) during ages 29–37 (Davies & Tanner, 2003). For males, suspension or being stopped had little effect; the more serious forms of contact, especially being jailed, had significant negative effects on average hours worked, status of job, and income. For women, suspension had negative effects; also, being sentenced and jailed had substantial impact, with incarceration resulting in a reduction in annual income of more than 50 percent! Another study of data from more than 3,600 men compared those who had been incarcerated with men who had not (Western, 2002); overall, incarceration reduced subsequent earnings by 19 percent, and also resulted in slower wage growth in later years. The impact on Black and Hispanic men was about twice as large as the impact on White men.

Discrimination in employment following formal labeling may occur because others perceive these persons as "delinquents," "ex-cons," or "crazies" and expect them to behave in ways consistent with the label. A study of the impact of a criminal record on decisions by employers used the audit method, which involves sending matched pairs of people (testers) to apply for real job openings. The pairs differ on some characteristic, and the researcher observes whether employers respond differently to the two people. Two young White men and two young Black men, matched on age, physical appearance, and style, applied for the same jobs 1 day apart. One man of each race had a criminal record—a felony drug conviction that resulted in 18 months in prison. Each tester posed as the convicted felon in alternate weeks. The dependent variable was whether the employer called back the young man for an interview. Having a record had a significant effect. For the White pair, the man without the record was called 34 percent of the time, whereas the man with the record was called 17 percent of the time. Among the Black testers, the percentages were 14 percent and 5 percent (Pager, 2003). Thus, the combination of being Black and having a criminal record makes it unlikely one will be called for an interview. This contributes to the high rates of unemployment in some minority communities.

Questionnaires and interviews with men hoping to transition from AIDS disability back to work focused on issues of identity. The men countered the stigma of living with AIDS by developing a romanticized anticipatory identity as worker/gay man/recovered. As they sought re-employment, they experienced discrimination and shame, and had to adjust to medical and other constraints, producing an actualized identity at odds with the anticipatory one. The transition back into the worker role was thus very stressful for some (Ghaziani, 2004).

Another approach to studying the long-term effects is to compare persons who have and have not been labeled. A study of psychiatrically disturbed persons compared the income and employment status of those who had been treated (labeled) with the income and status of those who had not been treated. Treatment was negatively associated with both income and employment (Link, 1982). The impact seemed to depend partly on whether occupational competence was developed before or after the onset of the illness. Men who had no history of competent work performance had more difficulty obtaining employment following hospitalization. Men who had a history of occupational competence usually kept their jobs, even during periods when their work performance was seriously affected.

Some persons turn a career history of deviance into an occupational asset by becoming a "professional ex-" (Brown, 1991a). Individuals with histories of alcohol or drug abuse or other problem behaviors sometimes become counselors, working with others who are involved in these behaviors. Professionalizing rather than giving up the deviant identity is another way of going straight.

A study of the long-term impact of being labeled as mentally ill suggests that it is not the label by itself that has impact but the label combined with changes in self-perception (Link, 1987). The study compared samples of residents and clinic patients from the same area of New York City. Three samples involved people who had been labeled: first-treatment contact patients, repeat-treatment contact patients, and formerly treated community residents. The other two groups were untreated "cases" (people with symptoms) and a sample of residents. All participants completed a scale that measured the belief that mental patients are stigmatized and discriminated against. High scores on the measure were associated with reduced income and unemployment in the labeled groups but not in the unlabeled ones. Later research shows that when people enter treatment, those who expect discrimination use strategies such as keeping their condition secret or withdrawing from interaction (Link, Cullen, Struening, Shrout, & Dohrenwend, 1989). This tends to cut them off from social support and interfere with their work performance.

A longitudinal study of recovery from mental illness obtained data from members of self-help groups ($N = 590$) and outpatients ($N = 90$) two times, 18 months apart. The results indicate that recovery is a complex process. As reported satisfaction with job status, income, place of residence, and time spent with family and friends increase, symptoms decrease. Decreases in symptoms over the 18-month period were associated with increases in self-esteem. In turn, we would expect increases in self-esteem to be associated with reduced symptoms and recovery (Markowitz, 2001). Thus, an important part of recovery is the quality of social, economic, and occupational roles available to the person.

The long-term effects of formal labeling on the reactions of others may be limited, because persons who have been labeled in the past engage in various tactics to prevent others from learning about their stigma. These tactics include selective concealment of past labeling, preventive disclosure to close friends, and various deception strategies (Miall, 1986). On the other hand, longitudinal research suggests that persons who have been publicly labeled and treated continue to anticipate rejection from others even though they no longer engage in the symptomatic behavior (Link, Struening, Rahav, Phelan, & Nuttbrock, 1997). A longitudinal study of 88 persons released following an average of 8 years of hospitalization measured experiences with rejection following release. Those who reported a larger number of such experiences subsequently attained low scores on mastery. Former patients' self-views appear to fluctuate, perhaps in response to alternating experiences of acceptance and stigma (Wright, Gronfein, & Owens, 2000). Thus, stigma may have lasting effects on a person's psychological well-being.

SUMMARY

Deviant behavior is any act that violates the social norms that apply in a given situation.

The Violation of Norms (1) Norms are local, subcultural, or societal in scope. The repercussions of deviant behavior depend on which type of norm an individual violates. (2) Anomie theory asserts that deviance occurs when persons do not have legitimate means available for attaining cultural success goals. Possible responses to anomie include ritualism, retreatism, innovation, and rebellion. General strain theory suggests that emotions link structural position and behavior. (3) Control theory states that deviance occurs when an individual is not responsive to the expectations of others. This responsiveness, or social bond, includes attachment to others, commitment to long-term goals, involvement in conventional activities, and a respect for law and authority. Research indicates that social integration is associated with reduced rates of deviance in adolescence and adulthood. (4) Differential association theory emphasizes the importance of learning through interaction with others. Individuals often learn the motives and actions that constitute deviant behavior just as they learn socially approved behavior. (5) The routine activities perspective calls attention to situations that facilitate the convergence of offenders and targets, in the absence of a guardian.

Reactions to Norm Violations Deviant behavior involves not only acts that violate social norms but also the societal reactions to these acts. (1) There are numerous possible responses to rule breaking. Very often, we ignore it. At other times, we deny that the act occurred, define the act as normal, excuse the perpetrator, or recognize the act but de-emphasize its significance. Only after an act is discovered and labeled "deviant" is it recognized as such. (2) Our reaction to rule breaking depends on the characteristics of the actor, the audience, and the situation. People often have a stereotyped image of deviant persons; these stereotypes influence how audiences react to rule violations. (3) The consequences of rule breaking depend on the reactions of the audience and the response of the rule breaker. If members of the audience reward the person, the deviance may become institutionalized. Alternatively, the person may decide to avoid further deviance, in spite of others' encouragement. If the person is punished, he or she may either give up the behavior or respond with additional rule violations.

Labeling and Secondary Deviance The process of labeling has two important consequences. (1) It leads members of an audience to change their perceptions of and behavior toward the actor. If they withdraw from the stigmatized person, they may create a self-fulfilling prophecy and elicit the behavior they expected from the actor. (2) Labeling often causes the actor to change his or her self-image and to come to define the self as deviant. This, in turn, may lead to secondary deviance—an open and active involvement in a lifestyle based on deviance. Such lifestyles are often embedded in deviant subcultures.

Formal Social Controls Every society gives certain agents the authority to respond to deviant behavior. (1) In U.S. society, the major formal social control agents are the criminal justice, juvenile justice, and mental health systems. These agencies select persons and identify them as deviant through a sequence of decisions. In the criminal justice system, the sequence includes the decisions to arrest, prosecute, and sentence the person. Various factors influence each step in decision making, including the strength of the evidence, the seriousness of the rule violation, and the individual's prior record, and sometimes gender and race. (2) Contrary to popular belief, people do not systematically stigmatize former deviant persons. Most families do not continue to stigmatize relatives following their release from mental hospitals, and most employers do not stigmatize ex-patients who have established competent work records. On the other hand, employers may stigmatize minority men with prison records, and stigma may have long-term effects on the ex-deviant person's psychological well-being.

LIST OF KEY TERMS AND CONCEPTS

anomie theory (p. 495)
control theory (p. 498)
deterrence hypothesis (p. 515)
deviant behavior (p. 494)
deviant subculture (p. 513)
differential association theory (p. 501)
formal social control (p. 514)
informal social control (p. 514)
institutionalization of deviance (p. 509)
labeling theory (p. 506)
learning structure (p. 496)
legitimate means (p. 495)
opportunity structure (p. 496)
routine activities perspective (p. 503)
rule breaking (p. 506)
secondary deviance (p. 513)
self-fulfilling prophecy (p. 511)

20

COLLECTIVE BEHAVIOR AND SOCIAL MOVEMENTS

Introduction

Collective Behavior

Crowds

Gatherings

Underlying Causes of Collective Behavior

Precipitating Incidents

Empirical Studies of Riots

Social Movements

The Development of a Movement

Social Movement Organizations

The Consequences of Social Movements

Summary

List of Key Terms and Concepts

INTRODUCTION

- After the Los Angeles Lakers won the 2000 NBA Championship, a crowd of fans swept through the streets of Los Angeles throwing rocks, setting fires, looting stores, and causing several hundred thousand dollars worth of damage.
- Rumors about the failure of a major South Korean bank set off waves of panicky selling on Wall Street and foreign stock markets, completely disrupting the flow of transactions on the New York Stock Exchange.
- In the wake of the shooting of a Black teenager by police officers, thousands of Blacks marched through the streets to city hall, and a series of speakers demanded changes in police practices.
- On February 13, 2000, more than 20,000 members of the Unification Church were married at a mass wedding ceremony in Seoul, South Korea.

Events such as these occur daily and sometimes receive national media coverage. In part because they occur frequently and often have serious consequences, they have been of interest to social scientists since the turn of the 20th century.

Collective behavior refers to two or more persons engaged in behavior judged common or concerted on one or more dimensions (McPhail, 1991). This is an intentionally broad definition, because a wide range of events have been studied by social scientists as examples of collective behavior, including the four just described.

Collective behavior has three dimensions: the spatial frame, the temporal frame, and the scale of social activity. With regard to space, collective behavior may occur at a single point (such as a street corner), at a larger site (such as a football stadium), or across an entire state or nation. The temporal duration of collective behavior can vary from a few minutes (such as a violent attack by a gang) to several hours (such as a victory celebration) to several days (such as the riot in Los Angeles that followed the Rodney King verdict in 1992).

Collective behavior also varies in terms of the third dimension—the scale of the activity. In fact, we usually only learn about large-scale events, because newspapers and television news programs tend to report only the largest rallies, demonstrations, riots, victory celebrations, or political campaigns (Oliver & Myers, 1999; Myers & Caniglia, 2004).

Most research on collective behavior focuses either on short-term, unorganized events—often referred to as crowds—or on long-term, relatively organized social movements.

The first part of this chapter is concerned with collective behavior. The second part discusses social movements. Specifically, this chapter addresses the following questions:

1. What social processes are involved in collective behavior?
2. What causes collective behavior? That is, what conditions facilitate it, and what conditions precipitate particular collective activities?
3. What factors influence the behavior of people when they gather in a crowd?
4. How do social movements develop? What are the processes by which social movements define issues and attract members?
5. How do social movement organizations mobilize supporters? How is their operation affected by processes inside and outside the organization?
6. How do social movements affect the larger society?

COLLECTIVE BEHAVIOR

Several years ago, the University of Wisconsin Badgers, playing a football game at home, beat the University of Michigan Wolverines, 13 to 10. When the game ended, fans in the stadium rushed onto the field in celebration of the win. In Sections O, P, and Q of the stadium, an estimated 12,000 persons attempted to move forward onto the field, seemingly in unison. But their progress was blocked by a 3-foot-high iron railing in front of the stands and by a 6-foot-high chain-link fence just beyond it. As the people in the back pressed forward, those in front were crushed against the barrier, and many fell and were trampled. The force ripped the railing out of its concrete moorings and flattened the fence. At least 68 people were injured, and 16 were hospitalized for one night or more.

Miraculously, no one was killed. The incident received nationwide publicity.

Can social psychology help us understand incidents like these? Since the very beginnings of the discipline, social psychologists have been concerned with crowd behavior. As we will see, however, the earliest, or classic, perspectives on the crowd were largely uninformed by thorough empirical study and in fact, most of these theoretical notions have been thoroughly debunked (McPhail, 1991). But these ideas about crowds and "mob psychology" live on in the popular press and the mind of the public. Therefore, we discuss them not only to provide an understanding of the historical development of thinking about collective behavior, but also to show that some of these popular ideas are, in fact, mistaken.

Crowds

A **crowd** is a temporary gathering of persons in close physical proximity, engaging in joint activity that is unconventional (Snow & Oliver, 1995). Participants may engage in one common activity (such as listening to a speech or spontaneously singing a song) or in concerted action (such as vandalizing cars or rescuing victims from a collapsed building) or in a large variety of activities (such as observing others in the crowd, milling, discussing actions with friends, looting, and running away from police). Crowd incidents often seem to be marked by high levels of emotion (Turner & Killian, 1972). For instance, in the surge at the Wisconsin football game, most persons in the stadium were Wisconsin fans and were elated at the unexpected victory over a powerful rival.

The classic perspective on crowds has its roots in the writings of Gustave Le Bon (1895) and focuses much of its attention on emotion in the crowd. According to Le Bon and other early writers, emotion in a crowd produces unity among its members and gives direction to the crowd's behavior (Locher, 2002; McPhail, 1991; Miller, 2000). Le Bon referred to this as "the mental unity of the crowd." This unanimity is then supposed to lead participants to think, feel, and act in ways that are different than if each member were alone. Thus, the elation shared by the Wisconsin fans led them to want to celebrate on the field, whereas they might not have tried to enter the field if it weren't for the influence of others. If these persons all had different feelings, the surge would not have occurred.

Deindividuation One influence on behavior in crowds is that the members may feel more anonymous in the large group than they usually do. This **deindividuation** can result in a temporary reduction in self-awareness and sense of personal responsibility (Festinger, Pepitone, & Newcomb, 1952). When an individual participates in a crowd with others—many of whom may be strangers—there can be a decline in the sense of responsibility for one's behavior. This can make it easier for the person to act on impulse and to engage in behavior that violates social norms—such as shoving hard against others, breaking windows, or overturning cars. Social psychologists have long thought that this effect arises from a reduced sense of self-awareness (Diener, 1980), increased arousal, and the diffusion of responsibility (Zimbardo, 1969; see also Chapter 11). More recently, however, some social psychologists have begun to recognize the possibility that behavior in deindividuated situations may actually be the result of conforming to norms that are specific to the situation (Postmes & Spears, 1998). This view is more consistent with sociological understandings of crowds such as the emergent norm theory discussed later. Furthermore, although the effects of deindividuation have been amply demonstrated in laboratory settings, it is considerably more controversial whether individuals in crowds experience marked degrees of deindividuation in the first place. In fact, most people in crowds attend as part of a group of friends or family members and are thus not particularly anonymous (McPhail, 1991).

Contagion People in general have a tendency to imitate the behavior of others. When many people are crowded into a relatively small area, imitation can spread a behavior very quickly. When one person provides a creative model for behavior, it can quickly be assessed and performed by others in the crowd. The outbreak of violence in the Super Bowl XXXII celebration that followed the Denver Broncos win

© John W. Gertz/zefa/Corbis

The behavior that occurs in groups and crowds often seems very different from behavior in our daily routines. Coordination of activity in a crowd can produce an innocuous, even fun, outcome, such as this water balloon attack. Or it can result in something much more serious, such as a destructive riot. Collective behavior scholars try to understand the dynamics that produce both kinds of action.

occurred when a group of people began kicking in the windows of an Athlete's Foot store. This behavior spread; others began to break windows in neighboring stores.

Le Bon's understanding of crowd behavior was deeply influenced by his understanding and observation of French politics (Miller, 2000). The tremendous interpersonal violence that accompanied the French Revolution and its long aftermath provided a frightening model of crowd behavior. Le Bon tried to understand how people could participate in such extraordinary fits of violence and revert to their normal routines only hours later. He concluded that a disease-like **contagion** was spreading through the crowd and infecting everyone present. Thus, an unconscious contagious effect transformed individuals in crowds into a unanimous mass. He thought the hypnotized mass was highly suggestible and thus could easily be turned to destructive behavior.

Myths About Crowds Although casual and unsystematic observation of crowds has a tendency to produce impressions similar to Le Bon's, social scientists have produced evidence that these notions of contagion and the accompanying view of mob psychology are simply wrong (McPhail, 1991; Miller, 2000). Almost 40 years ago, sociologist Carl Couch (1968) summarized research on crowds and identified a number of stereotypes of crowds held by social scientists and the broader public. These stereotypes have been so difficult to shake that collective behavior scholars now refer to them as myths! Recently, David Schweingruber and Ronald Wolstein (2005) summarized the seven dominant myths about crowds:

- Irrationality. Although people in crowds do things that to the outsider look irrational, research shows that people are no less rational in their decision-making processes when they are in crowd situations than they are at other times. Even in emergency evacuation situations where we might expect people to panic (for example, the evacuation of the World Trade Center during the September 11, 2001 attacks), people remain orderly and calm throughout (Tierney, 2002)

- Emotionality. People are engaged emotionally in crowd situations, but crowds do not cause individuals to supplant rationality with emotion nor is emotionality the exclusive domain of crowds by any means.
- Suggestibility. Despite how it may appear to an outsider, people in crowds are not particularly likely to obey the directives of others or to mindlessly imitate other behavior they see. If crowds are so easily suggestible, one might ask why they do not disperse immediately when asked to by the authorities! In the end, however, there is simply no research to substantiate this claim.
- Destructiveness. Crowd situations sometimes end up producing some kind of destruction or violence. But even though these violent episodes may be emphasized by the media, and thus become associated with crowds (Myers & Caniglia, 2004), violent crowds are very much the exception rather than the rule (McPhail, 1994). Furthermore, even in crowds that do produce violence, only a small fraction of the crowd is engaged in any kind of destructive act. Therefore, it does not seem that being in a crowd causes individuals to become violent.
- Spontaneity. Crowds are often thought to be spontaneous in producing action that is not thought through, not rational, and unpredictable. Once again, this myth seems to emerge from the outsider not being able to predict what the crowd will do rather than from those in the crowd engaging in unplanned action. Much crowd action, and in particular the protest crowds that are so often the focus of crowd psychology arguments, require a great deal of planning. Furthermore, those in crowds rely on well-established script and norms to guide their action much of the time (Tilly, 1995). Consider the emergence of the "wave" in a football stadium. Although the individual fan might not be able to predict when it will occur, the crowd at the game is responding to an action planned by the cheer squad and knows what to do as a result of prior experience seeing or being part of the wave. It is not a spontaneous act.
- Anonymity. As mentioned earlier, individuals are rarely anonymous in crowds. They assemble at the event with friends and family and usually stay with that group throughout the event.
- Unanimity. Although people in crowds are sometimes thought to all be doing the same thing at the same time, those who have actually systematically observed crowds have found that people in crowds are engaged in a huge variety of activities while in crowds (Turner & Killian, 1987) and rarely can one observe even near-unanimous activity—even at a rally where everyone is supposed to be paying attention to a speaker or praying together (Schweingruber & McPhail, 1999).

Some of these seven myths about crowds may seem very intuitively attractive to you, but as has often been shown to be the case throughout this textbook, systematic social psychology does not always verify our common sense understandings. To better illustrate various aspects (perhaps unexpected) of crowd behavior, Box 20.1 discusses a crowd rush at a rock concert, and how people's behavior in the rush was quite different than the portrayals of the event in the popular press suggested.

Emergent Norms Emergent norm theory, proposed by Turner and Killian (1972, 1987), was constructed to correct the mistakes of earlier crowd theorists. Emergent norm theory applies to collective behavior that occurs when people find themselves in an undefined or unanticipated situation. The situation may be novel, so there are few cultural norms to guide or direct action. For instance, in recent years there have been several incidents in which a person with a gun walked into a school and opened fire. On December 8, 1997, 14-year-old Michael Carneal walked into his high school in Paducah, Kentucky, carrying five guns. He stood in the hall waiting for a group of students to finish a prayer. Then he took a pistol from his backpack and fired 12 shots at the group, killing 3 girls and wounding 5 other students. At first, no one moved or attempted to stop Carneal. Such incidents are completely unexpected, and there are no behavioral guidelines; people don't know what to do. In other cases, the

BOX 20.1 "Stampede" at The Who Concert

In December of 1979, the rock group The Who were preparing to play a concert at Cincinnati, Ohio's Riverfront Coliseum. The concert promoters had arranged open seating for the concert, meaning that whoever got to a seat first claimed it for the concert. As a result, concertgoers began gathering at the Coliseum some 6 hours before the concert was scheduled to begin. By the time the doors opened, thousands of people were waiting to enter the concert. As the crowd surged forward to enter the Coliseum, the crowd became compacted and people near the front were caught in a massive crush. Some fell and were trampled by the crowd. In the end, 11 people were killed and many more injured.

The press and other observers reacted angrily to the incident, decrying the "mob psychology" that seemed to have produced incredibly callous behavior on the part of the concertgoers. The press called on traditional theories of crowd behavior to characterize what happened. They believed either that the crowd situation had transformed people into ruthless monsters who were willing to kill other people just to get a better seat at the concert or that the concert was a gathering of sociopaths who "stomped 11 persons to death [after] having numbed their brains on weeds, chemicals, and Southern Comfort . . ." (Ryoko, 1979).

Aware that the traditional theories of crowd behavior had serious shortcomings, sociologist Norris Johnson examined the crush at the Who concert to determine how accurate the press had been in characterizing the events, and if there was any empirical evidence for or against the traditional theories of crowds. In short, Johnson found that systematic, detailed investigation revealed a quite different set of behaviors than what had been reported by outsiders.

First, the concertgoers' behavior could not be described as "unregulated competition" for seats. People in the crowd near the area where the injuries occurred reported that everyone in the area was desperately trying to help others. When the first group of approximately 25 people fell, others tried to form a protective ring around them. Unfortunately, the immense pressure from behind them forced them either to walk across the fallen group or to fall into it themselves. Others reported trying to help or to get help for other people, but everyone around was trapped by the crowd. People reported not being able to move their arms, and others were carried along by the crowd, unable to touch the ground for long periods.

Second, social norms were not suspended during the surge. For example, one norm in our society is that men, by virtue of being stronger, should help women. During the crush, men did not abandon this norm and just attend to their own interests. Instead, they offered much more help than they received and offered the majority of their help to women.

In the end, it is clear that the structure of the situation had much more to do with the emergence of the crush than with transformative contagion or amoral, individualistic behavior. For one thing, communication in the crowd was extremely limited. The only ones who knew that people were being hurt were those close to the injured. The vast majority of the people in the crowd had no idea what was happening and were just moving forward in what they perceived as a routine situation. A second structural problem was that the crowd had been allowed to build up to such a large number with no place to go until the doors were opened. When the doors were opened, too few were opened to accommodate the large number of people who were waiting. Those at the back of the crowd could see that the doors were open—a signal to move forward—but they had no idea that they were moving forward faster than those in the front could move through the doors.

Source: Adapted from "Panic at 'The Who Concert Stampede'" by Johnson. Copyright © 1987 by The Society for the Study of Social Problems. Reprinted from *Social Problems, 34*(4), 362–373 by permission.

social structure may be temporarily disrupted by a natural disaster—such as a tornado—or by an event such as a citywide strike by police officers. Another possibility is that there may be conflicting definitions of how people should behave. To act in these situations, those present must develop a shared definition of the situation and the associated behavioral norms.

In all these circumstances, people want to find out what is going on or what they should do. Because they need information, conventional barriers to communication may break down. Strangers talk to each other or to members of groups they usually avoid. Furthermore, the usual standards of judgment and morality may be suspended (although research has shown that this is usually quite rare). **Rumor**—communication through informal and often novel channels that cannot be validated—can exert a major influence on the emerging definition of the situation (Knopf, 1975). In August 2002, a riot broke out in Indianapolis after a police bullet fired in a drug raid ricocheted and hit a young bystander in the arm. Rumors quickly spread that the police had aimed at the boy, producing a quite different understanding of the situation. In some incidents, rumors are broadcast by radio and TV stations, making them appear to be true. In other incidents, milling—the movement of persons within a setting, and the consequent exchange of information between crowd members—is the primary method through which rumor is transmitted. Persons who are not physically present may learn about the emerging situation from radio or television reports or through telephone calls (McPhail, 1991).

Diverse interpretations and action tendencies are usually present in a crowd situation (Turner & Killian, 1972). Someone initiates an act—perhaps in the belief that others will support him or her. Once a person initiates an act, the support of those nearby determines whether that person will persist in attempts to influence others. If enough people reinforce that person's position or behavior, a shared understanding will emerge. The definition of the situation that results from interaction in an initially ambiguous situation is termed an **emergent norm.** The emergent norm is usually not completely novel; it involves a modification or transformation of preexisting norms (Killian, 1984).

Once a definition of the situation develops, people are able to act purposively. In a crowd, behaviors consistent with the norms are encouraged, whereas behaviors inconsistent with the norms are discouraged. Thus, there are normative limits on the behavior of crowd participants. Crowds celebrating a football championship usually do not engage in looting. Conversely, crowds of looters in the inner city do not congregate in bars for several hours, drinking alcoholic beverages and chanting for their victorious team.

A distinctive image of crowds emerges from this perspective. People in crowds are viewed as emotional. Crowd activity reflects the rapid spread of a behavior—often one that violates social norms, such as looting stores—through the crowd. The spread is facilitated by anonymity; people can engage in deviance without fear of sanction, because the others present don't know them. But people are still making rational decisions in line with their current understanding of the situation. Nevertheless, when large numbers of people engage in the same behavior, unintended and undesirable outcomes can occur, such as the injuries suffered in the surge at the Wisconsin-Michigan game.

While the emergent norm perspective has been helpful in understanding certain kinds of actions in a limited set of collective settings, the emergent norm perspective has been criticized because emerging norms via negotiation are not unique or even especially relevant to crowd situations. Norms emerge in many social situations, including but hardly limited to crowds. Furthermore, norms of behavior that individuals have prior to being in a crowd are used and acted on within the crowd settings.

Gatherings

The traditional view of crowds as emotional is at best incomplete and at times has become misleading. In more recent years, an alternative perspective has been developed that calls our attention to other aspects of crowd incidents and collective behavior (McPhail, 1991, 1994). This perspective uses such concepts as the gathering, the phases of a gathering, and companion clusters to analyze collective behavior (McPhail, 1997).

According to this view, the social setting for many forms of collective behavior is a **gathering**—that is, a temporary collection of two or more persons occupying a common space and time frame (McPhail, 1991). Gatherings are the basis of collective behavior. People may gather for a variety of reasons. Some gatherings are for purposes of recreation or "hanging out," as in parks, theaters, swimming pools, or at the scene of a fire, accident, or arrest. Other gatherings are demonstrations that involve two or more people meeting in public to protest or celebrate some person, principle, or condition; these may be political or religious in nature, or involve an athletic event. Still other gatherings are ceremonies, intended to mark a change in status or a life course transition; these may be semipublic or private events.

Behavior in Gatherings The behaviors of persons in gatherings reflect their purposes. Many persons who attend share the stated purpose (say, to celebrate a victory). But others come with other purposes—to accompany a friend, to meet potential dates, or to pick someone's pocket. What occurs reflects two influences: (1) participants' purposes and (2) features of the situation. Consider again the surge at the Wisconsin-Michigan game. At the end of the game, some of those in the student section wanted to go onto the field to celebrate. Others wanted to leave the stadium. Others wanted to get something to eat or drink. Each of these required movement toward the lower level of the stands. A situational feature, unknown to most of them, was the iron railing; because it was only 3 feet high, it was not visible to those standing more than a few rows back. The interaction of the participants' purposes and the situational feature—the railing—caused the undesirable outcome—injuries to 68 people.

Gatherings have three phases: assembling, activities, and dispersal (McPhail, 1991). We will briefly examine each in turn.

Assembling Any gathering is the result of people coming together in a common space and time frame (McPhail, n.d.). This process may involve convergence, or it may reflect the ecology of the location at which the gathering occurs.

Convergence refers to the situation in which those present at a gathering share certain qualities. The spectators at a football game are there to see the game. They may have other, more idiosyncratic purposes as well, but they are fans, and this fact influences their behavior. Indeed, because the crowd has so many fans in it and because they have gathered because they are fans, their identity as fans may become more salient to them and contribute more to their behavior than it does in other circumstances. The surge at the Wisconsin-Michigan game occurred in part because many fans wanted to celebrate the unexpected victory. Convergence at a gathering is much more likely if the gathering has been publicized in advance. It is also more likely if the media broadcast news of it as it occurs.

More often, the composition of a gathering reflects the social ecology of the environment. Other factors being equal, the greater the density of an area, the larger the number of potential participants. Crowd events are much more likely to occur in central cities than in suburbs or rural areas. Organizers of demonstrations learned many years ago that they need to provide buses if sympathizers are not located near the site of the demonstration.

Research suggests that many individuals come to gatherings in small groups of two to five people (McPhail, 1991). These small groups are usually made up of acquaintances, friends, or family members. This group-within-a-group composition is important because the presence of others who know the person establishes some informal social control over his or her behavior. The classic perspective on crowds—which emphasized anonymity and the resulting lack of control over participants' behavior—was incorrect on this point. Although it is true that many participants in gatherings do not know each other, each participant is often part of a small group. These groups form the fundamental social unit in collective behavior situations—they discuss what is happening, decide jointly how to interpret the activities of others, and act together.

Activities The activities of participants in gatherings are not random. McPhail (1991) has identified the "elementary forms of collective action." The most common activity is simply staying together in

companion clusters—of family, friends, or acquaintances—throughout the gathering. A second activity form is the queue or line, as participants wait for admission, access, or service. A third form is arcs or rings of participants around performers, speakers, or fights. A fourth is the display of evaluation—oohs, aahs, whistles, boos, and applause.

Certain activities are common in particular types of gatherings. For instance, religious, concert, and sport gatherings involve celebration rituals—individual or collective chanting, singing, or praying, combined with symbolic gestures, such as the "wave" or holding up cigarette lighters to request an encore. Most members of the culture are familiar with these rituals because of childhood socialization or exposure via mass media. Thus, even a first-time participant in a religious service can act in unison with others. Ceremonial gatherings often involve singing, dancing, and musical performances; these are sometimes quite complex and require considerable advance planning and coordination. Retirement and farewell ceremonies and funerals often involve tributes to the person and have a typical form.

Dispersal Gatherings end or disperse in one of three ways: routine, coerced, or emergency. By far the most common but least studied is the routine type, in which those present leave the setting in an orderly fashion. People often queue as they leave an airplane, football stadium, or concert hall. They typically leave in the company of the same people they assembled with. When there are large numbers of persons or vehicles, officials may facilitate dispersal by directing traffic. In more open settings, such as a concert in a park, people may leave in clusters.

Coerced dispersal refers to the situation in which social control agents, such as police officers or firefighters, direct people to leave before the intended purpose of the gathering is achieved. This occurs when the authorities suspect that those gathered are in some danger, are causing others to be in danger, or because the assembled group is being too disruptive of others' life routines. An example is police directing citizens to leave a stadium or concert because of a bomb threat. Another type of coerced dispersal occurred in downtown Denver during the Super Bowl victory celebration, when the police used tear gas to force the revelers to leave the area.

The most frequently studied is emergency dispersal. This refers to situations where people have to deal with a suddenly disrupted or dangerous environment. The evacuation of the World Trade Center on September 11, 2001, after the first plane hit the tower is one example of an emergency dispersal. Studies have shown that even in such extraordinary circumstances, most people do not panic (Perry & Pugh, 1978). In fact, critical thinking and problem solving may even be enhanced in these dangerous situations.

Underlying Causes of Collective Behavior

Having considered the internal dynamics of gatherings, we turn now to the causes of collective behavior. In some instances, collective behavior is simply a response to some event, such as a natural disaster, an athletic victory, or an assassination. Other types of collective behavior—demonstrations, boycotts, lynchings, lootings, and epidemics—frequently are thought to involve not only a specific event but also more basic underlying conditions in the larger society. Three such conditions are strain, relative deprivation, and grievances.

Strain Society may be viewed as normally in a state of equilibrium, maintaining a relative balance between the emphasis on achieving society's goals and the provision of the means to achieve them—education and jobs (Merton, 1957). At times, however, social change may disrupt this equilibrium, so that one aspect of society is no longer in balance with other aspects. Advances in technology, for example, demand changes in occupational structure. Machines and robots have replaced many blue-collar workers in automobile plants. This change has contributed to high unemployment in cities like Detroit that depend heavily on the auto industry. Such change produces strains in society that cause some individuals to experience stress (see Chapter 18). Although those who are affected may not recognize the source (for example, automation), they experience stress or frustration, which has been theorized to contribute to collective behavior.

Historically, economic issues have often been the key grievances articulated by protestors (Rudé, 1964). Food riots to protest the lack of sufficient food, attacks on factories and businesses to prevent mechanization, and sabotage to disable machinery and other property are often economically motivated. These activities were common in preindustrial England and France (Hobsbawm & Rudé, 1968; Tilly, 1995). More recently, bank failures in Japan and Korea produced economic crises in several Southeast Asian countries, where currencies declined sharply in value. The reduced purchasing power that resulted led to widespread rioting in Indonesia in February 1998 ("Indonesians die," 1998). Rioters frequently targeted businesses and homes of ethnic Chinese, whom they blamed for soaring prices. These protests may reflect the strain caused by widespread unemployment and inadequate incomes.

The evidence that economic issues drive strain and produce collective action is mixed. Whereas economic grievances seem to be related to people's attitudes and their support of radical policies (Plutzer, 1987), other researchers have had great difficulty connecting economic conditions to actual collective behavior or protest (Myers, 1997; Shorter & Tilly, 1974; Spilerman, 1970, 1976). Useem (1998) argues that these ambiguous findings can be clarified if we make a distinction between routine (election rallies, peaceful protest) and nonroutine (riots, rebellion, violence) collective action. Whereas routine collective action cannot be well explained by strain, there is more evidence that strain does produce nonroutine collective action (Myers, 1997; Myers & Li, 2001; Olzak & Shanahan, 1996)

Relative Deprivation In the 18th century, the revolt against the feudal socioeconomic structure occurred first in France. Yet France had already lost many feudal characteristics by the time the French Revolution began in 1789. The French peasant was free to travel, to buy and sell goods, and to contract services. In Germany, however, the feudal social structure was still intact. Thus, based on objective conditions, we would have expected a revolution to occur in Germany before it did in France. Why didn't it? One analyst (de Tocqueville, 1856/1955) argued that the decline of medieval institutions in France caused peasants to become obsessed with the ownership of land. The improvement in their objective situation created subjective expectations for further improvement. Peasant participation in the French Revolution was motivated by the desire to fulfill subjective expectations—to obtain land—rather than by a desire to eliminate oppressive conditions.

This basic notion about the causes of revolutions was expanded into a more systematic view. According to the **J-curve theory** (Davies, 1962, 1971), the "state of mind" of citizens determines whether there is political stability or revolution. Based on external conditions, individuals develop expectations regarding the satisfaction of their needs. These expectations may be derived from one's own experience or from a comparison with the experiences of other groups. Under certain conditions, persons expect continuing improvement in the satisfaction of their needs. If these expectations are met, people are content, and political stability results. But if the gap between expectations and reality becomes too great, people can become frustrated and engage in protest and rebellious activity.

Some have posited that revolutions occur when the level of actual satisfaction declines following a period of rising expectations and their relative satisfaction (Davies, 1971). These relationships are summarized in Figure 20.1. Note the J shape of the actual need satisfaction curve; as satisfaction declines, an intolerable gap between expected need satisfaction and actual need satisfaction emerges.

Such a gap between one's desired level of need satisfaction and one's actual need satisfaction is called **relative deprivation.** Relative deprivation arises when people make a comparison between their own circumstances and some comparison situation. The comparison might be made by an individual comparing oneself or one's group to people from a different group or to an expected standard. For example, an African American person might experience relative deprivation by comparing his or her own salary to the salaries of Whites, by comparing the overall level of Black salaries to White salaries, or by comparing to what that individual thinks is a fair salary level. Thus, even if things are

FIGURE 20.1 THE J-CURVE MODEL

One theory of the causes of revolt is the J-curve theory. According to this model, rebellion occurs when there is an intolerable gap between people's expectations of need satisfaction and the actual level of satisfaction they experience. In response to improved economic and social conditions, people expect continuing improvement in the satisfaction of their needs. As long as they experience satisfaction, there is political stability, even if there is a gap between expected and actual satisfaction. If the level of actual satisfaction declines, the gap gets bigger; at some point it becomes intolerable, and collective action occurs.

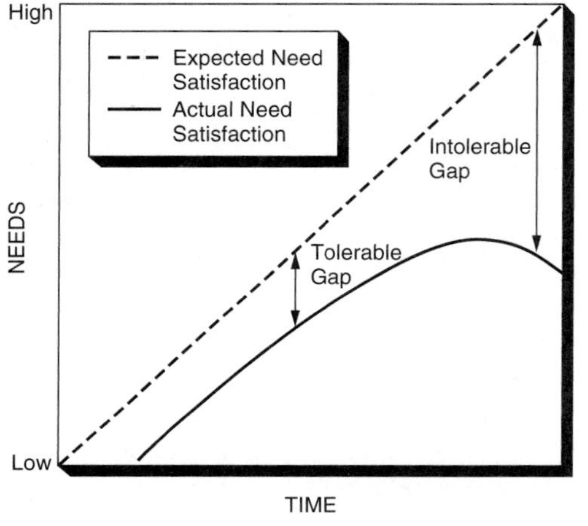

Source: Adapted from Davies, 1962.

improving, as was the case for the French peasants, relative deprivation can still occur—if the standard for comparison is improving even faster.

Although it has been hypothesized that increasing relative deprivation in a group should increase the chances that collective behavior will break out, a large number of studies have failed to find such a relationship (Gurney & Tierney, 1982). Other studies have tried to measure and analyze the relationship between the individual's level of frustration or deprivation and his or her participation in protest. These studies found no differences in frustration or relative deprivation between participants and nonparticipants (McPhail, 1994).

On the other hand, there have been successes for relative deprivation theory as well. Corning and Myers (2002), for example, measured relative deprivation among women by focusing on group comparisons and found that women with higher relative deprivation scores were more likely to have engaged in collective behavior and were more likely to expect to participate in the future. This and other research suggests that it is the feeling about one's group being deprived that is most important, rather than feelings about our own individual condition (Begley & Alker, 1982; Guimond & Dubé-Simard, 1983). In one laboratory experiment, individuals were led to believe they had an easy task or a hard task. Furthermore, either their individual ability or their group membership was made salient. Participants' evaluations of outcomes reflected the salience manipulation; those whose group identity was made salient interpreted their disadvantage (the hard task) in terms of their group membership (Smith & Spears, 1996). Another experiment manipulated whether deprivation (the loss of a promised $10 payment) was seen as due to individual failure or group membership. Participants who perceived it as due to group membership were more supportive of collective action (Foster & Matheson, 1995).

A common form of collective protest in the contemporary United States is the strike, in which union members collectively withhold their labor in the hope of improving their economic position. Can strikes be explained by relative deprivation? The answer is yes—under certain conditions. Strikes are more likely when economic conditions are good and there is a large gap between what workers expect and what management offers—that is, high relative deprivation (Snyder, 1975). Furthermore, strikes are only effective when labor unions are institutionalized—that is, when there is ongoing labor-management accommodation, when union membership is large and stable, and when organized labor has influence in the national political system (Rubin, 1986).

Grievances and Competition In any society, certain resources are highly valued but scarce. These resources include income or property, skills of certain types, and power and influence over others. Because of their scarcity, such resources are unequally distributed. Some groups have more access to a given resource than others. When one group has a grievance—discontent with the existing distribution of resources—collective behavior may occur to change that distribution (Oberschall, 1973). Attempts to change the existing arrangement frequently elicit responses by other groups that are designed to preserve the status quo. The result may be a series of actions by challengers and power holders.

There are three types of collective action (Tilly, Tilly, & Tilly, 1975). Competitive action involves conflict between communal groups, usually on a local scale. One example is the conflict and violence directed toward members of certain ethnic groups. Such incidents are more likely when members of the two groups are competing for low-wage jobs or where there are sharp increases in immigration (Olzak, 1989, 1992). The high rates of lynching of Blacks in the South in the 1890s is another example. From 1865 to 1880, Blacks enjoyed large gains in political influence. By 1890, however, Whites were attempting to regain political control. Between 1890 and 1900, several state legislatures discussed laws that would have taken the vote away from Blacks. During these years, the number of Blacks lynched in Alabama, Georgia, Louisiana, Mississippi, and South Carolina reached a peak (Wasserman, 1977). The lynching of Blacks also increased during economic downturns—for example, when the price of cotton was declining (Beck & Tolnay, 1990).

A second type of collective action, called reactive, involves a conflict between a local group and the agents of a national political system. Tax rebellions, draft resistance movements, and protests of governmental policy are reactive. Such behavior is a response to attempts by the state to enforce its rules (regarding military service, for example) or to extend its control (such as imposing a new tax). Thus, such events represent resistance to the centralization of authority.

A third type of collective action, called proactive, involves demands for material resources, rights, or power. Unlike reactive behavior, it is an attempt to influence rather than resist authority. Strikes by workers, demonstrations for equal rights or against abortion, and various nonviolent protest activities are all proactive. Most proactive situations involve broad coalitions rather than one or two locally based groups.

The three underlying conditions discussed in this section differ in their emphasis. The strain model emphasizes the individual's emotional state in explaining collective behavior. The relative deprivation view emphasizes the person's subjective assessment of need satisfaction. The grievance model suggests that collective behavior results from rational attempts to redistribute resources in society (Zurcher & Snow, 1990).

Precipitating Incidents

Conditions of strain, relative deprivation, and grievances may be present in a society over extended periods of time. By contrast, incidents of collective behavior are often sporadic. Frequently, there are warning signals that a group is frustrated or dissatisfied. Members of the dissatisfied group or third parties may attempt to convince those in power to make changes (Oberschall, 1973). If these changes are not made, members may increasingly perceive legitimate channels as ineffective, leading to marches, protests, or other activities. Eventually, an incident may occur that adversely affects members of the group and highlights the problem, triggering collective behavior by group members; such an incident is referred to as a precipitating event. Such incidents appear to sharply increase the dissatisfaction of those who may have had low levels of grievance prior to the event (Opp, 1988).

An incident is more likely to trigger collective behavior if it occurs in an area accessible to many members of the affected group; this facilitates the assembling process. It is also more likely to lead to collective action if it occurs in a location that has special significance to group members (Oberschall, 1973). An event that occurs in such a place may produce a stronger reaction than would the same incident in a less meaningful location.

In April 1992, a California jury acquitted four police officers charged with beating a Black motorist, Rodney King. Word of the acquittal was broadcast throughout Southern California. Within minutes, a crowd of young men gathered at the intersection of Florence and Normandie in mostly Black South Central Los Angeles. At first, some of the men shouted at and harassed passing motorists. As their numbers grew, others began stopping cars and beating the occupants. Many of those present merely observed these activities. Violence and looting spread rapidly. The ensuing disorder lasted 3 days, resulting in 53 deaths and the destruction of 10,000 businesses.

The acquittal of the officers symbolized for many Blacks their inferior position. The relations between Black citizens and White police officers in Los Angeles and other cities have been characterized by hostility for many years. The videotaped beating of Rodney King was a clear example of the mistreatment many Blacks had suffered. The verdict suggested that White police officers could abuse Black citizens without fear of punishment. This increased the frustration felt by large numbers of Blacks. Some of them acted, and others quickly joined in.

Empirical Studies of Riots

Because they are unpredictable, hostile crowd events such as the one in Los Angeles are difficult to study empirically. Nevertheless, extensive and sophisticated research has been conducted on past racial disturbances, such as occurred in many U.S. cities between 1965 and 1971. These studies have made careful attempts to examine many of the theories presented earlier in this chapter.

In the first 9 months of 1967, there were more than 160 serious racial disturbances. In response, President Lyndon Johnson appointed a commission to study the causes of these incidents. In its report (National Advisory Commission on Civil Disorders, 1968), the commission concluded that the racial disturbances were caused by the underlying social and economic conditions affecting Blacks in our society. The report pointed to the high rates of unemployment, poverty, and the poor health and sanitation conditions in Black ghettos; the exploitation of Blacks by retail merchants; and the experience of racial discrimination; all of which produced a sense of deprivation and frustration among Blacks.

The Commission studied 24 disorders in 23 cities in depth. It concluded that

> Disorder was generated out of an increasingly disturbed social atmosphere, in which typically a series of tension-heightening incidents over a period of weeks or months became linked in the minds of many in the Negro community with a reservoir of underlying grievances. At some point in the mounting tension, a further incident—in itself often routine or trivial—became the breaking point, and the tension spilled over into violence. Violence usually occurred almost immediately following the occurrence of the final precipitating incident, and then escalated rapidly. Disorder generally began with rock and bottle throwing and window breaking. Once store windows were broken, looting usually followed. (National Advisory Commission on Civil Disorders, 1968, p. 6)

The precipitating event frequently involved contacts between police officers and Blacks. In Tampa, Florida, a disturbance in 1967 began after a policeman shot a fleeing robbery suspect. A rumor quickly spread that the Black suspect was surrendering when the officer shot him. In other cities, disorder was triggered by incidents involving police attempts to disperse a crowd in a shopping district or to arrest predominantly Black patrons of a tavern selling alcoholic beverages after the legal closing time (Bergesen, 1982). To many Blacks, police officers symbolize White society and are therefore a readily available target for grievances and frustration. When a police officer arrests or injures a Black under ambiguous circumstances, it provides a concrete focus for discontent and a poignant reminder of grievances.

Severity of Disturbances In some cities, racial disorders involved a few dozen people, and there was little property damage. In other cities, they involved thousands of people, and millions of dollars worth of property was destroyed. What determined how severe a disorder was?

The Commission's report suggested that the deprivations experienced by Blacks fueled the disorders. Numerous researchers have studied this hypothesis. One question is whether absolute or relative deprivation is more influential. Are grievances greater only when unemployment, poor housing, and poor health are widespread, or are grievances greater when the conditions experienced by Blacks are poorer than the conditions experienced by Whites?

Measures of both absolute and relative deprivation were included in a series of studies of riots that occurred from 1961 to 1968 (Spilerman, 1970, 1976). The absolute level of deprivation was measured by the unemployment rate, the average income, and the average education of non-Whites in each city where a disturbance occurred. Relative deprivation was measured by the differences between White and non-White unemployment rates, average income, average education, and average occupational status. Spilerman examined both the frequency and the severity of rioting. The results were sobering for deprivation theorists. Spilerman found that although both severity and frequency of disturbances were associated with the size of the non-White population of a city and its location in the southern region of the United States, neither absolute nor relative deprivation was associated with the severity of disorders. This finding was challenged by subsequent work, and other analysts have located small effects of relative deprivation and strain-related variables (Carter, 1983, 1990; Myers, 1997; Olzak & Shanahan, 1996). Nevertheless, these explanations do not appear to be very important in predicting where and when rioting will occur.

The National Advisory Commission on Civil Disorders sponsored a survey in early 1968. A probability sample of 200 Blacks ages 15 to 65 was drawn in each of 15 major cities. This survey included several measures of relative deprivation, including perceived job discrimination, unresponsiveness of local government, and police abuse of Blacks. It also included objective measures of these three aspects of deprivation. The measures of relative deprivation (perceived conditions) did not correlate significantly with the objective measures of deprivation. Two measures of objective (economic) grievances were correlated with the severity of disturbance. But the variable that was most strongly related to the severity of disturbance was the size of the Black population of the city.

These results suggest that Black protests were not due to local community conditions but to general features of the society, such as increased Black consciousness, heightened racial awareness, and greater identification with other Blacks because of the civil rights movement. Television may have contributed to the disorders of the late 1960s by providing role models: Blacks in one city witnessed and later copied the actions of those in other cities (Myers, 2000). Blacks who were engaged in vandalism, looting, and other collective behavior served as a model for Blacks experiencing grievances and deprivation.

Temperature and Collective Violence It is often suggested that high temperatures contribute to large-scale racial disturbances. The Report of the National Advisory Commission on Civil Disorders (1968) noted that 60 percent of the 164 racial disorders that occurred in U.S. cities in 1967 took place in July, during hot weather. Of the 24 serious disturbances studied in detail, in most instances the temperature during the day on which violence first erupted was very high.

One study of the relationship between temperature and collective violence focused on 102 incidents that occurred between 1967 and 1971 (Baron & Ransberger, 1978). The results showed a strong relationship between temperature and the occurrence of violence. The incidents of collective violence were much more likely to have begun on days when the temperature was high (71–90 degrees). This relationship is depicted in Figure 20.2.

As Figure 20.2 indicates, there were few riots on days when the temperature was higher than 90 degrees. Does this indicate that it is too hot on such days, or that there are very few days when the temperature is that high? To answer this question, the researchers estimated the probability of a disturbance, controlling for the number of days in each temperature range. The results show a direct relationship (Carlsmith & Anderson, 1979). In other words, the higher the temperature, the more likely a disturbance is to occur. One interpretation of this relationship is that in high-density neighborhoods with little air conditioning, the number of people on the streets increases with the temperature. Large street crowds facilitate the transmission of rumors and increase the likelihood of supportive responses to acts initiated by an individual or a small group.

If high temperature is associated with the occurrence of collective violence, an obvious question is whether high temperature is related to other types of violent behavior (Anderson, 2001). One readily available measure of behavior is the rate of violent crime—of murder, sexual assault, and assault. One study analyzed the relationship between rates of violent crime

FIGURE 20.2 AMBIENT TEMPERATURE AND COLLECTIVE VIOLENCE

There is a strong relationship between mean temperature and the occurrence of collective disturbances. An analysis of 102 incidents between 1967 and 1971 generated this graph. As the temperature increased, the frequency of riots also increased. As the temperature increases, so does the number of people who are outside. Large numbers of people on the streets facilitate the development of a crowd.

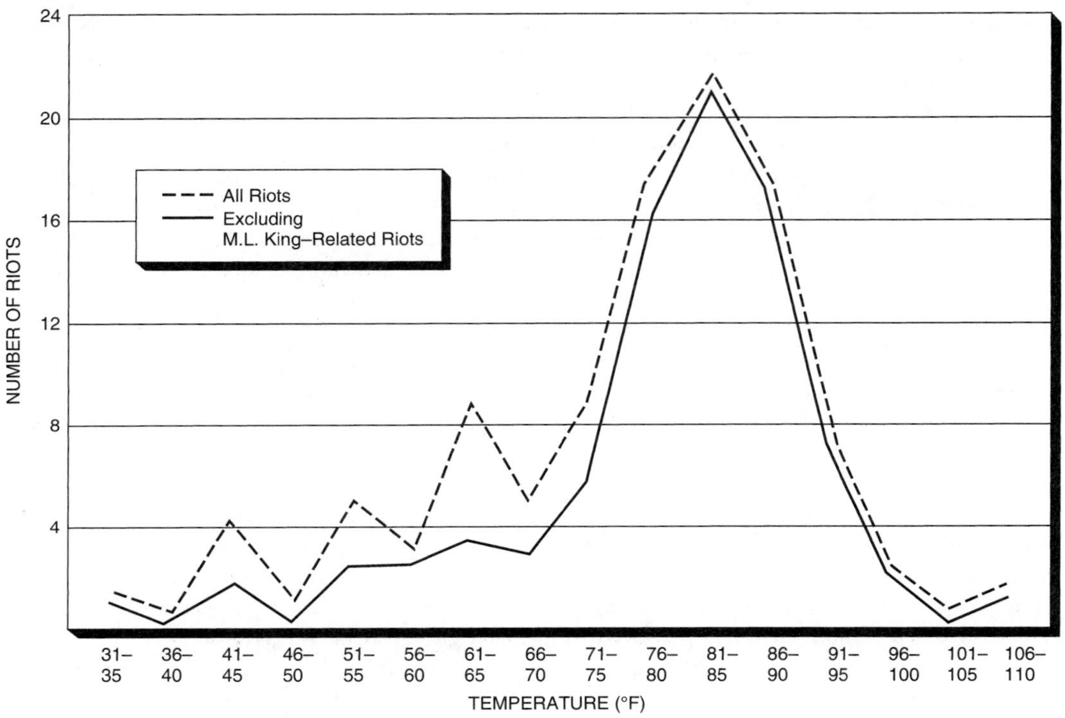

Source: Adapted from "Ambient Temperatures and Occurrence of Collective Violence" (1978) by Baron and Ransberger, *Journal of Personality and Social Psychology, 36,* 351–360. Copyright © 1978 by the American Psychological Association. Adapted with permission.

and average temperature in data from 260 cities for the year 1980 (Anderson, 1987). As expected, the higher the average temperature, the higher the rates of violent crime in a city were. Another study analyzed data from Dayton, Ohio, for a 2-year period (Rotton & Frey, 1985). The researchers looked at daily variations in the number of reports to police of assaults and family disturbances. The number of each was positively associated with the temperature. These results suggest that people are more irritable in hot weather and, thus, more likely to engage in aggressive or violent behavior (Berkowitz, 1993).

Some analysts have wondered if the relationship between violence and heat is merely an artifact of other processes related to heat, and not a product of increased aggressiveness. As in the case of riots, people may be outside, interacting in different kinds of circumstances, and providing themselves with more opportunities for violent behavior in the summer rather than in the winter. If violence is more likely to occur in places where people sell illegal drugs, for example, could it be that people are more willing to go and buy drugs in the summer, or more likely to spend more time in the area than they would in the winter?

Civil disorders, which occur periodically in American cities, often involve members of disadvantaged groups. Looting and vandalism are common, with businesses whose owners are disliked as the likely targets.

Although we do not know for certain how much these kinds of artifacts contribute to the heat-violence relationship, some field studies and laboratory studies confirm an increase in irritability and aggression, even when gathering processes are held constant (Anderson, Anderson, Dorr, DeNeve, & Flanagan, 2000). For example, Reifman, Larrick, and Fein (1991) found that baseball pitchers are more likely to hit batters with a pitch on hot days. This result suggests that aggression is partially a direct product of heat irritability.

Selection of Targets Looting during a civil disturbance does not occur randomly. During racial disorders, such as the one in Los Angeles in 1992, property damage is primarily to retail stores. Residences, public buildings such as schools, and medical facilities such as clinics and hospitals are usually unaffected. Moreover, the looting and vandalism of businesses is often selective. Some stores are cleaned out, whereas others in the same block are untouched.

According to one survey (Berk & Aldrich, 1972), the reason for this discrepancy is that businesses with higher average prices of merchandise (that is, more attractive merchandise) were more likely to be attacked. A second factor was familiarity with the interior of the store. The larger the percentage of Black customers, the more likely the store was to be looted. Retaliation was also a factor; stores whose owners refused to cash checks and give credit to Blacks were more likely to be attacked. White ownership by itself was the least important factor.

Thus, the selection of targets during a riot reflects the desire of participants to obtain expensive consumer goods and to retaliate against anti-Black owners. This may reflect the operation of social control within the crowd. Emergent norms may define some buildings and types of stores as appropriate targets and others as inappropriate targets. These norms are probably enforced by members of the crowd itself (Oberschall, 1973).

Even in deadly collective violence, assailants retain a clear sense of rationality as they choose whom they are going to kill. In a careful study of ethnic riots in which people were attacked and killed, Horowitz (2001) documents the great pains rioters often go to as they try to avoid what he calls "false positives." In other words, rioters work very hard to make sure that they do not kill people who do not belong to the group they wish to attack.

Social Control and Collective Behavior Social control agents such as police officers strongly influence the course of a collective incident once it begins. In some cases, the mere appearance of authorities at the scene of an incident sets off collective action.

The importance of control agents is especially clear in protest situations. Protesters usually enter a situation with (1) beliefs about the efficacy of violence, and (2) norms regarding the use of violence (Kritzer, 1977). If the participants' norms do not oppose violence, and if they believe violence may be effective, they are predisposed to choose violent tactics. Similarly, control agents enter a situation with (1) beliefs about what tactics the protesters are likely to use, and (2) informal norms regarding violence. If the police anticipate violence, they prepare by bringing specially trained personnel and special equipment. Based on these beliefs and expectations, either the control agents or the protesters may initiate violence. Violence by one group is likely to produce a violent response from the other. Data from 126 protest events support this view (Kritzer, 1977).

In some instances, the response of authorities determines the severity and the duration of disorders (Spiegel, 1969). In any disturbance, there are two critical points at which either undercontrol or overcontrol can cause a protest to escalate. The first is the authorities' response to the initial phase. Undercontrol by police in reaction to the initial disorder may be interpreted by protesters as an "invitation to act." It suggests that illegal behavior will not be punished. Overcontrol at this point—such as an unnecessary show of force or large numbers of arrests—may arouse moral indignation, which may attract new participants and increase violence. In Los Angeles in April 1992, the police initially maintained a low profile; officers assembled and remained in an area several blocks from Normandie and Florence Avenues.

The second critical point in a disturbance is the response to widespread disorder and looting during the second day. As the disorder progresses, participants gradually become physically exhausted. Undercontrol by authorities may facilitate the collapse of the protest. Overcontrol may result in incidents that fuel hostility, draw in new participants, and increase the intensity of the disturbance.

The response of authorities to one incident also may affect the severity of subsequent disorders in the same city (Spilerman, 1976). One study investigating incidents of collective violence in France, Germany, and Italy between 1830 and 1930 (Tilly et al., 1975) found that episodes involving violence were often preceded by nonviolent collective action. Moreover, a substantial amount of the violence consisted of the forcible reaction of authorities (often military or police forces employed by the government) to the nonviolent protests of citizens. Thus, violence was not necessarily associated with attempts to influence authority. It was equally associated with reactions to such attempts by the agents of authority. A study of supporters of the Irish Republican Army found that they supported the use of violence only when they viewed peaceful protest as ineffective and knew others who had experienced repressive acts by authorities (White, 1989). A survey of the death rate associated with political violence in 49 countries in the period from 1968 to 1977 found that the death rate was higher in countries with moderate scores on an index of regime repressiveness (Muller, 1985).

SOCIAL MOVEMENTS

The difference between collective behavior and social movements is in part one of degree. Both involve gatherings of people who engage in unconventional behavior—behavior that is inconsistent with some norms of society or is unconventional for the social space and time in which it occurs (Snow & Oliver, 1995). Both are caused by social conditions that generate strain, frustration, or grievances. The differences lie in their degree of organization. Crowd incidents are often unorganized; they occur spontaneously, with no widely recognized leaders and no specific goals.

A **social movement** is a collective activity that expresses a high level of concern about some issue (Zurcher & Snow, 1990). Its participants are people who feel strongly enough about an issue to act. Persons involved in a movement take a variety of actions—sign petitions, donate time or money, talk to family or friends, participate in rallies and marches, engage in civil disobedience, or campaign for candidates. These activities are drawn from the repertoire of ordinary political activities in society, but are often outside the conventional two-party structure. Within the movement, an organization may emerge—a group of persons with defined roles who engage in sustained activity to promote or resist social change (Turner & Killian, 1972).

In this part of the chapter, we first discuss the development of a social movement. Then we consider the movement organization and some influences on how it operates. Finally, we discuss the consequences of social movements.

The Development of a Movement

Preconditions By itself, strain or grievances cannot create a social movement. For a movement to appear, people must perceive their discontent as the result of controllable forces external to themselves (Ferree & Miller, 1985, McAdam, 1999). If they attribute their discontent to such internal forces as their own failings or bad luck, they are not predisposed to attempt to change their environment. Moreover, people must believe they have a right to the satisfaction of their unmet expectations (Oberschall, 1973). These attributions are often the result of interaction with others in similar circumstances. The moral principles used to legitimize their demand may be taken from the culture or from a specific ideology or philosophy (Snow, Rochford, Worden, & Benford, 1986). Thus, at the core of any social movement are beliefs rooted in the larger society.

A current social movement taking place in the United States involves abortion. In the late 1960s, an organized social movement developed that pressed for change in the laws restricting the availability of abortion. This movement culminated in the Supreme Court decision Roe v. Wade on January 22, 1973. This decision held that the state cannot interfere in an abortion decision by a woman and her physician during the first 3 months of pregnancy. The increasingly widespread availability of abortion created strain for others in our society. Many people view a fetus as a person and thus define abortion as murder. Drawing primarily on conservative Christian theology, these people gradually organized a movement in the mid-1970s to obtain legislation that would sharply restrict a woman's right to abortion.

In addition to perceiving unmet needs, people in a social movement must believe that the satisfaction of their needs cannot be achieved through established channels. This perception may be based on a lack of access to such channels (Graham & Hogan, 1990), or it may result from the failure of attempts to bring about change through those channels. At first, the anti-abortion movement emphasized lobbying and attempts to influence elections. As these activities did not produce the legislative or judicial action they sought, the movement increasingly adopted more aggressive tactics, including attempts to physically prevent women from entering abortion clinics. In the 1990s, some individuals have engaged in violent acts, including bombing clinics where abortions are performed and killing physicians who perform the procedure.

Another precondition may be a solution—action that people believe will ameliorate their discontent or redress their grievance (Wilmoth & Ball, 1995). A case study of the movement to control world population growth suggests that the development of birth control in the 1960s—especially oral contraceptives and surgical sterilization procedures—provided a feasible solution to the problem of overpopulation. As a result, population control efforts were more organized and successful in the 1970s than they had been in the 1950s.

Ideology and Framing As affected individuals interact, an ideology or generalized belief emerges. **Ideology** is a conception of reality that emphasizes certain values and justifies a movement (Turner & Killian, 1972; Zurcher & Snow, 1990). Ideologies are often developed by movement participants as the movement grows.

The anti-abortion, or pro-life, ideology rests on several assumptions. First, each conception is an act of God. Thus, abortion violates God's will. Second, the fetus is an individual who has a constitutional right to life. Third, every human life is unique and should be valued by every other human being. Pro-life forces view the status of abortion as temporary—a departure from the past when it was morally unacceptable. Persons and programs (such as sex education) are evaluated in terms of whether they support or undermine these beliefs. Any person or group who favors continued legal abortion is defined as immoral. In recent elections in many communities, a candidate's position on abortion has been a major political issue. Pro-life activists believe that by opposing people and programs that encourage abortion, they will cause a sharp decline in its availability and redefine it as illegitimate.

Such an ideology fulfills a variety of functions (Turner & Killian, 1972). First, it provides a way of identifying people and events and a set of beliefs regarding appropriate behavior toward them. Ideology is usually oversimplified because it emphasizes one or a few values at the expense of others. A second function of ideology is that it gives a movement a temporal perspective. It provides a history (what caused the present undesirable situation), an assessment of the present (what is wrong), and a conception of the future (what goals can be attained by the movement; Martin, Scully, & Levitt, 1990; Snow et al., 1986). Third, it defines group interests and gives preference to them. Finally, it creates villains. It identifies certain persons or aspects of society as responsible for the discontent. This latter function is essential because it provides the rationale for the activity designed to produce change (Oberschall, 1973).

A common element of the ideology of contemporary groups seeking to produce change is the rhetoric of "the public good" (Williams, 1995). Like other elements of ideology, this rhetoric is drawn from the larger culture and provides a resource that movements use in making their claims.

Once social movement groups have identified and committed to ideological positions, they must articulate and present their ideas to others in the broader social environment so that they can win support, recruit participants, and gather the resources they need to accomplish their goals. In essence, they must sell their ideas and causes to others. The processes of articulating their idea for the consumption of others is called social movement **framing** (Snow et al., 1986; Snow & Benford, 1992).

When activists frame their issues, they attempt to link them in some way with the values and ideologies potential recruits already hold. For example, in the United States, people commonly accept the notion that "all men are created equal" and that individuals, by their mere existence, have certain rights. These two general values, equality and basic rights, can be tapped into by activists to align others to their cause. For example, in a battle over gay rights in Cincinnati, those supporting the gay and lesbian cause argued that gay men and lesbian women were being denied basic rights and that all activists wanted was a guarantee of the same rights (to be free from employment discrimination, for example) that everyone should enjoy (Dugan, 2004). The opposition, however, argued that the gay and lesbian cause would provide special privileges to a select group, thereby violating equality notions: The rights of gay and lesbian people were equally protected by general principles that applied to everyone in Cincinnati—why should there be special treatment of this particular group? In this particular case, the frame chosen by the opposition resonated more with the population of Cincinnati—the framing attempts made by the gay and lesbian population were unsuccessful and were defeated.

These results and the many other studies of movement framing demonstrate the importance of social construction in protest. It is not enough for social problems to exist that are not being address by traditional government and institutional approaches. These problems have to diagnosed, the population has to be convinced that the problems are worthy of correction because they are unjust, and activists must also develop a corrective action plan that will be accepted (Gamson, 1992). Without these critical processes, social movements will fail to even get off the ground.

Recruitment The development and continuing existence of any movement depends on recruitment—the process of attracting supporters. Some people are attracted to a movement because they share some distinctive attributes (Zurcher & Snow, 1990). In many instances, these are persons who experience the discontent or grievances at the base of the movement.

A study comparing people who participated in the movement to prevent the reopening of Pennsylvania's Three Mile Island nuclear power plant with a group of nonparticipants found that the activists had opposed commercial and military uses of nuclear energy before the accident, and the accident had served to increase their discontent substantially (Walsh & Warland, 1983). There are limitations, however, to this grassroots view of recruitment (Turner & Killian, 1972). Many studies have found that the supporters of a movement are not the most deprived or frustrated. Also, the goals of a movement may not be aimed at removing the sources of the discontent. The content of the ideology reflects several influences—not just a desire to eliminate a particular source of frustration. Once the movement has developed, people may be attracted by the ideology who do not share the discontent.

Recruitment depends heavily on three catalysts: ideology, identity, and existing social networks (Zurcher & Snow, 1990). The content and framing of the ideology is what attracts supporters. The ideology spreads, at least in part, through existing social networks. Supporters communicate the ideology to their friends, families, and coworkers as they interact with them.

Research on recruitment into religious movements documents the importance of friendship and kinship ties (Stark & Bainbridge, 1980). Adherents to the Mormon religion, for example, are called to proselytize. They establish friendship ties with nonmembers, then gradually introduce their beliefs to their new friends. Likewise, the members of a doomsday cult who believed the Earth would soon be destroyed often were relatives of other members. Those with kinship ties were less likely to leave the cult.

A study of conversion to Catholic Pentecostalism compared 150 converts with a control group of non-Pentecostal Catholics (Heirich, 1977). The major difference between the two groups was in their social networks. Converts had been introduced to the "born again" movement by someone they trusted—a parent, priest, or teacher. Converts also reported positive or neutral reactions to their participation from family and friends.

Sometimes, entire groups are recruited all at once (Oberschall, 1973). The civil rights movement in the South in the 1950s is one example. Because of their religious views, Black ministers were predisposed to support a movement whose ideology emphasized freedom and equality. These ministers recruited their entire congregations and communicated the ideology to other ministers. As a result, the movement spread rapidly. More recently, the pro-life movement has grown by recruiting entire congregations of Catholics and Mormons. Such en bloc recruitment is much more efficient than recruiting individuals (Jenkins, 1983).

An alternative mechanism for recruitment is the mass media. On October 3, 1970, an estimated 15,000 to 20,000 persons participated in Reverend Carl McIntire's March for Victory in Washington. McIntire, a fundamentalist pastor who supported the Vietnam War, communicated his views in a weekly radio program carried by 600 stations and a weekly newsletter. Interviews with 201 march demonstrators revealed that most had come from outside the Washington area, and the overwhelming majority had learned of the march through McIntire's radio program or newsletter (Lin, 1974–1975).

Thus, the media play an important role in social movements. In the McIntire case, prior political beliefs appear to have led persons to seek specific information about the organizational program and the march. Media reports convey a movement's ideology and attract members by providing role models or by providing information about the time and place of activities. It is no accident that movement groups devote considerable effort to getting reporters and camera crews to cover their activities. Media attention—time on television and radio, space in newspapers and magazines—is a scarce resource. To get it, movement groups may have to engage in novel or dramatic acts (Holgartner & Bosk, 1988). Such actions may lead to confrontation and violence.

© Paul Conklin/PhotoEdit, Inc.

This demonstration is part of the social movement aimed at protecting workers' rights and the environment. Organizers of such events hope for coverage by the mass media, because reports of movement activities may attract additional supporters.

No matter what ideology drives a movement group and what kinds of social networks guide recruitment, successful mobilization is also dependent on a successful notion of collective identity. **Collective identity** is a shared understanding that a group of people has of who they are as a group—who is a member of the group, who is not, and what the boundary is that separates them (Taylor, 1989; Taylor & Whittier, 1992). Constructing the identity boundary for an activist group is a critical activity and often requires a great deal of work. If the group defines its boundaries too widely, it will not be able to present itself as an alternative to the status quo, and its members will not feel they are contributing to change. If the collective identity is too narrow and its boundaries are too rigidly drawn, the group risks alienating potential members, thwarting its chances of having a political impact, and being viewed as a fringe group with little relevance to the social and political realities of the day (McVeigh, Myers, & Sikkink, 2004).

Social Movement Organizations

If a movement is to have any impact, there must be some degree of organization to exert continuing pressure for—or resistance to—change. A group of persons with defined roles engaged in sustained activity that reflects a movement's ideology is called a **movement organization.** It develops through the mobilization of people and resources. Once developed, the organization is influenced by its environment and by its own internal processes.

An important influence on the development of a movement organization over time may be other movements (Meyer & Whittier, 1994). A case study of the impact of the women's movement on the peace movement found that the peace movement adopted parts of the ideology (such as the identification of masculine patriarchy as the villain), some of the tactics (such as nonviolent blockades), and some of the organizational structures (such as decentralized decision making) from the women's movement. This transmission from one movement to another occurred partly through overlapping communities of supporters and shared personnel.

Resource Mobilization Having attracted supporters, a movement must induce some of them to become committed members (Zurcher & Snow, 1990). Commitment involves the creation of links between the interests of an individual and those of the movement so that the individual will be willing to contribute actively to the achievement of movement goals. Committed members are necessary if the movement is to become active and self-sustaining. **Mobilization** is the process through which individuals surrender personal resources and commit them to the pursuit of group or organizational goals (Oberschall, 1978). Resources can be many things: money or other material goods, time and energy, leadership or other skills, or moral or political authority. From the individual's viewpoint, mobilization may involve a rational decision. The person weighs the costs and benefits of various actions. If the potential rewards outweigh the potential risks, a particular course of action is taken. Mobilization also involves group processes, such as collective rituals and democratic decision making, which increase the individual's commitment to movement actions (Hirsch, 1990).

Leadership is obviously essential to an organization. Taking a leadership role in a movement organization may be risky but potentially very rewarding (Oberschall, 1973). If the movement is successful, leaders may attain prestige, visibility, a permanent, well-paid position with the organization, and opportunities to interact with powerful, high-status members of society. Leaders of movements are often persons with substantial education (such as lawyers, writers, professors, and students) and at least moderate status in society (Weed, 1990). They are frequently persons whose skills cannot be confiscated by authorities, who can expect social support, and who will be dealt with leniently if arrested. Thus their risk/reward ratio is favorable for involvement.

For a movement to succeed, others also must be induced to work actively in the organization (Zurcher & Snow, 1990). One basis of commitment is moral—anchoring an individual's worldview in the movement's ideology. Members who are attracted by the ideology tend to see their own interests as furthered by the achievement of movement goals. Many women become involved in pro-abortion organizations because preserving freedom of choice for all women will benefit them. At the same time, movement adherents clarify, extend, and even reinterpret the ideology as they attempt to persuade others to commit resources to the movement (Snow et al., 1986). A second basis of commitment is a sense of belonging, which is facilitated by collective rituals in which members participate. One advantage of recruiting pre-existing groups, such as church congregations, is that this sense of belonging is already developed. A third basis of commitment is instrumental. If the organization has enough resources at its disposal, it can provide utilitarian rewards for committed members. These rewards may be distributed equally among members or selectively to members who make a particular contribution (Oliver, 1980).

What determines the relative success of a movement organization in mobilizing resources? One study compared local MADD (Mothers Against

Drunk Driving) organizations, assessing the effect of effort, strategy, organizational structure, and national affiliation (McCarthy & Wolfson, 1996). Effort (hours of work by officers, number of public appearances) predicted the mobilization of volunteer labor, revenue, and members. Organizational structure, as measured by the number of task committees, also predicted mobilization success.

Depending on its overall strategy, a movement organization can use moral, affective, or instrumental rewards as bases for building commitment. These are sufficient for most organizations to induce members to contribute time, materials, and other resources. Other organizations demand that members commit themselves to exclusive participation. They require that members renounce other roles and commitments and undergo **conversion**—the process through which a movement's ideology becomes the individual's fundamental perspective. This degree of commitment is required by some religious movements and by so-called utopian communities. Conversion involves persuasion and "consciousness raising"; it is an attempt to change the individual's worldview. Conversion is usually accomplished during a period of intensive interaction with other movement members. It is thus a very labor-intensive mobilization strategy (Ferree & Miller, 1985).

Movement efforts to mobilize resources vary according to what potential supporters are willing to contribute. If there is considerable latent support for a movement's goals but supporters are unwilling to make large contributions, mobilization efforts may focus on monetary donations. Another low-cost means of translating weak support into a resource is through using initiative or referendum ballots, asking supporters to vote for a movement's goal, such as clean water or a nuclear-free zone. Mobilization efforts also vary according to the resources being sought. An organization seeking volunteers for high-risk activities, such as those involving the risk of arrest, must use powerful inducements. A study of volunteers for a project to register voters in rural Mississippi in 1964 found that several characteristics distinguished participants from volunteers who decided not to go (McAdam, 1986). Participants had intense ideological commitment, previous experience with activism, and ties to other activists (see Figure 20.3).

The other side of mobilization is loss of members and support, or erosion. A study of a mobilization campaign carried out by the Dutch peace movement involved telephone interviews with random samples before (May 1985) and after (November 1985) the campaign (Oegema & Klandermans, 1994). The factors associated with erosion of support were—in May—moderate (compared to strong) willingness to participate, declining willingness to engage in activities requiring greater effort, and a social environment perceived as less supportive.

Organization-Environment Relations The social environment is another influence on the development of a movement organization. First, changes in the environment can increase or decrease the size of the movement and the probability of its success (McAdam, Tarrow, & Tilly, 2001). This is illustrated in a study comparing the evolution of the National Farm Labor Union (NFLU) movement from 1946 to 1955 and the United Farm Workers (UFW) movement from 1965 to 1972 (Jenkins & Perrow, 1977). Both movements had very similar ideologies and organizational structures. However, whereas the NFLU was generally unsuccessful, the UFW achieved major benefits for many farm workers. The difference was the amount of resource support from the environment due to a change in the political climate in the United States. An analysis of newspaper articles published during each period indicates that government, liberal political organizations, and organized labor gave much greater support to the UFW. Thus, challenges to the established order are more likely to succeed when there is sustained support from the environment and when organized opposition is absent.

Second, the growth and effectiveness of a movement organization depends, in part, on whether there are other organizations with similar goals. The National Association for the Advancement of Colored People (NAACP) was the principal Black civil rights organization before 1960. Its goal was racial integration, and it relied heavily on a combination of litigation and lobbying. From 1960 to 1973, other Black organizations developed with more radical goals, methods,

FIGURE 20.3 INFLUENCES ON RECRUITMENT TO HIGH-RISK ACTIVISM

Many social movement activities are low risk—involving little or no cost to participants. Such activities include donating money, distributing leaflets, and attending rallies. Sometimes, however, movement leaders decide to engage in activities that are high risk—involving the risk of injury or death. Political activism in the rural South in 1964 is an example of a high-risk activity. A study of the characteristics associated with the willingness to participate in this activity compared those who went with those who initially volunteered but decided not to go. Analyses suggested that a sequence of experiences, beginning with particular socialization experiences and including contact with other activists and prior activism, were associated with being available for and participating in the high-risk activity. Once involved, participation facilitated greater involvement.

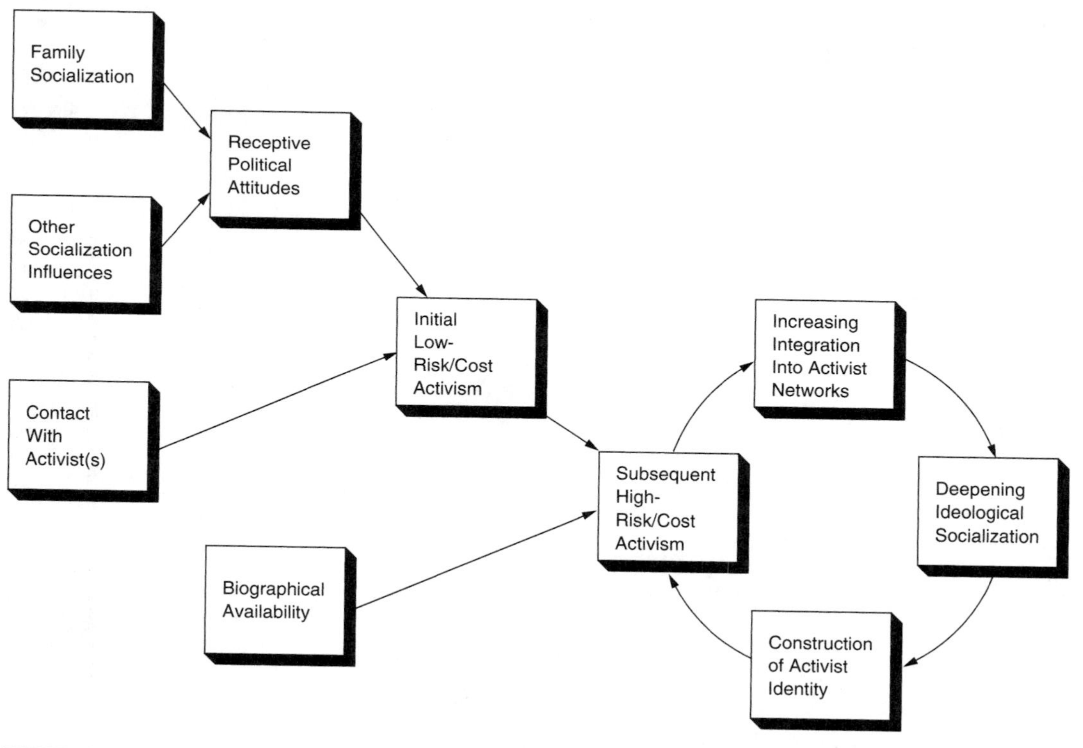

Source: From "Recruitment to High-Risk Activism: The Case of Freedom Summer" (1986) by McAdam, *American Journal of Sociology, 92,* 64–90. Reprinted by permission from the University of Chicago Press.

or both. The Congress of Racial Equality (CORE) and the Student Nonviolent Coordinating Committee (SNCC) favored direct action, and both later excluded Whites from membership. Membership in the NAACP remained stable, whereas the other organizations grew rapidly. The competition between these organizations for resources forced the NAACP to alter its short-term goals to give greater priority to economic advances. The NAACP also shifted from relying on members for financial support to relying on donations from foundations and corporations (Marger, 1984). Paradoxically, the more radical programs of CORE and SNCC made it easier for more moderate organizations like the NAACP to attract

BOX 20.2 Protest Against Pornography: The Vagaries of Collective Action

The underlying causes of collective behavior are present more or less continuously, yet protests and other collective acts occur only occasionally. Consider the example of pornography. Many communities have adult, or X-rated, bookstores and theaters that show X-rated movies. Many of these businesses have been operating for years. There are many people in these communities who find the goods and services sold by these businesses offensive. Yet public attention and protests focus on these businesses only once or twice a year (or less often) and for only a few days at a time.

One measure of the level of public concern about such businesses is the number of newspaper articles published per month or per year. Generally, collective action aimed at the availability of pornography is reported in local

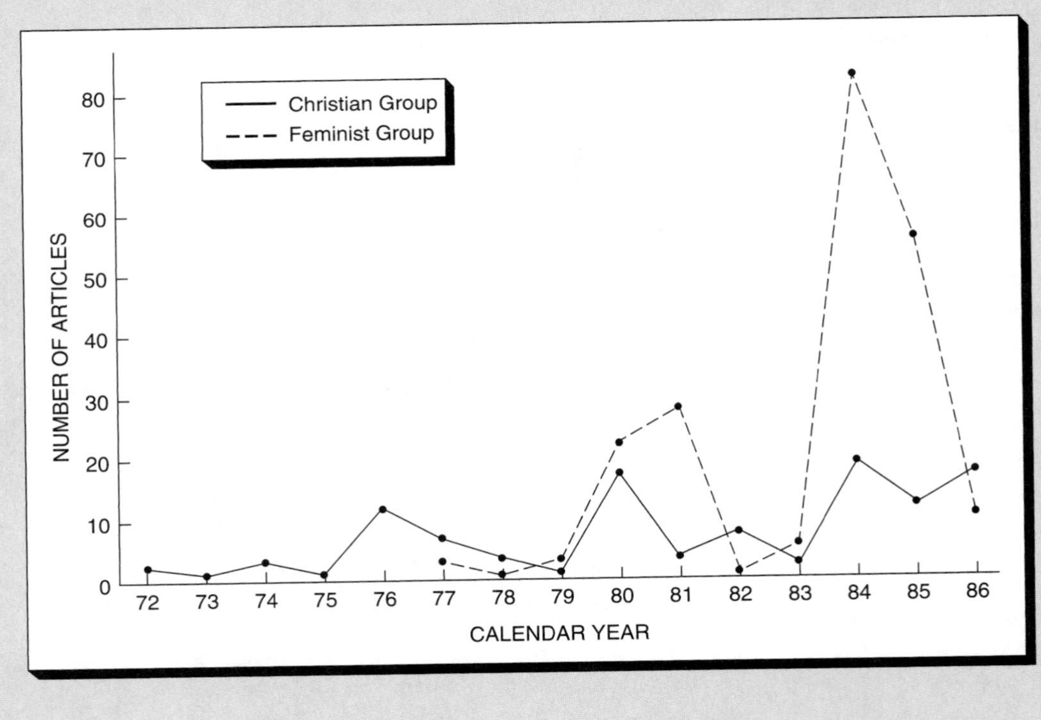

financial support from Whites, especially after the riots of 1967 and 1968 (Haines, 1984).

Under some circumstances, organizations with similar goals form coalitions. Coalitions are possible if organizations appear to have similar values and clearly defined common interests (Ferree & Miller, 1985). A study of cooperation and competition between movement organizations focused on 13 organizations and 6 coalitions involved in pro-abortion activity between 1966 and 1983 (Staggenborg, 1986). Coalitions were

BOX 20.2 Continued

newspapers. Madison, Wisconsin, is a community of 180,000 people. There are two daily newspapers, one more liberal in editorial policy and one more conservative. Thus, it is likely that collective action concerning pornography would be reported in at least one of these papers.

A case study of the anti-pornography movement in Madison included a count and an analysis of newspaper articles reporting local collective action (Duesterhoeft, 1987). The results of the count for the years 1972 to 1986 are displayed in the accompanying figure. The study identified two distinct groups that engaged in collective action—one composed of Christian women and one of feminists. In general, these two groups had been unable to form a coalition because of ideological differences. The Christian group defined as pornographic any depiction of sex outside of heterosexual, monogamous marriage. The feminists defined as pornographic any depiction that was degrading to women or children. The Christian group did not distinguish between acceptable portrayals ("erotica") and pornography, whereas the feminist group did. Because of these differences, the two groups engaged in separate activities. Most newspaper articles during the period of the study reported action by one group or the other but not of the two in coalition.

A look at the figure indicates that in 9 of the 16 years, there were fewer than 10 newspaper articles. These were years in which there was little protest activity. There were two periods in which numerous articles were published: 1980–1981 and 1984–1985. In the first of these periods, there were three small protests in response to the showing of notorious films (*Caligula* and *Windows*) and a single large protest march that focused on pornography and sexual assault.

The only instance of sustained protest activity during the 16-year period occurred in 1984–1985. In August 1984, local interest developed in passing a county ordinance that would define the sale or exhibition of pornography as a violation of civil rights, similar to legislation passed in Minneapolis. Protests, news conferences, forums, and other activities occurred regularly from August 1984 through 1985. The protests were aimed at the sale of magazines such as *Playboy* and *Penthouse* at both adult bookstores and at other businesses such as convenience stores. Thus, the introduction, debate, and, ultimately, action on the proposed ordinance increased public interest in the availability of pornography. Both the Christian and the feminist groups took advantage of this interest and engaged in a series of actions.

This case study illustrates several principles about collective behavior and social movements. First, although the underlying causes are always present, collective action occurs sporadically. Second, collective action is often a response to a precipitating incident, such as the showing of a notorious film. Third, to sustain collective action over time, there must be support from the community—for example, people willing to attend speeches and demonstrations. Finally, when there is more than one organization, differences in ideology may prevent the formation of a coalition and, thus, limit the effectiveness of the protest.

likely under conditions of exceptional opportunity or threat. However, abortion coalitions often dissolved because of differences over the use of civil disobedience or conflicts between the maintenance needs of the organizations.

Under other circumstances, organizations with similar goals may engage in sustained conflict. Organizations with conflicting ideologies may compete for influence over legislation (see Box 20.2). For example, between 1890 and 1920, both Catholic and

Protestant organizations endorsed the goal of providing shelters for neglected and homeless children. The organizations might have cooperated and built a single home in each city. However, each insisted that the children be raised in its own faith, which resulted in two or more homes in some cities (Sutton, 1990).

A third important aspect of the environment is countermobilization. The development of an organization that seeks change of some kind is likely to stimulate those who would be adversely affected by that change to mobilize as well. For instance, an analysis of the resistance by corporate managers to the union movement in the 1920s identified a wide range of tactics that were used to counter union efforts to mobilize workers (Griffin, Wallace, & Rubin, 1986). Managers tried to diminish the common interests of the workers by hiring immigrants and mechanizing factories. They attempted to reduce the power of the union movement by boycotting union goods and harassing other firms that recognized unions. Often, these tactics were successful.

The women's suffrage movement in the early 1900s and the Equal Rights Amendment (ERA) movement in the 1970s both provoked countermovements. The National Association Opposed to Women Suffrage (NAOWS) defended the homemaker's role and lifestyle and attempted to safeguard the material privileges enjoyed by well-to-do women with an "educational campaign" (Marshall, 1986). The anti-ERA movement in the 1970s used more overtly political strategies than the NAOWS, including rallies, marches, and parades. The leader of the anti-ERA movement, Phyllis Schlafly, had considerable political experience working in the National Federation of Republican Women, and she used her experience and contacts (Marshall, 1985).

Internal Processes Other influences on the development of a movement organization are internal processes. These include factionalization, professionalization, and radicalization.

Factionalization is the emergence of subgroups within an organization. It is more likely to occur in a democratic movement organization. For example, the Clamshell Alliance, an antinuclear protest group, was based on the ideology that all people are equal. This was an important element in mobilizing people to protest the activities of utilities and government agencies. As a result, every decision had to be unanimous; one dissenter could prevent the group from acting. Over a period of 5 years, sharp disagreements developed over strategy, leading to factionalism and reduced effectiveness (Downey, 1986).

Another internal process involves increasing reliance on paid professionals. Over time, as an organization develops, it is subject to increasing routinization. The organization develops an administrative structure. Roles are increasingly filled by secretaries, accountants, and lawyers rather than by volunteers committed to the movement, which may inhibit, substitute for, or facilitate volunteer activism (Kleidman, 1994). This change also may produce an increasingly conservative stance among the movement's leadership, which may alienate members who are more radical in orientation.

A study analyzed the effects of these changes in the Black protest movement from 1953 to 1980 (Jenkins & Eckert, 1986). The data indicate that as the flow of financial resources into movement organizations grew, there was increasing professionalization of organization staff. This process did not generate further growth and may have contributed to later decline. At the same time, professionalization does not appear to have produced changes in movement goals and tactics.

There are several other reasons why social movements often become more conservative over time (Myers, 1971). First, the leader-follower relationship is characterized by increasing distance and formality. Second, there is increased bureaucracy, including a role hierarchy, rules and procedures, and membership criteria. Third, there is a tendency for goal displacement to occur; leaders become increasingly concerned with maintaining the organization and less concerned with the goals of the broader movement.

Increasing conservatism is not inevitable. An organization committed to democracy (not necessarily equality) may adopt practices that allow members to retain control. These include a system of open elections, an open decision-making process, and regular communication between leaders and members (Nyden, 1985). A study of the United Steelworkers of

America documented one case in which a bureaucratized union organization was turned into a democratic one by a reform group that emerged from within.

Much rarer is a movement that undergoes increased radicalization, like the Committee of 100 in Great Britain (Myers, 1971). The Committee was formed to protest the increasing reliance of the British government on nuclear weapons for national defense. From the beginning, the group was committed to the use of civil disobedience to achieve its goals. This commitment led to greater radicalization rather than conservatism. In general, groups committed to illegal tactics attract members who are radical. This commitment also produces extreme demands on the time and energy of leaders, which leads to a predominance of younger leaders who have more energy and fewer competing commitments. Organizations of this type are less likely to achieve their goals because the use of illegal tactics tends to alienate the larger society and reduce the amount of support received. Such tactics may even elicit violence by control agents (Kritzer, 1977).

Other frequently overlooked sources of decline in a social movement are the affective and sexual relationships that develop between members (Goodwin, 1997). A case study of the Huk rebellion in the Philippines suggests that the intensity and isolation of the participants of this high-risk movement facilitated romantic and sexual relationships. These relationships weakened identification with the movement and competed with movement activity for participants' time and resources. Some members embezzled movement funds and stole other resources for these "families," and some of the women had babies and dropped out. The movement eventually disintegrated.

The Consequences of Social Movements

Once a movement has established an ideology, attracted supporters, and developed an organization that embodies its ideals, its ultimate success or failure often depends on the reaction of authorities outside the movement organization. Those in positions of power—whether political, economic, or religious—can use various strategies in dealing with a social movement (Oberschall, 1973).

One obvious strategy is to restrict a movement through the exercise of social control. For example, authorities can control access to social roles; they determine the educational, occupational, or religious roles available. A student protesting racism on campus may be expelled from school by administrators. In 2002, a group of Catholic nuns were excommunicated by the Pope when they challenged the church's doctrine on ordaining only men for the priesthood. Beyond this, authorities can manipulate material benefits and punishments, which may affect the risk-reward assessments of movement members (Oliver, 1984). Authorities also can use physical power in the form of police or troops. Of course, this method often results in violence.

A second general strategy is conciliation. Authorities may open channels of communication and negotiate with movement leaders in an attempt to resolve grievances. Conciliation is more likely when a movement is relatively powerful and has the support of large numbers or influential members of society. It is also more likely if both sides are highly organized, united internally, and have strong leadership.

By responding to conciliation from authorities, a movement may achieve at least some of its objectives. As a result of negotiation, there may be changes in the distribution of resources or in power relationships in society. A movement that succeeds in this sense may itself become institutionalized. The changes it brings may ensure a continuing flow of resources that perpetuate the movement organization. This is what happened to the labor, civil rights, and women's movements in the United States.

These consequences do not occur overnight. Many movements span decades as they attempt to bring about change. The movement's founders are likely to maintain their collective identity over time even as the movement ebbs and wanes (Rupp & Taylor, 1987; Taylor, 1989). But as succeeding cohorts join a movement over the years, they may have different identities, based on the environmental context at the time they join (Whittier, 1997). This process of replacement of members over time may lead to changes in the movement's ideology and strategy and influence whether the movement is successful.

SUMMARY

This chapter discusses both collective behavior and social movements.

Collective Behavior (1) The classic perspective on collective behavior focuses on the crowd. Early crowd theorists focused on ideas about crowd behavior that included unanimity of feeling, deindividuation, and contagion. Later analysts debunked some of these ideas and focused instead on how people in undefined situations, such as disasters, develop new norms and social organization. (2) An alternative perspective considers the gathering. Gatherings have a purpose; participants' behaviors reflect that purpose in interaction with features of the setting. (3) Three conditions have been studied as underlying causes of collective behavior: strain, relative deprivation, and grievances. (4) Collective behavior is often triggered by an event that adversely affects those already experiencing strain, deprivation, or grievances. (5) Empirical studies of riots suggest that the severity of a disturbance is influenced mainly by the number of potential participants and less so by differences in underlying social conditions. Once it begins, the course of a collective incident and the likelihood of future disorders are influenced by the behavior of police and other social control agents.

Social Movements A social movement is collective activity that expresses a high level of concern about some issue. (1) The development of a social movement rests on several factors. First, people must experience strain or deprivation, believe they have a right to satisfy their unmet needs, and believe that satisfaction cannot be achieved through established channels. Second, as participants interact, an ideology must emerge that justifies collective activity. Third, to sustain the movement, additional people must be recruited by spreading the ideology, often through existing social networks. (2) A movement organization is a group of persons engaged in sustained activity that reflects the movement ideology. Development of a movement organization depends on resource mobilization—getting individuals to commit personal resources to the group. The organization's effectiveness depends in part on its external environment, on whether there are cooperating or competing organizations, and on whether countermobilization occurs. Over time, organizations may experience such internal processes as factionalization, professionalization, or radicalization. (3) The ultimate success or failure of a social movement also depends on the reactions of authorities outside the movement—whether those in positions of power try to exercise control or to negotiate and attempt to ameliorate the sources of strain or deprivation.

LIST OF KEY TERMS AND CONCEPTS

collective behavior (p. 525)
collective identity (p. 544)
companion cluster (p. 532)
contagion (p. 527)
conversion (p. 546)
crowd (p. 526)
deindividuation (p. 526)
emergent norm (p. 530)
framing (p. 542)
gathering (p. 531)
ideology (p. 541)
J-curve theory (p. 533)
mobilization (p. 545)
movement organization (p. 545)
relative deprivation (p. 533)
rumor (p. 530)
social movement (p. 541)

GLOSSARY

A

Access display A signal (verbal or nonverbal) from one person indicating to another that further social interaction is permissible.

Accommodation A constructive response to a potentially destructive act by a partner in a romantic relationship.

Accounts Explanations people offer after they have performed acts that threaten their social identities. Accounts take two forms—excuses that minimize one's responsibility and justifications that redefine acts in a more socially acceptable manner.

Actor-observer difference The bias in attribution whereby actors tend to see their own behavior as due to characteristics of the external situation, whereas observers tend to attribute actors' behavior to the actors' internal, personal characteristics.

Additive task A type of unitary group task in which the group's performance is equal to the sum (or average) of the performances of its members. *See also* Conjunctive task, Disjunctive task.

Affect Any subjective positive or negative evaluation of an object.

Ageism Prejudice and discrimination against elderly persons based on negative beliefs about aging.

Aggravated assault An attack by one person on another with the intent of causing bodily injury.

Aggression Behavior that is intended to harm another person and that the other wants to avoid.

Aggressive pornography Explicit depiction—in film, video, photograph, or story—of sexual activity in which force is threatened or used to coerce a person to engage in sex. *See also* Nonaggressive pornography.

Alienation The sense that one is uninvolved in the social world or lacks control over it.

Aligning actions Actions people use to define their apparently questionable conduct as actually in line with cultural norms, thereby repairing social identities, restoring meaning to situations, and reestablishing smooth interaction.

Altercasting Tactics we use to impose roles and identities on others that produce outcomes to our advantage.

Altruism Actions performed voluntarily with the intention of helping someone else that entail no expectation of receiving a reward or benefit in return (except possibly an internal feeling of having done a good deed for someone).

Anomie theory The theory that deviant behavior arises

when people striving to achieve culturally valued goals find they do not have access to the legitimate means of attaining these goals.

Anticipatory socialization The process of learning the knowledge, skills, and values of a role that is not yet assumed. Unlike explicit training, anticipatory socialization is not intentionally designed as role preparation by socialization agents.

Arbitrator In situations of conflict, a neutral third-party who has the power to decide how a conflict will be resolved. *See also* Mediator.

Archival research A research method that involves the acquisition and analysis (or reanalysis) of existing information collected by others.

Arousal/cost-reward model A theory of help giving proposing that bystanders weigh the needs of the victim and their own needs and goals and then decide whether helping is too costly in the circumstances. Bystanders often take into account not only the cost of giving direct help but also the cost to themselves of not giving help.

Attachment A warm, close relationship with an adult that provides an infant with a sense of security and stimulation.

Attitude A predisposition to respond to a particular object in a generally favorable or unfavorable way.

Attitude change A change in a person's attitudes about some issue, person, or situation.

Attitude inoculation A process that helps a target person to resist persuasion attempts by exposing him or her to a weak version of the arguments.

Attitudinal similarity The sharing by two people of beliefs, opinions, likes, and dislikes.

Attractiveness stereotype The belief that "what is beautiful is good"; the assumption that an attractive person possesses other desirable qualities.

Attribution The process by which people make inferences about the causes of behavior or attitudes.

Authority The capacity of one group member to issue orders to others—that is, to direct or regulate the behavior of other members by invoking rights that are vested in his or her role.

Availables Those persons with whom we come into contact and who constitute the pool of potential friends and lovers.

Aversive affect Negative feelings that the individual seeks to reduce or eliminate—for example, anger or pain.

Aversive event In intergroup relations, a situation or event caused by or attributed to an outside group that produces negative or undesirable outcomes for members of the target group.

B

Back channel feedback The small vocal and visual comments a listener makes while a speaker is talking, without taking over the speaking turn. This includes responses such as "Yeah," "Huh?" "M-hmn," head nods, brief smiles, and completions of the speaker's words. Back channel feedback is crucial to coordinate conversation smoothly.

Back region A setting used to manage appearances. In back regions, people allow themselves to violate appearances while they prepare, rehearse, and rehash performances. Contrasts with front regions, where people carry out interaction performances and exert efforts to maintain appropriate appearances.

Balance theory A theory concerning the determinants of consistency in three-element cognitive systems.

Bilateral threat In bargaining, a situation where both bargainers can issue threats and inflict punishments on one another.

Birth cohort A group of people who were born during the same period of one or several years and who are therefore all exposed to particular historical events at approximately the same age.

Body language (kinesics) Communication through the silent motion of body parts—scowls, smiles, nods, gazes, gestures, leg movements, postural shifts, caresses, slaps, and so on. Because body language entails movement, it is also known as kinesics.

Borderwork Interaction across gender boundaries that is based on and strengthens such boundaries.

Brainstorming In groups, a procedure intended to generate a large number of high-quality, novel ideas in a brief period. Brainstorming is based on the principles that members should freely express any idea that comes to mind, withhold criticism and defer judgment until later, try to generate as many ideas a possible, and build on ideas suggested by others.

Bystander effect The tendency for bystanders in an emergency to help less often and less quickly as the number of bystanders present increases.

Bystander intervention In an emergency situation, a quick response by a person witnessing the emergency to help another who is endangered by events.

C

Career A sequence of roles—each role with its own set of activities—that a person enacts during his or her lifetime. People's most important careers are in the domains of family and friends, education, and work.

Categorization The tendency to perceive stimuli as

members of groups or classes rather than as isolated entities; the act of encoding stimuli as members of classes.

Catharsis The reduction of aggressive arousal by means of performing aggressive acts. The catharsis hypothesis states that we can purge ourselves of hostile emotions by intensely experiencing these emotions while performing aggression.

Cautious shift In group decision making, the tendency for decisions made in groups after discussion to be more cautious (less risky) than decisions made by individual members prior to discussion.

Cognition An element of cognitive structure. Cognitions include attitudes, beliefs, and perceptions of behavior.

Cognitive dissonance A state of psychological tension induced by dissonant relationships between cognitive elements.

Cognitive labeling theory A theory about emotional experience involving a three-step process including a physiological reaction, an explanation for the reaction, and a label derived from situational cues.

Cognitive processes The mental activities of an individual, including perception, memory, reasoning, problem solving, and decision making.

Cognitive structure Any form of organization among a person's concepts and beliefs.

Cognitive theory A theoretical perspective based on the premise that an individual's mental activities (perception, memory, and reasoning) are important determinants of behavior.

Collective behavior Emergent and extrainstitutional behavior that is often spontaneous and subject to norms created by the participants.

Collective identity A shared understanding that a group of people has of who they are as a group.

Communication The process through which people transmit information about their ideas and feelings to one another.

Communication accuracy The extent to which the message inferred by a listener from a communication matches the message intended by the speaker.

Communication-persuasion paradigm A research paradigm that conceptualizes persuasion attempts in terms of source, message, target, channel, and impact—that is, who says what to whom by what medium with what effect.

Communicator credibility In persuasion, the extent to which the communicator is perceived by the target audience as a believable source of information.

Companion cluster A group of family, friends, or acquaintances who remain together throughout a gathering.

Comparison level (CL) A standard used to evaluate the outcomes of a relationship, based on the average of the person's experience in past relevant relationships.

Comparison level for alternatives (CL_{alt}) A standard specifying the lowest level of outcomes a person will accept in light of available alternatives. The level of profit available to an individual in his or her best alternative relationship.

Compliance In social influence, adherence by the target to the source's requests or demands. Compliance may occur either with or without concomitant change in attitudes.

Conditioning A process of learning in which, if a person performs a particular response and if this response is then reinforced, the response is strengthened.

Conflict spiral A form of escalation occurring in bilateral threat situations. In a conflict spiral, the participants exchange successively larger threats and counterthreats to bolster their demands and end up inflicting more and more damage on one another.

Conformity Adherence by an individual to group norms, so that behavior lies within the range of tolerable behavior.

Conjunctive task A type of unitary group task in which the group's performance depends entirely on that of its weakest or slowest member. *See also* Additive task, Disjunctive task.

Conservative coalition In groups, a union of medium- and low-status members who support the existing leadership and status order against revolutionaries.

Contagion The rapid spread through a group of visible and often unusual symptoms or behavior.

Content analysis A research method that involves a systematic scrutiny of documents or messages to identify specific characteristics and then making inferences based on their occurrence.

Contingencies (of self-esteem) Characteristics of self or categories of outcomes on which a person stakes his or her self-esteem.

Contingency model of leadership effectiveness A middle-range theory of leadership effectiveness that maintains that group performance is a function of the interaction between a leader's style (task oriented or relationship oriented) and various situational factors such as the leader's personal relations with members, the degree of task structure, and the leader's position power.

Control theory The theory that an individual's tendency to engage in deviant behavior is influenced by his or her ties to other persons. There are four components of such ties: attachment, commitment, involvement, and belief.

Conversion The process through which the ideology of a social movement becomes the individual's fundamental perspective.

Cooling out A response to repeated or glaring failures that gently persuades an offender to accept a less desirable, though still reasonable, alternative identity.

Cooperative principle The assumption conversationalists ordinarily make that a speaker is behaving cooperatively by trying to be (1) informative, (2) truthful, (3) relevant to the aims of the ongoing conversation, and (4) clear.

Co-optation In groups, a strategy by which an existing leader singles out one or several lower status members for favored treatment; this is done to weaken the bonds among lower status members who might otherwise form revolutionary coalitions.

Correspondence The degree to which the action, context, target, and time in a measure of attitude is the same as those in a measure of behavior.

Crowd A substantial number of persons who engage in behavior recognized as unusual by participants and observers.

Cultural routines Recurrent and predictable activities that are basic to day-to-day social life.

D

Death instinct In Freud's theory, the innate urge we carry within us to destroy. This instinct constantly generates hostile impulses that demand release and is the basis for aggression.

Definition of the situation In symbolic interaction theory, a person's interpretation or construal of a situation and the objects in it. An agreement among persons about who they are, what actions are appropriate in the setting, and what their behaviors mean.

Deindividuation A temporary reduction of self-awareness and sense of personal responsibility; it may be brought on by such situational conditions as anonymity, a crowd, darkness, and consciousness-altering drugs.

Dependent variable In an experiment, the variable that is measured to determine whether it is affected by the manipulation of one or more other variables (independent variables).

Deterrence hypothesis The view that the arrest and punishment of some individuals for violation of laws deters other persons from committing the same violations.

Deviant behavior Behavior that violates the norms that apply in a given situation.

Deviant subculture A group of people whose norms encourage participation in a specific form of deviance and who regard positively those who engage in it.

Differential association theory The theory that deviant behavior occurs when people learn definitions favorable to the behavior through their associations with other persons.

Diffusion of responsibility The process of accepting less personal responsibility to act because the responsibility is shared with others. Diffusion of responsibility among bystanders in an emergency is one reason why they sometimes fail to help.

Disclaimer A verbal assertion intended to ward off any negative implications of impending actions by defining these actions as irrelevant to one's established social identity. By using disclaimers, a person suggests that although the impending acts may ordinarily imply a negative identity, his or hers is an extraordinary case.

Discrepant message In persuasion, a message advocating a position that is different from what the target believes.

Discrimination Overt acts, occurring without apparent justification, that treat members of certain out-groups in an unfair or disadvantageous manner.

Disinhibition In aggression, the reduction of ordinary internal controls against socially disapproved behavior.

Disjunctive task A type of unitary group task in which the group's performance depends entirely on that of its strongest or fastest member. *See also* Additive task, Conjunctive task.

Displaced aggression Aggression toward a target that exceeds what is justified by any provocation from that target.

Display rules Culture-specific norms for modifying facial expressions of emotion to make them fit the social situation.

Dispositional attribution A decision by an observer to attribute a behavior to the internal state(s) of the person who performed it rather than to factors in that person's environment. *See also* Situational attribution.

Dissonance effect The greater the reward or incentive for engaging in counterattitudinal behavior, the less the resulting attitude change.

Distributive justice principle A criterion in terms of which group members can judge the fairness and appropriateness of the distribution of rewards. Three of the most important are the equality principle, the equity principle, and the relative needs principle.

Dyadic withdrawal The process of increasing reliance on

one person for gratification and decreasing reliance on others.

E

Ebonics A variety of American English spoken by many Blacks, with distinctive pronunciation of some words; Black English.

Egoism Helping behavior motivated by a helper's own sense of self-gratification.

Elaboration likelihood model A model of persuasion maintaining that there are two basic routes through which a message may alter a target's existing attitudes—the central route and the peripheral route.

Embarrassment The feeling that people experience when interaction is disrupted because the identity they have claimed in an encounter is discredited.

Emergent norm The definition of the situation that results from interaction in an initially ambiguous situation.

Emotional intelligence A unique dimension of intelligence that describes an individual's ability to understand the emotional content in social interactions.

Emotions Reactions to a stimulus outside the individual that involve both physiological and cognitive responses.

Emotion work An individual's attempt to change the intensity or quality of his or her emotions in order to bring them into line with the requirements of the occasion. Emotion work may be used to evoke feelings that are not present but should be (psyching oneself up) or to suppress feelings that are present but should not be (calming oneself down).

Empathy An emotional response to others as if we ourselves were in that person's situation; feeling pleasure at another's pleasure or pain at another's pain.

Empathy-altruism model A theory proposing that adults can experience two distinct states of emotional arousal while witnessing another's suffering: distress and empathic concern. These states of emotional arousal give rise to different motivations for helping.

Encoder-decoder model A theory that views communication as a linear process in which the message is encoded by a transmitter, transmitted, and decoded by a receiver.

Endorsement An attitude held by a group member indicating the extent to which he or she supports the group's leader.

Equity In exchange theory, a state of affairs that prevails in a dyad or group when people receive rewards in proportion to the contributions they make toward the attainment of group goals.

Equitable relationship A relationship in which the outcomes received by each person are equivalent.

Ethnocentrism In intergroup relations, the tendency to regard one's own group as the center of everything and to evaluate other groups in reference to it. The tendency to regard one's in-group as superior to all out-groups.

Evaluation apprehension A state of concern about what others expect of us and how others will evaluate our behavior. Evaluation apprehension inhibits helping when bystanders fear others will view their intervention as foolish; it promotes helping when bystanders believe others expect them to help.

Evolutionary psychology A theoretical perspective positing that predispositions toward some social behaviors are passed genetically from generation to generation and shaped by the process of natural selection.

Expectation states theory A theory proposing that status characteristics cause group members to form expectations regarding one another's potential performance on the group's task; these performance expectations affect subsequent interaction among members.

Experiment A research method used to investigate cause-and-effect relations between one variable (the independent variable) and another (the dependent variable). In an experiment, the investigator manipulates the independent variable, randomly assigns participants to various levels of that variable, and measures the dependent variable.

External validity The extent to which it is possible to generalize the results of one study to other populations, settings, or times.

Extraneous variable A variable that is not explicitly included in a research hypothesis but has a causal impact on the dependent variable.

Extrinsically motivated behavior A behavior that results from the motivation to obtain a reward (food, praise) or avoid a punishment (spanking, criticism) controlled by someone else.

F

Fateful events Events that are beyond an individual's control, unpredictable, often life-threatening, often large in magnitude, and that disrupt peoples' usual activities.

Feeling rules Social rules that dictate what an individual with a particular public identity ought to feel in a given situation.

Field study An investigation that involves the collection of data about ongoing activity in everyday settings.

Flirting According to Moore, a class of nonverbal facial expressions and behavior exhibited by women that serves to attract the attention and elicit the approach of a man.

Focus of attention bias The tendency to overestimate the causal impact of whomever or whatever we focus our attention on.

Formal social control Agencies that are given responsibility for dealing with violations of rules or laws.

Frame A set of widely understood rules or conventions pertaining to a transient but repetitive social situation that indicate which roles should be enacted and which behaviors are proper.

Front region A setting used to manage appearances. In front regions, people carry out interaction performances and exert efforts to maintain appropriate appearances. Contrasts with back regions, where they allow themselves to violate appearances while they prepare, rehearse, and rehash performances.

Frustration The blocking of goal-directed activity. According to the frustration-aggression hypothesis, frustration leads to aggression.

Frustration-aggression hypothesis The hypothesis that every frustration leads to some form of aggression and every aggressive act is due to some prior frustration.

Fundamental attribution error The tendency to underestimate the importance of situational influences and to overestimate personal, dispositional factors as causes of behavior.

G

Game Mead's second stage of social experience, in which children enter organized activities, and learn to imagine the viewpoints of several others at the same time.

Gathering A temporary collection of two or more people occupying a common space and time frame.

Gender role The behavioral expectations associated with gender.

Generalized other A conception of the attitudes and expectations held in common by the members of the organized groups with whom one interacts.

Goal isomorphism In groups, a state in which group goals and individual goals held by a member are similar in the sense that actions leading to the attainment of group goals also lead simultaneously to the attainment of individual goals.

GRIT A strategy for reducing intergroup conflict whereby one side initiates de-escalatory steps in the hope that they will eventually be reciprocated by the other side; GRIT is an acronym for Graduated and Reciprocal Initiatives in Tension Reduction.

Group A social unit that consists of two or more persons and has the following characteristics: shared goal(s), interaction (communication and influence) among members, normative expectations (norms and roles), and identification of members with the unit.

Group cohesion A property of a group, specifically the degree to which members of a group desire to remain in that group and resist leaving it. A highly cohesive group will maintain a firm hold over its members' time, energy, loyalty, and commitment.

Group goal A desirable outcome that group members strive collectively to accomplish or bring about.

Group goal effect An empirical generalization regarding group productivity—namely, that if a group establishes explicit, demanding objectives with respect to the group's performance, and if the group's members are highly committed to those objectives, then the group will perform at a higher level than if it does not do these things.

Group polarization In group decision making, the tendency for group members to shift their opinions toward a position that is similar to but more extreme than the positions they held prior to group discussion; both the risky shift and the cautious shift are instances of group polarization.

Group productivity The level of a group's output (per unit of time) gauged relative to something else, such as the level of resources used by the group or the group's targeted objectives.

Group self-esteem An individual's evaluation of self as a member of a racial or ethnic group.

Groupthink A mode of thinking within a cohesive group whereby pressures for unanimity overwhelm the members' motivation to realistically appraise alternative courses of action.

H

Halo effect The tendency of our general or overall liking for a person to influence our assessment of more specific traits of that person. The halo effect can produce inaccuracy in our ratings of others' traits and performances.

Helping Any behavior that has the consequences of providing some benefit to or improving the well-being of another person.

Heuristic A type of mental shortcut that allows individuals to quickly select and apply schemas to new or ambiguous situations.

Hypothesis A conjectural statement of the relation between two or more variables. Some hypotheses are explicitly causal in nature, whereas others are noncausal.

I

Identity The categories people use to specify who they are—that is, to locate themselves relative to other people.

Identity control theory Proposes that an actor uses the social meaning of his/her identity as a reference point for assessing what is occurring in the situation.

Identity degradation A response to repeated or glaring failures that destroys the offender's current identity and transforms him or her into a "lower" social type.

Ideology In the study of social movements, a conception of reality that emphasizes certain values and justifies the movement.

Illusion of out-group homogeneity The tendency among in-group members to overestimate the extent to which out-group members are homogeneous or all alike.

Imitation A process of learning in which the learner watches another person's response and observes whether that person receives reinforcement.

Incentive effect The greater the reward or incentive for engaging in counterattitudinal behavior, the greater the resulting attitude change.

Independent variable In an experiment, the variable that is manipulated by the investigator to study the effects on one or more other (dependent) variables.

Informal social control The reactions of family, friends, and acquaintances to rule violations by individuals.

Informational influence In groups, a form of influence that occurs when a group member accepts information from others as valid evidence about reality. Influence of this type is particularly likely to occur in situations of uncertainty, or where there are no external or "objective" standards of reference.

Informed consent Voluntary consent by an individual to participate in a research project based on information received about what his or her participation will entail.

Ingratiation The deliberate use of deception to increase a target person's liking for us in hopes of gaining tangible benefits that the target person controls. Techniques such as flattery, expressing agreement with the target person's attitudes, and exaggerating one's own admirable qualities may be used.

Institutionalization of deviance The process by which members of a group come to expect and support deviance by another member over time.

Instrumental conditioning The process through which an individual learns a behavior in response to a stimulus to obtain a reward or avoid a punishment.

Intentionalist model A theory that views communication as the exchange of communicative intentions, and views messages transmitted as merely the means to this end.

Intergroup conflict A state of affairs in which groups having opposing interests take antagonistic actions toward one another to control some outcome important to them.

Intergroup contact hypothesis An hypothesis holding that in intergroup relations, increased interpersonal contact between groups will reduce stereotypes and prejudice and consequently reduce antagonism between groups.

Internalization The process through which initially external behavioral standards become internal and subsequently guide an individual's behavior.

Internal validity The extent to which research findings are free from contamination by extraneous variables.

Interpersonal attraction A positive attitude held by one person toward another person.

Interpersonal spacing (proxemics) Nonverbal communication involving the ways in which people position themselves at varying distances and angles from others. Because interpersonal spacing refers to the proximity of people, it is also known as proxemics.

Intersubjectivity The information that each participant in an interaction needs about the other participant(s) in order for communication to be successful.

Interview survey A method of research in which a person (that is, an interviewer) asks a series of questions and systematically records the answers from the respondents. *See also* Questionnaire survey.

Intrinsically motivated behavior A behavior that results from the motivation to achieve an internal state that an individual finds rewarding.

J

J-curve theory The theory that revolutions occur when there is an intolerable gap between people's expectations of need satisfaction and the actual level of satisfaction they experience.

K

Kin selection Recognition of those who share genes, especially for the purposes of targeting altruistic behavior.

L

Labeling theory The view that reactions by others to a rule violation are an essential element in deviance.

Leadership In groups, the process whereby one member influences and coordinates the behavior of other members in pursuit of group goals. The enactment of several functions necessary for successful group performance; these functions include planning, organizing, and controlling the activities of group members.

Learning structure An environment in which an individual can learn the information and skills required to enact a role.

Legitimate means Those ways of striving to achieve goals that are defined as acceptable by social norms.

Legitimate power Authority that is accepted as a normal part of a social role.

Life course An individual's progression through a series of socially defined, age-linked social roles.

Life event An episode marking a transition point in the life course that provokes coping and readjustment.

Likert scale A technique for measuring attitudes that asks a respondent to indicate the extent to which he or she agrees with each of a series of statements about an object.

Linguistic intergroup bias Subtle and systematic differences in the language we use to describe events as a function of our group membership and the group to which the actor or target belongs.

Loneliness An unpleasant, subjective experience that results from the lack of social relationships satisfying in either quantity or quality.

Looking-glass self The term coined by Cooley that describes the self-schema we create based on how we think we appear to others.

Love story A script about what love should be like; it has characters, plot, and theme.

M

Majority influence The process by which a group's majority pressures an individual to adopt a specific position on some issue.

Mass media Those channels of communication (TV, radio, newspapers, and the Internet) that enable a source to reach and influence a large audience.

Matching hypothesis The hypothesis that each person looks for someone to date who is of approximately the same level of social desirability.

Media campaign A systematic attempt by an influencing source to use the mass media to change attitudes and beliefs of a target audience.

Mediator A third party who helps groups in conflict to identify issues and agree on some resolution to the conflict. Mediators usually serve as advisors rather than as decision-makers. *See also* Arbitrator.

Mere exposure effect Repeated exposure to the same stimulus that produces a positive attitude toward it.

Meta-analysis A statistical technique that allows the researcher to combine the results from all previous studies of a question.

Methodology A set of systematic procedures used to conduct empirical research. Usually these procedures pertain to how data will be collected and analyzed.

Middle-range theory A narrow, focused theoretical framework that explains the conditions that produce some specific social behavior. *See also* Theoretical perspective.

Minority influence An attempt by an active minority within a group to persuade majority members to accept their viewpoint and adopt a new position.

Mobilization The process through which individuals surrender personal resources and commit them to the pursuit of group or organizational goals.

Mood A general psychological condition that characterizes our experience and emotional orientation for an extended period of time.

Moral development The process through which children become capable of making moral judgments.

Movement organization A group of persons with defined roles who engage in sustained activity that reflects the ideology of a social movement.

Mutual deterrence A standoff that ends conflict in a bilateral threat situation because both parties wish to avoid suffering and damages.

N

Negotiation Communication process between opposing sides in a conflict, whereby parties make offers and counteroffers in hopes of eventually reaching an agreement.

Nonaggressive pornography Explicit depictions—in film, video, photograph, or story—of adults engaging in consenting sexual activity. *See also* Aggressive pornography.

Nonstandard speech A style of speech characterized by limited vocabulary, improper pronunciation, and incorrect grammar. The use of this style is associated with low status and low power. *See also* Standard speech.

Norm In groups, a standard or rule that specifies how members are expected to behave under given circumstances. Expectations concerning which behaviors are acceptable and which are unacceptable for specific persons in specific situations.

Norm of homogamy A social norm requiring that friends, lovers, and spouses be characterized by similarity in age, race, religion, and socioeconomic status.

Norm of reciprocity A social norm stating that people should (1) help those who have previously helped them and (2) not help those who have denied them help for no legitimate reason.

Normative influence In groups, a form of influence that occurs when a member conforms to group norms in order to receive the rewards or avoid the punishments that are contingent on adherence to these norms.

Normative life stage A discrete period in the life course during which individuals are expected to perform the set of activities associated with a distinct age-related role.

Normative transition Socially expected changes made by all or most members of a defined population.

O

Observational learning The acquisition of behavior based on the observation of another person's behavior and of its consequences for that person. Also known as modeling.

Opportunity structure An environment in which an individual has opportunities to enact a role, which usually requires the assistance of those in complementary roles.

P

Panel study A method of research in which a given sample of respondents is surveyed at one point in time and then resurveyed at a later point (or several later points). Also known as a longitudinal survey.

Paralanguage All the vocal aspects of speech other than words, including loudness, pitch, speed of speaking, pauses, sighs, laughter, and so on.

Passionate love An state of intense longing for union with another and intense physiological arousal.

Perceived behavioral control The tendency of our behavior to be influenced not only by intentions but also by whether we are able to carry out those intentions.

Perspective-taking model A theory that views communication as the exchange of messages using symbols whose meaning is created by the interaction itself.

Persuasion An effort by a source to change the beliefs or attitudes of a target person through the use of information or argument.

Play Mead's first stage of social experience, in which young children imitate the activities of people around them.

Population A set of all people whose attitudes, behavior, or characteristics are of interest to the researcher.

Position A designated location in a social system.

Powerlessness The sense of having little or no control over events.

Prejudice A strong like or dislike for members of a specific group.

Primacy effect The tendency, when forming an impression, to be most influenced by the earliest information received. The primacy effect accounts for the fact that first impressions are especially powerful.

Primary group A small group in which there exist strong emotional ties among members. *See also* Secondary group.

Primary relationship An interpersonal relationship that is personal, emotionally involving, and of long duration.

Principle of cognitive consistency In cognitive theory, a principle maintaining that if a person holds several ideas that are incongruous or inconsistent with one another, he or she will experience discomfort or conflict and will subsequently change one or more of the ideas to render them consistent.

Principle of covariation A principle that attributes behavior to the potential cause that is present when the behavior occurs and absent when the behavior fails to occur.

Principle of determinism The principle that discoverable causes exist for all events in a science's domain of interest.

Production blocking In brainstorming groups, a phenomenon that inhibits the production of novel ideas. Production blocking occurs when participants in a brainstorming group are unable to express their ideas due to bottlenecks caused by turn taking among members.

Promise An influence technique that is a communication taking the general form, "If you do X (which I want), then I will do Y (which you want)." *See also* Threat.

Prosocial behavior A broad category of actions considered by society as being beneficial to others and as having positive social consequences. A wide variety of specific behaviors qualify as prosocial, including donation to charity, intervention in emergencies, cooperation, sharing, volunteering, sacrifice, and the like.

Prototype In person perception, an abstraction that represents the "typical" or quintessential instance of a class or group.

Punishment A painful or discomforting stimulus that reduces the frequency with which the target behavior occurs.

Q

Questionnaire survey A method of research in which a series of questions appear on a printed questionnaire, and the respondents read and answer them at their own pace. Usually no interviewer is present. *See also* Interview survey.

R

Random assignment In an experiment, the assignment of participants to experimental conditions on the basis of chance.

Rape myths Prejudicial, stereotyped, and false beliefs about rape, rape victims, and persons who commit rape.

Reactance Resistance to persuasion attempts that occurs when the persuasion attempt threatens the independence or freedom of the target.

Realistic group conflict theory A theory of intergroup conflict that explains the development and the resolution of conflict in terms of the goals of each group; its central hypothesis is that groups will engage in conflictive behavior when their goals involve opposition of interest.

Recency effect The tendency, when forming an impression, to be most influenced by the latest information received.

Referent power Social influence that occurs because individuals seek to be liked and accepted by valued social groups.

Reinforcement Any favorable outcome or consequence that results from a behavioral response by a person. Reinforcement strengthens the response—that is, it increases the probability it will be repeated.

Reinforcement theory A theoretical perspective based on the premise that social behavior is governed by external events, especially rewards and punishments.

Relative deprivation A gap between the expected level and the actual level of satisfaction of the individual's needs in which the level expected by the individual exceeds the level of need satisfaction experienced.

Reliability The degree to which a measuring instrument produces the same results each time it is employed under a set of specified conditions.

Response rate In a survey, the percentage of people contacted who complete the survey.

Return potential model A formal model of a social norm. This model treats norms as having two dimensions—the behavior dimension and the evaluation dimension. A norm is a function that specifies what level of evaluation will correspond to what behavior.

Revolutionary coalition In groups, a union of medium- and low-status members who oppose the existing leadership and wish to overturn it.

Ringelmann effect An empirical generalization with respect to group performance. On additive tasks, large groups produce a higher total output than smaller ones, but the output per member in the large groups is lower than that in the smaller groups.

Risk-benefit analysis A technique that weighs the potential risks to research participants against the anticipated benefits to participants and the importance of the knowledge that may result from the research.

Risky shift In group decision making, the tendency for decisions made in groups after discussion to be riskier than decisions made by individual members prior to discussion.

Role A set of functions to be performed by a person on behalf of a group of which he or she is a member. A cluster of rules indicating the set of duties to be performed by a member occupying a given position within a group. The set of expectations governing the behavior of an occupant of a specific position within a social structure.

Role differentiation The emergence of distinct roles within a group. The division of labor within a group.

Role discontinuity The discontinuity that results when the values and identities associated with a new role contradict those of earlier roles.

Role identity An individual's concept of self in a specific social role.

Role overload The condition in which the demands placed on a person by his or her roles exceed the amount of time, energy, and other resources available to meet those demands.

Role taking In symbolic interaction theory, the process of imaginatively occupying the position of another person and viewing the situation and the self from that person's perspective; the process of imagining the other's attitudes and anticipating that person's responses.

Role theory A theoretical perspective based on the premise that a substantial portion of observable, day-to-day social behavior is simply persons carrying out role expectations.

Romantic love ideal Five beliefs regarding love, including the belief (1) in love at first sight, (2) that there is one and only one true love for each person, (3) that love conquers all, (4) that our beloved is (nearly) perfect, and (5) that one should follow his or her heart.

Routine activities perspective A theory that considers how deviant behavior, such as crime and substance abuse, emerges from the routines of everyday life.

Rule breaking Behavior that violates social norms.

Rumor Communication via informal and often novel channels that cannot be validated.

S

Salience The relative importance of a specific role identity to the individual's self-schema. The salience hierarchy refers to the ordering of an individual's role identities according to their importance.

Schema A specific cognitive structure that organizes the processing of complex information about other persons, groups, and situations. Our schemas guide what we perceive in the environment, how we organize information in memory, and what inferences and judgments we make about people and things.

Secondary deviance Deviant behavior employed by a person as a means of defense or adjustment to the problems created by others' reactions to rule breaking by him or her.

Secondary group A group whose members have few emotional ties with one another and relate in terms of limited roles. Interaction in secondary groups is formal, impersonal, and nonspontaneous. *See also* Primary group.

Self The individual viewed as both the active source and the passive object of reflexive behavior.

Self-awareness A state in which we take the self as the object of our attention and focus on our own appearance, actions, and thoughts.

Self-disclosure The process of revealing personal information (aspects of our feelings and behaviors) to another person. Self-disclosure is sometimes used as an impression management tactic.

Self-discrepancy The state in which a component of the individual's actual self is the opposite of a component of the ideal self or the ought self.

Self-esteem The evaluative component of the self-concept. The positive and negative evaluations people have of themselves.

Self-estrangement The awareness that one is engaging in activities that are not rewarding in themselves.

Self-fulfilling prophecy When persons behave toward another person according to a label (impression) and cause the person to respond in ways that confirm the label.

Self-presentation All conscious and unconscious attempts by people to control the images of self they project in social interaction.

Self-reinforcement An individual's use of internalized standards to judge his or her own behavior and reward the self.

Self-schema The organized structure of information that people have about themselves; the primary influence on the processing of information about the self.

Self-serving bias In attribution, the tendency for people to take personal credit for acts that yield positive outcomes and to deflect blame for bad outcomes by attributing them to external causes.

Sentiments Socially significant feelings such as grief, love, jealousy, or indignation that arise out of enduring social relationships. Each sentiment is a pattern of sensations, emotions, actions, and cultural beliefs appropriate to a social relationship.

Sexual assault Sexual contact or intercourse without consent, accomplished by threat or use of force.

Shaping The learning process in which an agent initially reinforces any behavior that remotely resembles the desired response and subsequently requires increasing correspondence between the learner's behavior and the desired response before providing reinforcement.

Significant other A person whose views and attitudes are very important and worthy of consideration. The reflected views of a significant other have great influence on the individual's self-concept and self-regulation.

Simple random sample A sample of individuals selected from a population in such a way that everyone is equally likely to be selected. *See also* Stratified sample.

Situated identity A conception held by a person in a situation that indicates who he or she is in relation to the other people involved in that situation.

Situated self The subset of self-concepts that constitutes the self people recognize in a particular situation. Selected from the person's various identities, qualities, and self-evaluations, the situated self depends on the demands of the situation.

Situational attribution A decision by an observer to attribute a behavior to environmental forces facing the person who performed it rather than to that person's internal state. *See also* Dispositional attribution.

Situational constraint An influence on behavior due to the likelihood that other persons will learn about that behavior and respond positively or negatively to it.

Social emotion An emotion that involves an awareness of oneself in the social context, emerges out of interaction with at least one other actor, and is often experienced in reference to some kind of societal standard.

Social-emotional specialist In groups, a person who strives to keep emotional relationships pleasant among members; a person who initiates acts that ease the tension and soothe hurt feelings.

Social exchange theory A theoretical perspective, based on the principle of reinforcement, that assumes that people will choose whatever actions maximize rewards and minimize costs.

Social facilitation A phenomenon in which the mere presence of other individuals causes persons to perform better.

Social identity A definition of the self in terms of the defining characteristics of a social group.

Social identity theory of intergroup behavior A theory of intergroup relations based on the premise that people spontaneously categorize the social world into various groups (specifically, in-groups and out-groups) and experience high self-esteem to the extent that the in-groups to which they belong have more status than the out-groups.

Social impact theory A theoretical framework, applicable to both persuasion and obedience, stating that the impact of an influence attempt is a function of strength, immediacy, and number of sources that are present.

Social influence An interaction process in which one person's behavior causes another person to change an opinion or to perform an action that he or she would not otherwise do.

Socialization The process through which individuals learn skills, knowledge, values, motives, and roles appropriate to their positions in a group or society.

Social learning theory A theoretical perspective maintaining that one person (the learner) can acquire new responses without enacting them simply by observing the behavior of another person (the model). This learning process, called imitation, is distinguished by the fact that the learner neither performs a response nor receives any reinforcement.

Social loafing The tendency by group members to slack off and reduce their effort on additive tasks, which causes the group's output to fall short of its potential.

Social movement Collective activity that expresses a high level of concern about some issue; the activity may include participation in discussions, petition drives, demonstrations, or election campaigns.

Social network The sets of interpersonal relationships associated with the social positions a person occupies.

Social perception The process through which we construct an understanding of the social world out of the data we obtain through our senses. More narrowly defined, the processes through which we use available information to form impressions of people.

Social psychology The field that systematically studies the nature and causes of human social behavior.

Social responsibility norm A widely accepted social norm stating that individuals should help people who are dependent on them.

Social structure The ordered and persisting relationships among the positions in a social system.

Sociolinguistic competence Knowledge of the implicit rules for generating socially appropriate sentences that make sense because they fit the listeners' social knowledge.

Source In social influence, the person who intentionally engages in some behavior (persuasion, threat, promise) to cause another person to behave in a manner different from how he or she otherwise would. *See also* Target.

Speech act theory The theory that verbal utterances both state something and do something.

Spoken language A socially acquired system of sound patterns with meanings agreed on by the members of a group.

Standard speech A speech style characterized by diverse vocabulary, proper pronunciation, correct grammar, and abstract content. The use of this style is associated with high status and power. *See also* Nonstandard speech.

Status characteristic Any property of a person around which evaluations and beliefs about that person come to be organized; properties such as race, occupation, age, sex, ethnicity, education, and so on.

Status generalization A process through which differences in members' status characteristics lead to different performance expectations and hence affect patterns of interaction in groups. The tendency for a member's status inside a group to reflect his or her status outside that group.

Stereotype A fixed set of characteristics that is attributed to all the members of a group. A simplistic and rigid perception of members of one group that is widely shared by others.

Stereotype threat The suspicion a member of a group holds that he or she will be judged based on a common stereotype of the group.

Stigma Personal characteristics that others view as insurmountable handicaps preventing competent or morally trustworthy behavior.

Stratified sample In survey research, a sampling design whereby researchers subdivide the population into groups according to characteristics known or thought to be important, select a random sample of groups, and then draw a sample of units within each selected group. *See also* Simple random sample.

Stress The condition in which the demands made on the person exceed the individual's ability to cope with them.

Subjective expected value (SEV) With respect to threats, the product of a threat's credibility times its magnitude; with respect to promises, the product of a promise's credibility times its magnitude.

Subjective norm An individual's perception of others' beliefs about whether or not a behavior is appropriate in a specific situation.

Summons-answer sequence The most common verbal method for initiating a conversation, in which one person summons the other as with a question or greeting, and the other indicates his or her availability for conversation by responding. This sequence establishes the mutual obligation to speak and to listen that produces conversational turn taking.

Superordinate goal In intergroup conflict, an objective held in common by all conflicting groups that cannot be achieved by any one group without the supportive efforts of the others.

Supplication An impression management tactic that involves convincing a target person that you are needy and deserving.

Symbol A form used to represent ideas, feelings, thoughts, intentions, or any other object. Symbols represent our experiences in a way that others can perceive with their sensory organs—through sounds, gestures, pictures, and so on.

Symbolic interaction theory A theoretical perspective based on the premise that human nature and social order are products of communication among people.

T

Tactical impression management The selective use of self-presentation tactics by a person who wishes to manipulate the impressions that others form of him or her.

Target In social influence, the person who is affected by a social influence attempt from the source. In aggression, the person toward whom an aggressive act is directed. *See also* Source.

Task specialist In groups, a member who pushes the group toward the attainment of its goals; a person who contributes many ideas and suggestions to the group.

Theoretical perspective A theory that makes broad assumptions about human nature and offers general explanations of a wide range of diverse behaviors. *See also* Middle-range theory.

Theory A set of interrelated propositions that organizes and explains a set of observed facts; a network of hypotheses that may be used as a basis for prediction.

Theory of cognitive dissonance A theory concerning the sources and effects of inconsistency in cognitive systems with two or more elements.

Theory of reasoned action The theory that behavior is determined by behavioral intention, which in turn is determined by both attitude and subjective norm.

Theory of speech accommodation The theory that people express or reject intimacy with others by adjusting their speech behavior (accent, vocabulary, or language) during interaction. They make their own speech behavior more similar to their partner's to express liking, and more dissimilar to reject intimacy.

Threat An influence technique that is a communication taking the general form, "If you don't do X (which I want), then I will do Y (which you don't want)." *See also* Promise.

Trait centrality A personality trait has a high level of trait centrality when information about a person's standing on that trait has a large impact on the overall impression that others form of that person. The *warm-cold* trait, for instance, is highly central.

Transactional leadership Leadership in groups based on an exchange between the leader and other group members. The leader performs actions that move the group toward the attainment of its goals; in return, the leader receives support, endorsement, and rewards.

Transformational leadership Leadership that strengthens group performance by changing the way members view their group, its opportunities, and its mission. Leadership that conveys an extraordinary sense of mission to group members and arouses new ways of thinking within the group.

Trust The belief that a person is both honest and benevolent.

U

Ultimate attribution error A perceptual bias occurring in intergroup relations. Negative behaviors by out-group members are attributed to stable, internal factors such as undesirable personal traits or dispositions, but positive behaviors by out-group members are attributed to unstable, external factors such as situational pressures or luck. As a result, in-group observers will blame the out-group for negative outcomes but will not give it credit for positive outcomes.

W

Weapons effect An increase in aggression caused by the presence of a gun in a situation where an individual is already frustrated.

REFERENCES

Abbey, A., McAuslan, P., Zawacki, Clinton, & Buck (2001). Attitudinal, experiential, and situational predictors of sexual assault perpetration. *Journal of Interpersonal Violence, 16,* 784–807.

Abbey, A., Ross, L. T., McDuffie, D., & McAuslan, P. (1996). Alcohol and dating risk factors for sexual assault among college women. *Psychology of Women Quarterly, 20,* 147–169.

Abelson, R. P. (1981). The psychological status of the script concept. *American Psychologist, 36,* 715–729.

Aberson, C. L., Healy, M., & Romero, V. (2000). Ingroup bias and self-esteem: A meta-analysis. *Personality and Social Psychology Review, 4,* 157–173.

Abrams, D., Marques, J., Brown, N., & Henson, M. (2000). Pro-norm and anti-norm deviance within and between groups. *Journal of Personality and Social Psychology, 78,* 906–912.

Acock, A., & Scott, W. (1980). A model for predicting behavior: The effect of attitude and social class on high and low visibility political participation. *Social Psychology Quarterly, 43,* 59–72.

Adams, J. S. (1963). Toward an understanding of inequity. *Journal of Abnormal and Social Psychology, 67,* 422–436.

Adams, J. S., & Jacobsen, P. R. (1964). Effects of wage inequities on work quality. *Journal of Abnormal and Social Psychology, 69,* 19–25.

Adams, J. S., & Rosenbaum, W. B. (1962). The relationship of worker productivity to cognitive dissonance about wage inequities. *Journal of Applied Psychology, 46,* 161–164.

Adelman, R., & Verbrugge, L. (2000). Death makes news: The social impact of disease on newspaper coverage. *Journal of Health and Social Behavior, 41,* 347–367.

Aderman, D., & Berkowitz, L. (1983). Self-concern and the unwillingness to be helpful. *Social Psychology Quarterly, 46,* 293–301.

Adler, P. A., & Adler, P. (1995). Dynamics of inclusion and exclusion in preadolescent cliques. *Social Psychology Quarterly, 58,* 145–162.

Adlerfer, C. P. (1982). Problems in changing White males' behavior and beliefs concerning race relations. In P. Goodman (Ed.), *Change in organizations* (pp. 122–165). San Francisco: Jossey-Bass.

Agnew, R. (1992). Foundation for a general strain theory of crime and delinquency. *Criminology, 30,* 47–87.

Agnew, R., & White, H. R. (1992). An empirical test of general strain theory. *Criminology, 30,* 475–499.

Aiken, L. R. (1992). Some measures of interpersonal attraction and group cohesiveness. *Educational and Psychological Measurement, 52*(1), 63–67.

Ainsworth, M. (1979). Infant-mother attachment. *American Psychologist, 34,* 932–937.

Ajzen, I. (1982). On behaving in accordance with one's attitudes. In M. Zanna, E. Higgins, & C. Herman (Eds.), *Consistency in social behavior: The Ontario symposium* (Vol. 2). Hillsdale, NJ: Erlbaum.

Ajzen, I. (1985). From intentions to action: A theory of planned behavior. In J. Kuhl & J. Beckman (Eds.), *Action control: From cognition to behavior.* Heidelberg: Springer Verlag.

Ajzen, I., & Fishbein, M. (1977). Attitude-behavior relations: A theoretical analysis and review of research. *Psychological Bulletin, 84,* 888–918.

Ajzen, I., & Fishbein, M. (1980). *Understanding attitudes and predicting social behavior.* Englewood Cliffs, NJ: Prentice Hall.

Ajzen, I., & Holmes, W. H. (1976). Uniqueness of behavioral effects in causal attribution. *Journal of Personality, 44,* 98–108.

Akers, R. L., Krohn, M., Lanza-Kaduce, L., & Radosevich, M. (1979). Social learning and deviant behavior: A specific test of a general theory. *American Sociological Review, 44,* 636–655.

Akers, R. L., LaGreca, H.J., Cochran, J., & Sellers, C. (1989). Social learning theory and alcohol behavior among the elderly. *The Sociological Quarterly, 30,* 625–638.

Aldag, R. J., & Fuller, S. F. (1993). Beyond fiasco: A reappraisal of the groupthink phenomenon and a new model of group decision processes. *Psychological Bulletin, 113,* 533–552.

Aldous, J. (1987). New views on the family life of the elderly and the near-elderly. *Journal of Marriage and the Family, 49,* 227–234.

Aldwin, C., & Revenson, T. (1987). Does coping help? A reexamination of the relationship between coping and mental health. *Journal of Personality and Social Psychology, 53,* 337–348.

Alexander, C. N., Jr., & Rudd, J. (1984). Predicting behaviors from situated identities. *Social Psychology Quarterly, 47,* 172–177.

Alexander, C. N., Jr., & Wiley, M. G. (1981). Situated activity and identity formation. In M. Rosenberg & R. Turner (Eds.), *Social psychology: Sociological perspectives.* New York: Basic Books.

Allan, E. A., & Steffensmeier, D. J. (1989). Youth underemployment and property crime: Differential effects of job availability and job quality on juvenile and young adult arrest rates. *American Sociological Review, 54,* 107–123.

Allen, H. (1972). Bystander intervention and helping on the subway. In L. Bickman & T. Henchy (Eds.), *Beyond the laboratory: Field research in social psychology.* New York: McGraw-Hill.

Allen, V. L., & Levine, J. M. (1969). Consensus and conformity. *Journal of Experimental Social Psychology, 5,* 389–399.

Allen, V. L., & Levine, J. M. (1971). Social support and conformity: The role of independent assessment of reality. *Journal of Experimental Social Psychology, 7,* 48–58.

Allen, V. L., & Van de Vliert, E. (1982). A role theoretical perspective on transitional processes. In V. L. Allen & E. Van de Vliert (Eds.), *Role transitions: Explorations and explanations* (pp. 3–18). New York: Plenum.

Allison, S. T., Mackie, D. M., Muller, M. M., & Worth, L. T. (1993). Sequential correspondence biases and perceptions of change: The Castro studies revisited. *Personality and Social Psychology Bulletin, 19,* 151–157.

Allison, S. T., McQueen, L. R., & Schaerfl, L. M. (1992). Social decision making processes and the equal partitioning of shared resources. *Journal of Experimental Social Psychology, 28,* 23–42.

Allport, F. H. (1920). The influence of the group upon association and thought. *Journal of Experimental Psychology, 3,* 159–182.

Allport, F. H. (1924). *Social psychology.* Cambridge, MA: Houghton Mifflin.

Allport, G. W. (1935). Attitudes. In C. Murchison (Ed.), *Handbook of social psychology.* Worcester, MA: Clark University Press.

Allport, G. W. (1954). *The nature of prejudice.* Reading, MA: Addison-Wesley.

Allport, G. W. (1961). *Pattern and growth in personality.* New York: Holt, Rinehart and Winston.

Allport, G. W. (1985). The historical background of social psychology. In G. Lindzey & E. Aronson (Eds.), *Handbook of social psychology* (3rd. ed., Vol. I, pp. 1–46). New York: Random House.

Altman, I., & Taylor, D. A. (1973). *Social penetration: The development of interpersonal relationships.* New York: Holt, Rinehart and Winston.

Alwin, D. (1990). Cohort replacement and changes in parental socialization values. *Journal of Marriage and the Family, 52,* 347–360.

Alwin, D. F. (1986). Religion and parental child-rearing orientations: Evidence of Catholic-Protestant convergence. *American Journal of Sociology, 92,* 412–440.

Al-Zahrani, S. S. A., & Kaplowitz, S. A. (1993). Attributional biases in individualistic and collectivistic cultures: A comparison of Americans with Saudis. *Social Psychology Quarterly, 56,* 223–233.

Amato, P. R. (1983). Helping behavior in urban and rural environments: Field studies based on a taxonomic organization of helping episodes. *Journal of Personality and Social Psychology, 45,* 571–586.

Amato, P. R. (2001). The consequences of divorce for adults and children. In R. Milardo (Ed.), *Understanding families into the new millennium: A decade in review* (pp. 488–506). Minneapolis, MN: National Council on Family Relations.

Amato, P. R., & Booth, A. (2001). The legacy of parents' marital discord: Consequences for children's marital quality. *Journal of Personality and Social Psychology, 81,* 627–638.

Amato, P. R., & Cheadle, J. (2005). The long-reach of divorce: Divorce and child well-being across three generations. *Journal of Marriage and Family, 67,* 191–206.

Amato, P. R., & Fowler, F. (2002). Parenting practices, child adjustment, and family diversity. *Journal of Marriage and Family, 64,* 703–716.

Amato, P. R., & Rogers, S. J. (1997). A longitudinal study of marital problems and subsequent divorce. *Journal of Marriage and the Family, 59,* 612–624.

American Association of Retired Persons. (1999). *AARP/Modern maturity sexuality survey.* Washington, DC: Author.

American Association of University Women. (1992). *The AAUW report: How schools shortchange girls.* Washington, DC: Author.

American Psychiatric Association. (1981). *Diagnostic and statistical manual of mental disorders* (3rd ed.). Washington, DC: Author.

American Psychological Association. (1993). *Summary report of the APA Commission on Violence and Youth.* Washington, DC: Author.

Amir, Y. (1969). Contact hypothesis in ethnic relations. *Psychological Bulletin, 71,* 319–342.

Amir, Y. (1976). The role of intergroup contact in change of prejudice and ethnic relations. In P. A. Katz (Ed.), *Towards the elimination of racism* (pp. 245–308). New York: Pergamon.

Anderson, B. A., & Silver, B. D. (1987). The validity of survey responses: Insights from interviews of married couples in a survey of Soviet emigrants. *Social Forces, 66,* 537–554.

Anderson, C., Benjamin, A., & Bartholow, B. (1998). Does the gun pull the trigger? Automatic priming effects of weapon pictures and weapon names. *Psychological Science, 9,* 308–314.

Anderson, C., & Dill, K. (2000). Video games and aggressive thoughts, feelings, and behavior in the laboratory and in life. *Journal of Personality and Social Psychology, 78,* 772–790.

Anderson, C. A. (1987). Temperature and aggression: Effects on quarterly, yearly, and city rates of violent and nonviolent crime. *Journal of Personality and Social Psychology, 52,* 1161–1173.

Anderson, C. A. (2001). Heat and violence. *Current Directions in Psychological Science, 10,* 33–38.

Anderson, C. A., Anderson, K. B., & Denser, W. E. (1996). Examining an affective aggression framework: Weapon and temperature effects on aggressive thought, affect, and attitudes. *Personality and Social Psychology Bulletin, 22,* 366–376.

Anderson, C. A., Anderson, K. B., Dorr, N., DeNeve, K. M., & Flanagan, M. (2000). Temperature and aggression. In M. Zanna (Ed.), *Advances in experimental social psychology* (Vol. 32). New York: Academic Press.

Anderson, C. A., & Bushman, B. J. (2002). The effects of media violence on society. *Science, 295,* 2377–2380.

Anderson, C. A., & Sedikides, C. (1991). Thinking about people: Contribution of a typological alternative to associationistic and dimensional models of person perception. *Journal of Personality and Social Psychology, 60,* 203–217.

Anderson, K. B., Cooper, H., & Okamura, L. (1997). Individual differences and attitudes toward rape: A meta-analytic review. *Personality and Social Psychology Bulletin, 23,* 295–315.

Anderson, N. H. (1968). Likeableness ratings of 555 personality trait words. *Journal of Personality and Social Psychology, 9,* 272–279.

Anderson, N. H. (1981). *Foundations of information integration theory.* New York: Academic Press.

Andrews, B., & Brewin, C. R. (1990). Attributions of blame for marital violence: A study of antecedents and consequences. *Journal of Marriage and the Family, 52,* 757–767.

Andrews, I. R. (1967). Wage inequity and job performance: An experimental study. *Journal of Applied Psychology, 51,* 39–45.

Ansolabehere, S., & Iyengar, S. (1994). Of horseshoes and horse races: Experimental studies of the impact of poll results on electoral behavior. *Political Communication, 11,* 412–430.

Anson, O. (1989). Marital status and women's health revisited: The importance of a proximate adult. *Journal of Marriage and the Family, 51,* 185–194.

Appleton, W. (1981). *Fathers and daughters.* New York: Doubleday.

Apsler, R., & Sears, D. O. (1968). Warning, personal involvement, and attitude change. *Journal of Personality and Social Psychology, 9,* 162–166.

Aquino, K. F., & Reed, A., II. (2002). The self-importance of moral identity. *Journal of Personality and Social Psychology, 83,* 1423–1440.

Arbuthnot, J., & Wayner, M. (1982). Minority influence: Effects of size, conversion, and sex. *Journal of Psychology, 111,* 285–295.

Archer, J. (1991). Human sociobiology: Basic concepts and limitations. *Journal of Social Issues, 47*(3), 11–26.

Archer, R. L. (1980). Self-disclosure. In D. M. Wegner & R. R. Vallacher (Eds.), *The self in social psychology.* New York: Oxford University Press.

Archer, D., & Akert, R. (1977). Words and everything else: Verbal and nonverbal cues in social interpretation. *Journal of Personality and Social Psychology, 35,* 443–449.

Arditti, J. A. (1999). Rethinking relationships between divorced mothers and their children: Capitalizing on family strengths. *Family Relations, 48,* 109–119.

Arendt, H. (1965). *Eichmann in Jerusalem: A report on the banality of evil.* New York: Viking.

Aries, E. (1996). *Men and women in interaction.* New York: Oxford University Press.

Aristotle. (1952). On poetics, Book 6 (I. Bywater, Trans.). In Ross W. D. (Ed.), *The works of Aristotle* (Vol. 2, pp. 681–699). Chicago: Encyclopaedia Britannica. (Original work published 4th century BCE.)

Armeli, S., Carney, M. A., Tennen, H., Affleck, G., & O'Neil, T. P. (2000). Stress and alcohol use: A daily process examination of the stressor-vulnerability model. *Journal of Personality and Social Psychology, 78,* 979–994.

Arnett, J. J. (2004). *Emerging adulthood: The winding road from the late teens through the twenties.* Oxford/New York: Oxford University Press.

Aron, A., Norman, C., Aron, E., McKenna, C., & Heyman, R. (2000). Couples' shared participation in novel and arousing activities and experienced relationship quality. *Journal of Personality and Social Psychology, 78,* 273–284.

Aronfreed, J., & Reber, A. (1965). Internalized behavior suppression and the timing of social punishment. *Journal of Personality and Social Psychology, 1,* 3–16.

Aronoff, D., & Tedeschi, J. T. (1968). Original stakes and behavior in the prisoner's dilemma game. *Psychonomic Science, 12,* 79–80.

Aronson, E., Ellsworth, P. C., Carlsmith, J. M., & Gonzales, J. H. (1990). *Methods of research in social psychology* (2nd ed.). New York: McGraw-Hill.

Aronson, E., Turner, J. A., & Carlsmith, J. M. (1963). Communicator credibility and communication discrepancy as determinants of opinion change. *Journal of Abnormal and Social Psychology, 67,* 31–36.

Aronson, E., Wilson, T., & Brewer, M. (1998). Experimentation in social psychology. In D. Gilbert, S. Fiske, & G. Lindzey (Eds.), *The handbook of social psychology* (4th ed., Vol. 1, pp. 99–142). Boston: McGraw-Hill.

Asch, S. E. (1946). Forming impressions of personality. *Journal of Abnormal and Social Psychology, 41,* 258–290.

Asch, S. E. (1951). Effects of group pressure upon the modification and distortion of judgments. In H. Guetzkow (Ed.), *Groups, leadership, and men.* Pittsburgh, PA: Carnegie Press.

Asch, S. E. (1952). *Social psychology.* Englewood Cliffs, NJ: Prentice Hall.

Asch, S. E. (1955). Opinions and social pressure. *Scientific American, 193,* 31–35.

Asch, S. E. (1957). An experimental investigation of group influence. In *Symposium on preventive and social psychiatry.* Washington, DC: Walter Reed Army Institute of Research.

Asch, S. E., & Zukier, H. (1984). Thinking about persons. *Journal of Personality and Social Psychology, 46,* 1230–1240.

Aseltine, R., Jr., Gore, S., & Gordon, J. (2000). Life stress, anger and anxiety, and delinquency: An empirical test of general strain theory. *Journal of Health and Social Behavior, 41,* 256–275.

Asendorpf, J. B. (1987). Videotape reconstruction of emotions and cognitions related to shyness. *Journal of Personality and Social Psychology, 53,* 542–549.

Ashforth, B., & Lee, R. (1990). Defensive behavior in organizations: A preliminary model. *Human Relations, 43,* 621–648.

Ashmore, R. D. (1981). Sex stereotypes and implicit personality theory. In D. L. Hamilton (Ed.), *Cognitive processes in stereotyping and intergroup behavior* (pp. 37–81). Hillsdale, NJ: Erlbaum.

Ashmore, R. D., Solomon, M. R., & Longo, L. C. (1996). Thinking about fashion models' looks: A multidimensional approach to the structure of perceived physical attractiveness. *Personality and Social Psychology Bulletin, 22,* 1083–1104.

Astone, N., & McLanahan, S. S. (1991). Family structure, parental practices, and high school completion. *American Sociological Review, 56,* 309–320.

Atchley, R. C. (1980). *The social forces in later life.* Belmont, CA: Wadsworth.

Atkin, C. K. (1981). Mass media information campaign effectiveness. In R. E. Rice & W. J. Paisley (Eds.), *Public communication campaigns.* Beverly Hills, CA: Sage.

Attridge, M., Berscheid, E., & Simpson, J. A. (1995). Predicting relationship stability from both partners vs. one. *Journal of Personality and Social Psychology, 69,* 254–268.

Austin, W. (1980). Friendship and fairness: Effects of type of relationship and task performance on choice distribution rules. *Personality and Social Psychology Bulletin, 6,* 402–408.

Austin, W. G., & Hatfield, E. (1980). Equity theory, power, and social justice. In G. M. Kula (Ed.), *Justice and social interaction.* Bern, Switzerland: Hans Huber.

Austin, W., & Walster [Hatfield], E. (1974). Participants' reactions to "equity with the world." *Journal of Experimental Social Psychology, 10,* 528–548.

Averill, J. R. (1980). The emotions. In E. Staub (Ed.), *Personality: Basic aspects and current research* (pp. 134–199). Englewood Cliffs, NJ: Prentice Hall.

Averill, J. R. (1982). *Anger and aggression: An essay on emotion.* New York: Springer Verlag.

Ayers, L., Nacci, P., & Tedeschi, J. T. (1973). Attraction and reaction to noncontingent promises. *Bulletin of the Psychonomic Society, 1*(1B), 75–77.

Babcock, M. K., & Sabini, J. (1990). On differentiating embarrassment from shame. *European Journal of Social Psychology, 20,* 151–169.

Bacharach, S. B., & Lawler, E. J. (1981). *Bargaining: Power, tactics, and outcomes.* San Francisco: Jossey-Bass.

Backman, C. (1990). Attraction in interpersonal relationships. In M. Rosenberg & R. Turner (Eds.), *Social psychology: Sociological perspectives.* New Brunswick, NJ: Transaction.

Backman, C., & Secord, P. (1962). Liking, selective interaction, and misperception in congruent interpersonal relations. *Sociometry, 25,* 321–325.

Baer, F. C. (2003). *Creative quotations from Richard Armour (1906–1989).* Retrieved February 26, 2003, from http://creativequotations.com/one/862a.htm

Bagozzi, R. P. (1981). Attitudes, intentions, and behavior: A test of some key hypotheses. *Journal of Personality and Social Psychology, 41,* 607–627.

Bailey, W. C. (1990). Murder, capital punishment, and television: Execution publicity and homicide rates. *American Sociological Review, 55,* 628–633.

Bailey, J. M., Kim, P., Hills, A., & Linsenmeier, J. (1997). Butch, femme, or straight acting? Partner preferences of gay men and lesbians. *Journal of Personality and Social Psychology, 73,* 960–973.

Bales, R. F. (1953). The equilibrium problem in small groups. In T. Parsons, R. F. Bales, & E. A. Shils (Eds.), *Working papers in the theory of action.* New York: Free Press.

Bales, R. F. (1965). Task roles and social roles in problem solving groups. In I. D. Steiner & M. Fishbein (Eds.), *Current studies in social psychology* (pp. 321–333). New York: Holt, Rinehart and Winston.

Bales, R. F. (1970). *Personality and interpersonal behavior.* New York: Holt, Rinehart and Winston.

Balkwell, J. W. (1991). From expectations to behavior: An improved postulate for expectation states theory. *American Sociological Review, 56,* 355–369.

Balkwell, J. W. (1994). Status. In M. Foschi & E. J. Lawler (Eds.), *Group processes: Sociological analyses* (pp. 119–148). Chicago: Nelson-Hall.

Ball, D. W. (1976). Failure in sports. *American Sociological Review, 41,* 726–739.

Baller, R., & Richardson, K. (2002). Social integration, imitation and the geographic patterning of suicide. *American Sociological Review, 67,* 873–888.

Baltes, P., & Willis, S. (1982). Plasticity and enhancement of intellectual functioning in old age: Penn State's adult development and enrichment project. In F. Craik & S. Trehub (Eds.), *Aging and cognitive processes.* New York: Plenum.

Banaji, M., & Prentice, D. (1994). The self in social contexts. *Annual Review of Psychology, 45,* 297–332.

Bandura, A. (1965). Influence of models' reinforcement contingencies on the acquisition of imitative responses. *Journal of Personality and Social Psychology, 1,* 589–595.

Bandura, A. (1969). Social-learning theory of identificatory processes. In D. Goslin (Ed.), *Handbook of socialization theory and research.* Chicago: Rand McNally.

Bandura, A. (1973). *Aggression: A social learning analysis.* Englewood Cliffs, NJ: Prentice Hall.

Bandura, A. (1977). *Social learning theory.* Englewood Cliffs, NJ: Prentice Hall.

Bandura, A. (1982a). The psychology of chance encounters and life paths. *American Psychologist, 37,* 747–755.

Bandura, A. (1982b). Self-efficacy mechanism in human agency. *American Psychologist, 37,* 122–147.

Bandura, A. (1982c). The self and the mechanisms of agency. In J. Suls (Ed.), *Psychological perspectives on the self* (Vol. 1). Hillsdale, NJ: Erlbaum.

Bandura, A., Ross, D., & Ross, S. (1961). Transmission of aggression through imitation of aggressive models. *Journal of Abnormal and Social Psychology, 63,* 575–582.

Barber, J. J., & Grichting, W. L. (1990). Australia's media campaign against drug abuse. *International Journal of the Addictions, 25,* 693–708.

Bargh, J. A., & Thein, R. D. (1985). Individual construct accessibility, person memory, and the recall-judgment link: The case of information overload. *Journal of Personality and Social Psychology, 49,* 1129–1146.

Barker, R. G., Dembo, T., & Lewin, K. (1941). Frustration and aggression: An experiment with young children. *University of Iowa Studies in Child Welfare, 18,* 1–34.

Barnes, G., Farrell, M., & Cairns, A. (1986). Parental socialization factors and adolescent drinking behavior. *Journal of Marriage and the Family, 48,* 27–36.

Barnett, M. A. (1987). Empathy and related responses in children. In N. Eisenberg & J. Strayer (Eds.), *Empathy and its development* (pp. 146–162). New York: Cambridge University Press.

Barnett, R. C. (1994). Home-to-work spillover revisited: A study of full-time employed women in dual-earner couples. *Journal of Marriage and the Family, 56,* 647–656.

Baron, L., & Straus, M. A. (1984). Sexual stratification, pornography, and rape in the United States. In N. M. Malamuth & E. Donnerstein (Eds.), *Pornography and sexual aggression.* Orlando, FL: Academic Press.

Baron, R., & Ransberger, V. (1978). Ambient temperature and the occurrence of collective violence: The "long, hot summer" revisited. *Journal of Personality and Social Psychology, 36,* 351–360.

Baron, R. A. (1977). *Human aggression.* New York: Plenum.

Baron, R. A., & Kepner, C. R. (1970). Model's behavior and attraction toward the model as determinants of adult aggressive behavior. *Journal of Personality and Social Psychology, 14,* 335–344.

Baron, R. S., & Roper, G. (1976). A reaffirmation of a social comparison view of choice shifts, averaging, and extremity effects in autokinetic situations. *Journal of Personality and Social Psychology, 33,* 521–530.

Baron, R. S., Vandello, U. A., & Brunsman, B. (1996). The forgotten variable in conformity research: Impact of task importance on social influence. *Journal of Personality and Social Psychology, 71,* 915–927.

Barrett, K. C. (1995). A functionalist approach to shame and guilt. In J. P. Tangney & K. W. Fischer (Eds.), *Self-conscious emotions: The psychology of shame, guilt, embarrassment, and pride.* New York: Guilford Press.

Bar-Tal, D. (1976). *Prosocial behavior: Theory and research.* New York: Halsted.

Basic HHS Policy for Protection of Human Research Subjects. 45 C.F.R. Pt. 46.101–46.124 (1992).

Bass, B. M. (1990). *Bass & Stogdill's handbook of leadership: Theory, research, and managerial applications* (3rd ed.). New York: Free Press.

Bates, E., O'Connell, B., & Shore, C. (1987). Language and communication in infancy. In J. Osofsky (Ed.), *Handbook of infant competence* (2nd ed.). New York: Wiley.

Batson, C. D. (1987). Prosocial motivation: Is it ever truly altruistic? In L. Berkowitz (Ed.), *Advances in experimental social psychology* (Vol. 20, pp. 65–122). New York: Academic Press.

Batson, C. D., & Coke, J. S. (1981). Empathy: A source of altruistic motivation for helping? In J. P. Rushton & R. M. Sorrentino (Eds.), *Altruism and helping behavior.* Hillsdale, NJ: Erlbaum.

Batson, C. D., & Oleson, K. C. (1991). Current status of the empathy-altruism hypothesis. In M. S. Clark (Ed.), *Review of personality and social psychology: Vol. 12. Prosocial behavior* (pp. 62–85). Newbury Park, CA: Sage.

Batson, C. D., O'Quin, K., Fultz, J., Vanderplas, M., & Isen, A. M. (1983). Influence of self-reported distress and empathy on egoistic versus altruistic motivation to help. *Journal of Personality and Social Psychology, 45,* 706–718.

Bauer, R. (1964). The obstinate audience: The influence process from the point of view of social communication. *American Psychologist, 19,* 319–328.

Baum, A., Riess, M., & O'Hara, J. (1974). Architectural variants of reaction to spatial invasion. *Environment and Behavior, 6,* 91–100.

Baumeister, R. F. (1982). A self-presentational view of social phenomena. *Psychological Bulletin, 91,* 3–26.

Baumeister, R. F. (1998). The self. In D. Gilbert, S. Fiske, & G. Lindzey (Eds.), *The handbook of social psychology* (4th ed., Vol. 1, pp. 680–740). Boston: McGraw-Hill.

Baumeister, R. F., Stillwell, A. M., & Heatherton, T. F. (1994). Guilt: An interpersonal approach. *Psychological Bulletin 115,* 243–267.

Baumrind, D. (1980). New directions in socialization research. *American Psychologist, 35,* 639–652.

Bavelas, J., Coates, L., & Johnson, T. (2000). Listeners as co-narrators. *Journal of Personality and Social Psychology, 79,* 941–952.

Beall, A., & Sternberg, R. (1995). The social construction of love. *Journal of Social and Personal Relationships, 12,* 417–438.

Bearman, P., Moody, J., & Stovel, K. (2004). Chains of affection: The structure of adolescent romantic and sexual networks. *American Journal of Sociology, 110,* 44–91.

Beck, E., & Tolnay, S. E. (1990). The killing fields of the deep South: The market for cotton and the lynching of Blacks; 1882–1930. *American Sociological Review, 55,* 526–539.

Beck, S. B., Ward-Hull, C. I., & McLear, P. M. (1976). Variables related to women's somatic preferences of the male and female body. *Journal of Personality and Social Psychology, 34,* 1200–1210.

Becker, B. J. (1986). Influence again: Another look at studies of gender differences in social influence. In J. S. Hyde & M. C. Linn (Eds.), *The psychology of gender: Advances through meta-analysis.* Baltimore: Johns Hopkins University Press.

Becker, H. S. (1963). *Outsiders: Studies in the sociology of deviance.* New York: Free Press.

Becker, H. S. (1964). What do they really learn at college? *Trans-Action, 1,* 14–17.

Beckett, K. (1994). Setting the public agenda: "Street crime" and drug use in American politics. *Social Problems, 41,* 425–447.

Beebe, L. M., & Giles, H. (1984). Speech accommodation theories: A discussion in terms of second language learning. *International Journal of the Sociology of Language, 46,* 5–32.

Begley, T., & Alker, H. (1982). Anti-busing protest: Attitudes and actions. *Social Psychology Quarterly, 45,* 187–197.

Bell, R. (1979). Parent, child, and reciprocal influences. *American Psychologist, 34,* 821–826.

Bell, R. A., Zahn, C. J., & Hopper, R. (1984). Disclaiming: A test of two competing views. *Communication Quarterly, 32*(1), 28–36.

Bellavia, G., & Murray, S. (2003). Did I do that? Self-esteem related differences in reactions to romantic partners' moods. *Personal Relationships, 10,* 77–95.

Belsky, J. (1990). Parental and nonparental child care and children's socioemotional development: A decade in review. *Journal of Marriage and the Family, 52,* 885–903.

Belsky, J., Lang, M., & Rovine, M. (1985). Stability and change in marriage across the transition to parenthood: A second study. *Journal of Marriage and the Family, 47,* 855–865.

Bem, D. J. (1970). *Beliefs, attitudes, and human affairs.* Belmont, CA: Brooks/Cole.

Bem, S. L. (1981). Gender schema theory: A cognitive account of sex typing. *Psychological Review, 88,* 354–364.

Benedict, R. (1938). Continuities and discontinuities in cultural conditioning. *Psychiatry, 1,* 161–167.

Bennett, M. (1990). Children's understanding of the mitigating function of disclaimers. *Journal of Social Psychology, 130,* 29–37.

Bennett, M., & Dewberry, C. (1989). Embarrassment at others' failures: A test of the Semin and Manstead model. *Journal of Social Psychology, 129,* 557–559.

Benokraitis, N. V. (1997). *Subtle sexism: Current practice and prospects for change.* Thousand Oaks, CA: Sage.

Benokraitis, N., & Feagin, J. (1986). *Modern sexism.* Englewood Cliffs, NJ: Prentice Hall.

Bensley, L. S., & Wu, R. (1991). The role of psychological reactance in drinking following alcohol prevention messages. *Journal of Applied Social Psychology, 21,* 1111–1124.

Benson, M., & Fox, G. L. (2004). When violence hits home: How economics and neighborhood play a role. Washington, DC: National Institute of Justice, NCJ 20504.

Benson, M. L. (1989). The influence of class position on the formal and informal sanctioning of white-collar offenders. *The Sociological Quarterly, 30,* 465–479.

Benson, M. L., & Walker, E. (1988). Sentencing the white-collar offender. *American Sociological Review, 53,* 294–302.

Benson, P. L., Karabenick, S. A., & Lerner, R. M. (1973). Pretty pleases: The effect of physical attraction, race, and sex on receiving help. *Journal of Experimental Social Psychology, 12,* 409–415.

Ben Ze'ev, A. (2000). I only have eyes for you: The partiality of positive emotions. *Journal for the Theory of Social Behaviour, 30*(3), 341–351.

Bergen, D., & Williams, J. (1991). Sex stereotypes in the United States revisited: 1972–1988. *Sex Roles, 24,* 413–423.

Berger, J., Cohen, B. P., & Zelditch, M., Jr. (1972). Status characteristics and social interaction. *American Sociological Review, 37,* 241–255.

Berger, J., Conner, T. L., & Fisek, M. H. (Eds.). (1974). *Expectation states theory.* Cambridge, MA: Winthrop.

Berger, J., Rosenholtz, S. J., & Zelditch, M., Jr. (1980). Status organizing processes. In A. Inkeles, N. J. Smelser, & R. H. Turner (Eds.), *Annual review of sociology* (Vol. 6, pp. 479–508). Palo Alto, CA: Annual Reviews.

Berger, J., Webster, M., Jr., Ridgeway, C., & Rosenholtz, S. J. (1986). Status cues, expectations, and behavior. In E. J. Lawler (Ed.), *Advances in group processes* (Vol. 3, pp. 1–22). Greenwich, CT: JAI Press.

Bergesen, A. (1982). Race riots of 1967: An analysis of police violence in Detroit and Newark. *Journal of Black Studies, 12,* 261–274.

Berk, R. A., & Aldrich, H. (1972). Patterns of vandalism during civil disorders as an indicator of selection of targets. *American Sociological Review, 37,* 533–547.

Berkowitz, L. (1972). Frustrations, comparisons, and other sources of emotion arousal as contributors to social unrest. *Journal of Social Issues, 28*(1), 77–91.

Berkowitz, L. (1978a). Decreased helpfulness with increased group size through lessening the effects of the needy individual's dependency. *Journal of Personality, 46,* 299–310.

Berkowitz, L. (1978b). Whatever happened to the frustration-aggression hypothesis? *American Behavioral Scientist, 21,* 691–708.

Berkowitz, L. (1989). Frustration-aggression hypothesis: Examination and reformulation. *Psychological Bulletin, 106,* 59–73.

Berkowitz, L. (1993). *Aggression: Its cause, consequences, and control.* New York: McGraw Hill.

Berkowitz, L., Klanderman, S. B., & Harris, R. (1964). Effects of experimenter awareness and sex of subject and experimenter on reactions to dependency relationships. *Sociometry, 27,* 327–337.

Berkowitz, M. W., Mueller, C. W., Schnell, S. V., & Pudberg, M. T. (1986). Moral reasoning and judgments of aggression. *Journal of Personality and Social Psychology, 51,* 885–891.

Bernard, J. S. (1981). *The female world.* New York: Free Press.

Bernstein, I., Kelly, W., & Doyle, P. (1977). Societal reaction to deviants: The case of criminal defendants. *American Sociological Review, 42,* 743–755.

Bernstein, I., Kick, E., Leung, J., & Schulz, B. (1977). Charge reduction: An intermediary stage in the process of labelling criminal defendants. *Social Forces, 56,* 362–384.

Bernstein, W. M., Stephan, W. G., & Davis, M. H. (1979). Explaining attributions for achievement: A path analytic approach. *Journal of Personality and Social Psychology, 37,* 1810–1821.

Berry, D. S., & Landry, J. R. (1997). Facial maturity and daily social interaction. *Journal of Personality and Social Psychology, 72,* 570–580.

Berscheid, E. (1966). Opinion change and communicator-communicatee similarity and dissimilarity. *Journal of Personality and Social Psychology, 4,* 670–680.

Berscheid, E., Dion, K., Walster [Hatfield], E., & Walster, G. (1971). Physical attractiveness and dating choice: A test of the matching hypothesis. *Journal of Experimental Social Psychology, 7,* 173–189.

Berscheid, E., & Reis, H. (1998). Attraction and close relationships. In D. Gilbert, S. Fiske, & G. Lindzey (Eds.), *The handbook of social psychology* (4th ed., Vol. 2, pp. 193–281). Boston: McGraw-Hill.

Berscheid, E., & Walster [Hatfield], E. (1978). *Interpersonal attraction* (2nd ed.). Reading, MA: Addison-Wesley.

Bertenthal, B. I., & Fischer, K. (1978). Development of self-recognition in the infant. *Developmental Psychology, 14,* 44–50.

Bettencourt, B. A., Brewer, M. B., Croak, M. R., & Miller, N. (1992). Cooperation and the reduction of intergroup bias: The role of reward structure and social orientation. *Journal of Experimental Social Psychology, 28,* 301–319.

Bianchi, S., Milkie, M., Sayer, L., & Robinson, J. (2000). Is anyone doing the housework? Trends in the gender division of household labor. *Social Forces, 79,* 191–228.

Bianchi, S. M., & Robinson, J. (1997). What did you do today? Children's use of time, family composition, and the acquisition of social capital. *Journal of Marriage and the Family, 59,* 332–344.

Bickman, L. (1971). The effect of another bystander's ability to help on bystander intervention in an emergency. *Journal of Experimental Social Psychology, 7,* 367–379.

Biddle, B. J. (1979). *Role theory: Expectations, identities, and behaviors.* New York: Academic Press.

Biddle, B. J. (1986). Recent developments in role theory. In A. Inkeles, J. Coleman, & N. Smelser (Eds.), *Annual review of sociology* (Vol. 12, pp. 67–92). Palo Alto, CA: Annual Reviews.

Bielby, W., & Baron, J. (1986). Men and women at work: Sex segregation and statistical discrimination. *American Journal of Sociology, 91,* 759–799.

Bielby, D., & Bielby, W. (1988). She works hard for the money: Household responsibilities and the allocation of work effort. *American Journal of Sociology, 93,* 1031–1059.

Bielby, W., & Bielby, D. (1989). Balancing commitments to work and family in dual career households. *American Sociological Review, 54,* 776–789.

Bienenstock, E., & Bianchi, A. (2004). Activating performance expectations and status differences through gift exchange: Experimental results. *Social Psychology Quarterly, 67,* 310–318.

Bird, C. E. (1999). Gender, household labor, and psychological distress: The impact of the amount and division of housework. *Journal of Health and Social Behavior, 40,* 32–45.

Birdwhistell, R. L. (1970). *Kinesics in context: Essays on body motion communications.* Philadelphia: University of Pennsylvania Press.

Bittman, M., England, P., Sayer, L., Folbre, N., & Matheson, G. (2003). When does gender trump money? Bargaining and time in household work. *American Journal of Sociology, 109,* 186–214.

Black, D. (1980). *The manners and customs of the police.* New York: Academic Press.

Blackwell, D., & Lichter, D. (2004). Homogamy among dating, cohabiting, and married couples. *The Sociological Quarterly, 45,* 719–737.

Blake, R. R., Shepard, H. A., & Mouton, J. S. (1964). *Managing intergroup conflict in industry.* Houston, TX: Gulf.

Blanck, P. D., & Rosenthal, R. (1982). Developing strategies for decoding "leaky" messages. In R. S. Feldman (Ed.), *Development of nonverbal behavior in children.* New York: Springer Verlag.

Blascovich, J., Ginsburg, G. P., & Veach, T. L. (1975). A pluralistic explanation of choice shifts on the risk dimension. *Journal of Personality and Social Psychology, 31,* 422–429.

Blascovich, J., Mendes, W. B., Hunter, S., Lickel, B., & Kowai-Bell, N. (2001). Perceiver threat in social interactions with stigmatized others. *Journal of Personality and Social Psychology, 80,* 253–267.

Blass, T. (1991). Understanding behavior in the Milgram obedience experiment: The role of personality, situations, and their interactions. *Journal of Personality and Social Psychology, 60,* 398–413.

Blau, P. (1964). *Exchange and power in social life.* New York: Wiley.

Blauner, R. (1964). *Alienation and freedom.* Chicago: University of Chicago Press.

Blieszner, R., & Mancini, J. (1987). Enduring ties: Older adults' parental role and responsibilities. *Family Relations, 36,* 176–180.

Blom, J. P., & Gumperz, J. J. (1972). Social meaning and linguistic structure: Code-switching in Norway. In J. J. Gumperz & D. Hymes (Eds.), *Directions in sociolinguistics.* New York: Holt, Rinehart and Winston.

Blumenthal, M., Kahn, R. L., Andrews, F. M., & Head, K. B. (1972). *Justifying violence: Attitudes of American men.* Ann Arbor, MI: Institute for Social Research.

Blumer, H. (1962). Society and symbolic interactionism. In A. M. Rose (Ed.), *Human behavior and social processes.* Boston: Houghton Mifflin.

Blumer, H. (1969). *Symbolic interactionism: Perspective and method.* Englewood Cliffs, NJ: Prentice Hall.

Blumstein, P. W. (1974). The honoring of accounts. *American Sociological Review, 39,* 551–566.

Blumstein, P. W. (1975). Identity bargaining and self-conception. *Social Forces, 53,* 476–485.

Boardman, J., Finch, B. K., Ellison, C., Williams, D., & Jackson, J. (2001). *Journal of Health and Social Behavior, 42,* 151–165.

Bobo, L. (1983). Whites' opposition to busing: Symbolic racism or realistic group conflict. *Journal of Personality and Social Psychology, 45,* 1196–1210.

Bobo, L. (1999). Prejudice as group position: Micro-foundations of a sociological approach to racism and race relations. *Journal of Social Issues, 55,* 445–472.

Bobo, L. (2000). Race and beliefs about affirmative action: Assessing the effects of interests, group threat, ideology and racism. In D. O. Sears, J. Sidanius, & L. Bobo (Eds.), *Racialized politics: The debate about racism in America.* Chicago: University of Chicago Press.

Bochner, S., & Insko, C. A. (1966). Communicator discrepancy, source credibility, and opinion change. *Journal of Personality and Social Psychology, 4,* 614–621.

Bodenhausen, G. V., & Lichtenstein, M. (1987). Social stereotypes and information processing strategies: The impact of task complexity. *Journal of Personality and Social Psychology, 52,* 871–880.

Bodenhausen, G., & Wyer, R., Jr. (1985). Effects of stereotypes on decision making and information-processing strategies. *Journal of Personality and Social Psychology, 48,* 267–282.

Body rebuilding. (1997, January 26). *Wisconsin State Journal,* p. 1G.

Bohn, A. (2003). Familiar voices: Using Ebonics communication techniques in the primary classroom. *Urban Education, 38,* 688–707.

Bohra, K. A, & Pandey, J. (1984). Ingratiation toward strangers, friends, and bosses. *Journal of Social Psychology, 122,* 217–222.

Bolger, N., DeLongis, A., Kessler, R. C., & Wethington, E. (1989). The contagion of stress across multiple roles. *Journal of Marriage and the Family, 51,* 175–183.

Bollen, K. A. (1989). *Structural equations with latent variables.* New York: Wiley.

Bollen, K. A., & Hoyle, R. H. (1990). Perceived cohesion: A conceptual and empirical examination. *Social Forces, 69,* 479–504.

Bollen, K. A., & Phillips, D. P. (1981). Suicidal motor vehicle fatalities in Detroit: A replication. *American Journal of Sociology, 87,* 404–412.

Bolzendahl, C. I., & Myers, D. J. (2002, August). *Twenty-five years of feminist attitudes.* Paper presented at the Annual Meeting of the American Sociological Association, Chicago, IL.

Bond, R., & Smith, P. B. (1996). Culture and conformity: A meta-analysis of studies using Asch's line judgment task. *Psychological Bulletin, 119,* 111–137.

Bondurant, B., & Donat, P. (1999). Perceptions of women's sexual interest and acquaintance rape: The role of sexual overperception and affective attitude. *Psychology of Women Quarterly, 23,* 691–705.

Booth, A., Johnson, D., White, L., & Edwards, J. (1984). Women, outside employment, and marital instability. *American Journal of Sociology, 90,* 567–583.

Bord, R. J. (1976). The impact of imputed deviant identities in structuring evaluations and reactions. *Sociometry, 39,* 108–116.

Borkenau, P., & Ostendorf, F. (1987). Fact and fiction in implicit personality theory. *Journal of Personality, 55,* 415–443.

Bottomore, T. B. (Ed.). (1964). *Karl Marx' early writings.* New York: McGraw-Hill.

Bouchard, T. J., Barsaloux, J., & Drauden, G. (1974). Brainstorming procedure, group size, and sex as determinants of the problem-solving effectiveness of groups and individuals. *Journal of Applied Psychology, 59,* 135–138.

Bouchard, T. J., Jr., & Hare, M. (1970). Size, performance, and potential in brainstorming groups. *Journal of Applied Psychology, 54,* 51–55.

Boucher, J. D., & Ekman, P. (1975). Facial areas of emotional information. *Journal of Communication, 25,* 21–29.

Boulding, K. E. (1981). *Ecodynamics: A new theory of societal evolution.* Beverly Hills, CA: Sage.

Bourhis, R. Y., Giles, H., Leyens, J. P., & Tajfel, H. (1979). Psycholinguistic distinctiveness: Language diversity in Belgium. In H. Giles & R. N. St. Clair (Eds.), *Language and social psychology.* Oxford, UK: Blackwell.

Bowlby, J. (1965). Maternal care and mental health (1953). In J. Bowlby (Ed.), *Child care and the growth of love.* London: Penguin.

Braaten, L. J. (1991). Group cohesion: A new multidimensional model. *Group, 15,* 39–55.

Brackett, M. A., & Mayer, J. D. (2003). Convergent, discriminant, and incremental validity of competing measures of emotional intelligence. *Personality and Social Psychology Bulletin, 29,* 1147–1158.

Bradford, S., Feeney, J., & Campbell, L. (2002). Links between attachment orientations and dispositional and diary-based measures of disclosure in dating couples: A study of actor and partner effects. *Personal Relationships, 9,* 491–506.

Bradley, G. W. (1978). Self-serving biases in the attribution process: A reexamination of the fact or fiction question. *Journal of Personality and Social Psychology, 36,* 56–71.

Braesicke, K., Parkinson, J. A., Reekie, Y., Man, M. S., Hopewell, L., Pears, A., et al. (2005). Autonomic arousal in an appetitive context in primates: A behavioural and neural analysis. *European Journal of Neuroscience 21*(6), 1733–1740.

Brauer, M. (2001). Intergroup perception in the social context: The effects of social status and group membership on perceived out-group homogeneity and ethnocentrism. *Journal of Experimental Social Psychology, 37,* 15–31.

Bray, R. M., Kerr, N. L., & Atkin, R. S. (1978). Effects of group size, problem difficulty, and sex on group performance and member reactions. *Journal of Personality and Social Psychology, 36,* 1224–1240.

Brehm, J. W. (1956). Postdecision changes in the desirability of alternatives. *Journal of Abnormal and Social Psychology, 52,* 384–389.

Brehm, J. W., & Cohen, A. (1962). *Explorations in cognitive dissonance.* New York: Wiley.

Brehm, P. (1966). *A theory of psychological reactance.* New York: Academic Press.

Brewer, M. B. (1979). In-group bias in the minimal intergroup situation: A cognitive-motivational analysis. *Psychological Bulletin, 86,* 307–324.

Brewer, M. B. (1986). The role of ethnocentrism in intergroup conflict. In S. Worchel & W. G. Austin (Eds.), *Psychology of intergroup relations* (2nd ed., pp. 88–102). Chicago: Nelson-Hall.

Brewer, M. B., & Brown, R. J. (1998). Intergroup relations. In D. T. Gilbert, S. T. Fiske, & G. Lindzey (Eds.), *The handbook of social psychology* (4th ed.). New York: McGraw-Hill.

Brewer, M. B., & Campbell, D. T. (1976). *Ethnocentrism and intergroup attitudes: East African evidence.* New York: Halsted.

Brewer, M. B., & Kramer, R. M. (1985). The psychology of intergroup attitudes and behavior. *Annual Review of Psychology, 36,* 219–243.

Brewer, M. B., & Lui, L. (1984). Categorization of the elderly by the elderly: Effects of perceiver's category membership. *Personality and Social Psychology Bulletin, 10,* 585–595.

Brewer, M. B., & Miller, N. (1984). Beyond the contact hypothesis: Theoretical perspectives on desegregation. In N. Miller & M. B. Brewer (Eds.), *Groups in contact:*

The psychology of desegregation (pp. 281–302). Orlando, FL: Academic Press.

Brewer, M. B., & Pierce, K. P. (2005). Social identity complexity and outgroup tolerance. *Personality and Social Psychology Bulletin, 31,* 428–437.

Brickman, P., Rabinowitz, V. C., Karuza, J., Coates, D., Cohn, E., & Kidder, L. (1982). Models of helping and coping. *American Psychologist, 37,* 368–384.

Brickner, M. A., Harkins, S. G., & Ostrom, T. M. (1986). Effects of personal involvement: Thought provoking implications for social loafing. *Journal of Personality and Social Psychology, 51,* 763–769.

Briggs, J. L. (1970). *Never in anger.* Cambridge, MA: Harvard University Press.

Brim, O. G., Jr., & Ryff, C. (1980). On the properties of life events. In P. Baltes & O. G. Brim, Jr. (Eds.), *Life-span development and behavior* (Vol. 3). New York: Academic Press.

Brockmann, H., & Klein, T. (2002). *Love and death in Germany: The marital biography and its impact on health* (Working Paper 2002-15). Rostock, Germany: Max Planck Institute.

Brockner, J., Konovsky, M. A., Cooper-Schneider, R., Folger, R., Martin, C., & Bies, R. J. (1994). The interactive effects of procedural justice and outcome negativity on victims and survivors of job loss. *Academy of Management Journal, 37,* 397–409.

Brockner, J., & Wiesenfeld, B. M. (1996). An integrative framework for explaining reactions to a decision: The interactive effects of outcomes and procedures. *Psychological Bulletin, 120,* 189–208.

Brody, E. M. (1990). *Women in the middle: Their parent-care years.* New York: Springer Verlag.

Brody, L. (1999). *Gender, emotion, and the family.* Cambridge, MA: Harvard University Press.

Broman, C. (1991). Gender, work-family roles, and psychological well-being of Blacks. *Journal of Marriage and the Family, 53,* 509–520.

Bronfenbrenner, U. (1961). The mirror image in Soviet-American relations: A social psychologist's report. *Journal of Social Issues, 17*(3), 45–56.

Broverman, I., Vogel, S., Broverman, D., Clarkson, F., Rosenkrantz, P. (1972). Sex-role stereotypes: A current appraisal. *Journal of Social Issues, 28*(2), 59–78.

Brown, J. D. (1991a). The professional ex-: An alternative for exiting the deviant career. *The Sociological Quarterly, 32,* 219–230.

Brown, J. D. (1991b). Staying fit and staying well: Physical fitness as a moderator of life stress. *Journal of Personality and Social Psychology, 60,* 555–561.

Brown, J. D., Collins, R. L., & Schmidt, G. W. (1988). Self-esteem and direct vs. indirect forms of self-enhancement. *Journal of Personality and Social Psychology, 55,* 445–453.

Brown, P., & Elliott, R. (1965). Control of aggression in a nursery school class. *Journal of Experimental Child Psychology, 2,* 103–107.

Brown, R. (1964). The acquisition of language. In D. Rioch & E. Weinstein (Eds.), *Proceedings of the Association for Research in Nervous and Mental Disease, Vol. 42: Disorders of communication.* Baltimore: Williams and Wilkins.

Brown, R. (1965). *Social psychology.* Glencoe, IL: Free Press.

Brown, R., & Fraser, C. (1963). The acquisition of syntax. In C. Cofer & B. Musgrave (Eds.), *Verbal behavior and learning.* New York: McGraw-Hill.

Brown, R. J., & Turner, J. C. (1981). Interpersonal and intergroup behavior. In J. Turner & H. Giles (Eds.), *Intergroup behavior* (pp. 33–65). Chicago: University of Chicago Press.

Brown, S., Sessions, J., & Taylor, K. (2004). *What will I be when I grow up? An analysis of childhood expectations and career outcomes* (Working Paper 05/2). University of Leicester, Department of Economics.

Brown, S. L. (2000). The effect of union type on psychological well-being: Depression among cohabitors versus marrieds. *Journal of Health and Social Behavior, 41,* 241–255.

Browning, C. (2003). The span of collective efficacy: Extending social disorganization theory to partner violence. *Journal of Marriage and Family, 64,* 833–850.

Brubaker, T. H. (1990). Families in later life: A burgeoning research area. *Journal of Marriage and the Family, 52,* 959–981.

Brunstein, J. C., Dangelmayer, G., & Schultheiss, O. C. (1996). Personal goals and social support in close relationships: Effects on relationship mood and marital satisfaction. *Journal of Personality and Social Psychology, 71,* 1006–1019.

Bryan, J. H., & Davenport, M. (1968). *Donations to the needy: Correlates of financial contributions to the destitute* (Research Bulletin No. 68-1). Princeton, NJ: Educational Testing Service.

Bryan, J. H., & Test, M. (1967). Models and helping: Naturalistic studies in aiding behavior. *Journal of Personality and Social Psychology, 6,* 400–407.

Buck, R., & Ginsburg, B. (1991). Spontaneous communication and altruism: The communicative gene hypothesis. In M. S. Clark (Ed.), *Prosocial behavior* (pp. 149–175). Newbury Park, CA: Sage.

Bugenthal, D. E. (1974). Interpretations of naturally occurring discrepancies between words and intonation: Modes of inconsistency resolution. *Journal of Personality and Social Psychology, 30,* 125–133.

Bui, K.-V. T., Peplau, L. A., & Hill, C. T. (1996). Testing the Rusbult model of relationship commitment and stability in a 15-year study of heterosexual couples. *Personality and Social Psychology Bulletin, 22,* 1244–1257.

Bumpass, L. L., Sweet, J. S., & Cherlin, A. (1991). The role of cohabitation in declining rates of marriage. *Journal of Marriage and the Family, 53,* 913–927.

Burawoy, M. (2003). Revisits: An outline of a theory of reflexive ethnography. *American Sociological Review, 68,* 645–679.

Burger, J. M. (1986) Increasing compliance by improving the deal: The That's-Not-All technique. *Journal of Personality and Social Psychology, 51,* 277–283.

Burger, J. M. (1999). The foot-in-the-door compliance procedure: A multiple-process analysis and review. *Personality and Social Psychology Review, 3,* 303–325.

Burger, J. M., & Petty, R. E. (1981). The low-ball compliance technique: Task or person commitment? *Journal of Personality and Social Psychology, 40,* 492–500.

Burger, J. M., Soroka, S., Gonzago, K., Murphy, E., & Somervell, E. (2001). The effect of fleeting attraction on compliance to requests. *Personality and Social Psychology Bulletin, 27,* 1578–1586.

Burgess, R. L., & Akers, K. L. (1966). A differential association-reinforcement theory of criminal behavior. *Social Problems, 14,* 128–147.

Burke, P. (2004). Identities and social structure: The 2003 Cooley-Mead Award Address. *Social Psychology Quarterly, 67,* 5–15.

Burke, P. J., & Reitzes, D. (1981). The link between identity and role performance. *Social Psychology Quarterly, 44,* 83–92.

Burnstein, E. (1982). Persuasion as argument processing. In H. Brandstatter, J. H. Davis, & G. Stocher-Kreichgauer (Eds.), *Contemporary problems in group decision-making* (pp. 103–124). New York: Academic Press.

Burnstein, E., Crandall, C., & Kitayama, S. (1994). An evolved heuristic for altruism: Evidence for a human propensity to calculate inclusive fitness. *Journal of Personality and Social Psychology, 67,* 773–789.

Burnstein, E., & Vinokur, A. (1973). Testing two classes of theories about group-induced shifts in individual choice. *Journal of Experimental Social Psychology, 9,* 123–137.

Burt, M. R. (1980). Cultural myths and supports for rape. *Journal of Personality and Social Psychology, 38,* 217–230.

Bush, D., & Simmons, R. (1990). Socialization processes over the life course. In M. Rosenberg & R. Turner (Eds.), *Social psychology: Sociological perspectives.* New Brunswick, NJ: Transaction.

Bushman, B. J. (1988). The effects of apparel on compliance: A field experiment with a female authority figure. *Personality and Social Psychology Bulletin, 14,* 459–467.

Bushman, B. J. (1995). Moderating role of trait aggressiveness in the effect of violent media on aggression. *Journal of Personality and Social Psychology, 69,* 950–960.

Bushman, B., Baumeister, R., & Stack, A. (1999). Catharsis, aggression, and persuasive influence: Self-fulfilling or self-defeating prophecies. *Journal of Personality and Social Psychology, 76,* 367–376.

Bushman, B. J., & Stack, A. D. (1996). Forbidden fruit versus tainted fruit: Effects of warning labels on attraction to television violence. *Journal of Experimental Psychology: Applied, 2,* 207–226.

Buss, D. M. (1994). *The evolution of desire: Strategies of human mating.* New York: Basic Books.

Buss, D. M. (1999). *Evolutionary psychology.* Boston: Allyn & Bacon.

Buss, D. M., & Kenrick, D. T. (1998). Evolutionary social psychology. In D. T. Gilbert, S. T. Fiske, & G. Lindzey (Eds.), *The handbook of social psychology* (4th ed.). Boston: McGraw-Hill.

Butterfield, F. (2002, Aug. 21). Father steals best: Crime in an American family. *New York Times.*

Buunk, B., Dijkstra, P., Fetchenhauer, D., and Kenrick, D. (2002). Age and gender differences in mate selection criteria for various involvement levels. *Personal Relationships, 9,* 271–278.

Buunk, B. P., & van der Eijnden, R. (1997). Perceived prevalence, perceived superiority, and relationship satisfaction: Most relationships are good but ours is the best. *Personality and Social Psychology Bulletin, 23,* 219–228.

Byrne, D. (1961a). The influence of propinquity and opportunities for interaction on classroom relationships. *Human Relations, 14,* 63–69.

Byrne, D. (1961b). Interpersonal attraction and attitude similarity. *Journal of Abnormal and Social Psychology, 62,* 713–715.

Byrne, D. (1971). *The attraction paradigm.* New York: Academic Press.

Byrne, D., & Clore, G. L. (1970). A reinforcement model of evaluative responses. *Personality: An International Journal, 1,* 103–128.

Byrne, D., Ervin, C., & Lamberth, J. (1970). Continuity between the experimental study of attraction and real-life computer dating. *Journal of Personality and Social Psychology, 16,* 157–165.

Byrne, D., & Kelley, K. (1984). Introduction: Pornography and sex research. In N. M. Malamuth & E. Donnerstein (Eds.), *Pornography and sexual aggression.* Orlando, FL: Academic Press.

Byrne, D., & Nelson, D. (1965). Attraction as a linear function of proportion of positive reinforcements. *Journal of Personality and Social Psychology, 1,* 659–663.

Cacioppo, J. T., Petty, R. E., Feinstein, J. A., & Jarvis, W. B. G. (1996). Dispositional differences in cognitive motivation: The life and times of individuals varying in need for cognition. *Psychological Bulletin, 119,* 197–253.

Cadinu, M. R., & Rothbart, M. (1996). Self-anchoring and differentiation processes in minimal group settings. *Journal of Personality and Social Psychology, 70,* 661–677.

Cahill, S. (1987). Children and civility: Ceremonial deviance and the acquisition of ritual competence. *Social Psychology Quarterly, 50,* 312–321.

Cahill, S. E. (1999). Emotional capital and professional socialization: The case of mortuary science students (and me). *Social Psychology Quarterly, 62,* 101–116.

Calahan, D. (1970). *Problem drinkers.* San Francisco: Jossey-Bass.

Callaway, M. R., & Esser, J. K. (1984). Groupthink: Effects of cohesiveness and problem-solving procedures on group decision making. *Social Behavior and Personality, 12,* 157–164.

Cameron, D. (1998). Gender, language, and discourse: A review essay. *Signs: Journal of Women in Culture and Society, 23,* 945–973.

Campbell, D. T. (1967). Stereotypes in the perception of group differences. *American Psychologist, 22,* 817–829.

Campbell, J. D. (1990). Self-esteem and clarity of self-concept. *Journal of Personality and Social Psychology, 59,* 538–549.

Campbell, A., Converse, P., & Rodgers, W. (1976). *The quality of American life.* New York: Russell Sage Foundation.

Campbell, W. K., & Sedikides, C. (1999). Self-threat magnifies the self-serving bias: A meta-analysis integration. *Review of General Psychology, 3,* 23–43.

Campbell, D. T., & Stanley, J. C. (1963). *Experimental and quasi-experimental designs for research.* Chicago: Rand McNally.

Cano, I., Hopkins, N., & Islam, M. R. (1991). Memory for stereotype-related material: A replication study with real-life social groups. *European Journal of Social Psychology, 21,* 349–357.

Cantor, J., Alfonso, H., & Zillmann, D. (1976). The persuasive effectiveness of the peer appeal and a communicator's first hand experience. *Communication Research, 3,* 293–310.

Caplan, F. (1973). *The first twelve months of life.* New York: Grosset & Dunlap.

Caporeal, L. R. (2001). Evolutionary psychology: Toward a unifying theory and a hybrid science. *Annual Review of Psychology, 52,* 607–628.

Carli, L. L., LaFleur, S. J., & Loeber, C. C. (1995). Nonverbal behavior, gender, and influence. *Journal of Personality and Social Psychology, 68,* 1030–1041.

Carlsmith, J. M., & Anderson, C. (1979). Ambient temperature and the occurrence of collective violence. *Journal of Personality and Social Psychology, 37,* 337–344.

Carlson, M., Charlin, V., & Miller, N. (1988). Positive mood and helping behavior: A test of six hypotheses. *Journal of Personality and Social Psychology, 55,* 211–229.

Carlson, M., Marcus-Newhall, A., & Miller, N. (1990). Effects of situational aggression cues: A quantitative review. *Journal of Personality and Social Psychology, 58,* 622–633.

Carnevale, P. J., & Henry, R. A. (1989). Determinants of mediator behavior: A test of the strategic choice model. *Journal of Applied Social Psychology, 19,* 481–498.

Carnevale, P. J., & Pegnetter, R. (1985). The selection of mediation tactics in public sector disputes: A contingency analysis. *Journal of Social Issues, 41*(2), 65–81.

Carnevale, P. J., Pruitt, D. G., & Carrington, P. I. (1982). Effects of future dependence, liking, and repeated requests for help on helping behavior. *Social Psychology Quarterly, 45*(1), 9–14.

Carpenter, W. A., & Hollander, E. P. (1982). Overcoming hurdles to independence in groups. *Journal of Social Psychology, 117,* 237–241.

Carroll, J. M., & Russell, J. A. (1996). Do facial expressions signal specific emotions? Judging emotion from face in context. *Journal of Personality and Social Psychology, 70,* 205–218.

Carter, G. L. (1983). *Explaining the severity of the 1960's Black rioting.* Unpublished dissertation, Columbia University, Department of Sociology.

Carter, G. L. (1990). Black attitudes and the 1960s Black riots: An aggregate-level analysis of the Kerner Commission's "15 cities" data. *The Sociological Quarterly, 31,* 269–286.

Cartwright, D. (1968). The nature of group cohesiveness. In D. Cartwright & A. Zander (Eds.), *Group dynamics* (3rd ed., pp. 91–109). New York: Harper & Row.

Cartwright, D. (1971). Risk taking by individuals and groups: An assessment of research employing choice dilemmas. *Journal of Personality and Social Psychology, 20,* 361–378.

Cartwright, D., & Zander, A. (1968). Motivational processes in groups: Introduction. In D. Cartwright & A. Zander (Eds.), *Group dynamics* (3rd ed.). New York: Harper & Row.

Case, A., & Paxson, C. (2004). *Sex differences in morbidity and mortality* (Working Paper w10653). Cambridge, MA: National Bureau of Economic Research.

Casper, L. M. (1997). My daddy takes care of me: Fathers as care providers. *U.S. Bureau of the Census: Current Population Reports,* P70-59.

Caspi, A. (1992). *Surmounting childhood disadvantage: Turning points and pathways to change in the life course.*

Caspi, A., & Herbener, E. S. (1990). Continuity and change: Assortative marriage and the consistency of personality in adulthood. *Journal of Personality and Social Psychology, 58,* 250–258.

Catalano, R., Novaco, R, & McConnell, W. (1997). A model of the net effect of job loss on violence. *Journal of Personality and Social Psychology, 72,* 1440–1447.

Cate, R., Levin, L., & Richmond, L. (2002). Premarital relationship stability: A review of recent research. *Journal of Social and Personal Relationships, 19,* 261–284.

Catrambone, R., & Markus, H. (1987). The role of self-schemas in going beyond the information given. *Social Cognition, 5,* 349–368.

Center on Addiction and Substance Abuse. (2002). *Teen tipplers: America's underage drinking epidemic.* New York: Columbia University.

Centers for Disease Control. (2001). Fatal occupational injuries—United States, 1980–1997. *Morbidity and Mortality Weekly Report, 50,* 317–320.

Centers, R. (1975). Attitude similarity-dissimilarity as a correlate of heterosexual attraction and love. *Journal of Marriage and the Family, 37,* 305–312.

Chaffee, S. (1981). Mass media in political campaigns: An expanding role. In R. E. Rice & W. J. Paisley (Eds.), *Public communication campaigns.* Beverly Hills, CA: Sage.

Chaiken, S. (1980). Heuristic versus systematic information processing and the use of source versus message cues in persuasion. *Journal of Personality and Social Psychology, 39,* 752–766.

Chaiken, S. (1986). Physical appearance and social influence. In C. P. Herman, M. P. Zanna, & E. T. Higgins (Eds.), *Physical appearance, stigma, and social behavior: The Ontario Symposium* (Vol. 3, pp. 143–177). Hillsdale, NJ: Erlbaum.

Chaiken, S., & Maheswaran, D. (1994). Heuristic processing can bias systematic processing: Effects of source credibility, argument ambiguity, and task importance on attitude judgment. *Journal of Personality and Social Psychology, 66,* 460–473.

Chaiken, S., & Yates, S. (1985). Affective-cognitive consistency and thought-induced polarization. *Journal of Personality and Social Psychology, 49,* 1470–1481.

Chambliss, W. J. (1994). Policing the ghetto underclass: The politics of law and law enforcement. *Social Problems, 41,* 177–193.

Chao, R. (1994). Beyond parental control and authoritarian parenting style: Understanding Chinese parenting through the cultural notion of training. *Child Development, 65,* 1111–1119.

Chaplin, W., Phillips, J., Brown, J., Clanton, N., & Stein, J. (2000). Handshaking, gender, personality, and first impressions. *Journal of Personality and Social Psychology, 79,* 110–117.

Chapman, R. S., Streim, N. W., Crais, E. R., Salmon, D., Strand, E. A., & Negri, N. A. (1992). Child talk: Assumptions of a developmental process model for early language learning. In R. S. Chapman (Ed.), *Processes in language acquisition and disorder.* Chicago: Mosby/Year Book.

Charon, J. M. (1995). *Symbolic interactionism: An introduction, interpretation, and integration* (5th ed.). Englewood Cliffs, NJ: Prentice Hall.

Chebat, J.-C., Filiatrault, P., & Perrien, J. (1990). Limits of credibility: The case of political persuasion. *Journal of Social Psychology, 130,* 157–167.

Chekroun, P., & Brauer, M. (2002). The bystander effect and social control behavior: The effect of the presence of others on people's reactions to norm violations. *European Journal of Social Psychology, 32,* 853–867.

Chemers, M. M. (1983). Leadership theory and research: A systems-process integration. In P. B. Paulus (Ed.), *Basic group processes* (pp. 9–39). New York: Springer Verlag.

Chemers, M. M., & Skrzypek, G. J. (1972). An experimental test of the contingency model of leadership effectiveness. *Journal of Personality and Social Psychology, 24,* 172–177.

Chen, H. C., Reardon, R., Rea, C., & Moore, D. J. (1992). Forewarning of content and involvement: Consequences for persuasion and resistance to persuasion. *Journal of Experimental Social Psychology, 22,* 23–33.

Chen, Y. R., & Church, A. H. (1993). Reward allocation preferences in groups and organizations. *International Journal of Conflict Management, 4*(1), 25–59.

Cheng, P. W., & Novick, L. R. (1990). A probabilistic contrast model of causal induction. *Journal of Personality and Social Psychology, 58,* 545–567.

Cherlin, A. J. (1981). *Marriage, divorce, remarriage.* Cambridge, MA: Harvard University Press.

Cherlin, A. J., Chase-Lansdale, P. L., & McRae, C. (1998). Effects of divorce on mental health throughout the life course. *American Sociological Review, 63,* 239–249.

Chiricos, T., & Waldo, G. (1975). Socioeconomic status and criminal sentencing: An empirical assessment of a conflict proposition. *American Sociological Review, 40,* 753–772.

Choi, I., & Nisbett, R. E. (1998). Situational salience and cultural differences in the correspondence bias and in actor-observer bias. *Personality and Social Psychology Bulletin, 24,* 949–960.

Choi, I., Nisbett, R. E., & Norenzayan, A. (1999). Causal attribution across cultures: Variation and universality. *Psychological Bulletin, 125,* 47–63.

Christopher, F. S., & Roosa, M. (1991). Factors affecting sexual decisions in the premarital relationships of adolescents and young adults. In K. McKinney & S. Sprecher (Eds.), *Sexuality in close relationships* (pp. 111–133). Hillsdale, NJ: Erlbaum.

Cialdini, R. B. (1993). *Influence: Science and practice* (3rd ed.). New York: HarperCollins.

Cialdini, R. B. (2001). *Influence: Science and Practice* (4th ed.). Boston: Allyn & Bacon.

Cialdini, R. B., Borden, R., Thorne, A., Walker, M., & Freeman, S. (1976). Basking in reflected glory: Three (football) field studies. *Journal of Personality and Social Psychology, 34,* 366–375.

Cialdini, R. B., Cacioppo, J. T., Basset, R., & Miller, J. A. (1978). Low-ball procedure for producing compliance: Commitment, then cost. *Journal of Personality and Social Psychology, 36,* 463–476.

Cialdini, R. B., Darby, B. L., & Vincent, J. E. (1973). Transgression and altruism: A case for hedonism. *Journal of Experimental Social Psychology, 9,* 502–516.

Cialdini, R. B., Kendrick, D. T., & Baumann, D. J. (1982). Effects of mood on prosocial behavior in children and adults. In N. Eisenberg (Ed.), *The development of prosocial behavior* (pp. 339–359). New York: Academic Press.

Cialdini, R. B., & Trost, M. R. (1998). Social influence: Social norms, conformity, and compliance. In D. T. Gilbert, S. T. Fiske, & G. Lindzey (Eds.), *Handbook of social psychology.* Boston: McGraw-Hill.

Cialdini, R. B., Trost, M. R., & Newsom, J. T. (1995). Preference for consistency: The development of a valid measure and the discovery of surprising behavioral implications. *Journal of Personality and Social Psychology, 69,* 318–328.

Clark, E. V. (1976). From gesture to word: On the natural history of deixis in language acquisition. In J. S. Bruner & A. Gartner (Eds.), *Human growth and development.* Oxford, UK: Clarendon.

Clark, M. S., Gotay, C. C., & Mills, J. (1974). Acceptance of help as a function of the potential helper and opportunity to repay. *Journal of Applied Social Psychology, 4,* 224–229.

Clark, R. D., III, & Word, L. E. (1972). Why don't bystanders help? Because of ambiguity? *Journal of Personality and Social Psychology, 24,* 392–400.

Clausen, J. A. (1968). *Socialization and society.* Boston: Little, Brown.

Clausen, J. S. (1991). Adolescent competence and the shaping of the life course. *American Journal of Sociology, 96,* 805–842.

Clore, G. L., Bray, R. M., Itkin, S. M., & Murphy, P. (1978). Interracial attitudes and behavior at a summer camp. *Journal of Personality and Social Psychology, 36,* 107–116.

Cloward, R. (1959). Illegitimate means, anomie, and deviant behavior. *American Sociological Review, 24,* 164–176.

CNPAAEMI—Council of National Psychological Associations for the Advancement of Ethnic Minority Interests. (2000). *Guidelines for research in ethnic minority communities.* Washington, DC: American Psychological Association.

Cognard-Black, A. (2004). Will they stay, or will they go? Sex-atypical work among token men who teach. *The Sociological Quarterly, 45,* 113–139.

Cohen, A. (1966). *Deviance and control.* Englewood Cliffs, NJ: Prentice Hall.

Cohen, C. E. (1981). Person categories and social perception: Testing some boundaries of the processing effects of prior knowledge. *Journal of Personality and Social Psychology, 40,* 441–452.

Cohen, D. (1996). Law, social policy, and violence: The impact of regional cultures. *Journal of Personality and Social Psychology, 70,* 961–978.

Cohen, E. G. (1980). Design and redesign of the desegregated school: Problems of status, power, and conflict. In W. G. Stephan & J. Feagin (Eds.), *School desegregation*. New York: Academic Press.

Cohen, E. G. (1984). The desegregated school: Problems in status power and interethnic climate. In N. Miller & M. Brewer (Eds.), *Groups in contact: The psychology of desegregation* (pp. 77–96). Orlando, FL: Academic Press.

Cohen, L., & Kluegel, J. (1978). Determinants of juvenile court dispositions: Ascriptive and achieved factors in two metropolitan courts. *American Sociological Review, 43,* 162–176.

Cohen, D., Nisbett, R. E., Bowdle, B. F., & Schwarz, N. (1996). Insult, aggression, and the Southern culture of honor: An experimental ethnography. *Journal of Personality and Social Psychology, 70,* 945–960.

Cohen, E. G., & Roper, S. (1972). Modification of interracial interaction disability: An application of status characteristic theory. *American Sociological Review, 37,* 643–657.

Cohen, G., Steele, C. M., & Ross, L. D. (1999). The mentor's dilemma: Providing critical feedback across the racial divide. *Personality and Social Psychology Bulletin, 25,* 1302–1318.

Cohen, B. P., & Zhou, X. (1991). Status processes in enduring work groups. *American Sociological Review, 56,* 179–188.

Cohn, E. G., & Rotton, J. (1997). Assault as a function of time and temperature: A moderator-variable time-series analysis. *Journal of Personality and Social Psychology, 72,* 1322–1334.

Cohn, R. M. (1978). The effect of employment status change on self-attitudes. *Social Psychology, 41,* 81–93.

Cole, T. (2001). Lying to the one you love: The use of deception in romantic relationships. *Journal of Social and Personal Relationships, 18,* 107–129.

Coleman, J. W. (1987). Toward an integrated theory of white-collar crime. *American Journal of Sociology, 93,* 406–439.

Coleman, R. P., & Neugarten, B. (1971). *Social status in the city.* San Francisco: Jossey-Bass.

Collett, P. (1971). On training Englishmen in the nonverbal behavior of Arabs: An experiment in intercultural communication. *International Journal of Psychology, 6,* 209–215.

Collins, N. L., & Miller, L. C. (1994). Self-disclosure and liking: A meta-analytic review. *Psychological Bulletin, 116,* 457–475.

Comstock, G. (1984). Media influences on aggression. In A. Goldstein (Ed.), *Prevention and control of aggression: Principles, practices, and research.* New York: Pergamon.

Comstock, G. S., Chaffee, S., Katzman, N., McCombs, M., & Roberts, D. (1978). *Television and human behavior.* New York: Columbia University Press.

Condon, J. W., & Crano, W. D. (1988). Inferred evaluation and the relation between attitude similarity and interpersonal attraction. *Journal of Personality and Social Psychology, 54,* 789–797.

Condon, W. S., & Ogston, W. D. (1967). A segmentation of behavior. *Journal of Psychiatric Research, 5,* 221–235.

Conger, R., & Conger, K. (2002). Resilience in Midwestern families: Selected findings from the first decade of a prospective, longitudinal study. *Journal of Marriage and the Family, 64,* 361–373.

Conger, J. A., Kanungo, R. N., & Menon, S. T. (2000). Charismatic leadership and follower outcome effects. *Journal of Organizational Behavior, 21,* 747–767.

Conley, J. J. (1985). Longitudinal stability of personality traits: A multitrait-multimethod-multioccasion analysis. *Journal of Personality and Social Psychology, 49,* 1266–1282.

Cook, K. (Ed.). (1987). *Social exchange theory.* Newbury Park, CA: Sage.

Cook, S. W. (1984). The 1954 social science statement and school desegregation: A reply to Gerard. *American Psychologist, 39,* 819–832.

Cook, S. W. (1985). Experimenting on social issues: The case of school desegregation. *American Psychologist, 40,* 452–460.

Cooke, B., Rossmann, M., McCubbin, H., & Patterson, J. (1988). Examining the definition and measurement of social support: A resource for individuals and families. *Family Relations, 37,* 211–216.

Cooley, C. H. (1902). *Human nature and the social order.* New York: Charles Scribner's Sons.

Cooley, C. H. (1908). A study of the early use of self-words by a child. *Psychological Review, 15,* 339–357.

Cooley, C. H. (1909). *Social organization: A study of the large mind.* New York: Charles Scribner's Sons.

Cooper, H. M. (1979). Statistically combining independent studies: A meta-analysis of sex differences in conformity research. *Journal of Personality and Social Psychology, 37,* 131–146.

Cooper, W. H. (1981). Ubiquitous halo. *Psychological Bulletin, 90,* 218–244.

Cooper, J., & Fazio, R. H. (1986). The formation and persistence of attitudes that support intergroup conflict. In S. Worchel & W. G. Austin (Eds.), *Psychology of intergroup relations* (2nd ed., pp. 183–195). Chicago: Nelson-Hall.

Coopersmith, S. (1967). *The antecedents of self-esteem.* San Francisco: Freeman.

Corcoran, M. (1995). From rags to riches: Poverty and mobility in the United States. *Annual Review of Sociology, 21,* 237–267.

Corning, A. F., & Myers, D. J. (2002). Individual orientation toward engagement in social action. *Political Psychology, 23,* 703–729.

Correll, S. (2004). Constraints into preferences: Gender, status and emerging career aspirations. *American Sociological Review, 69,* 93–113.

Corsaro, W. A. (1992). Interpretive reproduction in children's peer cultures. *Social Psychology Quarterly, 55,* 160–177.

Corsaro, W. A., & Eder, D. (1995). Development and socialization of children and adolescents. In K. S. Cook, G. A. Fine, & J. S. House (Eds.), *Sociological perspectives on social psychology* (pp. 421–451). Boston: Allyn & Bacon.

Corsaro, W., & Fingerson, L. (2003). Development and socialization in childhood. In DeLamater, J. (Ed.), *Handbook of social psychology.* New York: Kluwer/Plenum.

Corsaro, W., & Molinari, L. (2000). Priming events and Italian children's transition from preschool to elementary school: Representations and actions. *Social Psychology Quarterly, 63,* 16–33.

Corsaro, W. A., & Rizzo, T. A. (1988). *Discussione* and friendship: Socialization processes in the peer culture of Italian nursery school children. *American Sociological Review, 53,* 879–894.

Coser, L. A. (1967). *Continuities in the study of social conflict.* New York: Free Press.

Costa, M., Dinsbach, W., and Manstead, A. (2001). Social presence, embarrassment, and nonverbal behavior. *Journal of Nonverbal Behavior, 25,* 225–240.

Cota, A. A., & Dion, K. L. (1986). Salience of gender and sex composition of ad hoc groups: An experimental test of distinctiveness theory. *Journal of Personality and Social Psychology, 50,* 770–776.

Cota, A. A., Evans, C. R., Dion, K. L., Kilik, L., & Longman, R. S. (1995). The structure of group cohesion. *Personality and Social Psychology Bulletin, 21,* 572–580.

Cotter, D., Hermsen, J., & Vanneman, R. (2004). *Gender inequality at work.* New York: Russell Sage Foundation.

Couch, C. J. (1968). Collective behavior: An examination of some stereotypes. *Social Problems, 15,* 310–322.

Court, J. H. (1984). Sex and violence: A ripple effect. In N. M. Malamuth & E. Donnerstein (Eds.), *Pornography and sexual aggression.* Orlando, FL: Academic Press.

Courtright, J. A. (1978). A laboratory investigation of group-think. *Communications Monographs, 45,* 229–246.

Cowan, P. A. (1991). Individual and family life transitions: A proposal for a new definition. In D. A. Cowan & M. Hetherington (Eds.), *Family transitions.* Hillsdale, NJ: Erlbaum.

Cox, C. L., Wexler, M. O., Rusbult, C. E., & Gaines, S. O., Jr. (1997). Prescriptive support and commitment processes in close relationships. *Social Psychology Quarterly, 60,* 79–90.

Cozby, P. C. (1972). Self-disclosure, reciprocity, and liking. *Sociometry, 35,* 151–160.

Cramer, R., McMaster, M., Bartell, P., & Dragna, M. (1988). Subject competence and minimization of the bystander effect. *Journal of Applied Social Psychology, 18,* 1132–1148.

Crandall, C. S., Tsang, J., Goldman, S., & Pennington, J. (1999). Newsworthy moral dilemmas: Justice, caring, and gender. *Sex Roles, 40,* 187–210.

Crano, W. (1997). Vested interests, symbolic politics, and attitude-behavior consistency. *Journal of Personality and Social Psychology, 72,* 485–491.

Crocker, J., & Major, B. (1989). Social stigma and self-esteem: The self-protective properties of stigma. *Psychological Review, 96,* 608–630.

Crocker, J., Thompson, L. L., McGraw, K. M., & Ingerman, C. (1987). Downward comparison, prejudice, and evaluations of others: Effects of self-esteem and threat. *Journal of Personality and Social Psychology, 52,* 907–916.

Crocker, J., Voelkl, K., Testa, M., & Major, B. (1991). Social stigma: The affective consequences of attributional ambiguity. *Journal of Personality and Social Psychology, 60,* 218–228.

Crocker, J., & Wolfe, C. (2001). Contingencies of self-worth. *Psychological Review, 108,* 593–623.

Crohan, S. E. (1996). Marital quality and conflict across the transition to parenthood in African-American and White couples. *Journal of Marriage and the Family, 58,* 933–944.

Cropanzano, R. (1993). *Justice in the workplace: Approaching fairness in human resource management.* Hillsdale, NJ: Erlbaum.

Crosbie, P. V. (Ed.). (1975). *Interaction in small groups.* New York: MacMillan.

Crosnoe, R., & Elder, G. (2002). Successful adaptation in the later years: A life course approach to aging. *Social Psychology Quarterly, 65,* 309–328.

Cummings, K. M., Sciandra, R., Davis, S., & Rimer, B. (1989). Response to anti-smoking campaign aimed at mothers with young children. *Health Education Research, 4,* 429–437.

Cunningham, J. D. (1981). Self-disclosure intimacy: Sex, sex-of-target, cross-national, and "generational" differences. *Personality and Social Psychology Bulletin, 7,* 314–319.

Cunningham, M. R. (1986). Measuring the physical in physical attractiveness: Quasi-experiments on the sociobiology of female facial beauty. *Journal of Personality and Social Psychology, 50,* 925–935.

Cunningham, M. R., Barbee, A. P., & Pike, C. L. (1990). What do women want? Facial metric assessment of multiple motives in the perception of male facial physical attractiveness. *Journal of Personality and Social Psychology, 59,* 61–72.

Cunningham, M. R., Steinberg, J., & Grev, R. (1980). Wanting to and having to help: Separate motivations for positive mood and guilt-induced helping. *Journal of Personality and Social Psychology, 38,* 181–192.

Cutrona, C., Russell, D., Abraham, W. T., Gardner, K., Melby, J., Bryant, C., & Conger, R. (2003). Neighborhood context and financial strain as predictors of marital interaction and marital quality in African American couples. *Personal Relationships, 10,* 389–409.

Cutrona, C. E. (1982). Transition to college: Loneliness and the process of social adjustment. In L. A. Peplau & D. Perlman (Eds.), *Loneliness: A resource book of current theory research and therapy.* New York: Wiley.

Dabbs, J. M., Jr., & Leventhal, H. (1966). Effects of varying the recommendations in a fear-arousing communication. *Journal of Personality and Social Psychology, 4,* 525–531.

Daher, D. M., & Banikiotes, P. G. (1976). Interpersonal attraction and rewarding aspects of disclosure content and level. *Journal of Personality and Social Psychology, 33,* 492–496.

Daly, K. (1987). Discrimination in the criminal courts: Family, gender, and the problem of equal treatment. *Social Forces, 66,* 152–175.

Daniels, A. (1987). Invisible work. *Social Problems, 34,* 403–415.

Darley, J. M., & Batson, C. D. (1973). From Jerusalem to Jericho: A study of situational and dispositional variables in helping behavior. *Journal of Personality and Social Psychology, 27,* 100–108.

Darley, J. M., & Fazio, R. H. (1980). Expectancy confirmation processes arising in the social interaction sequence. *American Psychologist, 35,* 867–881.

Darley, J. M., & Latané, B. (1968). Bystander intervention in emergencies: Diffusion of responsibility. *Journal of Personality and Social Psychology, 8,* 377–383.

Darley, J. M., Teger, A. I., & Lewis, L. D. (1973). Do groups always inhibit individuals' response to potential emergencies? *Journal of Personality and Social Psychology, 26,* 395–399.

Darwin, C. (1998). *The expression of the emotions in man and animals* (P. Ekman, Ed.). New York: Oxford University Press.

Davidson, A. R., & Jaccard, J. (1979). Variables that moderate the attitude-behavior relation: Results of a longitudinal survey. *Journal of Personality and Social Psychology, 37,* 1364–1376.

Davidson, A. R., Yantis, S., Norwood, M., & Montano, D. (1985). Amount of information about the attitude object and attitude-behavior consistency. *Journal of Personality and Social Psychology, 49,* 1184–1198.

Davidson, L. R., & Duberman, L. (1982). Friendship: Communication and interaction patterns in same sex dyads. *Sex Roles, 8,* 809–822.

Davies, J. C. (1962). Toward a theory of revolution. *American Sociological Review, 27,* 5–19.

Davies, J. C. (Ed.). (1971). *When men revolt—and why.* New York: Free Press.

Davies, S., & Tanner, J. (2003). The long arm of the law: Effects of labeling on employment. *The Sociological Quarterly, 44,* 385–404.

Davis, D., & Perkowitz, W. T. (1979). Consequences of responsiveness in dyadic interaction: Effects of probability of response and proportion of content-related responses on interpersonal attraction. *Journal of Personality and Social Psychology, 37,* 534–550.

Davis, F. (1961). Deviance disavowal: The management of strained interaction by the visibly handicapped. *Social Problems, 9,* 120–132.

Davis, J. A., Smith, T. W., & Marsden, P. V. (2000). *General social surveys, 1972–2000.* Ann Arbor, MI: Inter-University Consortium for Political and Social Research. Retrieved March 5, 2003, from http://www.icpsr.umich.edu:8080/ICPSR-STUDY/03197.xml

Davis, J. D. (1976). Self-disclosure in an acquaintance exercise: Responsibility for level of intimacy. *Journal of Personality and Social Psychology, 33,* 787–792.

Davis, K. (1947). Final note on a case of extreme isolation. *American Journal of Sociology, 52,* 432–437.

Davis, K. E. (1985, February). Near and dear: Friendship and love. *Psychology Today, 22,* 22–30.

Davis, M. H. (1996). *Empathy: A social psychological approach.* Boulder, CO: Westview Press.

Davis, M. H., & Franzoi, S. L. (1986). Adolescent loneliness, self-disclosure, and private self-consciousness: A longitudinal investigation. *Journal of Personality and Social Psychology, 51,* 595–608.

Dawes, R. M. (1998). Behavioral decision making and judgement. In D. Gilbert, S. Fiske, & G. Lindzey (Eds.), *The handbook of social psychology* (4th ed., Vol.1, pp. 497–548). New York: McGraw-Hill.

Dawkins, R. (1976). *The selfish gene.* Oxford, UK: Oxford University Press.

Dawkins, R. (1982). *The extended phenotype.* San Francisco: Freeman.

de Jong-Gierveld, J. (1987). Developing and testing a model of loneliness. *Journal of Personality and Social Psychology, 53,* 119–128.

de Tocqueville, A. (1955). *The old regime and the French Revolution* (Stuart Gilbert, Trans.). Garden City, NY: Doubleday. (Original work published 1856.)

De Wolff, M. S., & Van IJzendoorn, M. H. (1997). Sensitivity and attachment: A meta-analysis on parental antecedents of infant attachment. *Child Development, 68,* 571–591.

Deater-Deckard, K., & Dodge, K. (1997). Externalizing behavior problems and discipline revisited: Nonlinear effects and variation by culture, context, and gender. *Psychological Inquiry, 8,* 161–175.

Deaux, K., & Lewis, L. (1983). Components of gender stereotypes. *Psychological Documents, 13,* 25 (No. 2583).

Deaux, K., & Martin, D. (2003). Interpersonal networks and social categories: Specifying levels of context in identity processes. *Social Psychology Quarterly, 66,* 101–117.

DeBono, K. G., & Telesca, C. (1990). The influence of source physical attractiveness on advertising effectiveness: A functional perspective. *Journal of Applied Social Psychology, 20,* 1383–1395.

Deci, E. (1975). *Intrinsic motivation.* New York: Plenum.

DeDreu, C., & DeVries, N. (Eds.). (2001). *Group consensus and minority influence.* Oxford, UK: Blackwell.

Degelman, D., & Price, N. (2002). Tattoos and ratings of personal characteristics. *Psychological Reports, 90,* 507–514.

DeLamater, J., & Sill, M. (2005). Sexual desire in later life. *Journal of Sex Research, 42,* 138–149.

DeLongis, A., Folkman, J., & Lazarus, R. S. (1988). The impact of daily stress on health and mood: Psychological and social resources as mediators. *Journal of Personality and Social Psychology, 54,* 486–495.

Deluga, R. J. (1987). Corporate coups d'etat: What happens and why. *Leadership and Organizational Development Journal, 8,* 9–15.

DeMaris, A., Benson, M., Fox, G., Hill, T., & Van Wyk, J. (2003). Distal and proximal factors in domestic violence: A test of an integrated model. *Journal of Marriage and Family, 65,* 652–667.

Dembroski, T. M., Lasater, T. M., & Ramires, A. (1978). Communicator similarity, fear-arousing communications, and compliance with health care recommendations. *Journal of Applied Social Psychology, 8,* 254–269.

Demo, D. H. (1992). The self-concept over time: Research issues and directions. *Annual Review of Sociology, 18,* 303–326.

Demo, D. H., & Acock, A. C. (1988). The impact of divorce on children. *Journal of Marriage and the Family, 50,* 619–648.

Demo, D., & Cox, M. (2001). Families with young children: A review of research in the 1990s. In R. Milardo (Ed.), *Understanding families into the new millennium: A decade in review* (pp. 95–114). Minneapolis, MN: National Council on Family Relations.

Demo, D. H., & Hughes, M. (1990). Socialization and racial identity among Black Americans. *Social Psychology Quarterly, 53,* 364–374.

Dentler, R., & Erikson, K. (1959). The functions of deviance in groups. *Social Problems, 7,* 98–107.

Denton, K., & Krebs, D. (1990). From the scene to the crime: The effect of alcohol and social context on moral judgment. *Journal of Personality and Social Psychology, 59,* 242–248.

Denzin, N. (1977). *Childhood socialization: Studies in the development of language, social behavior, and identity.* San Francisco: Jossey-Bass.

Denzin, N. (1983). *On understanding emotion.* San Francisco: Jossey-Bass.

DePaulo, B. M. (1992). Nonverbal behavior and self-presentation. *Psychological Bulletin, 111,* 203–243.

DePaulo, B. M., & Fisher, J. D. (1980). The costs of asking for help. *Basic and Applied Social Psychology, 1,* 23–35.

DePaulo, B. M., Lassiter, G. D., & Stone, J. T. (1982). Attentional determinants of success at determining deception and truth. *Personality and Social Psychology Bulletin, 8,* 273–279.

DePaulo, B. M., Nadler, A., & Fisher, J. D. (1983). *New directions in helping: Help seeking* (Vol. 2). New York: Academic Press.

DePaulo, B. M., & Rosenthal, R. (1979). Ambivalence, discrepancy, and deception in nonverbal communication. In R. Rosenthal (Ed.), *Skill in nonverbal communication.* Cambridge, MA: Oelgeschlager, Gunn and Hain.

DePaulo, B. M., Rosenthal, R., Eisenstat, R. A., Rogers, P. L., & Finkelstein, S. (1978). Decoding discrepant non-verbal cues. *Journal of Personality and Social Psychology, 36,* 313–323.

DePaulo, B. M., Stone, J. I., & Lassiter, G. D. (1985). Deceiving and detecting deceit. In B. R. Schlenker (Ed.), *The self and social life* (pp. 323–370). New York: McGraw-Hill.

DePaulo, B. M., Zuckerman, M., & Rosenthal, R. (1980). Detecting deception: Modality effects. In L. Wheeler (Ed.), *Review of personality and social psychology* (Vol. 1, pp. 125–162). Beverly Hills, CA: Sage.

Der-Karabetian, A., & Smith, A. (1977). Sex-role stereotyping in the United States: Is it changing? *Sex Roles, 3,* 193–198.

Derlega, V. J., Durham, B., Gockel, B., & Sholis, D. (1981). Sex differences in self disclosure: Effects of topic content, friendship, and partner's sex. *Sex Roles, 7,* 433–447.

Derlega, V. J., & Grzelak, J. (1979). Appropriate self-disclosure. In G. J. Chelune et al. (Eds.), *Self-disclosure: Origins, patterns, and implications of openness in interpersonal relationships* (pp. 151–176). San Francisco: Jossey-Bass.

Derlega, V. J., Metts, S., Petronio, S., & Margulis, S. T. (1993). *Self-disclosure.* Newbury Park, CA: Sage.

Dervin, B. (1981). Mass communicating: Changing conceptions of the audience. In R. E. Rice & W. J. Paisley (Eds.), *Public communication campaigns.* Beverly Hills, CA: Sage.

Deutsch, M. (1985). *Distributive justice: A social psychological perspective.* New Haven, CT: Yale University Press.

Deutsch, M., & Gerard, H. B. (1955). A study of normative and informational social influences upon individual judgment. *Journal of Abnormal and Social Psychology, 51,* 629–636.

Deutsch, M., & Krauss, R. M. (1962). Studies of interpersonal bargaining. *Journal of Abnormal and Social Psychology, 61,* 181–189.

Devine, P. C. (1989). Stereotypes and prejudice: Their automatic and controlled components. *Journal of Personality and Social Psychology, 56,* 5–18.

Devine, J. A., Sheley, J. F., & Smith, M. D. (1988). Macroeconomic and social-control policy influences on crime-rate changes, 1948–1985. *American Sociological Review, 53,* 407–420.

Dickoff, H. (1961). *Reactions to evaluations by others as a function of self-evaluation and the interaction context.* Unpublished PhD dissertation, Duke University, Raleigh, NC.

Diehl, M., & Stroebe, W. (1987). Productivity loss in brainstorming groups: Toward the solution of a riddle. *Journal of Personality and Social Psychology, 53,* 497–509.

Diehl, M., & Stroebe, W. (1991). Productivity loss in idea-generating groups: Tracking down the blocking effect. *Journal of Personality and Social Psychology, 61,* 392–403.

Diekman, A. B., Eagly, A. H., Mladinic, A., & Ferreira, M. C. (2005). Dynamic stereotypes about women and men in Latin America and the United States. *Journal of Cross-Cultural Psychology, 36,* 209–226.

Diener, E. (1980). Deindividuation: The absence of self-awareness and self-regulation in group members. In P. B. Paulus (Ed.), *Psychology of group influence.* Hillsdale, NJ: Erlbaum.

Dietz, T. L. (1998). An examination of violence and gender-role portrayals in video games: Implications for gender socialization and aggressive behavior. *Sex Roles, 38,* 425–442.

DiMaggio, P., & Mohr, J. (1985). Cultural capital, educational attainment, and marital selection. *American Journal of Sociology, 90,* 1231–1257.

Dindia, K, & Allen, M. (1992). Sex differences in self-disclosure: A meta-analysis. *Psychological Bulletin, 112,* 106–124.

Dion, K. L. (1979). Intergroup conflict and intragroup cohesiveness. In W. G. Austin & S. Worchel (Eds.), *The social psychology of intergroup relations* (pp. 211–224). Monterey, CA: Brooks/Cole.

Dion, K., Baron, R., & Miller, N. (1970). Why do groups make riskier decisions than individuals? In L. Berkowitz (Ed.), *Advances in experimental social psychology* (Vol. 5). New York: Academic Press.

Dion, K., Berscheid, E., & Walster [Hatfield], E. (1972). What is beautiful is good. *Journal of Personality and Social Psychology, 24,* 285–290.

Dion, K. L., & Schuller, R. A. (1991). The Ms. stereotype: Its generality and its relation to managerial and marital status stereotypes. *Canadian Journal of Behavioural Science, 23,* 25–40.

Dixon, J., Gordon, C., & Khomusi, T. (1995). Sexual symmetry in psychiatric diagnosis. *Social Problems, 42,* 429–448.

Dodson, L., & Dickert, J. (2004). Girls' family labor in low-income households: A decade of qualitative research. *Journal of Marriage and Family, 66,* 318–332.

Dohrenwend, B. P. (1961). The social psychological nature of stress: A framework for causal inquiry. *Journal of Abnormal and Social Psychology, 62,* 294–302.

Dohrenwend, B. P. (2000). The role of adversity and stress in psychopathology: Some evidence and its implications for theory and research. *Journal of Health and Social Behavior, 41,* 1–19.

Dohrenwend, B. P., & Dohrenwend, B. S. (1974). Social and cultural influences on psychopathology. *Annual Review of Psychology, 25,* 417–452.

Dollard, J., Doob, J., Miller, N., Mowrer, O., & Sears, R. (1939). *Frustration and aggression.* New Haven, CT: Yale University Press.

Doms, M. (1984). The minority influence effect: an alternative approach. In S. Moscovici & W. Doise (Eds.), *Current issues in European social psychology* (pp. 1–31). Cambridge, UK: Cambridge University Press.

Doms, M., & Van Avermaet, E. (1980). Majority influence, minority influence, and conversion behavior: A replication. *Journal of Experimental Social Psychology, 16,* 283–293.

Donnerstein, E. (1984). Pornography: Its effect on violence against women. In N. M. Malamuth & E. Donnerstein (Eds.), *Pornography and sexual aggression.* Orlando, FL: Academic Press.

Donnerstein, E., & Barrett, G. (1978). The effects of erotic stimuli on male aggression toward females. *Journal of Personality and Social Psychology, 36,* 180–188.

Donnerstein, E., & Berkowitz, L. (1981). Victim reactions in aggressive erotic films as a factor in violence toward women. *Journal of Personality and Social Psychology, 41,* 710–724.

Donohue, W. A., Allen, M., & Burrell, N. (1988). Mediator communication competence. *Communication Monographs, 55,* 104–119.

Dooley, D., Prause, J., & Ham-Rowbottom, K. (2000). Underemployment and depression: Longitudinal relationships. *Journal of Health and Social Behavior, 41,* 421–436.

Dorfmann, L. (2002). Retirement and family relationships: An opportunity in later life. *Generations, 26,* 74–79.

Dovidio, J. F. (1984). Helping behavior and altruism: An empirical and conceptual overview. In L. Berkowitz (Ed.), *Advances in experimental social psychology* (Vol. 17, pp. 362–427). New York: Academic Press.

Dovidio, J. F., Allen, J. L., & Schroeder, D. A. (1990). Specificity of empathy-induced helping: Evidence for altruistic motivation. *Journal of Personality and Social Psychology, 59,* 249–260.

Dovidio, J., Brigham, J. C., Johnson, B., & Gaertner, S. (1996). Stereotyping, prejudice, and discrimination: Another look. In N. Macrea, M. Hewstone, & C. Stangor (Eds.), *Foundations of stereotypes and stereotyping.* New York: Guilford Press.

Dovidio, J. F., & Ellyson, S. L. (1982). Decoding visual dominance: Attributions of power based on relative percentages of looking while speaking and looking while listening. *Social Psychology Quarterly, 45,* 106–113.

Dovidio, J. F., & Gaertner, S. L. (1981). The effects of race, status, and ability on helping behavior. *Social Psychology Quarterly, 44,* 192–203.

Dovidio, J. F., & Gaertner, S. L. (1996). Affirmative action, unintentional racial biases, and intergroup relations. *Journal of Social Issues, 52*(4), 51–75.

Dovidio, J. F., Gaertner, S. L., Isen, A. M., & Lowrance, R. (1995). Group representations and intergroup bias: Positive affect, similarity, and group size. *Personality and Social Psychology Bulletin, 21,* 856–865.

Dovidio, J. F., Piliavin, J. A., Gaertner, S. L., Schroeder, D. A., & Clark, R. D., III (1991). The arousal/cost-reward model and the process of intervention: A review of the evidence. In M. S. Clark (Ed.), *Review of personality and social psychology: Vol. 12. Prosocial behavior* (pp. 86–118). Newbury Park, CA: Sage.

Downey, G. L. (1986). Ideology and the Clamshell identity: Organizational dilemmas in the anti-nuclear power movement. *Social Problems, 33,* 357–373.

Draghi-Lorenz, R., Reddy, V., & Costall, A. (2001). Rethinking the development of "nonbasic" emotions: A critical review of existing theories. *Developmental Review, 21,* 263–304.

Drass, K. A., & Spencer, J. W. (1987). Accounting for presentencing recommendations: Typologies and probation officers' theory of office. *Social Problems, 34,* 277–293.

Dreben, E. K., Fiske, S. T., & Hastie, R. (1979). The independence of evaluative and item information: Impression and recall order effects in behavior-based impression formation. *Journal of Personality and Social Psychology, 37,* 1758–1768.

Drew, P. (2003). Precision and exaggeration in interaction. *American Sociological Review, 68,* 917–938.

Drews, D., Allison, C., & Probst, J. (2000). Behavioral and self-concept differences in tattooed and nontattooed college students. *Psychological Reports, 86,* 475–481.

Dubé-Simard, L. (1983). Genesis of social categorization, threat to identity, and perceptions of social injustice: Their role in intergroup communication breakdown. *Journal of Language and Social Psychology, 2,* 183–206.

Duesterhoeft, D. (1987). *An unholy alliance? A case study comparison of religious and feminist anti-pornography activists.* Unpublished masters' thesis, University of Wisconsin.

Dugan, K. (2005). *The struggle over gay, lesbian, and bisexual rights: Facing off in Cincinnati.* New York: Routledge.

Duncan, S., Jr., & Fiske, D. W. (1977). *Face-to-face interaction: Research methods and theory.* Hillsdale, NJ: Erlbaum.

Duneier, M. (2001). *Sidewalk.* New York: Farrar, Straus, and Giroux.

Dunphy, D. (1972). *The primary group.* New York: Appleton-Century-Crofts.

Dura, J. A., & Kiecolt-Glaser, J. K. (1991). Family transitions, stress, and health. In D. A. Cowan & M. Hetherington (Eds.), *Family transitions.* Hillsdale, NJ: Erlbaum.

Duran, R. L., & Prusank, D. T. (1997). Relational themes in men's and women's popular nonfiction magazine articles. *Journal of Social and Personal Relationships, 14,* 165–189.

Dutton, D., & Aron, A. (1974). Some evidence for heightened sexual attraction under conditions of high anxiety. *Journal of Personality and Social Psychology, 30,* 510–517.

Dutton, D., & Lake, R. (1973). Threat of own prejudice and reverse discrimination in interracial situations. *Journal of Personality and Social Psychology, 28,* 94–100.

Dyck, R. J., & Rule, B. G. (1978). Effect on retaliation of causal attribution concerning attack. *Journal of Personality and Social Psychology, 36,* 521–529.

Dynes, R. R., & Quarantelli, E. L. (1980). Helping behavior in large-scale disasters. In D. H. Smith & J. Macaulay (Eds.), *Participation in social and political activities* (pp. 339–354). San Francisco: Jossey-Bass.

Eagly, A. (1987). *Sex differences in social behavior: A social-role interpretation.* Hillsdale, NJ: Erlbaum.

Eagly, A. H., Ashmore, R. D., Makhijani, M. G., & Longo, L. C. (1991). What is beautiful is good, but . . . : A meta-analytic review of research on the physical attractiveness stereotype. *Psychological Bulletin, 110,* 109–128.

Eagly, A. H., & Carli, L. L. (1981). Sex of researchers and sex-typed communications as determinants of sex differences in influenceability. *Psychological Bulletin, 90,* 1–20.

Eagly, A. H., & Chaiken, S. (1975). An attribution analysis of the effects of communicator characteristics on opinion change: The case of communicator attractiveness. *Journal of Personality and Social Psychology, 32,* 136–144.

Eagly, A. H., & Chrvala, C. (1986). Sex differences in conformity: Status and gender role interpretations. *Psychology of Women Quarterly, 10,* 203–220.

Eagly, A. H., & Crowley, M. (1986). Gender and helping behavior: A meta-analytic review of the social psychological literature. *Psychological Bulletin, 100,* 283–308.

Eagly, A., Karau, S., & Makhijani, M. (1995). Gender and the effectiveness of leaders: A meta-analysis. *Psychological Bulletin, 117,* 125–145.

Eagly, A., & Steffen, V. J. (1984). Gender stereotypes stem from distribution of women and men into social roles. *Journal of Personality and Social Psychology, 46,* 735–754.

Eagly, A. H., Wood, W., & Chaiken, S. (1978). Causal inferences about communicators and their effect on attitude change. *Journal of Personality and Social Psychology, 36,* 424–435.

East, P. L. (1998). Racial and ethnic differences in girls' sexual, marital, and birth expectations. *Journal of Marriage and the Family, 60,* 150–162.

Eaton, W. (1980). *The sociology of mental disorders.* New York: Praeger.

Ebbesen, E. B., & Bowers, R. J. (1974). Proportion of risky to conservative arguments in a group discussion and choice shift. *Journal of Personality and Social Psychology, 29,* 316–327.

Ebert, J. (2005). A broken heart harms your health. *Nature online.* Available at http://www.nature.com/news/2005/050207/full/050207-11.html

Eccles, J. S. (1987). Gender roles and women's achievement related decisions. *Psychology of Women Quarterly, 11,* 135–172.

Eckert, P., & McConnell-Ginet, S. (1999). New generalizations and explanations in language and gender research. *Language in Society, 28,* 185–201.

Eckes, T. (1995). Features of situations: A two-mode clustering study of situation prototypes. *Personality and Social Psychology Bulletin, 21,* 366–374.

Edelmann, R. J. (1985). Social embarrassment: An analysis of the process. *Journal of Social and Personal Relationships, 2,* 195–213.

Edelmann, R. J. (1987). *The psychology of embarrassment.* Chichester, UK: Wiley.

Edelmann, R. J., Evans, G., Pegg, I., & Tremain, M. (1983). Responses to physical stigma. *Perceptual and Motor Skills, 57,* 294.

Edelmann, R. J., & Iwawaki, S. (1987). Self-reported expression and consequences of embarrassment in the United Kingdom and Japan. *Psychologia, 30,* 205–216.

Eder, D. (with Evans, C. C., & Parker, S.). (1995). *School talk: Gender and adolescent culture.* New Brunswick, NJ: Rutgers University Press.

Editorial. (2004). True lies. *Nature, 428,* 679.

Edwards, C. P., Knoche, L., & Kumru, A. (2001). Play patterns and gender. In J. Worell (Ed.), *Encyclopedia of gender* (pp. 809–816). San Diego: Academic Press.

Edwards, K. (1990). The interplay of affect and cognition in attitude formation and change. *Journal of Personality and Social Psychology, 59,* 202–216.

Efran, M. G., & Cheyne, J. A. (1974). Affective concomitants of the invasion of shared space: Behavioral, physiological, and verbal indicators. *Journal of Personality and Social Psychology, 29,* 219–226.

Eibl-Eibesfeldt, I. (1979). Universals in human expressive behavior. In A. Wolfgang (Ed.), *Nonverbal behavior: Applications and cultural implications.* New York: Academic Press.

Eisenberg, N., & Fabes, R. A. (1991). Prosocial behavior and empathy: A multimethod developmental perspective. In M. S. Clark (Ed.), *Review of personality and social psychology: Vol. 12. Prosocial behavior* (pp. 34–61). Newbury Park, CA: Sage.

Eisenberg, N., Guthrie, I., Cumberland, A., Murphy, B. C., Shepard, S. A., Zhou, Q., & Carlo, G. (2002). Prosocial development in early adulthood: A longitudinal study. *Journal of Personality and Social Psychology, 82,* 993–1006.

Eisenberg, N., & Miller, P. A. (1987). The relation of empathy to prosocial and related behaviors. *Psychological Bulletin, 101,* 91–119.

Eiser, J. R. (Ed.). (1984). *Attitudinal judgment.* New York: Springer Verlag.

Eitle, D., & Turner, R. J. (2003). Stress exposure, race, and young adult male crime. *The Sociological Quarterly, 44,* 243–269.

Ekman, P. (1972). Universals and cultural differences in facial expression of emotion. In J. K. Cole (Ed.), *Nebraska symposium on motivation, 1971.* Lincoln: Nebraska University Press.

Ekman, P. (1984). Expression and nature of emotion. In K. Scherer & P. Ekman (Eds.), *Approaches to emotion.* Hillsdale, NJ: Erlbaum.

Ekman, P. (1992). An argument for basic emotions. In N. L. Stein & K. Oatley (Eds.), *Basic emotions* (pp. 169–200). Hillsdale, NJ: Erlbaum.

Ekman, P. (1993). Facial expression and emotion. *American Psychologist, 48,* 384–392.

Ekman, P., & Friesen, W. V. (1969). Nonverbal leakage and clues to deception. *Psychiatry, 32,* 88–106.

Ekman, P., & Friesen, W. V. (1974). Detecting deception from the body or face. *Journal of Personality and Social Psychology, 29,* 288–298.

Ekman, P., & Friesen, W. V. (1975). *Unmasking the face.* Englewood Cliffs, NJ: Prentice–Hall.

Ekman, P., Friesen, W. V., & O'Sullivan, M. (1988). Smiles when lying. *Journal of Personality and Social Psychology, 54,* 414–420.

Ekman, P., Friesen, W. V., & Scherer, K. R. (1976). Body movements and voice pitch in deceptive interaction. *Semiotica, 16,* 23–27.

Ekman, P., Friesen, W. V., & Tomkins, S. S. (1971). Facial affect scoring technique (FAST): A first validity study. *Semiotica, 3,* 37–58.

Ekman, P., Friesen, W. V., O'Sullivan, M., Chan, A., et al. (1987). Universals and cultural differences in the judgments of facial expressions of emotion. *Journal of Personality and Social Psychology, 53,* 712–717.

Ekman, P., & O'Sullivan, M. (1991). Who can catch a liar? *American Psychologist, 46,* 913–920.

Ekman, P., O'Sullivan, M., & Frank, M. (1999). A few can catch a liar. *Psychological Science, 10,* 263–266.

Ekman, P., Sorenson, E. R., & Friesen, W. V. (1969). Pan cultural elements in facial displays of emotions. *Science, 164,* 86–88.

Elder, G. H., Jr. (1975). Age differentiation and the life course. In A. Inkeles, J. Coleman, & N. Smelser (Eds.), *Annual Review of Sociology* (Vol. 1). Palo Alto, CA: Annual Reviews.

Elder, G. H., Jr., & O'Rand, A. M. (1995). Adult lives in a changing society. In K. S. Cook, G. A. Fine, & J. S. House (Eds.), *Sociological perspectives on social psychology* (pp. 452–475). Needham Heights, MA: Allyn & Bacon.

Elkin, F., & Handel, G. (1989). *The child and society: The process of socialization* (5th ed.). New York: Random House.

Elkin, R. A., & Leippe, M. (1986). Physiological arousal, dissonance, and attitude change: Evidence for a dissonance-arousal link and a "Don't remind me" effect. *Journal of Personality and Social Psychology, 51,* 55–65.

Ellard, J. H., & Bates, D. D. (1990). Evidence for the role of the justice motive in status generalization processes. *Social Justice Research, 4*(2), 115–134.

Ellemers, N., Van Rijswijk, W., Roefs, M., & Simons, C. (1997). Bias in intergroup perceptions: Balancing group identity with social reality. *Personality and Social Psychology Bulletin, 23,* 186–198.

Elliott, G. C., & Meeker, B. F. (1986). Achieving fairness in the face of competing concerns: The different effects of individual and group characteristics. *Journal of Personality and Social Psychology, 50,* 754–760.

Ellsworth, P. C., Carlsmith, J. M., & Henson, A. (1972). The stare as a stimulus to flight in human subjects. *Journal of Personality and Social Psychology, 21,* 302–311.

Emirbayer, M., & Mische, A. (1998). What is agency? *American Journal of Sociology, 103,* 962–1023.

Emmons, R. A., & Diener, E. (1986). Situation selection as a moderator of response consistency and stability. *Journal of Personality and Social Psychology, 51,* 1013–1019.

Emmons, R. A., Diener, E., & Larsen, R. J. (1986). Choice and avoidance of everyday situations and affect congruence: Two models of reciprocal interactionism. *Journal of Personality and Social Psychology, 51,* 815–826.

Emswiller, T., Deaux, K., & Willis, J. E. (1971). Similarity, sex, and requests for small favors. *Journal of Applied Social Psychology, 1,* 284–291.

Epstein, J. L. (1985). After the bus arrives: Resegregation in desegregated schools. *Journal of Social Issues, 41*(3), 23–44.

Erikson, E. H. (1968). *Identity: Youth and crisis.* New York: Norton.

Erikson, K. (1964). Notes on the sociology of deviance. In H. Becker (Ed.), *The other side.* New York: Free Press.

Erikson, K. (1966). *The wayward Puritans.* New York: Wiley.

Etaugh, C., & Bridges, J. (2001). Midlife transitions. In J. Worell (Ed.), *Encyclopedia of gender* (pp. 759–770). San Diego: Academic Press.

Ettinger, R. F., Marino, C. J., Endler, N. S., Geller, S. H., & Natziuk, T. (1971). Effects of agreement and correctness on relative competence and conformity. *Journal of Personality and Social Psychology, 19,* 204–212.

Etzioni, A. (1967). The Kennedy experiment. *Western Political Quarterly, 20,* 361–380.

Evans, C. R., & Dion, K. L. (1991). Group cohesion and performance: A meta-analysis. *Small Group Research, 22,* 175–186.

Evans, N. J., & Jarvis, P. A. (1980). Group cohesion: A review and reevaluation. *Small Group Behavior, 11,* 359–370.

Faley, T., & Tedeschi, J. T. (1971). Status and reactions to threats. *Journal of Personality and Social Psychology, 17,* 192–199.

Farhar-Pilgrim, B., & Shoemaker, F. F. (1981). Campaigns to affect energy behavior. In R. E. Rice & W. J. Paisley (Eds.), *Public communication campaigns.* Beverly Hills, CA: Sage.

Farina, A., Gliha, D., Boudreau, L. A., Allen, J. G., & Sherman, M. (1971). Mental illness and the impact of believing others know it. *Journal of Abnormal Psychology, 77,* 1–5.

Farnworth, M., & Leiber, M. J. (1989). Strain theory revisited: Economic goals, educational means, and delinquency. *American Sociological Review, 54,* 263–274.

Fazio, R. H. (1990). Multiple processes by which attitudes guide behavior: The MODE model as an integrative framework. In L. Berkowitz (Ed.), *Advances in experimental social psychology* (Vol. 23, pp. 75–109). New York: Academic Press.

Fazio, R. H., Powell, M., & Herr, P. (1983). Toward a process model of the attitude-behavior relation: Accessing one's attitude upon mere observation of the attitude object. *Journal of Personality and Social Psychology, 44,* 723–735.

Fazio, R. H., Sanbonmatsu, D. M., Powell, M. C., & Kardes, F. R. (1986). On the automatic activation of attitudes. *Journal of Personality and Social Psychology, 50,* 229–238.

Fazio, R. H., & Williams, C. J. (1986). Attitude accessibility as a moderator of the attitude-perception and attitude-behavior relations: An investigation of the 1984 presidential election. *Journal of Personality and Social Psychology, 51,* 505–514.

Fazio, R. H., & Zanna, M. (1981). Direct experience and attitude-behavior consistency. In L. Berkowitz (Ed.), *Advances in experimental social psychology* (Vol. 14). New York: Academic Press.

Feather, N. T. (1967). A structural balance approach to the analysis of communication effects. In L. Berkowitz (Ed.), *Advances in experimental social psychology* (Vol. 3). New York: Academic Press.

Feather, N. T. (1995). Values, valences, and choice: The influence of values on the perceived attractiveness and choice of alternatives. *Journal of Personality and Social Psychology, 68,* 1135–1151.

Feather, N. T., & Armstrong, D. J. (1967). Effects of variations in source attitude, receiver attitude, and communication stand on reactions to source and contents of communications. *Journal of Personality, 35,* 435–455.

Feeney, J. A., & Noller, P. (1990). Attachment style as a predictor of adult romantic relationships. *Journal of Personality and Social Psychology, 58,* 281–291.

Fejfar, M. C., & Hoyle, R. H. (2000). Effect of private self-awareness on negative affect and self-referent attribution. *Personality and Social Psychology Review, 4,* 132–142.

Feld, S. (1981). The focused organization of social ties. *American Journal of Sociology, 86,* 1015–1035.

Felipe, N. J., & Sommer, R. (1966). Invasions of personal space. *Social Problems, 14,* 206–214.

Felmlee, D., Sprecher, S., & Bassin, E. (1990). The dissolution of intimate relationships: A hazard model. *Social Psychology Quarterly, 53,* 13–30.

Felson, M. (1994). *Crime and everyday life: Insights and implications for society.* Thousand Oaks, CA: Pine Forge Press.

Felson, R. B. (1981). Ambiguity and bias in the self-concept. *Social Psychology Quarterly, 44,* 64–69.

Felson, R. B. (1985). Reflected appraisal and the development of self. *Social Psychology Quarterly, 48,* 71–78.

Felson, R. B. (1989). Parents and the reflected appraisal process: A longitudinal analysis. *Journal of Personality and Social Psychology, 56,* 965–971.

Felson, R. B., Liska, A. E., South, S. J., & McNulty, T. L. (1994). The subculture of violence and delinquency: Individual vs. school context effects. *Social Forces, 73,* 155–173.

Felson, R. B., & Reed, M. (1986). The effects of parents on the self-appraisals of children. *Social Psychology Quarterly, 49,* 302–308.

Felson, R. B., & Zielinski, M. A. (1989). Children's self-esteem and parental support. *Journal of Marriage and the Family, 51,* 727–735.

Fernandez-Dols, J. M., & Ruiz-Belda, M.-A. (1995). Are smiles a sign of happiness? Gold medal winners at the Olympic Games. *Journal of Personality and Social Psychology, 69,* 1113–1119.

Ferree, M. M., & Miller, F. D. (1985). Mobilization and meaning: Toward an integration of social psychological and resource perspectives on social movements. *Sociological Inquiry, 55,* 38–55.

Festinger, L. (1954). A theory of social comparison processes. *Human Relations, 7,* 117–140.

Festinger, L. (1957). *A theory of cognitive dissonance.* Stanford, CA: Stanford University Press.

Festinger, L., & Carlsmith, J. (1959). Cognitive consequences of forced compliance. *Journal of Abnormal and Social Psychology, 58,* 203–210.

Festinger, L., Pepitone, A., & Newcomb, T. (1952). Some consequences of de-individuation in a group. *Journal of Abnormal and Social Psychology, 47,* 382–389.

Festinger, L., Schachter, S., & Back, K. W. (1950). *Social pressures in informal groups.* New York: Harper & Row.

Fiedler, F. E. (1966). The effect of leadership and cultural heterogeneity on group performance: A test of the contingency model. *Journal of Experimental Social Psychology, 2,* 237–264.

Fiedler, F. E. (1978a). Recent developments in research on the contingency model. In L. Berkowitz (Ed.), *Group processes.* New York: Academic Press.

Fiedler, F. E. (1978b). The contingency model and the dynamics of the leadership process. In L. Berkowitz (Ed.), *Advances in experimental social psychology* (Vol. 11). New York: Academic Press.

Fiedler, F. E. (1981). Leadership effectiveness. *American Behavioral Scientist, 24,* 619–632.

Fields, J. (2003). *America's families and living arrangements: 2003.* Washington, DC: U.S. Bureau of the Census, Current Population Reports, P20-53.

Fife, B., & Wright, E. (2000). The dimensionality of stigma: A comparison of its impact on the self of persons with HIV/AIDS and cancer. *Journal of Health and Social Behavior, 41,* 50–67.

Fine, G. A. (2000). Games and truths: Learning to construct social problems in high school debate. *The Sociological Quarterly, 41,* 103–123.

Fink, E. L., Kaplowitz, S. A., & Bauer, C. L. (1983). Positional discrepancy, psychological discrepancy, and attitude change: Experimental tests of some mathematical models. *Communication Monographs, 50,* 413–430.

Finkel, S. E. (1993). Reexamining the 'minimal effects' model in recent presidential elections. *Journal of Politics, 55,* 1–21.

Fischer, C. S. (1976). *The urban experience.* New York: Harcourt Brace Jovanovich.

Fischer, C. S. (1984). *The urban experience* (2nd ed.). San Diego: Harcourt Brace Jovanovich.

Fisek, M. H. (1974). A model for the evolution of status structures in task-oriented discussion groups. In J. Berger, T. L. Conner, & M. H. Fisek (Eds.), *Expectation states theory.* Cambridge, MA: Winthrop.

Fishbein, M., & Ajzen, I. (1975). *Belief, attitude, intention, and behavior.* Reading, MA: Addison-Wesley.

Fisher, J., Nadler, D., & Whitcher-Alagna, S. (1982). Recipient reactions to aid. *Psychological Bulletin, 91,* 33–54.

Fishman, P. M. (1980). Conversational insecurity. In H. Giles & W. P. Robinson (Eds.), *Language: Social psychological perspectives.* New York: Pergamon.

Fishman, P. (1983). Interaction: The work women do. In B. Thorne, C. Kramerae, & N. Henley (Eds.), *Language, gender, and society* (pp. 89–101). Rowley, MA: Newbury House.

Fisicaro, S. A. (1988). A reexamination of the relation between halo error and accuracy. *Journal of Applied Psychology, 73,* 239–244.

Fiske, S. T. (1998). Stereotyping, prejudice, and discrimination. In D. T. Gilbert, S. T. Fiske, & G. Lindzey

(Eds.), *Handbook of social psychology* (4th ed.). New York: McGraw-Hill.

Fiske, S. T., Kinder, D. R., & Larter, W. M. (1983). The novice and the expert: Knowledge-based strategies in political cognition. *Journal of Experimental Social Psychology, 19,* 381–400.

Fiske, S. T., & Linville, P. (1980). What does the schema concept buy us? *Personality and Social Psychology Bulletin, 6,* 543–557.

Fiske, S. T., & Taylor, S. E. (1991). *Social cognition* (2nd ed.). New York: McGraw-Hill.

Flavell, J., Shipstead, S., & Croft, K. (1978). *What young children think you see when their eyes are closed.* Unpublished report, Stanford University.

Fleming, J. H., Darley, J. M., Hilton, J. L., & Kojetin, B. A. (1991). Multiple audience problem: A strategic communication perspective on social perception. *Journal of Personality and Social Psychology, 58,* 593–609.

Flora, J., & Segrin, C. (2000). Relationship development in dating couples: Implications for relational satisfaction and loneliness. *Journal of Social and Personal Relationships, 17,* 811–825.

Flowers, M. L. (1977). A laboratory test of some implications of Janis' groupthink hypothesis. *Journal of Personality and Social Psychology, 35,* 888–896.

Fodor, E. M., & Smith, T. (1982). The power motive as an influence on group decision making. *Journal of Personality and Social Psychology, 42,* 178–185.

Forbes, G. (2001). College students with tattoos and piercings: Motives, family experiences, personality factors, and perception by others. *Psychological Reports, 89,* 774–786.

Fordham, S. (1999). Dissin' "the standard": Ebonics as guerrilla warfare at Capital High. *Anthropology and Education Quarterly, 30,* 272–293.

Forgas, J. P., & Bond, M. H. (1985). Cultural influences on the perception of interaction episodes. *Personality and Social Psychology Bulletin, 11,* 75–88.

Form, W. H., & Nosow, S. (1958). *Community in disaster.* New York: Harper.

Forsyth, D. R. (1999). *Group dynamics* (3rd ed.). Belmont, CA: Wadsworth.

Forsyth, D. R., Berger, R. E., & Mitchell, T. (1981). The effects of self-serving vs. other-serving claims of responsibility on attraction and attribution in groups. *Social Psychology Quarterly, 44,* 59–64.

Forthofer, M. S., Markman, H. J., Cox, M., Stanley, S., & Kessler, R. C. (1996). Associations between marital distress and work loss in a national sample. *Journal of Marriage and the Family, 58,* 597–605.

Foster, C., & Campbell, W. K. (2005). The adversity of secret relationships. *Personal Relationships, 12,* 125–143.

Foster, M. D., & Matheson, K. (1995). Double relative deprivation: Combining the personal and political. *Personality and Social Psychology Bulletin, 21,* 1167–1177.

Fox, G. L., Benson, M. L., DeMaris, A. A., & Van Wyk, J. (2002). Economic distress and intimate violence: Testing family stress and resources theories. *Journal of Marriage and the Family, 64,* 793–807.

Francis, L. (1997). Ideology and interpersonal emotion management: Redefining identity in two support groups. *Social Psychology Quarterly, 60,* 153–171.

Franiuk, R., Cohen, D., & Pomerantz, E. (2002). Implicit theories of relationships: Implications for relationship satisfaction and longevity. *Personal Relationships, 9,* 345–367.

Frank, F., & Anderson, L. R. (1971). Effects of task and group size upon group productivity and member satisfaction. *Sociometry, 34,* 135–149.

Franks, D., & Marolla, J. (1976). Efficacious action and social approval as interacting dimensions of self-esteem. *Sociometry, 39,* 324–341.

Fraser, C., Gouge, C., & Billig, M. (1971). Risky shifts, cautious shifts, and group polarization. *European Journal of Social Psychology, 1,* 7–30.

Fredricks, A., & Dossett, D. (1983). Attitude-behavior relations: A comparison of the Fishbein-Ajzen and the Bentler-Speckart models. *Journal of Personality and Social Psychology, 45,* 501–512.

Freedman, J., & Fraser, S. (1966). Compliance without pressure: The foot-in-the-door technique. *Journal of Personality and Social Psychology, 4,* 195–202.

Freedman, J. L., & Sears, D. O. (1965). Warning, distraction, and resistance to influence. *Journal of Personality and Social Psychology, 1,* 262–266.

Freese, L. (1976). The generalization of specific performance expectations. *Sociometry, 39,* 194–200.

Freese, L., & Cohen, B. P. (1973). Eliminating status generalization. *Sociometry, 36,* 177–193.

French, J. R. P., & Raven, B. (1959). Bases of social power. In D. Cartwright (Ed.), *Studies in social power.* Ann Arbor: University of Michigan.

Freud, S. (1905). Fragment of an analysis of a case of hysteria. In *Collected papers* (Vol. 3). New York: Basic Books.

Freud, S. (1930). *Civilization and its discontents.* London: Hogarth Press.

Freud, S. (1950). Why war? In J. Strachey (Ed.), *Collected papers* (Vol. 5). London: Hogarth Press.

Frey, D. (1978). Reactions to success and failure in public and private conditions. *Journal of Experimental Social Psychology, 14,* 172–179.

Frey, K. S., & Ruble, D. N. (1985). What children say when the teacher is not around: Conflicting goals in social comparison and performance assessment in the classroom. *Journal of Personality and Social Psychology, 48,* 550–562.

Friedrich-Cofer, L., & Huston, A. C. (1986). Television violence and aggression: The debate continues. *Psychological Bulletin, 100,* 364–371.

Fries, A., & Pollak, S. (2004). Emotion understanding in postinstitutionalized Eastern European children. *Development and Psychopathology, 16,* 355–369.

Frieze, I. H., Parsons, J., Johnson, P., Ruble, D., & Zellman, G. (1978). *Women and sex roles: A social psychological perspective.* New York: Norton.

Frieze, I., & Weiner, B. (1971). Cue utilization and attributional judgments for success and failure. *Journal of Personality, 39,* 591–605.

Frijda, N. H., & Mequita, B. (1994). The social roles and functions of emotions. In S. Kitayama & H. Markus (Eds.), *Emotions and culture: Empirical studies of mutual influence* (pp. 51–87). Washington, DC: American Psychological Association.

Fritzsche, B. A., Finkelstein, M. A., & Penner, L. A. (2000). To help or not to help: Capturing individuals' decision policies. *Social Behavior and Personality, 28,* 561–578.

Funk, J., Flores, G., Buchman, D., & Germann, J. (1999). Rating electronic games: Violence is in the eye of the beholder. *Youth and Society, 30,* 283–312.

Furstenberg, F., Kennedy, S., McLoyd, V., Rumbaut, R., & Setersten, R. Growing up is harder to do. *Contexts, 3*(3), 33–41.

Furstenberg, F., & Kiernan, K. (2001). Delayed parental divorce: How much do children suffer? *Journal of Marriage and the Family, 63,* 446–457.

Gaertner, S. L., Dovidio, J. F., Anastasio, P. A., Bachman, B. A., & Rust, M. C. (1993). The common ingroup identity model: Recategorization and the reduction of intergroup bias. In W. Stroebe & H. Hewstone (Eds.), *European Review of Social Psychology, 4,* 1–26.

Gaertner, S. L., Dovidio, J. F., Rust, M. C., Nier, J. A., Banker, B., Ward, C. M., et al. (1999). Reducing intergroup bias: Elements of intergroup cooperation. *Journal of Personality and Social Psychology, 76,* 388–402.

Gaertner, S. L., Mann, J., Murrell, A., & Dovidio, J. F. (1989). Reducing intergroup bias: The benefits of recategorization. *Journal of Personality and Social Psychology, 57,* 239–249.

Gahagan, J. P., & Tedeschi, J. T. (1968). Strategy and the credibility of promises in the Prisoner's Dilemma game. *Journal of Conflict Resolution, 12,* 224–234.

Galton, M. (1987). An ORACLE chronicle: A decade of classroom research. *Teaching and Teacher Education, 3,* 299–313.

Gamson, W. A. (1992). Social psychology of collective action. In A. D. Morris & C. M. Mueller (Eds.), *Frontiers of social movement theory.* New Haven, CT: Yale University Press.

Garcia, S. M., Weaver, K., Darley, J. M., & Moskowitz, G. B. (2002). Crowded minds: The implicit bystander effect. *Journal of Personality and Social Psychology, 83,* 843–853.

Gardner, L., & Shoemaker, D. J. (1989). Social bonding and delinquency: A comparative analysis. *The Sociological Quarterly, 30,* 481–500.

Garfinkel, H. (1956). Conditions of successful degradation ceremonies. *American Sociological Review, 61,* 420–424.

Garfinkel, I., & McLanahan, S. S. (1986). *Single mothers and their children: A new American dilemma.* Washington, DC: Urban Institute.

Gartner, R. (1990). The victims of homicide: A temporal and cross-national comparison. *American Sociological Review, 55,* 92–106.

Gaskell, G., & Smith, P. (1986). Group membership and social attitudes of youth: An investigation of some implications of social identity theory. *Social Behaviour, 1,* 67–77.

Gates, G. S. (1924). The effect of an audience upon performance. *Journal of Abnormal and Social Psychology, 18,* 334–342.

Gauthier, A., & Furstenberg, F., Jr. (2002). *Historical trends in the patterns of time use of young adults in the United States and Canada.* Paper presented at the International Conference on Time Use, Waterloo, Ontario, Canada.

Gawronski, B., & Strack, F. (2004). On the propositional nature of cognitive consistency: Dissonance changes explicit, but not implicit attitudes. *Journal of Experimental Social Psychology, 40,* 535–542.

Gecas, V. (1990). Contexts of socialization. In M. Rosenberg & R. Turner (Eds.), *Social psychology: Sociological perspectives.* New Brunswick, NJ: Transaction.

Gecas, V., & Burke, P. J. (1995). Self and identity. In K. S. Cook, G. A. Fine, & J. S. House (Eds.), *Sociological perspectives on social psychology* (pp. 41–67). Needham Heights, MA: Allyn & Bacon.

Gecas, V., & Schwalbe, M. (1983). Beyond the looking-glass self: Social structure and efficacy-based self-esteem. *Social Psychology Quarterly, 46,* 77–88.

Geen, R. G. (1968). Effects of frustration, attack, and prior training in aggressiveness upon aggressive behavior. *Journal of Personality and Social Psychology, 9,* 316–321.

Geen, R. G., & Quanty, M. G. (1977). The catharsis of aggression: An analysis of a hypothesis. In L. Berkowitz (Ed.), *Advances in experimental social psychology* (Vol. 10). New York: Academic Press.

Geen, R. G., Stonner, L., & Shope, G. L. (1975). The facilitation of aggression by aggression: A study in response inhibition and disinhibition. *Journal of Personality and Social Psychology, 31,* 721–726.

Geis, M. L. (1995). *Speech acts and conversational interaction.* New York: Cambridge University Press.

Gelfand, D. M., & Hartmann, D. P. (1982). Response consequences and attributions: Two contributors to prosocial behavior. In N. Eisenberg (Ed.), *The development of prosocial behavior.* New York: Academic Press.

Geller, V. (1977). *The role of visual access in impression management and impression formation.* Unpublished doctoral dissertation, Columbia University, New York.

Gelles, R. J., & Cornell, C. P. (1990). *Intimate violence in families* (2nd ed.). Newbury Park, CA: Sage.

Gelles, R. J., & Strauss, M. A. (1988). *Intimate violence.* New York: Simon & Schuster.

Gerard, H. B. (1983). School desegregation: The social science role. *American Psychologist, 38,* 869–877.

Gerard, H. B., Wilhelmy, R. A., & Conolley, E. S. (1968). Conformity and group size. *Journal of Personality and Social Psychology, 8,* 79–82.

Gergen, K. J., Ellsworth, P., Maslach, C., & Siepel, M. (1975). Obligation, donor resources, and reactions to aid in three cultures. *Journal of Personality and Social Psychology, 31,* 390–400.

Gergen, K. J., Gergen, M. M., & Meter, K. (1972). Individual orientation to prosocial behavior. *Journal of Social Issues, 28*(1), 105–130.

Gergen, K. J., Morse, S. J., & Bode, K. A. (1974). Overpaid or overworked? Cognitive and behavioral reactions to inequitable rewards. *Journal of Applied Social Psychology, 4,* 259–274.

Gesell, A., & Ilg, F. (1943). *Infant and child in the culture of today.* New York: Harper & Row.

Ghaziani, A. (2004). Anticipatory and actualized identities: A cultural analysis of the transition from AIDS disability to work. *The Sociological Quarterly, 45,* 273–301.

Gibbons, F. X. (1990). Self-evaluation and self-perception: The role of attention in the experience of anxiety. *Anxiety Research, 2,* 153–163.

Gibbons, F. X., Smith, T. W., Ingram, R. E., Pearce, K., Brehm, S. S., & Schroeder, D. J. (1985). Self-awareness and self-confrontation: Effects of self-focussed attention on members of a clinical population. *Journal of Personality and Social Psychology, 48,* 662–675.

Gifford, R. (1982). Projected interpersonal distance and orientation choices: Personality, sex, and social situation. *Social Psychology Quarterly, 45,* 145–152.

Gilbert, T. F. (1978). *Human competence.* New York: McGraw-Hill.

Gilbert, D. T., & Malone, P. S. (1995). The correspondence bias. *Psychological Bulletin, 117,* 21–38.

Giles, H. (1980). Accommodation theory: Some new directions. In S. deSilva (Ed.), *Aspects of linguistic behavior.* York, UK: York University Press.

Giles, H., & Coupland, N. (1991). *Language: Contexts and consequences.* Pacific Grove, CA: Brooks/Cole.

Giles, H., Hewstone, M., & St. Clair, R. (1981). Speech as an independent and dependent variable of social situations: An introduction and new theoretical framework. In H. Giles & R. St. Clair (Eds.), *The social psychological significance of speech.* Hillsdale, NJ: Erlbaum.

Gill, V. T., & Maynard, D. W. (1995). On "labeling" in actual interactions: Delivering and receiving diagnoses of developmental disabilities. *Social Problems, 42,* 11–37.

Gilligan, C. (1982). *In a different voice.* Cambridge, MA: Harvard University Press.

Ginorio, A., Gutierrez, L., Cauce, A. M., & Acosta, M. (1995). Psychological issues for Latinas. In H. Landrine (Ed.), *Bringing cultural diversity to feminist psychology: Theory, research, and practice* (pp. 241–263). Washington, DC: American Psychological Association.

Giordano, P., Cernkovich, S., & Rudolph, J. (2002). Gender, crime, and desistance: Toward a theory of cognitive transformation. *American Journal of Sociology, 107,* 990–1064.

Glaser, B. G., & Strauss, A. (1971). *Status passage: A formal theory.* Chicago: Aldine.

Glass, J., Bengston, V. L., & Dunham, C. (1986). Attitude similarity in three-generation families: Status inheritance or reciprocal influence? *American Sociological Review, 51,* 685–698.

Glick, P. C. (1997). Demographic pictures of African-American families. In H. P. McAdoo (Ed.), *Black families* (3rd ed., pp. 118–137). Thousand Oaks, CA: Sage.

Global Forum for Health Research. (2002). *Mapping a global pandemic: Review of current literature on rape, sexual assault, and sexual harassment of women.* Retrieved April 1, 2002, from http://www.globalforumhealth.org

Goethals, G. R., & Zanna, M. P. (1979). The role of social comparison in choice shift. *Journal of Personality and Social Psychology, 37,* 1469–1476.

Goetting, A. (1986). The developmental tasks of siblingship over the life cycle. *Journal of Marriage and the Family, 48,* 703–714.

Goffman, E. (1952). Cooling the mark out: Some adaptations to failure. *Psychiatry, 15,* 451–463.

Goffman, E. (1959a). The moral career of the mental patient. *Psychiatry, 22,* 125–169.

Goffman, E. (1959b). *The presentation of self in everyday life.* Garden City, NY: Anchor/Doubleday.

Goffman, E. (1963). *Behavior in public places.* New York: Free Press.

Goffman, E. (1963). *Stigma: Notes on the management of spoiled identity.* Englewood Cliffs, NJ: Spectrum/Prentice Hall.

Goffman, E. (1967). *Interaction ritual.* Chicago: Aldine.

Goffman, E. (1974). *Frame analysis.* New York: Harper & Row.

Goffman, E. (1983). Felicity's condition. *American Journal of Sociology, 89,* 1–53.

Goldberg, C. (1974). Sex roles, task competence, and conformity. *Journal of Psychology, 86,* 157–164.

Goldberg, C. (1975). Conformity to majority type as a function of task and acceptance of sex-related stereotypes. *Journal of Psychology, 89,* 25–37.

Goldberg, L. R. (1981). Language and individual differences: The search for universals in personality lexicons. In L. Wheeler (Ed.), *Review of personality and social psychology* (Vol. 1, pp. 203–234). Hillsdale, NJ: Erlbaum.

Goldin, C., & Katz, L. (2002). The power of the pill: Oral contraceptives and women's career and marriage decisions. *Journal of Political Economy, 110,* 730–770.

Goldstein, B., & Oldham, J. (1979). *Children and work: A study of socialization.* New Brunswick, NJ: Transaction.

Goleman, D. (1995). *Emotional intelligence: Why it can matter more than IQ.* New York: Bantam Books.

Gonos, G. (1977). "Situation" vs. "frame": The "interactionist" and the "structuralist" analysis of everyday life. *American Sociological Review, 42,* 854–867.

Gonzales, M. H., Davis, J. M., Loney, G. L., Lukens, C. K., & Junghans, C. M. (1983). Interactional approach to interpersonal attraction. *Journal of Personality and Social Psychology, 44,* 1192–1197.

Gonzales, M., & Meyers, S. (1993). "Your mother would like me": Self-presentation in the personals ads of heterosexual and homosexual men and women. *Personality and Social Psychology Bulletin, 19,* 131–142.

Goodchilds, J. D., & Zellman, G. L. (1984). Sexual signaling and sexual aggression in adolescent relationships. In N. M. Malamuth & E. Donnerstein (Eds.), *Pornography and sexual aggression.* Orlando, FL: Academic Press.

Gooden, A. M., & Gooden, M. A. (2001). Gender representation in notable children's picture books: 1995–1999. *Sex Roles, 45,* 89–101.

Goodman, P. S., & Friedman, A. (1969). An examination of quantity and quality of performance under conditions of overpayment in piece rate. *Organizational Behavior and Human Performance, 4,* 365–374.

Goodman, P. S., & Friedman, A. (1971). An examination of Adams' theory of inequity. *Administrative Science Quarterly, 16,* 271–288.

Goodwin, C. (1987). Forgetfulness as an interactive resource. *Social Psychology Quarterly, 50,* 115–131.

Goodwin, J. (1997). The libidinal constitution of a high-risk social movement: Affectual ties and the Huk Rebellion, 1946 to 1954. *American Sociological Review, 62,* 53–69.

Goodwin, R. (1991). A re-examination of Rusbult's responses to dissatisfaction typology. *Journal of Social and Personal Relationships, 8,* 569–574.

Goodwin, S. A., Gubin, A., Fiske, S. T., & Yzerbyt, V. Y. (2000). Power can bias impression processes: Stereotyping subordinates by default and by design. *Group Processes & Intergroup Relations, 3,* 227–256.

Gordon, C. (1968). Self-conceptions: Configurations of content. In C. Gordon & K. J. Gergen (Eds.), *The self in social interaction I: Classic and contemporary perspectives* (pp. 115–136). New York: Wiley.

Gordon, P. (2004). Numerical cognition without words: Evidence from Amazonia. *Science, 306,* 496–499.

Gordon, R. A. (1996). Impact of ingratiation on judgments and evaluations: A meta-analytic investigation. *Journal of Personality and Social Psychology, 71,* 54–70.

Gordon, S. L. (1990). The sociology of sentiments and emotion. In M. Rosenberg & R. H. Turner (Eds.), *Social psychology: Sociological perspectives.* New Brunswick, NJ: Transaction.

Gorfein, D. S. (1964). The effects of a nonunanimous majority on attitude change. *Journal of Social Psychology, 63,* 333–338.

Gottlieb, J., & Carver, C. S. (1980). Anticipation of future interaction and the bystander effect. *Journal of Experimental Social Psychology, 16,* 253–260.

Gould, R. L. (1978). *Transformations.* New York: Simon & Schuster.

Gove, W., Hughes, M., & Geerken, M. R. (1980). Playing dumb: A form of impression management with undesirable effects. *Social Psychology Quarterly, 43,* 89–102.

Gowan, M., & Treviño, M. (1998). An examination of gender differences in Mexican-American attitudes toward family and career roles. *Sex Roles, 38,* 1079–1094.

Gowen, C. R. (1986). Managing work group performance by individual goals and group goals for an interdependent group task. *Journal of Organizational Behavior Management, 7*(3–4), 5–27.

Graham, L., & Hogan, R. (1990). Social class and tactics: Neighborhood opposition to group homes. *The Sociological Quarterly, 31,* 513–529.

Graham, T., & Ickes, W. (1997). When women's intuition isn't greater than men's. In W. Ickes (Ed.), *Empathic accuracy* (pp. 117–143). New York: Guilford Press.

Gramling, R., & Forsyth, C. J. (1987). Exploiting stigma. *Sociological Forum, 2,* 401–415.

Grammar, K., Kruck, K., Juette, A., & Fink, B. (2000). Non-verbal control as courtship signals: The role of control and choice in selecting partners. *Evolution and Human Behavior, 21,* 371–390.

Granberg, D. (1978). GRIT in the final quarter: Reversing the arms race through unilateral initiatives. *Bulletin of Peace Proposals, 9,* 210–221.

Granberg, D., & Holmberg, S. (1990). The intention-behavior relationship among U.S. and Swedish voters. *Social Psychology Quarterly, 53,* 44–54.

Granovetter, M. S. (1973). The strength of weak ties. *American Journal of Sociology, 78,* 1360–1380.

Grant, P. R., & Holmes, J. G. (1981). The integration of implicit personality schemas and stereotype images. *Social Psychology Quarterly, 44,* 107–115.

Grasmick, H. G., & Bryjak, G. J. (1980). The deterrent effect of perceived severity of punishment. *Social Forces, 59,* 471–491.

Gray, J. (1993). *Men are from Mars, women are from Venus: A practical guide for improving communication and getting what you want in your relationships.* New York: HarperCollins.

Gray-Little, B., & Hafdahl, A. (2000). Factors influencing racial comparisons of self-esteem: A quantitative review. *Psychological Bulletin, 126,* 26–54.

Grayshon, M. C. (1980). Social grammar, social psychology, and linguistics. In H. Giles, W. P. Robinson, & P. M. Smith (Eds.), *Language: Social psychological perspectives.* New York: Pergamon.

Green, A. (2002, July 13). Teen-agers are misunderstood. *Wisconsin State Journal,* p. D1.

Green, J. A. (1972). Attitudinal and situational determinants of intended behavior toward Blacks. *Journal of Personality and Social Psychology, 22,* 13–17.

Greenbaum, P., & Rosenfeld, H. (1978). Patterns of avoidance in response to interpersonal staring and proximity: Effects of bystanders on drivers at a traffic intersection. *Journal of Personality and Social Psychology, 36,* 575–587.

Greenberg, J. (1996). *The quest for justice on the job: Essays and experiments.* Thousand Oaks, CA: Sage.

Greenberg, M. (1980). A theory of indebtedness. In K. J. Gergen, M. S. Greenberg, & R. H. Willis (Eds.), *Social exchange: Advances in theory and research.* New York: Plenum.

Greenberg, J., & Cohen, R. L. (1982). *Equity and justice in social behavior.* New York: Academic Press.

Greenberg, M., & Frisch, D. (1972). Effect of intentionality on willingness to reciprocate a favor. *Journal of Experimental Social Psychology, 8,* 99–111.

Greenberg, J., & Scott, K. S. (1996). Why do workers bite the hands that feed them? Employee theft as a social exchange process. In B. M. Staw & L. L. Cummings (Eds.), *Research in organizational behavior* (Vol. 18, pp. 111–156). Greenwich, CT: JAI Press.

Greenberg, M. S., & Westcott, D. R. (1983). Indebtedness as a mediator of reactions to aid. In J. D. Fisher, A. Nadler, & B. M. DePaulo (Eds.), *New directions in helping: Vol. 1. Recipient reactions to aid* (pp. 85–112). San Diego: Academic Press.

Greenberger, E., & Steinberg, L. (1983). Sex differences in early labor force participation. *Social Forces, 62,* 467–486.

Greenfeld, L., Rand, M., Craven, D., et al. (1998). *Violence by intimates.* Washington, DC: U.S. Department of Justice, Bureau of Justice Statistics.

Greenley, J. (1979). Familial expectations, post-hospital adjustment, and the societal reaction perspective on mental illness. *Journal of Health and Social Behavior, 20,* 217–227.

Greenstein, T. N. (1995). Are the "most advantaged" children truly disadvantaged by early maternal employment? Effects on child cognitive outcomes. *Journal of Family Issues, 16,* 149–169.

Greenwald, A., & Farnham, S. (2000). Using the implicit association test to measure self-esteem and self-concept. *Journal of Personality and Social Psychology, 79,* 1022–1038.

Greenwald, A. G., & Pratkanis, A. (1984). The self. In R. S. Wyer & T. K. Srull (Eds.), *Handbook of social cognition.* Hillsdale, NJ: Erlbaum.

Gregory, S. W., Jr., & Webster, S. (1996). A nonverbal signal in voices of interview partners effectively predicts communication accommodation and social status perceptions. *Journal of Personality and Social Psychology, 70,* 1231–1240.

Grice, P. H. (1975). Logic and conversation. In P. Cole & J. L. Morgan (Eds.), *Syntax and semantics, Vol. 3: Speech acts.* New York: Academic Press.

Griffin, D., Gonzalez, R., & Varey, C. (2001). The heuristics and biases approach to judgment under uncertainty. In A. Tesser & N. Schwartz (Eds.), *Blackwell handbook of social psychology: Intraindividual processes.* Oxford, UK: Blackwell.

Griffin, L., Wallace, M., & Rubin, B. (1986). Capitalist resistance to the organization of labor before the New Deal: Why? How? Success? *American Sociological Review, 51,* 147–167.

Grimshaw, A. D. (1973). On language in society. Part 1. *Contemporary Sociology, 2,* 575–585.

Grimshaw, A. D. (1990). Talk and social control. In M. Rosenberg & R. H. Turner (Eds.), *Social psychology: Sociological perspectives.* New Brunswick, NJ: Transaction.

Gross, A. E., & Latané, J. G. (1974). Receiving help, reciprocation, and interpersonal attraction. *Journal of Applied Social Psychology, 4,* 210–223.

Gross, A. E., & McMullen, P. A. (1983). Models of the help seeking process. In B. M. DePaulo, A. Nadler, & J. D. Fisher (Eds.), *New directions in helping: Vol. 2. Help seeking* (pp. 47–73). New York: Academic Press.

Gross, E., & Stone, G. P. (1970). Embarrassment and the analysis of role requirements. In G. P. Stone & H. A. Farberman (Eds.), *Social psychology through symbolic interaction.* Waltham, MA: Ginn-Blaisdell.

Grzywacz, J., & Marks, N. (2000). Family, work, work-family spillover, and problem drinking during midlife. *Journal of Marriage and the Family, 62,* 336–348.

Guimond, S., & Dubé-Simard, L. (1983). Relative deprivation theory and the Quebec nationalist movement: The cognition-emotion distinction and the personal-group deprivation issue. *Journal of Personality and Social Psychology, 44,* 526–535.

Gully, S. M., Devine, D. J., & Whitney, D. J. (1995). A meta-analysis of cohesion and performance: Effects of level of analysis and task interdependence. *Small Group Research, 26,* 497–520.

Gumperz, J. J. (1976). The sociolinguistic significance of conversational code-switching. In J. Cook-Gumperz & J. J. Gumperz (Eds.), *Papers on language and context.* Berkeley, CA: University of California, Language Behavior Research Laboratory.

Gurevitch, Z. D. (1990). The embrace: On the element of non-distance in human relations. *The Sociological Quarterly, 31,* 187–201.

Gurin, G., Veroff, J., & Feld, S. (1960). *Americans view their mental health.* New York: Basic Books.

Gurney, J. F., & Tierney, K. J. (1982). Relative deprivation and social movements: A critical look at twenty years of theory and research. *The Sociological Quarterly, 23,* 33–47.

Gustafson, R. (1989). Frustration and successful vs. unsuccessful aggression: A test of Berkowitz's completion hypothesis. *Aggressive Behavior, 15*(1), 5–12.

Guttman, D. (1977). The cross-cultural perspective: Notes toward a comparative psychology of aging. In J. Birren & K. Schaie (Eds.), *Handbook of the psychology of aging.* New York: Van Nostrand Reinhold.

Haan, N. (1978). Two moralities in action contexts. Relationships to thought, ego regulation, and development. *Journal of Personality and Social Psychology, 36,* 286–305.

Haan, N. (1986). Systematic variability in the quality of moral action, as defined in two paradigms. *Journal of Personality and Social Psychology, 50,* 1271–1284.

Hacker, H. M. (1981). Blabbermouths and clams: Sex differences in self-disclosure in same-sex and cross-sex friendship dyads. *Psychology of Women Quarterly, 5,* 385–401.

Haidt, J., & Keltner, D. (1999). Culture and facial expression: Open-ended methods find more expressions and a gradient of recognition. *Cognition and Emotion, 13,* 225–266.

Haines, H. H. (1984). Black radicalization and the funding of civil-rights: 1957–1970. *Social Problems, 32,* 31–43.

Halberstam, D. (1979). *The powers that be.* New York: Knopf.

Hall, E. T. (1966). *The hidden dimension.* Garden City, NY: Doubleday.

Hall, J., & Friedman, G. (1999). Status, gender, and nonverbal behavior: A study of structured interactions between employees of a company. *Personality and Social Psychology Bulletin, 25,* 1082–1091.

Hamilton, D. L. (1979). A cognitive-attributional analysis of stereotyping. In L. Berkowitz (Ed.), *Advances in ex-*

perimental social psychology (Vol. 12). New York: Academic Press.

Hamilton, D. L. (1981). Stereotyping and intergroup behavior: Some thoughts on the cognitive approach. In D. L. Hamilton (Ed.), *Cognitive processes in stereotyping and intergroup behavior* (pp. 333–353). Hillsdale, NJ: Erlbaum.

Hamilton, W. (1964). The genetical evolution of social behavior, I & II. *Journal of Theoretical Biology, 7,* 1–52.

Hamilton, D. L., & Bishop, G. D. (1976). Attitudinal and behavioral effects of initial integration of White suburban neighborhoods. *Journal of Social Issues, 32*(2), 47–67.

Hamilton, D. L., & Zanna, M. P. (1972). Differential weighting of favorable and unfavorable attributes in impressions of personality. *Journal of Experimental Research in Personality, 6,* 204–212.

Han, G., & Lindskold, S. (1983). *A comparison of strategies with and without communication in a prisoner's dilemma game.* Unpublished manuscript, Ohio University.

Hanni, R. (1980). What is planned during speech pauses? In H. Giles, W. P. Robinson, & P. M. Smith (Eds.), *Language: Social psychological perspectives.* New York: Pergamon.

Harasty, A. (1997). The interpersonal nature of social stereotypes: Discussion patterns about in-groups and out-groups. *Personality and Social Psychology Bulletin, 23,* 270–284.

Hardy, R. C. (1975). A test of poor leader-member relations cells of the contingency model on elementary school children. *Child Development, 45,* 958–964.

Hardy, C., & Latané, B. (1986). Social loafing on a cheering task. *Social Science, 71,* 165–172.

Hardy, R. C., Sack, S., & Harpine, F. (1973). An experimental test of the contingency model on small classroom groups. *Journal of Psychology, 85,* 3–16.

Harkins, S. G., & Petty, R. E. (1981a). The effects of source magnification of cognitive effect on attitudes: An information processing view. *Journal of Personality and Social Psychology, 40,* 401–413.

Harkins, S. G., & Petty, R. E. (1981b). The multiple source effect in persuasion: The effects of distraction. *Personality and Social Psychology Bulletin, 4,* 627–635.

Harkins, S. G., & Petty, R. (1982). Effects of task difficulty and task uniqueness on social loafing. *Journal of Personality and Social Psychology, 43,* 1214–1229.

Harkins, S. G., & Petty, R. E. (1983). Social context effects in persuasion: The effects of multiple sources and multiple targets. In P. B. Paulus (Ed.), *Basic group processes.* New York: Springer Verlag.

Harkins, S. G., & Petty, R. E. (1987). Information utility and the multiple source effect. *Journal of Personality and Social Psychology, 52,* 260–268.

Harkins, S. G., & Szymanski, K. (1987). Social loafing and social facilitation. In C. Hendrick (Ed.), *Group processes and intergroup relations* (pp. 167–188). Newbury Park, CA: Sage.

Harlow, R. E., & Cantor, N. (1995). To whom do people turn when things go poorly? Task orientation and functional social contacts. *Journal of Personality and Social Psychology, 69,* 329–340.

Harlow, R. E., & Cantor, N. (1996). Still participating after all these years: A study of life task participation in later life. *Journal of Personality and Social Psychology, 71,* 1235–1249.

Harnish, J., Aseltine, R., Jr., & Gore, S. (2000). Resolution of stressful experiences as an indicator of coping effectiveness in young adults: An event history analysis. *Journal of Health and Social Behavior, 41,* 121–136.

Harris, L. (with associates). (1975). *The myth and the reality of aging in America.* Washington, DC: National Council on Aging.

Harris, L. (with associates). (1981). *Aging in the 80's.* Washington, DC: National Council on Aging.

Harris, M. B. (1974). Mediators between frustration and aggression in a field experiment. *Journal of Experimental Social Psychology, 10,* 561–571.

Harris, P., & Sachau, D. (2005). Is cleanliness next to godliness? The role of housekeeping in impression formation. *Environment and Behavior, 37,* 81–101.

Harris, R. J., Lee, D. J., Hensley, D. L., & Schoen, L. M. (1988). The effect of cultural script knowledge on memory for stories over time. *Discourse Processes, 11,* 413–431.

Harrison, A. (1977). Mere exposure. In L. Berkowitz (Ed.), *Advances in experimental social psychology* (Vol. 10). New York: Academic Press.

Harrison, A., & Saeed, L. (1977). Let's make a deal: An analysis of revelations and stipulations in lonely hearts advertisements. *Journal of Personality and Social Psychology, 35,* 257–264.

Harvey, E. (1999). Short-term and long-term effects of parental employment on children of the National Longitudinal Study of Youth. *Developmental Psychology, 35,* 445–459.

Harvey, J. H., Weber, A. L., & Orbuch, T. L. (1990). *Interpersonal accounts.* Oxford, UK: Blackwell.

Harvey, J. H., Yarkin, K. L., Lightner, J. M., & Tolin, J. P. (1980). Unsolicited interpretation and recall of interpersonal events. *Journal of Personality and Social Psychology, 38,* 551–568.

Haskell, M. R., & Yablonsky, L. (1983). *Criminology: Crime and Criminality* (3rd ed.). Boston: Houghton Mifflin.

Hass, R. G. (1981). Effects of source characteristics on cognitive responses in persuasion. In R. E. Petty, T. M. Ostrom, & T. C. Brock (Eds.), *Cognitive responses in persuasion* (pp. 141–172). Hillsdale, NJ: Erlbaum.

Hassin, R., & Trope, Y. (2000). Facing faces: Studies on the cognitive aspects of physiognomy. *Journal of Personality and Social Psychology, 78,* 837–852.

Hastorf, A. H., Wildfogel, J., & Cassman, T. (1979). Acknowledgment of a handicap as a tactic in social interaction. *Journal of Personality and Social Psychology, 31,* 1790–1797.

Hater, J. J., & Bass, B. M. (1988). Supervisors' evaluations and subordinates' perceptions of transformational and transactional leadership. *Journal of Applied Psychology, 73,* 695–702.

Hatfield, E. (1982). What do women and men want from love and sex? In E. R. Allgeier & N. B. McCormick (Eds.), *Changing boundaries: Gender roles and sexual behavior.* Palo Alto, CA: Mayfield.

Hatfield, E., & Sprecher, S. (1986). Measuring passionate love in intimate relationships. *Journal of Adolescence, 9,* 383–410.

Hatfield, E., & Walster, G. W. (1978). *A new look at love.* Lanham, MA: University Press of America.

Hauser, R. M., and Warren, J. R. (1997). Socioeconomic indexes for occupations: A review, update, and critique. *Sociological Methodology 1997* (pp. 177–298). Cambridge, MA: Basil Blackwell.

Haveman, R., Wolfe, B., & Spaulding, J. (1991). Childhood events and circumstances influencing high school completion. *Demography, 28,* 133–157.

Hawkes, D., Senn, C., & Thom, C. (2004). Factors that influence attitudes toward women with tattoos. *Sex Roles, 50,* 593–604.

Hawkins, A. (2004). Reflections on body hair. *Off our backs, 34,* 40–41.

Hayduk, L. A. (1978). Personal space: An evaluation and orienting review. *Psychological Bulletin, 85,* 117–134.

Haynie, D. (2001). Delinquent peers revisited: Does network structure matter? *American Journal of Sociology, 106,* 1013–1057.

Hazan, C., & Shaver, P. (1987). Romantic love conceptualized as an attachment process. *Journal of Personality and Social Psychology, 52,* 511–524.

Hazan, C., & Shaver, P. R. (1992). Broken attachments: Relationship loss from the perspective of attachment theory. In T. L. Orbuch (Ed.), *Close relationship loss: Theoretical approaches* (pp. 90–107). New York: Springer Verlag.

Hebl, M., & Dovidio, J. (2005). Promoting the "social" in the examination of social stigmas. *Personality and Social Psychology Review, 9,* 156–182.

Hechter, M., & Opp, K. (2001). *Social norms.* New York: Russell Sage Foundation.

Hedge, A., & Yousif, Y. H. (1992). Effects of urban size, urgency, and cost on helpfulness: A cross-cultural comparison between the United Kingdom and the Sudan. *Journal of Cross Cultural Psychology, 23*(1), 107–115.

Heider, E. R., & Olivier, D. (1972). The structure of the color space in naming and memory of two languages. *Cognitive Psychology, 3,* 337–354.

Heider, F. (1944). Social perception and phenomenal causality. *Psychological Review, 51,* 258–374.

Heider, F. (1958). *The psychology of interpersonal relations.* New York: Wiley.

Heimer, K., & Matsueda, R. (1994). Role-taking, role commitment, and delinquency: A theory of differential social control. *American Sociological Review, 59,* 365–390.

Heirich, M. (1977). Change of heart: A test of some widely held theories about religious conversion. *American Journal of Sociology, 83,* 653–680.

Heiss, D. (1979). *Understanding events: Affect and the construction of social action.* New York: Cambridge University Press.

Heiss, J. (1981). Social roles. In M. Rosenberg & R. H. Turner (Eds.), *Social psychology: Sociological perspectives.* New York: Basic Books.

Heiss, J. (1990). Social roles. In M. Rosenberg & R. Turner (Eds.), *Social psychology: Sociological perspectives.* New Brunswick, NJ: Transaction.

Heiss, D. A., & Calhan, C. (1995). Emotion norms in interpersonal events. *Social Psychology Quarterly, 53,* 223–240.

Heiss, J., & Owens, S. (1972). Self-evaluations of Blacks and Whites. *American Journal of Sociology, 78,* 360–370.

Hendrick, C., & Hendrick, S. (1989). Research on love: Does it measure up? *Journal of Personality and Social Psychology, 56,* 784–794.

Hendrick, S., & Hendrick, C. (1992). *Liking, loving, and relating* (2nd ed.). Pacific Grove, CA: Brooks/Cole.

Henley, J., Danziger, S., & Offer, S. (2005). The contribution of social support to the material well-being of low-income families. *Journal of Marriage and Family, 67,* 122–140.

Hennessy, D., & Wiesenthal, D. (2001). Gender, driver aggression, and driver violence: An applied evaluation. *Sex Roles, 44,* 661–676.

Henretta, J. C., Frazier, C., & Bishop, D. (1986). The effect of prior case outcome on juvenile justice decision-making. *Social Forces, 65,* 554–562.

Hensley, T. R., & Griffin, G. W. (1986). Victims of groupthink: The Kent State University board of trustees and the 1977 gymnasium controversy. *Journal of Conflict Resolution, 30,* 497–531.

Hepburn, C., & Locksley, A. (1983). Subjective awareness of stereotyping: Do we know when our judgments are prejudiced? *Social Psychology Quarterly, 45,* 311–318.

Herek, G. (1987). Can functions be measured? A new perspective on the functional approach to attitudes. *Social Psychology Quarterly, 50,* 285–303.

Herek, G. M., & Capitanio, J. P. (1996). "Some of my best friends": Intergroup contact, concealable stigma, and heterosexuals' attitudes toward gay men and lesbians. *Personality and Social Psychology Bulletin, 22,* 412–424.

Heritage, J., & Greatbatch, D. (1986). Generating applause: A study of rhetoric and response at party political conferences. *American Journal of Sociology, 92,* 110–157.

Hess, U., Banse, R., & Kappas, A. (1995). The intensity of facial affect is determined by underlying affective state and social situation. *Journal of Personality and Social Psychology, 69,* 280–288.

Hess, T. M., & Slaughter, S. J. (1990). Schematic knowledge influences on memory for scene information in young and older adults. *Developmental Psychology, 26,* 855–865.

Hetherington, E. M. (1999). Should we stay together for the sake of the children? In Hetherington, E. M. (Ed.), *Coping with divorce, single parenting, and remarriage: A risk and resiliency perspective* (pp. 93–116). Mahwah, NJ: Erlbaum.

Hewitt, J. P. (1997). *Self and society* (7th ed.). Boston: Allyn & Bacon.

Hewitt, J. P., & Stokes, R. (1975). Disclaimers. *American Sociological Review, 40,* 1–11.

Hewstone, M. (1990). The "ultimate attribution error"? A review of the literature on intergroup causal attribution. *European Journal of Social Psychology, 20,* 311–335.

Hewstone, M., & Brown, R. (1986). Contact is not enough: An intergroup perspective on the contact hypothesis. In M. Hewstone & R. Brown (Eds.), *Contact and conflict in intergroup encounters* (pp. 1–44). Oxford, UK: Blackwell.

Hewstone, M., & Giles, H. (1986). Social groups and social stereotypes in intergroup communication: A review and model of intergroup communication. In W. B. Gudykunst (Ed.), *Intergroup communication* (pp. 10–26). London: Edward Arnold.

Hewstone, M., & Jaspars, J. (1984). Social dimensions of attribution. In H. Tajfel (Ed.), *The social dimension* (Vol. 2). Cambridge, UK: Cambridge University Press.

Hewstone, M., & Jaspars, J. (1987). Covariation and causal attribution: A logical model of the intuitive analysis of variance. *Journal of Personality and Social Psychology, 53,* 663–672.

Heyl, B. S. (1977). The Madam as teacher: The training of house prostitutes. *Social Problems, 24,* 545–555.

Higbee, K. L. (1969). Fifteen years of fear arousal: Research on threat appeals, 1953–1968. *Psychological Bulletin, 72,* 426–444.

Higgins, C., & Judge, T. (2004). The effect of applicant influence tactics on recruiter perceptions of fit and hiring recommendations: A field study. *Journal of Applied Psychology, 89,* 622–632.

Higgins, E. T. (1989). Self-discrepancy theory: What patterns of self-beliefs cause people to suffer? In L. Berkowitz (Ed.), *Advances in experimental social psychology* (Vol. 22). New York: Academic Press.

Higgins, E. T., & Bargh, J. A. (1987). Social cognition and social perception. *Annual Review of Psychology, 38,* 369–425.

Higgins, E. T., & Bryant, S. L. (1982). Consensus information and the fundamental attribution error: The role of development and in-group versus out-group knowledge. *Journal of Personality and Social Psychology, 47,* 422–435.

Higgins, E. T., Klein, R., & Strauman, T. (1985). Self-concept discrepancy theory: A psychological model for distinguishing among different aspects of depression and anxiety. *Social Cognition, 3,* 51–76.

Higuchi, M., & Fukada, H. (2002). A comparison of four causal factors of embarrassment in public and private situations. *The Journal of Psychology, 136,* 399–406.

Hill, C., Rubin, Z., & Peplau, L. (1976). Breakups before marriage: The end of 103 affairs. *Journal of Social Issues, 32*(1), 147–168.

Hiltrop, J. M. (1985). Mediator behavior and the settlement of collective bargaining disputes in Britain. *Journal of Social Issues, 41*(2), 83–99.

Hinkle, S., & Schopler, J. (1986). Bias in the evaluation of in-group and out-group performance. In S. Worchel & W. G. Austin (Eds.), *Psychology of intergroup relations* (2nd ed., pp. 196–212). Chicago: Nelson-Hall.

Hirsch, E. L. (1990). Sacrifice for the cause: Group processes, recruitment, and commitment in a student social movement. *American Sociological Review, 55,* 243–254.

Hirschi, T. (1969). *Causes of delinquency.* Berkeley: University of California Press.

Hirt, E. R., & Markmann, K. D. (1995). Multiple explanation: A consider-an-alternative strategy for debiasing judgments. *Journal of Personality and Social Psychology, 69,* 1069–1086.

Hitlin, S. (2003). Values as the core of personal identity: Drawing links between two theories of self. *Social Psychology Quarterly, 66,* 118–137.

Hobsbawm, E., & Rudé, G. (1975). *Captain Swing: A social history of the great English agricultural uprising of 1830.* New York: Norton.

Hochschild, A. R. (1975). Disengagement theory: A critique and a proposal. *American Sociological Review, 40,* 553–569.

Hochschild, A. R. (1983). *The managed heart: Commercialization of human feeling.* Berkeley: University of California Press.

Hodson, R. (1996). Dignity in the workplace under participative management. *American Sociological Review, 61,* 719–738.

Hoelter, J. W. (1983). The effects of role evaluation and commitment on identity salience. *Social Psychology Quarterly, 46,* 140–147.

Hoelter, J. W. (1984). Relative effects of significant others on self-evaluation. *Social Psychology Quarterly, 47,* 255–262.

Hoelter, J. W. (1986). The relationship between specific and global evaluations of self: A comparison of several models. *Social Psychology Quarterly, 49,* 129–141.

Hofferth, S., & Sandberg, J. (2001). How American children spend their time. *Journal of Marriage and the Family, 63,* 295–308.

Hoffman, C., Lau, I., & Johnson, D. (1986). The linguistic relativity of person cognition: An English-Chinese comparison. *Journal of Personality and Social Psychology, 51,* 1097–1105.

Hoffman, M. L. (1998). Varieties of empathy-based guilt. In J. Bybee (Ed.), *Guilt and children* (pp. 91–112). New York: Academic Press.

Hoffmann, J. (2002). The community context of family structure and adolescent drug use. *Journal of Marriage and the Family, 64,* 314–330.

Hofling, C. K., Brotz, E., Dalrymple, S., Graves, N., & Pierce, C. M. (1966). An experimental study of nurse-physician relationships. *Journal of Nervous and Mental Disease, 143,* 171–180.

Hogan, D. P. (1981). *Transitions and social change.* New York: Academic Press.

Hogg, M. A., & Haines, S. C. (1996). Intergroup relations and group solidarity: Effects of group identification and social beliefs on depersonalized attraction. *Journal of Personality and Social Psychology, 70,* 25–39.

Hogg, M. A., Terry, D. J., & White, K. M. (1995). A tale of two theories: A critical comparison of identity theory with social identity theory. *Social Psychology Quarterly, 58,* 255–269.

Holahan, C. J. (1977). Effects of urban size and heterogeneity on judged appropriateness of altruistic responses: Situational vs. subject variables. *Sociometry, 40,* 378–382.

Holden, R. T. (1986). The contagiousness of aircraft hijacking. *American Journal of Sociology, 91,* 874–904.

Holgartner, S., & Bosk, C. L. (1988). The rise and fall of social problems: A public arenas model. *American Journal of Sociology, 94,* 53–78.

Hollander, E. P. (1975). Independence, conformity and civil liberties: Some implications from social psychological research. *Journal of Social Issues, 31*(2), 55–67.

Hollander, E. P. (1985). Leadership and power. In G. Lindzey & E. Aronson (Eds.), *Handbook of social psychology* (3rd ed, Vol. 2, pp. 485–537). New York: Random House.

Hollander, E. P., Fallon, B. J., & Edwards, M. T. (1977). Some aspects of influence and acceptability for appointed and elected group leaders. *Journal of Psychology, 95,* 289–296.

Hollander, E. P., & Julian, J. W. (1970). Studies in leader legitimacy, influence, and innovation. In L. Berkowitz (Ed.), *Advances in experimental social psychology* (Vol. 5, pp. 33–69). New York: Academic Press.

Hollinger, R. C., & Clark, J. P. (1983). Deterrence in the workplace: Perceived certainty, perceived severity, and employee theft. *Social Forces, 62,* 398–418.

Holmes, D. S. (1971). Compensation for ego threat: Two experiments. *Journal of Personality and Social Psychology, 18,* 234–237.

Holmes, J. G., & Grant, P. (1979). Ethnocentric reactions to social threats. In L. H. Strickland (Ed.), *Social psychology: East-West perspectives.* Oxford, UK: Pergamon.

Holohan, C., & Moos, R. (1987). Personal and contextual determinants of coping strategies. *Journal of Personality and Social Psychology, 52,* 946–955.

Holzer, H., & Offner, P. (2001). *Trends in employment outcomes of young black men, 1979–2000.* Washington, DC: Georgetown Public Policy Institute.

Homans, G. C. (1974). *Social behavior: Its elementary forms* (2nd ed.). New York: Harcourt Brace Jovanovich.

Homer, P. M., & Kahle, L. R. (1988). A structural-equation test of the value-attitude-behavior hierarchy. *Journal of Personality and Social Psychology, 54,* 638–646.

Hood, W. R., & Sherif, M. (1962). Verbal report and judgment of an unstructured stimulus. *Journal of Psychology, 54,* 121–130.

Hopkins, N., & Moore, C. (2001). Categorizing the neighbors: Identity, distance, and stereotyping. *Social Psychology Quarterly, 64,* 239–252.

Hopper, J. R., & Nielsen, J. M. (1991). Recycling as altruistic behavior: Normative and behavioral strategies to expand participation in a community recycling program. *Environment and Behavior, 23,* 195–220.

Horai, J., Naccari, N., & Fatoullah, E. (1974). The effects of expertise and physical attractiveness upon opinion agreement and liking. *Sociometry, 37,* 601–606.

Horai, J., & Tedeschi, J. T. (1969). The effects of threat credibility and magnitude of punishment upon compliance. *Journal of Personality and Social Psychology, 12,* 164–169.

Horne, C. (2001). The enforcement of norms: Group cohesion and meta-norms. *Social Psychology Quarterly, 64,* 253–266.

Horne, C. (2004). Collective benefits, exchange interests, and norm enforcement. *Social Forces, 82,* 1037–1062.

Horne, W. C., & Long, G. (1972). Effect of group discussion on universalistic-particularistic orientation. *Journal of Experimental Social Psychology, 8,* 236–246.

Horney, J., Osgood, D. W., & Marshall, I. H. (1995). Criminal careers in the short-term: Intra-individual variability in crime and its relation to local life circumstances. *American Sociological Review, 60,* 655–673.

Hornstein, G. (1985). Intimacy in conversational style as a function of the degree of closeness between members of a dyad. *Journal of Personality and Social Psychology, 49,* 671–681.

Hornstein, H. A. (1965). The effects of different magnitudes of threat upon interpersonal bargaining. *Journal of Experimental Social Psychology, 1,* 282–293.

Hornstein, H. A. (1978). Promotive tension and prosocial behavior: A Lewinian analysis. In L. Wispe (Ed.), *Altruism, sympathy, and helping.* New York: Academic Press.

Horowitz, D. L. (2001). *The deadly ethnic riot.* Berkeley: University of California Press.

Horowitz, R. (1987). Community tolerance of gang violence. *Social Problems, 34,* 437–450.

Horowitz, H. V., White, H. R., & Howell-White, S. (1996). Becoming married and mental health: A longitudinal study of a cohort of young adults. *Journal of Marriage and the Family, 58,* 895–907.

Hossain, Z., & Roopnarine, J. L. (1993). Division of household labor and child-care in dual-earner African-American families with infants. *Sex Roles, 29,* 571–583.

House, J. S. (1974). Occupational stress and coronary heart disease: A review and theoretical integration. *Journal of Health and Social Behavior, 15,* 17–21.

House, J. S. (1981). *Work stress and social support.* Reading, MA: Addison-Wesley.

House, J. S., Lepkowski, J. M., Kinney, A. M., Mero, R. P., Kessler, R. C., & Herzog, A. R. (1994). The social stratification of aging and health. *Journal of Health and Social Behavior, 35,* 213–234.

House, J. S., & Wolf, S. (1978). Effects of urban residence on interpersonal trust and helping behavior. *Journal of Personality and Social Psychology, 36,* 1029–1043.

Howard, J. (2000). Social psychology of identities. *Annual Review of Sociology, 26,* 367–393.

Hrdy, S. B. (1977). Infanticide as a primate reproductive strategy. *American Scientist, 65,* 40–49.

Hrdy, S. B. (1982). Positivist thinking encounters field primatology, resulting in agonistic behavior. *Social Science Information, 21,* 245–250.

Hue, C., & Erickson, J. R. (1991). Normative studies of sequence strength and scene structure of 30 scripts. *American Journal of Psychology, 104,* 229–240.

Huesmann, L. R. (1986). Psychological processes promoting the relation between exposure to media violence and aggressive behavior by the viewer. *Journal of Social Issues, 42*(3), 125–139.

Huesmann, L. R., & Moise, J. (1996). Media violence: A demonstrated public health threat to children. *Harvard Mental Health Letter, 12*(12), 5–7.

Huffine, C., & Clausen, J. (1979). Madness and work: Short and long-term effects of mental illness on occupational careers. *Social Forces, 57,* 1049–1062.

Hughes, M., & Demo, D. H. (1989). Self-perceptions of Black Americans: Self-esteem and personal efficacy. *American Journal of Sociology, 95,* 132–159.

Hughes, D., & Surra, C. (2000). The reported influence of research participation on premarital relationships. *Journal of Marriage and the Family, 62,* 822–832.

Hughes, M. E., & Waite, L. (2002). Health in household context: Living arrangements and health in late middle age. *Journal of Health and Social Behavior, 43,* 1–21.

Hull, C. L. (1943). *Principles of behavior.* New York: Appleton-Century-Crofts.

Hultsch, D., & Plemons, J. (1979). Life events and life span development. In P. Baltes & O. G. Brim, Jr. (Eds.),

Life-span development and behavior (Vol. 2). New York: Academic Press.

Hundleby, J. D., & Mercer, G. W. (1987). Family and friends as social environments and their relationship to young adolescents' use of alcohol, tobacco, and marijuana. *Journal of Marriage and the Family, 49,* 151–164.

Hunter, C. H. (1984). Aligning actions: Types and social distribution. *Symbolic Interaction, 7,* 155–174.

Hunter, J. A., Stringer, M., & Watson, R. P. (1991). Intergroup violence and intergroup attributions. *British Journal of Social Psychology, 30,* 261–266.

Huston, T. L., Ruggiero, M., Conner, R., & Geis, G. (1981). Bystander intervention into crime: A study based on naturally-occurring episodes. *Social Psychology Quarterly, 44*(1), 14–23.

Huston, A. C., & Wright, J. C. (1982). Effects of communication media on children. In C. B. Kopp & J. B. Krakow (Eds.), *The child: Development in a social context.* Reading, MA: Addison-Wesley.

Hyde, J. (2003). *Half the human experience* (6th ed.). Boston: Allyn & Bacon.

Hyde, J., DeLamater, J., Plant, A., & Byrd, J. (1996). Sexuality during pregnancy and the year post-partum. *Journal of Sex Research, 33,* 143–151.

Hyde, J., Fenema, E., & Lamon, S. (1990). Gender differences in mathematics performance: A meta-analysis. *Psychological Bulletin, 107,* 139–155.

Hyde, J., & Linn, M. (Eds.). (1986). *The psychology of gender: Advances through meta-analysis.* Baltimore: Johns Hopkins University Press.

Hyde, J. S. (1984). How large are gender differences in aggression? A developmental meta-analysis. *Developmental Psychology, 20,* 722–736.

Hymes, D. (1974). *Foundations in sociolinguistics.* London: Tavistock.

Ikle, F. C. (1971). *Every war must end.* New York: Columbia University Press.

Indonesians die as riots over price rises widen. (1998, Feb. 15). *New York Times,* p. A10.

Ingham, A. G., Levinger, G., Graves, J., & Peckham, V. (1974). The Ringelmann effect: Studies of group size and group performance. *Journal of Experimental Social Psychology, 10,* 371–384.

Insko, C. A., Arkoff, A., & Insko, V. M. (1965). Effects of high and low fear-arousing communications upon opinions toward smoking. *Journal of Experimental Social Psychology, 40,* 256–266.

Insko, C. A., Drenan, S., Solomon, M. R., Smith, R., & Wade, T. J. (1983). Conformity as a function of the consistency of positive self-evaluation with being liked and being right. *Journal of Experimental Social Psychology, 19,* 341–358.

Insko, C. A., Smith, R. H., Alicke, M. S., Wade, J., & Taylor, S. (1985). Conformity and group size: The concern with being right and the concern with being liked. *Personality and Social Psychology Bulletin, 11,* 41–50.

Isen, A. M. (1970). Success, failure, attention, and reaction to others: The warm glow of success. *Journal of Personality and Social Psychology, 15,* 294–301.

Isen, A. M., Clark, M., & Schwartz, M. F. (1976). Duration of the effect of good mood on helping: Footprints on the sands of time. *Journal of Personality and Social Psychology, 34,* 385–393.

Isen, A. M., & Levin, P. F. (1972). Effect of feeling good on helping. Cookies and kindness. *Journal of Personality and Social Psychology, 21,* 384–388.

Isen, A. M., & Simmons, S. F. (1978). The effect of feeling good on a helping task that is incompatible with good mood. *Social Psychology, 41,* 346–349.

Isenberg, D. J. (1986). Group polarization: A critical review and meta-analysis. *Journal of Personality and Social Psychology, 50,* 1141–1151.

Ishii, K., Reyes, J. A., & Kitayama, S. (2003). Spontaneous attention to word content versus emotional tone: Differences among three cultures. *Psychological Science, 14,* 39–46.

Isozaki, M. (1984). The effect of discussion on polarization of judgments. *Japanese Psychological Research, 26,* 187–193.

Jaccard, J. (1981). Toward theories of persuasion and belief change. *Journal of Personality and Social Psychology, 40,* 260–269.

Jackson, J. (1954). The adjustment of the family to the crisis of alcoholism. *Quarterly Journal of Studies on Alcohol, 15,* 564–586.

Jackson, J. (1965). Structural characteristics of norms. In I. D. Steiner & M. Fishbein (Eds.), *Current studies in social psychology.* New York: Holt, Rinehart and Winston.

Jackson, J. M. (1986). In defense of social impact theory: Comment on Mullen. *Journal of Personality and Social Psychology, 50,* 511–513.

Jackson, J. M. (1987). Social impact theory: A social forces model of influence. In B. Mullen & G. R. Goethals (Eds.), *Theories of group behavior* (pp. 111–124). New York: Springer Verlag.

Jackson, P. B. (1997). Role occupancy and minority mental health. *Journal of Health and Social Behavior, 38,* 237–255.

Jackson, L. A., Hunter, J. E., & Hodge, C. N. (1995). Physical attractiveness and intellectual competence: A meta-analytic review. *Social Psychology Quarterly, 58,* 108–122.

Jackson-Jacobs, C. (2004). Hard drugs in a soft context: Managing trouble and crack use on a college campus. *The Sociological Quarterly, 45,* 835–856.

Jacobson, N., & Gottman, J. (1998). *When men batter women: New insights into ending abusive relationships.* New York: Simon & Schuster.

Jacques, J. M., & Chason, K. (1977). Self-esteem and low status groups: A changing scene? *The Sociological Quarterly, 18,* 399–412.

Jaffee, S., & Hyde, J. S. (2000). Gender differences in moral orientation: A meta-analysis. *Psychological Bulletin, 126,* 703–726.

James, W. (1890). *Principles of psychology.* New York: Holt, Rinehart and Winston.

Janes, L., & Olson, J. M. (2000). Jeer pressure: The behavioral effects of observing ridicule of others. *Personality and Social Psychology Bulletin, 26,* 474–485.

Janis, I. L. (1982). *Groupthink* (2nd ed.). Boston: Houghton Mifflin.

Janis, I. L., & Mann, L. (1977). *Decision making: A psychological analysis of conflict, choice, and commitment.* New York: Free Press.

Jankowski, S., Videka-Sherman, L., & Laquidara-Dickinson, K. (1996). Social support networks of confidants to people with AIDS. *Social Work, 41,* 206–213.

Jekielek, S. M. (1998). Parental conflict, marital disruption, and children's emotional well-being. *Social Forces, 76,* 905–935.

Jellison, J. M., & Riskind, J. (1970). A social comparison of abilities interpretation of risk-taking behavior. *Journal of Personality and Social Psychology, 15,* 375–390.

Jemmott, J. B., & Magloire, K. (1988). Academic stress, social support, and secretory immunoglobulin A. *Journal of Personality and Social Psychology, 55,* 803–810.

Jenkins, J. C. (1983). Resource mobilization theory and the study of social movements. In R. H. Turner & J. F. Short (Eds.), *Annual review of sociology* (Vol. 9). Palo Alto, CA: Annual Reviews.

Jenkins, J. C., & Eckert, C. (1986). Channeling Black insurgency: Elite patronage and professional social movement organizations in the development of the Black movement. *American Sociological Review, 51,* 812–829.

Jenkins, J. C., & Perrow, C. (1977). Insurgency of the powerless: Farm worker movements 1946–1972. *American Sociological Review, 42,* 249–268.

Jenkins, C., Rosenman, R., & Zyzanski, S. (1974). Prediction of clinical coronary heart disease by a test for coronary prone behavior pattern. *The New England Journal of Medicine, 290,* 1271–1275.

Jensen, G., Erickson, M., & Gibbs, J. (1978). Perceived risk of punishment and self-reported delinquency. *Social Forces, 57,* 57–78.

Jessor, R., Costa, F., Jessor, L., & Donovan, J. (1983). Time of first intercourse: A prospective study. *Journal of Personality and Social Psychology, 44,* 608–626.

Job, R. F. S. (1988). Effective and ineffective use of fear in health promotion campaigns. *American Journal of Public Health, 78,* 163–167.

Johnson, B. T. (1991). Insight about attitudes: Meta-analytic perspectives. *Personality and Social Psychology Bulletin, 17,* 289–299.

Johnson, C. (1994). Gender, legitimate authority, and leader-subordinate conversations. *American Sociological Review, 59,* 122–135.

Johnson, N. R. (1987). Panic at "the Who Concert Stampede": An empirical assessment. *Social Problems, 34,* 362–373.

Johnson, R. E. (1979). *Juvenile delinquency and its origins.* New York: Cambridge University Press.

Johnson, D. L., & Andrews, I. R. (1971). The risky-shift hypothesis tested with consumer products as stimuli. *Journal of Personality and Social Psychology, 20,* 382–385.

Johnson, J. T., & Boyd, K. R. (1995). Dispositional traits versus the content of experience: Actor/observer differences in judgments of the "authentic self." *Personality and Social Psychology Bulletin, 21,* 375–383.

Johnson, J. G., Cohen, P., Smailes, E. M., Kasen, S., & Brooks, J. S. (2002). Television viewing and aggressive behavior during adolescence and adulthood. *Science, 295,* 2468–2471.

Johnson, B. T., & Eagly, A. H. (1989). Effects of involvement on persuasion: A meta-analysis. *Psychological Bulletin, 106,* 290–314.

Johnson, M., & Ferraro, K. (2001). Research on domestic violence in the 1990s: Making distinctions. In R. Milardo (Ed.), *Understanding families into the new millennium: A decade in review* (pp. 167–182). Minneapolis, MN: National Council on Family Relations.

Johnson, M., & Leslie, L. (1982). Couple involvement and network structure: A test of the dyadic withdrawal hypothesis. *Social Psychology Quarterly, 45,* 34–43.

Johnson, K. J., Lund, D. A., & Dimond, M. F. (1986). Stress, self-esteem, and coping during bereavement among the elderly. *Social Psychology Quarterly, 49,* 273–279.

Johnson, D. J., & Rusbult, C. E. (1989). Resisting temptation: Devaluation of alternative partners as a means of maintaining commitment in close relationships. *Journal of Personality and Social Psychology, 57,* 967–980.

Johnson, M. M., Stockard, J., Acker, J., & Naffziger, G. (1975). Expressiveness reevaluated. *School Review, 83,* 617–644.

Johnson-George, C., & Swap, W. (1982). Measurement of specific interpersonal trust: Construction and validation of a scale to assess trust in a specific other. *Journal of Personality and Social Psychology, 43,* 1306–1317.

Jones, E. E. (1964). *Ingratiation: A social psychology perspective.* New York: Appleton-Century-Crofts.

Jones, E. E. (1979). The rocky road from acts to dispositions. *American Psychologist, 34,* 107–117.

Jones, E. E. (1985). Major developments in social psychology during the past five decades. In G. Lindzey & E. Aronson (Eds.), *Handbook of social psychology* (3rd. ed, Vol. I, pp. 47–107). New York: Random House.

Jones, E. E., & Davis, K. E. (1965). From acts to dispositions. In L. Berkowitz (Ed.), *Advances in experimental social psychology* (Vol. 2). New York: Academic Press.

Jones, E. E., Davis, K. E., & Gergen, K. J. (1961). Role playing variations and their informational value for person perception. *Journal of Abnormal and Social Psychology, 63,* 302–310.

Jones, E. E., Farina, A., Hastorf, A. H., Markus, H., Miller, D. T., & Scott, R. A. (1984). *Social stigma.* New York: Freeman.

Jones, E. E., Gergen, K. J., Gumpert, P., & Thibaut, J. (1965). Some conditions affecting the use of ingratiation to influence performance evaluation. *Journal of Personality and Social Psychology, 1,* 613–626.

Jones, E. E., & Goethals, G. R. (1971). *Order effects in impression formation: Attribution context and the nature of the entity.* Morristown, NJ: General Learning Press.

Jones, E. E., & Harris, V. A. (1967). The attribution of attitudes. *Journal of Experimental Social Psychology, 3,* 1–24.

Jones, E. E., & McGillis, D. (1976). Correspondent inferences and the attribution cube: A comparative reappraisal. In J. H. Harvey, W. J. Ickes, & R. F. Kidd (Eds.), *New directions in attribution research* (Vol. 1). Hillsdale, NJ: Erlbaum.

Jones, E. E., & Nisbett, R. (1972). The actor and observer: Divergent perceptions of the causes of behavior. In E. E. Jones, D. E. Kanouse, H. H. Kelley, R. E. Nisbett, S. Valins, & B. W. Weiner (Eds.), *Attribution: Perceiving the causes of behavior.* Morristown, NJ: General Learning Press.

Jones, E. E., & Pittman, T. S. (1982). Toward a general theory of strategic self-presentation. In J. Suls (Ed.), *Psychological perspectives on the self* (Vol. 1). Hillsdale, NJ: Erlbaum.

Jones, E. E., Rock, L., Shaver, K. G., Goethals, G. R., & Ward, L. M. (1968). Pattern of performance and ability attribution: An unexpected primacy effect. *Journal of Personality and Social Psychology, 10,* 317–340.

Jones, E. E., & Wortman, C. (1973). *Ingratiation: An attributional approach.* Morristown, NJ: General Learning Press.

Jones, V. C. (1948). *The Hatfields and the McCoys.* Chapel Hill: University of North Carolina Press.

Jones, W. H., Briggs, S. R., & Smith, T. G. (1986). Shyness: Conceptualization and measurement. *Journal of Personality and Social Psychology, 51,* 629–639.

Jong, P. (1999). Communicative and remedial effects of social blushing. *Journal of Nonverbal Behavior, 23,* 197–217.

Jöreskög, K. G., & Sörbom, D. (1979). *Advances in factor analysis and structural equation models.* Cambridge, MA: Abt Books.

Joseph, N., & Alex, N. (1972). The uniform: A sociological perspective. *American Journal of Sociology, 77,* 719–730.

Joule, R., & Azdia, T. (2003). Cognitive dissonance, double forced compliance, and commitment. *European Journal of Social Psychology, 33,* 565–571.

Jourard, S. M. (1971). *Self-disclosure.* New York: Wiley.

Judd, C. M., Drake, R. A., Downing, J. W., & Krosnick, J. A. (1991). Some dynamic properties of attitude structures: Context-induced response facilitation and polarization. *Journal of Personality and Social Psychology, 60,* 193–202.

Judge, T. A., & Bono, J. E. (2000) Five-factor model of personality and transformational leadership. *Journal of Applied Psychology, 85,* 751–765.

Jussim, L., Coleman, L., & Nassau, S. (1987). The influence of self-esteem on perceptions of performance and feedback. *Social Psychology Quarterly, 50,* 95–99.

Kacmar, K. M., Carlson, D., & Bratton, V. (2004). Situational and dispositional factors as antecedents of ingratiatory behaviors in organizational settings. *Journal of Vocational Behavior, 65,* 309–331.

Kahneman, D., & Tversky, A. (1973). On the psychology of prediction. *Psychological Review, 80,* 237–251.

Kahn, A. S., Mathie, V. A., & Torgler, C. (1994). Rape scripts and rape acknowledgment. *Psychology of Women Quarterly, 18,* 53–66.

Kahne, M., & Schwartz, C. (1978). Negotiating trouble: The social construction and management of trouble in a psychiatric context. *Social Problems, 25,* 461–475.

Kalmijn, M. (2004). Marriage rituals as reinforcers of role transitions: An analysis of weddings in the Netherlands. *Journal of Marriage and Family, 66,* 582–594.

Kalven, H., Jr., & Zeisel, H. (1966). *The American jury.* Boston: Little, Brown.

Kanter, R. M. (1976). *Men and women of the corporation.* New York: Basic Books.

Kantrowitz, B., & Wingert, P. (2002, July 1). The group. *Newsweek,* pp. 52–53.

Kao, G., & Joyner, K. (2004). Do race and ethnicity matter among friends? Activities among interracial, interethnic, and intraethnic adolescent friends. *The Sociological Quarterly, 45,* 557–573.

Kaplan, H. B., Johnson, R., & Bailey, C. A. (1987). Deviant peers and deviant behavior: Further elaboration of a model. *Social Psychology Quarterly, 50,* 277–284.

Kaplan, H. B., Martin, S. S., & Johnson, R. J. (1986). Self-rejection and the explanation of deviance: Specification of the structure among latent constructs. *American Journal of Sociology, 92,* 384–411.

Kaplan, M. F. (1977). Discussion polarization effects in a modified jury discussion paradigm: informational influences. *Sociometry, 40,* 262–271.

Kaplan, M. F. (1987). The influencing process in group decision-making. In C. Hendrick (Ed.), *Group processes.* Newbury Park, CA: Sage.

Kaplan, M. F., & Miller, C. E. (1987). Group decision making and normative versus informational influence: Effects of type of issue and assigned decision rule. *Journal of Personality and Social Psychology, 53,* 306–313.

Karabenick, S. A. (1983). Sex-relevance of content and influenceability. *Personality and Social Psychology Bulletin, 9,* 243–252.

Karasek, R., Theorell, T., Schwartz, J., Schnall, P., Pieper, C., & Michela, J. (1988). Job characteristics in relation to the prevalence of myocardial infarction in the US Health Examination Survey (HES) and the Health and Nutrition Examination Survey (HANES). *American Journal of Public Health, 78,* 910–918.

Karau, S. J., & Williams, K. D. (1993). Social loafing: A meta-analytic review and theoretical integration. *Journal of Personality and Social Psychology, 65,* 681–706.

Karau, S. J., & Williams, K. D. (1995). Social loafing: Research findings, implications, and future directions. *Current Directions in Psychological Science, 4*(5), 134–140.

Karau, S. J., & Williams, K. D. (2001). Understanding individual motivation in groups: The collective effort model. In M. E. Turner (Ed.), *Groups at work: Advances in theory and research* (pp. 113–141). Mahwah, NJ: Erlbaum.

Karlins, M., & Abelson, H. I. (1970). *How opinions and attitudes are changed* (2nd ed.) New York: Springer Verlag.

Karlins, M., Coffman, T. L., & Walters, G. (1969). On the fading of social stereotypes: Studies on three generations of college students. *Journal of Personality and Social Psychology, 13,* 1–16.

Katz, D. (1960). The functional approach to the study of attitudes. *Public Opinion Quarterly, 24,* 163–204.

Katz, D., & Braly, K. (1933). Racial stereotypes of one hundred college students. *Journal of Abnormal and Social Psychology, 28,* 280–290.

Katz, I. (1970). Experimental study in Negro-White relationships. In L. Berkowitz (Ed.), *Advances in experimental social psychology* (Vol. 5). New York: Academic Press.

Katz, I. (1981). *Stigma: A social psychological analysis.* Hillsdale, NJ: Erlbaum.

Katz, I., Wackenhut, J., & Glass, D. C. (1986). An ambivalence-amplification theory of behavior toward the stigmatized. In S. Worchel & W. G. Austin (Eds.), *Psychology of intergroup relations* (2nd ed., pp. 103–117). Chicago: Nelson-Hall.

Katz, J. (1988). *Seductions of crime: Moral and sensual attractions in doing evil.* New York: Basic Books.

Katz, P. (1986). Gender identity: Development and consequences. In R. Ashmore & F. Del Boca (Eds.), *The social psychology of male-female relations* (pp. 21–67). Orlando, FL: Academic Press.

Katzner, K. (1995). *The languages of the world* (3rd ed.). London: Routledge.

Kauffman, D. R., & Steiner, I. D. (1968). Some variables affecting the use of conformity as an ingratiation technique. *Journal of Experimental Social Psychology, 4,* 400–414.

Kaufman, J., & Johnson, K. (2004). Stigmatized individuals and the process of identity. *The Sociological Quarterly, 45,* 807–833.

Keller, C. E., Hallahan, D. P., McShane, E. A., Crowley, E. P., & Blandford, B. J. (1990). The coverage of persons with disabilities in American newspapers. *The Journal of Special Education, 24,* 271–282.

Kelley, H. H. (1950). The warm-cold variable in first impressions. *Journal of Personality, 18,* 431–439.

Kelley, H. H. (1967). Attribution theory in social psychology. In D. Levine (Ed.), *Nebraska symposium in motivation, 1967.* Lincoln: University of Nebraska Press.

Kelley, H. H. (1972). Causal schemata and the attribution process. In E. Jones, D. Kanouse, H. Kelley, R. Nisbett, S. Valms, & B. Weiner (Eds.), *Attribution: Perceiving the causes of behavior.* Morristown, NJ: General Learning Press.

Kelley, H. H. (1973). The process of causal attribution. *American Psychologist, 28,* 107–128.

Kelley, H. H., Berscheid, E., Christensen, A., Harvey, H. H., Huston, T., Levinger, G., et al. (1983). *Close relationships.* New York: Freeman.

Kelley, H. H., & Michela, J. L. (1980). Attribution theory and research. *Annual Review of Psychology, 31,* 457–501.

Kelley, H. H., & Thibaut, J. W. (1978). *Interpersonal relations: A theory of interdependence.* New York: Wiley.

Kelley, K., & Byrne, D. (1976). Attraction and altruism: With a little help from my friends. *Journal of Research in Personality, 10,* 59–68.

Kelman, H. C. (1974). Attitudes are alive and well and gainfully employed in the sphere of action. *American Psychologist, 29,* 310–324.

Kelman, H. C., & Hamilton, V. L. (1989). *Crimes of obedience: Toward a social psychology of authority and responsibility.* New Haven, CT: Yale University Press.

Keltner, D. (1995). The signs of appeasement: Evidence for the distinct displays of embarrassment, amusement, and shame. *Journal of Personality and Social Psychology, 68,* 441–454.

Kemper, T. D. (1973). The fundamental dimensions of social relationship: A theoretical statement. *Acta Sociologica, 16,* 41–57.

Kendon, A. (1970). Movement coordination in social interaction. Some examples described. *Acta Psychologica, 32,* 100–125.

Kendon, A., Harris, R. M., & Key, M. R. (1975). *Organization of behavior in face-to-face interaction.* The Hague: Mouton.

Kendzierski, D., & Whitaker, D. J. (1997). The role of self-schema in linking intention with behavior. *Personality and Social Psychology Bulletin, 23,* 139–147.

Kenney, D., & La Voie, L. (1982). Reciprocity of interpersonal attraction: A confirmed hypothesis. *Social Psychology Quarterly, 45,* 54–58.

Kenrick, D. T. (1995). Evolutionary theory versus the confederacy of dunces. *Psychological Inquiry, 6,* 56–62.

Kenrick, D. T., McCreath, H. E., Govern, J., King, R., & Bordin, J. (1990). Person-environment intersections: Everyday settings and common trait dimensions. *Journal of Personality and Social Psychology, 58,* 685–698.

Kent, G., Davis, J., & Shapiro, D. (1978). Resources required in the construction and reconstruction of conversations. *Journal of Personality and Social Psychology, 36,* 13–22.

Kerber, K. W. (1984). The perception of nonemergency helping situations: Costs, rewards, and the altruistic personality. *Journal of Personality, 52,* 177–187.

Kerckhoff, A. C. (1974). The social context of interpersonal attraction. In T. Huston (Ed.), *Foundations of interpersonal attraction* (pp. 61–78). New York: Academic Press.

Kerpelman, J. L., & Schvaneveldt, P. (1999). Young adults' anticipated identity importance of career, marital, and parental roles: Comparisons of men and women with different role balance orientations. *Sex Roles, 41,* 189–218.

Kerr, N., & Bruun, S. (1981). Ringelmann revisited: Alternative explanations for the social loafing effect. *Personality and Social Psychology Bulletin, 7,* 224–231.

Kerr, N. L., & Bruun, S. (1983). The dispensability of member effort and group motivation losses: Free rider effects. *Personality and Social Psychology Bulletin, 44,* 78–94.

Kessler, R. C. (1982). A disaggregation of the relationship between socioeconomic status and psychological distress. *American Sociological Review, 47,* 752–764.

Kessler, R. C., & Cleary, P. (1980). Social class and psychological distress. *American Sociological Review, 45,* 463–478.

Ketelaar, T., & Ellis, B. J. (2000). Are evolutionary explanations unfalsifiable? Evolutionary psychology and the Lakatosian philosophy of science. *Psychological Inquiry, 11,* 1–21.

Khanna, N. (2004). The role of reflected appraisals in racial identity: The case of multiracial Asians. *Social Psychology Quarterly, 67,* 115–131.

Kiecolt-Glaser, J. K., & Newton, T. L. (2001). Marriage and health: His and hers. *Psychological Bulletin, 127,* 472–503.

Kiesler, C. A., & Corbin, L. H. (1965). Commitment, attraction, and conformity. *Journal of Personality and Social Psychology, 2,* 890–895.

Kiesler, C. A., & Kiesler, S. B. (1969). *Conformity.* Reading, MA: Addison-Wesley.

Kiesler, C. A., Zanna, M., & DeSalvo, J. (1966). Deviation and conformity: Opinion change as a function of commitment, attraction, and the presence of a deviate. *Journal of Personality and Social Psychology, 3,* 458–467.

Kiesling, S. (2004). Dude. *American Speech, 79,* 281–305.

Kilham, W., & Mann, L. (1974). Level of destructive obedience as a function of transmitter and executant roles in the Milgram obedience paradigm. *Journal of Personality and Social Psychology, 29,* 696–702.

Killian, L. (1984). Organization, rationality, and spontaneity in the civil rights movement. *American Sociological Review, 49,* 770–783.

Kimberly, J. C. (1984). Cognitive balance, inequality, and consensus: Interrelations among fundamental processes in groups. In E. J. Lawler (Ed.), *Advances in group processes* (Vol. 1). Greenwich, CT: JAI Press.

Kingston, P., & Nock, S. (1987). Time together among dual-earner couples. *American Sociological Review, 52,* 391–400.

Kiparsky, P. (1976). Historical linguistics and the origin of language. *Annals of the New York Academy of Sciences, 280,* 97–103.

Kirchmeyer, C. (1993). Nonwork-to-work spillover: A more balanced view of the experiences and coping of professional men and women. *Sex Roles, 28,* 531–552.

Kitayama, S., & Burnstein, E. (1988). Automaticity in conversations: A reexamination of the mindlessness hypothesis. *Journal of Personality and Social Psychology, 54,* 219–224.

Kitcher, P. (1985). *Vaulting ambition.* Cambridge, MA: MIT Press.

Kitsuse, J. (1964). Societal reaction to deviant behavior: Problems of theory and method. In H. Becker (Ed.), *The other side.* New York: Free Press.

Klapper, J. T. (1960). *The effects of mass communication.* New York: Free Press.

Kleck, R. E., & Strenta, A. (1980). Perceptions of the impact of negatively valued physical characteristics on social interaction. *Journal of Personality and Social Psychology, 39,* 861–873.

Kleidman, R. (1994). Volunteer activism and professionalism in social movement organizations. *Social Problems, 41,* 257–276.

Klein, O., Snyder, M., & Livingston, R. W. (2004). Prejudice on the stage: Self-monitoring and the public expression of group attitudes. *British Journal of Social Psychology, 43,* 299–314.

Kling, K. C., Hyde, J., Showers, C., & Buswell, B. (1999). Gender differences in self-esteem: A meta-analysis. *Psychological Bulletin, 125,* 470–500.

Klinger, L. J., Hamilton, J. A., & Cantrell, P. J. (2001). Children's perceptions of aggressive and gender-specific content in toy commercials. *Social Behavior and Personality, 29,* 11–20.

Klitzner, M., Gruenewald, P. J., & Bamberger, E. (1991). Cigarette advertising and adolescent experimentation with smoking. *British Journal of Addiction, 86,* 287–298.

Knapp, M. L., Stafford, L., & Daly, J. A. (1986). Regrettable messages: Things people wish they hadn't said. *Journal of Communication, 36,* 40–58.

Knopf, T. A. (1975). *Rumors, race, and riots.* New Brunswick, NJ: Transaction.

Knottnerus, J. D. (1988). A critique of expectation states theory: Theoretical assumptions and models of social cognition. *Sociological Perspectives, 31,* 420–445.

Knutson, M. C. (2002). Tattooing, piercing, and branding: A guide for clinicians. *Medscape.* Available at www.medscape.com/viewarticle/444871.

Koestner, R., & Wheeler, L. (1988). Self-presentation in personal advertisements: The influence of implicit notions of attraction and role expectations. *Journal of Social and Personal Relationships, 5,* 149–160.

Kogan, N., & Wallach, M. (1964). *Risk taking: A study in cognition and personality.* New York: Holt, Rinehart and Winston.

Kohl, W. L., Steers, R., & Terborg, J. (1995). The effects of transformational leadership on teacher attitudes and student performance in Singapore. *Journal of Organizational Behavior, 73,* 695–703.

Kohlberg, L. (1969). Stage and sequence: The cognitive-developmental approach to socialization. In D. Goslin (Ed.), *Handbook of socialization theory and research.* Chicago: Rand McNally.

Kohn, M. (1969). *Class and conformity: A study in values.* Homewood, IL: Dorsey Press.

Kohn, M. (1976). Occupational structure and alienation. *American Journal of Sociology, 82,* 111–130.

Kohn, M., Naoi, A., Schoenbach, C., Schooler, C., & Slomczynski, K. M. (1990). Position in the class structure and psychological functioning in the United States, Japan, and Poland. *American Journal of Sociology, 95,* 964–1008.

Kohn, M., & Schooler, C. (1973). Occupational experience and psychological functioning: An assessment of reciprocal effects. *American Sociological Review, 38,* 97–118.

Kohn, M., & Schooler, C. (1982). Job conditions and personality: A longitudinal assessment of their reciprocal effects. *American Journal of Sociology, 87,* 1257–1286.

Kohn, M., & Schooler, C. (with Miller, J., Miller, K., Schoenbach, S., & Schoenberg, R.). (1983). *Work and personality: An inquiry into the impact of social stratification.* Norwood, NJ: Ablex.

Konecni, V. J. (1979). The role of aversive events in the development of intergroup conflict. In W. G. Austin & S. Worchel (Eds.), *The social psychology of intergroup relations.* Monterey, CA: Brooks/Cole.

Korte, C., Ypma, I., & Toppen, A. (1975). Helpfulness in Dutch society as a function of urbanization and environmental input level. *Journal of Personality and Social Psychology, 32,* 996–1003.

Korten, D. C. (1962). Situational determinants of leadership structure. *Journal of Conflict Resolution, 6,* 222–235.

Kortenhaus, C. M., & Demarest, J. (1993). Gender role stereotyping in children's literature: An update. *Sex Roles, 28,* 219–232.

Koskinen, S., & Martelin, T. (1994). Why are socio-conomic mortality differences smaller among women than among men? *Social Science and Medicine, 38,* 1385–1396.

Koss, M. P., & Leonard, K. E. (1984). Sexually aggressive men: Empirical findings and theoretical implications. In N. M. Malamuth & E. Donnerstein (Eds.), *Pornography and sexual aggression.* Orlando, FL: Academic Press.

Koss, M. P., & Oros, C. J. (1982). Sexual experiences survey: A research instrument investigating sexual aggression and victimization. *Journal of Consulting and Clinical Psychology, 50,* 455–457.

Kothandapani, V. (1971). Validation of feeling, belief, and intention to act as three components of attitude and their contribution to prediction of contraceptive behavior. *Journal of Personality and Social Psychology, 19,* 321–333.

Kozielecki, J. (1984). Rodzaje samoprezentacji [Types of self-presentation]. *Psychologia Wychowawcza, 27,* 129–137.

Kraus, S. (Ed.). (1962). *The great debates.* Bloomington: Indiana University Press.

Krauss, R., & Chiu, C. (1998). Language and social behavior. In D. Gilbert, S. Fiske, & G. Lindzey (Eds.), *The handbook of social psychology* (4th ed., Vol. 2, 41–88). Boston: McGraw-Hill.

Krauss, R. M., & Fussell, S. K. (1996). Social psychological models of interpersonal communication. In E. T. Higgins & A. Kruglanski (Eds.), *Social psychology: Handbook of basic principles* (pp. 655–701). New York: Guilford Press.

Krauss, R. M., Morrel-Samuels, P., & Colasante, C. (1991). Do conversational hand gestures communicate? *Journal of Personality and Social Psychology, 61,* 743–754.

Kraut, R. E. (1980). Humans as lie-detectors: Some second thoughts. *Journal of Communication, 30,* 209–216.

Kraut, R. E., & Poe, D. (1980). Behavior roots of person perception: The deception judgments of customs inspectors and laymen. *Journal of Personality and Social Psychology, 39,* 784–798.

Kravitz, D. A., & Martin, B. (1986). Ringelmann rediscovered: The original article. *Journal of Personality and Social Psychology, 50,* 936–941.

Krebs, D. L. (1975). Empathy and altruism. *Journal of Personality and Social Psychology, 32,* 1134–1146.

Krebs, D. L. (1982). Psychological approaches to altruism: An evaluation. *Ethics, 92,* 147–158.

Krebs, D. L., & Miller, D. T. (1985). Altruism and aggression. In G. Lindzey & E. Aronson, (Eds.), *The handbook of social psychology* (3rd ed.). New York: Random House.

Kressel, K., & Pruitt, D. G. (1985). Themes in the mediation of social conflict. *Journal of Social Issues, 41*(2), 179–198.

Kriesberg, L. (1973). *The sociology of social conflicts.* Englewood Cliffs, NJ: Prentice Hall.

Kritzer, H. (1977). Political protest and political violence: A nonrecursive causal model. *Social Forces, 55,* 630–640.

Krohn, M. D. (1986). The web of conformity: A network approach to the explanation of delinquent behavior. *Social Problems, 33,* 581–593.

Krohn, M. D., Massey, J. L., & Zielinski, M. (1988). Role overlap, network multiplexity, and adolescent deviant behavior. *Social Psychology Quarterly, 51,* 346–356.

Krosnick, J. A., & Schuman, H. (1988). Attitude intensity, importance, and certainty and susceptibility to response effects. *Journal of Personality and Social Psychology, 54,* 940–952.

Kruger, D. J. (2001). Inclusive fitness and judgments of helping behaviors: Adaptations for kin directed altruism. *Social Behavior and Personality, 29,* 323–330.

Kruglanski, A. W., & Mackie, D. M. (1990). Majority and minority influence: A judgmental process analysis. In W. Stroebe & M. Hewstone (Eds.), *European review of social psychology* (Vol. 1, pp. 229–261). Chichester, UK: Wiley.

Krull, D. S., & Dill, J. C. (1996). On thinking first and responding fast: Flexibility in social inference processes. *Personality and Social Psychology Bulletin, 22,* 949–959.

Krull, D. S., Loy, M. H., Lin, J., Wang, C., Chen, S., & Zhao, X. (1999). The fundamental correspondence bias in individualist and collectivist cultures. *Personality and Social Psychology Bulletin, 25,* 1208–1219.

Kuhn, D., Langer, J., Kohlberg, L., & Haan, N. (1977). The development of formal operations in logical and moral judgement. *Genetic Psychology Monographs, 95,* 97–188.

Kuhn, M. H., & McPartland, T. (1954). An empirical investigation of self-attitudes. *American Sociological Review, 19,* 68–76.

Kulick, J. A., & Brown, R. (1979). Frustration, attribution of blame, and aggression. *Journal of Experimental Social Psychology, 15,* 183–194.

Kumkale, G. T., & Albarracín, D. (2004). The sleeper effect in persuasion: A meta-analytic review. *Psychological Bulletin, 130,* 143–172.

Kunkel, D. (1997). *Policy implications of the National Television Violence Study.* Paper presented at the American Psychological Association meeting, Chicago, IL.

Kurdek, L. A. (1991). The relation between reported well-being and divorce history, availability of proximate adult, and gender. *Journal of Marriage and the Family, 53,* 71–78.

Kurmeyer, S. L., & Biggers, K. (1988). Environmental demand and demand engendering behavior: An observational analysis of the Type A pattern. *Journal of Personality and Social Psychology, 54,* 997–1005.

Kurtines, M. M. (1986). Moral behavior as rule governed behavior: Person and situation effects on moral decision-making. *Journal of Personality and Social Psychology, 50,* 784–791.

Labov, W. (1972). *Sociolinguistic patterns.* Philadelphia: University of Pennsylvania Press.

Lachman, S. J., & Bass, A. R. (1985). A direct study of halo effect. *Journal of Psychology, 119,* 535–540.

LaFrance, M., & Mayo, C. (1978). *Moving bodies: Nonverbal communication in social relationships.* Monterey, CA: Brooks/Cole.

LaFree, G., & Drass, K. A. (1996). The effect of changes in intraracial income inequality and educational attainment on changes in arrest rates for African Americans and Whites, 1957–1990. *American Sociological Review, 61,* 614–634.

Lakoff, R. T. (1979). Women's language. In O. Buturff & E. L. Epstein (Eds.), *Women's language and style.* Akron, OH: University of Akron.

Lamb, M. E. (1979). Paternal influence and the father's role: A personal perspective. *American Psychologist, 34,* 938–993.

Lamm, H. (1988). A review of our research on group polarization: Eleven experiments on the effects of group discussion on risk acceptance, probability estimation, and negotiation positions. *Psychological Reports, 62,* 807–813.

Lamm, H., & Sauer, C. (1974). Discussion-induced shift toward higher demands in negotiation. *European Journal of Social Psychology, 4,* 85–88.

Lamm, H., & Schwinger, T. (1980). Norms concerning distributive justice: Are needs taken into consideration in allocation decisions? *Social Psychology Quarterly, 43,* 425–429.

Lamm, H., & Trommsdorff, G. (1973). Group versus individual performance on tasks requiring ideational proficiency (brainstorming): A review. *European Journal of Social Psychology, 3,* 361–388.

Land, K. C., McCall, P. L., & Cohen, L. E. (1990). Structural correlates of homicide rates: Are there invariances across time and social space? *American Journal of Sociology, 95,* 922–963.

Landis, D., & O'Shea, W., III. (2000). Cross-cultural aspects of passionate love: An individual differences analysis. *Journal of Cross-Cultural Psychology, 31,* 752–777.

Landrine, H., Klonoff, E., & Brown-Collins, A. (1995). Cultural diversity and methodology in feminist psychology: Critique, proposal, empirical example. In H. Landrine (Ed.), *Bringing cultural diversity to feminist psychology: Theory, research, and practice* (pp. 55–75). Washington, DC: American Psychological Association.

Lange, C. G. (1922). The emotions: A psychophysiological study. In C. G. Lange & W. James (Eds.), *The emotions* (pp. 33–90). Baltimore: Williams & Wilkins.

Lantz, H., Keyes, J., & Schultz, M. (1975). The American family in the preindustrial period: From base lines in history to change. *American Sociological Review, 40,* 21–36.

Lantz, H., Schultz, M., & O'Hara, M. (1977). The changing American family from the preindustrial to the industrial period: A final report. *American Sociological Review, 42,* 406–421.

Lantz, D., & Stefflre, V. (1964). Language and cognition revisited. *Journal of Abnormal and Social Psychology, 69,* 472–481.

LaPiere, R. (1934). Attitudes versus actions. *Social Forces, 13,* 230–237.

Larimer, M., Lydum, A., Anderson, B., & Turner, A. (1999). Male and female recipients of unwanted sexual contact in a college student sample. *Sex Roles, 40,* 295–308.

Larson, L. L., & Rowland, K. (1973). Leadership style, stress, and behavior in task performance. *Organizational Behavior and Human Performance, 9,* 407–421.

Larzelere, T., & Huston, T. (1980). The dyadic trust scale: Toward understanding interpersonal trust in close relationships. *Journal of Marriage and the Family, 42,* 595–604.

Latané, B. (1981). The psychology of social impact. *American Psychologist, 36,* 343–356.

Latané, B., & Darley, J. M. (1970). *The unresponsive bystander: Why doesn't he help?* New York: Appleton-Century-Crofts.

Latané, B., & Nida, S. (1981). Ten years of research on group size and helping. *Psychological Bulletin, 89,* 308–324.

Latané, B., Nida, S. A., & Wilson, D. W. (1981). The effects of group size on helping behavior. In J. P. Rushton & R. M. Sorrentino (Eds.), *Altruism and helping behavior: Social, personality, and developmental perspectives.* Hillsdale, NJ: Erlbaum.

Latané, B., & Rodin, J. (1969). A lady in distress: Inhibiting effects of friends and strangers on bystander intervention. *Journal of Experimental Social Psychology, 5,* 189–202.

Latané, B., Williams, K., & Harkins, S. (1979). Many hands make light the work: The causes and consequences of social loafing. *Journal of Personality and Social Psychology, 37,* 822–832.

Latané, B., & Wolf, S. (1981). The social impact of majorities and minorities. *Psychological Review, 88,* 438–453.

Lau, R. R. (1989). Individual and contextual influences on group identification. *Social Psychology Quarterly, 52,* 220–231.

Lau, R. R., & Russell, D. (1980). Attributions in the sports pages. *Journal of Personality and Social Psychology, 39,* 29–38.

Lauderdale, P. (1976). Deviance and moral boundaries. *American Sociological Review, 41,* 660–676.

Laughlin, P. R. (1980). Social combination processes of cooperative, problem-solving groups as verbal intellective tasks. In M. Fishbein (Ed.), *Progress in social psychology* (Vol. 1). Hillsdale, NJ: Erlbaum.

Laumann, E., Ellingson, S., Mahay, J., Paik, A., and Youm, Y. (Eds.). (2004). *The sexual organization of the city.* Chicago: University of Chicago Press.

Laumann, E. O., Gagnon, J. H., Michael, R. T., & Michaels, S. (1994). *The social organization of sexuality.* Chicago: University of Chicago Press.

Lavine, H., Thomsen, C. J., & Gonzales, M. H. (1997). The development of interattitudinal consistency: The shared-consequences model. *Journal of Personality and Social Psychology, 72,* 735–749.

Lavoie, F., Hebert, M., Tremblay, R., Vitaro, F., Vezina, L., & McDuff, P. (2002). History of family dysfunction and perpetration of dating violence by adolescent boys: A longitudinal study. *Journal of Adolescent Health, 30,* 375–383.

Lawler, E. J. (1975a). An experimental study of factors affecting the mobilization of revolutionary coalitions. *Sociometry, 38,* 163–179.

Lawler, E. J. (1975b). The impact of status differences on coalitional agreements: An experimental study. *Journal of Conflict Resolution, 19,* 271–285.

Lawler, E. J. (1983). Cooptation and threats as "divide and rule" tactics. *Social Psychology Quarterly, 46,* 89–98.

Lawler, E. J. (1986). Bilateral deterrence and conflict spiral: A theoretical analysis. In E. J. Lawler (Ed.), *Advances in group processes* (Vol. 3, pp. 107–130). Greenwich, CT: JAI Press.

Lawler, E. J., & Bacharach, S. B. (1987). Comparison of dependence and punitive forms of power. *Social Forces, 66,* 446–462.

Lawler, E. J., & Ford, R. (1995). Bargaining and influence in conflict situations. In K. S. Cook, G. A. Fine, & J. S. House (Eds.), *Sociological perspectives on social psychology* (pp. 236–256). Boston: Allyn & Bacon.

Lawler, E. J., Ford, R. S., & Blegen, M. A. (1988). Coercive capability in conflict: A test of bilateral deterrence versus conflict spiral theory. *Social Psychology Quarterly, 51,* 93–107.

Lawler, E. J., & O'Gara, P. W. (1967). Effects of inequity produced by underpayment on work output, work quality, and attitudes toward work. *Journal of Applied Psychology, 51,* 403–410.

Lawler, E. J., Youngs, J. A., Jr., & Lesh, M. D. (1978). Cooptation and coalition mobilization. *Journal of Applied Social Psychology, 8,* 199–214.

Lazarsfeld, P. F., Berelson, B., & Gaudet, H. (1948). *The people's choice: How the voter makes up his mind in a presidential campaign.* New York: Columbia University Press.

Lazarus, R. S. (1991). *Emotion and adaptation.* New York: Oxford University Press.

Le Bon, G. (1895). *Psychologie des foules* [Crowd psychology]. London: Unwin (1903).

Leana, C. R. (1985). A partial test of Janis' groupthink model: Effects of group cohesiveness and leader behavior on defective decision making. *Journal of Management, 11,* 5–17.

Leander, K. (2002). Silencing in classroom interaction: Producing and relating social spaces. *Discourse Processes, 34,* 193–235.

Leaper, C., Anderson, K., & Sanders, P. (1998). Moderators of gender effects on parents' talk to their children: A meta-analysis. *Developmental Psychology, 34,* 3–27.

Lear, D. (1997). *Sex and sexuality: Risk and relationships in the age of AIDS.* Thousand Oaks, CA: Sage.

Leary, A., & Tobin, T. (2005, March 18). In schools, violence starting at earlier ages. *St. Petersburg Times.*

Leary, M., Tchividjian, L., & Kraxberger, B. (1994). Self-presentation can be hazardous to your health: Impression management and health risk. *Health Psychology, 13,* 461–470.

Leary, M. R. (1995). *Self-presentation: Impression management and interpersonal behavior.* Madison, WI: Brown & Benchmark.

Leary, M. R., & Kowalski, R. M. (1990). Impression management: A literature review and two-component model. *Psychological Bulletin, 107,* 34–47.

Leary, M. R., Wheeler, D. S., & Jenkins, T. B. (1986). Aspects of identity and behavioral preference: Studies of occupational and recreational choice. *Social Psychology Quarterly, 49,* 11–18.

Ledvinka, J. (1971). Race of interviewer and the language elaboration of Black interviewees. *Journal of Social Issues, 27,* 185–197.

Lee, G. R., Seccombe, K., & Shehan, C. L. (1991). Marital status and personal happiness: An analysis of trend data. *Journal of Marriage and the Family, 53,* 839–844.

Lee, T. R., Mancini, J. A., & Maxwell, J. W. (1990). Sibling relationships in adulthood: Contact patterns and motivations. *Journal of Marriage and the Family, 52,* 431–440.

Leffler, A., Gillespie, D. L., & Conaty, J. C. (1982). The effects of status differentiation on nonverbal behavior. *Social Psychology Quarterly, 45,* 153–161.

Leigh, J. P., Markowitz, S. B., Fahs, M., Shin, C., & Landrigan, P. J. (1997). Occupational injury and illness in the United States: Estimates of costs, morbidity, and mortality. *Archives of Internal Medicine, 157,* 1557–1568.

Leippe, M. R., & Elkin, R. A. (1987). When motives clash: Issue involvement and response involvement as determinants of persuasion. *Journal of Personality and Social Psychology, 52,* 269–278.

Lemert, E. (1951). *Social pathology.* New York: McGraw-Hill.

Lemert, E. (1962). Paranoia and the dynamics of exclusion. *Sociometry, 25,* 2–20.

Lennington, S. (1981). Child abuse: The limits of sociobiology. *Ethology and Sociobiology, 2,* 17–29.

Lennon, M. C., Link, B. G., Marbach, J. J., & Dohrenwend, B. P. (1989). The stigma of chronic facial pain and its impact on social relationships. *Social Problems, 36,* 117–134.

Lepper, M., Greene, D., & Nisbett, R. (1973). Undermining children's intrinsic interest with extrinsic reward: A test of the "overjustification" hypothesis. *Journal of Personality and Social Psychology, 28,* 129–137.

Leventhal, G. S. (1979). Effects of external conflict on resource allocation and fairness within groups and organizations. In W. G. Austin & S. Worchel (Eds.), *The social psychology of intergroup relations.* Monterey, CA: Brooks/Cole.

Leventhal, H. (1970). Findings and theory in the study of fear communications. In L. Berkowitz (Ed.), *Advances in experimental social psychology* (Vol. 5). New York: Academic Press.

Leventhal, H. (1984). A perceptual motor theory of emotion. In K. R. Scherer & P. Ekman (Eds.), *Approaches to emotion.* Hillsdale, NJ: Erlbaum.

Leventhal, G. S., & Lane, D. W. (1970). Sex, age, and equity behavior. *Journal of Personality and Social Psychology, 15,* 312–316.

Leventhal, G. S., Michaels, J. W., & Sanford, C. (1972). Inequity and interpersonal conflict: Reward allocation and secrecy about reward as methods of preventing conflict. *Journal of Personality and Social Psychology, 23,* 88–102.

Leventhal, H., & Singer, R. P. (1966). Affect arousal and positioning of recommendations in persuasive comminications. *Journal of Personality and Social Psychology, 4,* 137–146.

Leventhal, G. S., Weiss, T., & Long, G. (1969). Equity, reciprocity, and reallocating the rewards in the dyad. *Journal of Personality and Social Psychology, 13,* 300–305.

Levin, J., & Arluke, A. (1982). Embarrassment and helping behavior. *Psychological Reports, 51,* 999–1002.

Levin, P. E., & Isen, A. M. (1975). Something you can still get for a dime: Further studies on the effects of feeling good on helping. *Sociometry, 38,* 141–147.

Levine, M., Prosser, A., Evans, D., & Reicher, S. (2005). Identity and emergency intervention: How social group membership and inclusiveness of group boundaries shape helping behavior. *Personality and Social Psychology Bulletin, 31,* 443–453.

Levine, P. (2000). *The sexual activity and birth control use of American teenagers* (Working Paper W7601). Cambridge, MA: National Bureau of Economic Research.

Levine, J. M., & Moreland, R. L. (1998). Small groups. In D. Gilbert, S. Fiske, & G. Lindzey (Eds.), *The handbook of social psychology* (4th ed., pp. 415–469). Boston: McGraw-Hill.

Levine, J. M., Saxe, L., & Ranelli, C. J. (1975). Extreme dissent, conformity reduction, and the bases of social influence. *Social Behavior and Personality, 3,* 117–126.

LeVine, R. A., & Campbell, D. T. (1972). *Ethnocentrism: Theories of conflict, ethnic attitudes, and group behavior.* New York: Wiley.

Levinger, G. (1974). A three-level approach to attraction: Toward an understanding of pair relatedness. In T. Huston (Ed.), *Foundations of interpersonal attraction.* New York: Academic Press.

Levinger, G. (1976). A social psychological perspective on marital dissolution. *Journal of Social Issues, 32*(1), 21–47.

Levinson, D. (1978). *The seasons of a man's life.* New York: Knopf.

Levinson, R., Powell, B., & Steelman, L. C. (1986). Social location, significant others, and body image among adolescents. *Social Psychology Quarterly, 49,* 330–337.

Levitin, T. E. (1975). Deviants are active participants in the labelling process: The visibly handicapped. *Social Problems, 22,* 548–557.

Lewicki, P. A. (1982). Social psychology as viewed by its practitioners: Survey of SESP members' opinions. *Personality and Social Psychology Bulletin, 8,* 409–416.

Lewin, D., Feuille, P., & Kochan, T. (1977). *Public sector labor relations: Analysis and readings.* Glen Springs, NJ: Thomas Horton.

Lewis, G. H. (1972). Role differentiation. *American Sociological Review, 37,* 424–434.

Lewis, K., Winsett, R., Cetingok, M., Martin, J., & Hathaway, D. (2000). Social network mapping with transplant recipients. *Progress in Transplantation, 10,* 262–266.

Lewis, M., & Brooks-Gunn, J. (1979). Toward a theory of social cognition: The development of self. In I. Uzgiris (Ed.), *Social interaction and communication during infancy. New directions for child development* (Vol. 4). San Francisco: Jossey-Bass.

Lewis, S. A., Langan, C. J., & Hollander, E. P. (1972). Expectation of future interaction and the choice of less desirable alternatives in conformity. *Sociometry, 35,* 440–447.

Lex, B. W. (1986). Measurement of alcohol consumption in fieldwork settings. *Medical Anthropology Quarterly, 17,* 95–98.

Lichter, D., and Qian, Z. (2004). *Marriage and family in a multiracial society.* New York: Russell Sage Foundation.

Lichterman, P. (1995). Piecing together multicultural community: Cultural differences in community building among grass-roots environmentalists. *Social Problems, 42,* 513–534.

Lickel, B., Hamilton, D. L., Wieczorkowski, G., Lewis, A., Sherman, S. J., & Uhles, A. N. (2000). Varieties of groups and the perception of group entiativity. *Journal of Personality and Social Psychology, 78,* 223–246.

Liden, R. C., & Mitchell, T. R. (1988). Ingratiatory behaviors in organizational settings. *The Academy of Management Review, 13,* 572–614.

Lieberman, P. (1975). *On the origins of human language: An introduction to the evolution of human speech.* New York: MacMillan.

Lieberman, S. (1965). The effect of changes of roles on the attitudes of role occupants. In H. Proshansky & B. Seidenberg (Eds.), *Basic studies in social psychology* (pp. 485–494). New York: Holt, Rinehart and Winston.

Lieberman, A., & Chaiken, S. (1992). Defensive processing of personally relevant health messages. *Personality and Social Psychology Bulletin, 18,* 669–679.

Lieberman, A., & Chaiken, S. (1996). The direct effect of personal relevance on attitudes. *Personality and Social Psychology Bulletin, 22,* 269–279.

Lightdale, J. R., & Prentice, D. A. (1994). Rethinking sex differences in aggression: Aggressive behavior in the absence of social roles. *Personality and Social Psychology Bulletin, 20,* 34–44.

Likert, R. (1932). A technique for the measurement of attitudes. *Archives of Psychology* (Whole no. 142).

Lin, N. (1974–1975). The McIntire march: A study of recruitment and commitment. *Public Opinion Quarterly, 38,* 562–573.

Lin, N., Ensel, W., & Vaughn, J. (1981). Social resources and strength of ties: Structural factors in occupational status attainment. *American Sociological Review, 46,* 393–405.

Lin, N., & Xie, W. (1988). Occupational prestige in urban China. *American Journal of Sociology, 93,* 793–832.

Lincoln, J., & Kalleberg, A. (1985). Work organization and workforce commitment: A study of plants and employees in the United States and Japan. *American Sociological Review, 50,* 738–760.

Linder, D. E., Cooper, J., & Jones, E. (1967). Decision freedom as a determinant of the role of incentive magnitude in attitude change. *Journal of Personality and Social Psychology, 6,* 245–254.

Lindskold, S. (1986). GRIT: Reducing distrust through carefully introduced conciliation. In S. Worchel & W. G. Austin (Eds.), *Psychology of intergroup relations* (2nd ed., pp. 305–323). Chicago: Nelson-Hall.

Lindskold, S., & Aronoff, J. R. (1980). Conciliatory strategies and relative power. *Journal of Experimental Social Psychology, 16,* 187–198.

Lindskold, S., Cullen, P., Gahagen, J., & Tedeschi, J. T. (1970). Developmental aspects of reaction to positive inducements. *Developmental Psychology, 3,* 277–284.

Lindskold, S., & Finch, M. L. (1981). Styles of announcing conciliation. *Journal of Conflict Resolution, 25,* 145–155.

Lindskold, S., Han, G., & Betz, B. (1986). The essential elements of communication in the GRIT strategy. *Personality and Social Psychology Bulletin, 12,* 179–186.

Lindskold, S., & Tedeschi, J. T. (1971). Reward power and attraction in interpersonal conflict. *Psychonomic Science, 22,* 211–213.

Link, B. G. (1982). Mental patient status, work, and income: An examination of the effects of a psychiatric label. *American Sociological Review, 47,* 202–215.

Link, B. G. (1987). Understanding labelling effects in the area of mental disorders: An assessment of the effects of expectations of rejection. *American Sociological Review, 52,* 96–112.

Link, B. G., Cullen, F. T., Struening, E., Shrout, P. E., & Dohrenwend, B. (1989). A modified labeling theory approach to mental disorders: An empirical assessment. *American Sociological Review, 54,* 400–423.

Link, B. G., & Phelan, J. C. (2001). Conceptualizing stigma. *Annual Review of Sociology, 27,* 363–385.

Link, B. G., Struening, E. L., Rahav, M., Phelan, J. C., & Nuttbrock, L. (1997). On stigma and its consequences: Evidence from a longitudinal study of men with dual diagnosis of mental illness and substance abuse. *Journal of Health and Social Behavior, 38,* 177–190.

Linkenhoker, B. A., & Knudsen, E. I. (2002). Incremental training increases the plasticity of the auditory space in adult barn owls. *Nature, 419,* 293–296.

Linville, P. W. (1982). The complexity-extremity effect and age-based stereotyping. *Journal of Personality and Social Psychology, 42,* 193–211.

Linville, P. W., Fischer, G. W., & Salovey, P. (1989). Perceived distributions of the characteristics of in-group and out-group members: Empirical evidence and a computer simulation. *Journal of Personality and Social Psychology, 57,* 165–188.

Linville, P. W., & Jones, E. E. (1980). Polarized appraisals of out-group members. *Journal of Personality and Social Psychology, 38,* 689–703.

Lipe, M. G. (1991). Counterfactual reasoning as a framework for attribution theories. *Psychological Bulletin, 109,* 456–471.

Lipscomb, T. J., Larrieu, J. A., McAllister, H. A., & Bregman, N. J. (1982). Modeling and children's generosity: A developmental perspective. *Merrill-Palmer Quarterly, 28,* 275–282.

Lipscomb, T. J., McAllister, H. A., & Bregman, N. J. (1985). A developmental inquiry into the effects of multiple models on children's generosity. *Merrill-Palmer Quarterly, 31,* 335–344.

Lisak, D., & Roth, S. (1988). Motivational factors in nonincarcerated sexually aggressive men. *Journal of Personality and Social Psychology, 55,* 795–802.

Liska, A. (1984). A critical examination of the causal structure of the Fishbein-Ajzen attitude-behavior model. *Social Psychology Quarterly, 47,* 61–74.

Littlepage, G. E. (1991). Effects of group size and task characteristics on group performance: A test of Steiner's model. *Personality and Social Psychology Bulletin, 17,* 449–456.

Littlepage, G., & Pineault, T. (1978). Verbal, facial, and paralinguistic cues to the detection of truth and lying. *Personality and Social Psychology Bulletin, 4,* 461–464.

Lloyd, S., & Emery, B. (2000). The context and dynamics of intimate aggression against women. *Journal of Social and Personal Relationships, 17,* 503–521.

Locher, D. A. (2002). *Collective behavior.* Upper Saddle River, NJ: Prentice Hall.

Locke, K. D., & Horowitz, L. M. (1990). Satisfaction in interpersonal interactions as a function of similarity in level of dysphoria. *Journal of Personality and Social Psychology, 58,* 823–831.

Lohr, J. M., & Staats, A. (1973). Attitude conditioning in Sino-Tibetan languages. *Journal of Personality and Social Psychology, 26,* 196–200.

Lois, J. (1999). Socialization to heroism: Individualism and collectivism in a voluntary search and rescue group. *Social Psychology Quarterly, 62,* 117–135.

Lopata, H. (1988). Support systems of American urban widowhood. *Journal of Social Issues, 44*(3), 113–128.

Lopes, P. N., Salovey, P., Cote, S., & Beers, M. (2005). Emotion regulation abilities and the quality of social interaction. *Emotion, 5,* 133–118.

Lord, C. G., Lepper, M. R., & Mackie, D. (1984). Attitude prototypes as determinants of attitude-behavior consistency. *Journal of Personality and Social Psychology, 46,* 1254–1266.

Lorence, J., & Mortimer, J. (1985). Job involvement through the life course: A panel study of three age groups. *American Sociological Review, 50,* 618–638.

Lorenz, K. (1966). *On aggression.* New York: Harcourt Brace Jovanovich.

Lorenz, K. (1974). *Civilized man's eight deadly sins.* New York: Harcourt Brace Jovanovich.

Lott, A., & Lott, B. (1965). Group cohesiveness as interpersonal attraction: A review of relationships with antecedent and consequent variables. *Psychological Bulletin, 64,* 259–309.

Lott, A., & Lott, B. (1974). The role of reward in the formation of positive interpersonal attitudes. In T. Huston (Ed.), *Foundations of interpersonal attraction.* New York: Academic Press.

Lottes, I. L. (1993). Nontraditional gender roles and the sexual experience of heterosexual college students. *Sex Roles, 29,* 645–669.

Lottes, I. L., & Kuriloff, P. J. (1994). The impact of college experience on political and social attitudes. *Sex Roles, 31,* 31–54.

Lowenthal, M. F., Thurnher, M., & Chiriboga, D. (1975). *Four stages of life: A comparative study of women and men facing transitions.* San Francisco: Jossey-Bass.

Luchins, A. S. (1957). Experimental attempts to minimize the impact of first impressions. In C. I. Hovland (Ed.), *The order of presentation in persuasion.* New Haven, CT: Yale University Press.

Luhman, R. (1990). Appalachian English stereotypes: Language attitudes in Kentucky. *Language in Society, 19,* 331–348.

Lundman, R. J. (1974). Routine police arrest practices: A commonweal perspective. *Social Problems, 22,* 127–141.

Lundman, R. J., Sykes, R. E., & Clark, J. P. (1978). Police control of juveniles: A replication. *Journal of Research in Crime and Delinquency, 15,* 74–91.

Lurigis, A. J., & Carroll, J. S. (1985). Probation officers' schemata of offenders: Content, development, and impact on treatment decisions. *Journal of Personality and Social Psychology, 48,* 1112–1126.

Lutfey, K., & Mortimer, J. (2003). Development and socialization through the adult life course. In J. DeLamater (Ed.), *Handbook of social psychology.* New York: Kluwer-Plenum.

Lynch, J. C., Jr., & Cohen, J. L. (1978). The use of subjective expected utility theory as an aid to understanding variables that influence helping behavior. *Journal of Personality and Social Psychology, 36,* 1138–1151.

Maass, A., & Arcuri, L. (1992). The role of language in the persistence of stereotypes. In G. R. Semin & K. Fiedler (Eds.), *Language, interaction, and social cognition* (pp. 129–143). Newbury Park, CA: Sage.

Maass, A., West, S. G., & Cialdini, R. B. (1987). Minority influence and conversion. In C. Hendrick (Ed.), *Group processes* (pp. 55–79). Newbury Park, CA: Sage.

MacEwen, K. E., & Barling, J. (1991). Effects of maternal employment experiences on children's behavior via mood, cognitive difficulties, and parenting behavior. *Journal of Marriage and the Family, 53,* 635–644.

Macintyre, S., Hunt, K., & Sweeting, H. (1996). Gender differences in health: Are things really as simple as they seem? *Social Science and Medicine, 42,* 617–624.

Mackey, R., & O'Brien, B. (1999). Adaptation in lasting marriages. *Families in Society, 80,* 587–596.

Mackie, M. (1983). The domestication of self: Gender comparisons of self-imagery and self-esteem. *Social Psychology Quarterly, 46,* 343–350.

Mackie, D. M., & Goethals, G. R. (1987). Individual and group goals. In C. Hendrick (Ed.), *Group processes* (pp. 144–166). Newbury Park, CA: Sage.

Macmillan, R. (2001). Violence and the life course: The consequences of victimization for personal and social development. *Annual Review of Sociology, 27,* 1–22.

Maddi, S., Bartone, P., & Puccetti, M. (1987). Stressful events are indeed a factor in physical illness: Reply to Schroeder and Costa (1984). *Journal of Personality and Social Psychology, 52,* 833–843.

Maddux, J. E., & Rogers, R. W. (1980). Effects of source expertness, physical attractiveness, and supporting arguments on persuasion: A case of brains over beauty. *Journal of Personality and Social Psychology, 39,* 235–244.

Maddux, J. E., & Rogers, R. W. (1983). Protection motivation and self-efficacy: A revised theory of fear appeals and attitude change. *Journal of Experimental Social Psychology, 19,* 469–479.

Madon, S., Jussim, L., & Eccles, J. (1997). In search of the powerful self-fulfilling prophecy. *Journal of Personality and Social Psychology, 72,* 791–809.

Madsen, D. B. (1978). Issue importance and choice shifts: A persuasive arguments approach. *Journal of Personality and Social Psychology, 36,* 1118–1127.

Magdol, L. H. (1989). *Variations in social support from friends and kin.* Unpublished master's thesis, University of Wisconsin.

Magner, N., Johnson, G., Sobery, J., & Welker, R. (2000). Enhancing procedural justice in local government budget and tax decision making. *Journal of Applied Social Psychology, 30,* 798–815.

Mahon, N. E. (1982). The relationship of self-disclosure, interpersonal dependency, and life changes to loneliness in young adults. *Nursing Research, 31,* 343–347.

Maines, D., & Hardesty, M. (1987). Temporality and gender: Young adults' career and family plans. *Social Forces, 66,* 102–120.

Major, B., Schmidlin, A. M., & Williams, L. (1990). Gender patterns in social touch: The impact of setting and age. *Journal of Personality and Social Psychology, 58,* 634–643.

Major, B., Zubek, J. M., Cooper, M. L., Cozarelli, C., & Richards, C. (1997). Mixed messages: Implications of social conflict and social support within close relationships for adjustment to a stressful life event. *Journal of Personality and Social Psychology, 72,* 1349–1363.

Malamuth, N. M. (1983). Factors associated with rape as predictors of laboratory aggression against women. *Journal of Personality and Social Psychology, 45,* 432–442.

Malamuth, N. M. (1984). Aggression against women: Cultural and individual causes. In N. M. Malamuth & E. Donnerstein (Eds.), *Pornography and sexual aggression.* Orlando, FL: Academic Press.

Malamuth, N. M. (1989). Sexually violent media, thought patterns, and antisocial behavior. *Public communication and behavior,* Vol. 2. Orlando, FL: Academic Press.

Malamuth, N. M., & Briere, J. (1986). Sexual violence in the media: Indirect effects on aggression against women. *Journal of Social Issues, 42*(3), 75–92.

Malamuth, N. M., & Check, J. V. P. (1981). The effects of mass media exposure on acceptance of violence against women: A field experiment. *Journal of Research in Personality, 15,* 436–446.

Malamuth, N. M., Heavey, C. L., & Linz, D. (1993). Predicting men's antisocial behavior against women: The interaction model of sexual aggression. In N. G. C. Hall & R. Hirschman (Eds.), *Sexual aggression: Issues in etiology and assessment, treatment and policy.* New York: Hemisphere.

Malamuth, N. M., Linz, D., Heavey, C. L., Barnes, G., & Acker, M. (1995). Using the confluence model of sexual aggression to predict men's conflict with women: A 10-year follow up study. *Journal of Personality and Social Psychology, 69,* 353–369.

Mallozzi, J., McDermott, V., & Kayson, W. A. (1990). Effects of sex, type of dress, and location on altruistic behavior. *Psychological Reports, 67,* 1103–1106.

Manis, M., Shedler, J., Jonides, J., & Nelson, T. E. (1993). Availability heuristic in judgments of set size and frequency of occurrence. *Journal of Personality and Social Personality, 65,* 448–457.

Mann, C. R. (1996). *When women kill.* Albany, NY: SUNY Press.

Mannheim, B. F. (1966). Reference groups, membership groups, and the self-image. *Sociometry, 29,* 265–279.

Mannheimer, D., & Williams, R. M., Jr. (1949). A note on Negro troops in combat. In S. A. Stouffer, E. A. Suchman, L. C. DeVinney, S. A. Star, & R. M. Williams, Jr. (Eds.), *The American soldier* (Vol. 1). Princeton, NJ: Princeton University Press.

Manning, P. K., & Cullum-Swan, B. (1992). Semiotics and framing: Examples. *Semiotica, 92,* 239–257.

Mannix, E. A., Neale, M.A., & Northcraft, G. B. (1995). Equity, equality, or need? The effects of organizational culture on the allocation of benefits and burdens. *Organizational Behavior and Human Decision Processes, 63,* 276–286.

Manstead, A. S. R., Proffitt, C., & Smart, J. L. (1983). Predicting and understanding mother's infant-feeding intentions and behavior: Testing the theory of reasoned action. *Journal of Personality and Social Psychology, 44,* 657–671.

Manucia, G. K., Baumann, D. J., & Cialdini, R. B. (1984). Mood influences on helping: Direct effects or side effects? *Journal of Personality and Social Psychology, 46,* 357–364.

Manz, C. C., & Sims, H. P. (1982). The potential for "groupthink" in autonomous work groups. *Human Relations, 35,* 773–784.

Marangoni, C., & Ickes, W. (1989). Loneliness: A theoretical review with implications for measurement. *Journal of Social and Personal Relationships, 6,* 93–128.

Marchman, V. A. (1991). The acquisition of language in normally developing children: Some basic strategies and approaches. In I. Pavao-Martins, A. Castro-Caldas, H. Van Dongen, & A. Van Hout (Eds.), *Acquired aphasia in children.* Dordrecht, The Netherlands: Kluwer.

Marcus, D. K., Wilson, J. R., & Miller, R. S. (1996). Are perceptions of emotion in the eye of the beholder? A social relations analysis of judgments of embarrassment. *Personality and Social Psychology Bulletin, 22,* 1220–1228.

Marcus-Newhall, A., Pedersen, W., Carlson, M., & Miller, N. (2000). Displaced aggression is alive and well: A meta-analytic review. *Journal of Personality and Social Psychology, 78,* 670–689.

Marger, M. N. (1984). Social movement organizations and response to environmental change: The NAACP, 1960–1973. *Social Problems, 32,* 16–30.

Markovsky, B., Smith, L. F., & Berger, J. (1984). Do status interventions persist? *American Sociological Review, 49,* 373–382.

Markowitz, F. (2001). Modeling processes in recovery from mental illness: Relationships between symptoms, life satisfaction, and self-concept. *Journal of Health and Social Behavior, 42,* 64–79.

Markowitz, F., & Felson, R. (1998). Social-demographic differences in attitudes and violence. *Criminology, 36,* 401–422.

Markus, H. (1977). Self-schemas and processing information about the self. *Journal of Personality and Social Psychology, 35,* 63–78.

Markus, H., & Kitayama, S. (1991). Culture and the self: Implications for cognition, emotion, and motivation. *Psychological Review, 98,* 224–253.

Markus, H. R., Smith, J., & Moreland, R. L. (1985). Role of the self-concept in the perception of others. *Journal of Personality and Social Personality, 49,* 1494–1512.

Markus, H., & Wurf, E. (1987). The dynamic self-concept: A social psychological perspective. *Annual Review of Psychology, 38,* 299–337.

Markus, H., & Zajonc, R. B. (1985). The cognitive perspective in social psychology. In G. Lindzey & E. Aronson (Eds.), *Handbook of social psychology* (3rd ed., Vol. I, pp. 137–230). New York: Random House.

Marmot, M. G., Bosma, H., Hemingway, H., Brunner, E., & Stansfeld, S. (1997). Contribution of job control and other risk factors to social variations in coronary heart disease. *The Lancet, 350,* 235–239.

Marsden, P., & Campbell, K. (1984). Measuring tie strength. *Social Forces, 63,* 482–501.

Marsh, H. W., Barnes, J., & Hocevar, D. (1985). Self-other agreement on multidimensional self-concept ratings: Factor analysis and multitrait-multimethod analysis. *Journal of Personality and Social Psychology, 49,* 1360–1377.

Marshall, S. E. (1985). Ladies against women: Mobilization dilemmas of antifeminist movements. *Social Problems, 32,* 348–362.

Marshall, S. E. (1986). In defense of separate spheres: Class and status politics in the antisuffrage movement. *Social Forces, 65,* 327–351.

Martin, C. L. (1987). A ratio measure of sex stereotyping. *Journal of Personality and Social Psychology, 52,* 489–499.

Martin, K., & Leary, M. (1999). Would you drink after a stranger? The influence of self-presentational motives on willingness to take a health risk. *Personality and Social Psychology Bulletin, 25,* 1092–1100.

Martin, K., Leary, M., & Rejeski, W. J. (2000). Self-presentational concerns in older adults: Implications for health and well-being. *Basic and Applied Social Psychology, 22,* 169–179.

Martin, J., Pescosolido, B., & Tuch, S. (2000). Of fear and loathing: The role of "disturbing behavior," labels, and causal attributions in shaping public attitudes toward people with mental illness. *Journal of Health and Social Behavior, 41,* 208–223.

Martin, J., Scully, M., & Levitt, B. (1990). Injustice and the legitimation of revolution: Damning the past, excusing the present, and neglecting the future. *Journal of Personality and Social Psychology, 58,* 281–290.

Martin, M. W., & Sell, J. (1985). The effect of equating status characteristics on the generalization process. *Social Psychology Quarterly, 48,* 178–182.

Martin, M. W., & Sell, J. (1986). Rejection of authority: The importance of type of distribution rule and extent of benefit. *Social Science Quarterly, 67,* 855–868.

Martinez, R., Jr. (1996). Latinos and lethal violence: The impact of poverty and inequality. *Social Problems, 43,* 131–145.

Marwell, G., Aiken, M. T., & Demerath, N. J., III. (1987). The persistence of political attitudes among 1960s civil rights activists. *Public Opinion Quarterly, 51,* 383–399.

Mathews, K. E., Jr., & Canon, L. K. (1975). Environmental noise level as a determinant of helping behavior. *Journal of Personality and Social Psychology, 32,* 571–577.

Matsueda, R. (1982). Testing control theory and differential association: A causal modeling approach. *American Sociological Review, 47,* 489–504.

Matsueda, R. L. (1992). Reflected appraisals, parental labeling, and delinquency: Specifying a symbolic interactionist theory. *American Journal of Sociology, 97,* 1577–1611.

Matsumoto, D. (1987). The role of facial response in the experience of emotion: More methodological problems and a meta-analysis. *Journal of Personality and Social Psychology* 52(4), 769–774.

Matsumoto, D., & Ekman, P. (1989). American-Japanese cultural differences in intensity ratings of facial expressions of emotion. *Motivation and Emotion, 13*(2), 143–157.

Matthews, S. H. (1977). *The social world of old women.* Beverly Hills, CA: Sage.

Matthews, L. S., Conger, R. D., & Wickrama, K. A. S. (1996). Work-family conflict and marital quality: Mediating processes. *Social Psychology Quarterly, 59,* 62–79.

Matthews, S., & Rosner, T. (1988). Shared filial responsibility: The family as primary care giver. *Journal of Marriage and the Family, 50,* 185–195.

Maurer, T., Pleck, J., & Rane, T. (2001). Parental identity and reflected-appraisals: Measurement and gender dynamics. *Journal of Marriage and the Family, 63,* 309–321.

Mayer, J. D., Rapp, H. C., & Williams, L. (1993). Individual differences in behavioral prediction: The acquisition of personal-action schemata. *Personality and Social Psychology Bulletin, 19,* 443–451.

Maynard, D. W. (1983). Social order and plea bargaining in the court. *The Sociological Quarterly, 24,* 215–233.

Maynard, D. W., & Whalen, M. R. (1995). Language, action, and social interaction. In K. S. Cook, G. A. Fine, & J. S. House (Eds.), *Sociological perspectives on social psychology* (pp. 149–175). Needham Heights, MA: Allyn & Bacon.

Mazur, J. E. (1998). *Learning and behavior* (4th ed.). Upper Saddle River, NJ: Prentice Hall.

McAdam, D. (1986). Recruitment to high-risk activism: The case of Freedom Summer. *American Journal of Sociology, 92,* 64–90.

McAdam, D. (1999). *Political process and the development of Black insurgency, 1930–1970* (2nd ed.). Chicago: University of Chicago Press.

McAdam, D., Tarrow, S., & Tilly, C. (2001). *Dynamics of contention.* New York: Cambridge University Press.

McAdoo, H. P. (1997). Upward mobility across generations of African-American families. In H. P. McAdoo (Ed.), *Black families* (3rd ed., pp. 139–162). Thousand Oaks, CA: Sage.

McArthur, L. Z. (1972). The how and what of why: Some determinants and consequences of causal attribution. *Journal of Personality and Social Psychology, 22,* 171–193.

McArthur, L. Z., & Post, D. L. (1977). Figural emphasis and person perception. *Journal of Experimental Social Psychology, 13,* 520–535.

McCall, G. J., & Simmons, J. L. (1978). *Identities and interactions.* New York: Free Press.

McCannell, K. (1988). Social networks and the transition to motherhood. In R. Milardo (Ed.), *Families and social networks.* Newbury Park, CA: Sage.

McCarthy, J. D., & Hoge, D. R. (1984). The dynamics of self-esteem and delinquency. *American Journal of Sociology, 90,* 396–410.

McCarthy, J. D., & Wolfson, M. (1996). Resource mobilization by local social movement organizations: Agency, strategy, and organization in the movement against drinking and driving. *American Sociological Review, 61,* 1070–1088.

McCauley, C. (1989). The nature of social influence in groupthink: compliance and internalization. *Journal of Personality and Social Psychology, 57,* 250–260.

McCauley, C., Stitt, C. L., & Segal, M. (1980). Stereotyping: From prejudice to prediction. *Psychological Bulletin, 87,* 195–208.

McConahay, J. B. (1983). Modern racism and modern discrimination: The effects of race, racial attitudes, and context on simulated hiring decisions. *Personality and Social Psychology Bulletin, 9,* 551–558.

McConahay, J. B. (1986). Modern racism, ambivalence, and the modern racism scale. In J. F. Dovidio & S. L. Gaertner (Eds.), *Prejudice, discrimination, and racism.* Orlando, FL: Academic Press.

McConahay, J. B., Hardee, B. B., & Batts, V. (1981). Has racism declined in America? It depends on who is asking and what is asked. *Journal of Conflict Resolution, 25,* 563–579.

McCrae, R. R., & Costa, P. T., Jr. (1987). Validation of the five-factor model of personality across instruments and observers. *Journal of Personality and Social Psychology, 52,* 81–90.

McFarland, C., & Ross, M. (1982). The impact of causal attributions on affective reactions to success and failure. *Journal of Personality and Social Psychology, 43,* 937–946.

McFarland, D. (2004). Resistance as a social drama: A study of change-oriented encounters. *American Journal of Sociology, 109,* 1249–1318.

McGrath, J. E. (1984). *Groups: Interaction and performance.* Englewood Cliffs, NJ: Prentice Hall.

McGuire, W. J. (1964). Inducing resistance to persuasion: Some contemporary approaches. In L. Berkowitz (Ed.), *Advances in experimental social psychology* (Vol. 1, pp. 191–229). New York: Academic Press.

McGuire, W. J. (1985). Attitude and attitude change. In G. Lindzey & E. Aronson (Eds.), *The handbook of social psychology* (3rd ed., Vol. II). New York: Random House.

McGuire, W. J., & McGuire, C. (1982). Significant others in self-space: Sex differences and developmental trends in the social self. In J. Suls (Ed.), *Psychological perspectives on the self* (Vol. 1). Hillsdale, NJ: Erlbaum.

McGuire, W. J., & McGuire, C. (1986). Differences in conceptualizing self versus conceptualizing other people as manifested in contrasting verb types used in natural speech. *Journal of Personality and Social Psychology, 51,* 1135–1143.

McGuire, W. J., & Padawer-Singer, A. (1976). Trait salience in the spontaneous self-concept. *Journal of Personality and Social Psychology, 33,* 743–754.

McGuire, W. J., & Papageorgis, D. (1961). The relative efficacy of various types of prior belief-defense in producing immunity against persuasion. *Journal of Abnormal and Social Psychology, 62,* 327–337.

McLanahan, S., & Booth, K. (1989). Mother-only families: Problems, prospects, and politics. *Journal of Marriage and the Family, 51,* 557–580.

McLanahan, S., & Sandefur, G. (1994). *Growing up with a single parent: What hurts, what helps?* Cambridge, MA: Harvard University Press.

McLeod, J. M., Price, K. O., & Harburg, E. (1966). Socialization, liking, and yielding of opinions in imbalanced situations. *Sociometry, 29,* 197–212.

McLoyd, V., Cauce, A. M., Takeuchi, D., & Wilson, L. (2001). Marital processes and parental socialization in families of color: A decade review of research. In R. Milardo (Ed.), *Understanding families into the new millennium: A decade in review.* Minneapolis, MN: National Council on Family Relations.

McPhail, C. (1991). *The myth of the madding crowd.* Hawthorne, NY: Aldine de Gruyter.

McPhail, C. (1994). Presidential address – The dark side of purpose: Individual and collective violence in riots. *The Sociological Quarterly, 35,* 1–32.

McPhail, C. (1997). Stereotypes of crowds and collective behavior: Looking backward, looking forward. *Studies in Symbolic Interactionism, 3,* 35–58.

McPhail, C. (n.d.). *Stereotypes of crowds and collective behavior: Looking backward, looking forward.* Unpublished manuscript, Department of Sociology, University of Illinois, Urbana-Champaign.

McPherson, M., Smith-Lovin, L., & Cook, J. (2001). Birds of a feather: Homophily in social networks. *Annual Review of Sociology, 27,* 415–444.

McVeigh, R., Myers, D. J., & Sikkink, D. (2004). Corn, klansmen, and Coolidge: Structure and framing in social movements. *Social Forces, 83,* 653–690.

McWorter, G. A., & Crain, R. L. (1967). Subcommunity gladiatorial competition: Civil rights leadership as a competitive process. *Social Forces, 46,* 8–21.

Mead, G. H. (1934). *Mind, self, and society.* Chicago: University of Chicago Press.

Meeker, B. F. (1981). Expectation states and interpersonal behavior. In M. Rosenberg & R. H. Turner (Eds.), *Social psychology: Sociological perspectives* (pp. 290–319). New York: Basic Books.

Mehrabian, A. (1972). *Nonverbal communication.* New York: Aldine-Atherton.

Mehrabian, A., & Ksionzky, S. (1970). Models for affiliative and conformity behavior. *Psychological Bulletin, 74,* 110–126.

Meier, R. F., & Johnson, W. T. (1977). Deterrence as social control: The legal and extralegal production of conformity. *American Sociological Review, 42,* 292–304.

Melzer, S. (2002). Gender, work, and intimate violence: Men's occupational violence spillover and compensatory violence. *Journal of Marriage and Family, 64,* 820–832.

Mendelsohn, H. (1973). Some reasons why information campaigns can succeed. *Public Opinion Quarterly, 37,* 50–61.

Mennino, S. F., Rubin, B., & Brayfield, A. (2005). Home-to-job and job-to-home spillover: The impact of company policies and workplace culture. *The Sociological Quarterly, 46,* 107–135.

Mequita, B., & Karasawa, M. (2002). Different emotional lives. *Cognition and Emotion, 16,* 127–141.

Merrens, M. R. (1973). Nonemergency helping behavior in various sized communities. *Journal of Social Psychology, 90,* 327–338.

Merton, R. (1957). *Social theory and social structure.* Glencoe, IL: Free Press.

Messner, S. F., & Krohn, M. D. (1990). Class, compliance structure, and delinquency: Assessing integrated structural-Marxist theory. *American Journal of Sociology, 96,* 300–328.

Metts, S., & Cupach, W. R. (1989). Situational influence on the use of remedial strategies in embarrassing predicaments. *Communication Monographs, 56,* 151–162.

Meyer, P. (2000). The sociobiology of human cooperation: The interplay of ultimate and proximate causes. In J. M. G. van der Dennen, D. Smillie, & D. R. Wilson, (Eds.), *The Darwinian heritage and sociobiology.* Westport, CT: Praeger.

Meyer, D. S., & Whittier, N. (1994). Social movement spillover. *Social Problems, 41,* 277–298.

Meyrowitz, J. (1985). *No sense of place: The impact of electronic media on social behavior.* New York: Oxford University Press.

Miall, C. E. (1986). The stigma of involuntary childlessness. *Social Problems, 33,* 268–282.

Michaels, J. W., Edwards, J. N., & Acock, A. C. (1984). Satisfaction in intimate relationships as a function of inequality, inequity, and outcomes. *Social Psychology Quarterly, 47,* 347–357.

Michener, H. A., & Burt, M. R. (1974). Legitimacy as a base of social influence. In J. T. Tedeschi (Ed.), *Perspectives on social power.* Chicago: Aldine-Atherton.

Michener, H. A., & Burt, M. R. (1975). Components of "authority" as determinants of compliance. *Journal of Personality and Social Psychology, 31,* 605–614.

Michener, H. A., & Cohen, E. D. (1973). Effects of punishment magnitude in the bilateral threat situation: Evidence for the deterrence hypothesis. *Journal of Personality and Social Psychology, 26,* 427–438.

Michener, H. A., & Lawler, E. J. (1971). Revolutionary coalition strength and collective failure as determinants of status reallocation. *Journal of Experimental Social Psychology, 7,* 448–460.

Michener, H. A., & Lawler, E. J. (1975). Endorsement of formal leaders: An integrative model. *Journal of Personality and Social Psychology, 31,* 216–223.

Michener, H. A., & Lyons, M. (1972). Perceived support and upward mobility as determinants of revolutionary coalitional behavior. *Journal of Experimental Social Psychology, 8,* 180–195.

Michener, H. A., Plazewski, J. G., & Vaske, J. J. (1979). Ingratiation tactics channeled by target values and threat capability. *Journal of Personality, 47,* 36–56.

Michener, H. A., & Tausig, M. (1971). Usurpation and perceived support as determinants of the endorsement accorded formal leaders. *Journal of Personality and Social Psychology, 18,* 364–372.

Michener, H. A., Vaske, J. J., Schleifer, S. L., Plazewski, J. G., & Chapman, L. J. (1975). Factors affecting concession rate and threat usage in bilateral conflict. *Sociometry, 38,* 62–80.

Michener, H. A., & Wasserman, M. (1995). Group decision making. In K. S. Cook, G. A. Fine, & J. S. House (Eds.), *Sociological perspectives on social psychology* (pp. 336–361). Boston: Allyn & Bacon.

Milardo, R. M. (1982). Friendship networks in developing relationships: Converging and diverging social environments. *Social Psychology Quarterly, 45,* 162–172.

Milardo, R. (1988). Families and social networks: An overview of theory and methodology. In R. Milardo (Ed.), *Families and social networks.* Newbury Park, CA: Sage.

Milavsky, J. R., Kessler, R., Stipp, H., & Rubens, W. (1983). *Television and aggression: A panel study.* New York: Academic Press.

Milburn, M. A. (1987). Ideological self-schemata and schematically induced attitude consistency. *Journal of Experimental Social Psychology, 23,* 383–398.

Milburn, M. A., Mather, R., & Conrad, S. (2000). The effects of viewing R-rated movie scenes that objectify women on perceptions of date rape. *Sex Roles, 43,* 645–664.

Miles, R. H. (1977). Role-set configuration as a predictor of role conflict and ambiguity in complex organizations. *Sociometry, 40,* 21–34.

Milgram, S. (1963). Behavioral study of obedience. *Journal of Abnormal and Social Psychology, 67,* 371–378.

Milgram, S. (1965a). Some conditions of obedience and disobedience to authority. *Human Relations, 18,* 57–76.

Milgram, S. (1965b). Liberating effects of group pressure. *Journal of Personality and Social Psychology, 1,* 127–134.

Milgram, S. (1970). The experience of living in cities. *Science, 167,* 1461–1468.

Milgram, S. (1974). *Obedience to authority.* New York: Harper & Row.

Milgram, S. (1976). Obedience to criminal orders: The compulsion to do evil. In T. Blass (Ed.), *Contemporary social psychology: Representative readings* (pp. 175–184). Itasca, IL: Peacock.

Milgram, S., Liberty, H. J., Toledo, R., & Wackenhut, J. (1986). Response to intrusion into waiting lines. *Journal of Personality and Social Psychology, 51,* 683–689.

Miller, A. G. (1976). Constraint and target effects on the attribution of attitudes. *Journal of Experimental Social Psychology, 12,* 325–339.

Miller, A. G. (Ed.). (1982). *In the eye of the beholder: Contemporary issues in stereotyping.* New York: Praeger.

Miller, A. G., Collins, B. E., & Brief, D. E. (1995). Perspectives on obedience to authority: The legacy of the Milgram experiments. *Journal of Social Issues, 51*(3), 1–19.

Miller, C. E., & Komorita, S. S. (1995). Reward allocation in task-performing groups. *Journal of Personality and Social Psychology, 69,* 80–90.

Miller, D. L. (2000). *Introduction to collective behavior and collective action.* Prospect Heights, IL: Waveland Press.

Miller, G. (1991). Family as excuse and extenuating circumstance: Social organization and use of family rhetoric in a work incentive program. *Journal of Marriage and the Family, 53,* 609–621.

Miller, J., Schooler, C., Kohn, M., & Miller, K. (1979). Women and work: The psychological effects of occupational conditions. *American Journal of Sociology, 85,* 66–94.

Miller, K., Kohn, M., & Schooler, C. (1986). Educational self-direction and personality. *American Sociological Review, 51,* 372–390.

Miller, L. K., & Hamblin, R. L. (1963). Interdependence, differential rewarding, and productivity. *American Sociological Review, 43,* 193–204.

Miller, M., & Megowen, K. R. (2000). The painful truth: Physicians are not invincible. *Southern Medical Journal, 93,* 966–972.

Miller, R. S. (1987). Empathic embarrassment: Situational and personal determinants of reactions to the embarrassment of another. *Journal of Personality and Social Psychology, 53,* 1061–1069.

Miller, R. S. (1992). The nature and severity of self-reported embarrassing circumstances. *Personality and Social Psychology Bulletin, 18,* 190–198.

Miller, S., Olson, M., & Fazio, R. (2004). Perceived reactions to interracial romantic relationships: When race is used as a cue to status. *Group Processes and Intergroup Relations, 7,* 354–369.

Miller-McPherson, J., & Smith-Lovin, L. (1982). Women and weak ties: Differences by sex in the size of voluntary organizations. *American Journal of Sociology, 87,* 883–904.

Minnigerode, F., & Lee, J. A. (1978). Young adults' perceptions of sex roles across the lifespan. *Sex Roles, 4,* 563–569.

Miranne, A. C., & Gray, L. N. (1987). Deterrence: A laboratory experiment. *Deviant Behavior, 8,* 191–203.

Mirowsky, J., & Ross, C. (1986). Social patterns of distress. In A. Inkeles, J. Coleman, & N. Smelser (Eds.), *Annual review of sociology* (Vol. 12). Palo Alto, CA: Annual Reviews.

Mirowsky, J., & Ross, C. E. (1995). Sex differences in distress: Real or artifact? *American Sociological Review, 60,* 449–468.

Mischel, W., & Liebert, R. (1966). Effects of discrepancies between deserved and imposed reward criteria on their acquisition and transmission. *Journal of Personality and Social Psychology, 3,* 45–53.

Miyamoto, S. F., & Dornbusch, S. (1956). A test of interactionist hypotheses of self-conception. *American Journal of Sociology, 61,* 399–403.

Mizrahi, T. (1984). Coping with patients: Subcultural adjustments to the conditions of work among internists-in-training. *Social Problems, 32,* 156–166.

Modigliani, A. (1971). Embarrassment, face-work, and eye contact: Testing a theory of embarrassment. *Journal of Personality and Social Psychology, 17,* 15–24.

Moede, W. (1927). Die Richtlinien der Leistungs-Psychologie [Guidelines for the psychology of achievement]. *Industrielle Psychotechnik, 4,* 193–209.

Moen, P., Erickson, M. A., & Dempster-McClain, D. (1997). Their mother's daughters? The intergenerational transmission of gender attitudes in a world of changing roles. *Journal of Marriage and the Family, 59,* 281–293.

Money, J., & Ehrhardt, A. (1972). *Man and woman, boy and girl.* Baltimore: Johns Hopkins University Press.

Moore, J. C., Jr. (1968). Status and influence in small group interaction. *Sociometry, 31,* 47–63.

Moore, M. M. (1985). Nonverbal courtship patterns in women: Context and consequences. *Ethology and Sociobiology, 6,* 237–247.

Moore, M. M. (1995). Courtship signaling and adolescents: "Girls just wanna have fun"? *Journal of Sex Research, 32,* 319–328.

Moore, M. M., & Butler, D. L. (1989). Predictive aspects of nonverbal courtship behavior in women. *Semiotica, 76,* 205–215.

Moore, R. (2004). Managing troubles in answering survey questions: Respondents' uses of projective reporting. *Social Psychology Quarterly, 67,* 50–69.

Moorhead, G., & Montanari, J. R. (1986). An empirical investigation of the groupthink phenomenon. *Human Relations, 39,* 399–410.

Moorhead, G., Ference, R., & Neck, C. P. (1991). Group decision fiascoes continue: Space shuttle Challenger and a revised groupthink framework. *Human Relations, 44,* 539–550.

Moran, G. (1966). Dyadic attraction and orientational consensus. *Journal of Personality and Social Psychology, 4,* 94–99.

Morgan, W., Alwin, D., & Griffin, L. (1979). Social origins, parental values, and the transmission of inequality. *American Journal of Sociology, 85,* 156–166.

Morgan, S., & Rindfuss, R. (1985). Marital disruption: Structural and temporal dimensions. *American Journal of Sociology, 90,* 1055–1077.

Moriarty, T. (1975). Crime, commitment, and the responsive bystander: Two field experiments. *Journal of Personality and Social Psychology, 31,* 370–376.

Morris, M. W., & Larrick, R. P. (1995). When one cause casts doubt on another: A normative analysis of discounting in causal attribution. *Psychological Review, 102,* 331–355.

Morris, W. N., & Miller, R. S. (1975). The effect of consensus-breaking and consensus-preempting partners on reduction of conformity. *Journal of Experimental Social Psychology, 11,* 215–223.

Morse, K., & Neuberg, S. (2004). How do holidays influence relationship processes and outcomes? Examining the instigating and catalytic effects of Valentine's Day. *Personal Relationships, 11,* 509–527.

Morse, S., & Gergen, K. (1970). Social comparison, self-consistency, and the concept of self. *Journal of Personality and Social Psychology, 16,* 148–156.

Mortimer, J. T., Finch, M., & Kumka, D. (1982). Persistence and change in development: The multidimensional self-concept. In P. Baltes & O. Brim, Jr. (Eds.), *Life span development and behavior* (Vol. 4). New York: Academic Press.

Mortimer, J. T., & Lorence, J. (1995). Social psychology of work. In K. S. Cook, G. A. Fine, & J. S. House (Eds.), *Sociological perspectives on social psychology* (pp. 497–523). Needham Heights, MA: Allyn & Bacon.

Mortimer, J. T., & Simmons, R. (1978). Adult socialization. In R. Turner, J. Coleman, & R. Fox (Eds.), *Annual Review of Sociology* (Vol. 4). Palo Alto, CA: Annual Reviews.

Moscovici, S. (1980). Toward a theory of conversion behavior. In L. Berkowitz (Ed.), *Advances in experimental social psychology* (Vol. 13, pp. 209–239). New York: Academic Press.

Moscovici, S. (1985a). Innovation and minority influence. In S. Moscovici, G. Mugny, & E. Van Avermaet (Eds.), *Perspectives on minority influence* (pp. 9–52). Cambridge, UK: Cambridge University Press.

Moscovici, S. (1985b). Social influence and conformity. In G. Lindzey & E. Aronson (Eds.), *Handbook of social psychology* (3rd ed., Vol. 2, pp. 347–412). New York: Random House.

Moscovici, S., & Lage, E. (1976). Studies in social influence III: Majority versus minority influence in a group. *European Journal of Social Psychology, 6,* 149–174.

Moscovici, S., Lage, E., & Naffrechoux, M. (1969). Influence of a consistent minority on the responses of a majority in a color perception task. *Sociometry, 32,* 365–379.

Muedeking, G. D. (1992). Authentic/inauthentic identities in the prison visiting room. *Symbolic Interaction, 15,* 227–236.

Mugny, G. (1982). *The power of minorities.* London: Academic Press.

Mugny, G. (1984). The influence of minorities: Ten years later. In H. Tajfel (Ed.), *The social dimension: European developments in social psychology* (Vol. 2, pp. 498–517). Cambridge, UK: Cambridge University Press.

Mulatu, M. S., & Schooler, C. (2002). Causal connections between socio-economic status and health: Reciprocal effects and mediating mechanisms. *Journal of Health and Social Behavior, 43,* 22–41.

Mullen, B. (1985). Strength and immediacy of sources: A meta-analytic evaluation of the forgotten elements of social impact theory. *Journal of Personality and Social Psychology, 48,* 1458–1466.

Mullen, B., & Copper, C. (1994). The relation between group cohesiveness and performance: An integration. *Psychological Bulletin, 115,* 210–227.

Mullen, B., & Hu, L. (1989). Perceptions of ingroup and outgroup variability: A meta-analytic integration. *Basic and Applied Social Psychology, 10,* 233–252.

Mullen, C. K., & Linz, D. (1995). Desensitization and resensitization to violence against women: Effects of exposure to sexually violent films on judgements of domestic violence victims. *Journal of Personality and Social Psychology, 69,* 449–459.

Muller, C., Ride, S., Fouke, J., Whitney, T., Denton, D., Cantor, N., et al. (2005, Feb. 18). Gender differences and performance in science. *Science, 307,* 1043.

Muller, E. (1985). Income inequality, regime repressiveness, and political violence. *American Sociological Review, 50,* 47–61.

Murray, S. L., Holmes, J. G., & Griffin, D. W. (1996a). The benefits of positive illusions: Idealization and the construction of satisfaction in close relationships. *Journal of Personality and Social Psychology, 70,* 79–98.

Murray, S. L., Holmes, J. G., & Griffin, D. W. (1996b). The self-fulfilling nature of positive illusions in romantic relationships: Love is not blind, but prescient. *Journal of Personality and Social Psychology, 71,* 1155–1180.

Murray, J. P., & Kippax, S. (1979). From the early window to the late night show: International trends in the study of television's impact on children and adults. In L. Berkowitz (Ed.), *Advances in experimental social psychology* (Vol. 12). New York: Academic Press.

Mussweiler, T., Strack, F., & Pfeiffer, T. (2000). Overcoming the inevitable anchoring effect: Considering the opposite compensates for selective accessibility. *Personality and Social Psychology Bulletin, 26,* 1142–1150.

Mutz, D. C., & Martin, P. S. (2001). Facilitating communication across lines of political difference: The role of mass media. *American Political Science Review, 95,* 97–114.

Myers, D. G. (1975). Discussion-induced attitude polarization. *Human Relations, 28,* 699–714.

Myers, D. G., Bruggink, J. B., Kersting, R. C., & Schlosser, B. A. (1980). Does learning others' opinions change one's opinion? *Personality and Social Psychology Bulletin, 6,* 253–260.

Myers, D. G., & Kaplan, M. F. (1976). Group-induced polarization in simulated juries. *Personality and Social Psychology Bulletin, 2,* 63–66.

Myers, D. G., & Lamm, H. (1976). The group polarization phenomenon. *Psychological Bulletin, 83,* 602–627.

Myers, D. J. (1997). Racial rioting in the 1960s: An event history analysis of local conditions. *American Sociological Review, 62,* 94–112.

Myers, D. J. (2000). The diffusion of collective violence: Infectiousness, susceptibility, and mass media networks. *American Journal of Sociology, 106,* 173–208.

Myers, D. J., & Caniglia, B. S. (2004). All the rioting that's fit to print: Selection effects in national newspaper coverage of civil disorders, 1968–1969. *American Sociological Review, 69,* 519–543.

Myers, D. J., & Li, Y. E. (2001, August). *City conditions and riot susceptibility reconsidered.* Paper presented at the Annual Meeting of the American Sociological Association, Anaheim, CA.

Myers, F. E. (1971). Civil disobedience and organization change: The British Committee of 100. *Political Science Quarterly, 86,* 92–112.

Myers, M. A., & Hagan, J. (1979). Private and public trouble: Prosecutors and the allocation of court resources. *Social Problems, 26,* 439–451.

Myers, M. A., & Talarico, S. M. (1986). The social contexts of racial discrimination in sentencing. *Social Problems, 33,* 236–251.

Nadler, A. (1987). Determinants of help seeking behaviour: The effects of helper's similarity, task centrality and recipient's self esteem. *European Journal of Social Psychology, 17*(1), 57–67.

Nadler, A. (1991). Help-seeking behavior: Psychological costs and instrumental benefits. In M. S. Clark (Ed.), *Review of personality and social psychology: Vol. 12. Prosocial behavior* (pp. 290–311). Newbury Park, CA: Sage.

Nadler, A., & Fisher, J. D. (1984a). Effects of donor-recipient relationships on recipients' reactions to aid. In E. Staub, D. Bar-Tal, J. Karylowski, & J. Reykowski (Eds.), *Development and maintenance of prosocial behavior: International perspectives on positive morality.* New York: Plenum.

Nadler, A., & Fisher, J. D. (1984b). The role of threat to self-esteem and perceived control in recipient reaction to aid. In L. Berkowitz (Ed.), *Advances in experimental social psychology* (Vol. 17). New York: Academic Press.

Nadler, A., & Fisher, J. D. (1986). The role of threat to self-esteem and perceived control in recipient reaction to help: Theory development and empirical validation. In L. Berkowitz (Ed.), *Advances in experimental social psychology* (Vol. 19, pp. 81–122). San Diego: Academic Press.

Nadler, A., Fisher, J. D., & Ben-Itzhak, S. (1983). With a little help from my friend: Effect of single or multiple act aid as a function of donor and task characteristics. *Journal of Personality and Social Psychology, 44,* 310–321.

Nadler, A., Mayseless, O., Peri, N., & Chemerinski, A. (1985). Effects of opportunity to reciprocate and self-esteem on help-seeking behavior. *Journal of Personality, 53,* 23–35.

Nagel, J. (1995). American Indian ethnic revival: Politics and the resurgence of identity. *American Sociological Review, 60,* 947–965.

Nail, P. R., Harton, H., & Decker, B. (2003). Political orientation and aversive versus modern racism: Tests of Dovidio and Gaertner's integrated model. *Journal of Personality and Social Psychology, 84,* 754–770.

Nakao, K., & Treas, J. (1994). Updating occupational prestige and socioeconomic scores: How the new measures measure up. In P. Marsden (Ed.), *Sociological methodology, 24,* 1–72.

National Advisory Commission on Civil Disorders. (1968). *Report of the National Advisory Commission on Civil Disorders.* New York: Bantam Books.

National Center for Education Statistics. (2004) *Digest of Education Statistics, 2003,* Table 185. Available at http://nces.ed.gov/programs/digest/d03/tables/dt185.asp

National Center for Health Statistics. (2004). *Health, United States, 2004.* Hyattsville, MD: National Center for Health Statistics.

National Institute of Occupational Safety and Health. (2002). *The effects of workplace hazards on male reproductive health* (NIOSH Publication No. 96-132). Washington, DC: Author.

National Public Radio. (2002, October 14). How pitchmen sell their wares at state fairs. *Morning Edition* [Program transcript]. Washington, DC: Public Broadcasting Service.

Navarro, M. (2005, April 24). When you contain multitudes. *New York Times,* pp. 1–2.

Neal, A. G., & Groat, H. (1974). Social class correlates of stability and change in levels of alienation: A longitudinal study. *The Sociological Quarterly, 15,* 548–558.

Neale, M., & Bazerman, M. (1991). *Cognition and rationality in bargaining.* New York: Free Press.

Nelson, L. J., & Klutas, K. (2000). The distinctiveness effect in social interaction: Creation of a self-fulfilling prophecy. *Personality and Social Psychology Bulletin, 26,* 126–135.

Nelson, T. D. (2002). *The psychology of prejudice.* Boston: Allyn & Bacon.

Nemeth, C. J. (1986). Differential contributions of majority and minority influence. *Psychological Review, 93,* 23–32.

Nemeth, C. J. (1995). Dissent as driving cognition, attitudes, and judgment. *Social Cognition, 13,* 273–291.

Nemeth, C. J., & Kwan, J. L. (1985). Originality of word associations as a function of majority vs. minority influence processes. *Social Psychological Quarterly, 48,* 277–282.

Nemeth, C. J., & Kwan, J. L. (1987). Minority influence, divergent thinking, and detection of correct solutions. *Journal of Applied Social Psychology, 17,* 788–799.

Nemeth, C. J., Swedlund, M., & Kanki, B. (1974). Patterning of the minority's responses and their influence on the majority. *European Journal of Social Psychology, 4,* 53–64.

Nemeth, C. J., Wachtler, J., & Endicott, J. (1977). Increasing the size of the minority: Some gains and some losses. *European Journal of Social Psychology, 7,* 15–27.

Neugarten, B. L., & Datan, N. (1973). Sociological perspectives on the life cycle. In P. Baltes & K. Schaie (Eds.), *Life-span developmental psychology: Personality and social processes.* New York: Academic Press.

Neville, B., & Parke, R. D. (1997). Waiting for paternity: Interpersonal and contextual implications of the timing of fatherhood. *Sex Roles, 37,* 45–60.

Newcomb, T. M. (1943). *Personality and social change.* New York: Dryden.

Newcomb, T. M. (1968). Interpersonal balance. In R. P. Abelson, W. J. McGuire, T. M. Newcomb, M. J. Rosenberg, & P.H. Tannenbaum (Eds.), *Theories of cognitive consistency: A sourcebook.* Chicago: Rand McNally.

Newcomb, T. M. (1971). Dyadic balance as a source of clues about interpersonal attraction. In B. Murstein (Ed.), *Theories of attraction and love.* New York: Springer Verlag.

Newspaper Advertising Bureau. (1980). *Mass media in the family setting: Social patterns in media availability and use by parents.* New York: Author.

NICHD Early Child Care Research Network. (1997a). Familial factors associated with the characteristics of nonmaternal care for infants. *Journal of Marriage and the Family, 59,* 389–408.

NICHD Early Child Care Research Network. (1997b). The effects of infant child care on mother-infant attachment security: Results of the NICHD Study of Early Child Care. *Child Development, 68,* 860–879.

NICHD Early Child Care Research Network. (2002). Early child care and children's development prior to school entry: Results from the NICHD Study of Early Child Care. *American Educational Research Journal, 39,* 133–164.

Nisbett, R. E., Caputo, C., Legant, P., & Maracek, J. (1973). Behavior as seen by the actor and as seen by the observer. *Journal of Personality and Social Psychology, 27,* 154–164.

Nizer, L. (1973). *The implosion conspiracy.* New York: Doubleday.

Norenzayan, A., & Nisbett, R. E. (2000). Culture and causal cognition. *Current Directions in Psychological Science, 9,* 132–135.

Norstrom, T. (1995). The impact of alcohol, divorce, and unemployment on suicide. *Social Forces, 74,* 293–314.

Norwood, A., Ursano, R., & Fullerton, C. (2002). *Disaster psychiatry: Principles and practice.* American Psychiatric Association. Retrieved March 7, 2003, from http://www.psych.org/pract_of_psych/principles_and_practice3201.cfm

Nyden, P. W. (1985). Democratizing organizations: A case study of a union reform movement. *American Journal of Sociology, 90,* 1179–1203.

Oakes, P. J., & Turner, J. C. (1980). Social categorization and intergroup behavior: Does minimal intergroup discrimination make social identity more positive? *European Journal of Social Psychology, 10,* 295–301.

Oberschall, A. (1973). *Social conflict and social movements.* Englewood Cliffs, NJ: Prentice Hall.

Oberschall, A. (1978). Theories of social conflict. In R. Turner, J. Coleman, & R. Fox (Eds.), *Annual review of sociology* (Vol. 4). Palo Alto, CA: Annual Reviews.

O'Bryant, S. (1988). Sibling support and older widows' well-being. *Journal of Marriage and the Family, 50,* 173–183.

Oegema, D., & Klandermans, B. (1994). Why social movement sympathizers don't participate: Erosion and nonconversion of support. *American Sociological Review, 59,* 703–722.

Ohbuchi, K., Kameda, M., & Agarie, N. (1989). Apology as aggression control: Its role in mediating appraisal of and response to harm. *Journal of Personality and Social Psychology, 56,* 219–227.

Okamoto, D. G., & Smith-Lovin, L. (2001). Changing the subject: Gender, status, and the dynamics of topic change. *American Sociological Review, 66,* 852–873.

O'Leary-Kelly, A. M., Martocchio, J. J., & Frink, D. D. (1994). A review of the influence of group goals on group performance. *Academy of Management Journal, 37,* 1285–1301.

Oliver, P. (1980). Rewards and punishments as selective incentives for collective action: Theoretical investigations. *American Journal of Sociology, 85,* 1356–1375.

Oliver, P. (1984). Rewards and punishments as selective incentives: An apex game. *Journal of Conflict Resolution, 28,* 123–148.

Oliver, P. E., & Myers, D. J. (1999). How events enter the public sphere: Conflict, location, and sponsorship in local newspaper coverage of public events. *American Journal of Sociology, 105,* 38–87.

Olver, R. (1961). *Developmental study of cognitive equivalence.* Unpublished doctoral dissertation, Radcliffe College, Cambridge, MA.

Olzak, S. (1989). Labor unrest, immigration, and ethnic conflict in urban America, 1880–1914. *American Journal of Sociology, 94,* 1303–1333.

Olzak, S. (1992). *The dynamics of ethnic competition and conflict.* Palo Alto, CA: Stanford University Press.

Olzak, S., & Shanahan, S. (1996). Deprivation race riots: An extension of Spilerman's analysis. *Social Forces, 74,* 931–961.

Olzak, S., Shanahan, S., & McEneaney, E. H. (1996). Poverty, segregation, and race riots: 1990 to 1993. *American Sociological Review, 61,* 590–613.

Opp, K.-D. (1988). Grievances and participation in social movements. *American Sociological Review, 53,* 853–864.

Oppenheimer, V. K. (1970). The female labor force in the United States. *Population Monograph Series* (No. 5). Berkeley, CA: Institute of International Studies.

Orbuch, T. L., & Custer, L. (1995). The social context of married women's work and its impact on Black husbands and White husbands. *Journal of Marriage and the Family, 57,* 333–345.

Orbuch, T. L., House, J. S., Mero, R. P., & Webster, P. S. (1996). Marital quality over the life course. *Social Psychology Quarterly, 59,* 162–171.

Orbuch, T., Veroff, J., Hassan, H., & Horrocks, J. (2002). Who will divorce? A 14-year longitudinal study of Black couples and White couples. *Journal of Social and Personal Relationships, 19,* 549–568.

Orcutt, J. (1975). Deviance as a situated phenomenon: Variations in the social interpretation of marijuana and alcohol use. *Social Problems, 22,* 346–356.

Orwell, G. (1949). *1984.* New York: Harcourt Brace Jovanovich.

Osborn, A. F. (1963). *Applied imagination* (3rd rev. ed.) New York: Scribner.

Osgood, C. E. (1962). *An alternative to war or surrender.* Urbana: University of Illinois Press.

Osgood, C. E. (1979). GRIT for MBFR: A proposal for unfreezing force-level postures in Europe. *Peace Research Review, 8,* 77–92.

Osgood, C. E. (1980, May). The GRIT strategy. *Bulletin of the Atomic Scientists,* 58–60.

Osgood, C. E., Suci, G., & Tannenbaum, P. (1957). *The measurement of meaning.* Urbana: University of Illinois Press.

Osgood, D. W., Wilson, J. R., O'Malley, P. M., Bachman, J. G., & Johnston, L. D. (1996). Routine activities and deviant behavior. *American Sociological Review, 61,* 635–655.

Oskamp, S. (1971). Effects of programmed strategies on cooperation in the Prisoner's Dilemma and other mixed-motive games. *Journal of Conflict Resolution, 15,* 225–259.

Oskamp, S. (1991). *Attitudes and opinions.* Englewood Cliffs, NJ: Prentice Hall.

O'Sullivan, M. (2003). The fundamental attribution error in detecting deception: The boy who cried wolf effect. *Personality and Social Psychology Bulletin, 29*(10), 1316–1327.

Otten, C. A., Penner, L. A., & Waugh, G. (1988). That's what friends are for: The determinants of psychological helping. *Journal of Social and Clinical Psychology, 7*(1), 34–41.

Padden, S. L., & Buehler, C. (1995). Coping with the dual-income lifestyle. *Journal of Marriage and the Family, 57,* 101–110.

Page, A. L., & Clelland, D. A. (1978). The Kanawha County textbook controversy: A study of the politics of lifestyle concern. *Social Forces, 57,* 265–281.

Pager, D. (2003). The mark of a criminal record. *American Journal of Sociology, 108,* 1249–1291.

Paicheler, G., & Bouchet, J. (1973). Attitude polarization, familiarization, and group process. *European Journal of Social Psychology, 3,* 83–90.

Palmore, E. (1981). *Social patterns in normal aging.* Durham, NC: Duke University Press.

Pandey, J. (1981). A note about social power through ingratiation among workers. *Journal of Occupational Psychology, 54*(1), 65–67.

Pantin, H. M., & Carver, C. S. (1982). Induced competence and the bystander effect. *Journal of Applied Social Psychology, 12,* 100–111.

Papastamou, S., & Mugny, G. (1985). Rigidity and minority influence: The influence of the social in social influence. In S. Moscovici, G. Mugny, & E. Van Avermaet (Ed.), *Perspectives on minority influence* (pp. 113–136). Cambridge, UK: Cambridge University Press.

Park, W.-W. (1990). A review of research on groupthink. *Journal of Behavioral Decision Making, 3,* 229–245.

Park, B., & Rothbart, M. (1982). Perception of out-group homogeneity and levels of social categorization: Memory for the subordinate attributes of in-group and out-group members. *Journal of Personality and Social Psychology, 42,* 1051–1068.

Parke, R. (1969). Effectiveness of punishment as an interaction of intensity, timing, agent nurturance, and cognitive structuring. *Child Development, 40,* 213–235.

Parke, R. (1970). The role of punishment in the socialization process. In R. Hoppe, G. Milton, & E. Simmel (Eds.), *Early experiences and the processes of socialization.* New York: Academic Press.

Parke, R. D. (1996). *Fatherhood.* Cambridge, MA: Harvard University Press.

Parkinson, B., Fischer, A. H., & Manstead, A. S. R. (2005). *Emotion in social relations: Cultural, group and interpersonal processes.* New York: Psychology Press.

Parrott, W. G., & Smith, S. F. (1991). Embarrassment: Actual vs. typical cases, classical vs. prototypical representations. *Cognition and Emotion, 5,* 467–488.

Patrick, S. L., & Jackson, J. J. (1991). Further examination of the equity sensitivity construct. *Perceptual and Motor Skills, 73,* 1091–1106.

Patterson, M. L., Mullens, S., & Romano, J. (1971). Compensatory reactions to spatial intrusion. *Sociometry, 34,* 114–121.

Patterson, R. J., & Neufeld, R. W. J. (1987). Clear danger: Situational determinants of the appraisal of threat. *Psychological Bulletin, 101,* 404–416.

Paulus, P. B. (1998). Developing consensus about groupthink after all these years. *Organizational Behavior and Human Decision Processes, 73,* 362–374.

Paulus, P. B., Larey, T. S., & Dzindolet, M. T. (2001). Creativity in groups and teams. In M. Turner (Ed.), *Groups at work: Advances in theory and research* (pp. 319–338). Hillsdale, NJ: Erlbaum.

Pavlidis, J., Eberhardt, N., & Levine, J. (2002). Seeing through the face of deception. *Nature, 415,* 35.

Paxton, P., & Moody, J. A. (2003). Structure and sentiment: Explaining emotional attachment to group. *Social Psychology Quarterly, 66,* 34–47.

Pearce, P. L. (1980). Strangers, travelers, and Greyhound terminals: A study of small-scale helping behaviors. *Journal of Personality and Social Psychology, 38,* 935–940.

Pearlin, L., & Johnson, J. (1977). Marital status, life-strains and depression. *American Sociological Review, 42,* 704–715.

Pearlin, L., & Radabaugh, C. (1976). Economic strains and the coping functions of alcohol. *American Journal of Sociology, 82,* 652–663.

Peirce, K. (1993). Socialization of teenage girls through teen-magazine fiction: The making of a new woman or an old lady? *Sex Roles, 29,* 59–68.

Pelham, B., Mirenberg, M., & Jones, J. (2002). Why Susie sells seashells by the seashore: Implicit egotism and major life decisions. *Journal of Personality and Social Psychology, 82,* 469–487.

Pendry, L., & Carrick, R. (2001). Doing what the mob do: Priming effects on conformity. *European Journal of Social Psychology, 31,* 83–92.

Penedo, F., & Dahn, J. (2005). Exercise and well-being: A review of mental and physical health benefits associated with physical activity. *Current Opinion in Psychiatry, 18,* 189–193.

Pennebaker, J. W. (1980). Self-perception of emotion and internal sensation. In D. W. Wegner & R. R. Vallacher (Eds.), *The self in social psychology.* New York: Oxford University Press.

Pennebaker, J., & Sanders, D. (1976). American graffiti: Effects of authority and reactance arousal. *Personality and Social Psychology Bulletin, 2,* 264–267.

Perlman, D. (1988). Loneliness: A life-span family perspective. In R. Milardo (Ed.), *Families and social networks.* Newbury Park, CA: Sage.

Perry, L. S. (1993). Effects of inequity on job satisfaction and self-evaluation in a national sample of African-American workers. *Journal of Social Psychology, 133,* 565–573.

Perry, J. B., & Pugh, M. D. (1978). *Collective behavior: Response to social stress.* St. Paul, MN: West.

Perry-Jenkins, M., Repetti, R., & Crouter, A. (2001). Work and family in the 1990s. In R. Milardo (Ed.), *Understanding families into the new millennium: A decade in review.* Minneapolis, MN: National Council on Family Relations.

Personnaz, B. (1981). Study in social influence using the spectrometer method: Dynamics of the phenomena of conversion and covertness in perceptual responses. *European Journal of Social Psychology, 11,* 431–438.

Pescosolido, A. T. (2001). Informal leaders and the development of group efficacy. *Small Group Research, 32,* 74–93.

Peters, L. H., Hartke, D. D., & Pohlmann, J. T. (1985). Fiedler's contingency theory of leadership: An application of the meta-analysis procedures of Schmidt and Hunter. *Psychological Bulletin, 97,* 274–285.

Petersen, T., & Morgan, C. A. (1995). Separate and unequal: Occupation-establishment sex segregation and the gender wage gap. *American Journal of Sociology, 101,* 329–365.

Petersen, T., Saporta, I., & Seidel, M. L. (2000). Offering a job: Meritocracy and social networks. *American Journal of Sociology, 106,* 763–816.

Petrunik, M., & Shearing, C. D. (1983). Fragile façades: Stuttering and the strategic manipulation of awareness. *Social Problems, 31,* 125–138.

Pettigrew, T. F. (1979). The ultimate attribution error: Extending Allport's cognitive analysis of prejudice. *Personality and Social Psychology Bulletin, 5,* 461–476.

Pettigrew, T. F. (1997). Generalized intergroup contact effects on prejudice. *Personality and Social Psychology Bulletin, 23,* 173–185.

Petty, R. E. (1995). Attitude change. In A. Tesser (Ed.), *Advanced social psychology* (pp. 195–255). New York: McGraw-Hill.

Petty, R. E., & Brock, T. C. (1981). Thought disruption and persuasion: Assessing the validity of attitude change experiments. In R. E. Petty, T. M. Ostrom, & T. C. Brock (Eds.), *Cognitive responses in persuasion.* Hillsdale, NJ: Erlbaum.

Petty, R. E., & Cacioppo, J. T. (1979a). Effects of forewarning of persuasive intent and involvement on cognitive responses and persuasion. *Personality and Social Psychology Bulletin, 5,* 173–176.

Petty, R. E., & Cacioppo, J. T. (1979b). Issue involvement can increase or decrease persuasion by enhancing message-relevant cognitive responses. *Journal of Personality and Social Psychology, 37,* 1915–1926.

Petty, R. E., & Cacioppo, J. T. (1986a). *Communication and persuasion: Central and peripheral routes to attitude change.* New York: Springer Verlag.

Petty, R. E., & Cacioppo, J. T. (1986b). The elaboration likelihood model of persuasion. In L. Berkowitz (Ed.), *Advances in experimental social psychology* (Vol. 19, pp. 207–249). New York: Academic Press.

Petty, R. E., & Cacioppo, J. T. (1990). Involvement and persuasion: Tradition versus integration. *Psychological Bulletin, 107,* 367–374.

Petty, R. E., Cacioppo, J. T., & Goldman, R. (1981). Personal involvement as a determinant of argument-based persuasion. *Journal of Personality and Social Psychology, 41,* 847–855.

Petty, R. E., Cacioppo, J. T., & Heesacker, M. (1981). Effects of rhetorical questions on persuasion: A cognitive response analysis. *Journal of Personality and Social Psychology, 40,* 432–440.

Petty, R. E., Cacioppo, J. T., Strathman, A. J., & Priester, J. R. (1994). To think or not to think: Exploring two routes to persuasion. In S. Shavitt & T. C. Brock (Eds.), *Persuasion: Psychological insights and perspectives* (pp. 113–147). Boston: Allyn & Bacon.

Petty, R. E., Wells, G. L., & Brock, T. C. (1976). Distraction can enhance or reduce yielding to propaganda: Thought disruption versus effort justification. *Journal of Personality and Social Psychology, 34,* 876–884.

Phillips, D. P. (1974). The influence of suggestion on suicide: Substantive and theoretical implications of the Werther effect. *American Sociological Review, 39,* 340–354.

Phillips, D. P. (1979). Suicide, motor vehicle fatalities, and the mass media: Evidence toward a theory of suggestion. *American Sociological Review, 84,* 1150–1174.

Piaget, J. (1954). *The construction of reality in the child.* New York: Basic Books.

Piaget, J. (1965). *The moral judgement of the child.* New York: Free Press.

Piliavin, I. M., & Briar, S. (1964). Police encounters with juveniles. *American Journal of Sociology, 70,* 206–214.

Piliavin, I. M., Piliavin, J. A., & Rodin, J. (1975). Cost, diffusion, and the stigmatized victim. *Journal of Personality and Social Psychology, 32,* 429–438.

Piliavin, I. M., Rodin, J., & Piliavin, J. A. (1969). Good Samaritanism: An underground phenomenon? *Journal of Personality and Social Psychology, 13,* 289–299.

Piliavin, I. M., Thornton, C., Gartner, R., & Matsueda, R. (1986). Crime, deterrence, and rational choice. *American Sociological Review, 51,* 101–119.

Piliavin, J. A., & Charng, H.-W. (1990). Altruism: A review of recent theory and research. *Annual Review of Sociology, 16,* 27–65.

Piliavin, J. A., Dovidio, J. F., Gaertner, S. L., & Clark, R. D., III. (1981). *Emergency intervention.* New York: Academic Press.

Piliavin, J. A., & LePore, P. C. (1995). Biology and social psychology: Beyond nature versus nurture. In K. S. Cook, G. A. Fine, & J. S. House (Eds.), *Sociological perspectives on social psychology* (pp. 9–40). Boston: Allyn & Bacon.

Piliavin, J. A., & Unger, R. K. (1985). The helpful but helpless female: Myth or reality? In V. O'Leary, R. K. Unger, & B. S. Wallston (Eds.), *Women, gender, and social psychology* (pp. 149–186). Hillsdale, NJ: Erlbaum.

Pillemer, K., & Suitor, J. J. (1991). "Will I ever escape my children's problems?" Effects of adult children's problems on elderly parents. *Journal of Marriage and the Family, 53,* 585–594.

Pinker, S. (1997). *How the mind works.* New York: Norton.

Pipher, M. (1994). *Reviving Ophelia: Saving the selves of adolescent girls.* New York: Ballantine Books.

Plant, E. A., Hyde, J., Keltner, D., & Devine, P. (2000).The gender stereotyping of emotions. *Psychology of Women Quarterly, 24,* 81–92.

Plant, E. A., Peruche, B. M., & Butz, D. A. (2005). Eliminating automatic racial bias: Making race non-diagnostic for responses to criminal suspects. *Journal of Experimental Social Psychology, 41*(2), 141–156.

Pleck, J. H. (1976). The male sex role: Definitions, problems, and sources of change. *Journal of Social Issues, 32*(3), 155–164.

Pleck, J. H. (1985). *Working wives/Working husbands.* Beverly Hills, CA: Sage.

Plutzer, E. (1987). Determinants of leftist radical belief in the United States: A test of competing theories. *Social Forces, 65,* 1002–1017.

Plutzer, E. (1988). Work life, family life, and women's support of feminism. *American Sociological Review, 53,* 640–649.

Podolny, J. M., & Baron, J. N. (1997). Resources and relationships: Social networks and mobility in the workplace. *American Sociological Review, 62,* 673–693.

Porter, J. R., & Washington, R. E. (1993). Minority identity and self-esteem. *Annual Review of Sociology, 19,* 139–161.

Postmes, T., & Spears, R. (1998). Deindividuation and antinormative behavior: A meta-analysis. *Psychological Bulletin, 123,* 238–259.

Powell, M. A., & Parcell, T. L. (1997). Effects of family structure on the earnings attainment process: Differences by gender. *Journal of Marriage and the Family, 59,* 419–433.

Powers, T. A., & Zuroff, D. C. (1988). Interpersonal consequences of overt self-criticism: A comparison with neutral and self-enhancing presentations of self. *Journal of Personality and Social Psychology, 54,* 1054–1062.

Poyatos, F. (1983). *New perspectives in nonverbal communication: Studies in cultural anthropology, social psychology, linguistics, literature, and semantics.* Oxford, UK: Pergamon.

Prager, I. G., & Cutler, B. L. (1990). Attributing traits to oneself and to others: The role of acquaintance level. *Personality and Social Psychology Bulletin, 16,* 309–319.

Pratkanis, A. R., & Greenwald, A. G. (1989). A sociocognitive model of attitude structure and function. In L. Berkowitz (Ed.), *Advances in experimental social psychology* (Vol. 22, pp. 245–285). New York: Academic Press.

Previti, D., & Amato, P. (2003). Why stay married? Rewards, barriers, and marital stability. *Journal of Marriage and Family, 65,* 561–573.

Price, K. O., Harburg, E., & Newcomb, T. M. (1966). Psychological balance in situations of negative interpersonal attitudes. *Journal of Personality and Social Psychology, 3,* 265–270.

Priest, R. T., & Sawyer, J. (1967). Proximity and peership: Bases of balance in interpersonal attraction. *American Journal of Sociology, 72,* 633–649.

Prince-Gibson, E., & Schwartz, S. H. (1998). Value priorities and gender. *Social Psychology Quarterly, 61,* 49–67.

Pritchard, R. D., Jones, S. D., Roth, P. L., Stuebing, K. K., & Ekeberg, S. (1988). Effects of group feedback, goal setting, and incentives on organizational productivity. *Journal of Applied Psychology, 73,* 337–358.

Pritchard, R. D., & Watson, M. D. (1992). Understanding and measuring group productivity. In S. Worchel, W. Wood, & J. A. Simpson (Eds.), *Group process and productivity* (pp. 251–275). Newbury Park, CA: Sage.

Pruitt, D. G. (1983). Achieving integrative agreements. In M. Bazerman & R. Lewicki (Eds.), *Negotiating in organizations* (pp. 35–50). Beverly Hills, CA: Sage.

Pruitt, D. G., & Carnevale, P. J. (1993). *Negotiation in social conflict.* Buckingham, UK: Open University Press.

Pruitt, D. G., & Insko, C. A. (1980). Extension of the Kelley attribution model: The role of comparison-object consensus, target-object consensus, distinctiveness, and consistency. *Journal of Personality and Social Psychology, 39,* 39–58.

Pryor, J. B., McDaniel, M. A., & Kott-Russo, T. (1986). The influence of the level of schema abstractness upon the processing of social information. *Journal of Experimental Social Psychology, 22,* 312–327.

Pugh, M. D., & Wahrman, R. (1983). Neutralizing sexism in mixed-sex groups: Do women have to be better than men? *American Journal of Sociology, 88,* 746–762.

Purdum, T. S. (1997, April 27). Legacy of riots in Los Angeles traces conflict. *New York Times, 1,* p. 16.

Quattrone, G. A. (1986). On the perceptions of a group's variability. In S. Worchel & W. G. Austin (Eds.), *Psychology of intergroup relations* (2nd ed., pp. 25–48). Chicago: Nelson-Hall.

Quattrone, G. A., & Jones, E. E. (1980). The perception of variability within in-groups and out-groups: Implications for the law of small numbers. *Journal of Personality and Social Psychology, 38,* 141–152.

Quillian, L., & Campbell, M. (2003). Beyond black and white: The present and future of multiracial friendship segregation. *American Sociological Review, 68,* 540–566.

Quinney, R. (1970). *The social reality of crime.* Boston: Little, Brown.

Rabbie, J. M., & Bekkers, F. (1978). Threatened leadership and intergroup competition. *European Journal of Social Psychology, 8,* 9–20.

Rabow, J., Neuman, C. A., & Hernandez, A. (1987). Cognitive consistency in attitudes, social support, and consumption of alcohol: Additive and interactive effects. *Social Psychology Quarterly, 50,* 56–63.

Rahn, J., & Mason, W. (1987). Political alienation, cohort size, and the Easterlin hypothesis. *American Sociological Review, 52,* 155–169.

Raley, R. K. (1996). A shortage of marriageable men? A note on the role of cohabitation in Black-White differences in marriage rates. *American Sociological Review, 61,* 973–983.

Rank, S. G., & Jacobson, C. K. (1977). Hospital nurses' compliance with medication overdose orders: A failure to replicate. *Journal of Health and Social Behavior, 18,* 188–193.

Rashotte, L. S. (2002). What does that smile mean? The meaning of nonverbal behaviors in social interaction. *Social Psychology Quarterly, 65,* 92–102.

Rasinski, K. A., Berktold, J., Smith, T. W., & Albertson, B. L. (2002). *America recovers: A follow-up to a national study of public responses to the September 11th terrorist attacks.* Chicago: National Opinion Research Center.

Ratzan, S. C. (1989). The real agenda setters: Pollsters in the 1988 presidential campaign. *American Behavioral Scientist, 32,* 451–463.

Raven, B. H. (1992). A power/interaction model of interpersonal influence: French and Raven thirty years later. *Journal of Social Behavior and Personality, 7,* 217–244.

Raven, B. H., & Kruglanski, A. W. (1970). Conflict and power. In P. Swingle (Ed.), *The structure of conflict.* New York: Academic Press.

Raven, B. H., & Rietsema, J. (1957). The effects of varied clarity of group goal and group path upon the individual and his relation to the group. *Human Relations, 10,* 29–44.

Ray, M. (1973). Marketing communication and the hierarchy of effects. In P. Clarke (Ed.), *New models for communication research.* Beverly Hills, CA: Sage.

Rea, L., & Parker, R. (1997). *Designing and conducting survey research: A comprehensive guide* (2nd ed.). San Francisco: Jossey-Bass.

Rees, C. R., & Segal, M. W. (1984). Role differentiation in groups: The relations between instrumental and expressive leadership. *Small Group Behavior, 15,* 109–123.

Regan, P. (2000). The role of sexual desire and sexual activity in dating relationships. *Social Behavior and Personality, 28,* 51–59.

Regan, D. T., & Fazio, R. (1977). On the consistency between attitudes and behavior: Look to the method of attitude formation. *Journal of Experimental Social Psychology, 35,* 21–30.

Reifenberg, R. J. (1986). The self-serving bias and the use of objective and subjective methods for measuring success and failure. *Journal of Social Psychology, 126,* 627–631.

Reifman, A. S., Larrick, R. P., & Fein, S. (1991). Temper and temperature on the diamond: The heat-aggression relationship in major-league baseball. *Personality and Social Psychology Bulletin, 17,* 580–585.

Reis, H. T., Senchak, M., & Solomon, B. (1985). Sex differences in the intimacy of social interaction: Further examination of potential explanations. *Journal of Personality and Social Psychology, 48,* 1204–1217.

Rempel, J. K., Holmes, J. G., & Zanna, M. P. (1985). Trust in a close relationship. *Journal of Personality and Social Psychology, 49,* 95–112.

Renfrow, D. (2004). A cartography of passing in everyday life. *Symbolic Interaction, 27,* 485–506.

Rennison, C. (2001). *Intimate partner violence and age of victim, 1993–1999.* Washington, DC: U.S. Department of Justice, Bureau of Justice Statistics.

Repetti, R. (1987). Individual and common components of the social environment at work and psychological well-being. *Journal of Personality and Social Psychology, 52,* 710–720.

Repetti, R. (1989). Effects of daily workload on subsequent behavior during marital interaction: The roles of social withdrawal and spouse support. *Journal of Personality and Social Psychology, 57,* 651–659.

Reskin, B., & Hartmann, H. (Eds.). (1986). *Women's work, men's work: Sex segregation on the job.* Washington, DC: National Academy Press.

Reskin, B., & Padavic, I. (1994). *Women and men at work.* Thousand Oaks, CA: Pine Forge Press.

Rexroat, C., & Shehan, C. (1987). The family life cycle and spouses' time in housework. *Journal of Marriage and the Family, 49,* 737–750.

Reynolds, J. R. (1997). The effects of industrial employment conditions on job-related distress. *Journal of Health and Social Behavior, 38,* 105–116.

Reynolds, P. D. (1984). Leaders never quit: Talking, silence, and influence in interpersonal groups. *Small Group Behavior, 15,* 404–413.

Rhine, R. J., & Severance, L. J. (1970). Ego-involvement, discrepancy, source credibility, and attitude change. *Journal of Personality and Social Psychology, 16,* 175–190.

Rice, R. W., Marwick, N. J., Chemers, M. M., & Bentley, J. C. (1982). Task performance and satisfaction: Least preferred coworker (LPC) as a moderator. *Personality and Social Psychology Bulletin, 8,* 534–541.

Richeson, J. A., & Ambady, N. (2003). Effects of situational power on automatic racial prejudice. *Journal of Experimental Social Psychology, 39,* 177–183.

Ridgeway, C. L. (1982). Status in groups: The importance of motivation. *American Sociological Review, 47,* 76–88.

Ridgeway, C. L. (1987). Nonverbal behavior, dominance, and the basis of status in task groups. *American Sociological Review, 52,* 683–694.

Ridley, M., & Dawkins, R. (1981). The natural selection of altruism. In J. P. Rushton & R. M. Sorrentino (Eds.),

Altruism and helping behavior: Social, personality, and developmental perspectives (pp. 19–39). Hillsdale, NJ: Erlbaum.

Riggio, R. E., & Friedman, H. S. (1983). Individual differences and cues to deception. *Journal of Personality and Social Psychology, 45,* 899–915.

Rigney, J. (1962). *A developmental study of cognitive equivalence transformations and their use in the acquisition and processing of information.* Unpublished honors thesis, Radcliffe College, Cambridge, MA.

Riketta, M. (2005). Cognitive differentiation between self, ingroup, and outgroup: The roles of identification and perceived intergroup conflict. *European Journal of Social Psychology, 35,* 97–106.

Riley, M. (1987). On the significance of age in sociology. *American Sociological Review, 52,* 1–14.

Riley, A., & Burke, P. J. (1995). Identities and self-verification in the small group. *Social Psychology Quarterly, 58,* 61–73.

Rindfuss, R., Swicegood, C. G., & Rosenfeld, R. (1987). Disorder in the life course: How common and does it matter? *American Sociological Review, 52,* 785–801.

Rinehart, A. J., & Dunwoody, P. T. (2005, April). *Groupthink in the Bush administration's decision for Operation Iraqi Freedom.* Poster session presented at the annual Western Pennsylvania Undergraduate Psychology Conference, Chatham College, Pittsburgh, PA.

Ring, K., & Kelley, H. H. (1963). A comparison of augmentation and reduction as modes of influence. *Journal of Abnormal and Social Psychology, 66,* 95–102.

Ringelmann, M. (1913). Recherches sur les moteurs animés: Travail de l'homme [Research on animate sources of power: The work of man]. *Annales de l' Institut National Agronomique, 2e Série, 12,* 1–40.

Riordan, C. (1978). Equal-status interracial contact: A review and revision of the concept. *International Journal of Intercultural Relations, 2,* 161–185.

Riordan, C. A., Marlin, N. A., & Kellogg, R. T. (1983). The effectiveness of accounts following transgression. *Social Psychology Quarterly, 46,* 213–219.

Riordan, C., & Ruggiero, J. A. (1980). Producing equal-status interracial interaction: A replication. *Social Psychology Quarterly, 43,* 131–136.

Robertson, J. F., & Simons, R. L. (1989). Family factors, self-esteem, and adolescent depression. *Journal of Marriage and the Family, 51,* 125–138.

Robinson, J. W., Jr., & Preston, J. D. (1976). Equal-status contact and the modification of racial prejudice: A reexamination of the contact hypothesis. *Social Forces, 54,* 911–924.

Robinson, D. T., Smith-Lovin, L., & Tsoudis, O. (1994). Heinous crime or unfortunate accident: The effects of remorse on responses to mock criminal confessions. *Social Problems, 73,* 175–190.

Robson, P. (1982). Patterns of mobility and activity among the elderly. In E. Warnes (Ed.), *Geographical perspectives on the elderly.* New York: Wiley.

Roethlisberger, F. J., & Dickson, W. J. (1939). *Management and the worker.* Cambridge, MA: Harvard University Press.

Rogers, M., Miller, N., Mayer, F. S., & Duvall, S. (1982). Personal responsibility and salience of the request for help: Determinants of the relation between negative affect and helping behavior. *Journal of Personality and Social Psychology, 43,* 956–970.

Rogers, R. G. (1995). Marriage, sex, and mortality. *Journal of Marriage and the Family, 57,* 515–526.

Rogers, R. W. (1980). Expressions of aggression: Aggression-inhibiting effects of anonymity to authority and threatened retaliation. *Personality and Social Psychology Bulletin, 6,* 315–320.

Rogers, S., & May, D. (2003). Spillover between marital quality and job satisfaction: Long-term patterns and gender differences. *Journal of Marriage and Family, 65,* 482–495.

Rogers, T. B. (1977). Self-reference in memory: Recognition of personality items. *Journal of Research in Personality, 11,* 295–305.

Rohrer, J. H., Baron, S. H., Hoffman, E. L., & Swander, D. V. (1954). The stability of autokinetic judgments. *Journal of Abnormal and Social Psychology, 49,* 595–597.

Rokeach, M. (1973). *The nature of human values.* New York: Free Press.

Rommetveit, R. (1955). *Social norms and roles.* Minneapolis: University of Minnesota Press.

Ronis, D. L., & Lipinski, E. R. (1985). Value and uncertainty as weighting factors in impression formation. *Journal of Experimental Social Psychology, 21,* 47–60.

Root, M. (1995). The psychology of Asian American women. In H. Landrine (Ed.), *Bringing cultural diversity to feminist psychology: Theory, research, and practice* (pp. 241–263). Washington, DC: American Psychological Association.

Rose, H., & Rose, S. (2000). *Alas poor Darwin: Arguments against evolutionary psychology.* New York: Harmony Books.

Rose, S., & Frieze, I. H. (1993). Young singles' contemporary dating scripts. *Sex Roles, 28,* 499–509.

Rosen, B., & D'Andrade, R. (1959). The psychological origins of achievement motivation. *Sociometry, 22,* 185–218.

Rosen, S. (1984). Some paradoxical status implications of helping and being helped. In E. Staub, D. Bar-Tal, J. Karylowski, & J. Reykowski (Eds.), *Development and maintenance of prosocial behavior: International perspectives on positive morality.* New York: Plenum.

Rosenbaum, M. E. (1986). The repulsion hypothesis: On the nondevelopment of relationships. *Journal of Personality and Social Psychology, 51,* 1156–1166.

Rosenberg, L. A. (1961). Group size, prior experience, and conformity. *Journal of Abnormal and Social Psychology, 63,* 436–437.

Rosenberg, M. (1965). *Society and the adolescent self-image.* Princeton, NJ: Princeton University Press.

Rosenberg, M. (1973). Which significant others? *American Behavioral Scientist, 16,* 829–860.

Rosenberg, M. (1990). The self-concept: Social product and social force. In M. Rosenberg & R. H. Turner (Eds.), *Social psychology: Sociological perspectives.* New Brunswick, NJ: Transaction.

Rosenberg, M. J., & Abelson, R. (1960). An analysis of cognitive balancing. In C. Hovland & M. Rosenberg (Eds.), *Attitude organization and change.* New Haven, CT: Yale University Press.

Rosenberg, S. V., Nelson, C., & Vivekananthan, P. S. (1968). A multidimensional approach to the structure of personality impressions. *Journal of Personality and Social Psychology, 9,* 283–294.

Rosenberg, M., & Pearlin, L. (1978). Social class and self-esteem among children and adults. *American Journal of Sociology, 84,* 53–77.

Rosenberg, M., Schooler, C., & Schoenbach, C. (1989). Self-esteem and adolescent problems: Modeling reciprocal effects. *American Sociological Review, 54,* 1004–1018.

Rosenberg, M., Schooler, C., Schoenbach, C., & Rosenberg, F. (1995). Global self-esteem and specific self-esteem: Different concepts, different outcomes. *American Sociological Review, 60,* 141–156.

Rosenberg, S. V., & Sedlak, A. (1972). Structural representations in implicit personality theory. In L. Berkowitz (Ed.), *Advances in experimental social psychology* (Vol. 6). New York: Academic Press.

Rosenberg, M., & Simmons, R. (1972). *Black and White self-esteem: The urban school child.* Washington, DC: American Sociological Association.

Rosenblatt, A., & Greenberg, J. (1988). Depression and interpersonal attraction: The role of perceived similarity. *Journal of Personality and Social Psychology, 55,* 112–119.

Rosenblatt, A., & Greenberg, J. (1991). Examining the world of the depressed: Do depressed people prefer others who are depressed? *Journal of Personality and Social Psychology, 60,* 620–629.

Rosenhan, D. L. (1973). On being sane in insane places. *Science, 179,* 250–258.

Rosenhan, D. L., Salovey, P., & Hargis, K. (1981). The joys of helping: Focus of attention mediates the impact of positive affect on altruism. *Journal of Personality and Social Psychology, 40,* 899–905.

Rosenthal, R. (1966). *Experimenter effects in behavioral research.* New York: Appleton-Century-Crofts.

Rosenthal, R. (1980). Replicability and experimenter influence: Experimenter effects in behavioral research. *Parapsychology, 11,* 5–11.

Roskos-Ewoldsen, D. R., Bichsel, J., & Hoffman, K. (2002). The influence of accessibility of source likability on persuasion. *Journal of Experimental Social Psychology, 38,* 137–143.

Rosow, I. (1974). *Socialization to old age.* Berkeley: University of California Press.

Ross, A. S. (1971). Effect of increased responsibility on bystander intervention: The presence of children. *Journal of Personality and Social Psychology, 19,* 306–310.

Ross, L. (1977). The intuitive psychologist and his shortcomings: Distortion in the attribution process. In L. Berkowitz (Ed.), *Advances in experimental social psychology* (Vol. 10). New York: Academic Press.

Ross, L. D. (2001). Getting down to fundamentals: Lay dispositionism and the attributions of psychologists. *Psychological Inquiry, 12,* 37–40.

Ross, M., & Fletcher, G. (1985). Attribution and social perception. In G. Lindzey & E. Aronson (Eds.), *The handbook of social psychology* (3rd ed.). Reading, MA: Addison-Wesley.

Ross, M., & Lumsden, H. (1982). Attributions of responsibility in sports settings: It's not how you play the game but whether you win or lose. In H. Hiebsch, H. Brandstatter, & H. H. Kelley (Eds.), *Social psychology.* East Berlin: Deutscher Verlag der Wissenschaften.

Ross, C. E., Mirowsky, J., & Goldsteen, K. (1990). The impact of the family on health: The decade in review. *Journal of Marriage and the Family, 52,* 1059–1078.

Ross, M., Thibaut, J., & Evenbeck, S. (1971). Some determinants of the intensity of social protests. *Journal of Experimental Social Psychology, 7,* 401–418.

Ross, C. E., & Van Willigen, M. (1996). Gender, parenthood, and anger. *Journal of Marriage and the Family, 58,* 572–584.

Ross, C. E., & Van Willigen, M. (1997). Education and the subjective quality of life. *Journal of Health and Social Behavior, 38,* 275–297.

Ross, L., & Ward, A. (1995). Psychological barriers to dispute resolution. In M. P. Zanna (Ed.), *Advances in experimental social psychology* (Vol. 27, pp. 255–304). San Diego: Academic Press.

Ross, C. E., & Wu, C. (1995). The links between education and health. *American Sociological Review, 60,* 719–745.

Rossi, A. S. (1980). Parenthood in the middle years. In P. Baltes & O. Brim, Jr. (Eds.), *Life-span development and behavior* (Vol. 3). New York: Academic Press.

Rotenberg, K. J., & Mann, L. (1986). The development of the norm of the reciprocity of self-disclosure and its function in children's attraction to peers. *Child Development, 57,* 1349–1357.

Roth, D. L., Wiebe, D. J., Fillingian, R. B., & Shay, K. A. (1989). Life events, fitness, hardiness, and health: A simultaneous analysis of proposed stress-resistance effects. *Journal of Personality and Social Psychology, 57,* 136–142.

Rothbart, M., Dawes, R., & Park, B. (1984). Stereotypes and sampling biases in intergroup perception. In J. R. Eiser (Ed.), *Attitudinal judgment* (pp. 109–134). New York: Springer Verlag.

Rothbart, M., Fulero, S., Jensen, C., Howard, J., & Birrell, B. (1978). From individual to group impressions: Availability heuristics in stereotype formation. *Journal of Experimental Social Psychology, 14,* 237–255.

Rothbart, M., & John, O. P. (1985). Social categorization and behavioral episodes: A cognitive analysis of the effects of intergroup contact. *Journal of Social Issues, 41*(3), 81–104.

Rotheram-Borus, M. J. (1990). Adolescents' reference group choices, self-esteem, and adjustment. *Journal of Personality and Social Psychology, 59,* 1075–1081.

Rotton, J., & Frey, J. (1985). Air pollution, weather, and violent crimes: Concomitant time-series analyses of archival data. *Journal of Personality and Social Psychology, 49,* 1207–1220.

Ruback, R. (1987). Deserted (and nondeserted) aisles: Territorial intrusion can produce persistence, not flight. *Social Psychology Quarterly, 50,* 270–276.

Ruback, R. B., Pape, K. D., & Doriot, P. (1989). Waiting for a phone: Intrusion on callers leads to territorial defense. *Social Psychology Quarterly, 52,* 232–241.

Rubin, B. A. (1986). Class struggle, American style: Unions, strikes, and wages. *American Sociological Review, 51,* 618–631.

Rubin, Z. (1970). Measurement of romantic love. *Journal of Personality and Social Psychology, 16,* 265–273.

Rubin, Z. (1974). From liking to loving: Patterns of attraction in dating relationships. In T. Huston (Ed.), *Foundations of interpersonal attraction.* New York: Academic Press.

Rubin, J. Z., & Brown, B. (1975). *The social psychology of bargaining and negotiation.* New York: Academic Press.

Rubin, M., & Hewstone, M. (1998). Social identity theory's self-esteem hypothesis: A review and some suggestions for clarification. *Personality and Social Psychology Review, 2,* 40–62.

Rubin, Z., Hill, C., Peplau, L., & Dunkel-Scheker, C. (1980). Self-disclosure in dating couples: Sex roles and the ethic of openness. *Journal of Marriage and the Family, 42,* 305–317.

Rubin, J. Z., & Lewecki, R. J. (1973). A three-factor experimental analysis of promises and threats. *Journal of Applied Social Psychology, 3,* 240–257.

Rudé, G. (1964). *The crowd in history.* New York: Wiley.

Ruiter, R. A. C., Kok, G., Verplanken, B., & van Eersel, G. (2003). Strengthening the persuasive impact of fear appeals: The role of action framing. *The Journal of Social Psychology, 143,* 397–400.

Rupp, L., & Taylor, V. (1987). *Survival in the doldrums: The American women's rights movement, 1945 to the 1960s.* New York: Oxford University Press.

Rusbult, C. E. (1983). A longitudinal test of the investment model: The development (and deterioration) of satisfaction and commitment in heterosexual involvements. *Journal of Personality and Social Psychology, 45,* 101–117.

Rusbult, C. E., Johnson, D. J., & Morrow, G. D. (1986). Predicting satisfaction and commitment in adult romantic involvements: An assessment of the generalizability of the investment model. *Social Psychology Quarterly, 49,* 81–89.

Rusbult, C. E., Verette, J., Whitney, G. A., Slovik, L. A., & Lipkus, I. (1991). Accommodation processes in close relationships: Theory and preliminary empirical evidence. *Journal of Personality and Social Psychology, 60,* 53–78.

Rusbult, C. E., Zembrodt, I. M., & Gunn, L. K. (1982). Exit, voice, loyalty, and neglect: Responses to dissatisfaction in romantic involvement. *Journal of Personality and Social Psychology, 43,* 1230–1242.

Rushton, J. (1978). Urban density and altruism: Helping strangers in a Canadian city, suburb, and small town. *Psychological Reports, 43,* 987–990.

Rushton, J. P., & Campbell, A. C. (1977). Modeling, vicarious reinforcement and extraversion on blood donating in adults: Immediate and long-term effects. *European Journal of Social Psychology, 7,* 297–306.

Rushton, J. P., Russell, R. J., & Wells, P. A. (1984). Genetic similarity theory: Beyond kin selection. *Behaviour Genetics, 14,* 179–193.

Ryder, N. B. (1965). The cohort as a concept in the study of social change. *American Sociological Review, 30,* 843–861.

Ryen, A. H., & Kahn, A. (1975). The effects of intergroup orientation on group attitudes and proxemic behavior. *Journal of Personality and Social Psychology, 31,* 302–310.

Ryoko, M. (1979, December 4). The new barbarians: A glimpse of the future. *Cincinnati Post.*

Sabini, J., Siepmann, M., & Stein, J. (2001). The really fundamental attribution error in social psychological research. *Psychological Inquiry, 12,* 1–15.

Sacks, H., Schegloff, E., & Jefferson, G. (1978). A simplest systematics for the organization of turn-taking in conversations. In J. Schenkein (Ed.), *Studies in the organization of conversational interaction.* New York: Academic Press.

Saegert, S. C., Swap, W., & Zajonc, R. B. (1973). Exposure, context, and interpersonal attraction. *Journal of Personality and Social Psychology, 25,* 234–242.

Sagarin, E. (1975). *Deviants and deviance.* New York: Praeger.

Saito, Y. (1988). Situational characteristics as the determinants of adopting distributive justice principles: II. *Japanese Journal of Experimental Social Psychology, 27,* 131–138.

Sakurai, M. M. (1975). Small group cohesiveness and detrimental conformity. *Sociometry, 38,* 340–357.

Sales, S. (1969). Organizational roles as a risk factor in coronary heart disease. *Administrative Science Quarterly, 14,* 325–336.

Salovey, P., Mayer, J. D., & Rosenhan, D. L. (1991). Mood and helping: Mood as a motivator of helping and helping as a regulator of mood. In M. S. Clark (Ed.), *Review of personality and social psychology: Vol. 12. Prosocial behavior* (pp. 215–237). Newbury Park, CA: Sage.

Sammon, S., Reznikoff, M., & Geisinger, K. (1985). Psychosocial development and stressful life events among religious professionals. *Journal of Personality and Social Psychology, 48,* 676–687.

Sample, J., & Warland, R. (1973). Attitude and the prediction of behavior. *Social Forces, 51,* 292–304.

Sampson, R. J., & Laub, J. H. (1990). Crime and deviance over the life course: The salience of adult social bonds. *American Sociological Review, 55,* 609–627.

Sampson, H., Messinger, S., Towne, R., Russ, D., Livson, F., Bowers, M., et al. (1964). The mental hospital and marital family ties. In H. Becker (Ed.), *The other side.* New York: Free Press.

Samuels, F. (1970). The intra- and inter-competitive group. *The Sociological Quarterly, 11,* 390–396.

Sanday, P. R. (1981). The socio-cultural context of rape: A cross-cultural study. *Journal of Social Issues, 37*(4), 5–27.

Sande, G. N., Goethals, G. R., & Radloff, C. E. (1988). Perceiving one's own traits and others': The multifaceted self. *Journal of Personality and Social Psychology, 54,* 13–20.

Sanders, C. (1988). Risk factors in bereavement outcome. *Journal of Social Issues, 44*(3), 97–111.

Sanitioso, R., Kunda, Z., & Fong, G. T. (1990). Motivated recruitment of autobiographical memories. *Journal of Personality and Social Psychology, 59,* 229–241.

Sarbin, T., & Rosenberg, B. (1955). Contributions to role-taking theory IV: A method for obtaining a qualitative estimate of the self. *Journal of Social Psychology, 42,* 71–81.

Sarkisian, N., & Gerstel, N. (2004). Kin support among blacks and whites: Race and family organization. *American Sociological Review, 69,* 812–837.

Sawyer, A. (1973). The effects of repetition of refutational and supportive advertising appeals. *Journal of Marketing Research, 10,* 23–33.

Sayer, L., Bianchi, S., and Robinson, J. (2004). Are parents investing less in children? Trends in mothers' and fathers' time with children. *American Journal of Sociology, 110,* 1–43.

Schachter, S. (1964). The interaction of cognitive and physiological determinants of emotional state. In L. Berkowitz (Ed.), *Advances in experimental social psychology* (Vol. 1). New York: Academic Press.

Schachter, S., & Singer, J. (1962). Cognitive, social, and physiological determinants of emotional state. *Psychological Review, 69,* 379–399.

Schaller, M., & Cialdini, R. B. (1988). The economics of empathic helping: Support for a mood management motive. *Journal of Experimental Social Psychology, 24,* 163–181.

Schank, R. C., & Abelson, R. P. (1977). *Scripts, plans, goals and understanding.* Hillsdale, NJ: Erlbaum.

Scheff, T. (1966). *Being mentally ill.* Chicago: Aldine.

Scheff, T. J., & Retzinger, S. M. (1991). *Emotions and violence: Shame and rage in destructive conflicts.* Lexington, MA: Lexington Books/D.C. Heath.

Schegloff, E. (1968). Sequencing in conversational openings. *American Anthropologist, 70,* 1075–1095.

Scheier, M. F., & Carver, C. (1981). Public and private aspects of the self. In L. Wheeler (Ed.), *Review of personality and social psychology* (Vol. 2). Beverly Hills, CA: Sage.

Scher, S. J. (1997). Measuring the consequences of injustice. *Personality and Social Psychology Bulletin, 23,* 482–497.

Scherer, K. R. (1979). Nonlinguistic indicators of emotion and psychopathology. In C. E. Izard (Ed.), *Emotions in personality and psychopathology.* New York: Plenum.

Scherer, S. E. (1974). Proxemic behavior of primary school children as a function of their socioeconomic class and subculture. *Journal of Personality and Social Psychology, 29,* 800–805.

Schiffenbauer, A., & Schiavo, R. S. (1976). Physical distance and attraction: An intensification effect. *Journal of Experimental Social Psychology, 12,* 274–282.

Schiffrin, D. (1977). Opening encounters. *American Sociological Review, 42,* 679–691.

Schifter, D. E., & Ajzen, I. (1985). Intention, perceived control, and weight loss: An application of the theory of planned behavior. *Journal of Personality and Social Psychology, 45,* 843–851.

Schlenker, B. R. (1975). Self-presentation. Managing the impression of consistency when reality interferes with self-enhancement. *Journal of Personality and Social Psychology, 32,* 1030–1037.

Schlenker, B. R., Helm, B., & Tedeschi, J. T. (1973). The effects of personality and situational variables on behavioral trust. *Journal of Personality and Social Psychology, 25,* 419–427.

Schlenker, B. R., & Weigold, M. F. (1992). Interpersonal processes involving impression regulation and management. *Annual Review of Psychology, 43,* 133–168.

Schlenker, B. R., Weigold, M. E., & Hallam, J. K. (1990). Self-serving attributions in social context: Effects of self-esteem and social pressure. *Journal of Personality and Social Psychology, 58,* 855–863.

Schoenborn, C. (2004, Dec. 15). Marital status and health: United States, 1999–2002. *Advance Data from Vital and Health Statistics,* No. 351. Hyattsville, MD: National Center for Health Statistics.

Schmitt, D. P. (2004). Patterns and universals of mate poaching across 53 nations: The effects of sex, culture, and personality on romantically attracting another person's partner. *Journal of Personality and Social Psychology, 86*(4), 560–584.

Schmitt, D. P., & Buss, D. M. (1996). Strategic self-promotion and competitor derogation: Sex and context effects in the perceived effectiveness of mate attraction tactics. *Journal of Personality and Social Psychology, 70,* 1185–1204.

Schmitt, D. P., & Buss, D. M. (2001). Human mate poaching: Tactics and temptations for infiltrating existing mateships. *Journal of Personality and Social Psychology, 80*(6), 894–917.

Schmitt, D. R., & Marwell, G. (1972). Withdrawal and reward reallocation as responses to inequity. *Journal of Experimental Social Psychology, 8,* 207–221.

Schmitt, D. P., & Shackelford, T. K. (2003). Nifty ways to leave your lover: The tactics people use to entice and disguise the process of human mate poaching. *Personality and Social Psychology Bulletin, 29*(8), 1018–1035.

Schnittker, J. (2000). Gender and reactions to psychological problems: An examination of social tolerance and perceived dangerousness. *Journal of Health and Social Behavior, 41,* 224–240.

Schnittker, J. (2002). The self-esteem of Chinese immigrants. *Social Psychology Quarterly, 65,* 56–76.

Schonbach, P. (1980). A category system for account phrases. *European Journal of Social Psychology, 10,* 195–200.

Schooler, C. (1996). Cultural and social-structural explanations of cross-national psychological differences. *Annual Review of Sociology, 22,* 323–349.

Schrauger, J. S., & Schoeneman, T. (1979). Symbolic interactionist view of self-concept: Through the looking glass darkly. *Psychological Bulletin, 86,* 549–573.

Schroeder, D. A., Penner, L. A., Dovidio, J. F., & Piliavin, J. A. (1995). *The psychology of helping and altruism: Problems and puzzles.* New York: McGraw-Hill.

Schrum, W., & Creek, N. A., Jr. (1987). Social structure during the school years: Onset of the degrouping process. *American Sociological Review, 52,* 218–223.

Schulz, A., Williams, D., Israel, B., Becker, A., Parker, E., James, S., & Jackson, J. (2000). Unfair treatment, neighborhood effects, and mental health in the Detroit metropolitan area. *Journal of Health and Social Behavior, 41,* 314–332.

Schuman, H., & Johnson, M. (1976). Attitudes and behavior. In A. Inkeles, J. Coleman, & N. Smelser (Eds.), *Annual review of sociology* (Vol. 2). Palo Alto, CA: Annual Reviews.

Schutte, J., & Light, J. (1978). The relative importance of proximity and status for friendship choices in social hierarchies. *Social Psychology, 41,* 260–264.

Schutte, N. S., Kendrick, D. T., & Sadalla, E. K. (1985). The search for predictable settings: Situational prototypes, constraint, and behavioral variation. *Journal of Personality and Social Psychology, 49,* 121–128.

Schwartz, S. H. (1978). Temporal instability as a moderator of the attitude-behavior relationship. *Journal of Personality and Social Psychology, 36,* 715–724.

Schwartz, S. H. (1992). Universals in the content and structure of values: Theoretical advances and empirical tests in 20 countries. In M. Zanna (Ed.), *Advances in experimental social psychology* (Vol. 25, pp. 1–65). Orlando, FL: Academic Press.

Schwartz, S. H. (1994). Are there universal aspects in the content and structure of values? *Journal of Social Issues, 50,* 19–45.

Schwartz, S., & Ames, R. (1977). Positive and negative referent others as sources of influence: A case of helping. *Sociometry, 40,* 12–20.

Schwartz, S. H., & Clausen, G. T. (1970). Responsibility, norms, and helping in an emergency. *Journal of Personality and Social Psychology, 16,* 299–310.

Schwartz, S. H., & Fleishman, J. (1978). Personal norms and the mediation of legitimacy effects on helping. *Social Psychology, 41,* 306–315.

Schwartz, S. H., & Gottlieb, A. (1976). Bystander reactions to a violent theft: Crime in Jerusalem. *Journal of Personality and Social Psychology, 34,* 1188–1199.

Schwartz, S. H., & Gottlieb, A. (1980). Bystander anonymity and reactions to emergencies. *Journal of Personality and Social Psychology, 39,* 418–430.

Schwartz, S. H., & Howard, J. A. (1980). Explanations of the moderating effect of responsibility denial on personal norm-behavior relationship. *Social Psychology Quarterly, 43,* 441–446.

Schwartz, S. H., & Howard, J. A. (1981). A normative decision-making model of altruism. In J. P. Rushton & R. M. Sorrentino (Eds.), *Altruism and helping behavior.* Hillsdale, NJ: Erlbaum.

Schwartz, S. H., & Howard, J. A. (1982). Helping and cooperation: A self-based motivational model. In V. J. Derlega & J. Grzelak (Eds.), *Cooperation and helping behavior: Theories and research* (pp. 327–353). New York: Academic Press.

Schwartz, S. H., & Howard, J. A. (1984). Internalized values as motivators of altruism. In E. Staub, E. Bar-Tal, J. Karylowski, & J. Reykowski (Eds.), *Development and maintenance of prosocial behavior: International perspectives on positive morality.* New York: Plenum.

Schwarz, N., Groves, R., & Schuman, H. (1998). Survey methods. In D. Gilbert, S. Fiske, & G. Lindzey (Eds.), *The handbook of social psychology* (4th ed., Vol. 1, pp. 143–179). Boston: McGraw-Hill.

Schweingruber, D., & McPhail, C. (1999). A method for systematically observing and recording collective action. *Sociological Methods and Research, 27*(4), 451–498.

Schweingruber, D., & Wohlstein, R. (2005) The madding crowd goes to school: Myths about crowds in introductory sociology textbooks. *Teaching Sociology, 33,* 136–153.

Scott, R. (1976). Deviance, sanctions, and social integration in small-scale societies. *Social Forces, 54,* 604–620.

Scott, M., & Lyman, S. (1968). Accounts. *American Sociological Review, 33,* 46–62.

Scotton, C. M. (1983). The negotiation of identities in conversation. *International Journal of the Sociology of Language, 44,* 115–136.

Searle, J. R. (1979). *Expression and meaning: Studies in the theory of speech acts.* Cambridge, UK: Cambridge University Press.

Sears, D., Fu, M., Henry, P. J., & Bui, K. (2003). The origins and persistence of ethnic identity among the "New Immigrant" Groups. *Social Psychology Quarterly, 66,* 419–437.

Sears, D. O., & Freedman, J. L. (1967). Selective exposure to information: A critical review. *Public Opinion Quarterly, 31,* 194–213.

Sears, D. O., & Whitney, R. E. (1973). Political persuasion. In I. de S. Pool, W. Schramm, N. Maccoby, & E. Parker (Eds.), *Handbook of communication.* Chicago: Rand McNally.

Sedikides, C., & Jackson, J. M. (1990). Social impact theory: A field test of source strength, source immediacy, and number of targets. *Basic and Applied Social Psychology, 11,* 273–281.

Seedman, A. A., & Hellman, P. (1975). *Chief.* New York: Avon.

Seeman, M. (1975). Alienation studies. In A. Inkeles, J. Coleman, & N. Smelser (Eds.), *Annual Review of Sociology* (Vol. 1). Palo Alto, CA: Annual Reviews.

Seeman, M., Seeman, M., & Sayles, M. (1985). Social networks and health status: A longitudinal analysis. *Social Psychology Quarterly, 48,* 237–248.

Segal, B. E. (1965). Contact, compliance, and distance among Jewish and non-Jewish undergraduates. *Social Problems, 13,* 66–74.

Select Committee on Intelligence, United States Senate. (2004). *Report on the U.S. intelligence community's prewar intelligence assessments on Iraq.* Retrieved

March 31, 2006 from http://intelligence.senate.gov/conclusions.pdf

Semin, G. R., & Manstead, A. S. R. (1982). The social implications of embarrassment displays and restitution behaviour. *European Journal of Social Psychology, 12,* 367–377.

Semin, G. R., & Manstead, A. S. R. (1983). *The accountability of conduct: A social psychological analysis.* London: Academic Press.

Seplaki, C., Goldman, N., Weinstein, M., & Lin, Y.-H. (2003). *Before and after the 1999 Chi-Chi earthquake: Traumatic events and depressive symptoms in an older population* (Working Paper 2003-02). Princeton University: Office of Population Research.

Serbin, L., & O'Leary, K. (1975). How nursery schools teach girls to shut up. *Psychology Today, 9,* 56–58.

Serpe, R. T. (1987). Stability and change in self: A structural symbolic interactionist explanation. *Social Psychology Quarterly, 50,* 44–55.

Sewell, W. H., & Hauser, R. M. (1975). *Education, occupation, and earnings: Achievement in the early career.* New York: Academic Press.

Sewell, W. H., & Hauser, R. M. (1980). The Wisconsin longitudinal study of social and psychological factors in aspirations and achievements. *Research in Sociology of Education and Socialization, 1,* 59–99.

Sewell, W. H., Hauser, R. M., & Wolf, W. (1980). Sex, schooling, and occupational status. *American Journal of Sociology, 86,* 551–583.

Shanas, E. (1979). The family as a support system in old age. *Gerontologist, 19,* 169–174.

Shapiro, S. P. (1990). Collaring the crime, not the criminal: Considering the concept of white-collar crime. *American Sociological Review, 55,* 346–365.

Sharkey, W. F., & Stafford, L. (1990). Responses to embarrassment. *Human Communication Research, 17,* 315–342.

Shaver, P. R., Schwartz, J. C., Kirson, D., & O'Connor, C. (1987). Emotion knowledge: Further exploration of a prototype approach. *Journal of Personality and Social Psychology, 52,* 1061–1086.

Shaver, P. R., Wu, S., & Schwartz, J. C. (1992). Cross-cultural similarities and differences in emotion and its representation: A prototype approach. In M. S. Clark (Ed.), *Review of personality and social psychology* (Vol. 13, *Emotion,* pp. 175–212). Newbury Park, CA: Sage.

Shaw, M., & Costanzo, P. (1982). *Theories of social psychology* (2nd ed.). New York: McGraw-Hill.

Shaw, M. E., & Shaw, L. M. (1962). Some effects of sociometric grouping upon learning in a second grade classroom. *Journal of Social Psychology, 57,* 453–458.

Shell, R. M., & Eisenberg, N. (1992). A developmental model of recipients' reactions to aid. *Psychological Bulletin, 111,* 413–433.

Shelton, B. A., & John, D. (1996). The division of household labor. *Annual Review of Sociology, 22,* 299–322.

Sherif, M. (1935). A study of some social factors in perception. *Archives of Psychology, 27*(no. 187).

Sherif, M. (1936). *The psychology of social norms.* New York: Harper & Row.

Sherif, M. (1966). *In common predicament.* Boston: Houghton Mifflin.

Sherif, M., Harvey, O. J., White, B. J., Hood, W. R., & Sherif, C. W. (1961). *Intergroup cooperation and competition: The Robbers Cave experiment.* Norman, OK: University Book Exchange.

Sherif, M., & Sherif, C. (1964). *Exploration into conformity and deviation of adolescents.* New York: Harper & Row.

Sherif, M., & Sherif, C. W. (1982). Production of intergroup conflict and its resolution—Robbers Cave experiment. In J. W. Reich (Ed.), *Experimenting in society: Issues and examples in applied social psychology.* Glenview, IL: Scott, Foresman.

Sherman, P. (1980). The limits of ground squirrel nepotism. In G. Barlow & J. Silverberg (Eds.), *Sociobiology: Beyond nature/nurture?* Boulder, CO: Westview.

Sherman, S. J. (1970). Effects of choice and incentive on attitude change in a discrepant behavior situation. *Journal of Personality and Social Psychology, 15,* 245–252.

Sherman, S. J., Judd, C. M., & Park, B. (1989). Social cognition. *Annual Review of Psychology, 40,* 281–326.

Sherrod, D. R., & Downs, R. (1974). Environmental determinants of altruism: The effects of stimulus overload and perceived control on helping. *Journal of Experimental Social Psychology, 10,* 468–479.

Sherwood, J. J. (1965). Self-identity and referent others. *Sociometry, 28,* 66–81.

Shibutani, T. (1961). *Society and personality.* Englewood Cliffs, NJ: Prentice Hall.

Shihadeh, E. S. (1991). The prevalence of husband-centered migration: Employment consequences for married mothers. *Journal of Marriage and the Family, 53,* 431–444.

Shorter, E., & Tilly, C. (1974). *Strikes in France: 1830–1968.* New York: Cambridge University Press.

Shotland, R. L., & Huston, T. L. (1979). Emergencies: What are they and do they influence bystanders to intervene? *Journal of Personality and Social Psychology, 37,* 1822–1834.

Shotland, R. L., & Stebbins, C. A. (1983). Emergency and cost as determinants of helping behavior and the slow accumulation of social psychological knowledge. *Social Psychology Quarterly, 46,* 36–46.

Shotland, R. L., & Straw, M. K. (1976). Bystander response to an assault: When a man attacks a woman. *Journal of Personality and Social Psychology, 34,* 990–999.

Shott, S. (1979). Emotion and social life: A symbolic interactionist analysis. *American Journal of Sociology, 84,* 1317–1334.

Shover, N., Novland, S., James, J., & Thornton, W. (1979). Gender roles and delinquency. *Social Forces, 58,* 162–175.

Shweder, R. A. (1977). Likeness and likelihood in everyday thought: Magical thinking in judgments about personality. *Current Anthropology, 18,* 637–658.

Sigall, H., & Landy, D. (1973). Radiating beauty: The effects of having a physically attractive partner on person perception. *Journal of Personality and Social Psychology, 28,* 218–224.

Silberman, M. (1976). Toward a theory of criminal deterrence. *American Sociological Review, 41,* 442–461.

Sime, J. D. (1983). Affiliative behavior during escape to building exits. *Journal of Environmental Psychology, 3,* 21–41.

Simmons, R. G. (1991). Presidential address on altruism and sociology. *The Sociological Quarterly, 32,* 1–22.

Simon, R. (1997). The meanings individuals attach to role-identities and their implications for mental health. *Journal of Health and Social Behavior, 38,* 256–274.

Simon, R., & Marcussen, K. (1999). Marital transitions, marital beliefs, and mental health. *Journal of Health and Social Behavior, 40,* 111–125.

Simons, R., Lin, K.-H., Gordon, L., Brody, G., Murry, V., & Conger, R. (2002). Community differences in the association between parenting practices and child conduct problems. *Journal of Marriage and Family, 64,* 331–345.

Simpson, J. A. (1987). The dissolution of romantic relationships: Factors involved in emotional stability and emotional distress. *Journal of Personality and Social Psychology, 53,* 683–692.

Simpson, J. A. (1990). Influence of attachment styles on romantic relationships. *Journal of Personality and Social Psychology, 59,* 971–980.

Simpson, J. A., Gangestad, S. W., & Lerum, M. (1990). Perception of physical attractiveness: Mechanisms involved in maintenance of romantic relationships. *Journal of Personality and Social Psychology, 59,* 1192–1201.

Sinclair, S., Dunn, E., & Lowery, B. S. (2005). The relationship between parental racial attitudes and children's implicit prejudice. *Journal of Experimental Social Psychology, 41,* 283–289.

Singer, J. L., & Singer, D. G. (1981). *Television, imagination and aggression: A study of preschoolers.* Hillsdale, NJ: Erlbaum.

Singer, J. L., & Singer, D. G. (1983). Psychologists look at television: Cognitive, developmental, personality, and social policy implications. *American Psychologist, 38,* 826–834.

Sistrunk, F., & McDavid, J. W. (1971). Sex variable in conforming behavior. *Journal of Personality and Social Psychology, 17,* 200–207.

Skinner, B. F. (1953). *Science and human behavior.* New York: MacMillan.

Skinner, B. F. (1957). *Verbal behavior.* New York: Appleton-Century-Crofts.

Skinner, B. F. (1971). *Beyond freedom and dignity.* New York: Knopf.

Skowronski, J. J., & Carlston, D. E. (1989). Negativity and extremity biases in impression formation: A review of explanations. *Psychological Bulletin, 105,* 131–142.

Slater, P. (1963). On social regression. *American Sociological Review, 28,* 339–364.

Slater, P. E. (1955). Role differentiation in small groups. *American Sociological Review, 20,* 300–310.

Sloane, D., & Potrin, R. H. (1986). Religion and delinquency: Cutting through the maze. *Social Forces, 65,* 87–105.

Slomczynski, K. M., Miller, J., & Kohn, M. (1981). Stratification, work, and values: A Polish-United States comparison. *American Sociological Review, 46,* 720–744.

Small, K. H., & Peterson, J. (1981). The divergent perceptions of actors and observers. *Journal of Social Psychology, 113,* 123–132.

Smith, D. A. (1987). Police response to interpersonal violence: Defining the parameters of legal control. *Social Forces, 65,* 767–782.

Smith, W. P., & Anderson, A. (1975). Threats, communication, and bargaining. *Journal of Personality and Social Psychology, 32,* 76–82.

Smith, E. R., Fazio, R. H., & Cejka, M. A. (1996). Accessible attitudes influence categorization of multiply categorizable objects. *Journal of Personality and Social Psychology, 71,* 888–898.

Smith, E. R., & Henry, S. (1996). An in-group becomes part of the self: Response time evidence. *Personality and Social Psychology Bulletin, 22,* 635–642.

Smith, E. S., Powers, A. S., & Suarez, G. (2005). If Bill Clinton were a woman: The effectiveness of male and female politicians' account strategies following alleged transgressions. *Political Psychology, 26,* 115–134.

Smith, W. P., & Leginski, W. A. (1970). Magnitude and precision of punitive power in bargaining strategy. *Journal of Experimental Social Psychology, 6,* 57–76.

Smith, H. J., & Spears, R. (1996). Ability and outcome evaluations as a function of personal and collective (dis)advantage: A group escape from individual bias. *Personality and Social Psychology Bulletin, 22,* 690–704.

Smith, C. J., Tindale, R. S., & Dugoni, B. L. (1996). Minority and majority influence in freely interacting groups: Qualitative versus quantitative differences. *British Journal of Social Psychology, 35,* 137–149.

Smith-Lovin, L. (1990). Emotions as the confirmation and disconfirmation of identity: The affect control model. In T. D. Kemper (Ed.), *Research agendas in the sociology of emotion.* Albany, NY: SUNY Press.

Smith-Lovin, L. (1995). The sociology of affect and emotion. In K. S. Cook, G. A. Fine, & J. S. House (Eds.), *Sociological perspectives on social psychology* (pp. 118–148). Boston: Allyn & Bacon.

Smith-Lovin, L., & Robinson, D. (2006). Control theories of identity, action and emotion: In search of testable differences between affect control theory and identity control theory. In K. A. McClelland & T. J. Fararo (Eds.), *Purpose, meaning, and action: Control systems theories in sociology.* New York: Palgrave Macmillan.

Sniezek, J. A., & May, D. R. (1990). Conflict of interests and commitment in groups. *Journal of Applied Social Psychology, 20,* 1150–1165.

Snow, D. A., & Benford, R. D. (1992). Master frames and cycles of protest. In A. D. Morris & C. M. Mueller (Eds.), *Frontiers in social movement theory.* New Haven, CT: Yale University Press.

Snow, D. A., & Oliver, P. E. (1995). Social movements and collective behavior: Social psychological dimensions and considerations. In K. S. Cook, G. A. Fine, & J. S. House (Eds.), *Sociological perspectives on social psychology* (pp. 571–599). Boston: Allyn & Bacon.

Snow, D. A., Rochford, E., Jr., Worden, S., & Benford, R. (1986). Frame alignment processes, micromobilization, and movement participation. *American Sociological Review, 51,* 464–481.

Snyder, D. (1975). Institutional setting and industrial conflict: Comparative analysis of France, Italy, and the United States. *American Sociological Review, 40,* 259–278.

Snyder, M. (1981). On the self-perpetuating nature of social stereotypes. In D. L. Hamilton (Ed.), *Cognitive processes in stereotyping and intergroup behavior.* Hillsdale, NJ: Erlbaum.

Snyder, C. R., Higgins, R. L., & Stucky, R. J. (1983). *Excuses: Masquerades in search of grace.* New York: Wiley.

Snyder, C. R., Lassegard, M. A., & Ford, C. E. (1986). Distancing after group success and failure: Basking in reflected glory and cutting off reflected failure. *Journal of Personality and Social Psychology, 51,* 382–388.

Snyder, M., & Swann, W. B., Jr. (1978). Hypothesis-testing processes in social interaction. *Journal of Personality and Social Psychology, 36,* 1202–1212.

Snyder, M., Tanke, E. D., & Berscheid, E. (1977). Social perception and interpersonal behavior: On the self-fulfilling nature of social stereotypes. *Journal of Personality and Social Psychology, 35,* 656–666.

Solano, C. H., Batten, P. G., & Parish, E. A. (1982). Loneliness and patterns of self-disclosure. *Journal of Personality and Social Psychology, 43,* 524–531.

Solomon, D. S. (1982). Mass media campaigns for health promotion. *Prevention in Human Services, 2,* 115–123.

Sommer, B. (2001). Menopause. In J. Worell (Ed.), *Encyclopedia of women and gender* (Vol. II, pp. 729–738). San Diego: Academic Press.

Sommer, R. (1969). *Personal space.* Englewood Cliffs, NJ: Prentice Hall.

Sommers-Flanagan, R., Sommers-Flanagan, J., & Davis, B. (1993). What's happening on Music Television? A gender role content analysis. *Sex Roles, 28,* 745–753.

Sorrentino, R. M., & Field, N. (1986). Emergent leadership over time: The functional value of positive motivation. *Journal of Personality and Social Psychology, 50,* 1091–1099.

South, S., & Spitze, G. (1994). Housework in marital and nonmarital households. *American Sociological Review, 59,* 327–347.

South, S., Trent, K., & Shen, Y. (2001). Changing partners: Toward a macrostructural-opportunity theory of marital dissolution. *Journal of Marriage and the Family, 63,* 743–754.

Spence, J. T., & Buckner, C. E. (2000). Instrumental and expressive traits, trait stereotypes, and sexist attitudes. *Psychology of Women Quarterly, 24,* 44–62.

Spencer, J. W. (1987). Self-work in social interaction: Negotiating role-identities. *Social Psychology Quarterly, 50,* 131–142.

Spencer, S. J., Steele, C. M, & Quinn, D. M. (1999). Stereotype threat and women's math performance. *Journal of Experimental Social Psychology, 35,* 4–28.

Spiegel, J. P. (1969). Hostility, aggression, and violence. In A. Grimshaw (Ed.), *Patterns in American racial violence.* Chicago: Aldine.

Spilerman, S. (1970). The causes of racial disturbances: A comparison of alternative explanations. *American Sociological Review, 35,* 627–649.

Spilerman, S. (1976). Structural characteristics of cities and severity of racial disorders. *American Sociological Review, 41,* 771–793.

Spitz, R. (1945). Hospitalism. *The Psychoanalytic Study of the Child, 1,* 53–72.

Spitz, R. (1946). Hospitalism: A follow-up report. *The Psychoanalytic Study of the Child, 2,* 113–117.

Sprecher, S., & Metts, S. (1989). Development of the "Romantic Beliefs Scale" and examination of the effects of gender and gender role orientation. *Journal of Social and Personal Relationships, 6,* 387–411.

Sprecher, S., & Regan, P. (1998). Passionate and companionate love in courting and young married couples. *Sociological Inquiry, 68,* 163–185.

Squier, W. (2002, April 15). My son goofs around, and I think it's great. *Newsweek,* p. 10.

St. C. Oates, G. (2004). The color of the undergraduate experience and the Black self-concept: Evidence from longitudinal data. *Social Psychology Quarterly, 67,* 16–32.

Staats, A. W., & Staats, C. (1958). Attitudes established by classical conditioning. *Journal of Abnormal and Social Psychology, 57,* 37–40.

Stack, S. (1987). Celebrities and suicide: A taxonomy and analysis, 1948–1983. *American Sociological Review, 52,* 401–412.

Stafford, L., & Canary, D. J., (1991). Maintenance strategies by romantic relationship type, gender, and relational characteristics. *Journal of Social and Personal Relationships, 8,* 217–242.

Stafford, M. C., Gray, L. N., Menke, B. A., & Ward, D. A. (1986). Modelling the deterrent effects of punishment. *Social Psychology Quarterly, 49,* 338–347.

Staggenborg, S. (1986). Coalition work in the pro-choice movement: Organizational and environmental opportunities and obstacles. *Social Problems, 33,* 374–390.

Stang, D. J. (1972). Conformity, ability, and self-esteem. *Representative Research in Social Psychology, 3,* 97–103.

Stanhope, P. D. (1989). *Lord Chesterfield's letters to his son.* London: Transworld. (Original work published 1774.)

Stark, R., & Bainbridge, W. S. (1980). Networks of faith: Interpersonal bonds and recruitment in cults and sects. *American Journal of Sociology, 85,* 1376–1395.

Starrels, M. E. (1994). Husband's involvement in female gender-typed household chores. *Sex Roles, 31,* 473–491.

Stasser, G. (1992). Pooling of unshared information during group discussion. In S. Worchel, W. Wood, & J. Simpson (Eds.), *Group process and productivity* (pp.48–67). Newbury Park, CA: Sage.

Stasser, G., & Titus, W. (1987). Effects of information load and percentage of shared information on the dissemination of unshared information during group discussion. *Journal of Personality and Social Psychology, 53,* 81–93.

Staub, E. (1974). Helping a distressed person: Social, personality, and stimulus determinants. In L. Berkowitz (Ed.), *Advances in experimental social psychology* (Vol. 7). New York: Academic Press.

Steblay, N. M. (1987). Helping behavior in rural and urban environments: A meta-analysis. *Psychological Bulletin, 102,* 346–356.

Steele, C. M. (1997). A threat in the air: How stereotypes shape the intellectual identities and performance of women and African-Americans. *American Psychologist, 52,* 613–629.

Steele, C. M. (1999, August). Thin ice: "Stereotype threat" and Black college students. *The Atlantic Monthly, 284*(2), 44–47, 50–54.

Steele, C. M., & Aronson, J. (1995). Stereotype threat and the intellectual test performance of African-Americans. *Journal of Personality and Social Psychology, 69,* 797–811.

Steffensmeier, D., & Demuth, S. (2000). Ethnicity and sentencing outcomes in U.S. federal courts: Who is punished more harshly? *American Sociological Review, 65,* 705–729.

Steffensmeier, D. J., & Allan, E. (1996). Gender and crime: Toward a gendered theory of female offending. *Annual Review of Sociology, 22,* 459–487.

Steffensmeier, D. J., Allan, E. A., Haver, M. D., & Streifel, C. (1989). Age and the distribution of crime. *American Journal of Sociology, 94,* 803–831.

Steffensmeier, D. J., & Terry, R. M. (1973). Deviance and respectability: An observational study of reactions to shoplifting. *Social Forces, 51,* 417–426.

Steffy, L. (2005, Jan. 12). CEOs are playing dumb in attempts to get off hook. *Houston Chronicle.* Available at http://www.chron.com/disp/story.mpl/business/steffy/2988866.html

Stein, J. A., Newcomb, M. D., & Bentler, P. M. (1987). An eight-year study of multiple influences on drug use and drug use consequences. *Journal of Personality and Social Psychology, 53,* 1094–1105.

Steiner, D. D., & Rain, J. S. (1989). Immediate and delayed primacy and recency effects in performance evaluation. *Journal of Applied Psychology, 74,* 136–142.

Steiner, I. D. (1972). *Group process and productivity.* New York: Academic Press.

Steiner, I. D., & Rogers, E. (1963). Alternative responses to dissonance. *Journal of Abnormal and Social Psychology, 66,* 128–136.

Stemp, P., Turner, R., & Noh, S. (1986). Psychological distress in the postpartum period: The significance of social support. *Journal of Marriage and the Family, 48,* 271–277.

Stephan, W. G. (1987). The contact hypothesis in intergroup relations. In C. Hendrick (Ed.), *Group processes and intergroup relations* (pp. 13–40). Beverly Hills, CA: Sage.

Stephenson, W. (1953). *The study of behavior.* Chicago: University of Chicago Press.

Sternberg, R. J. (1985). Implicit theories of intelligence, creativity, and wisdom. *Journal of Personality and Social Psychology, 49,* 607–627.

Sternberg, R. (1996). Love stories. *Personal Relationships, 3,* 59–79.

Sternberg, R. (1998). *Love is a story: A new theory of relationships.* New York: Oxford University Press.

Sternthal, B., Dholakia, R., & Leavitt, C. (1978). The persuasive effect of source credibility: A test of cognitive response analysis. *Journal of Consumer Research, 4,* 252–260.

Stets, J. E. (2003). Emotions and sentiments. In J. Delamater (Ed.), *Handbook of social psychology* (pp. 309–335). New York: Kluwer.

Stets, J. E., & Burke, P. (2000). Identity theory and social identity theory. *Social Psychology Quarterly, 63,* 224–237.

Stevenson, M. B., Ver Hoeve, J. N., Roach, M. A., & Leavitt, L. A. (1986). The beginning of conversation: Early patterns of mother-infant vocal responsiveness. *Infant Behavior and Development, 9,* 423–440.

Stewart, A. (2000). *Uses of the past: Toward a psychology of generations.* Paper presented at the University of Michigan, Ann Arbor, MI.

Stewart, A. (2002). *Gender, race, and generation in the psychology of women.* Paper presented at the Annual Meeting of the American Psychological Association, Chicago, IL.

Stewart, A. J., Henderson-King, D., Henderson-King, E., & Winter, D. G. (2000). *Work and family values: Life scripts in the Midwest in the 1950s.* Paper presented at Conference on Work and Families, San Francisco, CA.

Stewart, R. H. (1965). Effect of continuous responding on the order effect in personality impression formation. *Journal of Personality and Social Psychology, 1,* 161–165.

Stewart, S., Stinnett, H., & Rosenfeld, L. (2000). Sex differences in desired characteristics of short-term and long-term relationship partners. *Journal of Social and Personal Relationships, 17,* 843–853.

Stiff, J. B. (1986). Cognitive processing of persuasive message cues: A meta-analytic review of the effects of supporting information on attitudes. *Communication Monographs, 53,* 75–89.

Stiles, W., Orth, J., Scherwitz, L., Hennrikus, D., & Vallbona, C. (1984). Role behaviors in routine medical interviews with hypertensive patients: A repertoire of verbal exchanges. *Social Psychology Quarterly, 47,* 244–254.

Stokes, J. P. (1985). The relation of social network and individual difference variables to loneliness. *Journal of Personality and Social Psychology, 48,* 981–990.

Stokes, J., & Levin, I. (1986). Gender differences in predicting loneliness from social network characteristics. *Journal of Personality and Social Psychology, 51,* 1069–1074.

Stone, G. P. (1962). Appearances and the self. In A. Rose (Ed.), *Human behavior and social processes.* Boston: Houghton Mifflin.

Stoner, J. A. F. (1968). Risky and cautious shifts in group decisions: The influence of widely held values. *Journal of Experimental Social Psychology, 4,* 442–459.

Storms, M. D. (1973). Videotape and attribution process: Reversing actors' and observers' points of view. *Journal of Personality and Social Psychology, 27,* 165–175.

Strauman, T. J., Vookles, J., Berenstein, V., Chaiken, S., & Higgins, E. T. (1991). Self-discrepancies and vulnerability to body dissatisfaction and disordered eating. *Journal of Personality and Social Psychology, 61,* 946–956.

Straus, M., & Field, C. (2003). Psychological aggression by American parents: National data on prevalence, chronicity, and severity. *Journal of Marriage and Family, 65,* 795–808.

Straus, M., & Stewart, J. (1999). Corporal punishment by American parents: National data on prevalence, chronicity, severity, and duration, in relation to child and family characteristics. *Clinical Child and Family Psychology Review, 2,* 55–70.

Straus, M., Sugarman, D., & Giles-Sims, J. (1997). Spanking by parents and subsequent antisocial behavior of children. *Archives of Pediatric and Adolescent Medicine, 151,* 761–767.

Stricker, L. J., Jacobs, P. I., & Kogan, N. (1974). Trait interrelations in implicit personality theories and questionnaire data. *Journal of Personality and Social Psychology, 30,* 198–207.

Strodtbeck, F. L., Simon, R. J., & Hawkins, C. (1965). Social status in jury deliberations. In I. D. Steiner & M. Fishbein (Eds.), *Current studies in social psychology.* New York: Holt, Rinehart and Winston.

Stroebe, W., Thompson, V., Insko, C., & Reisman, S. R. (1970). Balance and differentiation in the evaluation of linked attitude objects. *Journal of Personality and Social Psychology, 16,* 38–47.

Strube, M. J., & Garcia, J. E. (1981). A meta-analytic investigation of Fiedler's contingency model of leadership effectiveness. *Psychological Bulletin, 90,* 307–321.

Struch, N., & Schwartz, S. H. (1989). Intergroup aggression: Its predictors and distinctness from in-group bias. *Journal of Personality and Social Psychology, 56,* 364–373.

Stryker, S. (1980). *Symbolic interactionism: A social structural version.* Menlo Park, CA: Benjamin/Cummings.

Stryker, S. (1987). The vitalization of symbolic interactionism. *Social Psychology Quarterly, 50,* 83–94.

Stryker, S., & Gottlieb, A. (1981). Attribution theory and symbolic interactionism: A comparison. In J. H. Hawes, W. Ickes, & R. F. Kidd (Eds.), *New directions in attribution theory* (Vol. 3). Hillsdale, NJ: Erlbaum.

Stryker, S., & Serpe, R. (1981). Commitment, identity salience and role behavior: Theory and research example. In W. Ickes & E. Knowles (Eds.), *Personality, roles and social behavior.* New York: Springer Verlag.

Stryker, S., & Serpe, R. (1982). Towards a theory of family influence in the socialization of children. In A. Kerckhoff (Ed.), *Research in sociology of education and socialization* (Vol. 4). Greenwich, CT: JAI Press.

Sudman, S., & Bradburn, N. M. (1974). *Response effects in surveys.* Chicago: Aldine.

Sueda, K., & Wiseman, R. L. (1992). Embarrassment remediation in Japan and the United States. *International Journal of Intercultural Relations, 16*(2), 159–173.

Suls, J., Lemos, K., & Stewart, H. L. (2002). Self-esteem, construal, and comparisons with self, friends, and others. *Journal of Personality and Social Psychology, 82,* 252–261.

Suls, J. M., & Miller, R. (Eds.). (1977). *Social comparison processes: Theoretical and empirical perspectives.* New York: Wiley.

Summers-Effler, E. (2005). The emotional significance of solidarity for social movement communities: Sustaining Catholic worker community and service. In H. Flam & D. King (Eds.), *Emotions and social movements* (pp. 133–149). London and New York: Routledge.

Sumner, W. G. (1906). *Folkways.* New York: Ginn.

Sun, Y., & Li, Y. (2002). Children's well-being during parents' marital disruption process: A pooled time-series analysis. *Journal of Marriage and Family, 64,* 472–488.

Surra, C. (1990). Research and theory on mate selection and premarital relationships in the 1980s. *Journal of Marriage and the Family, 52,* 844–865.

Surra, C., & Longstreth, M. (1990). Similarity of outcomes, interdependence, and conflict in dating relationships. *Journal of Personality and Social Psychology, 59,* 501–516.

Suskind, R. (2004). *The price of loyalty: George W. Bush, the White House, and the education of Paul O'Neill.* New York: Simon & Schuster.

Sussman, N. M., & Rosenfeld, H. M. (1982). Influence of culture, language, and sex on conversational distance. *Journal of Personality and Social Psychology, 42,* 66–74.

Sutherland, E., Cressey, D., & Luckenbill, D. (1992). *Principles of criminology* (11th ed.). Dix Hills, NY: General Hall.

Suttles, G. (1968). *The social order of the slum.* Chicago: University of Chicago Press.

Sutton, J. R. (1990). Bureaucrats and entrepreneurs: Institutional responses to deviant children in the United States, 1890–1920s. *American Journal of Sociology, 95,* 1367–1400.

Swann, W. B., Jr. (1987). Identity negotiation: Where two roads meet. *Journal of Personality and Social Psychology, 53,* 1038–1051.

Swann, W. B., Jr., & Predmore, S. C. (1985). Intimates as agents of social support: Sources of consolation or despair? *Journal of Personality and Social Psychology, 49,* 1609–1617.

Swann, W. B., Jr., & Schroeder, D. G. (1995). The search for beauty and truth: A framework for understanding reactions to evaluations. *Personality and Social Psychology Bulletin, 21,* 1307–1318.

Swanson, D. L. (Ed.). (1979). The uses and gratifications approach to mass communications research. *Communication Research, 3,* 3–111.

Sweeney, P. D. (1990). Distributive justice and pay satisfaction: A field test of an equity theory prediction. *Journal of Business and Psychology, 4,* 329–341.

Sweet, J. A., & Bumpass, L. (1987). *American families and households.* New York: Russell Sage Foundation.

Swigert, V., & Farrell, R. (1977). Normal homicides and the law. *American Sociological Review, 42,* 16–32.

Swim, J. K., Aikin, K. J., Hall, W. S., & Hunter, B. A. (1995). Sexism and racism: Old fashioned and modern prejudices. *Journal of Personality and Social Psychology, 68,* 199–214.

Syme, S. L., & Berkman, L. (1976). Social class, susceptibility, and sickness. *American Journal of Epidemiology, 104,* 1–8.

Symons, D. (1992). On the use and misuse of Darwinism in the study of human behavior. In J. Barkow, L. Cosmides, & J. Tooby (Eds.), *The adapted mind: Evolutionary psychology and the generation of culture* (pp. 137–159). New York: Oxford University Press.

Szinovacz, M. (2000). Changes in housework following retirement. *Journal of Marriage and the Family, 62,* 78–92.

Tajfel, H. (1981). *Human groups and social categories: Studies in social psychology.* Cambridge, UK: Cambridge University Press.

Tajfel, H. (1982a). *Social identity and intergroup relations.* Cambridge, UK: Cambridge University Press.

Tajfel, H. (1982b). Social psychology of intergroup relations. *Annual Review of Psychology, 33,* 1–39.

Tajfel, H., & Billig, M. (1974). Familiarity and categorization in intergroup behavior. *Journal of Experimental Social Psychology, 10,* 159–170.

Tajfel, H., Billig, M. G., Bundy, R. P., & Flament, C. (1971). Social categorization and intergroup behaviour. *European Journal of Social Psychology, 1,* 149–178.

Tajfel, H., and Turner, J. C. (1979). An integrative theory of intergroup conflict. In W. G. Austin & S. Worchel, *The social psychology of intergroup relations* (pp. 33–47). Monterey, CA: Brooks/Cole.

Tajfel, H., & Turner, J. (1986). The social identity theory of intergroup behavior. In S. Worchel & W. G. Austin (Eds.), *Psychology of intergroup relations* (2nd ed., pp. 7–24). Chicago: Nelson-Hall.

Takooshian, H., Haber, S., & Lucido, D. J. (1977). Who wouldn't help a lost child? You, maybe. *Psychology Today, 10,* 67–68.

Tanford, S., & Penrod, S. (1984). Social influence model: A formal integration of research on majority and minority influence processes. *Psychological Bulletin, 95,* 189–225.

Tangney, J. P. (1992). Situational determinants of shame and guilt in young adulthood. *Personality and Social Psychology Bulletin, 18,* 199–206.

Tangney, J. P. (1995). Shame and guilt in interpersonal relationships. In J. P. Tangney & K. W. Fischer (Eds.), *Self-conscious emotions: The psychology of shame, guilt, embarrassment, and pride* (pp. 114–139). New York: Guilford Press.

Tannen, D. (1991). *You just don't understand: Women and men in conversation.* New York: Morrow.

Taylor, D. A., & Belgrave, F. Z. (1986). The effects of perceived intimacy and valence on self-disclosure reciprocity. *Personality and Social Psychology Bulletin, 12,* 247–255.

Taylor, D. M., & Jaggi, V. (1974). Ethnocentrism and causal attribution in a South Indian context. *Journal of Cross-Cultural Psychology, 5,* 162–171.

Taylor, D. M., & Moghaddam, F. M. (1987). *Theories of intergroup relations: International social psychological perspectives.* New York: Praeger.

Taylor, D. M., & Royer, L. (1980). Group processes affecting anticipated language choice in intergroup relations. In H. Giles, W. P. Robinson, & P. M. Smith (Eds.), *Language: Social psychological perspectives.* New York: Pergamon.

Taylor, D. W., Berry, P. C., & Block, C. (1958). Does group participation when using brainstorming facilitate creative thinking? *Administrative Science Quarterly, 3,* 23–47.

Taylor, R. J., Chatters, L. M., Tucker, M. B., & Lewis, E. (1990). Developments in research on Black families: A decade review. *Journal of Marriage and the Family, 52,* 993–1014.

Taylor, S. E. (1981). A categorization approach to stereotyping. In D. L. Hamilton (Ed.), *Cognitive processes in stereotyping and intergroup behavior* (pp. 83–114). Hillsdale, NJ: Erlbaum.

Taylor, S. E., & Crocker, J. (1981). Schematic bases of social information processing. In E. T. Higgins, C. P. Herman, & M. P. Zanna (Eds.), *Social cognition: The Ontario symposium* (Vol. 1). Hillsdale, NJ: Erlbaum.

Taylor, S. E., & Fiske, S. T. (1978). Salience, attention, and attribution: Top of the head phenomena. In L. Berkowitz (Ed.), *Advances in experimental social psychology* (Vol. 11). New York: Academic Press.

Taylor, S. E., Fiske, S. T., Etcoff, N. L., & Ruderman, A. J. (1978). The categorical and contextual bases of person memory and stereotyping. *Journal of Personality and Social Psychology, 36,* 778–793.

Taylor, S. P. (1967). Aggressive behavior and physiological arousal as a function of provocation and the tendency to inhibit aggression. *Journal of Personality, 35,* 297–310.

Taylor, V. (1989). Social movement continuity: The women's movement in abeyance. *American Sociological Review, 54,* 761–775.

Taylor, V., & Whittier, N. (1992). Collective identity in social movement communities: Lesbian feminist mobilization. In A. D. Morris & C. M. Mueller (Eds.),

Frontiers in social movement theory. New Haven, CT: Yale University Press.

Teachman, J. (1987). Family background, educational resources, and educational attainment. *American Sociological Review, 52,* 548–557.

Tedeschi, J. T. (Ed.). (1981). *Impression management theory and social psychological research.* New York: Academic Press.

Tedeschi, J. T., Bonoma, T. V., & Schlenker, B. R. (1972). Influence, decision, and compliance. In J. T. Tedeschi (Ed.), *The social influence processes.* Chicago: Aldine-Atherton.

Tedeschi, J. T., Schlenker, B. R., & Lindskold, S. (1972). The exercise of power and influence: The source of influence. In J. T. Tedeschi (Ed.), *The social influence processes.* Chicago: Aldine-Atherton.

Terry, D. J., & Hogg, M. A. (1996). Group norms and the attitude-behavior relationship: A role for group identification. *Personality and Social Psychology Bulletin, 22,* 776–793.

Tesser, A., & Campbell, J. (1983). Self-definition and self-evaluation maintenance. In J. Suls & A. G. Greenwald (Eds.), *Psychological perspectives on the self* (Vol. 2). Hillsdale, NJ: Erlbaum.

Teti, D. M., & Lamb, M. E. (1989). Socioeconomic and marital outcomes of adolescent marriage, adolescent childbirth, and their co-occurence. *Journal of Marriage and the Family, 51,* 203–212.

Teti, D., Lamb, M., & Elster, A. (1987). Long-range socioeconomic and marital consequences of adolescent marriage in three cohorts of adult males. *Journal of Marriage and the Family, 49,* 499–506.

Tetlock, P. E. (1981). The influence of self-presentational goals on attributional reports. *Social Psychology Quarterly, 44,* 300–311.

Tetlock, P. E. (1986). A value pluralism model of ideological reasoning. *Journal of Personality and Social Psychology, 50,* 819–827.

Tetlock, P. E., & Manstead, A. S. R. (1985). Impression management versus intrapsychic explanations in social psychology: A useful dichotomy? *Psychological Review, 92,* 59–77.

Thakerar, J. N., Giles, H., & Cheshire, J. (1982). Psychological and linguistic parameters of speech accommodation theory. In C. Fraser & K. R. Scherer (Eds.), *Advances in the social psychology of language.* Cambridge, UK: Cambridge University Press.

Thibaut, J., & Kelley, H. (1959). *The social psychology of groups.* New York: Wiley.

Thoennes, N. A., & Pearson, J. (1985). Predicting outcomes in divorce mediation: The influence of people and process. *Journal of Social Issues, 41*(2), 115–126.

Thoits, P. A. (1985). Self-labelling processes in mental illness: The role of emotional deviance. *American Journal of Sociology, 91,* 221–249.

Thoits, P. A. (1990). Emotional deviance: Research agendas. In T. D. Kemper (Ed.), *Research agendas in the sociology of emotions* (pp. 180–203). Albany, NY: SUNY Press.

Thomas, W. I., & Znaniecki, F. (1918). *The Polish peasant in Europe and America* (Vol. 1). Boston: Badger.

Thompson, L. (1990). Negotiation behavior and outcomes: Empirical evidence and theoretical issues. *Psychological Bulletin, 108,* 515–532.

Thompson, L. (1995). They saw a negotiation: Partisanship and involvement. *Journal of Personality and Social Psychology, 68,* 839–853.

Thompson, L., & Walker, A. J. (1989). Gender in families: Women and men in marriage, work, and parenthood. *Journal of Marriage and the Family, 51,* 845–871.

Thompson, T. L., & Zerbinos, E. (1995). Gender roles in animated cartoons: Has the picture changed in 20 years? *Sex Roles, 32,* 651–673.

Thomsen, C. T., & Borgida, E. (1996). Throwing out the baby with the bathwater? Let's not overstate the overselling of the base rate fallacy. *Behavioral and Brain Sciences, 19,* 39–40.

Thorlundsson, T. (1987). Bernstein's sociolinguistics: An empirical test in Iceland. *Social Forces, 65,* 695–718.

Thornberry, T. D., & Christenson, R. L. (1984). Juvenile justice decision-making as a longitudinal process. *Social Forces, 63,* 433–444.

Thorndike, E. L. (1920). A constant error in psychological ratings. *Journal of Applied Psychology, 4,* 25–29.

Thorne, B. (1993). *Gender play: Girls and boys in school.* New Brunswick, NJ: Rutgers University Press.

Thornhill, R., & Grammar, K. (1999). The body and face of woman: One ornament that signals quality? *Evolution and Human Behavior, 20,* 105–120.

Thornton, A. (1984). Changing attitudes toward separation and divorce: Causes and consequences. *American Journal of Sociology, 90,* 856–872.

Thornton, R., & Nardi, P. (1975). The dynamics of role acquisition. *American Journal of Sociology, 80,* 870–885.

Thorton, A., Alwin, D. F., & Camburn, D. (1983). Causes and consequences of sex-role attitudes and attitude change. *American Sociological Review, 48,* 211–227.

Tierney, K. (2002). *Strength of a city: A disaster research perspective on the World Trade Center attack*. Retrieved March 31, 2006, from www.ssrc.org/sept11/essays/tierney.htm

Tilker, H. A. (1970). Socially responsible behavior as a function of observer responsibility and victim feedback. *Journal of Personality and Social Psychology, 14,* 95–100.

Tilly, C. (1995). *Popular contention in Great Britain: 1758–1834.* Cambridge, MA: Harvard University Press.

Tilly, C., Tilly, L., & Tilly, R. (1975). *The rebellious century, 1830–1930.* Cambridge, MA: Harvard University Press.

Tobin, S. (1980). Institutionalization of the aged. In N. Datan & N. Lohman (Eds.), *Transitions of aging.* New York: Academic Press.

Toch, H. (1969). *Violent men: An inquiry into the psychology of violence.* Chicago: Aldine.

Toi, M., & Batson, C. D. (1982). More evidence that empathy is a source of altruism. *Journal of Personality and Social Psychology, 43,* 289–292.

Touhey, J. (1979). Sex-role stereotyping and individual differences in liking for the physically attractive. *Social Psychology Quarterly, 42,* 285–289.

Triandis, H. C. (1980). Values, attitudes, and interpersonal behavior. In H. Howe & M. Page (Eds.), *Nebraska symposium on motivation* (Vol. 27). Lincoln: University of Nebraska Press.

Triandis, H. C. (1989). The self and social behavior in differing cultural contexts. *Psychological Review, 96,* 506–520.

Triandis, H. C. (1995). *Individualism and collectivism.* Boulder, CO: Westview Press.

Triandis, H. C., McCusker, C., & Hui, C. H. (1990). Multimethod probes of individualism and collectivism. *Journal of Personality and Social Psychology, 59,* 1006–1020.

Triplett, N. (1898). The dynamogenic factors in pacemaking and competition. *American Journal of Psychology, 9,* 507–533.

Trivers, R. L. (1971). The evolution of reciprocal altruism. *Quarterly Review of Biology, 46,* 35–57.

Trivers, R. L. (1983). The evolution of cooperation. In D. L. Bridgeman (Ed.), *The nature of prosocial behavior.* New York: Academic Press.

Trope, Y., & Cohen, O. (1989). Perceptual and inferential determinants of behavior-correspondent attributions. *Journal of Experimental Social Psychology, 25,* 142–158.

Trope, Y., Cohen, O., & Maoz, Y. (1988). The perceptual and inferential effects of situational inducements on dispositional attribution. *Journal of Personality and Social Psychology, 55,* 165–177.

Turner, J. C. (1981). The experimental social psychology of intergroup behavior. In J. C. Turner & H. Giles (Eds.), *Intergroup behavior* (pp. 66–101). Oxford, UK: Blackwell.

Turner, J. C. (1991). *Social influence.* Pacific Grove, CA: Brooks/Cole.

Turner, R. H. (1978). The role and the person. *American Journal of Sociology, 84,* 1–23.

Turner, R. H. (1990). Role change. In W. R. Scott & J. Blake (Eds.), *Annual review of sociology* (Vol. 16, pp. 87–110). Palo Alto, CA: Annual Reviews.

Turner, R. H., & Killian, L. M. (1972). *Collective behavior* (2nd ed.). Englewood Cliffs, NJ: Prentice Hall.

Turner, R. H., & Killian, L. M. (1987). *Collective behavior* (3rd ed.). Englewood Cliffs, NJ: Prentice Hall.

Turner, J. C., Wetherell, M. S., & Hogg, M. A. (1989). Referent informational influence and group polarization. *British Journal of Social Psychology, 28,* 135–147.

Tversky, A., & Kahneman, D. (1974). Judgement under uncertainty: Heuristics and biases. *Science, 185,* 1124–1131.

Twenge, J. M. (1999). Mapping gender: The multifactorial approach and the organization of gender-related attributes. *Psychology of Women Quarterly, 23,* 485–502.

Twenge, J., & Crocker, J. (2002). Race and self-esteem: Meta-analyses comparing Whites, Blacks, Hispanics, Asians, and American Indians and comment on Gray-Little and Hafdahl (2000). *Psychological Bulletin, 128,* 371–408.

Tyler, T. R., & Caine, A. (1981). The influence of outcomes and procedures on satisfaction with formal leaders. *Journal of Personality and Social Psychology, 41,* 642–655.

Tyler, T. R., & Folger, R. (1980). Distributional and procedural aspects of satisfaction with citizen-police encounters. *Basic and Applied Social Psychology, 1,* 281–292.

Tyler, T. R., & Sears, D. O. (1977). Coming to like obnoxious people when we must live with them. *Journal of Personality and Social Psychology, 35,* 200–211.

Tziner, A. (1982). Differential effects of group cohesiveness types: A clarifying overview. *Social Behavior and Personality, 10,* 227–239.

Uggen, C., & Thompson, M. (2003). The socioeconomic determinants of ill-gotten gains: Within-person changes in drug use and illegal earnings. *American Journal of Sociology, 109,* 146–185.

U.S. Bureau of Labor Statistics. (1989). *Handbook of labor statistics* (Bulletin No. 2340). Washington, DC: U.S. Government Printing Office.

U.S. Bureau of Labor Statistics. (2001). Workplace injuries and illnesses in 1999. *Workplace Injury and Illness Summary* (00-357). Washington, DC: U.S. Department of Labor.

U.S. Bureau of Labor Statistics. (2004). *Fatal occupational injuries by event or exposure, 1999–2004.* Available at http://www.bls.gov/news.release/cfoi.t01.htm

U.S. Bureau of Labor Statistics. (2005). Average hours per day spent in primary activities, Table 6. Available at www.bls.gov/news.release/atus.t06.htm

U.S. Bureau of the Census. (1992). *Statistical abstract of the United States: 1992* (112th ed.). Washington, DC: U.S. Government Printing Office.

U.S. Bureau of the Census. (1996). *Statistical abstract of the United States: 1996* (116th ed.). Washington, DC: U.S. Government Printing Office.

U.S. Bureau of the Census. (2000). *Statistical abstract of the United States: 2000* (120th ed.). Washington, DC: U.S. Government Printing Office.

U.S. Bureau of the Census. (2005). Facts for features: Father's Day. Available at www.census.gov/Press-Release/www/releases/archives/facts_for_features_special_editions/004711.html

Useem, B. (1998). Breakdown theories of collective action. *Annual Review of Sociology, 24,* 215–238.

Valle, V. A., & Frieze, I. H. (1976). Stability of causal attributions as a mediator in changing expectations for success. *Journal of Personality and Social Psychology, 35,* 579–589.

van Baaren, R., Holland, R., Kawakami, K., & Van Knippenberg, A. (2004). Mimicry and prosocial behavior. *Psychological Science, 15,* 71–74.

van Baaren, R., Holland, R., Steenaert, B., and Van Knippenberg, A. (2003). Mimicry for money: Behavioral consequences of imitation. *Journal of Experimental Social Psychology, 39,* 393–398.

Van Gennep, A. (1960). *The rites of passage.* Chicago: University of Chicago Press. (Original work published 1908.)

Van Maanen, J. (1976). Breaking in: Socialization to work. In R. Dubin (Ed.), *Handbook of work, organization and society.* Chicago: Rand McNally.

Van Yperen, N. W., Hagedoorn, M., & Geurts, S. A. E. (1996). Intent to leave and absenteeism as reactions to perceived inequity: The role of psychological and social constraints. *Journal of Occupational and Organizational Psychology, 69,* 367–372.

Vandewater, E. A., Ostrove, J. M., & Stewart, A. (1997). Predicting women's well-being in midlife: The importance of personality development and social role involvements. *Journal of Personality and Social Psychology, 72,* 1147–1160.

VanLaningham, J., Johnson, D. R., & Amato, P. (2001). Marital happiness, marital duration, and the U-shaped curve: Evidence from a five-wave panel study. *Social Forces, 79,* 1313–1341.

Vega, W. A., & Rumbaut, R. G. (1991). Ethnic minorities and mental health. *Annual Review of Sociology, 17,* 351–383.

Verbrugge, L. (1979). Marital status and health. *Journal of Marriage and the Family, 41,* 267–285.

Verplanck, W. S. (1955). The control of the content of conversation: Reinforcement of statements of opinion. *Journal of Abnormal and Social Psychology, 51,* 668–676.

Vidmar, N. (1974). Effects of group discussion on category width judgments. *Journal of Personality and Social Psychology, 29,* 187–195.

Vinokur, A., & Burnstein, E. (1978). Novel argumentation and attitude change: The case of polarization following group discussion. *European Journal of Social Psychology, 8,* 335–348.

Vinokur, A. D., Price, R. H., & Caplan, R. D. (1996). Hard times and hurtful partners: How financial strain affects depression and relationship satisfaction of unemployed persons and their spouses. *Journal of Personality and Social Psychology, 71,* 166–179.

Violence in our culture. (1991, April 1). *Newsweek,* pp. 46ff.

Vittengl, J., & Holt, C. S. (2000). Getting acquainted: The relationship of self-disclosure and social attraction to positive affect. *Journal of Social and Personal Relationships, 17,* 53–66.

Voissem, N. H., & Sistrunk, F. (1971). Communication schedule and cooperative game behavior. *Journal of Personality and Social Psychology, 19,* 160–167.

Volling, B. L., & Belsky, J. (1991). Multiple determinants of father involvement during infancy in dual career and single-earner families. *Journal of Marriage and the Family, 53,* 461–474.

von Baeyer, C. L., Sherk, D. L., & Zanna, M. P. (1981). Impression management in the job interview: When the female applicant meets the male (chauvinist) interviewer. *Personality and Social Psychology Bulletin, 7,* 45–51.

von Hippel, W., Hawkins, C., & Schooler, J. (2001). Stereotype distinctiveness: How counter stereotypic behavior shapes the self-concept. (2001). *Journal of Personality and Social Psychology, 81,* 193–205.

Vonk, R. (2002). Self-serving interpretations of flattery: Why ingratiation works. *Journal of Personality and Social Psychology, 82,* 515–526.

Voss, M., Nylén, L., Floderus, B., Diderichsen, F., & Terry, P. (2004). Unemployment and early cause-specific mortality: A study based on the Swedish Twin Registry. *American Journal of Public Health, 94,* 2155–2161.

Voydanoff, P. (1990). Economic distress and family relations: A review of the eighties. *Journal of Marriage and the Family, 52,* 1099–1115.

Vygotsky, L. S. (1962). *Thought and language.* Cambridge, MA: MIT Press.

Waite, L., Haggstrom, G., & Kanouse, D. (1986). The effects of parenthood on the career orientations and job characteristics of young adults. *Social Forces, 65,* 43–73.

Waldman, D. A., Ramirez, G. G., House, R. J., & Puranam, P. (2001). Does leadership matter? CEO leadership attributes and profitability under conditions of perceived environmental uncertainty. *Academy of Management Journal, 44,* 134–143.

Waldron, I. (1976). Why do women live longer than men? *Social Science and Medicine, 10,* 349–362.

Walker, H. A., & Simpson, B. 2000. Equating characteristics and status-organizing processes. *Social Psychology Quarterly, 63,* 175–185.

Walker, H. A., Thomas, G. M., & Zelditch, M., Jr. (1986). Legitimation, endorsement, and stability. *Social Forces, 64,* 620–643.

Walker, L. L., & Heyns, R. W. (1962). *An anatomy for conformity.* Englewood Cliffs, NJ: Prentice Hall.

Wallin, P. (1950). Cultural contradictions and sex roles: A repeat study. *American Sociological Review, 15,* 288–293.

Walsh, E. J., & Taylor, M. (1982). Occupational correlates of multidimensional self-esteem: Comparisons among garbage collectors, bartenders, professors, and other workers. *Sociology and Social Research, 66,* 252–258.

Walsh, E. J., & Warland, R. H. (1983). Social movement involvement in the wake of a nuclear accident: Activists and free riders in the TMI area. *American Sociological Review, 48,* 764–780.

Walster [Hatfield], E., Aronson, E., & Abrahams, D. (1966). On increasing the persuasiveness of a low prestige communicator. *Journal of Experimental Social Psychology, 2,* 325–342.

Walster [Hatfield], E., Aronson, V., Abrahams, D., & Rottman, L. (1966). The importance of physical attractiveness in dating behavior. *Journal of Personality and Social Psychology, 4,* 508–516.

Walster [Hatfield], E., Berscheid, E., & Walster, G. W. (1973). New directions in equity research. *Journal of Personality and Social Psychology, 25,* 151–176.

Walster [Hatfield], E., Walster, G. W., & Berscheid, E. (1978). *Equity: Theory and research.* Boston: Allyn & Bacon.

Walster [Hatfield], E., Walster, G. W., & Traupmann, J. (1978). Equity and premarital sex. *Journal of Personality and Social Psychology, 36,* 82–92.

Walters, J., & Walters, L. (1980). Parent–child relationships: A review, 1970–1979. *Journal of Marriage and the Family, 42,* 807–822.

Warner, L. C., & DeFleur, M. (1969). Attitude as an interactional concept: Social constraint and social distance as intervening variables between attitudes and action. *American Sociological Review, 34,* 153–169.

Wasserman, I. (1977). Southern violence and the political process. *American Sociological Review, 42,* 359–362.

Watkins, P. L., & Whaley, D. (2000). Gender role stressors and women's health. In R. Eisler & M. Hersen (Eds.), *Handbook of gender, culture, and health* (pp. 43–62). Mahwah, NJ: Erlbaum.

Watson, D. (1982). The actor and the observer: How are their perceptions of causality divergent? *Psychological Bulletin, 92,* 682–700.

Watts, B. L., Messe, L. A., & Vallacher, R. R. (1982). Toward understanding sex differences in pay allocation: Agency, communion, and reward distribution behavior. *Sex Roles, 8,* 1175–1187.

Webb, E. J., Campbell, D. J., Schwartz, R. D., & Sechrest, L. (1981). *Unobtrusive measures: Nonreactive research in the social sciences* (2nd ed.). Chicago: Rand McNally.

Weber, R., & Crocker, J. (1983). Cognitive processes in the revision of stereotypic beliefs. *Journal of Personality and Social Psychology, 45,* 961–977.

Webster, M., Jr., & Driskell, J. E., Jr. (1978). Status generalization: A review and some new data. *American Sociological Review, 43,* 220–236.

Webster, M., Jr., & Foschi, M. (1988). *Status generalization: New theory and research.* Stanford, CA: Stanford University Press.

Webster, P. S., Orbuch, T. L., & House, J. S. (1995). Effects of childhood family background on adult marital quality and perceived stability. *American Journal of Sociology, 101,* 404–432.

Webster, M., & Smith, L. F. (1978). Justice and revolutionary coalitions: A test of two theories. *American Journal of Sociology, 84,* 267–292.

Weed, F. J. (1990). The victim-activist role in the anti-drunk driving movement. *The Sociological Quarterly, 31,* 459–473.

Wegner, D. M., Lane, J. D., & Dimitri, S. (1994). The allure of secret relationships. *Journal of Personality and Social Psychology, 66,* 287–300.

Weigand, B. (1994). Black money in Belize: The ethnicity and social structure of black-market crime. *Social Forces, 73,* 135–154.

Weigel, D., & Murray, C. (2000). The paradox of stability and change in relationships: What does chaos theory offer for the study of romantic relationships? *Journal of Social and Personal Relationships, 17,* 425–449.

Weigel, R. H., & Newman, L. (1976). Increasing attitude-behavior correspondence by broadening the scope of the behavioral measure. *Journal of Personality and Social Psychology, 33,* 793–802.

Weinberg, M. (1976). The nudist management of respectability. In M. Weinberg (Ed.), *Sex research: Studies from the Kinsey Institute.* New York: Oxford University Press.

Weiner, B. (1985). An attributional theory of achievement motivation and emotion. *Psychological Review, 92,* 548–573.

Weiner, B. (1986). *An attributional theory of motivation and emotion.* New York: Springer Verlag.

Weiner, B., Amirkhan, J., Folkes, V. S., & Verette, J. A. (1987). An attributional analysis of excuse giving: Studies of a naive theory of emotion. *Journal of Personality and Social Psychology, 52,* 316–324.

Weiner, B., Frieze, I., Kukla, A., Reed, L., Rest, B., & Rosenbaum, R. M. (1971). *Perceiving the causes of success and failure.* Morristown, NJ: General Learning Press.

Weiner, B., Heckhausen, H., Meyer, W. U., & Cook, R. E. (1972). Causal ascriptions and achievement behavior: A conceptual analysis of effort and reanalysis of locus of control. *Journal of Personality and Social Psychology, 21,* 239–248.

Weiner, B., Perry, R. P., & Magnusson, J. (1988). An attributional analysis of reactions to stigmas. *Journal of Personality and Social Psychology, 55,* 738–748.

Weinstein, E. A., & Deutschberger, P. (1963). Some dimensions of altercasting. *Sociometry, 26,* 454–466.

Weiss, R. S. (1973). *Loneliness: The experience of emotional and social isolation.* Cambridge, MA: MIT Press.

Weitzer, R. (1991). Prostitute's rights in the United States: The failure of a movement. *The Sociological Quarterly, 32,* 23–41.

Weitzman, L. J., Eifler, D., Hokada, E., & Ross, K. (1972). Sex role socialization in picture books for pre-school children. *American Journal of Sociology, 77,* 1125–1150.

Weldon, E., & Weingart, L. R. (1993). Group goals and group performance. *British Journal of Social Psychology, 32,* 307–334.

Wellman, B., & Worley, S. (1990). Different strokes from different folks: Community ties and social support. *American Journal of Sociology, 96,* 558–588.

Wellman, H. M., Phillips, A. T., & Rodriguez, T. (2000). Young children's understanding of perception, desire, and emotion. *Child Development 71*(4), 895–912.

Werner, C., & Parmelee, P. (1979). Similarity of activity preferences among friends: Those who play together stay together. *Social Psychology Quarterly, 42,* 62–66.

Western, B. (2002). The impact of incarceration on wage mobility and inequality. *American Sociological Review, 67,* 526–546.

Weyant, J. (1996). Application of compliance techniques to direct-mail requests for charitable donations. *Psychology and Marketing, 13,* 157–170.

Whalen, M. R., & Zimmerman, D. H. (1987). Sequential and institutional contexts in calls for help. *Social Psychology Quarterly, 50,* 172–185.

White, C., Bushnell, N., & Regnemer, J. (1978). Moral development in Bahamian school children: A 3-year examination of Kohlberg's stages of moral development. *Developmental Psychology, 14,* 58–65.

White, G. (1980). Physical attractiveness and courtship progress. *Journal of Personality and Social Psychology, 39,* 660–668.

White, J. W., & Gruber, K. J. (1982). Instigative aggression as a function of past experience and target characteristics. *Journal of Personality and Social Psychology, 42,* 1069–1075.

White, L., & Edwards, J. N. (1990). Emptying the nest and parental well-being: An analysis of national panel data. *American Sociological Review, 55,* 235–242.

White, R. W. (1989). From peaceful protest to guerilla war: Micromobilization of the Provisional Irish Republican Army. *American Journal of Sociology, 94,* 1277–1302.

Whitt, H. P., & Meile, R. L. (1985). Alignment, magnification, and snowballing: Processes in the definition of symptoms of mental illness. *Social Forces, 63,* 682–697.

Whittier, N. (1997). Political generations, micro-cohorts, and the transformation of social movements. *American Sociological Review, 62,* 760–778.

Whorf, B. L. (1956). In J. B. Carroll (Ed.), *Language, thought, and reality.* Cambridge, MA: MIT Press.

Wicker, A. W. (1969). Attitudes versus actions: The relationship of verbal and overt behavioral responses to attitude objects. *Journal of Social Issues, 25*(4), 41–78.

Wicklund, R. A. (1975). Objective self-awareness. In L. Berkowitz (Ed.), *Advances in experimental social psychology* (Vol. 8). New York: Academic Press.

Wicklund, R. A., & Brehm, J. (1976). *Perspectives on cognitive dissonance.* Hillsdale, NJ: Erlbaum.

Wicklund, R. A., & Frey, D. (1980). Self-awareness theory: When the self makes a difference. In D. M. Wegner & R. R. Vallacher (Eds.), *The self in social psychology.* New York: Oxford University Press.

Wickrama, K., Conger, R. D., Lorenz, F. O., & Matthews, L. (1995). Role identity, role satisfaction, and perceived physical health. *Social Psychology Quarterly, 58,* 270–283.

Wickrama, K. A. S., Conger, R., Wallace, L. E., & Elder, G. H., Jr. (1999). The inter-generational transmission of health-risk behaviors: Adolescent lifestyles and gender moderating effects. *Journal of Health and Social Behavior, 40,* 258–272.

Wickrama, K. A. S., Lorenz, F. O., Conger, R. D., & Elder, G. H., Jr. (1997). Marital quality and physical illness: A latent growth curve analysis. *Journal of Marriage and the Family, 59,* 143–155.

Wigboldus, D., Semin, G., & Spears, R. (2000). How do we communicate stereotypes? Linguistic biases and inferential consequences. *Journal of Personality and Social Psychology, 78,* 5–18.

Wiggins, J. S., Wiggins, N., & Conger, J. C. (1968). Correlates of heterosexual somatic preference. *Journal of Personality and Social Psychology, 10,* 82–90.

Wilcox, C., & Williams, L. (1990). Taking stock of schema theory. *Social Science Journal, 27,* 373–393.

Wilder, D. A. (1981). Perceiving persons as a group: Categorization and intergroup relations. In D. L. Hamilton (Ed.), *Cognitive processes in stereotyping and intergroup behavior* (pp. 213–257). Hillsdale, NJ: Erlbaum.

Wilder, D. A., & Thompson, J. E. (1980). Intergroup contact with independent manipulations on in-group and out-group interaction. *Journal of Personality and Social Psychology, 38,* 589–603.

Wiley, M. G. (1973). Sex roles in games. *Sociometry, 36,* 526–541.

Wilke, H. A. M. (1985). Coalition formation from a socio-psychological point of view. In H. A. M. Wilke (Ed.), *Coalition formation* (Chapter 3). Amsterdam: Elsevier.

Wilke, H., & Lanzetta, J. T. (1982). The obligation to help: Factors affecting response to help received. *European Journal of Social Psychology, 12,* 315–319.

Wilke, H. A. M., Van Knippenberg, A. F. M., & Bruins, J. (1986). Conservative coalitions: An expectation states approach. *European Journal of Social Psychology, 16,* 51–63.

Williams, D. R. (1990). Socioeconomic differentials in health: A review and redirection. *Social Psychology Quarterly, 53,* 81–99.

Williams, L. S. (2002). Trying on gender, gender regimes, and the process of becoming women. *Gender and Society, 16,* 29–52.

Williams, R. H. (1995). Constructing the public good: Social movements and cultural resources. *Social Problems, 42,* 124–144.

Williams, R. M., Jr. (1977). *Mutual accommodation: Ethnic conflict and cooperation.* Minneapolis: University of Minnesota Press.

Williams, K. D., Harkins, S. G., & Latané, B. (1981). Identifiability as a deterrent to social loafing: Two cheering experiments. *Journal of Personality and Social Psychology, 40,* 303–311.

Willis, F. N., & Carlson, R. A. (1993). Singles ads: Gender, social class, and time. *Sex Roles, 29,* 387–404.

Wills, T. A. (1992). The helping process in the context of personal relationships. In S. Spacapan & S. Oskamp (Eds.), *Helping and being helped: Naturalistic studies* (pp. 17–48). Newbury Park, CA: Sage.

Wilmoth, J. K., & Ball, P. (1995). Arguments and action in the life of a social problem: A case study of "overpopulation," 1946–1990. *Social Problems, 42,* 318–343.

Wilson, E. O. (1971). *The insect societies.* Cambridge, MA: Belknap.

Wilson, E. O. (1975). *Sociobiology: The new synthesis.* Cambridge, MA: Harvard University Press.

Wilson, E. O. (1978). *On human nature.* Cambridge, MA: Harvard University Press.

Wilson, T. D., Houston, C. E., Etling, K. M., & Brekke, N. C. (1996). A new look at anchoring effects: Basic anchoring and its antecedents. *Journal of Experimental Psychology: General, 125,* 387–402.

Windle, M., & Dumenci, L. (1997). Parental and occupational stress as predictors of depressive symptoms among dual-income couples: A multilevel modeling approach. *Journal of Marriage and the Family, 59,* 625–634.

Winschild, P. D., & Wells, G. L. (1997). Behavioral consensus information affects people's inferences about population traits. *Personality and Social Psychology Bulletin, 23,* 148–156.

Wit, A., & Wilke, H. (1988). Subordinates' endorsement of an allocating leader in a commons dilemma: An equity theoretical approach. *Journal of Economic Psychology, 9,* 151–168.

Wit, A. P., Wilke, H. A. M., & Van Dijk, E. (1989). Attribution of leadership in a resource management situation. *European Journal of Social Psychology, 19,* 327–338.

Wittenbaum, G. M., Vaughan, S. I., & Stasser, G. (1998). Coordination in task-performing groups. In R. S. Tindale, L. Heath, J. D. Edwards, E. J. Posavac, F. B. Bryant, Y. Suarez-Balcazar, et al. (Eds.), *Social psychological applications to social issues: Applications of theory and research on groups* (pp. 177–204). New York: Plenum.

Wittenbrink, B., Judd, C. M., & Park, B. (2001). Evaluative versus conceptual judgments in automatic stereotyping and prejudice. *Journal of Experimental Social Psychology, 37*(3), 244–252.

Wolf, S. (1985). Manifest and latent influence of majorities and minorities. *Journal of Personality and Social Psychology, 48,* 899–908.

Wolf, S., & Bugaj, A. M. (1990). The social impact of courtroom witnesses. *Social Behaviour, 5*(1), 1–13.

Wolf, S., & Latané, B. (1983). Majority and minority influences on restaurant preferences. *Journal of Personality and Social Psychology, 45,* 282–292.

Woll, S. B., & Young, P. (1989). Looking for Mr. or Ms. Right: Self-presentation in videodating. *Journal of Marriage and the Family, 51,* 483–488.

Won-Doornink, M. J. (1979). On getting to know you: The association between the stage of a relationship and the reciprocity of self-disclosure. *Journal of Experimental Social Psychology, 15,* 229–241.

Won-Doornink, M. J. (1985). Self-disclosure and reciprocity in conversation: A cross-national study. *Social Psychology Quarterly, 48,* 97–107.

Wood, W., Lundgren, S., Ouellette, J. A., Busceme, S., & Blackstone, T. (1994). Minority influence: A meta-analytic review of social influence processes. *Psychological Bulletin, 115,* 323–345.

Worchel, S. (1986). The role of cooperation in reducing intergroup conflict. In S. Worchel & W. G. Austin (Eds.), *Psychology of intergroup relations* (2nd ed., pp. 288–304). Chicago: Nelson-Hall.

Worchel, S., & Norvell, N. (1980). Effect of perceived environmental conditions during cooperation on intergroup attraction. *Journal of Personality and Social Psychology, 38,* 764–772.

Worchel, S., Lind, E., & Kaufman, K. (1975). Evaluations of group products as a function of expectations of group longevity, outcome of competition, and publicity of evaluations. *Journal of Personality and Social Psychology, 31,* 1089–1097.

Wortman, C. B., & Linsenmeier, J. A. (1977). Interpersonal attraction and techniques of ingratiation in organizational settings. In B. M. Staw & G. R. Salancik (Eds.), *New directions in organizational behavior* (pp. 133–178). Chicago: St Clair Press.

Wright, E. O., Costello, C., Hachen, D., & Sprague, J. (1982). The American class structure. *American Sociological Review, 47,* 709–726.

Wright, E., Gronfein, W., & Owens, T. (2000). Deinstitutionalization, social rejection, and the self-esteem of former mental patients. *Journal of Health and Social Behavior, 41,* 68–90.

Wu, X., & DeMaris, A. (1996). Gender and marital status differences in effects of chronic strain. *Sex Roles, 34,* 299–319.

Wulfert, E., & Wan, C. K. (1995). Safer sex intentions and condom use viewed from a health belief, reasoned action, and social cognitive perspective. *Journal of Sex Research, 32,* 299–311.

Wyer, R. S., Jr. (1966). Effects of incentive to perform well, group attraction, and group acceptance on conformity in a judgmental task. *Journal of Personality and Social Psychology, 4,* 21–26.

Wyer, R. S., Jr., & Srull, T. K. (Eds.). (1984). *Handbook of social cognition* (Vols. 1–3). Hillsdale, NJ: Erlbaum.

Wylie, R. C. (1979). *The self-concept: Theory and research on selected topics* (Rev. ed., Vol. 2). Lincoln: University of Nebraska Press.

Yarrow, M., Schwartz, C., Murphy, H., & Deasy, L. (1955). The psychological meaning of mental illness in the family. *Journal of Social Issues, 11,* 12–24.

Youngs, G. A., Jr. (1986). Patterns of threat and punishment reciprocity in a conflict setting. *Journal of Personality and Social Psychology, 51,* 541–546.

Yukl, G. (1981). *Leadership in organizations.* Englewood Cliffs, NJ: Prentice Hall.

Zaccaro, S. J. (1984). Social loafing: The role of task attractiveness. *Personality and Social Psychology Bulletin, 10,* 99–106.

Zajonc, R. B. (1965). Social faciliation. *Science, 149,* 269–274.

Zajonc, R. B. (1968). The attitudinal effects of mere exposure. *Journal of Personality and Social Psychology, 9* (Monograph Suppl. no. 2, Pt. 2), 1–27.

Zajonc, R. B., Heingartner, A., & Herman, E. M. (1969). Social enhancement and impairment of performance in the cockroach. *Journal of Personality and Social Psychology, 13,* 83–92.

Zajonc, R. B., & McIntosh, D. N. (1992). Emotions research: Some promising questions and some questionable promises. *Psychological Science, 3,* 70–74.

Zander, A. (1985). *The purposes of groups and organizations.* San Francisco: Jossey-Bass.

Zanna, M., & Fazio, R. (1982). The attitude-behavior relation: Moving toward a third generation of research. In M. Zanna, E. Higgins, & C. Herman (Eds.), *Consistency in social behavior: The Ontario symposium* (Vol. 2). Hillsdale, NJ: Erlbaum.

Zanna, M. P., & Hamilton, D. L. (1977). Further evidence for meaning change in impression formation. *Journal of Experimental Social Psychology, 13,* 224–238.

Zantra, A. J., Reich, J. W., & Guarnaccia, C. A. (1990). Some everyday life consequences of disability and bereavement for older adults. *Journal of Personality and Social Psychology, 59,* 550–561.

Zartman, I. W., & Touval, S. (1985). International mediation: Conflict resolution and power politics. *Journal of Social Issues, 41*(2), 27–45.

Zebrowitz, L. A., Andreoletti, C., Collins, M. A., Lee, S. Y., & Blumenthal, J. (1998). Bright, bad, babyfaced boys: Appearance stereotypes do not always yield self-fulfilling prophecy effects. *Journal of Personality and Social Psychology, 75,* 1300–1320.

Zebrowitz, L., Voinescu, L., & Collins, M. A. (1996). "Wide-eyed" and "crooked-faced": Determinants of perceived and real honesty across the life span. *Personality and Social Psychology Bulletin, 22,* 1258–1269.

Zelditch, M., Jr. (1972). Authority and performance expectations in bureaucratic organizations. In C. G. McClintock (Ed.), *Experimental social psychology.* New York: Holt, Rinehart and Winston.

Zelditch, M., Jr., Lauderdale, P., & Stublarec, S. (1980). How are inconsistencies between status and disability resolved? *Social Forces, 58,* 1025–1043.

Zelditch, M., Jr., & Walker, H. A. (1984). Legitimacy and the stability of authority. In E. J. Lawler (Ed.), *Advances in group processes* (Vol. 1, pp. 1–26). Greenwich, CT: JAI Press.

Zey, M. (1993). *Banking on fraud: Junk bonds and buyouts.* New York: Aldine De Gruyter.

Zillman, D. (1978). Attribution and misattribution of excitatory reactions. In J. Harvey, W. Ickes, & R. F. Kidd (Eds.), *New directions in attribution research* (Vol. 2). Hillsdale, NJ: Erlbaum.

Zillman, D. (1979). *Hostility and aggression.* Hillsdale, NJ: Erlbaum.

Zimbardo, P. (1969). The human choice: Individuation, reason, and order versus deindividuation, impulse, and chaos. In W. J. Arnold & D. Levine (Eds.), *Nebraska symposium on motivation* (Vol. 17). Lincoln: University of Nebraska Press.

Zimmerman, D. H., & West, C. (1975). Sex roles, interruptions, and silences in conversations. In B. Thorne & N. Henley (Eds.), *Language and sex: Difference and dominance.* Rowley, MA: Newbury House.

Zipf, S. G. (1960). Resistance and conformity under reward and punishment. *Journal of Abnormal and Social Psychology, 61,* 102–109.

Zollar, A., & Williams, J. (1987). The contribution of marriage to the life satisfaction of Black adults. *Journal of Marriage and the Family, 49,* 87–92.

Zuckerman, E., Kim, T.-Y., Ukanwa, K., and von Rittman, J. (2003). Robust identities or nonentities? Typecasting in the feature-film labor market. *American Journal of Sociology, 108,* 1018–1074.

Zuckerman, M., DePaulo, B. M., & Rosenthal, R. (1981). Verbal and nonverbal communication of deception. In L. Berkowitz (Ed.), *Advances in experimental social psychology* (Vol. 14). New York: Academic Press.

Zuckerman, M., Koestner, R., & Alton, A. O. (1984). Learning to detect deception. *Journal of Personality and Social Psychology, 46,* 519–528.

Zurcher, L. A., & Snow, D. A. (1990). Collective behavior and social movements. In M. Rosenberg & R. H. Turner (Eds.), *Social psychology: Sociological perspectives.* New Brunswick, NJ: Transaction.

Zvonkovic, A. M., Greaves, K. M., Schmiege, C. J., & Hull, C. D. (1996). The marital construction of gender through work and family decisions: A qualitative analysis. *Journal of Marriage and the Family, 58,* 91–100.

CREDITS

This page constitutes an extension of the copyright page. We have made every effort to trace the ownership of all copyrighted material and to secure permission from copyright holders. In the event of any question arising as to the use of any material, we will be pleased to make the necessary corrections in future printings. Thanks are due to the following authors, publishers, and agents for permission to use the material indicated.

Chapter 1. **9:** Mark Richards/PhotoEdit, Inc. **13:** Bob Daemmrich/Stock, Boston

Chapter 2. **30:** David Frazier **37:** Michael Kagan **42:** Cary Wolinsky/Stock, Boston

Chapter 3. **54:** Michael Newman/PhotoEdit, Inc. **65:** David Shaefer/Jeroboam **69:** David Sams/Stock, Boston **71:** David Woo/Stock, Boston **75:** Michael Newman/PhotoEdit, Inc.

Chapter 4. **86:** Elizabeth Crews **88:** Tony Freeman/PhotoEdit, Inc. **98:** A. Ramey/PhotoEdit, Inc. **103:** Brian Bahr/Getty Images

Chapter 5. **117:** Judy S. Gelles/Stock, Boston **125:** Roger Allyn Lee/SuperStock **134:** Glyn Kirk/Getty Images

Chapter 6. **152, top:** Tony Freeman/PhotoEdit, Inc. **152, bottom:** Bill Aron/PhotoEdit, Inc. **156:** Sylvia Johnson/Woodfin Camp

Chapter 7. **167:** Myrleen Ferguson Cate/PhotoEdit, Inc. **177:** J. Applewhite/AP/Wide World Photo **179:** Jane Scherr/Jeroboam **182:** Richard Lord/The Image Works **191:** Katherine Karnow/Corbis

Chapter 8. **201:** Francisco Cruz/SuperStock **205:** Bill Aron/PhotoEdit, Inc. **214:** V.C.L./Chris Ryan/Getty Images **218:** Bob Mahoney/The Image Works

Chapter 9. **228:** Mug Shots/Corbis **230:** Oscar Palmquist/Lightwave **232:** Roger Garwood/Corbis **238:** Dynamic Graphics, Inc./Jupiterimages **243:** Georges DeKeerle/Getty Images

Chapter 10. **253:** From *The Expression of the Emotions in Man and Animals,* Charles Darwin, p. 118, drawing by Mr. Wood **254:** Paul Ekman/Human Interaction Laboratory **260:** Ali Al-Saadi/AFP/Getty Images **266:** Purestock/SuperStock **271:** Mark Peterson/Corbis

Chapter 11. **276:** B. P./Taxi/Getty Images **285:** Keith Horan/Stock, Boston **295:** David Maialetti/Philadelphia Daily News

Chapter 12. **303:** Richard Hutchings/PhotoEdit, Inc. **312:** Spencer Grant/Stock, Boston **321:** Rick Kopstein

Chapter 13. **328:** PhotoDisc/Getty Images **330, left:** Ronnie Kaufman/Corbis **330, right:** Gerhard Steiner/Corbis **335:** PhotoDisc/Getty Images

Chapter 14. **361:** Tony Freeman/PhotoEdit, Inc. **364:** John Neubauer/PhotoEdit, Inc. **366, top:** Bill Bachmann/PhotoEdit, Inc. **366, bottom:** PhotoDisc/Getty Images

Chapter 15. **386, top:** Phyllis Graber Jensen/Stock, Boston **386, bottom:** Courtesy of Catherine Comet, Conductor of the Grand Rapids, Michigan, Symphony and Shaw Concerts, Inc. **391, top:** Jean-Claude Lejeune/Stock, Boston **391, bottom:** Cheryl Maeder/Getty Images **394:** Rose Skytta/Jeroboam

Chapter 16. **409:** AP/Wide World Photo **415:** David McNew/Getty Images **417 left and right:** Michael Siluk **423:** Roger Leo/Index Stock Imagery/Jupiterimages **428:** Mark Richards/PhotoEdit, Inc.

Chapter 17. **437:** Myrleen Ferguson/PhotoEdit, Inc. **451:** Bob Daemmrich/The Image Works **456:** Kathy McLaughlin/The Image Works **458:** PhotoDisc/Getty Images

Chapter 18. **468, left:** Tim Barnwell/Stock, Boston **468, right:** PhotoDisc/Getty Images **473:** Dana White/PhotoEdit, Inc. **486:** A. Ramey/PhotoEdit, Inc. **490:** Robert A. Ginn/PhotoEdit, Inc.

Chapter 19. **496:** PhotoDisc/Getty Images **499:** Catherine Karnow/Woodfin Camp **507:** David Young-Wolff/PhotoEdit, Inc. **511:** Fred Frames/Woodfin Camp **515:** The Boston Globe **516:** Bob Daemmrich **519:** Jeremy Woodhouse/SuperStock

Chapter 20. **527:** John W. Gertz/zefa/Corbis **539:** Lester Sloan/Woodfin Camp **544:** Paul Conklin/PhotoEdit, Inc.

NAME INDEX

A

Abbey, A., 318
Abelson, H. I., 208
Abelson, R., 147, 148
Abelson, R. P., 11 1
Aberson, C. L., 414
Abrahams, D., 202, 330
Abrams, D., 508
Acker, M., 316
Ackerman, P., 276
Acock, A., 161
Acock, A. C., 62, 347
Acosta, M., 70, 455
Acquino, K. F., 419
Adams, J. S., 13, 397, 398
Adelman, R., 40
Aderman, D., 287
Adler, P., 62
Adler, P. A., 62
Adlerfer, C. P., 426
Affleck, G., 481
Agarie, N., 308
Agnew, R., 498
Aiken, L. R., 356
Aiken, M. T., 157
Aikin, K. J., 160
Ainsworth, M., 57, 58
Ajzen, I., 129, 142, 158, 161, 162, 163
Akers, K. L., 501
Akers, R. L., 502
Akert, R., 179
Albarracin, D., 201
Albertson, B. L., 482
Aldag, R. J., 402
Aldous, J., 452
Aldrich, H., 539
Aldwin, C., 486
Alex, N., 176
Alexander, C. N., Jr., 228
Alfonso, H., 202
Alicke, M. S., 370
Alker, H., 534
Allan, E., 503
Allan, E. A., 497, 500
Allen, H., 294
Allen, J. G., 247
Allen, J. L., 276
Allen, M., 229, 427–28
Allen, V. L., 10, 371, 373
Allison, S. T., 135, 394
Allison, C., 231
Allport, F. H., 6, 11, 387
Allport, G. W., 6, 7, 86, 142, 427
Altman, I., 339
Alton, A. O., 242
Alwin, D., 61, 78
Alwin, D. F., 70, 160
Al-Zahrani, S. S. A., 139
Amato, P. R., 61, 350, 443, 448
Ambady, N., 120
Ames, R., 296
Amir, Y., 425, 427
Amirkhan, J., 236
Anastasio, P. A., 424–25
Anderson, A., 216
Anderson, B., 315
Anderson, B. A., 32
Anderson, C., 311, 324, 537
Anderson, C. A., 114, 303, 311, 323, 537, 538, 539
Anderson, K., 70, 73
Anderson, K. B., 303, 317, 539
Anderson, L. R., 392
Anderson, N. H., 123
Andreoletti, C., 122
Andrews, B., 308

Andrews, F. M., 310
Andrews, I. R., 397, 405
Ansolabehere, S., 207
Anson, O., 477
Appleton, W., 70
Apsler, R., 222
Arbuthnot, J., 375
Archer, D., 179
Archer, J., 278
Archer, R. L., 229
Arcuri, L., 174
Arditti, J. A., 61
Arendt, H., 219
Aries, E., 446, 447
Aristotle, 314
Arkoff, A., 204
Arluke, A., 244
Armeli, S., 481
Armour, Richard, 192
Armstrong, D. J., 373
Arnett, J. J., 440
Aron, A., 261, 338
Aron, E., 338
Aronfreed, J., 66
Aronoff, D., 430
Aronoff, J. R., 430
Aronson, E., 41, 45, 202, 203
Aronson, V., 330
Asch, S. E., 122, 123, 368, 369, 370, 371, 372
Asch, Solomon, 7
Aseltine, R., Jr., 486, 498
Asendorpf, J. B., 351
Ashforth, B., 234
Ashmore, R. D., 119, 332
Astone, N., 62
Atchley, R. C., 453
Atkin, C. K., 206, 207
Atkin, R. S., 392
Attridge, M., 348
Austin, W., 394, 397
Austin, W. G., 398
Averill, J. R., 260, 303
Ayers, L., 430
Azdia, T., 150

B

Babcock, M. K., 265
Bacharach, S. B., 216, 217
Bachman, B. A., 424–25
Bachman, J. G., 503
Back, K. W., 328
Backman, C., 105, 333, 338, 339, 341, 342
Bagozzi, R. P., 163
Bailey, C. A., 512
Bailey, J. M., 40
Bailey, W. C., 516
Bainbridge, W. S., 543
Bales, R. F., 359
Bales, Robert Freed, 7
Balkwell, J. W., 356, 361
Ball, D. W., 245
Ball, P., 541
Baller, R., 505
Baltes, P., 436
Bamberger, E., 207
Banaji, M., 92
Bandura, A., 12, 68, 69, 85, 102, 304, 309, 435
Banikiotes, P. G., 229
Banse, R., 257
Barbee, A. P., 288, 332
Barber, J. J., 206
Bargh, J. A., 114
Barker, R. G., 301
Barling, J., 58
Barnes, G., 316, 499
Barnes, J., 91
Barnett, M. A., 276
Barnett, R. C., 484
Baron, J., 469
Baron, J. N., 471
Baron, L., 320
Baron, R., 311, 405, 537, 538
Baron, R. A., 313
Baron, R. S., 370, 405
Baron, S. H., 371
Barrett, G., 319
Barrett, K. C., 264, 265
Barsaloux, J., 399
Bar-Tal, D., 283
Bartell, P., 290
Bartholow, B., 311
Bartone, P., 475
Bass, A. R., 115
Bass, B. M., 381, 383
Basset, R., 153
Bassin, E., 350
Bates, D. D., 361
Bates, E., 72
Batson, C. D., 276, 277, 283, 289, 294
Batten, P. G., 229–30
Batts, V., 160
Bauer, C. L., 203
Bauer, R., 206
Baum, A., 190
Baumann, D. J., 287, 288
Baumeister, R., 103, 228, 234, 235, 310
Baumeister, R. F., 226, 265
Baumrind, D., 70
Bavelas, J., 193
Bazerman, M., 216
Beall, A., 346
Bearman, P., 329
Beck, E., 535
Beck, S. B., 332
Becker, H. S., 441, 506
Beckett, K., 508
Beebe, L. M., 187
Beers, M., 268
Begley, T., 534
Bekkers, F., 422
Belden, Timothy N., 509
Belgrave, F. Z., 229
Bell, R., 57
Bell, R. A., 236
Bellavia, G., 105
Belsky, J., 60, 447
Bem, D. J., 145
Bem, S. L., 121
Benedict, R., 79
Benford, R., 541
Benford, R. D., 542
Bengston, V. L., 143
Ben-Itzhak, S., 297
Benjamin, A., 311
Bennett, M., 236, 242
Benokraitis, N. V., 160
Bensley, L. S., 222
Benson, M. L., 311, 520
Benson, P. L., 280
Bentler, P. M., 502
Bentley, J. C., 384
Ben Ze'ev, A., 265
Berelson, B., 206
Berenstein, V., 101
Bergen, D., 121
Berger, J., 360, 361, 363, 364, 365
Berger, R. E., 235
Bergesen, A., 536
Berk, R. A., 539
Berkman, L., 487
Berkowitz, L., 282, 287, 294, 302, 303, 311, 320, 415, 416, 538
Berkowitz, M. W., 77
Berktold, J., 482
Bernstein, I., 520
Bernstein, W. M., 138
Berry, D. S., 186
Berry, P. C., 399
Berscheid, E., 13, 126, 202, 331, 332, 336, 348, 395
Bertenthal, B. I., 86
Bettencourt, B. A., 424
Betz, B., 430
Beyer, Jennifer, 274

Bianchi, A., 275
Bianchi, S., 60, 444
Bianchi, S. M., 465
Bichsel, J., 202
Bickman, L., 292
Biddle, B. J., 9, 462
Bielby, D., 442, 445
Bielby, W., 442, 445, 469
Bienenstock, E., 275
Biggers, K., 476
Billig, M., 405, 413
Billig, M. G., 413
Birch, K., 276
Bird, C. E., 444
Birdwhistell, R. L., 175, 182
Birrell, B., 119
Bishop, D., 520
Bishop, G. D., 426
Bitman, M., 444
Black, D., 38, 519
Blackstone, T., 375
Blackwell, D., 329
Blake, R. R., 424
Blanck, P. D., 240
Blandford, B. J., 40
Blascovich, J., 246, 405
Blass, T., 220
Blau, P., 333
Blauner, R., 489
Blegen, M. A., 216
Blieszner, R., 452
Block, C., 399
Blom, J. P., 185, 187
Blumenthal, J., 122
Blumenthal, M., 310
Blumer, H., 227
Blumstein, P. W., 236, 237
Boardman, J., 497
Bobo, L., 412
Bochner, S., 203
Bode, K. A., 398
Bodenhausen, G., 144
Bodenhausen, G. V., 119
Bohn, A., 184
Bohra, K. A., 233
Bolger, N., 484
Bollen, K. A., 35, 356, 505
Bolzendahl, C. I., 160
Bond, M. H., 114
Bond, R., 371
Bondurant, B., 318
Bono, J. E., 383
Bonoma, T. V., 213, 214
Booth, A., 443, 448
Booth, K., 62
Bord, R. J., 246
Borden, R., 106
Bordin, J., 96
Borgida, E., 126
Borkenau, P., 116
Bosk, C. L., 543
Bosma, H., 476
Bottomore, T. B., 490
Bouchard, T. J., 399
Bouchard, T. J., Jr., 399
Boucher, J. D., 255
Bouchet, J., 405
Boudreau, L. A., 247
Boulding, K. E., 211
Bourhis, R. Y., 187
Bowdle, B. F., 307
Bowers, R. J., 406
Bowlby, J., 58
Boyd, K. R., 136
Braaten, L. J., 356
Brackett, M. A., 268
Bradburn, N. M., 32
Bradford, S., 340
Bradley, G. W., 138
Braesicke, K., 258, 260
Braly, K., 160
Bratton, V., 233
Brauer, M., 293, 417
Bray, R. M., 392, 426
Brayfield, A., 485
Bregman, N. J., 286
Brehm, J., 150, 151
Brehm, J. W., 151
Brehm, P., 222
Brekke, N. C., 127
Brewer, M., 41
Brewer, M. B., 412–13, 414, 417, 420, 424, 425
Brewin, C. R., 308
Briar, S., 229
Brickman, P., 280
Brickner, M. A., 390
Bridges, J., 78, 450, 451, 452, 453
Brief, D. E., 219
Briere, J., 320
Briggs, J. L, 257, 258, 260
Briggs, S. R., 351
Brigham, J. C., 160
Brim, O. G., Jr., 435
Brock, T. C., 210
Brockmann, H., 477
Brockner, J., 394, 395
Brody, E. M., 286
Broman, C., 449
Bronfenbrenner, U., 418
Brook, J. S., 322, 323
Brooks-Gunn, J., 69, 70, 73
Brotz, E., 219
Broverman, D., 70
Broverman, I., 70, 120, 121
Brown, B., 72, 217
Brown, J., 176
Brown, J. D., 106, 475, 522
Brown, N., 508
Brown, P., 309
Brown, R., 172, 302, 303, 425
Brown, R. J., 414, 425
Brown, S., 78
Brown-Collins, A., 47
Browning, C., 311
Brubaker, T. H., 454
Bruggink, J. B., 405
Bruins, J., 363
Brunner, E., 476
Brunsman, B., 370
Brunstein, J. C., 350
Bruun, S., 390
Bryan, J. H., 280, 285
Bryant, S. L., 135
Bryjak, G. J., 214
Buchman, D., 323
Buck, P. O., 318
Buck, R., 279
Buckley, T., 276
Buehler, C., 485
Bugaj, A. M., 202
Bui, K., 93
Bui, K.-V. T., 350, 352
Bumpass, L., 441, 456
Bumpass, L. L., 443
Bundy, McGeorge, 400
Bundy, R. P., 413
Bundy, Ted, 319
Burger, J. M., 153, 202
Burgess, R. L., 501
Burke, P., 90, 91, 92, 95
Burke, P. J., 85, 89, 93, 94, 101
Burnstein, E., 170, 279, 405, 406
Burawoy, M., 39
Burrell, N., 427–28
Burt, M. R., 218, 220, 221
Busceme, S., 375
Bush, D., 79
Bush, G. W., 177, 402, 403
Bushman, B., 310, 314, 324
Bushman, B. J., 221, 222, 322
Bushnell, N., 76
Buss, D. M., 18, 19, 278, 279, 333
Buswell, B., 102
Butler, D. L., 182
Butterfield, F., 499
Butz, D. A., 110
Buunk, B., 333

Buunk, B. P., 345
Byrne, D., 233, 280, 319, 327, 336, 337, 338

C

Cacioppo, J. T., 8, 153, 199, 201, 208, 209, 210, 222
Cadinu, M. R., 127
Cahill, S., 64
Cahill, S. E., 230
Caine, A., 382
Cairns, A., 499
Calhan, C., 270
Callaway, M. R., 402
Callero, P. L., 96
Calley, Lieutenant, 219
Camburn, D., 160
Cameron, D., 181
Campbell, A., 442, 447, 448, 450
Campbell, A. C., 285
Campbell, D. J., 38
Campbell, D. T., 412, 413
Campbell, J., 106
Campbell, J. D., 103
Campbell, K., 471
Campbell, L., 340
Campbell, M., 64
Campbell, W. K., 138, 338
Canary, D. J., 338
Caniglia, B. S., 144, 525, 528
Cano, I., 113
Cantor, J., 202
Cantor, N., 454, 485
Cantrell, P. J., 121
Capitanio, J. P., 425
Caplan, F., 55
Caplan, R. D., 481
Caporeal, L. R., 20, 279
Caputo, C., 136
Carli, L. L., 186
Carlsmith, J., 150
Carlsmith, J. M., 45, 190, 203, 537
Carlson, D., 233
Carlson, M., 287, 308, 311
Carlson, R. A., 331
Carlston, D. E., 115
Carneal, Michael, 525
Carnevale, P. J., 216, 283, 427, 428
Carney, M. A., 481
Carrick, R., 372
Carrington, P. I., 283
Carroll, J. M., 180
Carroll, J. S., 518
Carter, G. L., 537
Carter, Jimmy, 207
Cartwright, D., 356, 405
Carver, C., 100
Carver, C. S., 292, 294
Case, A., 476
Casper, L. M., 443
Caspi, A., 439, 441
Cassman, T., 218
Castro, Fidel, 134–35, 400, 401
Catalano, R., 302
Cate, R., 349
Catrambone, R., 111
Cauce, A. M., 60, 70, 455
Cejka, M. A., 154
Cenkovich, S., 500
Centers, R., 343
Cetingok, M., 487
Chaffee, S., 206, 207
Chaiken, S., 101, 143, 155, 157, 199, 202, 205, 208, 210
Chambliss, W. J., 515, 519
Chao, R., 61
Chaplin, W., 176
Chapman, L. J., 217
Chapman, R. S., 174
Charlin, V., 287
Charng, H.-W., 275
Charon, J. M., 16, 227
Chase-Lansdale, P. L., 61
Chason, K., 104
Chatters, L. M., 448–49
Cheadle, J., 61
Chebat, J.-C., 200
Check, J. V. P., 319–20
Chekroun, P., 293
Chemerinski, A., 283
Chemers, M. M., 384, 385
Chen, H. C., 222
Chen, Y. R., 393
Cheney, Richard, 202, 403
Cheng, P. W., 131
Cherlin, A., 443
Cherlin, A. J., 61
Cheshire, J., 187
Chesterfield, Lord, 233
Cheyne, J. A., 190
Chiriboga, D., 442
Chiricos, T., 520
Chiu, C., 186
Choi, I., 139
Christenson, R. L., 520
Christopher, F. S., 343
Church, A. H., 393
Cialdini, R. B., 106, 153, 202, 287, 288, 369, 376
Clanton, N., 176
Clark, E. V., 87
Clark, J. P., 38, 516
Clark, M., 286, 287
Clark, M. S., 296
Clark, R. D., III, 275, 279, 291
Clarkson, F., 70
Clausen, G. T., 292
Clausen, J. A., 79
Clausen, J. S., 441
Cleary, P., 487
Clelland, D. A., 410
Clinton, A. M., 318
Clinton, Bill, 207, 237, 515
Clinton, Hillary, 202
Clore, G. L., 338, 426, 427
Cloward, R., 496, 497
Coates, L., 193
Cochran, J., 502
Coffman, T. L., 117
Cognard-Black, A., 469
Cohen, A., 151, 513, 515, 517
Cohen, B. P., 360, 361, 365
Cohen, C. E., 112
Cohen, D., 307, 347, 501
Cohen, E. D., 216, 217, 430
Cohen, E. G., 365, 426
Cohen, G., 118
Cohen, J. L., 275
Cohen, L., 520
Cohen, L. E., 312
Cohen, O., 128
Cohen, P., 322, 323
Cohen, R. L., 395
Cohn, E. G., 311
Cohn, R. M., 103
Coke, J. S., 277
Colasante, C., 178
Cole, T., 239
Coleman, J. W., 496
Coleman, L., 105
Coleman, R. P., 464
Collett, P., 189
Collins, B. E., 219
Collins, M. A., 122, 186
Collins, N. L., 229, 425
Collins, R. L., 106
Comstock, G., 322
Comstock, G. S., 206
Conaty, J. C., 186, 447
Condon, J. W., 337
Condon, W. S., 193
Conger, J. A., 383
Conger, J. C., 332
Conger, K., 488
Conger, R., 477, 488
Conger, R. D., 435, 484
Conley, J. J., 115

Conner, T. L., 363
Connor, R., 286
Conolley, E. S., 371
Conrad, S., 319
Converse, P., 442
Cook, J., 329
Cook, K., 12
Cook, R. E., 133
Cook, S. W., 425
Cooke, B., 485
Cooley, Charles Horton, 6, 87, 252, 462
Cooper, H., 317
Cooper, J., 151, 418
Cooper, M. L., 485
Cooper, W. H., 115
Coopersmith, S., 101
Copper, C., 356, 357, 358
Corbin, L. H., 373
Corcoran, M., 467
Cornell, C. P., 304
Corning, A. F., 534
Correll, S., 469
Corsaro, W., 56, 64
Corsaro, W. A., 62, 64
Coser, L. A., 421
Costa, F., 465
Costa, M., 243
Costa, P. T., Jr., 115
Costall, A., 266
Costanzo, P., 68
Cota, A. A., 93, 356
Cote, S., 268
Cotter, D., 456
Couch, C. J., 527
Coupland, N., 169, 174, 181, 184, 421
Court, J. H., 320
Courtright, J. A., 402
Cowan, P. A., 437
Cox, C. L., 350
Cox, M., 58, 60, 61
Cozarelli, C., 485
Cozby, P. C., 229
Crain, R. L., 422
Cramer, R., 290
Crandall, C., 279
Crandall, C. S., 77
Crano, W., 157
Crano, W. D., 337
Creek, N. A., Jr., 64
Cressey, D., 501
Croak, M. R., 424
Crocker, J., 101, 104, 111, 121, 144, 218, 425
Croft, K., 86
Crohan, S. E., 443
Cropanzano, R., 397
Crosbie, P. V., 382
Crosnoe, R., 454
Crouter, A., 58
Crowley, E. P., 40
Crowley, M., 286
Cullen, F. T., 522
Cullum-Swan, B., 228
Cummings, K. M., 206
Cunningham, J. D., 229
Cunningham, M. R., 287, 288, 332
Cupach, W. R., 244
Custer, L., 445, 449
Cutler, B. L., 137
Cutrona, C. E., 351
Cutrona, C., 449

D

Dabbs, J. M., Jr., 205
Daher, D. M., 229
Dahn, J., 476
Dalrymple, S., 219
Daly, J. A., 244
Daly, K., 507
Dangelmayer, G., 350
Daniels, A., 446
Danziger, S., 487
Darby, B. L., 288
Darley, J. M., 42, 126, 174, 283, 289, 290, 291, 292, 294
Darwin, Charles, 18, 253, 278
Datan, N., 436
Davenport, M., 280
Davidson, A. R., 156, 158
Davies, J. C., 533, 534
Davies, S., 521
Davis, B., 71
Davis, D., 229
Davis, F., 247–48
Davis, J., 192
Davis, J. A., 455
Davis, J. D., 229, 339
Davis, J. M., 336
Davis, K., 56
Davis, K. E., 129, 130, 343
Davis, M. H., 138, 266, 267, 351
Davis, S., 206
Dawes, R., 417
Dawes, R. M., 126
Dawkins, R., 19, 278
Deasy, L., 506
Deater-Deckard, K., 61
Deaux, K., 90, 121, 280
DeBono, K. G., 202
Deci, E., 68
Decker, B., 160
DeDreu, C., 374
Deeg, Colin, 274
DeFleur, M., 159
Degelman, D., 231
de Jong-Gierveld, J., 351
DeLamater, J. D., 28
DeLongis, A., 475, 484
Deluga, R. J., 382
DeMaris, A. A., 311
Dembo, T., 301
Dembroski, T. M., 204
Demerath, N. J., III, 157
Demo, D., 58, 60, 61, 62
Demo, D. H., 92, 104
Dempster-McClain, D., 441
Demuth, S., 520
DeNeve, K. M., 539
Denser, W. E., 303
Dentler, R., 509
Denton, K., 77
Denzin, N., 64, 86
DePaulo, B. M., 179, 180, 231, 240, 241, 242, 296, 297
Der-Karabetian, A., 121
Derlega, V. J., 229
DeSalvo, J., 373
de Tocqueville, A., 533
Deutsch, M., 216, 369, 393
Deutschberger, P., 237
Devine, D. J., 357
Devine, J. A., 497
Devine, P. C., 154
DeVries, N., 374
Dewberry, C., 242
De Wolff, M. S., 58
Dholakia, R., 201
Diallo, Amadou, 109–10, 113
Dickert, J., 63
Dickoff, H., 240
Dickson, W. J., 367
Diderichsen, F., 481
Diehl, M., 399
Diekmann, A. B., 117
Diener, E., 92, 526
Dietz, T. L., 323
Dijkstra, P., 333
Dill, J. C., 136
Dill, K., 324
Dillon, Douglas, 400
DiMaggio, P., 465
Dimitri, S., 338
Dimond, M. F., 103
Dindia, K., 229
Dinsbach, W., 243
Dion, K., 331, 332, 405

Dion, K. L., 93, 119, 356, 357, 358, 421
Dixon, J., 508
Dodge, K., 61
Dodson, L., 63
Dohrenwend, B., 522
Dohrenwend, B. P., 247, 435, 482, 488
Dohrenwend, B. S., 488
Dollard, J., 301, 314
Doms, M., 375, 376
Donat, P., 318
Donnerstein, E., 319, 320
Donohue, W. A., 427–28
Donovan, J., 465
Doob, J., 301
Dooley, D., 488
Dorfmann, L., 452, 453
Doriot, P., 189
Dornbusch, S., 84, 91
Dorr, N., 539
Dossett, D., 155
Dovidio, J., 160, 246
Dovidio, J. F., 119, 186, 275, 276, 279, 280, 290, 292, 424–25
Downey, G. L., 550
Downing, J. W., 145
Doyle, P., 520
Draghi-Lorenz, R., 266
Dragna, M., 290
Drake, R. A., 145
Drass, K. A., 497, 518
Drauden, G., 399
Dreben, E. K., 125
Drenan, S., 369
Drew, P., 193
Drews, D., 231
Driskell, J. E., Jr., 363
Duberman, L., 229
Dubé-Simard, L., 421, 534
Duesterhoeft, D., 549
Dugan, K., 542
Dugoni, B. L., 376
Dumenci, L., 484
Duncan, B., 276
Duncan, S., Jr., 192
Duneier, M., 38
Dunham, C., 143
Dunkel-Scheker, C., 339
Dunn, E., 143
Dunwoody, P., 402–3
Dura, J. A., 451
Duran, R. L., 335
Durham, B., 229
Dutton, D., 98, 261
Duvall, S., 287
Dyck, R. J., 308
Dynes, R. R., 279
Dzindolet, M. T., 399

E

Eagly, A., 119
Eagly, A. H., 117, 202, 208, 286, 332, 333
East, P. L., 436
Eaton, W., 488
Ebbesen, E. B., 406
Ebert, J., 475
Eccles, J., 513
Eccles, J. S., 441
Eckert, C., 550
Eckert, P., 181
Eckes, T., 111
Edelmann, R. J., 242, 243, 246
Eder, D., 38, 62, 63
Edwards, C. P., 70
Edwards, J., 448
Edwards, J. N., 347, 450
Edwards, K., 157
Edwards, M. T., 381
Efran, M. G., 190
Ehrhardt, A., 70
Eibl-Eibesfeldt, I., 255
Eifler, D., 71
Eisenberg, N., 276, 279, 286, 296
Eisenstat, R. A., 180
Eiser, J. R, 418
Eitle, D., 498
Ekeberg, S., 392
Ekman, P., 240, 241, 242, 252, 255–56, 257, 258
Elder, G. H., Jr., 433, 434, 435, 438, 439, 454, 477
Elkin, F., 54, 74
Elkin, R. A., 149, 208
Ellard, J. H., 361
Ellemers, N., 90
Ellingson, S., 443
Elliott, G. C., 393
Elliott, R., 309
Ellis, B. J., 278
Ellison, C., 497
Ellsworth, P., 283
Ellsworth, P. C., 45, 190
Ellyson, S. L., 186
Elster, A., 437
Emery, B., 319
Emirbayer, M., 85
Emmons, R. A., 92
Emswiller, T., 280
Endicott, J., 375
Endler, N. S., 374
England, P., 444
Epstein, J. L., 426
Erickson, J. R., 111
Erickson, M., 515
Erickson, M. A., 441
Erikson, E. H., 92, 441, 449
Erikson, K., 504, 509, 515, 517, 521
Ervin, C., 336
Esser, J. K., 402
Etaugh, C., 78, 450, 451, 452, 453
Etcoff, N. L., 119
Etling, K. M., 127
Ettinger, R. F., 374
Etzioni, A., 430
Evans, C. R., 356, 357
Evans, D., 281
Evans, G., 246
Evans, N. J., 356
Evenbeck, S., 382

F

Fabes, R. A., 286
Fahs, M., 472
Faley, T., 214
Fallon, B. J., 381
Farhar-Pilgrim, B., 206
Farina, A., 247
Farmer, Denise, 300, 304, 312
Farnham, S., 101
Farnworth, M., 497
Farrell, M., 499
Farrell, R., 508
Fatoullah, E., 202
Fazio, R., 154, 155, 330
Fazio, R. H., 126, 154, 155, 161, 418
Feagin, J., 160
Feather, N. T., 148, 373, 473
Feeney, J., 340
Feeney, J. A., 57
Fein, S., 539
Feinstein, J. A., 210
Fejfar, M. C., 136
Feld, S., 471
Felipe, N. J., 189, 190
Felmlee, D., 350
Felson, M., 503, 509
Felson, R., 310
Felson, R. B., 91, 92, 102, 106, 501
Fenema, E., 470
Ference, R., 402
Fernandez-Dols, J. M., 256, 257
Ferraro, K., 305, 317
Ferree, M. M., 541, 546, 548
Ferreira, M. C., 117
Festinger, L., 102, 106, 149, 150, 328, 337, 526

Festinger, Leon, 7
Fetchenhauer, D., 333
Feuille, P., 429
Fiedler, F. E., 384, 385, 387
Field, N., 359
Field, C., 66, 67
Fields, J., 444
Filiatrault, P., 200
Fillingian, R. B., 475
Finch, B. K., 497
Finch, M. L., 430
Fine, G. A., 64
Fingerson, L., 56, 64
Fink, B., 334
Fink, E. L., 203
Finkel, S. E., 206
Finkelstein, M. A., 290
Finkelstein, S., 180
Fischer, C. S., 327
Fischer, G. W., 417
Fischer, A. H., 258
Fischer, K., 86
Fisek, M. H., 359, 363
Fishbein, M., 158, 161, 162
Fisher, J., 283
Fisher, J. D., 296, 297
Fishman, P. M., 181, 193
Fisicaro, S. A., 115
Fiske, D. W., 192
Fiske, S. T., 13, 111, 113, 119, 120, 121, 125, 135, 160
Flament, C., 413
Flanagan, M., 539
Flavell, J., 86
Fleishman, J., 280
Fleming, J. H., 174
Fletcher, G., 127, 138
Floderus, B., 481
Flora, J., 351
Flores, G., 323
Flowers, M. L., 402
Fodor, E. M., 402
Folbre, N., 444
Folger, R., 382
Folkes, V. S., 236
Folkman, J., 475
Fong, G. T., 99
Forbes, G., 231
Ford, C. E., 106
Ford, R., 216
Ford, R. S., 216
Fordham, S., 184
Forgas, J. P., 114
Form, W. H., 279
Forsyth, C. J., 247
Forsyth, D. R., 235, 364, 373
Fortenbach, V. A., 276
Forthofer, M. S., 484
Foschi, M., 361
Foster, C., 338
Foster, M. D., 534
Fowler, F., 61
Fox, G. L., 311
Francis, L., 231
Franiuk, R., 347
Frank, F., 392
Frank, M., 242
Franks, D., 102
Franzoi, S. L., 351
Fraser, C., 72, 405
Fraser, S., 153
Frazier, C., 520
Fredricks, A., 155
Freedman, J., 153
Freedman, J. L., 222
Freeman, S., 106
Freese, L., 363, 365
French, J. R. P., 210, 212
Freud, Sigmund, 57, 240, 253, 259, 301, 314
Frey, D., 99, 235
Frey, D. L., 280
Frey, J., 538
Frey, K. S., 65
Friedman, A., 397, 398
Friedman, G., 181
Friedman, H. S., 241
Friedrich-Cofer, L., 322
Friesan, M., 255
Friesen, W. V., 240, 241, 255
Fries, A., 58
Frieze, I., 133–34
Frieze, I. H., 111–12, 335
Frijda, N. H., 252
Frink, D. D., 392
Frisch, D., 283
Fritzsche, B. A., 290
Fu, M., 93
Fukada, H., 243
Fulero, S., 119
Fuller, S. F., 402
Fullerton, C., 482
Fultz, J., 276, 277
Funk, J., 323
Furstenberg, F., 61, 440
Furstenberg, F., Jr., 443
Fussell, S. K., 168, 170

G

Gaertner, S., 160
Gaertner, S. L., 119, 275, 279, 280, 424–25
Gagnon, J. H., 327
Gaines, S. O., 350
Gallup, George, 6
Galton, M., 37
Gamson, W. A., 542
Gandhi, Mahatma, 313
Gangestad, S. W., 352
Garcia, J. E., 385
Garcia, S. M., 292
Gardner, L., 499
Garfinkel, H., 245
Gartner, R., 312
Gates, G. S., 387
Gaudet, H., 206
Gauthier, A., 443
Gawronski, B., 161
Gecas, V., 64, 85, 89, 101, 102
Geen, R. G., 308, 314
Geerken, M. R., 234
Geis, G., 286
Geis, M. L., 171, 175
Geisinger, K., 435
Gelfand, D. M., 275
Geller, S. H., 374
Geller, V., 242
Gelles, R. J., 304, 305
Genovese, Catherine (Kitty), 288–89, 292
Gerard, H. B., 369, 371, 425
Gergen, K., 103
Gergen, K. J., 130, 232–33, 276, 283, 398
Gergen, M. M., 276
Germann, J., 323
Gerstel, N., 487
Gesell, A., 54, 55
Geurts, S. A. E., 398
Ghaziani, A., 521
Gibbons, F. X., 99, 100
Gibbs, J., 515
Gifford, R., 189
Gilbert, D. T., 136
Gilbert, T. F., 379
Giles, H., 168, 169, 174, 181, 184, 187, 421
Giles-Sims, J., 67
Gill, V. T., 518
Gillespie, D. L., 186, 447
Gilligan, C., 77
Ginorio, A., 70, 455
Ginsburg, B., 279
Ginsburg, G. P., 405
Giordano, P., 500
Glaser, B. G., 80
Glass, D. C., 246
Glass, J., 143
Glick, P. C., 442
Gliha, D., 247

Gockel, B., 229
Goethals, G. R., 86, 125, 358, 405
Goetting, A., 450, 453
Goffman, E., 171, 190, 227–28, 232, 243, 244, 245, 255, 512
Goldberg, L. R., 115
Goldin, C., 443
Goldman, N., 483
Goldman, R., 208, 209
Goldman, S., 77
Goldsteen, K., 477
Goldstein, B., 77
Goleman, D., 268
Gonos, G., 228
Gonzales, J. H., 45
Gonzales, M., 331
Gonzales, M. H., 151, 336
Gonzalez, R., 127
Goodchilds, J. D., 318
Gooden, A. M., 71
Gooden, M. A., 71
Goodman, P. S., 397, 398
Goodwin, C., 193
Goodwin, J., 551
Goodwin, R., 350
Goodwin, S. A., 120
Gordon, C., 84, 508
Gordon, J., 498
Gordon, P., 173
Gordon, R. A., 153, 233, 240
Gordon, Randy, 300
Gordon, S. L., 252, 262
Gore, S., 486, 498
Gorfein, D. S., 371
Gotay, C. C., 296
Gottlieb, A., 227, 292
Gottlieb, J., 294
Gottman, J., 318
Gouge, C., 405
Gould, R. L., 442, 449
Gove, W., 234
Govern, J., 96
Gowan, M., 444
Gowen, C. R., 392
Graf, Cyndy, 274
Graham, L., 541
Graham, T., 267
Gramling, R., 247
Grammar, K., 333, 334
Granberg, D., 163, 429
Granovetter, M. S., 471
Grant, P., 422
Grant, P. R., 114
Grasmick, H. G., 214
Graves, J., 390
Graves, N., 219
Gray, J., 181, 446
Gray, L. N., 212, 213
Gray-Little, B., 104
Grayshon, M. C., 178
Greatbatch, D., 40, 193
Greaves, K. M., 445
Green, A., 239
Green, J. A., 157
Greenbaum, P., 190
Greenberg, J., 337, 395, 397
Greenberg, M., 283, 296
Greenberg, M. S., 295
Greenberger, E., 469
Greene, D., 68
Greenfeld, L., 305
Greenstein, T. N., 448
Greenwald, A., 101
Greenwald, A. G., 99, 142, 144, 159
Gregory, S. W., Jr., 188
Grev, R., 287
Grice, P. H., 171
Grichting, W. L., 206
Griffin, D., 127
Griffin, D. W., 345
Griffin, G. W., 402
Griffin, L., 78, 550
Grimshaw, A. D., 175, 184
Groat, H., 490
Gronfein, W., 522
Gross, A. E., 296
Gross, E., 243
Groves, R., 32
Gruber, K. J., 308
Gruenewald, P. J., 207
Grzelak, J., 229
Grzywacz, J., 485
Guarnaccia, C. A., 454
Gubin, A., 120
Guimond, S., 534
Gully, S. M., 357
Gumpert, P., 232–33
Gumperz, J. J., 185, 187
Gunn, L. K., 350
Gurevitch, Z. D., 188
Gurney, J. F., 534
Gustafson, R., 416
Gutierrez, L., 70, 455
Guttman, D., 452

H

Haan, N., 76, 77
Hachen, D., 464
Hacker, H. M., 229
Hackman, J. R., 388
Hafdahl, A., 104
Hagan, J., 520
Hagedoorn, M., 398
Haggstrom, G., 443
Haidt, J., 256
Haines, H. H., 548
Haines, S. C., 356
Halberstam, D., 30
Hall, E. T., 188
Hall, J., 181
Hall, W. S., 160
Hallahan, D. P., 40
Hallam, J. K., 105
Hamblin, R. L., 396
Hamilton, D. L., 111, 138, 426
Hamilton, J. A., 121
Hamilton, V. L., 218
Hamilton, W., 278
Ham-Rowbottom, K., 488
Han, G., 430
Handel, G., 54, 74
Hanni, R., 192
Harasty, A., 91
Harburg, E., 147, 373
Hardee, B. B., 160
Hardesty, M., 445
Hardy, C., 390
Hardy, R. C., 387
Hare, M., 399
Hargis, K., 287
Harkins, S., 390
Harkins, S. G., 202, 203, 390
Harlow, R. E., 454, 485
Harnish, J., 486
Harpine, F., 387
Harris, L., 452, 453
Harris, M. B., 302
Harris, P., 232
Harris, R., 282
Harris, R. J., 114
Harris, R. M., 192
Harris, V. A., 134
Harrison, A., 329, 331
Hartke, D. D., 385
Hartmann, D. P., 275
Hartmann, H., 449
Harton, H., 160
Harvey, E., 58
Harvey, J. H., 236, 276
Harvey, O. J., 411–12
Haskell, M. R., 507
Hass, R. G., 200, 201
Hassan, H., 36

Hassin, R., 122, 178
Hastie, R., 125
Hastorf, A. H., 218
Hatch, Milton, 289
Hater, J. J., 383
Hatfield, E., 340, 343, 344, 398
Hatfield, Ellison, 416
Hatfield, Floyd, 416
Hatfield, Tennis, 416
Hathaway, D., 487
Hauser, R. M., 449, 465, 466
Haveman, R., 58
Haver, M. D., 500
Hawkes, D., 231
Hawkins, A., 231
Hawkins, C., 95, 361, 362
Hayduk, L. A., 189
Haynie, D., 502
Hazan, C., 57
Hazen, C., 258
Head, K. B., 310
Healy, M., 414
Heatherton, T. F., 265
Heavey, C. L., 315, 316
Hebl, M., 246
Hechter, M., 365
Heckhausen, H., 133
Heesacker, M., 208
Heider, E. R., 172
Heider, F., 15, 127–28, 132, 135, 147
Heider, Fritz, 7
Heimer, K., 501
Heingartner, A., 387
Heirich, M., 543
Heiss, D., 34
Heiss, D. A., 270
Heiss, J., 9, 79, 106
Hellman, P., 289
Helm, B., 430
Hemingway, H., 476
Henderson-King, D., 458
Henderson-King, E., 458
Hendrick, C., 330, 344
Hendrick, S., 330, 344
Henley, J., 487
Hennessy, D., 302
Hennrikus, D., 185
Henretta, J. C., 520
Henry, R. A., 428
Henry, S., 90
Henry, P. J., 93
Hensley, D. L., 114
Hensley, T. R., 402
Henson, A., 190
Henson, M., 508
Hepburn, C., 117
Herbener, E. S., 439
Herek, G. M., 425
Heritage, J., 40, 193
Herman, E. M., 387
Hermsen, J., 456
Hernandez, A., 161
Herr, P., 154
Hess, T. M., 112
Hess, U., 257
Hetherington, E. M., 61, 62
Hewitt, J. P., 87, 93, 236
Hewstone, M., 131, 138, 168, 414, 418, 421, 425
Heyl, B. S., 496
Heyman, R., 338
Heyns, R. W., 373
Higbee, K. L., 204
Higgins, C., 235
Higgins, E. T., 86, 87, 100, 101, 114, 135
Higgins, R. L., 236
Higuchi, M., 243
Hill, C., 339, 349
Hill, C. T., 350
Hill, T., 311
Hills, A., 40
Hiltrop, J. M., 428, 429
Hinkle, S., 420
Hirsch, E. L., 545
Hirschi, T., 498
Hirt, E. R., 404
Hitler, Adolf, 219
Hitlin, S., 472
Hobsbaum, E., 533
Hocevar, D., 91
Hochschild, A. R., 230, 270, 271, 453
Hodge, C. N., 332
Hodson, R., 489
Hoelter, J. W., 96, 102, 106
Hofferth, S., 28
Hoffman, C., 173
Hoffman, E. L., 371
Hoffman, J., 499
Hoffman, K., 202
Hoffman, M. L., 264
Hofling, C. K., 219
Hogan, D. P., 440
Hogan, R., 541
Hoge, D. R., 513
Hogg, M. A., 90, 91, 161, 356, 405
Hojjat, M., 347
Hokada, E., 71
Holgartner, S., 543
Holland, R., 188
Hollander, E. P., 373, 381, 383, 507
Hollinger, R. C., 516
Holmberg, S., 163
Holmes, D. S., 244
Holmes, J. G., 114, 341, 345, 422
Holmes, W. H., 129
Holohan, C., 486
Holt, C. S., 339
Holzer, H., 441
Homans, G. C., 12, 333, 383, 395
Homans, George, 7
Homer, P. M., 473
Hood, W. R., 371, 411–12
Hopkins, N., 113, 414, 470
Hopper, J. R., 284
Hopper, R., 236
Horai, J., 202, 215
Horne, C., 370, 508
Horne, W. C., 405
Horney, J., 499
Hornstein, G., 187
Hornstein, H. A., 217, 280
Horowitz, D. L., 540
Horowitz, H. V., 484
Horowitz, L. M., 337
Horowitz, R., 509
Horrocks, J., 36
Hossain, Z., 445
House, J. S., 443, 450, 476, 479
House, R. J., 383
Houston, C. E., 127
Hovland, Carl, 7
Howard, J., 90, 119
Howard, J. A., 283, 284–85
Howell-White, S., 484
Hoyle, R. H., 136, 356
Hrdy, S. B., 278
Hu, L., 417
Hue, C., 111
Huesmann, L. R., 317, 321, 322
Hughes, D., 50
Hughes, M., 104, 234
Hui, C. H., 91
Hull, C. D., 445
Hull, C. L., 11
Hultsch, D., 435
Hundelby, J. D., 499
Hunt, K., 476, 481
Hunter, B. A., 160
Hunter, C. H., 235
Hunter, J. A., 418
Hunter, J. E., 332
Hunter, S., 246
Huston, A. C., 321, 322
Huston, T., 340, 341, 342

Huston, T. L., 286, 289
Hyde, J., 102, 440, 441, 470
Hyde, J. S., 70, 77
Hymes, D., 175

I

Ickes, W., 267, 351
Ikle, F. C., 421
Ilg, F., 54, 55
Ingerman, C., 144
Ingham, A. G., 390
Insko, C., 147
Insko, C. A., 131, 203, 204, 370
Insko, V. M., 204
Isen, A. M., 277, 286, 287, 424
Isenberg, D. J., 405, 406
Ishii, K., 258, 260
Islam, M. R., 113
Isozaki, M., 405
Itkin, S. M., 426
Iwawaki, S., 242
Iyengar, S., 207

J

Jaccard, J., 158, 203
Jackson, J., 367, 497, 506
Jackson, J. J., 397
Jackson, J. M., 202, 221
Jackson, L. A., 332, 360
Jackson, P. B., 484
Jackson-Jacobs, C., 505
Jacobs, P. I., 116
Jacobsen, P. R., 397
Jacobson, C. K., 219
Jacobson, N., 318
Jacques, J. M., 104
Jaffee, S., 77
Jaggi, V., 418
James, J., 499
James, W., 85, 90, 253
Janes, L., 369
Janis, I. L., 400–404, 420
Jankowski, S., 246
Jarvis, P. A., 356
Jarvis, W. B. G., 210
Jaspars, J., 131, 138
Jefferson, G., 192
Jekielek, S. M., 61
Jellison, J. M., 405
Jenkins, C., 476
Jenkins, J. C., 543, 550
Jenkins, T. B., 94
Jensen, C., 119
Jensen, G., 515
Jessor, L., 465
Jessor, R., 465
Job, R. F. S., 208
John, D., 444, 445
John, O. P., 8
Johnson, B., 160, 205
Johnson, B. T., 208
Johnson, C., 186
Johnson, D., 173, 448
Johnson, D. J., 13, 352
Johnson, D. L., 405
Johnson, D. R., 448
Johnson, G., 394
Johnson, J., 481
Johnson, J. G., 322, 323
Johnson, J. T., 136
Johnson, K., 245, 248
Johnson, K. J., 103
Johnson, Lyndon, 536
Johnson, M., 156, 157, 305, 317, 341
Johnson, N. R., 529
Johnson, R., 512
Johnson, R. E., 499
Johnson, R. J., 512
Johnson, T., 193
Johnson, W. T., 516
Johnson-George, C., 341
Johnston, L. D., 503
Jones, E., 151
Jones, E. E., 6, 113, 125, 129, 130, 134, 135, 136, 153, 230, 232–33, 245, 417, 418
Jones, J., 94
Jones, S. D., 392
Jones, V. C., 416
Jones, W. H., 351
Jong, P., 244
Jonides, J., 126
Jöreskög, K. G., 35
Joseph, N., 176
Joule, R., 150
Jourard, S. M., 229
Joyner, K., 329
Judd, C. M., 112, 115, 145
Judge, T., 235
Judge, T. A., 383
Juette, A., 334
Julian, J. W., 381, 383, 507
Junghans, C. M., 336
Jussim, L., 105, 513

K

Kacmar, K. M., 233
Kahle, L. R., 473
Kahn, A., 421
Kahn, A. S., 318
Kahn, R. L., 310
Kahne, M., 508
Kahneman, D., 126, 127
Kalleberg, A., 489
Kalmijn, M., 80
Kalven, H., Jr., 374
Kameda, M., 308
Kanki, B., 375
Kanouse, D., 443
Kanter, R. M., 446
Kantrowitz, B., 446
Kanungo, R. N., 383
Kao, G., 329
Kaplan, H. B., 512
Kaplan, M. F., 369, 370, 405, 406
Kaplowitz, S. A., 139, 203
Kappas, A., 257
Karabenick, S. A., 280
Karasawa, M., 258, 260
Karasek, R., 476
Karau, S. J., 390
Kardes, F. R., 154
Karlins, M., 117, 208
Kasen, S., 322, 323
Katz, D., 144, 160
Katz, I., 246, 365
Katz, J., 509
Katz, L., 443
Katz, P., 121
Katzman, N., 206
Katzner, K., 166
Kauffman, D. R., 240
Kaufman, K., 420
Kaufman, J., 245, 248
Kawakami, K., 188
Kayson, W. A., 280
Keller, C. E., 40
Kelley, H. H., 7, 12, 122, 123, 125, 127, 131–32, 215, 261, 334, 341
Kelley, K., 280, 319
Kellogg, R. T., 236
Kelly, W., 520
Kelman, H. C., 155, 218
Keltner, D., 256
Kemper, T. D., 185
Kendon, A., 192
Kendrick, D. T., 161, 287
Kendzierski, D., 163

Kennedy, John F., 400, 401, 430
Kennedy, Robert, 400
Kennedy, S., 440
Kenney, D., 338
Kenrick, D. T., 18, 20, 96, 278, 279
Kenrick, D., 333
Kent, G., 192
Kepner, C. R., 313
Kerber, K. W., 275, 276
Kerckhoff, A. C., 327, 329
Kerpelman, J. L., 441
Kerr, N., 390
Kerr, N. L., 392
Kersting, R. C., 405
Kessler, R., 321
Kessler, R. C., 484, 487
Ketelaar, T., 278
Key, M. R., 192
Keyes, J., 345, 346
Khanna, N., 93
Khomusi, T., 508
Khrushchev, Nikita, 430
Kick, E., 520
Kiecolt-Glaser, J. K., 451, 477
Kiernan, K., 61
Kiesler, C. A., 373
Kiesling, S., 188
Kiesler, S. B., 373
Kilham, W., 221
Kilik, L., 356
Killian, L. M., 526, 528, 530, 541, 542, 543
Kim, P., 40
Kim, T.-Y., 97
King, Larry, 188
King, R., 96
King, Rodney, 309, 525, 535–36
Kingston, P., 448
Kiparsky, P., 167
Kippax, S., 322
Kirchmeyer, C., 485
Kirson, D., 256
Kitayama, S., 170, 258, 260, 279
Kitcher, P., 279
Kitsuse, J., 511
Klanderman, S. B., 282
Klandermans, B., 546
Klapper, J. T., 206
Kleck, R. E., 246, 247
Kleidman, R., 550
Klein, Harold, 289
Klein, O., 159
Klein, R., 100
Klein, T., 477
Kling, K. C., 102
Klinger, L. J., 121
Klitzner, M., 207
Klonoff, E., 47
Kluegel, J., 520
Knapp, M. L., 244
Knoche, L., 70
Knopf, T. A., 530
Knottnerus, J. D., 363
Knudsen, E. I., 452
Knutson, M. C., 231
Kochan, T., 429
Koestner, R., 242, 331
Koffka, K., 14
Kogan, N., 116
Kohl, W. L., 383
Kohlberg, L., 76–77
Kohler, Wolfgang, 14
Kohn, M., 77, 78, 473, 474, 490
Kojetin, B. A., 174
Komorita, S. S., 395
Konecni, V. J., 415
Korten, D. C., 423
Koskinen, S., 479
Koss, M. P., 316, 317, 318
Kothandapani, V., 143
Kowai-Bell, N., 246
Kowalski, R. M., 226
Kozielecki, J., 226
Kramer, R. M., 425
Kraus, S., 207
Krauss, R., 186
Krauss, R. M., 168, 170, 178, 216
Kraut, R. E., 240, 241
Kravitz, D. A., 389, 390
Kraxberger, B., 238
Krebs, D., 77
Krebs, D. L., 276, 278, 300
Kressel, K., 427, 428, 429
Kriesberg, L., 422
Kritzer, H., 540, 551
Krohn, M., 502
Krohn, M. D., 499, 502
Krosnick, J. A., 145
Kruck, K., 334
Kruger, D. J., 279
Kruglanski, A. W., 218, 376
Krull, D. S., 136, 139
Ksionzky, S., 373
Kuhn, D., 76
Kuhn, M. H., 84
Kulick, J. A., 302, 303
Kumkale, G. T., 201
Kumru, A., 70
Kunda, Z., 99
Kunkel, D., 323
Kurdek, L. A., 484
Kuriloff, P. J., 121
Kurmeyer, S. L., 476
Kurtines, M. M., 77
Kwan, J. L., 376

L

Labov, W., 184
Lachman, S. J., 115
LaFleur, S. J., 186
LaFrance, M., 189, 447
LaFree, G., 497
Lage, E., 375
LaGreca, H. J., 502
Lake, R., 98
Lakoff, R. T., 446
Lamb, M., 437
Lamb, M. E., 70, 437
Lamberth, J., 336
Lamm, H., 393, 399, 405, 406
Lamon, S., 470
Land, K. C., 312
Landis, D., 345
Landrigan, P. J., 472
Landrine, H., 47
Landry, J. R., 186
Landy, D., 332
Lane, D. W., 393–94
Lane, J. D., 338
Lang, M., 447
Langan, C. J., 373
Lange, C. G., 253
Lantz, D., 168
Lantz, H., 345, 346
Lanza-Kaduce, L., 502
Lanzetta, J. T., 283
LaPiere, R., 154, 158
LaPiere, Richard T., 6
Laquidara-Dickinson, K., 246
Larey, T. S., 399
Larimer, M., 315
Larrick, R. P., 132, 539
Larrieu, J. A., 286
Larsen, R. J., 92
Larson, L. L., 385
Larzelere, T., 340, 341, 342
Lasater, T. M., 204
Lasrewski, Jeff, 274
Lassegard, M. A., 106
Lassiter, G. D., 240, 242

Latané, B., 42, 202, 289, 290, 291, 375, 390
Latané, J. G., 296
Lau, I., 173
Lau, R. R., 91, 138
Laub, J. H., 499
Lauderdale, P., 364, 517
Laughlin, P. R., 388
Laumann, E., 443
Laumann, E. O., 327, 329
Lavine, H., 151
La Voie, L., 338
Lavoie, F., 304
Lawler, E. J., 216, 217, 381, 382, 383, 397
Lazarsfeld, P. F., 206
Lazarus, R. S., 472, 475
Leana, C. R., 402
Leander, K., 186
Leaper, C., 70, 73
Lear, D., 342
Leary, M., 238, 239
Leary, M. R., 94, 226
Leary, A., 75
Leavitt, C., 201
Leavitt, L. A., 72
Le Bon, G., 526, 527
Ledvinka, J., 184
Lee, D. J., 114
Lee, G. R., 447
Lee, J. A., 119
Lee, R., 234
Lee, S. Y., 122
Lee, T. R., 450
Leffler, A., 186, 189, 447
Legant, P., 136
Leginski, W. A., 217
Leiber, M. J., 497
Leigh, J. P., 472
Leippe, M., 149
Leippe, M. R., 208
Lemert, E., 511, 513
Lemos, K., 106
Lennington, S., 19
Lennon, M. C., 247
Leonard, K. E., 317, 318
LePore, P. C., 19
Lepper, M., 68
Lepper, M. R., 120–21
Lerum, M., 352
Lesh, M. D., 383
Leslie, L., 341
Leung, J., 520
Leventhal, G. S., 393–94, 397, 423
Leventhal, H., 204, 205, 261
Levin, I., 351
Levin, J., 244
Levin, L., 349
Levin, P. F., 287
Levine, J. M., 371, 373, 405
Levine, M., 281
Levine, P., 30
LeVine, R. A., 412, 413
Levinger, G., 326, 350, 390
Levinson, D., 449
Levinson, R., 92
Levitin, T. E., 247
Levitt, B., 542
Lewecki, R. J., 215
Lewicki, P. A., 24
Lewin, D., 429
Lewin, K., 301
Lewin, Kurt, 7
Lewis, E., 448–49
Lewis, G. H., 360
Lewis, K., 487
Lewis, L., 121
Lewis, L. D., 291
Lewis, M., 69, 70, 73
Lewis, S. A., 373
Lex, B. W., 39
Leyens, J. P., 187
Li, Y., 61
Li, Y. E., 533
Liberty, H. J., 509
Lichtenstein, M., 119
Lichter, D., 60, 329, 442, 443
Lichterman, P., 114
Lickel, B., 246, 355
Liden, R. C., 153
Lieberman, A., 157, 205
Lieberman, P., 167
Lieberman, S., 10
Liebert, R., 67
Light, J., 328
Lightner, J. M., 276
Likert, R., 34
Lin, N., 464, 543
Lin, Y.-H., 483
Lincoln, Abraham, 114
Lincoln, J., 489
Lind, E., 420
Linder, D. E., 151
Lindskold, S., 197, 211, 215, 430
Link, B. G., 247, 510, 521, 522
Linkenhoker, B. A., 452
Linsenmeier, J., 40
Linsenmeier, J. A., 232
Linville, P., 111
Linville, P. W., 113, 417, 418
Linz, D., 315, 316, 320
Lipe, M. G., 127
Lipkus, I., 350
Lippitt, Ronald, 7
Lipscomb, T. J., 286
Lisak, D., 317
Liska, A., 163
Liska, A. E., 501
Littlepage, G., 242
Littlepage, G. E., 392
Livingston, R. W., 159
Lloyd, S., 319
Locher, D. A., 526
Locke, K. D., 337
Locksley, A., 117
Loeber, C. C., 186
Lohr, J. M., 144
Lois, J., 38, 91
Loney, G. L., 336
Long, G., 397, 405
Longman, R. S., 356
Longo, L. C., 332
Longstreth, M., 342
Lopata, H., 452
Lopes, P. N., 268
Lord, C. G., 120–21
Lorence, J., 488, 489, 490
Lorenz, F. O., 435, 477
Lorenz, K., 301
Lott, A., 337, 356, 357
Lott, B., 337, 356, 357
Lottes, I. L., 121, 335
Lowenthal, M. F., 442, 449
Lowery, B. S., 143
Lowrance, R., 424
Luckenbill, D., 501
Luhman, R., 184
Lui, L., 417
Lukens, C. K., 336
Lumsden, H., 138
Lund, D. A., 103
Lundgren, S., 375
Lundman, R. J., 38, 517, 518
Lurigis, A. J., 518
Lutfey, K., 437
Lydum, A., 315
Lyman, S., 236
Lynch, J. C., Jr., 275
Lyons, M., 382

M

Maass, A., 174, 376
MacEwen, K. E., 58
Macintyre, S., 476, 481
Mackey, R., 450

Mackie, D., 120–21
Mackie, D. M., 135, 358, 376
Mackie, M., 446
Mackovic, John, 382
Macmillan, R., 499
Maddi, S., 475
Maddux, J. E., 200, 208
Madon, S., 513
Madsen, D. B., 406
Magner, N., 394
Magnusson, J., 280
Mahay, J., 443
Maheswaran, D., 210
Mahon, N. E., 229
Maines, D., 445
Major, B., 104, 218, 447, 485
Makhijani, M. G., 332
Malamuth, N. M., 315, 316, 319–20
Mallozzi, J., 280
Malone, P. S., 136
Mancini, J., 452
Mancini, J. A., 450
Manis, M., 126
Mann, C. R., 520
Mann, J., 425
Mann, L., 221, 229, 400
Mannheim, B. F., 92
Mannheimer, D., 426
Manning, P. K., 228
Mannix, E. A., 393
Manstead, A., 243
Manstead, A. S. R., 163, 230, 236, 242, 258
Manucia, G. K., 288
Manz, C. C., 402
Maoz, Y., 128
Maracek, J., 136
Marangoni, C., 351
Marbach, J. J., 247
Marchman, V. A., 72
Marcus, D. K., 267
Marcus-Newhall, A., 308, 311, 312
Marcussen, K., 484
Marger, M. N., 547
Margulis, S. T., 229
Marino, C. J., 374
Markman, H. J., 484
Markmann, K. D., 404
Markovsky, B., 364
Markowitz, F., 310, 522
Markowitz, S. B., 472
Marks, N., 485
Markus, H., 13, 85, 93, 99, 111
Markus, H. R., 127
Marlin, N. A., 236
Marmot, M. G., 476
Marolla, J., 102
Marques, J., 508
Marsden, P., 471
Marsden, P. V., 455
Marsh, H., 91
Marshall, I. H., 499
Marshall, S. E., 550
Martelin, T., 479
Martin, B., 389, 390
Martin, C. L., 119
Martin, D., 90
Martin, J., 487, 507, 542
Martin, K., 238, 239
Martin, M. W., 220, 365
Martin, P. S., 207
Martin, S. S., 512
Martinez, R., Jr., 497
Martocchio, J. J., 392
Marwell, G., 157, 398
Marwick, N. J., 384
Marx, Karl, 394, 490
Maslach, C., 283
Mason, W., 491
Massey, J. L., 502
Mather, R., 319
Matheson, G., 444
Matheson, K., 534
Mathie, V. A., 318
Matsueda, R., 501
Matsueda, R. L., 92
Matsumoto, D., 258
Matthews, L., 477
Matthews, L. S., 484
Matthews, S., 450
Maurer, T., 92
Maxwell, J. W., 450
May, D., 484
May, D. R., 358
Mayer, F. S., 287
Mayer, J. D., 112, 268, 286
Maynard, D. W., 37, 169, 518
Mayo, C., 186, 189, 447
Mayseless, O., 283
Mazur, J. E., 11
McAdam, D., 262, 541, 546, 547
McAdoo, H. P., 467
McAllister, H. A., 286
McArthur, L. Z., 131, 136
McAuslan, P., 318
McCall, G. J., 89, 95, 105, 227
McCall, P. L., 312
McCannell, K., 447
McCarthy, J. D., 513, 546
McCarthy, P. M., 276
McCauley, C., 117, 402
McConahay, J. B., 160
McConnell, W., 302
McConnell-Ginet, S., 181
McCoy, Jim, 416
McCoy, Randolph, 416
McCrae, R. R., 115
McCreath, H. E., 96
McCubbin, H., 485
McCusker, C., 91
McDermott, V., 280
McDougall, William, 6
McDuffie, D., 318
McEneaney, E. H., 309
McFarland, C., 133–34
McFarland, D., 62
McGillis, D., 130
McGrath, J. E., 388
McGraw, K. M., 144
McGuire, C., 84, 86, 87, 93, 94
McGuire, W. J., 84, 86, 87, 93, 94, 198, 221, 222
McIntire, Carl, 543
McIntosh, D. N., 256
McKenna, C., 338
McLanahan, S., 62
McLanahan, S. S., 62
McLear, P. M., 332
McLeod, J. M., 373
McLoyd, V., 60, 440
McMaster, M., 290
McMullen, P. A., 296
McNamara, Robert, 400
McNulty, T. L., 501
McPartland, T., 84
McPhail, C., 525, 526, 527, 528, 530–32, 534
McPherson, M., 329
McQueen, L. R., 394
McRae, C., 61
McVeigh, R., 544
McWorter, G. A., 422
Mead, G. H., 6, 85, 88, 99
Meeker, B. F., 361, 393
Megowen, K. R., 481
Mehrabian, A., 180, 188, 373
Meier, R. F., 516
Meile, R. L., 507
Melzer, S., 485
Mendelsohn, H., 206
Mendes, W. B., 246
Menke, B. A., 213
Mennino, S. F., 485
Menon, S. T., 383
Mequita, B., 252, 260

Mercer, G. W., 499
Mero, R. P., 450
Merton, R., 495, 511, 532
Mesquita, B., 258
Messe, L. A., 393–94
Messner, S. F., 499
Meter, K., 276
Metts, S., 229, 244, 345
Meyer, D. S., 545
Meyer, P., 278
Meyers, S., 331
Meyrowitz, J., 232
Miall, C. E., 522
Michael, R. T., 327
Michaels, J. W., 347, 393
Michaels, S., 327
Michela, J. L., 127, 261
Michener, H. A., 216, 217, 218, 220, 221, 233, 381, 382, 402, 430
Milardo, R., 462
Milardo, R. M., 342
Milavsky, J. R., 321
Milburn, M. A., 319
Miles, R. H., 10
Milgram, S., 219, 220, 221, 509
Milkie, M., 444
Miller, A. G., 129, 219
Miller, C. E., 370, 395
Miller, D. L., 526, 527
Miller, D. T., 278
Miller, F. D., 541, 546, 548
Miller, G., 512
Miller, J., 474
Miller, J. A., 153
Miller, K., 474
Miller, L. C., 229, 425
Miller, L. K., 396
Miller, M., 481
Miller, N., 287, 301, 308, 311, 405, 424
Miller, P. A., 276
Miller, R., 106
Miller, R. S., 242, 243, 267, 371
Miller, S., 330
Miller-McPherson, J., 471
Mills, J., 296
Minnigerode, F., 119
Miranne, A. C., 212
Mirenberg, M., 94
Mirowsky, J., 477, 481, 487
Mische, A., 85
Mischel, W., 67
Mitchell, T., 235
Mitchell, T. R., 153
Miyamoto, S. F., 84, 91
Mizrahi, T., 486
Mladinic, A., 117
Modigliani, A., 244
Moede, W., 389, 390
Moen, P., 441
Moghaddam, F. M., 411
Mohr, J., 465
Moise, J., 321, 322
Molinari, L., 64
Mondale, Walter, 155
Money, J., 70
Montanari, J. R., 402
Montano, D., 156
Moody, J., 329
Moody, J. A., 357
Moore, C., 414
Moore, D. J., 222
Moore, J. C., Jr., 362
Moore, M. M., 182, 183
Moore, R., 31
Moorhead, G., 402
Moos, R., 486
Moran, G., 356
Moreland, R. L., 127, 405
Moreno, J. L., 6–7
Morgan, S., 443
Morgan, W., 78
Moriarty, T., 290
Morrel-Samuels, P., 178
Morris, C. G., 388
Morris, M. W., 132
Morris, W. N., 371
Morrow, G. D., 13
Morse, K., 350
Morse, S., 103
Morse, S. J., 398
Mortimer, J., 437, 490
Mortimer, J. T., 79, 441, 488, 489
Moscovici, S., 8, 374, 375, 376
Moskowitz, G. B., 292
Mouton, J. S., 424
Mowrer, O., 301
Muedeking, G. D., 228
Mueller, C. W., 77
Mugny, G., 375
Mulatu, M. S., 487
Mullen, B., 202, 356, 357, 358, 417
Mullen, C. K., 320
Mullens, S., 189
Muller, C., 470
Muller, E., 540
Muller, M. M., 135
Murphy, H., 506
Murphy, P., 426
Murray, C., 347
Murray, J. P., 322
Murray, S., 105
Murray, S. L., 345
Murrell, A., 425
Mussweiler, T., 127
Mutz, D. C., 207
Myers, D. G., 405, 528
Myers, D. J., 144, 160, 309, 525, 533, 534, 537, 544
Myers, F. E., 550, 551
Myers, M. A., 520

N

Naccari, N., 202
Nacci, P., 430
Nadler, A., 283, 296, 297
Nadler, D., 283
Naffrechoux, M., 375
Nagel, J., 104
Nail, P. R., 160
Nakao, K., 464
Naoi, A., 78
Nardi, P., 79
Nassau, S., 105
Natziuk, T., 374
Navarro, M., 93
Neal, A. G., 490
Neale, M., 216
Neale, M. A., 393
Neck, C. P., 402
Nelson, C., 115, 116
Nelson, D., 336
Nelson, T. D., 160
Nelson, T. E., 126
Nemeth, C. J., 8, 375, 376
Neuberg, S., 350
Neufeld, R. W. J., 208
Neugarten, B., 464
Neugarten, B. L., 436
Neuman, C. A., 161
Neville, B., 444
Newcomb, M. D., 502
Newcomb, T., 526
Newcomb, T. M., 15, 143, 147, 337
Newcomb, Theodore, 7
Newman, L., 158
Newsom, J. T., 153
Newton, T. L., 477
Nida, S., 291
Nida, S. A., 291
Nielsen, J. M., 284
Nisbett, R., 68, 136
Nisbett, R. E., 138, 139, 307
Nock, S., 448

Noh, S., 447
Noller, P., 57
Norenzayan, A., 138, 139
Norman, C., 338
Norstrom, T., 499
Northcraft, G. B., 393
Norvell, N., 421
Norwood, A., 156, 482
Nosow, S., 279
Novaco, R., 302
Novick, L. R., 131
Novland, S., 499
Nuttbrock, L., 522
Nyden, P. W., 550
Nylen, L., 481

O

Oakes, P. J., 414
Oberschall, A., 535, 539, 541, 542, 543, 551
O'Brien, B., 450
O'Bryant, S., 452
O'Connell, B., 72
O'Connor, C., 256
Oegema, D., 546
Offer, S., 487
Offner, P., 441
O'Gara, P. W., 397
Ogston, W. D., 193
O'Hara, J., 190
O'Hara, M., 345
Ohbuchi, K., 308
Okamoto, D. G., 181
Okamura, L., 317
Oldham, J., 77
O'Leary, K., 71
O'Leary-Kelly, A. M., 392
Oleson, K. C., 277
Oliver, P., 551
Oliver, P. E., 525, 526, 540
Olivier, D., 172
Olson, J. M., 369
Olson, M., 330
Olver, R., 73
Olzak, S., 309, 412, 533, 535, 537
O'Malley, P. M., 503
O'Neil, T. P., 481
O'Neill, Paul, 403
Opp, K., 365
Opp, K.-D., 535
Oppenheimer, V. K., 457
O'Quin, K., 277
O'Rand, A. M., 433, 439
Orbuch, T., 36, 47, 350
Orbuch, T. L., 236, 443, 445, 449, 450
Orcutt, J., 509
Oros, C. J., 316
Orth, J., 185
Orwell, George, 172
Osborn, A. F., 398–99
Osgood, C. E., 34, 84, 95, 252, 429, 430
Osgood, D. W., 499, 503
O'Shea, W., III, 345
Oskamp, S., 160, 430
Ostendorf, F., 116
Ostrom, T. M., 390
Ostrove, J. M., 451
O'Sullivan, M., 137, 241, 242
Otten, C. A., 286
Ouellette, J. A., 375
Owens, S., 106
Owens, T., 522

P

Padavic, I., 469
Padden, S. L., 485
Page, A. L., 410
Pager, D., 521
Paicheler, G., 405
Paik, A., 443
Palmore, E., 453
Pandey, J., 233
Pantin, H. M., 292
Papageorgis, D., 222
Papastamou, S., 375
Parcell, T. L., 467
Parish, E. A., 229–30
Park, B., 112, 115, 417
Park, W.-W., 402
Parke, R., 66, 67
Parke, R. D., 60, 70, 444
Parker, R., 32
Parkinson, B., 258
Parmelee, P., 338
Parrott, W. G., 242
Patrick, S. L., 397
Patterson, J., 485
Patterson, M. L., 189
Patterson, R. J., 208
Paulus, P. B., 399, 402
Pavlidis, J., 241
Pavlov, I., 11
Paxson, C., 476
Paxton, P., 357
Pearce, P. L., 280
Pearlin, L., 104, 481, 487
Pearson, J., 427–28
Peckham, V., 390
Pedersen, W., 308
Pegg, I., 246
Pegnetter, R., 427, 428
Peirce, K., 71
Pelham, B., 94
Pendry, L., 372
Penedo, F., 476
Pennebaker, J., 222
Pennebaker, J. W., 261
Penner, L. A., 275, 286, 290
Pennington, J., 77
Penrod, S., 203, 375
Pepitone, A., 526
Peplau, L., 339, 349
Peplau, L. A., 350
Peri, N., 283
Perkowitz, W. T., 229
Perlman, D., 351
Perrien, J., 200
Perry, J. B., 532
Perry, L. S., 397
Perry, R. P., 280
Perry-Jenkins, M., 58
Personnaz, B., 376
Peruche, B. M., 110
Pescosolido, A. T., 380
Pescosolido, B., 507
Peters, L. H., 385
Petersen, T., 471
Peterson, J., 135
Petronio, S., 229
Petrunik, M., 247
Pettigrew, T. F., 418, 425
Petty, R., 390
Petty, R. E., 8, 153, 199, 201, 202, 203, 204, 208, 209, 210, 222
Pfeiffer, T., 127
Phelan, J. C., 510, 522
Phillips, A. T., 258, 260
Phillips, D. P., 504, 505
Phillips, J., 176
Piaget, J., 76, 86
Pierce, C. M., 219
Pierce, K. P., 412–13
Pike, C. L., 288, 332
Piliavin, I. M., 229, 280, 294, 313
Piliavin, J. A., 19, 275, 279, 280, 286, 291, 292, 294
Pillemer, K., 454
Pineault, T., 242
Pinker, S., 278
Pipher, M., 102
Pittman, T. S., 230
Plant, E. A., 110

Plazewski, J. G., 217, 233
Pleck, J., 92
Pleck, J. H., 121, 450
Plemons, J., 435
Plutzer, E., 160, 533
Podolny, J. M., 471
Poe, D., 241
Pohlmann, J. T., 385
Pollak, S., 58
Pomerantz, E., 347
Porter, J. R., 104
Post, D. L., 136
Postmes, T., 526
Potrin, R. H., 499
Powell, B., 92
Powell, M., 154
Powell, M. A., 467
Powell, M. C., 154
Power, Emil, 289
Powers, A. S., 237
Powers, T. A., 235
Poyatos, F., 178
Prager, I. G., 137
Pratkanis, A., 99, 142
Pratkanis, A. R., 144, 159
Prause, J., 488
Predmore, S. C., 105
Prentice, D., 92
Preston, J. D., 425
Previti, D., 350
Price, K. O., 147, 373
Price, R. H., 481
Price, N., 231
Priest, R. T., 328
Priester, J. R., 199
Prince, Morton, 6
Prince-Gibson, E., 30, 472
Pritchard, R. D., 379, 392
Probst, J., 231
Proffitt, C., 163
Prosser, A., 281
Pruitt, D. G., 131, 216, 283, 427, 428, 429
Prusank, D. T., 335
Puccetti, M., 475
Pudberg, M. T., 77
Pugh, M. D., 364, 365, 532
Puranam, P., 383
Purdum, T. S., 310

Q

Qian, Z., 60, 442, 443
Quanty, M. G., 314
Quarantelli, E. L., 279
Quattrone, G. A., 417
Quillian, L., 64
Quinn, D. M., 118
Quinney, R., 518

R

Rabbie, J. M., 422
Rabow, J., 161
Radabaugh, C., 487
Radloff, C. E., 86
Radosevich, M., 502
Rahav, M., 522
Rahn, J., 491
Rain, J. S., 125
Raley, R. K., 442
Ramires, A., 204
Ramirez, G. G., 383
Rane, T., 92
Ranelli, C. J., 371
Rank, S. G., 219
Ransberger, V., 311, 537, 538
Rapp, H. C., 112
Rashotte, L. S., 177
Rasinski, K. A., 482
Ratzan, S. C., 30
Raven, B., 210, 212
Raven, B. H., 210, 212, 218, 356
Ray, M., 207
Rea, C., 222
Rea, L., 32
Reagan, Ronald, 155
Reardon, R., 222
Reber, A., 66
Reddy, V., 266
Reed, A., II, 419
Reed, M., 92
Rees, C. R., 360
Regan, D. T., 155
Regan, P., 344
Regnemer, J., 76
Reich, J. W., 454
Reicher, S., 281
Reifenberg, R. J., 138
Reifman, A. S., 539
Reis, H., 332
Reis, H. T., 339
Reisman, S. R., 147
Reitzes, D., 93, 94
Rejeski, W. J., 238
Rempel, J. K., 341
Renfrow, D., 248
Rennison, C., 305
Repetti, R., 58, 484, 490
Reskin, B., 449, 469
Retzinger, S. M., 265
Revenson, T., 486
Reyes, J. A., 258, 260
Reynolds, J. R., 481
Reynolds, P. D., 359
Reznikoff, M., 435
Rhine, R. J., 201, 203
Rice, R. W., 384
Richards, C., 485
Richardson, K., 505
Richeson, J. A., 120
Richmond, L., 349
Ride, Sally, 470
Ridgeway, C., 363
Ridgeway, C. L., 186, 364
Ridley, M., 278
Riess, M., 190
Rietsema, J., 356
Riggio, R. E., 241
Rigney, J., 73
Riketta, M., 418
Riley, A., 92
Riley, M., 436
Rimer, B., 206
Rindfuss, R., 436, 443
Rinehart, A., 402–3
Ring, K., 215
Ringelmann, M., 389, 390
Riordan, C., 365, 425
Riordan, C. A., 236
Riskind, J., 405
Rizzo, T. A., 64
Roach, M. A., 72
Roberts, D., 206
Robertson, J. F., 101
Robinson, D., 95
Robinson, D. T., 510
Robinson, J., 60, 444, 465
Robinson, J. W., Jr., 425
Robson, P., 453
Rochford, E., Jr., 541
Rock, L., 125
Rodgers, W., 442
Rodin, J., 280, 290, 294
Rodriguez, T., 258, 260
Roefs, M., 90
Roethlisberger, F. J., 367
Rogers, E., 147
Rogers, M., 287
Rogers, P. L., 180
Rogers, R. G., 478
Rogers, R. W., 200, 208, 308
Rogers, S., 484
Rogers, S. J., 443
Rohrer, J. H., 371
Rokeach, M., 472
Romano, J., 189

Romero, V., 414
Rommetveit, R., 462
Roopnarine, J. L., 445
Roosa, M., 343
Root, M., 71, 455–56
Roper, G., 405
Roper, S., 365
Rose, H., 20
Rose, S., 20, 111–12, 335
Rosen, S., 296
Rosenbaum, M. E., 337
Rosenbaum, W. B., 398
Rosenberg, B., 84
Rosenberg, F., 103
Rosenberg, L. A., 371
Rosenberg, M., 101, 102, 103, 104, 105
Rosenberg, M. J., 147, 148, 149
Rosenberg, S. V., 115, 116
Rosenblatt, A., 337
Rosenfeld, H., 190
Rosenfeld, H. M., 189
Rosenfeld, L., 333
Rosenfeld, R., 436
Rosenhan, D. L., 109, 286, 287, 288
Rosenholtz, S. J., 361, 363
Rosenkrantz, P., 70
Rosenmann, R., 476
Rosenthal, R., 45, 179, 180, 240
Roskos-Ewoldsen, D. R., 202
Rosner, T., 450
Ross, A. S., 292, 294
Ross, C., 487
Ross, C. E., 447, 477, 479, 481, 485, 487
Ross, D., 304
Ross, E. H., 6
Ross, K., 71
Ross, L., 135, 216
Ross, L. D., 118, 135
Ross, L. T., 318
Ross, M., 127, 133–34, 138, 382
Ross, S., 304
Rossi, A. S., 450
Rossmann, M., 485
Rotenberg, K. J., 229
Roth, D. L., 475
Roth, P. L., 392
Roth, S., 317
Rothbart, M., 8, 119, 127, 417
Rotheram-Borus, M. J., 104
Rottman, L., 330
Rotton, J., 311, 538
Rovine, M., 447
Rowland, K., 385
Royer, L., 187
Ruback, R., 189
Ruback, R. B., 189
Rubens, W., 321
Rubin, B., 485, 550
Rubin, B. A., 534
Rubin, J. Z., 215, 217
Rubin, M., 414
Rubin, Z., 188, 339, 343, 349
Ruble, D. N., 65
Rudd, J., 228
Rudé, G., 533
Ruderman, A. J., 119
Rudolph, J., 500
Ruggiero, J. A., 365
Ruggiero, M., 286
Ruiter, R. A. C., 204
Ruiz-Belda, M.-A., 256, 257
Rule, B. G., 308
Rumbaut, R. G., 440, 487
Rumsfeld, Donald, 403
Rupp, L., 551
Rusbult, C. E., 13, 350, 352
Rushton, J. P., 279, 285
Rusk, Dean, 400
Russell, D., 138
Russell, J. A., 180
Russell, R. J., 279
Rust, M. C., 424–25
Ryder, N. B., 438
Ryen, A. H., 421
Ryff, C., 435
Ryoko, M., 529

S

Sabini, J., 135, 265
Sachau, D., 232
Sack, S., 387
Sacks, H., 192
Sadalla, E. K., 161
Saeed, L., 331
Saegert, S. C., 329
Sagarin, E., 510
Saito, Y., 393
Sakurai, M. M., 357, 422
Sales, S., 476
Salovey, P., 268, 286, 287, 288, 417
Sammon, S., 435
Sample, J., 156
Sampson, R. J., 499
Samuels, F., 422
Sanbonmatsu, D. M., 154
Sanday, P. R., 315
Sandberg, J., 28
Sande, G. N., 86
Sandefur, G., 62
Sanders, C., 452
Sanders, D., 222
Sanders, P., 70, 73
Sanford, C., 393
Sanitioso, R., 99
Saporta, I., 471
Sarbin, T., 84
Sarkisian, N., 487
Sauer, C., 405
Sawyer, A., 208
Sawyer, J., 328
Saxe, L., 371
Sayer, L., 60, 444
Sayles, M., 486
Schachter, S., 261, 328
Schaerfl, L. M., 394
Schafly, Phyllis, 550
Schaller, M., 288
Schank, R. C., 111
Scheff, T., 245
Scheff, T. J., 265
Schegloff, E., 190, 192
Scheier, M. F., 100
Scher, S. J., 397
Scherer, K. R., 176, 240
Scherer, S. E., 189
Scherwitz, L., 185
Schiavo, R. S., 188
Schiffenbauer, A., 188
Schiffren, D., 335
Schifter, D. E., 163
Schleifer, S. L., 217
Schlenker, B. R., 105, 197, 211, 213, 214, 226, 235, 430
Schlesinger, Arthur, Jr., 400
Schlosser, B. A., 405
Schmidlin, A. M., 447
Schmidt, G. W., 106
Schmiege, C. J., 445
Schmitt, D. P., 19, 333
Schmitt, D. R., 398
Schnell, S. V., 77
Schnittker, J., 103, 507
Schoen, L. M., 114
Schoenbach, C., 78, 103, 105
Schoenborn, C., 477
Schoeneman, T., 92
Schooler, C., 78, 95, 103, 105, 473, 474, 487
Schopler, J., 420
Schrauger, J. S., 92
Schroeder, D. A., 275, 276, 279, 283
Schroeder, D. G., 98
Schrum, W., 64
Schuller, R. A., 119
Schultheiss, O. C., 350
Schultz, M., 345, 346

Schulz, A., 497
Schulz, B., 520
Schuman, H., 32, 156, 157
Schutte, J., 328
Schutte, N. S., 161
Schvaneveldt, P., 441
Schwalbe, M., 102
Schwartz, C., 506, 508
Schwartz, J. C., 256, 257, 260
Schwartz, M. F., 286, 287
Schwartz, R. D., 38
Schwartz, S., 296
Schwartz, S. H., 30, 157, 280, 283, 284–85, 292, 418, 472
Schwarz, N., 32, 307
Schweingruber, D., 528
Schwinger, T., 393
Sciandra, R., 206
Scott, K. S., 397
Scott, M., 236
Scott, R., 517
Scott, W., 161
Scotton, C. M., 185
Scully, M., 542
Searle, J. R., 170
Sears, D., 93
Sears, D. O., 147, 207, 222
Sears, R., 301
Seccombe, K., 447
Sechrest, L., 38
Secord, P., 105
Sedikides, C., 114, 138, 202, 221
Seedman, A. A., 289
Seeman, M., 486, 488, 490–91
Segal, B. E., 425
Segal, M. W., 360
Segrin, C., 351
Seidel, M. L., 471
Sell, J., 220, 365
Sellers, C., 502
Semin, G., 175
Semin, G. R., 236, 242
Senchak, M., 339
Senn, C., 231
Seplaki, C., 483
Serbin, L., 71
Serpe, R., 64, 70, 95, 96
Serpe, R. T., 92
Sessions, J., 78
Setersten, R., Jr., 440
Severance, L. J., 201, 203
Sewell, W. H., 449, 465, 468
Shackelford, T. K., 19
Shanahan, S., 309, 533, 537
Shanas, E., 453
Shapiro, D., 192
Shapiro, S. P., 497
Sharkey, W. F., 243
Shaver, K. G., 125
Shaver, P., 57
Shaver, P. R., 256, 257, 258, 260
Shaw, L. M., 357
Shaw, M., 68
Shaw, M. E., 357
Shay, K. A., 475
Shearing, C. D., 247
Shedler, J., 126
Shehan, C., 444
Shehan, C. L., 447
Sheley, J. F., 497
Shell, R. M., 296
Shelton, B. A., 444, 445
Shen, Y., 352
Shepard, H. A., 424
Sherif, C., 367
Sherif, C. W., 411–12
Sherif, M., 367, 370, 371, 372, 411–12, 424
Sherif, Muzafer, 6
Sherk, D. L., 231
Sherman, M., 247
Sherman, P., 278
Sherman, S. J., 112, 151
Sherwood, J. J., 101
Shibutani, T., 17, 56, 72, 100
Shihadeh, E. S., 468
Shin, C., 472
Shipstead, S., 86
Shoemaker, D. J., 499
Shoemaker, F. F., 206
Sholis, D., 229
Shope, G. L., 314
Shore, C., 72
Shorter, E., 533
Shotland, R. L., 276, 282, 289, 294
Shott, S., 266, 276
Shover, N., 499
Showers, C., 102
Shrout, P. E., 522
Shweder, R. A., 116
Siepel, M., 283
Siepmann, M., 135
Sigall, H., 332
Sikkink, D., 544
Silberman, M., 516
Sill, M., 28
Silver, B. D., 32
Sime, J. D., 279
Simmons, J. L., 89, 95, 105, 227
Simmons, R., 79, 102, 437
Simmons, R. G., 275, 282
Simmons, S. F., 287
Simon, R., 481, 484
Simon, R. J., 361, 362
Simons, C., 90
Simons, R., 68
Simons, R. L., 102
Simpson, B., 363
Simpson, J. A., 57, 348, 349
Sims, H. P., 402
Sinclair, S., 143
Singer, D. G., 322
Singer, J., 261
Singer, J. L., 322
Singer, Kevin, 300
Singer, R. P., 204
Sistrunk, F., 430
Skinner, B. F., 11, 65
Skowronski, J. J., 115
Skrzypek, G. J., 385
Slater, P., 341
Slater, P. E., 359
Slaughter, S. J., 112
Sloane, D., 499
Slomczynski, K. M., 78, 474
Slovik, L. A., 350
Smailes, E. M., 322, 323
Small, K. H., 135
Smart, J. L., 163
Smith, A., 121
Smith, C. J., 376
Smith, D. A., 519
Smith, E. R., 90, 154
Smith, E. S., 237
Smith, H. J., 534
Smith, J., 127
Smith, L. F., 364, 382
Smith, M. D., 497
Smith, P. B., 371
Smith, R., 370
Smith, R. H., 370
Smith, S. F., 242
Smith, T., 402
Smith, T. G., 351
Smith, T. W., 455, 482
Smith, W. P., 216, 217
Smith-Lovin, L., 34, 95, 181, 251, 329, 471, 510
Sniezek, J. A., 358
Snow, D. A., 526, 535, 540, 541, 542, 543, 545
Snyder, C. R., 106, 236
Snyder, D., 534
Snyder, M., 114, 126, 159, 332
Sobery, J., 394
Solano, C. H., 229–30
Solomon, B., 339
Solomon, D. S., 206
Solomon, M. R., 332, 370

Sommer, B., 436, 450
Sommer, R., 189, 190
Sommers-Flanagan, J., 71
Sommers-Flanagan, R., 71
Sontag, Deborah, 307
Sörbom, D., 35
Sorenson, E. R., 255
Sorrentino, R. M., 359
South, S., 352, 444, 445
South, S. J., 501
Spaulding, J., 58
Spears, R., 175, 526, 534
Spencer, J. W., 235, 518
Spencer, S. J., 118
Spiegel, J. P., 540
Spilerman, S., 533, 537, 540
Spitz, R., 57
Spitze, G., 444, 445
Sprague, J., 464
Sprecher, S., 343, 344, 345, 350
Squier, W., 53
Srull, T. K., 13
St. C. Oates, G., 103
St. Clair, R., 168
Staats, A., 144
Staats, A. W., 144
Stack, A., 310
Stack, A. D., 222
Stack, F., 161
Stack, S., 505
Stafford, L., 243, 244, 338
Stafford, M. C., 213, 214
Staggenborg, S., 548
Stang, D. J., 374
Stanley, S., 484
Stansfeld, S., 476
Stark, R., 543
Starrels, M. E., 444
Stasser, G., 392, 400
Staub, E., 294
Stebbins, C. A., 276
Steele, C. M., 117, 118
Steelman, L. C., 92
Steenaert, B., 188
Steers, R., 383
Steffen, V. J., 119
Steffensmeier, D., 520
Steffensmeier, D. J., 497, 500, 503, 507
Stefflre, V., 168
Steffy, L., 234
Stein, J., 176
Stein, J. A., 502
Steinberg, J., 287
Steinberg, L., 469
Steiner, D. D., 125
Steiner, I. D., 147, 240, 388, 392
Stemp, P., 447
Stephan, W. G., 138, 425
Stephenson, W., 84
Sternberg, R., 345, 346, 347
Sternberg, R. J., 114
Sternthal, B., 201
Stets, J. E., 91, 252, 264, 267
Stevenson, M. B., 72
Stewart, A., 439, 451, 458
Stewart, A. J., 458
Stewart, H. L., 106
Stewart, J., 66, 67
Stewart, R. H., 125
Stewart, S., 333
Stiff, J. B., 201
Stiles, W., 185
Stillwell, A. M., 265
Stinnett, H., 333
Stipp, H., 321
Stitt, C. L., 117
Stokes, J., 351
Stokes, J. P., 351
Stokes, R., 236
Stone, G. P., 231, 243
Stone, J. I., 240
Stone, J. T., 242
Stoner, J. A. F., 404, 405
Stonner, L., 314
Storms, M. D., 136
Stovel, K., 329
Strack, F., 127
Strathman, A. J., 199
Strauman, T., 100
Strauman, T. J., 101
Straus, M., 66, 67
Straus, M. A., 320
Strauss, A., 80
Strauss, M. A., 305
Straw, M. K., 282, 294
Streifel, C., 500
Strenta, A., 246, 247
Stricker, L. J., 116
Stringer, M., 418
Strodtbeck, F. L., 361, 362
Stroebe, W., 147, 399
Strube, M. J., 385
Struch, N., 418
Struening, E., 522
Struening, E. L., 522
Stryker, S., 16, 64, 70, 89, 95, 96, 227
Stublarec, S., 364
Stucky, R. J., 236
Stuebing, K. K., 392
Suarez, G., 237
Suci, G., 34, 252
Sudman, S., 32
Sueda, K., 244
Sugarman, D., 67
Suitor, J. J., 454
Suls, J., 106
Suls, J. M., 106
Summers, L., 470
Summers-Effler, E., 262–63
Sumner, W. G., 412
Sun, Y., 61
Surra, C., 50, 342, 348
Suskin, R., 403
Sussman, N. M., 189
Sutherland, E., 501
Suttles, G., 494
Sutton, J. R., 550
Swander, D. V., 371
Swann, W. B., Jr., 98, 99, 105, 114, 226
Swap, W., 329, 341
Swedlund, M., 375
Sweeney, P. D., 397
Sweet, J. A., 441, 443, 456
Sweet, J. S., 443
Sweeting, H., 476, 481
Swicegood, C. G., 436
Swigert, V., 508
Swim, J. K., 160
Sykes, R. E., 38
Syme, S. L., 487
Symons, D., 18
Szinovacz, M., 452
Szymanski, K., 390

T

Tajfel, H., 90, 187, 413, 414
Takeuchi, D., 60
Talarico, S. M., 520
Tanford, S., 203, 375
Tangney, J. P., 265, 288
Tanke, E. D., 126
Tannen, D., 180, 446
Tannenbaum, P., 34, 252
Tanner, J., 521
Tarrow, S., 546
Tausig, M., 381
Taylor, D. A., 229, 339
Taylor, D. M., 187, 411, 418
Taylor, D. W., 399
Taylor, K., 78
Taylor, M., 106
Taylor, R. J., 448–49
Taylor, S., 370
Taylor, S. E., 13, 111, 113, 117, 119, 121, 135
Taylor, S. P., 310

Taylor, V., 544, 551
Tchividjian, L., 238
Teachman, J., 465
Tedeschi, J. T., 197, 211, 213, 214, 215, 430
Teger, A. I., 291
Telesca, C., 202
Tennen, H., 481
Terborg, J., 383
Terry, D. J., 90, 161
Terry, P., 481
Terry, R. M., 507
Tesser, A., 106
Test, M., 285
Testa, M., 104
Teti, D., 437
Teti, D. M., 437
Tetlock, P. E., 230, 473
Thakerar, J. N., 187
Thibaut, J., 7, 232–33, 334, 382
Thibaut, J. W., 12, 334
Thoennes, N. A., 427–28
Thoits, P. A., 252, 513
Thom, C., 231
Thomas, E., 462
Thomas, G. M., 381
Thomas, W. I., 6
Thompson, J. E., 425
Thompson, L., 216, 444
Thompson, L. L., 144
Thompson, M., 497
Thompson, T. L., 71
Thompson, V., 147
Thomsen, C. J., 151
Thomsen, C. T., 126
Thorlundsson, T., 184
Thornberry, T. D., 520
Thorndike, E. L., 11, 115
Thorne, A., 106
Thorne, B., 37, 62, 63
Thornhill, R., 333
Thornton, A., 143, 157
Thornton, R., 79
Thornton, W., 499
Thorton, A., 160
Thurnher, M., 442
Thurstone, L. L., 6
Tierney, K., 528
Tierney, K. J., 534
Tilker, H. A., 221
Tilly, C., 528, 533, 535, 540, 546
Tilly, L., 535
Tilly, R., 535
Tindale, R. S., 376
Titus, W., 400
Tobin, S., 453
Tobin, T., 75
Toch, H., 310
Toi, M., 276
Toledo, R., 509
Tolin, J. P., 276
Tolnay, S. E., 535
Tomkins, S. S., 255
Torgler, C., 318
Touhey, J., 333
Touval, S., 428
Traupmann, J., 348
Treas, J., 464
Tremain, M., 246
Trent, K., 352
Treviño, M., 444
Triandis, H. C., 91, 138, 151
Triplett, Norman, 6, 387
Trivers, R. L., 278, 283
Trommsdorff, G., 399
Trope, Y., 122, 128, 178
Trost, M. R., 153, 369
Tsang, J., 77
Tsoudis, O., 510
Tuch, S., 507
Tucker, M. B., 448–49
Turner, A., 315
Turner, J., 414
Turner, J. A., 203
Turner, J. C., 90, 358, 369, 405, 414, 425
Turner, R., 447
Turner, R. H., 9, 90, 526, 528, 530, 541, 542, 543
Turner, R. J., 498
Tversky, A., 126, 127
Twenge, J., 104
Tyler, T. R., 147, 382
Tziner, A., 356

U

Uggen, C., 497
Ukanwa, K., 97
Unger, R. K., 286
Ursano, R., 482
Useem, B., 533

V

Vallacher, R. R., 393–94
Valle, V. A., 133–34
Van Avermaet, E., 376
van Baaren, R., 188
van de Eijnden, R., 345
Vandello, U. A., 370
Vanderplas, M., 277
Van de Vliert, E., 10
Vandewater, E. A., 451
Van Dijk, E., 381
Van Gennep, A., 80
Van IJzendoorn, M. H., 58
van Knippenberg, A., 188
Van Knippenberg, A. F. M., 363
VanLaningham, J., 448
Van Maanen, J., 445
Vanneman, R., 456
Van Rijswijk, W., 90
Van Willigen, M., 447, 487
Van Wyk, J., 311
Van Yperen, N. W., 398
Varey, C., 127
Varney, L. L., 276
Vaske, J. J., 217, 233
Vaughan, S. I., 392
Veach, T. L., 405
Vega, W. A., 487
Verbrugge, L., 40, 477
Verette, J., 350
Verette, J. A., 236
Ver Hoeve, J. N., 72
Veroff, J., 36
Verplanck, W. S., 11
Videka-Sherman, L., 246
Vidmar, N., 405
Vincent, J. E., 288
Vinokur, A., 405, 406
Vinokur, A. D., 481
Vittengl, J., 339
Vivekananthan, P. S., 115, 116
Voelkl, K., 104
Vogel, S., 70
Voinescu, L., 186
Voissem, N. H., 430
Volling, B. L., 60
von Baeyer, C. L., 231
von Hippel, W., 95
Vonk, R., 234
von Rittman, J., 97
Vookles, J., 101
Voss, M., 481
Voydanoff, P., 481
Vygotsky, L. S., 86

W

Wachtler, J., 375
Wackenhut, J., 246, 509
Wade, J., 370
Wade, T. J., 370
Wagler, Evelyn, 320, 321
Wahrman, R., 364, 365

Waite, L., 443
Waldman, D. A., 383
Waldo, G., 520
Waldron, I., 477
Walker, A. J., 444
Walker, E., 520
Walker, H. A., 363, 381
Walker, L. L., 373
Walker, M., 106
Wallace, L. E., 477
Wallace, M., 550
Wallin, P., 234
Walsh, E. J., 106, 543
Walster, G., 331
Walster, G. W., 13, 343, 348, 395
Walster [Hatfield], E., 13, 202, 330, 331, 332, 336, 348, 395, 397
Walters, G., 117
Walters, J., 70
Walters, L., 70
Wan, C. K., 164
Ward, A., 216
Ward, D. A., 213
Ward, L. M., 125
Ward-Hull, C. I., 332
Warland, R., 156
Warland, R. H., 543
Warner, L. C., 159
Washington, R. E., 104
Wasserman, I., 535
Wasserman, M., 402
Watkins, P. L., 477, 481
Watson, D., 136
Watson, M. D., 379
Watson, R. P., 418
Watts, B. L., 393–94
Waugh, G., 286
Wayner, M., 375
Weaver, K., 292
Webb, E. J., 38
Weber, A. L., 236
Weber, R., 121, 425
Webster, M., 382
Webster, M., Jr., 361, 363
Webster, P. S., 443, 450
Webster, S., 188
Weed, F. J., 545
Wegner, D. M., 338
Weigand, B., 496
Weigel, D., 347
Weigel, R. H., 158
Weigold, M. E., 105
Weigold, M. F., 226
Weinberg, M., 514
Weiner, B., 132, 133–34, 236, 280
Weingart, L. R., 393
Weinstein, E. A., 237
Weinstein, M., 483
Weiss, R. S., 351
Weiss, T., 397
Weitzer, R., 513
Weitzman, L. J., 71
Weldon, E., 393
Welker, R., 394
Wellman, B., 485, 487
Wellman, H. M., 258, 260
Wells, G. L., 131, 210
Wells, P. A., 279
Werner, C., 338
West, C., 37, 180, 446
West, S. G., 376
Westcott, D. R., 295
Western, B., 521
Wetherell, M. S., 405
Wethington, E., 484
Wexler, M. O., 350
Weyant, J., 153
Whalen, M. R., 38, 169, 191–92
Whaley, D., 477, 481
Wheeler, D. S., 94
Wheeler, L., 331
Whitaker, D. J., 163
Whitcher-Alagna, S., 283
White, B. J., 411–12
White, C., 76
White, G., 348
White, H. R., 484, 498
White, J. W., 308
White, K. M., 90
White, L., 448, 450
White, Ralph, 7
White, R. W., 540
White, William Foote, 7
Whitney, D. J., 357
Whitney, G. A., 350
Whitney, R. E., 207
Whitt, H. P., 507
Whittier, N., 544, 545, 551
Whorf, B. L., 172
Wicker, A. W., 154
Wicklund, R. A., 99, 100, 151
Wickrama, K. A. S., 435, 477, 484
Wiebe, D. J., 475
Wiesenfeld, B. M., 395
Wiesenthal, D., 302
Wigboldus, D., 175
Wiggins, J. S., 332
Wiggins, N., 332
Wilcox, C., 111
Wilder, D. A., 412, 425
Wildfogel, J., 218
Wiley, M. G., 228
Wilhelmy, R. A., 371
Wilke, H., 283, 382
Wilke, H. A. M., 363, 381, 382
Wilkes, Michael, 306–7
Williams, C. J., 155
Williams, Charles, 300, 303, 315
Williams, D., 497
Williams, D. R., 477, 478
Williams, J., 121, 448–49
Williams, K., 390
Williams, K. D., 390
Williams, L., 111, 112, 447
Williams, L. S., 71
Williams, R. H., 542
Williams, R. M., Jr., 426
Willis, F. N., 331
Willis, J. E., 280
Willis, S., 436
Wills, T. A., 296
Wilmoth, J. K., 541
Wilson, D. W., 267, 291
Wilson, E. O., 7, 18, 278
Wilson, J. R., 503
Wilson, L., 60
Wilson, T., 41
Wilson, T. D., 127
Windle, M., 484
Winfrey, Oprah, 114
Wingert, P., 446
Winschild, P. D., 131
Winsett, R., 487
Winter, D. G., 458
Wiseman, R. L., 244
Wit, A., 382
Wit, A. P., 381
Wittenbaum, G. M., 392
Wittenbrink, B., 115
Wohlstein, R., 528
Wolf, S., 202, 375, 376
Wolf, W., 449
Wolfe, B., 58
Wolfe, C., 101
Wolfowitz, Paul, 403
Wolfson, M., 546
Woll, S. B., 331
Won-Doornink, M. J., 229, 339, 340
Wood, W., 202, 375, 376
Worchel, S., 420, 421, 424
Word, L. E., 291
Worden, S., 541
Worley, S., 485, 487
Wortman, C. B., 232
Wright, E., 522
Wright, E. O., 464, 490
Wright, J. C., 321
Wu, C., 479

Wu, R., 222
Wu, S., 257, 260
Wulfert, E., 164
Wurf, E., 85, 93, 99
Wyer, R., Jr., 144
Wyer, R. S., Jr., 13
Wylie, R. C., 104

X

Xie, W., 464

Y

Yablonsky, L., 507
Yantis, S., 156
Yarkin, K. L., 276
Yarrow, M., 506
Yates, S., 143, 155
Youm, Y., 443
Young, Carrie, 300
Young, P., 331
Young, Steven, 300
Youngs, G. A., Jr., 217
Youngs, J. A., Jr., 383
Yukl, G., 383
Yzerbyt, V. Y., 120

Z

Zaccaro, S. J., 390
Zahn, C. J., 236
Zajonc, R. B., 13, 256, 328, 329, 387, 388
Zander, A., 392
Zanna, M., 154, 155, 373
Zanna, M. P., 231, 341, 405
Zantra, A. J., 454
Zartman, I. W., 428
Zawacki, T., 318
Zebrowitz, L., 186
Zebrowitz, L. A., 122
Zeisel, H., 374
Zelditch, M., Jr., 218, 360, 361, 364, 381
Zellman, G. L., 318
Zembrodt, I. M., 350
Zerbinos, E., 71
Zey, M., 509
Zhou, X., 361
Zielinski, M., 502
Zielinski, M. A., 102
Zillman, D., 202, 261, 302
Zimbardo, P., 526
Zimmerman, D. H., 37, 38, 180, 191–92, 446
Zipf, S. G., 215
Znaniecki, F., 6
Zollar, A., 448–49
Zubek, J. M., 485
Zuckerman, E., 97
Zuckerman, M., 179, 240, 241, 242
Zuckor, Bryan, 300
Zukier, H., 123
Zurcher, L. A., 535, 541, 542, 543, 545
Zuroff, D. C., 235
Zvonkovic, A. M., 445
Zyzanski, S., 476

SUBJECT INDEX

Note: page numbers in *italics* refer to figures and tables

A

abortion, 485, 541, 548–549
access displays, 335
accessibility of attitudes, 154–155
accommodation, 350
accounts, 236–237, 244
acquaintanceship and helping, 279–280
activation of an attitude, 154–155
activism, social. *See* social movements
actor-observer difference, 136–137
actors and reaction to deviance, 506–508
additive model, 123, *124*
additive tasks, 389–390
address, forms of, 185
adolescence and self-esteem, 102
adulthood, 78–79, 436, 440. *See also* careers and work; life course; parenting
adversary, stereotyping of the, 401
affect, 251–252
affective aggression, 303
affective-cognitive consistency, 155
African Americans: and child-rearing, 60–61; and civil disorders, 536–537; and civil rights movement, 458–459; and crime, 497, 498; and employment discrimination, 521; lynching of, 535; and marriage, 442–443, 445, 448–449; and mental health, 487; and modern vs. old-fashioned racism, 160; and movement organizations, 546–548; and nonstandard English, 184–185; and representativeness heuristic, 126; and research methods, 47–48; and schematic judgment, 113; and self-esteem, 104; stereotyping of, 115, 118, 126–127. *See also* race and ethnicity
age grading, 436–437
ageism, 452
agency, self as source of, 85
aggravated assault, 306
aggression and violence: affective, 303; attribution for attack, 308; definition of aggression, 300; displaced, 308; domestic violence, 305, *305,* 306–307, 308, 311; and empathy, 267; frustration-aggression hypothesis, 301–302, 416; gang violence, 509; and gender, 305–306; and heat, 537–539, *538;* and helping, 278; and media, 320–324, *323;* motivations for, 300–304; pornography and violence, 319–320; and race, 305; reducing, 312–315; riots, 309–310, 535–540; sexual assault, 315–319, *318;* situational impacts on, 309–312; and social norms, 306–308; subculture of violence, 501; targets of, 304–308. *See also* intergroup conflict
aggressive cues, 311–312
aggressive pornography, 319–320
aging, biological, 436, 450, 451, *454,* 454–455
AIDS and stigma, 246, 521
alcohol consumption, 484–485
alienation, 488–491, *489*
aligning actions, 235–237
altercasting, 237
altruism. *See* prosocial behavior, helping, and altruism
anchoring effects, 127
anger, 257, *258,* 259, 265
anomie theory, 495–498
anonymity in crowds, 526, 528, 530
anticipatory socialization, 79
anxious/ambivalent style of attachment, 59

apologies, 244, 308
appearance. *See* physical appearance, management of; physical attractiveness
applause, 40, 193–194
arbitrators, 426
archival research, 39–41, *46*
arousal, 204–208, 322, 343
arousal/cost-reward model, 292, *294*
Asch conformity paradigm, *368,* 368–369, 370
Asian Americans, 61, 70–71, 103. *See also* race and ethnicity
assembly for gatherings, 531
assembly-line work, 489
attachment: to groups, 357; and mothers, 58; and social bond, 498, 499; styles of, 59
attention, focus of, *135,* 135–136
attenuation, 506
attitude change: and communication-persuasion paradigm, 198–199, *200;* definition of, 197; and media campaigns, 206–207; and messages, 203–208; processing of persuasive messages, 198–199; sources of persuasion, 200–203; targets of persuasion, 208–210, *209. See also* social influence and persuasion
attitude inoculation, 221–222
attitudes: activation and accessibility of, 154–155; attitudes-behavior correspondence, 157–158, *158;* behavior, relationship with, 154–161; and cognitive consistency, 15, 146–153, 155; cognitive structure of, 144–146, *146;* components of, 142–143; definition of, 142; and deviance, 501; and direct experience, 155–156; formation of, 143–144; functions of, 144; and individuals' impact on individuals, 3–4; reasoned action model of, 161–164, *162, 163;* in role theory, 10; and schemas, 73–74; and sexual assault, 315–316; similarity and liking, 336–337; situational constraints on, *159,* 159–161; strength and certainty of, 156–157, 159; surveys for measuring, 32–34; and women's employment, 455–456. *See also* beliefs; prejudice
attitudinal similarity, 336–337
attraction to a group, 373
attraction-to-a-stranger paradigm, 336, 337
attractiveness. *See* interpersonal attraction; physical attractiveness
attractiveness stereotype, 332–333
attribution: of aggression, 308; bias and error in, 134–138, 418–419; covariation model of, 131–132; cultural basis of, 138–139; definition of, 109, 127; of deservingness for help, 280; dispositional vs. situational, 127–128, 132–133; inferring dispositions from acts, 128–131; and social movements, 541; for success and failure, 132–134, *133,* 138
audience and reaction to deviance, 508–509
authentic impression management, 226, 227, 229
authority: definition of, 217; in formal groups, 380; and influence, 198; obedience to, 213, 217–221
auto accidents, 504–505
autokinetic effect experiment, 370–371
availability heuristic, 126
availables, 327–329, 350–352
averaging model, 123, *124*
aversive affect, 303
aversive events, 415–416
avoidant style of attachment, 59

B

back channel feedback, 193, 447
back regions, 232
backtracking, 510
balance theory, 147–148, *148*
balancing, 506
bargaining over identities, 237
Bay of Pigs invasion, 400–401
behavior: and attitudes, 154–161; in cognitive theory, 21–22; as concern of social psychology, 3; counterattitudinal, 150–151; in evolutionary theory, 18–19, 21–22; extrinsically vs. intrinsically motivated, 68; and identities, 93–95; reflexive, 85; in reinforcement theory, 11, 21; in role theory, 10–11, 21; self-awareness, influence of, 99–100; and self-esteem, 103–105; situational constraints on, 159–161; in symbolic interaction theory, 16, 21–22. *See also* deviant behavior
behavioral change, 22
behavioral control, perceived, 163–164
behavioral dimension of norms, 367
behavioral intentions, 161–163
behavioral style and dissenting minorities, 375
beliefs: and attribution bias, 138; about gender and math skills, 470; primitive or fundamental, 145; in reasoned action model, 161–163; and social bond, 498; and social movements, 541; and violence, 540. *See also* attitudes
belonging and social movements, 545
bias: in attributions, 134–139, 418–419; cultural, 138–139; linguistic intergroup bias, 174; in perception of outgroups, 417–420; and reward equity, 398; self-serving, 138; in social control, 518–520
bilateral threat, 216–217
biological aging, 436, 450, 451, *454,* 454–455
birth cohorts, 438–439
Blacks. *See* African Americans
bluffing, 212–213
blushing, 244
bodily movement, development of, *55*
body language (kinesics), 176, *176,* 186, 188
borderwork, 63
brainstorming, 398–400
burden of proof and status, 363, 365
bureaucratic organizations, 489
bystander effect, 290–292, *291*
bystander intervention, 42–43, 280–282, 288–294, *290, 291*

C

careers and work: in achieving independence stage, 441; in adult role performance stage, 449–450; anomie and employment, 496–497; in coping with loss stage, 452–453; definition of, 434; in family-work balancing stage, 442–445; health and occupational roles, 475–476; labeling and discrimination, 521–522; maternal employment, effects on children, 58; mental health and occupational stress, 479–481; occupational segregation, 469–471; occupational status, 463–464, *464;* and role acquisition, 78–79; self-estrangement and job satisfaction, 488–490, *489;* socialization and work orientations, 77–78; upward mobility, *167,* 465–471, *466;* values and occupational roles, 473–474; women's employment, 455–457, *457,* 468, 469–471; work-family conflict, 484–485
caregiving, 58–60, 444, 450

caring, morality of, 77
categorization: and cognitive competence, 73; definition of, 110; and intergroup conflict, 413; and schemas, 110–111; of self, 91
catharsis, 313–315, *314*
Catholic worker houses, 262–263
CATI (computer assisted telephone interviewing), 31
causal analysis of survey data, 35–36
causal hypotheses, 28
cause, attribution of. *See* attribution
cautious shifts, 405
celebration rituals, 532
Census data, 39
central route persuasion, 199, 210
chaos theory, 347
charismatic leadership, 383
child care, 58–60, 444
children: and aggression, 304; developmental processes, 54–55, *55,* 56; divorce, effects of, 61–63; and gender role socialization, 70; grandchildren and grandparents, 453; living arrangements of, *60;* and maternal employment, 58; mother, relationship with, 57–58; and self-esteem, 101–102; and social learning, 55–56. *See also* parenting; socialization
Chinese Americans, 103
choice dilemmas, 404–405
civil disorders, 536–540
civil rights movement, 458–459
CL (comparison level), 334, 347–348
CL_{alt} (comparison level for alternatives), 334, 348
Clamshell Alliance, 550
class. *See* social class
classical conditioning, 143–144
coalitions, 374, 382–383, 548–549
codability, 168–169
coerced dispersal, 532
cognition needs of targets, 210
cognitions, 142, 146
cognitive conflict tasks, 388–389
cognitive consistency: overview, 146–147; and attitude similarity, 337; and balance theory, 147–148, *148;* definition of, 15; reasons for, 151–153; and theory of cognitive dissonance, 149–151
cognitive dissonance: counterattitudinal behavior, 150–151; definition of, 149; and Festinger, 7; and intergroup contact, 425–426; postdecision dissonance, *149,* 149–150; sales techniques using, 153
cognitive labeling theory, 261–263
cognitive processes, 14. *See also* information processing
cognitive structure, 14, 144–146, *146*
cognitive theory, 14–16, 20–22, 36
cohesion of groups. *See* group cohesion
cohorts, 438–439
collective behavior: crowds, 526–530; definition and dimensions of, 525; gatherings, 530–532; grievances, 534–535; precipitating events, 535–536; riots, empirical studies of, 536–540; routine vs. unroutine, 533; underlying causes of, 532–535. *See also* social movements
collective identity, 544
collective rationalization, 401, 403
collective rituals, 545
collectivist cultures, 91, 138–139, 257–260, *258*
college, 441
commitment, 349–350, 545–546
Committee of 100 (Great Britain), 551
commonality of effects, 129
common in-group identity model, 424–425
commonsense knowledge, 2
communication: breakdowns in, 421; conversational analysis, 190–194; definition of terms, 166; encoder-decoder model, 168–170, *169;* and gender, 180–181; intentionalist model, 170–172; and intimacy, 185, 187–188; linguistic communication, 167–168; linguistic relativity hypothesis, 172–173; and mediation, 427–428; and normative distances, 188–190; perspective-taking model, 172–175; roles in group communication, 358–359, *359;* rumor, 530; social stratification and standard vs. nonstandard speech, 181–185; and status, 184, 185–187; verbal insults, 509. *See also* language; nonverbal communication
communication accuracy, 168–169, 170–171, 173–174, 178
communication-persuasion paradigm, 199–200, *200*
communicator credibility, 200–202, 203–204, *204*
companion clusters, 531–532
comparison, social, 65, 102–103, 106
competence and group conformity, 373–374
competitive collective action, 535
compliance: definition of, 197, 210; obedience to authority, 217–221; and social power, 212–213; with threats and promises, 210–215, *215*
compromise, 216
computer-based causal analysis, 35
concessions, 216
conciliation, 551
conciliatory initiatives, unilateral, 429–430
conditioning, 11–12, 65–68, 143–144
confidentiality, breach of, 48–49
conflict, intergroup. *See* intergroup conflict
conflict resolution programs, 313
conflict spirals, 216–217
conformity: Asch conformity paradigm, *368,* 368–369; and impact of groups on individuals, 4; increase factors, 371–374, *373;* and intergroup conflict, 422–423; and role theory, 10
conjunctive tasks, 390–392, *392*
connotative meaning, 34
consensus, 16, 18, 131
conservatism in movement organizations, 550–551
conservative coalitions, 382–383
consistency, 97–98, 131, 143. *See also* cognitive consistency
construct validity, 32
contact, intergroup, 425–426
contagion, 526–527
content analysis, 40
contest/battle tasks, 389
contingencies of self-esteem, 101
contingency model of leadership effectiveness, 383–387, *385*
contingent reinforcement, 215
control, social, 514–522, *518,* 540, 551
control theory, 498–500, *500*
convergence, 531
conversational analysis, 190–194
conversational scripts, 169
conversion, 546
conversion theory, 374–375
cooling out, 244–245
cooperation, 16, 18
cooperative principle, 171–172
CORE (Congress of Racial Equality), 547
coronary heart disease, 476
coronary prone behavior patterns, 476
corporal punishment, 66, *67,* 304
correspondence, 158
cost-reward motivation, 275–276
costs and bystander intervention, 292–294, *294*

counterattitudinal behavior, 150–151
countermobilization, 550
courtship signaling, 182–183, *183*
covariation, principle of, 131–132
creativity tasks, 388
credibility, 200–202, 203–204, *204,* 212–215
crime. *See* deviant behavior
criminal justice system, 313, 514–515, *518,* 519–520. *See also* labeling of deviance
criterion validity, 32
crowds, 526–530
Cuban missile crisis, 134–135
cultural routines, 56
culture: and cognitive consistency, 153; and communication, 168, 177, 184; and emotions, *255,* 255–260, *258;* and gender, 70–71, 121; and identity adoption, 91; individualist vs. collectivist cultures, 91, 138–139, 257–260, *258;* interpretive reproduction by children, 56–57; and normative distances, 188, 189; and representative samples, 47; schemas as cultural elements, 114; and self-esteem, 103; and sexual assault, 315, 318

D

data, archival, 39–40
date rape, 318
dating. *See* interpersonal relationships and dating
death and grief, *258,* 259–260, 451, 452, 454
death instinct, 301
deception, 137, 239–242
decision making in groups, 400–406
decision making tasks, 388
defense, refutational, 222
deficit theories, 184
definition of the situation, 227–229, 530
deindividuation, 526
delinquency. *See* deviant behavior
demand characteristics, 45
denial, 506
dependent variables, 28–29. *See also* variables
depression, 337, 454. *See also* mental health
deprivation, absolute, 536–537
deprivation, relative, 533–534, 536–537
description, 28, 86–87, *87*
desensitization to violence, 322
deservingness and helping, 280–282
destructiveness and crowds, 528
determinism, principle of, 22, 24
deterrence, mutual, 216–217
deterrence hypothesis, 515–516
development: cognitive, 73–74; linguistic, 72–73; moral, 74–77, *76;* processes of, 54–55, *55,* 56; self-differentiation, 85–86
developmental perspective on socialization, 54–55
deviant behavior: definition of, 494; differential association theory, 501–502; formal social controls, 514–522, *518;* institutionalization of, 509–510; legitimate means and anomie theory, 495–498; reactions and labeling, 503–510; and role theory, 10–11; routine activities perspective, 503; secondary deviance and subcultures, 513–514; social bonds and control theory, 498–500, *500;* societal reaction and stigmatized identity, 510–513, *512;* suicide and power of suggestion, 504–505; unanticipated, 510. *See also* aggression and violence
deviant subcultures, 513–514
diagnosis of mental illness, 109
dialogue, 73, 85. *See also* language
differential association theory, 501–502
differential rewarding, *396,* 396–397
diffuse status characteristics, 360, 364–365
diffusion of responsibility, 292
direct experience, attitudes based on, 155–156
disclaimers, 235–236
discrepant messages, 203–204, *204*
discrimination: ageism, 452; definition of, 413; intergroup, 413–414; and labeling, 521–522; and modern vs. old-fashioned racism and sexism, 160; and schemas, 15. *See also* prejudice; race and ethnicity
disinhibition, 314–315
disjunctive tasks, 390–392, *392*
dispersal, 532
displaced aggression, 308
display rules, 257
dispositional attributions, 127–128, 132–133
dissent, 374–375, 401, 403, 423. *See also* conformity
dissonance effect, 151
dissonance theory. *See* cognitive dissonance
distinctiveness and covariation, 131
distraction, 210
distributive justice principles, 393–394
diversity and minorities, 47–48, 60–61, 104. *See also* prejudice; race and ethnicity
divorce, 36, 61–63, 443
dolls, 68–69
domestic violence, 305, *305,* 306–307, 308, 311
dominant communication styles, 186
drug use, 502
dual-process model, 376
"Dude," 188
dyadic withdrawal, 341–342

E

ebonics, 184–185
ECI (Emotional Competence Inventory), 268
economic frustration, 312–313
economic strain, 533
economic uncertainty, 481
education, 465–467, *467,* 474, 496–497. *See also* schools
effective social control, 510
ego centrality and self-esteem threat, 296
egoism, 275–276
elaboration likelihood model, 199
embarrassment, 242–244, 267
emergency dispersal, 532
emergency intervention by bystanders, 42–43, 280–282, 288–294, *290, 291*
emergent norms, 528–530
emerging adulthood, 440
emotional appeals, 204–205
emotional cohesion, 357
emotional intelligence, 268–269
emotional loneliness, 351
emotional support, 485
emotions: and crowds, 528, 530; cultural differences in display of, 257–260, *258;* definition of terms, 251–252; emotional intelligence, 268–269; emotion work and feeling rules, 270–272, *271;* facial expressions of, 253–256, *255,* 257; and impression management, 230–231; origins, classical theories on, 252–253; physiological reactions and cognitive labeling, 260–263; social, 264–270; and social movements, 262–263
emotion work, 270–272, *271*
empathy, 264, 266–267, 276–277

empathy-altruism model, 277
empirical research, 27–29
employment. *See* careers and work
encoder-decoder model, 168–170, *169*
endorsement, 381–382
equality principle, 393–394
Equal Rights Amendment (ERA), 550
equal-status contact, 425–426, 427
equitable relationships, 348
equity, 13, 393–398
equity principle, 393–394
equity theory, 348, 394–395
ERA (Equal Rights Amendment), 550
erosion of support, 546
Eskimo culture, 257, 259
esteem support, 485
ethical issues in research, 48–50
ethnic identity, 93
ethnicity. *See* race and ethnicity
ethnic stereotypes, 117–119
ethnocentrism, 412–413, *413*
European (White) Americans: and child rearing, 60–61; and competitive collective action, 535; and employment discrimination, 521; and gender roles, 47; and housework by men, 445; and mental health, 487; and research methods, 47–48; and schematic judgment, 113; and self-esteem, 104; stereotyping of Blacks by, 115, 126–127. *See also* race and ethnicity
evaluation: and attitudes, 142; and attitudinal similarity, 337; group productivity and evaluation biases, 420; and mental maps, 123; of relationships, 348; of self, 89, 100, 105–106, 441–442
evaluation apprehension, 291–292
evaluative dimension of norms, 367
events, precipitating, 535–536
event schemas (scripts), 111–112, *112,* 169, 334–335, 346
evolutionary theory: and attractiveness, 333; behavior, evolutionary foundations of, 18–19; in comparison, 20–22; and helping, 277–279; limitations of, 20; as unifying theory, 20
exchange processes and attraction, 333–335
exchange theory. *See* social exchange theory
excuses, 236–237
expansion of issues, 420–421
expectation states theory, 363–364
experience, personal, 2
experimental realism, 45
experimenter effects, 45
experiments, 41–45, *46*
expertise, 200–201, 213
Expression of the Emotions in Man and Animals, The (Darwin), 253
external attribution, 132–133
external validity, 29, 39, 45
extraneous variables, 29, 41, 43
extrinsically motivated behavior, 68
eye contact, 190

F

face saving, 244
face validity, 32
facial features and expressions: attractiveness of, 332; emotional, 253–256, *255,* 257; and nonverbal communication, 177–178, 180
facial maturity, 186
factionalization, 550
failures: attribution for, 132–134, *133,* 138; and leader endorsement, 381; and self-efficacy, 68
falling in love, 346
familiarity, 328–329
family: and norms, 74; peer groups vs., 63–64; and self-esteem, 101–102; sibling role, 450; and socialization, 57–63; work vs., 442–449, 484–485. *See also* parenting
fateful events, 482–483
fathers, 60, 443–444. *See also* parenting
fear arousal, 204–208
feedback: back channel, 193, 447; in conversation, 192–194; negative, 48; self-confirming, 97–98; and self-esteem, 102, 105; and self-schema, 99
feeling rules, 270
feminism, 549
feudalism, 533
field experiments, 41–42, *46*
field studies, 37–39, *46*
first date scripts, 335
first impressions, 124–125
fitness, 475–476
fixed-interval reinforcement schedules, 66
flattery, 233
flight attendants, 270–271, *271*
flirting, 182–183, *183*
focus of attention bias, *135,* 135–136
football games, 525–527, 531, 532
foot-in-the-door technique, 153
forewarning and persuasion, 222
formal social controls, 514–522, *518*
frames, 227–228
framing, 542
French Revolution, 533
friends, choosing, 333–334. *See also* interpersonal relationships and dating
front regions, 232
frustration, 301–302, *302,* 312–313, 534
frustration-aggression hypothesis, 301–302, 416
fundamental attribution error, 135, 136, 137, 138–139
fundamental beliefs, 145

G

game stage, 88
gang violence, 509
gatherings, 530–532
gay rights movement, 542
gender roles and gender differences: in achieving independence stage, 441, 442; in adult role performance stage, 449; and aggression, *305,* 305–306, 315–319; and communication, 180–181, 446–447; in coping with loss stage, 452; and dating, 112; and deviance, 503, 507–508; and empathy, 267; in family-work balancing stage, 443–446; and health, 476–477, 481; and helping, 286; and language socialization, 73; and life course expectations, 440; and mate selection, 18, 19; and math skills, 470; modern vs. old-fashioned sexism, 160; and moral development, 77; and parenting, 58; and peer groups, 62–63; and personal identity, 441; and playing dumb, 234; and self-disclosure, 229, 339; and self-esteem, 102; and situated self, 93; socialization of, 69–71; and status, 361–362, 467–469; stereotypes and gender schemas, 118, 119, 120–121; and stress and satisfaction, 442; and women's employment, 455–457, *457,* 468, 469–471
generalized other, 88–89
general strain theory, 498
genetic fitness and helping, 278–279
genetics and emotions, 253. *See also* evolutionary theory
gestalt psychology, 14
global self-esteem, 101
goal isomorphism, 358

goals: commitment to long-term goals, 498, 499; and deviance, 495; and emotions, 252; group goal effect, 392–393; and group structure, 358
Graduated Reciprocated Initiatives in Tension Reduction (GRIT), 429–430
Graduate Record Examination (GRE), 118
graffiti, 293
grammaticization, 72
grandchildren and grandparents, 453
GRE (Graduate Record Examination), 118
grief, *258*, 259–260, 451
grievances, 534–535
group cohesion: consequences of, 356–358; definition of, 355–356; and deviant behavior, 508; and intergroup conflict, 421–422; and labeling, 517; nature and types of, 356
group goal effect, 392–393
group leadership. *See* leadership
group polarization, 405
group productivity: and brainstorming, 398–400; decision making, 388, 400–406; definition of, 379; evaluation biases, 420; group goal effect, 392–393; and reward distribution, 393–398; and social facilitation, 387–388; and task interdependence, 395–397, *396;* task types, 388–389; and transformational leadership, 383
groups: communication in, 181, 186; conformity in, 368–374; definition of, 355; formal vs. informal, 380; in gatherings, 531; and helping, 281–282; impact of groups on, 4–5; impact of individuals on, 4; impact on individuals, 4; linguistic intergroup bias, 174; minority influence in, 374–376; norms in, 365–368; in role theory, 9–10; and social identities, 90–91; status in, 360–365, *362. See also* intergroup conflict; peer groups
group schemas. *See* stereotypes (group schemas)
group self-esteem, 104
group structure: and goals, 358, 379; hierarchical, 379; and intergroup conflict, 421; and productivity, 387–393; size, 388–392, *390, 392;* and status generalization, 361; status structure in a work group, *380. See also* leadership
groupthink, 400–404, *404*
GSR (galvanic skin response), 149
guilt, *258*, 264–265, 288

H

hair, 231
halo effect, 115
handicaps and stigma, 247
handshakes, 176
happiness, *258*, 259
harm from research, risk of, 48, 50
Hatfield-McCoy feud, 416
Hay Group, 268
health, mental, 179–185, 490
health, physical, 238–239, 454, 475–479, *478*
heart disease, 476
heat and violence, 537–539, *538*
hedonism, 12
helping. *See* prosocial behavior, helping, and altruism
heuristic function of attitudes, 144
heuristics, 126–127
hierarchical structure, 379, 381
high-risk activism, *547*
Hispanics, 70. *See also* race and ethnicity
history and the life course, 437–439, *439,* 455–459
Holocaust, 219
homogamy norm, 329–330
homogeneity and outgroups, 417–418
horizontal structure, 145–146, *146*
household work, 444–445, *445*
human nature, 22
Human Subjects Committee, 50
humor, 263
hypotheses, 28–29

I

ideal impression management, 226
identity and identities: in achieving independence stage, 441; in adult role performance stage, 450; aligning actions to repair, 235–237; altercasting of, 237; bargaining over, 237; choosing among, 95–97; collective, 544; consistency in, 97–98; cooling out and identity degradation, 244–245; in coping with loss stage, 453; definition of, 93; enactment of, 93–98; in family-work balancing stage, 445–446; and group norms, 367; and intergroup conflict, 412–415; and life course, 434; multiracial, 92–93; and normative transitions, 437; and peer associations, 64; role identities, 89–90; salience hierarchy of, 95–96, 97; selective commitment to, 106; situated identities, 228–229; and situated self, 93; social, 90–91; spoiled or discredited, 242–248; stigmatized, 511–512; superordinate social identity, 424; support, need for, 96. *See also* self
identity control theory, 95
identity crisis, 97
identity degradation, 245
ideology and social movements, 541–542, 543, 545
illegitimate means, 496
illusion of out-group homogeneity, 417–418
illusions of invulnerability, 401, 403
illusions of morality, 401, 403
imbalanced systems, 147, *148*
imitation, 12, 304, 321, 526–527
implicational relationship, 151
implicit personality theories, 114–116
impression formation, 122–125
impression management. *See* self-presentation and impression management
incentive effect, 151
inconsistency in communication, 179–180
independence, achieving, 440–442
independent variables, 28–29. *See also* variables
independent verification, 27
individualist cultures, 91, 138–139; and emotions, 257–260, *258*
individuals, 3–4, 17
inference, schematic, 113
influence. *See* social influence and persuasion
informal social control, 514
information, 136–137, 213
informational influence, 370–371
information integration and impressions, 123–125
information processing, 99, 105–106, 161
informed consent, 39, 50
ingratiation, 232–235, 240
in-groups and out-groups: and aversive events, 416; and helping, 281–282; identification and ethnocentrism, 412–414, *413;* and linguistic intergroup bias, 174; and moral identity, 419; perception biases, 417–420; and silencing, 187. *See also* intergroup conflict
inner-directedness, 130, *130*
innovation, 4, 496
inoculation against persuasion, 221–222
instinct theories of aggression, 301
institutionalization of deviance, 509–510

institutionalization of social movements, 551
Institutional Review Board, 50
instrumental commitment, 545
instrumental conditioning, 65–68, 143
instrumental function of attitudes, 144
instrumental support, 485
insults, 509
integrative solutions, 216
intellective tasks, 388
intelligence, emotional, 268–269
intentionalist model, 170–172
intentions, behavioral, 161–163
interdependence, 341–343, 395–397, *396*
intergenerational mobility, *167,* 465–471, *466*
intergenerational similarity and socialization, 53
intergroup conflict: definition of, 410; development of, 411–416; group processes, impact on, 421–423; and impact of groups on groups, 4–5; overview, 409–411; persistence of, 416–421; resolution of, 424–430
intergroup contact hypothesis, 425–426
internal attribution, 132–133
internal dialogue, 85
internalization, 69
internal validity, 29, 43, 44–45
interpersonal attraction: availability, 327–329; definition of, 326; desirability, 329–335; liking, determinants of, 335–338. *See also* physical attractiveness
interpersonal context, 174–175, 184
interpersonal development, *55*
interpersonal relationships and dating: breakups, 347–352; deception in, 239; degree of mutuality, 338–339; equitable relationships, 348; exchange processes, 333–335; first date event schema, 111–112, *112;* flirting, 182–183; identity bargaining, 237; interdependence in, 341–343; jealousy in, 265–266; and loving, 343–347; normative distance, 188–190; primary relationships, 462; and self-disclosure, 339–340; and self-esteem, 105; singles ads, 331; and social exchange theory, 12–13; in social exchange theory, 12–13; and trust, 340–341, *341, 342*
interpersonal spacing (proxemics), 176, *176,* 188
interpretive perspective on socialization, 56–57
interruptions, 446
intersubjectivity, 173, 175
interview surveys, 30–31
intimacy, 185, 187–188, 339–340, *340,* 342–343
intimate distance, 189
intrinsically motivated behavior, 68
involvement, 208–210, *209,* 498, 499
invulnerability, illusions of, 401, 403
Iraq War, 402–403
irrationality and crowds, 527–528

J

Japanese culture, 258–259
J-curve theory, 533, *534*
jealousy, 265–266
judgment, moral, 76–77
judgment, schematic, 113
justice, morality of, 77
justification and media violence, 322
justifications, 236–237
juvenile justice system, 515, 520

K

Kanawha County textbook controversy (West Virginia), 409–410, 416, 420–421
kinesics (body language), 176, *176,* 186, 188
kin selection, 278

L

labeling of deviance: consequences of, 509–510; formal labeling and creation of deviance, 515–520, *518;* long-term effects of, 520–521; as redefinition of social status, 506; and stereotypes, 508
labeling of emotions, cognitive, 261–263
labeling theory, 506, 521
laboratory experiments, 41–42, *46*
labor strikes, 534
language: advantages of use of, 168; components of, 167–168; conversation, initiation of, 190–192; development and socialization of, 72–73; disclaimers, 235–236; encoder-decoder model, 168–170, *169;* feedback and coordination, 192–194; intentionalist model, 170–172; internal dialogue, 85; linguistic relativity hypothesis, 172–173; norms, linguistic, 74; and person schemas, 116; perspective-taking model, 172–175; and role taking, 87; and self-differentiation, 86; and social learning, 56; social stratification and standard vs. nonstandard speech, 181–185; speech accommodation theory, 187–188; speech repertoires and status, 185; spoken language, defined, 167; turn-taking in conversation, 192; verbal and nonverbal communication, combining, 178–180. *See also* communication
latency of responses, 145–146, 155
Latinos, 70. *See also* race and ethnicity
leadership: activities of leaders, 383; authority in formal groups, 380–381; competition and militancy, 422; contingency model of effectiveness, 383–387, *385;* definition of, 383; endorsement of formal leaders, 381–382; and groupthink, 401, 403; and impact of individuals on groups, 4; and movement organizations, 545; perceptions of, in small groups, 92; promotional, 401; and revolutionary and conservative coalitions, 382–383; and roles, 359; task specialists and socio-emotional specialists, 359–360; transactional and transformational, 383
learning: instrumental, 65–68; observational, 68–69; reinforcement in, 11–12; schools, 64–65, 71, 75; social learning theory, 12, 20, 55–56, 304. *See also* education; socialization
learning structure, 496
lectures and feedback, 193
legitimate means, 495
legitimate power, 212–213
legitimation and media violence, 322
life course: achieving independence (stage I), 440–441; balancing family and work commitments (stage II), 442–449; and biological aging, 435–436; components of, 434–435; coping with loss (stage IV), 451–455; definition of, 433; and history, 437–439, *439,* 455–459; and life events, 435; performing adult roles (stage III), 449–451; and social age grading, 436–437
life events, 435
Likert scales, 33, 34
liking: and familiarity, 329; and helping, 279–280; and persuasion, 202; reciprocal, 338; and self-disclosure, 229; and shared activities, 337–338; and similarity, 335–337. *See also* interpersonal attraction

linguistic intergroup bias, 174
linguistic relativity hypothesis, 172–173
littering, 293
local norms, 494–495
loneliness, 229–230, 351
longitudinal surveys, 36
looking-glass self, 87
Los Angeles riots, 309–310, 535–536, 539
loss, life stage of coping with, 451–455
love and loving: as emotion, 267, 270; liking vs., 343; passionate love, 343–345; research questions on, 44; as story, 345–347
love stories, 345–347
low-balling technique, 153
lying, 137, 240–242

M

majority influence, 368–369, 371, 375–376
manipulation, experimental, 48
manipulative influence, 197
manual dexterity, development of, *55*
marriage: death of spouse, 452, 454; delay in, *442,* 442–443; divorce, 36, 61–63, 443; and mental health, 481–485; and physical health, 477; and retirement, 452; stress and satisfaction in, 447–449, *448,* 450, 477, 484; transition to, 436–437
mass media: and aggression, 320–324, *323;* and attitude formation, 144; definition of, 206; and emotions, universality of, 255–256; and front vs. back regions, 232; and gender role socialization, 71; media campaigns and persuasion, 206–207; and social movements, 543, 548–549; suicide and power of suggestion, 504–505
mate poaching, 19
mate selection in evolutionary theory, 18
mating. *See* interpersonal relationships and dating
meanings: connotative, 34; and identities, 94–95; in intentionist model of communication, 170–171; negotiated, 16
media. *See* mass media
media campaigns, 206–207
mediators, 426–429
membership, 90, 91. *See also* groups
memory, 99, 106, 112–113
men. *See* gender roles and gender differences
menopause, 450
mental health, 179–185, 490
mental illness: depression, 337, 454; diagnosis of, 109, 507–508; impact of labeling, 522; reactions to, 506–507, 512. *See also* labeling of deviance
mental maps, 115, *116,* 123
mere exposure effect, 328
message discrepancy, 203–204, *204*
messages: discrepant, 203–204, *204;* and encoder-decoder model, 168, 170; one-sided vs. two-sided, 208; and persuasion, 198–199, 200, *201,* 203–208
meta-analysis, 46–47
methodology, 23, 27. *See also* research methods
middle-range theories, 8
militancy, 422
mindguarding, 401, 403
minimal risk, 50
minorities. *See* diversity and minorities; race and ethnicity
minority influence, 374–376
mixed motive tasks, 389
mobility, intergenerational, *167,* 465–471, *466*
mobilization, 544, 545–546, *547*
mob psychology, 527, 529
modeling, behavioral, 285–286, 309–310. *See also* observational learning
mood, 252, 286–288, *287,* 337
moral commitment, 545
moral development, 74–77, *76*
moral identity, 419
morality, illusions of, 401, 403
morphology, 167–168
mothers, 57–58, 444, 447–448. *See also* parenting
motivational attribution biases, 137–138
motives, ulterior, 239–240
mourning, 260
movement organizations, 545–551
MTV, 71
multiple-source effect, 202–203
mundane realism, 45
mutual deterrence, 216–217
mutuality, degrees of, 338–339

N

NAACP (National Association for the Advancement of Colored People), 546–548
NAOWS (National Association Opposed to Women Suffrage), 550
National Advisory Commission on Civil Disorders, 536, 537
naturalistic observation, 37–39
negative feedback, 48
negative reinforcement, 66
negotiated meanings, 16
negotiation, 216, 342
network multiplicity, 502
networks, social. *See* social networks
New Guinea, 256
newspaper singles ads, 331
NFLU (National Farm Labor Union), 546
Nineteen Eighty-Four (Orwell), 172–173
nonaggressive models, 313
nonaggressive pornography, 319
noncausal hypotheses, 28–29
nonroutine collective action, 533
nonstandard speech, 181–185
nonverbal communication: and culture, 177; deception cues, 240–242; embarrassment indicators, 243–244; facial features and expressions, 177–178; flirting, 182–183, *183;* and gender, 447; multiple cues, 178–180; types of, 175–177, *176*
normalization, 506
"normals" and stigma, 245–248
normative distances, 188–190
normative influence, 369–370
normative life stages, 436. *See also* life course
normativeness of effects, 129–130
normative transitions, 437, 439
norms: and aggression, 306–308; of communicative behavior, 174; definition of, 74, 365; and deviant behavior, 494–495; emergent, 528–530; emotional expression and culture, 257; group norms, 365–368; and helping, 282–285; homogamy norm and desirability, 329–330; and intergroup conflict, 422–423; local, subcultural, and societal, 494–495; and moral development, 74–76; personal, 283–285, *284;* privacy and embarrassment, 243; of reciprocity, 229, 239–240, 283, 295–296; in role theory, 10; of self-reliance, 295; situational, 164; of social responsibility, 282–283; subjective norm, 161–163, *162, 163;* and violence, 540. *See also* deviant behavior
Northern Ireland, 418–420
nudists, 513–514

O

Oak Valley school and civil rights, 458–459
observation, 22–23, 37–38
observational learning, 12, 68–69, 144. *See also* modeling, behavioral
occupational roles. *See* careers and work
occupational segregation, 469–471
occupational status, 463–464, *464*
one-sided messages, 208
open influence, 197
opinion conformity, 233
opportunity structure, 496
organization-environment relations, 546–550
other, generalized, 88–89
other-directedness, 130, *130*
other enhancement, 233–234
outcomes and relationships, 333–334, 347–349
out-groups. *See* in-groups and out-groups
overgeneralization, 117, 119
overreward, responses to, 397–398

P

pair relatedness, 326
panel studies, 36
paralanguage: definition of, 175–176, *176;* and multiple cues, 178, 180; and speech accommodation, 187; and status, 186
parenting: and aggression, 304; and gender roles, 58; and punishment, 66–68, *67;* and self-direction, 78; and self-esteem, 101–102; and stress, 447–448; transition to, 443
participant observation, 38
passing, 248, 414
passionate love, 343–345, *346*
peace movement, 545, 546
peer groups: and attitude formation, 143; attitudes and social support, congruence of, 161; and deviance, 502, 503; and norms, 74–75; and socialization, 62–64
perceived behavioral control, 163–164
perceptual distortion, 398
performance feedback and self-esteem, 102. *See also* failures; successes
performance of groups. *See* group productivity
performance/psychomotor tasks, 389
peripheral route persuasion, 199
personal distance, 189–190, *190*
personal effects and nonverbal communication, *176,* 176–177
personality disorders, 507–508
personality theories, implicit, 114–116
personality traits, 122–123, 210, 450–451
personal norms, 283–285, *284*
person schemas, 111, 114–116, 228. *See also* stereotypes (group schemas)
perspective-taking model, 172–175
persuasion, defined, 198. *See also* social influence and persuasion
persuasive argument theory, 405–406
phonetics, 167–168
physical appearance, management of, 231–232
physical attractiveness: attractiveness stereotype, 332–333; desirability and matching hypothesis, 330–332; and evaluation of relationships, 348; and mate selection, 18, 19; and persuasion, 202; and self-appraisals, 92. *See also* interpersonal attraction
physical health, 238–239, 454, 475–479, *478*
physiognomy, 178
planning tasks, 388
playing dumb, 234
play stage, 88
plea bargaining, 520
polarization, 405, 421
police-citizen encounters, 518
political alienation, 491
political scandals, 236–237
populations and sample selection, 34, 35
pornography, 319–320, 548–549
position, 462
positive reinforcement, 66
postdecision dissonance, *149,* 149–150
power, 212–213, 217
powerlessness, 490–491
precipitating events, 535–536
prediction, 23–24, 154
predisposition, behavioral, 142–143
prejudice: ageism, 452; definition of, 144; and intergroup contact, 425–426; modern vs. old-fashioned racism and sexism, 160, 161; and stereotyping, 144; and survey sampling, 34
prestige, 464. *See also* status
pride, 266
primacy effect, 124–125
primary relationships, 462
priming, 320, 321–322, 372
primitive belief, 145
privacy and embarrassment, 243
private speech, 72–73
proactive collective action, 535
production blocking, 399–400
productivity of groups. *See* group productivity
professionalization, 550
pro-life movement, 542, 543
promises, 198, 210–215, *215*
promotional leadership, 401
props, 231–232
prosocial behavior, helping, and altruism: bystander emergency intervention, 42–43, 280–282, 288–294, *290, 291;* definition of terms, 274–285; and empathy, 267; in evolutionary theory, 18–19; motivation to help, 275–279; needy, characteristics of, 279–282; normative factors, 282–285; personal and situational factors, 285–288, *287;* seeking and receiving help, 295–297
prototypes, 110
proxemics (interpersonal spacing), 176, *176,* 188
proximity and attraction, 327–328
psychoanalytic theory, 57
psychology and social psychology, 5
PTSD (post-traumatic stress disorder), 482–483
public distance, 189
public opinion polls, 30, 35
punishment: and aggression, 304, 313; deterrence hypothesis, 515–516; and instrumental conditioning, 66–68, *67;* in role theory, 10. *See also* threats
pyramid structure, 381

Q

questionnaire surveys, 31

R

race and ethnicity: and aggression, 305; and attribution, 110, 138; and child rearing, 60–61; and civil disturbance, 539; and civil rights movement, 458–459; and competitive collective action, 535; and crime, 497–498; and education, 467; and employment discrimination, 521; equal-status contact and interracial attitudes, 426, 427; and homogamy norm, 329–330; and housework, 445; and marriage delay, 442–443; and mental health, 484; modern

race and ethnicity (*continued*)
vs. old-fashioned racism, 160; multiracial heritage and identity, 92–93; and occupational attainment, 471; and representativeness heuristic, 126–127; and research methods, 47–48; and schematic judgment, 113; and self-esteem, 104; and situated self, 93; and social control bias, 520; and social support, 487; and status generalization, 361–362, 365; and stereotyping, 115, 117–119
radicalization, 551
random assignment, 41
rape, 306, 315–319, *318*
rape-prone societies, 315
rational appeals, 204
rationality, 17
rationalization, collective, 401, 403
reactance, 222
reactive collective action, 535
realism, mundane vs. experimental, 45
realistic group conflict theory, 411–412
reasoned action, theory of, 161–164, *162, 163*
rebellion and anomie, 496
recency effect, 125
reciprocal liking, 338
reciprocity: and intimacy, 339, *340;* norm of, 229, 239, 283, 295–296
recruitment for social movements, 543, *547*
referent power, 212
reflected appraisals, 91–92, 99
reflexive behavior, 85
refutational defense, 222
regions, front and back, 232
reinforcement: and aggression, 304, 309; and attitude similarity, 337; contingent, 215; definition of, 11; and differential association, 501–502
reinforcement schedules, 66
reinforcement theory, 11–13, 20–22
relationship-oriented leaders, 384–387, *385*
relative deprivation, 533–534, 536–537
relative needs principle, 393–394
relevance, personal, 293
relevance of attitudes, 157
reliability, 31, 341
religion and gender role socialization, 70
religious movements, 543
representativeness heuristic, 126
representative samples, 47
reproduction and health, 476
research, empirical, 27–29
Research Ethics Committee, 50
research in diverse populations, 45–50
research methods: archival research and content analysis, 39–41; comparison of, 45–46, *46;* field studies and naturalistic observation, 37–39; surveys, 30–37
resilience, 488
resource mobilization, 545–546, *547*
resource scarcity, 534–535
response latency, 145–146, 155
response rates, 31, 35
responsibility, 290, 292, 526
retaliation, 244, 308
retirement, 79, 97, 452, 453
retreatism, 495–496, 504
return potential model, 367
revolutionary coalitions, 382
revolutions, 533
rewards: for aggression, 309; arousal/cost-reward model, 292, *294;* equity of distribution in groups, 393–398; group allocation of, 381–382; in role theory, 10. *See also* promises
rhetorical forms of speech, 40
riots, 309–310, 535–540
risk-benefit analysis, 50
risky shifts, 405
rites of passage, 80. *See also* life course
ritual, 263, 532, 545
ritualism, 495
road rage, 302
Robbers Cave study, 411–412, 420, 421, 424
Roe v. Wade, 541
role acquisition, 78–79
role differentiation, 360
role discontinuity, 79–80
role identities, 89–90, 91, 96. *See also* identity and identities
role overload, 476
roles: age-graded, 436–437; altercasting of, 237; and anticipatory socialization, 79; definition of, 9–10, 462; in groups, 358–360, 380; and life course, 434; out-of-role behaviors and inference, 130; parental vs. occupational, 443–444; and social structure, 462; and stereotypes, 119
role schemas, 111
role taking, 16, 87, 88, 266
role theory, 8–11, 20–22
romantic love ideal, 345, *346*
romantic relationships. *See* interpersonal relationships and dating
routine activities and attraction, 327
routine activities perspective, 503
routine collective action, 533
routine dispersal, 532
routines, cultural, 56
rule breaking, 506. *See also* deviant behavior
rules, social, 74–76
rumor, 530

S

sales techniques and cognitive dissonance, 153
salience hierarchy of identities, 95–96, 97
samples, 34–35, 47
Sapir-Whorf hypothesis, 172–173
sarcasm, 171–172
satisfaction: in achieving independence stage, 442; in coping with loss stage, 453–455, *454;* in family-work balancing stage, 448–449; and life course, 435
scapegoating, 516–517
schemas: and categorization, 110–111; and cognitive development, 73–74; and cognitive theory, 14–15; as cultural elements, 114; definition of, 14, 111; and formal labeling, 518; and heuristics, 126–127; impression formation, 122–127; person schemas, 111, 114–116; schematic processing, 112–114; types of, 111–112. *See also* stereotypes (group schemas)
schematic function of attitudes, 144
schematic inference, 113
schematic judgment, 113
schematic memory, 112–113
schools, 64–65, 71, 75. *See also* education
science, 22–24
scripts, 111–112, *112,* 169, 334–335, 346
secondary deviance, 513–514
secure attachment, 58, 59
segregation, occupational, 469–471
selective self-presentation, 235
self: nature and genesis of, 85–89; and private speech, 72–73; situated, 93; social origins of, *87,* 87–89; as source and object of action, 85; in symbolic interaction theory, 17. *See also* identity and identities
self-awareness, 99–100
self-categorization, 91
self-censorship, 401, 403
self-concept. *See* self-schema or self-concept

self-deprecation, 235
self-description, 86–87, *87*
self-differentiation, 85–87
self-direction, 78
self-disclosure, 229–230, 339–340, 425
self-discrepancies, 100–101
self-efficacy, 68
self-enhancement, 105, 235
self-esteem: in achieving independence stage, 441–442; in adult role performance stage, 450; assessment of, 101; and attitudes, 144; and behavior, 103–105; in coping with loss stage, 453; definition of, 101; in family-work balancing stage, 446; group self-esteem, 104; and life course, 434–435; protecting and maintaining, 105–106; receiving help as threat to, 296–297; sources of, 101–103
self-estrangement, 488–490, *489*
self-evaluation, 89, 100, 105–106, 441–442
self-fulfilling prophecies: and deviance labeling, 511, 513; on girls and math, 470; impressions as, 125–126; and perspective-taking model of communication, 174
self-identity, 16–17
selfishness and helping, 278
self-presentation and impression management: cooling-out and identity degradation, 244–245; and deception, detection of, 239–242; definition of, 226; embarrassment and saving face, 242–244; in everyday situations, 227–230; and risky behavior, 238–239; stigma, 245–248, *248;* tactical, 230–237; types of, 226–227
self-promotion, 235
self-reinforcement, 68
self-rejection, *512,* 512–513
self-reliance and receiving help, 295, 297
self-reports, 36–37
self-sacrificing behavior. *See* prosocial behavior, helping, and altruism
self-schema or self-concept: and attitudes, 144; consistency in, 97; definition of, 83, 111; and deviant status, 511–512, 513–514; information processing, influence on, 99; measurement of, 83–84; mirroring of, 101; and reflected appraisals, 91–92; and role identities, 89–90, 91; and self-discrepancies, 100–101; and shame, 265; and social identities, 90–91; and social identity theory of intergroup behavior, 414; social origins of, *87,* 87–89; social vs. personal, 94, *94. See also* identity and identities
self-serving bias, 138
self-verification strategies, 97–98
semantic differential scales, 33, 34
semantics, 167–168
sentiment, 252, 263
sentiment relations, 147
September 11 terrorist attacks, 482–483, 532
SEV (subjective expected value) model, 213–215
sex, casual, 19, 20
sex roles and differences. *See* gender roles and gender differences
sexual assault, 315–319, *318*
sexual gratification, 342–343
shame, *258,* 259, 265
shame-anger cycle, 265
shaping, 65–66
shared activities, 337–338, 342
shifts, risky vs. cautious, 405
shoplifting, 494, 496, 498–499, 501, 507
shyness, 351
sibling role, 450
significant others, 17, 87, *87,* 92, 102
silencing, 186–187
similarity, 280, 296, 336–337
simple random samples, 35
single-item scales, 33–34
single-process model, 375–376
singles ads, 331
situated identities, 228–229
situated self, 93
situation: and aggression, 309–312; and crowds, 529, 530; definition of the, 227–229, 530; and emotions, 257, 262–263; and reaction to deviance, 509
situational attributions, 127–128, 132–133
situational constraints, *159,* 159–161
situational identity choices, 96–97
size of groups, 388–392, *390, 392*
skin cancer, 238–239
SNCC (Student Nonviolent Coordinating Committee), 547
social age grading, 436–437
social bond and control theory, 498–500
social change and cohorts, 438–439
social class: and anomie, 496–498; and occupational status, 464; and physical health, 477–479; and powerlessness, 491; and social control bias, 518–519, 520; and stress, 487–488; and work orientations, 78
social cohesion, 356
social communication styles, 186
social comparison theory, 405
social control, 514–522, *518,* 540, 551
social desirability of effects, 129
social distance, 189
social emotions, 264–270
social exchange theory: definition of, 12; and evolutionary theory, 20; and household work, 444; and interpersonal attraction, 333–334; and reinforcement, 12–13
social facilitation, 387–388
social identities, 90–91. *See also* identity and identities
social identity theory of intergroup behavior, 414–415, 508
social impact theory, 202–203
social influence and persuasion: and attitude change, 198–210; and authority, 198; communication-persuasion paradigm, 199–200, *200;* compliance with threats and promises, 198, 210–215, *215;* definition of terms, 197, 198; forms of, 197–198; and group conformity, 368–373; in high-cohesion groups, 357; and individuals' impact on individuals, 3; informational influence, 370–371; majority influence, 368–369, 371, 375–376; and mass media, 206–207; messages, 200, *201,* 203–208; minority influence, 374–376; negotiation and bilateral threat, 216–217; normative influence, 369–370; obedience to authority, 217–221; resistance, 221–222; sources, 197, 200–203, *201;* targets, 197, 200, *201,* 208–210, *209*
socialization: adult, 78–80; anticipatory, 79; and attitude formation, 143; childhood agents of, 57–65; and cognitive development, 73–74; and emotions, management of, 230; empathy and gender, 267; and family, 57–63; and gender roles, 69–71; girls and math, 470; and impact of groups on individuals, 4; instrumental conditioning, 65–68; internalization, 69; and language development, 72–73; and moral development, 74–77, *76;* observational learning, 68–69; overview, 53–54; perspectives on, 54–57; role acquisition, 78–79; role discontinuity, 79–80; and role transitions, 437; and romantic love ideal, 345; and work orientations, 77–78

social learning theory, 12, 20, 55–56, 304
social loafing, 390
social loneliness, 351
social movements: consequences of, 551–552; definition of, 540–541; development of, 541–544; emotions and activism, 262–263; movement organizations, 545–551, *547. See also* collective behavior
social networks: and differential association, 502; and occupational attainment, 471; and role identities, 96; and social movements, 543; and social structure, 462, *463;* support from, 485–487
social perception: and attribution theory, 127–134; definition of, 109; impression formation, 122–127; overview, 109–110; and stereotypes, 119. *See also* schemas
social psychology, 3–7, *4,* 22–24
social responsibility norm, 282–283
social rules, 74–76
social situation. *See* situation
social stratification, 181–185
social structure: and alienation, 488–491, *489;* and communication, 180–190; definition of terms, 462; feudal, 533; historical events, impact of, 458–459; and mental health, 479–488; and physical health, 475–479, *478;* and socialization, 57; and status attainment, 463–471; and values, 471–474, *472*
societal norms, 495
socioeconomic background and occupational attainment, 465–467. *See also* status
socioeconomic status. *See* status
socio-emotional specialists, 359–360
sociolinguistic competence, 175
sociology and social psychology, 5
sociometry, 6
solidarity, 517
sororities, 357
sources and persuasion, 197, 200–203, *201*
South, culture of honor in, 307–308
space violations, 189–190, *190*
specific status characteristics, 360–361
speech accommodation, theory of, 187–188
speech act theory, 170–171
speech expansion, 72
speech perception and production, 72
spillover, 484–485
split-half method, 31
spoken language, defined, 167. *See also* language
spontaneity and crowds, 528
stability and instability: of attitudes, 157; of attributions, 132–133, *133;* of relationships, 347–349, 352
standard speech, 181–185
staring, 190
status: and coalitions, 382–383; and communication, 184, 185–187; equal-status contact, 425–426, 427; and gender, 181; and impression management, 233, 234; intergenerational mobility, *167,* 465–471, *466;* and interpersonal spacing, 189; and labeling, 506, 510–513; occupational, 463–464, *464;* and physical health, 477–479; and social control bias, 518–519; and social structure, 462; status characteristics, 360–361, 363–364; status generalization, 361–363, *362,* 364–365; and stereotyping, 119–120; and stress, 487–488. *See also* leadership
stereotypes (group schemas): of adversaries, 401; attractiveness stereotype, 332–333; and attribution bias, 138; common types of, 117–119; of crowds, 527; definition of, 111, 116–117; errors caused by, 119–122; in intergroup conflict, 418; and intergroup contact, 425–426; and labeling of deviance, 508; and mental maps, 115; origins of, 119; and overgeneralization, 117; and perspective-taking model of communication, 173–174; and prejudice, 144; and schemas, 15; and situated identities, 228; and social identities, 91, 95; and stigma, 248
stereotype threat, 117, 118
stigma, 245–248, *248,* 510–513, 521
stimuli, 11, 14
stimulus discrimination, 12
STIs (sexually transmitted infections), 238
strain and collective behavior, 532–533
stratified samples, 35
stress: in achieving independence stage, 442; in adult role performance stage, 450–451; definition of, 435, 480; in family-work balancing stage, 447–448; and gender, 481; and mental health, 479; and physical health, 475, 477; post-traumatic stress disorder (PTSD), 482–483; responses to, 480; support from social networks, 485–487; work-related, 479–481, 484
strikes, 534
subcultural norms, 495
subculture of violence, 501
subcultures, deviant, 513–514
subject effects, 45
subjective expected value (SEV) model, 213–215
subjective norm, 161–163, *162, 163*
submissive communication styles, 186
subtractive rule, 128, 130, 135
successes, 68, 132–134, *133,* 138
suggestibility and crowds, 528
suicide, 504–505
summons-answer sequence, 190–191
superordinate goals, 424–425
supplication, 234–235
support from social networks, 485–487
surveillance, 215, 221, 369
surveys: attitudes, measuring, 32–34; causal analysis of data, 35–36; compared to other methods, *46;* definition of, 30; and informed consent, 50; measurement reliability and validity, 31–32; purpose of, 30; questions, formulating, 32; samples, 34–35; strengths and weaknesses, 36–37; types of, 30–31
symbolic communication. *See* communication; language
symbolic interaction theory: in comparison, 20–22; definition of, 16; limitations of, 17–18; negotiated meanings in, 16; and reflexive behavior and the self, 85; self in relationship to others, 16–17; and surveys, 36
symbols, 166, 221
syntax, 167–168
systems of production, 489

T

tactical impression management: aligning actions, 235–237; altercasting, 237; and appearance management, 231–232; deception, detection of, 239–242; definition of, 226, 230; and emotions, 230–231; failures of, 242; ingratiation, 232–235, 240; playing dumb, 234; reasons for, 230
tag questions, 446–447
tape recording, 38
targets: of aggression, 304–308; of ingratiation and deception, 232–234, 239–

240; of persuasion, 197, 200, *201,* 208–210, *209;* of riots, 539–540
task cohesion, 356, 357
task interdependence, 395–397, *396*
task-oriented communication styles, 186
task-oriented leaders, 384–387, *385*
task specialists, 359–360
tattoos, 231
teachers and socialization, 64
telephone interviews, 31
television, 71, 320–323. *See also* mass media
temperature and violence, 537–539, *538*
temporal stability of attitudes, 157
terrorist attacks (9/11/01), 482–483, 532
test-retest method, 31
theoretical perspectives: cognitive theory, 14–16; comparison of, 20–22; definition of, 8; evolutionary theory, 18–20; middle-range theories vs., 8; reinforcement theory, 11–13; role theory, 8–11; symbolic interactionism, 16–18
theory, 23, 28, 47
therapy and stigma, 248
threats: bilateral, 216–217; definition of, 211; and influence, 198, 210–215, *215;* and social power, 212
trait centrality, 122–123
transactional leadership, 383
transformational leadership, 383
transitions in life. *See* life course
trust, 340–341, *341, 342,* 427
trustworthiness of a source, 201–202
turn taking in conversation, 192
two-sided messages, 208

U

UFW (United Farm Workers), 546
ulterior motives, 239–240
ultimate attribution error, 418–419
unanimity: apparent, 401, 403; and conformity, 371–373, *373;* in crowds, 526, 528
unanticipated deviance, 510
underreward, responses to, 397, 398
understatement, 171
unemployment, 481, 497
unilateral conciliatory initiatives, 429–430
unit relations, 147
universalistic norms, 75–76
universality of emotions, 255–256
unobtrusive measures, 38–39
upward mobility, *167,* 465–471, *466*
utterances, significance of, 170–171

V

Valentine's Day, 349–350
validity, 29, 32, 39, 40–41
values and social structure, 471–474, *472*
variable-interval reinforcement schedules, 66
variables, 28–29, 41–45
verbal communication. *See* communication; language
verification, independent, 27
vertical structure, 145, *146*
videodating services, 331
video games, violent, 323–324
Vietnam War, 219
violence. *See* aggression and violence
visual activity, development of, *55*
visual perspective and actor-observer difference, 136
vocal activity, development of, *55*

W

warning against persuasion, 222
weapons effect, 311
weight, self-appraisals of, 92
weighted averaging model, 124
Whites. *See* European (White) Americans
widows and widowers, 452, 454
witch hunt, 517
women. *See* gender roles and gender differences
women's movement, 545, 550
work. *See* careers and work
work relations in social exchange theory, 12–13